YANG-BAXTER EQUATION IN INTEGRABLE SYSTEMS

ADVANCED SERIES IN MATHEMATICAL PHYSICS

Also in the Series

Volume 1: Mathematical Aspects of String Theory
edited by S-T Yau (*Harvard*)

Volume 2: Bombay Lectures on Highest Weight Representations of Infinite Dimensional Lie Algebras
by V G Kac (*MIT*) & A K Raina (*Tata*)

Volume 3: Kac-Moody and Virasoro Algebras: A Reprint Volume for Physicists
edited by P Goddard (*Cambridge*) & D Olive (*Imperial College*)

Volume 4: Harmonic Mappings, Twistors and σ-Models
edited by P Gauduchon (*CNRS, Paris*)

Volume 5: Geometric Phases in Physics
edited by A Shapere & F Wilczek (*IAS, Princeton*)

Volume 7: Infinite Dimensional Lie Algebras and Groups
edited by V Kac (*MIT*)

Volume 8: Introduction to String Field Theory
by W Siegel (*SUNY*)

Volume 9: Braid Group, Knot Theory and Statistical Mechanics
edited by C N Yang (*SUNY*) & M L Ge (*Nankai Inst*)

Advanced Series in Mathematical Physics

Vol. 10

YANG-BAXTER EQUATION IN INTEGRABLE SYSTEMS

Editor

Michio Jimbo

Kyoto University

World Scientific

Singapore • New Jersey • London • Hong Kong

Published by

World Scientific Publishing Co. Pte. Ltd.,
P O Box 128, Farrer Road, Singapore 9128
USA office: 687 Hartwell Street, Teaneck, NJ 07666
UK office: 73 Lynton Mead, Totteridge, London N20 8DH

The editor and publisher are grateful to the authors and the following publishers for their assistance and permission to reproduce the reprinted papers found in these volumes:

Academic Press (*Ann. Phys.*, *Algebraic Analysis*)
American Mathematical Society (*Soviet Math. Doklady*)
American Physical Society (*Phys. Rev. Lett.*, *Phys. Rev.*)
Elsevier Science Publishers B.V (*Phys. Lett.*, *Nucl. Phys.*, *Adv. Stud. Pure Math*)
Kluwer Academic Publishers (*Lett. Math. Phys.*)
Plenum Publishing Corporation (*J. Stat. Phys.*, *J. Soviet Math.*, *Funct. Anal. Appl.*)
Springer-Verlag (*Comm. Math. Phys.*)
The Royal Society (*Phil. Trans. Roy. Soc. London*)

While every effort has been made to contact the publishers of reprinted papers prior to publication, we have not been successful in a few cases. Where we could not contact the publishers, we have acknowledged the source of the material. Proper credit will be given to these publishers in future editions of this work after permission is granted.

Library of Congress Cataloging-in-Publication data is available.
Printed in Singapore by Kim Hup Lee Printing Co. Pte. Ltd.

Preface

The present volume is a collection of reprints related to the Yang-Baxter equation. The papers are grouped into 7 chapters, each headed by a brief guide about their contents and related works. A bibliography consisting of some 270 papers is given at the end of the book.

A few words are in order on the nature of this volume. Firstly, there exist in the literature several other ways to call the equation— the *star-triangle relation*, the *triangle equation* and the *factorization equation*. We have chosen here the name *Yang-Baxter equation* as it seems to be comparatively more in common use. Secondly, and more importantly, this is a subject related to many other branches. To name a few: the Bethe Ansatz, solvable lattice models, factorized S-matrices, quantum inverse method, quantum groups, link invariants and conformal field theory. Some of them are so intimately connected with the Yang-Baxter equation — in fact contain the latter as a part of the strucutre— that it might be senseless to separate this specific topic from them at all. Even so we have limited our selection of reprints to only those papers that are directly related to the Yang-Baxter equation in order to keep the book within a reasonable size. As for conformal field theory the reprint volume of Itzykson, Saleur and Zuber is available ("Conformal Invariance and Applications to Statistical Mechanics", World Scientific 1988). A reprint volume on knot theory is also being prepared by T. Kohno. We hope the other topics could be covered by future publications. Thirdly, we have not always attempted to single out the very first papers that deal with a specific problem. We have rather tried to include readable and informative account among the papers of similar nature. About this point the reader is referred to the comments at the beginning of each chapter. We have also tried to avoid overlaps with the reprint volumes mentioned above. Finally, despite our effort we feel that the bibliography is still very incomplete. We apologize to the authors whose papers happened to be out of the list because of our ignorance.

In preparing this volume we have benefited from conversations with V. V. Bazhanov, E. Date, V. E. Korepin, T. Miwa, Y. Nakagami and J. H. H. Perk, to whom we are most grateful. We hope this volume is useful for people interested in this rich field.

November 1989

M. J.

Preface

CONTENTS

YANG-BAXTER EQUATION
IN
INTEGRABLE SYSTEMS

PIONEERING WORKS

1. Pioneering works

At an early stage the Yang-Baxter equation (YBE) appeared in several different guises in the literature, and sometimes its solutions have preceded the equation. One can trace basically three streams of ideas from which YBE has emerged: the Bethe Ansatz, commuting transfer matrices in statistical mechanics, and factorizable S matrices in field theory. For general reference about the first two topics the reader is referred to the books of Gaudin[96)] and of Baxter[20)] (see also [19)]).

One of the first occurrences of YBE can be found in the study of a one-dimensional quantum mechanical many-body problem with δ function interaction. By building the Bethe-type wavefunctions, McGuire and others[44-45),49),182)] discovered that the N-particle S-matrix factorized into the product of two-particle ones. Yang's reprints [1,2] treat the case of arbitrary statistics of particles by introducing the 'nested Bethe Ansatz'. The YBE appears here in the present form as the consistency condition for the factorization.

The significance of YBE in statistical mechanics lies in that it implies the existence of a commuting family of transfer matrices (Onsager[192-193)]). Baxter's celebrated solution of the eight vertex model (reprint [3]) utilizes this property to derive equations that determine the eigenvalues of the transfer matrix. In the reprint [4] Baxter investigated further the role of YBE and its implications for the eight vertex model on an arbitrary irregular lattice. He found that the partition function is left unchanged upon parallel displacement of the lines forming the lattice — a property which he called Z invariance.

Nearly a decade after the works [182),265-266)], the theory of factorized S matrix has been resumed in the relativistic setting by Zamolodchikov, the Berlin group and others [127-130),226),267-269),274-276)]. The reprint of Zamolodchikov and Zamolodchikov [5] is a detailed exposition of how the S matrix is determined by requiring the factorization property along with unitarity and crossing symmetry. The review [269)] is also a readable account of this theory. Subsequently Zamolodchikov[268)] pointed out that the algebraic mechanism of factorization is precisely the same as in Baxter's Z-invarinace.

Having these works on the one hand, and the development in soliton theory on the other, Sklyanin, Takhtajan and Faddeev[234-235)] proposed the quantum inverse scattering method (QISM) as a synthesis of classical and quantum integrable systems. The traditional Bethe Ansatz found an algebraisation here in the framework of commutation relations of operators that are derived from YBE. Sklyanin's reprint [6] reviews QISM on the basic example of the quantum and classical non-linear Schrödinger equation — the case treated earlier by McGuire, Yang and others. It is stressed that the transition between the quantum and classical systems can be done by preserving integrability. Through the correspondence principle there naturally comes about the notion of the classical Yang-Baxter equation (CYBE), first introduced by Sklyanin in [228)]. For general reviews of QISM see also [83–84),246)].

In the reprint [7] Kulish and Sklyanin gives the first survey devoted to YBE from QISM point of view, describing basic concepts, properties and many examples of solutions that have been known upto 1980. The paper [145)] also contains a detailed exposition on QISM and YBE. For a more recent review on YBE, see [116)].

VOLUME 19, NUMBER 23 PHYSICAL REVIEW LETTERS 4 DECEMBER 1967

SOME EXACT RESULTS FOR THE MANY-BODY PROBLEM IN ONE DIMENSION WITH REPULSIVE DELTA-FUNCTION INTERACTION*

C. N. Yang
Institute for Theoretical Physics, State University of New York, Stony Brook, New York
(Received 2 November 1967)

The repulsive δ interaction problem in one dimension for N particles is reduced, through the use of Bethe's hypothesis, to an eigenvalue problem of matrices of the same sizes as the irreducible representations R of the permutation group S_N. For some R's this eigenvalue problem itself is solved by a second use of Bethe's hypothesis, in a generalized form. In particular, the ground-state problem of spin-$\frac{1}{2}$ fermions is reduced to a generalized Fredholm equation.

(1) Consider the one-dimensional N-body problem

$$H=-\sum_1^N \partial^2/\partial x_i^2+2c\sum_{i<j}\delta(x_i-x_j),\quad c>0, \tag{1}$$

with no limitation on the symmetry of the wave function ψ. For a given irreducible representation R_ψ of the permutation group S_N of the N coordinates x_i, we want to determine the wave function ψ. Assume Bethe's hypothesis[1] to be valid: Let $p_1,\cdots,p_N$ = a set of unequal numbers. For $0<x_{Q1}<x_{Q2}<\cdots<x_{QN}<L$,

$$\psi=\sum_P[Q,P]\exp i[p_{P1}x_{Q1}+\cdots+p_{PN}x_{QN}], \tag{2}$$

where $P=[P1,P2,\cdots,PN]$ and $Q=[Q1,Q2,\cdots,QN]$ are two permutations of the integers $1,2,\cdots,N$. $[Q,P]$ can be arranged as a $N!\times N!$ matrix. Denote the columns of this matrix by ξ_P. To satisfy the continuity of ψ and the proper discontinuity of its derivative as required by (1) at $x_{Q3}=x_{Q4}$, it is sufficient to have

$$\xi_{\ldots ij\ldots}=Y_{ji}^{34}\xi_{\ldots ji\ldots}, \tag{3}$$

where the subscripts for ξ on the two sides represent any two permutation P and P' so that $P1=P'1$, $P2=P'2$, $P3=i=P'4$, $P4=j=P'3$, etc. The operator Y is defined by

$$Y_{ij}^{34}=(y_{ij}^{-1}-1)+y_{ij}^{-1}P_{34}=Y_{ij}^{43}, \tag{4}$$

where

$$y_{ij}=1+x_{ij}, \tag{5}$$

$$x_{jk}=ic(p_j-p_k)^{-1}=-x_{kj}, \tag{6}$$

and P_{34} = the permutation operator on ξ so that it interchanges $Q3$ and $Q4$. Altogether there are $N!(N-1)$ equations of the form (3). Are they mutually consistent? The answer is yes for any set of unequal p's. This can be seen with the aid of the following identities:

$$Y_{ij}^{ab}Y_{ji}^{ab}=1, \tag{7}$$

and

$$Y_{jk}^{ab}Y_{ik}^{bc}Y_{ij}^{ab}=Y_{ij}^{bc}Y_{ik}^{ab}Y_{jk}^{bc}, \tag{8}$$

which are easily verified. Thus given a set of unequal p's, and $\xi_0=\xi_P$ for P = identity, all ξ_P's are determined.

(2) The imposition of the periodic boundary conditions leads to equations which, upon expressing ξ_P in terms of ξ_0, become

$$\lambda_j\xi_0=X_{(j+1)j}\times X_{(j+2)j}\cdots X_{Nj}X_{1j}X_{2j}\cdots X_{(j-1)j}\xi_0, \tag{9}$$
$$j=1,\cdots,N,$$

where

$$\lambda_j=\exp(ip_jL), \tag{10}$$

and

$$X_{ij}=P_{ij}Y_{ij}^{ij}=(1-P_{ij}x_{ij})(1+x_{ij})^{-1}. \tag{11}$$

The N Eqs. (9) say that ξ_0 is simultaneously an eigenvector of N operators. These N operators can be shown to commute with each other, using

$$X_{ij}X_{ji}=1,\quad X_{jk}X_{ik}X_{ij}=X_{ij}X_{ik}X_{jk}\ \ (\text{sic: }X_{jk}X_{ik}X_{ij}X_{kj}X_{ki}X_{ji}=1),$$

$$X_{ij}X_{kl}=X_{kl}X_{ij};\quad i,j,k,\text{ and } l \text{ all unequal}. \tag{12}$$

(3) The operators P_{ij} on ξ form a $N!\times N!$ representation of S_N. To find the eigenfunctions ξ_0 in (9) we can first reduce this representation to irreducible ones. Choosing one specific irreducible representation R reduces the

eigenvalue problem (9) to one of smaller dimensions. It can be shown that the resultant wave function (2) would have a permutation symmetry R_ψ which is the same as R. For example, if R = identity representation = $[N]$, then $P_{ij}=1$, and (9) becomes 1×1 matrix equations and the result is precisely the well-known boson result.[2] If R = antisymmetric representation = $[1^N]$, then $P_{ij}=-1$, and $X_{ij}=1$, so that (9) and (10) reduce to $\exp(ip_jL)=1$, showing there is no interaction, a result to be expected for the antisymmetrical wave function.

(4) The λ_j's are functions of the p's, c, and R. It is easily seen (that R and $\bar{R}$ being conjugate representations)

$$\lambda_j(p;c;R)=\prod_{i\neq j}\left(\frac{1-x_{ij}}{1+x_{ij}}\right)\lambda_j(p;-c;\bar{R}). \tag{13}$$

(5) Define $\mu_j(p;c;R)$ by

$$\mu_j\Phi=X_{(j+1)j}{}'X_{(j+2)j}{}'\cdots \times X_{Nj}{}'X_{1j}{}'X_{2j}{}'\cdots X_{(j-1)j}{}'\Phi, \tag{14}$$

where

$$X_{ij}{}'=(1+P_{ij}x_{ij})(1+x_{ij})^{-1}. \tag{15}$$

Clearly

$$\mu_j(p;c;\bar{R})=\lambda_j(p;c;R). \tag{16}$$

(6) We now evaluate λ_j for $R_\psi=R=[2^{M}1^{N-2M}]$. By (16) we need to find $\mu_j(p;c;[N-M,M])$. To do this we first define a convenient representation for P_{ij} of (15):

Consider N spin-$\frac{1}{2}$ particles, and consider the spin wave functions Φ for total z spin $=\frac{1}{2}(N-2M)$. These spin wave functions transform under S_N according to a sum of irreducible representations,

$$[N]+[N-1,1]+[N-2,2]+\cdots+[N-M,M]. \tag{17}$$

We consider the P_{ij}'s of (15) as operating on these spin wave functions Φ. The eigenvalue equations (14) for μ_j are then to be solved for a Φ that belongs to the symmetry $[N-M,M]$.

(7) Consider the N spins as forming a cyclic chain. The wave function Φ has $C_M{}^N$ components [$N-M$ spins up, M spins down]. The eigenvalue problem (14) can be solved with a generalized Bethe's hypothesis:

$$\Phi=\sum_P A_P F(\Lambda_{P1},y_1) \times F(\Lambda_{P2},y_2)\cdots F(\Lambda_{PM},y_M), \tag{18}$$

where $y_1<y_2<\cdots<y_M$ are the "coordinates," along the chain, of the M down spins, and Λ_1, Λ_2, $\cdots$, Λ_M are a set of unequal numbers. With this hypothesis, one finds

$$F(\Lambda,y)=\prod_{j=1}^{y-1}\frac{ip_j-i\Lambda-c'}{ip_{j+1}-i\Lambda+c'}\quad(c'=\tfrac{1}{2}c); \tag{19}$$

$$-\prod_j\frac{ip_j-i\Lambda_\alpha-c'}{ip_j-i\Lambda_\alpha+c'}=\prod_\beta\frac{-i\Lambda_\beta+i\Lambda_\alpha+c}{-i\Lambda_\beta+i\Lambda_\alpha-c}; \tag{20}$$

and

$$\mu_j(p;c;[N-M,M])=\prod_\beta\frac{ip_j-i\Lambda_\beta-c'}{ip_j-i\Lambda_\beta+c'}. \tag{21}$$

(8) Thus for the $R_\psi=[2^{M}1^{N-2M}]$ symmetry, we need to solve

$$\exp(ip_jL)=\text{right-hand side of (21)}, \tag{22}$$

together with (20). In taking the logarithm of (20) and (22) care must be taken to add terms $2\pi i$(integer). The value of the integer can be determined by going to the limit $c\to+\infty$. One obtains, for the ground state with the symmetry $R_\psi=[2^{M}1^{N-2M}]$, for the case N = even, M = odd,

$$-\sum_p\theta(2\Lambda-2p)=2\pi J_\Lambda-\sum_{\Lambda'}\theta(\Lambda-\Lambda'), \tag{23a}$$

$$Lp=2\pi I_p+\sum_\Lambda\theta(2p-2\Lambda), \tag{23b}$$

where the p's are a set of N ascending real numbers, the Λ's a set of M ascending real numbers,

$$\theta(p)=-2\tan^{-1}(p/c)\quad(-\pi\leq\theta<\pi), \tag{24}$$

and

J_Λ = successive integers from

$$-\tfrac{1}{2}(M-1)\text{ to }+\tfrac{1}{2}(M-1), \tag{24a}$$

$\frac{1}{2}+I_p$ = successive integers from

$$1-\tfrac{1}{2}N\text{ to }\tfrac{1}{2}N. \tag{24b}$$

Equation (23a) differs from that given in a recent paper,[3] in the definition of θ and our introduction of J_Λ. The present equation allows for a natural discussion of the limit $c\to+\infty$ (not $c\to 0$!) and hence the values of J_Λ.

(9) We can now approach the limit $N\to\infty$, $M=\infty$, $L\to\infty$ proportionally, obtaining

$$-\int_{-Q}^{Q}\theta(2\Lambda-2p)\rho(p)dp=2\pi g-\int_{-B}^{B}\theta(\Lambda-\Lambda')\sigma(\Lambda')d\Lambda', \quad (25a)$$

$$p=2\pi f+\int_{-B}^{B}\theta(2p-2\Lambda)\sigma(\Lambda)d\Lambda, \quad (25b)$$

$$dg/d\Lambda=\sigma,\quad df/dp=\rho. \quad (25c)$$

Or, after differentiation,

$$2\pi\sigma=-\int_{-B}^{B}\frac{2c\sigma(\Lambda')d\Lambda'}{c^2+(\Lambda-\Lambda')^2}+\int_{-Q}^{Q}\frac{4c\rho dp}{c^2+4(p-\Lambda)^2}, \quad (26a)$$

$$2\pi\rho=1+\int_{-B}^{B}\frac{4c\sigma d\Lambda}{c^2+4(p-\Lambda)^2}, \quad (26b)$$

$$N/L=\int_{-Q}^{Q}\rho dp,\quad M/L=\int_{-B}^{B}\sigma d\Lambda, \quad (27a)$$

and

$$E/L=\int_{-Q}^{Q}p^2\rho(p)dp. \quad (27b)$$

(10) Equations (26) are generalized Fredholm equations with a symmetrical kernel. It is easy to show that the equations are nonsingular by first studying the eigenvalues of the kernel in the limit $B=Q=\infty$.

(11) Equations (26) and (27) yield the ground-state energy per particle for spatial wave functions with the symmetry $[2^M 1^{N-2M}]$, at a given density N/L. For N fermions with spin $\frac{1}{2}$ interacting through the Hamiltonian (1), this spatial wave function is coupled to a spin wave function of conjugate symmetry $[N-M, M]$, i.e., the total spin of the system is $\frac{1}{2}N-M$.

(12) For $B=\infty$, integration of (26a) over all Λ yields $N=2M$. Thus for the fermion problem with spin $\frac{1}{2}$, $B=\infty$ gives the ground state for states with total spin $=0$. This state is also the absolute ground state for the problem, by a theorem due to Lieb and Mattis.[4]

(13) For the case $B\cong 0$, M/L is proportional to B. One can readily expand all quantities in powers of B, obtaining, for fixed $r=N/L$,

$$\frac{E}{L}=\text{const.}+\frac{M}{L}\left[cr-\left(\frac{c^2}{2\pi}+2\pi r^2\right)\tan^{-1}\frac{2\pi r}{c}\right]+\cdots. \quad (28)$$

This result is in agreement with results already obtained by McGuire[5] for the case $M=1$ and by Flicker and Lieb[6] for the case $M=2$.

(14) For each symmetry R_ψ of spatial wave function ψ, the excited states near the ground state can be obtained in a similar way as in the boson case.[7] More quantum numbers are, however, necessary to designate the excitations than in the boson case, because of the existence of the integers J_Λ (which are in fact quantum numbers). Details will be published elsewhere.

(15) For the boson problem the thermodynamics and excitations for finite T were treated by Yang and Yang.[8] Extension to the present problem presents no difficulty. Details will be published elsewhere.

(16) Using (13) one could generalize all the considerations above to the case of $R_\psi=[N-M, M]$. Details will be published elsewhere. The main change is that while all Eqs. (26) and (27) remain the same, (26b) is replaced by

$$2\pi\rho=1-\int_{-B}^{B}\frac{4c\sigma d\Lambda}{c^2+4(p-\Lambda)^2}+\int_{-Q}^{Q}\frac{2c\rho(p')dp'}{c^2+(p-p')^2}. \quad (26b')$$

It is a pleasure to acknowledge useful discussions with J. B. McGuire in 1963 at University of California, Los Angeles, with T. T. Wu in 1964 at Brookhaven National Laboratory, with C. P. Yang in 1966 in Princeton, and with B. Sutherland in 1967 in Stony Brook.

---

*Research partly supported by U. S. Atomic Energy Commission under Contract No. AT(30-1)-3668B.

[1]H. A. Bethe, Z. Physik 71, 205 (1931), first used the hypothesis for the spin-wave problem. E. Lieb and W. Linger, Phys. Rev. 130, 1605 (1963), and J. B. McGuire, J. Math. Phys. 5, 622 (1964), first used the same hypothesis for the δ-function interaction problem. The present author believes that for all states (excited as well as the ground state) with periodic boundary condition the hypothesis is valid. Justification of this belief for some special cases is found in C. N. Yang and C. P. Yang, Phys. Rev. 150, 321 (1966), and to be published.

[2]Lieb and Linger, Ref. 1.

[3]M. Gaudin, Phys. Letters 24A, 55 (1967).

[4]E. Lieb and D. Mattis, Phys. Rev. 125, 164 (1962).

[5]J. B. McGuire, J. Math. Phys. 6, 432 (1965), and 7,

Reprinted from THE PHYSICAL REVIEW, Vol. 168, No. 5, 1920–1923, 25 April 1968
Printed in U. S. A.

S Matrix for the One-Dimensional N-Body Problem with Repulsive or Attractive δ-Function Interaction

C. N. YANG
Institute for Theoretical Physics, State University of New York, Stony Brook, New York 11790
(Received 14 December 1967)

For N particles with equal mass, interacting with repulsive or attractive δ-function interaction of the same strength, the S matrix is explicitly given and shown to be symmetrical and unitary. The incoming and outgoing states may consist of bound compounds as well as single particles. The momenta of the particles and compounds are not changed in the scattering, but particles are exchanged, such as $ABC+DE \to ADC+BE$. Only distinguishable particles are considered.

1. INTRODUCTION

FOR the one-dimensional N-body problem

$$H=-\sum_1^N \partial^2/\partial x_i^2+2c\sum_{i<j}\delta(x_i-x_j)\,, \tag{1}$$

with positive or negative c, the S matrix was discussed by McGuire[1] and by Zinn-Justine and Brezin.[2] (*Note added in proof.* K. Hepp kindly informed the author that F. A. Berezin and V. N. Sushko, Zh. Eksperim. i Teor. Fiz. **48**, 1293 (1965) [English transl.: Soviet Phys.—JETP **21**, 865 (1965)] have also discussed this problem.) We give in this paper a complete explicit expression for S. Only distinguishable particles are considered.

2. METHOD

The method used follows that of Sec. 1 of a recent paper.[3] We observe that all formulas there are also applicable to the case $c<0$.

If boundary conditions are not imposed, it is clear that all solutions of the Schrödinger equation are superpositions of solutions of the type ($Y2$). In other words, Bethe's hypothesis is proved in such a case.

3. INCOMING AND OUTGOING STATES

To construct scattering states, we need real values of the p's. Let us choose them so that

$$p_1<p_2<\cdots<p_N\,. \tag{2}$$

A term in ($Y2$) that has $P=$identity permutation$=I$, then, represents an *outgoing* wave. [A wave packet constructed out of such a term would have the left-most particle (at X_{Q1}) travel with velocity $2p_1$; the second left-most particle (at X_{Q2}) travel with velocity $2p_2$, etc. Thus the wave packet in *future* movement develops no collisions, meaning it is an outgoing wave packet.] A term in ($Y2$) that has $P=[N,N-1,\cdots,1]=I'$, i.e., the "reversed" permutation, represents an *incoming* wave.

Now each permutation Q represents a definite ordering of the coordinates and represents a scattering channel. A scattering state $Q_i\to Q_0$ is obtained if there are only incoming waves in channel Q_i:

$$[Q_i,I']=1\,,\qquad [Q,I']=0\quad\text{for}\quad Q\neq Q_i\,. \tag{3}$$

In other words,

$$\langle Q_i|\xi_{I'}\rangle=1\,,\qquad \langle Q|\xi_{I'}\rangle=0\quad\text{for}\quad Q\neq Q_i\,. \tag{4}$$

The amplitudes of the outgoing waves are the elements of ξ_I. Now ξ_I can be related to $\xi_{I'}$ through repeated use of ($Y2$):

$$\xi_I=[Y_{21}{}^{12}Y_{31}{}^{23}Y_{41}{}^{34}\cdots Y_{N1}{}^{(N-1)N}]\times[Y_{32}{}^{12}Y_{42}{}^{23}\cdots Y_{N2}{}^{(N-2)(N-1)}]\cdots[Y_{N(N-1)}{}^{12}]\xi_{I'}\,. \tag{5}$$

Thus the scattering amplitude for $Q_i\to Q_0$ is

$$\langle Q_0|S'|Q_i\rangle\,, \tag{6}$$

where S' is the right-hand side of (5) with $\xi_{I'}$ deleted.

4. OPERATOR: {ij}

We did not call the matrix S' in (6) the S matrix because it differs from the usual one in that the labeling of the columns is not in accordance with the usual rules. This is so because the incoming wave in Q_i, represented by the $[Q_i,I']$ term, describes particle $Q1$ with momentum p_N, $Q2$ with momentum p_{N-1}, etc. Thus the correct S matrix is

$$S=S'[P^{N1}P^{(N-1)2}\cdots]=S'[P^{12}][P^{23}P^{12}][P^{34}P^{23}P^{12}]\cdots[P^{(N-1)N}\cdots P^{12}]\,. \tag{7}$$

If in (7) one explicitly writes S', as given in (5), one observed that the superscripts for the Y's are the same as those for the P's, but in reverse order. One now permutes the last factor P^{12} through to just behind the first factor $Y_{21}{}^{12}$; then the new last factor P^{23} through to just behind the second factor $Y_{31}{}^{23}$, etc. The final

[1] J. B. McGuire, J. Math. Phys. 5, 622 (1964). This is a very interesting paper in which by geometrical construction many of the results of the present paper were obtained.

[2] E. Brezin and J. Zinn-Justine, Compt. Rend. Acad. Sci. Paris B263, 670 (1966).

[3] C. N. Yang, Phys. Rev. Letters, 19, 1312 (1967). Formula (m) of this paper will be called (Ym) in the present paper.

result is

$$S=[\{21\}\{31\}\{41\}\cdots\{N1\}]\times[\{32\}\{42\}\cdots\{N2\}]\cdots[\{N(N-1)\}], \quad (8)$$

where

$$\{ij\}\equiv X_{ij}=P^{ij}Y_{ij}{}^{ij}=(1-P^{ij}x_{ij})(1+x_{ij})^{-1}. \quad (9)$$

5. S MATRIX

In (8) we have *an explicit formula for the S matrix* (for both $c\geq 0$ and $c<0$, $p_1<p_2<p_3\cdots<p_N$ being all real). S is an $N!\times N!$ matrix. The scattering only exchanges particle momenta. The elements of S have the following meaning:

$\langle A'B'C'\cdots|S|ABC\cdots\rangle$
= matrix element of S for
[State: particle A with p_1, B with p_2, etc.] →
[State: particle A' with p_1, B' with p_2, etc.].

In (9) the permutation operator P^{ij} is defined so that, e.g.,

$$P^{31}|CDBA\rangle=|BDCA\rangle=P^{41}|ADCB\rangle.$$

It is easy to verify that each $\{ij\}$ is unitary. Hence S is unitary. S is a symmetrical matrix, as required by the time-reversal invariance of the interaction we have, because each $\{ij\}$ is symmetrical and the order of the operators $\{ij\}$ in (8) can be reversed by repeated application of Eq. (Y12). For example, for $N=4$,

$$\begin{aligned} S&=\{21\}\{31\}\{41\}\{32\}\{42\}\{43\}\\ &=\{21\}\{31\}\{41\}\{43\}\{42\}\{32\}\\ &=\{21\}\{43\}\{41\}\{31\}\{42\}\{32\}\\ &=\{43\}\{21\}\{41\}\{42\}\{31\}\{32\}\\ &=\{43\}\{42\}\{41\}\{21\}\{31\}\{32\}\\ &=\{43\}\{42\}\{41\}\{32\}\{31\}\{21\}\\ &=\{43\}\{42\}\{32\}\{41\}\{31\}\{21\}=\tilde{S}. \end{aligned}$$

6. ATTRACTIVE CASE

For the case $c<0$, there are bound states[1] for the system of N particles. The wave function for the bound state is

$$\psi=\exp[\tfrac{1}{2}c\sum_{i<j}|x_i-x_j|]. \quad (10)$$

It is easy to show directly that (10) satisfies the Schrödinger equation.

It is clear that (10) is of Bethe's form (Y2) with

$$p_1=\tfrac{1}{2}ic(N-1),\quad p_2=\tfrac{1}{2}ic(N-3),\ \cdots,\quad p_N=-\tfrac{1}{2}ic(N-1), \quad (11)$$

and with

$$\xi_I=(\text{a column with all elements equal}), \quad (12a)$$

$$\xi_P=0 \quad \text{for all } P\neq I. \quad (12b)$$

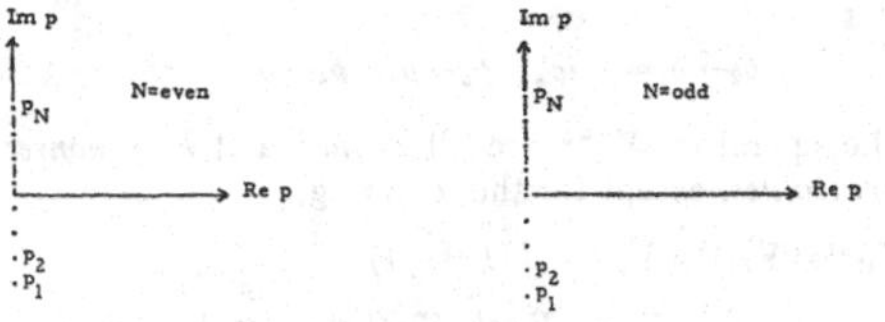

FIG. 1. The p's for the N-particle bound state. The p's are pure imaginary, and the difference between successive p's is $-ic$.

The numbers $p_1,\ p_2,\ \cdots,\ p_N$ are plotted in Fig. 1. Equation (12a) can also be written as

$$p^{ab}\xi_I=\xi_I \quad \text{for any } a \text{ and } b. \quad (12c)$$

The energy of this bound state is

$$E=\sum_j p_j{}^2=-c^2N(N^2-1)/12, \quad (13)$$

a result already given by[1] McGuire.

It can be shown that for the N-particle problem, (10) gives the *only* bound state. This fact was already noted by McGuire.[1]

7. S MATRIX FOR BOUND STATES

If one multiplies the wave function (10) by $\exp(ik\sum x)$, one obtains a new one describing the bound state moving with a momentum Nk (k = real). The wave function is again of the form (Y2) with the p's equal to those of Fig. 1 displaced by k along the real p axis.

Would such bound particles scatter each other? To study this problem, we evidently need to fuse the considerations of Secs. 3 and 4 with those of Sec. 6.

Consider as an example the scattering of a two-particle bound state by a three-particle bound state. The p's for such a problem are plotted in Fig. 2(a). Note

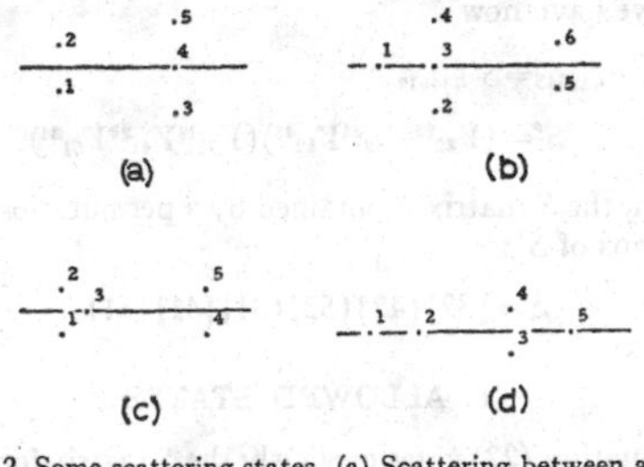

FIG. 2. Some scattering states. (a) Scattering between a bound doublet with momentum p_1+p_2 and a bound triplet with momentum $p_3+p_4+p_5$. (b) Scattering between a particle with momentum p_1, a bound triplet with momentum $p_2+p_3+p_4$, and a bound doublet with momentum p_5+p_6. (c) Scattering between a doublet of momentum p_1+p_2, a particle with momentum p_3, and a doublet of momentum p_4+p_5. (d) Scattering between three particles of momenta p_1, p_2, and p_5 and a doublet of momentum p_3+p_4. The difference between two successive p's in any vertical column is $-ic$. Note that if the p at the top of the left-most column is p_a, then the S matrix is $S=[\{\cdot a\}\{\cdot a\}\{\cdot a\}\cdots\{\cdot a\}]\cdots$.

that

$$p_2-p_1=-ic,\quad p_5-p_4=p_4-p_3=-ic. \tag{14}$$

The operators Y_{ij}^{ab} are all *defined* and *have nonzero eigenvalues*, except for the following:

$$Y_{12}^{ab}=Y_{34}^{ab}=Y_{45}^{ab}=\tfrac{1}{2}(P^{ab}-1);$$

$$Y_{21}^{ab},\ Y_{43}^{ab},\ Y_{54}^{ab} \text{ are not defined}, \tag{15}$$

$$y_{21}=y_{43}=y_{54}=0. \tag{16}$$

For the wave function ($Y2$) to be bounded, such columns as $\xi_2\cdots$ must be zero; for otherwise as $x_{Q1}\to\infty$, the terms in ($Y2$) with the elements of $\xi_2\cdots$ as coefficients will diverge exponentially. Considerations like this and a reexamination of ($Y2$), which remains valid except for the cases where Y_{ij}^{ab} is not defined, finally lead to

$$\xi_P\neq 0 \quad \text{if } P \text{ is of type } A, \tag{17a}$$

$$\xi_P=0 \quad \text{if } P \text{ is not of type } A, \tag{17b}$$

where P is defined to be of type A if in

$$P=[P1,P2,P3,P4,P5]$$

1 and 2 are in that order and 3, 4, 5 are in that order (e.g., [23145] is not in A, [34152] is in A). Furthermore,

$$\xi_I=\xi_{12345}$$

satisfies

$$\xi_I=P^{12}\xi_I=P^{34}\xi_I=P^{45}\xi_I. \tag{18}$$

Because of (18), we have, e.g.,

$$\xi_{21345}=Y_{12}{}^{12}\xi_{12345}=\tfrac{1}{2}(P^{12}-1)\xi_{12345}=0,$$
$$\xi_{54321}=Y_{45}{}^{12}Y_{35}{}^{23}\cdots Y_{12}{}^{12}\xi_{12345}=0. \tag{19}$$

ξ_{12345} still gives the outgoing waves, but the incoming waves are not given by ξ_{54321}, which is zero by (19). Instead, it is given by ξ_{34512}. Thus, instead of the S' of (6), we have now

$$\xi_{12345}=S'\xi_{34512}, \tag{20}$$

$$S'=(Y_{32}{}^{23}Y_{42}{}^{34}Y_{52}{}^{45})(Y_{31}{}^{12}Y_{41}{}^{23}Y_{51}{}^{34}). \tag{21}$$

Again, the S matrix is obtained by a permutation of the columns of S':

$$S=\{32\}\{42\}\{52\}\{31\}\{41\}\{51\}. \tag{22}$$

8. ALLOWED STATES

Equation (22) gives explicitly the S matrix for a two-particle bound state scattered by a three-particle bound state. Because of (18), S should only operate between states Φ, satisfying

$$\Phi=P^{12}\Phi=P^{34}\Phi=P^{45}\Phi. \tag{23}$$

We shall call such states "allowed" states. Among the $5!=120$ components of the column Φ, there are only $5!/2!3!=10$ independent allowed ones. For example,

$$\begin{aligned}\langle ABCDE|\Phi\rangle&=\langle ABCED|\Phi\rangle=\langle ABDCE|\Phi\rangle\\&=\text{etc.}=\langle ABEDC|\Phi\rangle\\&=\langle BACDE|\Phi\rangle=\langle BACED|\Phi\rangle\\&=\langle BADCE|\Phi\rangle=\text{etc.}\\&=\langle BAEDC|\Phi\rangle=1/(12)^{1/2}\end{aligned}$$

together describe the allowed incoming state

$$AB+CDE, \tag{24}$$

where AB is the symmetrical bound state of A and B with momentum p_1+p_2, and CDE is the symmetrical bound state of C, D, and E with momentum $p_3+p_4+p_5$.

9. SOME IDENTITIES

We shall prove in Sec. 10 three important properties of the S of Eq. (22). A few mathematical preliminaries will be given in this section.

We note that

$$\{12\}=\tfrac{1}{2}(1-P^{12}),\quad \{34\}=\tfrac{1}{2}(1-P^{34}),$$
$$\{45\}=\tfrac{1}{2}(1-P^{45}), \tag{25}$$

so that Eq. (23) is equivalent to

$$0=\{12\}\Phi=\{34\}\Phi=\{45\}\Phi. \tag{26}$$

($Y12$) remains valid, or rather the following hold true:

$$\{ij\}\{ji\}=1, \tag{27a}$$

$$\{ij\}\{kj\}\{ki\}=\{ki\}\{kj\}\{ij\}, \tag{27b}$$

$$\{ij\}\{kl\}=\{kl\}\{ij\} \quad \text{if } i,j,k,l \text{ are all different}, \tag{27c}$$

provided the undefined $\{21\}$, $\{43\}$, and $\{54\}$ do not appear. [Note that $\{35\}$ and $\{53\}$ are defined.]

Although $\{21\}$ is not defined, we can try to define y_{21} ($\{21\}$) so that $y_{21}=0$ does not appear any more in the denominator. In other words, we define

$$\begin{aligned}\{21'\}&=(1-y_{21})P^{21}+1=P^{21}+1,\\\{43'\}&=P^{43}+1,\\\{54'\}&=P^{54}+1.\end{aligned} \tag{28}$$

With this definition, (27b) is true also for those cases where $\{21\}$, $\{43\}$, and/or $\{54\}$ appear, provided we replace them by $\{21'\}$, $\{43'\}$, and $\{54'\}$. For example,

$$\{43'\}\{53\}\{54'\}=\{54'\}\{53\}\{43'\},$$
$$\{21'\}\{51\}\{52\}=\{52\}\{51\}\{21'\}. \tag{29}$$

Φ is allowed if, and only if,

$$2\Phi=\{21'\}\Phi=\{43'\}\Phi=\{54'\}\Phi. \tag{30}$$

Equations (23), (26), and (30) are equivalent.

10. UNITARITY AND SYMMETRY OF S

We now first prove that if Φ is allowed, so is $S\Phi$. This follows from

$$\{12\}S\Phi=\{12\}(\{32\}\{31\})(\{42\}\{41\})(\{52\}\{51\})\Phi$$
$$=(\{31\}\{32\})(\{41\}\{42\})(\{51\}\{52\})\{12\}\Phi$$

and

$$\{34\}S\Phi=\{45\}S\Phi=0.$$

Next we shall prove that S is symmetrical for allowed states, i.e., if Φ_1 and Φ_2 are both allowed,

$$\Phi_1{}^{\dagger}S\Phi_2=\Phi_1{}^{\dagger}\tilde{S}\Phi_2. \tag{31}$$

To prove this, we note that $\{ij\}$ is symmetrical. Thus

$$\tilde{S}=\{51\}\{41\}\{31\}\{52\}\{42\}\{32\}. \tag{23}$$

Now

$$S\{21'\}=(\{32\}\{31\})(\{42\}\{41\})(\{52\}\{51\})\{21'\}=\{21'\}(\{31\}\{32\})(\{41\}\{42\})(\{51\}\{52\}),$$
$$S\{21'\}\{54'\}=\{21'\}\{54'\}(\{31\}\{32\})(\{51\}\{41\})(\{52\}\{42\}),$$
$$S\{21'\}\{54'\}\{53\}=\{21'\}\{54'\}\{53\}(\{51\}\{31\})(\{52\}\{32\})(\{41\}\{42\}), \tag{33}$$
$$S\{21'\}\{54'\}\{53\}\{43'\}=\{21'\}\{54'\}\{53\}\{43'\}(\{51\}\{52\})(\{41\}\{31\})(\{42\}\{32\})=\{21'\}\{54'\}\{53\}\{43'\}\tilde{S}.$$

But

$$\{21'\}\{54'\}\{53\}\{43'\}\Phi_2=8(2y_{53}{}^{-1}-1)\Phi_2=24\Phi_2,$$
$$\Phi_1{}^{\dagger}\{21'\}\{54'\}\{53\}\{43'\}=8(2y_{53}{}^{-1}-1)\Phi_1{}^{\dagger}=24\Phi_1{}^{\dagger}.$$

Thus (33) yields directly (31).

Last we shall prove that S is unitary for allowed states, i.e., if Φ_2 is allowed,

$$\Phi_2{}^{\dagger}S^{\dagger}S\Phi_2=\Phi_2{}^{\dagger}\Phi_2. \tag{34}$$

To prove this we find that

$$S^{\dagger}=\{51\}^{\dagger}\{41\}^{\dagger}\{31\}^{\dagger}\{52\}^{\dagger}\{42\}^{\dagger}\{32\}^{\dagger}.$$

Now

$$P^{12}P^{53}\{51\}^{\dagger}P^{53}P^{12}=\{23\}\text{ etc.}$$

Thus

$$S^{\dagger}P^{53}P^{12}=P^{53}P^{12}\{23\}\{24\}\{25\}\{13\}\{14\}\{15\}.$$

By (32),

$$S^{\dagger}P^{53}P^{12}\tilde{S}=P^{53}P^{12}.$$

Thus

$$\Phi_2{}^{\dagger}S^{\dagger}P^{53}P^{12}\tilde{S}\Phi_2=\Phi_2{}^{\dagger}\Phi_2. \tag{35}$$

Put

$$\Phi_1=S\Phi_2.$$

Thus Φ_1 is allowed, and $\Phi_1=P^{12}P^{53}\Phi_1$. Equations (35) and (31) together give

$$\Phi_2{}^{\dagger}\Phi_2=\Phi_1{}^{\dagger}\tilde{S}\Phi_2=\Phi_1{}^{\dagger}S\Phi_2=\Phi_2{}^{\dagger}S^{\dagger}S\Phi_2.$$

11. GENERAL CASE

The results of Secs. 7–10 can be generalized in a straightforward way to the scattering between any number of particles or compounds, each of which may be a bound state of any number of particles. The S matrix can be easily written down. For example, we write down the S matrix for a scattering between a single particle of momentum p_1, a bound triplet of momentum $p_2+p_3+p_4$, and a bound doublet of momentum p_5+p_6. These p's are plotted in Fig. 2(b). We have, like Eq. (22),

$$S=(\{21\}\{31\}\{41\}\{51\}\{61\})(\{54\}\{64\})\times(\{53\}\{63\})(\{52\}\{62\}). \tag{36}$$

For the case where the p's are given by Fig. 2(c), we have

$$S=(\{32\}\{42\}\{52\})(\{31\}\{41\}\{51\})(\{43\}\{53\}). \tag{37}$$

For the case where the p's are given by Fig. 2(d), we have

$$S=(\{21\}\{31\}\{41\}\{51\})(\{32\}\{42\}\{52\})\times(\{54\})(\{53\}). \tag{38}$$

All these S matrices are unitary and symmetrical for the allowed Φ in each case.

12. REDUNDANT POLES

The S matrix discussed above has evidently matrix elements that are rational functions of the relative momenta of the particles involved. For real values of these relative momenta, S is regular. But for complex values of these relative momenta, S may have poles. For example, in the reaction $AB+CDE$ discussed in Secs. 7–10, for which the S matrix is given by (22), there are poles when y_{32}, y_{42}, y_{52}, y_{31}, y_{41}, or y_{51} vanishes. However, only the pole $y_{32}=0$ corresponds to a bound state (the 5-particle bound state). The others are *redundant poles.* This point was already realized by McGuire.[1]

ACKNOWLDGMENTS

It is a pleasure to acknowledge useful discussions with T. T. Wu, M. Moshinsky, and J. Groeneveld.

Partition Function of the Eight-Vertex Lattice Model

RODNEY J. BAXTER

Research School of Physical Sciences, The Australian National University, Canberra, A.C.T. 2600, Australia

Received May 20, 1971

The partition function of the zero-field "Eight-Vertex" model on a square M by N lattice is calculated exactly in the limit of M, N large. This model includes the dimer, ice and zero-field Ising, F and KDP models as special cases. In general the free energy has a branch point singularity at a phase transition, with an irrational exponent.

1. INTRODUCTION

There are very few models in statistical mechanics for which the partition function has been calculated exactly. The only models of multidimensional interacting systems that have been solved are certain two-dimensional lattice models. These can be classified into two types,

(a) those whose partition function can be expressed as a Pfaffian, notably the Ising, dimer and "free-fermion" models [1];

(b) the "ice-type" models which can be solved by a Bethe-type ansatz for the eigenvectors of the transfer matrix [2–4].

For the square lattice all of these (except for the ferroelectric models in the presence of electric fields) can be regarded as special cases of a more general zero-field "Eight-Vertex" model (c.f. [1], [5], and Appendix A of this paper). As previously reported [6], we have calculated the partition function of this model exactly in the limit of a large lattice. In this paper we present this calculation. The method is new, but is inspired by the results of the Bethe ansatz.

The model is defined in Section 2 and certain symmetry relations stated. The principal results are given in Sections 7 and 8. As far as possible, detailed working is left to the appendices.

2. DEFINITION OF THE MODEL

Consider a lattice of M rows (labelled $I = 1,..., M$) and N columns (labelled $J = 1,..., N$), with toroidal boundary conditions. Place arrows on the bonds of the

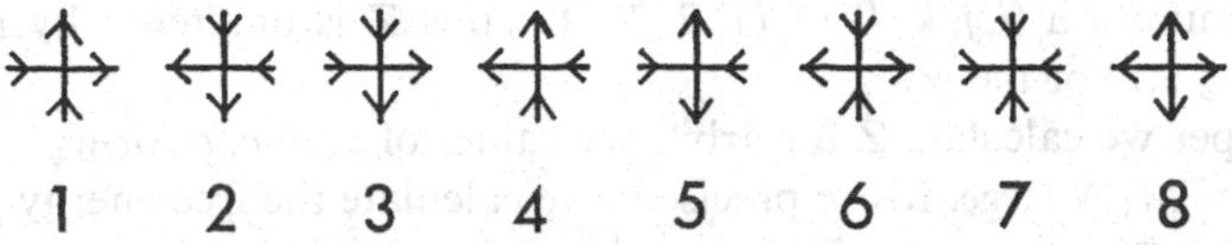

FIG. 1. The eight-arrow configurations allowed at a vertex.

lattice and allow only those configurations with an even number of arrows pointing into each vertex. Then there are eight possible configurations of arrows at each vertex (hence our name for the model), as shown in Fig. 1. Associating energies $\epsilon_1, \ldots, \epsilon_8$ with these vertex configurations, the problem is to evaluate the partition function

$$Z = \sum \exp\left(-\beta \sum_{j=1}^{8} N_j \epsilon_j\right), \tag{2.1}$$

where the summation is over all allowed configurations of arrows on the lattice and N_j is the number of vertices of type j.

Clearly vertices of type 7 and 8 occur in pairs, being sources and sinks of arrows. Similarly, reversing all horizontal arrows we see that vertices of type 5 and 6 occur in pairs. Thus there is no loss of generality in setting

$$\epsilon_5 = \epsilon_6\,, \qquad \epsilon_7 = \epsilon_8\,. \tag{2.2a}$$

We further require that

$$\epsilon_1 = \epsilon_2\,, \qquad \epsilon_3 = \epsilon_4\,, \tag{2.2b}$$

so that the model is unchanged by reversing all arrows (in ferroelectric terminology this implies no electric fields). We can then write the vertex weights

$$\omega_j = \exp(-\beta\epsilon_j) \tag{2.3}$$

as

$$\begin{aligned} \omega_1 = \omega_2 = a, &\qquad \omega_3 = \omega_4 = b \\ \omega_5 = \omega_6 = c, &\qquad \omega_7 = \omega_8 = d. \end{aligned} \tag{2.4}$$

A related set of quantities that we shall use are

$$\begin{aligned} w_1 = \tfrac{1}{2}(c + d), &\qquad w_2 = \tfrac{1}{2}(c - d), \\ w_3 = \tfrac{1}{2}(a - b), &\qquad w_4 = \tfrac{1}{2}(a + b). \end{aligned} \tag{2.5}$$

It has been pointed out [1] that Z satisfies a number of symmetry relations. These are particularly simple if we think of Z as a function of w_1, w_2, w_3, w_4, for then they become (taking M, N to be even)

$$Z(w_1, w_2, w_3, w_4) = Z(\pm w_i, \pm w_j, \pm w_k, \pm w_l) \tag{2.6}$$

for any permutation (i, j, k, l) of $(1, 2, 3, 4)$. Thus Z is unaltered by negating or interchanging any of the w's.

In this paper we calculate Z for arbitrary values of a, b, c, d (or w_1, w_2, w_3, w_4) in the limit of M, N large. More precisely, we calculate the free energy per vertex f of an infinite lattice, given by

$$-\beta f = \lim_{M\to\infty} \lim_{N\to\infty} (MN)^{-1} \ln Z. \tag{2.7}$$

3. Transfer Matrix

We use the transfer matrix method (c.f. [2–4]). Look at some particular row of vertical bonds in the lattice and let $\alpha_J = +$ or $-$ according as whether there is an up or down arrow in column J. Let $\boldsymbol{\alpha}$ denote the set $\{\alpha_1, ..., \alpha_N\}$, so that $\boldsymbol{\alpha}$ defines the configuration of arrows on the whole row of vertical bonds and has 2^N possible values. Suppose $\boldsymbol{\alpha}$, $\boldsymbol{\alpha}'$ correspond to the configurations of two successive rows and introduce the 2^N by 2^N transfer matrix $\mathbf{T}$, with elements

$$T_{\boldsymbol{\alpha}|\boldsymbol{\alpha}'} = \sum \exp\left(-\beta \sum_{j=1}^{8} n_j \epsilon_j\right), \tag{3.1}$$

where the sum is over allowed arrangements of arrows on the intervening row of horizontal bonds and n_j is the number of vertices of type j in this row. The partition function is then

$$\begin{aligned} Z &= \sum_{\boldsymbol{\alpha}_1} \cdots \sum_{\boldsymbol{\alpha}_M} T_{\boldsymbol{\alpha}_1|\boldsymbol{\alpha}_2} T_{\boldsymbol{\alpha}_2|\boldsymbol{\alpha}_3} \cdots T_{\boldsymbol{\alpha}_M|\boldsymbol{\alpha}_1}, \\ &= \mathrm{Tr}\{\mathbf{T}^M\}. \end{aligned} \tag{3.2}$$

Let $\lambda_J = +$ or $-$ according as whether there is a right- or left-pointing arrow on the horizontal bond between columns $J - 1$ and J. Then (3.1) can be written more explicitly as

$$T_{\boldsymbol{\alpha}|\boldsymbol{\alpha}'} = \sum_{\lambda_1} \cdots \sum_{\lambda_N} \prod_{J=1}^{N} R(\alpha_J, \alpha_J' \mid \lambda_J, \lambda_{J+1}), \tag{3.3}$$

where if α, α', λ, λ' specify the arrow configurations round a vertex as indicated in Fig. 2, then

$$R(\alpha, \alpha' \mid \lambda, \lambda') = 0 \quad \text{or} \quad \omega_j \tag{3.4}$$

according as whether this vertex configuration is not allowed, or is of type j.

Regarding the λ's as indices, (3.3) can in turn be written as

$$T_{\boldsymbol{\alpha}|\boldsymbol{\alpha}'} = \mathrm{Tr}\{\mathbf{R}(\alpha_1, \alpha_1')\, \mathbf{R}(\alpha_2, \alpha_2') \cdots \mathbf{R}(\alpha_N, \alpha_N')\}, \tag{3.5}$$

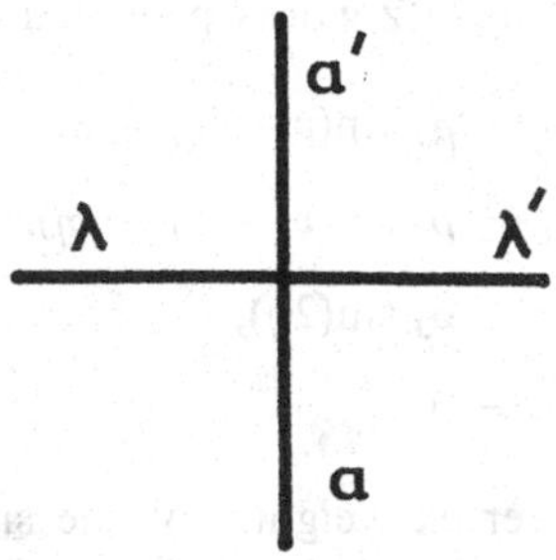

FIG. 2. The arrangement of the bond parameters α, α', λ, λ' round a vertex.

where the $\mathbf{R}(\alpha, \alpha')$ are 2 by 2 matrices

$$\mathbf{R}(+,+) = \begin{pmatrix} a & 0 \\ 0 & b \end{pmatrix}, \quad \mathbf{R}(+,-) = \begin{pmatrix} 0 & d \\ c & 0 \end{pmatrix},$$
$$\mathbf{R}(-,+) = \begin{pmatrix} 0 & c \\ d & 0 \end{pmatrix}, \quad \mathbf{R}(-,-) = \begin{pmatrix} b & 0 \\ 0 & a \end{pmatrix}. \tag{3.6}$$

An alternative formulation of R is to introduce the Pauli and unit matrices

$$\sigma^1 = \begin{pmatrix} 0 & 1 \\ 1 & 0 \end{pmatrix}, \quad \sigma^2 = \begin{pmatrix} 0 & -i \\ i & 0 \end{pmatrix},$$
$$\sigma^3 = \begin{pmatrix} 1 & 0 \\ 0 & -1 \end{pmatrix}, \quad \sigma^4 = \begin{pmatrix} 1 & 0 \\ 0 & 1 \end{pmatrix}. \tag{3.7}$$

From (2.4) and (3.6) we then find that

$$R(\alpha, \alpha' \mid \lambda, \lambda') = \sum_{j=1}^{4} w_j \sigma^j_{\alpha,\alpha'} \sigma^j_{\lambda,\lambda'}. \tag{3.8}$$

This formulation gives some insight into the symmetry properties (2.6).

4. ICE MODEL RESULTS

We now seek to diagonalize the transfer matrix $\mathbf{T}$, thereby making it a simple matter to calculate the partition function from (3.2). To do this we are guided by some recent results [7] for an inhomogeneous system satisfying the "ice" condition $d = 0$. With fairly extensive changes in notation, these results can be summarized as follows:

(i) A generalized Bethe ansatz works provided on each row

$$\begin{aligned} a_J &= \rho_J \sin(v - v_J'' + \eta), \\ b_J &= \rho_J \sin(v - v_J'' - \eta), \\ c_J &= \rho_J \sin(2\eta), \\ d_J &= 0, \end{aligned} \tag{4.1}$$

where a_J, b_J, c_J, d_J are the vertex weights for the site on column J, the ρ_J are arbitrary normalization factors, η and $v_1'',\ldots, v_N''$ are the same for each row, but v can vary from row to row. Thus we can regard η and the ρ_J, v_J'' as constants, v as a variable, and write the transfer matrix associated with such a row as $\mathbf{T}(v)$.

(ii) When these conditions are satisfied the transfer matrices for any two rows commute, i.e., $\mathbf{T}(u)$, $\mathbf{T}(v)$ commute for all values of u and v.

(iii) The eigenvalues $T(v)$ of $\mathbf{T}(v)$ are entire functions of v and are given by the identity

$$T(v)\, Q(v) = \phi(v - \eta)\, Q(v + 2\eta) + \phi(v + \eta)\, Q(v - 2\eta), \tag{4.2}$$

where

$$\phi(v) = \prod_{J=1}^{N} [\rho_J \sin(v - v_J'')] \tag{4.3}$$

and $Q(v)$ can be written in the form

$$Q(v) = \prod_{j=1}^{n} \sin(v - v_j), \tag{4.4}$$

where n is the number of down arrows in each row of vertical bonds [this quantity is conserved in each row, so $\mathbf{T}(v)$ breaks up into diagonal blocks corresponding to $n = 0, 1,\ldots, N$].

The identity (4.2) is sufficient to determine $T(v)$, since on setting $v = v_1,\ldots, v_n$ the l.h.s. vanishes, giving n equations for $v_1,\ldots, v_n$. These can in principle be solved (there will be many solutions, corresponding to the different eigenvalues). $Q(v)$ can then be calculated from (4.4), and then $T(v)$ from (4.2).

Equation (4.2) was obtained using the Bethe ansatz for the eigenvectors of $\mathbf{T}(v)$. Note however that as $\mathbf{T}(u)$, $\mathbf{T}(v)$ commute, there exists a representation (independent of v) in which $\mathbf{T}(v)$ is diagonal for all v. Let $\mathbf{Q}(v)$ be a matrix which is also diagonal in this representation, with corresponding diagonal elements $Q(v)$. Then (4.2) can be thought of as a matrix equation in this diagonal representation. Returning to the original representation, we see that there exists a matrix $\mathbf{Q}(v)$ which commutes with $\mathbf{T}(v)$ and whose elements are entire functions of v, such that

$$\mathbf{T}(v)\,\mathbf{Q}(v) = \phi(v-\eta)\,\mathbf{Q}(v+2\eta) + \phi(v+\eta)\,\mathbf{Q}(v-2\eta). \tag{4.5}$$

Thus instead of using the Bethe ansatz, we could attempt to construct $\mathbf{Q}(v)$ directly. This can be done: more generally, it can be done when $d \neq 0$ and the Bethe ansatz is not applicable. This is the method we use to solve the generalized problem.

For clarity we return to considering an homogeneous system (no variation of vertex weights from site to site) in the following working. The allowed extensions to inhomogeneous systems are straightforward and are outlined in Section 11.

5. Commuting Transfer Matrices

Guided by the results given in Section 4, we first consider under what circumstances two transfer matrices given by (3.3) and (3.8) (with different w's) commute. In Appendix B we show that they do so if

$$(w_j^2 - w_k^2)(w_l^2 - w_m^2) \tag{5.1}$$

is the same for both matrices, for all values 1, 2, 3, 4 of j, k, l and m. Thus a simple parametrization would be to set

$$w_j^2 = p(\xi - s_j), \qquad j = 1, 2, 3, 4. \tag{5.2}$$

All transfer matrices with the same values of s_1, s_2, s_3, s_4 would then commute, even if their values of p and ξ were different.

A more useful parametrization is derived in a natural way in Appendix B, namely,

$$w_1 : w_2 : w_3 : w_4 = \frac{\text{cn}(V, l)}{\text{cn}(\zeta, l)} : \frac{\text{dn}(V, l)}{\text{dn}(\zeta, l)} : 1 : \frac{\text{sn}(V, l)}{\text{sn}(\zeta, l)}, \tag{5.3}$$

where $\text{sn}(u, l)$, $\text{cn}(u, l)$, $\text{dn}(u, l)$ are the Jacobian elliptic functions of modulus l defined in §8.14 of Gradshteyn and Ryzhik [8], hereafter referred to as GR. (At this stage we need only consider ratios of the w's, and of a, b, c, d, since multiplying them by a normalization factor p simply multiplies the transfer matrix by p^N.)

Using the formulae

$$\begin{aligned} \text{sn}^2(u, l) + \text{cn}^2(u, l) &= 1, \\ l^2\,\text{sn}^2(u, l) + \text{dn}^2(u, l) &= 1, \end{aligned} \tag{5.4}$$

it is easy to verify that all the expressions (5.1) depend only on ζ and l. Thus if we regard ζ, l as constants and V as a variable, all transfer matrices with w's satisfying (5.3) commute for all values of V.

For the purposes of the next two sections it is convenient to work with elliptic functions of modulus

$$k = (1 - l)/(1 + l). \tag{5.5}$$

Defining v, η by

$$v = iV/(1 + k), \qquad \eta = i\zeta/(1 + k), \tag{5.6}$$

and using §§8.152, 8.153 of GR [8], we can verify from (2.4) and (5.3) that

$$\begin{aligned} a : b : c : d = {} & \mathrm{sn}(v + \eta, k) : \mathrm{sn}(v - \eta, k) : \mathrm{sn}(2\eta, k) : \\ & k\, \mathrm{sn}(2\eta, k)\, \mathrm{sn}(v - \eta, k)\, \mathrm{sn}(v + \eta, k). \end{aligned} \tag{5.7}$$

Note that when $k = 0$, (5.7) becomes the ice model parametrization (4.1) (with $v_J'' = 0$).

6. Matrix Equation for $\mathbf{T}(v)$

Throughout this section and in Appendices C and D, we work with elliptic functions of modulus k, given by (5.5). We therefore adopt the convention that any elliptic function or integral is to be interpreted as of this modulus, unless explicitly shown or stated to be otherwise.

From §8.191 of GR [8] we have the formula

$$\mathrm{sn}\, u = k^{-1/2} H(u)/\Theta(u), \tag{6.1}$$

where $H(u)$, $\Theta(u)$ are the elliptic theta functions. Thus from (5.7) we can write

$$\begin{aligned} a &= \rho\Theta(2\eta)\, \Theta(v - \eta)\, H(v + \eta), \\ b &= \rho\Theta(2\eta)\, H(v - \eta)\, \Theta(v + \eta), \\ c &= \rho H(2\eta)\, \Theta(v - \eta)\, \Theta(v + \eta), \\ d &= \rho H(2\eta)\, H(v - \eta)\, H(v + \eta), \end{aligned} \tag{6.2}$$

where ρ is some normalization constant.

Regarding ρ, k, η as constants and v as a variable, we can write the transfer matrix associated with the vertex weights (6.2) as $\mathbf{T}(v)$. Then from Section 5 we see that two matrices $\mathbf{T}(u)$, $\mathbf{T}(v)$ commute for any values of u and v. Further, since the theta functions are entire, all the elements of $\mathbf{T}(v)$ must be entire functions of v.

Thus far we have generalized part (ii) of Section 4. We now look for a matrix equation of the type (4.5).

In Appendix C we construct a 2^N by 2^N matrix $\mathbf{Q}(v)$ such that (4.5) is satisfied, only now we have

$$\phi(v) = [\rho\Theta(0)\, H(v)\, \Theta(v)]^N. \tag{6.3}$$

We also find that any two matrices $\mathbf{Q}(u)$, $\mathbf{Q}(v)$ commute, and that $\mathbf{Q}(v)$ commutes with $\mathbf{T}(v)$. Thus there exists a representation in which all these matrices are diagonal (for all v). Writing the diagonal elements of $\mathbf{T}(v)$, $\mathbf{Q}(v)$ in this representation (i.e., their eigenvalues) as $T(v)$, $Q(v)$, we regain the scalar equation (4.2).

We still need to generalize (4.4). To do this, first note that there are two elementary ways in which the matrix $\mathbf{T}(v)$, and hence $\mathbf{Q}(v)$, can be broken up into diagonal blocks or subspaces. We associate "quantum numbers" with these subspaces as follows:

$\nu' = 0$ or 1 if the number of down arrows in each row of vertical bonds is even or odd, respectively; (6.4)

$\nu'' = 0$ or 1 in the subspace symmetric or antisymmetric, respectively, with respect to reversing all arrows. (6.5)

Let K be the complete elliptic integral of the first kind of modulus k, and K' the same integral of the complementary modulus $k' = (1 - k^2)^{1/2}$ (§§8.110–8.112 of GR [8]). Let

$$q = \exp(-\pi K'/K). \tag{6.6}$$

Then the elliptic theta functions satisfy the quasiperiodic conditions

$$H(u + 2K) = -H(u), \qquad \Theta(u + 2K) = \Theta(u), \tag{6.7}$$

$$\begin{aligned} H(u + iK') &= iq^{-1/4} \exp(-\tfrac{1}{2}i\pi u/K)\, \Theta(u), \\ \Theta(u + iK') &= iq^{-1/4} \exp(-\tfrac{1}{2}i\pi u/K)\, H(u). \end{aligned} \tag{6.8}$$

Using these, it follows from Eqs. (C2), (C17), (C19) of Appendix C that (taking N to be even)

$$\begin{aligned} \mathbf{Q}(v + 2K) &= (-1)^{\nu'}\, \mathbf{Q}(v), \\ \mathbf{Q}(v + iK') &= (-1)^{\nu''}\, q^{-N/4} \exp(-\tfrac{1}{2}iN\pi v/K)\, \mathbf{Q}(v). \end{aligned} \tag{6.9}$$

It is also apparent from Appendix C that the elements of $\mathbf{Q}(v)$ are entire functions of v. Hence its eigenvalues are also entire functions and satisfy the conditions (6.9).

It follows that it must be possible to factorize them in the form

$$Q(v) = \exp(-\tfrac{1}{2}i\nu\pi v/K) \prod_{j=1}^{\frac{1}{2}N} \{H(v - v_j)\, \Theta(v - v_j)\}, \tag{6.10}$$

where the integer ν and the parameters $v_1, \ldots, v_{N/2}$ satisfy

$$\nu + \nu' + \tfrac{1}{2}N = (\text{even integer}), \tag{6.11}$$

$$K^{-1}\left\{\tfrac{1}{2}i\nu K' - \sum_{j=1}^{\frac{1}{2}N} v_j\right\} = \nu'' + \tfrac{1}{2}N + (\text{even integer}). \tag{6.12}$$

We can now in principle use (4.2), (6.3) and (6.10) to calculate the eigenvalues of the transfer matrix in the same way as for the ice-type models. Namely, we set $v = v_1, \ldots, v_{N/2}$ in (4.2): the l.h.s. vanishes from (6.10), giving $N/2$ equations for $v_1, \ldots, v_{N/2}$. If these can be solved then $Q(v)$ is given by (6.10) and $T(v)$ by (4.2).

It is interesting to note that if $4\eta = 2m_1K + im_2K'$, where m_1 and m_2 are integers, then the equations for $v_1, \ldots, v_{N/2}$ split up into $N/2$ independent equations and are easy to solve. This is the situation for the Ising, dimer and free-fermion models, all of which can alternatively be solved by the Pfaffian method.

7. Free Energy

From (2.7) and (3.2) we see that the free energy f of the infinite lattice is given by

$$-\beta f = \lim_{N\to\infty} N^{-1} \ln[T(v)]_{\max}, \tag{7.1}$$

where $[T(v)]_{\max}$ is the maximum eigenvalue of the transfer matrix $\mathbf{T}(v)$.

We consider the regime

$$w_1 > w_2 > w_3 > |\, w_4 \,|. \tag{7.2}$$

[Strictly speaking this regime is unphysical, since from (2.5) it implies that b is negative. However, negating b does not alter the partition function or the following discussion.] From (5.3) or Appendix B the parameters l, ζ, V are given by

$$l = [(w_1^2 - w_4^2)(w_2^2 - w_3^2)/(w_1^2 - w_3^2)(w_2^2 - w_4^2)]^{1/2}, \tag{7.3}$$

$$\mathrm{sn}(\zeta, l) = [(w_1^2 - w_3^2)/(w_1^2 - w_4^2)]^{1/2}, \tag{7.4}$$

$$\mathrm{sn}(V, l) = w_3^{-1} w_4\, \mathrm{sn}(\zeta, l). \tag{7.5}$$

Thus when (7.2) is satisfied we can choose l, ζ, V to be real and such that

$$0 < l < 1, \qquad |V| < \zeta < K_l\,, \tag{7.6}$$

where K_l is the complete elliptic integral of the first kind of modulus l.

When this is so we show in Appendix D that we can calculate the maximum eigenvalue of $\mathbf{T}(v)$ from the equation (4.2), using a perturbation-type technique that neglects certain terms that become relatively exponentially small as N becomes large. We find that there are actually two numerically largest eigenvalues, equal in magnitude but positive or negative according to whether the associated eigenvector is symmetric or antisymmetric with respect to reversing all arrows on the lattice (i.e., whether $\nu'' = 0$ or 1). This indicates that we are considering an ordered state of the antiferroelectric type, which agrees with the observation that in the regime (7.2) c is the numerically largest vertex weight. (Vertices of type 5 and 6 are favoured, as in the F-model of Lieb [3].)

Provided M is even this choice of sign does not affect (7.1), so using the result of Appendix D we obtain

$$-\beta f = \ln(w_1 + w_2) + 2 \sum_{n=1}^{\infty} \frac{\sinh^2[(\tau - \lambda)\, n]\{\cosh(n\lambda) - \cosh(n\alpha)\}}{n \sinh(2n\tau) \cosh(n\lambda)}\,, \tag{7.7}$$

where

$$\tau = \pi K_l / K_l', \qquad \lambda = \pi \zeta / K_l', \qquad \alpha = \pi V / K_l', \tag{7.8}$$

and K_l' is the complete elliptic integral of the first kind of modulus $l' = (1 - l^2)^{1/2}$.

For tunately there is no need to repeat the working for any other cases, since from the symmetry relations (2.6) we can always arrange w_1, w_2, w_3, w_4 so as to lie in the regime (7.2) (or its boundaries: as f is continuous these can be handled by taking an appropriate limit). Thus we have solved the general problem.

8. Phase Transitions

Suppose now that the vertex configuration energies $\epsilon_1, \ldots, \epsilon_8$ are fixed and satisfy (2.2), and we vary the temperature T from 0 to ∞. Then w_1, w_2, w_3, w_4, given by (2.5) and negated or arranged where necessary to satisfy (7.2), are entire functions of T. Further, so long as (7.2) is satisfied, the free energy f given by (7.7) is an entire function of the w's, and hence of T. A phase transition (i.e., a singularity in the function $f(T)$) can therefore occur only when the w's cross a boundary of the regime (7.2).

In general this will correspond to just two of the w's becoming numerically equal (if three are equal, or two pairs are equal, we have a more complicated situation which we do not discuss here: the F and KDP models [3, 4] are of this type).

Across the $w_1 = w_2$ or $w_3 = | w_4 |$ boundaries of the regime (7.2) we find that f is the same analytic function of T on either side, so there is no phase transition.

The behaviour when $w_2 - w_3$ becomes zero is discussed in Appendix E [and alternative expressions for f given which are more suitable near this boundary than (7.7)]. In general this will correspond to $w_2 - w_3$ having a simple zero at some value T_c of the temperature T. We find that f can be written as the sum of an analytic and a singular function of T. The analytic part is the same on both sides of T_c, while near T_c the singular part is proportional to

$$\cot(\tfrac{1}{2}\pi^2/\mu) \mid T - T_c \mid^{\pi/\mu}, \tag{8.1a}$$

or if $\pi/(2\mu) = m =$ integer, to

$$\pi^{-1}2(T - T_c)^{2m} \ln \mid T - T_c \mid, \tag{8.1b}$$

where

$$\mu = \pi\zeta/K_l\,. \tag{8.2}$$

The constant of proportionality which multiplies (8.1) is positive and is the same on both sides of T_c. From (7.6) we see that

$$0 < \mu < \pi. \tag{8.3}$$

Also, from (7.3) we see that $l = 0$ when $w_2 = w_3$, so $K_l = \pi/2$, $\mathrm{sn}(u, l) = \sin u$, and from (7.4) we find that at the transition temperature

$$\cos \mu = (2w_2^2 - w_1^2 - w_4^2)/(w_1^2 - w_4^2). \tag{8.4}$$

Thus the 8-vertex model undergoes a phase transition if the middle two w's, arranged in numerically decreasing order, cross one another as the temperature T is varied. (Unless $\pi/\mu =$ odd integer greater than one, in which case all singularities in f disappear.) This verifies the conjecture of Sutherland [5]. At the critical temperature the free energy has a branch point singularity, and it is interesting to note that the exponent π/μ of this singularity can be varied continuously (by appropriate choices of the interaction energies) from one to infinity. Thus it is not confined to any simple set of rational values. Rather it is the special cases previously solved that correspond to such special values of the exponent.

To see this, define three intermediate parameters A, B, C such that

$$\begin{aligned} A : B : C &= (-w_1^2 + w_2^2 + w_3^2 - w_4^2) : \\ &\quad (w_1^2 - w_2^2 + w_3^2 - w_4^2) : (w_1^2 + w_2^2 - w_3^2 - w_4^2) \\ &= -ab - cd : cd - ab : \tfrac{1}{2}(c^2 + d^2 - a^2 - b^2) \end{aligned} \tag{8.5}$$

[using (2.5)]. From (7.3) and (7.4) we see that

$$A : B : C = \operatorname{cn}(2\zeta, l) : \operatorname{dn}(2\zeta, l) : 1, \tag{8.6}$$

and when the restrictions (7.2) are satisfied

$$C/B > 1 > | A/B |. \tag{8.7}$$

Interchanging the w's simply rearranges A, B, C, and possibly negates two of them. Their ratios occur fairly naturally in the above working [e.g., in Eq. (C11)], and can be thought of as generalizations of the parameter Δ used by Lieb [3, 4] for the F and KDP models.

When one of the terms on the r.h.s. of (8.5) vanishes the 8-vertex model can be solved by Pfaffians. This is the case for the dimer, Ising and free-fermion models. Thus for such models one of the parameters A, B, C must be zero, and after rearranging the w's to satisfy (7.2) we see from (8.7) that we must have $A = 0$. From (8.6) it follows that $\operatorname{cn}(2\zeta, l) = 0$ and, using the restriction (7.6), $\zeta = \frac{1}{2}K_l$. From (8.2) we therefore see that $\mu = \pi/2$, so the singularity in the free energy is therefore of the type (8.1b) with $m = 1$. Clearly this is a very special case.

From (7.8) we also see that for the Pfaffian models $\lambda = \frac{1}{2}\tau$, so the denominator in the series in (7.7) simplifies. One can verify (after a rather lengthy calculation) that the expression (7.7) for the free energy is then the same as those previously obtained (c.f. [1, 9]).

For the ice-type models $d = 0$, so (8.5) can be written

$$A : B : C = 1 : 1 : \Delta, \tag{8.8}$$

where

$$\Delta = (a^2 + b^2 - c^2)/(2ab) \tag{8.9}$$

(this is the Δ used by Lieb). In addition we define

$$\Delta_1 = [(a^2 - b^2)^2 - 4c^2(a^2 + b^2)]/(8abc^2) \tag{8.10}$$

[this is obtained by inverting the w's in the definition (8.5), (8.8) of Δ, which from (5.3) is equivalent to interchanging V and ζ].

When $d = 0$ we see from (2.5) that $w_1 = w_2$, so it is impossible to arrange the w's to satisfy (7.2). Nevertheless we can arrange them to lie on a boundary of the region (7.2), and then take an appropriate limit in the formula (7.7) for the free energy. We find there are only three cases to consider, so for completeness we list the results,

(i) $$\Delta_1 > \Delta > 1 : l = 1, \qquad \tau = \lambda = \alpha = \infty,$$
$$\mu = U = \pi, \qquad \Delta = \cosh(\tau - \lambda),$$
$$\Delta_1 = \cosh(\tau - \alpha), \qquad -\beta f = \ln a. \tag{8.11}$$

(ii) $$-1 \leqslant \Delta_1 < \Delta < 1 : l = 0, \qquad \tau = 0,$$

$$\lambda = \alpha = 0, \qquad \Delta = -\cos \mu, \qquad \Delta_1 = -\cos U,$$

$$-\beta f = \ln \left\{ \frac{a+b+c}{2} \right\} + \int_{-\infty}^{\infty} dx \, \frac{\sinh^2[(\pi - \mu)\, x][\cosh(\mu x) - \cosh(Ux)]}{x \sinh(2\pi x) \cosh(\mu x)}. \tag{8.12}$$

(iii) $$-1 \geqslant \Delta_1 > \Delta : l = 1, \qquad \tau = \infty,$$

$$\mu = U = 0, \qquad \Delta = -\cosh \lambda, \qquad \Delta_1 = -\cosh \alpha,$$

$$-\beta f = \ln c + \sum_{n=1}^{\infty} \frac{e^{-2n\lambda}[\cosh(n\lambda) - \cosh(n\alpha)]}{n \cosh(n\lambda)}. \tag{8.13}$$

(The parameter U is the one used in Appendix E, and λ, α, μ, U must satisfy the inequalities $|\alpha| \leqslant \lambda \leqslant \tau$, $|U| \leqslant \mu \leqslant \pi$.) These results agree with those of Lieb [3, 4] for the F and KDP models.

Phase transitions occur when $\Delta = \pm 1$. Since more than two w's are then numerically equal, the formulae (8.1) no longer apply (see Lieb [3, 4] for discussions of the transition behaviour). Nevertheless it is interesting to note that $\mu = 0$ or π at such a transition, which again correspond to very special values of π/μ. Indeed for the F-model transition $\Delta = -1$, $\mu = 0$, and we can think of this as a limiting case in which the singularity (8.1) becomes of infinitely high order.

9. Generalized F-Model

As an example we consider a generalized F-model in which the interaction energies have the values

$$\begin{gathered} \epsilon_1 = \epsilon_2 = \epsilon_3 = \epsilon_4 = \epsilon > 0, \\ \epsilon_5 = \epsilon_6 = 0, \qquad \epsilon_7 = \epsilon_8 = \epsilon' > 0. \end{gathered} \tag{9.1}$$

Let

$$p = p(T) = \tfrac{1}{2}(1 + e^{-\epsilon'/kT}), \tag{9.2}$$

$$q = q(T) = \tfrac{1}{2}(1 - e^{-\epsilon'/kT}), \tag{9.3}$$

$$r = r(T) = e^{-\epsilon/kT}. \tag{9.4}$$

Then from (2.5) we see that

$$w_1, w_2, w_3, w_4 = p, q, 0, r. \tag{9.5}$$

Now arrange w_1, w_2, w_3, w_4 in numerically decreasing order. Since $p > q > 0$, there are at most three cases to consider, according as whether $r > p$, $p > r > q$, or $q > r$.

When $\epsilon' > 2\epsilon$ all three cases occur as T varies from 0 to ∞. Define two temperatures T_1, T_2 by

$$r(T_1) = q(T_1), \tag{9.6}$$

$$r(T_2) = p(T_2). \tag{9.7}$$

Then $0 < T_1 < T_2$ and the three cases are

(i) $0 < T < T_1 : q > r$,

$$w_1, w_2, w_3, w_4 = p, q, r, 0. \tag{9.8}$$

(ii) $T_1 < T < T_2 : p > r > q$,

$$w_1, w_2, w_3, w_4 = p, r, q, 0. \tag{9.9}$$

(iii) $T > T_2 : r > p$,

$$w_1, w_2, w_3, w_4 = r, p, q, 0. \tag{9.10}$$

In each case the restrictions (7.2) are satisfied and the free energy $f(T)$ can be calculated from Eqs. (7.3)–(7.7). The function $f(T)$ is analytic inside each region and across the $T = T_2$ boundary (largest two w's crossing). At $T = T_1$ the middle two w's cross, so there is a phase transition and $f(T)$ has a branch point singularity of the form (8.1), with exponent π/μ. At this temperature T_1 we see from (8.4) that μ is given by

$$\cos \mu = (2q^2 - p^2)/p^2. \tag{9.11}$$

When $\epsilon' \leqslant 2\epsilon$, $p > r$ at all temperatures, so the region (iii) no longer occurs. Nevertheless there is still a transition at the temperature T_1 defined by (9.6), with μ given by (9.11).

When $\epsilon' = \infty$ we regain the original F-model [3]. If we vary ϵ' from 0 to ∞, then T_1 decreases from ∞ to the F-model transition temperature, while μ decreases from π to 0. Thus again we see that we can think of the F-model transition as a limiting case in which $\mu \to 0$, giving an infinitely high-order singularity. When $\epsilon' = 2\epsilon$ we find that μ has the Pfaffian value $\pi/2$, and indeed in this case the model becomes the modified F-model of Wu [9], which is soluble by the Pfaffian method.

It is intriguing to note that for certain special values of ϵ', namely, those corresponding to π/μ being an odd integer greater than one, the phase transition disappears, $f(T)$ being analytic at T_1.

10. Inhomogeneous Systems

As a final point we mention that we can handle a restricted class of inhomogeneous 8-vertex models by the above methods (in the same way as the Bethe ansatz can be used for a restricted class of ice-type models).

Suppose a, b, c, d, and hence w_1, w_2, w_3, w_4, can vary from site to site of the lattice. At each site define l, ζ, V, and hence k, η, v, by (5.3), (5.7) or (6.2). Note that the interaction between columns of the lattice enters the above working only via the matrices **R**, **P**, **X**, **Y** of Appendices B and C, and that these depend only on l, ζ (or k, η) and the difference between two V's or v's. Using this, we find that we can generalize the results of these Appendices provided

(i) The parameters l, ζ, and hence k, η, are the same for each site of the lattice.

(ii) There exist parameters $V_1',\ldots, V_M'$, $V_1'',\ldots, V_N''$ such that at the site (I, J) V has the value

$$V = V_I' - V_J'', \tag{10.1}$$

and similarly, using (5.6),

$$v = v_I' - v_J''. \tag{10.2}$$

The normalization factors ρ in (6.2) can be varied arbitrarily from site to site. However, the effect of this is a trivial variation in the normalization of the partition function, so without loss of generality we can regard ρ as a constant. Then V_I', or v_I', is the only parameter that can vary from row to row, so we can write the transfer matrix of row I as $\mathbf{T}(v_I')$.

All such transfer matrices commute, and again we can construct **Q** so that (4.5) is satisfied, only now

$$\phi(v) = \prod_{J=1}^{N} \{\rho\Theta(0)\, H(v - v_J'')\, \Theta(v - v_J'')\}. \tag{10.3}$$

We can still use the technique of Appendix D to sum a perturbation expansion about the purely ordered F-model state in the limit of M and N large. We find that the total free energy F of the lattice is given by

$$\begin{aligned} -\beta F &= \ln Z, \\ &= -\beta \sum_{I=1}^{M} \sum_{J=1}^{N} f_{I,J}, \end{aligned} \tag{10.4}$$

where $f_{I,J}$ is given by (7.7), using the parameters w_1, w_2, α appropriate to the site (I, J). This is a remarkably simple result, and it is certainly true when each α is not large compared with unity and λ, $\tau - \lambda$ are sufficiently large (i.e., the system is sufficiently close to the purely ordered state). However, since $z_1, \ldots, z_r$ of Appendix D no longer lie on the unit circle it is not clear under what precise conditions the perturbation expansion is convergent and the result (10.4) is valid.

11. Summary

In the absence of fields the interaction energies of the 8-vertex model can be chosen to satisfy (2.2). The free energy f in the limit of a large lattice is then given by the following procedure:

(i) Calculate w_1, w_2, w_3, w_4 from (2.3)–(2.5).

(ii) Replace each w_j by its absolute value and arrange them in nonincreasing order, so that

$$w_1 \geqslant w_2 \geqslant w_3 \geqslant w_4 \geqslant 0.$$

(iii) Calculate l, ζ, V from (7.3)–(7.6).

(iv) Calculate τ, λ, α from (7.8).

(v) Calculate f from (7.7). Alternatively, if the series in (7.7) converges only slowly it may be more convenient to calculate f from equation (E6) of Appendix E.

In general there is a phase transition when, and only when, the middle two w's cross. At this transition the free energy has the branch point singularity (8.1), with exponent π/μ defined by (8.4). This exponent can be varied continuously (by varying the interaction energies) from one to infinity. It is not confined to any simple set of rational values.

It would be very interesting to solve the 8-vertex model in the presence of fields, i.e., to remove the restriction (2.2b), since then one could solve the staggered F-model [10] and the Ising model with first- and second-neighbour interactions. Unfortunately we have not been able to do this. In this respect our method appears at a disadvantage compared with the Pfaffian and Bethe ansatz techniques, which when applicable can be used in the presence of fields.

Appendix A

Clearly the zero-field F, KDP and free-fermion models are special cases of the zero-field eight-vertex model. The same is true of the nearest-neighbour Ising model and the dimer model, as we now show.

The required transformation of the Ising model is discussed by Wu [9]. For completeness we also give the argument here. Consider a square lattice with spins $\sigma_{i,j} = \pm 1$ on sites (i,j), with total energy

$$E = -\sum_i \sum_j \{J\sigma_{i,j}\sigma_{i+1,j+1} + J'\sigma_{i,j}\sigma_{i+1,j-1}\}. \tag{A1}$$

Thus we allow only second-neighbour interactions (crossed bonds). However, as there is no interaction between sites on the A sublattice ($i+j$ even) and sites on the B sublattice ($i+j$ odd), the partition function factorizes into the product of the partition functions of the A and B sublattices. On each of these sublattices we have an Ising system with *first*-neighbour interactions J, J', so that in the limit of a large lattice

$$f_{\text{crossed bonds}} = f_{\text{nearest neighbour}}\,, \tag{A2}$$

where f is the free energy per spin.

Now transform to the dual of the complete lattice, so the spins are associated with faces, rather than sites. Draw right- or up-pointing arrows on bonds between adjacent faces with the same spin, and left- or down-pointing arrows between adjacent faces with opposite spins. Then only the arrow configurations shown in Fig. 1 are allowed at a vertex. With each such vertex configuration we associate the interaction energy of the spins on the faces surrounding the vertex. The ϵ_j of Section 2 are then given by

$$\begin{aligned} \epsilon_1 = \epsilon_2 = -J - J', \qquad \epsilon_3 = \epsilon_4 = J + J', \\ \epsilon_5 = \epsilon_6 = J' - J, \qquad \epsilon_7 = \epsilon_8 = J - J'. \end{aligned} \tag{A3}$$

Noting that there is a simple 2 to 1 correspondence between spin orientations on the faces of the lattice and allowed configurations of arrows on the bonds, for a large lattice it follows that

$$f = f_{\text{Ising}}\,, \tag{A4}$$

where f is the free energy per vertex of the zero-field 8-vertex model specified by (A3), and f_{Ising} is the free energy per spin of an Ising model with first-neighbour interactions J, J'.

To express the close-packed dimer problem on the square lattice as an 8-vertex model, draw the lattice diagonally and add to it "superbonds" as shown in Fig. 3. At each vertex of the original lattice draw an arrow on the superbond passing through it, pointing right or left (up or down) according to whether the dimer at that vertex lies right or left of (above or below) that vertex.

Each vertex of the superbond lattice is surrounded by four bonds of the original lattice. Considering possible dimer arrangements on these four bonds, we see that only the arrow configurations 1, 2, 3, 4, 7, 8 of Fig. 1 can occur at a superbond

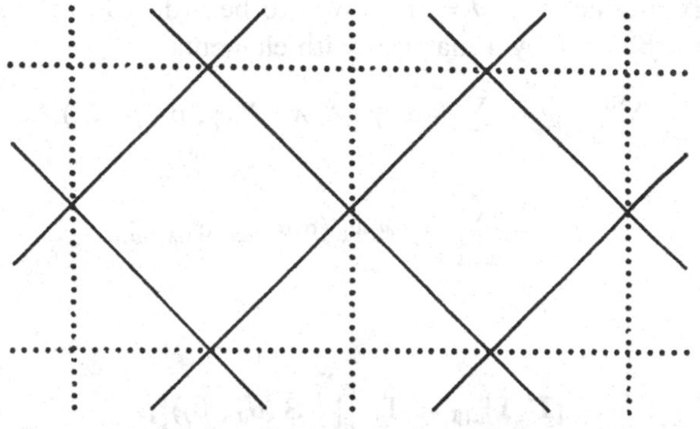

FIG. 3. The lattice transformation of Appendix A. A dimer covering of the original lattice (solid lines) is equivalent to an eight-vertex model on the superbond lattice (dotted lines).

vertex. With each such configuration we associate as a weight the product of the activities of the dimers on the surrounding four bonds. Noting that there are two dimer arrangements corresponding to configuration 7, this gives

$$\omega_1 = \omega_2 = z, \qquad \omega_3 = \omega_4 = z', \tag{A5}$$

$$\omega_5 = \omega_6 = 0, \tag{A6}$$

$$\omega_7 = z^2 + z'^2, \qquad \omega_8 = 1, \tag{A7}$$

where z, z' are the activities of dimers on the horizontal and vertical bonds, respectively, of the original lattice.

Clearly vertices of type 7 and 8 occur in pairs (being sinks and sources of arrows). Thus the partition function is unaffected if we replace (A7) by

$$\omega_7 = \omega_8 = (z^2 + z'^2)^{1/2}. \tag{A8}$$

Summing over allowed arrow configurations on the superbond lattice with these weights is equivalent to summing over dimer coverings of the original lattice, so we have reduced the dimer problem to a zero-field 8-vertex model.

APPENDIX B

Here we obtain the conditions under which two transfer matrices commute. Let $\mathbf{T}$ be the transfer matrix defined by (3.3) and (3.8), and let $\mathbf{T}'$ be similarly defined, but with the w_j replaced by w_j'. Then from (3.3)

$$[\mathbf{T}\,\mathbf{T}']_{\alpha|\beta} = \mathrm{Tr}\left\{\prod_{J=1}^{N} \mathbf{S}(\alpha_J, \beta_J)\right\}, \tag{B1}$$

where the matrix product over $J = 1, \ldots, N$ is to be ordered in the same way as in (3.5), and the $\mathbf{S}(\alpha, \beta)$ are 4 by 4 matrices with elements

$$\begin{aligned} S^{\alpha,\beta}_{\lambda,\mu|\lambda',\mu'} &= \sum_{\gamma} R(\alpha, \gamma \mid \lambda, \lambda')\, R'(\gamma, \beta \mid \mu, \mu') \\ &= \sum_{j=1}^{4} \sum_{k=1}^{4} w_j w_k' (\boldsymbol{\sigma}^j \boldsymbol{\sigma}^k)_{\alpha,\beta}\, \sigma^j_{\lambda,\lambda'} \sigma^k_{\mu,\mu'} \end{aligned} \tag{B2}$$

[using (3.8)].

Similarly,

$$[\mathbf{T}'\,\mathbf{T}]_{\alpha|\beta} = \operatorname{Tr} \left\{ \prod_{J=1}^{N} \mathbf{S}'(\alpha_J, \beta_J) \right\}, \tag{B3}$$

where the $\mathbf{S}'(\alpha, \beta)$ are given by interchanging the primed and unprimed w's in (B2). If we also interchange λ with μ, and λ' with μ' [this has no effect on (B3)], and j with k, the elements of $\mathbf{S}'(\alpha, \beta)$ are

$$\sum_{j=1}^{4} \sum_{k=1}^{4} w_j w_k' (\boldsymbol{\sigma}^k \boldsymbol{\sigma}^j)_{\alpha,\beta}\, \sigma^j_{\lambda,\lambda'} \sigma^k_{\mu,\mu'} . \tag{B4}$$

For $\mathbf{T}$ and $\mathbf{T}'$ to commute, the right-hand sides of (B1), (B3) must be the same. This will be so if there exists a 4 by 4 nonsingular matrix $\mathbf{R}$ such that

$$\mathbf{S}'(\alpha, \beta) = \mathbf{R}\mathbf{S}(\alpha, \beta)\, \mathbf{R}^{-1}, \tag{B5}$$

or, more conveniently,

$$\mathbf{S}'(\alpha, \beta)\mathbf{R} = \mathbf{R}\mathbf{S}(\alpha, \beta), \tag{B6}$$

for $\alpha = \pm$ and $\beta = \pm$.

Some inspection shows that if such a matrix $\mathbf{R}$ exists it must have elements of the form

$$R_{\lambda,\mu|\lambda',\mu'} = \sum_{j=1}^{4} x_j \sigma^j_{\lambda,\lambda'} \sigma^j_{\mu,\mu'} . \tag{B7}$$

Substituting the forms (B2), (B4), (B7) into (B6), performing the matrix multiplications, we find that (B6) is satisfied provided

$$w_m w_l' x_j - w_l w_m' x_k + w_k w_j' x_l - w_j w_k' x_m = 0 \tag{B8}$$

for all permutations (j, k, l, m) of (1, 2, 3, 4).

There are six such equations (B8). Regarding them as linear homogeneous equations for x_1, x_2, x_3, x_4, we find that they have a nontrivial solution provided

$$(w_j^2 - w_k^2)/(w_l^2 - w_m^2) = (w_j'^2 - w_k'^2)/(w_l'^2 - w_m'^2) \tag{B9}$$

for all values 1, 2, 3, 4 of j, k, l and m. Thus these are the conditions for **T** and **T**′ to commute. There are only two such independent conditions, and an obvious parametrization that satisfies (B9) is to set

$$w_j^2 = p(\xi - s_j), \qquad w_j'^2 = p'(\xi' - s_j) \tag{B10}$$

for $j = 1,\ldots, 4$.

When (B9) is satisfied, we can verify from (B8) that

$$(x_j^2 - x_k^2)/(x_l^2 - x_m^2) = (w_j^2 - w_k^2)/(w_l^2 - w_m^2) \tag{B11}$$

for all values of j, k, l, m. Thus we can also introduce p'', ξ'' such that

$$x_j^2 = p''(\xi'' - s_j). \tag{B12}$$

Clearly p, p', p'' are arbitrary normalization factors that cancel out of (B8). Regarding s_1, s_2, s_3, s_4 as constants, any of the equations (B8) can be regarded as defining ξ'' as a function of ξ and ξ'. Differentiating, we find that

$$\frac{\partial \xi''}{\partial \xi} \Big/ \frac{\partial \xi''}{\partial \xi'} = \frac{-g(\xi)}{g(\xi')}, \tag{B13}$$

where

$$g(\xi) = \text{constant} \times \prod_{j=1}^{4} (\xi - s_j)^{-1/2}. \tag{B14}$$

The factorization of the r.h.s. of (B13) is interesting, since it implies that if we transform from ξ, ξ' to new variables V, V' such that

$$dV/d\xi = g(\xi), \qquad dV'/d\xi' = g(\xi'), \tag{B15}$$

then ξ'' will be a function only of $V - V'$. (This transformation is related to the transformation to a difference kernel that is used in the Bethe ansatz for the ice models.) The integrations in (B15) can be performed using elliptic functions (§3.147 of GR [8]), and with an appropriate choice of the constant in (B14) we find that

$$\xi = [(s_3 - s_1)\, s_4 - s_3(s_4 - s_1)\, \mathrm{sn}^2(V, l)]/[s_3 - s_1 - (s_4 - s_1)\, \mathrm{sn}^2(V, l)], \tag{B16}$$

where $\mathrm{sn}(u, l)$ is the elliptic sine-amplitude function (§8.143 of GR [8]) of modulus

$$l = [(s_3 - s_2)(s_4 - s_1)/(s_4 - s_2)(s_3 - s_1)]^{1/2}. \tag{B17}$$

(We implicitly consider the case $\xi > s_4 > s_3 > s_2 > s_1$, when V is real and $0 < l < 1$. However, the parametrization is not limited to this region.) Adding primes to ξ and V in (B16), we also obtain the relation between ξ' and V'.

Substituting the expression (B16) into (B10) and defining a parameter ζ such that

$$\operatorname{sn}(\zeta, l) = [(s_3 - s_1)/(s_4 - s_1)]^{1/2}, \tag{B18}$$

it follows that we can choose the w's so that

$$w_1 : w_2 : w_3 : w_4 = \frac{\operatorname{cn}(V, l)}{\operatorname{cn}(\zeta, l)} : \frac{\operatorname{dn}(V, l)}{\operatorname{dn}(\zeta, l)} : 1 : \frac{\operatorname{sn}(V, l)}{\operatorname{sn}(\zeta, l)}, \tag{B19}$$

where $\operatorname{cn}(u, l)$, $\operatorname{dn}(u, l)$ are the other Jacobian elliptic functions (§8.14 of GR [8]). This is the parametrization (5.3) quoted in the text.

Note that l, ζ are defined by s_1, s_2, s_3, s_4, but V depends also on ξ. Thus any two transfer matrices with the same values of ζ and l, but different values of V, must commute.

APPENDIX C

Using the parametrization (5.7), or more explicitly (6.2), of the vertex weights, we look for a matrix $\mathbf{Q}(v)$ which satisfies an equation of the form (4.5).

As a first step, we attempt to construct a 2^N by 2^N matrix $\mathbf{Q}_R$ such that

$$\mathbf{T}\mathbf{Q}_R = \text{sum of two } Q\text{-type matrices}. \tag{C1}$$

From the form (3.5) of $\mathbf{T}$, we see that the most general matrix $\mathbf{Q}_R$ that we can conveniently handle is one with elements

$$[\mathbf{Q}_R]_{\alpha|\beta} = \operatorname{Tr}\left\{\prod_{J=1}^{N} \mathbf{S}(\alpha_J, \beta_J)\right\}, \tag{C2}$$

where the $\mathbf{S}(\alpha, \beta)$ are matrices of some dimension L, and the matrix product over $J = 1,\ldots, N$ is to be ordered in the same way as in (3.5).

From (3.5), (3.6) and (C2), we see that

$$[\mathbf{T}\,\mathbf{Q}_R]_{\alpha|\beta} = \operatorname{Tr}\left\{\prod_{J=1}^{N} \mathbf{U}(\alpha_J, \beta_J)\right\}, \tag{C3}$$

where the $\mathbf{U}(\alpha, \beta)$ are $2L$ by $2L$ matrices and

$$\begin{aligned} \mathbf{U}(+, \beta) &= \begin{pmatrix} a\mathbf{S}(+, \beta) & d\mathbf{S}(-, \beta) \\ c\mathbf{S}(-, \beta) & b\mathbf{S}(+, \beta) \end{pmatrix}, \\ \mathbf{U}(-, \beta) &= \begin{pmatrix} b\mathbf{S}(-, \beta) & c\mathbf{S}(+, \beta) \\ d\mathbf{S}(+, \beta) & a\mathbf{S}(-, \beta) \end{pmatrix}. \end{aligned} \tag{C4}$$

The r.h.s. of (C3) will decompose into the sum of two matrices if we can find a $2L$ by $2L$ matrix **M** (independent of α and β) such that

$$\mathbf{M}^{-1}\mathbf{U}(\alpha, \beta)\,\mathbf{M} = \begin{pmatrix} \mathbf{A}(\alpha, \beta) & \mathbf{0} \\ \mathbf{C}(\alpha, \beta) & \mathbf{B}(\alpha, \beta) \end{pmatrix}, \tag{C5}$$

since substituting this form for $\mathbf{U}(\alpha, \beta)$ into (C3) gives

$$\mathbf{TQ}_R = \mathbf{H}_1 + \mathbf{H}_2, \tag{C6}$$

where

$$[\mathbf{H}_1]_{\alpha|\beta} = \mathrm{Tr}\left\{\prod_{J=1}^{N} \mathbf{A}(\alpha_J, \beta_J)\right\},$$

$$[\mathbf{H}_2]_{\alpha|\beta} = \mathrm{Tr}\left\{\prod_{J=1}^{N} \mathbf{B}(\alpha_J, \beta_J)\right\}. \tag{C7}$$

The form of the requirement (C5) is unaffected by postmultiplying **M** by a lower blocktriangular matrix, so we can in general choose

$$\mathbf{M} = \begin{pmatrix} \mathbf{E} & \mathbf{P} \\ \mathbf{0} & \mathbf{E} \end{pmatrix}, \tag{C8}$$

where **E** is the identity matrix and **P** is some L by L matrix. Further, (C2) and the form of (C5) are unaffected by applying a similarity transformation to the $\mathbf{S}(\alpha, \beta)$ and **P**, so we can in general choose **P** to be diagonal, with elements

$$P_{m,n} = p_m \delta_{m,n} \qquad (m, n = 1, \ldots, L). \tag{C9}$$

Writing the elements of $\mathbf{S}(\alpha, \beta)$ as $s_{m,n}^{\alpha,\beta}$, using (C4), (C8), (C9), we find that the upper right block of the matrix on the l.h.s. of (C5) vanishes provided

$$\begin{aligned} (ap_n - bp_m)\, s_{m,n}^{+,\beta} + (d - cp_m p_n)\, s_{m,n}^{-,\beta} &= 0, \\ (c - dp_m p_n)\, s_{m,n}^{+,\beta} + (bp_n - ap_m)\, s_{m,n}^{-,\beta} &= 0 \end{aligned} \tag{C10}$$

for $m = 1, \ldots, L$ and $n = 1, \ldots, L$.

For given values of m, n, β, (C10) is a pair of homogeneous linear equations. They have a nontrivial solution provided the determinant of the coefficients vanishes, i.e.,

$$(a^2 + b^2 - c^2 - d^2)\, p_m p_n = ab(p_m^2 + p_n^2) - cd(1 + p_m^2 p_n^2). \tag{C11}$$

This can only happen for certain values of m and n. For all other values we must have

$$s_{m,n}^{\alpha,\beta} = 0. \tag{C12}$$

Using the parametrization (5.7) of a, b, c, d we find that

$$\frac{a^2+b^2-c^2-d^2}{ab} = 2\,\mathrm{cn}(2\eta)\,\mathrm{dn}(2\eta), \qquad cd/ab = k\,\mathrm{sn}^2(2\eta) \tag{C13}$$

(note that these quantities are independent of v). It follows that if $p_m = k^{1/2}\,\mathrm{sn}\,u$, then (C11) can be solved to give $p_n = k^{1/2}\,\mathrm{sn}(u \pm 2\eta)$. This suggests a very natural ordering of $p_1, \ldots, p_L$, namely,

$$p_m = k^{1/2}\,\mathrm{sn}[K + (2m-1)\eta], \qquad m = 1,\ldots, L, \tag{C14}$$

where K is the complete elliptic integral of the first kind of modulus k.

We have chosen (C14) so that (formally) $p_0 = p_1$. Thus (C11) is satisfied for $m = n = 1$ as well as $n = m \pm 1$. Hence the matrices $\mathbf{S}(\alpha, \beta)$ can have nonzero elements in the upper left diagonal and one-off diagonal positions. We can also require that the bottom right diagonal element be nonzero, which implies that p_{L+1} given by (C14) be such that $p_{L+1} = p_L$. Using the periodicity and oddness properties of sn u (§8. 151of GR [8]) this will be so if there exist integers m_1, m_2 such that

$$2L\eta = 2m_1K + im_2K', \tag{C15}$$

where K' is the complete elliptic integral of the first kind of modulus $k' = (1-k^2)^{1/2}$.

Clearly (C15) is a restriction on the possible values of η. However, as the positive integer L is arbitrary, we can approach arbitrarily close to any desired value.

We can now solve (C10) and (C5) for $\mathbf{S}(\alpha, \beta)$, $\mathbf{A}(\alpha, \beta)$, $\mathbf{B}(\alpha, \beta)$. All these matrices are of the form

$$\begin{pmatrix} z_0 & z_{-1} & 0 & 0 & \cdot & 0 \\ z_1 & 0 & z_{-2} & 0 & \cdot & \cdot \\ 0 & z_2 & 0 & \cdot & \cdot & \cdot \\ 0 & 0 & \cdot & \cdot & \cdot & 0 \\ \cdot & \cdot & \cdot & \cdot & 0 & z_{1-L} \\ 0 & \cdot & \cdot & 0 & z_{L-1} & z_L \end{pmatrix}. \tag{C16}$$

Using (6.2), we find the following values of z_m ($m = 1-L,\ldots, L$):

$$\begin{aligned} \mathbf{S}(\alpha,\beta) &: z_m = q(\alpha, \beta, m \mid v), \\ \mathbf{A}(\alpha,\beta) &: z_m = \rho\Theta(0)\,H(v-\eta)\,\Theta(v-\eta)(x_m/x_{m+1})\,q(\alpha,\beta,m \mid v+2\eta), \\ \mathbf{B}(\alpha,\beta) &: z_m = \rho\Theta(0)\,H(v+\eta)\,\Theta(v+\eta)(x_{m+1}/x_m)\,q(\alpha,\beta,m \mid v-2\eta), \end{aligned} \tag{C17}$$

where

$$x_m = \Theta[K + (2m - 1)\eta] \tag{C18}$$

and the functions $q(\alpha, \beta, m \mid v)$ of v are defined by

$$\begin{aligned} q(+, \beta, m \mid v) &= H(v + K + 2m\eta)\, \tau_{\beta,m}\,, \\ q(-, \beta, m \mid v) &= \Theta(v + K + 2m\eta)\, \tau_{\beta,m}\,, \end{aligned} \tag{C19}$$

the parameters $\tau_{\beta,m}$ being arbitrary. To derive (C17) we have used (6.1), the fact that sn u and $H(u)$ are odd functions while $\Theta(u)$ is an even function, and the general formulae

$$\begin{aligned} \operatorname{sn} A \operatorname{sn} B - \operatorname{sn} C \operatorname{sn} D &= \frac{\Theta(0)\, \Theta(A + B)\, H(A - D)\, H(B - D)}{k\Theta(A)\, \Theta(B)\, \Theta(C)\, \Theta(D)}\,, \\ 1 - k^2 \operatorname{sn} A \operatorname{sn} B \operatorname{sn} C \operatorname{sn} D &= \frac{\Theta(0)\, \Theta(A + B)\, \Theta(A - D)\, \Theta(B - D)}{\Theta(A)\, \Theta(B)\, \Theta(C)\, \Theta(D)}\,, \end{aligned} \tag{C20}$$

when $A + B = C + D$.

From (C17) it is apparent that we can regard $\mathbf{S}(\alpha, \beta)$ as a function of v. Exhibiting this dependence explicitly by writing the matrix as $\mathbf{S}(\alpha, \beta \mid v)$, we see that

$$\begin{aligned} \mathbf{A}(\alpha, \beta) &= \rho\Theta(0)\, H(v - \eta)\, \Theta(v - \eta)\, \mathbf{X}^{-1}\mathbf{S}(\alpha, \beta \mid v + 2\eta)\, \mathbf{X}, \\ \mathbf{B}(\alpha, \beta) &= \rho\Theta(0)\, H(v + \eta)\, \Theta(v + \eta)\, \mathbf{X}\mathbf{S}(\alpha, \beta \mid v + 2\eta)\, \mathbf{X}^{-1}, \end{aligned} \tag{C21}$$

where $\mathbf{X}$ is an L by L diagonal matrix with elements $x_m \delta_{m,n}$.

Clearly $\mathbf{T}$ and $\mathbf{Q}_R$ are also functions of v. Exhibiting this dependence explicitly, substituting the expressions (C21) into (C7) and using (C2), the equation (C6) becomes

$$\mathbf{T}(v)\, \mathbf{Q}_R(v) = \phi(v - \eta)\, \mathbf{Q}_R(v + 2\eta) + \phi(v + \eta)\, \mathbf{Q}_R(v - 2\eta), \tag{C22}$$

where

$$\phi(v) = \{\rho\Theta(0)\, H(v)\, \Theta(v)\}^N. \tag{C23}$$

[The matrices $\mathbf{X}$ cancel out of (C7).]

Commutation Relations

We have obviously come a long way towards finding the generalization of (4.5). It remains to construct a matrix $\mathbf{Q}(v)$ which satisfies (C22) and also commutes with $\mathbf{T}(v)$.

To do this we first repeat the above working, replacing $\mathbf{Q}_R$ by $\mathbf{Q}_L$ and forming the product $\mathbf{Q}_L\mathbf{T}$ in (C3). The effect of this is to interchange c and d in (C4) and (C10), but the matrix $\mathbf{P}$ remains the same. We find that

$$\mathbf{Q}_L(v)\,\mathbf{T}(v) = \phi(v-\eta)\,\mathbf{Q}_L(v+2\eta) + \phi(v+\eta)\,\mathbf{Q}_L(v-2\eta), \tag{C24}$$

where

$$[Q_L(v)]_{\alpha|\beta} = \mathrm{Tr}\left\{\prod_{J=1}^{N} \mathbf{S}'(\alpha_J, \beta_J \mid v)\right\} \tag{C25}$$

and the $\mathbf{S}'(\alpha, \beta \mid v)$ are L by L matrices of the form (C16) with

$$\begin{aligned} \mathbf{S}'(\alpha, + \mid v) : z_m &= \Theta(v - K - 2m\eta)\,\tau'_{\alpha,m}\,, \\ \mathbf{S}'(\alpha, - \mid v) : z_m &= H(v - K - 2\eta)\,\tau'_{\alpha,m}\,, \end{aligned} \tag{C26}$$

the parameters $\tau'_{\alpha,m}$ being arbitrary.

Now note that (C22), (C24) are unaffected by post- and pre-multiplying $\mathbf{Q}_R(v)$, $\mathbf{Q}_L(v)$, respectively, by constant matrices. Thus to construct a matrix $\mathbf{Q}(v)$ which commutes with $\mathbf{T}(v)$, it is sufficient to show that there exist nonsingular constant matrices $\mathbf{F}$, $\mathbf{G}$ such that

$$\mathbf{Q}(v) = \mathbf{Q}_R(v)\,\mathbf{F} = \mathbf{G}\mathbf{Q}_L(v). \tag{C27}$$

We also expect any two matrices $\mathbf{Q}(u)$, $\mathbf{Q}(v)$ to commute. From (C27) this will be so if

$$\mathbf{Q}_L(u)\,\mathbf{Q}_R(v) = \mathbf{Q}_L(v)\,\mathbf{Q}_R(u). \tag{C28}$$

We can prove this relation. From (C2) and (C25) we see that

$$[\mathbf{Q}_L(u)\,\mathbf{Q}_R(v)]_{\alpha|\beta} = \mathrm{Tr}\left\{\prod_{J=1}^{N} \mathbf{W}(\alpha_J, \beta_J \mid u, v)\right\}, \tag{C29}$$

where the $\mathbf{W}(\alpha, \beta \mid u, v)$ are L^2 by L^2 matrices, with elements

$$[\mathbf{W}(\alpha, \beta \mid u, v)]_{m,m'|n,n'} = \sum_{\gamma=\pm} [\mathbf{S}'(\alpha, \gamma \mid u)]_{m,n}\,[\mathbf{S}(\gamma, \beta \mid v)]_{m',n'}\,. \tag{C30}$$

Using (C17), (C19), (C26) and the identity

$$H(A)\,\Theta(B) + \Theta(A)\,H(B) = f(A+B)\,g(A-B), \tag{C31}$$

where $f(u)$, $g(u)$ are functions such that

$$g(u) = g(-u) = -g(u+4K), \tag{C32}$$

we find that

$$[\mathbf{W}(\alpha, \beta \mid u, v)]_{m,m'|n,n'} = y_{m,m'}[\mathbf{W}(\alpha, \beta \mid v, u)]_{m,m'|n,n'}/y_{n,n'}, \tag{C33}$$

where

$$y_{m,m'} = t_{m+m'}t_{m-m'+1} \tag{C34}$$

and the t_m are defined by the recurrence relation

$$t_m/t_{m+2} = g(u - v + 2K + 2m\eta)/g(v - u + 2K + 2m\eta). \tag{C35}$$

Equation (C33) can be written in matrix form as

$$\mathbf{W}(\alpha, \beta \mid u, v) = \mathbf{Y}\mathbf{W}(\alpha, \beta \mid v, u)\, \mathbf{Y}^{-1}, \tag{C36}$$

where $\mathbf{Y}$ is a diagonal L^2 by L^2 matrix with elements $y_{m,m'}\delta_{m,n}\delta_{m',n'}$. Substituting the expression (C36) into (C29), the matrices $\mathbf{Y}$ cancel and we obtain the required identity (C28).

Although we have not been able to construct a proof, it appears that the matrices $\mathbf{Q}_R(v)$, $\mathbf{Q}_L(v)$ are in general nonsingular (certainly this is so for $N = 1$ and 2). Assuming this, it follows from (C28) that we can define

$$\mathbf{Q}(v) = \mathbf{Q}_R(v)\, \mathbf{Q}_R^{-1}(v_0) = \mathbf{Q}_L^{-1}(v_0)\, \mathbf{Q}_L(v), \tag{C37}$$

where v_0 is some fixed value of v. From (C22), (C24) and (C28) it then follows that

$$\begin{aligned} \mathbf{T}(v)\,\mathbf{Q}(v) &= \phi(v-\eta)\,\mathbf{Q}(v+2\eta) + \phi(v+\eta)\,\mathbf{Q}(v-2\eta), \\ \mathbf{T}(v)\,\mathbf{Q}(v) &= \mathbf{Q}(v)\,\mathbf{T}(v), \\ \mathbf{Q}(u)\,\mathbf{Q}(v) &= \mathbf{Q}(v)\,\mathbf{Q}(u), \end{aligned} \tag{C38}$$

for all values of v and u. These are the required generalizations of the equations for the ice-type models.

Note that if the eigenvalues of $\mathbf{Q}(v)$ are nondegenerate (as appears to be the case), then (C38) is sufficient to establish that any two transfer matrices $\mathbf{T}(u)$, $\mathbf{T}(v)$ commute. Thus in this sense Appendix B is redundant. Nevertheless we have included it, partly for completeness but also because it provides a natural way of starting the calculation.

Appendix D

We here use (4.2), (6.3), (6.10) to obtain the maximum eigenvalue $T(v)$ of $\mathbf{T}(v)$ in the limit of N large. We could employ the method used by Yang [11] and Lieb [2-4] for the Bethe ansatz, namely to look at the equations for $v_1, \ldots, v_{N/2}$, assume that they tend to a continuous distribution along some line segment in the complex plane, and thus obtain an integral equation for the distribution function. Instead we adopt a perturbation expansion approach similar to that used by Baxter [7], which we believe makes some of the properties of the equation (4.2) more transparent. In addition, it can be applied to the inhomogeneous system discussed at the end of this paper.

We consider the regime

$$w_1 > w_2 > w_3 > |w_4|. \tag{D1}$$

In this case it follows from (5.3) (or Appendix B) that we can choose l, V, ζ so that

$$0 < l < 1, \qquad |V| < \zeta < K_l, \tag{D2}$$

K_l being the complete elliptic integral of the first kind of modulus l. Using (5.5), (5.6) and the formulae (§8.126 of GR [8])

$$K_l = \tfrac{1}{2}(1+k)\,K_k{}', \qquad K_l{}' = (1+k)\,K_k, \tag{D3}$$

we can therefore define real parameters τ, λ, α such that

$$\tau = \pi K_k{}'/(2K_k) = \pi K_l/K_l{}', \tag{D4}$$

$$\lambda = -i\pi\eta/K_k = \pi\zeta/K_l{}', \tag{D5}$$

$$\alpha = -i\pi v/K_k = \pi V/K_l{}', \tag{D6}$$

and

$$0 < k < 1, \qquad |\alpha| < \lambda < \tau. \tag{D7}$$

We further define

$$q = e^{-2\tau}, \qquad x = e^{-\lambda}, \qquad z = e^{-\alpha}, \tag{D8}$$

so from (D7) we see that

$$q < x^2 < 1, \tag{D9}$$

$$x < z < x^{-1}, \tag{D10}$$

These parameters occur quite naturally when we expand the elliptic theta functions of modulus k in (6.3) and (6.10) as infinite products, using the general formulae (§8.181 of GR [8])

$$H(v) = 2i\gamma q^{1/4}\sinh(\tfrac{1}{2}\alpha)\prod_{n=1}^{\infty}(1-q^{2n}z)(1-q^{2n}z^{-1}),$$

$$\Theta(v) = \gamma\prod_{n=1}^{\infty}(1-q^{2n-1}z)(1-q^{2n-1}z^{-1}), \tag{D11}$$

where α, q, z are defined in terms of k and v by (D6) and (D8), and

$$\gamma = \prod_{n=1}^{\infty} (1 - q^{2n}). \tag{D12}$$

We find that it is convenient also to define the quantities

$$\xi = i\rho\Theta(0)\, \gamma^2 q^{1/4} x^{-1}, \tag{D13}$$

$$r = \tfrac{1}{2}N, \tag{D14}$$

$$z_j = \exp(i\pi v_j/K_k), \qquad j = 1,\ldots, r. \tag{D15}$$

(We take N to be even, so r is an integer.)

First Order Approximation

As a first step consider the extreme case $q \ll x^2 \ll 1$ and suppose that $|\, z_j \,| \sim 1$ for $j = 1,\ldots, r$. Then the condition (6.12) implies that the integer ν is zero and gives the *exact* equation

$$z_1 \cdots z_r = (-1)^{r+\nu''}. \tag{D16}$$

Although v must be pure imaginary for the vertex weights a, b, c, d to be real, the equation (4.2) is an algebraic identity, valid for all complex values of v, and hence z. Using (6.3), (6.10), (D11) to expand the terms in (4.2) as infinite products, we find that when $|\, z \,| \sim 1$ each of the products on the r.h.s. is dominated by a single term. Retaining only such dominant terms, the equation (4.2) becomes

$$T(v) \prod_{j=1}^{r} (z - z_j) \simeq \xi^N \{(-1)^r z_1 \cdots z_r z^r + 1\}. \tag{D17}$$

This equation is an identity, true for all values of z close to the unit circle. As $T(v)$ is an entire function, it follows that the r.h.s. is a polynomial of degree r, with zeros $z = z_1 ,\ldots, z_r$. Thus

$$\prod_{j=1}^{r} (z - z_j) \simeq z^r + (-1)^r/(z_1 \cdots z_r). \tag{D18}$$

Equating final coefficients on both sides of (D18), we see that

$$(z_1 \cdots z_r)^2 = 1. \tag{D19}$$

This agrees with the exact equation (D16). There are two solutions, corresponding to which square root of (D19) we take, i.e., to whether $\nu'' = 0$ or 1 in (D16). We also note that (D18) implies that $z_1 ,\ldots, z_r$ lie on the unit circle, as was assumed.

The polynomials on both sides of (D17) therefore cancel. Using (D16) we obtain

$$T(v) \simeq (-1)^{\nu''} \xi^N. \tag{D20}$$

Having located $z_1, \ldots, z_r$, we can now let v, and hence z, assume any value in (4.2). There are many cases to be considered, corresponding to which terms dominate in the various product expansions, but in particular we find that $T(v)$ is given by (D20) when

$$\max[x^2, q^{1/2}] < |z| < \min[x^{-2}, q^{-1/2}]. \tag{D21}$$

Restricting attention to real values of a, b, c, d, and therefore z, we also find that when (D10) is satisfied, $c \simeq \xi$ and c is the dominant vertex weight. The energetically favoured configurations of the lattice are therefore the two completely ordered F-model states, in which vertices on one sublattice are of type 5, those on the other of type 6. Thus in the region (D10), which is contained in (D21), the above working does indeed give the two largest eigenvalues (D20) of $\mathbf{T}(v)$, equal in magnitude and opposite in sign.

Outside (D10) the maximum eigenvalue is not given by (D20). Since all matrices $\mathbf{T}(v)$ commute, and hence have the same eigenvectors, it follows that the eigenvectors which correspond to the largest eigenvalues of $\mathbf{T}(v)$ inside (D10) do not do so outside, and vice versa. This does not violate any theorems of the Frobenius type, since at the boundaries of (D10) $v = \pm\eta$. Thus from (5.7) we see that two of the weights a, b, c, d become zero and $\mathbf{T}(v)$ becomes a matrix with mostly zero elements, breaking up into many diagonal blocks. The restriction (D10) on the validity of the formula for the maximum eigenvalue is therefore a very natural one.

Perturbation Expansion

We now implicitly consider a perturbation expansion about the solution (D20) and show that if we neglect certain terms that tend to zero as N becomes large, then we can solve (4.2) for $T(v)$.

To do this we define the functions

$$A(z) = \prod_{n=0}^{\infty} (1 - q^n z)^N, \tag{D22}$$

$$F(z) = \prod_{j=1}^{r} \prod_{n=0}^{\infty} (1 - q^n z/z_j), \tag{D23}$$

$$G(z) = \prod_{j=1}^{r} \prod_{n=0}^{\infty} (1 - q^n z_j/z). \tag{D24}$$

[Thus $A(z)$ is a known function. The functions $F(z)$, $G(z)$ are to be determined.] Using the expansions (D11), we can then write (4.2) exactly as

$$(-1)^r (z_1 \cdots z_r)^{-1} T(v) F(qz) G(q^{-1}z) \prod_{j=1}^{r} (z - z_j)$$
$$= \xi^N A(x^{-1}qz) A(x^{-1}q/z) F(x^2 z) G(x^{-2} z) \Gamma(z), \tag{D25}$$

where

$$\Gamma(z) = z^r A(x/z)\, G(q^{-1}x^2z)/[A(x^{-1}q/z)\, G(x^{-2}z)]$$
$$+ (-1)^r (z_1 \cdots z_r)^{-1} A(xz)\, F(x^{-2}qz)/[A(x^{-1}qz)\, F(x^2z)]. \tag{D26}$$

We assume again that $|z_j| = 1$, $j = 1,..., r$. Then when $|z| = 1$ and the restrictions (D9) are satisfied, we can develop convergent expansions of the functions on the r.h.s. of (D26) in increasing powers of x and $x^{-2}q$. Neglecting terms of order x^r, $x^{-2r}q^r$, or smaller, we find that $\Gamma(z)$ is a polynomial of degree r, with leading coefficient unity. Further, from (D25) we see that $\Gamma(z)$ has zeros $z = z_1 ,..., z_r$, so to this order of approximation we have the identity

$$\prod_{j=1}^{r} (z - z_j) \equiv \Gamma(z). \tag{D27}$$

Thus the parameters $z_1 ,..., z_r$ can in principle be calculated by equating coefficients on both sides of (D27). In particular, the last such equation gives (D19), in agreement with the exact result (D16). We therefore again have two solutions, corresponding to whether we take $\nu'' = 0$ or 1 in (D16).

From (D27) we see that the last terms on both sides of (D25) cancel. Using (D16), we obtain

$$T(v) = (-1)^{\nu''} \xi^N [A(x^{-1}qz)\, A(x^{-1}q/z)\, F(x^2z)\, G(x^{-2}z)/F(qz)\, G(q^{-1}z)]. \tag{D28}$$

Rather than evaluate $z_1 ,..., z_r$ directly, we see from (D28) that it is sufficient to determine $F(z)$ for $|z| \leqslant x^2$ and $G(z)$ for $|z| \geqslant x^{-2}$. We can do this from (D26) and (D27).

Since (D27) is an identity, it must be true for all values of z, so long as we replace $\Gamma(z)$ by its truncated expansion, or ensure that any terms neglected, or extra terms introduced, are negligible. When $|z| \geqslant x^{-2}$, to order x^N only the first term on the r.h.s. of (D26) contributes to this expansion. Using (D24), we can then write (D27) as

$$G(z)/G(q^{-1}z) = A(x/z)\, G(q^{-1}x^2z)/[A(x^{-1}q/z)\, G(x^{-2}z)]. \tag{D29}$$

This can be thought of as a recurrence relation for $G(z)$. It is greatly simplified if we note that it is equivalent to

$$H(z) = H(q^{-1}x^2z), \tag{D30}$$

where

$$H(z) = G(z)\, G(x^{-2}z)/A(x/z). \tag{D31}$$

The function $H(z)$ is Laurent expandable in inverse powers of z with leading term unity, and $q^{-1}x^2 > 1$. Expanding both sides of (D30) and equating coefficients, it follows that

$$H(z) = 1. \tag{D32}$$

Thus from (D31) the recurrence relation for $G(z)$ simplifies to

$$G(z) = A(x/z)/G(x^{-2}z). \tag{D33}$$

Solving this by iteration, using the fact that $G(z) \to 1$ as $z \to \infty$, we obtain

$$G(z) = \prod_{m=0}^{\infty} \frac{A(x^{4m+1}/z)}{A(x^{4m+3}/z)} \tag{D34}$$

provided $| z | \geqslant x^{-2}$.

Similarly, when $| z | \leqslant x^2$ only the second term on the r.h.s. of (D26) is significant and from (D23) and (D27) we can deduce that

$$F(z) = \prod_{m=0}^{\infty} \frac{A(x^{4m+1}z)}{A(x^{4m+3}z)} \cdot \tag{D35}$$

Substituting these forms for the functions $G(z)$ and $F(z)$ into (D28), using (D22) and noting from (6.1), (6.2), (D11) and (D13) that

$$\begin{aligned} c/\xi = \prod_{n=1}^{\infty} \{&(1 - q^{2n-1})^2 (1 - q^{2n-2}x^2)(1 - q^{2n}x^{-2}) \\ &\times (1 - q^{2n-1}xz)(1 - q^{2n-1}x^{-1}z)(1 - q^{2n-1}xz^{-1})(1 - q^{2n-1}x^{-1}z^{-1})\}, \end{aligned} \tag{D36}$$

we find that $(-1)^{\nu''} c^{-N}T(v)$ is a product of terms of the form $1 - \alpha$, where $| \alpha | < 1$. Taking logarithms and Taylor expanding each $\ln(1 - \alpha)$ term, the various m and n summations can be performed to give

$$N^{-1} \ln[(-1)^{\nu''} T(v)] = \ln c + \sum_{j=1}^{\infty} \frac{x^{-j}(x^{2j} - q^j)^2 (x^j + x^{-j} - z^j - z^{-j})}{j(1 - q^{2j})(1 + x^{2j})} . \tag{D37}$$

Using the definitions (D8), this is the result quoted in the text.

We must verify the assumption that $z_1, \ldots, z_r$ lie in the unit circle. To do this, note that each z_j is a zero of $\Gamma(z)$. Using the forms (D34) and (D35) of $G(z)$ and $F(z)$ in (D26), it follows that $z_1, \ldots, z_r$ are roots of the equation

$$Z^r = -(-1)^{\nu''} \prod_{m=0}^{\infty} \left(\frac{(1 - x^{4m+1}z)(1 - x^{4m+3}z^{-1})}{(1 - x^{4m+3}z)(1 - x^{4m+1}z^{-1})} \right)^N . \tag{D38}$$

(All terms explicitly involving q cancel out.) This equation can be written quite nearly in terms of elliptic functions of modulus k_1, with associated elliptic integrals K_1, K_1' such that

$$x = \exp(-\pi K_1'/K_1). \tag{D39}$$

Define u so that

$$z = \exp(i\pi u/K_1), \tag{D40}$$

then (D38) becomes

$$\exp\{iN \operatorname{am}(u, k_1)\} = -(-1)^{\nu''}, \tag{D41}$$

where $\operatorname{am}(u, k_1)$ is the elliptic amplitude function of argument u and modulus k_1 (§8.141 of GR [8]). Thus when $z = z_j$, $\operatorname{am}(u, k_1)$ is real. Hence u is real and z_j lies on the unit circle, as assumed.

In deriving (D37) we have neglected only terms of order x^r, $x^{-2r}q^r$, or smaller terms. Thus in the limit of N and r large we expect it to be exact, provided (D9) is satisfied. Also, although we took $|z| = 1$ in obtaining (D37), having located $z_1, \ldots, z_r$ this condition can be relaxed and we find that $T(v)$ is given by (D37) throughout the region (D21). However, from the above discussion we expect it to give the largest eigenvalues of $\mathbf{T}(v)$ only in the restricted region (D10).

To summarize: the numerically largest eigenvalues $T(v)$ of the transfer matrix $\mathbf{T}(v)$ are given by (D37), provided the restrictions (D9) and (D10) are satisfied. These restrictions are equivalent to (D7), to (D2), and to (D1).

Appendix E

We here discuss the behaviour of the free energy f, given by (7.7), across the boundary $w_2 = w_3$ of the regime (7.2).

When $w_2 \to w_3$ we see from (7.3)–(7.5) that $l \to 0$, so that $K_l \to \pi/2$, $K_l' \to +\infty$, and ζ and V tend in general to finite nonzero values. From (7.8) it follows that $\tau, \lambda, \alpha \to 0$, and hence in this limit the summation in (7.7) becomes an integral.

This suggests that to consider the behaviour about this limit we should make a Poisson transformation of the series in (7.7). Define

$$\mu = \pi\zeta/K_l, \qquad U = \pi V/K_l, \tag{E1}$$

$$g(x) = \sinh^2[(\pi - \mu)\, x]\{\cosh(\mu x) - \cosh(Ux)\}/x \sinh(2\pi x) \cosh(\mu x), \tag{E2}$$

$$h(s) = \int_{-\infty}^{\infty} e^{isx} g(x)\, dx. \tag{E3}$$

Noting that $g(x)$ is an even function and $g(0) = 0$, the equation (7.7) can be written as

$$-\beta f = \ln(w_1 + w_2) + \pi^{-1}\tau \sum_{n=-\infty}^{\infty} g(\pi^{-1}\tau n). \tag{E4}$$

Expressing $g(x)$ in terms of its Fourier transform $h(s)$ by inverting the relation (E3) and substituting the result in (E4), the summation and integration can be interchanged. The summation can then be performed to give a periodic delta function. Performing the integration then gives

$$\begin{aligned} -\beta f &= \ln(w_1 + w_2) + \sum_{m=-\infty}^{\infty} h(2\pi^2\tau^{-1}m) \\ &= \ln(w_1 + w_2) + h(0) + 2\sum_{m=1}^{\infty} h(2\pi K_l'm/K_l) \end{aligned} \tag{E5}$$

[using (7.8) and the evenness of $h(s)$].

When s is positive the integration in (E3) can be closed round the upper halfplane, giving $h(s)$ as a sum of residues. Substituting the result into (E5), the summation over m can be performed for each residue, giving

$$\begin{aligned} -\beta f = \psi + h(0) &- 2\sum_{n=1}^{\infty} \frac{q^{2n}}{1-q^{2n}} \frac{\sin^2(n\mu)[\cos(n\mu) - \cos(nU)]}{n\cos(n\mu)} \\ &+ 4\sum_{n=1}^{\infty} \frac{(-)^n\, q^{(2n-1)\pi/\mu}}{1-q^{(2n-1)\pi/\mu}} \frac{1}{2n-1} \\ &\times \cot\left(\frac{(2n-1)\,\pi^2}{2\mu}\right) \cos\left\{\frac{(2n-1)\,\pi U}{2\mu}\right\}, \end{aligned} \tag{E6}$$

where

$$q = \exp(-\pi K_l'/K_l) \tag{E7}$$

and

$$\begin{aligned} \psi = \ln(w_1 + w_2) &+ 4\sum_{n=1}^{\infty} \frac{q^{2n-1}}{(2n-1)[1-q^{2n-1}]} \cos\left(\frac{2n-1}{2}\mu\right) \\ &\times \left\{\cos\left(\frac{2n-1}{2}\mu\right) - \cos\left(\frac{2n-1}{2}U\right)\right\}. \end{aligned} \tag{E8}$$

(Note that this q is *not* the same as the q of Appendix D, being defined in terms of the modulus l, rather than k.)

The series in (E8) can be summed, using the Fourier series for ln[sn u/cn u dn u] (§8.146 of GR [8]). After some manipulation, using (5.3), we find that

$$\psi = \frac{1}{2}\ln\left\{\frac{\pi}{2l'^{1/2}K_l}\tan\left(\frac{1}{2}\mu\right)\frac{\cos(\frac{1}{2}U)+\cos(\frac{1}{2}\mu)}{\cos(\frac{1}{2}U)-\cos(\frac{1}{2}\mu)}\right\}$$

$$+\frac{1}{8}\ln\{(w_1^2-w_2^2)(w_1^2-w_3^2)(w_2^2-w_4^2)(w_3^2-w_4^2)\}. \qquad \text{(E9)}$$

As $q \to 0$ when $w_2 \to w_3$, the remaining series in (E6) are uniformly convergent in some neighbourhood of $w_2 = w_3$. Also, from the definitions (7.3)–(7.5) we can verify that l', K_l, q, μ, U are analytic functions of the w's in some such neighbourhood, that q has a simple zero at $w_2 = w_3$, but the other parameters are nonzero. Using (E3) and (E9), it follows that the first three terms on the r.h.s. of (E6) are analytic in this neighbourhood, but the last is a sum of functions with branch-point singularities, the dominant singular contribution to βf near $w_2 = w_3$ being

$$(\beta f)_{\text{sing}} = 4\cos(\pi U/2\mu)\cot(\pi^2/2\mu)\, q^{\pi/\mu}. \qquad \text{(E10)}$$

As q has a simple zero at $w_2 = w_3$, this singular term is proportional to

$$(w_2 - w_3)^{\pi/\mu}. \qquad \text{(E11)}$$

This result breaks down if $\pi/\mu = 2m$, where m is an integer. In this case some of the poles of the integrand of (E3) coincide and we find that

$$(\beta f)_{\text{sing}} = \pi^{-1}\, 8\cos(mU)\, q^{2m}\ln q, \qquad \text{(E12)}$$

which is proportional to

$$(w_2 - w_3)^{2m}\ln(w_2 - w_3). \qquad \text{(E13)}$$

To summarize the working so far: the free energy f is given by the above equations when the restrictions (7.2) are satisfied, in particular when $w_2 > w_3$. The above equations also define an analytic continuation of f across the $w_2 = w_3$ boundary, and we see that this function has a branch-point singularity.

However, when $w_2 < w_3$ the true free energy is not given by the above equations. Rather we see from the symmetry relations (2.6) that it is given by the above equations when w_2 and w_3 are interchanged in the definitions (7.3)–(7.5). Let us denote any variable obtained by interchanging w_2 and w_3 in these equations by a suffix 1. Then we must still compare the analytic continuation of f into the region $w_2 < w_3$ with the true free energy f_1 in this region. For instance the analytic parts of these functions might be different, leading to a discontinuity in the free energy or its derivatives at $w_2 = w_3$. Such a discontinuity might be more important in the critical region than the branch point singularity already found.

From (7.3) we see that $l' = (1 - l^2)^{1/2}$ is given by

$$l' = [(w_1^2 - w_2^2)(w_3^2 - w_4^2)/(w_1^2 - w_3^2)(w_2^2 - w_4^2)]^{1/2}. \tag{E14}$$

Interchanging w_2 and w_3 simply inverts the r.h.s. of this equation. Thus

$$l_1' = 1/l'. \tag{E15}$$

From §§8.126, 8.128 of GR [8], it follows that

$$K_{l_1} = l'K_l, \tag{E16}$$

$$K'_{l_1} = l'(K_l' + iK_l), \tag{E17}$$

and hence from (E7) that

$$q_1 = -q. \tag{E18}$$

From §§8.152.1, 8.153 of GR [8], we can also deduce the transformation formula

$$\mathrm{sn}(l'u, l_1) = l' \,\mathrm{sn}(u, l)/\mathrm{dn}(u, l). \tag{E19}$$

From (7.3) and (7.4),

$$\begin{aligned} l' \,\mathrm{sn}(\zeta, l)/\mathrm{dn}(\zeta, l) &= [(w_1^2 - w_2^2)/(w_1^2 - w_4^2)]^{1/2}, \\ &= \mathrm{sn}(\zeta_1, l_1). \end{aligned} \tag{E20}$$

Hence

$$\zeta_1 = l'\zeta, \tag{E21}$$

and similarly we can show that

$$V_1 = l'V. \tag{E22}$$

From these results and (E1) it follows that

$$l_1'^{1/2}K_{l_1} = l'^{1/2}K_l, \qquad \mu_1 = \mu, \qquad U_1 = U. \tag{E23}$$

Using (E18) and (E23), together with (E2), (E3), (E9), we see that the first three terms on the r.h.s. of (E6) are unaltered by interchanging w_2 and w_3. Thus the analytic part of the true free energy is in fact the same function of w_1, w_2, w_3, w_4 on both sides of the $w_2 = w_3$ boundary and there are no discontinuities of the type mentioned above.

We also see from (E18) and (E23) that the dominant singular contribution to the free energy on *either* side of the $w_2 = w_3$ boundary is given by (E10) or (E12), with q replaced by $|q|$. Thus if $w_2 - w_3$, and hence q, has a simple zero at some value T_c of the temperature T, the singularity is of the form (8.1) of the text.

Acknowledgments

The author thanks Professor E. H. Lieb for stimulating his interest in the above problem, Professor K. J. Le Couteur for his encouragement in this work, and Professor F. Y. Wu for drawing the author's attention to the paper by B. Sutherland.

References

1. C. Fan and F. Y. Wu, *Phys. Rev.* **B2** (1970), 723–733.
2. E. H. Lieb, *Phys. Rev.* **162** (1967), 162–172.
3. E. H. Lieb, *Phys. Rev. Lett.* **18** (1967), 1046–1048.
4. E. H. Lieb, *Phys. Rev. Lett.* **19** (1967), 108–110.
5. B. Sutherland, *J. Math. Phys.* **11** (1970), 3183–3186.
6. R. J. Baxter, *Phys. Rev. Lett.* **26** (1971), 832–834.
7. R. J. Baxter, *Stud. Appl. Math.* (Mass. Inst. of Technology) **50** (1971), 51–69.
8. I. S. Gradshteyn and I. M. Ryzhik, "Tables of Integrals, Series and Products," pp. 241–242 and 909–925, Academic Press, New York, 1965.
9. F. Y. Wu, *Phys. Rev.* **183** (1969), 604–607.
10. J. F. Nagle, *J. Chem. Phys.* **50** (1969), 2813–2818.
11. C. N. Yang and C. P. Yang, *Phys. Rev.* **150** (1966), 327–339.

[315]

SOLVABLE EIGHT-VERTEX MODEL ON AN ARBITRARY PLANAR LATTICE

By R. J. BAXTER
Research School of Physical Sciences, The Australian National University, Canberra ACT. 2600 Australia

(*Communicated by C. Domb, F.R.S. – Received* 23 *September* 1977)

CONTENTS

Any planar set of intersecting straight lines forms a four-coordinated graph, or 'lattice', provided no three lines intersect at a point. For any such lattice an eight-vertex model can be constructed. Provided the interactions satisfy certain constraints (which are in general temperature-dependent), the model can be solved exactly in the thermodynamic limit, its local properties at a particular site being those of a related square lattice.

A particular case is a solvable model on the Kagomé lattice. It is shown that this model includes as special cases many of the models in statistical mechanics that have been solved exactly, notably the square, triangular and honeycomb Ising models, and the square eight-vertex model.

Some remarkable equivalences between correlations on different lattices are also established.

1. Introduction

There are a number of two-dimensional statistical mechanical models of interacting systems for which the free energy has been evaluated exactly in the thermodynamic limit. In particular, the following models have been solved in the absence of magnetic or electric fields:

(i) the translation-invariant Ising model on the square lattice (Onsager 1944);

(ii) the translation-invariant Ising model on the triangular or honeycomb lattice (Houtappel 1950; Husimi & Syozi 1950; Wannier 1950; Stephenson 1964);

(iii) the ice-type ferroelectric models on the square lattice (Lieb 1967; Sutherland 1967);

(iv) the eight-vertex model on the square lattice (Baxter 1972);

(v) the three-spin model on the triangular lattice (Wood & Griffiths 1972; Baxter & Wu 1974);

(vi) an ice-type model on the triangular lattice satisfying certain temperature-dependent restrictions (Baxter 1969; Kelland 1974); this has recently been shown to be equivalent to a restricted ice-type model on the Kagomé lattice (Baxter, Temperley & Ashley 1977). Both (iii) and (vi) are also equivalent to Potts models at their transition temperatures, on the square and triangular lattices, respectively (see also Baxter, Kelland & Wu 1976).

From a mathematical point of view, all these models have two common features. One is that their solution leads sooner or later to the introduction of elliptic functions. (In the ice-type models these functions occur in the distribution of the wavenumbers $k_1, \ldots, k_n$ (Baxter 1971, appendix)). The other common feature is that they can all be solved by some appropriately generalized Bethe ansatz (see for example, Baxter 1973).

This suggests that there may be some more general model which includes all of (i) to (vi) as special cases. Considerable progress has already been made in this direction, for by construction (iv) contains (i) and (iii) as special cases, and has also been shown to include (v) (Baxter & Enting 1976). However, it does not include (ii) or (vi).

In this paper a general model including (i)–(vi) is presented. If some reasonable assumptions are made concerning the thermodynamic limit, then the properties of the model can be obtained directly from the known or conjectured results for (iv).

It should be noted that there are very few exactly solved two-dimensional models that are not included in the list (i) to (vi). The only ones that come to mind are the spherical model, non-translation-invariant Ising models, ice-type models in the presence of direct electric fields, and colouring problems on the honeycomb and square lattices (Baxter 1970*a*, *b*). There are basic mathematical differences between the solutions of the first three and those of models (i) to (vi), so in this sense they are 'exceptional' models.

The two solved colouring problems share the same common mathematical features as (i) to (vi) and should probably be listed with them. However, they are defined very differently and it is not yet obvious that they are equivalent to any special case of the general model discussed in this paper.

2. The general model

A model that includes (i) to (vi) can be constructed on the Kagomé lattice, as is shown in § 8. However, it is actually simpler and more illuminating to consider a yet more general model, namely a restricted eight-vertex model on an arbitrary lattice of intersecting straight lines.

Consider some simply connected convex planar region, such as the interior of a circle, and draw N straight lines within it, starting and ending at the boundary. No three lines are allowed to intersect at a common point.

Two typical sets of such lines are shown in figure 1. The intersections of these lines from the sites of a graph, or 'lattice' $\mathscr{L}$. The line segments between sites form the edges of $\mathscr{L}$. Each site is the endpoint of four edges.

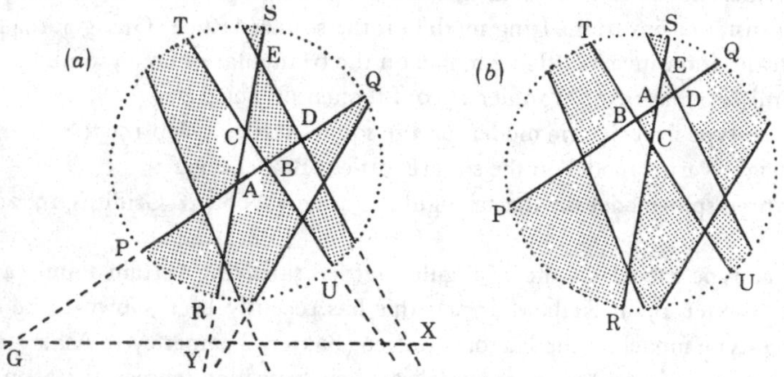

Figure 1. Typical irregular straight-line graphs, or 'lattices' $\mathscr{L}$. The second differs from the first only in that the line PQ has been shifted upwards. The broken lines in the first figure are a possible base line, with the lattice lines extended to cross it.

Consider two intersecting lines, such as PQ and RS in figure 1. Let A be their point of intersection. To the site A assign an interaction coefficient K''_A, to the angle PAR a coefficient K_{PAR}, and to the angle RAQ another coefficient K_{RAQ}. Make no distinction between opposite angles at an intersection, or between the senses of an angle, so that for example K_{PAR}, K_{RAP}, K_{QAS}, K_{SAQ} are all identical.

Do this for every intersecting pair of lines. Thus if all lines intersect there will be $\frac{1}{2}N(N-1)$ interaction coefficients associated with sites, and $N(N-1)$ coefficients associated with angles.

We now construct an eight-vertex model on $\mathscr{L}$, with K''_A, K_{PAR}, etc. as interaction coefficients. To do this we first label the *faces* of $\mathscr{L}$ in some way, and with each face l we associate a spin σ_l. Each such spin has values $+1$ or -1.

Define an Hamiltonian $\mathscr{H}$ by

$$-\beta\mathscr{H} = \Sigma[K_{PAR}\,\sigma_l\sigma_n + K_{RAQ}\,\sigma_m\sigma_p + K''_A\,\sigma_l\sigma_m\sigma_n\sigma_p], \tag{2.1}$$

where the summation is over all sites A of $\mathscr{L}$, and for each term in the summation l, m, n, p are the four faces surrounding the site, arranged as in figure 2 so that l and n are opposite, the angle PAR is a corner of either face l or face n, RAQ is a corner of either face m or face p.

The object of statistical mechanics is to calculate the partition function

$$Z = \sum_{\sigma} e^{-\beta \mathscr{H}} \tag{2.2}$$

(the σ-summation being over all values of all spins in the lattice) and various thermodynamic averages, such as the two-spin correlation

$$\langle \sigma_l \sigma_m \rangle = Z^{-1} \sum_{\sigma} \sigma_l \sigma_m e^{-\beta \mathscr{H}}. \tag{2.3}$$

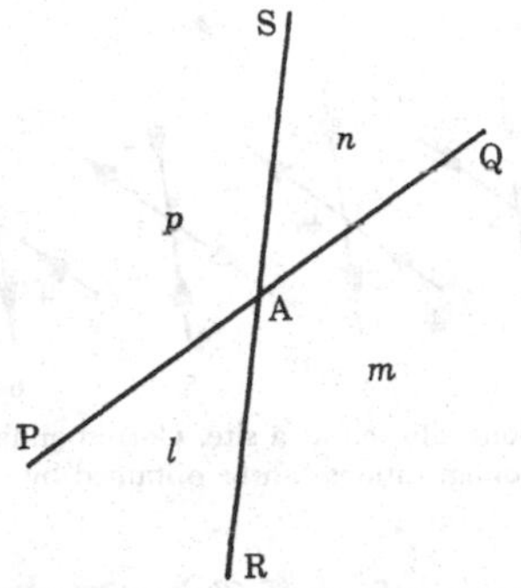

FIGURE 2. A typical site A of $\mathscr{L}$, showing the surrounding faces l, m, n, p, ordered as in equation (2.1).

As is indicated in figure 1 the faces of $\mathscr{L}$ can be grouped into two classes X and Y (shaded and unshaded) so that no two faces of the same class have an edge in common. (An equivalent statement is that the dual lattice of $\mathscr{L}$ is bipartite.) The two-spin interactions in (2.1) link only faces of the same class, so (2.1) is the sum of three terms:

(*a*) a nearest-neighbour two-spin Ising Hamiltonian defined on the class X faces,

(*b*) a similar Hamiltonian defined on the class Y faces,

(*c*) a purely four-spin Hamiltonian coupling the X and Y spins, with coefficients K''_A.

In particular, if the K''_A are all zero, then the model reduces to two independent ordinary Ising models, one on the X spins, the other on the Y spins.

It follows that (2.1) is a fairly obvious generalization to a rather arbitrary planar lattice of the eight-vertex model, which was originally defined on the square lattice (Fan & Wu 1970).

It would be marvellous to calculate Z for any choice of the lattice coefficients K''_A, K_{PAR}, etc., but the author knows of no way to do this. What can be done is to calculate Z (in the thermodynamic limit when N becomes infinite) if the parameters satisfy the conditions given in §4.

3. Formulation as an eight-vertex model

In this paper the above Ising-type formulation will mostly be used. However, to connect with previous results it is sometimes desirable to regard (2.2) as the partition function of a ferroelectric-type model.

This can be done by using an argument due to Wu (1971) and Kadanoff & Wegner (1971). Every edge of $\mathscr{L}$ lies between a shaded (class X) and an unshaded (class Y) face. If the spins on the two faces are alike, draw an arrow on the edge so that an observer following the arrow has the shaded face on his left. If the spins are different, draw the arrow so that the observer has the shaded face on his right.

Now consider a site j of the lattice $\mathscr{L}$. There are 16 possible choices of the four surrounding spins. To each choice there corresponds a configuration of arrows on the four edges. Each arrow configuration corresponds to two spin configurations, one being obtained from the other by reversing all spins. Thus there are eight arrow configurations, as shown in figure 3. In each arrow configuration there are an even number of arrows pointing into the site (or vertex). We call this the *eight-vertex condition*. An arrow covering of the edges of $\mathscr{L}$ is 'allowed' if the eight-vertex condition is satisfied at every site.

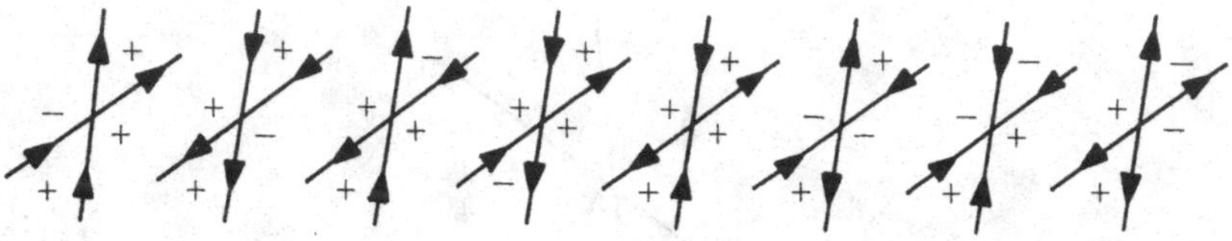

FIGURE 3. The eight arrow configurations allowed at a site. Corresponding spin configurations are also shown. The other eight spin configurations can be obtained by reversing all four spins.

To every configuration of spins on the faces of $\mathscr{L}$ there corresponds an allowed arrow covering of the edges. To every allowed arrow covering there correspond two spin configurations, one being obtained from the other by reversing all spins. Since the Hamiltonian (2.1) is unchanged by reversing all spins, it follows from (2.2) that for a lattice of M sites

$$Z = 2\sum_C \omega_1 \omega_2 \ldots \omega_M, \tag{3.1}$$

where the sum is over all allowed arrow configurations C and ω_j is the Boltzmann weight of site j for configuration C.

Consider the site A, or j, shown in figure 2, and for brevity set

$$K_j = K_{\mathrm{RAQ}}, \quad K'_j = K_{\mathrm{PAR}}, \quad K'_j = K''_{\mathrm{A}}. \tag{3.2}$$

Then the Boltzmann weight ω_j is given by

$$\omega_j = \exp(K_j \sigma_m \sigma_p + K'_j \sigma_l \sigma_n + K''_j \sigma_l \sigma_m \sigma_n \sigma_p). \tag{3.3}$$

From figure 3 it follows that

$$\omega_j = a_j(b_j, c_j, d_j) \text{ if the arrows at site } j \text{ are in configuration 1 or 2 (3 or 4, 5 or 6, 7 or 8)}, \tag{3.4}$$

where

$$\begin{aligned} a_j, b_j, c_j, d_j = \exp(K_j - K'_j - K''_j), \quad & \exp(-K_j + K'_j - K''_j), \\ \exp(K_j + K'_j + K''_j), \quad & \exp(-K_j - K'_j + K''_j). \end{aligned} \tag{3.5}$$

In figure 3 the top-right and bottom-left faces have been regarded as shaded, but reversing the shading merely reverses the arrows, which leaves the weights unchanged. Thus ω_j is always given by the above rule (3.4).

Duality

From (3.5), the site weights satisfy the normalization condition $a_j b_j c_j d_j = 1$. For some purposes it is convenient to ignore this requirement and regard a_j, b_j, c_j, d_j as independent variables.

(This is equivalent to re-defining the energy zero of the Hamiltonian (2.1).) Then Z is given by (3.1) and (3.4) and is a linear function of $a_1, a_2, \ldots, d_M$. Using the same weak-graph symmetry argument as that employed by Fan & Wu (1970), and Wegner (1973), for the square lattice, one can establish for the general lattice $\mathscr{L}$ that

$$Z(\{a_j^*, b_j^*, c_j^*, d_j^*\}) = Z(\{a_j, b_j, c_j, d_j\}), \tag{3.6}$$

where

$$\left.\begin{aligned} a_j^* &= \tfrac{1}{2}(a_j - b_j + c_j - d_j),\\ b_j^* &= \tfrac{1}{2}(-a_j + b_j + c_j - d_j),\\ c_j^* &= \tfrac{1}{2}(a_j + b_j + c_j + d_j),\\ d_j^* &= \tfrac{1}{2}(-a_j - b_j + c_j + d_j).\end{aligned}\right\} \tag{3.7}$$

This is a duality relation, taking a low-temperature system to a high-temperature one.

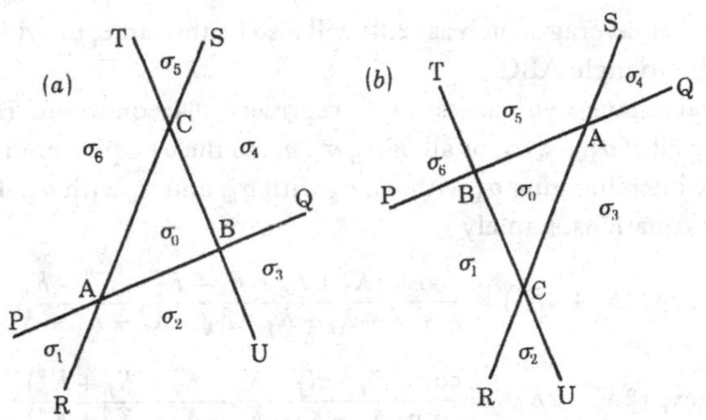

FIGURE 4. The triangles ABC in figures 1 *a* and *b*, showing the positions of the interior spin σ_0 and the surrounding spins $\sigma_1, \ldots, \sigma_6$.

4. STAR-TRIANGLE CONDITIONS

Consider the two lattices shown in figure 1. They differ only in that the line PQ has been shifted from one side of site C to the other. This changes the triangle ABC, but leaves the rest of the lattice unaltered. In both lattices A is the intersection of the lines PQ and RS. Similarly B is the intersection of PQ and TU, C is the intersection of RS and TU.

Construct eight-vertex models as above on both lattices, using the same boundary conditions (e.g. all boundary spins up) and the same coefficients K''_{A}, K_{PAR}, etc.

For each model, let $\sigma_1, \ldots, \sigma_6$ be the six spins on the faces surrounding the triangle ABC, and σ_0 the spin inside the triangle, as indicated in figure 4. For brevity set

$$\left.\begin{aligned} K_1 &= K_{\mathrm{SAQ}}, & K_2 &= K_{\mathrm{PBT}}, & K_3 &= K_{\mathrm{UCR}},\\ K_1' &= K_{\mathrm{RAQ}}, & K_2' &= K_{\mathrm{QBT}}, & K_3' &= K_{\mathrm{TCR}},\\ K_1'' &= K_{\mathrm{A}}'', & K_2'' &= K_{\mathrm{B}}'', & K_3'' &= K_{\mathrm{C}}''.\end{aligned}\right\} \tag{4.1}$$

Thus K_1, K_2, K_3 are the coefficients of the *interior* angles of the triangle ABC, while K_1', K_2', K_3' are the coefficients of the *exterior* angles.

The summand in (2.2) is the same for both models, except the factors coming from the interactions round sites A, B, C. The centre spin σ_0 occurs in only these factors, so the summation over

σ_0 may in each case be performed to give a total Boltzmann weight for the triangle ABC. Doing this, the weight for the first model is

$$W_1 = 2\exp(K_1'\sigma_6\sigma_2 + K_2'\sigma_2\sigma_4 + K_3'\sigma_4\sigma_6) \\ \times \cosh(K_1\sigma_1 + K_2\sigma_3 + K_3\sigma_5 + K_1''\sigma_1\sigma_6\sigma_2 + K_2''\sigma_3\sigma_2\sigma_4 + K_3''\sigma_5\sigma_4\sigma_6). \quad (4.2a)$$

while the weight for the second model is

$$W_2 = 2\exp(K_1'\sigma_3\sigma_5 + K_2'\sigma_5\sigma_1 + K_3'\sigma_1\sigma_3) \\ \times \cosh(K_1\sigma_4 + K_2\sigma_6 + K_3\sigma_2 + K_1''\sigma_4\sigma_3\sigma_5 + K_2''\sigma_6\sigma_5\sigma_1 + K_3''\sigma_2\sigma_1\sigma_3). \quad (4.2b)$$

Both weights are functions of $\sigma_1, \ldots, \sigma_6$. The other factors in the summand of (2.2) are identical, for all values of $\sigma_1, \ldots, \sigma_6$, so the two partition functions will be the same if

$$W_1 = W_2 \quad (4.3)$$

for all $\sigma_1, \ldots, \sigma_6$. Further, averages such as (2.3) will also be the same, provided neither spin σ_l nor spin σ_m lies inside the triangle ABC.

Since $\sigma_1, \ldots, \sigma_6$ each have two values, (4.3) represents 64 equations. However, (4.3) is unchanged by negating all of $\sigma_1, \sigma_3, \sigma_5$, or all of $\sigma_2, \sigma_4, \sigma_6$, so the 64 equations reduce to 16. Further, (4.3) is unaltered by interchanging σ_1 with σ_4, σ_3 with σ_6, and σ_5 with σ_2. This means that there are only six distinct equations, namely

$$\exp(2K_j' + 2K_k') = \frac{\cosh(K_1 + K_2 + K_3 + K_i'' - K_j'' - K_k'')}{\cosh(-K_i + K_j + K_k - K_i'' + K_j'' + K_k'')} \quad (4.4)$$

$$\exp(2K_j' - 2K_k') = \frac{\cosh(K_i - K_j + K_k - K_i'' + K_j'' + K_k'')}{\cosh(K_i + K_j - K_k - K_i'' + K_j'' + K_k'')} \quad (4.5)$$

where (i, j, k) is any permutation of (1, 2, 3). There are three distinct equations of the form (4.4), three of form (4.5).

(It is tempting to try to generalize the present work by allowing $K_1, \ldots, K_3''$ to be different for the two lattices, but the interchange symmetry of (4.3) is then destroyed, leaving 16 apparently distinct equations. Thus one gains nine more degrees of freedom, but ten more apparent restrictions, and it appears that no such generalization is possible.)

The equations (4.4) and (4.5) have been reported before: they are equivalent to the commutation conditions for the square lattice eight-vertex model (Baxter 1972, eqn B8), and a special case has been discussed for the six-vertex models (Baxter, Temperley & Ashley 1977, eqn 111).

The six equations (4.4) and (4.5) are not independent. The K_1', K_2', K_3' can all be eliminated by taking ratios and products, giving three apparently distinct equations which can be regarded as defining K_1'', K_2'', K_3'' as functions of K_1, K_2, K_3. However, these three equations are identically satisfied if $K_1'' = K_2'' = K_3''$, so cannot be independent.

It appears that there is in general no other solution of these three equations, so a corollary of (4.4) and (4.5) is, taking K'' to be the common value,

$$K_1'' = K_2'' = K_3'' = K''. \quad (4.6)$$

If $K'' = 0$, then the equations (4.4) become the star-triangle relation between an Ising model on the honeycomb lattice, with interactions K_1, K_2, K_3, and an equivalent Ising model on the triangular lattice with interactions K_1', K_2', K_3' (Houtappel 1950, eqn 23). This follows directly from their derivation, since if $K'' = 0$ the weights W_1 and W_2 each factor into a function of $\sigma_1, \sigma_3, \sigma_5$ and a function of $\sigma_2, \sigma_4, \sigma_6$, and (4.3) factors into two independent star-triangle transformations with identical coefficients. Thus for $K'' \neq 0$ (4.3) can be regarded as a generalized star-triangle transformation.

By using (4.6), the equations (4.5) can be obtained from (4.4) by taking ratios. Thus (4.4) and (4.6) ensure that (4.3) is satisfied, i.e. Z is unchanged by shifting the line PQ across the site C. Remembering that $K_1, \ldots, K_3''$ are defined by (4.1) and figure 4, we call equations (4.4) and (4.6) the *star-triangle relations for the triangle* ABC. Remember that in these equations K_1, K_2, K_3 are the interaction coefficients assigned to the inside angles of the triangle ABC, while K_1', K_2', K_3' are assigned to the corresponding supplementary exterior angles.

Corollaries

Various corollaries of (4.4) and (4.6) can be obtained by eliminating two of the angle coefficients $K_1, \ldots, K_3'$. In particular, eliminating K_i between equations (4.4) and (4.5) as written, using (4.6), gives

$$\Delta_j = \Delta_k, \tag{4.7}$$

where

$$\Delta_j = -\sinh 2K_j \sinh 2K_j' - \tanh 2K'' \cosh 2K_j \cosh 2K_j'. \tag{4.8}$$

Since Δ_j is a symmetric function of the angle coefficients K_j, K_j' of a single site, it can be thought of as a 'site parameter' similar to K_j''. Since Δ_1 corresponds to the site A in figure 4, it can alternatively be written as Δ_A. Similarly, Δ_2 and Δ_3 can be written as Δ_B and Δ_C, respectively. From (4.7), they have a common value Δ. By using (4.1), it follows that (4.6) and (4.7) can be written as

$$K_A'' = K_B'' = K_C'' = K'', \quad \Delta_A = \Delta_B = \Delta_C = \Delta. \tag{4.9}$$

In the limit when K_j, K_j', $-K''$ all tend to plus infinity, their differences remaining constant, the eight-vertex model reduces to an 'ice-type' six-vertex model. The Δ defined above is then the same as that used by Lieb (1967).

Also, eliminating K_2' and K_3' gives the equations

$$-\cosh 2K_2 \cosh 2K_3 + \coth 2K_1' \sinh 2K_2 \sinh 2K_3 = \cosh 2K_1 \cosh 2K'' + \coth 2K_1' \sinh 2K_1 \sinh 2K''. \tag{4.10}$$

Two similar equations can be obtained by permuting the suffixes 1, 2, 3.

Quadrilateral theorem

Consider the quadrilateral shown in figure 5. Let K_A'', K_{DFB}, etc. be the interaction coefficients assigned to the six sites and twelve angles. Suppose that these satisfy the star-triangle relations (4.4) and (4.6) for each of the triangles AEF, BFD, CDE. Then it follows that they also satisfy the star-triangle relations for the triangle ABC.

A brute-force proof of this theorem is given in appendix A, by using only elementary algebra. A much neater proof is given in § 6, but this makes use of elliptic functions. From the form of the equations (4.8) and (4.10), it seems likely that a third proof, of intermediate difficulty, could be obtained by using hyperbolic trigonometry (Onsager 1944, p. 135; Coxeter 1947, p. 238).

Note in particular that if the star-triangle relations are satisfied for AEF, BFD and CDE, then

$$K_A'' = K_B'' = K_C'' = K_D'' = K_E'' = K_F''$$
$$\Delta_A = \Delta_B = \Delta_C = \Delta_D = \Delta_E = \Delta_F, \tag{4.11}$$

i.e. the site parameters K'', Δ are the same for each site.

Conditions for model to be solvable

Consider a connected lattice $\mathscr{L}$ and extend the lines until they each cross some base line, as indicated by the broken lines in figure 1*a*. Assign site and angle coefficients to the sites and angles on the base line.

Every site A of $\mathscr{L}$ is then a vertex of a triangle consisting of the two lines through A and the base line, such as AGY in figure 1*a*. Call this the *basic triangle* with vertex A. In the next section we shall show that the eight-vertex model on $\mathscr{L}$ is exactly solvable (in the thermodynamic limit) provided the site and angle coefficients on the base line can be chosen to that the star-triangle conditions are satisfied for every basic triangle. If this can be done we say that the model is *Z-invariant.*

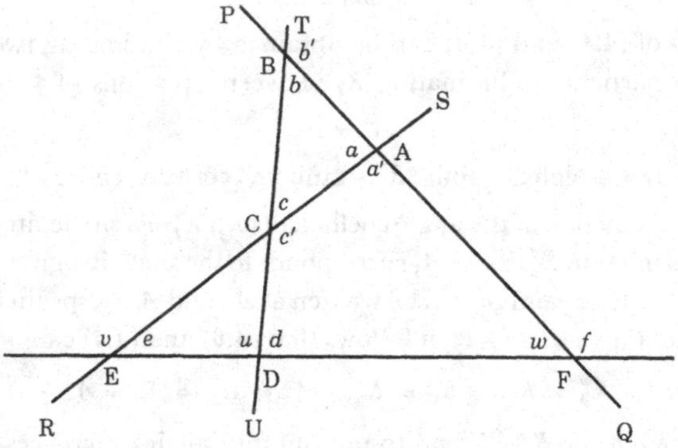

FIGURE 5. The quadrilateral ABCDEF. The lower-case letters denote the angle weights used in appendix A, e.g. $w = \exp(1K_{\mathrm{EFB}})$.

From (4.9), this implies that K'' must have the same value for all sites of $\mathscr{L}$ and all sites on the base line. So must Δ. Also, any triangle in $\mathscr{L}$ (not necessarily a face of $\mathscr{L}$) is part of a quadrilateral, with fourth side the base line. The other three triangles in this quadrilateral are basic, so from the quadrilateral theorem the star-triangle conditions must be satisfied *for every triangle in* $\mathscr{L}$. For example, they must be satisfied for the triangles ADE in figure 1*a* and *b*.

It follows that equation (4.3) is satisfied whenever a line of $\mathscr{L}$ is shifted across a vertex. Hence Z is unchanged by shifting the lines, so long as their order at the boundary is preserved. The correlation $\langle \sigma_l \sigma_m \rangle$ is also unchanged, provided no line is shifted across face l or face m.

Such models certainly exist. They can be constructed by choosing a set of site and angle coefficients for the base line, such that K'' and Δ have the same value at each such site. As is explicity shown in equation (A 3)–(A 7) of appendix A, the star-triangle conditions then determine the coefficients at the third vertex of every basic triangle, i.e. at every site of $\mathscr{L}$.

Since K'' and Δ are fixed, there is only one degree of freedom in choosing the angle coefficients of a site formed by the intersection of a lattice line with the base line. Ift here are N lattice lines, it follows that there are $N+2$ disposable parameters in the model, namely K'', Δ and N 'line parameters'.

Extended lattices

A lattice $\mathscr{L}$ can be extended either by moving the convex boundary outwards and extending the lattice lines to the new boundary, or by adding new lines, or both. If the model on $\mathscr{L}$ is Z-invariant, then the coefficients of the new sites and angles can be chosen so that the model on the extended lattice $\mathscr{L}'$ is also Z-invariant.

To see this, first consider a new site formed by extending two lines of $\mathscr{L}$ until they intersect. This new site is a vertex of a basic triangle, so its coefficients can be obtained from those of the sites on the base line by the star-triangle relations, and the Z-invariance condition remains satisfied.

If a new line is added, select an intersection A of this with a previous line. Assign site and angle coefficients to A such that K'', Δ have the same values as on the previous sites of $\mathscr{L}$.

Extend the new line to cross the base line at a point P. Let Q be the intersection of the old line through A with the base line. From figure 6 it is apparent that PQA is a basic triangle. The coefficients at P can be obtained from the star-triangle relations for PQA.

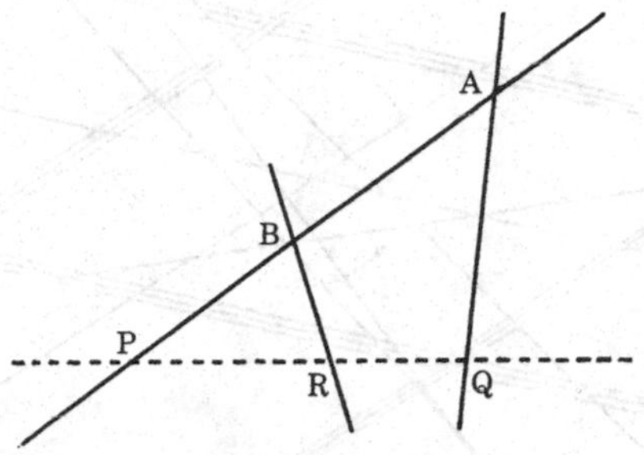

FIGURE 6. Assignment of coefficients to a new line ABP in $\mathscr{L}$: PRQ is the base line, AQ and BR are already existing lines. The coefficients at Q, R are given. If those of A are also given, then those at P, B are determined by the star-triangle relations for PQA, PRB.

Every other site B on the new line is a vertex of a basic triangle, for example BPR in figure 6. The coefficients at B can therefore be obtained from the star-triangle relations for BPR, thereby maintaining Z-invariance.

Note that if such a line crosses all previous lines of $\mathscr{L}$, it can be regarded as an alternative base line. From the quadrilateral theorem, the star-triangle conditions are satisfied for all triangles, in particular for those with the new line as a side. Hence if a lattice model is Z-invariant with respect to one base line, it is also Z-invariant with respect to any other.

5. Local thermodynamic properties

Consider a site A near the centre of $\mathscr{L}$. Let PAQ, RAS be the two lines through A, as in figure 1. Let l, m, n, p be the four faces round A, as in figure 2.

Extend $\mathscr{L}$ as follows: draw $2M$ lines parallel to PAQ, M of them being close together on one side of $\mathscr{L}$, the other M being close together on the other side. Similarly, draw $2M$ lines parallel to RAS.

This creates a parallelogram 'frame' for $\mathscr{L}$, each side being made up of M lines close together.

Suppose that no other line of $\mathscr{L}$ is parallel to either PAQ or RAS (if one is, rotate it slightly). Extend every such line to cross all the $4M$ framing lines. Move the convex boundary outwards to enclose all these intersections, and extend all lines to the boundary.

This creates an extended lattice $\mathscr{L}'$, as in figure 7. Assign coefficients to the new sites and angles in $\mathscr{L}'$ according to the rules given above, choosing the angle coefficients between PAQ (and RAS) and the framing lines to be K_{PAR} and K_{RAQ}, necessarily equal angles having equal coefficients. Then by considering triangles it is straightforward to verify that the angle coefficients for any intersection of two framing lines are also K_{PAR} and K_{RAQ}, necessarily equal angles having equal coefficients.

Suppose that originally the boundary spins in $\mathscr{L}$ (i.e. those on faces adjacent to the boundary) were fixed to be up. Do the same for $\mathscr{L}'$ and consider correlations between the spins round A, such as $\langle\sigma_l\rangle$, $\langle\sigma_l\sigma_n\rangle$.

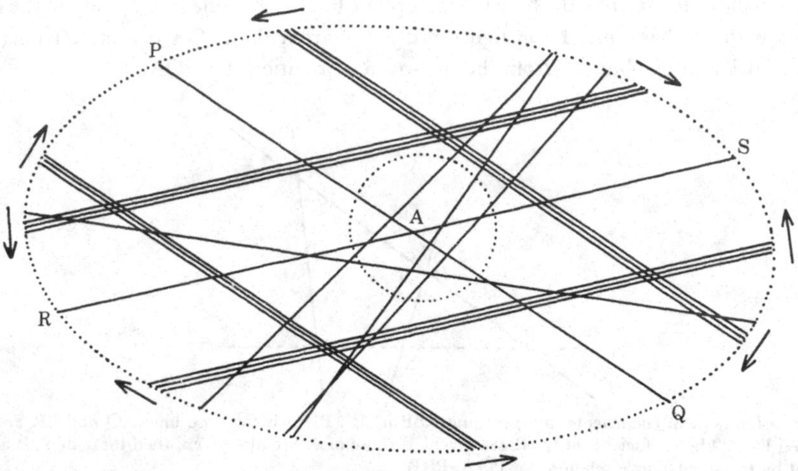

FIGURE 7. The lattice of figure 1*a*, extended as in §5. The inner dotted line is the old boundary, the outer one is the new boundary. The correlations of the four spins round A are unchanged by making parallel shifts of the framing lines inwards towards PQ and RS.

These are of course *not* the same for the model defined on $\mathscr{L}$ as for the model defined on $\mathscr{L}'$. However, if A is originally deep in the centre of $\mathscr{L}$ (i.e. any path through faces from A to the boundary crosses a large number of edges), then we expect the local correlations to be insensitive to the position of the boundary, and in the limit of $\mathscr{L}$ large we do expect them to be the same for $\mathscr{L}$ and $\mathscr{L}'$.

There is a problem here: if the Boltzmann weights are not all positive then local correlations may be sensitive to boundary conditions. From the construction used to extend $\mathscr{L}$ it is not obvious that all the new weights generated will be positive. However, this is probably not a serious difficulty. In particular, if $\mathscr{L}$ originally has a periodic structure (e.g. the Kagomé lattice discussed later), no new weights are generated by extending $\mathscr{L}$ to $\mathscr{L}'$, so if they were originally all positive for $\mathscr{L}$, then they are also all positive for $\mathscr{L}'$.

Now shift the framing lines of $\mathscr{L}'$ inwards towards A, keeping them parallel, until all the original sites of $\mathscr{L}$ (other than A) lie outside the frame. This does not alter the order of the lines at the boundary, and no line crosses A, so $\langle\sigma_l\rangle$, $\langle\sigma_l\sigma_m\rangle$, etc. are left unchanged, because of the Z-invariance of $\mathscr{L}'$.

Now, however, the picture is quite different: A is at the centre of a regular parallelogram lattice of $(2M+1)$ by $(2M+1)$ lines. Outside this lies the original lattice $\mathscr{L}$. We still expect the local correlations to be insensitive to what is going on many sites away from A, so if M is large we expect the local correlations to be the same as those of this regular lattice.

But this lattice is just the regular square lattice eight-vertex model, with interaction coefficients K_{PAR}, K_{RAQ}, K''. Thus the correlations $\langle\sigma_l\rangle$, $\langle\sigma_l\sigma_n\rangle$, etc. *are the same as those of this regular model.*

In particular, the local magnetization, polarization and internal energy U_{A} of the lattice $\mathscr{L}$ are given by

$$\langle\sigma_l\rangle \quad = M(\Delta, K'') \tag{5.1}$$

$$\langle\sigma_l\sigma_m\rangle = P(\Delta, K'') \tag{5.2}$$

$$-\beta U_{\mathrm{A}} = K_{\mathrm{PAR}}\langle\sigma_l\sigma_n\rangle + K_{\mathrm{RAQ}}\langle\sigma_m\sigma_p\rangle + K''\langle\sigma_l\sigma_m\sigma_n\sigma_p\rangle = u(K_{\mathrm{PAR}}, K_{\mathrm{RAQ}}, K''), \tag{5.3}$$

where M, P, $-\beta^{-1}u$ are the magnetization, polarization and internal energy per site of the regular square-lattice eight-vertex model with interaction coefficients K_{PAR}, K_{RAQ}, K''. The first two of these are known to depend on K_{PAR}, K_{RAQ} only via Δ (Barber & Baxter 1973; Baxter & Kelland 1974). (This can also be established by the present methods.)

Note that $\langle\sigma_l\rangle$ is therefore the same (in the thermodynamic limit) for every face of $\mathscr{L}$. Similarly, $\langle\sigma_l\sigma_m\rangle$ is the same for every edge.

Free energy

Let $F = -\beta^{-1}\ln Z$ be the total free energy of $\mathscr{L}$. Increment all the site and angle coefficients by infinitesimal amounts $\delta K''$, δK_{PAR}, etc. Then from (2.1), (2.2) and (2.3) the increment induced in $-\beta F$ is

$$-\delta(\beta F) = \Sigma[\delta K_{\mathrm{PAR}}\langle\sigma_l\sigma_n\rangle + \delta K_{\mathrm{RAQ}}\langle\sigma_m\sigma_p\rangle + \delta K''\langle\sigma_l\sigma_m\sigma_n\sigma_p\rangle], \tag{5.4}$$

where the summation is over all sites A of $\mathscr{L}$, as in (1).

Let $f_{\mathrm{A}} = f(K_{\mathrm{PAR}}, K_{\mathrm{RAQ}}, K'')$ be the free energy per site of the regular square lattice model, with coefficients K_{PAR}, K_{RAQ}, K''. Then from the above remarks

$$\left.\begin{aligned}\langle\sigma_l\sigma_n\rangle &= -\partial(\beta f_{\mathrm{A}})/\partial K_{\mathrm{PAR}},\\ \langle\sigma_m\sigma_p\rangle &= -\partial(\beta f_{\mathrm{A}})/\partial K_{\mathrm{RAQ}},\\ \langle\sigma_l\sigma_m\sigma_n\sigma_p\rangle &= -\partial(\beta f_{\mathrm{A}})/\partial K'',\end{aligned}\right\} \tag{5.5}$$

provided A is deep inside $\mathscr{L}$.

Assuming that we can ignore the contribution to the sum in (5.4) of sites that are near the boundary, it follows that

$$\delta(\beta F) = \Sigma\,\delta(\beta f_{\mathrm{A}}), \tag{5.6}$$

where $\delta(\beta f_{\mathrm{A}})$ is the increment induced in βf_{A}. Thus

$$\beta[F - \Sigma f_{\mathrm{A}}]$$

is stationary with respect to variations in the interaction coefficients, provided the model on $\mathscr{L}$ is Z-invariant.

However, while keeping the model Z-invariant, one can continuously vary the interactions until all the coefficients are large and positive, when it is trivially true that $F - \Sigma f_{\mathrm{A}}$ is zero. Thus for any Z-invariant model one must have

$$F = \Sigma f(K_{\mathrm{PAR}}, K_{\mathrm{RAQ}}, K''), \tag{5.7}$$

the summation being over all sites A of $\mathscr{L}$.

This is the key result of this paper. The free energy of any Z-invariant lattice model is the sum of site free energies, the site energies being those of the regular square lattice model.

There are previous results that have suggested this: notably an inhomogeneous square lattice model (Baxter 1972, eqn 10.4) and the six-vertex model on the Kagomé lattice (Baxter, Temperley & Ashley 1977, eqn 47).

6. Elliptic function parametrization

The star-triangle relations and the Z-invariance conditions can be written very simply by introducing elliptic functions.

The four-spin interaction coefficient K'' must have the same value at each site of $\mathscr{L}$, and so must Δ. If these two values are given, then (4.8) is a relation between the two angle coefficients K_j and K_j' at site j.

Define two site-independent parameters ω, Ω by

$$\coth 2K'' = \cosh \Omega$$

$$-\Delta \coth 2K'' = \cosh \omega. \tag{6.1}$$

Then (4.8) can be written as

$$\cosh \omega = \cosh 2K_j \cosh 2K_j' + \cosh \Omega \sinh 2K_j \sinh 2K_j'. \tag{6.2}$$

Although this relation concerns only a single site of $\mathscr{L}$, formally it is the same as that between the sides ω, $2K_j$, $2K_j'$ of an hyperbolic triangle, with angle $\pi + \mathrm{i}\Omega$ between the sides $2K_j$, $2K_j'$ (Onsager 1944, p. 135; Coxeter 1947). This is the same as that for a spherical triangle with sides of pure imaginary length.

It is well known that this relation can be simplified by introducing elliptic functions of modulus

$$k = \sinh \Omega / \sinh \omega \tag{6.3}$$

(Greenhill 1892, §129). Onsager (1944, p. 144) refers to this as a uniformizing substitution.

From (6.1) and (6.3) it follows that

$$k^{-2} = \Delta^2 \cosh^2 2K'' - \sinh^2 2K'', \tag{6.4}$$

and from (4.8) that

$$k^{-2} = \frac{16(1+vv'v'')(v+v'v'')(v'+v''v)(v''+vv')}{(1-v^2)^2(1-v'^2)^2(1-v''^2)}, \tag{6.5}$$

where

$$v = \tanh K_j, \quad v' = \tanh K_j', \quad v'' = \tanh K''. \tag{6.6}$$

It is interesting, but probably quite irrelevant, to note that this k is the same as the elliptic modulus which occurs in the solution of a triangular two-spin Ising model with interaction coefficients K_j, K_j', K'' (Green 1963; Stephenson 1964).

Let $\mathscr{K}$, $\mathscr{K}'$ be the complete elliptic integrals of the first kind of moduli, k, $k' = (1-k^2)^{\frac{1}{2}}$, respectively. They are *not* to be confused with any of the interaction coefficients used above, notably with K_j and K_j'.

Following Greenhill (1892), but making a few notational changes, we define α_j, β_j, λ by

$$\left.\begin{aligned} 2K_j' &= -\mathrm{i}\,\mathrm{am}\,[\mathrm{i}(\mathscr{K}'-\alpha_j)],\\ 2K_j &= -\mathrm{i}\,\mathrm{am}\,[\mathrm{i}(\mathscr{K}'-\beta_j)],\\ 2K'' &= \mathrm{i}\,\mathrm{am}\,[\mathrm{i}(\mathscr{K}'-\lambda)]. \end{aligned}\right\} \tag{6.7}$$

Then from (6.1), (6.3) and various standard relations for the Jacobi elliptic functions (Gradshteyn & Ryzhik 1965, §§ 8.143 and 8.151.2)

$$\left.\begin{aligned} \cosh 2K_j' &= \mathrm{i}k^{-1}\,\mathrm{ds}\,(\mathrm{i}\alpha_j),\\ \sinh 2K_j' &= \mathrm{i}/[k\,\mathrm{sn}\,(\mathrm{i}\alpha_j)],\\ \cosh 2K_j &= \mathrm{i}k^{-1}\,\mathrm{ds}\,(\mathrm{i}\beta_j),\\ \sinh 2K_j &= \mathrm{i}/[k\,\mathrm{sn}\,(\mathrm{i}\beta_j)],\\ \coth 2K'' &= \cosh\Omega = -\,\mathrm{dn}\,(\mathrm{i}\lambda),\\ \sinh 2K'' &= -\,\mathrm{i}/[k\,\mathrm{sn}\,(\mathrm{i}\lambda)],\\ \omega &= \mathrm{i}\,\mathrm{am}\,[\mathrm{i}(2\mathscr{K}'-\lambda)],\\ \Delta &= -\,\mathrm{cn}(\mathrm{i}\lambda)/\mathrm{dn}(\mathrm{i}\lambda), \end{aligned}\right\} \tag{6.8}$$

where $\mathrm{ds}(u) = \mathrm{dn}(u)/\mathrm{sn}(u)$.

Substituting these expressions into (6.2) and using the elliptic function identity

$$k^2\,\mathrm{cn}\,(u+v)\,\mathrm{sn}\,u\,\mathrm{sn}\,v = \mathrm{dn}\,u\,\mathrm{dn}\,v - \mathrm{dn}\,(u+v), \tag{6.9}$$

we find that (4.8) is satisfied if $\qquad \alpha_j+\beta_j = \lambda. \qquad$ (6.10)

Also, by using the relations (6.7) and (6.8) in the equation (4.10), and using (6.9) and (6.10) to simplify the right hand side, (4.10) becomes

$$[\mathrm{dn}\,(\mathrm{i}\beta_2)\,\mathrm{dn}\,(\mathrm{i}\beta_3) - \mathrm{dn}\,(\mathrm{i}\alpha_1)]/[k^2\,\mathrm{sn}\,(\mathrm{i}\beta_2)\,\mathrm{sn}\,(\mathrm{i}\beta_3)] = \mathrm{cn}\,(\mathrm{i}\alpha_1). \tag{6.11}$$

From (6.9) this is clearly satisfied if

$$\alpha_1 = \beta_2+\beta_3 \tag{6.12}$$

or, by using (6.10), if $\qquad \alpha_1+\alpha_2+\alpha_3 = 2\lambda, \quad \beta_1+\beta_2+\beta_3 = \lambda. \qquad$ (6.13)

Since these relations are unchanged by permuting the suffixes 1, 2, 3, the other two equations that can thereby be obtained from (4.10) are also satisfied.

The relations (4.7) and (4.10) imply (4.4), so the original star-triangle relations are satisfied by the definitions (6.4) and (6.7), and the relations (6.10) and (6.13).

So far we have made no restriction on the values of the interaction coefficients, other than those imposed by the star-triangle relations. From (6.4), k may be greater or less than one, or may be imaginary. Whatever the value of k, there will always be parameters α_j, β_j, λ that satisfy (6.7), (6.10) and (6.13), but they may be complex.

To fix our ideas, and to give a real single-valued definition of α_j, β_j, λ, it is desirable to focus attention on the case when at every site j of $\mathscr{L}$ the two angle coefficients K_j and K_j' satisfy

$$\sinh\,(K_j+K_j') > \mathrm{e}^{-2K''}\cosh\,(K_j-K_j'). \tag{6.14}$$

From (3.5) this is equivalent to the condition $c_j > a_j+b_j+d_j$. This is the ferromagnetically ordered phase of the eight-vertex model.

From (4.8), this implies that $\qquad \Delta < -1, \qquad$ (6.15)

so from (6.4), $k^{-2} > 1$ and we can choose k so that

$$0 < k < 1. \tag{6.16}$$

The obvious solution of (6.7), together with the definition of $\mathcal{K}$, $\mathcal{K}'$, is then

$$\alpha_j = \int_{2K'_j}^{\infty} (1+k^2\sinh^2\phi)^{-\frac{1}{2}}\,\mathrm{d}\phi, \tag{6.17a}$$

$$\beta_j = \int_{2K_j}^{\infty} (1+k^2\sinh^2\phi)^{-\frac{1}{2}}\,\mathrm{d}\phi, \tag{6.17b}$$

$$\lambda = \int_{-2K''}^{\infty} (1+k^2\sinh^2\phi)^{-\frac{1}{2}}\,\mathrm{d}\phi, \tag{6.17c}$$

$$\mathcal{K}' = \int_{0}^{\infty} (1+k^2\sinh^2\phi)^{-\frac{1}{2}}\,\mathrm{d}\phi, \tag{6.17d}$$

$$\mathcal{K} = \int_{0}^{\frac{1}{2}\pi} (1-k^2\sin^2\theta)^{-\frac{1}{2}}\,\mathrm{d}\theta, \tag{6.17e}$$

so that α_j, β_j, λ are real and positive. For this case this solution is indeed the one that satisfies (6.10) and (6.13), so

$$0 < \alpha_j, \beta_j < \lambda < 2\mathcal{K}'. \tag{6.18}$$

Note that K'_1, K'_2, K'_3, α_1, α_2, α_3 are associated with the exterior angles of the triangle ABC, while K_1, K_2, K_3, β_1, β_2, β_3 are associated with the interior angles.

Z-invariance conditions in terms of the elliptic angle parameters

In the above equations (6.2) to (6.18) we have considered a triangle with vertices 1, 2, 3, interior angle coefficients K_1, K_2, K_3, and exterior angle coefficients K'_1, K'_2, K'_3. To develop a notation appropriate to the whole lattice, we note that K'', Δ, k, λ are the same for all sites, to an angle PAR is assigned an interaction coefficient K_{PAR}, and, from (6.17*a* or *b*), an elliptic angle parameter

$$\alpha_{\mathrm{PAR}} = \int_{2K_{\mathrm{PAR}}}^{\infty} (1+k^2\sinh^2\phi)^{-\frac{1}{2}}\,\mathrm{d}\phi. \tag{6.19}$$

Then a lattice $\mathcal{L}$ is Z-invariant if the relations (6.10) and (6.13) are satisfied for every basic triangle formed by two lattice lines and the base line, i.e. if

(i) the sum of the two elliptic angle parameters at every site is λ,

(ii) the sum of the elliptic parameters of the interior (exterior) angles of every basic triangle is λ (2λ).

General formulae for all the elliptic angle parameters can now be given. Suppose that all sites of $\mathcal{L}$ lie on the same side of the base line and rotate the lattice until the base line is horizontal and below $\mathcal{L}$, as in figure 1. Label the lattice lines 1, ..., N. Let X be a point on the base line to the right of all the lattice lines.

Consider a line r. Let A be a lattice site on r and E its intersection with the base line, as in figure 5 (regarding EDFX as the base line). Define

$$\alpha_{\mathrm{r}} = \alpha_{\mathrm{AEX}}. \tag{6.20}$$

There are N such 'line angle parameters'.

Now consider a typical site of $\mathcal{L}$, such as A in figure 5. Let r be the line AE, s be the line AF, E and F lying on the base line, E to the left of F.

Since AEF is a basic triangle, it follows from (i) and (ii) that

$$\left.\begin{aligned}\alpha_{\mathrm{EAF}} &= \alpha_{\mathrm{s}}-\alpha_{\mathrm{r}},\\ \alpha_{\mathrm{EAB}} &= \lambda+\alpha_{\mathrm{r}}-\alpha_{\mathrm{s}}.\end{aligned}\right\} \tag{6.21}$$

Thus the elliptic parameters of the angles of intersection of lines r and s can be simply written in terms of $\alpha_{\mathrm{s}}-\alpha_{\mathrm{r}}$ and λ.

The quadrilateral theorem is now trivial. Let t be the line BCD in figure 5. Then the star-triangle relations (i) and (ii) for AEF, BFD, CDE imply (6.21), i.e.

$$\left.\begin{aligned}\alpha_{\mathrm{BAC}} &= \lambda+\alpha_{\mathrm{r}}-\alpha_{\mathrm{s}},\\ \alpha_{\mathrm{ACB}} &= \alpha_{\mathrm{t}}-\alpha_{\mathrm{r}},\\ \alpha_{\mathrm{CBA}} &= \alpha_{\mathrm{s}}-\alpha_{\mathrm{t}}.\end{aligned}\right\} \tag{6.22}$$

Adding these equations gives the required star-triangle relation for ABC, namely

$$\alpha_{\mathrm{BAC}}+\alpha_{\mathrm{ACB}}+\alpha_{\mathrm{CBA}} = \lambda. \tag{6.23}$$

Thus the sum of the elliptic parameters of the interior angles must be 2λ.

More generally, one can readily prove:

(iii) the sum of the elliptic parameters of the exterior angles of any polygon in $\mathscr{L}$ must be 2λ.

This is a useful observation, since if (i) and (iii) are satisfied, then the model on $\mathscr{L}$ is Z-invariant. This provides an alternative definition of Z-invariance that does not require the artificial introduction of a base line.

Geometric model

This formulation with elliptic functions and integrals makes it quite clear that the Z-invariance conditions have a geometric interpretation. In fact, for a given lattice $\mathscr{L}$ and given values of $k, \lambda(K'', \Delta)$ there is a particularly obvious choice of the angle parameters, namely that for any angle PAR

$$\alpha_{\mathrm{PAR}} = (\lambda/\pi)\times\text{the angle PAR (radians)}. \tag{6.24}$$

The conditions (i), (ii), (iii) are then automatically satisfied. We call this the 'geometric' model. Many models, notably the anisotropic square and Kagomé lattice models, can be converted to geometric models by rotating some of the lattice lines (e.g. so as to convert the square lattice into a parallelogram lattice).

The condition (6.14) can be somewhat relaxed without introducing complex elliptic angle parameters. It can be replaced by the requirement that the v, v', v'' defined by (6.6) satisfy

$$v+v'v'',\quad v'+v''v,\quad v''+vv' > 0 \tag{6.25}$$

for every site of $\mathscr{L}$. This is automatically satisfied if the model is ferromagnetic, i.e. all the interaction coefficients are positive. From (6.5) it implies only that $k^2 > 0$, whereas (6.14) implies $0 < k^2 < 1$. Real positive parameters λ, α_{PAR} may still be defined by (6.17*c*) and (6.19), and the Z-invariance conditions are still equivalent to the conditions (i) and (iii).

Alternatively, if (6.25) is satisfied but (6.14) is not, then $k^2 > 1$ and it is natural to use elliptic functions of modulus $k^* = k^{-1}$. Specific formulae for doing this are given in equations (9.3)–(9.8).

Connection with Onsager's parametrization: the case $K'' = 0$

As was remarked in section 2, if $K'' = 0$ the eight-vertex model factors into two independent two-spin Ising models. From (4.8) and (6.4),

$$k = 1/|\sinh 2K_j \sinh 2K_j'|. \tag{6.26}$$

For the square lattice this is the elliptic modulus used by Onsager for the low-temperature case (Onsager 1944, eqn 2.1 *a*). Onsager's parameters a, $\mathscr{K}'-a$ are our parameters $\mathscr{K}'-\alpha_j$, $\mathscr{K}'-\beta_j$. From (6.17 *c*)

$$\lambda = \mathscr{K}' \tag{6.27}$$

so from (6.10) the sum of $\mathscr{K}'-\alpha_j$ and $\mathscr{K}'-\beta_j$ is $\mathscr{K}'$, in agreement with Onsager.

7. Expressions for f, M, P

The regular square-lattice function f in (5.7) has been obtained (Baxter 1972), using the ferroelectric formulation of the model described in § 3. The result can be summarized as follows (negating b in the 1972 paper).

Let K_j, K_j', K'' $(= K_j'')$ be the interaction coefficients (the same for every site in the regular square lattice model). Let a, b, c, d be the Boltzmann weights defined by (3.5) (temporarily dropping the suffix j). Define

$$\left.\begin{aligned} w_1 &= \tfrac{1}{2}(c+d), \quad w_2 = \tfrac{1}{2}(c-d), \\ w_3 &= \tfrac{1}{2}(a+b), \quad w_4 = \tfrac{1}{2}(a-b). \end{aligned}\right\} \tag{7.1}$$

Rearrange and negate (if necessary) $w_1, \ldots, w_4$ until they satisfy

$$w_1 > w_2 > w_3 > |w_4|. \tag{7.2}$$

Now define new weights a, b, c, d so that (7.1) is again satisfied. Call these a', b', c', d'. From (7.2) they are positive and satisfy

$$c' > a'+b'+d'. \tag{7.3}$$

(This procedure maps the model into the ordered ferromagnetic phase, while leaving the partition function unchanged.)

Define an elliptic modulus k_{I} and parameters η, v such that

$$\left.\begin{aligned} a':b':c':d' &= \operatorname{sn}(\eta+v, k_{\mathrm{I}}):\operatorname{sn}(\eta-v, k_{\mathrm{I}}): \\ &\operatorname{sn}(2\eta, k_{\mathrm{I}}): -k_{\mathrm{I}}\operatorname{sn}(2\eta, k_{\mathrm{I}})\operatorname{sn}(\eta+v, k_{\mathrm{I}})\operatorname{sn}(\eta-v, k_{\mathrm{I}}). \end{aligned}\right\} \tag{7.4}$$

These can and are to be chosen so that k_{I} is real, satisfying

$$0 < k_{\mathrm{I}} < 1, \tag{7.5}$$

while η and v are pure imaginary, satisfying

$$|\mathrm{Im}\,(v)| < \mathrm{Im}\,(\eta) < \tfrac{1}{2}\mathscr{K}_{\mathrm{I}}', \tag{7.6}$$

where in this section $\mathscr{K}_{\mathrm{I}}$ and $\mathscr{K}_{\mathrm{I}}'$ are the complete elliptic integrals of the first kind of moduli k_{I}, $k_{\mathrm{I}}' = (1-k_{\mathrm{I}}^2)^{\frac{1}{2}}$, respectively. They are *not* interaction coefficients.

Define q, x, z by (Baxter 1972, eqns D 4–D 8)

$$q = \exp\,(-\pi\mathscr{K}_{\mathrm{I}}'/\mathscr{K}_{\mathrm{I}}),\ x = \exp\,(\mathrm{i}\pi\eta/\mathscr{K}_{\mathrm{I}}),\ z = \exp\,(\mathrm{i}\pi v/\mathscr{K}_{\mathrm{I}}). \tag{7.7}$$

Then q, x, z are positive real, satisfying

$$0 < q < x^2 < 1, \quad x < z < x^{-1}, \tag{7.8}$$

and f, the free energy per site, is given by (Baxter 1972, eqn D 37)

$$-\beta f(K_j, K_j', K'') = \ln c' + \sum_{n=1}^{\infty} \frac{x^{-n}(x^{2n}-q^n)^2(x^n+x^{-n}-z^n-z^{-n})}{n(1-q^{2n})(1+x^{2n})}. \tag{7.9}$$

Also, Barber & Baxter (1973) and Baxter & Kelland (1974) conjecture that in the ordered phases M and P are given by

$$M = \prod_{n=1}^{\infty} \frac{1-x^{4n-2}}{1+x^{4n-2}} \tag{7.10}$$

$$P = \prod_{n=1}^{\infty} \left[\frac{1+q^n}{1-q^n}\frac{1-x^{2n}}{1+x^{2n}}\right]^2. \tag{7.11}$$

They are certainly independent of z.

Relation between the elliptic parametrizations in the ordered ferromagnetic phase

From (3.5) and (4.8), K'' and Δ may be expressed as functions of a, b, c, d:

$$\exp(4K'') = cd/ab \tag{7.12}$$

$$\Delta = \tfrac{1}{2}(a^2+b^2-c^2-d^2)/(ab+cd). \tag{7.13}$$

Suppose that (6.14) is satisfied, i.e. that $c > a+b+d$. Then the a', b', c', d' in (7.4) are the original weights a, b, c, d. Substituting (7.4) into (7.12) and (7.13) gives

$$\exp(2K'') = -ik_{\mathrm{I}}^{\frac{1}{2}} \operatorname{sn}(2\eta, k_{\mathrm{I}}) \tag{7.14}$$

$$\Delta = -\frac{\operatorname{cn}(2\eta, k_{\mathrm{I}}) \operatorname{dn}(2\eta, k_{\mathrm{I}})}{1-k_{\mathrm{I}} \operatorname{sn}^2(2\eta, k_{\mathrm{I}})}. \tag{7.15}$$

Substituting these expressions into (6.4) gives

$$k = 2k_{\mathrm{I}}^{\frac{1}{2}}/(1+k_{\mathrm{I}}). \tag{7.16}$$

Thus the elliptic moduli k, k_{I} are related by a Landen transformation. By using §8.152 of Gradshteyn & Ryzhik (1965), it follows from (7.14) or (7.15) and (6.8) that

$$i\lambda = (1+k_{\mathrm{I}})\, 2\eta. \tag{7.17}$$

Also, from (3.5) $$\sinh 2K_j' = (bc-ad)/[2(abcd)^{\frac{1}{2}}], \tag{7.18}$$

so, by using (7.4) $$\sinh 2K_j' = \tfrac{1}{2}ik_{\mathrm{I}}^{-\frac{1}{2}}[1+k_{\mathrm{I}} \operatorname{sn}^2(\eta+v, k_{\mathrm{I}})]/\operatorname{sn}(\eta+v, k_{\mathrm{I}}). \tag{7.19}$$

From §8.152 of Gradshteyn & Ryzhik (1965) and (6.8), it follows that

$$i\alpha_j = (1+k_{\mathrm{I}})(\eta+v), \tag{7.20}$$

and, noting that negating v is equivalent to interchanging K_j and K_j',

$$i\beta_j = (1+k_{\mathrm{I}})(\eta-v). \tag{7.21}$$

Note that (7.17), (7.20) and (7.21) imply the relation (6.10).

The elliptic integrals $\mathscr{K}, \mathscr{K}', \mathscr{K}_{\mathrm{I}}, \mathscr{K}_{\mathrm{I}}'$ are related by

$$\mathscr{K} = (1+k_{\mathrm{I}})\mathscr{K}_{\mathrm{I}}, \quad \mathscr{K}' = \tfrac{1}{2}(1+k_{\mathrm{I}})\mathscr{K}_{\mathrm{I}}'. \tag{7.22}$$

By eliminating $\eta, v, \mathscr{K}_{\mathrm{I}}, \mathscr{K}_{\mathrm{I}}'$ between (7.7), (7.17) and (7.20)–(7.22), and exhibiting the site dependence of z, it follows that

$$q = \exp(-2\pi\mathscr{K}'/\mathscr{K}_{\mathrm{I}}), \quad x = \exp(-\pi\lambda/2\mathscr{K}),$$

$$z = z_j = \exp[-\pi(\alpha_j-\beta_j)/2\mathscr{K}]. \tag{7.23}$$

Together with (6.17), (6.4) and (4.8), this provides an explicit real definition of q, x, z_j for the ordered ferromagnetic phase.

Note that q and x depend only on k and λ, i.e. on K'' and Δ. Thus they are the same for all sites of the lattice $\mathscr{L}$, while z varies from site to site. Its value z_j at site j is given by (7.23), α_j and β_j being the elliptic parameters of the two angles at j. The order of the two angles is irrelevant here, since (7.9) is unchanged by inverting z.

Phase boundaries

The free energy function $f(K_j, K_j', K'')$ defined by (3.5) and (7.1)–(7.9) is analytic except when the middle two w's, in numerically decreasing order, are equal, i.e. when

$$a = b+c+d, b = a+c+d, c = a+b+d \quad \text{or} \quad d = a+b+c. \tag{7.24}$$

At these surfaces the correlation length goes to infinity (Johnson, Krinsky & McCoy 1973), so they are surfaces of critical points.

From (7.12), (7.13), (6.4) and (3.7),

$$k^{-2} = a^*b^*c^*d^*/abcd, \tag{7.25}$$

$$\frac{1-k^2}{k^2} = \frac{(a-b+c+d)(-a+b+c+d)(a+b+c-d))(-a-b+c-d)}{16\,abcd}. \tag{7.26}$$

Thus $k^2 = 1$ if, and only if, the system is critical.

The system is in an ordered phase if one of a, b, c, d is greater than the sum of the other three. From (7.26) and (6.4) this implies that

$$0 < k^2 < 1(|\Delta| > 1), \tag{7.27}$$

and vice versa.

The system is disordered if each of a, b, c, d is less than the sum of the other three, i.e. if

$$k^{-2} < 1\,(|\Delta| < 1). \tag{7.28}$$

Although a, b, c, d may vary from site to site, k (and Δ) does not. Thus for any lattice $\mathscr{L}$ we expect a Z-invariant model to be ordered if (7.27) is satisfied, disordered if (7.28) is satisfied, and to be critical if $k^2 = 1$ (and $\Delta = \pm 1$).

The definition (7.1)–(7.9) of f is somewhat cumbersome to use in the disordered phase, there being eight different cases to consider. This is related to the fact that k^2 can be either positive (greater than one), or negative. Nevertheless, f is analytic throughout the disordered phase.

8. Kagomé lattice eight-vertex model

So far, we have considered a very general 'lattice' $\mathscr{L}$, made up of almost any set of intersecting straight lines. The advantage of doing this is that it brings out very clearly the trigonometric character of the star-triangle conditions on the interaction coefficients of the eight-vertex model. It is, however, unnecessarily general for the purpose outlined in § 1.

In this section we specialize to a regular eight-vertex model on the Kagomé lattice shown in figure 8. This can be divided into three equivalent sub-lattices, labelled 1, 2 and 3 in figure 8.

Associate spins with the *faces* of the lattice. On every up-triangle assign the same set of two-spin interaction coefficients K_1, K_2, K_3, K_1', K_2', K_3' as indicated in figure 9. Also assign the same four-spin interaction coefficient K'' to every set of four spins round a site of the Kagomé lattice. The

Hamiltonian is then given by (2.1), with K_{PAR} and K_{RAQ} equal to the appropriate two-spin coefficient, and $K''_A = K''$.

The spins on the triangular faces form a honeycomb lattice, interacting with their nearest neighbours with coefficients K_1, K_2, K_3, as indicated in figure 9. The spins on the hexagonal faces form a triangular lattice, with interaction coefficients K'_1, K'_2, K'_3.

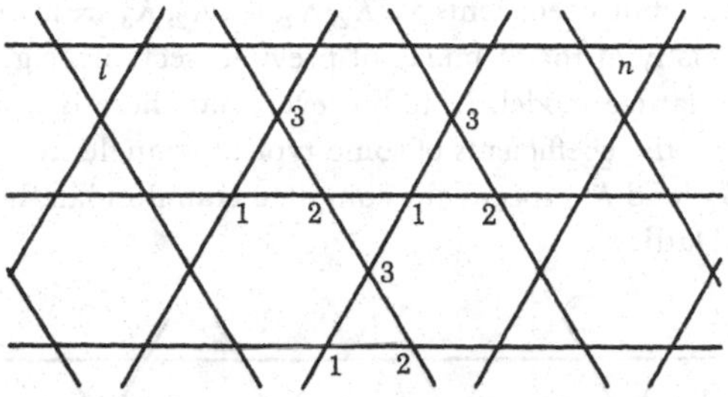

FIGURE 8. The Kagomé lattice, with sites divided into three equivalent classes 1, 2, 3.

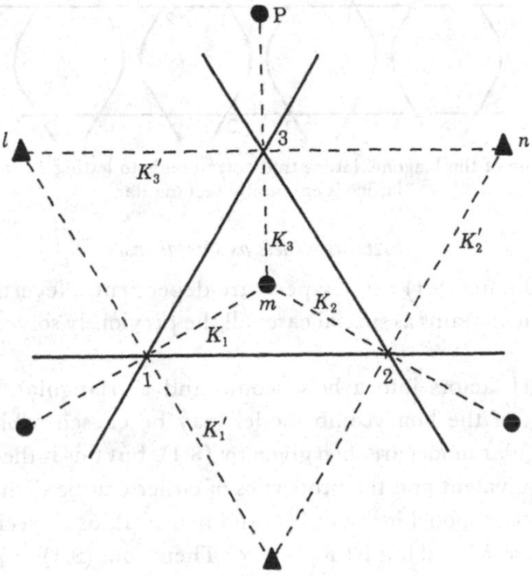

FIGURE 9. An up-triangle of the Kagomé lattice. The solid lines are lattice edges, while the circles and triangles denote the spins associated with the lattice faces. Broken lines represent two-spin intersections, the corresponding coefficients K_1, ..., K'_3 being shown.

Thus this eight-vertex model consists of an honeycomb and a triangular Ising model, with four-spin interactions between them.

We require that the star-triangle relations (4.4) be satisfied for every up-triangle of the type shown in figure 9, i.e.

$$\left.\begin{aligned} \exp(2K'_2+2K'_3) &= \cosh(K_1+K_2+K_3-K'')/\cosh(-K_1+K_2+K_3+K''), \\ \exp(2K'_3+2K'_1) &= \cosh(K_1+K_2+K_3-K'')/\cosh(K_1-K_2+K_3+K''), \\ \exp(2K'_1+2K'_2) &= \cosh(K_1+K_2+K_3-K'')/\cosh(K_1+K_2-K_3+K''). \end{aligned}\right\} \quad (8.1)$$

The model is then Z-invariant, so the thermodynamic properties are given by (5.1), (5.2) and (5.7). In particular, the mean free energy per site of the Kagomé lattice model is given by

$$f_{\text{Kagomé}} = \tfrac{1}{3}[f(K_1, K_1', K'') + f(K_2, K_2', K'') + f(K_3, K_3', K'')], \tag{8.2}$$

where $f(K_j, K_j', K'')$ is the free energy per site of a regular square lattice model with coefficients K_j, K_j', K'', given by equations (3.5) and (7.1)–(7.9).

Since the two-spin interaction coefficients $K_1, K_2, K_3, K_1', K_2', K_3'$ are arranged in the same way in figure 9 as in figure 4, many of the formulae of previous sections, e.g. (4.10), can be applied directly to this Kagomé lattice model. Note however that there is a difference in viewpoint: previously $K_1, \ldots, K_3'$ were the coefficients of some typical triangle in $\mathscr{L}$, different for different triangles. Here $K_1, \ldots, K_3'$ (and K'') specify the complete Hamiltonian, being the same for all up-triangles in the Kagomé lattice.

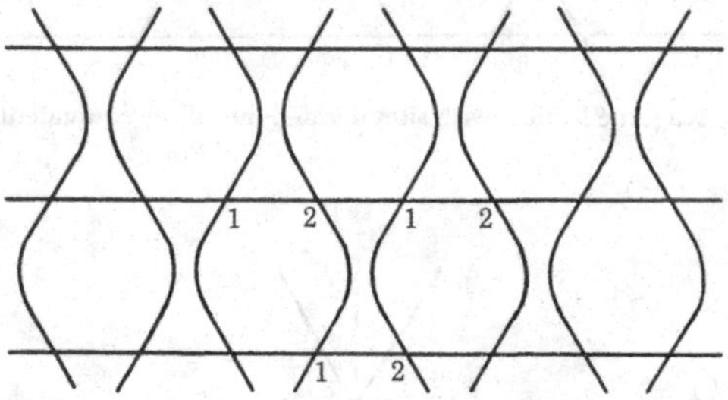

FIGURE 10. The deformation of the Kagomé lattice that corresponds to letting K_3 become infinite. The resulting lattice is essentially rectangular.

Previous models as special cases

In general the conditions (8.1) are temperature-dependent. Nevertheless, the model is still very interesting, since it contains as special cases all the previously solved models (i) to (vi) listed in §1.

If $K'' = 0$, the model factors into a honeycomb and a triangular Ising model. The three coefficients K_1, K_2, K_3 of the honeycomb model may be chosen arbitrarily. The coefficients K_1', K_2', K_3' of the triangular model are then given by (8.1), but this is the star-triangle relation, so the two models are equivalent and the properties of either can be deduced from the properties of their product. Thus this model includes (ii), and hence (i), as a special case.

Alternatively, suppose $K'' \neq 0$ but let $K_3 \to +\infty$. Then from (8.1)

$$K_1' = K_2, \quad K_2' = K_1, \tag{8.3a}$$

$$K_3' = -K''. \tag{8.3b}$$

Consider the interactions between the four spins on faces l, m, n, p round a site of type 3, as in figure 9. Since $K_3 \to +\infty$, σ_m and σ_p must be equal. The remaining interactions are

$$K_3' \sigma_l \sigma_n + K'' \sigma_l \sigma_n \sigma_m \sigma_p. \tag{8.4}$$

Since $\sigma_m = \sigma_p = \pm 1$, it follows that $\sigma_m \sigma_p = 1$. From (8.3b) the interaction (8.4) therefore vanishes.

Thus in the limit $K_3' \to \infty$ the faces m and p can be identified and the faces l and n separated. Graphically this is equivalent to deforming the Kagomé lattice of figure 8 to the lattice of figure 10.

The latter is simply a square lattice. Using (8.3*a*) we find that the model is now a regular square lattice eight-vertex model, with two-spin interactions K_1 and K_2, and four-spin interaction K''. Thus (iv), and hence (iii), are special cases of the Kagomé lattice model.

An interesting isotropic case is when $K_1 = K_2 = K_3 = K''$, which from (8.1) implies that $K_1' = K_2' = K_3' = 0$. Let $\sigma_1, \ldots, \sigma_6$ be the six spins round an up-triangle, as in figure 4*a*. Then from (4.2*a*) the contribution of the triangle to the partition function is

$$W_1 = 2\cosh[K''(\sigma_1+\sigma_3+\sigma_5+\sigma_1\sigma_6\sigma_2+\sigma_3\sigma_2\sigma_4+\sigma_5\sigma_4\sigma_6)]. \tag{8.5}$$

Using $\sigma_i^2 = 1$, one can verify that this is the same as

$$W_1 = 2\cosh[K''(\sigma_1\sigma_2+\sigma_2\sigma_3+\sigma_3\sigma_4+\sigma_4\sigma_5+\sigma_5\sigma_6+\sigma_6\sigma_1)]. \tag{8.6}$$

(An easy way to do this is to verify that the squares of the bracketted expressions in (8.5) and (8.6) are the same.)

From (8.6), W_1 can be written

$$W_1 = \sum_{\sigma_0} \exp[K''\sigma_0(\sigma_1\sigma_2+\ldots+\sigma_6\sigma_1)], \tag{8.7}$$

where σ_0 can be regarded as the spin inside the up-triangle, as in figure 4*a*. But this is just the triangle contribution to the partition function of a system with Hamiltonian

$$-\beta\mathscr{H} = K''\Sigma\,\sigma_i\sigma_j\sigma_k, \tag{8.8}$$

the summation being over all triplets of spins consisting of one spin inside an up-triangle and two surrounding spins that are adjacent to one another. If two spins are regarded as 'neighbours' if they both lie in such a triplet, then they form a triangular lattice, and the sum in (8.8) is over all faces of this lattice. Thus (8.8) is then the Hamiltonian of the three-spin model (v), which is therefore also a special case of the Kagomé lattice eight-vertex model.

As was remarked in § 1, (v) can also be transformed to a square-lattice eight-vertex model (Baxter & Enting 1976).

Finally, by using the vertex formulation of § 3 and remembering that K'' is site-independent, (3.5) and (8.1) imply that the vertex weights must satisfy

$$c_1 d_1/a_1 b_1 = c_2 d_2/a_2 b_2 = c_3 d_3/a_3 b_3, \tag{8.9}$$

$$b_i/c_i = (a_j a_k - b_j b_k)/(c_j c_k - d_j d_k), \tag{8.10}$$

for all permutations (i, j, k) of (1, 2, 3).

Setting $d_1 = d_2 = d_3 = 0$, the model becomes the six-vertex model solved by Baxter, Temperley & Ashley (1977), (8.10) becoming their restriction (6).

Classification of phases

Returning to the general Kagomé lattice model, the coefficients K_1, K_2, K_3, K'' can be regarded as independent real parameters. Then K_1', K_2', K_3' are uniquely defined by (8.1) and are also real.

The elliptic modulus k is given by (6.4) and (4.8). Using (8.1) to eliminate K_j', and setting

$$\left.\begin{aligned} L_1 &= -K_1+K_2+K_3-K'', & L_2 &= K_1-K_2+K_3-K'',\\ L_3 &= K_1+K_2-K_3-K'', & L_4 &= K_1+K_2+K_3+K'', \end{aligned}\right\} \tag{8.11a}$$

$$\left.\begin{aligned} M_1 &= -K_1+K_2+K_3+K'', & M_2 &= K_1-K_2+K_3+K'',\\ M_3 &= K_1+K_2-K_3+K'', & M_4 &= K_1+K_2+K_3-K'', \end{aligned}\right\} \tag{8.11b}$$

$$l = \cosh L_1, \quad m = \cosh L_2, \quad n = \cosh L_3, \quad p = \cosh L_4, \tag{8.12}$$

we find after some lengthy algebra that

$$\Delta = -\tfrac{1}{4}[\cosh M_1 \cosh M_2 \cosh M_3 \cosh M_4]^{-\frac{1}{2}}$$
$$\times \{\sinh 2K'' + 2\tanh 2K'' \cosh 2K_1 \cosh 2K_2 \cosh 2K_3 + 2\sinh 2K_1 \sinh 2K_2 \sinh 2K_3\}, \quad (8.13)$$

$$\frac{1-k^2}{k^2} = \frac{(-l+m+n+p)(l-m+n+p)(l+m-n+p)(-l-m-n+p)}{16\cosh M_1 \cosh M_2 \cosh M_3 \cosh M_4}. \quad (8.14)$$

[Also, k^{-2} is given by the right hand side of (8.14), but with each cosh in (8.12) replaced by sinh.]

Using the argument of §7, we expect the model to be in an ordered phase if $0 < k^2 < 1$, disordered otherwise. From (8.11)–(8.14) it follows that there are eight domains in (K_1, K_2, K_3, K'') space in which the system is ordered, namely those in which one of l, m, n, p is greater than the sum of the other three, the corresponding L_j being either positive or negative. There is one domain in which the system is disordered, namely when each of l, m, n, p is less than the sum of the other three.

The archetypal ordered phase is

$$p > l+m+n, \quad L_4 > 0. \quad (8.15)$$

In this domain the spins on each sub-lattice are ferromagnetically ordered. If $\alpha_1, \alpha_2, \alpha_3, \beta_1, \beta_2, \beta_3$ are defined by (6.17), then (6.10) and (6.13) are satisfied. Parameters q, x, z_1, z_2, z_3 can then be defined by (7.23). By using (7.9) and (3.5), the free energy function in (8.2) is then given by

$$-\beta f(K_j, K_j', K'') = K_j + K_j' + K'' + \sum_{n=1}^{\infty} \frac{x^{-n}(x^{2n}-q^n)^2 (x^n + x^{-n} - z_j^n - z_j^{-n})}{n(1-q^{2n})(1+x^{2n})}. \quad (8.16)$$

If the conjectures of Barber & Baxter (1973) and Baxter & Kelland (1974) are valid, then the spontaneous magnetization and polarization are given by (7.10) and (7.11).

The other seven ordered phases can all be mapped to (8.15) by reversing some of the lattice spins. For instance, reversing the spin inside each up-triangle is equivalent to negating K_1, K_2, K_3, K'', while leaving K_1', K_2', K_3' unchanged. This negates $L_1, \ldots, M_4$ and leaves k unchanged. Thus it maps the domain (8.15) to $p > l+m+n$, $L_4 < 0$; and vice versa.

Also, reversing the spins between alternate pairs of the horizontal lines in figure 8 is equivalent to negating K_1, K_2, K_1', K_2', while leaving K_3, K_3', K'' unchanged. From (8.11)–(8.14) this leaves unchanged but maps (8.15) to the domain $n > l+m+p$, $L_3 < 0$; and vice versa.

Similarly, one can reverse all the spins between alternate pairs of parallel diagonal lines in figure 8, thereby mapping (8.15) to either $l > m+n+p$, $L_1 < 0$, or to $m > l+n+p$, $L_2 < 0$. These domains can then be mapped to the domains with corresponding $L_j > 0$ by further reversing the spins inside each up-triangle.

The disordered domain is

$$l, m, n, p < \tfrac{1}{2}(l+m+n+p).$$

(Note that these inequalities are certainly satisfied if the interaction coefficients are all small.) In this domain M and P must be zero. The free energy can be obtained from (8.2), by using the definitions (3.5), (7.1)–(7.9) of the function $-\beta f(K_j, K_j', K'')$.

Critical behaviour

Since the restrictions (8.1) can in general only be satisfied for a few discrete values of the temperature (if any), we cannot discuss the temperature dependence of the model. Nevertheless, we can define a parameter which plays the same rôle, namely

$$t = (k^2-1)/k^2. \tag{8.17}$$

This is positive for the disordered phase, negative for ordered ones, and vanishes linearly on a path in (K_1, K_2, K_3, K'') space as a critical surface is crossed non-tangentially.

Suppose one starts in the ordered ferromagnetic phase and approaches the critical surface $p = l+m+n$, $L_4 > 0$. Then k^2 tends to one from below. From (6.17), $\alpha_1, \alpha_2, \alpha_3, \beta_1, \beta_2, \beta_3, \lambda$ and $\mathcal{K}'$ all remain finite and analytic, while $\mathcal{K}$ diverges, being given by

$$\mathcal{K} = \tfrac{1}{2}\ln[-16/t] + \text{vanishing terms}. \tag{8.18}$$

From (7.23), q, x, z_1, z_2, z_3 therefore all tend to one, and the expressions (7.9)–(7.11) for f, M, P become very slowly convergent. It is then appropriate to make a Poisson transform of these series and products (as is explained in Baxter (1972), Barber & Baxter (1973), Baxter & Kelland (1974)) Doing this, the dominant singular contribution to $-\beta f$ is found to be given by

$$(-\beta f)_{\text{sing}} \propto \exp(-2\pi\mathcal{K}/\lambda), \tag{8.19}$$

while M and P are given by

$$M \simeq 2^{\frac{1}{2}} \exp(-\pi\mathcal{K}/8\lambda), \tag{8.20}$$

$$P \simeq (2\mathcal{K}'/\lambda)\exp[-\pi\mathcal{K}(2\mathcal{K}'-\lambda)/4\lambda\mathcal{K}']. \tag{8.21}$$

From (6.17*d*), $\mathcal{K}' = \frac{1}{2}\pi$ at criticality. From (8.18) it follows that

$$(-\beta f)_{\text{sing}} \propto (-t)^{\pi/\lambda}, \tag{8.22}$$

$$M \propto (-t)^{\pi/16\lambda}, \tag{8.23}$$

$$P \propto (-t)^{(\pi-\lambda)/4\lambda}. \tag{8.24}$$

Thus the critical exponents are the same as those for a corresponding square lattice, namely

$$\alpha = 2-\pi/\lambda, \quad \beta_M = \pi/16\lambda, \quad \beta_e = (\pi-\lambda)/4\lambda, \tag{8.25}$$

λ having the same value at criticality as the parameter μ or $\bar{\mu}$ used in earlier papers. It is defined by (6.17*c*). At criticality $k^2 = 1$, so λ is then given by

$$\cos\lambda = -\tanh 2K'', \quad 0 < \lambda < \pi. \tag{8.26}$$

For the Ising model case, $K'' = 0$ and $\lambda = \mathcal{K}' = \pi/2$.

This formula applies on the critical surface $p = l+m+n$, $L_4 > 0$, when $\Delta = -1$, so can be replaced by

$$\cos\lambda = \Delta^{-1}\tanh 2K'', \quad 0 < \lambda < \pi. \tag{8.27}$$

From (8.13), $\Delta^{-1}\tanh 2K''$ is unchanged by the various mappings between the eight ordered states described above, so (8.27) and (8.25) are valid on all eight critical surfaces of the Kagomé lattice model. By using (7.12) and (7.13), (8.27) can be written in terms of the Boltzmann weights of a site of type j as

$$\cos\lambda = 2(c_j d_j - a_j b_j)/(a_j^2+b_j^2-c_j^2-d_j^2), \quad 0 < \lambda < \pi. \tag{8.28}$$

On a critical surface, one of a_j, b_j, c_j, d_j is equal to the sum of the other three.

9. Two-spin correlations

Arguments similar to those of §5 can be used to establish remarkable equivalences between the two-spin correlations of Z-invariant lattice models.

Consider any two faces l and n of an arbitrary lattice $\mathscr{L}$, as in figure 11. Construct a Z-invariant eight-vertex model on $\mathscr{L}$, with given values of K'' and Δ, and consider the correlation $\langle \sigma_l \sigma_n \rangle$. Suppose that l and n are deep within $\mathscr{L}$, so that boundary conditions are irrelevant.

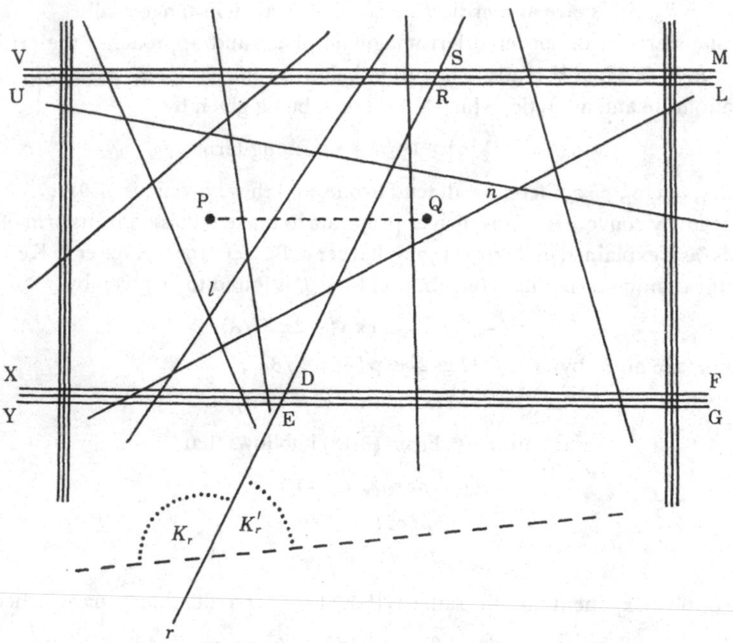

Figure 11. A lattice $\mathscr{L}$ extended by adding a rectangular frame aligned to the line PQ between faces l and m. The broken line is a base line of $\mathscr{L}$.

Let P be a point inside l, Q a point inside m, chosen so that the line PQ does not pass through any lattice site. Orient the lattice so that PQ is horizontal. The line segment PQ intercepts some of the lattice lines. Label these 1, 2, ..., m. Let K_r', K_r be the angle coefficients of the intersection of line r with some base line, ordered as in figure 11. Let α_r, β_r be the corresponding elliptic angle parameters, defined in the ferromagnetic phase by (6.17), (6.5) and (6.6). Thus $\alpha_r + \beta_r = \lambda$, and α_r is the 'line angle parameter' (6.20).

We shall show that $\langle \sigma_l \sigma_n \rangle$ is a function only of m, K'', Δ and $K_1', \ldots, K_m'$, $K_1, \ldots, K_m$, the function being *independent of the structure of* $\mathscr{L}$. In particular, it is unchanged by simultaneously interchanging K_i' with K_j', K_i with K_j.

By using the elliptic parameters of §6, this implies that

$$\langle \sigma_l \sigma_n \rangle = g_m(k, \lambda; \alpha_1, \ldots, \alpha_m), \tag{9.1}$$

where the function g_m is the same for all Z-invariant models, and is a symmetric function of $\alpha_1, \ldots, \alpha_m$. Adding the same constant to each of $\alpha_1, \ldots, \alpha_m$ re-defines the coefficients on the base line, but from (6.21) leaves the coefficients at the sites of $\mathscr{L}$ unchanged. Thus g_m must be a function only of the differences of $\alpha_1, ..., \alpha_m$.

Note that g_1 is the correlation between two adjacent spins, i.e. the polarization. We have already seen that this is a function only of k and λ.

To establish these results, first extend $\mathscr{L}$ by adding $2M$ lines parallel to PQ, M being close together and above $\mathscr{L}$, the others being below $\mathscr{L}$, as in figure 11. Choose the angle coefficients of the intersections of these lines with lattice line r to be K'_r, K_r, for $r = 1, ..., m$. For example, in figure 11

$$K_{\mathrm{RDF}} = K_{\mathrm{REG}} = K_{\mathrm{DRU}} = K_{\mathrm{DSV}} = K'_r,$$

$$K_{\mathrm{RDX}} = K_{\mathrm{REY}} = K_{\mathrm{DRL}} = K_{\mathrm{DSM}} = K_r.$$

This is consistent with Z-invariance.

Further extend $\mathscr{L}$ by adding $2M$ vertical lines, M to the left of $\mathscr{L}$ and M to the right of $\mathscr{L}$. Thus there are $4M$ lines forming a rectangular 'frame' around $\mathscr{L}$. At all intersections of framing lines assign the same coefficient K_0 to the top-left and bottom-right angles, K'_0 to the other two angles, choosing K_0, K'_0 to satisfy (4.8), with $\Delta_j = \Delta$.

Extend all lattice lines (rotating them slightly if necessary) to cross all the framing lines. Extend the convex boundary outwards to Include all these intersections. Assign coefficients to new intersections according to the rules of §4.

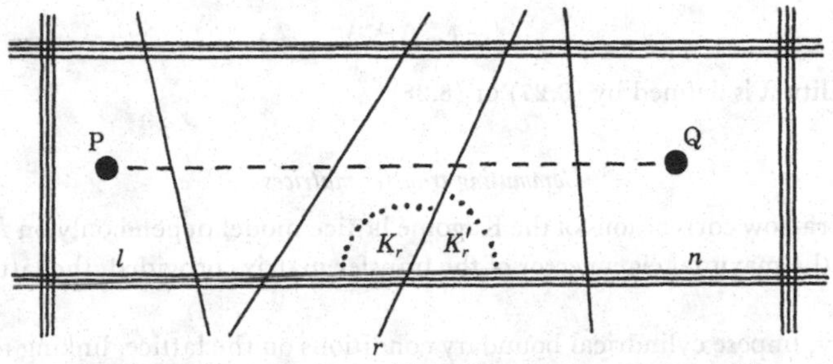

FIGURE 12. The irregular rectangular lattice obtained by shifting the framing lines of figure 11 inwards towards P and Q, and then neglecting sites outside the framing lines.

Suppose this can be done without introducing non-positive Boltzmann weights (for the geometric model of §6 this is certainly so). Then $\langle \sigma_l \sigma_n \rangle$ should be the same for this extended lattice as for the original lattice $\mathscr{L}$, provided l and n were originally deep within $\mathscr{L}$.

The extended lattice is by construction Z-invariant. Shift the framing lines inwards towards PQ until all sites of $\mathscr{L}$ lie outside the framing lines, as indicated in figure 12. This does not change the order of the lines at the boundary, nor does any line cross face l or face n. From §4 it follows that $\langle \sigma_l \sigma_n \rangle$ is unchanged.

However, the picture is now quite different. From figure 12, l and n lie in the same row of a rectangular lattice of $2M$ rows and $2M$ columns. The coefficient of the top-left (top-right) angle is K_r (K'_l) for a site on column and between P and Q. It is K_0 (K'_0) for a site on a column to the left of P, or to the right of Q. The four-spin coefficient K'' and the parameter Δ are of course the same as in the original lattice $\mathscr{L}$.

If M is large, we expect $\langle \sigma_l \sigma_n \rangle$ to be unchanged by deleting all sites outside this rectangular lattice, so $\langle \sigma_l \sigma_n \rangle$ is the correlation between two spins in the same row of a rectangular lattice model. The model is not regular, since the two-spin coefficients vary from column to column.

Further, in the extended lattice we could have also made parallel shifts of lines 1, ..., m to re-order their intersections with the line PQ. This also leaves the boundary order unchanged, and no line crosses face l or face n, so $\langle\sigma_l\sigma_n\rangle$ is unchanged. Hence $\langle\sigma_l\sigma_n\rangle$ is independent of the order of columns 1, ..., m in figure 12. This establishes the assertions made above, in particular equation (9.1).

Intra-row correlations in the Kagomé lattice

In specific cases it may be possible to simplify the above argument. For instance, consider the correlation $\langle\sigma_l\sigma_n\rangle$, where l and n are the faces of the Kagomé lattice shown in figure 8. By considering a finite, but large, lattice with extended external edges, it becomes apparent that the horizontal lines above (below) l and n can be shifted far upwards (downwards), while leaving $\langle\sigma_l\sigma_n\rangle$ unchanged. Thus $\langle\sigma_l\sigma_n\rangle$, and any correlation between spins in the same row, is the same as if all horizontal lines in figure 8 were deleted. This leaves a regular square lattice drawn diagonally, with coefficients K_3, K_3', K''. Thus intra-row correlations for the Kagomé lattice are the same as those of this square lattice.

For the six-vertex models this has been established directly by Baxter, Temperley & Ashley (1977, §2.3) using the results of Kelland (1974).

Since the correlation length ξ of the square lattice model diverges when $k^2 = 1$ (Johnson, Krinsky & McCoy 1973), that of the Kagomé lattice model of section 8 must do so in the same way, i.e.

$$\xi \propto (-t)^{-\pi/2\lambda}, \tag{9.2}$$

where at criticality λ is defined by (8.27) or (8.28).

Commuting transfer matrices

Just as the intra-row correlations of the Kagomé lattice model depend only on K_3, K_3', K'' so do the elements of the maximal eigenvector of the transfer matrix, provided, the lattice is infinitely large.

More strongly, impose cylindrical boundary conditions on the lattice, linking the right side to the left, and consider a finite Kagomé lattice model where K_1, K_1', K_2, K_2' can vary from row to row, but K_3, K_3', K'' are constant and the star-triangle relations (8.1) are satisfied for all up-triangles. Then the model is still Z-invariant, and one can establish that interchanging two horizontal lines (together with their associated values of K_1, K_1', K_2, K_2') leaves Z unchanged. It also leaves unchanged all correlations not involving the spins inbetween the two lines.

This implies that the transfer matrices associated with the two lines commute. In particular, they commute when K_3 becomes infinite and the model becomes a square lattice model, as explained in the text following equation (8.3).

Establishing this commutation property was the first step in the original solution of the square-lattice eight-vertex model (Sutherland 1970; Baxter 1972).

Disordered ferromagnetic phase

A particularly important case of the disordered phase, for any lattice $\mathscr{L}$, is when at each site the inequality (6.14) is *not* valid, but (6.25) is. In terms of the vertex weights (3.5) this is the domain

$$a_j + b_j + d_j > c_j > a_j + b_j - d_j, \quad c_j > |a_j - b_j| + d_j \tag{9.3}$$

i.e. w_1, ..., w_4 in (7.1) are ordered so that $w_1 > w_3 > w_2 > |w_4|$.

If the interaction coefficients are all positive, then this is the only disordered case that can occur. From (7.13), (7.25) and (7.26), in this case

$$-1 < \Delta < 0, \quad k^2 > 1, \tag{9.4}$$

where Δ, k are defined by (4.8) and (6.4).

Interchanging w_2 and w_3 is equivalent to making the duality transformation (3.7). Thus if we first make this transformation and then re-define k, α_j, β_j, λ, $\mathscr{K}'$, $\mathscr{K}$ by (6.4), and (6.8) or (6.17), then the relations (7.23) will still be valid, and the free energy function will be given by (7.9), with $c' = c^* = \frac{1}{2}(a+b+c+d)$.

Let k^*, α_j^*, β_j^*, $\mathscr{K}^{*\prime}$, $\mathscr{K}^*$ be the new elliptic parameters defined by this procedure. The duality transformation (3.7) inverts k, while leaving $k^{\frac{1}{2}}\sinh 2K''$, $k^{\frac{1}{2}}\sinh 2K_j'$, $k^{\frac{1}{2}}\sinh 2K_j$ unchanged. From (6.4) and (6.8) it follows that

$$k^* = \{\Delta^2\cosh^2 2K'' - \sinh^2 2K''\}^{\frac{1}{2}}, \tag{9.5}$$

$$\left.\begin{aligned} \sinh 2K'' &= -\mathrm{i}/\mathrm{sn}\,(\mathrm{i}\lambda^*, k^*),\\ \sinh 2K_j' &= \mathrm{i}/\mathrm{sn}\,(\mathrm{i}\alpha_j^*, k^*),\\ \sinh 2K_j &= \mathrm{i}/\mathrm{sn}\,(\mathrm{i}\beta_j^*, k^*), \end{aligned}\right\} \tag{9.6}$$

where $0 < k^* < 1$ and λ^*, α_j^*, β_j^* are all real, lying in the interval $(0, 2\mathscr{K}^{*\prime})$.

By noting that $k^* = k^{-1}$ and $\mathrm{sn}\,(u, k^{-1}) = k\,\mathrm{sn}\,(k^{-1}u, k)$ (Gradshteyn & Ryzhik 1965, §8.152), and comparing (9.6) with (6.8), it follows that

$$\lambda^* = k\lambda, \quad \alpha_j^* = k\alpha_j, \quad \beta_j^* = k\beta_j, \tag{9.7}$$

where λ, α_j, β_j are defined by (6.17). Thus the elliptic parameters in the ferromagnetic disordered phase are proportional to the analytic continuation of those in the ordered phase. The Z-invariance conditions (i), (ii) and (iii) are therefore unaltered. In particular, at each site j of $\mathscr{L}$ we must have

$$\alpha_j^* + \beta_j^* = \lambda. \tag{9.8}$$

Ising model case: $K'' = 0$

Unfortunately the correlation functions g_m have not yet been evaluated for general values of K'' (apart from g_2, which can in principle be obtained by differentiating the free energy).

They can be obtained when $K'' = 0$, since the model then factors into two independent Ising models. For even m the functions g_m can be obtained by generalizing the Pfaffian method of Montroll, Potts & Ward (1963). In particular, in the high-temperature disordered phase

$$g_2(k;\alpha_1,\alpha_2) = \frac{4\pi}{k^*\mathscr{K}^*}\sum_{n=1}^{\infty}\frac{p^{2n-1}\cosh\left[(n-\frac{1}{2})\,\pi(\alpha_1^*-\alpha_2^*)/\mathscr{K}^*\right]}{(1+p^{2n-1})^2}, \tag{9.9}$$

where

$$p = \exp\left(-\pi\mathscr{K}^{*\prime}/\mathscr{K}^*\right), \tag{9.10}$$

k^*, $\mathscr{K}^*$, $\mathscr{K}^{*\prime}$, are the elliptic parameters defined by (9.5) and (4.8), i.e. (using $K'' = 0$)

$$k^* = \sinh 2K_j \sinh 2K_j', \tag{9.11}$$

and α_1^*, α_2^* are the elliptic line parameters defined by (6.20) (with an asterisk on each α). For $m = 2$ there are two lines between spins σ_l and σ_n, and $|\alpha_1^* - \alpha_2^*|$ is the elliptic parameter of the angle between these lines that includes neither face l nor face n.

When $K'' = 0$ we see from (9.6) that $\lambda^* = \mathscr{K}^{*\prime}$, so the dependence of g_m on λ can be suppressed, and α_1^*, α_2^* must lie in the interval $(0, \mathscr{K}^{*\prime})$.

For the rest of this section we consider elliptic functions of modulus k^*. We omit the asterisks on $k, k', \mathscr{K}, \mathscr{K}', \alpha_j$. The formula (9.9) can be written as

$$g_2(k; \alpha_1, \alpha_2) = -\frac{2\mathscr{K}}{\pi}\left(\frac{k'}{k}\right)^{\frac{1}{2}} \frac{\Theta_1'(\mathrm{i}\alpha_1 - \mathrm{i}\alpha_2)}{H(\mathrm{i}\alpha_1 - \mathrm{i}\alpha_2)}, \tag{9.12}$$

$$= -\frac{k'^{\frac{1}{2}}\Theta^2(0)}{2\pi}\int_{-\mathscr{K}}^{\mathscr{K}} \frac{\Theta(\mathrm{i}\alpha_1 - \mathrm{i}\alpha_2)\,\Theta_1(2s - \mathrm{i}\alpha_1 - \mathrm{i}\alpha_2)}{h(s-\mathrm{i}\alpha_1)\,h(s-\mathrm{i}\alpha_2)}\,\mathrm{d}s, \tag{9.13}$$

where H, Θ, H_1, Θ_1 are the Jacobi theta functions (Gradshteyn & Ryzhik 1965, §§ 8.191 and 8.192), and

$$h(u) = H(u)\,\Theta(u). \tag{9.14}$$

For general even m, the integral form (9.13) can be generalized to

$$g_m(k; \alpha_1, \ldots, \alpha_m) = \frac{1}{(m/2)!\,H_1(0)}\left[-\frac{k'^{\frac{1}{2}}\Theta^3(0)}{2\pi}\right]^{m/2} \int_{-\mathscr{K}}^{\mathscr{K}} \cdots \int \frac{\prod\limits_{1 \leqslant j < l \leqslant m} \Theta(\mathrm{i}\alpha_j - \mathrm{i}\alpha_l) \prod\limits_{1 \leqslant j < l \leqslant \frac{1}{2}m} h^2(s_j - s_l)}{\prod\limits_{r=1}^{m}\prod\limits_{j=1}^{\frac{1}{2}m} h(s_j - \mathrm{i}\alpha_r)}$$

$$\times\, \psi[2(s_1 + \ldots + s_{\frac{1}{2}m}) - \mathrm{i}\alpha_1 - \ldots - \mathrm{i}\alpha_m]\,\mathrm{d}s_1 \ldots \mathrm{d}s_{\frac{1}{2}m}, \tag{9.15}$$

where

$$\psi(u) = \Theta_1(u) \quad \text{if } \tfrac{1}{2}m \text{ odd},$$
$$= H_1(u) \quad \text{if } \tfrac{1}{2}m \text{ even}. \tag{9.16}$$

This is a rather unwieldy formula, but it does explicitly exhibit the fact that g_m is a symmetric function of $\alpha_1, \ldots, \alpha_m$. A useful recurrence relation is

$$g_m(k; \alpha_1, \ldots, \alpha_{m-2}, \alpha_{m-1}, \alpha_{m-1} + \mathscr{K}') = g_{m-2}(k; \alpha_1, \ldots, \alpha_{m-2}), \quad g_2(k; \alpha, \alpha + \mathscr{K}') = 1. \tag{9.17}$$

Note that *any* two-spin correlation for any Z-invariant lattice Ising model (e.g. the triangular Ising model) must be of the form (9.15), with an appropriate choice of the parameters $\alpha_1, \ldots, \alpha_m$ of the intermediate lines.

References

Barber, M. N. & Baxter, R. J. 1973 *J. Phys.* C: *Solid St. Phys.* **6**, 2913–2921.
Baxter, R. J. 1969 *J. Math. Phys.* **10**, 1211–1216.
Baxter, R. J. 1970*a* *J. Math. Phys.* **11**, 784–789.
Baxter, R. J. 1970*b* *J. Math. Phys.* **11**, 3116–3124.
Baxter, R. J. 1971 *Stud. appl. Maths* (Massachusetts Institute of Technology) **50**, 51–69.
Baxter, R. J. 1972 *Annls Phys.* **70**, 193–228.
Baxter, R. J. 1973 *Annls Phys.* **76**, 1–71.
Baxter, R. J. & Enting, I. G. 1976 *J. Phys.* A: *Math. Gen.* **9**, L149–L152.
Baxter, R. J. & Kelland, S. B. 1974 *J. Phys.* C: *Solid St. Phys.* **7**, L403–L406.
Baxter, R. J., Kelland, S. B. & Wu, F. Y. 1976 *J. Phys.* A: *Math. Gen.* **9**, 397–406.
Baxter, R. J., Temperley, H. N. V. & Ashley, S. E. 1977 Triangular Potts model at its transition temperature, and related models, *Proc. R. Soc. Lond.* A
Baxter, R. J. & Wu, F. Y. 1974 *Aust. J. Phys.* **27**, 357–381.
Coxeter, H. S. M. 1947 *'Non-Euclidean geometry*. Toronto: University Press.
Fan, C. & Wu, F. Y. 1970 *Phys. Rev.* B **2**, 723–733.
Gradshteyn, I. S. & Ryzhik, I. M. 1965 *Table of integrals, series and products*. New York: Academic Press.
Green, H. S. 1963 *Z. Phys.* **171**, 129–148.
Greenhill, A. G. 1892 *Applications of elliptic functions* (Reprinted 1959). New York: Dover.
Houtappel, R. M. F. 1950 *Physica* **16**, 425–455.
Husimi, K. & Syozi, I. 1950 *Progr. theor. Phys.* **5**, 177–186 and 341–351.

Johnson, J. D., Krinsky, S. & McCoy, B. M. 1973 *Phys. Rev.* A **8**, 2526–2547.
Kadanoff, L. P. & Wegner, F. J. 1971 *Phys. Rev.* B **4**, 3989–3993.
Kelland, S. B. 1974 *Aust. J. Phys.* **27**, 813–829.
Lieb, E. H. 1967 *Phys. Rev.* **162**, 162–172; *Phys. Rev. Lett.* **18**, 1046–1048; **19**, 108–110.
Montroll, E. W., Potts, R. B. & Ward, J. C. 1963 *J. Math. Phys.* **4**, 308–322.
Onsager, L. 1944 *Phys. Rev.* **65**, 117–149.
Stephenson, J. 1964 *J. Math. Phys.* **5**, 1009–1024.
Sutherland, B. 1967 *Phys. Rev. Lett.* **19**, 103–104.
Sutherland, B. 1970 *J. Math. Phys.* **11**, 3183–3186.
Wannier, G. H. 1950 *Phys. Rev.* **79**, 357–364.
Wegner, F. J. 1973 *Physica* **68**, 570–578.
Wood, D. W. & Griffiths, H. P. 1972 *J. Phys.* C: *Solid St. Phys.* **5**, L253–L255.
Wu, F. Y. 1971 *Phys. Rev.* B **4**, 2312–2314.

Appendix A. Proof of quadrilateral theorem

First consider a typical triangle ABC as in figure 5. Use the notation of equation (4.1) and define using (4.6),

$$\left.\begin{aligned} a &= e^{2K_1}, \quad b = e^{2K_2}, \quad c = e^{2K_3},\\ a' &= e^{2K_1'}, \quad b' = e^{2K_2'}, \quad c' = e^{2K_3'},\\ t &= e^{-2K''}. \end{aligned}\right\} \tag{A 1}$$

(These a, b, c, a', b', c' are *not* the Boltzmann weights used in the text.) Then the star-triangle relations (4.4) are

$$b'c' = (1+tabc)/(bc+ta), \tag{A 2a}$$

$$c'a' = (1+tabc)/(ca+tb), \tag{A 2b}$$

$$a'b' = (1+tabc)/(ab+tc). \tag{A 2c}$$

Solving (A 2a) for a gives

$$a = t^{-1}(bcb'c'-1)/(bc-b'c'). \tag{A 3}$$

Alternatively, taking ratios of (A 2b) and (A 2c) to eliminate a', then solving for a, gives

$$a = t(bc'-b'c)/(bb'-cc'). \tag{A 4}$$

Eliminating a between (A 3) and (A 4) gives

$$\Delta(b,b') = \Delta(c,c'), \tag{A 5}$$

where

$$\Delta(b,b') = \tfrac{1}{2}[t^2(b^2+b'^2)-1-b^2b'^2]/[(1+t^2)\,bb']. \tag{A 6}$$

This is the equation $\Delta_B = \Delta_C$ of equation (4.7) and (4.8) of the text. Note that if it is satisfied, then (A 2a) is a corollary of (A 2b) and (A 2c).

Substituting the expression (A 3) for a into the numerator on the right hand side of (A 2b), and the expression (A 4) into the denominator, gives

$$a' = t^{-1}(bb'-cc')\,(b^2c^2-1)/[(bc-b'c')\,(b^2-c^2)]. \tag{A 7}$$

To summarize so far: the a, b, c, a', b', c' are Boltzmann weights associated with the angles of the triangle ABC; a, b, c are associated with the interior angles, while a', b', c' are associated with the complementary exterior angles. If the star-triangle relations (A 2) are satisfied, and b, b', c, c', t are known, then a can be obtained from either (A 3) or (A 4), and a' from (A 7). The b, b', c, c', t must satisfy (A 6).

Now consider the quadrilateral ABCDEF shown in figure 5. Let a, b, c, a', b', c', d, e, f, u, v, w be the weights associated with the indicated angles, e.g. $w = \exp(2K_{\mathrm{EFB}})$.

Suppose the star-triangle relations are satisfied for the triangles AEF, BFD, CDE. Then from (4.6), $K''_A, \ldots, K''_F$ all have the same value K'', so t is a constant weight. For the triangle AEF the relations (A 3), (A 4), (A 7) give

$$a' = t^{-1}(efvw-1)/(ew-fv), \tag{A 8}$$

$$a' = t(ef-vw)/(ev-fw), \tag{A 9}$$

$$a = t^{-1}(ev-fw)\,(e^2w^2-1)/[(ew-fv)\,(e^2-w^2)]. \tag{A 10}$$

For the triangle BFD they give

$$b = t^{-1}(dfuw-1)/(dw-fu), \tag{A 11}$$

$$b = t(df-uw)/(du-fw), \tag{A 12}$$

$$b' = t^{-1}(du-fw)\,(d^2w^2-1)/[(dw-fu)\,(d^2-w^2)]. \tag{A 13}$$

For CDE, the relation (A 4) gives

$$c = t(de-uv)/(ev-du). \tag{A 14}$$

We want to prove that the star-triangle relations are necessarily satisfied for the triangle ABC. In particular, we want to establish the relation (A 2c), i.e.

$$tc(ab-a'b') = aba'b'-1. \tag{A 15}$$

To do this, substitute the expressions (A 8), (A 10), (A 11), (A 13), (A 14) for a', a, b, b', c into the left hand side, and the expressions (A 9), (A 10), (A 12), (A 13) for a', a, b, b' into the right hand side. Multiplying out all denominator factors, (A 15) will be satisfied if $J = 0$, where J is the expression

$$\begin{aligned} J \equiv (de-uv)\,&[(ev-fw)\,(dfuw-1)\,(d^2-w^2)\,(e^2w^2-1) \\ &-(du-fw)\,(efvw-1)\,(e^2-w^2)\,(d^2w^2-1)] \\ &-(ev-du)\,[(ef-vw)\,(df-uw)\,(d^2w^2-1)\,(e^2w^2-1) \\ &-(ew-fv)\,(dw-fu)\,(d^2-w^2)\,(e^2-w^2)]. \end{aligned} \tag{A 16}$$

Note that J does not explicitly depend on t, which is a slight simplification. The choice of whether to use (A 8) or (A 9), (A 11) or (A 12) has been made to ensure this.

The expression J is a polynomial in w of degree six. Setting $w^2 = \pm 1$, we find that J vanishes, so it contains a factor w^4-1. Now it is not too difficult to verify that

$$J \equiv de\,(w^4-1)\,L, \tag{A 17}$$

where

$$\begin{aligned} L = du\,&[(e^2+v^2)\,(1+f^2w^2)-(f^2+w^2)\,(1+e^2v^2)] \\ +ev\,&[(f^2+w^2)\,(1+d^2u^2)-(d^2+u^2)\,(1+f^2w^2)] \\ +fw\,&[(d^2+u^2)\,(1+e^2v^2)-(e^2+v^2)\,(1+d^2u^2)]. \end{aligned} \tag{A 18}$$

The d, e, f, u, v, w are not independent, since from the corollary (A 5) of the star-triangle relations, applied to the triangles AEF, BFD, CDE, they must satisfy:

$$\Delta(d,u) = \Delta(e,v) = \Delta(f,w). \tag{A 19}$$

Using the form (A 6) of the function $\Delta(b,b')$, the equations (A 19) are linear in t^2. Eliminating t^2 gives the equation

$$L = 0, \tag{A 20}$$

where L is defined by (A 18). From (A 17) it follows that J does vanish, and hence the relation (A 2c) is satisfied for the triangle ABC in figure 4.

Now interchange d with u, e with w, and f with v in the above working. This leaves (A 19) and (A 20) still satisfied, and from the star-triangle relations for AEF, BFD, CDE, the right hand sides of equations (A 8)–(A 14) become a', a', a, c, c, c', b, respectively. Thus we have also established that

$$tb(ac - a'c') = aca'c' - 1, \tag{A 21}$$

which is the relation (A 2b).

Finally, note from (4.9) that the star-triangle relations for AEF, BFD, CDE imply that Δ has the same value at all points of the quadrilateral, and in particular that $\Delta_B = \Delta_C$. From the observation made after equation (A 6), it follows that the relation (A 2a) must also be satisfied.

Thus the star-triangle relations for the triangles AEF, BFD, CDE imply that the star-triangle relations are also satisfied for the triangle ABC, which is the required theorem.

ANNALS OF PHYSICS 120, 253–291 (1979)

Factorized S-Matrices in Two Dimensions as the Exact Solutions of Certain Relativistic Quantum Field Theory Models

ALEXANDER B. ZAMOLODCHIKOV

Institute of Theoretical and Experimental Physics, Moscow, USSR

AND

ALEXEY B. ZAMOLODCHIKOV

Joint Institute for Nuclear Research, Dubna, USSR

Received August 1, 1978

The general properties of the factorized S-matrix in two-dimensional space-time are considered. The relation between the factorization property of the scattering theory and the infinite number of conservation laws of the underlying field theory is discussed. The factorization of the total S-matrix is shown to impose hard restrictions on two-particle matrix elements: they should satisfy special identities, the so-called factorization equations. The general solution of the unitarity, crossing and factorization equations is found for the S-matrices having isotopic $O(N)$-symmetry. The solution turns out to have different properties for the cases $N = 2$ and $N \geqslant 3$. For $N = 2$ the general solution depends on one parameter (of coupling constant type), whereas the solution for $N \geqslant 3$ has no parameters but depends analytically on N. The solution for $N = 2$ is shown to be an exact soliton S-matrix of the sine-Gordon model (equivalently the massive Thirring model). The total S-matrix of the model is constructed. In the case of $N \geqslant 3$ there are two "minimum" solutions, i.e., those having a minimum set of singularities. One of them is shown to be an exact S matrix of the quantum $O(N)$-symmetric nonlinear σ-model, the other is argued to describe the scattering of elementary particles of the Gross-Neveu model.

1. INTRODUCTION

The general two-dimensional relativistic S-matrix (not to mention higher space-time dimensionalities) is a very complicated object. In two space-time dimensions, however, a situation is possible in which the total S-matrix being nontrivial is simplified drastically. This is the case of factorized scattering. Generally, the factorization of a two-dimensional S-matrix means a special structure of the multiparticle S-matrix element: it is factorized into the product of a number of two-particle ones as if an arbitrary process of multiparticle scattering would be a succession of space-time separated elastic two-particle collisions, the movement of the particles in between being free.

The factorized S-matrix has been first discovered in the nonrelativistic problem of one-dimensional scattering of particles interacting through the δ-function pair potential [1-3]. Furthermore, the factorization is typical for the scattering of solitons of the nonlinear classical field equations completely integrable by the inverse scattering method [4-6]. Note, that all the dynamical systems leading to the factorized S-matrix possess, as a common feature, an infinite set of "close to free" conservation laws.[1] This set of conservation laws is considered to be a necessary and sufficient condition for the S-matrix factorization [7-11]. Some speculations about this point are presented in Section 2.

The expressibility of the multiparticle S-matrix in terms of two-particle ones provides an essential simplification and enables one to construct in many cases the total S-matrix up to the explicit calculation of the two-particle matrix elements themselves. In the present paper we construct a certain class of the relativistic factorized S-matrices being invariant under $O(N)$ isotopic transformations. We use the method first suggested by Karowski, Thun, Truong and Weisz [12] (in the sine-Gordon context). The selfconsistency of the factorized structure of the total S-matrix turns out to impose special cubic equations (the factorization equations in what follows) on the two-particle S-matrix elements (see Section 2). Therefore, the factorization, unitarity and crossing symmetry provide a nontrivial system of equations which is basic for the method mentioned above. The general solution of these equations has an ambiguity of CDD type: there is a "minimum solution" (i.e., the solution having minimum set of singularities); one obtains the general one adding an arbitrary number of auxiliary CDD poles.

Are there any two-dimensional quantum field theory (QFT) models that lead to these S-matrices? Most of the nonlinear classical field equations have evident QFT versions. The problem of factorizing quantum S-matrices of these models (which is closely connected with that of "surviving" classical conservation laws under quantization) is nontrivial and requires special investigation in each case. In this paper we consider three models, the aim is to show that they lead to $O(N)$-symmetric factorized S-matrices.

(1) The quantum sine-Gordon model, i.e., the model of a single scalar field $\phi(x)$, which is defined by the Lagrangian density:

$$\mathscr{L}_{SG} = \frac{1}{2}(\partial_\mu \phi)^2 + \frac{m_0^2}{\beta^2}\cos(\beta\phi), \tag{1.1}$$

where m_0 is a mass-like parameter and β is a coupling constant.

It is well-known that the classical sine-Gordon equation is completely integrable [6]. The structure of the quantum theory has been also studied in detail. The mass

[1] The meaning of this term is as follows. In the asymptotic states where all particles are far enough from each other these conservation laws tend to those of the theory of free particles. The latter laws lead to conservation of the individual momentum of each particle and can be formulated, e.g., as the conservation of sums of the entire powers of all particle momenta $\Sigma_a p_a^n$; $n = 1, 2, \ldots$ [7].

spectrum of this model has been found by a quasiclassical method [13–15]. It contains particles carrying the so-called "topological charge"[2]—quantum solitons and corresponding antisolitons—and a number of neutral particles (quantum doublets) which can be thought of as the soliton-antisoliton bound states; the "elementary particle" corresponding to field ϕ turns out to be one of these bound states. Some of the quasiclassical results (the mass formula for the doublets) proved to be exact [15].

The other exact result has been obtained by Coleman [16] (see also Refs. [17, 19]). The quantum sine-Gordon model is equivalent to the massive Thirring model, i.e., the model of charged fermion field, defined by the Lagrangian density

$$\mathscr{L}_{MTM} = i\bar{\psi}\gamma_\mu\partial_\mu\psi - m\bar{\psi}\psi - g/2(\bar{\psi}\gamma_\mu\psi)^2 \tag{1.2}$$

provided the coupling constants are connected by

$$g/\pi = 4\pi/\beta^2 - 1. \tag{1.3}$$

Fundamental fermions of (1.2) are identical to quantum solitons of (1.1).

There is a considerable amount of results in support of the factorization of the quantum sine-Gordon S-matrix; these results are mentioned in Section 4.

(2) The quantum chiral field on the sphere S^{N-1} ($N = 3, 4, \ldots$) ($O(N)$ symmetric nonlinear σ-model) defined by the Lagrangian density and the constraint

$$\mathscr{L}_{CF} = \frac{1}{2g_0}\sum_{i=1}^{N}(\partial_\mu n_i)^2; \qquad \sum_{i=1}^{N} n_i^2 = 1, \tag{1.4}$$

where g_0 is a (bare) coupling constant. This model is $O(N)$ symmetric, renormalizable and asymptotically free [20, 12]. The infrared charge singularity of this model seems to cause the disintegration of the Goldstone vacuum [22]. True vacuum is $O(N)$ symmetric and nondegenerate; all particles of the model are massive and form $O(N)$-multiplets. This situation is surely the case when N is large enough [23, 24] and we suppose it is valid for $N \geqslant 3$.

In Section 5 some arguments in favour of S-matrix factorization in the model (1.4) are presented. The first evidence of this phenomenon is based on the properties of the $1/N$ expansion of the model [23, 24]. Namely, the absence of $2 \to 4$ production amplitude and the factorization of $3 \to 3$ amplitude can be shown to the leading order in $1/N$ [25]. A more rigorous proof of the σ-model S-matrix factorization follows from the recently discovered infinite set of quantum conservation laws [26, 27]. In Section 5 we review briefly the results of Ref. [26].

[2] In model (1.1) the topologic charge q is connected with the asymptotic behaviour of the field $\phi(x, t)$ as $x \to \pm\infty$:

$$q = \beta/2\pi\int_{-\infty}^{\infty}\frac{d\phi}{dx}\,dx = \beta/2\pi[\phi(\infty) - \phi(-\infty)].$$

(3) The Gross–Neveu model, i.e., the model of N-component self-conjugated Fermi-field $\psi_i(x)$; $i = 1, 2,..., N$ ($N \geqslant 3$) with four-fermion interaction

$$\mathscr{L}_{GN} = \frac{i}{2} \sum_{i=1}^{N} \bar{\psi}_i \gamma_\mu \partial_\mu \psi_i + \frac{g_0}{8} \left[\sum_{i=1}^{N} \bar{\psi}_i \psi_i \right]^2, \tag{1.5}$$

where $\bar{\psi}_i = \psi_i \gamma_0$. Like the chiral field this model is renormalizable, asymptotically free and explicitly $O(N)$ symmetric.

Model (1.5) has been studied by Gross and Neveu in the limit of $N \to \infty$ [28]. They have found a spontaneous breakdown of discrete γ_5-symmetry (the field $\sum_{i=1}^{N} \bar{\psi}_i \psi_i$ acquires a nonzero vacuum expectation value) leading to the dynamical mass transmutation. Using quasiclassical method Dashen, Hasslacher and Neveu [29] have studied the model in the same $N \to \infty$ limit. These authors have found a rich spectrum of bound states of the fundamental fermions of this model and determinded their masses.

We support the factorization of the Gross–Neveu S-matrix by arguments which are quite analogous to those for model (1.4).

The paper is arranged as follows. In Section 2 general properties of factorized scattering are considered, factorization equations are introduced and their meaning is cleared up. Furthermore, a convenient algebraic representation of the factorized S-matrix is suggested. Sect. 3 contains the general solution of analyticity, unitarity and factorization equations for the S-matrix having $O(N)$ isotopic symmetry. The "minimum" solutions of these equations turn out to be essentially different for $N = 2$ and $N \geqslant 3$. The solution for $N = 2$ depends on one parameter of coupling constant type. As it is shown in Section 4 this solution turns out to be the exact S-matrix of quantum sine-Gordon solitons. In this Section we construct the total sine-Gordon S-matrix too, which includes all bound states (doublets). For the case $N \geqslant 3$ "minimum" solutions of Section 3 do not depend on any free parameters. They correspond to asymptotically free field theories with the dynamical mass transmutation. In Sections 5 and 6 one of these solutions is shown to be the S-matrix of model (1.4) and the other—to be an exact one of elementary fermions of (1.5).

2. Factorized Scattering, General Properties, Factorization Equations

Consider a two-dimensional scattering theory and suppose that the underlying dynamical theory is governed by the infinite set of conservation laws, the corresponding conserving charges Q_n ; $n = 1, 2,..., \infty$ being diagonal in one-particle states:

$$Q_n \mid p^{(a)} \rangle = \omega_n^{(a)}(p) \mid p^{(a)} \rangle. \tag{2.1}$$

In (2.1) p is the particle momentum and (a) letters the kind of particle (if the theory contains more than one kind). Suppose, furthermore, that the eigenvalues $\omega_n(p)$ form the set of independent functions. In fact, all the known systems with the infinite number of conservation laws permit such a choice of the set Q_n that $\omega_n(p)$, speaking roughly, should be the entire powers of the momentum p. E.g., for the sine-Gordon case they are:

$$\omega_{2n+1}^{(a)}(p) = p^{2n+1}; \qquad \omega_{2n}^{(a)}(p) = (p^2 + m_a^2)^{1/2}\, p^{2n}, \tag{2.2}$$

where m_a is the mass of the particle (a). The laws described above are said to be "deformation of free laws". If the theory is governed by laws of this type the corresponding scattering theory satisfies strong selection rules (first noticed by Polyakov). Namely:

(i) Let $\{m_a\}$ be the mass spectrum of the theory. Then the number of particles of the same mass m_a remains unchanged after collision.

(ii) The final set of the two-momenta of particles is the same as the initial one.[3]

These two selection rules become evident if one takes into account that:

(a) $$Q_n \mid p_1^{(a_1)}, p_2^{(a_2)}, \ldots, p_k^{(a_k)}, \text{in(out)}\rangle = [\omega_n^{(a_1)}(p_1) + \cdots + \omega_n^{(a_k)}(p_k)]\; p_1^{(a_1)}, \ldots, p_k^{(a_k)} \cdot \text{in(out)}\rangle. \tag{2.3}$$

(b) $dQ_n/dt = 0$

and hence

$$\sum_{j \in \text{in}} \omega_n^{(a_j)}(p_j) = \sum_{j \in \text{out}} \omega_n^{(a_j)}(p_j). \tag{2.4}$$

Note that all the intermediate states where the particles are far enough from each other should satisfy both selection rules (i) and (ii) too. This note, together with the special properties of the two-dimensional kinematics, gives an impression that if the theory is governed by an infinite set of conservation laws the multiparticle S-matrix elements can be expressed in terms of two-particle ones.

To clarify this point consider the space of configurations of the system of N particles. There are $N!$ disconnected domains in this space where all the particles are far enough from each other and one can neglect interaction between them. Let $\{x\} = \{x_1, \ldots, x_N\}$ be the coordinates of the particles and R—the interaction range (we suppose the latter to be finite). Then each domain can be identified by the succession of inequalities $x_{P1} < x_{P2} < \cdots < x_{PN}$ where $|x_{Pi+1} - x_{Pi}| \gg R$ and P is any permutation of the integers $1, 2, \ldots, N$. We denote this domain by X_P.

The free motion of particles in these domains can be described in terms of the wave function $\Psi(x_1, \ldots, x_N)$; $\{x\} \in X_P$ Selection rules (i) and (ii) mean now that if the incident particles are of momenta $p_1 > p_2 > \cdots > p_N$, the wave function in each domain

[3] It may appear that (i) and (ii) mean that the S-matrix is diagonal in the momentum representation. It is not true if the theory contains different particles (having different internal quantum numbers, e.g., particle and antiparticle) of the same mass. In this case the exchanges of momenta between these particles and other nondiagonal processes are possible (see Sect. 3).

should be a superposition of waves, the set of wave vectors being selected by these rules:

$$\Psi_P(x_1, ..., x_N) = \sum_{P'} C(P, P') \exp\{ip_{P1}x_{P'1} + \cdots + ip_{PN}x_{P'N}\}; \qquad \{x\} \in X_P. \quad (2.5)$$

Here summation is carried out over all permutation P' of $p_1, ..., p_N$, permitted by (i) and (ii). Symmetrization (antisymmetrization) in the coordinates of identical particles is implied in (2.5). The coefficients $C(P, P')$ are functions of the domain X_P and of the permutation P'. In particular, the coefficient $C(P, I)$ describes the inci-incident wave in the domain X_P; to obtain the scattering wave function one puts $C(P, I) = 0$ if $P \neq I$ and $C(I, I) = 1$ (here I is identical transposition). The coefficients $C(P, \tilde{I})$ ($\tilde{I}$ is the inverse transposition $\tilde{I}(1, 2, ..., N) = (N, N-1, ..., 1)$) describe outgoing waves in these domains and thus they are elements of the N-particle S-matrix. For example, in the case of two particles of the same mass the wave function becomes

$$\begin{aligned} \Psi_{x_1 \ll x_2}(x_1, x_2) &= e^{ip_1x_1}e^{ip_2x_2} + S_R(p_1, p_2)\, e^{ip_2x_1}e^{ip_1x_2}; \\ \Psi_{x_1 \gg x_2}(x_1, x_2) &= S_T(p_1, p_2)\, e^{ip_1x_1}e^{ip_2x_2}. \end{aligned} \qquad (2.6)$$

In Eq. (2.6) S_R and S_T are two-particle S-matrix elements corresponding to backward scattering (reflection) and forward scattering (transition).

It is convenient to picture the situation as the scattering of the N-dimensional plane wave in the system of semipenetrable hypersurfaces $x_i = x_j$ (for any i and j). Far enough from these hypersurfaces the wave is described by (2.5); near them the motion is more complicated because of the interaction between the particles. Moreover, if the relativistic problem is taken into consideration the motion in the interaction region cannot be treated in terms of the wave function of a finite number of variables (because the virtual pair creation is possible). The determination of the coefficients $C(P, P')$ in (2.5) requires the extrapolation of the wave function from one domain of free motion to another through the boundary between them, where the particles are in interaction. The solution of the problem of interacting particles is, in general, a very complicated task. Note, however, that the extrapolation of the wave function can pass through the region of the boundary, where two particles are close and others are arbitrary far from them and each other (e.g., $|x_1 - x_2| \lesssim R$, $|x_i - x_1| \gg R$, $|x_2 - x_i| \gg R$ and $|x_i - x_j| \gg R$; $i, j = 3, 4, ...$). These regions describe two-particle collisions and there the extrapolation conditions are the same as in the two-particle problem. Therefore, in this case the knowledge of the two-particle S-matrix elements provides one with a sufficient information to determine all the coefficients $C(P, P')$ and, therefore, to obtain the multiparticle S-matrix. N-particle S-matrix element turns out to be a product of $N(N-1)/2$ two particle ones. Such a structure is spoken about as the factorized S-matrix. Note that the possibility of this structure is due to the fact that the wave function in each domain X_P is a superposition of a finite number of waves, the latter being a consequence of the infinite set of conservation laws.

Of course, this consideration connecting the factorization and the ėxistance of infinite number of conservation laws is not a rigorous proof; a complete evidence can be found in a recent paper [11]. All the considerations presented in the above paragraph are of exact sense in the case of the one dimensional problem of nonrelativistic particles interacting via the δ-function potential [1–3].

The factorized S-matrix corresponds to the following simple scattering picture. In the infinite past particles of momenta $p_1 > p_2 > \cdots > p_N$ were spatially arranged in the opposite order: $x_1 < x_2 < \cdots < x_N$. In the interaction region the particles successively collide in pairs; they move as free real (not virtual) particles in between. The set of momenta of particles is conserved in each pair collision; if the particles are of different mass the transition is possible only, the collision of particles of the same mass may result in the reflection too. After $N(N-1)/2$ pair collisions the particles are arranged along the x axis in the order of momenta increasing. This corresponds to the final state of scattering—outgoing particles.

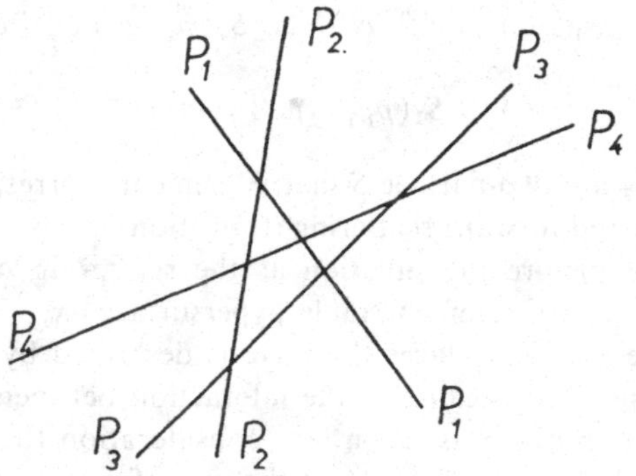

FIG. 1. The space-time picture illustrating the multiparticle factorized scattering.

The space-time picture of the multiparticle factorized scattering can be represented by a spacial diagram; an example is drawn in Fig. 1. Each straight line in the diagram corresponds to any value of momentum, obviously connected with the slope of the line (in this diagram time is assumed to flow up). Two-particle collisions are represented by the vertices where the lines cross each other; the corresponding two particle amplitude should be attached to each cross. The total multiparticle S-matrix element of the process drawn in the diagram is given by a sum of products of all the $N(N-1)/2$ two-particle amplitudes corresponding to each vertex. The summation mentioned above should be carried out over all possible kinds of particles flowing through the internal lines of the diagram and resulting in a given final state.

The following is to be mentioned. The same scattering process can be represented by a number of different diagrams in which some of the lines are translated in parallel (e.g., see Figs. 2a and 2b). The amplitudes of these diagrams should not be added in the multiparticle S-matrix element. In terms of the wave function in sectors X_P amplitudes drawn in Figs. 2a and 2b correspond to different semifronts of the same

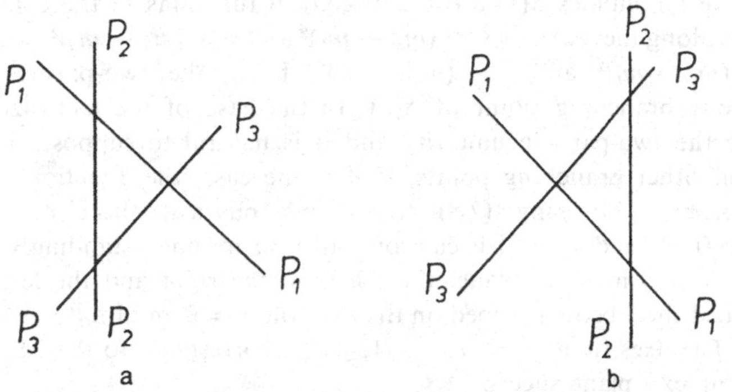

FIG. 2. Two possible ways of the three-particle scattering.

outgoing wave. Both should have the same amplitudes and phases (because of (i) and (ii)), i.e., be coherent. This requirement makes two-particle matrix elements satisfy special cibic equations, the latter being necessary conditions of the factorization. In what follows these equations play an essential role and we shall call them the factorization equations.[4]

In the present paper the relativistic scattering is mainly considered. The following notations are convenient in this case. We shall use rapidities θ_a instead of momenta p_a of particles (of mass m_a)

$$p_a^0 = m_a \operatorname{ch} \theta_a ; \qquad p_a^1 = m_a \operatorname{sh} \theta_a \tag{2.7}$$

Two-particle amplitudes $S(p_a, p_b)$ become functions of the rapidity difference of colliding particles $\theta_{ab} = \theta_a - \theta_b$, the latter being simply connected with the s-channel invariant $s_{ab} = (p_a^\mu + p_b^\mu)^2$

$$s_{ab} = m_a^2 + m_b^2 + 2m_a m_b \operatorname{ch} \theta_{ab} \tag{2.8}$$

(m_a and m_b are masses of the particles).

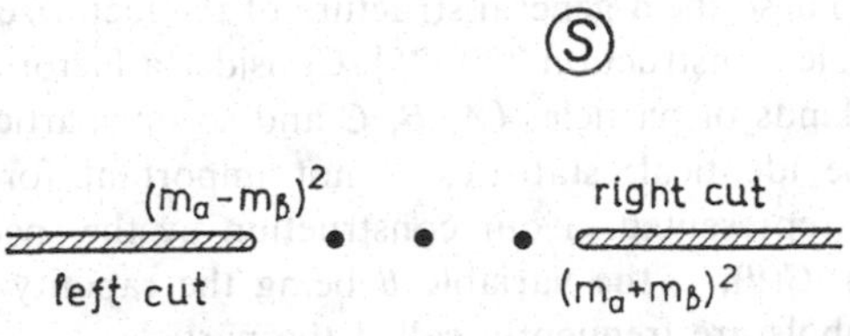

FIG. 3. The analytical structure of two-particle amplitudes in the physical sheet of the s-plane.

[4] Factorization equations and their physical sense in the problem of nonrelativistic particles interacting via the δ-function potential have been considered in Ref. [2]; in the case of the sine-Gordon problem they were obtained in Refs. [30, 31] and used in Ref. [12].

Two-particle amplitudes $S(s)$ are the analytical functions in the complex s-plane with two cuts along the real axis $s \leqslant (m_a - m_b)^2$ and $s \geqslant (m_a + m_b)^2$ (see Fig. 3). The points $s = (m_a - m_b)^2$ and $s = (m_a + m_b)^2$, being the two-particle thresholds, are square root branching point of $S(s)$. In the case of the factorized scattering there is only the two-particle unitarity and it is natural to suppose functions $S(s)$ not to exibit other branching points. If it is the case, the functions $S(\theta)$ should be meromorphic. Mapping (2.8) transforms physical sheet of the s-plane into the strip $0 < \mathrm{Im}\,\theta < \pi$ (if it cannot lead to a misunderstanding we shall drop subindices $\theta \equiv \theta_{ab}$) in the θ-plane, the edges of the right and the left cuts of the s-plane physical sheet being mapped on the axes $\mathrm{Im}\,\theta = 0$ and $\mathrm{Im}\,\theta = \pi$, respectively (see Fig. 4). The axes $\mathrm{Im}\,\theta = l\pi$; $l = -1, \pm 2, \ldots$ correspond to the edges of cuts of the other complex s-plane sheets.

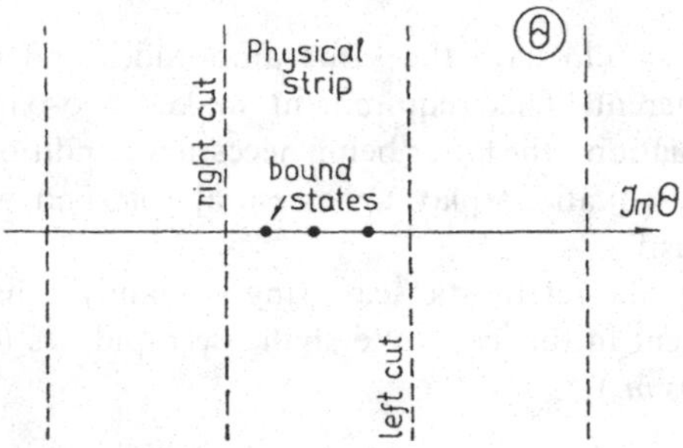

FIG. 4. The structure of the θ-plane.

The functions $S(\theta)$ are real at the imaginary axis of the θ-plane (real analyticity). In particular, at $\mathrm{Im}\,\theta = 0$ the relation $S(-\theta) = S^*(\theta)$ is valid. Crossing symmetry transformation $s \to 2m_a^2 + 2m_b^2 - S$ corresponds in terms of the variable θ to substitution $\theta \to i\pi - \theta$.

In the nonrelativitic limit $p_a^1 \ll m_a$ rapidities can be replaced by the nonrelativistic velocities $\theta_a \to v_a = p_a/m_a$. All the following expressions (except those connected with the crossing relations) can be applied to the case of nonrelativistic S-matrices after replacement $\theta_a \to v_a$, $\theta_b \to v_b$, $\theta_{ab} \to v_a - v_b$.

It is convenient to describe a general structure of the factorized S-matrix by means of a special algebraic construction [30, 25]. Consider a factorized scattering theory containing several kinds of particles (A, B, C and so on: particles of the same kind are supposed to be identical; statistics is not important for our consideration). These particles are represented in our construction by the special noncommutative symbols $A(\theta)$, $B(\theta)$, $C(\theta)$,..., the variable θ being the rapidity of the correponding particle. These symbols are frequently called the particles.

The scattering theory is stated as follows. Identify asymptotical states of the scattering theory with the products of all the particles in the state. The arrangement of the symbols in the product corresponds to that of particles along the spatial axis x: in-states should be identified with the products arranged in the order of decreasing rapidities of particles while out-states with those arranged in the order of increasing rapidities. For example, in-state of three particles A, A and B having rapidities θ_1, θ_2 and θ_3, respectively ($\theta_1 > \theta_2 > \theta_3$), acquires the form $A(\theta_1)\, A(\theta_2)\, B(\theta_3)$.

Any product can be rearranged by means of a number of subsequent commutations of neighbour particles (the associativity of the symbol multiplication is supposed). Each commutation corresponds to the certain two-particle collision; this leads to commutation rules for the symbols $A(\theta)$, $B(\theta)$,.... For example, if particles A and B are of different mass, one writes

$$A(\theta_1)\, B(\theta_2) = S_T^{AB}(\theta_{12})\, B(\theta_2)\, A(\theta_1), \tag{2.9}$$

where $S_T^{AB}(\theta_{12})$ is the transition amplitude for the reaction $AB \to AB$ (remind, that in the case of different masses, reflection is forbidden by (ii)). If particles of differerent kinds (say A and C) but of the same mass are under consideration the reflection is permitted and we should write

$$A(\theta_1)\, C(\theta_2) = S_T^{AC}(\theta_{12})\, C(\theta_2)\, A(\theta_1) + S_R^{AC}(\theta_{12})\, A(\theta_2)\, C(\theta_1). \tag{2.10}$$

Reflection and transition are indistinguishable in the case of identical particles, therefore

$$A(\theta_1)\, A(\theta_2) = S^{AA}(\theta_{12})\, A(\theta_2)\, A(\theta_1). \tag{2.11}$$

As it was mentioned above (see footnote 3) if there are different particles of the same mass one of them is permitted to turn into the other in the process of two-particle scattering. It means that additional channels in the two-particle scattering are open and, hence, corresponding terms should be added into the right hand sides of Eqs. (2.9), (2.10) and (2.11). We shall not discuss this point here, there are some examples of such situation in the next section.

The consistency of the commutation relations of type (2.9), (2.10), (2.11) in the calculation of symbols $A(\theta)$, $B(\theta)$ and so on and their associativity requires certain equations for the two-particle amplitudes to be satisfied. The latter are of two kinds. The identities of the first kind arise when one performs the opposite transposition of symbols after the direct one, and requires the result to be equal to the initial combination; these identities coincide with the two-particle unitarity relations. The multiparticle in-states may be rearranged into out-states in many possible successions of pair commutations but the result should be the same. This leads to the identities of the second kind. Clearly, it is sufficient to consider three particle states only and require the same result of permutations in two possible successions. One obtains all the required identities which coincide, of course, with the conditions ensuring the equality of triangle diagrams (see Figs. 2a, 2b), and so they are the factorization equations.

If identities of both kinds are satisfied the commutation relations permit one to rearrange unambiguously any in-state into a superposition of out-states and then this construction represents the total factorized S-matrix. Its unitarity is trivial. One obtains the matrix S^{-1} after the rearrangement of out-states into in-states; it differs from the S-matrix in the signs of the arguments of all two-particle amplitudes $\theta_{ab} \to -\theta_{ab}$. This change of signs leads to the complex conjugation of the two-particle matrix elements. Taking into account the symmetry of the S-matrix one obtains $S^{+} = S^{-1}$.

3. Relativistic S-Matrix with O(N)-Isosymmetry. General Solution

Following the general consideration of the previous section we treat now the class of relativistic factorized S-matrices characterized by the isotopic $O(N)$ symmetry. To introduce the $O(N)$ symmetry we assume the existance of isovector N-plet of particles A_i ; $i = 1, 2,..., N$ with equal masses m and require the $O(N)$ symmetry of the two-particle scattering (this ensures $O(N)$ symmetry of the total S-matrix due to the factorization). Namely, we assume for two-particle S-matrix the form:

$$\begin{aligned} {}_{ik}S_{jl} &= \langle A_j(p_1') \, A_l(p_2'), \text{out} \mid A_i(p_1) \, A_k(p_2), \text{in}\rangle \\ &= \delta(p_1 - p_1') \, \delta(p_2 - p_2')[\delta_{ik}\delta_{jl}S_1(s) + \delta_{ij}\delta_{kl}S_2(s) + \delta_{il}\delta_{jk}S_3(s)] \\ &\quad \pm (i \leftrightarrow k, \, p_1 \leftrightarrow p_2), \end{aligned} \tag{3.1}$$

where $s = (p_1^{\mu} + p_2^{\mu})^2$ and the $+(-)$ refers to bosons (fermions). The functions $S_2(s)$ and $S_3(s)$ are the transition and reflection amplitudes, respectively, while $S_1(s)$ describes the "annihilation" type processes: $A_i + A_i \to A_j + A_j$ $(i \neq j)$.

The S-matrix (3.1) will be cross-symmetric provided the amplitudes $S(s)$ satisfy equations $S_2(s) = S_2(4m^2 - s)$ and $S_1(s) = S_3(4m^2 - s)$. After introducing the rapidity variables (2.7), (2.8) we deal with meromorphic functions $S_1(\theta)$, $S_2(\theta)$ and $S_3(\theta)$, where $s = 4m^2 \operatorname{ch}^2(\theta/2)$, and the cross-symmetry relations become

$$S_2(\theta) = S_2(i\pi - \theta), \tag{3.2a}$$

$$S_1(\theta) = S_3(i\pi - \theta). \tag{3.2b}$$

To describe now the factorized total S-matrix let us introduce, following the general method of Section 2, symbols $A_i(\theta)$; $i = 1, 2,..., N$. The commutation rules corresponding to (3.1) are

$$\begin{aligned} A_i(\theta_1) \, A_j(\theta_2) &= \delta_{ij} S_1(\theta_{12}) \sum_{k=1}^{N} A_k(\theta_2) \, A_k(\theta_1) \\ &\quad + S_2(\theta_{12}) \, A_j(\theta_2) \, A_i(\theta_1) + S_3(\theta_{12}) \, A_i(\theta_2) \, A_j(\theta_1). \end{aligned} \tag{3.3}$$

It is straightforward to obtain the unitarity conditions for two-particle S-matrix (3.1)

$$S_2(\theta) \, S_2(-\theta) + S_3(\theta) \, S_3(-\theta) = 1, \tag{3.4a}$$

$$S_2(\theta) \, S_3(-\theta) + S_2(-\theta) \, S_3(\theta) = 0, \tag{3.4b}$$

$$\begin{aligned} &NS_1(\theta) \, S_1(-\theta) + S_1(\theta) \, S_2(-\theta) + S_1(\theta) S_3(-\theta) \\ &\quad + S_2(\theta) \, S_1(-\theta) + S_3(\theta) \, S_1(-\theta) = 0. \end{aligned} \tag{3.4c}$$

Obviously, Eqs. (3.2) and (3.4) are not sufficient to determine the functions $S(\theta)$. Further restrictions arise from the factorization equations (see Section 2). One obtains the factorization equations considering all possible three-particle in-products $A_i(\theta_1)\,A_j(\theta_2)\,A_k(\theta_3)$, reordering them to get out-products by means of (3.3) and requiring the results obtained in two possible successions of two-particle commutations to be equal. The equations arising are evidently different for the cases $N = 2$ and $N \geqslant 3$ (fewer different three-particle products are possible at $N = 2$). Therefore it is convenient to make a notational distinction between these two cases. Dealing with the case $N \geqslant 3$ we redenote the amplitudes S_1, S_2 and S_3 by σ_1, σ_2 and σ_3, respectively, reserving the original notations for the case $N = 2$.

The factorization equations have the form (the derivation is straightforward but somewhat cumbersome)

$$S_2S_1S_3 + S_2S_3S_3 + S_3S_3S_2 = S_3S_2S_3 + S_1S_2S_3 + S_1S_1S_2\,, \tag{3.5a}$$

$$\begin{aligned} S_3S_1S_3 + S_3S_2S_3 &= S_3S_3S_1 + S_3S_3S_2 + S_2S_3S_1 \\ &+ S_2S_3S_3 + 2S_1S_3S_1 + S_1S_3S_2 + S_1S_3S_3 + S_1S_2S_1 + S_1S_1S_1 \end{aligned} \tag{3.5b}$$

for $N = 2$ and

$$\sigma_2\sigma_3\sigma_3 + \sigma_3\sigma_3\sigma_2 = \sigma_3\sigma_2\sigma_3\,, \tag{3.6a}$$

$$\sigma_2\sigma_1\sigma_1 + \sigma_3\sigma_2\sigma_1 = \sigma_3\sigma_1\sigma_2\,, \tag{3.6b}$$

$$\begin{aligned} &N\sigma_1\sigma_3\sigma_1 + \sigma_1\sigma_3\sigma_2 + \sigma_1\sigma_3\sigma_3 + \sigma_1\sigma_2\sigma_1 \\ &\quad + \sigma_2\sigma_3\sigma_1 + \sigma_3\sigma_3\sigma_1 + \sigma_1\sigma_1\sigma_1 = \sigma_3\sigma_1\sigma_3 \end{aligned} \tag{3.6c}$$

for $N \geqslant 3$. For each term in (3.5) and (3.6) the argument of the first, the second and the third S (σ in (3.6)) is implied to be θ, $\theta + \theta'$ and θ', respectively.

The factorization equations turn out to be rather restrictive. They allow one to express explicitly all the amplitudes in terms of one function.

General solutions for both systems (3.5) and (3.6) satisfying the real-analyticity condition (all the amplitudes are real if θ is purely imaginary) are derived in Appendix A. For system (3.5) (i.e., for $N = 2$) this solution is

$$S_3(\theta) = i\,\mathrm{ctg}\left(\frac{4\pi\delta}{\gamma}\right)\mathrm{cth}\left(\frac{4\pi\theta}{\gamma}\right)S_2(\theta) \tag{3.7a}$$

$$S_1(\theta) = i\,\mathrm{ctg}\left(\frac{4\pi\delta}{\gamma}\right)\mathrm{cth}\left(\frac{4\pi(i\delta-\theta)}{\gamma}\right)S_2(\theta) \tag{3.7b}$$

with arbitrary real γ and δ. The general solution for (3.6) contains only one free parameter λ and have the form:

$$\sigma_3(\theta) = -\frac{i\lambda}{\theta}\,\sigma_2(\theta) \tag{3.8a}$$

$$\sigma_1(\theta) = -\frac{i\lambda}{i[(N-2)/2]\,\lambda - \theta}\,\sigma_2(\theta). \tag{3.8b}$$

The restrictions on the amplitudes $S_2(\theta)$ and $\sigma_2(\theta)$ come from the unitarity conditions (3.4). The equations (3.4b) and (3.4c) are satisfied by (3.7) and (3.8) identically, while equation (3.4a) gives

$$S_2(\theta)\, S_2(-\theta) = \frac{\sin^2\left(\frac{4\pi\delta}{\gamma}\right) \mathrm{sh}^2\left(\frac{4\pi\theta}{\gamma}\right)}{\sin^2\left(\frac{4\pi\delta}{\gamma}\right) \mathrm{sh}^2\left(\frac{4\pi\theta}{\gamma}\right) + \cos^2\left(\frac{4\pi\delta}{\gamma}\right) \mathrm{ch}^2\left(\frac{4\pi\theta}{\gamma}\right)} \tag{3.9}$$

for $N = 2$ and

$$\sigma_2(\theta)\, \sigma_2(-\theta) = \frac{\theta^2}{\theta^2 + \lambda^2} \tag{3.10}$$

for $N \geqslant 3$.

Until now we have deliberately avoided the use of the cross-symmetry relations (3.2). Although the above consideration concerns the relativistic case, the unitarity conditions (3.4) and factorization equations (3.5), (3.6) are valid for any nonrelativistic $O(N)$ symmetric factorized S-matrix as well, under the substitution:

$$\theta \to \frac{k}{m} = \frac{k_1 - k_2}{m}, \tag{3.11}$$

where k_1 and k_2 are momenta of the colliding particles. Therefore, the general solutions (3.7), (3.9) and (3.8), (3.10) are still valid (after the substitution (3.11)) in a nonrelativistic case. This will be used at the end of Section 4.

Equations (3.2) are especially relativistic. They turn out to give restrictions on free parameters in (3.7) and (3.8). It is easy to see that (3.2) is satisfied only if

$$\delta = \pi \tag{3.12}$$

in (3.7), (3.9) and

$$\lambda = \frac{2\pi}{N-2} \tag{3.13}$$

in (3.8), (3.10). Thus, the formulas for $N \geqslant 3$ do not actually contain any free parameter. This circumstance will be important in Section 5.

Equation (3.2a) (which is certainly valid for $\sigma_2(\theta)$ as well as for $S_2(\theta)$) together with (3.9) and (3.10) will be used to determine $S_2(\theta)$ and $\sigma_2(\theta)$. In both cases $N = 2$ and $N \geqslant 3$ the solution admits the CDD-ambiguity only [32]: an arbitrary solution can be obtained multiplying some "minimum" solution by a meromorphic function of the type

$$f(\theta) = \prod_{k=1}^{L} \frac{\mathrm{sh}\,\theta + i \sin \alpha_k}{\mathrm{sh}\,\theta - i \sin \alpha_k}, \tag{3.14}$$

where $\alpha_1, \alpha_2, ..., \alpha_L$ are arbitrary real[5] numbers. It is the "minimum" solutions, i.e., the solutions having a minimum set of singularities in the θ plane, that will be of most interest below. For $N = 2$ such a solution can be represented in the form

$$S_2(\theta) = \frac{2}{\pi} \sin\left(\frac{4\pi^2}{\gamma}\right) \operatorname{sh}\left(\frac{4\pi\theta}{\gamma}\right) \operatorname{sh}\left[\frac{4\pi(i\pi - \theta)}{\gamma}\right] U(\theta), \tag{3.15}$$

where

$$U(\theta) = \Gamma\left(\frac{8\pi}{\gamma}\right) \Gamma\left(1 + i\frac{8\theta}{\gamma}\right) \Gamma\left(1 - \frac{8\pi}{\gamma} - i\frac{8\theta}{\gamma}\right) \prod_{n=1}^{\infty} \frac{R_n(\theta)\, R_n(i\pi - \theta)}{R_n(0)\, R_n(i\pi)}, \tag{3.16}$$

$$R_n(\theta) = \frac{\Gamma\left(2n\dfrac{8\pi}{\gamma} + i\dfrac{8\theta}{\gamma}\right) \Gamma\left(1 + 2n\dfrac{8\pi}{\gamma} + i\dfrac{8\theta}{\gamma}\right)}{\Gamma\left((2n+1)\dfrac{8\pi}{\gamma} + i\dfrac{8\theta}{\gamma}\right) \Gamma\left(1 + (2n-1)\dfrac{8\pi}{\gamma} + i\dfrac{8\theta}{\gamma}\right)}.$$

In the case $N \geqslant 3$ there are, in general, two different "minimum" solutions (the exceptional cases are $N = 3$ and $N = 4$, when these two solutions coincide). We denote these solutions $\sigma_2^{(+)}(\theta)$ and $\sigma_2^{(-)}(\theta)$; they can be written in the form

$$\sigma_2^{(\pm)}(\theta) = Q^{(\pm)}(\theta)\, Q^{(\pm)}(i\pi - \theta), \tag{3.17}$$

where

$$Q^{(\pm)}(\theta) = \frac{\Gamma\left(\pm\dfrac{\lambda}{2\pi} - i\dfrac{\theta}{2\pi}\right) \Gamma\left(\dfrac{1}{2} - i\dfrac{\theta}{2\pi}\right)}{\Gamma\left(\dfrac{1}{2} \pm \dfrac{\lambda}{2\pi} - i\dfrac{\theta}{2\pi}\right) \Gamma\left(-i\dfrac{\theta}{2\pi}\right)}. \tag{3.18}$$

The difference between these two solutions is of the CDD type (3.14)

$$\sigma_2^{(-)}(\theta) = \frac{\operatorname{sh}\theta + i\sin\lambda}{\operatorname{sh}\theta - i\sin\lambda}\, \sigma_2^{(+)}(\theta). \tag{3.19}$$

In the following Sections we point out the relation between the solutions (3.15) and (3.17) and certain two-dimensional quantum field theory models. Namely, we show that (3.15) together with (3.7a, b) is an exact S-matrix of quantum sine-Gordon solitons, while the solutions $\sigma_2^{(+)}(\theta)$ and $\sigma_2^{(-)}(\theta)$ for $N \geqslant 3$ give the exact S-matrices for the quantum chiral field (1.4) and for the "fundamental" fermions of Gross–Neveu model (1.5), respectively.

[5] We consider the solutions having singularities on the imaginary θ axis only, i.e., the solutions exhibiting bound and virtual states only.

4. Exact S-Matrix of the Quantum Sine-Gordon Model

Quantum sine-Gordon model (1.1) is the most known example of relativistic quantum field theory leading to the factorized scattering. There are various results ensuring the factorization of the sine-Gordon S-matrix. Complete integrability of the classical sine-Gordon equation [6] means the existance of an infinite number of conservation laws in the classical theory which are "deformation of free ones". The analogous set of conservation laws is also present in the "classical" massive Thirring model (which corresponds to the "tree" approximation for the Lagrangian (1.2) [33, 34]). The important problem of the conservation laws in quantum theory has been treated in Ref. [34] where such conservation laws were shown to survive after quantization by the perturbation theory approach (in all perturbational orders). The absence of particle production and factorization of multiparticle quantum S-matrix which are the consequenses of conservation laws has been previously demonstrated applying the direct sine-Gordon perturbative calculations by Arefyeva and Korepin [8].[6] The same result can be obtained in perturbation theory of massive Thirring model, i.e., for the soliton scattering [35, 36]. The semiclassical arguments for the soliton S-matrix factorization are also possible [14]. We use here the results mentioned above and treat the total sine-Gordon S-matrix as a factorized one.

The bound states of quantum solitons (the quantum doublets) and the soliton scattering have been investigated by a semiclassical approach in Refs. [13–15, 37, 38]. We represent here some semiclassical formulas which will be necessary below.

The two-particle scattering amplitude $S(\theta)$ for the solitons of the same sign and the transition amplitude $S_T(\theta)$ for soiton-antisoliton scattering calculated in the main semiclassical approximation have the form [13, 14, 37]

$$\begin{aligned} S_T^{(\mathrm{sem})}(\theta) &= S^{(\mathrm{sem})}(\theta)\exp\left\{i\,\frac{8\pi}{\beta^2}\right\}, \\ S^{(\mathrm{sem})}(\theta) &= \exp\left\{\frac{8}{\beta^2}\int_0^\pi \ln\left[\frac{e^{\theta-i\eta}+1}{e^\theta+e^{-i\eta}}\right]d\eta\right\} \end{aligned} \tag{4.1}$$

(where β is introduced in (1.1)) which is connected in a simple way with parameters of the classical soliton scattering. Calculation of one-loop correction leads to the change $\beta^2 \to \gamma'$ in (4.1) (see Ref. [46]) where[7]

$$\gamma' = \beta^2\left[1-\frac{\beta^2}{8\pi}\right]^{-1}.$$

[6] Besides the explicit expressions for the factorized sine-Gordon "elementary" particle S-matrix has been first proposed in [8] (see also [9, 10]).

[7] The singularity of the sine-Gordon theory at $\beta^2 = 8\pi$ has been discussed by Coleman [16]. As shown in [16], the Hamiltonian of the theory becomes unbounded from below at $\beta^2 \geqslant 8\pi$, provided the standard renormalization technique is used; the phenomenon is of the ultraviolet nature. This scarcely means the failure of the theory with $\beta^2 \geqslant 8\pi$, but rather indicates a lack of superrenormalizability property and suggests that another renormalization prescription is necessary at $\beta^2 \geqslant 8\pi$. Throughout this paper we restrict our consideration to the case $\beta^2 < 8\pi$.

The semiclassical soliton-antisoliton reflection amplitude (which takes into account an imaginary time classical trajectory, see Ref. [38]) is

$$S_R^{(\mathrm{sem})}(\theta) = i \sin\left(\frac{8\pi^2}{\gamma'}\right) \exp\left(-\frac{8\pi}{\gamma'} \mid \theta \mid\right) S_T^{(\mathrm{sem})}(\theta). \tag{4.2}$$

The derivation of the semiclassical mass spectrum of the quantum doublets was carried out in papers [13–15]. In the first two semiclassical approximations it is

$$m_n^{(\mathrm{sem})} = 2m \sin\left(\frac{n\gamma'}{16}\right); \quad n = 1, 2,\ldots < \frac{8\pi}{\gamma'}, \tag{4.3}$$

where m is a soliton mass. The authors of [15] have presented some arguments for formula (4.3) to be not only semiclassical but exact. Independent supports for this hypothesis have been given in [10, 39, 40]. The exact solution for the S-matrix which is derived in this section also confirms the exactness of spectrum (4.3).

We begin constructing the quantum sine-Gordon S-matrix stressing that the model exhibits an $O(2)$ isotopic symmetry. In terms of the massive Thirring fields ψ this symmetry is quite obvious; it corresponds to the phase invariance $\psi \to e^{i\alpha}\psi$ of (1.2). From the view-point of the sine-Gordon Lagrangian $O(2)$ symmetry is of a more delicate nature; it is the rotational symmetry of the disorder parameter (see Ref. [41] for the concept of the disorder parameter). The detailed discussion of the last point is beyond the scope of this paper. For our purpose it is sufficient to note only that the soliton and antisoliton of model (1.1) can be incorporated into an isovector $O(2)$ doublet. Following the convention of Section 3 we denote real components of this doublet by symbols $A_i(\theta)$; $i = 1, 2$. Then the soliton and antisoliton themselves will be the combinations

$$A(\theta) = A_1(\theta) + iA_2(\theta); \qquad \bar{A}(\theta) = A_1(\theta) - iA_2(\theta). \tag{4.4}$$

In terms of the particles $A(\theta)$ and $\bar{A}(\theta)$ commutation rules (3.3) take the form

$$\begin{aligned} A(\theta_1)\, \bar{A}(\theta_2) &= S_T(\theta_{12})\, \bar{A}(\theta_2)\, A(\theta_1) + S_R(\theta_{12})\, A(\theta_2)\, \bar{A}(\theta_1), \\ A(\theta_1)\, A(\theta_2) &= S(\theta_{12}) A(\theta_2)\, A(\theta_1), \\ \bar{A}(\theta_1)\, \bar{A}(\theta_2) &= S(\theta_{12})\, \bar{A}(\theta_2)\, \bar{A}(\theta_1). \end{aligned} \tag{4.5}$$

In (4.5) $S_T(\theta)$ and $S_R(\theta)$ are transition and reflection amplitudes for the soliton-antisoliton scattering while $S(\theta)$ is the scattering amplitude for identical solitons. They are connected in a simple way with amplitudes $S_1(\theta)$, $S_2(\theta)$ and $S_3(\theta)$ from (3.3)

$$\begin{aligned} S(\theta) &= S_3(\theta) + S_2(\theta), \\ S_T(\theta) &= S_1(\theta) + S_2(\theta), \\ S_R(\theta) &= S_1(\theta) + S_3(\theta). \end{aligned} \tag{4.6}$$

It is seen from (3.2) that

$$S(\theta) = S_T(i\pi - \theta); \qquad S_R(\theta) = S_R(i\pi - \theta). \tag{4.7}$$

The factorization and $O(2)$ symmetry of the sine-Gordon soliton S-matrix allows one to apply immediately the results of the previous section. It follows from (3.71, b), (3.12) and (3.15) that

$$S_T(\theta) = -i \frac{\operatorname{sh}\left(\frac{8\pi}{\gamma}\theta\right)}{\sin\left(\frac{8\pi^2}{\gamma}\right)} S_R(\theta), \tag{4.8a}$$

$$S(\theta) = -i \frac{\operatorname{sh}\left(\frac{8\pi}{\gamma}(i\pi - \theta)\right)}{\sin\left(\frac{8\pi^2}{\gamma}\right)} S_R(\theta), \tag{4.8b}$$

where (with arbitrariness of the CDD-type (3.14) only)

$$S_R(\theta) = \frac{1}{\pi} \sin\left(\frac{8\pi^2}{\gamma}\right) U(\theta) \tag{4.9}$$

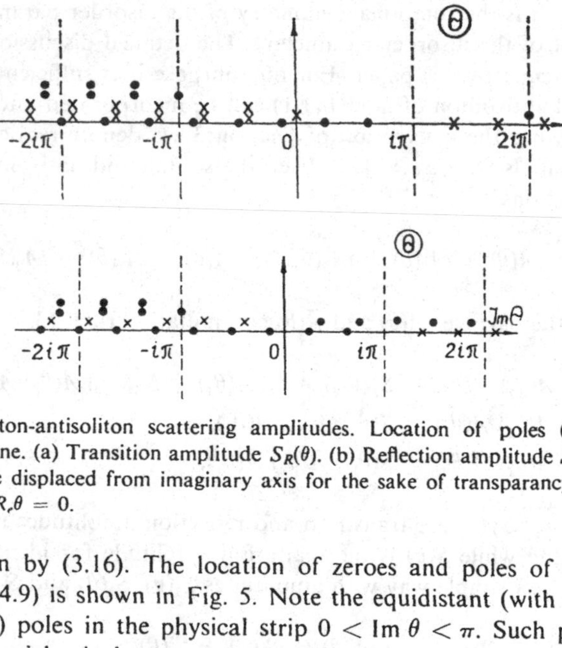

FIG. 5. The soliton-antisoliton scattering amplitudes. Location of poles (dots) and zeroes (crosses) in the θ-plane. (a) Transition amplitude $S_R(\theta)$. (b) Reflection amplitude $S_T(\theta)$. Some of the dots and crosses are displaced from imaginary axis for the sake of transparancy; actually all the singularities are at $R,\theta = 0$.

and $U(\theta)$ is given by (3.16). The location of zeroes and poles of functions $S_T(\theta)$ and $S_R(\theta)$ (4.8), (4.9) is shown in Fig. 5. Note the equidistant (with separation $\gamma/8$) positions of $S_T(\theta)$ poles in the physical strip $0 < \operatorname{Im}\theta < \pi$. Such positions are in accord with the semiclassical mass spectrum (4.3). The correspondence is exact if

$$\gamma = \gamma'. \tag{4.10}$$

Therefore, the whole bound state spectrum (4.3) is already contained in the "minimum" solution (4.9), (3.16) and CDD poles need not be added. This solution automatically stisfies also another necessary requirement for the exact sine-Gordon S-matrix. If $\gamma = 8\pi(\beta^2 = 4\pi)$ the massive Thirring-model coupling vanished and the S-matrix should become unity. In fact when $\gamma = 8\pi$ one has from (4.8), (4.9) and (3.16)

$$S_T(\theta) \equiv S(\theta) \equiv 1; \qquad S_R(\theta) \equiv 0.$$

These two remarkable properties of the "minimum" solution (together with its obvious aesthetic appeal) may serve as initial arguments to choose it as the exact S-matrix of quantum sin-Gordon solitons. We present below a number of checks which confirm such a choice.

If $\gamma \to 8\pi$ the formulas (4.8), (4.9) and (3.16) can be expanded in powers of $2g/\pi = 8\pi/\gamma - 1$ and the expansion coefficients can be compared with the results of diagrammatic calculus in massive Thirring model (1.2). Such a comparison has been carried out in [42] up to g^3 and the coincidence has been found.

Another check is a comparison with semiclassical formulas (4.1), (4.2). The semiclassical limit for Lagrangian (1.1) corresponds to $\beta^2 \to 0$. At θ fixed and $\beta^2 \to 0$ exact relation (4.8a) converts into semiclassical one (4.2). Furthermore, it can be easily verified that asymptotics of the exact amplitudes $S(\theta)$ and $S_T(\theta)$ as $\gamma \to 0$ coincide with (4.1). To do this one represents the exact $S_T(\theta)$ from (4.8a), (4.9) in the form

$$S_T(\theta) = \prod_{l=1}^{\infty} \frac{\Gamma\left(\frac{l\gamma}{16\pi} - i\frac{\theta}{2\pi}\right)\Gamma\left(\frac{l-1}{16\pi}\gamma - i\frac{\theta}{2\pi}\right)}{\Gamma\left(\frac{1}{2} + \frac{l\gamma}{16\pi} - i\frac{\theta}{2\pi}\right)\Gamma\left(-\frac{1}{2} + \frac{l-1}{16\pi}\gamma - i\frac{\theta}{2\pi}\right)} \times \frac{\Gamma\left(\frac{3}{2} + \frac{l\gamma}{16\pi} + i\frac{\theta}{2\pi}\right)\Gamma\left(\frac{1}{2} + \frac{l-1}{16\pi}\gamma + i\frac{\theta}{2\pi}\right)}{\Gamma\left(1 + \frac{l\gamma}{16\pi} + i\frac{\theta}{2\pi}\right)\Gamma\left(1 + \frac{l-1}{16\pi}\gamma + i\frac{\theta}{2\pi}\right)} \tag{4.11}$$

Changing in (4.4) the infinite product by a sum in the exponent and then replacing at $\gamma \to 0$ the summation by the integration one reproduced (4.1) exactly.

The larger the coupling parameter, the larger the mass of each bound state (4.3) (in units of the soliton mass). The n-th bound state acquires the soliton-antisoliton threshold when $\gamma = 8\pi/n$, and when $\gamma \geqslant 8\pi/n$ it disappears from the spectrum converting into the virtual state. At $\gamma \geqslant 8\pi$ all bound states (4.3) including the "elementary" particle of sine-Gordon Lagrangian (1.1) become unbound (remind that "elementary" particle is one of states (4.3), corresponding to $n = 1$ [15, 14]). Thus, at $\gamma \geqslant 8\pi$ the spectrum contains soliton and antisoliton only. The values $\gamma \geqslant 8\pi$ correspond to $g \leqslant 0$ in (1.2), i.e., to the repulsion between soliton and antisoliton.

Note that at $\gamma = 8\pi/n$ the reflection amplitude (4.9) vanishes identically (this property appears already in semi-classical formula (4.2)) while transition apmplitude $S_T(\theta)$ acquires, as a result of special cancellation of poles and zeroes in Fig. 5, a simple form:

$$S_T(\theta) = e^{in\pi} \prod_{k=1}^{n-1} \frac{e^{\theta - i(\pi k/n)} + 1}{e^{\theta} + e^{-i(\pi k/n)}} \tag{41.2}$$

This expression together with the hypothesis of its exact nature at $\gamma = 8\pi/n$ has been first presented by Korepin and Faddeev [14]. The general formulas (4.8), (4.9), (3.16) for arbitrary γ were given in [43][8].

Commutation rules (4.5) together with explicit expressions (4.8), (4.9) and (3.16) for the two-particle amplitudes represent S-matrix for an arbitrary number of solitons and antisolitions. To obtain the total sine-Gordon S-matrix one should supplement it with elements describing the scattering of any number of solitons and bound states (4.3). We denote the latter particles as B_n ; $n = 1, 2, \ldots < 8\pi/\gamma$.

Particles B_n with even (odd) values of n turn out to have positive (negative) C-parity. This can be seen if one considers the soliton-antisoliton ammplitudes with the definite s-channel C-parity:

$$S_+(\theta) = \tfrac{1}{2}[S_T(\theta) + S_R(\theta)], \tag{4.14a}$$

$$S_-(\theta) = \tfrac{1}{2}[S_T(\theta) - S_R(\theta)]. \tag{4.14b}$$

The amplitude $S_+(\theta)$ has even subset $n = 2, 4, \ldots$ of bound state poles $\theta = i\pi - in(\gamma/8)$ only, while $S_-(\theta)$ exhibits only odd subset $n = 1, 3, \ldots$ (these poles of $S_\pm(\theta)$ have positive residues as it should be). In particular, the sine-Gordon "elementary" particle B_1 is C-odd.

Since the particles B_n appear as poles of soliton-antisoliton amplitudes, an arbitrary S-matrix element involving these particles can be calculated as a residue of an appropriate multiparticle soliton amplitude. In terms of the algebraic formalism described in Section 2 algebra (3.5) of particles $A(\theta)$ and $\bar{A}(\theta)$ should be supplemented with new symbols $B_n(\theta)$; $n = 1, 2, \ldots < 8\pi/\gamma$ and commutation rules of B_n with A and $\bar{A}$ and of B_n with B_m should be specified. The procedure for the residue calculation mentioned above corresponds to the following definition of symbols $B_n(\theta)$ in terms of A and $\bar{A}$

$$B_n\left(\frac{\theta_1 + \theta_2}{2}\right) = \lim_{\theta_1 - \theta_2 \to i(n\gamma/8)} [A(\theta_2)\, \bar{A}(\theta_1) + \bar{A}(\theta_2)\, A(\theta_1)] \tag{4.15a}$$

for n even, and

$$B_n\left(\frac{\theta_1 + \theta_2}{2}\right) = \lim_{\theta_1 - \theta_2 \to i(n\gamma/8)} [A(\theta_2)\, \bar{A}(\theta_1) - \bar{A}(\theta_2)\, A(\theta_1)] \tag{4.15b}$$

for n odd.

[8] The derivation presented in [43] is based on certain special assumptions such as exactness of mass spectrum (4.3) and vanishing of reflection at $\gamma = 8\pi/n$. A derivation relying on the factorization equations and not referring to these assumptions was first given in [12].

This definition has a formal character and should be used to derive the rules for commutation of B_n with A and of B_n with B_m. Considering, for instance, in-product $A(\theta_1)\,\bar{A}(\theta_2)\,A(\theta_3)$, using (4.5) and taking the limit $\theta_1 - \theta_2 \to in(\gamma/8)$ by means of (4.15) one obtains

$$\begin{aligned} A(\theta_1)\,B_n(\theta_2) &= S^{(n)}(\theta_{12})\,B_n(\theta_2)\,A(\theta_1), \\ \bar{A}(\theta_1)\,B_n(\theta_2) &= S^{(n)}(\theta_{12})\,B_n(\theta_2)\,\bar{A}(\theta_1), \end{aligned} \tag{4.16}$$

where

$$S^{(n)}(\theta) = \frac{\operatorname{sh}\theta + i\cos\frac{n\gamma}{16}}{\operatorname{sh}\theta - i\cos\frac{n\gamma}{16}} \prod_{l=1}^{n-1} \frac{\sin^2\left(\frac{n-2l}{32}\gamma - \frac{\pi}{4} + i\frac{\theta}{2}\right)}{\sin^2\left(\frac{n-2l}{32}\gamma - \frac{\pi}{4} - i\frac{\theta}{2}\right)} \tag{4.17}$$

is the amplitude of two-particle scattering $A + B_n \to A + B_n$. Analogous consideration leads to commutation rules

$$B_n(\theta_1)\,B_m(\theta_2) = S^{(n,m)}(\theta_{12})\,B_m(\theta_2)\,B_n(\theta_1) \tag{4.18}$$

where $S^{(n,m)}(\theta)$ is the two-particle amplitude for $B_n + B_m \to B_n + B_m$ scattering. Its explicit form is

$$\begin{aligned} S^{(n,m)}(\theta) = {} & \frac{\operatorname{sh}\theta + i\sin\left(\frac{n+m}{16}\gamma\right)}{\operatorname{sh}\theta - i\sin\left(\frac{n+m}{16}\gamma\right)} \frac{\operatorname{sh}\theta + i\sin\left(\frac{n-m}{16}\gamma\right)}{\operatorname{sh}\theta - i\sin\left(\frac{n-m}{16}\gamma\right)} \\ & \times \prod_{l=1}^{m-1} \frac{\sin^2\left(\frac{m-n-2l}{32}\gamma + i\frac{\theta}{2}\right)\cos^2\left(\frac{m+n-2l}{32}\gamma + i\frac{\theta}{2}\right)}{\sin^2\left(\frac{m-n-2l}{32}\gamma - i\frac{\theta}{2}\right)\cos^2\left(\frac{m+n-2l}{32}\gamma - i\frac{\theta}{2}\right)}; \\ & n \geqslant m. \end{aligned} \tag{4.19}$$

Amplitudes (4.17) and (4.18) turn out to be $2\pi i$-periodic functions of θ (in fact, this property is dictated by the cross-symmetry and the two-particle unitarity of $S^{(n)}(\theta)$ and $S^{(n,m)}(\theta)$). The location of poles and zeroes of these amplitudes is shown in Fig. 6. Note the set of double poles $\theta_l = i(\pi/2) + [(2l - n)/16]\,\gamma$; $l = 1, 2, \ldots, n - 1$ of $S^{(n)}(\theta)$ for $n \geqslant 2$; these "redundant" poles do not correspond to any bound states. Single poles $\theta = i(\pi/2) + in(\gamma/16)$ and $\theta = i(\pi/2) - in(\gamma/16)$ are the s-channel and u-channel soliton poles, respectively: in the s-plane these poles are at $s = m^2$ and $u = m^2$ (m is the soliton mass).

In amplitude $S^{(n,m)}(\theta)$ only the poles $\theta = i[(n + m)/16]\,\gamma$ and $\theta = i\pi - i[(n + m)/16]\,\gamma$ correspond to the real particle B_{n+m}, all other poles are redundant. The appearance of poles B_{n+m} in the amplitude $S^{(n,m)}(\theta)$ allows one to interpret

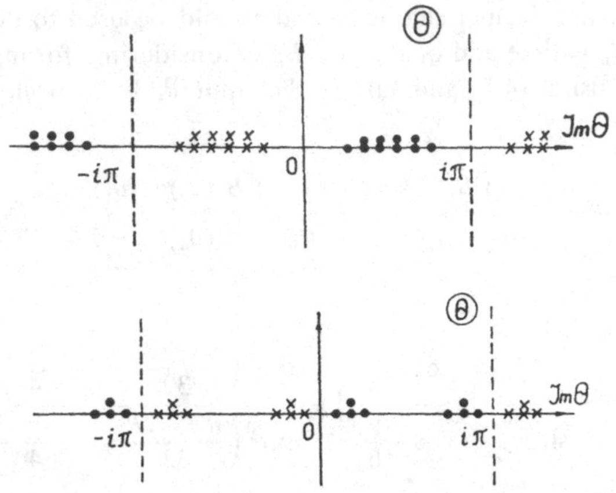

FIG. 6. Poles and zeroes of the soliton-bound state and the bound state — bound state scattering amplitudes. (a) $S^{(n)}(\theta)$ amplitude for $n = 5$. (b) $S^{(n,m)}(\theta)$ amplitude for $n = 4$, $m = 2$.

any particle B_l for $l \geqslant 2$ as a bound state $B_n + B_m$ with $n + m = l$[9] and, consequently, to interpret B_l as a bound state of l "elementary" particles B_1. A possibility of such interpretation was mentioned in Ref. [15].

In the case $m = n = 1$ Eq. (4.19) gives the two-particle amplitude of "elementary" particles

$$S^{(1,1)}(\theta) = \frac{\operatorname{sh}\theta + i\sin(\gamma/8)}{\operatorname{sh}\theta - i\sin(\gamma/8)}. \tag{4.20}$$

This expression can be expanded in powers of β^2 and compared with β^2-perturbation theory results for Lagrangian (1.1). Formula (4.20) together with its perturbation verification was presented in [8–10] as a solution of analyticity and unitarity for particles B_1.

Formulas (4.16–4.19) solve the bound state problem of the sine-Gordon model. Together with (4.5), (4.8–4.10) and (3.16) they represent the total quantum sine-Gordon *S*-matrix.

To conclude this section let us consider the nonrelativistic version of $O(2)$-symmetric *S*-matrix. After substitution (3.11) the general solution of the factorization equations becomes, instead of (4.8), (see Appendix A)

[9] One can verify that definition

$$B_{n+m}\left(\frac{\theta_1+\theta_2}{2}\right) = \lim_{\theta_1-\theta_2\to i[(n+m/16)]\gamma} B_m(\theta_2)B_n(\theta_1)$$

is consistent with (4.14) and completely self-consistent [31].

$$S_T(k) = -i \frac{\operatorname{sh}\left(\frac{8\pi k}{\gamma m}\right)}{\sin(\pi\kappa)} S_R(k), \tag{4.21a}$$

$$S(\kappa) = -i \frac{\operatorname{sh}\left(i\pi\kappa - \frac{8\pi k}{\gamma m}\right)}{\sin(\pi\kappa)} S_R(k), \tag{4.21b}$$

where $\kappa = 8\delta/\gamma$. Since we do not require any crossing-symmetry now, γ and κ are independent parameters. The unitarity condition, then, gives the following equation

$$S_R(k)\, S_R(-k) = \frac{\sin^2(\pi\kappa)}{\operatorname{sh}^2\left(\frac{8\pi k}{\gamma m}\right) + \sin^2(\pi\kappa)}. \tag{4.22}$$

The "minimum" solution of (4.22) is

$$S_R(k) = \frac{\sin(\pi\kappa)}{i\operatorname{sh}\left(\frac{8\pi k}{\gamma m}\right)} \frac{\Gamma\left(-i\frac{8k}{\gamma m} - \kappa\right)\Gamma\left(-i\frac{8k}{\gamma m} + \kappa + 1\right)}{\Gamma\left(-i\frac{8k}{\gamma m}\right)\Gamma\left(1 - i\frac{8k}{\gamma m}\right)}. \tag{4.23}$$

Formulas (4.21–4.23) clearly give the nonrelativistic limit of the sine-Gordon soliton S-matrix. Furthermore, amplitudes (4.23) and (4.21a) are just reflection and transition ones for the scattering on the potential

$$V_{A\bar{A}}(x) = -\frac{m}{64} \frac{\gamma^2 G}{\operatorname{ch}^2\left(\frac{m\gamma}{8} x\right)} \tag{4.24}$$

while amplitude (4.21b) describes the scattering on the potential

$$V_{AA}(x) = \frac{m}{64} \frac{\gamma^2 G}{\operatorname{sh}^2\left(\frac{m\gamma}{8} x\right)} \tag{4.25}$$

where $G = \kappa^2 - \kappa + \frac{3}{4}$. It is known that a system of $N + M$ nonrelativistic particles of two different kinds described by the Hamiltonian

$$H = -\frac{1}{2m}\sum_{i=1}^{N} \frac{d^2}{dx_i^2} - \frac{1}{2m}\sum_{j=1}^{M} \frac{d^2}{dy_j^2} + \sum_{i<i'}^{N} 2V_{AA}(x_i - x_{i'})$$
$$+ \sum_{j<j'}^{M} 2V_{AA}(y_j - y_{j'}) + \sum_{i=1}^{N}\sum_{j=1}^{M} 2V_{A\bar{A}}(x_i - y_j) \tag{4.26}$$

is completely integrable, i.e., it possesses an infinite number of conservation laws, and its S-matrix is factorized [44, 45]. So, system (4.26) describes just the nonrelativistic dynamics of quantum sine-Gordon solitons. The analogy between the sine-Gordon soliton scattering and the one on potential $\sim -1/ch^2x$ was noticed in Ref. [38].

5. $O(N)$-Symmetric Nonlinear σ-Model with $N \geqslant 3$

Now we consider the problem what type of quantum field theory can serve as a dynamical background of the $O(N)$-symmetric factorized S-matrix of Section 3 with $N \geqslant 3$.

First of all, remind the essential difference between a general solution of Section 3 with $N \geqslant 3$ and that with $N = 2$. For $N = 2$ the solution depends on a free parameter which could be interpreted as a coupling constant (in particular, a "weak coupling regime" can be achieved by a special choice of this parameter), while for $N \geqslant 3$ no free parameter enters the solution (the only umbiguity is (3.14)), but it depends analytically on the symmetry group rank N[10] and can be expanded in powers of $1/N$.

The coupling constant independence of all observable properties but the overall mass scale is the phenomenon characteristic for asymptotically free theories with dynamical mass transmutation (actually, it is a consequence of renormalizability [28]). One can believe, therefore, that a theory of this very type describes the dynamics of factorized scattering of Section 3 with $N \geqslant 3$.

To ensure the fact that a certain quantum field theory really leads to the S-matrix of Section 3 one should reveal its following properties:

(a) The model is one of the massive particles with $O(N)$-symmetric spectrum. The spectrum contains the isovector N-plet.

(b) The total S-matrix of this theory is factorized.

If these properties turn out to be true the problem of CDD-umbiguity (3.14) should be solved to obtain an exact S-matrix of the theory.

Consider $O(N)$-symmetric chiral field model (1.4). The usual g-perturbation theory of (1.4) is based on the Goldstone vacuum and leads in two dimensions to infrared catastrophe. Therefore, it is unlikely applicable to elucidate observable properties such as the spectrum and the S-matrix. However, there is another powerful approach to this model, namely, the $1/N$-expansion. This method for the model (1.4) has been developed in Refs. [23, 24] (see Appendix B). It is based on the exact solution at $N \to \infty$ which obviously satisfies the requirement (a): at $N \to \infty$ the model contains an isovector N-plet of free massive particles only.[11] The interaction of these particles is of the order of $1/N$ and $1/N$ expansion is just the perturbation theory in this interaction. The property (a) is still valid in any order of this perturbation theory. Thus, it is obviously true as N is sufficiently large. It is not ultimately clear whether the

[10] In this case the weak coupling limit is achieved as $N \to \infty$.

[11] The absence of other particles at large N is shown in [48].

situation is the same if N becomes not large, say $N = 3, 4$. However, we shall assume that the situation characteristic of large N is still valid at all $N \geqslant 3$. The result of this section confirms this assumption to some extent.

Let us turn to the scattering properties of this model. The existence of an infinite number of conservation laws for classical model (1.4) has been discovered by Pohlmeyer [47]. However, since the quantum vacuum of the model appears to be crucially different from the classical one, the relation between the classical conservation laws and quantum ones cannot be straightforward. In particular, the conformal invariance of the classical theory which is of essential use in Pohlmeyer's derivation is surely broken in a quantum case due to coupling constant renormalization.

The presence of higher conservation laws in quantum model (1.4) has been shown by Polyakov [26]. Here we present briefly Polyakov's derivation.

The equations of motion corresponding to Lagrangian (1.4) are:

$$n^i_{,\sigma\tau} + \omega n^i = 0; \qquad \sum_{i=1}^{N} (n^i)^2 = 1, \tag{5.1}$$

where ω is a Lagrange field (see Appendix B) and the indices σ and τ mean derivation with respect to

$$\sigma = x^0 + x^1; \qquad \tau = x^0 - x^1. \tag{5.2}$$

Equations (5.1) in the classical theory imply

$$\left[\sum_{i=1}^{N} (n^i_{,\sigma})^2\right]_\tau = \left[\sum_{i=1}^{N} (n^i_{,\tau})^2\right]_\sigma = 0. \tag{5.3}$$

These equations, which are of essential use in Pohlmeyer's derivation [47], mean both the conservation of energy-momentum and the conformal invariance of classical theory. In the quantum case the conformal symmetry is broken by stress-energy tensor anomaly and instead of (5.3) one has

$$\begin{aligned} \left[\sum_{i=1}^{N} (n^i_{,\sigma})^2\right]_\tau &= b\omega_{,\sigma}, \\ \left[\sum_{i=1}^{N} (n^i_{,\tau})^2\right]_\sigma &= b\omega_{,\tau}, \end{aligned} \tag{5.4}$$

where b is a constant which can be easily related to the Gell-Mann–Low function. Of course, relations (5.4) imply the energy-momentum conservation in the quantum theory since they are just of the divergence-zero type.

To obtain the next conservation law let us consider, following Polyakov, the derivatives $[\sum_{i=1}^{N} (n^i_{,\sigma\sigma})^2]_\tau$ and $[\sum_{i=1}^{N} (n^i_{,\tau\tau})^2]_\sigma$. In a classical theory one has, for instance,

$$\left[\sum_{i=1}^{N} (n^i_{,\sigma\sigma})^2\right]_\tau = -\left[\sum_{i=1}^{N} (n^i_{,\sigma})^2 \omega\right]_\sigma + 3 \sum_{i=1}^{N} (n^i_{,\sigma})^2 \omega_{,\sigma}. \tag{5.5}$$

In the quantum case this relation is deformed by anomalies. It is easy to see that the most general quantum variant of (5.5) is

$$\left[\sum_{i=1}^{N} (n^i_{,\sigma\sigma})^2\right]_{,\tau} = (3+\alpha)\left[\sum_{i=1}^{N} (n^i_{,\sigma})^2\right]\omega_{,\sigma} + (\cdots)_{\sigma}, \tag{5.6}$$

where the term proportional to α is the anomaly. Furthermore, one can consider the derivative $[\sum_{i=1}^{N} (n^i_{,\sigma})^2]_{,\tau}$ and obtain in a quantum case, quite analogously to (5.6),

$$\left[\sum_{i=1}^{N} (n^i_{,\sigma})^2\right]_{,\tau} = (2b+\alpha')\left[\sum_{i=1}^{N} (n^i_{,\sigma})^2\right]\omega_{,\sigma} + (\cdots)_{\sigma}. \tag{5.7}$$

Now it is straightforward to construct a new conservation law

$$\left[\sum_{i=1}^{N} (n^i_{,\sigma\sigma})^2 - \frac{3+\alpha}{2b+\alpha'}\left[\sum_{i=1}^{N} (n^i_{,\sigma})^2\right]^2\right]_{,\tau} = (\cdots)_{,\sigma}. \tag{5.8}$$

No difficulties arise in constructing a further conservation law. The equation

$$\left[\sum_{i=1}^{N} (n^i_{,\sigma\sigma\sigma})^2 + a_1\left[\sum_{i=1}^{N} (n^i_{,\sigma\sigma})^2\right]\left[\sum_{i=1}^{N} (n^i_{,\sigma})^2\right]\right.$$
$$\left. + a_2\left[\sum_{i=1}^{N} (n^i_{,\sigma})^2\right]^3 + a_3\left[\sum_{i=1}^{N} (n^i_{,\sigma\sigma}n^i_{,\sigma})\right]^2\right]_{,\tau} = (\cdots)_{,\sigma} \tag{5.9}$$

can be satisfied, in the same manner as (5.8), by an appropriate choice of parameters a_1, a_2 and a_3. Higher conservation laws of the infinite set require more delicate investigation; they have been constructed in a recent paper [27]. We do not discuss all the infinite set here. As shown in [26], the first two conservation laws (5.4) and (5.8) are already sufficient to restrict the S-matrix to the processes satisfying the selection rules (i) and (ii) pointed in Section 2. According to the general consideration of Section 2, this implies the S-matrix factorization for model (1.4).

It is instructive to observe the last property of the chiral field S-matrix in the $1/N$-perturbation theory [25]. Let us do this in the order of $1/N^2$ (in the order of $1/N$ these properties are trivial, being determined by kinematics). In this order we are

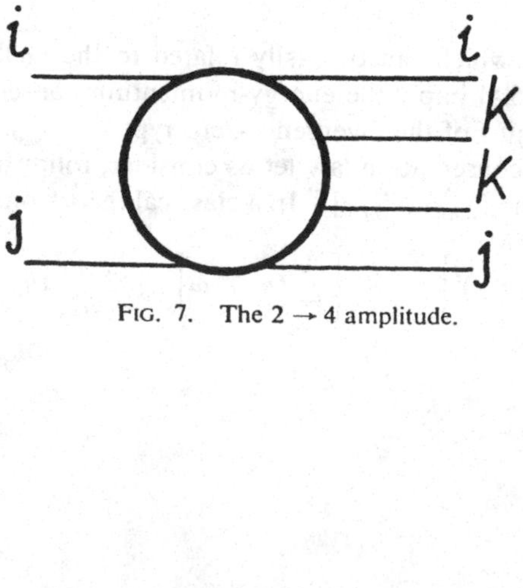

FIG. 7. The 2 → 4 amplitude.

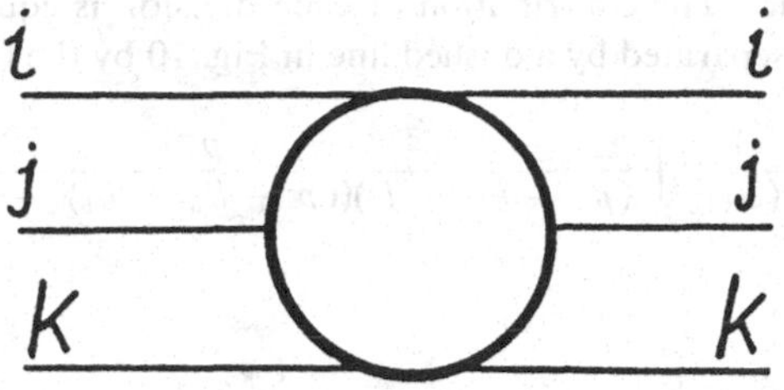

FIG. 8. The 3 → 3 amplitude.

interested in the amplitudes 2 → 4 (Fig. 7) and in connected amplitudes 3 → 3 (Fig. 8).

Using the diagrammatic technique of $1/N$-expansion (see Appendix B) one can represent the amplitudes 2 → 4 by a sum of diagrams shown in Fig. 9 (we consider only the case $i \neq j \neq k \neq i$ for simplicity; a general case includes more diagrams, but the result is the same). To demonstrate the total cancellation among the diagrams in Fig. 9 it is useful to take into account the explicit expression for arbitrary two dimensional one-loop diagram [8]. The extession is shown schemtatically in Fig. 10: an arbitrary boson loop is the sum of terms, each corresponding to any division of the

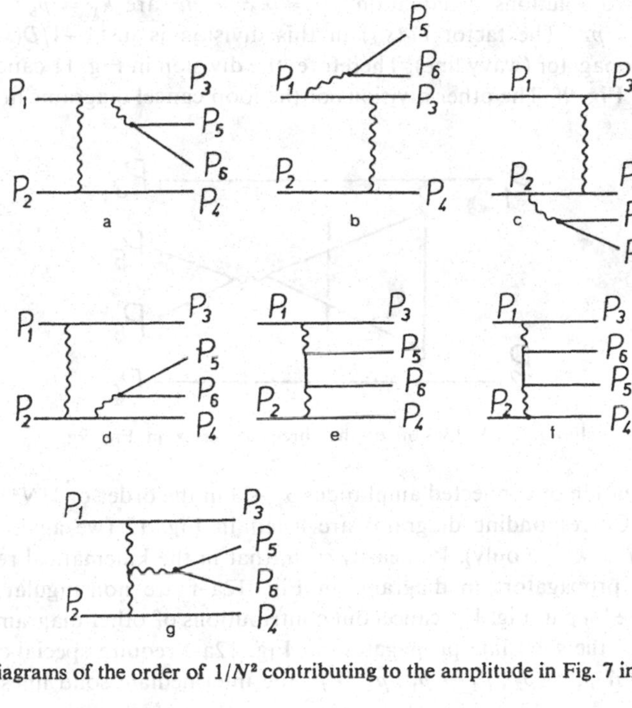

FIG. 9. Diagrams of the order of $1/N^2$ contributing to the amplitude in Fig. 7 in the case $i \neq j \neq k \neq i$.

loop through two lines. The contribution of each division is equal to the product of two "tree" diagrams separated by a dashed line in Fig. 10 by the function

$$i\Phi(s_{ab}) = \frac{1}{(2\pi)^2} \int \frac{d^2p}{(p^2 - m^2 + i\epsilon)((p + k_a + k_b)^2 - m^2 + i\epsilon)} \tag{5.10}$$

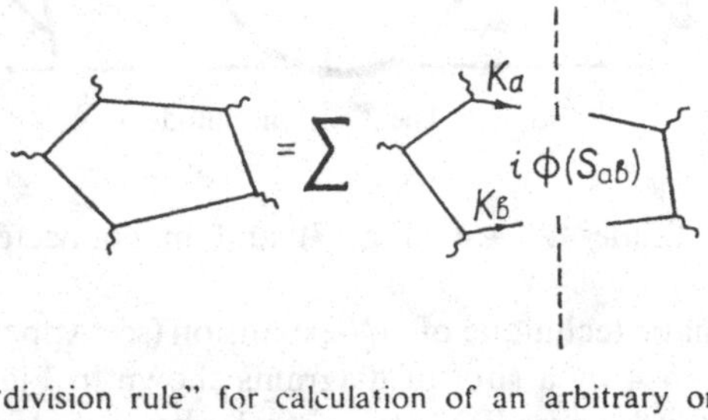

FIG. 10. The "division rule" for calculation of an arbitrary one-loop diagram.

where $s_{ab} = (k_a + k_b)^2$, the momenta k_a and k_b of cut lines being determined by the condition $K_a{}^2 = k_b{}^2 = m^2$. At s_{ab} fixed this equation has two solutions connected by the exchange $k_a \leftrightarrow k_b$, both should be taken into account in Fig. 10.

It is easy to see that all possible divisions of triangle loop in diagram in Fig. 9g cancel exactly the other diagrams in Fig. 9. Consider, for example, the division shown in Fig. 11. Two solutions of equation $k_1{}^2 = k_2{}^2 = m^2$ are $k_1 = p_5$; $k_2 = p_6$ and $k_1 = p_6$; $k_2 = p_5$. The factor $i\Phi(s_{56})$ in this division is just $-1/D(s_{56})$, where D is a ω' field propagator (wavy line). Therefore, the division in Fig. 11 cancels diagrams in Fig. 9e and Fig. 9f. The other divisions of the loop cancel diagrams in Fig. 9a–d.

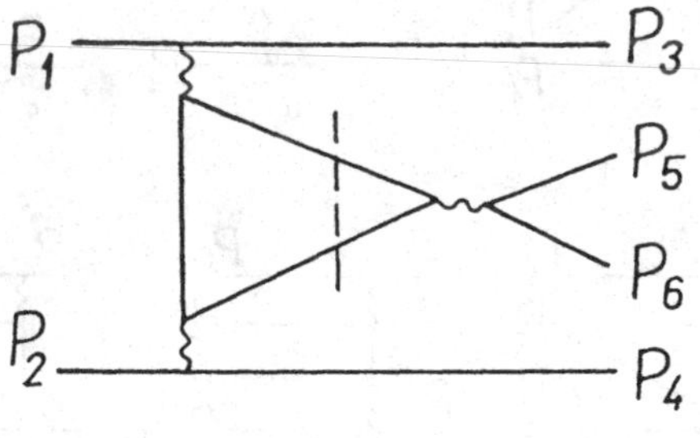

FIG. 11. A division of the three ω' vertex in Fig. 9g.

The factorization of connected amplitudes 3 → 3 in the order of $1/N^2$ can be shown analogously. Corresponding diagrams are listed in Fig. 12 (we again consider the the case $i \neq j \neq k \neq i$ only). It is easily seen, that in the kinematical regions, where the solid-line propagators in diagrams in Fig. 12a–f are nonsingular, all possible divisions of the loop in Fig. 12g cancel out contributions of other diagrams. Mass-shell singularities of the solid-line propagators in Fig. 12a–f require special consideration. For example, if $p_1' \to p_3$, $p_2' \to p_1$, $p_3' \to p_2$ the intermediate solid lines of diagrams

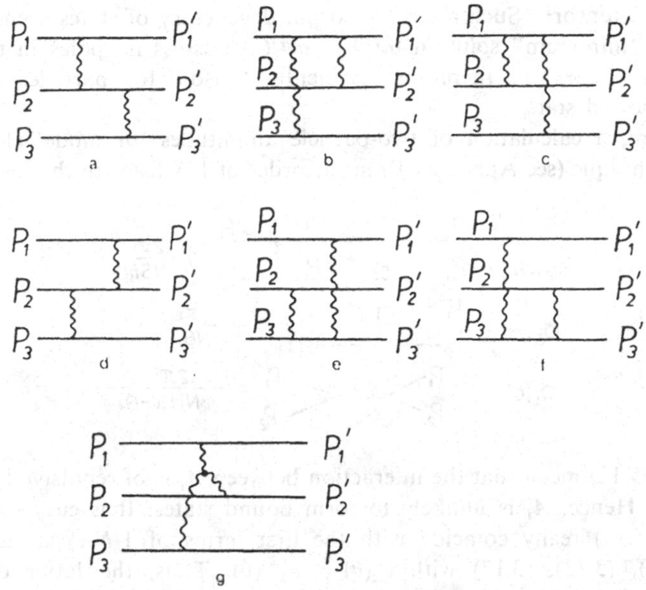

FIG. 12. Diagrams of the order of $1/N^2$ contributing to the amplitude in Fig. 8 in the case $i \neq j \neq k \neq i$.

in Figs. 12c, d and f acquire their mass shell poles. These propagators can be written in the form

$$\frac{1}{p^2 - m^2 + i\epsilon} = P\,\frac{1}{p^2 - m^2} + i\pi\,\delta(p^2 - m^2) \tag{5.11}$$

It can be shown that the principal parts of these singularities are cancelled among three diagrams in Figs. 12c, d and f, leaving the δ-function terms only, corresponding to intermediate particles on the mass shell. The diagram 12g cannot cancel these δ-function terms, being nonsingular in the region under consideration (where all the momenta transferred are space-like). So, only the terms with mass-shell δ-functions remain in the sum of diagrams in Fig. 12. These δ-functions ensure the factorized structure of the amplitude in Fig. 8.

We have shown that quantum model (1.4) satisfies properties (a) and (b). Hence, one of the solutions (3.8), (3.13), (3.17) specified by a choice of CDD poles (3.14) can be used to describe the scattering of this model. We do not know any way to remove the CDD-ambiguity rigorously, but the choice of "minimum" solution, namely $\sigma_2(\theta) = \sigma_2^{(+)}(\theta)$ from (3.17), appears to be the most natural. Below we present some arguments in support of this choice.

At first let us note that CDD-poles (3.14), if added, in general, result in additional poles in all three channels of two-particle scattering: isoscalar, antisymmetric-tensor and symmetric tensor[12]. Such a strong isospin degenerasy of states seems to be unnatural. The "minimum" solution $\sigma_2(\theta) = \sigma_2^{(+)}(\theta)$ possess no poles in the physical strip $0 < \operatorname{Im} \theta < \pi$ and therefore "elementary" isovector particles A_i of (1.4) produce no bound states.

Furthermore, a calculation of two-particle amplitudes for model (1.4) by $1/N$-expansion technique (see Appendix B) in the order of $1/N$ leads to the result:

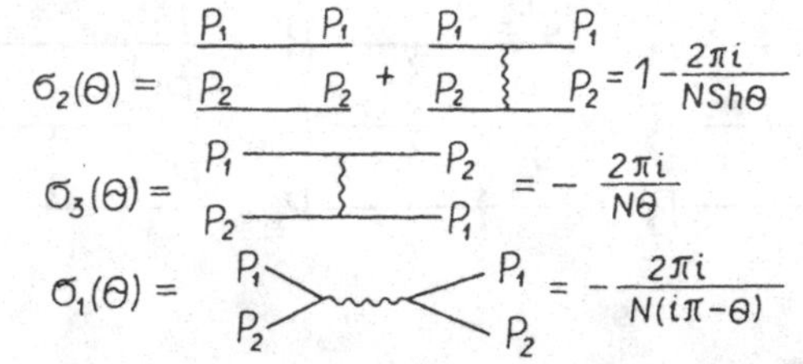

$$\sigma_2(\theta) = \begin{array}{cc} P_1 \quad P_1 \\ P_2 \quad P_2 \end{array} + \begin{array}{cc} P_1 \quad P_1 \\ P_2 \quad P_2 \end{array} = 1 - \frac{2\pi i}{N \operatorname{Sh}\theta}$$

$$\sigma_3(\theta) = \begin{array}{cc} P_1 \quad\quad P_2 \\ P_2 \quad\quad P_1 \end{array} = -\frac{2\pi i}{N\theta} \qquad (5.12)$$

$$\sigma_1(\theta) = \begin{array}{cc} P_1 \quad P_1 \\ P_2 \quad P_2 \end{array} = -\frac{2\pi i}{N(i\pi - \theta)}$$

The signs in (5.12) mean that the interaction between A_i is of repulsive type (at least for large N). Hence, A_i is unlikely to form bound states. It is easy to verify that expressions (5.12) really coincide with the first terms of $1/N$-expansion of exact solution (3.8), (3.13), (3.17) with $\sigma_2(\theta) = \sigma_2^{(+)}(\theta)$. Thus, the latter choice is in accordance with $1/N$-expansion of (1.4).

It is interesting to compare the solution of Section 3 with the results of the ordinary g-perturbation theory of model (1.4). Adopting the S-matrix (3.8), (3.17) to correspond to some renormalizable asymptotically free field theory, one can expand the scattering amplitudes, which are the functions of variable

$$\ln \frac{s}{m^2} = \ln \frac{s}{\mu^2} + \int^{g(\mu)} \frac{dg}{\beta(g)} \qquad (5.13)$$

in the asymptotic series in powers of $g(\mu)$. Using the first term of σ-model Gell–Mann–Low function $\beta(g)$ [20]

$$\beta(g) = -\frac{N-2}{4\pi} g^2 + O(g^3) \qquad (5.14)$$

one obtains up to g^2 ($g = g(\mu)$)

$$\sigma_2(s) = 1 - i\frac{g^2}{8} + O(g^3),$$

$$\sigma_3(s) = -i\frac{g}{2} + i\frac{N-2}{8\pi} g^2 \ln \frac{s}{\mu^2} + O(g^3), \qquad (5.15)$$

$$\sigma_1(s) = i\frac{g}{2} - i\frac{N-2}{8\pi} g^2 \ln \frac{s}{\mu^2} - \frac{N-2}{8} g^2 + O(g^3).$$

[12] The only exceptional case is that of single CDD-pole $\alpha_1 = \lambda$ added to $\sigma_2^{(+)}(\theta)$, where bound states appear in isoscalar and antisymmetric tensor channels only. This case corresponds to $\sigma_2(\theta) = \sigma_2^{(-)}(\theta)$ (see (3.19)) and is under consideration in the further Section.

In (5.15) the asymptotics $s \to \infty$ is written down and power terms in s are dropped.

The usual perturbation theory (expansion in g) is based on Goldstone vacuum and deals with $N - 1$ plet of Goldstone particles instead of N-plet of massive particles A_i. In this perturbation theory the loop diagram calculation leads to infrared divergencies. However, all the infrared divergencies are cancelled among the diagrams contributing to the scattering amplitude of Goldstone particles in the order of g^2. Hence, one can believe that calculation of these diagrams results in the correct ultraviolet asymptotics of the reeal A_i particle scattering amplitude. This calculation is straightforward and does lead to (5.15).

6. S-MATRIX OF THE GROSS-NEVEU "ELEMENTARY" FERMIONS

Another example of an asymptotically free field theory exhibiting the properties (a) and (b) of Section 5 and therefore leading to the factorized S-matrix of Section 3 with $N \geqslant 3$ is Gross-Neveu model (1.5).

An infinite set of nontrivial conservation laws for a classical version of model (1.5) has been found in a recent paper [49]. These classical conservation laws are quite analogous in their structure to those of the nonlinear σ-model found by Pohlmeyer. Again the conformal invariance of the classical theory (1.5) (which is broken in a quantum case) plays the crucial role in the derivation of these conserved currents. However, it is natural to expect that higher conservation laws are present in quantum theory (1.5) as well. Dashen, Hasslacher and Neveu [29] have investigated the classical field equation which determine the stationary phase points of the effective action (B.5′) (see Appendix B). They have been able to find out explicitly a series of time-dependent solutions. It means almost surely the complete integrability of the system determined by these equations.

To make sure that quantum theory (1.5) really possesses higher conservation laws let us derive the first nontrivial law following Polyakov's method [26]. All the considerations will be quite parallel to those applied in Section 5 to the case of the σ-model.

It is convenient to use the motion equations of (1.5) explicitly in terms of right and left-handed components of Majorana "bispinors" $\psi_i(x) = (\psi_i^r(x), \psi_i^l(x))$

$$\begin{aligned} i\psi^r_{i,\tau} &= \omega\psi_i^l, \\ i\psi^l_{i,\sigma} &= -\omega\psi_i^r; \qquad i = 1, 2, \ldots, N, \end{aligned} \tag{6.1}$$

where $\omega = g_0 \sum_{i=1}^N \psi_i^r\psi_i^l$. The momentum-energy conservation and conformal invariance of equations in classical case imply analogously to (5.3)

$$\left(\sum_{i=1}^N \psi_i^r\psi^r_{i,\sigma}\right)_{,\tau} = \left(\sum_{i=1}^N \psi_i^l\psi^l_{i,\tau}\right)_{,\sigma} = 0 \tag{6.2}$$

which should be, of course, replaced in the quantum theory

$$\left(\sum_{i=1}^{N} \psi_i{}^r \psi_{i,\sigma}^r\right)_{,\tau} = -b\omega_{,\sigma}^2,$$
$$\left(\sum_{i=1}^{N} \psi_i{}^l \psi_{i,\tau}^l\right)_{,\sigma} = -b\omega_{,\tau}^2. \tag{6.3}$$

Using these quantum equations one can easily verify that the following equations can be satisfied by an appropriate choice of the parameter C

$$\left[\sum_{i=1}^{N} \psi_{i,\sigma}^r \psi_{i,\sigma\sigma}^r + C\left(\sum_{i=1}^{N} \psi_i{}^r \psi_{i,\sigma}^r\right)^2\right]_{,\tau} = (\cdots)_{,\sigma},$$
$$\left[\sum_{i=1}^{N} \psi_{i,\tau}^l \psi_{i,\tau\tau}^l + C\left(\sum_{i=1}^{N} \psi_i{}^l \psi_{i,\tau}^l\right)^2\right]_{,\sigma} = (\cdots)_{,\tau}. \tag{6.4}$$

These equations are just the first nontrivial conservation law of the Gross–Neveu model.

The existence of higher conservation laws implies that the Gross–Neveu S-matrix satisfies property (b) of Section 5 [26].

Alternatively one could discover this property of the Gross–Neveu S-matrix in $1/N$-expansion [50]. The $1/N$-expansion technique for this model has been developed in Ref. [28] (it is described briefly in Appendix B). It is similar in the main to that used in the case of the nonlinear σ-model. In particular, the diagrammatic consideration of previous section can be repeated word for word in the Gross–Neveu case.

There is an important difference, however, between the σ-model $1/N$ technique and that of the Gross-Neveu model. Note the additional minus sign in (B.9′) against (B.9) which is connected with the fermion nature of ψ_i-fields (see Appendix B). This leads to the significant difference between two scattering theories. For instance, for model (1.5) one has, instead of (5.12),

$$\sigma_2^{GN}(\theta) = 1 + \frac{2\pi i}{N \operatorname{sh} \theta} + O\left(\frac{1}{N^2}\right);$$
$$\sigma_3^{GN}(\theta) = \frac{2\pi i}{N\theta} + O\left(\frac{1}{N^2}\right); \qquad \sigma_1^{GN}(\theta) = \frac{2\pi i}{N(i\pi - \theta)} + O\left(\frac{1}{N^2}\right). \tag{6.5}$$

The signs in Eq. (6.5) correspond to attractive interaction of "elementary" fermions (which we denote again by A_i; $i = 1, 2, \ldots, N$). Therefore, bound states of A_i should exist.

The bound state problem in model (1.5) has been investigated by a semiclassical large N method in Ref. [29]. The rich spectrum of $O(N)$-multiplets of bound states has been found. There are isoscalar, isovector multiplets and a number of higher rank antisymmetric-tensor ones. The semiclassical spectrum possesses a strong isospin degenracy: different isospin multiplets are gathered into supermultiplets defined by the "main quantum number" n which corresponds to the number of bounded "elementary" fermions. Semiclassical masses depend on this number only

$$m_n^{(\text{sem})} = m \frac{\sin\left(\frac{\pi n}{N}\right)}{\sin\left(\frac{\pi}{N}\right)}; \qquad n = 1, 2, \ldots < \frac{N}{2}, \tag{6.6}$$

where m is the mass of A_i.

There are other particles apart from A_i—bound states—"kinks" of the $\omega(x)$ field [29]. Their semiclassical masses are

$$M_{\text{kink}} = \frac{m}{2 \sin\left(\frac{\pi}{N}\right)} \tag{6.7}$$

and the existence of these "kinks" is just an explanation of the upper bound for n in Eq. (6.6).

The qualitative structure of semiclassical bound state spectrum (which becomes exact as $N \to \infty$) makes one choose solution (3.8), (3.13), (3.17) with $\sigma_2(\theta) = \sigma_2^{(-)}(\theta)$ as an exact S-matrix of "elementary" Gross–Neveu fermions. The $1/N$-expansion of this solution turns out to coincide with (6.5). Furthermore, let us consider two particle amplitudes of A_i-scattering with the definite s-channel isospin

$$\sigma_{(\text{isocalar})} = N\sigma_1 + \sigma_2 + \sigma_3 = -\frac{(\theta + i\lambda)(\theta + i\pi)}{\theta(i\pi - \theta)} \sigma_2^{(-)}(\theta), \tag{6.8a}$$

$$\sigma_{(\text{antisymm})} = \sigma_2 - \sigma_3 = \frac{\theta + i\lambda}{\theta} \sigma_2^{(-)}(\theta), \tag{6.8b}$$

$$\sigma_{(\text{symm})} = \sigma_2 + \sigma_3 = \frac{\theta - i\lambda}{\theta} \sigma_2^{(-)}(\theta). \tag{6.8c}$$

As it is seen from (6.8), bound states exist only in isoscalar and antisymmetric-tensor channels. We denote these particles B and B_{ij}. Their masses are

$$m_B = m_{B_{kj}} \equiv m_2 = m \sin\left(\frac{2\pi}{N-2}\right)\left[\sin\left(\frac{\pi}{N-2}\right)\right]^{-1}. \tag{6.9}$$

Higher bound states appear as poles in multiparticle amplitudes. The investigation of these poles (quite parallel to that made in Ref. [10] for the bound states of sine-Gordon "elementary" particles) leads to the spectrum of multiparticle A_i-bound states which agrees qualitatively with the semiclassical spectrum of Dashen, Hasslacher and Neveu [29]. Semiclassical isospin degeneracy turns out to be exact while the exact mass formula is

$$m_n = m \sin\left(\frac{\pi n}{N-2}\right)\left[\sin\left(\frac{\pi}{N-2}\right)\right]^{-1}; \qquad n = 1, 2, \ldots < \frac{N-2}{2}, \tag{6.10}$$

which differs from the semiclassical one (6.6) by substitution $N \to N-2$ only. It is natural to suppose that all the qualitative picture of semiclassical spectrum remains unchanged in the exact solution provided the substituion $N \to N-2$ is made. In particular, there are "kink" particles at any N and formula (6.10) in terms of kink mass becomes

$$m_n = 2M_{\text{kink}} \sin\left(\frac{\pi n}{N-2}\right); \qquad n = 1, 2, \ldots < \frac{N-2}{2}. \tag{6.11}$$

It follows from Eq. (6.11) that the bound states subsequently disappear from the spectrum with the decrease of N and no particles but the "kinks" remain in the system at $N \leqslant 4$. In particular, there are no "fundamental" fermions A_i at $N = 3, 4$. Therefore, the exact S-matrix of Gross–Neveu "fundamental" fermions presented above has a direct physical meaning at $N > 4$ only being quite fictitious at $N = 3, 4$.

To construct the total Gross–Neveu S-matrix for any $N \geqslant 3$ one should calculate the factorized S-matrix for the "kinks". The essential problem arises in this way: what representation of internal symmetry group do the "kinks" belong to? There are some arguments that these particles form $O(N)$ isospinor multiplets (E. Witten, Private communication). In any event the problem of obtaining the total Gross–Neveu S-matrix remains open.

For the further development of the subject reviewed in this paper see Refs. [51–54].

Appendix A

In this Appendix we derive solutions of Eqs. (3.5) and (3.6) [12, 25].

(1) Consider system (3.5). It is convenient to introduce the ratios

$$h(\theta) = \frac{S_2(\theta)}{S_3(\theta)}; \qquad g(\theta) = \frac{S_1(\theta)}{S_3(\theta)}. \tag{A.1}$$

Then Eqs. (3.5a, b) become

$$\begin{aligned} h(\theta) + h(\theta') - h(\theta+\theta') &= g(\theta')\, h(\theta+\theta') \\ &\quad - h(\theta')\, g(\theta+\theta') + g(\theta')\, g(\theta+\theta')\, h(\theta), \end{aligned} \tag{A.2}$$

$$\begin{aligned} [1 + h(\theta+\theta') + g(\theta+\theta')][1 - g(\theta)\, g(\theta')] + h(\theta)\, h(\theta') \\ = (1 + g(\theta) + h(\theta))(1 + g(\theta') + h(\theta')). \end{aligned} \tag{A.3}$$

Substituting $\theta = 0$ or $\theta' = 0$ into (A.3) and (A.2) one obtains the following relations

$$\begin{aligned} [1 - g^2(\theta)]\, h(0) &= 0, \\ [1 + g(\theta)][(1 + g(\theta) + h(\theta))\, g(0) + h(0)] &= 0, \\ [1 + g(\theta)][h(0) - g(0)\, h(\theta)] &= 0. \end{aligned}$$

These equations can be satisfied in three possible ways: (a) $g(\theta) \equiv 1$; $h(\theta) \equiv -1$ (b) $g(\theta) \equiv -1$; $h(\theta)$ is arbitrary, (c) $g(0) = h(0) = 0$. The first two cases are not interesting for us since possibility (a) is against unitarity (3.4b) and possibility (b) cannot satisfy crossing relation (3.2b). Therefore,

$$g(0) = h(0) = 0. \tag{A.4}$$

Differentiating (A.2) and (A.3) with respect to θ' and then setting $\theta' = 0$ one gets

$$h'(\theta) = (1 + g(\theta)((\alpha - \beta h(\theta)); \tag{A.5}$$

$$h'(\theta) + g'(\theta) = (1 + g(\theta))[\alpha + \beta h(\theta) + \beta(1 + g(\theta))], \tag{A.6}$$

where $\alpha = h'(0)$ and $\beta = g'(0)$. These equations can be easily turned to the form

$$g(\theta) = \beta h(\theta) \frac{h(\theta) + 1}{\alpha - \beta h(\theta)}; \tag{A.7}$$

$$h'(\theta) = \beta h^2(\theta) + \alpha. \tag{A.8}$$

The solution of (A.8) and (A.7) is

$$h(\theta) = -i \operatorname{tg}\left(\frac{4\pi\delta}{\gamma}\right) \operatorname{th}\left(\frac{4\pi\theta}{\gamma}\right); \tag{A.9}$$

$$g(\theta) = \operatorname{th}\left(\frac{4\pi\theta}{\gamma}\right) \operatorname{cth}\left[\frac{4\pi}{\gamma}(i\delta - \theta)\right], \tag{A.10}$$

where $\alpha = -i(4\pi/\gamma)\operatorname{tg}(4\pi\delta/\gamma)$; $\beta = -i(4\pi/\gamma)\operatorname{ctg}(4\pi\delta/\gamma)$. The real analyticity condition for the scattering amplitudes requires $h(\theta)$ and $g(\theta)$ to be real at $\operatorname{Re}\theta = 0$. Hence, γ and δ are real parameters. Formulas (A.9) and (A.10) are equivalent to (3.7).

(2) Let us turn to the system (3.6). Using the notation $h(\theta) = \sigma_2(\theta)/\sigma_3(\theta)$ one reduces (3.6a) to the form

$$h(\theta) + h(\theta') = h(\theta + \theta'). \tag{A.11}$$

Hence

$$\sigma_3(\theta) = -i\frac{\lambda}{\theta}\sigma_2(\theta), \tag{A.12}$$

where λ is an arbitrary real parameter. Substitution of (A.12) into (3.6b) leads to the following equation

$$\rho(\theta + \theta')\,\rho(\theta') = \frac{i\lambda}{\theta}[\rho(\theta') - \rho(\theta + \theta')], \tag{A.13}$$

where $\rho(\theta) = \sigma_1(\theta)/\sigma_2(\theta)$. The solution of (A.13) is

$$\rho(\theta) = -\frac{i\lambda}{i\kappa - \theta}, \tag{A.14}$$

where κ is the other real parameter. Now Eq. (3.6c) leads to the restriction

$$\kappa = i\frac{N-2}{2}\lambda \tag{A.15}$$

and we obtain (3.8).

Appendix B

This Appendix is intended for the derivation of the diagrammatic technique of $1/N$ expansion of models (1.4) and (1.5) [23, 24, 28].

All the following calculations will be performed for both models (1.4) and (1.5) simultaneously. To avoid any confusion, relative variables corresponding to models (1.4) and (1.5) are marked by subindices CF (chiral field) and GN (Gross–Neveu). Furthermore, all the numbers of formulas relating to model (1.5) are primed.

Following [23, 28] introduce the auxiliary Lagrange field and write

$$\mathscr{L}'_{CF} = \frac{1}{2g_0}\sum_{i=1}^{N}[(\partial_\mu n_i)^2 + \omega(x)\, n_i{}^2] - \frac{\omega(x)}{2g_0}; \tag{B.1}$$

$$\mathscr{L}'_{GN} = \frac{1}{2}\sum_{i=1}^{N}[i\bar\psi_i\, \partial_\mu\gamma^\mu\psi_i - \omega(x)\, \bar\psi_i\psi_i] - \frac{\omega^2(x)}{2g_0}. \tag{B.1'}$$

The generating functional for the Green functions of the field $n_i(x)(\psi_i(x)$ in the case of (1.5)) can be written in the form

$$Z[J] = I[J]/I[0], \tag{B.2}$$

where

$$I_{CF}[J] = \int \prod_x \left[d\omega(x)\prod_i dn_i(x)\right]$$
$$\times \exp\left\{i\int d^2x \left[\mathscr{L}'_{CF}[n_i, \omega] + (g_0)^{1/2}\sum_{i=1}^{N} J_i(x)\, n_i(x)\right]\right\}, \tag{B.3}$$

$$I_{GN}[J] = \int \prod_x \left[d\omega(x)\prod_i d\psi_i(x)\right]$$
$$\times \exp\left\{i\int d^2x \left[\mathscr{L}'_{GN}[\psi_i, \omega] + \sum_{i=1}^{N} \bar J_i(x)\, \psi_i(x)\right]\right\}. \tag{B.3'}$$

The integration over $n_i(x)$ in expression (B.3) (over $\psi_i(x)$ in (B.3′)) may be carried out explicitly, and results in (irrelevant factor which is cancelled in (B.2) is dropped):

$$I_{CF}[J_i] = \int \prod_x d\omega(x) \times \exp\left\{iS_{CF}^{(\mathrm{eff})}[\omega] + \frac{i}{2}\int d^2x\, d^2x'\, J_i(x)\, J_i(x')\, G_{CF}(x, x' \mid \omega)\right\}; \quad \text{(B.4)}$$

$$I_{GN}[J_i] = \int \prod_x d\omega(x) \times \exp\left\{iS_{GN}^{(\mathrm{eff})}[\omega] + \frac{i}{2}\int d^2x\, d^2x'\, \bar{J}_i(x)\, G_{GN}(x, x' \mid \omega)\, J_i(x')\right\}, \quad \text{(B.4′)}$$

where

$$S_{CF}^{(\mathrm{eff})} = i\,\frac{N}{2}\,\mathrm{tr}\,\ln[\partial_\mu{}^2 - \omega(x)] - \int d^2x\,\frac{\omega(x)}{2g_0}; \quad \text{(B.5)}$$

$$S_{GN}^{(\mathrm{eff})} = -i\,\frac{N}{2}\,\mathrm{tr}\,\ln[\gamma_\mu\partial^\mu - \omega(x)] - \int d^2x\,\frac{\omega^2(x)}{2g_0}; \quad \text{(B.5′)}$$

and $C_{CF}(x, x' \mid \omega)$ and $G_{GN}(x, x' \mid \omega)$ are Green functions of differential operators

$$\partial_\mu{}^2 - \omega(x), \quad \text{(B.6)}$$

$$\gamma_\mu\partial^\mu - \omega(x), \quad \text{(B.6′)}$$

respectively.

One obtains the $1/N$-series of model (1.4)((1.5)) calculating integral (B.4)((B.4′)) perturbatively. The stationary phase point of this integral at $\omega = \tilde{\omega}$

$$\tilde{\omega}_{CF} = m_{CF}^2 = \Lambda^2 \exp\left\{-\frac{4\pi}{Ng_0}\right\}, \quad \text{(B.7)}$$

$$\tilde{\omega}_{GN} = m_{GN} = \Lambda \exp\left\{-\frac{2\pi}{Ng_0}\right\} \quad \text{(B.7′)}$$

should be taken into account and functionals $S_{CF}^{(\mathrm{eff})}[\omega]$ and $G_{CF}(x, x' \mid \omega)(S_{GN}^{(\mathrm{eff})}[\omega]$ and $G_{GN}(x, x' \mid \omega))$ should be expanded in $\omega' = \omega - \tilde{\omega}$.[13]

It is easy to follow from the integrals (B.4) and (B.4′) to the simple diagrammatic technique with elements drawn in Fig. 13a, b, where the sign $+(-)$ in multileg vertices (Fig. 13b) corresponds to the case of the chiral field (Gross–Neveu) model Constructing any diagram from these elements one should not draw closed solidline loops since they are already taken into account by multileg vertices in Fig. 13b.

[13] In fact, there are two symmetrical stationary phase points $\tilde{\omega} = \pm m_{GN}$ in (B.4′). The system is settled in one of them by the Higgs effect. This corresponds to spontaneous breakdown of the discrete γ_5-symmetry [28].

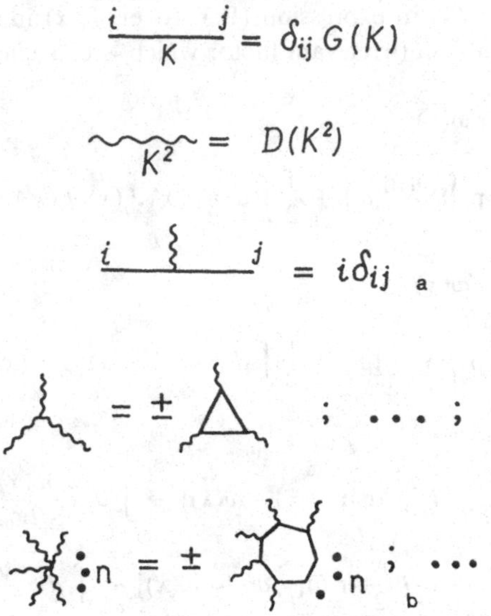

FIG. 13. Elements of the $1/N$-diagrammatic technique for chiral field and Gross-Neveu models.

Functions corresponding to the solid an wavy lines in Fig. 13 are different for the cases of (1.4) and (1.5):

$$G_{CF}(k) = \frac{i}{k^2 - m^2 + i\epsilon}, \tag{B.8}$$

$$G_{GN}(k) = i\,\frac{\hat{k} + m}{k^2 - m^2 + i\epsilon}, \tag{B.8$'$}$$

and

$$[D_{CF}(k^2)]^{-1} = \frac{1}{(2\pi)^2} \int \frac{d^2p}{(p^2 - m^2 + i\epsilon)((p + k)^2 - m^2 + i\epsilon)}; \tag{B.9}$$

$$[D_{GN}(k^2)]^{-1} = -\frac{1}{(2\pi)^2}\,\mathrm{tr} \int \Big[\frac{d^2p}{(\hat{p} - m + i\epsilon)(\hat{p} + \hat{k} - m + i\epsilon)} - \frac{d^2p}{(\hat{p} - m + i\epsilon)^2}\Big]. \tag{B.9$'$}$$

In formulas (B.9) and (B.9′) and in the general part of the paper indices CF and GN near the masses of particles are dropped.

ACKNOWLEDGMENTS

We are obliged to L. D. Faddeev, M. S. Marinov, A. M. Polyakov and K. A. Ter-Martirosyan for interesting discussions and helpful suggestions.

References

1. F. A. Berezin, G. P. Pokhil, and V. M. Finkelberg, *Vestn. Mosk. Univ. Mat. Mekh.* **1** (1964), 21.
2. J. B. McCuire, *J. Math. Phys.* **5** (1964), 622.
4. C. S. Gardner, T. M. Green, M. D. Kruskal, and R. M. Miura, *Phys. Rev. Lett.* **19** (1967), 1095.
5. V. E. Zakharov, *Zh. Exp. Teor. Fiz.* **60** (1970), 993. V. E. Zakharov and A. B. Shabat, *Zh. Exp. Teor. Fiz.* **61** (1971), 118.
6. L. A. Takhtadjyan and L. D. Faddeev, *Teor. Mat. Fiz.* **21** (1974), 160. M. J. Ablowitz, D. J. Kaup, A. C. Newell, and H. Segur, *Phys. Rev. Lett.* **31** (1975), 125; D. McLaughlin, *J. Math. Phys.* **16** (1975), 96.
7. P. P. Kulish, *Teor. Mat. Fiz.* **26** (1976), 188.
8. I. Ya. Arefyeva and V. E. Korepin, *Pis'ma Zh. Eksp. Teor. Fiz.* **20** (1974), 680.
9. S. N. Vergeles and V. M. Gryanik, *Yyad. Fiz.* **23** (1976), 1324.
10. B. Schroer, T. T. Truong, and P. H. Weisz, *Phys. Lett. B* **63** (1976), 422.
11. D. Iagolnitzer, Saclay, preprint DPh-T/77-130, 1977.
12. M. Karowski, H.-J Thun, T. T. Truong, and P. H. Weisz, *Phys. Lett. B* **67** (1977), 321.
13. V. E. Korepin, P. P. Kulish, and L. D. Faddeev, *Pis'ma Zh. Eksp. Teor. Fiz.* **21** (1975), 302.
14. V. E. Korepin and L. D. Faddeev, *Teor. Mat. Fiz.* **25** (1975), 147.
15. R. Dashen, B. Hasslacher, and A. Neveu, *Phys. Rev. D* **11** (1975), 3424.
16. S. Coleman, *Phys. Rev. D* **11** (1975), 2088.
17. S. Mandelstam, *Phys. Rev. D* **11** (1975), 3026.
18. B. Schroer, *Phys. Rep. C* **23**, No. 3 (1976).
19. B. Schroer and T. T. Truong, *Phys. Rev. D* **15** (1977), 1684.
20. A. M. Polyakov, *Phys. Lett. B* **59** (1975), 87.
21. A. A. Migdal, *Zh. Exp. Teor. Fiz.* **69** (1975), 1457.
22. E. Brezin, J. Zinn-Justin, and J. C. Le Guillou, *Phys. Rev. D* **14** (1976), 2615.
23. E. Brezin and J. Zinn-Justin, *Phys. Rev. B* **14** (1976), 3110.
24. W. A. Bardeen, B. W. Lee, and R. E. Shrock, *Phys. Rev. D* **14** (1976), 985.
25. A. B. Zamolodchikov and Al. B. Zamolodchikov, preprint JINR E2-10857, 1977.
26. A. M. Polyakov, preprint ICTP, 112, 1977.
27. I. Ya. Arefyeva, P. P. Kulish, E. R. Nisimov, and S. J. Pacheva, LOMI preprint E-I-1978.
28. D. Gross and A. Neveu, *Phys. Rev. D* **10** (1974), 3235.
29. R. Dashen, B. Hasslacher, and A. Neveu, *Phys. Rev. D* **12** (1975), 2443.
30. A. B. Zamolodchikov, preprint ITEP-12, 1977.
31. M. Karowski and H.-J. Thun, *Nucl. Phys. B* (1977), 295.
32. L. Castillejo, R. H. Dalitz, and F. J. Dyson, *Phys. Rev.* **101** (1956), 453.
33. I. Ya. Arefyeva, *Teor. Mat. Fiz.* **26** (1976), 306.
34. P. P. Kulish and E. R. Nisimov, *Teor. Mat. Fiz.* **29** (1976), 161.
35. B. Berg, M. Karowski, and H.-J. Thun, *Phys. Lett. B* (1976), 63, 187.
36. B. Berg, *Nuovo Cimento* **41** (1977), 58.
37. R. Jackiw and G. Woo, *Phys. Rev. D* **12** (1976), 1643.
38. V. E. Korepin, *Pis'ma Zh. Eksp. Teor. Fiz.* **23** (1976), 224; *Teor. Mat. Fiz.* **34** (1978), 3.
39. A. Luter, *Phys. Rev. B* **14** (1976), 2153.
40. S. Nussinov, *Phys. Rev. D* **14** (1976), 647.
41. L. Kadanoff and H. Ceva, *Phys. Rev. B* **3** (1970), 3918.
42. P. H. Weisz, *Nucl. Phys. B* **112** (1977), 1.
43. A. B. Zamolodchikov, *Commun. Math. Phys.* **55** (1977), 183.
44. F. Calogero, *Lett. Nuovo Cimento* **13** (1975), 411.
45. M. A. Olshanetsky and A. M. Perelomov, *Lett. Math. Phys.* **1** (1976), 187; *Invent. Math.* **37** (1976), 94; preprint ITEP-172, 1976.

46. M. T. Jaekel, *Nucl. Phys. B* **118** (1977), 506.
47. K. Pohlmeyer, *Commun. Math. Phys.* **46** (1976), 2u7.
48. T. Banks and A. Zaks, *Nucl. Phys. B* **128** (1977), 333.
49. A. Neveu and N. Papanicolaou, IAS Princeton preprint COO-2220-116, 1977.
50. A. B. Zamolodchikov and Al. B. Zamolodchikov, *Phys. Lett. B* **72** (1978), 481.
51. R. Shankar and E. Witten, Harvard preprint 77/A076, 1977.
52. B. Berg, M. Karowski, V. Kurak, and P. Weisz, preprint FUB/HEP 26/77, 1977.
53. B. Berg, M. Karowski, V. Kurak, and P. Weisz, DESY preprint 78/09, 1978.
54. M. Luscher, Copenhagen preprint, NBI-HE-77-44, 1977.

QUANTUM VERSION OF THE METHOD OF INVERSE SCATTERING PROBLEM

E. K. Sklyanin

UDC 517.43+530.145

This paper is a developed and consecutive account of a quantum version of the method of the inverse scattering problem on the example of the nonlinear Schrödinger equation. The method of R-matrices developed by the author is given basic consideration. The generating functions of quantum integrals of motion and action-angle variables for the quantum nonlinear Schrödinger equation are obtained. There is also described a classical version of the method of R-matrices.

Introduction

1. The last decade was marked by a sharp increase in interest in completely integrable systems of classical and quantum mechanics. Although exactly solvable problems are always of interest for physics and mathematics and models for the study of general laws of behavior of complicated nonlinear systems or as sources of "zeroth" approximations to nonintegrable equations, only comparatively recently were powerful methods for their study available, which allowed the essential extension of the class of completely integrable systems. We are concerned here, on the one hand, with the method of the inverse scattering problem [1], which has its origin in Gardner, Greene, Kruskal, and Miura [2], Lax [3], Zakharov and Faddeev [4] and allows one to investigate completely integrable models of classical mechanics. On the other hand, there exists a tradition of studying exactly solvable models of quantum mechanics and statistical physics, going back to Bethe [5] and Onsager [6] and achieving its highest development in Baxter [7-8]. (We intentionally schematize the situation here, leaving aside for example, the group-theoretic methods of investigation of classical and quantum completely integrable models.)

After the complete integrability of certain relativistically invariant models was proved: by the method of the inverse scattering problem sin-Gordon equations [9], chiral fields [10], etc. there arose the question, doesn't the corresponding quantum models also turn out to be completely integrable. A positive solution of this question would be of great interest for the quantum theory of fields, since it would give a nontrivial example of an exactly solvable quantum model. This circumstance gave rise to a series of attempts at the quasiclassical quantization of completely integrable models, e.g., [11, 12], which, however, were not completely satisfactory, since one could give only an approximate and not an exact answer.

Thus, there remains the actual problem of synthesis of the two approaches indicated above, the classical and the quantum, into one method of investigation of quantum fields of completely integrable systems, which one could call the "quantum method of the inverse scattering problem."

The first steps in this direction were undertaken in [13] by Faddeev and the author and in [14]. In [13] there was formulated a program for generalizing the method of the inverse scattering problem to the quantum case. In [14] this program was successfully realized for the nonlinear Schrödinger equation. However, the small size of [14] did not allow the inclusion in it of all the necessary proofs. The present paper, written on the basis of the author's dissertation [15], contains a developed and consecutive account of the quantum method of the inverse scattering problem for the nonlinear Schrödinger equation taking account of the achievements of the 2-yr period of development of this method [16-28]. In addition, we have included here a brief outline of the theory of the classical nonlinear Schrödinger equation, based on the method proposed by the author in [29] of the classical r-matrix.

Translated from Zapiski Nauchnykh Seminarov Leningradskogo Otdeleniya Matematicheskogo Instituta im. V. A. Steklova AN SSSR, Vol. 95, pp. 55-128, 1980.

The author profoundly thanks Academician L. D. Faddeev for formulating the problem and constant interest in the work and also all the collaborators in the laboratory of mathematical problems of physics of the Leningrad Branch of the Mathematics Institute (LOMI) for many fruitful discussions.

2. We consider in more detail the basic lines of the approach to the quantum generalized method of the inverse scattering problem, proposed in [13, 14] and developed in the present paper. First of all we compare the formulation of the problem in classical and quantum mechanics. If in classical mechanics one is usually interested in the evolution of the initial data in time, and the classical method of the inverse scattering problem is usually aimed at finding the law of evolution in time of the scattering data of an auxiliary linear problem, then for quantum mechanics the "stationary" approach is more characteristic, in which the greatest interest is in the spectrum of the Hamiltonian and S-matrix. However, despite such an apparent dissimilarity in the formulation of the problem, there nevertheless exists an approach to the classical method of the inverse scattering problem, completely analogous to the "stationary" quantum mechanical approach. We are concerned here with the Hamiltonian interpretation of the method of the inverse scattering problem, developed in [4, 30]. The study of the nonlinear evolution equation with such an approach is aimed at constructing from the scattering data of the auxiliary linear problem variables of action-angle type for the equation considered and thus proving its complete integrability, determining in passing the spectrum of the elementary excitations of the system. Now the question of the time of evolution is from this point of view of secondary interest. Precisely this Hamiltonian approach was taken as the starting point for the quantum mechanical generalization of the method of the inverse scattering problem in [13, 14] and in the present paper.

The choice of the nonlinear Schrödinger equation (n.S.e.)

$$i\Psi_t = -\Psi_{xx} + 2\varkappa|\Psi|^2\Psi \tag{1}$$

as the object of study was dictated by the following considerations:

1) The quantum version of the nonlinear Schrödinger equation (with zero boundary conditions at infinity) describes a one-dimensional system of Bose particles with pointwise interaction. Thus, the problem reduces to the quantum mechanics of a finite number of particles and does not contain difficulties specific for the quantum field theory (nonfocal representations of commuting relations, divergence, etc.).

2) The nonlinear Schrödinger equation is studied in detail in the classical as well as the quantum case. The classical n.S.e. admits the application of the method of the inverse scattering problem [30-32], in the quantum case there is a complete description of the spectrum and eigenfunctions of the Hamiltonian [33, 34]. The latter circumstance is valuable in that it allows one to compare the results obtained by the new method with the exact quantum answer.

Now one can give a concrete formulation of the problem. An important role in the method of the inverse scattering problem is played by the transition matrix $T(\lambda)$ (see below Sec. 1.1)

$$T(\lambda) = \begin{pmatrix} a(\lambda), & \varkappa\bar{b}(\lambda) \\ b(\lambda), & \bar{a}(\lambda) \end{pmatrix}, \quad \lambda = \bar{\lambda}, \tag{2}$$

which is defined by means of the auxiliary linear problem, and whose matrix elements are functionals of the fields $\bar{\Psi}(x)$, $\Psi(x)$.

As is known [30], (1) describes the dynamics of a Hamiltonian system with Hamiltonian

$$H = \int_{-\infty}^{\infty} dx \left(|\Psi_x|^2 + \varkappa|\Psi|^4\right)$$

and Poisson bracket, defined by the relations

$$\{\Psi(x), \Psi(y)\} = \{\overline{\Psi}(x), \overline{\Psi}(y)\} = 0 ,$$
$$\{\Psi(x), \overline{\Psi}(y)\} = i\delta(x-y) .$$

The method of the inverse scattering problem allows one to calculate the Poisson brackets between the matrix elements of the matrix $T(\lambda)$. In particular:

$$\{a(\lambda), a(\mu)\} = \{\bar{b}(\lambda), \bar{b}(\mu)\} = 0 , \tag{3}$$

$$\{a(\lambda), \bar{b}(\mu)\} = -\frac{\varkappa}{\lambda-\mu+i0}\, a(\lambda)\, \bar{b}(\mu) . \tag{4}$$

In turns out [31, 32], that $\ln a(\lambda)$ is the generating function of the local integrals of motion of (1), and from the quantities $b(\lambda)$ and $\bar{b}(\lambda)$ one can construct variables of action-angle type.

In [13] the problem of generalizing the method of the inverse problem to the quantum case was formulated as the problem of constructing quantum operators $A(\lambda)$ and $B^+(\lambda)$, which would have the following properties:

1) In the classical limit the operators $A(\lambda)$ and $B^+(\lambda)$ should go respectively into the quantities $a(\lambda)$ and $b(\lambda)$.

2) Between the operators $A(\lambda)$ and $B^+(\mu)$ there should be the following commutation relations:

$$[A(\lambda), A(\mu)] = [B^+(\lambda), B^+(\mu)] = 0, \tag{5}$$

$$A(\lambda) B^+(\mu) = c(\lambda,\mu) B^+(\mu) A(\lambda), \tag{6}$$

where $c(\lambda,\mu)$ is some numerical function of λ and μ.

We note that such a formulation makes sense not only for the n.S.e., but also for many other completely integrable equations [16].

The quantization of (1) is carried out in terms of the annihilation and birth operators $\Psi(x)$ and $\Psi^+(x)$, satisfying the canonical commutation relations

$$[\Psi(x), \Psi^+(y)] = [\Psi^+(x), \Psi^+(y)] = 0, \tag{7}$$
$$[\Psi(x), \Psi^+(y)] = \delta(x-y) .$$

Here the problem of constructing the operators $A(\lambda)$ and $B^+(\lambda)$ reduces, essentially, to the question of choosing a proper ordering of the operators Ψ and Ψ^+, i.e., an ordering such that condition 2), formulated above, should be satisfied. As will be proved later, for the n.S.e. the normal (Wick) ordering is proper, i.e., the operators $A(\lambda)$ and $B^+(\lambda)$ are defined as the operators whose normal symbols are respectively the classical functionals $a(\lambda; \Psi, \overline{\Psi})$ and $\bar{b}(\lambda; \Psi, \overline{\Psi})$.

The basic technical difficulty of the approach considered consists in proving the commutation relations (5, 6) and calculating the coefficients $c(\lambda,\mu)$. This problem can be solved by the method of R-matrices proposed by the author in [14], which makes it possible to write down the commutation relations between the matrix elements of the quantum transition matrix in compact matrix form and to reduce their proof to the verification of simple infinitesimal relations. The idea of this method was suggested to the author by the papers of Baxter [7, 8].

Calculation of the coefficient $c(\lambda,\mu)$ by the method of R-matrices gives the following result:

$$c(\lambda,\mu)=\frac{\lambda-\mu+i\varkappa}{\lambda-\mu}\ . \tag{8}$$

We list the main results which will be proved in the basic text as consequences of the commutation relations (5, 6):

1) The operator-valued function $\ln A(\lambda)$, as in the classical case, is the generating function of locally pairwise commuting integrals of motion $\mathcal{J}_m$ for the quantum n.S.e.

2) States $|k_1,\ldots,k_N\rangle_B$, obtained by the action in a vacuum of the operators $B^+(k_j)$ $(j=1,\ldots,N)$

$$|k_1,\ldots,k_N\rangle_B=B^+(k_1)\ldots B^+(k_N)|0\rangle\ , \tag{9}$$

are eigenvectors of the quantum Hamiltonian $\mathbb{H}$ and all integrals of motion $\mathcal{J}_m$, where the corresponding eigenvalues are additive in the momenta k_j:

$$\mathbb{H}\,|k_1,\ldots,k_N\rangle_B=\sum_{j=1}^{N}k_j^2\,|k_1,\ldots,k_N\rangle_B\ . \tag{10}$$

The wave functions of the states $k_1,\ldots,k_N\rangle_B$ coincide with the wave functions obtained by the Bethe substitution method [33, 34].

3) The operators $\varphi(\lambda),\varphi^+(\lambda)$, defined by the formulas

$$\varphi^+(\lambda)=B^+(\lambda)\left(2\pi A^+(\lambda)A(\lambda)\right)^{-1/2},\ \varphi(\lambda)=\varphi^+(\lambda)^+, \tag{11}$$

satisfy the canonical commutation relations

$$\begin{gathered}[\varphi(\lambda),\varphi(\mu)]=[\varphi^+(\lambda),\varphi^+(\mu)]=0\ ,\\ [\varphi(\lambda),\varphi^+(\mu)]=\delta(\lambda-\mu)\ .\end{gathered} \tag{12}$$

and allow (in the case $\varkappa>0$) the explicit diagonalization of the Hamiltonian

$$\mathbb{H}=\int_{-\infty}^{\infty}d\mu\,\mu^2\,\varphi^+(\mu)\,\varphi(\mu) \tag{13}$$

and all other integrals of motion $\mathcal{J}_m$. On this basis the operators $\varphi(\lambda),\varphi^+(\lambda)$ can be called the quantum analogs of action-angle variables.

3. The basic text of the paper consists of two chapters, a Conclusion and a Supplement. In Chap. I we consider the classical nonlinear Schrödinger equation, in Chap. II the quantum one. Here the consideration of the classical case is carried out so that all the results obtained have direct analogs in the quantum case.

The composition of the paper is also subordinate to this idea: to each section of Chap. I corresponds an analogous section of Chap. II.

Chapter I consists of five sections. In Sec. 1.1 the basic concepts and notation are introduced. In Sec. 1.2 the Poisson brackets between matricial elements of the transition matrix $T_{x_1}^{x_2}(\lambda)$ on a finite interval are computed. In Sec. 1.3 the cases of semi-infinite and infinite intervals are considered. In Sec. 1.4, which has an auxiliary character, known results about the integrals of motion for the n.S.e. are collected and action-angle variables are constructed. In Sec. 1.5, with the help of the method of R-matrices, the generating functions of the M-operators for the n.S.e. are constructed.

Chapter II also consists of five sections. In Sec. 2.1 the known results for the quantum n.S.e. are listed, the quantum transition matrix $\mathbb{T}_{x_1}^{x_2}(\lambda)$ is introduced. In Sec. 2.2 the commutation relations between the matricial elements of the quantum transition matrix $\mathbb{T}_{x_1}^{x_2}(\lambda)$ are calculated. In Sec. 2.3 the cases of semi-infinite and infinite intervals are considered. In Sec. 2.4, the results obtained are summarized, the question of construction of quantum action-angle variables is considered. In Sec. 2.5 the question of the quantum M-operator is studied.

In the Conclusion the basic derivations and results of the paper are summarized, a brief survey is given of unsolved problems in the domain of quantum completely integrable systems, the future developments in this direction are discussed.

In the Supplement a summary is given of the classical and quantum commutation relations between the matricial elements of the transition matrix for finite, semi-infinite, and infinite intervals.

CHAPTER I
CLASSICAL NONLINEAR SCHRÖDINGER EQUATIONS

1.1. Transition Matrix

In the present section we introduce the basic notation and list some results, basically known in [30-32], for the classical nonlinear Schrödinger equation. We allow ourselves to deviate somewhat from the notation and formulations of the original papers [30-32], giving them a form more convenient for our goals.

The nonlinear Schrödinger equation, as was indicated in the Introduction, has the form

$$i\Psi_t = -\Psi_{xx} + 2\varkappa|\Psi|^2\Psi . \qquad (1.1.1)$$

The complex-valued function $\Psi(x,t)$ will be assumed to be infinitely differentiable in both arguments and for any t, decreasing in x faster than any power of x.

The study of (1.1.1) by the method of the inverse scattering problem reduces, as was shown in [31, 32], to the study of the spectral characteristics of the sheaf of linear differential operators $\frac{d}{dx} - L(x,\lambda)$, where

$$L(x,\lambda) = \begin{pmatrix} -i\frac{\lambda}{2} & , i\varkappa\overline{\psi}(x) \\ -i\psi(x), & i\frac{\lambda}{2} \end{pmatrix} = -i\frac{\lambda}{2}\sigma_3 + i\varkappa\overline{\psi}(x)\sigma_+ - i\psi(x)\sigma_- . \qquad (1.1.2)$$

$$\sigma_3 = \begin{pmatrix} 1 & 0 \\ 0 & -1 \end{pmatrix}, \quad \sigma_+ = \begin{pmatrix} 0 & 1 \\ 0 & 0 \end{pmatrix}, \quad \sigma_- = \begin{pmatrix} 0 & 0 \\ 1 & 0 \end{pmatrix}.$$

Starting here and up to Sec. 1.5 we shall consider the moment of time t fixed.

In contrast with [30-32], we have chosen the matrix L with nonsymmetric occurrence of the connection constant $\varkappa$, which allows us to consider in a uniform way both the case of repulsion $\varkappa > 0$, and that of attraction $\varkappa < 0$.

We introduce the transition matrix $T_{x_1}^{x_2}(\lambda)$ on the finite interval $[x_1, x_2]$ as the solution of the differential equation

$$\frac{\partial}{\partial x_2} T_{x_1}^{x_2}(\lambda) = L(x_2,\lambda) T_{x_1}^{x_2}(\lambda) \qquad (1.1.3)$$

with the initial condition

$$T_x^x(\lambda) = \begin{pmatrix} 1 & 0 \\ 0 & 1 \end{pmatrix} = I. \qquad (1.1.4)$$

We list some properties of the matrix $T_{x_1}^{x_2}(\lambda)$:

1) $$\left(T_{x_1}^{x_2}(\lambda)\right)^{-1} = T_{x_2}^{x_1}(\lambda), \qquad (1.1.5)$$

2) $$T_{x_2}^{x_3}(\lambda)\, T_{x_1}^{x_2}(\lambda) = T_{x_1}^{x_3}(\lambda), \qquad (1.1.6)$$

3) $$\frac{\partial}{\partial x_1} T_{x_1}^{x_2}(\lambda) = -T_{x_1}^{x_2}(\lambda) L(x_1,\lambda), \qquad (1.1.7)$$

4) $$\overline{T_{x_1}^{x_2}(\lambda)} = K\, T_{x_1}^{x_2}(\bar{\lambda}) K, \qquad (1.1.8)$$

where

$$K = \begin{pmatrix} 0 & , \varkappa^{1/2} \\ \varkappa^{-1/2}, & 0 \end{pmatrix}, \quad K^2 = I,$$

and the line over a matrix denotes elementwise complex conjugation.

5) $$\det T_{x_1}^{x_2}(\lambda) = 1. \qquad (1.1.9)$$

Properties 1)-3) follow directly from the definition of $T_{x_1}^{x_2}(\lambda)$. The symmetry property 4) follows from the analogous property of the L-operator

$$\overline{L(x,\lambda)} = K L(x,\bar{\lambda}) K, \qquad (1.1.10)$$

which can be verified directly. We note that (1.1.8) means that the matrix $T_{x_1}^{x_2}(\lambda)$ has the form

$$T_{x_1}^{x_2}(\lambda)=\begin{pmatrix} a_{x_1}^{x_2}(\lambda), & \varkappa\overline{b_{x_1}^{x_2}(\bar\lambda)} \\ b_{x_1}^{x_2}(\lambda), & \overline{a_{x_1}^{x_2}(\bar\lambda)}\end{pmatrix} \qquad (1.1.11)$$

Finally, property 5) follows from the equation $\operatorname{tr} L(x,\lambda)=0$.

Substituting (1.1.11) in (1.1.9) we get for real λ the important relation of "unitariness"

$$|a_{x_1}^{x_2}(\lambda)|^2-\varkappa|b_{x_1}^{x_2}(\lambda)|^2=1,\ \lambda=\bar\lambda. \qquad (1.1.12)$$

Now we define the transition matrices $T_-(x,\lambda), T_+(x,\lambda), T(\lambda)$ for semi-infinite intervals $(-\infty,x]$ and $[x,\infty)$ and the infinite interval $(-\infty,\infty)$, respectively, as the following limits:

$$T_-(x,\lambda)=\lim_{x_1\to-\infty} T_{x_1}^{x}(\lambda)\,e(x_1,\lambda), \qquad (1.1.13)$$

$$T_+(x,\lambda)=\lim_{x_2\to\infty} e(-x_2,\lambda)\,T_{x}^{x_2}(\lambda), \qquad (1.1.14)$$

$$T(\lambda)=\lim_{\substack{x_1\to-\infty\\ x_2\to+\infty}} e(-x_2,\lambda)\,T_{x_1}^{x_2}(\lambda)\,e(x_1,\lambda), \qquad (1.1.15)$$

where we have introduced the notation

$$e(x,\lambda)=\exp\left(-i\frac{\lambda}{2}\sigma_3 x\right).$$

It follows from (1.1.13) that $T_-(x,\lambda)$ satisfies with respect to the variables x the differential equation (1.1.3) with the boundary condition as $x\to-\infty$

$$T_-(x,\lambda)-e(x,\lambda)\xrightarrow[x\to-\infty]{}0. \qquad (1.1.16)$$

Analogously, $T_+(x,\lambda)$ satisfies with respect to x the differential equation (1.1.7) with the boundary condition as $x\to+\infty$

$$T_+(x,\lambda)-e(-x,\lambda)\xrightarrow[x\to+\infty]{}0. \qquad (1.1.17)$$

It follows from (1.1.6) that for any x

$$T(\lambda)=T_+(x,\lambda)\,T_-(x,\lambda). \qquad (1.1.18)$$

Analogously, to the case of a finite interval, for $T_\pm(x,\lambda)$ and $T(\lambda)$ one proves the symmetry property (1.1.8) and the "unitariness" property (1.1.12).

We list analytic properties of the matricial functions $T_{x_1}^{x_2}(\lambda)$, $T_\pm(x,\lambda)$, and $T(\lambda)$ with respect to the spectral parameter λ. $T_{x_1}^{x_2}(\lambda)$ is a holomorphic function on the entire complex plane λ. The matrix elements $a_-(x,\lambda), b_-(x,\lambda), a_+(x,\lambda), \overline{b_+(x,\bar\lambda)}, a(\lambda)$ can be analytically extended to the half-plane $\operatorname{Im}\lambda>0$, then as matrix elements $\overline{a_-(x,\bar\lambda)}, \overline{b_-(x,\bar\lambda)}, b_+(x,\lambda), \overline{a_+(x,\bar\lambda)}, \overline{a(\bar\lambda)}$ can be analytically extended to the half-plane $\operatorname{Im}\lambda<0$. The matrix elements $b(\lambda)$ and $\overline{b(\bar\lambda)}$, in general, are defined only for real λ. (The notation for the matrix elements of the matrices $T_\pm(x,\lambda)$ and $T(\lambda)$ is given in the Supplement.)

The proof of the analytic properties listed above and also of the existence of the limits (1.1.13-17) is carried out in the standard way [30-32] using integral equations and integral representations for $T^{x_2}_{x_1}(\lambda)$, $T_\pm(x,\lambda)$, and $T(\lambda)$, and we now start in on its consideration.

The Cauchy problems (1.1.3-4) and (1.1.7)-(1.1.4) are equivalent respectively with the Volterra integral equations

$$T^{x_2}_{x_1}(\lambda) = I + \int_{x_1}^{x_2} dx\, L(x,\lambda) T^{x}_{x_1}(\lambda) \tag{1.1.19}$$

and

$$T^{x_2}_{x_1}(\lambda) = I + \int_{x_1}^{x_2} dx\, T^{x_2}_{x}(\lambda) L(x,\lambda) . \tag{1.1.20}$$

Extracting from $L(x,\lambda)$ the potential $V(x) = i\varkappa\overline{\Psi}(x)\sigma_+ - i\Psi(x)\sigma_-$, we get the following integral equations for $T^{x_2}_{x_1}(\lambda)$:

$$T^{x_2}_{x_1}(\lambda) = e(x_2 - x_1, \lambda) + \int_{x_1}^{x_2} dx\, e(x_2 - x, \lambda) V(x) T^{x}_{x_1}(\lambda) \tag{1.1.21}$$

or

$$T^{x_2}_{x_1}(\lambda) = e(x_2 - x_1, \lambda) + \int_{x_1}^{x_2} dx\, T^{x_2}_{x}(\lambda) V(x) e(x - x_1, \lambda) . \tag{1.1.22}$$

From (1.1.13), (1.1.21), and (1.1.22) follow the integral equation

$$T_-(x,\lambda) = e(x,\lambda) + \int_{-\infty}^{x} d\xi\, e(x-\xi,\lambda) V(\xi) T_-(\xi,\lambda) \tag{1.1.23}$$

and the integral representation

$$T_-(x,\lambda) = e(x,\lambda) + \int_{-\infty}^{x} d\xi\, T^{x}_{\xi}(\lambda) V(\xi) e(\xi,\lambda) \tag{1.1.24}$$

for $T_-(x,\lambda)$. Analogously, from (1.1.14), (1.1.21), and (1.1.22) one derives the integral equation

$$T_+(x,\lambda) = e(-x,\lambda) + \int_{x}^{\infty} d\xi\, T_+(\xi,\lambda) V(\xi) e(\xi - x, \lambda) \tag{1.1.25}$$

and the integral representation

$$T_+(x;\lambda) = e(-x,\lambda) + \int_{x}^{\infty} d\xi\, e(-\xi,\lambda) V(\xi) T^{\xi}_{x}(\lambda) \tag{1.1.26}$$

for $T_+(x,\lambda)$. For $T(\lambda)$, from (1.1.15), (1.1.21), (1.1.22), we get the two integral representations:

$$T(\lambda) = I + \int_{-\infty}^{\infty} dx\, e(-x,\lambda) V(x) T_-(x,\lambda) \tag{1.1.27}$$

and

$$T(\lambda) = I + \int_{-\infty}^{\infty} dx\, T_+(x,\lambda) V(x) e(x,\lambda) . \tag{1.1.28}$$

Iterating the integral equation (1.1.21) or (1.1.22), we get the following expansions of the matricial elements of $T^{x_2}_{x_1}(\lambda)$ in power series in $\varkappa$:

$$a^{x_2}_{x_1}(\lambda)=e^{-i\frac{\lambda}{2}(x_2-x_1)}\left[1+\sum_{n=1}^{\infty}\varkappa^n\int\limits_{x_2>\xi_n>\eta_n>\xi_{n-1}>\ldots>\eta_1>x_1}\!\!\!\ldots\int d\xi_1\ldots d\xi_n d\eta_1\ldots d\eta_n\, e^{i\lambda(\xi_1+\ldots+\xi_n-\eta_1-\ldots-\eta_n)}\,\overline{\Psi}(\xi_1)\ldots\overline{\Psi}(\xi_n)\Psi(\eta_1)\ldots\Psi(\eta_n)\right], \tag{1.1.29}$$

$$b^{x_2}_{x_1}(\lambda)=-ie^{i\frac{\lambda}{2}(x_1+x_2)}\left[\sum_{n=0}^{\infty}\varkappa^n\int\limits_{x_2>\eta_{n+1}>\xi_n>\ldots>\eta_1>x_1}\!\!\!\ldots\int d\xi_1\ldots d\xi_n d\eta_1\ldots d\eta_{n+1}\times\right.$$

$$\left.\times e^{i\lambda(\xi_1+\ldots+\xi_n-\eta_1-\ldots-\eta_{n+1})}\,\overline{\Psi}(\xi_1)\ldots\overline{\Psi}(\xi_n)\Psi(\eta_1)\ldots\Psi(\eta_{n+1})\right]. \tag{1.1.30}$$

Expansions for $\overline{a^{x_2}_{x_1}(\bar\lambda)}$ and $\overline{b^{x_2}_{x_1}(\bar\lambda)}$ are obtained by complex conjugation. Analogous expansions for the matricial elements of $T_-(x,\lambda)$, $T_+(x,\lambda)$, and $T(\lambda)$ are obtained from (1.1.29) and (1.1.30) by cancellation, respectively, of x_1, x_2, or x_1 and x_2.

To conclude this section, some words on the discrete spectrum. As shown in [31], for $\varkappa>0$ the function $a(\lambda)$ has no zeros in $\operatorname{Im}\lambda>0$, but for $\varkappa<0$ can have in the upper half-plane a finite number of zeros:

$$a(\lambda_j)=0\,,\ \operatorname{Im}\lambda_j>0\,,\ j=1,\ldots,N\,. \tag{1.1.31}$$

This property of the coefficient $a(\lambda)$ is closely connected with the existence for $\varkappa<0$ of soliton solutions of (1.1.1), and in the quantum case, as we shall see later, the connected states of the basic particles.

1.2. Poisson Brackets. r-Matrix

As is known [30], (1.1.1) describes the dynamics of a Hamiltonian system with Hamiltonian

$$H=\int_{-\infty}^{\infty}dx\left(|\Psi_x|^2+\varkappa|\Psi|^4\right) \tag{1.2.1}$$

and Poisson brackets

$$\begin{gathered}\{\Psi(x),\Psi(y)\}=\{\overline{\Psi}(x),\overline{\Psi}(y)\}=0\,,\\ \{\Psi(x),\overline{\Psi}(y)\}=i\delta(x-y)\,.\end{gathered} \tag{1.2.2}$$

In other words, (1.1.1) can be represented in the following form:

$$\Psi_t=\{H,\Psi\}\,. \tag{1.2.3}$$

In [30] from the matrix elements of the matrix $T(\lambda)$ (1.1.15) there were constructed action-angle variables for (1.1.1). The basis for this construction includes the calculation of the Poisson brackets between the matrix elements of $T(\lambda)$ as functionals of the fields $\Psi(x)$ and $\overline{\Psi}(x)$. Below we shall calculate these Poisson brackets by a new method, proposed by the author in [29]. This method is based on using the so-called r-matrix and has the advantage over the traditional methods [4, 30], that it admits direct generalization to the quantum case, and also allows one to appreciably simplify calculations.

In what follows it will be convenient for us to use the following notation. With each (2×2) matrix

$$T = \begin{pmatrix} t_{11} & t_{12} \\ t_{21} & t_{22} \end{pmatrix}$$

we associate two (4×4) matrices $\tilde{T}$ and $\tilde{\tilde{T}}$:

$$\tilde{T} = T \otimes I_2 = \begin{pmatrix} t_{11} & 0 & t_{12} & 0 \\ 0 & t_{11} & 0 & t_{12} \\ t_{21} & 0 & t_{22} & 0 \\ 0 & t_{21} & 0 & t_{22} \end{pmatrix}, \tag{1.2.4}$$

and

$$\tilde{\tilde{T}} = I_2 \otimes T = \begin{pmatrix} t_{11} & t_{12} & 0 & 0 \\ t_{21} & t_{22} & 0 & 0 \\ 0 & 0 & t_{11} & t_{12} \\ 0 & 0 & t_{21} & t_{22} \end{pmatrix}. \tag{1.2.5}$$

Thus, the matrix

$$\{\tilde{T}(\lambda), \tilde{\tilde{T}}(\mu)\} = \begin{pmatrix} \{a(\lambda), a(\mu)\}, \{a(\lambda), \varkappa\overline{b(\bar{\mu})}\}, \{\varkappa\overline{b(\bar{\lambda})}, a(\mu)\}, \{\varkappa\overline{b(\bar{\lambda})}, \varkappa\overline{b(\bar{\mu})}\} \\ \{a(\lambda), b(\mu)\}, \{a(\lambda), \overline{a(\bar{\mu})}\}, \{\varkappa\overline{b(\bar{\lambda})}, b(\mu)\}, \{\varkappa\overline{b(\bar{\lambda})}, \overline{a(\bar{\mu})}\} \\ \{b(\lambda), a(\mu)\}, \{b(\lambda), \varkappa\overline{b(\bar{\mu})}\}, \{\overline{a(\bar{\lambda})}, a(\mu)\}, \{\overline{a(\bar{\lambda})}, \varkappa\overline{b(\bar{\mu})}\} \\ \{b(\lambda), b(\mu)\}, \{b(\lambda), \overline{a(\bar{\mu})}\}, \{\overline{a(\bar{\lambda})}, b(\mu)\}, \{\overline{a(\bar{\lambda})}, \overline{a(\bar{\mu})}\} \end{pmatrix} \tag{1.2.6}$$

will contain all 16 possible Poisson brackets between the matrix elements of the matrices $T(\lambda)$ and $T(\mu)$. We note that the matrices $T(\lambda)$ and $T(\mu)$ commute with one another

$$\tilde{T}\tilde{\tilde{T}} = \tilde{\tilde{T}}\tilde{T}. \tag{1.2.7}$$

Later we shall also need the permutation matrix $\mathcal{P}$

$$\mathcal{P} = \begin{pmatrix} 1 & 0 & 0 & 0 \\ 0 & 0 & 1 & 0 \\ 0 & 1 & 0 & 0 \\ 0 & 0 & 0 & 1 \end{pmatrix}, \quad \mathcal{P}^2 = I_4 \tag{1.2.8}$$

and the following easily verifiable relations

$$\mathcal{P}\tilde{T}\mathcal{P} = \tilde{\tilde{T}}, \quad \mathcal{P}\tilde{\tilde{T}}\mathcal{P} = \tilde{T}, \tag{1.2.9}$$

$$\mathcal{P}\tilde{T}\tilde{\tilde{T}}\mathcal{P} = \tilde{T}\tilde{\tilde{T}}. \tag{1.2.10}$$

The fundamental result of the present section is the following theorem.

THEOREM 1. The matrix of Poisson brackets $\{\tilde{T}_{x_1}^{x_2}(\lambda), \tilde{\tilde{T}}_{x_1}^{x_2}(\mu)\} = \mathcal{P}_{x_1}^{x_2}(\lambda,\mu)$ admits the following representation for $x_2 > x_1$

$$\mathcal{P}_{x_1}^{x_2}(\lambda,\mu) = \left[r(\lambda-\mu), \tilde{T}_{x_1}^{x_2}(\lambda)\, \tilde{\tilde{T}}_{x_1}^{x_2}(\mu) \right], \tag{1.2.11}$$

where $[\ ,\ ]$ denotes the commutator of matrices, and the 4×4 matrix $r(\lambda-\mu)$ has the form

$$\tau(\lambda-\mu) = -\frac{\varkappa}{\lambda-\mu}\,\mathcal{P}\,. \tag{1.2.12}$$

For $x_2 < x_1$, as follows from (1.1.5), (1.2.11) goes into

$$\mathcal{P}^{x_2}_{x_1}(\lambda,\mu) = -\left[\tau(\lambda-\mu), \tilde{T}^{x_2}_{x_1}(\lambda)\tilde{\tilde{T}}^{x_2}_{x_1}(\mu)\right].$$

<u>Proof.</u> For the proof it suffices to see that the right and left sides of (1.2.11) satisfy the same differential equation with the same initial conditions.

Differentiating the right side of (1.2.11) with respect to x_2 we get for $\mathcal{P}^{x_2}_{x_1}(\lambda,\mu)$ the following differential equation:

$$\frac{\partial}{\partial x_2}\mathcal{P}^{x_2}_{x_1}(\lambda,\mu) = \left[\tau(\lambda-\mu), \tilde{L}(x_2,\lambda)+\tilde{\tilde{L}}(x_2,\mu)\right]\tilde{T}^{x_2}_{x_1}(\lambda)\tilde{\tilde{T}}^{x_2}_{x_1}(\mu) + \left(\tilde{L}(x_2,\lambda)+\tilde{\tilde{L}}(x_2,\mu)\right)\mathcal{P}^{x_2}_{x_1}(\lambda,\mu) \tag{1.2.13}$$

with the initial condition

$$\mathcal{P}^{x}_{x}(\lambda,\mu) = 0\,. \tag{1.2.14}$$

In order to calculate the derivative with respect to x_2 of the left side of (1.2.11), we represent it in the form:

$$\frac{\partial}{\partial x_2}\left\{\tilde{T}^{x_2}_{x_1}(\lambda), \tilde{\tilde{T}}^{x_2}_{x_1}(\mu)\right\} = \lim_{\delta\to 0}\left(\mathcal{P}^{x_2+\delta}_{x_1}(\lambda,\mu) - \mathcal{P}^{x_2}_{x_1}(\lambda,\mu)\right) =$$

$$= \lim_{\delta\to 0}\frac{1}{\delta}\left\{\tilde{T}^{x_2+\delta}_{x_2}(\lambda), \tilde{\tilde{T}}^{x_2+\delta}_{x_2}(\mu)\right\}\tilde{T}^{x_2}_{x_1}(\lambda)\tilde{\tilde{T}}^{x_2}_{x_1}(\mu) + \lim_{\delta\to 0}\frac{1}{\delta}\left(\tilde{T}^{x_2+\delta}_{x_2}(\lambda)\tilde{\tilde{T}}^{x_2+\delta}_{x_2}(\mu) - I_4\right)\left\{\tilde{T}^{x_2}_{x_1}(\lambda), \tilde{\tilde{T}}^{x_2}_{x_1}(\mu)\right\}\,. \tag{1.2.15}$$

We study the first summand of the expression obtained. Substituting in $\mathcal{P}^{x_2+\delta}_{x_2}(\lambda,\mu)$ the expression

$$T^{x_2+\delta}_{x_2} = I_2 + \int_{x_2}^{x_2+\delta} dx\, L(x) + O(\delta^2)\,,$$

following from (1.1.19), we get

$$\left\{\tilde{T}^{x_2+\delta}_{x_2}(\lambda), \tilde{\tilde{T}}^{x_2+\delta}_{x_2}(\mu)\right\} = \int_{x_2}^{x_2+\delta} dx \int_{x_2}^{x_2+\delta} dy \left\{\tilde{L}(x,\lambda), \tilde{\tilde{L}}(y,\mu)\right\} + O(\delta^2) =$$

$$= \int_{x_2}^{x_2+\delta} dx \int_{x_2}^{x_2+\delta} dy \left\{\tilde{L}(x,\lambda), \tilde{\tilde{L}}(x,\mu)\right\}'\delta(x-y) + O(\delta^2) = \int_{x_2}^{x_2+\delta} dx \left\{\tilde{L}(x,\lambda), \tilde{\tilde{L}}(x,\mu)\right\}' + O(\delta^2)\,.$$

Here $\{,\}'$ denotes the "local" Poisson bracket

$$\left\{\Psi(x), \Psi(x)\right\}' = \left\{\overline{\Psi}(x), \overline{\Psi}(x)\right\}' = 0\,,$$
$$\left\{\Psi(x), \overline{\Psi}(x)\right\}' = i\,. \tag{1.2.16}$$

Finally, passing on (1.2.15) to the limit as $\delta \to 0$, we get

$$\frac{\partial}{\partial x_2}\mathcal{P}^{x_2}_{x_1}(\lambda,\mu) = \left\{\tilde{L}(x_2,\lambda), \tilde{\tilde{L}}(x_2,\mu)\right\}'\tilde{T}^{x_2}_{x_1}(\lambda)\tilde{\tilde{T}}^{x_2}_{x_1}(\mu) + \left(\tilde{L}(x_2,\lambda)+\tilde{\tilde{L}}(x_2,\mu)\right)\mathcal{P}^{x_2}_{x_1}(\lambda,\mu)\,. \tag{1.2.17}$$

In order to identify (1.2.13) and (1.2.17), it suffices to note that one has

$$\left\{\tilde{L}(x,\lambda), \tilde{\tilde{L}}(x,\mu)\right\}' = \left[\tau(\lambda-\mu), \tilde{L}(x,\lambda)+\tilde{\tilde{L}}(x,\mu)\right] \tag{12.18}$$

or

$$\{\tilde{L}(x,\lambda),\tilde{\tilde{L}}(y,\mu)\} = [r(\lambda\ \mu),\tilde{L}(x,\lambda)+\tilde{\tilde{L}}(y,\mu)]\,\delta(x-y), \tag{1.2.19}$$

which is easily verified directly, taking into account (1.1.2), (1.2.2), (1.2.12).

In order to complete the proof of Theorem 1, it remains to note that $\{\tilde{T}^{x_2}_{x_1}(\lambda),\tilde{\tilde{T}}^{x_2}_{x_1}(\mu)\}$ satisfies the initial condition (1.2.14).

A complete summary of all Poisson brackets between matrix elements of $T^{x_2}_{x_1}(\lambda)$ following from (1.2.11) is given in the Supplement (Eqs. (S1-6)).

We note that the domain of application of the method of the r-matrix is not restricted to (1.1.1). The existence of a representation of the Poisson bracket in the form (1.2.11) is based, as follows from the proof given above, only on (1.2.19), which also holds for the sin-Gordon equation and the Landau–Lifshits equations [29]. The matrix r for these equations has a more complicated form than (2.2.12).

1.3. Passage to an Infinite Interval

To achieve our ultimate goal, the construction of action-angle variables, we need to calculate the Poisson brackets for the transition matrix on the interval $(-\infty,\infty)$. But first we concern ourselves with the calculation of the Poisson bracket $\{\tilde{T}_-(x,\lambda),\tilde{\tilde{T}}_-(x,\mu)\}$, which it is convenient to denote by $\mathcal{P}_-(x;\lambda,\mu)$.

Repeating word for word the argument of the preceding section, one can see that $\mathcal{P}_-(x;\lambda,\mu)$ satisfies with respect to the variable x, the differential equation (1.2.13). On the other hand, the same differential is satisfied by the expression $[r(\lambda-\mu),\ \tilde{T}_-(x,\lambda)\tilde{\tilde{T}}_-(x,\mu)]$. Consequently, their difference satisfies the corresponding homogeneous differential equation, whose general solution we can write as $\tilde{T}_-(x,\lambda)\tilde{\tilde{T}}_-(x,\mu)\,C_-(\lambda,\mu)$. Thus, we get the following representation for $\mathcal{P}_-(x;\lambda,\mu)$:

$$\mathcal{P}_-(x;\lambda,\mu)=[r(\lambda-\mu),\tilde{T}_-(x,\lambda)\tilde{\tilde{T}}_-(x,\mu)]-\tilde{T}_-(x,\lambda)\tilde{\tilde{T}}_-(x,\mu)\,C_-(\lambda,\mu). \tag{1.3.1}$$

The (4×4) matrix $C_-(\lambda,\mu)$ is determined from comparison with the asymptotics of (1.3.1) as $x\to-\infty$.

It is easy to get the asymptotic behavior of the right side of (1.3.1), using (1.1.16):

$$\mathcal{P}_-(x;\lambda,\mu)\underset{x\to-\infty}{\simeq}[r(\lambda\ \mu),E(x;\lambda,\mu)]-E(x;\lambda,\mu)\,C_-(\lambda,\mu), \tag{1.3.2}$$

where

$$E(x;\lambda,\mu)=\tilde{e}(x,\lambda)\tilde{\tilde{e}}(x,\mu)=e^{-\frac{i}{2}(\lambda\tilde{\sigma}_3+\mu\tilde{\tilde{\sigma}}_3)x}. \tag{1.3.3}$$

To calculate the asymptotics of $\{\tilde{T}_-(x,\lambda),\tilde{\tilde{T}}_-(x,\mu)\}$ as $x\to-\infty$ we use the integral representation (1.1.24). We have:

$$\{\tilde{T}_-(x,\lambda),\tilde{\tilde{T}}_-(x,\mu)\}=\int_{-\infty}^{x}d\xi\int_{-\infty}^{x}d\eta\,\{\tilde{T}^{x}_{\xi}(\lambda)\tilde{V}(\xi),\tilde{\tilde{T}}^{x}_{\eta}(\mu)\tilde{\tilde{V}}(\eta)\}\,\tilde{e}(\xi,\lambda)\tilde{\tilde{e}}(\eta,\mu). \tag{1.3.4}$$

Calculating the Poisson bracket standing in (1.3.4) under the integral

$$\left\{\widetilde{T}^{x}_{\xi}(\lambda)\widetilde{V}(\xi),\overset{\approx}{T}^{x}_{\eta}(\mu)\overset{\approx}{V}(\eta)\right\}=\widetilde{T}^{x}_{\xi}(\lambda)\overset{\approx}{T}^{x}_{\eta}(\mu)\left\{\widetilde{V}(\xi),\overset{\approx}{V}(\eta)\right\}+\left\{\widetilde{T}^{x}_{\xi}(\lambda),\overset{\approx}{T}^{x}_{\eta}(\mu)\right\}\widetilde{V}(\xi)\overset{\approx}{V}(\eta)+$$
$$+\overset{\approx}{T}^{x}_{\eta}(\mu)\left\{\widetilde{T}^{x}_{\xi}(\lambda),\overset{\approx}{V}(\eta)\right\}\widetilde{V}(\xi)+\widetilde{T}^{x}_{\xi}(\lambda)\left\{\widetilde{V}(\xi),\overset{\approx}{T}^{x}_{\eta}(\mu)\right\}\overset{\approx}{V}(\eta), \qquad (1.3.5)$$

we see that only the summand

$$\widetilde{T}^{x}_{\xi}(\lambda)\overset{\approx}{T}^{x}_{\eta}(\mu)\left\{\widetilde{V}(\xi),\overset{\approx}{V}(\eta)\right\}=\widetilde{T}^{x}_{\xi}(\lambda)\overset{\approx}{T}^{x}_{\eta}(\mu)\, i\varkappa\left(\widetilde{\sigma}_{-}\overset{\approx}{\sigma}_{+}-\widetilde{\sigma}_{+}\overset{\approx}{\sigma}_{-}\right)\delta(\xi-\eta) \qquad (1.3.6)$$

gives a nondecreasing contribution as $x\to-\infty$ in (1.3.4). Substituting (1.3.5) and (1.3.6) in (1.3.4), and considering that as follows from (1.1.21),

$$T^{x}_{\xi}(\lambda)-e(x-\xi,\lambda)\to 0,\quad x,\xi\to-\infty,$$

we get, that as $x\to-\infty$,

$$\mathcal{P}_{-}(x;\lambda,\mu)\sim\int_{-\infty}^{x}E(x-\xi;\lambda,\mu)\, i\varkappa\left(\widetilde{\sigma}_{-}\overset{\approx}{\sigma}_{+}-\widetilde{\sigma}_{+}\overset{\approx}{\sigma}_{-}\right)E(\xi;\lambda,\mu)\,d\xi. \qquad (1.3.7)$$

It is easy to calculate the integral in (1.3.7), and comparing the asymptotics of (1.3.7) with (1.3.2), we get

$$C_{-}(\lambda,\mu)=\varkappa\begin{pmatrix}0&0&0&0\\0&0&\frac{1}{\lambda-\mu-i0}&0\\0&\frac{1}{\lambda-\mu+i0}&0&0\\0&0&0&0\end{pmatrix}. \qquad (1.3.8)$$

Completely analogously one calculates the Poisson bracket $\left\{\widetilde{T}_{+}(x,\lambda),\overset{\approx}{T}(x,\mu)\right\}=\mathcal{P}_{+}(x;\lambda,\mu)$:

$$\mathcal{P}_{+}(x;\lambda,\mu)=\left[r(\lambda-\mu),\widetilde{T}_{+}(x,\lambda)\overset{\approx}{T}_{+}(x,\mu)\right]+C_{+}(\lambda,\mu)\widetilde{T}_{+}(x,\lambda)\overset{\approx}{T}_{+}(x,\mu), \qquad (1.3.9)$$

where

$$C_{+}(\lambda,\mu)=\varkappa\begin{pmatrix}0&0&0&0\\0&0&\frac{1}{\lambda-\mu+i0}&0\\0&\frac{1}{\lambda-\mu-i0}&0&0\\0&0&0&0\end{pmatrix}. \qquad (1.3.10)$$

It is interesting to note that although (1.3.1) and (1.3.9) should be understood in the sense of generalized functions, the summands in (1.3.1) and (1.3.9) containing $r(\lambda-\mu)$, are not needed in the regularization for $\lambda=\mu$, since the corresponding numerator $\left[\mathcal{P},\widetilde{T}_{\pm}(x,\lambda)\overset{\approx}{T}_{\pm}(x,\mu)\right]$ vanishes for $\lambda=\mu$ by virtue of (1.2.10).

However, the choice of a definite regularization of $r(\lambda-\mu)$, e.g., $r(\lambda-\mu)=-\varkappa\mathcal{P}\,\mathrm{v.p.}\frac{1}{\lambda-\mu}$, allows one to write (1.3.1) and (1.3.9) in compact form:

$$\mathcal{P}_{-}(x;\lambda,\mu)=r(\lambda-\mu)\widetilde{T}_{-}(x,\lambda)\overset{\approx}{T}_{-}(x,\mu)-\widetilde{T}_{-}(x,\lambda)\overset{\approx}{T}_{-}(x,\mu)\,r_{-}(\lambda-\mu) \qquad (1.3.11)$$

and

$$\mathcal{P}_+(x;\lambda,\mu) = r_+(\lambda-\mu)\mathrm{T}_+(x,\lambda)\tilde{\mathrm{T}}_+(x,\mu) - \mathrm{T}_+(x,\lambda)\tilde{\mathrm{T}}_+(x,\mu)\, r(\lambda-\mu)\ , \tag{1.3.12}$$

where

$$r_\pm(\lambda-\mu) = r(\lambda-\mu) + C_\pm(\lambda,\mu) = -\varkappa \begin{pmatrix} \mathrm{v.p.}\frac{1}{\lambda-\mu}, & 0\ , & 0\ , \\ 0\ , & 0\ , & \pm\pi i\delta(\lambda-\mu), \\ 0\ , & \mp\pi i\delta(\lambda-\mu), & 0\ , \\ 0\ , & 0\ , & 0\ , & \mathrm{v.p.}\frac{1}{\lambda-\mu} \end{pmatrix} . \tag{1.3.13}$$

Now everything is ready to calculate the Poisson bracket $\{\mathrm{T}(\lambda), \tilde{\mathrm{T}}(\mu)\} = \mathcal{P}(\lambda,\mu)$. Keeping in mind (1.1.18), (1.3.11), and (1.3.12), we get

$$\mathcal{P}(\lambda,\mu) = r_+(\lambda-\mu)\mathrm{T}(\lambda)\tilde{\mathrm{T}}(\mu) - \mathrm{T}(\lambda)\tilde{\mathrm{T}}(\mu)\, r_-(\lambda-\mu)\ . \tag{1.3.14}$$

The results proved above allow us to formulate the following theorem.

THEOREM 2. The Poisson brackets between matrix elements of transition matrices for semi-infinite and infinite intervals are given by (1.3.11-14).

A complete summary of all Poisson brackets is given in the Supplement (Eqs. (S7-24)).

1.4. Integrals of Motion. Action-Angle Variables

In the present section, which has an auxiliary character, we list results basically known from [30-32] for the nonlinear Schrödinger equation, which will be useful later for comparison with the quantum case.

As shown in [31, 32], $\ln a(\lambda)$ is the generating function for the local integrals of motion $\mathcal{I}_m$ for (1.1.1), i.e., the coefficients $\mathcal{I}_m$ of the expansion of $\ln a(\lambda)$ in an asymptotic series in powers of λ^{-1}

$$\ln a(\lambda) = i\varkappa \sum_{m=1}^{\infty} \mathcal{I}_m \lambda^{-m} \tag{1.4.1}$$

are the integrals of the local densities with respect to $\Psi(x)$ and $\overline{\Psi}(x)$

$$\mathcal{I}_m = \int_{-\infty}^{\infty} dx\, \overline{\Psi}(x) \chi_m(x)\ , \tag{1.4.2}$$

where $\chi_m(x)$ are determined from the recursion relation

$$\chi_{m+1}(x) = -i\frac{d}{dx}\chi_m(x) + \varkappa\overline{\Psi}(x)\sum_{k=1}^{m-1}\chi_k(x)\chi_{m-k}(x) \tag{1.4.3}$$

with the initial condition

$$\chi_1(x) = \Psi(x)\ . \tag{1.4.4}$$

By virtue of (S19) the quantities $\mathcal{I}_m$ are in involution with respect to the Poisson bracket (1.2.2). We shall write down the first few of the integrals $\mathcal{I}_m$:

$$\mathcal{I}_1 = N = \int_{-\infty}^{\infty} dx\, |\Psi(x)|^2, \tag{1.4.5}$$

$$\mathcal{I}_2 = P = \frac{i}{2}\int_{-\infty}^{\infty} dx\left(\overline{\psi}_x\psi - \overline{\psi}\psi_x\right), \tag{1.4.6}$$

$$\mathcal{I}_3 = H = \int_{-\infty}^{\infty} dx\left(|\psi_x|^2 + \varkappa|\psi|^4\right). \tag{1.4.7}$$

The quantities N, P, and H are called, respectively, the number of particles, momentum, and energy. Since the Hamiltonian $H = \mathcal{I}_3$ is included among the $\mathcal{I}_m$, the quantities $\mathcal{I}_m$ are integrals of motion for (1.1.1).

We proceed now to the construction of action-angle variables for (1.1.1). The concept of "action-angle variables" we shall treat here broadly, calling such any canonical variables in which the Hamiltonian H can be written as a quadratic form (and the equations of motion, correspondingly, become linear).

We introduce quantities $\varphi(\lambda)$ and $\overline{\varphi}(\lambda)$ by the formulas

$$\varphi(\lambda) = \frac{b(\lambda)}{|b(\lambda)|}\sqrt{\frac{\ln|a(\lambda)|}{\pi\varkappa}},$$
$$\overline{\varphi}(\lambda) = \frac{\overline{b}(\lambda)}{|b(\lambda)|}\sqrt{\frac{\ln|a(\lambda)|}{\pi\varkappa}} \tag{1.4.8}$$

In (1.4.8) it is necessary to take the positive value of the root. The expression under the radical sign here remains positive for any value of $\varkappa$, since by (1.1.12), $|a(\lambda)|>1$ for $\varkappa>0$ and $|a(\lambda)|<1$ for $\varkappa<0$, when λ runs through the real axis.

The quantities $\varphi(\lambda)$ and $\overline{\varphi(\lambda)}$ satisfy all the requirements listed above for action-angle variables. In fact, using the Poisson brackets (Eqs. (S9-24)), it is easy to verify that $\varphi(\lambda)$ and $\overline{\varphi}(\lambda)$ are canonical conjugate variables:

$$\{\varphi(\lambda), \varphi(\mu)\} = \{\overline{\varphi}(\lambda), \overline{\varphi}(\mu)\} = 0,$$
$$\{\varphi(\lambda), \overline{\varphi}(\mu)\} = i\delta(\lambda-\mu). \tag{1.4.9}$$

Further, for $\varkappa>0$, the generating function of the integrals of motion $\ln a(\lambda)$ has, as proved in [31, 32], the following integral representation:

$$\ln a(\lambda) = \frac{i}{\pi}\int_{-\infty}^{\infty} d\mu\, \frac{\ln|a(\mu)|}{\lambda-\mu},$$

which can be rewritten, using (1.4.8), as

$$\ln a(\lambda) = i\varkappa\int_{-\infty}^{\infty} d\mu\, \frac{|\varphi(\mu)|^2}{\lambda-\mu}. \tag{1.4.10}$$

Decomposing (1.4.10) into powers of λ^{-1}, we get

$$\mathcal{I}_m = \int_{-\infty}^{\infty} d\mu\, \mu^{m-1}\, |\varphi(\mu)|^2. \tag{1.4.11}$$

In particular,

$$N = \int_{-\infty}^{\infty} d\mu\, |\varphi(\mu)|^2, \tag{1.4.12}$$

$$P=\int_{-\infty}^{\infty} d\mu\,\mu\,|\varphi(\mu)|^2, \tag{1.4.13}$$

$$H=\int_{-\infty}^{\infty} d\mu\,\mu^2\,|\varphi(\mu)|^2 . \tag{1.4.14}$$

Thus, all the integrals of motion $\mathcal{I}_m$ in the variables $\bar{\varphi}(\mu)$ are quadratic, and correspondingly, the equations of motion

$$\frac{d}{dt}\varphi(t;\mu)=\{\mathcal{I}_m,\varphi(t;\mu)\}=-i\mu^{m-1}\varphi(t;\mu) \tag{1.4.15}$$

are linear in $\varphi,\bar{\varphi}$, which allows us to call the variables $\varphi(\mu),\bar{\varphi}(\mu)$ variables of action-angle type for (1.1.1) in the sense of the definition given above. The action-angle variables introduced by us differ from the traditional ones [30], but have the advantage that they admit a quantum-mechanical generalization.

The case $\varkappa<0$ requires the calculation of the discrete spectrum. We shall not write down here the corresponding variables of action-angle type, since they, apparently, have no reasonable analogs in the quantum case, and we restrict ourselves to indicating how in this case one generalizes (1.4.10-14):

$$\ln a(\lambda)=\sum_{j=1}^{N}\ln\frac{\lambda-\lambda_j}{\lambda-\bar{\lambda}_j}+i\varkappa\int_{-\infty}^{\infty}\frac{|\varphi(\mu)|^2}{\lambda-\mu}\,d\mu , \tag{1.4.10'}$$

$$\mathcal{I}_m=\frac{2}{m|\varkappa|}\sum_{j=1}^{N}\mathrm{Im}(\lambda_j^m)+\int_{-\infty}^{\infty}d\mu\,\mu^{m-1}|\varphi(\mu)|^2 , \tag{1.4.11'}$$

$$N=\frac{2}{|\varkappa|}\sum_{j=1}^{N}\mathrm{Im}\,\lambda_j+\int_{-\infty}^{\infty}d\mu\,|\varphi(\mu)|^2 , \tag{1.4.12'}$$

$$P=\frac{2}{|\varkappa|}\sum_{j=1}^{N}\mathrm{Re}\,\lambda_j\cdot\mathrm{Im}\,\lambda_j+\int_{-\infty}^{\infty}d\mu\,\mu\,|\varphi(\mu)|^2 , \tag{1.4.13'}$$

$$H=\frac{2}{3|\varkappa|}\sum\left(3\mathrm{Re}^2\lambda_j\cdot\mathrm{Im}\,\lambda_j-\mathrm{Im}^3\lambda_j\right)+\int_{-\infty}^{\infty}d\mu\,\mu^2|\varphi(\mu)|^2 . \tag{1.4.14'}$$

1.5. M -Operator

All the arguments of the preceding sections were based on the study of the operator $L(x,\lambda)$ at a fixed moment of time t. This did not prevent us from proving the complete integrability of (1.1.1) and finding the spectrum of its elementary excitations. For the traditional approach [31], however, the consideration of the time evolution from the very start is characteristic. The initial point here is the representation of the equation of motion (1.1.1) as the commutativity condition of two differential operators:

$$\left[\frac{\partial}{\partial x}-L(x,t;\lambda),\frac{\partial}{\partial t}-M(x,t;\lambda)\right]=0 , \tag{1.5.1}$$

or

$$L_t=M_x+[M,L] , \tag{1.5.2}$$

where

$$M(x,t;\lambda) = \begin{pmatrix} i\frac{\lambda^2}{2} + i\varkappa|\Psi(x,t)|^2, & \varkappa\overline{\Psi}_x(x,t) - i\varkappa\lambda\overline{\Psi}(x,t) \\ \Psi_x(x,t) + i\lambda\Psi(x,t), & -i\frac{\lambda^2}{2} - i\varkappa|\Psi(x,t)|^2 \end{pmatrix}. \tag{1.5.3}$$

Although, we stress again, the approach developed in the present paper is in principle not necessary in introducing the operator M both in the classical and the quantum case, the study of the question of the M-operator is of definite systematic interest and allows us to demonstrate again the possibility of the method of the τ-matrix.

We raise the question in the following way. Let the time evolution of the fields $\Psi(x,t)$ and $\overline{\Psi}(x,t)$ be given by the m-th local integral of motion:

$$\frac{\partial}{\partial t}\Psi(x,t) = \{J_m, \Psi(x,t)\}. \tag{1.5.4}$$

How can we find under these conditions a matrix $M_m(x,t)$, satisfying (1.5.2)? Although the answer to this question is known [35], the method of the τ-matrix allows us to express it in a simple and compact form.

We consider first the case of periodic boundary conditions on the interval $[x_1, x_2]$

$$\Psi(x + x_2 - x_1) = \Psi(x) \tag{1.5.5}$$

and we formulate the following proposition:

Proposition 1.5.1. One has the equation

$$\{\mathrm{tr}\, T_{x_1}^{x_2}(\lambda), L(x,\mu)\} = \frac{\partial}{\partial x} M_{x_1}^{x_2}(x;\lambda,\mu) + [M_{x_1}^{x_2}(x;\lambda,\mu), L(x,\mu)], \tag{1.5.6}$$

where

$$M_{x_1}^{x_2}(x;\lambda,\mu) = \widetilde{\mathrm{tr}}\left(\widetilde{T}_x^{x_2}(\lambda)\,\tau(\lambda-\mu)\,\widetilde{T}_{x_1}^{x}(\lambda)\right). \tag{1.5.7}$$

The operation $\widetilde{\mathrm{tr}}$ introduced by us here is the convolution operator in $\mathbb{C}^2 \otimes \mathbb{C}^2$ in the indices relating to the first factor. The result, thus, is a (2 × 2) matrix. In particular,

$$\widetilde{\mathrm{tr}}(A \otimes B) = \widetilde{\mathrm{tr}}(\widetilde{A}\widetilde{\widetilde{B}}) = (\mathrm{tr} A)B; \quad A, B \in \mathrm{Mat}(2,2).$$

The proof of Proposition 1.5.1. is based on the following lemma:

LEMMA 1.5.1. For any functional $X(\psi,\overline{\psi})$ of the fields $\Psi(x)$ and $\overline{\Psi}(x)$ one has the following relation:

$$\{T_{x_1}^{x_2}(\lambda), X\} = \int_{x_1}^{x_2} dx\, T_x^{x_2}(\lambda)\{L(x,\lambda), X\} T_{x_1}^{x}(\lambda). \tag{1.5.8}$$

To prove Lemma 1.5.1 we introduce the notation $\mathcal{P}_{x_1}^{x_2}(\lambda, X) = \{T_{x_1}^{x_2}(\lambda), X\}$. Differentiating the left and right sides of (1.5.8) with respect to x_2, it is easy to see that they satisfy the same differential equation

$$\frac{\partial}{\partial x_2}\mathcal{P}_{x_1}^{x_2}(\lambda; X) = L(x_2,\lambda)\mathcal{P}_{x_1}^{x_2}(\lambda, X) + \{L(x_2,\lambda), X\} T_{x_1}^{x_2}(\lambda) \tag{1.5.9}$$

with the same initial condition

$$\mathcal{P}_{x_1}^{x_1}(\lambda;X)=0\ . \tag{1.5.10}$$

Reference to the corresponding uniqueness theorem completes the proof.

We begin now the proof of Proposition 1.5.1. First, using the notation $\widetilde{\mathrm{tr}}$, we have introduced, we transform the left side of (1.5.6):

$$\{\mathrm{tr}\,T_{x_1}^{x_2}(\lambda),L(x,\mu)\}=\widetilde{\mathrm{tr}}\{\widetilde{T}_{x_1}^{x_2}(\lambda),\widetilde{L}(x,\mu)\}\ . \tag{1.5.11}$$

To calculate the right side of (1.5.11), we use Lemma 1.5.1, which we have just proved. Substituting in (1.5.8) $X=\widetilde{L}(x,\mu)$, we get

$$\{\widetilde{T}_{x_1}^{x_2}(\lambda),\widetilde{L}(x,\mu)\}=\int_{x_1}^{x_2}d\xi\,\widetilde{T}_{\xi}^{x_2}(\lambda)\{\widetilde{L}(\xi,\lambda),\widetilde{L}(x,\mu)\}\widetilde{T}_{x_1}^{\xi}(\lambda), \tag{1.5.12}$$

or, by virtue of (1.2.19),

$$\{\widetilde{T}_{x_1}^{x_2}(\lambda),\widetilde{L}(x,\mu)\}=\widetilde{T}_{x}^{x_2}(\lambda)\left[r(\lambda-\mu),\widetilde{L}(x,\lambda)+\widetilde{\widetilde{L}}(x,\mu)\right]\widetilde{T}_{x_1}^{x}(\lambda)\ . \tag{1.5.13}$$

Substituting (1.5.13) in (1.5.11) and using (1.1.3) and (1.1.7), we get

$$\begin{aligned}\{\mathrm{tr}\,T_{x_1}^{x_2}(\lambda),L(x,\mu)\}&=\widetilde{\mathrm{tr}}\,\widetilde{T}_{x}^{x_2}(\lambda)\left[r(\lambda-\mu),\widetilde{L}(x,\lambda)\right]\widetilde{T}_{x_1}^{x}(\lambda)+\\ &+\widetilde{\mathrm{tr}}\,\widetilde{T}_{x}^{x_2}(\lambda)\left[r(\lambda-\mu),\widetilde{\widetilde{L}}(x,\lambda)\right]\widetilde{T}_{x_1}^{x}(\lambda)=\frac{\partial}{\partial x}\widetilde{\mathrm{tr}}\,\widetilde{T}_{x}^{x_2}(\lambda)\,r(\lambda-\mu)\widetilde{T}_{x_1}^{x}(\lambda)+\\ &+\left[\widetilde{\mathrm{tr}}\,\widetilde{T}_{x}^{x_2}\,r(\lambda-\mu)\widetilde{T}_{x_1}^{x}(\lambda),L(x,\mu)\right]=\frac{\partial}{\partial x}M_{x_1}^{x_2}(x;\lambda,\mu)+\left[M_{x_1}^{x_2}(x;\lambda,\mu),L(x,\mu)\right],\end{aligned} \tag{1.5.14}$$

where $M_{x_1}^{x_2}(x;\lambda,\mu)$ is given by (1.5.7), which is what had to be proved.

From (1.2.11) follows

$$\{\mathrm{tr}\,T_{x_1}^{x_2}(\lambda),\mathrm{tr}\,T_{x_1}^{x_2}(\mu)\}=0\ , \tag{1.5.15}$$

which allows us to consider $\ln\mathrm{tr}\,T_{x_1}^{x_2}(\lambda)$, as the generating function of the integrals of motion of (1.1.1) with periodic boundary conditions.

In order to find the analog of (1.5.6) for an infinite interval, it is necessary to divide both sides of (1.5.6) by $\mathrm{tr}\,T_{x_1}^{x_2}(\lambda)$ and pass to the limit as $x_1\to-\infty$, $x_2\to+\infty$. The answer depends, obviously, on the sign of $\mathrm{Im}\,\lambda$ and is given by the following formula:

$$\{\ln a^{(\pm)}(\lambda),L(x,\mu)\}=M_x^{(\pm)}(x;\lambda,\mu)+\left[M^{(\pm)}(x;\lambda,\mu),L(x,\mu)\right], \tag{1.5.16}$$

where

$$M^{(\pm)}(x;\lambda,\mu)=\widetilde{\mathrm{tr}}\left(\widetilde{P}^{(\pm)}\,\widetilde{T}_{+}(x,\lambda)\,r(\lambda-\mu)\,\widetilde{T}_{-}(x,\lambda)\widetilde{P}^{(\pm)}\right)\ . \tag{1.5.17}$$

We explain the notation. The upper sign corresponds to $\mathrm{Im}\,\lambda>0$, the lower to $\mathrm{Im}\,\lambda<0$. $a^{(+)}(\lambda)=a(\lambda)$, $a^{(-)}(\lambda)=\overline{a(\bar{\lambda})}$. The projectors $P^{(\pm)}$ onto the eigenspaces $e(x,\lambda)$ have the form

$$P^{(+)}=\begin{pmatrix}1&0\\0&0\end{pmatrix}\quad,\quad P^{(-)}=\begin{pmatrix}0&0\\0&1\end{pmatrix}.$$

The proof of (1.5.16) and (1.5.17) becomes obvious if one notes that as $x_1\to-\infty$, $x_2\to+\infty$

$$\operatorname{tr}T_{x_1}^{x_2}(\lambda)\sim\operatorname{tr}\left(e(x_2,\lambda)T(\lambda)e(-x_1,\lambda)\right)\sim e^{\mp i\frac{\lambda}{2}(x_2-x_1)}\operatorname{tr}P^{(\pm)}T(\lambda)P^{(\pm)}=e^{\mp i\frac{\lambda}{2}(x_2-x_1)}a^{(\pm)}(\lambda).$$

We consider the results obtained. Since $\ln a(\lambda)$ (or $\ln\overline{a(\bar\lambda)}$) is, as discussed in Sec. 1.4 the generating function of the local integrals of motion $\mathcal{I}_m$, (1.5.16) means that $M^{(\pm)}(x;\lambda,\mu)$ is the generating function of the corresponding M-operators:

$$M^{(\pm)}(x;\lambda,\mu)=\pm i\varkappa\sum_{m=1}^{\infty}\lambda^{-m}M_m(x,\mu). \tag{1.5.18}$$

In particular,

$$M_1(x,\lambda)=\frac{i}{2}\sigma_3, \tag{1.5.19}$$

$$M_2(x,\lambda)=-L(x,\lambda)=i\frac{\lambda}{2}\sigma_3-i\varkappa\overline{\Psi}(x)\sigma_++i\Psi(x)\sigma_-, \tag{1.5.20}$$

$$M_3(x,\lambda)=\left(i\frac{\lambda^2}{2}+i\varkappa|\Psi(x)|^2\right)\sigma_3+\varkappa\left(\overline{\Psi}_x(x)-i\lambda\overline{\Psi}(x)\right)\sigma_++\left(\Psi_x(x)+i\lambda\Psi(x)\right)\sigma_-. \tag{1.5.21}$$

We note that the specific structure of the r-matrix for the nonlinear Schrödinger equation (1.2.12) allows one to simplify the expression (1.5.17) for $M^{(\pm)}(x;\lambda,\mu)$:

$$M^{(+)}(x;\lambda,\mu)=-\frac{\varkappa}{\lambda-\mu}\frac{1}{a(\lambda)}\begin{pmatrix}a_-(x,\lambda)\\ b_-(x,\lambda)\end{pmatrix}\left(a_+(x,\lambda),\varkappa\overline{b_+(x,\bar\lambda)}\right), \tag{1.5.22}$$

$$M^{(-)}(x;\lambda,\mu)=-\frac{\varkappa}{\lambda-\mu}\frac{1}{\overline{a(\bar\lambda)}}\begin{pmatrix}\varkappa\overline{b_-(x,\bar\lambda)}\\ \overline{a_-(x,\bar\lambda)}\end{pmatrix}\left(b_+(x,\lambda),\overline{a_+(x,\bar\lambda)}\right), \tag{1.5.23}$$

thus reproducing a known result [35], stating that the generating function of the M-operators for the n.S.e. is proportional to the diagonal of the kernel of the resolvent operator $\frac{\partial}{\partial x}-L$.

We stress, however, that such a simplification makes essential use of the specifics of the nonlinear Schrödinger equation, at the same time that (1.5.17) carries a universal character and is suitable for any completely integrable models, whose L-operators have a r-matrix (e.g., the sin-Gordon equation, the Landau-Lifshits equation [29]).

To conclude this section, we introduce a series of formulas, describing the time evolution of the transition matrices $T_{x_1}^{x_2}$, $T_\pm$, and T.

<u>Proposition 1.5.2.</u> The transition matrices $T_{x_1}^{x_2}(t;\lambda)$, $T_\pm(x,t;\lambda)$, and $T(t;\lambda)$ satisfy the following differential equations:

$$\frac{\partial}{\partial t}T_{x_1}^{x_2}(t;\lambda)=M(x_2,t;\lambda)T_{x_1}^{x_2}(t;\lambda)-T_{x_1}^{x_2}(t,\lambda)M(x_1;t,\lambda), \tag{1.5.24}$$

$$\frac{\partial}{\partial t}T_-(x,t;\lambda)=M(x,t;\lambda)T_-(x,t;\lambda)-T_-(x,t;\lambda)\,i\frac{\lambda^2}{2}\sigma_3, \tag{1.5.25}$$

$$\frac{\partial}{\partial t}T_+(x,t;\lambda)=i\frac{\lambda^2}{2}\sigma_3 T_+(x,t;\lambda)-T_+(x,t;\lambda)M(x,t;\lambda)\,, \tag{1.5.26}$$

$$\frac{\partial}{\partial t}T(t,\lambda)=\left[i\frac{\lambda^2}{2}\sigma_3\,,\,T(t,\lambda)\right]. \tag{1.5.27}$$

Proof. We introduce the notation $\frac{\partial}{\partial t}T^{x_2}_{x_1}(t,\lambda)=\mathcal{M}^{x_2}_{x_1}(t;\lambda)$. Differentiating (1.5.24) with respect to x_2 and using (1.1.3) and (1.5.2), it is easy to see that both sides of (1.5.24) satisfy the same differential equation

$$\frac{\partial}{\partial x_2}\mathcal{M}^{x_2}_{x_1}(t;\lambda)=L(x_2,t;\lambda)\mathcal{M}^{x_2}_{x_1}(t;\lambda)+\left(M_x(x_2,t;\lambda)+\left[L(x_2,t;\lambda),M(x_2,t;\lambda)\right]\right)T^{x_2}_{x_1}(t;\lambda) \tag{1.5.28}$$

with the same initial condition

$$\mathcal{M}^{x_1}_{x_1}(t,\lambda)=0\,, \tag{1.5.29}$$

which proves (1.5.24). Equations (1.5.25-27) are now obtained from (1.5.24) by passage to the limit according to (1.1.13-15), taking into account

$$\lim_{|x|\to\infty}M(x,\lambda)=i\frac{\lambda^2}{2}\sigma_3\,, \tag{1.5.30}$$

which follows directly from (1.5.3) and the boundary condition $\Psi(x)\xrightarrow[|x|\to\infty]{}0$.

We note that (1.5.27) has the obvious general solution:

$$T(t;\lambda)=e^{i\frac{\lambda^2}{2}\sigma_3 t}\,T(0,\lambda)e^{-i\frac{\lambda^2}{2}\sigma_3 t}. \tag{1.5.31}$$

Writing out the matrix elements of (1.5.31), we get the well-known result [31]:

$$a(t,\lambda)=a(0,\lambda)\,, \tag{1.5.32}$$

$$\bar{a}(t,\lambda)=\bar{a}(0,\lambda), \tag{1.5.33}$$

$$b(t,\lambda)=e^{-i\lambda^2 t}\,b(0,\lambda) \tag{1.5.34}$$

$$\bar{b}(t,\lambda)=e^{i\lambda^2 t}\,\bar{b}(0,\lambda). \tag{1.5.35}$$

CHAPTER II

QUANTUM NONLINEAR SCHRÖDINGER EQUATION

2.1. Quantization

In this section we enter upon the study of the quantum version of the nonlinear Schrödinger equation, which constitutes the basic object of study of the present paper. As already noted in the Introduction, the nonlinear Schrödinger equation admits detailed description in the classical as well as the quantum case, which stipulated its choice as the object of the first application of the quantum method of the inverse problem.

We shall briefly describe the quantum system corresponding to the classical equation (1.1.1). The Hilbert space of states of the system is the Focke space F for Bose particles in one dimension [36, 37].

The elements of the space F are columns of the form

$$f=\begin{pmatrix} f_0 \\ f_1(x_1) \\ \cdots\cdots \\ f_N(x_1,\ldots,x_N) \\ \cdots\cdots\cdots \end{pmatrix}, \tag{2.1.1}$$

where $f_0 \in \mathbb{C}$, and $f_N(x_1,\dots,x_N)$ is a complex-valued symmetric square-integrable function of N real variables. We define the scalar product in the space F by the formula

$$\langle f,g\rangle = \bar{f}_0 g_0 + \sum_{N=1}^{\infty} \int_{-\infty}^{\infty}\dots\int_{-\infty}^{\infty} dx_1\dots dx_N \bar{f}_N(x_1,\dots,x_N)\, g_N(x_1,\dots,x_N).$$

The Focke space F splits into the orthogonal sum of N partial subspaces F_N:

$$F = \sum_{N=0}^{\infty}{}^{\oplus} F_N \quad ; \quad f_N(x_1,\dots,x_N) \in F_N .$$

A vector of the form

$$\begin{pmatrix} 1 \\ 0 \\ \vdots \\ 0 \\ \vdots \end{pmatrix} = |0\rangle \; , \quad \langle 0|0\rangle = 1,$$

will be called a vacuum and will be denoted by $|0\rangle$.

Let the generalized operator-valued functions $\Psi(x)$ and $\Psi^+(x)$ define the standard [36, 37] representation of the canonical commutation relations

$$[\Psi(x),\Psi(y)] = [\Psi^+(x),\Psi^+(y)] = 0 \; ,$$

$$[\Psi(x),\Psi^+(y)] = \delta(x-y) \tag{2.1.2}$$

in the space F (we set Planck's constant $\hbar$ equal to 1), having the property

$$\Psi(x)|0\rangle = 0 \; . \tag{2.1.3}$$

Here any element f [Eq. (2.1.1)] of the space F can be represented in the form

$$f = f_0|0\rangle + \sum_{N=1}^{\infty}\frac{1}{\sqrt{N!}}\int_{-\infty}^{\infty}\dots\int_{-\infty}^{\infty} dx_1\dots dx_N f_N(x_1,\dots,x_N)\Psi^+(x_1)\dots\Psi^+(x_N)|0\rangle \; . \tag{2.1.4}$$

In what follows we shall use as representative both (2.1.1) and (2.1.4).

The classical Hamiltonian H (1.2.1) corresponds in the quantum case to the self-adjoint operator $\mathbb{H}$ in the space F, defined by the expression

$$\mathbb{H} = \int_{-\infty}^{\infty} dx \left(\Psi_x^+\Psi_x + \varkappa\Psi^+\Psi^+\Psi\Psi\right). \tag{2.1.5}$$

The Heisenberg equation of motion for the operator $\Psi(x,t)$, generated by the Hamiltonian $\mathbb{H}$, has the form

$$i\Psi_t = [\Psi,\mathbb{H}] = -\Psi_{xx} + 2\varkappa\Psi^+\Psi\Psi \; . \tag{2.1.6}$$

Using the commutation relations (2.1.2), it is easy to verify that the Hamiltonian $\mathbb{H}$ commutes with the operators of the number of particles

$$\mathbb{N} = \int_{-\infty}^{\infty} dx\, \Psi^+\Psi \tag{2.1.7}$$

and momentum

$$\mathbb{P} = \frac{i}{2}\int_{-\infty}^{\infty} dx\left(\Psi_x^+\Psi - \Psi^+\Psi_x\right). \tag{2.1.8}$$

On vectors $f_N(x_1,\ldots,x_N)$ from the N-partial subspace F_N, the Hamiltonian $\mathbb{H}$ acts as a differential operator [33, 34, 38]:

$$\mathbb{H} f_N(x_1,\ldots,x_N) = \left(-\sum_{j=1}^{N} \frac{\partial^2}{\partial x_j^2} + 2\varkappa \sum_{i<j} \delta(x_i - x_j)\right) f_N(x_1,\ldots,x_N), \tag{2.1.9}$$

having the form of a multipartial Schrödinger operator with twin pointwise interaction. It is easy to give strict meaning to the singular potential $\delta(x_i - x_j)$ in (2.1.9), replacing it by the boundary condition [33, 34, 38]

$$\frac{\partial}{\partial x_i} f_N(x_1,\ldots,x_N)\Big|_{x_i = x_j - 0}^{x_i = x_j + 0} = \varkappa f_N(x_1,\ldots,x_N)\Big|_{x_i = x_j}, \quad i \neq j. \tag{2.1.10}$$

The eigenfunctions of the operator $\mathbb{H}$ admit a simple description [33, 34]. Namely, the eigenfunction $f_N(x_1,\ldots,x_N|k_1,\ldots,k_N)$, corresponding to the eigenvalue $\sum_{j=1}^{N} k_j^2$ and describing the state of scattering of N particles with momenta $k_1,\ldots,k_N$ $(\operatorname{Im} k_j = 0; j = 1,\ldots,N)$, has for $x_1 < x_2 < \ldots < x_N$ the following form

$$f_N(x_1,\ldots,x_N|k_1,\ldots,k_N) = \frac{(i)^N}{(2\pi)^{N/2}\sqrt{N!}} \sum C_{\ell_1\ldots\ell_N} e^{i(k_{\ell_1}x_1 + \ldots + k_{\ell_N}x_N)} \tag{2.1.11}$$

(to other values of x the function f_N is extended by symmetry). A substitution of the form (2.1.11) for an eigenfunction is usually called a Bethe substitution in honor of Bethe, who in [5] first proposed such a substitution, studying a lattice model of a ferromagnet.

The summation in (2.1.11) is taken over all permutations $(\ell_1,\ldots,\ell_N)$ of $(1,\ldots,N)$, and the coefficients $C_{\ell_1\ldots\ell_N}$ must satisfy the condition

$$\frac{C_{\ell_1\ldots\ell_r\ldots\ell_s\ldots\ell_N}}{C_{\ell_1\ldots\ell_s\ldots\ell_r\ldots\ell_N}} = \frac{k_{\ell_r} - k_{\ell_s} + i\varkappa}{k_{\ell_r} - k_{\ell_s} - i\varkappa}. \tag{2.1.12}$$

Following [38], we choose a solution of (2.1.12) in the form

$$C^{(\mathrm{norm})}_{\ell_1\ldots\ell_N} = \prod_{r<s} \sqrt{\frac{k_{\ell_r} - k_{\ell_s} + i\varkappa}{k_{\ell_r} - k_{\ell_s} - i\varkappa}}. \tag{2.1.13}$$

For such a choice of $C^{(\mathrm{norm})}_{\ell_1\ldots\ell_N}$ the system of functions $f_N^{(\mathrm{norm})}(x_1,\ldots,x_N|k_1,\ldots,k_N)$ is orthonormalized:

$$\int_{-\infty}^{\infty}\!\!\ldots\!\int_{-\infty}^{\infty} dx_1\ldots dx_N\, \overline{f_N^{(\mathrm{norm})}(x_1,\ldots,x_N|k_1,\ldots,k_N)}\, f_N^{(\mathrm{norm})}(x_1,\ldots,x_N|p_1,\ldots,p_N) = \sum_{(\ell_1,\ldots,\ell_N)} \prod_{j=1}^{N} \delta(k_j - p_{\ell_j}). \tag{2.1.14}$$

In the case of repulsion $\varkappa > 0$, the system of functions $f_N^{(\mathrm{norm})}(x_1,\ldots,x_N|k_1,\ldots,k_N)$ $(N = 0, 1, 2\ldots)$, moreover, is complete in $\mathbb{F}$. In the case of attraction $\varkappa < 0$, we must consider connected states. It turns out that for each $N = 2, 3, \ldots$ there is only one N-particle connected state obtained from (2.1.11) by analytic continuation with respect to the momenta k_j:

$$k_j = \frac{P}{N} + i|\varkappa|\left(j - \frac{N+1}{2}\right); \quad j = 1, 2, \ldots, N, \tag{2.1.15}$$

where P is the total momentum of the connected state. The corresponding normalized wave functions are given in [38].

The proofs of all the assertions given above are contained in [33, 34, 38].

Concluding the list of results obtained for the quantum n.S.e. by the method of Bethe substitution, we can enter upon the consecutive account of the quantum version of the method of the inverse problem. We note right away that in contrast with the method of Bethe substitution, our method allows us to find the spectrum of the Hamiltonian and other integrals of motion of (2.1.6), without using the explicit form of the eigenfunctions (2.1.11).

The main object of our investigation will be, as noted in the Introduction, the quantum analog of the fundamental solution $T_{x_1}^{x_2}(\lambda)$ of the auxiliary linear problem (1.1.3). Generally speaking, there exist many methods of associating with a given functional of classical canonical variables $\Psi(x)$ and $\overline{\Psi}(x)$ a quantum operator (for example, Wick, anti-Wick, Weyl quantization [39]). We dwell here on Wick (normal) quantization. The advantage of such a choice is indicated, for example, by the result of [38] in which it is proved that the integrals of motion $\mathcal{I}_m$ (1.4.2) for the classical equation (1.1.1) after Wick quantization go into quantum integrals of motion for (2.1.6).

Thus, we define the quantum transition matrix $\mathbb{T}_{x_1}^{x_2}(\lambda)$ by

$$\mathbb{T}_{x_1}^{x_2}(\lambda) = :T_{x_1}^{x_2}(\lambda): \quad . \tag{2.1.16}$$

The colons :: in (2.1.16) denote Wick quantization. In other words, the matrix elements of the matrix $\mathbb{T}_{x_1}^{x_2}(\lambda)$ are defined as quantum operators whose Wick symbols are the corresponding elements of $T_{x_1}^{x_2}(\lambda)$.

There immediately arises the question of the propriety of such a definition, i.e., of the existence of such quantum operators in the space $\mathcal{F}$. The clarification of this question we postpone to the end of the section, and meanwhile we list the properties of the quantum transition matrix, formally following from the definition (2.1.16).

$$1) \quad \mathbb{T}_{x_1}^{x_3}(\lambda) = \mathbb{T}_{x_2}^{x_3}(\lambda)\,\mathbb{T}_{x_1}^{x_2}(\lambda) \quad \text{for } x_1 < x_2 < x_3 \quad \text{or} \quad x_1 > x_2 > x_3 \tag{2.1.17}$$

2) $\mathbb{T}_{x_1}^{x_2}(\lambda)$ has the form

$$\mathbb{T}_{x_1}^{x_2}(\lambda) = \begin{pmatrix} A_{x_1}^{x_2}(\lambda), & \varkappa B^{+\,x_2}_{\;\;x_1}(\lambda) \\ B_{x_1}^{x_2}(\lambda), & A^{+\,x_2}_{\;\;x_1}(\lambda) \end{pmatrix}, \tag{2.1.18}$$

where the superscript + denotes Hermitian conjugation, and $A_{x_1}^{x_2}(\lambda) = :a_{x_1}^{x_2}(\lambda):$, $B_{x_1}^{x_2}(\lambda) = :b_{x_1}^{x_2}(\lambda):$, $A^{+\,x_2}_{\;\;x_1}(\lambda) = (A_{x_1}^{x_2}(\bar{\lambda}))^+$, $B^{+\,x_2}_{\;\;x_1}(\lambda) = (B_{x_1}^{x_2}(\bar{\lambda}))^+$.

3) $\mathbb{T}_{x_1}^{x_2}(\lambda)$ satisfies the differential equations

$$\frac{\partial}{\partial x_2}\mathbb{T}_{x_1}^{x_2}(\lambda) = :L(x_2,\lambda)\,T_{x_1}^{x_2}(\lambda): = \left(-i\frac{\lambda}{2}\sigma_3 + i\varkappa\Psi^+(x_2)\sigma_+\right)\mathbb{T}_{x_1}^{x_2}(\lambda) - i\sigma_-\mathbb{T}_{x_1}^{x_2}(\lambda)\,\Psi(x_2) \tag{2.1.19}$$

and

$$\frac{\partial}{\partial x_1}\mathbb{T}_{x_1}^{x_2}(\lambda) = -:T_{x_1}^{x_2}(\lambda)\,L(x_1,\lambda): = -i\varkappa\Psi^+(x_1)\,\mathbb{T}_{x_1}^{x_2}(\lambda)\sigma_+ - \mathbb{T}_{x_1}^{x_2}(\lambda)\left(-i\frac{\lambda}{2}\sigma_3 - i\sigma_-\Psi(x_1)\right) \tag{2.1.20}$$

with the initial condition

$$\mathbb{T}_x^x(\lambda) = I \,. \tag{2.1.21}$$

Property 1) follows from the analogous property (1.1.6) for the classical transition matrix and the commutativity of the operators $\Psi(x)$ and $\Psi^+(x)$ on disjoint intervals. We stress that in contrast with the classical case, in the quantum case the condition $x_1 < x_2 < x_3$ or $x_1 > x_2 > x_3$ is essential.

Property 2) follows from (1.1.11) for the classical transition matrix and the obvious property of Wick quantization that the complex conjugate of the Wick symbol corresponds to the Hermitian conjugate operator.

In order to prove property 3), we formulate the following simple assertion. Let $X(\psi,\bar\psi)$ be a functional of the fields $\psi(x)$ and $\bar\psi(x)$. Then one has

$$:\bar\psi X: = \Psi^+ :X:\,, \qquad :X\psi: = :X:\Psi\,. \tag{2.1.22}$$

The proof is obvious.

Now, differentiating $\mathbb{T}_{x_1}^{x_2}(\lambda)$ with respect to x_2 or x_1, and using (2.1.16), (1.1.3), (1.1.7), and (2.1.22), we get (2.1.19-21).

In order to write (2.1.19) and (2.1.20) more compactly, we introduce the sign of normal arrangement of operator factors $\vdots\,\vdots$. The sign $\vdots\,\vdots$ should not be confused with the sign for Wick quantization ::, applied here only to classical functionals, the sign $\vdots\,\vdots$, applied to the product of several operator factors (including Ψ and Ψ^+), guarantees the arrangement of all Ψ^+ on the left, and all Ψ on the right, without altering the order of the remaining factors. For example,

$$\vdots\, X\Psi\Psi^+ Y \,\vdots = \Psi^+ X Y \Psi\,. \tag{2.1.23}$$

Using the notation introduced, (2.1.19) and (2.1.20) can be rewritten, respectively, in the form

$$\frac{\partial}{\partial x_2}\mathbb{T}_{x_1}^{x_2}(\lambda) = \vdots\, \mathbb{L}(x_2,\lambda)\,\mathbb{T}_{x_1}^{x_2}(\lambda)\,\vdots \tag{2.1.24}$$

$$\frac{\partial}{\partial x_1}\mathbb{T}_{x_1}^{x_2}(\lambda) = -\vdots\, \mathbb{T}_{x_1}^{x_2}(\lambda)\,\mathbb{L}(x_1,\lambda)\,\vdots\,, \tag{2.1.25}$$

where $\mathbb{L}(x,\lambda)$ is the quantum $\mathbb{L}$-operator

$$\mathbb{L}(x,\lambda) = :L(x,\lambda): \; = -i\frac{\lambda}{2}\sigma_3 - i\varkappa\sigma_-\Psi(x) + i\varkappa\sigma_+\Psi^+(x) = \begin{pmatrix} -i\frac{\lambda}{2}\,, & i\varkappa\Psi^+(x) \\ -i\Psi(x)\,, & i\frac{\lambda}{2} \end{pmatrix}. \tag{2.1.26}$$

As also in the classical case, the differential equations (2.1.24) and (2.1.25) with initial condition (2.1.21) are equivalent with the Volterra integral equations

$$\mathbb{T}_{x_1}^{x_2}(\lambda) = I + \int_{x_1}^{x_2} dx\, \vdots\, \mathbb{L}(x,\lambda)\,\mathbb{T}_{x_1}^{x}(\lambda)\,\vdots \tag{2.1.27}$$

and

$$\mathbb{T}_{x_1}^{x_2}(\lambda) = I + \int_{x_1}^{x_2} dx \, : \mathbb{T}_{x}^{x_2}(\lambda) \, \mathbb{L}(x,\lambda) : \tag{2.1.28}$$

or

$$\mathbb{T}_{x_1}^{x_2}(\lambda) = e(x_2 - x_1, \lambda) + \int_{x_1}^{x_2} dx \, e(x_2 - x, \lambda) : \mathbb{V}(x) \mathbb{T}_{x_1}^{x}(\lambda) : \tag{2.1.29}$$

and

$$\mathbb{T}_{x_1}^{x_2}(\lambda) = e(x_2 - x_1, \lambda) + \int_{x_1}^{x_2} dx \, : \mathbb{T}_{x}^{x_2}(\lambda) \mathbb{V}(x) : e(x - x_1, \lambda) , \tag{2.1.30}$$

where

$$\mathbb{V}(x) = \, : \mathbb{V}(x) : \, = i\varkappa \sigma_+ \Psi^+(x) - i\sigma_- \Psi(x) . \tag{2.1.31}$$

Iterating (2.1.29) and (2.1.30), we get for the matrix elements of $\mathbb{T}_{x_1}^{x_2}(\lambda)$ [Eq. (2.1.18)] the following expansions:

$$A_{x_1}^{x_2}(\lambda) = e^{-i\frac{\lambda}{2}(x_2 - x_1)} \Big[1 + \sum_{n=1}^{\infty} \varkappa^n \underset{x_2 > \xi_n > \eta_n > \ldots > \eta_1 > x_1}{\int \ldots \int} d\xi_1 \ldots d\xi_n d\eta_1 \ldots d\eta_n \; e^{i\lambda(\xi_1 + \ldots + \xi_n - \eta_1 - \ldots - \eta_n)} \Psi^+(\xi_1) \ldots \Psi^+(\xi_n) \Psi(\eta_1) \ldots \Psi(\eta_n) \Big] , \tag{2.1.32}$$

$$A^{+}{}_{x_1}^{x_2}(\lambda) = e^{i\lambda(x_2 - x_1)} \Big[1 + \sum_{n=1}^{\infty} \varkappa^n \underset{x_2 > \eta_n > \xi_n > \ldots > \xi_1 > x_1}{\int \ldots \int} d\xi_1 \ldots d\xi_n d\eta_1 \ldots d\eta_n \; e^{i\lambda(\xi_1 + \ldots + \xi_n - \eta_1 - \ldots - \eta_n)} \Psi^+(\xi_1) \ldots \Psi^+(\xi_n) \Psi(\eta_1) \ldots \Psi(\eta_n) \Big] , \tag{2.1.33}$$

$$B_{x_1}^{x_2}(\lambda) = -ie^{i\frac{\lambda}{2}(x_1 + x_2)} \sum_{n=0}^{\infty} \varkappa^n \underset{x_2 > \eta_{n+1} > \xi_n > \ldots > \eta_1 > x_1}{\int \ldots \int} d\xi_1 \ldots d\xi_n d\eta_1 \ldots d\eta_{n+1} \; e^{i\lambda(\xi_1 + \ldots + \xi_n - \eta_1 - \ldots - \eta_{n+1})} \Psi^+(\xi_1) \ldots \Psi^+(\xi_n) \Psi(\eta_1) \ldots \Psi(\eta_{n+1}) , \tag{2.1.34}$$

$$B^{+}{}_{x_1}^{x_2}(\lambda) = ie^{-i\frac{\lambda}{2}(x_1 + x_2)} \sum_{n=0}^{\infty} \varkappa^n \underset{x_2 > \xi_{n+1} > \eta_n > \ldots > \xi_1 > x_1}{\int \ldots \int} d\xi_1 \ldots d\xi_{n+1} d\eta_1 \ldots d\eta_n \; e^{i\lambda(\xi_1 + \ldots + \xi_{n+1} - \eta_1 - \ldots - \eta_{n+1})} \Psi^+(\xi_1) \ldots \Psi^+(\xi_{n+1}) \Psi(\eta_1) \ldots \Psi(\eta_n) , \tag{2.1.35}$$

analogous to the classical expansions (1.1.29) and (1.1.30).

To conclude this section we turn to the question of the propriety of the definition (2.1.16) of the quantum transition matrix $\mathbb{T}_{x_1}^{x_2}(\lambda)$. Unfortunately, the general theorems contained, e.g., in [40], do not give an answer to this question, since too restrictive conditions are imposed on the Wick symbol of the operator (of the type of decreasing as $|\Psi| \to \infty$ or summability). However, the specific construction of $\mathbb{T}_{x_1}^{x_2}(\lambda)$ as a functional of $\Psi(x)$ and $\overline{\Psi}(x)$ essentially simplifies the situation.

We take as a basis the expansions (2.1.32-35). Analysis of (2.1.32-35) shows that the operators $A_{x_1}^{x_2}(\lambda)$ and $A^{+}{}_{x_1}^{x_2}(\lambda)$ do not change the number of particles, $B^{+}{}_{x_1}^{x_2}(\lambda)$ increases it by 1, and lowers it by 1 (annihilating vacuum). Here, in order to define the action of any of these four operators on an $\mathcal{N}$-particle state, it suffices to know only a finite number of first terms of the series (2.1.32-35) (up to terms containing $\mathcal{N}$ annihilating operators, inclusive). Thus, on functions from $F_{\mathcal{N}}$ the matrix elements of $\mathbb{T}_{x_1}^{x_2}(\lambda)$ act as certain integral operators.

We shall not discuss here such properties of these operators as the domain of definition, range of values, etc. For our purposes (i.e., for calculating commutation relations between matrix elements $T^{x_2}_{x_1}(\lambda)$) it will suffice to consider the operators $A^{x_2}_{x_1}(\lambda)$, $A^{+x_2}_{x_1}(\lambda)$, $B^{x_2}_{x_1}(\lambda)$, and $B^{+x_2}_{x_1}(\lambda)$, as formal series (2.1.32-35). The integral equations (2.1.27-30) we shall consider here as compact notation for the series (2.1.32-35).

Some additional information on the properties of the matrix elements of $T^{x_2}_{x_1}(\lambda)$ as operators in the Focke space will be given in Sec. 2.4.

2.2. R-Matrix

As in the classical case too, our final goal is the definition of the quantum transition matrix $T(\lambda)$ for an infinite interval and the calculation of the commutation relations between its elements. An important intermediate stage here is the calculation of the commutation relations between the matrix elements of $T^{x_2}_{x_1}(\lambda)$. Analogously to Sec. 1.2, it is convenient to introduce the matrices $\widetilde{T}^{x_2}_{x_1}(\lambda)$ and $\widetilde{\widetilde{T}}^{x_2}_{x_1}(\mu)$ by (1.2.4-5). The basic result of the present section is the following

THEOREM 3. The commutation relations $T^{x_2}_{x_1}(\lambda)$ and $T^{x_2}_{x_1}(\mu)$ can be written compactly in the form

$$R(\lambda-\mu)\widetilde{T}^{x_2}_{x_1}(\lambda)\widetilde{\widetilde{T}}^{x_2}_{x_1}(\mu)=\widetilde{\widetilde{T}}^{x_2}_{x_1}(\mu)\widetilde{T}^{x_2}_{x_1}(\lambda)R(\lambda-\mu), \tag{2.2.1}$$

where

$$R(\lambda)=I_4+ir(\lambda)=I_4-\frac{i\varkappa}{\lambda}P . \tag{2.2.2}$$

As also in Sec. 1.2, the proof of Theorem 3 is based on the verification of (2.2.1) in infinitesimal form. Here the following lemma will be useful.

LEMMA 2.2.1. The products $\widetilde{T}^{x_2}_{x_1}(\lambda)\widetilde{\widetilde{T}}^{x_2}_{x_1}(\mu)$ and $\widetilde{\widetilde{T}}^{x_2}_{x_1}(\mu)\widetilde{T}^{x_2}_{x_1}(\lambda)$ satisfy the following differential equations:

$$\frac{\partial}{\partial x_2}\left(\widetilde{T}^{x_2}_{x_1}(\lambda)\widetilde{\widetilde{T}}^{x_2}_{x_1}(\mu)\right)=:\mathcal{L}(x_2;\lambda,\mu)\widetilde{T}^{x_2}_{x_1}(\lambda)\widetilde{\widetilde{T}}^{x_2}_{x_1}(\mu): \tag{2.2.3}$$

and

$$\frac{\partial}{\partial x_2}\left(\widetilde{\widetilde{T}}^{x_2}_{x_1}(\mu)\widetilde{T}^{x_2}_{x_1}(\lambda)\right)=:\mathcal{L}'(x_2;\lambda,\mu)\widetilde{\widetilde{T}}^{x_2}_{x_1}(\mu)\widetilde{T}^{x_2}_{x_1}: \tag{2.2.4}$$

and initial condition

$$\widetilde{T}^{x}_{x}(\lambda)\widetilde{\widetilde{T}}^{x}_{x}(\mu)=\widetilde{\widetilde{T}}^{x}_{x}(\mu)\widetilde{T}^{x}_{x}(\lambda)=I . \tag{2.2.5}$$

The operators $\mathcal{L}(x;\lambda,\mu)$ and $\mathcal{L}'(x;\lambda,\mu)$ in (2.2.3) and (2.2.4) have the following form

$$\mathcal{L}(x;\lambda,\mu)=\widetilde{L}(x,\lambda)+\widetilde{\widetilde{L}}(x,\mu)+\varkappa\widetilde{\sigma}_-\widetilde{\widetilde{\sigma}}_+=\begin{pmatrix}-i\frac{\lambda+\mu}{2}, & i\varkappa\Psi^+(x), & i\varkappa\Psi^+(x), & 0\\ -i\Psi(x), & i\frac{\mu-\lambda}{2}, & 0, & i\varkappa\Psi^+(x)\\ -i\Psi(x), & \varkappa, & i\frac{\lambda-\mu}{2}, & i\varkappa\Psi^+(x)\\ 0, & -i\Psi(x), & -i\Psi(x), & i\frac{\lambda+\mu}{2}\end{pmatrix}, \tag{2.2.6}$$

$$\mathcal{L}'(x;\lambda,\mu)=\mathbb{L}(x,\lambda)+\tilde{\mathbb{L}}(x,\mu)+\varkappa\tilde{\sigma}_+\tilde{\sigma}_- = \begin{pmatrix} -i\frac{\lambda+\mu}{2}, & i\varkappa\Psi^+(x), & i\varkappa\Psi^+(x), & 0 \\ -i\Psi(x), & i\frac{\mu-\lambda}{2}, & \varkappa, & i\varkappa\Psi^+(x) \\ -i\Psi(x), & 0, & i\frac{\lambda-\mu}{2}, & i\varkappa\Psi^+(x) \\ 0, & -i\Psi(x), & -i\Psi(x), & i\frac{\lambda+\mu}{2} \end{pmatrix}. \tag{2.2.7}$$

We give two proofs of Lemma 2.2.1.

Proof 1. Let ε be an arbitrary positive number. We differentiate the product $\tilde{\mathbb{T}}_{x_1}^{x_2+\varepsilon}(\lambda)\tilde{\tilde{\mathbb{T}}}_{x_1}^{x_2}(\mu)$ with respect to x_2. Using (2.1.19), we get

$$\frac{\partial}{\partial x_2}\left(\tilde{\mathbb{T}}_{x_1}^{x_2+\varepsilon}(\lambda)\tilde{\tilde{\mathbb{T}}}_{x_1}^{x_2}(\mu)\right)=\left(-i\frac{\lambda}{2}\tilde{\sigma}_3-i\frac{\mu}{2}\tilde{\tilde{\sigma}}_3+i\varkappa\tilde{\sigma}_+\Psi^+(x_2+\varepsilon)\right)\tilde{\mathbb{T}}_{x_1}^{x_2+\varepsilon}(\lambda)\tilde{\tilde{\mathbb{T}}}_{x_1}^{x_2}(\mu)-i\tilde{\tilde{\sigma}}_-\tilde{\mathbb{T}}_{x_1}^{x_2+\varepsilon}(\lambda)\tilde{\tilde{\mathbb{T}}}_{x_1}^{x_2}(\mu)\Psi(x_2)+$$
$$+i\varkappa\tilde{\tilde{\sigma}}_+\tilde{\mathbb{T}}_{x_1}^{x_2+\varepsilon}(\lambda)\Psi^+(x_2)\tilde{\tilde{\mathbb{T}}}_{x_1}^{x_2}(\mu)-i\tilde{\sigma}_-\tilde{\mathbb{T}}_{x_1}^{x_2+\varepsilon}(\lambda)\Psi(x_2+\varepsilon)\tilde{\tilde{\mathbb{T}}}_{x_1}^{x_2}(\mu)=:\left(\tilde{\mathbb{L}}(x_2+\varepsilon,\lambda)+\tilde{\tilde{\mathbb{L}}}(x_2,\mu)\right)\tilde{\mathbb{T}}_{x_1}^{x_2+\varepsilon}(\lambda)\tilde{\tilde{\mathbb{T}}}_{x_1}^{x_2}(\mu):-$$
$$-i\tilde{\sigma}_-\tilde{\mathbb{T}}_{x_1}^{x_2+\varepsilon}(\lambda)\left[\Psi(x_2+\varepsilon),\tilde{\tilde{\mathbb{T}}}_{x_1}^{x_2}(\mu)\right]+i\varkappa\tilde{\tilde{\sigma}}_+\left[\tilde{\mathbb{T}}_{x_1}^{x_2+\varepsilon}(\lambda),\Psi^+(x_2)\right]\tilde{\tilde{\mathbb{T}}}_{x_1}^{x_2}(\mu). \tag{2.2.8}$$

In order to get the final answer (2.2.3), we must calculate the commutators in the last two terms on the right side of (2.2.8) and pass to the limit as $\varepsilon\to 0$. The commutator $[\Psi(x_2+\varepsilon),$ $\tilde{\tilde{\mathbb{T}}}_{x_1}^{x_2}(\mu)]$ obviously vanishes, since $\Psi(x_2+\varepsilon)$ commutes with all operators $\Psi(x),\Psi^+(x)$ for $x\in[x_1,x_2]$ since $\varepsilon>0$. To calculate the second commutator in (2.2.8), we use Lemma 1.5.1 and the following easily verifiable equation,

$$\left[:X(\psi,\bar{\psi}):,\Psi^+(x)\right]=:\frac{1}{i}\left\{X(\psi,\bar{\psi}),\bar{\psi}(x)\right\}:, \tag{2.2.9}$$

which is valid for any functional $X(\psi,\bar{\psi})$ of the fields $\psi(x)$, $\bar{\psi}(x)$. As a result we get

$$\left[\tilde{\mathbb{T}}_{x_1}^{x_2+\varepsilon}(\lambda),\Psi^+(x_2)\right]=:\int_{x_1}^{x_2+\varepsilon}dx\,\tilde{\mathbb{T}}_{x}^{x_2+\varepsilon}(\lambda)\frac{1}{i}\left\{\tilde{\mathbb{L}}(x,\lambda),\bar{\Psi}(x_2)\right\}\tilde{\mathbb{T}}_{x_1}^{x}(\lambda):\ \tilde{\mathbb{T}}_{x_2}^{x_2+\varepsilon}(\lambda)\,\varkappa\tilde{\sigma}_-\tilde{\mathbb{T}}_{x_1}^{x_2}(\lambda). \tag{2.2.10}$$

Substituting (2.2.10) in (2.2.8) and letting $\varepsilon\to 0$, we get the answer required. We note that the result is independent of the sign of ε. For $\varepsilon<0$ the second commutator in (2.2.8) vanishes, and the first gives the needed summand in (2.2.3). One proves (2.2.4) analogously.

The method of "extension" used in the proof given above we borrow from [30].* This method allows us to avoid consideration of indeterminate expressions of the form $[\mathbb{T}_{x_1}^{x_2}(\lambda),$ $\Psi^+(x_2)]$ (containing indeterminacies of the type of the product of a function and a discontinuous one, as is easy to see, using, for example, the expansions (1.1.29-30)). However, here one uses implicitly an unproved, generally speaking, proposition about the continuous dependence of $\frac{\partial}{\partial x_2}\left(\tilde{\mathbb{T}}_{x_1}^{x_2+\varepsilon}(\lambda)\tilde{\tilde{\mathbb{T}}}_{x_1}^{x_2}(\mu)\right)$ on ε. Hence we give a second proof of Lemma 2.2.1, more straightforward, although also more complicated.

Proof 2. We substitute into the product $\tilde{\mathbb{T}}_{x_1}^{x_2}(\lambda)\tilde{\tilde{\mathbb{T}}}_{x_1}^{x_2}(\mu)$ the integral equations for $\tilde{\mathbb{T}}_{x_1}^{x_2}(\lambda)$ and $\tilde{\tilde{\mathbb{T}}}_{x_1}^{x_2}(\mu)$ of the form (2.1.27). As a result, we get:

*The author thanks S. V. Manakov for indicating the possibility of using the method of "extension" in the quantum case.

$$\tilde{\mathbb{T}}_{x_1}^{x_2}(\lambda)\tilde{\tilde{\mathbb{T}}}_{x_1}^{x_2}(\mu)=I+\int_{x_1}^{x_2}d\xi\left[\left(-i\frac{\lambda}{2}\tilde{\sigma}_3+i\varkappa\Psi^+(\xi)\tilde{\sigma}_+\right)\tilde{\mathbb{T}}_{x_1}^{\xi}(\lambda)-i\tilde{\sigma}_-\tilde{\mathbb{T}}_{x_1}^{\xi}(\lambda)\Psi(\xi)\right]+$$
$$+\int_{x_1}^{x_2}d\eta\left[\left(-i\frac{\mu}{2}\tilde{\tilde{\sigma}}_3+i\varkappa\Psi^+(\eta)\tilde{\tilde{\sigma}}_+\right)\tilde{\tilde{\mathbb{T}}}_{x_1}^{\eta}(\mu)-i\tilde{\tilde{\sigma}}_-\tilde{\tilde{\mathbb{T}}}_{x_1}^{\eta}(\mu)\Psi(\eta)\right]+$$
$$+\int_{x_1}^{x_2}d\xi\int_{x_1}^{x_2}d\eta\left[\left(-i\frac{\lambda}{2}\tilde{\sigma}_3+i\varkappa\Psi^+(\xi)\tilde{\sigma}_+\right)\tilde{\mathbb{T}}_{x_1}^{\xi}(\lambda)-i\tilde{\sigma}_-\tilde{\mathbb{T}}_{x_1}^{\xi}(\lambda)\Psi(\xi)\right]\times$$
$$\times\left[\left(-i\frac{\mu}{2}\tilde{\tilde{\sigma}}_3+i\varkappa\Psi^+(\eta)\tilde{\tilde{\sigma}}_+\right)\tilde{\tilde{\mathbb{T}}}_{x_1}^{\eta}(\mu)-i\tilde{\tilde{\sigma}}_-\tilde{\tilde{\mathbb{T}}}_{x_1}^{\eta}(\mu)\Psi(\eta)\right]. \qquad (2.2.11)$$

Subsequently, complicated, but in terms of ideas entirely transparent, calculations produce as their ultimate goal the reduction, using the commutation relations (2.1.2), of equation (2.2.11) to a Volterra integral equation, equivalent with the differential equation (2.2.3) with initial condition (2.2.5).

We transform the fourth summand in (2.2.11), opening the brackets in the integrand and using the commutation relation (2.1.2), after which it assumes the form

$$\int_{x_1}^{x_2}d\xi\int_{x_1}^{x_2}d\eta\left[\left(-i\frac{\lambda}{2}\tilde{\sigma}_3+i\varkappa\Psi^+(\xi)\tilde{\sigma}_+\right)\tilde{\mathbb{T}}_{x_1}^{\xi}(\lambda)\left(-i\frac{\mu}{2}\tilde{\tilde{\sigma}}_3+i\varkappa\Psi^+(\eta)\tilde{\tilde{\sigma}}_+\right)\tilde{\tilde{\mathbb{T}}}_{x_1}^{\eta}(\mu)+\right.$$
$$\left(-i\frac{\lambda}{2}\tilde{\sigma}_3+i\varkappa\Psi^+(\xi)\tilde{\sigma}_+\right)\left(-i\tilde{\tilde{\sigma}}_-\right)\tilde{\mathbb{T}}_{x_1}^{\xi}(\lambda)\tilde{\tilde{\mathbb{T}}}_{x_1}^{\eta}(\mu)\Psi(\eta)-\tilde{\sigma}_-\tilde{\tilde{\sigma}}_-\tilde{\mathbb{T}}_{x_1}^{\xi}(\lambda)\Psi(\xi)\tilde{\tilde{\mathbb{T}}}_{x_1}^{\eta}(\mu)\Psi(\eta)+$$
$$\left.-i\tilde{\sigma}_-\tilde{\mathbb{T}}_{x_1}^{\xi}(\lambda)\left(-i\frac{\mu}{2}\tilde{\tilde{\sigma}}_3+i\varkappa\Psi^+(\eta)\tilde{\tilde{\sigma}}_+\right)\Psi(\xi)\tilde{\tilde{\mathbb{T}}}_{x_1}^{\eta}(\mu)+\varkappa\tilde{\sigma}_-\tilde{\tilde{\sigma}}_+\delta(\xi-\eta)\tilde{\mathbb{T}}_{x_1}^{\xi}(\lambda)\tilde{\tilde{\mathbb{T}}}_{x_1}^{\eta}(\mu)\right]. \qquad (2.2.12)$$

We transform (2.2.12) in the following way. Firstly, we integrate the δ-function in the fifth summand. Secondly, we divide the domain of integration in the remaining summands into two parts: $x_1<\xi<\eta<x_2$ and $x_1<\eta<\xi<x_2$. Then, for $\xi<\eta$, in the first and fourth summands we transfer $\Psi^+(\eta)$ to the left, using the fact that $\tilde{\mathbb{T}}_{x_1}^{\xi}(\lambda)$ commutes with $\Psi^+(\eta)$ for the indicated relation of ξ and η, and for $\xi>\eta$ analogously we transfer $\Psi(\xi)$ to the right in the third and fourth summands. Then using (2.1.19), we can rewrite (2.2.12) in the form

$$\int_{x_1}^{x_2}d\eta\int_{x_1}^{\eta}d\xi\left[\left(-i\frac{\mu}{2}\tilde{\tilde{\sigma}}_3+i\varkappa\Psi^+(\eta)\tilde{\tilde{\sigma}}_+\right)\left(\frac{\partial}{\partial\xi}\tilde{\mathbb{T}}_{x_1}^{\xi}(\lambda)\right)\tilde{\tilde{\mathbb{T}}}_{x_1}^{\eta}(\mu)-i\tilde{\tilde{\sigma}}_-\left(\frac{\partial}{\partial\xi}\tilde{\mathbb{T}}_{x_1}^{\xi}(\lambda)\right)\tilde{\tilde{\mathbb{T}}}_{x_1}^{\eta}(\mu)\Psi(\eta)\right]+$$
$$+\int_{x_1}^{x_2}d\xi\int_{x_1}^{\xi}d\eta\left[\left(-i\frac{\lambda}{2}\tilde{\sigma}_3+i\varkappa\Psi^+(\xi)\tilde{\sigma}_+\right)\tilde{\mathbb{T}}_{x_1}^{\xi}(\lambda)\left(\frac{\partial}{\partial\eta}\tilde{\tilde{\mathbb{T}}}_{x_1}^{\eta}(\mu)\right)-\right.$$
$$\left.-i\tilde{\sigma}_-\tilde{\mathbb{T}}_{x_1}^{\xi}(\lambda)\left(\frac{\partial}{\partial\eta}\tilde{\tilde{\mathbb{T}}}_{x_1}^{\eta}(\mu)\right)\Psi(\xi)\right]+\varkappa\int_{x_1}^{x_2}dx\,\tilde{\sigma}_-\tilde{\tilde{\sigma}}_+\tilde{\mathbb{T}}_{x_1}^{x}(\lambda)\tilde{\tilde{\mathbb{T}}}_{x_1}^{x}(\mu). \qquad (2.2.13)$$

Carrying out integration of the total derivatives in (2.2.13) and substituting the result in (2.2.11), we get for $\tilde{\mathbb{T}}_{x_1}^{x_2}(\lambda)\tilde{\tilde{\mathbb{T}}}_{x_1}^{x_2}(\mu)$ the following integral equation

$$\tilde{\mathbb{T}}_{x_1}^{x_2}(\lambda)\tilde{\tilde{\mathbb{T}}}_{x_1}^{x_2}(\mu)=I+\int_{x_1}^{x_2}dx:\left(\tilde{\mathbb{L}}(x,\lambda)+\tilde{\tilde{\mathbb{L}}}(x,\mu)+\varkappa\tilde{\sigma}_-\tilde{\tilde{\sigma}}_+\right)\tilde{\mathbb{T}}_{x_1}^{x}(\lambda)\tilde{\tilde{\mathbb{T}}}_{x_1}^{x}(\mu):, \qquad (2.2.14)$$

which, obviously, is equivalent with the Cauchy problem (2.2.5) for (2.2.3). The investigation of the product $\tilde{\tilde{\mathbb{T}}}_{x_1}^{x_2}(\mu)\tilde{\mathbb{T}}_{x_1}^{x_2}(\lambda)$ is carried out analogously. Thus, Lemma 2.2.1 is proved.

The proved lemma allows the reduction of the proof of Theorem 3 to the verification of the equation

$$R(\lambda-\mu)\mathcal{L}(x;\lambda,\mu)=\mathcal{L}'(x;\lambda,\mu)R(\lambda-\mu)\ . \tag{2.2.15}$$

In fact, by virtue of (2.2.3-5) and (2.2.15), the quantities $\tilde{\mathbb{T}}^{x_2}_{x_1}(\mu)\mathbb{T}^{x_2}_{x_1}(\lambda)$ and $R(\lambda-\mu)\mathbb{T}^{x_2}_{x_1}(\lambda)\tilde{\mathbb{T}}^{x_2}_{x_1}(\mu)\cdot R^{-1}(\lambda-\mu)$ satisfy the same differential equation with the same initial condition. We note that in establishing this fact it is extraordinarily important that R is a numerical matrix, whose matrix elements commute with the matrix elements of $\mathbb{T}^{x_2}_{x_1}$.

Equation (2.2.15), to which the proof of Theorem 3 is reduced, is easily verified directly.

We discuss in the conclusion of this section the connection of the formula (2.2.1) which we have obtained with the result (1.2.11) of Theorem 1. For this it is convenient to introduce in the commutation relations (2.1.2) the Planck constant $\hbar$:

$$[\Psi(x),\Psi^{+}(y)]=\hbar\,\delta(x-y)\ . \tag{2.2.16}$$

Then the R-matrix assumes the form

$$R(\lambda)=I-\frac{i\varkappa\hbar}{\lambda}P=I+i\hbar r(\lambda)\ . \tag{2.2.17}$$

We shall show that in the quasiclassical limit $\hbar\to 0$ (2.2.1) goes into (1.2.7). In fact, in view of (2.2.17), (1.2.1) can be written in the form

$$[\mathbb{T}^{x_2}_{x_1}(\lambda),\tilde{\mathbb{T}}^{x_2}_{x_1}(\mu)]=-i\hbar\left(r(\lambda-\mu)\mathbb{T}^{x_2}_{x_1}(\lambda)\tilde{\mathbb{T}}^{x_2}_{x_1}(\mu)-\tilde{\mathbb{T}}^{x_2}_{x_1}(\mu)\mathbb{T}^{x_2}_{x_1}(\lambda)r(\lambda-\mu)\right). \tag{2.2.18}$$

Using the fact that as $\hbar\to 0$, $\mathbb{T}^{x_2}_{x_1}(\lambda)$ goes into the classical transition matrix $T^{x_2}_{x_1}(\lambda)$, and the commutator goes into the Poisson bracket

$$[\ ,\ ]\longrightarrow -i\hbar\{\ ,\ \} \tag{2.2.19}$$

and retaining in (2.2.18) terms of order $\hbar$, we arrive at (1.2.11).

We note that this result is also valid for R-matrices of more general form, for which (2.2.17) is false (see [29]). In the general case it is replaced by the relation

$$R(\lambda)=I+i\hbar r(\lambda)+O(\hbar^2) \tag{2.2.20}$$

or

$$r(\lambda)=-i\frac{\partial}{\partial\hbar}\Big|_{\hbar=0}R(\lambda,\hbar)\ . \tag{2.2.21}$$

2.3. Passage to an Infinite Interval

This section is devoted to the derivation of the most important result of the present paper, the commutation relations between the matrix elements of the quantum transition matrix for an infinite interval.

Analogously to the way the quantum transition matrix $\mathbb{T}^{x_2}_{x_1}(\lambda)$ was introduced in Sec. 2.1 for a finite interval, we define quantum transition matrices $\mathbb{T}_-(x,\lambda)$, $\mathbb{T}_+(x,\lambda)$ for the semi-infinite and $\mathbb{T}(\lambda)$ for the infinite intervals by the formulas

$$\mathbb{T}_\pm(x,\lambda)=:\mathbb{T}_\pm(x,\lambda):\ , \tag{2.3.1}$$

$$\mathbb{T}(\lambda)=:\mathbb{T}(\lambda):\ .$$

The properties of the matrices $\mathbb{T}_\pm(x,\lambda)$ and $\mathbb{T}(\lambda)$ follow directly from (2.3.1), completely analogously to the corresponding properties of the matrix $\mathbb{T}_{x_1}^{x_2}(\lambda)$, established in Sec. 2.1. The matrices $\mathbb{T}_\pm(x,\lambda)$ and $\mathbb{T}(\lambda)$ have the same symmetry as $\mathbb{T}_{x_1}^{x_2}(\lambda)$ (for the notation for the matrix elements $\mathbb{T}_\pm(x,\lambda)$ and $\mathbb{T}(\lambda)$, see Points 6-8 in the Supplement) $\mathbb{T}_-(x,\lambda)$ satisfies the differential equation

$$\frac{\partial}{\partial x}\mathbb{T}_-(x,\lambda) = :\mathbb{L}(x,\lambda)\,\mathbb{T}_-(x,\lambda):, \tag{2.3.2}$$

and $\mathbb{T}_+(x,\lambda)$ the equation

$$\frac{\partial}{\partial x}\mathbb{T}_+(x,\lambda) = -:\mathbb{T}_+(x,\lambda)\,\mathbb{L}(x,\lambda):. \tag{2.3.3}$$

For $\mathbb{T}_\pm(x,\lambda)$ and $\mathbb{T}(\lambda)$ the quantum analogs of the integral equations (1.1.23-28) are valid, which we shall write down later insofar as they are needed.

We recall again that for now we are considering the matrix elements of $\mathbb{T}_\pm(x,\lambda)$ and $\mathbb{T}(\lambda)$ which are formal series of the form (2.1.32-35) (we shall not write these series down here, since they are obtained from (2.1.32-35) by eliminating x_1 and/or x_2).

Now we can formulate the basic result of the present paper, Theorem 4.

THEOREM 4. The commutation relations between the matrix elements of the quantum transition matrices $\mathbb{T}_\pm(x,\lambda)$ and $\mathbb{T}(\lambda)$ can be written for real λ and μ in the following form:

$$R(\lambda-\mu)\tilde{\mathbb{T}}_-(x,\lambda)\tilde{\tilde{\mathbb{T}}}_-(x,\mu)\Big(1-\frac{i\varkappa}{\lambda-\mu+i0}\tilde{\sigma}_-\tilde{\tilde{\sigma}}_+\Big) = \tilde{\tilde{\mathbb{T}}}_-(x,\mu)\tilde{\mathbb{T}}_-(x,\lambda)\Big(1+\frac{i\varkappa}{\lambda-\mu-i0}\tilde{\sigma}_+\tilde{\tilde{\sigma}}_-\Big)R(\lambda-\mu), \tag{2.3.4}$$

$$R(\lambda-\mu)\Big(1+\frac{i\varkappa}{\lambda-\mu-i0}\tilde{\sigma}_-\tilde{\tilde{\sigma}}_+\Big)\tilde{\mathbb{T}}_+(x,\lambda)\tilde{\tilde{\mathbb{T}}}_+(x,\mu) = \Big(1-\frac{i\varkappa}{\lambda-\mu+i0}\tilde{\sigma}_+\tilde{\tilde{\sigma}}_-\Big)\tilde{\tilde{\mathbb{T}}}_+(x,\mu)\tilde{\mathbb{T}}_+(x,\lambda)R(\lambda-\mu), \tag{2.3.5}$$

$$R(\lambda-\mu)\Big(1+\frac{i\varkappa}{\lambda-\mu-i0}\tilde{\sigma}_-\tilde{\tilde{\sigma}}_+\Big)\tilde{\mathbb{T}}(\lambda)\tilde{\tilde{\mathbb{T}}}(\mu)\Big(1-\frac{i\varkappa}{\lambda-\mu+i0}\tilde{\sigma}_-\tilde{\tilde{\sigma}}_+\Big) =$$

$$= \Big(1-\frac{i\varkappa}{\lambda-\mu+i0}\tilde{\sigma}_+\tilde{\tilde{\sigma}}_-\Big)\tilde{\tilde{\mathbb{T}}}(\mu)\tilde{\mathbb{T}}(\lambda)\Big(1+\frac{i\varkappa}{\lambda-\mu-i0}\tilde{\sigma}_+\tilde{\tilde{\sigma}}_-\Big)R(\lambda-\mu). \tag{2.3.6}$$

Proof. First we prove (2.3.4). The proof will be based on the study of the asymptotic behavior of the products $\tilde{\mathbb{T}}_{x_1}^{x_2}(\lambda)\tilde{\tilde{\mathbb{T}}}_{x_1}^{x_2}(\mu)$ and $\tilde{\tilde{\mathbb{T}}}_{x_1}^{x_2}(\mu)\tilde{\mathbb{T}}_{x_1}^{x_2}(\lambda)$ as $x_1 \to -\infty$. Here we shall devote basic attention to the formal-algebraic side, not going into the analytic justifications of our calculations and making it our goal to give as simply and rapidly as possible a method of calculating the desired commutation relations.

It was proved in Sec. 2.2 that the products $\tilde{\mathbb{T}}_{x_1}^{x_2}(\lambda)\tilde{\tilde{\mathbb{T}}}_{x_1}^{x_2}(\mu)$ and $\tilde{\tilde{\mathbb{T}}}_{x_1}^{x_2}(\mu)\tilde{\mathbb{T}}_{x_1}^{x_2}(\lambda)$ satisfy, respectively, the differential equations (2.2.3) and (2.2.4). The operators $\mathcal{L}$ and $\mathcal{L}'$ (2.2.6-7) figuring in (2.2.3) and (2.2.4) do not coincide with the sum of the operators $\tilde{\mathbb{L}}(x,\lambda)$ and $\tilde{\tilde{\mathbb{L}}}(x,\mu)$, as would be so in the classical case, but differ from it by the summands $\varkappa\tilde{\sigma}_-\tilde{\tilde{\sigma}}_+$ and $\varkappa\tilde{\sigma}_+\tilde{\tilde{\sigma}}_-$, respectively, arising from the noncommutativity of the quantum operators. In connection with this, in the quantum case in describing the asymptotic behavior of the products $\tilde{\mathbb{T}}_{x_1}^{x_2}(\lambda)\tilde{\tilde{\mathbb{T}}}_{x_1}^{x_2}(\mu)$ and $\tilde{\tilde{\mathbb{T}}}_{x_1}^{x_2}(\mu)\tilde{\mathbb{T}}_{x_1}^{x_2}(\lambda)$ as $x_1 \to -\infty$ or $x_2 \to +\infty$ the role of the classical matrix $E(x;\lambda,\mu)$ (see Sec. 1.3) will be played, respectively, by the matrices $\mathcal{E}(x;\lambda,\mu)$ and $\mathcal{E}'(x;\lambda,\mu)$:

$$\mathcal{E}(x;\lambda,\mu) = \exp \mathcal{L}_0(\lambda,\mu)\, x = \begin{pmatrix} e^{-i\frac{\lambda+\mu}{2}x}, & 0, & 0, & 0 \\ 0, & e^{i\frac{\mu-\lambda}{2}x}, & 0, & 0 \\ 0, & 2\varkappa\frac{\sin\frac{\lambda-\mu}{2}x}{\lambda-\mu}, & e^{i\frac{\lambda-\mu}{2}x}, & 0 \\ 0, & 0, & 0, & e^{i\frac{\lambda+\mu}{2}x} \end{pmatrix}, \tag{2.3.7}$$

$$\mathcal{E}'(x;\lambda,\mu) = \exp \mathcal{L}_0'(\lambda,\mu)\, x = \begin{pmatrix} e^{-i\frac{\lambda+\mu}{2}x}, & 0, & 0, & 0 \\ 0, & e^{i\frac{\mu-\lambda}{2}x}, & 2\varkappa\frac{\sin\frac{\lambda-\mu}{2}x}{\lambda-\mu}, & 0 \\ 0, & 0, & e^{i\frac{\lambda-\mu}{2}x}, & 0 \\ 0, & 0, & 0, & e^{i\frac{\lambda+\mu}{2}x} \end{pmatrix}, \tag{2.3.8}$$

where $\mathcal{L}_0(\lambda,\mu)$ and $\mathcal{L}_0'(\lambda,\mu)$ are the "asymptotic" (as $|x|\to\infty$) values of the operators $\mathcal{L}(x;\lambda,$ and $\mathcal{L}'(x;\lambda,\mu)$, respectively,

$$\mathcal{L}_0(\lambda,\mu) = -i\frac{\lambda}{2}\tilde{\sigma}_3 - i\frac{\mu}{2}\tilde{\tilde{\sigma}}_3 + \varkappa\tilde{\sigma}_-\tilde{\tilde{\sigma}}_+ , \tag{2.3.9}$$

$$\mathcal{L}_0'(\lambda,\mu) = -i\frac{\lambda}{2}\tilde{\sigma}_3 - i\frac{\mu}{2}\tilde{\tilde{\sigma}}_3 + \varkappa\tilde{\sigma}_+\tilde{\tilde{\sigma}}_- . \tag{2.3.10}$$

We note that by virtue of (2.2.15) one has

$$R(\lambda-\mu)\,\mathcal{L}_0(\lambda,\mu) = \mathcal{L}_0'(\lambda,\mu)\,R(\lambda-\mu). \tag{2.3.11}$$

From (2.3.11) in combination with (2.3.7-8) follows the analogous equation for the matrices $\mathcal{E}(x;\lambda,\mu)$ and $\mathcal{E}'(x;\lambda,\mu)$:

$$R(\lambda-\mu)\,\mathcal{E}(x;\lambda,\mu) = \mathcal{E}'(x;\lambda,\mu)\,R(\lambda-\mu). \tag{2.3.12}$$

Now we concern ourselves with the investigation of the asymptotic behavior of the product $\tilde{\mathbb{T}}_{x_1}^{x_2}(\lambda)\tilde{\tilde{\mathbb{T}}}_{x_1}^{x_2}(\mu)$ as $x_1\to-\infty$. We note, first of all, that the differential equation (2.2.3), which, one can rewrite using the notation (2.3.9-10) and (2.1.31) in the form

$$\frac{\partial}{\partial x_2}\left(\tilde{\mathbb{T}}_{x_1}^{x_2}(\lambda)\tilde{\tilde{\mathbb{T}}}_{x_1}^{x_2}(\mu)\right) = :\left(\mathcal{L}_0(\lambda,\mu) + \tilde{\mathbb{V}}(x_2) + \tilde{\tilde{\mathbb{V}}}(x_2)\right)\left(\tilde{\mathbb{T}}_{x_1}^{x_2}(\lambda)\tilde{\tilde{\mathbb{T}}}_{x_1}^{x_2}(\mu)\right): , \tag{2.3.13}$$

is equivalent under the initial condition (2.2.5) with the Volterra integral equation

$$\left(\tilde{\mathbb{T}}_{x_1}^{x_2}(\lambda)\tilde{\tilde{\mathbb{T}}}_{x_1}^{x_2}(\mu)\right) = \mathcal{E}(x_2-x_1;\lambda,\mu) + \int_{x_1}^{x_2} dx :\left(\tilde{\mathbb{T}}_{x}^{x_2}(\lambda)\tilde{\tilde{\mathbb{T}}}_{x}^{x_2}(\mu)\right)\left(\tilde{\mathbb{V}}(x) + \tilde{\tilde{\mathbb{V}}}(x)\right): \mathcal{E}(x-x_1;\lambda,\mu). \tag{2.3.14}$$

We introduce into consideration the limit

$$\mathcal{T}(x;\lambda,\mu) = \lim_{x_1\to-\infty}\left(\tilde{\mathbb{T}}_{x_1}^{x}(\lambda)\tilde{\tilde{\mathbb{T}}}_{x_1}^{x}(\mu)\right)\mathcal{E}(x_1;\lambda,\mu). \tag{2.3.15}$$

Substituting (2.3.14) in (2.3.15), we get for $\mathcal{T}(x;\lambda,\mu)$ the following integral representation

$$\mathcal{T}(x;\lambda,\mu) = \mathcal{E}(x;\lambda,\mu) + \int_{-\infty}^{x} d\eta :\left(\tilde{\mathbb{T}}_{\eta}^{x}(\lambda)\tilde{\tilde{\mathbb{T}}}_{\eta}^{x}(\mu)\right)\left(\tilde{\mathbb{V}}(\eta) + \tilde{\tilde{\mathbb{V}}}(\eta)\right): \mathcal{E}(\eta;\lambda,\mu). \tag{2.3.16}$$

We note that $\mathcal{T}(x;\lambda,\mu)$ as before satisfies the differential equation (2.2.3):

$$\frac{d}{dx}\mathcal{T}(x;\lambda,\mu) = \vdots\, \mathcal{L}(x;\lambda,\mu)\,\mathcal{T}(x;\lambda,\mu) \vdots\, . \qquad (2.3.17)$$

On the other hand, arguing exactly as in the proof of Lemma 2.2.1 in Sec. 2.2, one can see that the product $\widetilde{\mathbb{T}}_-(x,\lambda)\widetilde{\widetilde{\mathbb{T}}}_-(x,\mu)$, which we denote by $\mathcal{T}_-(x;\lambda,\mu)$, satisfies the same differential equation

$$\frac{d}{dx}\mathcal{T}_-(x;\lambda,\mu) = \vdots\, \mathcal{L}(x;\lambda,\mu)\,\mathcal{T}_-(x;\lambda,\mu) \vdots\, . \qquad (2.3.18)$$

Consequently, the quantities $\mathcal{T}(x;\lambda,\mu)$ and $\mathcal{T}_-(x;\lambda,\mu)$ can differ only by some matrix factor $\mathbb{C}(\lambda,\mu)$:

$$\mathcal{T}_-(x;\lambda,\mu) = \mathcal{T}(x;\lambda,\mu)\,\mathbb{C}(\lambda,\mu)\, . \qquad (2.3.19)$$

We note the obvious similarity of our arguments with the arguments made in proving Theorem 2 (Sec. 1.3). As also in Sec. 1.3, we find the matrix $\mathbb{C}(\lambda,\mu)$, comparing the asymptotics of $\mathcal{T}(x;\lambda,\mu)$ and $\mathcal{T}_-(x;\lambda,\mu)$ as $x\to-\infty$. The asymptotics of $\mathcal{T}(x;\lambda,\mu)$ as $x\to-\infty$ are easily determined from the integral representation (2.3.16):

$$\mathcal{T}(x;\lambda,\mu) \underset{x\to-\infty}{\sim} \mathcal{E}(x;\lambda,\mu)\, . \qquad (2.3.20)$$

It remains to investigate the asymptotics of $\mathcal{T}_-(x;\lambda,\mu)$. For this we use the quantum analog of the integral representation (1.1.24):

$$\widetilde{\mathbb{T}}_-(x,\lambda) = \tilde{e}(x,\lambda) + \int_{-\infty}^{x} d\eta \,\vdots\, \widetilde{\mathbb{T}}_\eta^x(\lambda)\widetilde{\mathbb{V}}(\eta) \,\vdots\, \tilde{e}(\eta,\lambda) \qquad (2.3.21)$$

or

$$\widetilde{\widetilde{\mathbb{T}}}_-(x,\mu) = \tilde{\tilde{e}}(x,\mu) + \int_{-\infty}^{x} d\eta \,\vdots\, \widetilde{\widetilde{\mathbb{T}}}_\eta^x(\mu)\widetilde{\widetilde{\mathbb{V}}}(\eta) \,\vdots\, \tilde{\tilde{e}}(\eta,\mu)\, . \qquad (2.3.22)$$

Substituting (2.3.21) and (2.3.22) in the product $\widetilde{\mathbb{T}}_-(x,\lambda)\widetilde{\widetilde{\mathbb{T}}}_-(x,\mu)$, we get

$$\mathcal{T}_-(x;\lambda,\mu) = \mathrm{E}(x;\lambda,\mu) + \int_{-\infty}^{x} d\eta \,\vdots\, \widetilde{\mathbb{T}}_\eta^x(\lambda)\widetilde{\mathbb{V}}(\eta) \,\vdots\, \tilde{e}(\eta,\lambda)\tilde{\tilde{e}}(x,\mu) +$$

$$+\int_{-\infty}^{x} d\eta \,\vdots\, \widetilde{\widetilde{\mathbb{T}}}_\eta^x(\mu)\widetilde{\widetilde{\mathbb{V}}}(\eta) \,\vdots\, \tilde{e}(x,\lambda)\tilde{\tilde{e}}(\eta,\mu) + \int_{-\infty}^{x} d\eta_1 \int_{-\infty}^{x} d\eta_2 \left[i\varkappa \Psi^+(\eta_1)\widetilde{\mathbb{T}}_{\eta_1}^x(\lambda)\tilde{\sigma}_+ - i\widetilde{\mathbb{T}}_{\eta_1}^x(\lambda)\tilde{\sigma}_-\Psi(\eta_1)\right] \times$$

$$\times\left[i\varkappa \Psi^+(\eta_2)\widetilde{\widetilde{\mathbb{T}}}_{\eta_2}^x(\mu)\tilde{\tilde{\sigma}}_+ - i\widetilde{\widetilde{\mathbb{T}}}_{\eta_2}^x(\mu)\tilde{\tilde{\sigma}}_-\Psi(\eta_2)\right]\tilde{e}(\eta_1,\lambda)\tilde{\tilde{e}}(\eta_2,\mu)\, . \qquad (2.3.23)$$

Here we again use the notation introduced in Sec. 1.3 $\mathrm{E}(x;\lambda,\mu) = \tilde{e}(x,\lambda)\tilde{\tilde{e}}(x,\mu)$.

The fourth summand in (2.3.23) can be transformed completely analogously to the way the corresponding summand in (2.2.11) was transformed. Omitting the corresponding calculations, which coincide almost identically with the chain of calculations (2.2.11-14), we give only the final result:

$$\mathcal{T}_-(x;\lambda,\mu) = \mathrm{E}(x;\lambda,\mu) + \int_{-\infty}^{x} d\eta \,\vdots\, (\widetilde{\mathbb{T}}_\eta^x(\lambda)\widetilde{\widetilde{\mathbb{T}}}_\eta^x(\mu))(\widetilde{\mathbb{V}}(\eta)+\widetilde{\widetilde{\mathbb{V}}}(\eta)+\varkappa\tilde{\sigma}_-\tilde{\tilde{\sigma}}_+) \,\vdots\, \mathrm{E}(\eta;\lambda,\mu)\, . \qquad (2.3.24)$$

In order to find the asymptotics of $\mathcal{T}_-(x;\lambda,\mu)$ as $x\to-\infty$, we note that the product $\widetilde{\mathbb{T}}_\eta^x(\lambda)\widetilde{\widetilde{\mathbb{T}}}_\eta^x(\mu)$ has the following asymptotics:

$$\widetilde{\mathbb{T}}^{x}_{\eta}(\lambda)\,\widetilde{\widetilde{\mathbb{T}}}^{x}_{\eta}(\mu)\underset{x,\eta\to-\infty}{\sim}\mathcal{E}(x-\eta;\lambda,\mu) \tag{2.3.25}$$

as $x,\eta\to-\infty$. Formula (2.3.25) follows from (2.3.14). Substituting (2.3.25) in (2.3.24) and discarding terms which decrease as $x\to-\infty$, we get

$$\mathcal{T}_{-}(x;\lambda,\mu)\underset{x\to-\infty}{\sim}E(x;\lambda,\mu)+\int_{-\infty}^{x}d\eta\,\mathcal{E}(x-\eta;\lambda,\mu)\,\varkappa\,\tilde{\sigma}_{-}\,\tilde{\tilde{\sigma}}_{+}\,E(\eta;\lambda,\mu). \tag{2.3.26}$$

Calculating the integral in (2.3.26), we arrive at the following result:

$$\mathcal{I}(x;\lambda,\mu)\underset{x\to-\infty}{\sim}E(x;\lambda,\mu)+\frac{i\varkappa}{\lambda-\mu+i0}e^{i\frac{\mu-\lambda}{2}x}\tilde{\sigma}_{-}\tilde{\tilde{\sigma}}_{+}=\mathcal{E}(x;\lambda,\mu)\,\mathbb{C}(\lambda,\mu), \tag{2.3.27}$$

where

$$\mathbb{C}(\lambda,\mu)=I_4+\frac{i\varkappa}{\lambda-\mu+i0}\tilde{\sigma}_{-}\tilde{\tilde{\sigma}}_{+}. \tag{2.3.28}$$

Rewriting (2.3.19) in the form

$$\mathcal{T}(x;\lambda,\mu)=\mathcal{I}(x;\lambda,\mu)\,\mathbb{C}^{-1}(\lambda,\mu) \tag{2.3.29}$$

and using the fact that

$$\mathbb{C}^{-1}(\lambda,\mu)=I_4-\frac{i\varkappa}{\lambda-\mu-i0}\tilde{\sigma}_{-}\tilde{\tilde{\sigma}}_{+} \tag{2.3.30}$$

(since $\sigma_-^2=\sigma_+^2=0$), and recalling the definition of $\mathcal{T}(x;\lambda,\mu)$ and $\mathcal{I}(x;\lambda,\mu)$, we get finally

$$\lim_{x_1\to-\infty}\left(\widetilde{\mathbb{T}}^{x}_{x_1}(\lambda)\widetilde{\widetilde{\mathbb{T}}}^{x}_{x_1}(\mu)\right)\mathcal{E}(x;\lambda,\mu)=\widetilde{\mathbb{T}}_{-}(x,\lambda)\widetilde{\widetilde{\mathbb{T}}}_{-}(x,\mu)\left(I_4-\frac{i\varkappa}{\lambda-\mu+i0}\tilde{\sigma}_{-}\tilde{\tilde{\sigma}}_{+}\right). \tag{2.3.31}$$

The analogous formula for $\widetilde{\widetilde{\mathbb{T}}}^{x}_{x_1}(\mu)\widetilde{\mathbb{T}}^{x}_{x_1}(\lambda)$ is obtained by interchanging in (2.3.31) $\lambda\leftrightarrow\mu$ and $\sim\;\leftrightarrow\;\approx$:

$$\lim_{x_1\to-\infty}\left(\widetilde{\widetilde{\mathbb{T}}}^{x}_{x_1}(\mu)\widetilde{\mathbb{T}}^{x}_{x_1}(\lambda)\right)\mathcal{E}'(x;\lambda,\mu)=\widetilde{\widetilde{\mathbb{T}}}_{-}(x,\mu)\widetilde{\mathbb{T}}_{-}(x,\lambda)\left(I_4+\frac{i\varkappa}{\lambda-\mu-i0}\tilde{\sigma}_{+}\tilde{\tilde{\sigma}}_{-}\right). \tag{2.3.32}$$

Now everything is ready for getting the commutation relation (2.3.4). For this, we multiply (2.2.1) on the right by $\mathcal{E}(x_1;\lambda,\mu)$ and use (2.3.12), and we get

$$R(\lambda-\mu)\widetilde{\mathbb{T}}^{x}_{x_1}(\lambda)\widetilde{\widetilde{\mathbb{T}}}^{x}_{x_1}(\mu)\mathcal{E}(x_1;\lambda,\mu)=\widetilde{\widetilde{\mathbb{T}}}^{x}_{x_1}(\mu)\widetilde{\mathbb{T}}^{x}_{x_1}(\lambda)\mathcal{E}'(x_1;\lambda,\mu)R(\lambda-\mu). \tag{2.3.33}$$

Passing in (2.3.33) to the limit as $x_1\to-\infty$, according to (2.3.31) and (2.3.32) we get (2.3.4).

Equation (2.3.5) is proved completely analogously. Combining (2.3.4) and (2.3.5) and using the obvious equation

$$\mathbb{T}(\lambda)=\mathbb{T}_{+}(x,\lambda)\mathbb{T}_{-}(x,\lambda), \tag{2.3.34}$$

we get (2.3.6), thus completing the proof of Theorem 4.

We proceed to discuss the results obtained. We note first of all that a calculation, completely analogous to that given at the end of Sec. 2.2, allows us to get in the classical limit of (2.3.4) the formula (1.3.1) and analogous formulas for T_+ and T. Thus, the results of Theorem 4 generalize the results of Theorem 2 to the quantum case.

A summary of the commutation relations between the matrix elements of the matrices $\mathbb{T}_{\pm}(x,\lambda)$ and $\mathbb{T}(\lambda)$ is given in the Supplement (formulas (S31-48)). In order to show how one gets these formulas, we calculate, for example, the commutation relation between the operators $A_-(x,\mu)$ and $B_-^+(x,\lambda)$ (formula (S34)). For this we write in (2.3.4) the matrix element found at the intersection of the first row and third column:

$$\left(1-\frac{i\varkappa}{\lambda-\mu}\right)B_-^+(x,\lambda)A_-(x,\mu)=-\frac{i\varkappa}{\lambda-\mu}B_-^+(x,\mu)A_-(x,\lambda)+\frac{i\varkappa}{\lambda-\mu-i0}B_-^+(x,\mu)A_-(x,\lambda)+A_-(x,\mu)B_-^+(x,\lambda). \quad (2.3.35)$$

Regrouping terms, we get

$$A_-(x,\mu)B_-^+(x,\lambda)=B_-^+(x,\lambda)A_-(x,\mu)-\frac{i\varkappa}{\lambda-\mu-i0}B_-^+(x,\mu)A_-(x,\lambda)+\frac{i\varkappa}{\lambda-\mu}\left[B_-^+(x,\mu)A_-(x,\lambda)-B_-^+(x,\lambda)A_-(x,\mu)\right]. \quad (2.3.36)$$

We note that in the denominator of the third term on the right side of (2.3.36) it is unnecessary to regularize for $\lambda=\mu$, since the numerator here vanishes. This means that we can choose the regularization of the denominator arbitrarily, in particular, replace $(\lambda-\mu)^{-1}$ by $(\lambda-\mu-i0)^{-1}$ (see the analogous argument in Sec. 1.3 in connection with (1.3.11)). Then terms containing the product $B_-^+(x,\mu)A_-(x,\lambda)$, are preserved, and we get (S34).

Analogously one also gets the remaining formulas (S31-48). The calculations here, however, turn out to be rather complicated. It turns out that if one is interested in commutation relations only for $\lambda\neq\mu$ then (2.3.4-6) can be essentially simplified.

In fact, for $\lambda\neq\mu$ the regularization of $\pm i0$ in the denominator $(\lambda-\mu)$ is inessential, and we can divide, for example, (2.3.4) on the right by $\left(I-\frac{i\varkappa}{\lambda-\mu+i0}\tilde{\sigma}_-\tilde{\tilde{\sigma}}_+\right)$, obtaining here the following equation:

$$R(\lambda-\mu)\tilde{\mathbb{T}}_-(x,\lambda)\tilde{\tilde{\mathbb{T}}}_-(x,\mu)=\tilde{\tilde{\mathbb{T}}}_-(x,\mu)\tilde{\mathbb{T}}_-(x,\lambda)R_0(\lambda-\mu), \quad (2.3.37)$$

where

$$R_0(\lambda)=\left(1+\frac{i\varkappa}{\lambda}\tilde{\sigma}_+\tilde{\tilde{\sigma}}_-\right)R(\lambda)\left(1+\frac{i\varkappa}{\lambda}\tilde{\sigma}_-\tilde{\tilde{\sigma}}_+\right)=\begin{pmatrix}1-\frac{i\varkappa}{\lambda}, & 0, & 0, & 0\\ 0, & 1+\frac{\varkappa^2}{\lambda^2}, & 0, & 0\\ 0, & 0, & 1, & 0\\ 0, & 0, & 0, & 1-\frac{i\varkappa}{\lambda}\end{pmatrix}. \quad (2.3.38)$$

We note that without making any preliminary statement about $\lambda\neq\mu$, we would get in (2.3.38) a meaningless product of generalized functions of the form $(\lambda-\mu-i0)^{-1}(\lambda-\mu+i0)^{-1}$. Analogously from (2.3.5) and (2.3.6) one gets

$$R_0(\lambda-\mu)\tilde{\mathbb{T}}_+(x,\lambda)\tilde{\tilde{\mathbb{T}}}_+(x,\mu)=\tilde{\tilde{\mathbb{T}}}_+(x,\mu)\tilde{\mathbb{T}}_+(x,\lambda)R(\lambda-\mu), \quad (2.3.39)$$

$$R_0(\lambda-\mu)\tilde{\mathbb{T}}(\lambda)\tilde{\tilde{\mathbb{T}}}(\mu)=\tilde{\tilde{\mathbb{T}}}(\mu)\tilde{\mathbb{T}}(\lambda)R_0(\lambda-\mu). \quad (2.3.40)$$

Equation (2.3.40) reproduces a result obtained by Faddeev in [16].

We note one interesting thing. Although the commutation relations for the matrix elements of $T_{\pm}(x,\lambda)$, obtained from (2.3.37) and (2.3.39), are only defined for $\lambda \neq \mu$, we can, using the analytic properties of the matrix elements of the matrices $T_{\pm}(x,\lambda)$, extend the corresponding commutation relations to the real axis and find thus their proper regularizations for $\lambda = \mu$. We clarify what has been said with an example.

We write the matrix element lying at the intersection of the first row and second column in (2.3.37):

$$\left(1-\frac{i\varkappa}{\lambda-\mu}\right) B_-^+(x,\lambda) A_-(x,\mu) = A_-(x,\mu) B_-^+(x,\lambda). \tag{2.3.41}$$

By virtue of (2.3.1) the operator functions $A_-(x,\mu)$ and $B_-^+(x,\lambda)$ have the same analytic properties as the corresponding classical quantities $a_-(x,\mu)$ and $\bar{b}_-(x,\lambda)$. Thus, (2.3.41) is initially defined for $\mathrm{Im}\,\mu > 0$ and $\mathrm{Im}\,\lambda < 0$. When λ and μ leave the real axis we must regularize the denominator $(\lambda-\mu)$ in (2.3.41) in the following way:

$$A_-(x,\mu) B_-^+(x,\lambda) = \left(1-\frac{i\varkappa}{\lambda-\mu-i0}\right) B_-^+(x,\lambda) A_-(x,\mu), \tag{2.3.42}$$

thus getting the proper commutation relation

In the same way from (2.3.37) and (2.3.39) one can reproduce all the commutation relations (S31-42). For the quantum transition matrix $T(\lambda)$ on the infinite interval, analogously from (2.3.40) one can reproduce the commutation relations (S43-46), i.e., those commutation relations, in which at least one factor admits analytic continuation to the real axis. Exceptions are the commutation relations (S47-48), since the functions $B(\lambda)$ and $B^+(\lambda)$ are defined only for real λ. These commutation relations can be obtained only from (2.3.6).

The commutation relation (S48) between $B(\lambda)$ and $B^+(\mu)$ deserves special commentary. We write it separately:

$$B(\lambda)B^+(\mu) = \left(1-\frac{i\varkappa}{\lambda-\mu+i0}\right)\left(1-\frac{i\varkappa}{\lambda-\mu-i0}\right)B^+(\mu)B(\lambda) + 2\pi\delta(\lambda-\mu)A^+(\lambda)A(\lambda). \tag{2.3.43}$$

On the right side of (2.3.43) we see, generally speaking, the undefined product of generalized functions $(\lambda-\mu+i0)^{-1}(\lambda-\mu-i0)^{-1}$. This indicates the highly singular operator character of $B(\lambda)$ and $B^+(\mu)$. It turns out, however, that one can be saved from the singularities in the ratio (2.3.46) by regularizing the operators $B(\lambda)$ and $B^+(\mu)$ in a definite way.

Namely, we define operators $\Phi(\lambda)$ and $\Phi^+(\lambda)$ by

$$\Phi(\lambda) = (2\pi A^+(\lambda)A(\lambda))^{-1/2} B(\lambda),$$
$$\Phi^+(\lambda) = B^+(\lambda)(2\pi A^+(\lambda)A(\lambda))^{-1/2}, \tag{2.3.44}$$

and we formulate the following proposition.

Proposition 2.3.1. The operators $\Phi(\lambda)$ and $\Phi^+(\lambda)$ introduced by (2.3.44) satisfy the canonical commutation relations:

$$[\Phi(\lambda), \Phi(\mu)] = [\Phi^+(\lambda), \Phi^+(\mu)] = 0, \quad [\Phi(\lambda), \Phi^+(\mu)] = \delta(\lambda-\mu). \tag{2.3.45}$$

Proof. We note first of all that from (S46)

$$A(\lambda)B^{+}(\mu) = \left(1+\frac{i\varkappa}{\lambda-\mu+i0}\right)B^{+}(\mu)A(\lambda) \tag{2.3.46}$$

and from (S43) follows the analogous relation

$$f(A(\lambda))\,B^{+}(\mu) = B^{+}(\mu)\,f\left[\left(1+\frac{i\varkappa}{\lambda-\mu+i0}\right)A(\lambda)\right] \tag{2.3.47}$$

for any analytic function $f(\lambda)$. In fact, from (2.3.46) the validity of (2.3.47) follows directly for polynomial functions $f(\lambda)$, and consequently also for analytic functions $f(\lambda)$, considered as infinite power series in λ (we recall that all our arguments carry formal algebraic character). We can also extend (2.3.47) to functions $f(\lambda)$ of the form $f(\lambda)=\lambda^{-1/2}$, by decomposing them into power series near any point $\lambda=\lambda_0\neq 0$.

Analogous arguments allow us to get from (S45) the following relation

$$B(\lambda)\,f(A(\mu)) = f\left[\left(1-\frac{i\varkappa}{\lambda-\mu-i0}\right)A(\mu)\right]B(\lambda). \tag{2.3.48}$$

Now one can enter upon the proof of Proposition 2.3.1. We derive, e.g., the commutation relation between $\Phi(\lambda)$ and $\Phi^{+}(\mu)$. For this we substitute in the product $\Phi(\lambda)\Phi^{+}(\mu)$ the expression (2.3.44). We get

$$\Phi(\lambda)\Phi^{+}(\mu) = (2\pi)^{-1}\left(A^{+}(\lambda)A(\lambda)\right)^{-1/2}B(\lambda)B^{+}(\mu)\left(A^{+}(\mu)A(\mu)\right)^{-1/2}. \tag{2.3.49}$$

Now using (2.3.43),

$$\Phi(\lambda)\Phi^{+}(\mu) = (2\pi)^{-1}\left(A^{+}(\lambda)A(\lambda)\right)^{-\frac{1}{2}}B(\mu)B(\lambda)\left(A^{+}(\mu)A(\mu)\right)^{-\frac{1}{2}}\left(1+\frac{i\varkappa}{\lambda-\mu+i0}\right)\left(1-\frac{i\varkappa}{\lambda-\mu-i0}\right) +$$

$$+(2\pi)^{-1}\left(A^{+}(\lambda)A(\lambda)\right)^{-\frac{1}{2}}(2\pi)A^{+}(\lambda)A(\lambda)\,\delta(\lambda-\mu)\left(A^{+}(\mu)A(\mu)\right)^{-1/2}. \tag{2.3.50}$$

In order to get the answer needed, it remains to transform (2.3.50), using (2.3.47-48) and the commutativity of $A(\lambda)$ and $A^{+}(\mu)$ (Eqs. (S43-44)):

$$\Phi(\lambda)\Phi^{+}(\mu) = (2\pi)^{-1}B^{+}(\mu)\left(A^{+}(\mu)A(\mu)A^{+}(\lambda)A(\lambda)\right)^{-1/2}B(\lambda)+\delta(\lambda-\mu) = \Phi^{+}(\mu)\Phi(\lambda)+\delta(\lambda-\mu).$$

The remaining relations from (2.3.45) are obtained analogously.

We note to conclude this section that the quantum operators $\Phi(\lambda)$ and $\Phi^{+}(\lambda)$ correspond to the classical variables of action-angle type $\varphi(\lambda)$ and $\overline{\varphi}(\lambda)$ (Eqs. (1.4.8)).

2.4. Spectral Decomposition

In the present section we shall show how, using the commutation relations between the matrix elements of the transition matrix $\mathbb{T}(\lambda)$, one can study the spectra of the integrals of motion of the quantum n.S.e.

First, however, we consider the connection of the quantum method of the inverse problem and the Bethe substitution method. This connection is given by the following proposition.

Proposition 2.4.1. The wave function of an N-particle state

$$|k_1,\ldots,k_N\rangle_B = B^{+}(k_1)\ldots B^{+}(k_N)|0\rangle$$

coincides with the wave function defined by (2.1.11) for the following choice of coefficients $C_{\ell_1\ldots\ell_N}$:

$$C^{(B)}_{\ell_1\ldots\ell_N} = (2\pi)^{N/2} \prod_{r<s} \frac{k_{\ell_r}-k_{\ell_s}+i\varkappa}{k_{\ell_r}-k_{\ell_s}} \tag{2.4.1}$$

The assertion just formulated was first announced in the author's paper [14]. The proof of this fact available to the author is quite complicated and reduces, essentially, to the direct calculation of the result of the action of a segment of the series (2.1.34) on a wave function of the form (2.1.11). An analogous proof was recently published in [27]. Almost simultaneously with [27] there appeared [23], containing an elegant and short proof of an assertion, equivalent with Proposition 2.4.1. Hence we shall not give here the proof of Proposition 2.4.1, but proceed directly to the discussion of the consequences following from it.

Comparing (2.4.1) with (2.1.13), we see that one has

$$|k_1,\ldots,k_N\rangle_B = (2\pi)^{N/2} \prod_{r<s} \left|\frac{k_r-k_s+i\varkappa}{k_r-k_s}\right| |k_1,\ldots,k_N\rangle_{norm}\,, \tag{2.4.2}$$

where $|k_1,\ldots,k_N\rangle_{norm}$ is the N-particle state which is associated by (2.1.4) with the wave function $f_N^{(norm)}(x_1,\ldots,x_N|k_1,\ldots,k_N)$ (Eq. (2.1.11)). Equation (2.4.2) shows that the wave functions generated by operators $B^+(k_j)$ are not normalized on the δ-function. Moreover, the denominators $(k_r-k_s)^{-1}$ make the normalizations of these wave functions so singular that $B^+(\lambda)$ cannot be defined even as a generalized operator-valued function.* This fact allows us to clarify the singular commutation relations (2.4.43).

Now we consider the N-particle state generated by the normalized operators $\Phi^+(k_j)$ (Eqs. (2.3.44)):

$$|k_1,\ldots,k_N\rangle = \Phi^+(k_1)\ldots\Phi^+(k_j)|0\rangle\,. \tag{2.4.3}$$

Substituting (2.3.44) in (2.4.3), we get

$$|k_1,\ldots,k_N\rangle = B^+(k_1)(2\pi A^+(k_1)A(k_1))^{-\frac{1}{2}}\ldots B^+(k_N)(2\pi A^+(k_N)A(k_N))^{-\frac{1}{2}}|0\rangle\,. \tag{2.4.4}$$

With the help of (2.3.47-48) moving the factor $(2\pi A^+(k_j)A(k_j))^{-1/2}$ to the right in (2.4.4) and using the equation

$$A(k_j)|0\rangle = A^+(k_j)|0\rangle = |0\rangle\,, \tag{2.4.5}$$

which follows directly from (2.1.32-33), we arrive at

$$|k_1,\ldots,k_N\rangle = \frac{1}{(2\pi)^{N/2}} \prod_{r<s} \left|\frac{k_r-k_s}{k_r-k_s+i\varkappa}\right| |k_1,\ldots,k_N\rangle_B\,, \tag{2.4.6}$$

or by virtue of (2.4.2),

$$|k_1,\ldots,k_N\rangle = |k_1,\ldots,k_N\rangle_{norm}$$

*A. K. Pogrebkov pointed this out to the author.

Thus, the operators $\Phi^{+}(k_j)$ give birth to the normalized eigenfunctions of the Hamiltonian (2.1.5).

We proceed now to the consideration of a circle of questions connected with the quantum integrals of motion for (2.1.6).

We shall show that, analogously to the classical case, the role of the generating function of the quantum integrals of motion is played by $\ln A(\lambda)$. In fact, from (S43) follows the commutation relation

$$[\ln A(\lambda), \ln A(\mu)] = 0 . \tag{2.4.7}$$

Moreover, substituting in (2.3.47) $f(\lambda) = \ln\lambda$, we get

$$(\ln A(\lambda))B^{+}(\mu) = B^{+}(\mu)\left[\ln A(\lambda) + \ln\left(1 + \frac{i\varkappa}{\lambda - \mu + i0}\right)\right] \tag{2.4.8}$$

or

$$[\ln A(\lambda), \Phi^{+}(\mu)] = \Phi^{+}(\mu) \ln\left(1 + \frac{i\varkappa}{\lambda - \mu + i0}\right) . \tag{2.4.9}$$

We let the operator $\ln A(\lambda)$ act on the N-particle state $|k_1, \ldots, k_N\rangle$

$$\ln A(\lambda) \Phi^{+}(k_1) \ldots \Phi^{+}(k_N) |0\rangle . \tag{2.4.10}$$

Using (2.4.9), we can move $\ln A(\lambda)$ in (2.4.10) to the right. Noting, in addition, that by virtue of (2.4.5) one has

$$\ln A(\lambda) |0\rangle = 0 , \tag{2.4.11}$$

we arrive at the following result. The state $|k_1, \ldots, k_N\rangle$ is an eigenfunction of the operator $\ln A(\lambda)$:

$$\ln A(\lambda) |k_1, \ldots, k_N\rangle = \sum_{j=1}^{N} \ln\left(1 + \frac{i\varkappa}{\lambda - k_j + i0}\right) |k_1, \ldots, k_N\rangle , \tag{2.4.12}$$

where the corresponding eigenvalue is additive with respect to the momenta k_j;

Decomposing both sides of (2.4.12) in powers of λ^{-1}, we get that the state $|k_1, \ldots, k_N\rangle$ is proper also for the operators A_m, defined as coefficients of the expansion

$$\ln A(\lambda) = i\varkappa \sum_{m=1}^{\infty} A_m \lambda^{-m} \tag{2.4.13}$$

The corresponding eigenvalues $c_m(k)$

$$A_m |k_1, \ldots, k_N\rangle = \sum_{j=1}^{N} c_m(k_j) |k_1, \ldots, k_N\rangle \tag{2.4.14}$$

are defined from the expansion

$$\ln\left(1 + \frac{i\varkappa}{\lambda - k}\right) = i\varkappa \sum_{m=1}^{\infty} c_m(k) \lambda^{-m} \tag{2.4.15}$$

and have the form

$$c_m(k) = \frac{k^m - (k - i\varkappa)^m}{im\varkappa} = \sum_{s=1}^{m} \frac{(m-1)!}{s!(m-s)!} (-i\varkappa)^{s-1} k^{m-s+1} . \tag{2.4.16}$$

Unfortunately, for the quantum case a method of calculating the operators A_m, analogous to the method of the Riccati equation (1.4.2-4) in the classical case, is still unknown. Hence, in order to connect the operators A_m with the local integrals of motion for (2.1.6), we have to use a result of Tsvetkov [38]. As shown in [38], the classical integrals of motion $\mathcal{I}_m$ (1.4.2) for (1.1.1) after Wick quantization become quantum self-adjoint operators $\mathcal{I}_m$ in the space $\mathbb{F}$:

$$\mathcal{I}_m = \,:\mathcal{I}_m: \tag{2.4.17}$$

commuting with the Hamiltonian $\mathbb{H}$. In particular,

$$\mathcal{I}_1 = \,:N: = \mathbb{N}\,, \tag{2.4.18}$$

$$\mathcal{I}_2 = \,:P: = \mathbb{P}\,, \tag{2.4.19}$$

$$\mathcal{I}_3 = \,:H: = \mathbb{H}\,. \tag{2.4.20}$$

(For the definition of $\mathbb{N}, \mathbb{P}$, and $\mathbb{H}$, see Eqs. (2.1.5, 7, 8).)

The eigenvalues of the operators $\mathcal{I}_m$ on the states $|k_1,\ldots,k_N\rangle$ have the form

$$\mathcal{I}_m |k_1,\ldots,k_N\rangle = \sum_{j=1}^{N} k_j^m \,|k_1,\ldots,k_N\rangle\,. \tag{2.4.21}$$

Comparing (2.4.21) and (2.4.14), we get the relation

$$A_m = \sum_{s=1}^{m} \frac{(m-1)!}{s!(m-s)!} (-i\varkappa)^{s-1} \mathcal{I}_{m-s+1}\,. \tag{2.4.22}$$

In particular,

$$A_1 = \mathcal{I}_1\,, \tag{2.4.23}$$

$$A_2 = \mathcal{I}_2 - \frac{i\varkappa}{2}\mathcal{I}_1\,, \tag{2.4.24}$$

$$A_3 = \mathcal{I}_3 - i\varkappa \mathcal{I}_2 - \frac{\varkappa^2}{3}\mathcal{I}_1\,. \tag{2.4.25}$$

Formulas (2.4.23-25) allow one to express $\mathbb{N}, \mathbb{P}$, and $\mathbb{H}$ in terms of A_1, A_2, and A_3:

$$\mathbb{N} = A_1\,, \tag{2.4.26}$$

$$\mathbb{P} = A_2 + \frac{i\varkappa}{2}A_1\,, \tag{2.4.27}$$

$$\mathbb{H} = A_3 + i\varkappa A_2 - \frac{\varkappa^2}{6}A_1\,. \tag{2.4.28}$$

For positive values of the connection constant $\varkappa$ (the case of repulsion), the states $|k_1,\ldots,k_N\rangle$ $(N=0,1,2\ldots)$, as indicated in Sec. 2.1, form a complete system of eigenfunctions of $\mathbb{H}$ in the space $\mathbb{F}$. This fact, and also the additivity of the eigenvalues of $\ln A(\lambda)$ in (2.4.12), allow us to write for the generating function of the quantum integrals of motion $\mathcal{I}_m$ the following spectral decomposition:

$$\ln A(\lambda) = \int_{-\infty}^{\infty} d\mu \ln\left(1 + \frac{i\varkappa}{\lambda-\mu}\right) \Phi^+(\mu)\Phi(\mu), \quad \operatorname{Im}\lambda > 0. \tag{2.4.29}$$

The analogous decompositions for $\mathbb{N}, \mathbb{P}$, and $\mathbb{H}$ have the form

$$\mathbb{N}=\int_{-\infty}^{\infty} d\mu\, \Phi^{+}(\mu)\Phi(\mu), \tag{2.4.30}$$

$$\mathbb{P}=\int_{-\infty}^{\infty} d\mu \cdot \mu\, \Phi^{+}(\mu)\Phi(\mu), \tag{2.4.31}$$

$$\mathbb{H}=\int_{-\infty}^{\infty} d\mu\, \mu^{2}\Phi^{+}(\mu)\Phi(\mu). \tag{2.4.32}$$

Equations (2.4.30-32) show that $\Phi^{+}(\mu)$ and $\Phi(\mu)$ are operators of birth and annihilation of elementary particles with momentum μ and energy μ^2. The operator $\Phi^{+}(\mu)\Phi(\mu)$ can here be interpreted as the operator of the density of the number of particles with momentum μ. We note the obvious similarity of (2.4.29-32) with (1.4.10, 12-14) for the classical n.S.e.

To conclude this section, we discuss the case of attraction ($\varkappa<0$). Here, as noted in Sec. 2.1, in the spectrum there appear connected states, which can be obtained from the scattering states $|k_1,\ldots,k_N\rangle$ by analytic continuation with respect to the momenta (2.1.15).

We calculate the eigenvalues of the integrals of motion for the connected states. This, of course, can be done by simply substituting (2.1.15) in (2.4.21), but we choose another method of calculation, which allows us at the same time to get interesting deductions of general character.

First, we find the eigenvalue of the operator $A(\lambda)$ on the state $|k_1,\ldots,k_N\rangle$. This is easy to do, letting the operator $A(\lambda)$ act on the expression $\Phi^{+}(k_1)\ldots\Phi^{+}(k_N)|0\rangle$ and moving $A(\lambda)$ to the right with the help of (S46). As a result we have

$$A(\lambda)|k_1,\ldots,k_N\rangle=\prod_{j=1}^{N}\frac{\lambda-k_j+i\varkappa}{\lambda-k_j}|k_1,\ldots,k_N\rangle,\ \mathrm{Im}\,\lambda>0\ . \tag{2.4.33}$$

We note two things in connection with (2.4.33). Firstly, the eigenvalues of $A(\lambda)$ are multiplicative (hence additive as eigenvalues of $\ln A(\lambda)$) in the momenta k_j. Secondly, the eigenvalue $\prod_{j=1}^{N}\frac{\lambda-k_j+i\varkappa}{\lambda-k_j}$ has in the upper half-plane with respect to λ exactly N zeros $\lambda=k_j-i\varkappa=k_j+i|\varkappa|$ (we recall that we are considering the case $\varkappa<0$).

An eigenvalue of the operator $A(\lambda)$ on a connected state of N particles is obtained from (2.4.33) by analytic continuation of (2.1.15) with respect to the momenta k_j. Here in the product

$$\prod_{j=1}^{N}\frac{\lambda-k_j-i|\varkappa|}{\lambda-k_j}=\frac{\lambda-\frac{P}{N}+i|\varkappa|\frac{N-3}{2}}{\lambda-\frac{P}{N}+i|\varkappa|\frac{N-1}{2}}\cdot\frac{\lambda-\frac{P}{N}+i|\varkappa|\frac{N-5}{2}}{\lambda-\frac{P}{N}+i|\varkappa|\frac{N-3}{2}}\cdot\ldots\cdot\frac{\lambda-\frac{P}{N}-i|\varkappa|\frac{N+1}{2}}{\lambda-\frac{P}{N}-i|\varkappa|\frac{N-1}{2}} \tag{2.4.34}$$

there occurs consecutive cancellation of numerators and denominators, and as a result there remains the factor

$$\frac{\lambda-\frac{P}{N}-i|\varkappa|\frac{N+1}{2}}{\lambda-\frac{P}{N}+i|\varkappa|\frac{N-1}{2}} \tag{2.4.35}$$

having a unique zero in the upper half-plane with respect to λ at the point $\lambda=\frac{P}{N}+i|\varkappa|\frac{N+1}{2}$. It is interesting to note that on the other hand (2.1.15) can be obtained, by requiring that the eigenvalue $A(\lambda)$ have a unique zero in the half-plane $\mathrm{Im}\,\lambda>0$ and that the momenta k_j be distributed symmetrically with respect to the real axis. In fact, the condition of cancellation of numerators and denominators in (2.4.34) leads to the requirement of equidistance of momenta $k: k_{j+1}-k_j = i|\varkappa|$, which in combination with the requirement of symmetry $k_N = \bar{k}_1$ gives (2.1.15).

This result has interesting analogs in the theory of the classical nonlinear Schrödinger equation. It is known [11, 41], that connected states of quantum particles correspond in the classical limit to solutions for (1.1.1). The classical coefficient of passage $a(\lambda)$ to a one-soliton solution, characterized by the momentum P and the number of particles N, has the form

$$a(\lambda)=\frac{\lambda-\frac{P}{N}-i|\varkappa|\frac{N}{2}}{\lambda-\frac{P}{N}+i|\varkappa|\frac{N}{2}}\;. \tag{2.4.36}$$

Comparing (2.4.36) and (2.4.35), we see that they coincide up to the translation $\lambda\to\lambda-i\frac{|\varkappa|}{2}$, which in the quasiclassical limit is inessential.

The eigenvalues $c_m^{(N)}(p)$ of the integrals of motion A_m on an N-particle connected state are obtained, as earlier, by the expansion of the generating function

$$\ln\frac{\lambda-\frac{P}{N}-i|\varkappa|\frac{N+1}{2}}{\lambda-\frac{P}{N}+i|\varkappa|\frac{N-1}{2}}=-i|\varkappa|\sum_{m=1}^{\infty}c_m^{(N)}(p)\,\lambda^{-m}, \tag{2.4.37}$$

$$c_m^{(N)}(p)=\frac{i}{m|\varkappa|}\left[\left(\frac{P}{N}-i|\varkappa|\frac{N-1}{2}\right)^m-\left(\frac{P}{N}+i|\varkappa|\frac{N+1}{2}\right)^m\right], \tag{2.4.38}$$

$$c_m^{(1)}(p)=c_m(p)\,.$$

Using (2.4.26-28) it is easy to get the eigenvalues of the integrals of motion $\mathbf{N}$, $\mathbf{P}$, and $\mathbf{H}$ on an N-particle connected state $|p,N\rangle$:

$$|p,N\rangle=|k_1,\dots,k_N\rangle,\quad k_j=\frac{P}{N}+i|\varkappa|\left(j-\frac{N+1}{2}\right),\quad j=1,\dots,N\,. \tag{2.4.39}$$

These eigenvalues have the form

$$\mathbf{N}|p,N\rangle=N|p,N\rangle, \tag{2.4.40}$$

$$\mathbf{P}|p,N\rangle=p|p,N\rangle, \tag{2.4.41}$$

$$\mathbf{H}|p,N\rangle=\left(\frac{p^2}{N}-\frac{\varkappa^2}{12}(N^3-N)\right)|p,N\rangle\,. \tag{2.4.42}$$

Unfortunately, we still do not have available a method based on the quantum method of the inverse problem for constructing the canonical operators $\Phi_N^{+}(p)$ and $\Phi_N(p)$ of birth and annihilation of normalized connected states of N particles with total momentum p. If, however, one admits that such operators are constructed, then the proper generalization of the spectral decompositions (2.4.29-32) to the case $\varkappa<0$ must assume the form:

$$\ln A(\lambda)=\sum_{N=1}^{\infty}\int_{-\infty}^{\infty}dp\,\ln\frac{\lambda-\frac{P}{N}-i|\varkappa|\frac{N+1}{2}}{\lambda-\frac{P}{N}+i|\varkappa|\frac{N-1}{2}}\,\Phi_N^{+}(p)\Phi_N(p), \tag{2.4.29'}$$

$$\mathbb{N}=\sum_{\mathcal{N}=1}^{\infty}\int_{-\infty}^{\infty} dp\, \mathcal{N}\, \Phi_{\mathcal{N}}^{+}(p)\Phi_{\mathcal{N}}(p)\,, \tag{2.4.30'}$$

$$\mathbb{P}=\sum_{\mathcal{N}=1}^{\infty}\int_{-\infty}^{\infty} dp\cdot p\, \Phi_{\mathcal{N}}^{+}(p)\Phi_{\mathcal{N}}(p)\,, \tag{2.4.31'}$$

$$\mathbb{H}=\sum_{\mathcal{N}=1}^{\infty}\int_{-\infty}^{\infty} dp\left(\frac{p^2}{\mathcal{N}}-\frac{\varkappa^2}{12}(\mathcal{N}^3-\mathcal{N})\right)\Phi_{\mathcal{N}}^{+}(p)\Phi_{\mathcal{N}}(p). \tag{2.4.32'}$$

2.5. Quantum $\mathbb{M}$-Operator

All arguments of the preceding sections were carried out for a fixed moment of time t. The introduction of temporal evolution does not present any difficulty in quantum mechanics. In fact, the solution of the Heisenberg equation of motion

$$X_t = i\,[X, \mathbb{H}] \tag{2.5.1}$$

for any observable quantity X is given by the formula

$$X(t) = e^{i\mathbb{H}t} X(0) e^{-i\mathbb{H}t}\,. \tag{2.5.2}$$

In particular, for matrix elements of the quantum transition matrix $\mathbb{T}(\lambda)$, we get, using the commutation relations

$$[A(\lambda), \mathbb{H}] = 0\,, \tag{2.5.3}$$

$$[B^{+}(\lambda), \mathbb{H}] = \lambda^2 B^{+}(\lambda) \tag{2.5.4}$$

the following result

$$A(t,\lambda) = A(0,\lambda)\,, \tag{2.5.5}$$

$$B^{+}(t,\lambda) = e^{i\lambda^2 t} B^{+}(0,\lambda)\,. \tag{2.5.6}$$

Nevertheless, there is definite methodological interest in the following question: Does there exist in the quantum case an operator $\mathbb{M}(x,\lambda)$, allowing one to describe the temporal evolution of the transition matrix $\mathbb{T}_{x_1}^{x_2}(\lambda)$, analogous to the operator M in the classical case (Sec. 1.5)?

A positive answer to this question is given by the following proposition.

Proposition 2.5.1. The Heisenberg equation of motion for the quantum transition matrix on a finite interval

$$\frac{d}{dt}\mathbb{T}_{x_1}^{x_2}(\lambda) = i\left[\mathbb{H}, \mathbb{T}_{x_1}^{x_2}(\lambda)\right]$$

can be represented in the form analogous to (1.5.24):

$$\frac{d}{dt}\mathbb{T}_{x_1}^{x_2}(\lambda) = :\mathbb{M}(x_2,\lambda)\mathbb{T}_{x_1}^{x_2}(\lambda) - \mathbb{T}_{x_1}^{x_2}(\lambda)\mathbb{M}(x_1,\lambda):\;, \tag{2.5.7}$$

where the operator $\mathbb{M}(x,\lambda)$ has the form

$$\mathbb{M}(x,\lambda) = :M(x,\lambda): \quad \left(i\frac{\lambda^2}{2} + i\varkappa\Psi^{+}(x)\Psi(x)\right)\sigma_3 + \varkappa\left(\Psi_x^{+}(x) - i\lambda\Psi^{+}(x)\right)\sigma_+ + \left(\Psi_x(x) + i\lambda\Psi(x)\right)\sigma_-\,.$$

Proof. We denote the commutator $i\left[\mathbb{H}, \mathbb{T}_{x_1}^{x_2}(\lambda)\right]$ by the symbol $\mathcal{M}_{x_1}^{x_2}(\lambda)$ and we find the differential equation with respect to the variable x_2, to which this quantity is subordinate.

For this we differentiate $\mathcal{M}_{x_1}^{x_2}(\lambda)$ with respect to x_2, using (2.1.24). We get

$$\frac{\partial}{\partial x_2}\mathcal{M}_{x_1}^{x_2}(\lambda)=i\left[\mathbb{H},\,:\mathbb{L}(x_2,\lambda)\mathbb{T}_{x_1}^{x_2}(\lambda):\right]=$$
$$=:\mathbb{L}(x_2,\lambda)\mathcal{M}_{x_1}^{x_2}(\lambda):-i\sigma_-\mathbb{T}_{x_1}^{x_2}(\lambda)\left[i\mathbb{H},\Psi(x_2)\right]+i\varkappa\sigma_+\left[i\mathbb{H},\Psi^+(x_2)\right]\mathbb{T}_{x_1}^{x_2}(\lambda). \tag{2.5.8}$$

Using the equation of motion (2.1.6) for $\Psi(x)$ and its conjugate equation, we reduce (2.5.8) to the following form

$$\frac{\partial}{\partial x_2}\mathcal{M}_{x_1}^{x_2}(\lambda)=:\mathbb{L}(x_2,\lambda)\mathcal{M}_{x_1}^{x_2}(\lambda):\;-i\sigma_-\mathbb{T}_{x_1}^{x_2}(\lambda)\left(i\Psi_{xx}(x_2)-2i\varkappa\Psi^+(x_2)\Psi(x_2)\Psi(x_2)\right)+$$
$$+i\varkappa\sigma_+\left(-i\Psi^+_{xx}(x_2)+2i\varkappa\Psi^+(x_2)\Psi^+(x_2)\Psi(x_2)\right)\mathbb{T}_{x_1}^{x_2}(\lambda). \tag{2.5.9}$$

Before moving further, we formulate the following lemma.

LEMMA 2.5.1. One has the equation:

$$\sigma_-\left[\mathbb{T}_{x_1}^{x_2}(\lambda),\Psi^+(x_2)\right]=\sigma_+\left[\Psi(x_2),\mathbb{T}_{x_1}^{x_2}(\lambda)\right]=0. \tag{2.5.10}$$

We shall not give the proof of Lemma 2.5.1, since it is carried out with the help of the same method of "extension" which was used in Proof 1 of Lemma 2.2.1 in Sec. 2.2. Here by virtue of the equations $\sigma_-^2=\sigma_+^2=0$ the result, as also in Sec. 2.2, is independent of the sign.

Using Lemma 2.5.1, we transform (2.5.9), moving $\Psi^+(x_2)$ to the right, and $\Psi(x_2)$ to the left. We get:

$$\frac{\partial}{\partial x_2}\mathcal{M}_{x_1}^{x_2}(\lambda)=:\mathbb{L}(x_2,\lambda)\mathcal{M}_{x_1}^{x_2}(\lambda):+\sigma_-\mathbb{T}_{x_1}^{x_2}(\lambda)\Psi_{xx}(x_2)-2\varkappa\sigma_-\Psi^+(x_2)\mathbb{T}_{x_1}^{x_2}(\lambda)\Psi(x_2)\Psi(x_2)+$$
$$+\varkappa\sigma_+\Psi^+_{xx}(x_2)\mathbb{T}_{x_1}^{x_2}(\lambda)-2\varkappa^2\sigma_+\Psi^+(x_2)\Psi^+(x_2)\mathbb{T}_{x_1}^{x_2}(\lambda)\Psi(x_2)=$$
$$=:\mathbb{L}(x_2,\lambda)\mathcal{M}_{x_1}^{x_2}(\lambda):+:\left(\mathbb{M}_x(x_2,\lambda)+\left[\mathbb{M}(x_2,\lambda),\mathbb{L}(x_2,\lambda)\right]\right)\mathbb{T}_{x_1}^{x_2}(\lambda): \tag{2.5.11}$$

On the other hand, the right side of (2.5.7), which we denote by $\mathcal{M}'^{x_2}_{x_1}(\lambda)$, satisfies exactly the same differential equation. In fact, differentiating the right side of (2.5.7) with respect to x_2 and using (1.5.28), we get

$$\frac{\partial}{\partial x_2}\mathcal{M}'^{x_2}_{x_1}(\lambda)=\frac{\partial}{\partial x_2}:\mathbb{M}(x_2,\lambda)\mathbb{T}_{x_1}^{x_2}(\lambda)-\mathbb{T}_{x_1}^{x_2}(\lambda)\mathbb{M}(x_1,\lambda):= \tag{2.5.12}$$
$$=\frac{\partial}{\partial x_2}:\mathbb{M}(x_2,\lambda)\mathbb{T}_{x_1}^{x_2}(\lambda)-\mathbb{T}_{x_1}^{x_2}(\lambda)\mathbb{M}(x_1,\lambda):\;=:\mathbb{L}(x_2,\lambda)\mathcal{M}'^{x_2}_{x_1}(\lambda):+:\left(\mathbb{M}_x(x_2,\lambda)+\left[\mathbb{M}(x_2,\lambda),\mathbb{L}(x_2,\lambda)\right]\right)\mathbb{T}_{x_1}^{x_2}(\lambda):$$

Since the quantities $\mathcal{M}_{x_1}^{x_2}(\lambda)$ and $\mathcal{M}'^{x_2}_{x_1}(\lambda)$ satisfy the same differential equation and the same initial condition

$$\mathcal{M}_{x_1}^{x_1}(\lambda)=\mathcal{M}'^{x_1}_{x_1}(\lambda)=0, \tag{2.5.13}$$

we conclude that they in fact coincide, which is what had to be proved.

Analogously, one can prove the quantum analogs of (1.5.25-27).

CONCLUSIONS

We summarize. In the present paper, for the example of the nonlinear Schrödinger equation we developed a new method of exact quantization of completely integrable field-theoretic models. This method allowed not only the reproduction of known results for the quantum nonlinear Schrödinger equation, obtained earlier with the help of the Bethe substitution, but also getting a series of new results, namely, constructing a generating function for the quantum integrals of motion and operators of birth-annihilation of elementary excitations. In comparison with the method of Bethe substitution our method has the advantage that it allows one to construct and study eigenvectors of the Hamiltonian by a purely algebraic method, without writing down the explicitly corresponding wave functions in a coordinate representation.

A central role in the method we propose is played, as we saw, by the R-matrix, which gives its name to the method. The use of the R-matrix allows us compactly and effectively to calculate the commutation relations between the matrix elements of the quantum transition matrix, without resorting to infinite series, as was done, e.g., in [22, 26], which appeared after the author's paper [14].

We list some problems concerning the quantum nonlinear Schrödinger equation, which still remain unsolved:

1) It would be desirable to find an effective method of construction of quantum integrals of motion analogous to the Riccati equation in the classical case. This would allow one to definitively free oneself in studying quantum integrals of motion from references to results obtained with the help of Bethe substitution.

2) To construct in the realms of the method of the R-matrix operators of birth and annihilation of connected states of $\mathcal{N}$ particles $\Phi_{\mathcal{N}}^{+}(k)$ and $\Phi_{\mathcal{N}}(k)$.

3) To construct the generating function of the quantum M-operators analogous to the way this was done for the classical case in Sec. 1.5.

After the publication of [13, 14], problems connected with the quantum generalization of the method of the inverse problem attracted the attention of a large number of investigators, both in the USSR and abroad. In the Soviet Union work on the quantum method of the inverse problem was conducted at the Leningrad Branch of the Mathematics Institute (LOMI) under the direction of Faddeev [13-21]. Of the foreign authors one should single out Thacker (USA, Batavia) [22-25] and Honerkamp (GFR, Freiburg) [26, 27].

We list the basic directions in which the quantum method of the inverse problem is developing at the present time:

1) Quantum relativistically invariant completely integrable models [17], in which the method of the R-matrix was successfully applied to the quantization of the sin-Gordon equation.

2) The study of completely integrable lattice spin models, such as the Heisenberg ferromagnet [18] and the XYZ-model [19].

3) The investigation of models with several kinds of particles, having isotopic symmetry [20, 21].

4) And, finally, the very long-range direction, intensively developed recently — the attempt to solve the inverse scattering problem for the auxiliary linear equation, i.e., to express the field operators, for example, $\Psi(x)$ and $\Psi^{+}(x)$, for the n.S.e. in terms of the

scattering data $A(\lambda)$ and $B^{+}(\lambda)$. The solution of this problem is of great interest for quantum field theory, since it would allow the effective study of Green's functions of completely integrable quantum field systems. Some results in this direction are obtained in [23-25, 28] for the nonlinear Schrödinger equation.

In conclusion, it is necessary to mention the classical version of the R-matrix method developed in Chap. I of the present paper. This method allowed one not only to simplify the calculations connected with the computation of the Poisson brackets, but also to get a new result such as the expression for the generating function of the M-operators. Thus, in the theory of classical completely integrable equations there arises a new object, the r-matrix. The place which the r-matrix occupies in the method of the inverse problem is still not entirely clear. One does not know, e.g., the precise class of L-operators, which have a r-matrix. In connection with this there is great interest in the problem of generalizing the method of the r-matrix to nonultralocal L-operators in the terminology of Faddeev [16], i.e., L-operators, the Poisson brackets between whose matrix elements contain derivatives of the δ-function.

SUPPLEMENT

In the Supplement we gather together the Poisson brackets (in the classical case) and commutation relations (in the quantum case) between the matrix elements of the transition matrices for finite, semi-infinite, and infinite intervals. All the formulas are written for real values of λ and μ.

1. Summary of Poisson brackets between matrix elements of the classical transition matrix $T_{x_1}^{x_2}(\lambda)$ for the finite interval $[x_1, x_2]$.

We recall that the matrix $T_{x_1}^{x_2}(\lambda)$ has the form (1.1.11):

$$T_{x_1}^{x_2}(\lambda)=\begin{pmatrix} a_{x_1}^{x_2}(\lambda) , & \varkappa \bar{b}_{x_1}^{x_2}(\lambda) \\ b_{x_1}^{x_2}(\lambda) , & \bar{a}_{x_1}^{x_2}(\lambda) \end{pmatrix}, \qquad \lambda=\bar{\lambda} .$$

The desired Poisson brackets are given by (1.2.11):

$$\{\tilde{T}_{x_1}^{x_2}(\lambda), \tilde{\tilde{T}}_{x_1}^{x_2}(\mu)\} = [\, r(\lambda-\mu), \tilde{T}_{x_1}^{x_2}(\lambda)\tilde{\tilde{T}}_{x_1}^{x_2}(\mu)] .$$

Below are written six independent matrix elements of (1.2.11) of the 16 possible ones.

$$\{a_{x_1}^{x_2}(\lambda), a_{x_1}^{x_2}(\mu)\} = 0 , \tag{S1}$$

$$\{a_{x_1}^{x_2}(\lambda), \bar{a}_{x_1}^{x_2}(\mu)\} = \frac{\varkappa^2}{\lambda-\mu}\left(\bar{b}_{x_1}^{x_2}(\lambda) b_{x_1}^{x_2}(\mu) - b_{x_1}^{x_2}(\lambda)\bar{b}_{x_1}^{x_2}(\mu)\right), \tag{S2}$$

$$\{a_{x_1}^{x_2}(\lambda), b_{x_1}^{x_2}(\mu)\} = \frac{\varkappa}{\lambda-\mu}\left(a_{x_1}^{x_2}(\lambda) b_{x_1}^{x_2}(\mu) - b_{x_1}^{x_2}(\lambda) a_{x_1}^{x_2}(\mu)\right), \tag{S3}$$

$$\{a_{x_1}^{x_2}(\lambda), \bar{b}_{x_1}^{x_2}(\mu)\} = \frac{\varkappa}{\lambda-\mu}\left(\bar{b}_{x_1}^{x_2}(\lambda) a_{x_1}^{x_2}(\mu) - a_{x_1}^{x_2}(\lambda)\bar{b}_{x_1}^{x_2}(\mu)\right), \tag{S4}$$

$$\{b_{x_1}^{x_2}(\lambda), b_{x_1}^{x_2}(\mu)\} = 0 , \tag{S5}$$

$$\{b_{x_1}^{x_2}(\lambda), \bar{b}_{x_1}^{x_2}(\mu)\} = \frac{1}{\lambda-\mu}\left(\bar{a}_{x_1}^{x_2}(\lambda) a_{x_1}^{x_2}(\mu) - a_{x_1}^{x_2}(\lambda)\bar{a}_{x_1}^{x_2}(\mu)\right). \tag{S6}$$

The remaining 10 relations are obtained from the ones listed by complex conjugation, interchange of λ and μ, and the use of the antisymmetry of the Poisson brackets.

2. Summary of Poisson brackets between matrix elements of the classical transition matrix $T_-(x,\lambda)$ for the semi-infinite interval $(-\infty, x]$.

The matrix $T_-(x,\lambda)$ has the form

$$T_-(x,\lambda) = \begin{pmatrix} a_-(x,\lambda), & \varkappa \bar{b}_-(x,\lambda) \\ b_-(x,\lambda), & \bar{a}_-(x,\lambda) \end{pmatrix}, \quad \lambda = \bar{\lambda}.$$

We recall that the matrix elements $a(x,\lambda)$ and $b_-(x,\lambda)$ admit analytic continuation with respect to λ to the upper half-plane, and $\bar{b}_-(x,\lambda)$ and $\bar{a}_-(x,\lambda)$ to the lower (see Sec. 1.1).

Original formula for Poisson brackets (1.3.11):

$$\{\tilde{T}_-(x,\lambda), \tilde{\tilde{T}}_-(x,\mu)\} = r(\lambda-\mu)\tilde{T}_-(x,\lambda)\tilde{\tilde{T}}_-(x,\mu) - \tilde{T}_-(x,\lambda)\tilde{\tilde{T}}_-(x,\mu)r_-(\lambda-\mu).$$

Independent matrix elements:

$$\{a_-(x,\lambda), a_-(x,\mu)\} = 0, \tag{S7}$$

$$\{a_-(x,\lambda), \bar{a}_-(x,\mu)\} = -\frac{\varkappa^2}{\lambda-\mu+i0} b_-(x,\lambda)\bar{b}_-(x,\mu), \tag{S8}$$

$$\{a_-(x,\lambda), b_-(x,\mu)\} = \frac{\varkappa}{\lambda-\mu}(a_-(x,\lambda)b_-(x,\mu) - b_-(x,\lambda)a_-(x,\mu)), \tag{S9}$$

$$\{a_-(x,\lambda), \bar{b}_-(x,\mu)\} = -\frac{\varkappa}{\lambda-\mu+i0} a_-(x,\lambda)\bar{b}_-(x,\mu), \tag{S10}$$

$$\{b_-(x,\lambda), b_-(x,\mu)\} = 0, \tag{S11}$$

$$\{b_-(x,\lambda), \bar{b}_-(x,\mu)\} = -\frac{1}{\lambda-\mu+i0} a_-(x,\lambda)\bar{a}_-(x,\mu). \tag{S12}$$

In formula (S9) regularization of the denominator is not necessary, since the numerator vanishes for $\lambda = \mu$. Here the Poisson bracket admits analytic continuation to the same half-plane with respect to λ and μ.

3. Summary of Poisson brackets between matrix elements of the classical transition matrix $T_+(x,\lambda)$ for the semi-infinite interval $[x, +\infty)$.

The matrix $T_+(x,\lambda)$ has the form:

$$T_+(x,\lambda) = \begin{pmatrix} a_+(x,\lambda), & \varkappa \bar{b}_+(x,\lambda) \\ b_+(x,\lambda), & \bar{a}_+(x,\lambda) \end{pmatrix}, \quad \lambda = \bar{\lambda}.$$

The matrix elements $a_+(x,\lambda)$ and $\bar{b}_+(x,\lambda)$ admit analytic continuation with respect to λ to the upper half-plane, and $b_+(x,\lambda)$ and $\bar{a}_+(x,\lambda)$ to the lower (see Sec. 1.1).

Original formula for Poisson brackets (1.3.12):

$$\{\tilde{T}_+(x,\lambda), \tilde{\tilde{T}}_+(x,\mu)\} = r_+(\lambda-\mu)\tilde{T}_+(x,\lambda)\tilde{\tilde{T}}_+(x,\mu) - \tilde{T}_+(x,\lambda)\tilde{\tilde{T}}_+(x,\mu)r(\lambda-\mu).$$

Independent matrix elements:

$$\{a_+(x,\lambda), a_+(x,\mu)\} = 0, \tag{S13}$$

$$\{a_+(x,\lambda), \bar{a}_+(x,\mu)\} = \frac{\varkappa^2}{\lambda-\mu+i0} \bar{b}_+(x,\lambda)b_+(x,\mu), \tag{S14}$$

$$\{a_+(x,\lambda), b_+(x,\mu)\} = \frac{\varkappa}{\lambda-\mu+i0} a_+(x,\lambda)b_+(x,\mu), \tag{S15}$$

$$\{a_+(x,\lambda), \bar{b}_+(x,\mu)\} = \frac{\varkappa}{\lambda-\mu}(\bar{b}_+(x,\lambda)a_+(x,\mu) - a_+(x,\lambda)\bar{b}_+(x,\mu)), \tag{S16}$$

$$\{ b_+(x,\lambda), b_+(x,\mu)\} = 0\,, \tag{S17}$$

$$\{ b_+(x,\lambda), \bar{b}_+(x,\mu)\} = \frac{1}{\lambda-\mu+i0}\, \bar{a}_+(x,\lambda)\, a_+(x,\mu)\,. \tag{S18}$$

In connection with (S16) one can make a remark analogous to that made in Paragraph 2 about (S9).

4. Summary of Poisson brackets between matrix elements of the classical transition matrix $T(\lambda)$ on the infinite interval $(-\infty,\infty)$.

The matrix $T(\lambda)$ has the form

$$T(\lambda) = \begin{pmatrix} a(\lambda)\,, & \varkappa \bar{b}(\lambda) \\ b(\lambda)\,, & \bar{a}(\lambda) \end{pmatrix}, \quad \lambda = \bar{\lambda}\,.$$

The matrix element $a(\lambda)$ admits analytic continuation with respect to λ to the upper half-plane, $\bar{a}(\lambda)$ to the lower. The matrix elements $b(\lambda)$ and $\bar{b}(\lambda)$, generally speaking, do not admit analytic continuation (see Sec. 1.1).

Original formula for Poisson brackets (1.3.14):

$$\{ \tilde{T}(\lambda), \tilde{\tilde{T}}(\mu)\} = r_+(\lambda-\mu)\tilde{T}(\lambda)\tilde{\tilde{T}}(\mu) - \tilde{T}(\lambda)\tilde{\tilde{T}}(\mu)\, r_-(\lambda-\mu)\,.$$

Independent matrix elements:

$$\{ a(\lambda),\ a(\mu)\} = 0, \tag{S19}$$

$$\{ a(\lambda),\ \bar{a}(\mu)\} = 0, \tag{S20}$$

$$\{ a(\lambda), b(\mu)\} = \frac{\varkappa}{\lambda-\mu+i0}\, a(\lambda) b(\mu), \tag{S21}$$

$$\{ a(\lambda), \bar{b}(\mu)\} = -\frac{\varkappa}{\lambda-\mu+i0}\, a(\lambda)\bar{b}(\mu), \tag{S22}$$

$$\{ b(\lambda), b(\mu)\} = 0, \tag{S23}$$

$$\{ b(\lambda), \bar{b}(\mu)\} = 2\pi i\, |a(\lambda)|^2 \delta(\lambda-\mu)\,. \tag{S24}$$

5. Summary of commutation relations between matrix elements of the quantum transition matrix $\mathbb{T}_{x_1}^{x_2}(\lambda)$ for the finite interval $[x_1, x_2]$.

The matrix $\mathbb{T}_{x_1}^{x_2}(\lambda)$ has the form (2.1.18)

$$\mathbb{T}_{x_1}^{x_2}(\lambda) = \begin{pmatrix} A_{x_1}^{x_2}(\lambda)\,, & \varkappa B_{x_1}^{+x_2}(\lambda) \\ B_{x_1}^{x_2}(\lambda)\,, & A_{x_1}^{+x_2}(\lambda) \end{pmatrix}, \quad \lambda = \bar{\lambda}\,.$$

Original formula (2.2.1):

$$R(\lambda-\mu)\tilde{\mathbb{T}}_{x_1}^{x_2}(\lambda)\tilde{\tilde{\mathbb{T}}}_{x_1}^{x_2}(\mu) = \tilde{\tilde{\mathbb{T}}}_{x_1}^{x_2}(\mu)\tilde{\mathbb{T}}_{x_1}^{x_2}(\lambda) R(\lambda-\mu)\,.$$

Independent commutation relations:

$$A_{x_1}^{x_2}(\lambda) A_{x_1}^{x_2}(\mu) = A_{x_1}^{x_2}(\mu) A_{x_1}^{x_2}(\lambda)\,, \tag{S25}$$

$$A_{x_1}^{x_2}(\lambda) A_{x_1}^{+x_2}(\mu) = A_{x_1}^{+x_2}(\mu) A_{x_1}^{x_2}(\lambda) + \frac{i\varkappa^2}{\lambda-\mu}\left(B_{x_1}^{+x_2}(\mu) B_{x_1}^{x_2}(\lambda) - B_{x_1}^{+x_2}(\lambda) B_{x_1}^{x_2}(\mu)\right), \tag{S26}$$

$$B_{x_1}^{x_2}(\mu)A_{x_1}^{x_2}(\lambda)=\left(1+\frac{i\varkappa}{\lambda-\mu}\right)A_{x_1}^{x_2}(\lambda)B_{x_1}^{x_2}(\mu)-\frac{i\varkappa}{\lambda-\mu}A_{x_1}^{x_2}(\mu)B_{x_1}^{x_2}(\lambda), \tag{S27}$$

$$A_{x_1}^{x_2}(\lambda)B_{x_1}^{+x_2}(\mu)=\left(1+\frac{i\varkappa}{\lambda-\mu}\right)B_{x_1}^{+x_2}(\mu)A_{x_1}^{x_2}(\lambda)-\frac{i\varkappa}{\lambda-\mu}B_{x_1}^{+x_2}(\lambda)A_{x_1}^{x_2}(\mu), \tag{S28}$$

$$B_{x_1}^{x_2}(\lambda)B_{x_1}^{x_2}(\mu)=B_{x_1}^{x_2}(\mu)B_{x_1}^{x_2}(\lambda), \tag{S29}$$

$$B_{x_1}^{x_2}(\lambda)B_{x_1}^{+x_2}(\mu)=B_{x_1}^{+x_2}(\mu)B_{x_1}^{x_2}(\lambda)+\frac{i}{\lambda-\mu}\left(A_{x_1}^{+x_2}(\mu)A_{x_1}^{x_2}(\lambda)-A_{x_1}^{+x_2}(\lambda)A_{x_1}^{x_2}(\mu)\right), \tag{S30}$$

All other commutation relations are obtained by Hermitian conjugation and interchange of λ and μ.

Here and later all commutation relations are given in one form: on the left side is the product of two matrix elements, on the right side is a linear combination of matrix elements with coefficients having $(\lambda-\mu)$ in the denominator.

6. Summary of commutation relations between matrix elements of the quantum transition matrix $\mathbb{T}_-(x,\lambda)$ for the semi-infinite segment $(-\infty,x]$.

The matrix $\mathbb{T}_-(x,\lambda)$ has the form:

$$\mathbb{T}_-(x,\lambda)=\begin{pmatrix} A_-(x;\lambda), & \varkappa B_-^+(x,\lambda) \\ B_-(x,\lambda), & A_-^+(x,\lambda) \end{pmatrix}, \quad \lambda=\bar{\lambda}.$$

The analytic properties of the matrix elements of $\mathbb{T}_-(x,\lambda)$ are the same as in Paragraph 2.

Original formula (2.3.4):

$$R(\lambda-\mu)\tilde{\mathbb{T}}_-(x,\lambda)\tilde{\tilde{\mathbb{T}}}_-(x,\mu)\left(1-\frac{i\varkappa}{\lambda-\mu+i0}\tilde{\sigma}_-\tilde{\tilde{\sigma}}_+\right)=\tilde{\tilde{\mathbb{T}}}_-(x,\mu)\tilde{\mathbb{T}}_-(x,\lambda)\left(1+\frac{i\varkappa}{\lambda-\mu-i0}\tilde{\sigma}_+\tilde{\tilde{\sigma}}_-\right)R(\lambda-\mu).$$

Independent commutation relations:

$$A_-(x,\lambda)A_-(x,\mu)=A_-(x,\mu)A_-(x,\lambda), \tag{S31}$$

$$A_-(x,\lambda)A_-^+(x,\mu)=A_-^+(x,\mu)A_-(x,\lambda)+\frac{i\varkappa^2}{\lambda-\mu+i0}B_-^+(x,\mu)B_-(x,\lambda), \tag{S32}$$

$$B_-(x,\mu)A_-(x,\lambda)=\left(1+\frac{i\varkappa}{\lambda-\mu}\right)A_-(x,\lambda)B_-(x,\mu)-\frac{i\varkappa}{\lambda-\mu}A_-(x,\mu)B_-(x,\lambda), \tag{S33}$$

$$A_-(x,\lambda)B_-^+(x,\mu)=\left(1+\frac{i\varkappa}{\lambda-\mu+i0}\right)B_-^+(x,\mu)A_-(x,\lambda). \tag{S34}$$

$$B_-(x,\lambda)B_-(x,\mu)=B_-(x,\mu)B_-(x,\lambda). \tag{S35}$$

$$B_-(x,\lambda)B_-^+(x,\mu)=B_-^+(x,\mu)B_-(x,\lambda)+\frac{i}{\lambda-\mu+i0}A_-^+(x,\mu)A_-(x,\lambda). \tag{S36}$$

7. Summary of commutation relations between matrix elements of the quantum transition matrix $\mathbb{T}_+(x,\lambda)$ for the semi-infinite interval $[x,+\infty)$.

The matrix $\mathbb{T}_+(x,\lambda)$ has the form:

$$\mathbb{T}_+(x,\lambda)=\begin{pmatrix} A_+(x,\lambda), & \varkappa B_+^+(x,\lambda) \\ B_+(x,\lambda), & A_+^+(x,\lambda) \end{pmatrix}, \quad \lambda=\bar{\lambda}.$$

The analytic properties of the matrix elements are the same as in Paragraph 3.

Original formula (2.3.5):

$$R(\lambda-\mu)\left(1+\frac{i\varkappa}{\lambda-\mu-i0}\tilde{\sigma}_-\tilde{\tilde{\sigma}}_+\right)\tilde{\mathbb{T}}_+(x,\lambda)\tilde{\tilde{\mathbb{T}}}_+(x,\mu)=\left(1-\frac{i\varkappa}{\lambda-\mu+i0}\tilde{\sigma}_+\tilde{\tilde{\sigma}}_-\right)\tilde{\tilde{\mathbb{T}}}_+(x,\mu)\tilde{\mathbb{T}}_+(x,\lambda)R(\lambda-\mu).$$

Independent commutation relations:

$$A_+(x,\lambda)A_+(x,\mu)=A_+(x,\mu)A_+(x,\lambda), \tag{S37}$$

$$A_+(x,\lambda)A_+^+(x,\mu)=A_+^+(x,\mu)A_+(x,\lambda)-\frac{i\varkappa^2}{\lambda-\mu+i0}B_+^+(x,\lambda)B_+(x,\mu), \tag{S38}$$

$$B_+(x,\mu)A_+(x,\lambda)=\left(1+\frac{i\varkappa}{\lambda-\mu+i0}\right)A_+(x,\lambda)B_+(x,\mu), \tag{S39}$$

$$A_+(x,\lambda)B_+^+(x,\mu)=\left(1+\frac{i\varkappa}{\lambda-\mu}\right)B_+^+(x,\mu)A_+(x,\lambda)-\frac{i\varkappa}{\lambda-\mu}B_+^+(x,\lambda)A_+(x,\mu), \tag{S40}$$

$$B_+(x,\lambda)B_+(x,\mu)=B_+(x,\mu)B_+(x,\lambda), \tag{S41}$$

$$B_+(x,\lambda)B_+^+(x,\mu)=B_+^+(x,\mu)B_+(x,\lambda)-\frac{i}{\lambda-\mu-i0}A_+^+(x,\lambda)A_+(x,\mu). \tag{S42}$$

8. Summary of commutation relations between matrix elements of the quantum transition matrix $\mathbb{T}(\lambda)$ for the infinite interval $(-\infty,\infty)$.

The matrix $\mathbb{T}(\lambda)$ has the form:

$$\mathbb{T}(\lambda)=\begin{pmatrix} A(\lambda), & \varkappa B^+(\lambda) \\ B(\lambda), & A^+(\lambda)\end{pmatrix}, \quad \lambda=\bar{\lambda}.$$

The analytic properties of the matrix elements are the same as in Paragraph 4.

Original formula (2.3.6)

$$R(\lambda-\mu)\left(1+\frac{i\varkappa}{\lambda-\mu-i0}\tilde{\sigma}_-\tilde{\tilde{\sigma}}_+\right)\tilde{\mathbb{T}}(\lambda)\tilde{\tilde{\mathbb{T}}}(\mu)\left(1-\frac{i\varkappa}{\lambda-\mu+i0}\tilde{\sigma}_-\tilde{\tilde{\sigma}}_+\right)=\left(1-\frac{i\varkappa}{\lambda-\mu+i0}\tilde{\sigma}_+\tilde{\tilde{\sigma}}_-\right)\tilde{\tilde{\mathbb{T}}}(\mu)\tilde{\mathbb{T}}(\lambda)\left(1+\frac{i\varkappa}{\lambda-\mu-i0}\tilde{\sigma}_+\tilde{\tilde{\sigma}}_-\right)R(\lambda-\mu).$$

Independent commutation relations:

$$A(\lambda)A(\mu)=A(\mu)A(\lambda), \tag{S43}$$

$$A(\lambda)A^+(\mu)=A^+(\mu)A(\lambda), \tag{S44}$$

$$B(\mu)A(\lambda)=\left(1+\frac{i\varkappa}{\lambda-\mu+i0}\right)A(\lambda)B(\mu), \tag{S45}$$

$$A(\lambda)B^+(\mu)=\left(1+\frac{i\varkappa}{\lambda-\mu+i0}\right)B^+(\mu)A(\lambda), \tag{S46}$$

$$B(\lambda)B(\mu)=B(\mu)B(\lambda), \tag{S47}$$

$$B(\lambda)B^+(\mu)=\left(1+\frac{i\varkappa}{\lambda-\mu+i0}\right)\left(1-\frac{i\varkappa}{\lambda-\mu-i0}\right)B^+(\mu)B(\lambda)+2\pi A^+(\lambda)A(\lambda)\delta(\lambda-\mu). \tag{S48}$$

LITERATURE CITED

1. V. E. Zakharov, S. V. Manakov, S. P. Novikov, and L. P. Pitaevskii, Theory of Solitons: Method of the Inverse Problem [in Russian], Nauka, Moscow (1980).
2. C. S. Gardner, J. M. Greene, M. D. Kruskal, and R. M. Miura, "Method for solving the Korteweg—de Vries equation," Phys. Rev. Lett., 19, 1095-1097 (1967).

3. P. D. Lax, "Integrals of nonlinear evolutional equations and isolated waves," Matematika (collection of translations of foreign papers), 13, No. 5, 128-150 (1967).
4. V. E. Zakharov and L. D. Faddeev, "Korteweg-de Vries equation — completely integrable Hamiltonian system," Funkts. Anal. Prilozhen., 5, No. 4, 18-27 (1971).
5. H. Bethe, "Zur Theorie der Metalle. I. Eigenwerte und Eigenfunctionen der linearen Atomkette," Z. Phys., 71, 205-226 (1931).
6. L. Onsager, "Crystal statistics. I. A two-dimensional model with an order—disorder transition," Phys. Rev., 65 (1944).
7. R. J. Baxter, "Partition function of the eight-vertex lattice model," Ann. Phys., 70, No. 1, 193-288 (1972).
8. R. J. Baxter, "One-dimensional anisotropic Heisenberg chain," Ann. Phys., 70, No. 2, 323-337 (1972).
9. V. E. Zahkarov, L. A. Takhtadzhyan, and L. D. Faddeev, "Complete description of solutions of the 'sin-Gordon' equation," Dokl. Akad. Nauk SSSR, 219, No. 6, 1334-1337 (1974).
10. V. E. Zakharov and A. V. Mikhailov, "Relativistically invariant two-dimensional models of the theory of fields, integrable by the method of the inverse problem," Zh. Eksp. Teor. Fiz., 74, No. 6, 1953-1973 (1978).
11. P. P. Kulish, S. V. Manakov, and L. D. Faddeev, "Comparison of exact quantum and quasiclassical answers for the nonlinear Schrödinger equation," Teor. Mat. Fiz., 28, No. 1, 38-45 (1976).
12. L. D. Faddeev and V. E. Korepin, "Quantum theory of solitons," Phys. Reports, 42C, No. 1, 1-87 (1978).
13. E. K. Sklyanin and L. D. Faddeev, "Quantum mechanical approach to completely integrable models of field theory," Dokl. Akad. Nauk SSSR, 243, No. 6, 1430-1433 (1978).
14. E. K. Sklyanin, "Method of the inverse scattering problem and quantum nonlinear Schrödinger equation," Dokl. Akad. Nauk SSSR, 244, No. 6, 1337-1341 (1978).
15. E. K. Sklyanin, "Quantum version of the method of the inverse scattering problem," Candidates Dissertation, Leningrad Branch of the Mathematics Institute (1980).
16. L. D. Faddeev, "Quantum completely integrable models of field theory," in: Problems of Quantum Field Theory (Memoirs of the International Conference on Nonlocal Field Theories, Alushta, 1979) [in Russian], Dubna (1979), pp. 249-299.
17. E. K. Sklyanin, L. A. Takhtadzhyan, and L. D. Faddeev, "Quantum method of the inverse problem I," Teor. Mat. Fiz., 40, No. 2, 194-220 (1979).
18. P. P. Kulish and E. K. Sklyanin, "Quantum inverse scattering method and the Heisenberg ferromagnet," Phys. Lett. A, 70, Nos. 5-6, 461-463 (1979).
19. L. A. Takhtadzhyan and L. D. Faddeev, "Quantum method of the inverse problem and the XYZ-model of Heisenberg," Usp. Mat. Nauk, 34, No. 5, 13-63 (1979).
20. P. P. Kulish, "Generalized Bethe Ansatz and the quantum method of the inverse problem," Preprint Leningrad Branch of the Mathematics Institute P-3-79, Leningrad (1979).
21. P. P. Kulish and N. Yu. Reshetikhin, "Generalized Heisenberg ferromagnet and the Groos—Neveu model," Preprint Leningrad Branch of the Mathematics Institute E-4-79, Leningrad (1979).
22. H. B. Thacker and D. Wilkinson, "The inverse scattering transform as an operator method in quantum field theory," Phys. Rev., D19, No. 12, 3660-3665 (1979).
23. D. B. Creamer, H. B. Thacker, and D. Wilkinson, "Gelfand—Levitan method for operator fields," Preprint Fermilab-Pub-79/75-THY, September, 1979.
24. D. B. Creamer, H. B. Thacker, and D. Wilkinson, "Quantum Gelfand—Levitan method as a generalized Jordan—Wigner transformation," Fermilab-Pub-80/17-THY, January, 1980.
25. D. B. Creamer, H. B. Thacker, and D. Wilkinson, "Statistical mechanics of an exactly integrable system," Fermilab-Pub-80/25-THY, February, 1980.
26. J. Honerkamp, P. Weber, and A. Wiesler, "On the connection between the inverse transform method and the exact quantum eigenstates," Nucl. Phys., B152, No. 2, 266 (1979).
27. A. Wiesler, "Rigorous proof of "Bethe's hypothesis" from inverse scattering transformation," Preprint. Univ. Freiburg THEP 79/8, October, 1979.
28. H. Grosse, "On the construction of Möller operators for the nonlinear Schrödinger equation," Phys. Lett., 86B, Nos. 3-4, 267-271 (1979).
29. E. K. Sklyanin, "On complete integrability of the Landau—Lifshits equation," Preprint, Leningrad Branch of the Mathematics Institute, E-3-79, Leningrad (1979).
30. V. E. Zakharov and S. V. Manakov, "Complete integrability of the nonlinear Schrödinger equation," Teor. Mat. Fiz., 19, No. 3, 332-343 (1974).

31. V. E. Zakharov and A. B. Shabat, "Exact theory of two-dimensional self-focusing and one-dimensional automodulation of waves in nonlinear media," Zh. Eksp. Teor. Fiz., 61, No. 1, 118-134 (1971).
32. L. A. Takhtadzhyan, "Hamiltonian systems connected with Dirac's equation," Zap. Nauchn. Sem. Leningr. Otd. Mat. Inst., 37, 66-76 (1973).
33. F. A. Berezin, G. P. Pokhil, and V. M. Finkel'berg, "Schrödinger's equation for systems of one-dimensional particles with pointwise interactions," Vestn. Mosk. Gos. Univ., Ser. Mat., Mekh., No. 1, 21-28 (1964).
34. J. B. McGuire, "Study of exactly soluble one-dimensional N-body problem," J. Math. Phys., 5, No. 5, 622 (1964).
35. I. M. Gel'fand and L. A. Dikii, "Resolvent and Hamiltonian systems," Funkts. Anal., 11, No. 2, 11-27 (1977).
36. F. A. Berezin, Method of Secondary Quantization [in Russian], Nauka, Moscow (1965).
37. A. S. Shvarts, Mathematical Foundations of Quantum Field Theory [in Russian], Atomizdat, Moscow (1975).
38. A. A. Tsvetkov, "Integrals of motion of a system of bosons with pointwise interactions," Vestn. Mosk. Gos. Univ., Ser. Mat., Mekh., No. 4, 61-69 (1977).
39. F. A. Berezin, "Quantization," Izv. Akad. Nauk SSSR, Ser. Mat., 38, No. 5, 1116-1175 (1974).
40. F. A. Berezin, "Wick and anti-Wick symbols of operators," Mat. Sb., 86 (128), No. 4 (12), 578-610 (1971).
41. Yu. S. Tyupkin, V. A. Fateev, and A. S. Shvarts, "Connection of particle-similar solutions of classical equations with quantum particles," Yad. Fiz., 22, No. 3, 622-631 (1975).

SOLUTIONS OF THE YANG—BAXTER EQUATION

P. P. Kulish and E. K. Sklyanin UDC 517.43+530.145

We give the basic definitions connected with the Yang—Baxter equation (factorization condition for a multiparticle S-matrix) and formulate the problem of classifying its solutions. We list the known methods of solution of the Y—B equation, and also various applications of this equation to the theory of completely integrable quantum and classical systems. A generalization of the Y—B equation to the case of Z_2-graduation is obtained, a possible connection with the theory of representations is noted. The supplement contains about 20 explicit solutions.

0. By the Yang—Baxter equation [1, 2] is meant the following functional equation:

$$ {}_{\alpha\alpha'}R_{\gamma\gamma'}(u-v)\,{}_{\gamma\alpha''}R_{\beta\gamma''}(u)\,{}_{\gamma'\gamma''}R_{\beta'\beta''}(v) = {}_{\alpha'\alpha''}R_{\gamma'\gamma''}(v)\,{}_{\alpha\gamma''}R_{\gamma\beta''}(u)\,{}_{\gamma\gamma'}R_{\beta\beta'}(u-v) \tag{1} $$

for a collection of functions ${}_{\alpha\beta}R_{\gamma\delta}(u)$ of a complex parameter u, depending on four indices $\alpha, \beta, \gamma, \delta$, running through values from 1 to some natural number N. In (1) and later we understand summation over repeated indices.

Equation (1), which first appeared in [1, 2], has many applications to the theory of completely integrable quantum and classical systems and exactly solvable models of statistical physics. In recent years it has undergone intensive study. Here the profound connection of (1) with such areas of mathematics as group theory and algebraic geometry has become more and more apparent.

The present paper is an (apparently the first) attempt to give a systematic survey of the facts accumulated at the time it is written relating to the solutions of (1). The account is structured in the following way. In Sec. 1 we give the basic definitions and we

Translated from Zapiski Nauchnykh Seminarov Leningradskogo Otdeleniya Matematicheskogo Instituta im. V. A. Steklova AN SSSR, Vol. 95, pp. 129-160, 1980.

formulate the problem of the classification of the solutions of (1). In Sec. 2 we list various applications of (1) to the theory of completely integrable systems. In Sec. 3 we discuss known methods of solution of (1) and we formulate some assertions about properties of its solutions. In Sec. 4 we consider generalizations of the Yang–Baxter equation, and finally, in the Supplement we give a summary of the known solutions of (1).

The authors hope that the present paper will be useful to specialists in the theory of completely integrable systems and the method of the inverse scattering problem, and also helps to attract attention of mathematicians who are specialists in group theory and algebraic geometry to a new promising object of investigation, the Yang–Baxter equation.

The authors thank L. D. Faddeev, the initiator of studies on the quantum method of the inverse problem, V. E. Korepin, A. G. Reiman, M. A. Semenov-Tyan-Shanskii, L. A. Takhtadzhyan, N. Yu. Reshetikhin, and S. A. Tsyplyaev for many helpful discussions. We are grateful to A. A. Belavin, A. B. Zamolodchikov, and V. A. Fateev for giving us a series of solutions of the Yang–Baxter equation.

1. Since the systematic study of the Yang–Baxter equation (Y–B) has only just begun, there is still no generally accepted terminology in this area. In this section we make an attempt to propose a system of terms and definitions for the theory of solutions of the Y–B equation. Practice will show how successful this attempt is.

First of all, we discuss a series of equivalent ways of writing the Y–B equation (1). For this we note that the four-indexed quantities ${}_{\alpha\beta}R_{\gamma\delta}(u)$ can constitute a linear operator $\mathbb{R}(u)$ in the tensor product of two N-dimensional complex spaces $V\otimes V\,(V=\mathbb{C}^N)$. The action of this operator on the basis vector $e_\gamma\otimes e_\delta$ is given by the following formula:

$$\mathbb{R}(e_\gamma\otimes e_\delta)=(e_\alpha\otimes e_\beta)\,{}_{\alpha\beta}R_{\gamma\delta}\,. \tag{2}$$

The tensor ${}_{\alpha\beta}R_{\gamma\delta}$ can also constitute three operators $\mathbb{R}_{12}$, $\mathbb{R}_{13}$, $\mathbb{R}_{23}$ in the tensor product $V\otimes V\otimes V$, corresponding to the three ways of imbedding the space $V\otimes V$ in $V\otimes V\otimes V$:

$$\begin{aligned}
&\mathbb{R}_{12}(e_\gamma\otimes e_{\gamma'}\otimes e_{\gamma''})=(e_\alpha\otimes e_{\alpha'}\otimes e_{\gamma''})\,{}_{\alpha\alpha'}R_{\gamma\gamma'},\\
&\mathbb{R}_{13}(e_\gamma\otimes e_{\gamma'}\otimes e_{\gamma''})=(e_\alpha\otimes e_{\gamma'}\otimes e_{\alpha''})\,{}_{\alpha\alpha''}R_{\gamma\gamma''}\,,\\
&\mathbb{R}_{23}(e_\gamma\otimes e_{\gamma'}\otimes e_{\gamma''})=(e_\gamma\otimes e_{\alpha'}\otimes e_{\alpha''})\,{}_{\alpha'\alpha''}R_{\gamma'\gamma''}.
\end{aligned} \tag{3}$$

Notation (3) we have introduced allows us to write the Yang–Baxter equation (1) as an operator equation:

$$\mathbb{R}_{12}(u-v)\,\mathbb{R}_{13}(u)\,\mathbb{R}_{23}(v)=\mathbb{R}_{23}(v)\,\mathbb{R}_{13}(u)\,\mathbb{R}_{12}(u-v)\,. \tag{4}$$

In order to avoid misunderstandings, it is necessary to note that there also exists another system of notation, used, e.g., in [3-5]. In these papers, instead of the operator $\mathbb{R}$, introduced above, there is used an operator $\check{\mathbb{R}}$, differing from $\mathbb{R}$ by multiplication by the permutation operator $\mathbb{P}$:

$$\check{\mathbb{R}}=\mathbb{P}\mathbb{R}\,, \tag{5}$$

where

$${}_{\alpha\beta}P_{\gamma\delta}=\delta_{\alpha\delta}\delta_{\beta\gamma}\,. \tag{6}$$

Here the Yang–Baxter equation assumes the form:

$$(I\otimes\check{\mathbb{R}}(u-v))(\check{\mathbb{R}}(u)\otimes I)(I\otimes\check{\mathbb{R}}(v))=(\check{\mathbb{R}}(v)\otimes I)(I\otimes\check{\mathbb{R}}(u))(\check{\mathbb{R}}(u-v)\otimes I)\,. \tag{7}$$

This notation for the Y–B equation is interesting in that it preserves its form in the graduated case too (see Sec. 4).

Now we introduce a series of concepts which we need later. A solution $\mathbb{R}(u)$ of the Yang–Baxter equation (4) will be called a Yang–Baxter sheaf. The natural number N (dimension of the space V) will be called the dimension of the sheaf. The variable u, figuring in the Yang–Baxter equation (4), will be called the spectral parameter, in contrast with the other parameters $\xi, \eta, \zeta \ldots$, on which the sheaf $\mathbb{R}(u,\xi,\eta,\ldots)$ possibly depends, and which we shall call connection constants. We shall call the Yang–Baxter sheaf $\mathbb{R}(u)$ regular if for $u=0$ the operator $\mathbb{R}(u)$ is equal to the permutation operator $\mathbb{P}$ (6), which obviously satisfies (4):

$$\mathbb{P}_{12}\mathbb{P}_{13}\mathbb{P}_{23} = \mathbb{P}_{23}\mathbb{P}_{13}\mathbb{P}_{12}. \tag{8}$$

It is often useful to consider not the isolated sheaf $\mathbb{R}(u)$, but a family of Yang–Baxter sheaves $\mathbb{R}(u,\eta)$, depending on the connection constant η. We call the family $\mathbb{R}(u,\eta)$ quasiclassical if for some value of the parameter $\eta=\eta_0$ (one usually chooses the normalization $\eta_0=0$, which we shall also do in what follows) one has, identically in u,

$$\mathbb{R}(u,\eta)\big|_{\eta-\eta_0} = I, \tag{9}$$

where I is the identity operator in the space $V\otimes V$:

$${}_{\alpha\beta}I_{\gamma\delta} = \delta_{\alpha\gamma}\delta_{\beta\delta}. \tag{10}$$

If, moreover, for each η the sheaf $\mathbb{R}(u,\eta)$ is regular in the sense of the definition given above, such a family of sheaves will be called canonical.

Equation (4) admits a series of obvious transformations, leaving it invariant:

1) Multiplication of the solution $\mathbb{R}(u)$ by an arbitrary scalar function $f(u)$ again gives a solution of (4):

$$\mathbb{R}'(u) = f(u)\mathbb{R}(u). \tag{11}$$

Sheaves $\mathbb{R}(u)$ and $\mathbb{R}'(u)$, connected by (11), will be called homothetic.

2) Similarity transformation. Let T be a nondegenerate operator in the space V. Then, as one verifies easily, the sheaf

$$\mathbb{R}'(u) = (T\otimes T)\,\mathbb{R}(u)(T\otimes T)^{-1} \tag{12}$$

satisfies the Yang–Baxter equation (4). Sheaves $\mathbb{R}(u)$ and $\mathbb{R}'(u)$, connected by (12), will be called similar. We note that a similarity transformation preserves the properties of regularity, quasiclassicism, and canonicity of a Y–B sheaf or family of sheaves. Two sheaves connected by a similarity transformation and homothetic will be called equivalent.

If in the space V the representation $T(g)$ of some group G acts, then we shall call the sheaf $\mathbb{R}(u)$ invariant with respect to the representation $T(g)$, if for any $g\in G$ one has

$$\mathbb{R}(u)(T(g)\otimes T(g)) = (T(g)\otimes T(g))\,\mathbb{R}(u). \tag{13}$$

Let $\mathbb{R}^{(1)}(u)$ and $\mathbb{R}^{(2)}(u)$ be two solutions of (4) of dimensions N_1 and N_2, respectively. By the tensor product of the sheaves $\mathbb{R}^{(1)}(u)$ and $\mathbb{R}^{(2)}(u)$ we shall mean the sheaf $(\mathbb{R}^{(1)}\otimes\mathbb{R}^{(2)})(u)$ of dimension $N_1\times N_2$ defined by

$$(\mathbb{R}^{(1)}\otimes\mathbb{R}^{(2)})(u)=\mathbb{R}^{(1)}(u)\otimes\mathbb{R}^{(2)}(u)\,. \tag{14}$$

By the direct sum of the sheaves $\mathbb{R}^{(1)}(u)$ and $\mathbb{R}^{(2)}(u)$ we shall mean the sheaf $(\mathbb{R}^{(1)}+\mathbb{R}^{(2)})(u)$ of dimension N_1+N_2, defined in the following way. The operator $(\mathbb{R}^{(1)}+\mathbb{R}^{(2)})(u)$ acts on the basis vectors of the form $e^{(i)}_{\alpha_i}\otimes e^{(k)}_{\alpha_k}$ $(i,k=1,2;\ \alpha_i=1,2,\ldots,N_i;\ e^{(i)}_{\alpha_i}\in V_i)$ of the space $(V_1+V_2)\otimes(V_1+V_2)$ by:

$$\begin{aligned}(\mathbb{R}^{(1)}+\mathbb{R}^{(2)})(u)(e^{(1)}_{\alpha_1}\otimes e^{(1)}_{\beta_1})&=\mathbb{R}^{(1)}(u)(e^{(1)}_{\alpha_1}\otimes e^{(1)}_{\beta_1}),\\(\mathbb{R}^{(1)}+\mathbb{R}^{(2)})(u)(e^{(1)}_{\alpha_1}\otimes e^{(2)}_{\beta_2})&=e^{(1)}_{\alpha_1}\otimes e^{(2)}_{\beta_2},\\(\mathbb{R}^{(1)}+\mathbb{R}^{(2)})(u)(e^{(2)}_{\alpha_2}\otimes e^{(1)}_{\beta_1})&=e^{(2)}_{\alpha_2}\otimes e^{(1)}_{\beta_1},\\(\mathbb{R}^{(1)}+\mathbb{R}^{(2)})(u)(e^{(2)}_{\alpha_2}\otimes e^{(2)}_{\beta_2})&=\mathbb{R}^{(2)}(u)(e^{(2)}_{\alpha_2}\otimes e^{(2)}_{\beta_2}).\end{aligned} \tag{15}$$

Direct verification shows that $(\mathbb{R}^{(1)}\otimes\mathbb{R}^{(2)})(u)$ and $(\mathbb{R}^{(1)}+\mathbb{R}^{(2)})(u)$ actually satisfy (4). It is also obvious that the operation of tensor multiplication of Y–B sheaves preserves the property of regularity of families of sheaves. Now the operation of addition, on the contrary, preserves only the property of quasiclassicism of sheaves. In addition, it follows from (15) that the direct sum of two sheaves is never a regular sheaf.

If the space V admits a decomposition into a direct sum of two subspaces V_1 and V_2, such that the action of the operator $\mathbb{R}(u)$ on the basis vectors of the form $e^{(i)}_{\alpha_i}\otimes e^{(k)}_{\beta_k}$ (notation is the same as in (15)) has the following property

$$\mathbb{R}(u)(e^{(i)}_{\alpha_i}\otimes e^{(k)}_{\alpha_k})\in V_i\otimes V_k\,,\quad i,k=1,2\,; \tag{16}$$

then the sheaf $\mathbb{R}(u)$ is called reducible. In particular, a reducible sheaf is always the direct sum of two sheaves. If no such decomposition exists, we shall call such a sheaf irreducible. It is easy to prove that for a reducible sheaf $\mathbb{R}(u)$ the operators $\mathbb{R}^{(1)}(u)$ and $\mathbb{R}^{(2)}(u)$, acting in the spaces $V_1\otimes V_1$ and $V_2\otimes V_2$, respectively, according to the formula

$$\mathbb{R}^{(i)}(u)(e^{(i)}_{\alpha_i}\otimes e^{(i)}_{\beta_i})=\mathbb{R}(u)(e^{(i)}_{\alpha_i}\otimes e^{(i)}_{\beta_i}),\quad i=1,2, \tag{17}$$

will also be Yang–Baxter sheaves. It is also obvious that a reducible sheaf cannot be regular.

As we shall see later, Y–B sheaves play a large role in the theory of quantum completely integrable systems. The analogous role in the theory of classical completely integrable systems is played by the classical Yang–Baxter sheaf, whose definition we shall now give. Let $\mathbb{R}(u,\eta)$ be a quasiclassical family of Yang–Baxter sheaves, depending smoothly on the parameter η. Then, differentiating (4) with respect to η and setting $\eta=0$, we get, keeping (9) in mind, for the quantities

$$r(u)=\frac{\partial}{\partial\eta}\mathbb{R}(u,\eta)\Big|_{\eta=0} \tag{18}$$

the following equation

$$r_{12}(u-v)\,r_{13}(u)+r_{12}(u-v)\,r_{23}(v)+r_{13}(u)\,r_{23}(v) = r_{23}(v)\,r_{13}(u)+r_{23}(v)\,r_{12}(u-v)+r_{13}(u)\,r_{12}(u-v), \qquad (19)$$

which can be rewritten in the following commutator form:

$$\left[r_{12}(u-v),\, r_{13}(u)+r_{23}(v)\right]+\left[r_{13}(u),\, r_{23}(v)\right] = 0\ . \qquad (20)$$

By a classical Yang–Baxter sheaf will be meant any solution of the functional equation (20). The calculation given above shows that for a known quasiclassical family of Y–B sheaves one can always construct a classical Yang–Baxter sheaf. Whether the converse is valid, i.e., whether one can for any classical Y–B sheaf construct a corresponding quasiclassical family of Y–B sheaves, is still unknown. Using (18), one can transfer almost all concepts we have introduced for Y–B sheaves to the case of classical Y–B sheaves. In particular, instead of invariance with respect to homothety transformations (11), classical Y–B sheaves are invariant with respect to the translation transformation

$$r'(u) = r(u)+f(u)\,I\ . \qquad (21)$$

The definitions of similarity transformation (12) and group invariance (13) carry over to classical Y–B sheaves without change. We shall call two classical Y–B sheaves equivalent if one of them can be turned into the other by a similarity transformation and a translation.

We shall call a classical Y–B sheaf $r(u)$ canonical if for it one has the following equation:

$$r(u) = -\mathbb{P}\,r(-u)\,\mathbb{P}\ . \qquad (22)$$

The connection between the concepts of canonicity of classical and quantum Y–B sheaves is established by the following theorem.

THEOREM. Any canonical family of Y–B sheaves $\mathbb{R}(u,\eta)$ generates by (18) a classical Y–B sheaf $r(u)$, equivalent with a canonical one.

Proof. We differentiate (4) with respect to η and we set $u=0$ $\eta=0$. Multiplying the result obtained

$$r_{12}(-v)\,\mathbb{P}_{13}+\mathbb{P}_{13}\,r_{23}(v) = \mathbb{P}_{13}\,r_{12}(-v)+r_{23}(v)\,\mathbb{P}_{13} \qquad (23)$$

on the right by $\mathbb{P}_{13}$ and using the obvious equations

$$\mathbb{P}_{13}\,r_{21}(v)\,\mathbb{P}_{12} = r_{21}(v) = \mathbb{P}_{12}\,r_{12}(v)\,\mathbb{P}_{12}\ ,$$
$$\mathbb{P}_{13}\,r_{12}(-v)\,\mathbb{P}_{13} = r_{32}(-v) = \mathbb{P}_{23}\,r_{23}(-v)\,\mathbb{P}_{23}\ , \qquad (24)$$

we arrive at the equation

$$r_{12}(-v)+r_{21}(v) = r_{32}(-v)+r_{23}(v)\ . \qquad (25)$$

By virtue of the obvious symmetry of the Y–B equation with respect to permutation of the spaces V_1, V_2, V_3 , one also has

$$r_{12}(-v)+r_{21}(v) = r_{13}(-v)+r_{31}(v)\ . \qquad (26)$$

Comparing (25) and (26), we arrive at the inference that the operator $r_{12}(-v)+r_{21}(v)$ acts trivially on all three spaces V_1, V_2 , V_3 , i.e.,

$$r_{12}(-v)+r_{12}(v) = \varphi(v) I_1 \otimes I_2 \otimes I_3 ,$$

where the scalar function $\varphi(v)$ must be even $\varphi(v)=\varphi(-v)$. Redefining $r(v)\to r(v)+\frac{1}{2}\varphi(v)$, we get

$$r_{12}(-v) = -r_{21}(v) ,$$

equivalent with (22), which is what had to be proved.

The tensor product $(r^{(1)}\otimes r^{(2)})(u)$ of the classical Y–B sheaves $r^{(1)}(u)$ and $r^{(2)}(u)$ is given by:

$$(r^{(1)}\otimes r^{(2)})(u) = r^{(1)}(u)\otimes I^{(2)} + I^{(1)}\otimes r^{(2)}(u) , \tag{27}$$

and the direct sum by $(r^{(1)}+r^{(2)})(u)$

$$(r^{(1)}+r^{(2)})(u) = r^{(1)}(u)+r^{(2)}(u). \tag{28}$$

The definitions of reducible and irreducible sheaves are carried over to the case of classical Y–B sheaves unchanged.

We formulate to conclude this section a series of unsolved problems, standing in front of the theory of quantum and classical Yang–Baxter sheaves:

1) List all solutions of the Yang–Baxter equation (4) of a given dimension N up to equivalence. The analogous problem is of interest for regular, quasiclassical, and canonical sheaves, and families of Y–B sheaves, and also for Y–B sheaves having group invariance. The problem of listing all constant solutions of the Y–B equation, i.e., those independent of the spectral parameter, is also interesting:

$$\mathbb{R}_{12}\mathbb{R}_{13}\mathbb{R}_{23} = \mathbb{R}_{23}\mathbb{R}_{13}\mathbb{R}_{12} . \tag{29}$$

It is easy to see that (29) is satisfied, in particular, by the permutation operator $\mathbb{P}$ (6, 8) and the identity operator I (10). Setting in (4) $u=v=0$, we get that for any Y–B sheaf $\mathbb{R}(u)$, its value for $u=0$ also satisfies (29).

2) The same problems are naturally formulated also for classical Yang–Baxter sheaves. In addition to constant solutions, here there is also interest in solutions of the classical Yang–Baxter equation (20) of the form

$$r(u) = \frac{r}{u} , \tag{30}$$

where the operator r must by virtue of (20) satisfy

$$[r_{12}, r_{13}+r_{23}] = 0, \quad [r_{12}+r_{13}, r_{23}] = 0 . \tag{31}$$

(For canonical sheaves $r=\mathbb{P}r\mathbb{P}$, and the equations in (31) are equivalent.)

It is easy to construct a wide class of solutions of (31). In fact, let $\mathfrak{g}$ be an arbitrary semisimple Lie algebra, J_α $(\alpha=1,2,\dots,N)$ be a basis of its generators in an arbitrary representation, $k^{\alpha\beta}$ be the matrix inverse to the matrix of the Killing form of the Lie algebra $\mathfrak{g}$ in the basis of generators J_α . Then, as is easy to verify, the operator r, defined by the formula

$$r = k^{\alpha\beta} J_\alpha \otimes J_\beta , \tag{32}$$

satisfies (31). This solution was also obtained in [6], where there is the assertion that (32) gives a complete description of the solutions of (31).

3) Can one associate with any classical Y–B sheaf $r(u)$ a classical family of Y–B sheaves such that (18) holds?

2. Now we discuss applications of the Yang–Baxter equation to the theory of quantum and classical completely integrable systems.

1) We shall show, first of all, that to any regular Yang–Baxter sheaf one can associate a quantum completely integrable system with locally mutually commuting integrals of motion. In fact, let $\mathbb{R}(u)$ be a regular Y–B sheaf of dimension N. We take as state space $\mathcal{H}$ of the quantum system sought the space $\mathcal{H}=V_1\otimes V_2\ldots\otimes V_M$ ($V_n=\mathbb{C}^N$; $n=1,2,\ldots,M$), where M is an arbitrary natural number ≥ 2, and we define in this space a Hamiltonian $\mathbb{H}$ by the formula:

$$\mathbb{H}=\sum_{n=1}^{M-1}\mathbb{H}_{n+1,n}+\mathbb{H}_{1,M}, \tag{33}$$

where the local density of the Hamiltonian $\mathbb{H}_{n+1,n}$ is given by

$$\mathbb{H}_{n+1,n}=\left(\frac{d}{du}\mathbb{R}_{n+1,n}(u)\Big|_{u=0}\right)\mathbb{P}_{n+1,n}. \tag{34}$$

We explain the notation. The operator $\mathbb{R}_{n+1,n}(u)$ acts in the space $V_{n+1}\otimes V_n$, as the corresponding Yang–Baxter sheaf $\mathbb{R}(u)$, and on the remaining components of the tensor product $V_1\otimes\ldots\otimes V_M$ it acts as the identity operator. The same relates to the permutation operator $\mathbb{P}_{n+1,n}$. It is convenient to represent the quantum system we have constructed as a ring of M "atoms," each of which has N quantum states, where only the closest neighbors interact. We note that although the Hamiltonian $\mathbb{H}$, defined above, generally speaking need not be a self-adjoint operator (which, however, is not reflected in the following calculations), in practice, in the majority of cases it can be made self-adjoint by multiplying by a suitable constant.

A sequence of operators commuting with $\mathbb{H}$ is constructed in the following way. We extend our space $\mathcal{H}$ to the space $\hat{\mathcal{H}}=Q\otimes Q'\otimes\mathcal{H}$, introducing two auxiliary spaces Q and Q' isomorphic with $\mathbb{C}^N$. We define the transition operator $\mathbb{T}_1^M(u)$ by

$$\mathbb{T}_1^M(u)=\mathbb{L}_M(u)\mathbb{L}_{M-1}(u)\ldots\mathbb{L}_1(u), \tag{35}$$

where $\mathbb{L}_n(u)=\mathbb{R}_{qn}(u)$ (the notation is the same as above, the index n relates to the space V_n, the index q to the space Q). Analogously, replacing Q by Q' one defines operators $\mathbb{L}'_n(u)$ and $\mathbb{T}_1'^M(u)$.

Using the notation introduced, one can rewrite (4) in the form

$$\mathbb{R}_{qq'}(u-v)\mathbb{L}_n(u)\mathbb{L}'_n(v)=\mathbb{L}'_n(v)\mathbb{L}_n(u)\mathbb{R}_{qq'}(u-v). \tag{36}$$

Starting from (36) one can prove [2–4] the following remarkable equation:

$$\mathbb{R}_{qq'}(u-v)\mathbb{T}_1^M(u)\mathbb{T}_1'^M(v)=\mathbb{T}_1'^M(v)\mathbb{T}_1^M(u)\,\mathbb{R}_{qq'}(u-v). \tag{37}$$

The generating function $t(u)$ of the integrals of motion of the quantum system considered is defined as the trace of the transition operator $\mathbb{T}_1^M(u)$, taken with respect to the auxiliary space Q

$$t(u) = \mathrm{tr}_q \mathbb{T}_1^M(u) . \tag{38}$$

It follows [2-4] from (37) that $t(u)$ is a family of mutually commuting operators in $\mathcal{H}$:

$$[t(u), t(v)] = 0 . \tag{39}$$

As shown in [7], $\ln(t^{-1}(0)t(u))$ is the generating functional of the local integrals of motion $\mathcal{J}_n$ for the Hamiltonian $\mathbb{H}$:

$$\mathcal{J}_n = \frac{d^n}{du^n} \ln(t^{-1}(0)t(u))\Big|_{u=0} \tag{40}$$

(locality means that we can represent the operator $\mathcal{J}_n$ as the sum of operators, each of which acts nontrivially at no more than $n+1$ neighboring nodes of the lattice). In particular, for $n=1$, (40) gives the Hamiltonian $\mathbb{H} = \mathcal{J}_1$.

The question of completeness of the system of integrals of motion $\mathcal{J}_n$ in the space $\mathcal{H}$ has only been weakly studied so far (completeness is strictly proved only for one of the simplest models — the Heisenberg ferromagnet [8]). The conjecture on the completeness of the integrals of motion (40) for the known Y—B sheaves of dimension $N=2$ is quite plausible. On the other hand, for dimensions $N>2$, the completeness of the system $\mathcal{J}_n$ is automatically false as comparison with the corresponding classical completely integrable equations shows [9, 10]. The problem of constructing the missing integrals of motion in this case is still unsolved.

It is not excluded that there exist methods of constructing from a given Y—B sheaf other completely integrable quantum models too. For example, in the recent paper [11] there is constructed a relativistically invariant model of the quantum field theory, closely connected with the Y—B XYZ-sheaf (S8). Probably this result can be generalized to the case of an arbitrary Y—B sheaf.

2) The construction given above of the integrals of motion for a quantum model on a lattice was based on (37). This equation plays a most important role in the quantum method of the inverse problem [3, 4]. Besides the construction of integrals of motion, it allows one to find, in many cases, the eigenfunctions of the Hamiltonian $\mathbb{H}$ and its spectrum [3, 4]

If the Yang—Baxter sheaf $\mathbb{R}$, on which one constructs by the method described above a quantum completely integrable model on a lattice, depends also on additional parameters η, ξ,..., then it often turns out to be possible to perform the passage to the limit, as a result of which one now gets a continuous completely integrable model of quantum field theory on the line. For example, from the XXZ model one gets in this way the nonlinear Schrödinger equation [12], from the XYZ-model one gets the Tirring model [7] and the quantum sin-Gordon equation.

The limit passage mentioned is usually realized in the following way. Depending on the character of the model, one chooses "critical" points of the spectral parameter and connection constant and performs a scaling limit passage with respect to the size of the lattice $a \to 0$, so that $an = x$ remains fixed (n is the number of nodes of the lattice). For the choice of "critical" values and scaling parameter, one considers the natural requirements of decomposability of $\mathbb{L}_n(u)$ from (36) in a series with respect to the scaling parameter and simplicity (diagonality or multiplicity 1) of the highest term

$$\mathbb{L}_n(u,\eta,\ldots) \sim \mathbb{L}^{(0)}(\alpha,\gamma,\ldots) + a\mathbb{L}^{(1)}(\alpha,\gamma,\ldots) + O(a^2)\,. \tag{41}$$

It is important to emphasize that the quantities $\mathbb{R}_{qq'}(\alpha,\gamma,\ldots)$ and $\mathbb{L}(x,a) = \mathbb{L}^{(0)} + a\mathbb{L}^{(1)}$ in (36) now get (after passage to the limit) a completely different interpretation in contrast with the lattice case, where $\mathbb{R}_{qq'}(u)$ and $\mathbb{L}_n(u)$ represent one and the same Yang—Baxter sheaf of dimension N. At the same time that $\mathbb{R}_{qq'}(\alpha,\gamma,\ldots)$ remains a numerical $N^2 \times N^2$ matrix, the operator $\mathbb{L}(x,a,\ldots)$ is interpreted as an $N \times N$ matrix, whose elements are operator-valued functions on the line, e.g., $\psi(x), \psi^*(x)$ for the nonlinear Schrödinger equation [12, 13] and $\pi(x)$, $\exp(\pm i\varphi(x))$ for the quantum sin-Gordon equation [14].

We choose the following parametrization of the Y—B sheaf (S8), connected with the XYZ-model [2],

$$\mathbb{R}(u,\eta,k) = \sum_{j=1}^{4} w_j(u,\eta,k)\,\sigma_j \otimes \sigma_j'\,,$$

where $\sigma_j, \sigma_j', j=1,2,3$ are Pauli matrices acting in $V = V' = \mathbb{C}^2$, and σ_4, σ_4' are identity matrices in these spaces,

$$w_1 = \frac{\operatorname{sn}\eta}{\operatorname{sn}u}\,, \quad w_2 = \frac{\operatorname{sn}\eta\,\operatorname{dn}u}{\operatorname{sn}u\,\operatorname{dn}\eta}\,, \quad w_3 = \frac{\operatorname{sn}\eta\,\operatorname{cn}u}{\operatorname{sn}u\,\operatorname{cn}\eta}\,, \quad w_4 = 1\,.$$

The coefficients w_j can be expressed in terms of an analytic function of modulus k. Making the substitutions (K, K' are complete elliptic integrals of the first kind of moduli k and $k' = \sqrt{1-k^2}$)

$$u = i\alpha - iK'\,, \quad \eta = \gamma + \frac{\pi}{2} - K$$

and letting $k \to 0\,(k \sim a)$, we get the $\mathbb{L}$-operator of the quantum sin-Gordon equation [14]

$$\mathbb{L}(x,\alpha,\gamma,\ldots) \sim \begin{pmatrix} e^{i\gamma p} & 0 \\ 0 & e^{-i\gamma p} \end{pmatrix} + a \begin{pmatrix} 0 & e^{-\alpha}u^- - e^{\alpha}u^+ \\ e^{\alpha}u^- - e^{-\alpha}u^+ & 0 \end{pmatrix},$$

$$p = \int_x^{x+a} \pi(y)\,dy\,, \quad u^{\pm} = \exp\Big(\pm i\frac{1}{a}\int_x^{x+a}\varphi(y)\,dy\Big)\,.$$

The linear problem for the quantum nonlinear Schrödinger equation [13] is obtained in the following way. First we carry out the degeneration of the Y—B sheaf (S8) in the sheaf of the XXZ-model on a lattice (S9) (the modulus of the elliptic functions $k=0$). Then we proceed to the scaling limit in (S9) [12]

$$\eta = i\sqrt{2\gamma a}\ , \qquad u = \frac{\pi}{2} + i(\alpha - \frac{\gamma}{2})\sqrt{\frac{a}{2\gamma}}\ ,$$

where γ is the connection constant of the nonlinear Schrödinger equation, α is the spectral parameter of the linear problem. As a result, we get $(\sigma_n^{\pm}/2\sqrt{a} \longrightarrow \psi^{\pm}(x))$

$$\mathbb{L}_n(u,\eta) \sim \begin{pmatrix} 1 + i\frac{\alpha}{2}a\ , & i\sqrt{2\gamma}\int_x^{x+a} \psi^{-}(y)dy \\ i\sqrt{2\gamma}\int_x^{x+a} \psi^{+}(y)dy\ , & 1 - i\frac{\alpha}{2}a \end{pmatrix} .$$

Equation (36), in which $\mathbb{R}_{qq'}(u)$ is replaced by the limit of the matrices, and $\mathbb{L}_n$ is replaced by the approximate transition operator on the interval $(x, x+a)$, is satisfied only up to a. Now (37) remains valid after limit passage too, which allows one to apply for models on the line the quantum method of the inverse problem. The corresponding matrix $\mathbb{R}$, realizing the similarity of the tensor products of transition matrices of the quantum linear problems $\mathbb{T}(\lambda)\otimes\mathbb{T}(\mu)$ and $\mathbb{T}(\mu)\otimes\mathbb{T}(\lambda)$, will be called the quantum $\mathbb{R}$-matrix.

3) Most of the results of the first two parts of this section relating to quantum completely integrable systems can be carried over to the classical case too. The role of quantum Yang–Baxter sheaves will be played here by the classical sheaves. For example, analogously to the way for a quantum Yang–Baxter sheaf there was constructed a quantum completely integrable system, with any classical Yang–Baxter sheaf one can in a canonical way associate a classical completely integrable system of Heisenberg ferromagnet type [15]. Without describing this construction in detail, we note only that the quantum equations (36) and (37) correspond here to the classical equations

$$\{ L'(x,u), L''(y,v) \} = -i\left[r(u-v), L'(x,u) + L''(y,v) \right]\ , \tag{42}$$

$$\{ T'^{x}_{y}(u), T''^{x}_{y}(v) \} = -i\left[r(u-v), T'^{x}_{y}(u) + T''^{x}_{y}(v) \right]\ . \tag{43}$$

Equation (42) reproduces (20) with this difference that in the decomposition of $r_{13}(u)$ and $r_{23}(v)$ in terms of generators of some Lie algebra $\mathcal{J}_\alpha^{(i)}$, $i=1,2,3$, the generators $\mathcal{J}_\alpha^{(3)}$ are replaced by functions $S_\alpha(x)$, of the Poisson brackets for which the commutation relations for the generators $\mathcal{J}_\alpha$ are reproduced and the commutator $[r_{13}(u), r_{23}(v)]$ is replaced by the Poisson brackets $L'(x,u)$ and $L''(y,v)$. Thus, $L(x,u)$ is an $N\times N$ matrix whose matrix elements are functions on the phase space of the dynamical system with Poisson bracket

$$\{ S_\alpha(x), S_\beta(y) \} = -c^{\gamma}_{\alpha\beta} S_\gamma(x)\delta(x-y),\ ([\mathcal{J}_\alpha, \mathcal{J}_\beta] = i c^{\gamma}_{\alpha\beta}\mathcal{J}_\gamma)\ .$$

The classical transition matrix is determined as a fundamental solution of the differential equation

$$\frac{\partial}{\partial x} T^{x}_{y}(u) = L(x,u) T^{x}_{y}(u), \qquad T^{x}_{x}(u) = I\ , \tag{44}$$

and the matrices $L'(x,u)$ and $L''(y,v)$ are given by

$$L'(x,u) = L(x,u)\otimes I\ , \quad L''(y,v) = I \otimes L(y,v)\ . \tag{45}$$

In order to get (43) from (37), it is necessary to expand (37) in powers of the quasiclassical parameter η as $\eta \to 0$ and to use the relation $[\cdot,\cdot] \to -i\hbar\{\cdot,\cdot\}$ between the quantum commutator and the classical Poisson bracket, retaining in (37) terms of order η. The parameter η plays here the role of Planck's constant $\hbar$ [15].

As in the quantum case too, (43) allows one to calculate the Poisson brackets between matrix elements of the transition matrix and to construct commuting integrals of motion and action-angle variables [15]. The matrix figuring in (42), (43) will be called the classical r-matrix.

In the scheme described above there is contained a large number of completely integrable classical models, e.g., the nonlinear Schrödinger equation [13], the sin-Gordon equation, the Landau–Lifshits equation [15], Toda chain [3, 16], etc. [30]. It is essential here, however, that the Poisson brackets between dynamical variables are ultralocal in the terminology of [3], i.e., do not contain derivatives of the δ-function. There is interest in the problem of generalizing this scheme to equations with nonultralocal Poisson brackets, e.g., the Korteweg–de Vries equation. The first steps in this direction were taken by S. A. Tsyplyaev, who proved that for the sin-Gordon equation the classical r-matrices in the laboratory system $(\{\pi(x),\varphi(y)\}=\delta(x-y))$ and in the light cone system $(\{\varphi(x),\varphi(y)\}=\delta'(x-y))$ coincide.

4) To conclude this section, we note that there exists another important interpretation of the Yang–Baxter equation (1), as the condition for factorization of multiparticle S-matrices. The functions ${}_{\alpha\beta}R_{\gamma\delta}(\theta_1-\theta_2)$ are interpreted here as the scattering matrix of two particles of identical mass with relativistic speeds θ_1 and θ_2 and having N states ("polarizations") each, which are given by the indices $\alpha, \beta, \gamma, \delta$. Equation (1) here is the condition of reducibility of any multiparticle collision to a two-particle one (property of factorizability of multiparticle S-matrices). In this context (1) was first obtained by Yang [1], as a property of two-particle S-matrices of nonrelativistic one-dimensional Bose particles with exact interaction. Later, when the close connection was established between complete integrability of a model and the factorizability of its S-matrix [17], (1) was situated at the foundation of the method of calculation of factorized relativistic S-matrices (see [19] and the references in it). Here, in addition to (1) there are imposed on the S-matrix conditions of unitariness, analyticity, and crossing-symmetry. In the realms of this approach a large collection of S-matrices have been calculated. The fact is encouraging that the S-matrices found in a series of models of dynamics [20, 30] in the realms of the quantum Hamiltonian approach coincided with the answers obtained earlier by the method of factorization of S-matrices.

Using the S-matrix treatment of the Yang–Baxter equation one can give an intuitive interpretation of the operations on Y–B sheaves introduced in Sec. 1. The tensor product of sheaves describes the construction of the S-matrix for composite particles. The direct sum corresponds to the possibility of dividing the particles considered according to isotropic indices into two kinds such that particles of different kinds do not interact with one another. Reducibility means the possibility of dividing the particles into groups such that for particles from different groups the scattering is without reflections.

Besides the applications listed above, the Yang–Baxter equation is also used in the theory of exactly solvable models of statistical physics. The first paper in this area is Baxter [2]. Not having the possibility of discussing this interesting direction, we refer interested readers to [2, 4, 21].

3. In this section we list the basic methods of finding solutions of the Yang–Baxter equation, known from the literature.

1) The most straightforward method of solving (1) consists of choosing some more or less successful substitution for the matrix ${}_{\alpha\beta}R_{\gamma\delta}(u)$ (e.g., to impose some symmetry condition) and write down the so obtained cubical system of functional equations for the matrix elements ${}_{\alpha\beta}R_{\gamma\delta}(u)$. The system of equations obtained can either be solved directly, as was done by Baxter in [2], or one can differentiate it with respect to v and set $v=0$, getting a system of differential equations, which with some luck and skill can be solved. Although this method is rather awkward and success is not guaranteed, the overwhelming majority of known Yang–Baxter sheaves were obtained in precisely this way. In view of the large volume of calculations in verifying the Yang–Baxter equation for concrete sheaves, it can turn out to be helpful here to use a computer, in particular, programming languages allowing one to make analytic calculations [22].

2) An important improvement in the preceding method is connected with the algebra of Zamolodchikov [19]. We consider the algebra generated by elements $A_\alpha(\theta)$ and the commutation relations

$$A_\alpha(u)A_\beta(v) = {}_{\alpha\beta}R_{\gamma\delta}(u-v)A_\delta(v)A_\gamma(u). \tag{46}$$

The Yang–Baxter equation is the condition for associativity of this algebra under the assumption of the linear independence of the monomials of the third degree in A_α. The use of the Zamolodchikov algebra in practical calculations allows one easily to write down the matrix elements of the Yang–Baxter equation considering monomials of the form $A_\alpha(u)A_{\alpha'}(v)A_{\alpha''}(w)$ and performing commutations according to (46).

Cherednik constructed a realization of the operators $A_\alpha(u)$ for the XYZ- sheaf (S8) of dimension 2 [23] in the form of compositions of operators of multiplication and translation in the space of functions on an elliptic curves. Analogous relations of the Zamolodchikov algebra for $N>2$ were obtained in [5, 24]. Although Cherednik's method allows one to get new Yang–Baxter sheaves, for the corresponding realizations of the Zamolodchikov algebra for $N>2$ one does not have the independence of the third-degree monomials, and (1) must be verified independently each time.

3) And, finally, the third method of solving the Yang–Baxter equation consists of seeking a Y–B sheaf $\mathbb{R}(u)$ as an R-matrix $\mathbb{R}_{qq'}(u)$, involved in relations of type (36), (37) (or (42), (43) in the classical case) for some completely integrable model. The advantage of this method is that in the classical case the operator $L(u)$ is, as a rule, known in advance

from the classical method of the inverse scattering problem, and the problem reduces simply to the definition of $\tau(u)$ from the (it is true, redefined) system of linear equations (42). Knowledge of the classical L-operator essentially facilitates the search for the quantum L-operator for the corresponding quantum problem (which need not at all coincide with the classical L-operator [14]).

Of course, as in the Cherednik realization, the Yang–Baxter equation (4) follows from (37) only under the condition of the independence of the monomials of third degree in $\mathbb{T}(u)$, which it is difficult to verify. Hence to get the sheaves $\mathbb{R}(u)$ it is necessary to verify (4) directly each time.

4. In this section we consider some generalizations of the Yang–Baxter equation.

One of the natural generalizations of the Yang–Baxter equation is connected with the introduction of a graduation [25]. In particular, the classical and quantum equations which can be solved by the method of the inverse problem and which along with the usual functions contain functions with anticommuting values, and in quantum theory, Fermi fields, lead to this generalization [26-28]. In what follows we shall speak only of $\mathbb{Z}_2$-graduation with the symbol $\mathbb{Z}_2$ frequently omitted.

A vector space V is called $\mathbb{Z}_2$-graded if it is decomposed into a direct sum of two subspaces $V_0 \oplus V_1$. Elements of V, having zero projection onto one of these subspaces, are called homogeneous. For the homogeneous elements x there is defined a function p(x) with values in the group $\mathbb{Z}_2$:

$$p(x) = 0, \text{ if } x \in V_0 \text{ (even elements);}$$
$$p(x) = 1, \text{ if } x \in V_1 \text{ (odd elements).}$$

If the dimensions of the spaces V_0 and V_1 are equal to n and m, respectively, then one writes the dimension of the graded space thus: $\dim V = (n, m)$.

An algebra $\mathcal{A}$ is called graded if it is graded as a vector space $\mathcal{A} = \mathcal{A}_0 \oplus \mathcal{A}_1$ and for any homogeneous elements $a_\alpha \in \mathcal{A}_\alpha$ one has the property: $a_\alpha a_\beta \in \mathcal{A}_{\alpha+\beta}$, i.e., $p(a_\alpha a_\beta) = p(a_\alpha) + p(a_\beta)$ (addition of α and β reduced mod 2). $\mathbb{Z}_2$-graduations of an algebra are also called superalgebras. If for homogeneous elements of the graded algebra $\mathcal{A}$ one has the relation $ab = (-1)^{p(a)p(b)} ba$, then $\mathcal{A}$ is called a commutative superalgebra. An example of such an algebra is the Grassman algebra $\mathcal{G}$ [29].

We choose in the space $V = V_0 \oplus V_1$ a basis of homogeneous elements $e_1, \ldots, e_n \in V_0$ and $e_{n+1}, \ldots, e_{n+m} \in V_1$. The coefficients of the expansion of a vector $x \in V$ belong to the Grassman algebra $x = \sum_{i=1}^{n+m} e_i x_i$, $x_i \in \mathcal{G}$ (V is a right $\mathcal{G}$-module). Right linear operators in V can be represented in the chosen basis in the form of matrices

$$F(x) = F(e_i x_i) = F(e_i) x_i = e_j F_{ji} x_i .$$

Such matrices F_{ij} are graded – to their rows and columns one can ascribe parity: $p_r(i) = p(e_i)$, $p_c(j) = p(e_j)$, $i, j = 1, \ldots, n+m$. A graduation is also introduced in the linear space of such matrices. To matrix F one ascribes a definite parity $p(F)$ if the expression

$$p(F) \equiv p(i)+p(j)+p(F_{ij})$$

is independent of i and j (the matrix elements $F_{ij} \in \mathcal{Y}$ and $p(F_{ij})$ is its parity as an element of the Grassman algebra). In what follows we shall be interested only in matrices of parity zero, for which $p(F_{ij}) = p(i)+p(j)$.

We consider the graded Zamolodchikov algebra with generators $A_\alpha(u)$, some of which are even $p(A_\alpha(u)) \equiv p(\alpha)=0$, and the others odd $p(A_\beta(u)) \equiv p(\beta)=1$. The basic commutation relation (46) we rewrite in the form

$$A_\alpha(u)A_\beta(v) = (-1)^{p(\alpha)p(\beta)}\,{}_{\alpha\beta}R_{\gamma\delta}(u-v)A_\delta(v)A_\gamma(u), \tag{47}$$

assuming that all ${}_{\alpha\beta}R_{\gamma\delta}$ are even elements of $\mathcal{Y}$. If ${}_{\alpha\beta}R_{\gamma\delta}=\delta_{\alpha\gamma}\delta_{\beta\delta}$, then the algebra introduced becomes a commutative superalgebra. Just as for an ordinary algebra (46) the condition for associativity under the assumption of the independence of monomials of third degree in the generators $A_\alpha(u)$, will be the relation

$${}_{\alpha\alpha'}R_{\gamma\gamma'}(u-v)\,{}_{\gamma\alpha''}R_{\beta\gamma''}(u)\,{}_{\gamma'\gamma''}R_{\beta'\beta''}(v)(-1)^{p(\gamma')(p(\alpha'')+p(\gamma''))} = {}_{\alpha'\alpha''}R_{\gamma'\gamma''}(v)\,{}_{\alpha\gamma''}R_{\gamma\beta''}(u)\,{}_{\gamma\gamma'}R_{\beta\beta'}(u-v)(-1)^{p(\gamma')(p(\gamma'')+p(\beta''))}. \tag{48}$$

Between the solutions of the Yang—Baxter equation (1) and its graded analog (48) there is a one-to-one correspondence. In fact, we overdetermine the coefficients in (48)

$${}_{\alpha\beta}\tilde{R}_{\gamma\delta}(u) = (-1)^{p(\alpha)p(\beta)}\,{}_{\alpha\beta}R_{\gamma\delta}(u). \tag{49}$$

Then for $\tilde{\mathbb{R}}(u)$ we get (1). Here it is essential that $p({}_{\alpha\beta}R_{\gamma\delta})=0$ for any nonzero ${}_{\alpha\beta}R_{\gamma\delta}$ and as a matrix $\mathbb{R}$ has null parity $p(\mathbb{R})=0$.

$\mathbb{R}(u)$ can be considered as a matrix in the tensor product of two graded spaces $V\otimes V$, $\dim V=(n,m)$. For compact notation for (48), we need the permutation operator in the tensor product of graded spaces and the operation of tensor product of graded matrices.

As a basis in the tensor product of two spaces $V\otimes W$ we take $v_i \otimes w_j$ (v_i, w_j are homogeneous elements). The components of the vector $x\otimes y$ in this basis are equal to $x_i y_j (-1)^{p(x_i)p(j)}$:

$$x\otimes y = (v_i x_i)\otimes(w_j y_j) = (v_i \otimes w_j)\, x_i y_j (-1)^{p(x_i)p(j)}.$$

We define the action of the (right) linear operator $F\otimes G$ in the space $V\otimes W$ as $(F\otimes G)(x\otimes y) = F(x)\otimes G(y)$ (we consider operators of null parity $p(F)=p(G)=0$, otherwise there arises an additional factor $(-1)^{p(x)p(G)}$). As a result the matrix element of the tensor product of even matrices $\{F_{ij}\}$, $\{G_{ab}\}$ has the form

$${}_{ia}(F\otimes G)_{jb} = F_{ij}G_{ab}(-1)^{p(a)(p(i)+p(j))}. \tag{50}$$

The permutation operator $\mathbb{P}$ in $V\otimes V$, defined by its action on the product of homogeneous elements x, y has the form

$$\mathbb{P}(x\otimes y) = (-1)^{p(x)p(y)}\, y\otimes x, \quad {}_{ab}P_{cd} = \delta_{ad}\delta_{bc}(-1)^{p(c)p(d)}. \tag{51}$$

Just as the ordinary permutation operator satisfies (1), the operator (51) satisfies (48). Using this operator one can write one of the solutions of (48) for arbitrary dimension of the space V and graduation

$$R(u)=a(u)+b(u)P, \quad a(u)=1-b(u)=u/u+1 \ . \tag{52}$$

We note that the correspondence described above between $R(u)$ and $\tilde{R}(u)$ (see (49)) preserves the property of regularity of Y—B sheaves, but not quasiclassicity.

We make use of the definition of the tensor product (50) and the graded permutation operator (51). Then for the operator

$$\check{R}(u)=PR(u)$$

(48) assumes the same form as in Sec. 1 (7)

$$(I\otimes\check{R}(u-v))(\check{R}(u)\otimes I)(I\otimes\check{R}(v))=(\check{R}(v)\otimes I)(I\otimes\check{R}(u))(\check{R}(u-v)\otimes I) \ . \tag{53}$$

Just as in the nongraduated case, for a quasiclassical family of Y—B sheaves ($R(u,\eta)|_{\eta=0}=I$) one introduces a classical Y—B sheaf:

$$r(u)=\frac{d}{d\eta}R(u,\eta)\Big|_{\eta=0} \ . \tag{54}$$

It is convenient to get the equation for $r(u)$ from the following form of (48):

$$R_{12}(u-v)\bar{R}_{13}(u)R_{23}(v)=R_{23}(v)\bar{R}_{13}(u)R_{23}(u-v) \ , \tag{55}$$

where the lower indices indicate in which of the three spaces $V_1\otimes V_2\otimes V_3$ the Y—B sheaf acts nontrivially, and

$$\bar{R}_{13}(u)=P_{23}R_{12}(u)P_{23}=P_{12}R_{23}(u)P_{12} \ . \tag{56}$$

Differentiating (55) twice with respect to η and setting $\eta=0$, we get

$$[r_{12}(u-v),\bar{r}_{13}(u)+r_{23}(v)]+[\bar{r}_{13}(u),r_{23}(v)]=0 \ . \tag{57}$$

The arbitrariness connected with multiplication of $R(u)$ by an arbitrary function leads to additive arbitrariness in r: if $r(u)$ is a solution of (57), then $r(u)+\varphi(u)I$ is also a solution. One can make use of this arbitrariness and choose $\varphi(u)$ so that for $r(u)$ one has

$$r(u)=-Pr(-u)P \ . \tag{58}$$

We shall say a few words about applications. Graduated Y—B sheaves, just like ordinary ones, about which we spoke in Sec. 2, arise if one applies the quantum method of the inverse problem to equations in which anticommuting quantum fields enter. Such are, e.g., the matrix nonlinear Schrödinger equation with Bose and Fermi fields, the massive model of Tirring with anticommuting fields [26], the supersymmetric sin-Gordon equation [27], etc. If $R(u-v)$ interlaces the quantum transition matrices $T(u)$, $T(v)$, then $r(u-v)$ defines the Poisson brackets of the matricial elements of $T(u), T(v)$ in the classical theory. Equations (57) and (58) are the reflections, respectively, of the Jacobi identity and the antisymmetry property of the Poisson brackets.

Another possible generalization of the Yang—Baxter equation consists of considering (4) as a functional equation for three different operators $R_{12}(u-v), R_{13}(u)$, and $R_{23}(v)$ acting in the product of three different spaces $V_1\otimes V_2\otimes V_3$ with dimensions N_1, N_2, N_3, respectively, where the operator $R_{\alpha\beta}(u)$ acts nontrivially only in $V_\alpha\otimes V_\beta$. In the language of the S-matrix interpretation, discussed at the end of Sec. 2, this means that we consider scattering of three kinds of particles, where N_α is the number of internal states of the α-th kind of particles ($\alpha=1,2,3$). The coefficient of reflection for the scattering of particles of different kinds here must be equal to zero.

One gets an especially simple such generalization for classical Y—B sheaves. In fact, any classical Y—B sheaf of dimension N can be decomposed as an operator in $C^N \otimes C^N$ with respect to basis elements of the form $\mathcal{J}_\alpha \otimes \mathcal{J}_\beta$ ($\alpha,\beta=1,2,\ldots,N^2$), where $\mathcal{J}_\alpha$ is any basis in the Lie algebra $gl(N,C)$:

$$r(u) = \sum_{\alpha,\beta=1}^{N^2} r_{\alpha\beta}(u)\, \mathcal{J}_\alpha \otimes \mathcal{J}_\beta \,. \tag{59}$$

But since in the classical Yang—Baxter equation (20) only commutators appear, it can be considered as an equation on the Lie algebra $gl(N,C)$ and one can take as $\mathcal{J}_\alpha$ the generators of $gl(N,C)$ in any other representation besides the fundamental one. If the generators $\mathcal{J}_\alpha$, appearing in (59) with nonzero coefficients $r_{\alpha\beta}(u)$, form a subalgebra $\mathcal{A} \subset gl(N,C)$, then the same arguments work for the Lie algebra $\mathcal{A}$. Thus, for any classical Y—B sheaf one can construct an infinite series of classical Y—B sheaves of any dimensions, and also solutions of the generalized in the above sense Y—B equation.

If the conjecture discussed in Sec. 1 that to any classical Y—B sheaf corresponds a quantum quasiclassical Y—B sheaf is valid, then analogous results should be expected also in the quantum case. However, since the quantum Yang—Baxter equation (4) is not expressed in terms of commutators, the problem of extending a given Y—B sheaf to higher representations of $gl(N,C)$ (or other Lie algebras) becomes much more complicated than in the classical case. It is reasonable to assume that analogous to the way the classical Y—B equation (20) can be considered as an equation on a Lie algebra, the quantum Y—B equation (4) can be considered on the universal enveloping Lie algebra and one can get finite-dimensional Y—B sheaves by reducing a "universal" sheaf. The validity of this conjecture is verified by one of the authors (E.K.S.) for the simplest $gl(2,C)$ invariant sheaf (see Supplement, formula (S3)).* Using the generalized quantum linear problem for the sin-Gordon equation on higher representations with respect to an auxiliary space and applying the third method (Sec. 3) of finding solutions of the Y—B equation as quantum R-matrices, one can get a generalization of the XXZ-sheaf (S9) in terms of the universal enveloping algebra of $gl(2,c)$ [44].

SUPPLEMENT

In the Supplement we give a summary of known solutions of the quantum Yang—Baxter equation. Since in a series of cases the verification of (1) requires long and tiresome calculations, not all the sheaves given below will be reexamined. In the majority of examples, we indicate the authors to whom the assertion that the given sheaf satisfies (1) is due.

*As V. A. Fateev informed us, he, together with A. B. Zamolodchikov, obtained an analogous result.

It will be convenient for us to represent the tensor ${}_{\alpha\beta}R_{\gamma\delta}$ in the form of a square block-matrix, considering the index α as the number of the block-row, β as the number of the row in block α, γ as the number of the block-column, and δ as the number of the column in block γ.

The spectral parameter is denoted everywhere by u, the quasiclassical parameter (for quasiclassical sheaves) by η. The normalization of quasiclassical sheaves is chosen so that $R(u,\eta)=I$ for $\eta=0$. For quasiclassical sheaves we give the corresponding classical sheaf $r(u)=\frac{d}{d\eta}R(u,\eta)|_{\eta=0}$ (18).

1) GL(N,C)-Invariant Sheaf (Yang [1])

$$R(u)=\frac{u}{u+\eta}I+\frac{\eta}{u+\eta}P,\quad \dim R=N\geq 2\,. \tag{S1}$$

It is obvious that I and P are the unique operators in $C^N\otimes C^N$, invariant with respect to the group $GL(N,C)$ in the sense of (13). The sheaf (S1) is regular and quasiclassical, the corresponding classical sheaf $r(u)$ has the form

$$r(u)=\frac{1}{u}P+\varphi(u)I\,. \tag{S2}$$

The sheaf $R(u)$ is widely used in the quantum method of the inverse problem [3]. For $N=2$ it arises in studying the quantum nonlinear Schrödinger equation [13], the XXX-model (Heisenberg ferromagnet) [3, 12], Toda chains [3, 16]. Sheaves with $N\geq 3$ are used in considering multicomponent analogs of the equations mentioned: vector and matrix nonlinear Schrödinger equations [10], generalized Heisenberg ferromagnet [30], non-Abelian Toda chains [16], systems of n-waves (Lie model) [31]. In addition, the sheaf (S1) is used as S-matrix for nonrelativistic particles with pointwise interaction [1].

As noted at the end of Sec. 4, for $N=2$ there are known analogs of the sheaf (S1) for any finite-dimensional irreducible representation of the group $GL(2,C)$. Let there act in the space $V=C^\ell$ an ℓ-dimensional irreducible representation $\mathcal{D}_\ell(q)$ of the group $GL(2,C)$. Then the generalized sheaf (S1) has the form:

$$R(u)=\sum_{j=0}^{2\ell}\prod_{k=1}^{j}\frac{u-k\eta}{u+k\eta}P_j\,, \tag{S3}$$

where P_j is the projector onto the space of j-dimensional irreducible representations in the decomposition of the tensor product $\mathcal{D}_\ell\otimes\mathcal{D}_\ell$ into irreducible representations. The sheaf obtained is canonical. The proof of (S3) will be published separately.

2) SO(N,C) Invariant Sheaves

In the space $C^N\otimes C^N$ there are in all three operators, invariant with respect to the action of the group $SO(N,C)$. We have already met this with the operators I and P, and, in addition, the projector K:

$${}_{\alpha\beta}K_{\gamma\delta}=\delta_{\alpha\beta}\,\delta_{\gamma\delta}\,. \tag{S4}$$

An invariant Y–B sheaf corresponding to $SO(N,C)$ was found in [18] as S-matrix for the Gross–Neveu model:

$$\mathbb{R}(u)=\frac{u}{u+\eta}\mathbb{I}+\frac{\eta}{u+\eta}\mathbb{P}+\frac{u\eta}{(u+\eta)(\eta g-u)}\mathbb{K},\ g=-\frac{N-2}{2}\ . \tag{S5}$$

The sheaf (S5) is canonical. The corresponding classical sheaf:

$$r(u)=\frac{1}{u}\mathbb{P}-\frac{1}{u}\mathbb{K}+\varphi(u)\mathbb{I}\ . \tag{S6}$$

It is interesting to note that if one takes a linear combination of only $\mathbb{I}$ and $\mathbb{P}$, we get a $GL(N,C)$-invariant solution of (S1), for combinations of $\mathbb{P}$ and $\mathbb{K}$ we get a new sheaf

$$\mathbb{R}(u)=\mathbb{P}-\frac{\operatorname{sh}(u)}{\operatorname{sh}(u+\gamma)}\mathbb{K},\ e^{\gamma}=\frac{1}{2}(N-\sqrt{N^2-4}), \tag{S7}$$

which is regular, but not quasiclassical, and finally, for combinations of only $\mathbb{I}$ and $\mathbb{K}$ a solution does not exist.

3) XYZ -Sheaf (Baxter [2])

$$\mathbb{R}(u,\eta,k)=\begin{vmatrix} a & 0 & 0 & d \\ 0 & b & c & 0 \\ 0 & c & b & 0 \\ d & 0 & 0 & a \end{vmatrix}. \tag{S8}$$

TABLE 1

	R	$u=0$	$\eta=0$	$\partial/\partial\eta\cdot\vert_{\eta=0}$
a	1	1	1	0
b	$\frac{\operatorname{sn} u}{\operatorname{sn}(u+\eta)}$	0	1	$-\frac{\operatorname{cn} u\,\operatorname{dn} u}{\operatorname{sn} u}$
c	$\frac{\operatorname{sn}\eta}{\operatorname{sn}(u+\eta)}$	1	0	$\frac{1}{\operatorname{sn} u}$
d	$k\operatorname{sn} u\operatorname{sn}\eta$	0	0	$k\operatorname{sn} u$

In the formulas of Table 1 all elliptic functions have modulus k. The XYZ-sheaf is canonical. To the quantum XYZ-sheaf corresponds a completely integrable lattice XYZ-model [4, 32]. The corresponding classical sheaf corresponds to the Landau–Lifshits equation [15]. The presence in the XYZ-sheaf of two parameters: η and k allows one to get different degenerate cases, e.g., XXZ - and XXX-sheaves (see below), and quantum models on the line (sin-Gordon equation, nonlinear Schrödinger equation).

4) XXZ -Sheaf

$$\mathbb{R}(u,\eta)=\begin{vmatrix} a & 0 & 0 & 0 \\ 0 & b & c & 0 \\ 0 & c & b & 0 \\ 0 & 0 & 0 & a \end{vmatrix}. \tag{S9}$$

The XXZ-sheaf is obtained from the XYZ-sheaf (Table 1) as the limit as $k\to 0$. Like the XYZ-sheaf, the XXZ-sheaf is canonical. To it corresponds a quantum lattice XXZ-model, which was considered in the realms of the quantum method of the inverse problem (QMIP) in

TABLE 2

	R	$u=0$	$\eta=0$	$\partial/\partial\eta \cdot \|_{\eta=0}$
a	1	1	1	0
b	$\frac{\sin u}{\sin(u+\eta)}$	0	1	$-\operatorname{ctg} u$
c	$\frac{\sin\eta}{\sin(u+\eta)}$	1	0	$\frac{1}{\sin u}$

TABLE 3

	R		R
s_1	$\mathcal{J}(\operatorname{sn} u \operatorname{cn} 2\eta + \frac{\operatorname{sn}\eta \operatorname{sn} 2\eta}{\operatorname{sn}(u+\eta)}$	t	$\varepsilon_3 \mathcal{J} \operatorname{sn} u \operatorname{cn} 2\eta$
s_2	$\mathcal{J}(\operatorname{sn} u(\operatorname{cn} 2\eta + \operatorname{dn} 2\eta - 1) + \frac{\operatorname{sn}\eta \operatorname{sn} 2\eta}{\operatorname{sn}(u+\eta)})$	T	$\varepsilon_4 \mathcal{J} \operatorname{sn} u$
s_3	$\mathcal{J}(\operatorname{sn} u \operatorname{dn} 2\eta + \frac{\operatorname{sn}\eta \operatorname{sn} 2\eta}{\operatorname{sn}(u+\eta)})$	R	$\mathcal{J} \operatorname{sn} 2\eta$
α	$\varepsilon_1 \mathcal{J} \operatorname{sn} u \operatorname{cn}(u+\eta) \operatorname{sn} 2\eta/\operatorname{sn}(u+\eta)$	a	$\varepsilon_5 \mathcal{J} \operatorname{sn} u \operatorname{dn} 2\eta$
β	$\varepsilon_2 \mathcal{J} \operatorname{sn} u \operatorname{sn} 2\eta/\operatorname{sn}(u+\eta)$	r	$\mathcal{J} \operatorname{cn} u \operatorname{sn} 2\eta$
γ	$-\varepsilon_1\varepsilon_2 \mathcal{J} \operatorname{sn} u \operatorname{sn} 2\eta/\operatorname{sn}(u+\eta)$	q	$\mathcal{J} \operatorname{dn} u \operatorname{sn} 2\eta$

$$\mathcal{J} = 1/\operatorname{sn}(u+2\eta), \quad \varepsilon_i = \pm 1 .$$

TABLE 4

	R	$u=0$	$\eta=0$	$\partial/\partial\eta \cdot \|_{\eta=0}$
s	1	1	1	0
t	$\mathcal{J} \operatorname{sh} u$	0	1	$-2 \operatorname{cth} u$
r	$\mathcal{J} \operatorname{sh} 2\eta$	1	0	$\frac{2}{\operatorname{sh} u}$
a	$\varepsilon \mathcal{J} \frac{\operatorname{sh} u \operatorname{sh} 2\eta}{\operatorname{sh}(u+\eta)}$	0	0	$\frac{2\varepsilon}{\operatorname{sh} u}$
R	$\mathcal{J} \frac{\operatorname{sh}\eta \operatorname{sh} 2\eta}{\operatorname{sh}(u+\eta)}$	1	0	0
T	$\mathcal{J} \frac{\operatorname{sh} u \operatorname{sh}(u-\eta)}{\operatorname{sh}(u+\eta)}$	0	1	$-4 \operatorname{cth} u$
σ	$t+R$	1	1	$-2 \operatorname{cth} u$

$$\mathcal{J} = 1/\operatorname{sh}(u+2\eta) .$$

TABLE 5

	R	$u=0$	$\eta=0$	$\partial/\partial\eta \cdot \|_{\eta=0}$
a	1	1	1	0
b	$g(\eta) \frac{\operatorname{sh} u}{\operatorname{sh}(u+\eta)}$	0	$g(0)$	$g'(0) - g(0) \operatorname{cth} u$
$\bar{b}$	$g^{-1}(\eta) \frac{\operatorname{sh} u}{\operatorname{sh}(u+\eta)}$	0	$g^{-1}(0)$	$\frac{-g'(0)}{g^2(0)} - g^{-1}(0) \operatorname{cth} u$
c	$e^{-\frac{1}{3}u} \frac{\operatorname{sh}\eta}{\operatorname{sh}(u+\eta)}$	1	0	$e^{-\frac{1}{3}u}/\operatorname{sh} u$
$\bar{c}$	$e^{-\frac{1}{3}u} \frac{\operatorname{sh}\eta}{\operatorname{sh}(u+\eta)}$	1	0	$e^{\frac{1}{3}u}/\operatorname{sh} u$

[12]. This sheaf also arises in the quantization of the sin-Gordon equation by means of QMIP [14].

For the following degeneration $u=\varepsilon v$, $\eta=\varepsilon\gamma$, $\varepsilon\to 0$ the XXZ-sheaf degenerates into an XXX-sheaf corresponding with the sheaf (S1) for $N=2$.

5) XYZ-Sheaf for Spin 1 (Fateev [33])

$$\mathbb{R}(u,\eta,k)=\left|\begin{array}{ccc|ccc|ccc} s_1 & & & & \alpha & & & & \beta \\ & t & & r & & & & & \\ & & T & & & & R & & \\ \hline & r & & t & & & & & \\ \alpha & & & & s_2 & & & & \gamma \\ & & & & & a & & q & \\ \hline & & R & & & & T & & \\ & & & & & q & & a & \\ \beta & & & & \gamma & & & & s_3 \end{array}\right| \tag{S10}$$

As in Table 1, the elliptic functions in Table 3 have modulus k. The sheaf (S10) is canonical. The corresponding classical sheaf, which we shall not write down due to its complexity, is a sheaf (S8) rewritten in a basis of generators of a three-dimensional irreducible representation of $gl(2,C)$, which allows one to consider the sheaf (S10) as a generalization of the sheaf (S8) to a higher representation (see end of Sec. 4).

6) XXZ-Sheaf for Spin 1 (Zamolodchikov and Fateev [34])

$$\mathbb{R}(u,\eta)=\left|\begin{array}{ccc|ccc|ccc} s & & & & & & & & \\ & t & & r & & & & & \\ & & T & & a & & R & & \\ \hline & r & & t & & & & & \\ & & a & & \sigma & & a & & \\ & & & & & t & & r & \\ \hline & & R & & a & & T & & \\ & & & & & r & & t & \\ & & & & & & & & s \end{array}\right| \tag{S11}$$

In Table 4, $\varepsilon=\pm 1$. The sheaf (S11) for $\varepsilon=1$ is obtained from (S10) with $\varepsilon_i=1$ after passage to the limit as $k\to 0$ and a similarity transformation (12) with matrix

$$T=\frac{1}{\sqrt{2}}\left|\begin{array}{ccc} 1 & -i & 0 \\ 0 & 0 & \sqrt{2} \\ -i & 1 & 0 \end{array}\right|.$$

The property of canonicity of (S10) is also preserved for (S11).

7) Z_N-Invariant Sheaf (Cherednik [24])

For $N=3$

$$\mathbb{R}(u,\eta)=\left|\begin{array}{ccc|ccc|ccc} a & & & & & & & & \\ & b & & c & & & & & \\ & & \bar{b} & & & & \bar{c} & & \\ \hline & \bar{c} & & \bar{b} & & & & & \\ & & & & a & & & & \\ & & & & & b & & c & \\ \hline & & c & & & & b & & \\ & & & & & \bar{c} & & \bar{b} & \\ & & & & & & & & a \end{array}\right| \tag{S12}$$

TABLE 6

	R	$\partial/\partial\eta \cdot \vert_{\eta=0}$
a	$\mathcal{J}(\mathrm{sh}(u-3\eta)-\mathrm{sh}\,5\eta+\mathrm{sh}\,3\eta+\mathrm{sh}\,\eta)$	$X(\mathrm{ch}\,u-1)$
b	$\mathcal{J}(\mathrm{sh}(u-3\eta)+\mathrm{sh}\,3\eta)$	$X(\mathrm{ch}\,u+1)$
c	1	0
d	$\mathcal{J}(\mathrm{sh}(u-\eta)+\mathrm{sh}\,\eta)$	$2X\,\mathrm{ch}\,u$
e	$-2\mathcal{J}e^{-\frac{1}{2}u}\mathrm{sh}\,2\eta\,\mathrm{ch}(\frac{1}{2}u-3\eta)$	$-X(1+e^{-u})$
$\bar{e}$	$-2\mathcal{J}e^{\frac{1}{2}u}\mathrm{sh}\,2\eta\,\mathrm{ch}(\frac{1}{2}u-3\eta)$	$-X(1+e^{u})$
f	$-\mathcal{J}(2e^{-u}e^{2\eta}\mathrm{sh}\,\eta\,\mathrm{sh}\,2\eta+e^{-\eta}\mathrm{sh}\,4\eta)$	$-2X$
$\bar{f}$	$\mathcal{J}(2e^{u}e^{-2\eta}\mathrm{sh}\,\eta\,\mathrm{sh}\,2\eta-e^{\eta}\mathrm{sh}\,4\eta)$	$-2X$
g	$\mathcal{J}(e^{-\frac{1}{2}u}e^{2\eta}2\,\mathrm{sh}\frac{1}{2}u\,\mathrm{sh}\,2\eta)$	$X(1-e^{-u})$
$\bar{g}$	$-\mathcal{J}(e^{\frac{1}{2}u}e^{-2\eta}2\,\mathrm{sh}\frac{1}{2}u\,\mathrm{sh}\,2\eta)$	$-X(1-e^{u})$

$$\mathcal{J}=(\mathrm{sh}(u-5\eta)+\mathrm{sh}\,\eta)^{-1}, \qquad X=2/\mathrm{sh}\,u\,.$$

TABLE 7

	a_1	a_2	a_3	b_1	b_2	b_3	c_1	c_2	c_3	g
I	$\frac{u}{u+\eta}$	$\frac{\eta}{u+\eta}$	0	$\frac{u}{u+\eta}$	0	$\frac{u\eta}{(u+\eta)(\eta g-u)}$	0	0	0	$-\frac{N}{2}$
II	$\frac{u}{u+\eta}$	$\frac{\eta}{u+\eta}$	0	$\frac{u}{u+\eta}$	0	$\frac{u\eta}{(u+\eta)(\eta g-u)}$	0	$\frac{\eta}{u+\eta}$	$\frac{u\eta}{(u+\eta)(\eta g-u)}$	$-(N-1)$
III	$\frac{u}{u+\eta}$	$\frac{\eta}{u+\eta}$	0	$\frac{-u}{u+\eta}$	0	$\frac{u\eta}{(u+\eta)(\eta g+u)}$	0	$\frac{\eta}{u+\eta}$	$\frac{u\eta}{(u+\eta)(\eta g+u)}$	$1+N$
IV	0	1	0	0	0	$\frac{-\mathrm{sh}(u+\mu)}{\mathrm{sh}\,u}$	0	1	$\frac{-\mathrm{sh}(u+\mu)}{\mathrm{sh}\,u}$	$g=\mathrm{ch}\,\mu=$ $=N$
V	0	e^{u}	0	0	0	$-e^{-u-\mu}\frac{\mathrm{sh}(u+\mu)}{\mathrm{sh}\,u}$	0	1	$\frac{-\mathrm{sh}(u+\mu)}{\mathrm{sh}\,u}$	$g=e^{\mu}=$ $=N$

TABLE 8

	R	$u=0$	$\eta=0$	$\partial/\partial\eta\cdot\vert_{\eta=0}$
a	1	1	1	0
b	$\mathcal{J}\,\mathrm{sh}\,u\,\mathrm{ch}(u-\eta)/\mathrm{ch}(u+\eta)$	0	1	$-4\,\mathrm{cth}\,2u$
c	$\mathcal{J}\,\mathrm{sh}\,u$	0	1	$-2\,\mathrm{cth}\,u$
d	$\mathcal{J}(\mathrm{sh}\,u-\mathrm{sh}\,2\eta\,\mathrm{ch}\,\eta/\mathrm{ch}(u+\eta))$	-1	1	$-2\frac{\mathrm{ch}\,2u+3}{\mathrm{sh}\,2u}$
r	$\mathcal{J}\,\mathrm{sh}\,2\eta\,\mathrm{ch}\,\eta/\mathrm{ch}(u+\eta)$	1	0	$\frac{4}{\mathrm{sh}\,2u}$
x	$\mathcal{J}\,\mathrm{sh}\,2\eta$	1	0	$\frac{2}{\mathrm{sh}\,u}$
y	$\mathcal{J}\,\mathrm{sh}\,u\,\mathrm{sh}\,2\eta/\mathrm{ch}(u+\eta)$	0	0	$\frac{2}{\mathrm{ch}\,u}$

$$\mathcal{J}=1/\mathrm{sh}(u+2\eta)$$

The sheaf (S12) is regular, and if $g(0)=1$, then also quasiclassical. In addition, the sheaf is invariant with respect to a similarity transformation (12) with matrix T.

$$T=\begin{vmatrix} 0 & 0 & 1 \\ 1 & 0 & 0 \\ 0 & 1 & 0 \end{vmatrix}\,.$$

TABLE 9

	R	$u=0$	$\eta=0$	$\partial/\partial\eta\cdot\|_{\eta=0}$
a	$\mathcal{I}(\operatorname{sh}u+\operatorname{sh}\eta/\operatorname{ch}u)$	1	1	$-\operatorname{cth}u+2/\operatorname{sh}u$
$\bar{a}$	$\mathcal{I}(\operatorname{sh}u-\operatorname{sh}\eta/\operatorname{ch}u)$	-1	1	$-\operatorname{cth}u-2/\operatorname{sh}u$
b	$\mathcal{I}\operatorname{sh}u$	0	1	$-\operatorname{cth}u$
	R	$u=0$	$\eta=0$	$\partial/\partial\eta\cdot\|_{\eta=0}$
c	$\mathcal{I}\operatorname{sh}\eta$	1	0	$1/\operatorname{sh}u$
d	$\mathcal{I}\operatorname{sh}u\operatorname{sh}\eta/\operatorname{ch}u$	0	0	$1/\operatorname{ch}u$

$$\mathcal{I}=1/\operatorname{sh}(u+\eta).$$

TABLE 10

	R	$u=0$	$\eta=0$	$\partial/\partial\eta\cdot\|_{\eta=0}$
a	$\mathcal{I}\operatorname{sh}(u-\eta)$	-1	1	$-2\operatorname{cth}u$
b	$\mathcal{I}\operatorname{sh}u$	0	1	$-\operatorname{cth}u$
c	$\mathcal{I}\operatorname{sh}\eta$	1	0	$1/\operatorname{sh}u$

$$\mathcal{I}=1/\operatorname{sh}(u+\eta).$$

In view of the identity $T^3=1$, the corresponding group of transformations is isomorphic with Z_3. For $g(\eta)=1$ the sheaf (S12) arises upon quantization of the doubles Toda chain [36] with the help of QMIP (Reshetikhin).

Analogously, one constructs N-dimensional Z_N-invariant sheaves for any $N\geqslant 3$ [24]:

$$\begin{aligned}
&{}_{\alpha\beta}R_{\gamma\delta}=0 \quad \text{for } \alpha+\beta\neq\gamma+\delta \ (\mathrm{mod}\, N);\ \alpha,\beta,\gamma,\delta=0,1,\dots,N-1,\\
&{}_{\alpha\alpha}R_{\alpha\alpha}=1,\\
&{}_{\alpha\beta}R_{\beta\alpha}=\exp\Big(2u\frac{\alpha-\beta}{N}-\operatorname{sign}(\alpha-\beta)\Big)\frac{\operatorname{sh}\eta}{\operatorname{sh}(u+\eta)},\\
&{}_{\alpha\beta}R_{\alpha\beta}=\exp\Big(2\eta\frac{\beta-\alpha}{N}-\operatorname{sign}(\beta-\alpha)\Big)\frac{\operatorname{sh}u}{\operatorname{sh}(u+\eta)}.
\end{aligned}\tag{S13}$$

In algebraic terms a system of roots of the given sheaf, as we noted, is the quantum R-matrix for the relativistic field-theoretic model corresponding to the system of roots A_{N-1} [36]. It is not hard to calculate the classical r-matrix for the remaining root systems $(B_N, C_N, \dots, E_8)$, and it is interesting to find the corresponding quantum R-matrices.

8) Sheaf of Dimension 3 (Izergin and Korepin [5])

$$R(u,\eta)=\left|\begin{array}{ccc|ccc|ccc}
a & & & & & \bar{g} & & g & \\
 & b & & \bar{e} & & & & & \\
 & & b & & & & e & & \\
\hline
 & e & & b & & & & & \\
 & & & & c & & & & \\
g & & & & & d & & f & \\
\hline
 & & \bar{e} & & & & b & & \\
 & & & & & \bar{f} & & d & \\
\bar{g} & & & & & & & & c
\end{array}\right|.\tag{S14}$$

The sheaf (S14) is canonical — it is the $\mathbb{R}$-matrix for the quantum relativistic model of Mikhailov—Shabat [5].

9) Block $O(N)$-Invariant Sheaves [35]

$$\mathbb{R}(u) = \begin{vmatrix} A & & & \\ & B & C & \\ & C & B & \\ & & & A \end{vmatrix}, \tag{S15}$$

A, B, C are $N^2 \times N^2$ blocks of the form $A = a_1 I + a_2 P + a_3 K$ and analogously for B and C, where K is a projector (S4).

These sheaves were obtained in [35], as examples of factorized $SU(N)$-invariant S-matrices. This does not contradict our definition of them as $O(N)$-invariant, since the action of a group of transformations on an S-matrix is defined differently than we did it in (13).

The sheaf II is canonical, I is quasiclassical, III is regular.

Now we give some examples of Z_2-graded Yang—Baxter sheaves, satisfying (48).

10) $GL(n,m,C)$-Invariant Sheaf

$GL(n,m,C)$ is the analog of the group $GL(N,C)$ [25] in the graded space C^{n+m} with graduation (n,m). The sheaf has the form

$$\mathbb{R}(u) = \frac{u}{u+\eta} I + \frac{\eta}{u+\eta} P_{n,m}, \tag{S16}$$

where $P_{n,m}$ is the graded permutation operator (51). The sheaf given is a natural generalization to the graded case of the sheaf (S1). It is used in applying QMIP to equations containing Fermi fields [26, 27, 30].

11) Sheaf, Connected with the Tirring Massive Model for Fermi Fields

The auxiliary linear problem for the Tirring model [26] is defined by a (3 × 3) matrix differential operator of the first order. The graduation is equal to (2, 1). In the basis where $p(1) = p(2) = 0$, $p(3) = 1$, the $\mathbb{R}$-matrix, interlacing the monodromy operators for the auxiliary linear problem (37), has the form:

$$\mathbb{R}(u,\eta) = \left|\begin{array}{ccc|ccc|ccc} a & & & & & & & & y \\ & b & & r & & & & & \\ & & c & & & & x & & \\ \hline & r & & b & & & & & y \\ & & & & a & & & & \\ & & & & & c & & x & \\ \hline & & x & & & & c & & \\ & & & & & x & & c & \\ & y & & y & & & & & d \end{array}\right|. \tag{S17}$$

12) Graded Analog of XYZ-Sheaf (but in Hyperbolic Functions)

$$\mathbb{R}(u,\eta) = \begin{vmatrix} a & 0 & 0 & d \\ 0 & b & c & 0 \\ 0 & c & b & 0 \\ d & 0 & 0 & \bar{a} \end{vmatrix}. \tag{S18}$$

This sheaf arises as part of the interlacing R-matrix for the quantum supersymmetric sin-Gordon equation [27], corresponding to a majorizing field.

13) Graded Analog of the XXZ-Sheaf

$$R(u,\eta)=\begin{vmatrix} 1 & 0 & 0 & 0 \\ 0 & b & c & 0 \\ 0 & c & b & 0 \\ 0 & 0 & 0 & a \end{vmatrix}. \tag{S19}$$

LITERATURE CITED

1. C. N. Yang, Phys. Rev. Lett., 19, No. 23, 1312-1314 (1967).
2. R. J. Baxter, Ann. Phys., 70, No. 1, 193-228 (1972).
3. L. D. Faddeev, in: Problems of Quantum Field Theory (Memoirs of the International Conference on Nonlocal Field Theory, Alushta, 1979) [in Russian], Dubna (1979), pp. 249-299.
4. L. A. Takhtadzhyan and L. D. Faddeev, Usp. Mat. Nauk, 34, 13-63 (1979).
5. A. G. Izergin and V. E. Korepin, Commun. Math. Phys. (to be published); Preprint, Leningrad Branch of the Mathematical Institute, E-3-80, Leningrad (1980), pp. 3-28.
6. A. A. Belavin, Commun. Math. Phys. (to be published); Landau Institute for Theor. Physics, Preprint (1980), pp. 1-15.
7. M. Lüscher, Nucl. Phys., B117, No. 2, 475-492 (1976).
8. D. Babbitt and L. Thomas, Preprint Univ. Virginia (1978). V. V. Anmelevich, Teor. Mat. Fiz., 43, No. 1, 107-110 (1980).
9. S. V. Manakov, Zh. Eksp. Teor. Fiz., 65, No. 2, 505-516 (1973).
10. P. P. Kulish, Preprint Leningrad Branch of the Mathematical Institute, P-3-79, Leningrad (1979).
11. V. N. Dutyshev, Zh. Eksp. Teor. Fiz., 78, No. 4, 1332-1342 (1980).
12. P. P. Kulish and E. K. Sklyanin, Phys. Lett., 70A, Nos. 5-6, 461-463 (1979).
13. E. K. Sklyanin, Dokl. Akad. Nauk SSSR, 244, No. 6, 1337-1341 (1979). E. K. Sklyanin, Zap. Nauchn. Sem. Leningr. Otd. Mat. Inst., 95, 57-132 (1980).
14. E. K. Sklyanin, L. A. Takhtadzhyan, and L. D. Faddeev, Teor. Mat. Fiz., 40, No. 2, 194-220 (1979).
15. E. K. Sklyanin, Preprint Leningrad Branch of the Mathematics Institute, E-3-79, Leningrad (1979).
16. V. E. Korepin, Zap. Nauchn. Sem. Leningr. Otd. Mat. Inst., 101, 79-90 (1980).
17. P. P. Kulish, Teor. Mat. Fiz., 26, No. 2, 198-205 (1976). I. Ya. Aref'eva and V. E. Korepin, Pis'ma Zh. Eksp. Teor. Fiz., 20, No. 5, 680-683 (1974). B. Schroer et al., Phys. Lett., B63, 422-425 (1976).
18. A. B. Zamolodchikov and Al. B. Zamolodchikov, Nucl. Phys., B133, No. 3, 525-535 (1978).
19. A. B. Zamolodchikov and Al. B. Zamolodchikov, Ann. Phys., 120, No. 2, 253-291.
20. V. E. Korepin, Teor. Mat. Fiz., 41, No. 2, 169-189 (1979). N. Andrie and J. H. Lowenstein, Phys. Lett., 91B, Nos. 3-4, 401-405 (1980).
21. A. B. Zamolodchikov, Sov. Sci. Rev.; Phys. Rev., 2, 3-50 (1980).
22. V. P. Gerdt, O. V. Tarasov, and D. V. Shirokov, Usp. Fiz. Nauk, 130, No. 1, 113-148 (1980).
23. I. V. Cherednik, Dokl. Akad. Nauk SSSR, 249, No. 5, 1095-1098 (1979).
24. I. V. Cherednik, Teor. Mat. Fiz., 43, No. 1, 117-119 (1980).
25. F. A. Berezin, Yad. Fiz., 29, No. 6, 1670-1687 (1979). D. A. Leites, Usp. Mat. Nauk, 35, No. 1, 3-57 (1980).
26. A. G. Izergin and P. P. Kulish, Zap. Nauchn. Sem. Leningr. Otd. Mat. Inst., 77, 76-83 (1978).
27. M. Chaichian and P. Kulish, Phys. Lett., 78B, No. 4, 413-416 (1978). P. P. Kulish and S. A. Tsyplyaev, Teor. Mat. Fiz. (in press).
28. A. V. Mikhailov, Pis'ma Zh. Eksp. Teor. Fiz., 28, No. 8, 554-558 (1978).
29. F. A. Berezin, Method of Secondary Quantization [in Russian], Moscow (1965).
30. P. P. Kulish and N. Yu. Reshetikhin, Preprint Leningrad Branch of the Mathematical Institute, E-4-79, Leningrad (1979).
31. S. V. Manakov, Teor. Mat. Fiz., 28, No. 2, 172-179 (1976).
32. R. J. Baxter, Ann. Phys., 70, No. 2, 323-337 (1972).
33. V. A. Fateev, Yad. Fiz. (in press).

34. A. B. Zamolodchikov and V. A. Fateev, Yad. Fiz., 32, 587 (1980).
35. B. Berg, M. Karowski, P. Weisz, and V. Kurak, Nucl. Phys., B134, No. 1, 125-132 (1978).
36. A. Mikhailov, M. Olshanetsky, and A. Perelomov, preprint ITEP-64 (1980), pp. 1-26.
37. S. A. Bulgadaev, Landau Institute for Theor. Physics, Preprint (1980), pp. 1-7.
38. Yu. A. Bashilov and S. V. Pokrovsky, Commun. Math. Phys. (to be published); Landau Institute for Theor. Physics, Preprint (1980), pp. 1-23.
39. B. Berg and P. Weisz, Preprint FUB-HEP-21-78 (1978), pp. 1-15.
40. R. D. Pisarski, Princeton Univ., Preprint (1979), pp. 1-15.
41. A. V. Mikhailov, Pis'ma Zh. Eksp. Teor. Fiz., 30, No. 7, 443-448 (1979).
42. R. Z. Bariev, Pis'ma Zh. Eksp. Teor. Fiz., 32, No. 1, 10-14 (1980).
43. L. A. Takhtadzhyan, Zap. Nauchn. Sem. Leningr. Otd. Mat. Inst., 101, 121-150 (1980).
44. P. P. Kulish and N. Yu. Reshetikhin, Zap. Nauchn. Sem. Leningr. Otd. Mat. Inst., 101, 71-93 (1980).

ERRATA

The text around formula (A3) should read as follows ... Let the (2l+1)-dimensional representation $\mathcal{D}_l(g)$ of the group GL(2,C) act in the space $V = C^{2l+1}$. Then the generalization of (A1) reads:

$$\ldots \tag{A3}$$

where P_i is the projector onto the space of (2j+1)-dimensional irreducible representation ...

In Table 3 the value of γ should be replaced by

$$\gamma = -\varepsilon_1\varepsilon_2 J \,\mathrm{sn}\, u \,\mathrm{dn}(u+\eta)\,\mathrm{sn}\, 2\eta/\mathrm{sn}(u+\eta) .$$

The formula for T (immediately after Table 4) should read as follows

$$T = \frac{1}{2^{1/2}} \begin{pmatrix} 1 & 1 & 0 \\ 0 & 0 & 2^{1/2} \\ -1 & 1 & 1 \end{pmatrix} .$$

The last line of the Table 5 should read as follows

$$\bar{c} \mid e^{-u/3} \frac{\mathrm{sh}\,\eta}{\mathrm{sh}(u+\eta)} \mid 1 \mid 0 \mid e^{-u/3}/\mathrm{sh}\, u .$$

The lower-left block of the matrix $\mathbb{R}(u,\eta)$ in the formula (A14) should read as follows

$$\begin{array}{ccc|} \hline 0 & 0 & \bar{e} \\ \bar{g} & 0 & 0 \\ 0 & 0 & 0 \end{array}$$

The following factors should be inserted into the expressions for the off-diagonal elements of $\mathbb{R}(u,\eta)$ in the Table 6: $e^{u/3}$ for e and g, $e^{-u/3}$ for $\bar{e}$ and $\bar{g}$, $e^{2u/3}$ for f, $e^{-2u/3}$ for $\bar{f}$.

The upper-right block of the matrix $\mathbb{R}(u,\eta)$ in the formula (A17) should read as follows

$$\begin{array}{|ccc} 0 & 0 & 0 \\ 0 & 0 & y \\ x & 0 & 0 \\ \hline \end{array}$$

CLASSICAL YANG-BAXTER EQUATION

2. Classical Yang-Baxter equation

Compared to (quantum) YBE, an important and simplifying feature of CYBE is that it can be formulated in the realm of Lie algebras, independently of the way it is represented by matrices. In the reprint [8] (see also [42]), Belavin and Drinfel'd gave a detailed study of solutons of CYBE associated with complex simple Lie algebras. They succeeded in classifying all the non-degenerate elliptic and trigonometric solutions using the data from the Dynkin diagram. The paper [43] is a readable review of their theory. Analogous classification for Lie superalgebras is given in [155].

Schematically, each solution to YBE (often referred to as an R matrix) gives rise to integrable quantum systems; in the same way, with each classical r matrix (= a solution to CYBE) one can associate integrable classical systems, or soliton equations . A basic viewpoint of the QISM is to regard them as infinite dimensional Hamiltonian systems. The papers [58],[99–100] address the issue of describing the Hamiltonian structure of soliton equations in the framework of classical r matrices. What is characteristic here is that one can introduce more than one natural Hamiltonian structures.

As it turns out, these classical systems possess a natural action of groups (usually infinite dimensional), which are themselves Hamiltonian manifolds. In the reprint [9] Drinfel'd shows that a classical r matrix induces a Poisson-Lie group structure, namely a Poisson bracket on the corresponding Lie group such that the group multiplication law is a Hamiltonian map. The reprint of Semenov-Tyan-Shanskii [10] gives a similar but slightly different interpretation, relating CYBE to a second Lie algebra structure on the underlying Lie algebra. He showed that the group action on the associated Hamiltonian systems are again Hamiltonian[224–225]. Detailed exposition of these topics can be found in [81],[258].

SOLUTIONS OF THE CLASSICAL YANG – BAXTER EQUATION FOR SIMPLE LIE ALGEBRAS

A. A. Belavin and V. G. Drinfel'd

UDC 517.43+519.46

1. Introduction

1.1. By the classical Yang–Baxter equation is meant the functional equation

$$[X^{12}(u_1, u_2), X^{13}(u_1, u_3)] + [X^{12}(u_1, u_2), X^{23}(u_2, u_3)] + [X^{13}(u_1, u_3), X^{23}(u_2, u_3)] = 0 \tag{1.1}$$

with respect to the function $X(u_1, u_2)$, assuming values in $\mathfrak{g} \otimes \mathfrak{g}$, where $\mathfrak{g}$ is a Lie algebra. We explain the meaning of notation of the type $X^{13}(u_1, u_3)$. We fix an associative algebra A with unit containing $\mathfrak{g}$. $X^{13}(u_1, u_3)$, by definition, is the image of $X(u_1, u_3)$ under the linear map $\varphi_{13}: \mathfrak{g} \times \mathfrak{g} \to A \otimes A \otimes A$, defined by the formula $\varphi_{13}(a \otimes b) = a \otimes 1 \otimes b$. The notations $X^{12}(u_1, u_2)$ and $X^{23}(u_2, u_3)$ have the analogous meaning [we note only that $\varphi_{12}(a \otimes b) \stackrel{\text{def}}{=} a \otimes b \otimes 1$, $\varphi_{23}(a \otimes b) \stackrel{\text{def}}{=} 1 \otimes a \otimes b$]. It is easy to see that each of the three summands of the left side of (1.1) plays an important role in the theory of classical and quantum integrable systems (cf. [1, 5]).

We note that if $X(u_1, u_2)$ is a solution of (1.1) and $\varphi(u)$ is a function with values in Aut $\mathfrak{g}$, then $\tilde{X}(u_1, u_2) \stackrel{\text{def}}{=} (\varphi(u_1) \otimes \varphi(u_2)) X(u_1, u_2)$ is also a solution of (1.1). We shall call the solutions X and $\tilde{X}$ equivalent. Before formulating another method or propagation of solutions of (1.1), we introduce the following definition.

<u>Definition.</u> The function $X(u_1, u_2)$ is said to be invariant with respect to $g \in \text{Aut}\,\mathfrak{g}$, if $(g \otimes g)X(u_1, u_2) = X(u_1, u_2)$. The set of all such g is called the invariance group of the function $X(u_1, u_2)$. The function $X(u_1, u_2)$ is said to be invariant with respect to $h \in \mathfrak{g}$, if $[h \otimes 1 + 1 \otimes h, X(u_1, u_2)] = 0$ (i.e., if it is invariant with respect to $e^{t \cdot \text{ad} h}$ for any t).

The second method of propagation of solutions of (1.1) is the following: if $X(u_1, u_2)$ is a solution of (1.1), invariant with respect to the subalgebra $\mathfrak{h} \subset \mathfrak{g}$, and a tensor of $\mathfrak{h} \otimes \mathfrak{h}$ satisfies

$$[r^{12}, r^{13}] + [r^{12}, r^{23}] + [r^{13}, r^{23}] = 0, \tag{1.2}$$

$$r^{21} = -r^{12}, \tag{1.3}$$

then the function $\tilde{X}(u_1, u_2) \stackrel{\text{def}}{=} X(u_1, u_2) + r$ also is a solution of (1.1). It is easy to see this by direct calculation. We note that if the algebra $\mathfrak{h}$ is Abelian, then (1.2) holds automatically.

One often imposes the following additional conditions on the solutions of (1.1):

a) the so-called unitary solution $X^{12}(u_1, u_2) = -X^{21}(u_2, u_1)$,

b) the requirement that the function $X(u_1, u_2)$ depend only on $u_1 - u_2$ [in this case, taking some liberty we shall write $X(u_1 - u_2)$ instead of $X(u_1, u_2)$].

It is clear that property a) is preserved under both of the methods of propagation of solutions considered above, and property b) is preserved under the second method, but not always under the first. If $X(u_1 - u_2)$ is a solution of (1.1), then, in general it is not clear whether there exists a nonconstant function $\varphi(u)$ with values in Aut $\mathfrak{g}$ such that the function $\tilde{X}(u_1, u_2) \stackrel{\text{def}}{=} (\varphi(u_1) \otimes \varphi(u_2)) X(u_1 - u_2)$ depends only on $u_1 - u_2$. If, however, the invariance group G of the solution $X(u_1 - u_2)$ is nondiscrete, then one can set $\varphi(u) = e^{uP}$, where P is any element of the Lie algebra of the group G. For example, if the solution $X(u_1 - u_2)$ is invariant with respect to $h \in \mathfrak{g}$, then one can set $\varphi(u) = e^{u \cdot \text{ad} h}$.

We note that for functions $X(u_1, u_2)$, depending only on $u_1 - u_2$, (1.1) can be written in the form

$$[X^{12}(u), X^{13}(u+v)] + [X^{12}(u), X^{23}(v)] + [X^{13}(u+v), X^{23}(v)] = 0, \tag{1.4}$$

L. D. Landau Institute of Theoretical Physics, Academy of Sciences of the USSR. Physicotechnical Institute of Low Temperatures, Academy of Sciences of the Ukrainian SSR. Translated from Funktsional'nyi Analiz i Ego Prilozheniya, Vol. 16, No. 3, pp. 1-29, July-September, 1982. Original article submitted December 24, 1981.

and the unitary condition can be written in the form $X^{12}(u) = -X^{21}(-u)$.

1.2. In the present paper we investigate (1.4) under the assumption that $\mathfrak{g}$ is a finite-dimensional simple Lie algebra over C. Moreover, we shall seek a solution X(u) in the class of meromorphic functions, defined in some disk $U \subset \mathbf{C}$ with center at zero and satisfying one of the following three conditions, whose equivalence will be proved in Sec. 2:

A) The determinant of the matrix formed by the coordinates of the tensor X(u), is not identically equal to zero;

B) the function X(u) has at least one pole, and there does not exist a Lie subalgebra $\mathfrak{g}' \subset \mathfrak{g}$ such that $X(u) \in \mathfrak{g}' \otimes \mathfrak{g}'$ for any u;

C) the function X(u) has for u = 0 a pole of the first order with residue of the form $\sum_\mu I_\mu \otimes I_\mu$, where $c \in \mathbf{C}$, $\{I_\mu\}$ with respect to the Killing form.

Such a solution of (1.4) will be called nondegenerate. Our first basic result is the following.

THEOREM 1.1. Any nondegenerate solution X(u) of (1.4) also satisfies the unitary condition and extends meromorphically to the entire complex plane. All poles of X(u) are simple. They form a discrete subgroup $\Gamma \subset \mathbf{C}$. There exists a homomorphism $A: \Gamma \to \mathrm{Aut}\, \mathfrak{g}$ such that for any $u \in \mathbf{C}$, $\gamma \in \Gamma$ $X(u+\gamma) = (A(\gamma) \otimes 1) X(u) = (1 \otimes A(\gamma)^{-1}) X(u)$. If Γ has rank 2, then the restriction of A to some subgroup of finite index is trivial, so that X(u) is an elliptic function. If Γ has rank 1, then X(u) is equivalent to a solution $\tilde{X}(u)$ of the form $f(e^{ku})$, where f is a rational function. If $\Gamma = 0$, then X(u) is equivalent to a rational solution.

This theorem is proved in Sec. 4 with the help of the analog of the classical Weierstrass theorem on functions having an algebraic addition theorem obtained in Sec. 3. In Sec. 5 it is proved that nondegenerate solutions of (1.4) in elliptic functions exist only for $\mathfrak{g} = sl(n)$ and that all of them exhaust the solutions found in [1]. In Sec. 6 we find all nondegenerate solutions of (1.4) of the form $X(u) = f(e^{ku})$, where f is a rational function (we call such solutions trigonometric). It turns out that up to the methods of propagation of solutions described in point 1.1 and such trivial transformations as multiplication of a solution by a number and replacement of u by cu, the number of nondegenerate trigonometric solutions of (1.4) is finite. Moreover, we shall show that the simplest trigonometric solutions are the classical r-matrices (in the sense of [5], p. 141), corresponding to Toda–Bogoyavlenskii chains [10].

Unfortunately, we have not succeeded in getting essential results on rational solutions of (1.4). Even the problem of finding rational solutions not having a pole at infinity seems rather difficult. We have succeeded in finding only certain methods of constructing such solutions. These methods are given in Sec. 7.

2. Equivalence of Three Definitions of Nondegeneracy

2.1. We recall that $\mathfrak{g}$ denotes a finite-dimensional simple Lie algebra over C. We fix a nondegenerate invariant bilinear form on $\mathfrak{g}$. We choose in $\mathfrak{g}$ a basis $\{I_\mu\}$, orthonormal with respect to this form, and we set $t = I_\mu \otimes I_\mu$ (here and below we understand summation over identical indices). It is easy to see that t does not depend on the choice of $\{I_\mu\}$. Let X(u) be a meromorphic solution of (1.4), defined in some disk $U \subset \mathbf{C}$, containing 0.

Proposition 2.1. Let us assume that 1) the function X(u) has at least one pole, 2) there does not exist a Lie subalgebra $\mathfrak{g}' \subset \mathfrak{g}$, different from $\mathfrak{g}$ and such that $X(u) \in \mathfrak{g}' \otimes \mathfrak{g}'$ for any u. Then a) all poles of X(u) are simple, b) the function X(u) has a pole for u = 0 with residue of the form ct, $c \in \mathbf{C} \setminus \{0\}$.

Proof. Let X(u) have for $u = \gamma$ a pole of order k, and we set $\tau = \lim_{k \to \gamma} (u-\gamma)^k X(u)$. Multiplying both sides of (1.4) by $(v-\gamma)^k$ and letting v tend to γ, we get

$$[X^{12}(u), \tau^{23}] + [X^{13}(u+\gamma), \tau^{23}] = 0. \tag{2.1}$$

In exactly the same way, letting u tend to γ in (1.4), we get

$$[\tau^{12}, X^{13}(v+\gamma)] + [\tau^{12}, X^{23}(v)] = 0. \tag{2.2}$$

LEMMA. $[\tau^{12}, \tau^{13}] \neq 0$.

Proof. Let $V \subset \mathfrak{g}$ be the smallest vector space such that $\tau \in V \otimes \mathfrak{g}$. We set $\mathfrak{g}' = \{x \in \mathfrak{g} \mid [x, V] \subset V\}$. It is clear that $\mathfrak{g}' \subset \mathfrak{g}$ is a Lie subalgebra. Since $[X^{13}(u+\gamma), \tau^{23}] \in \mathfrak{g} \otimes V \otimes \mathfrak{g}$, it follows from (2.1) that $[X^{12}, \tau^{23}] \in \mathfrak{g} \otimes V \otimes \mathfrak{g}$, i.e., $X(u) \in \mathfrak{g} \otimes \mathfrak{g}'$. Exactly the same way one deduces from (2.2) that $X(v+\gamma) \in \mathfrak{g}' \otimes \mathfrak{g}$

for any v. Thus, $X(u) \in \mathfrak{g}' \otimes \mathfrak{g}'$ for any u. Consequently, $\mathfrak{g}' = \mathfrak{g}$, i.e., $[\mathfrak{g}, V] \subset V$. Whence, and from the simplicity of $\mathfrak{g}$ it follows that $V = \mathfrak{g}$. Hence $[\tau^{12}, \tau^{13}] \neq 0$.

It follows from the lemma that the function X(u) has, for u = 0, a pole of order not less than k: otherwise, letting v tend to 0 in (2.2), we would have $[\tau^{12}, \tau^{13}] = 0$. It remains to prove that the order of the pole of X(u) for u = 0 does not exceed one and $\lim_{u \to 0} uX(u) = ct$.

Let

$$X(u) = \frac{\theta}{u^l} + \frac{A}{u^{l-1}} + \sum_{i=2-l}^{\infty} X_i u^i, \quad \theta \neq 0.$$

If $l > 1$, then fixing v and comparing the coefficient of u^{1-l} in the Laurent series at the point u = 0 of the left side of (1.4) to zero, we get

$$[A^{12}, X^{13}(v) + X^{23}(v)] + \left[\theta^{12}, \frac{dX^{13}(v)}{dv}\right] = 0.$$

Now letting v tend to zero, we get $[\theta^{12}, \theta^{13}] = 0$, which contradicts the lemma. Thus, $l = 1$.

In (2.1), setting $\gamma = 0$, $\tau = \theta$, we get

$$[X^{12}(u) + X^{13}(u), \theta^{23}] = 0. \tag{2.3}$$

In exactly the same way, from (2.2) it follows that

$$[\theta^{(2)}, X^{13}(u) + X^{23}(u)] = 0. \tag{2.4}$$

We set $\mathfrak{g}' = \{x \in \mathfrak{g} \mid [x \otimes 1 + 1 \otimes x, \theta] = 0\}$, $\mathfrak{g}'$ is a Lie subalgebra in $\mathfrak{g}$. Solutions (2.3) and (2.4) mean that $X(u) \in \mathfrak{g}' \otimes \mathfrak{g}'$ for any u. Hence $\mathfrak{g}' = \mathfrak{g}$, i.e., $[x \otimes 1 + 1 \otimes x, \theta] = 0$ for any $x \in \mathfrak{g}$. Whence it follows that θ is proportional to t. The lemma is proved.

Thus, we have proved that from condition B), formulated in point 1.2, follows C). It is clear that from C), A) and B) follow. Hence to prove the equivalence of all three conditions, it remains to prove that there does not exist a solution X(u) of (1.4), holomorphic in U, such that for some u the tensor X(u) is nondegenerate. This will be done in the rest of this section.

2.2. Proposition 2.2. Let the solution X(u) of (1.4) be holomorphic in U and let there exist a $u_0 \in U$ such that the tensor $X(u_0)$ is nondegenerate. Then the tensor X(0) is also nondegenerate.

Proof. Setting v = 0 in (1.4), we get

$$[X^{12}(u), X^{13}(u)] + [X^{12}(u) + X^{13}(u), X^{23}(0)] = 0.$$

Let $X(u) = K^{\mu} \otimes I_{\mu}$. Then

$$[K^{\mu}(u), K^{\nu}(u)] \otimes I_{\mu} \otimes I_{\nu} + K^{\lambda}(u) \otimes [I_{\lambda} \otimes 1 + 1 \otimes I_{\lambda}, X(0)] = 0,$$

whence,

$$[K^{\mu}(u), K^{\nu}(u)] = C_{\lambda}^{\mu\nu} K^{\lambda}(u), \tag{2.5}$$

where $C_{\lambda}^{\mu\nu}$ are found from the relation $C_{\lambda}^{\mu\nu} I_{\mu} \otimes I_{\nu} = [X(0), I_{\lambda} \otimes 1 \otimes I_{\lambda}]$. By hypothesis, the vectors $K^{\mu}(u_0)$ form a basis in $\mathfrak{g}$. Hence for any $u \in U$ there exists exactly one linear operator $\varphi_u\colon \mathfrak{g} \to \mathfrak{g}$ such that $\varphi_u(K^{\mu}(u_0)) = K^{\mu}(u)$. Here φ_u depends holomorphically on u. It is necessary to prove that $\det \varphi_0 \neq 0$. From (2.5) it follows that φ_u is an endomorphism of $\mathfrak{g}$ as a Lie algebra.

LEMMA. Let φ be an endomorphism of $\mathfrak{g}$ as a Lie algebra. Then $\det \varphi \in \{0, 1, -1\}$.

Proof. Let us assume that $\varphi \neq 0$. From the simplicity of $\mathfrak{g}$ it follows that then φ is an automorphism and consequently preserves the Killing form. Hence $\det \varphi = \pm 1$. The lemma is proved.

Since $\varphi_u = 1$ and φ_u depends holomorphically on u, it follows from the lemma that $\det \varphi_u = 1$ for any u. In particular, $\det \varphi_0 = 1$.

It is clear that if X(u) is a solution of (1.4), holomorphic for u = 0, then the tensor $r \overset{\text{def}}{=} X(0)$ satisfies (1.2).

Proposition 2.3. Let $r \in \mathfrak{g} \otimes \mathfrak{g}$ be a nondegenerate solution of (1.2). Then r also satisfies (1.3).

<u>Proof.</u> Let $r = K^\mu \otimes I_\mu = I_\mu \otimes L^\mu$. The nondegeneracy of r means that $\{K^\mu\}$ and $\{L^\mu\}$ are bases in $\mathfrak{g}$. We define $C_\lambda^{\mu\nu}$ from the relation $C_\lambda^{\mu\nu} I_\mu \otimes I_\nu = [r, I_\lambda \otimes 1 + 1 \otimes I_\lambda]$. Arguing exactly as in the proof of the preceding proposition, we get

$$[K^\mu, K^\nu] = C_\lambda^{\mu\nu} K^\lambda, \tag{2.6}$$

$$[L^\mu, L^\nu] = -C_\lambda^{\mu\nu} L^\lambda. \tag{2.7}$$

From (2.6) it follows that $C_\lambda^{\mu\nu} + C_\lambda^{\nu\mu} = 0$, whence $[r^{12} + r^{21}, I_\lambda \otimes 1 + 1 \otimes I_\lambda] = 0$. Hence $r^{12} + r^{21} = at$, $a \in C$. This means that

$$K^\mu + L^\mu = aI_\mu. \tag{2.8}$$

Let φ: $\mathfrak{g} \to \mathfrak{g}$ be a linear operator such that $\varphi(K^\lambda) = -L^\lambda$. From (2.6) and (2.7) it follows that φ is an automorphism of $\mathfrak{g}$ as a Lie algebra. Solution (2.8) can be rewritten in the form

$$(1 - \varphi)K^\mu = aI_\mu. \tag{2.9}$$

We need to prove that $a = 0$. If $a \neq 0$, then it would follow from (2.9) that $\det(\varphi - 1) \neq 0$. As a matter of fact, for any $\varphi \in \operatorname{Aut} \mathfrak{g}$ there exists a nonzero $x \in \mathfrak{g}$ such that $\varphi(x) = x$. In fact, if φ has finite order, then this follows from Lemma 1 of [3]. Now if φ has infinite order, then it is necessary to apply to the cyclic subgroup, generated by φ, the following lemma.

<u>LEMMA.</u> Let $H \subset \operatorname{Aut} \mathfrak{g}$ be an infinite Abelian subgroup. Then there exists a nonzero $x \in \mathfrak{g}$ such that $gx = x$ for any $g \in H$.

<u>Proof.</u> We denote by $\bar{H}$ the smallest algebraic subgroup in $\operatorname{Aut} \mathfrak{g}$, containing H, and by $\mathfrak{h}$ the Lie algebra of the group $\bar{H}$. Since $|\bar{H}| = \infty$, one has $\mathfrak{h} \neq 0$. The Lie algebra of the group $\operatorname{Aut} \mathfrak{g}$ coincides with $\mathfrak{g}$, so $\mathfrak{h}$ can be considered as a subalgebra in $\mathfrak{g}$. As x one can take any nonzero element of $\mathfrak{h}$. The lemma is proved.

<u>2.3.</u> It remains to prove that the system of equations (1.2), (1.3) has no nondegenerate solutions.

<u>Proposition 2.4.</u> Let $r = r^{\mu\nu} I_\mu \otimes I_\nu$ be a nondegenerate skew symmetric tensor, (S_{kl}) be the matrix inverse to $(r^{\mu\nu})$, B be a bilinear form on $\mathfrak{g}$ with matrix (S_{kl}). In order that (1.2) hold it is necessary and sufficient that the form B be a 2-cocycle, i.e., that one have the identity

$$B([x, y], z) + B([y, z], x) + B([z, x], y) = 0, \quad x, y, z \in \mathfrak{g}. \tag{2.10}$$

<u>Proof.</u> Solution (1.2) is equivalent with

$$C_{ij}^\alpha r^{i\beta} r^{j\gamma} + C_{ij}^\beta r^{\alpha i} r^{j\gamma} + C_{ij}^\gamma r^{\alpha i} r^{\beta j} = 0,$$

where C_{ij}^α are the structural constants of $\mathfrak{g}$. This equation, by virtue of the skew symmetry of r, can be rewritten in the form

$$C_{ij}^\alpha r^{i\beta} r^{j\gamma} + C_{ij}^\beta r^{i\alpha} r^{i\gamma} + C_{ij}^\gamma r^{i\alpha} r^{j\beta} = 0. \tag{2.11}$$

Multiplying both sides of (2.11) by $S_{\alpha k} S_{\beta l} S_{\gamma m}$, we get

$$C_{lm}^\alpha S_{\alpha k} + C_{mk}^\beta S_{\beta l} + C_{kl}^\gamma S_{\gamma m} = 0,$$

which is equivalent with (2.10).

Now we shall show that a bilinear skew symmetric form B on $\mathfrak{g}$, which is a 2-cocycle, is degenerate. In fact, since $\mathfrak{g}$ is simple, any cocycle is a coboundary, i.e., $B(x, y) = l([x, y])$, where $l \in \mathfrak{g}^*$. The image of l under the isomorphism $\mathfrak{g}^* \rightleftarrows \mathfrak{g}$, defined by the Killing form, we denote by z. It is easy to see that z belongs to the kernel of B.

The equivalence of conditions A)-C) is proved.

3. Weierstrass-Type Theorem

The classical Weierstrass theorem asserts that if the function f(u) is meromorphic on the entire complex plane and satisfies a functional equation of the form

$$P(f(u), f(v), f(u + v)) = 0, \tag{3.1}$$

where P is a nonzero polynomial, then the function f is either elliptic or rational or has the form $\varphi(e^{kz})$, where φ is a rational function. Let us assume now that the function f is defined only in some neighborhood of zero $U \subset C$ and assumes vector values, and the polynomial P in (3.1) is also vector valued. We shall show that then under certain additional assumptions the function f has the form $f(u) = \bar{f}(ua)$, where $\bar{f}$ is a quasi-Abelian function on C^n (i.e., either an Abelian function or a degeneration of Abelian ones), $a \in C^n$.

We proceed to a precise formulation. We recall that a meromorphic function φ on C^n is called Abelian, if it has 2n periods, linearly independent over R.

Definition. A meromorphic function φ on an n-dimensional complex vector space L is called quasi-Abelian, if there exists a system of coordinates $z_1, \ldots, z_n$ in the space L, integers p, q, $r \geq 0$, $p + q + r = n$ and vectors $\gamma_1, \ldots, \gamma_{2r} \in L$ such that

1) for fixed $z_{p+q+1}, \ldots, z_n$, $\varphi(z_1, \ldots, z_n)$ is a rational function of $z_1, \ldots, z_p, e^{z_{p+1}}, \ldots, e^{z_{p+q}}$;

2) the vectors γ_i are periods of φ;

3) the vectors $\tilde{\gamma}_i \in C^r$, formed by the last r coordinates of the vectors γ_i, are linearly independent over R.

Let f(u) be a meromorphic function with values in C^m, defined in some disk $U \subset C$ with center at zero. We denote by U' the complement of the set of poles of f. Let us assume that the following identity holds,

$$P_j(f(u), f(v), f(u+v)) = 0, \quad j = 1, 2, \ldots, N,$$

where P_j are polynomials in 3n variables. We denote by S the set of points $(u, v) \in U' \times U'$ such that the system of equations

$$P_j(f(u), f(v), x) = 0, \quad j = 1, 2, \ldots, N$$

with respect to the unknown $x \in C^m$ has no more than one solution (it is clear that if $u + v \in U'$, then at least one solution of this system exists). We denote by T the set of points $(u, w) \in U' \times U'$ such that the system of equations

$$P_j(x, f(v), f(w)) = 0, \quad j = 1, 2, \ldots, N$$

has no more than one solution.

THEOREM 2.1. If S and T have nonempty interiors, then there exist a natural number n, a vector $a \in C^n$ and a quasi-Abelian function $\bar{f}$ on C^n such that

a) $f(u) = \bar{f}(ua)$,

b) one has the identities $P_j(\bar{f}(u), \bar{f}(v), \bar{f}(u+v)) = 0$, $j = 1, 2, \ldots, N$.

Proof. Let $X \in C^m$ be the Zariski closure of the set of points of the form $f(u)$, $u \in U'$. Let $\Gamma \subset X \times X \times X$ be the Zariski closure of the set of points of the form $(f(u), f(v), f(u+v))$, where $u, v, u+v \in U'$. It is clear that the varieties X and Γ are irreducible.

LEMMA. All three projections $\Gamma \to X \times X$ are birational isomorphisms.

Proof. We consider, for example, the projection π_{12} of the set Γ onto the product of the first two factors. Let $W \subset X^2$ be a nonempty Zariski open subset such that the fibers of the map π_{12} over points of W have identical cardinality k. We set $A = \{(u, v) \in U' \times U' \mid (f(u), f(v)) \in W\}$. It is clear that A is everywhere dense in $U' \times U'$. Hence $A \cap S \neq \phi$, whence $k \leq 1$. On the other hand, A contains at least one point (u, v) such that $u + v \in U'$. Hence $k \geq 1$. Thus π_{12} is a birational isomorphism. For the other two projections the proof is analogous. The lemma is proved.

Since π_{12} is a birational isomorphism, Γ is the graph of a rational map $\mu: X \times X \to X$, which can be considered as an "operation" on X. It is clear that this "operation" is commutative. Since π_{13} and π_{23} are birational isomorphisms, there exists an inverse "operation" for it. We shall show that the "operation" μ is associative. We denote by V the set of points $(x_1, x_2, x_3) \in X^3$, for which the expressions $\mu(x_1, x_2)$, $\mu(\mu(x_1, x_2)x_3)$, $\mu(x_2, x_3)$, $\mu(x_1, \mu(x_2, x_3))$ make sense. We set $R \stackrel{\text{def}}{=} \{(f(u_1), f(u_2), f(u_3)) \mid u_1, u_2, u_3, u_1 + u_2, u_2 + u_3, u_1 + u_2 + u_3 \in U'\}$. It is easy to see that if $(x_1, x_2, x_3) \in V \cap R$, then $\mu(\mu(x_1, x_2), x_3) = \mu(x_1, \mu(x_2, x_3))$. Since $V \cap R \subset X$ is Zariski everywhere dense, the associativity of μ follows from this.

Thus, X is a "birational group" in the sense of Weil. It is known ([16, 17]) that such a group is birationally isomorphic with a real algebraic group which is uniquely defined. Thus, we have proved that there exist a connected commutative algebraic group G, a rational function $\bar{f}: G \to C^m$ and a meromorphic map $\varphi: U \to G$ such that

$$P_j(\bar{f}(g_1), \bar{f}(g_2), \bar{f}(g_1+g_2)) = 0, \quad j = 1, 2, \ldots, N,$$
$$\varphi(u+v) = \varphi(u) + \varphi(v). \tag{3.2}$$

From (3.2) it is easy to deduce that φ is holomorphic and moreover extends to a holomorphic homomorphism $\mathbf{C} \to G$.

By Chevalley's theorem, any connected commutative algebraic group over C is an extension of an Abelian variety by a direct product of a finite number of additive and multiplicative groups. Hence the universal covering group for G is isomorphic with $\mathbf{C}^n$, and rational functions on G go into quasi-Abelian functions on $\mathbf{C}^n$. We denote by $\bar{f}$ the quasi-Abelian function on $\mathbf{C}^n$, corresponding to $\bar{f}$. The homomorphism $\varphi: \mathbf{C} \to G$ lifts uniquely to a holomorphic homomorphism $\bar{\varphi}: \mathbf{C} \to \mathbf{C}^n$. $\bar{\varphi}$ is defined by a formula of the form $\bar{\varphi}(u) = ua$, $a \in \mathbf{C}^n$. The $\bar{f}$ and a constructed in this way are the ones sought. The lemma is proved.

4. Properties of Nondegenerate Solutions

4.1. Let X(u) be a nondegenerate solution of (1.4), defined in some disk $U \subset \mathbf{C}$, containing 0. We shall always assume that $\lim_{u\to 0} uX(u) = t$ [according to Proposition 2.1, this can always be achieved by multiplying X(u) by a suitable number].

Proposition 4.1. X(u) satisfies the unitary condition.

Proof. We have:

$$[X^{12}(u_1-u_2), X^{13}(u_1-u_3)] + [X^{12}(u_1-u_2), X^{23}(u_2-u_3)] + [X^{13}(u_1-u_3), X^{23}(u_2-u_3)] = 0. \tag{4.1}$$

Interchanging the places of u_1 and u_2, and also of the first and second factors in the tensor product $\mathfrak{g} \otimes \mathfrak{g} \otimes \mathfrak{g}$, we get

$$[X^{21}(u_2-u_1), X^{23}(u_2-u_3)] + [X^{21}(u_2-u_1), X^{13}(u_1-u_3)] + [X^{23}(u_2-u_3), X^{13}(u_1-u_3)] = 0. \tag{4.2}$$

Adding (4.1) and (4.2), we arrive at the identity

$$[X^{12}(u_1-u_2) + X^{21}(u_2-u_1), X^{13}(u_1-u_3) + X^{23}(u_2-u_3)] = 0.$$

If now, fixing u_1 and u_2, we let u_3 tend to u_2, then we get $[X^{12}(u_1-u_2) + X^{21}(u_2-u_1), t^{23}] = 0$. Whence it is easy to deduce that $X^{12}(u_1-u_2) + X^{21}(u_2-u_2) = 0$.

Proposition 4.2. There exist a natural number n, a vector $a \to \mathbf{C}^n$, and a quasi-Abelian function: $\bar{X}: \mathbf{C}^n \to \mathfrak{g} \otimes \mathfrak{g}$, satisfying (1.4), such that $X(u) = \bar{X}(ua)$.

Proof. We set $U' = U \setminus \{0\}$. One can assume that the function X(u) is holomorphic in U'. According to Theorem 2.1, it is sufficient to show that the sets S and T have nonempty interiors, where S is the set of points $(u, v) \in U' \times U'$ such that the equation

$$[X^{12}(u) - X^{23}(v), Z^{13}] = 0 \tag{4.3}$$

with respect to the unknown $Z \in \mathfrak{g} \otimes \mathfrak{g}$ has only the zero solution, T is the set of points $(u, w) \in U' \times U'$ such that the equation $[Z^{12}, X^{23}(v) + X^{13}(w)] = 0$ has only the zero solution. Since X(u) satisfies the unitary condition, one has $[Z^{12}, X^{23}(v) + X^{13}(w)] = 0$. Since S is open, it suffices to prove that $S \neq \emptyset$. We shall show that if $u \neq 0$ is sufficiently small, then $(u, v) \in S$. For $v = u \neq 0$, (4.3) can be written in the form

$$[uX^{12}(u) - uX^{23}(u), Z^{13}] = 0. \tag{4.4}$$

For u = 0, (4.4) assumes the form

$$[t^{12} - t^{23}, Z^{13}] = 0. \tag{4.5}$$

We shall show that (4.5) has only the zero solution. Whence it will follow that for all sufficiently small u, (4.4) has only the zero solution.

Solution (4.5) means that for any μ

$$[I_\mu \otimes 1 - 1 \otimes I_\mu, Z] = 0. \tag{4.6}$$

Whence it follows that

$$[[I_\mu, I_\nu] \otimes 1 + 1 \otimes [I_\mu \otimes 1 - 1 \otimes I_\mu, I_\nu \otimes 1 - 1 \otimes I_\nu], Z] = 0. \tag{4.7}$$

Since elements of the form $[I_\mu, I_\nu]$ generate $\mathfrak{g}$ as a vector space, it follows from (4.7) that $[I_\mu \otimes 1 + 1 \otimes I_\mu, T] = 0$ for any μ. Whence and from (4.6) it follows that $[I_\mu \otimes 1, Z] = 0$, and consequently, $Z = 0$.

From Proposition 4.2, in particular, it follows that $X(u)$ extends to a meromorphic function on all of $\mathbf{C}$. We denote by Γ the set of its poles. According to Proposition 2.1, all of them are simple.

Proposition 4.3. Let $\gamma \in \Gamma$. Then there exists an $A_\gamma \in \operatorname{Aut} \mathfrak{g}$ such that

$$X(u+\gamma) = (A_\gamma \otimes 1)\, X(u). \tag{4.7a}$$

Proof. We set $\tau = \lim_{u\to\gamma}(u-\gamma)\, X(u)$. Let $A_\gamma\colon \mathfrak{g} \to \mathfrak{g}$ be a linear operator such that $\tau = A_\gamma(I_\mu) \otimes I_\mu$. From (2.2) and the identity $[t^{12}, r^{13} + r^{23}] = 0$, valid for any $r \in \mathfrak{g} \otimes \mathfrak{g}$, it follows that

$$[\tau^{12}, X^{13}(v+\gamma)] = -(A_\gamma \otimes 1 \otimes 1)([t^{12}, X^{23}(v)]) = (A_\gamma \otimes 1 \otimes 1)([t^{12}, X^{13}(v)]). \tag{4.8}$$

Comparing the residues of both sides of (4.8) for $v = 0$, we get $[\tau^{12}, \tau^{13}] = (A_\gamma \otimes 1 \otimes 1)([t^{12}, t^{13}])$, i.e., $[A_\gamma(I_\mu), A_\gamma(I_\nu)] \otimes I_\mu \otimes I_\nu = A_\gamma([I_\mu, I_\nu]) \otimes I_\mu \otimes I_\nu$. This means that $A_\gamma([I_\mu, I_\nu]) = [A_\gamma(I_\mu), A_\gamma(I_\nu)]$, i.e., A_γ is an endomorphism of the Lie algebra $\mathfrak{g}$. Since $A_\gamma \neq 0$, and the algebra $\mathfrak{g}$ is simple, A_γ is an automorphism. Applying to both sides of (4.8) the map $A_\gamma^{-1} \otimes 1 \otimes 1$, and using the fact that A_γ^{-1} is an automorphism of $\mathfrak{g}$ as a Lie algebra, we get the equation $[t^{12}, (A_\gamma^{-1} \otimes 1)X^{13}(v+\gamma) - X^{13}(v)] = 0$, whence $(A_\gamma^{-1} \otimes 1)X(v+\gamma) = X(v)$. The proposition is proved.

Proposition 4.4. 1) Γ is a discrete subgroup in $\mathbf{C}$. 2) $A_{\gamma_1+\gamma_2} = A_{\gamma_1}A_{\gamma_2}$ for any $\gamma_1, \gamma_2 \in \Gamma$; 3) $X(u+\gamma) = (1 \otimes A_\gamma^{-1})\, X(u)$, $u \in \mathbf{C}$, $\gamma \in \Gamma$; 4) $(A_\gamma \otimes A_\gamma)X(u)$, $u \in \mathbf{C}$, $\gamma \in \Gamma$.

Proof. Let $\gamma, \gamma' \in \Gamma$. The right side of (4.7) has a pole for $u = \gamma'$. Hence the left side has the same property, i.e., $\gamma + \gamma' \in \Gamma$. Since $X(u)$ satisfies the unitary condition, $\gamma \in \Gamma \Rightarrow -\gamma \in \Gamma$. Thus, Γ is a subgroup in $\mathbf{C}$. The discreteness of Γ and assertion 2) are obvious. Assertion 3) is equivalent with the equation $X^{21}(u+\gamma) = (A_\gamma^{-1} \otimes 1)$, $X^{21}(u)$, which follows from (4.7) and the unitary condition. Assertion 4) follows from 3) and (4.7). The proposition is proved.

4.2. Proposition 4.5. Let Γ have rank 2. Then

a) there does not exist a nonzero $x \in \mathfrak{g}$ such that $A_\gamma(x) = x$ for any $\gamma \in \Gamma$;

b) there exists a subgroup of finite index $\Gamma' \in \Gamma$ such that $A_\gamma = 1$ for $\gamma \in \Gamma'$.

Proof. a) Let us assume that $x \in \mathfrak{g}$, $x \neq 0$, $A_\gamma(x) = x$ for $\gamma \in \Gamma$. Let $X(u) = X_{\mu\nu}(u) I_\mu \otimes I_\nu$. We define a meromorphic function $\varphi\colon \mathbf{C} \to \mathfrak{g}$ by the formula $\varphi(u) = X_{\mu\nu}(I_\mu, x)\cdot I_\nu$. It is easy to see that the function φ is Γ-periodic, has, for $u = 0$, a simple pole, and does not have other poles in the period parallelogram. The contradiction obtained proves assertion a).

b) We set $H = \{A_\gamma \mid \gamma \in \Gamma\}$. The lemma from the proof of Proposition 2.3 and assertion a) which has been proved already show that $|H| < \infty$, whence follows b). The proposition is proved.

COROLLARY. If the rank of Γ is equal to 2, then $X(u)$ is an elliptic function.

4.3. In this section we shall prove the assertions of Theorem 1.1, concerning the case when the rank of Γ is equal to 0 or 1. Let n and $\overline{X}$ denote the same things as in Proposition 4.2.

Proposition 4.6. There exist an $(n-1)$-dimensional vector subspace $V \subset \mathbf{C}^n$ and a holomorphic homomorphism $\varphi\colon V \to \operatorname{Aut} \mathfrak{g}$ such that for any $z \in \mathbf{C}^n$, $h \in V$

$$\overline{X}(z+h) = (\varphi(h) \otimes 1)\, \overline{X}(z), \tag{4.9}$$

$$(\varphi(h) \otimes \varphi(h))\, \overline{X}(z) = \overline{X}(z). \tag{4.10}$$

Proof. Let $\overline{X}(z) = Y(z)/f(z)$, where Y and f are entire functions. Without loss of generality one can assume that the set $S \overset{\text{def}}{=} \{z \in \mathbf{C} \mid f(z) = 0,\ Y(z) \neq 0\}$ is nonempty. Let $h \in S$.

LEMMA. 1) There exist $L(h) \in \operatorname{Aut} \mathfrak{g}$ and $c(h) \in \mathbf{C} \mid \{0\}$ such that

$$Y(h) = (c(h) L(h) I_\mu) \otimes I_\mu \tag{4.11}$$

2) $$\overline{X}(\lambda + h) = (L(h) \otimes 1)\overline{X}(z),\quad z \in \mathbf{C}^n. \tag{4.12}$$

Proof. Let $\overline{X}(z) = K_\mu(z) \otimes I_\mu$. The same arguments as in the derivation of (2.2) show that $[Y^{12}(h), \overline{X}^{13}(z+h) + \overline{X}^{23}(z)] = 0$ and consequently,

$$[Y(h), K_\mu(Z+h)\otimes 1 + 1\otimes K_\mu(Z)] = 0. \tag{4.13}$$

We denote by W the set of those $Z \in C^n$ such that a) the function $\bar{X}$ is holomorphic at the points z and z + h, b) the tensors $\bar{X}(Z)$ and $\bar{X}(z+h)$ are nondegenerate. Let $z \in W$. We denote by $\mathfrak{a}$ the subalgebra in $\mathfrak{g}\otimes\mathfrak{g}$, generated by elements $K_\mu(z+h)\otimes 1 + 1\otimes K_\mu(z)$. Then $[Y(h), a] = 0$ for any $a \in \mathfrak{a}$. Since the vectors $K_\mu(z)$ and $K_\mu(z+h)$ form bases in $\mathfrak{g}$, both projections $\mathfrak{a}\to\mathfrak{g}$ are surjective. Whence and from the simplicity of $\mathfrak{g}$ it follows that either $\mathfrak{a} = \mathfrak{g}\times\mathfrak{g}$ or there exists an $L \in \mathrm{Aut}\,\mathfrak{g}$ such that $\mathfrak{a} = \{Lx\otimes 1 + 1\otimes x \mid x \in \mathfrak{g}\}$. The first case is impossible, since $[Y(h), \mathfrak{a}] = 0$, $Y(h) \neq 0$. Thus, we have proved the existence of an $L(z,h) \in \mathrm{Aut}\,\mathfrak{g}$ such that

$$K_\mu(z+h) = L(z,h)K_\mu(z). \tag{4.14}$$

From (4.13) and (4.14) it follows that $[(L(z,h)^{-1}\otimes 1)Y(h), K_\mu(z)\otimes 1 + 1\otimes K_\mu(Z)] = 0$, whence

$$Y(h) = (c(z,h)L(z,h)I_\mu)\otimes I_\mu. \tag{4.15}$$

From (4.15) it follows that c(z, h) and L(z, h) do not depend on z. From (4.14) it follows that (4.12) holds for $z \in W$, and hence also for any $z \in C^n$.

Let $H \subset C^n$ be the subgroup generated by S. From the lemma it follows that there exists a homeomorphism φ: $H \to \mathrm{Aut}\,\mathfrak{g}$ such that (4.9) holds for any $h \in H$, $z \in C^n$. The set of poles of the function $\bar{X}$ goes into itself under translations by elements of H. Hence $H \neq C^n$. Since H is generated by an analytic subset $S \subset C^n$ of codimension 1, S is an open subset in the union of a finite or countable number of mutually parallel affine hyperplanes, and H contains an (n − 1)-dimensional vector subspace $V \subset C^n$, parallel to these hyperplanes. From (4.11) it follows that L(h) depends holomorphically on h. Hence the map φ: $V \to \mathrm{Aut}\,\mathfrak{g}$ is holomorphic.

The same arguments as in the proof of Proposition 4.1 show that $\bar{X}$ satisfies the unitary condition. Whence and from (4.9) follows (4.10) (cf. the proof of Proposition 4.4). The corollary is proved.

LEMMA 4.1. Let a denote the same thing as in Proposition 4.2. If $\tilde{a} - a \in V$, then the function $\tilde{X}(u) \overset{\text{def}}{=} \bar{X}(u\tilde{a})$ is a solution of (1.4), equivalent with X(u).

Proof. From (4.9) and (4.10) it follows that $\tilde{X}(u_1 - u_2) = (\varphi(u_1, h)\otimes\varphi(u_2, h))X(u_1 - u_2)$, where $h = \tilde{a} - a$, φ denotes the same thing as in Proposition 4.6.

Proposition 4.7. If Γ has rank 1, then X(u) is equivalent with a solution $\tilde{X}(u)$ of the form $f(e^{ku})$, where f is a rational function. If $\Gamma = 0$, then X(u) is equivalent with a rational solution.

Proof. Let p, q, r, $z_1, \ldots, z_n$, $\gamma_1, \ldots, \gamma_{2r}$ denote the same things as in the definition of quasi-Abelianness [in our situation $\varphi(z) = \bar{X}(z)$]. We denote by $e_1, \ldots, e_n$ basis vectors in C^n, corresponding to the system of coordinates $z_1, \ldots, z_n$, and by W the subspace in C^n, defined by the equations $z_{p+q+1} = \ldots = z_n = 0$. We represent γ_i in the form $\delta_i a + h_i$, $\delta_i \in C$, $h_i \in V$. It is clear that $\delta_i \in \Gamma$.

Let us assume that the rank of Γ is equal to 0 or 1. Then $W \not\subset V$. In fact, if one had $W \subset V$, then the vectors $\gamma_1, \ldots, \gamma_{2r}$ would generate C^n/V as a vector space over R, so that $\delta_1, \ldots, \delta_{2r}$ would generate C as a vector space over R, and this is impossible, since $\delta_1, \ldots, \delta_{2r} \in \Gamma$. Since $W \not\subset V$, there exists an $i \le p + q$ such that $e_i \notin V$. Then a can be represented in the form $ke_i + h$, $k \in C$, $h \in V$. We set $\tilde{X}(u) = \bar{X}(uke_i)$. According to Lemma 4.1, $\tilde{X}(u)$ is a solution of (1.4), equivalent with X(u). It is clear that if i < p, then $\tilde{X}(u)$ is a rational function, and if i > p, then X(u) has the form $f(e^{ku})$, where f is rational.

The set of poles of $\tilde{X}(u)$ is equal to Γ. Hence if the rank of Γ is equal to 1, then the function $\tilde{X}(u)$ cannot be rational, and if $\Gamma = \{0\}$, then $\tilde{X}(u)$ cannot have the form $f(e^{ku})$, where f is rational. The proposition is proved.

Theorem 1.1 is completely proved. Solutions of the form $f(e^{ku})$, where f is a rational function, we shall call trigonometric.

4.4. It was already noted that any nondegenerate solution of (1.4), meromorphic in a neighborhood of zero, extends to a meromorphic function on all of C. One can also show that any formal solution of (1.4) of the form $X(u) = \frac{t}{u} + \sum_{i=0}^{\infty} X_i u^i$ converges for sufficiently small $u \neq 0$.

5. Elliptic Solutions

5.1. Let $\Gamma \subset C$ be a discrete subgroup of rank 2, ω_1 and ω_2 be generators of it. Let X(u) be a nondegenerate solution of (1.4) with set of poles Γ. We set $A_1 = A_{\omega_1}$, $A_2 = A_{\omega}$ (in connection with the notation A_γ, cf. Proposition 4.3). It is clear that $A_1A_2 = A_2A_1 = A_{\omega_1+\omega_2}$. According to Proposition 4.5, the automorphisms A_1 and A_2 have finite order, while there does not exist a nonzero $x \in \mathfrak{g}$ such that $A_1(x) = A_2(x) = x$. In this point we shall prove that if $A_1, A_2 \in \mathrm{Aut}\,\mathfrak{g}$ commute, have finite order, and do not have common fixed nonzero vectors, then to the pair (A_1, A_2) corresponds exactly one nondegenerate solution of (1.4) with set of poles Γ. As a preliminary we prove a lemma, valid for a discrete subgroup $\Gamma \subset \mathbf{C}$ of any rank.

LEMMA 5.1. Let $A : \Gamma \to \mathrm{Aut}\,\mathfrak{g}$ be a homomorphism, $X(u)$ be a meromorphic function on the complex plane with values in $\mathfrak{g} \otimes \mathfrak{g}$ such that a) $X(u + \gamma) = (A_\gamma \otimes 1)X(u)$ for $u \in \mathbf{C}$, $\gamma \in \Gamma$ [here $A_\gamma \overset{\mathrm{def}}{=} A(\gamma)$]; b) $X^{21}(u) = X^{12}(-u)$; c) $\lim_{u\to 0} u\,X(u) = t$; d) $X(u)$ has no poles for $u \notin \Gamma$. We set

$$Y(u_1, u_2, u_3) = [X^{12}(u_1 - u_2),\ X^{13}(u_1 - u_3)] + [X^{12}(u_1 - u_2), X^{23}(u_2, u_3)] + [X^{13}(u_1 - u_3),\ X^{23}(u_2 - u_3)]. \quad (5.1)$$

Then 1) the function $Y(u_1, u_2, u_3)$ has no poles; 2) for any $\gamma \in \Gamma$

$$Y(u_1 + \gamma, u_2, u_3) = (A_\gamma \otimes 1 \otimes 1)Y(u_1, u_2, u_3), \quad (5.2)$$

$$Y(u_1, u_2, u_3 + \gamma) = (1 \otimes 1 \otimes A_\gamma^{-1})Y(u_1, u_2, u_3). \quad (5.3)$$

Proof. Formula (5.2) can be verified directly. Solution (5.3) follows from (5.2) and the identity

$$Y^{321}(u_3, u_2, u_1) = -Y^{123}(u_1, u_2, u_3), \quad (5.4)$$

following from the unitary condition. If the set P of poles of the function Y is not empty, then it is the union of certain planes of the form $u_i - u_j = \gamma$, $\gamma \in \Gamma$. It is necessary to show that no such plane is contained in P. In view of (5.2)-(5.4) and the equation $Y^{213}(u_2, u_1, u_3) = -Y^{123}(u_1, u_2, u_3)$, it suffices to show that the plane $u_1 = u_2$ is not contained in P. In fact, for fixed u_2, u_3 we have $\lim_{u_1\to u_1} (u_1 - u_2)Y(u_1, u_2, u_3) = [t^{12}, X^{13}(u_2 - u_3) + X^{23}(u_2 - u_3)] = 0$. The lemma is proved.

Proposition 5.1. Let A_1, A_2 be commuting automorphisms of $\mathfrak{g}$ of finite order, not having common fixed nonzero vectors. Then there exists exactly one meromorphic function $X : \mathbf{C} \to \mathfrak{g} \otimes \mathfrak{g}$ such that 1) $\lim_{u\to 0} u\,X(u) = t$; 2) $X(u + \omega_i) = (A_i \otimes 1)X(u)$, $i = 1, 2$; 3) $X(u)$ has no poles for $u \notin \Gamma$. This function is a solution of (1.4).

Proof. Let $A_1^n = A_2^n = 1$. We have: $\mathfrak{g} = \bigoplus_{k,l \in \mathbf{Z}/n\mathbf{Z}} \mathfrak{g}_{kl}$, where $\mathfrak{g}_{kl} = \{x \in \mathfrak{g} \mid A_1(x) = \zeta^k x,\ A_2(x) = \zeta^l x\}$, $\zeta = e^{2\pi i/n}$. By hypothesis, $\mathfrak{g}_{00} = 0$. Since $(A_1 \otimes A_1)t = (A_2 \otimes A_2)t = t$, one has $t \in \bigoplus_{k,l}(\mathfrak{g}_{kl} \otimes \mathfrak{g}_{-k,-l})$. The projection of t on $\mathfrak{g}_{kl} \otimes \mathfrak{g}_{-k,-l}$ we denote by t_{kl}.

If the function $X(u)$ sought exists, then $(A_1 \otimes A_1)X(u) = (A_2 \otimes A_2)X(u) = X(u)$ (cf. Proposition 4.4). Hence $X(u)$ must be sought in the form

$$X(u) = \sum_{\substack{k,l \in \mathbf{Z}/n\mathbf{Z} \\ (k,l) \neq (0,0)}} X_{k,l}(u),\ X_{kl}(u) \in \mathfrak{g}_{kl} \otimes \mathfrak{g}_{-k,-l}.$$

In order that the function $X(u)$ have properties 1)-3), it is necessary and sufficient that the functions $X_{kl}(u)$ satisfy the conditions 1') $\lim_{u\to 0} u X_{kl}(u) = t_{kl}$; 2') $X_{kl}(u + \omega_1) = \zeta^k X_{kl}(u)$, $X_{kl}(u_1 + \omega_2) = \zeta^l X_{kl}(u)$; 3') $X_{kl}(u)$ has no poles for $u \notin \Gamma$. Since $(k, l) \neq (0, 0)$, there exists exactly one meromorphic function φ_{kl} such that $\lim_{u\to 0} u\varphi_{kl}(u) = 1$, $\varphi_{kl}(u + \omega_1) = \zeta^k\varphi_{kl}(u)$, $\varphi_{kl}(u + \omega_2) = \zeta^l\varphi_{kl}(u)$, $\varphi_{kl}(u)$ has no poles for $u \in \Gamma$. Hence there exists exactly one function X_{kl}, having properties 1')-3'), namely, $X_{kl}(u) = \varphi_{kl}(u) \cdot t_{kl}$.

The function $X(u) \overset{\mathrm{def}}{=} \sum_{k,l} \varphi_{kl}(u)t_{kl}$ satisfies the hypotheses of Lemma 5.1 [for example, the unitary condition follows from the equations $\varphi_{kl}(u) = -\varphi_{-k,-l}(-u)$, $\sigma(t_{kl}) = t_{-k,-l}$, where σ: $\mathfrak{g} \otimes \mathfrak{g} \to \mathfrak{g} \otimes \mathfrak{g}$ is permutation of the factors]. Hence the function Y, defined by (5.1), is a bounded entire function, and hence constant [the boundedness follows from (5.2), (5.3) and the obvious identity $Y(u_1 + u, u_2 + u, u_3 + u) = Y(u_1, u_2, u_3)$]. Let $Y(u_1, u_2, u_3) = y$. Let $(A_1 \otimes 1 \otimes 1)y = (A_2 \otimes 1 \otimes 1)y = y$, whence $y = 0$. The proposition is proved.

5.2. Thus, finding nondegenerate elliptic solutions of (1.4) reduces to describing triples $(\mathfrak{g}, A_1, A_2)$, where $\mathfrak{g}$ is a simple Lie algebra, A_1 and A_2 are commuting automorphisms of $\mathfrak{g}$ of finite order, not having common fixed nonzero vectors. An example of such a triple: $\mathfrak{g} = sl(n)$, A_1 and A_2 are inner automorphisms corresponding to the matrices

$$T_1=\begin{pmatrix}1 & & & 0\\ & \zeta & & \\ & & \ddots & \\ 0 & & & \zeta^{n-1}\end{pmatrix},\quad T_2=\begin{pmatrix}0 & 1 & & & 0\\ \vdots & 0 & 1 & & \\ \vdots & & 0 & 1 \ddots & \\ \vdots & & & \ddots & 1\\ 1 & \dots & \dots & \dots & 0\end{pmatrix}, \tag{5.5}$$

where ζ is a primitive root of degree n of unity. The corresponding solutions of (1.4) were found in [1]. The following proposition shows that there do not exist other nondegenerate elliptic solutions of (1.4).

Proposition 5.2. Let A_1 and A_2 be commuting automorphisms of $\mathfrak{g}$ of finite order, where there does not exist a nonzero $x\in\mathfrak{g}$ such that $A_1(x) = A_2(x) = x$. Then there exists an isomorphism $\mathfrak{g}\approx sl(n)$, under which A_1 and A_2 go into inner automorphisms corresponding to the matrices (5.5).

Proof. We set $\mathfrak{g}_0=\{x\in\mathfrak{g}\mid A_1(x)=x\}$.

LEMMA 1. The algebra $\mathfrak{g}_0$ is Abelian.

Proof. It is known that if σ is an automorphism of finite order of a semisimple Lie algebra $\mathfrak{a}$, $\mathfrak{a}^\sigma \stackrel{\text{def}}{=} \{x\in\mathfrak{a}\mid\sigma(x)=x\}$, then 1) $\mathfrak{a}^\sigma\neq 0$, 2) $\mathfrak{a}^\sigma$ is the direct product of semisimple and Abelian algebras (cf. [3], Lemma 1). If the algebra $\mathfrak{g}_0$ were non-Abelian, then taking as $\mathfrak{g}_0$ the semisimple part of σ, and as σ the restriction of A_2 to $\mathfrak{a}$, we would get that there exists a nonzero $x\in\mathfrak{g}_0$ such that $A_2(x) = x$, and this is impossible. The proposition is proved.

In [3], with each pair $(\mathfrak{a},\sigma)$, where $\mathfrak{a}$ is a simple Lie algebra, $\sigma:\mathfrak{a}\to\mathfrak{a}$ is an automorphism of finite order, there is associated a graph, called the Dynkin diagram of the pair $(\mathfrak{a},\sigma)$. Here if the algebra $\mathfrak{a}^\sigma\stackrel{\text{def}}{=}\{x\in\mathfrak{a}\mid\sigma(x)=x\}$ is Abelian, then any automorphism $\tau:\mathfrak{g}\to\mathfrak{g}$, commuting with σ, induces an automorphism of this graph. Let Δ be the Dynkin diagram of the pair $(\mathfrak{g}, A_1)$, φ be the automorphism of Δ, induced by the automorphism $A_2:\mathfrak{g}\to\mathfrak{g}$, H be the subgroup in Aut Δ, generated by φ.

LEMMA 2. The action of H on the set of vertices of Δ is transitive.

Proof. According to [3], to each vertex δ of the graph Δ there corresponds in a canonical way an element $h_\delta\in\mathfrak{g}_0$; here the vectors h_δ generate $\mathfrak{g}_0$ and the number of these vectors is equal to $\dim\mathfrak{g}_0+1$. Let us assume that the set of vertices of Δ can be represented as the union of H-invariant subsets S_1 and S_2, where $S_1\cap S_2=\phi$, $S_1\neq\phi$, $S_2\neq\phi$. We set $x_i=\sum_{\delta\in S_i}h_\delta$, $i=1,2$. Since S_i is invariant, one has $A_2(x_i) = x_i$. Since $x_i\in\mathfrak{g}_0$, one has $A_1(x_i) = x_i$. Hence $x_i = 0$. Thus, $\sum_{\delta\in S_1}h_\delta=\sum_{\delta\in S_2}h_\delta=0$, i.e., we have obtained two independent linear relations between the h_δ, and this is impossible.

From Lemma 2 it follows that the group Aut Δ acts transitively on the set of vertices of Δ. Hence Δ has type $A^{(1)}_{n-1}$ (cf. the tables from [3]). Whence it follows (cf. [3], Theorem 2), that $\mathfrak{g}\approx sl(n)$, and the automorphism A_1 is inner. Since A_1 and A_2 play identical roles, the automorphism A_2 is also inner.

Let $A_1: sl(n)\to sl(n)$, $A_2: sl(n)\to sl(n)$ be the inner automorphisms corresponding to matrices P_1, P_2 SL(n). Since $A_1A_2 = A_2A_1$, one has that $P_1P_2P_1^{-1}P_2^{-1}$ is a scalar matrix. Thus, associating with the element $(i,j)\in\mathbf{Z}^2$ the operator $p_1^ip_2^j$, we get a projective representation of $\mathbf{Z}^2$ on the space $\mathbf{C}^n$. This representation is irreducible: otherwise there would exist a nonzero matrix $B\in sl(n)$, commuting with P_1 and P_2, and then one would have the equations $A_1(B) = A_2(B) = B$. To prove the proposition it remains to use the well-known theorem that any n-dimensional irreducible projective representation of the group Z is equivalent with a representation under which the element $(i,j)\in\mathbf{Z}^2$ corresponds to the operator $T_1^iT_2^j$, where T_1 and T_2 are defined by (5.5). The lemma is proved.

6. Trigonometric Solutions

In describing trigonometric solutions important roles are played by the concepts of Coxeter automorphism and simple weights. These concepts are introduced in points 6.1 and 6.2. In point 6.3 there is given a formula for the simplest trigonometric solution and its connection with Toda–Bogoyavlenskii chains is clarified. In point 6.4 there is formulated the basic theorem, describing all nondegenerate trigonometric solutions. The rest of the section is devoted to the proof of this theorem.

6.1. We recall that $\mathfrak{g}$ denotes a simple finite-dimensional Lie algebra over C. We denote by $\mathrm{Aut}^0\,\mathfrak{g}$ the connected component of the identity of the group Aut $\mathfrak{g}$. The elements of $\mathrm{Aut}^0\mathfrak{g}$ are called inner automorphisms. It is known that $\mathrm{Aut}\,\mathfrak{g}/\mathrm{Aut}^0\,\mathfrak{g}$, where Δ is the Dynkin diagram of $\mathfrak{g}$. In particular, the order of the group $\mathrm{Aut}\,\mathfrak{g}/\mathrm{Aut}^0\,\mathfrak{g}$ can be equal to 1, 2, or 6, where the last possibility is realized only for $\mathfrak{g}=O(8)$ (in this case $\mathrm{Aut}\,\mathfrak{g}/\mathrm{Aut}^0\,\mathfrak{g}\simeq S_3$). Let $\sigma\in\mathrm{Aut}\,\Delta$, K_σ be the corresponding coset of the group Aut $\mathfrak{g}$ by the subgroup $\mathrm{Aut}^0\,\mathfrak{g}$.

Definition. The automorphism $A \in K_\sigma$ is called a Coxeter automorphism, if the following conditions hold:

a) the algebra $\mathfrak{g}^A \overset{\text{def}}{=} \{x \in \mathfrak{g} \mid Ax = x\}$ is Abelian;

b) A has smallest order among automorphisms $A' \in K_\sigma$ such that the algebra $\mathfrak{g}^{A'}$ is Abelian.

It follows from the results of [3] that for any pair $(\mathfrak{g}, \sigma)$ there is a Coxeter automorphism C which is unique up to conjugation by inner automorphisms (in terms of [3] the Coxeter automorphism corresponds to graduation of type (1, 1, . . . , 1)). The order h of the automorphism C is called the Coxeter number of the pair $(\mathfrak{g}, \sigma)$. We give a table for Coxeter numbers, taken from [11].

Type $(\mathfrak{g},\sigma)$	$A_n^{(1)}$	$A_{2n}^{(2)}$	$A_{2n+1}^{(2)}$	$B_n^{(1)}$	$C_n^{(1)}$	$D_n^{(1)}$	$D_n^{(2)}$	$D_3^{(4)}$	$E_6^{(1)}$	$E_6^{(2)}$	$E_7^{(1)}$	$E_8^{(1)}$	$F_4^{(1)}$	$G_2^{(1)}$
h	$n+1$	$4n+2$	$4n+2$	$2n$	$2n$	$2n-2$	$2n$	12	12	18	18	30	12	6

In this table it is meant that $(\mathfrak{g}, \sigma)$ has, say, type $D_n^{(2)}$ if $\mathfrak{g}$ has type D_n, and the order of σ is equal to 2 (we note that σ is determined by its order, uniquely up to conjugation).

We give a method of constructing a Coxeter automorphism. We choose in $\mathfrak{g}$ a system of Weyl generators $\{X_i, Y_i, H_i\}$, where i runs through the set of vertices of Δ (cf. [8], part III, Chap. VI, Sec. 4). We denote by C the automorphism of $\mathfrak{g}$ such that $C(H_i) = H\sigma(i)$, $C(X_i) = e^{2\pi i/h}X_{\sigma(i)}$, $C(Y_i) = e^{-2\pi i/h}Y_{\sigma(i)}$. It follows from the results of [3] that the automorphism C is Coxeter.

Finally, we give an explicit form for Coxeter automorphisms of the classical algebras. Notation: C is a Coxeter automorphism, m is the order of σ, S is the matrix with ones on the auxiliary diagonal and zeros everywhere else, $\omega = e^{2\pi i/h}$, where h is the Coxeter number of the pair $(\mathfrak{g}, \sigma)$. For the algebras $o(n)$ and $sp(n)$ one uses realizations which are not completely standard, namely: $o(h) \overset{\text{def}}{=} \{X \in \mathrm{Mat}(n, \mathbb{C}) \mid X^t = -SXS^{-1}\}$, $sp = \{X \in \mathrm{Mat}(n, \mathbb{C}) \mid X^t = -BXB^{-1}\}$, where $B = (b_{ij})$, $b_{ij} = -b_{ij}$, $b_{ij} = 0$ for $i + j \neq 2n + 1$, $b_{ij} \neq 0$ for $i + j = 2n + 1$. The Coxeter automorphisms of the classical Lie algebras are the following:

1) if $\mathfrak{g} = sl(n)$, m = 1, then h = n, $C(X) = TXT^{-1}$, where $T = \mathrm{diag}(1, \omega, \ldots, \omega^{n-1})$;

2) if $\mathfrak{g} = sl(2n+1)$, m = 2, then h = 4n + 2, $C(X) = -TX^tT^{-1}$, where $T = S \cdot \mathrm{diag}(1, \omega, \ldots, \omega^{2n})$;

3) if $\mathfrak{g} = sl(2n)$, m = 2, then h = 4n − 2, $C(X) = -TX^tT^{-1}$, where $T = S \cdot \mathrm{diag}(1, \omega, \ldots, \omega^{n-2}, \omega^{n-1}, \omega^n, \ldots, \omega^{2n-2})$;

4) if $\mathfrak{g} = sp(2n)$, then h = 2n, $C(X) = TXT^{-1}$, where $T = \mathrm{diag}(1, \omega, \ldots, \omega^{2n-1})$;

5) if $\mathfrak{g} = o(2n+1)$, then h = 2n, $C(x) = TXT^{-1}$, where $T = \mathrm{diag}(1, \omega, \ldots, \omega^{2n-1}, 1)$;

6) if $\mathfrak{g} = O(2n)$, m = 1, then h = 2n − 2, $C(X) = TXT^{-1}$, where $T = \mathrm{diag}(1, \omega, \ldots, \omega^{n-2}, \omega^{n-1}, \omega^n, \ldots, \omega^{2n-3}, 1)$;

7) if $\mathfrak{g} = o(2n)$, m = 2, then h = 2n, $C(X) = TXT^{-1}$, where

$$T = \begin{pmatrix} \omega & & & & & & & 0 \\ & \omega^2 & & & & & & \\ & & \ddots & & & & & \\ & & & \omega^{n-1} & & & & \\ & & & & 0 & 1 & & \\ & & & & 1 & 0 & & \\ & & & & & & \omega^{n+1} & \\ & & & & & & & \ddots \\ 0 & & & & & & & & \omega^{2n-1} \end{pmatrix}.$$

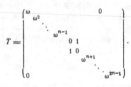

6.2. We fix $\sigma \in \mathrm{Aut}\Delta$ and a Coxeter automorphism $C \in K_\sigma$. We set $\mathfrak{h} = \{x \in \mathfrak{g} \mid Cx = x\}$, $\mathfrak{h}$ is an Abelian subalgebra in $\mathfrak{g}$. We set $\omega = e^{2\pi i/h}$, where h is the Coxeter number of the pair $(\mathfrak{g}, \sigma)$. We expand $\mathfrak{g}$ with respect to the eigenvalues of C: $\mathfrak{g} = \bigoplus_{j \in \mathbb{Z}/h\mathbb{Z}} \mathfrak{g}_j$, where $\mathfrak{g}_j \overset{\text{def}}{=} \{x \in \mathfrak{g} \mid Cx = \omega^j x\}$ (in particular, $\mathfrak{g}_0 = \mathfrak{h}$). For any $\alpha \in \mathfrak{h}^*$ we denote by $\mathfrak{g}_j^k$ the set of those $x \in \mathfrak{g}$, such that $[a, x] = \alpha(a)x$ for $a \in \mathfrak{h}$. According to [3], $\mathfrak{g}_j = \bigoplus_\alpha \mathfrak{g}_j^\alpha$ and $\dim \mathfrak{g}_j^\alpha \leqslant 1$ for $\alpha \neq 0$. We set $\Gamma = \{\alpha \in \mathfrak{h}^* \mid \mathfrak{g}_1^\alpha \neq 0\}$. The elements of Γ are called simple weights (since C is a Coxeter automorphism, it follows from the results of [3] that this definition of simple weights is equivalent with the definition given in [3]). According to [3], $0 \notin \Gamma$, so that for any $\alpha \in \Gamma$ $\dim \mathfrak{g}_1^\alpha = 1$. The mutual disposition of the simple weights is conveniently described with the help of the Dynkin diagram. The Dynkin diagram of the pair $(\mathfrak{g}, C)$ = is the graph, whose vertices are in one-to-one correspondence with the simple weights, and the character of the connection of the vertices A and B, corresponding to simple weights α and β, is determined by the following rules: a) the number of segments joining α and β is equal to $4(\alpha, \beta)^2/(\alpha, \alpha)(\beta, \beta)$; b) if $(\alpha, \alpha)/(\beta, \beta) > 1$, then these segments are provided with arrows pointing to B. (We note that the isomorphism $\mathfrak{h} \to \mathfrak{h}^*$ defined by the scalar product in $\mathfrak{h}$, allows us to transfer this scalar product to $\mathfrak{h}^*$.) The Dynkin diagrams of all pairs $(\mathfrak{g}, C)$ are given in [3].

6.3. Since $(C \otimes C)t = t$, one has $t \in \bigoplus_{j \in Z/hZ} (\mathfrak{g}_j \otimes \mathfrak{g}_{-j})$. The projection of t on $\mathfrak{g}_j \otimes \mathfrak{g}_{-j}$ we denote by t_j. We set

$$\xi(\lambda) = \frac{t_0}{2} + \frac{1}{\lambda^h - 1} \sum_{j=0}^{h-1} t_j \lambda^j, \quad X(u) = \xi(e^{u/h}). \tag{6.1}$$

Proposition 6.1. The function X(u), defined by (6.1), is a solution of (1.4) with set of poles $2\pi i Z$ and residue t at zero.

Proof. It is easy to verify that $X(u + 2\pi i) = (C \otimes 1)X(u)$, $X^{21}(u) = -X^{12}(-u)$, $\lim_{u \to 0} uX(u) = t$; the set of poles of X(u) is equal to $2\pi i Z$. Thus, it follows from Lemma 5.1 that the function

$$Z(\lambda, \mu) \overset{\text{def}}{=} [\xi^{12}(\lambda), \xi^{13}(\lambda\mu)] + [\xi^{12}(\lambda), \xi^{23}(\mu)] + [\xi^{13}(\lambda\mu), \xi^{33}(\mu)]$$

has no poles for $\lambda \neq 0, \infty$, $\mu \neq 0, \infty$. Since $\xi(\lambda)$ has no poles for $\lambda = 0, \infty$, one has that $Z(\lambda, \mu)$ also has no poles for $\lambda = 0, \infty$ and for $\mu = 0, \infty$. Hence $Z(\lambda, \mu)$ is constant. On the other hand, $\lim_{\lambda, \mu \to \infty} Z(\lambda, \mu) = 0$, since $t_0 \in \mathfrak{h} \otimes \mathfrak{h}$, and the algebra $\mathfrak{h}$ is Abelian. The proposition is proved.

Remarks. 1) It is easy to see that the invariance group G of the solution (6.1) consists of precisely those automorphisms of $\mathfrak{g}$, which commute with C. It is clear that G contains the subgroup H, generated by C and the automorphisms $e^{\operatorname{ad} a}$, $a \in \mathfrak{h}$. One can show that G(H) is the group of automorphisms of the Dynkin diagram of $(\mathfrak{g}, C)$.

2) Proposition 6.1 remains valid if C is replaced by any automorphism of finite order A such that the algebra $\mathfrak{g}^A \overset{\text{def}}{=} \{x \in \mathfrak{g} \mid Ax = x\}$ is Abelian. It turns out however that the solution corresponding to any such A is equivalent with the solution corresponding to C.

In [10] there was studied the equation

$$\ddot{\varphi} = -(\operatorname{grad} U)(\varphi), \quad \varphi(t) \in \mathfrak{h}, \quad U(\varphi) = \sum_{\alpha \in \Gamma} e^{2\alpha(\varphi)}. \tag{6.2}$$

In particular, for it there was found an (L, A)-pair of the form

$$L(\lambda) = \dot{\varphi} + \lambda e^{\operatorname{ad}\varphi} I + \lambda^{-1} e^{-\operatorname{ad}\varphi} J, \quad A(\lambda) = \lambda^{-1} e^{-\operatorname{ad}\varphi} J - \lambda e^{\operatorname{ad}\varphi} I, \tag{6.3}$$

where $I \in \mathfrak{g}$, $J \in \mathfrak{g}_{-1}$. The following proposition shows that the solution (6.1) of (1.4) is the classical r-matrix ([5], p. 141), corresponding to the operator L of the form (6.3).

Proposition 6.2. $\{L(\lambda), L(\mu)\} = 2[L(\lambda) \otimes 1 + 1 \otimes L(\mu), \xi(\lambda/\mu)]$, where ξ is defined by (6.1).

We explain that L(λ) and L(μ) are considered as $\mathfrak{g}$-valued functions of φ and $\dot{\varphi}$, and their Poisson bracket is a function of φ and $\dot{\varphi}$ with values in $\mathfrak{g} \otimes \mathfrak{g}$.

Proof. Let $I = \sum_{\alpha \in \Gamma} I_\alpha$, $J = \sum_{\alpha \in \Gamma} J_\alpha$, where $I_\alpha \in \mathfrak{g}_1^\alpha$, $J_\alpha \in \mathfrak{g}_{-1}^{-\alpha}$ (from the invariance of the scalar product in $\mathfrak{g}$ it follows that $\mathfrak{g}_{-1} = \bigotimes_{\alpha \in \Gamma} \mathfrak{g}_{-1}^{-\alpha}$). We denote by α^* the image of α under the isomorphism $\mathfrak{h}^* \rightleftarrows \mathfrak{h}$, defined by the scalar product in $\mathfrak{h}$. We have

$$\begin{aligned}\{L(\lambda), L(\mu)\} &= \{\dot{\varphi}, \mu e^{\operatorname{ad}\varphi} I + \mu^{-1} e^{\operatorname{ad}\varphi} J\} + \{\lambda e^{\operatorname{ad}\varphi} I + \lambda^{-1} e^{-\operatorname{ad}\varphi} J, \dot{\varphi}\} \\ &= \sum_{\alpha \in \Gamma} (\{\dot{\varphi}, \mu e^{\alpha(\varphi)} I_\alpha + \mu^{-1} e^{\alpha(\varphi)} I_\alpha\} + \{\lambda e^{\alpha(\varphi)} I_\alpha + \lambda^{-1} e^{\alpha(\varphi)} J_\alpha, \dot{\varphi}\}) \\ &= \sum_{\alpha \in \Gamma} e^{\alpha(\varphi)} (\mu \alpha^* \otimes I_\alpha + \mu^{-} / \alpha^* \otimes J_\alpha - \lambda I_\alpha \otimes \alpha^* - \lambda^{-1} J_\alpha \otimes \alpha^*).\end{aligned}$$

On the other hand, since $[\dot{\varphi} \otimes 1 + 1 \otimes \dot{\varphi}, \xi(\lambda/\mu)] = 0$, one has

$$\left[L(\lambda)\otimes 1+1\otimes L(\mu),\xi\left(\frac{\lambda}{\mu}\right)\right]=\sum_{\alpha\in\Gamma}e^{\alpha(\varphi)}\left[\lambda I_\alpha\otimes 1+1\otimes\mu I_\alpha+\lambda^{-1}J_\alpha\otimes+1\otimes\mu^{-1}J_\alpha,\xi\left(\frac{\lambda}{\mu}\right)\right].$$

It remains to verify that

$$\left[\lambda I_\alpha\otimes 1+1\otimes\mu I_\alpha,\xi\left(\frac{\lambda}{\mu}\right)\right]=\frac{1}{2}(\mu\alpha^*\otimes I_\alpha-\lambda I_\alpha\otimes\alpha^*), \tag{6.4}$$

$$\left[\lambda^{-1}J_\alpha\otimes 1+1\otimes\mu^{-1}J_\alpha,\xi\left(\frac{\lambda}{\mu}\right)\right]=\frac{1}{2}(\mu^{-1}\alpha^*\otimes J_\alpha-\lambda^{-1}J_\alpha\otimes\alpha^*), \tag{6.5}$$

We prove (6.4). From the equation $[I_\alpha\otimes 1+1\otimes I_\alpha, t]=0$ it follows that $[1\otimes I_\alpha, t_j]+[I_\alpha\otimes 1, t_{j-1}]=0$, $j\in Z/hZ$. Hence

$$\left[\lambda I_\alpha\otimes 1\otimes\mu I_\alpha,\xi\left(\frac{\lambda}{\mu}\right)\right]=\left[\lambda I_\alpha\otimes 1+1\otimes\mu I_\alpha,\frac{t_0}{2}\right]-[1\otimes\mu I_\alpha,t_0]$$
$$=\frac{\mu}{2}[t_0,1\otimes I_\alpha]-\frac{\lambda}{2}[t_0,I_\alpha\otimes 1]=\frac{\mu}{2}\alpha^*\otimes I_\alpha-\frac{\lambda}{2}I_\alpha\otimes\alpha^*.$$

(6.5) is proved in exactly the same way. The proposition is proved.

The same arguments as in the proof of Proposition 6.2 show that the solution (6.1) of (1.4) is the classical r-matrix corresponding to the two-dimensional generalization of (6.2) (cf. [6, 12]).

6.4. Let $X(u)$ be a nondegenerate trigonometric solution of (1.4). Without loss of generality one can assume that the set of poles of $X(u)$ is $2\pi i Z$. Let A be an automorphism of $\mathfrak{g}$ such that $X(u+2\pi i)=(A\otimes 1)X(u)$. We denote by σ the automorphism of the Dynkin diagram Δ of the algebra $\mathfrak{g}$ defined by A. In this situation we shall say that the solution $X(u)$ corresponds to σ. We note that if $X(u)$ is replaced by an equivalent solution, then A is replaced by $T_1AT_2^{-1}$, where T_1 and T_2 belong to the same connected component of $\mathrm{Aut}\,\mathfrak{g}$, and hence the conjugacy class of σ is unchanged.

We proceed to the description of the general form of trigonometric solutions, corresponding to a fixed $\sigma\in\mathrm{Aut}\Delta$. We fix a Coxeter automorphism $C\in K_\sigma$. Let h, Γ, $t_j,\ldots$ denote the same things as in points 6.1-6.3. The triple (Γ_1,Γ_2,τ), where $\Gamma_1,\Gamma_2\subset\Gamma$, τ is a one-to-one map of Γ_1 onto Γ_2 such that a) for any $\alpha,\beta\in\Gamma$ $(\tau(\alpha),\tau(\beta))=(\alpha,\beta)$, b) for any $\alpha\in\Gamma_1$ there exists a natural number k such that $\tau^k(\alpha)\notin\Gamma_1$ serves as a discrete parameter on which the solution depends. We note that the expression $\tau^k(\alpha)$ makes sense only if α, $\tau(\alpha),\ldots,\tau^{k-1}_{(\alpha)}\in\Gamma_1$. Hence condition b) actually means that the expression $\tau^k(\alpha)$ makes no sense for sufficiently large k. A triple (Γ_1,Γ_2,τ), satisfying conditions a) and b) will be called admissible.

Let (Γ_1,Γ_2,τ) be an admissible triple. The tensor $r\in\mathfrak{h}\otimes\mathfrak{h}$, satisfying the system of equations

$$r^{12}+r^{21}=t_0, \tag{6.6}$$

$$(\tau\alpha\otimes 1)(r)+(1\otimes\alpha)(r)=0,\ \alpha\in\Gamma_1, \tag{6.7}$$

serves as the continuous parameter on which the solution depends. We explain that if $r=\sum_{i=1}^k h_i\otimes h_i',\ h_i, h_i'\in\mathfrak{h}$, $\alpha\in\mathfrak{h}^*$, then

$$(\alpha\otimes 1)(r)\overset{\mathrm{def}}{=}\sum_{i=1}^k\alpha(h_i)h_i',\ (1\otimes\alpha)(r)\overset{\mathrm{def}}{=}\sum_{i=1}^k\alpha(h_i')h_i.$$

<u>LEMMA 6.1.</u> The system of equations (6.6), (6.7) is consistent. The skew symmetric tensors from $\mathfrak{h}_0\otimes\mathfrak{h}_0$, where $\mathfrak{h}_0\overset{\mathrm{def}}{=}\{a\in\mathfrak{h}\mid\forall\alpha\in\Gamma_1,\ \alpha(a)=(\tau\alpha)(a)\}$ and only these tensors are solutions of the corresponding homogeneous systems.

The proof of this lemma as well as those of Lemmas 6.2-6.4 will be given in point 6.6.

We denote by $\mathfrak{a}_i$ (i = 1, 2) the subalgebra in $\mathfrak{g}$, generated by the subspaces $\mathfrak{g}_1^\alpha$, $\alpha\in\Gamma_i$. We recall that $\mathfrak{g}=\bigoplus_{j,\alpha}\mathfrak{g}_j^\alpha$.

<u>LEMMA 6.2.</u> $\mathfrak{a}_i$ is the sum of some of the subspaces $\mathfrak{a}_j^\alpha$.

According to Lemma 6.2, there exists a unique projector $P:\mathfrak{g}\to\mathfrak{a}_1$ such that $P(\mathfrak{g}_j^\alpha)=0$, if $\mathfrak{g}_j^\alpha\not\subset\mathfrak{a}_1$. For any $\alpha\in\Gamma_1$ we fix an isomorphism of vector spaces $\mathfrak{g}_1^\alpha\simeq\mathfrak{g}_1^{\tau(\alpha)}$ (we recall that $\dim\mathfrak{g}_1^\alpha=\dim\mathfrak{g}_1^{\tau(\alpha)}=1$).

<u>LEMMA 6.3.</u> The isomorphisms $\mathfrak{g}_1^\alpha\simeq\mathfrak{g}_1^{\tau(\alpha)}$, $\alpha\in\Gamma_1$, extend to an isomorphism of Lie algebras $\theta:\mathfrak{a}_1\rightleftarrows\mathfrak{a}_2$.

We define a linear operator $\tilde{\theta}:\mathfrak{g}\to\mathfrak{g}$ by the formula $\theta(x)=\theta(P(x))$.

LEMMA 6.4. The operator $\tilde{\theta}$ is nilpotent.

We set $\psi = \tilde{\theta}/(1-\tilde{\theta}) = \tilde{\theta} + \tilde{\theta}^2 + \ldots$.

THEOREM 6.1. 1) Let $r \in \mathfrak{h} \otimes \mathfrak{h}$ satisfy the system of equations (6.6), (6.7). Then the function

$$X(u) = r + \frac{1}{e^u - 1}\sum_{j=0}^{h-1} e^{ju/h} t_j - \sum_{j=1}^{h-1} e^{ju/h}(\psi \otimes 1) t_j + \sum_{j=1}^{h-1} e^{-ju/h}(1 \otimes \psi) t_{-j} \tag{6.8}$$

is a solution of (1.4) with set of poles $2\pi i\mathbf{Z}$ and residue t at zero. In addition $X(u + 2\pi i) = (C \otimes 1)X(u)$.

2) Any trigonometric solution of (1.4) with set of poles $2\pi i\mathbf{Z}$ and residue t at zero, corresponding to the automorphism $\sigma \in \mathrm{Aut}\, \mathbf{Z}$, is equivalent with a solution of the form (6.8).

Points 6.5-6.7 are devoted to the proof of this theorem.

Remark. 1) The solution (6.1) corresponds to the case when $\Gamma_1 = \Gamma_2 = \emptyset$, $r = t_0/2$.

2) It is easy to see that the solution (6.8) is $\mathfrak{h}_0$-invariant, where $\mathfrak{h}_0$ denotes the same thing as in Lemma 6.1. Hence, adding to this solution any skew-symmetric tensor from $\mathfrak{h}_0 \otimes \mathfrak{h}_0$, we get a new solution of (1.4) (cf. point 1.1). According to Lemma 6.1, one can get by this method all solutions corresponding to a fixed triple $(\Gamma_1, \Gamma_2, \tau)$, starting from one solution. Further, it is easy to show that θ, ψ, and hence $X(u)$ depend on the choice of the isomorphisms $\mathfrak{g}_1^{\alpha} \simeq \mathfrak{g}_1^{\tau(\alpha)}$, $\alpha \in \Gamma_1$ and change of them leads to the replacement of $X(u)$ by $(e^{\mathrm{ad}\, a} \otimes e^{\mathrm{ad}\, a})\, X(u)$, $a \in \mathfrak{h}$. Thus, from Theorem 6.1 it follows that up to the methods of propagation of solutions described in point 1.1 and such trivial transformations as multiplication of a solution by a number and replacement of u by cu, the number of nondegenerate trigonometric solutions of (1.4) is finite.

3) One can show that if the solutions $X(u)$ and $\tilde{X}(u)$ of the form (6.8) are equivalent, then $\tilde{X}(u) = (g \otimes g) \times X(u)$, $g \in G$, where G denotes the same thing as in Remark 1 after Proposition 6.1.

4) From the preceding remark and Remark 1 after Proposition 6.1, it follows that a) if the solutions $X(u)$ and $\tilde{X}(u)$ of the form (6.8) corresponding to triples $(\Gamma_1, \Gamma_2, \tau)$ and $(\tilde{\Gamma}_1, \tilde{\Gamma}_2, \tilde{\tau})$ are equivalent, then $(\tilde{\Gamma}_1, \tilde{\Gamma}_2, \tilde{\tau})$ is obtained by applying to $(\Gamma_1, \Gamma_2, \tau)$ some automorphism of the Dynkin diagram of the pair $(\mathfrak{g}, C)$; b) if $(\tilde{\Gamma}_1, \tilde{\Gamma}_2, \tilde{\tau})$ is obtained by applying to $(\Gamma_1, \Gamma_2, \tau)$ an automorphism of the Dynkin diagram of the pair $(\mathfrak{g}, C)$, then any solution of the form (6.8), corresponding to $(\Gamma_1, \Gamma_2, \tau)$, is equivalent with some solution, corresponding to $(\tilde{\Gamma}_1, \tilde{\Gamma}_2, \tilde{\tau})$.

Examples. 1) $\mathfrak{g} = sl(2)$. The Dynkin diagram of $\mathfrak{g}$ has only the identity automorphism:

$$h = 2, \mathfrak{h} = \left\{\begin{pmatrix} a & 0 \\ 0 & -a\end{pmatrix} \middle| a \in \mathbf{C}\right\}, \quad \mathfrak{g}_1 = \left\{\begin{pmatrix} 0 & x \\ y & 0\end{pmatrix} \middle| x, y \in \mathbf{C}\right\}, \quad \Gamma = \{\alpha_1, \alpha_2\}.$$

$$C(X) = \begin{pmatrix} 1 & 0 \\ 0 & -1\end{pmatrix} X \begin{pmatrix} 1 & 0 \\ 0 & -1\end{pmatrix},$$

where $\alpha_1(e_{22} - e_{11}) = 2$, $\alpha_2 = -\alpha_1$ (here and later e_{ij} denotes the matrix which has a one at the intersection of the i-th row and j-th column and zeros in all other places). We have: $\mathfrak{g}_1^{\alpha_1} = Ce_{21}$, $\mathfrak{g}_1^{\alpha_2} = Ce_{12}$. The Dynkin diagram of $(\mathfrak{g}, C)$ has the form

There exist two essentially different admissible triples $(\Gamma_1, \Gamma_2, \tau)$: a) $\Gamma_1 = \Gamma_2 = \emptyset$, b) $\Gamma_1 = \{\alpha_1\}$, $\Gamma_2 = \{\alpha_2\}$, $\tau(\alpha_1) = \alpha_2$ [the case when $\Gamma_1 = \{\alpha_2\}$, $\Gamma_2 = \{\alpha_1\}$, need not be considered in view of Remark 4)]. We have: $t_0 = {}^1/_2(e_{11} - e_{22}) \otimes (e_{11} - e_{22})$, $t_1 = e_{12} \otimes e_{21} + e_{21} \otimes e_{12}$. The system of equations (6.6), (6.7) has both in case a) and case b) the unique solution $r = t_0/2$. In case b) we have $\mathfrak{a}_1 = Ce_{21}$, $\mathfrak{a}_2 = Ce_{22}$; θ can be chosen so that $\tilde{\theta}(e_{21}) = e_{12}$), so then $\tilde{\theta}(e_{21}) = e_{12}$, $\tilde{\theta}(e_{12}) = \tilde{\theta}(e_{11} - e_{22}) = 0$, $\psi = \tilde{\theta}$. Thus, we get two solutions:

a) $$X_1(u) = \frac{e^u + 1}{4(e^u - 1)}(e_{11} - e_{22}) \otimes (e_{11} - e_{22}) + \frac{e_{12} \otimes e_{21} + e_{21} \otimes e_{12}}{e^{u/2} - e^{-u/2}}, \tag{6.9}$$

b) $$X_2(u) = X_1(u) + (e^{-u/2} - e^{u/2})(e_{12} \otimes e_{12}). \tag{6.10}$$

Both solutions are well known. Moreover, the corresponding solutions of the quantum Yang–Baxter equation are known: (6.9) corresponds to the trigonometric degenerate solution of Baxter (cf. the supplement to [5], Eq. p9), and (6.10) corresponds to the solution found in [9, p. 118, case a].

2) $\mathfrak{g} = sl(3)$. The Dynkin diagram of $\mathfrak{g}$ has two automorphisms. The corresponding Coxeter automorphisms are as follows:

$$C_1(X) = \begin{pmatrix} 1 & 0 & 0 \\ 0 & e^{2\pi i/3} & 0 \\ 0 & 0 & e^{2\pi i/3} \end{pmatrix} X \begin{pmatrix} 1 & 0 & 0 \\ 0 & e^{2\pi i/3} & 0 \\ 0 & 0 & e^{4\pi i/3} \end{pmatrix}^{-1},$$

$$C_2(X) = -\begin{pmatrix} 0 & 0 & e^{2\pi i/3} \\ 0 & e^{\pi i/3} & 0 \\ 1 & 0 & 0 \end{pmatrix} X^t \begin{pmatrix} 0 & 0 & e^{2\pi i/3} \\ 0 & e^{2\pi i/3} & 0 \\ 0 & 0 & 0 \end{pmatrix}.$$

The Dynkin diagrams of the pairs $(\mathfrak{g}, C_1)$ and $(\mathfrak{g}, C_2)$ have the form

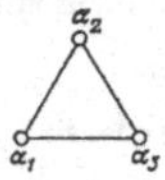

In the second case there is a unique admissible triple: $\Gamma_1 = \Gamma_2 = \emptyset$ [the fact is that $(\beta_1, \beta_1) \neq (\beta_2, \beta_2)$]. The corresponding solution is the classical r-matrix for the Zhiber–Shabat equation [2]. It is equivalent with the solution given in the supplement to [5] (Eq. PII). We write down the solution corresponding to C_1. In this case h = 3, $\mathfrak{g}_j$ is the set of matrices (a_{kl}) of $sl(3)$ such that $a_{kl} = 0$ for $k - l \not\equiv j \pmod 3$. In particular, $\mathfrak{h} = \mathfrak{g}_0$ is the set of diagonal matrices. We have $t = \sum\limits_{k-l\equiv j(\mathrm{mod}\,3)} e_{kl} \otimes e_{lk}$ for $j \neq 0$, $t_0 = \frac{1}{3}\sum\limits_{i<j}(e_{ii} - e_{jj}) \otimes (e_{ii} - e_{jj})$. The simple weights $\alpha_1, \alpha_2, \alpha_3$ on the matrix diag (a_1, a_2, a_3) assume values equal to $a_2 - a_1$, $a_3 - a_2$, $a_1 - a_3$. Here $\mathfrak{g}_1^{\alpha_1} = Ce_{21}$, $\mathfrak{g}_1^{\alpha_2} = Ce_{32}$, $\mathfrak{g}_1^{\alpha_3} = Ce_{13}$. Admissible triples: a) $\Gamma_1 = \Gamma_2 = \emptyset$; b) $\Gamma_1 = \{\alpha_1\}$, $\Gamma_2 = \alpha_2$, $\tau(\alpha_1) = \alpha_2$; c) $\Gamma_1 = \{\alpha_1, \alpha_2\}$, $\Gamma_2 = \{\alpha_2, \alpha_3\}$, $\tau(\alpha_1) = \alpha_2$, $\tau(\alpha_2) = \alpha_3$.

We consider case c). In this case

$$\mathfrak{a}_1 = \left\{\begin{pmatrix} 0 & 0 & 0 \\ a & 0 & 0 \\ c & b & 0 \end{pmatrix} \middle| a, b, c \in \mathbf{C}\right\}, \quad \mathfrak{a}_2 = \left\{\begin{pmatrix} 0 & c & b \\ 0 & 0 & 0 \\ 0 & a & 0 \end{pmatrix} \middle| a, b, c \in \mathbf{C}\right\},$$

$$\tilde{\theta}\begin{pmatrix} a_{11} & a_{12} & a_{13} \\ a_{21} & a_{22} & a_{23} \\ a_{31} & a_{32} & a_{33} \end{pmatrix} = \begin{pmatrix} 0 & a_{31} & a_{32} \\ 0 & 0 & 0 \\ 0 & a_{21} & 0 \end{pmatrix}, \quad \psi\begin{pmatrix} a_{11} & a_{12} & a_{13} \\ a_{21} & a_{22} & a_{23} \\ a_{31} & a_{32} & a_{33} \end{pmatrix} = \begin{pmatrix} 0 & a_{31} & a_{32} + a_{21} \\ 0 & 0 & 0 \\ 0 & 0_{21} & 0 \end{pmatrix}, \tag{6.11}$$

$$r = \frac{1}{3}\sum_{i,j=1}^{3} r_{ij} e_{ii} \otimes e_{jj}, \quad (r_{ij}) = \begin{pmatrix} 1 & 0 & -1 \\ -1 & 1 & 0 \\ 0 & -1 & 1 \end{pmatrix}.$$

One considers cases a) and b) analogously. Answers:

a) $X_1(u) = \sum\limits_{i,j=1}^{3} \rho_{ij} e_{ii} \otimes e_{jj} + Y(u)$, for

$$Y(u) = \frac{1}{e^u - 1}\left[\frac{1}{3}\sum_{i<j}(e_{ii} - e_{jj}) \otimes (e_{ii} - e_{jj}) + e^{u/3}\sum_{i-j\equiv 1(\mathrm{mod}\,3)} e_{ij} \otimes e_{ji} + e^{2u/3}\sum_{i-j\equiv 2(\mathrm{mod}\,3)} e_{ij} \otimes e_{ji}\right],$$

$$(\rho_{ij}) = \begin{pmatrix} 1/3 & a & b \\ b & 1/3 & a \\ a & b & 1/3 \end{pmatrix}, \quad a + b = -\frac{1}{3};$$

b) $X_2(u) = r + Y(u) - e^{u/3} e_{32} \otimes e_{12} + e_{12}^{-u/3} e_{12} \otimes e_{32}$, where r is defined by (6.11);

c) $X_3(u) = X_2(u) - e^{u/3} e_{13} \otimes (e_{12} + e_{23}) - e^{2u/3} e_{12} \otimes e_{13} + e^{-u/3}(e_{12} + e_{23}) \otimes e_{13} + e^{-2u/3} e_{13} \otimes e_{12}$.

6.5. As the first step toward proving Theorem 6.1, we translate the problem of classification of trigonometric solutions into different language.

Let X(u) be a trigonometric solution of (1.4) with set of poles $2\pi i\mathbf{Z}$ and residue t at zero (such a solution will be called normalized). We have

$$X(u + 2\pi i) = (A \otimes 1)X(u) = (1 \otimes A^{-1})X(u), \quad A \in \mathrm{Aut}\,\mathfrak{g}. \tag{6.12}$$

Since there exists a k such that X(u) is a rational function of e^{ku}, A has finite order m. Then $X(u) = \varphi(e^{u/m}) \times I_\mu \otimes I_\mu$, where $\{I_\mu\}$ is an orthonormal basis in $\mathfrak{g}$, φ is a rational function with values in the space of linear operators $\mathfrak{g} \to \mathfrak{g}$. We expand $\varphi(z)$ in a Laurent series in a neighborhood of the point $Z=\infty$: $\varphi(z)=\sum_{i=-\infty}^{N}\varphi_i z^i$. We denote by $\mathfrak{g}[z, z^{-1}]$ the algebra of polynomials of the form $\sum_{i=l}^{n} x_i z^i$, $x_i \in \mathfrak{g}$. We define the operator $\Phi: \mathfrak{g}[z, z^{-1}] \to \mathfrak{g}[z, z^{-1}]$ by the formula $\Phi(\sum_i x z^i)=\sum_i \varphi_i(x_i) z^i$. We define in $\mathfrak{g}[z, z^{-1}]$ an invariant scalar product by the formula $(\sum_i x_i z^i, \sum_j y_j z^j)=\sum_i (x_i, y_{-i})$. We set $\zeta = e^{2\pi i/m}$. We have: $\mathfrak{g}=\bigoplus_{i\in Z/mZ} \mathfrak{g}_i$, where $\mathfrak{g}_i \overset{\text{def}}{=} \{x\in\mathfrak{g} \mid Ax=\zeta^i x\}$. Let $\Pi_i: \mathfrak{g}\to\mathfrak{g}$ be the projector onto $\mathfrak{g}_i$. We define a projector $\Pi: \mathfrak{g}[z, z^{-1}] \to \mathfrak{g}[z, z^{-1}]$ by the formula $\Pi(\sum_i x_i z^i)=\sum_i \Pi_i(x_i) z^i$.

LEMMA 6.5. 1) The operator Φ has the following properties:

a) $\Phi(\mathfrak{g}z^i) \subset \mathfrak{g}z^i$,

b) $\Phi(\mathfrak{g}z^i)=0$ for $i \gg 0$,

c) $\Phi = \Pi\Phi\Pi$,

d) $\Phi + \Phi^* = \Pi$,

e) $[\Phi(w_1), \Phi(w_2)] = \Phi([w_1, \Phi(w_2)] + [\Phi(w_1), w_2] - [\Pi w_1, w_2])$, $w_1, w_2 \in \mathfrak{g}[z, z^{-1}]$.

2) The map constructed from the set of normalized trigonometric solutions of (1.4), satisfying (6.12), into the set of linear operators $\Phi: \mathfrak{g}[z, z^{-1}] \to \mathfrak{g}[z, z^{-1}]$, having properties a)-e), is bijective.

Proof. 1) Properties a) and b) are obvious. From (6.12) it follows that $\varphi(z, \zeta) = A\varphi(z) = \varphi(z)A$, so $\varphi_i = \Pi_i\varphi_i = \varphi_i\Pi_i$, whence follows c). To prove d), we use the equation

$$\varphi_i = -\operatorname*{res}_{z=\infty} z^{-i-1}\varphi(z) = \operatorname*{res}_{z=0} z^{-i-1}\varphi(z) + \sum_{k=0}^{m-1}\operatorname*{res}_{z=\zeta^k} z^{-i-1}\varphi(z).$$

Since $\operatorname*{res}_{u=0} X(u)=t$, one has $\operatorname*{res}_{u=2\pi k} X(u) = (A^k\otimes 1)\ t$, whence $\operatorname*{res}_{z=\zeta^k}\ \varphi\ (z) = 1/mA^{m}\cdot\zeta^k$. From the unitary condition it follows that $\varphi(z) = -\varphi(z^{-1})^*$. Thus,

$$\varphi_i = -\operatorname*{res}_{z=0} z^{-i-1}\varphi(z-1)^* + \frac{1}{m}\sum_{k=0}^{m-1}\zeta^{-ik}A^k = \Pi_i - \varphi^*_{-i} \tag{6.13}$$

whence follows d). To prove e) we use (1.4). We have:

$$[X^{12}(u), X^{13}(u+v)] = [\varphi(z_1)I_\mu, \varphi(z_1, z_2)I_\nu] \otimes I_\mu \otimes I_\nu,$$

where $z_1 = e^{u/m}$, $z_2 = e^{v/m}$. Further,

$$[X^{12}(u), X^{23}(v)] = (\varphi(z_1)\otimes 1\otimes 1)[t^{12}, X^{23}(v)] =$$
$$= -(\varphi(z_1)\otimes 1\otimes 1)[t^{12}, X^{13}(v)] = -\varphi(z_1)[I_\mu, \varphi(z_2)I_\mu]\otimes I_\mu\otimes I_\nu,$$
$$[X^{13}(u+v), X^{23}(v)] = (\varphi(z_1, z_2)\otimes 1\otimes 1)[t^{13}, X^{23}(v)] = -(\varphi(z, z_2)\otimes 1\otimes 1)[t^{13}, X^{21}(v)] =$$
$$= -(\varphi(z, z_2)\otimes 1\otimes 1)[t^{13}, \varphi(z_2)^* I_\mu\otimes I_\mu\otimes 1] = \varphi(z, z_2)[\varphi(z_2)^* I_\mu, I_\nu]\otimes I_\mu\otimes I_\nu.$$

Hence from (1.4) it follows that

$$[\varphi(z_1)I_\mu, \varphi(z, z_2)I_\nu] = \varphi(z_1)[I_\mu, \varphi(z_2)I_\nu] - \varphi(z, z_2)[\varphi(z_2)^* I_\mu I_\nu].$$

Expanding both parts of this equation in a Laurent series in a neighborhood of the point $z_1 = z_2 = \infty$, comparing the coefficients of $z_1^{i+j}z_2^j$, and using (6.13), we get

$$\varphi_i I_\mu, \varphi_j I_\nu] = \varphi_{i+j}[I_\mu, \varphi_j I_\nu] - \varphi_{i+j}[\varphi^*_{-j}I_\mu, I_\nu] = \varphi_{i+j}([I_\mu, \varphi_j I_\nu] + [\varphi_j I_\mu, I_\nu] - [\Pi_j I_\mu, I_\nu]),$$

whence follows e).

2) From the operator Φ, one uniquely reconstitutes φ_i, $\varphi(z)$ and finally $X(u)$. Reversing the argument used to prove assertion 1), we get that if Φ has properties a)- e), then $X(u)$ is a normalized trigonometric solution of (1.4). The lemma is proved.

We set $G_i = \mathfrak{g}_i z^i$, $G = \bigoplus_{i\in Z} G_i$. It is clear that G is a graded Lie subalgebra in $\mathfrak{g}[z, z^{-1}]$, and the scalar product in G is nondegenerate. Let Φ have properties a)-e) (cf. Lemma 6.5). From property c) it follows that

$\Phi(G) \subset G$. Let $f: G \to G$ be the restriction of Φ to G. Then

$$f(G_i) \subset G_i, \tag{6.14}$$

$$f(G_i) = 0 \text{ when } i > 0, \tag{6.15}$$

$$f + f^* = 1, \tag{6.16}$$

$$[f(w_1), f(w_2)] = f([w_1, f(w_2)] + [f(w_1), w_2] - [w_1, w_2]), \ w_1, w_2 \in G. \tag{6.17}$$

Conversely, any linear operator $f: G \to G$, having properties (6.14)-(6.17), extends uniquely to an operator Φ: $g[z, z^{-1}] \to g[z, z^{-1}]$, having properties a)-e).

LEMMA 6.6. Let the linear operator $f: G \to G$ satisfy (6.16). We set $C_1 = \mathrm{Im}(f-1)$, $C_2 = \mathrm{Im} f$. Then

a) $C_1^\perp = \mathrm{Ker} f \subset C_1$, $C_2^\perp = \mathrm{Ker}(f-1) \subset C_2$;

b) the map $\theta: C_1/C_1^\perp \to C_2/C_2^\perp$, carrying the coset $(f-1)w$ into the coset fw, is well-defined and is an orthogonal isomorphism;

c) for f to satisfy (6.17) it is necessary and sufficient that C_1 and C_2 be subalgebras in G, $C_1^\perp$ and $C_2^\perp$ be ideals in C_1 and C_2, and θ be an isomorphism of Lie algebras.

Proof. Assertions a) and b) are verified directly. Let f satisfy (6.17). Then

$$[(f-1)w_1, (f-1)w_2] = (f-1)([w_1, f(w_2)] + [f(w_1), w_2] - [w_1, w_2]). \tag{6.18}$$

From (6.18) it follows that C_1 is a subalgebra, and from (6.17) it follows that $[C_1, C_1^\perp] \subset C_1^\perp$, $[C_2, C_2^\perp] \subset C_2^\perp$ is a subalgebra. From the invariance of the scalar product in G it follows that θ is an isomorphism of Lie algebras.

Let C_1 and C_2 be subalgebras in G (so $C_1^\perp$ and $C_2^\perp$ are ideals in C_1 and C_2), and θ be an isomorphism of Lie algebras. Then for any $w_1, w_2 \in G$ there exist $u \in G$, $v \in \mathrm{Ker} f$ such that

$$[f(w_1), f(w_2)] = f(u), \tag{6.19}$$

$$[(f-1)w_1, (f-1)w_2] = (f-1)u + v. \tag{6.20}$$

Subtracting (6.20) from (6.19), we get

$$[f(w_1), w_2] + [w_1, f(w_2)] - [w_1, w_2] = u - v. \tag{6.21}$$

Applying the operator f to both sides of (6.21) and using the fact that $v \in \mathrm{Ker} f$, we get (6.17).

We note that passing from f to θ is a generalization of the Cayley transformation connecting skew symmetric and orthogonal operators.

6.6. In this point we shall prove Lemmas 6.1-6.4 and assertion 1) of Theorem 6.1. Let $(\Gamma_1, \Gamma_2, \tau)$ be an admissible triple.

LEMMA 6.7. The vectors $\tau\alpha - \alpha$, $\alpha \in \Gamma_1$ are linearly independent.

Proof. Let $\sum_{\alpha \in \Gamma_1} \lambda_\alpha (\tau\alpha - \alpha) = 0$. We write this relation in the form $\sum_{\alpha \in \Gamma} \mu_\alpha \alpha = 0$. Then $\sum_{\alpha \in \Gamma} \mu_\alpha = 0$. On the other hand, it is known [3] that there is exactly one linear relation among simple weights, where the corresponding coefficients have identical signs. Hence $\mu_\alpha = 0$ for any $\alpha \in \Gamma$. Whence it follows that $\tau(S) = S$, where $S \overset{\text{def}}{=} \{\alpha \in \Gamma_1 \mid \lambda_\alpha \neq 0\}$. From condition b) in the definition of an admissible triple and the equation $\tau(S) = S$ it follows that $S = \emptyset$.

LEMMA 6.8. Let V be a finite-dimensional vector space, provided with a nondegenerate symmetric bilinear form. Let $e_1, \ldots, e_k, f_1, \ldots, f_k \in V$, where the vectors $e_1, \ldots, e_k$ are linearly independent. In order that there exist a linear operator $R: V \to V$ such that $R + R^* = 1$ and $Re_i = f_i$ for $i = 1, 2, \ldots, k$, it is necessary and sufficient that $(e_i, f_j) + (e_j, f_i) = (e_i, e_j)$ for any $i, j \in \{1, 2, \ldots, k\}$.

Proof of Lemma 6.1. For any $\alpha \in \mathfrak{h}^*$ we denote by α^* the image of α under the canonical isomorphism $\mathfrak{h}^* \xrightarrow{\sim} \mathfrak{h}$. We set $r = (R \otimes 1)t_0$, $R: \mathfrak{h} \to \mathfrak{h}$. Then (6.6) and (6.7) can be rewritten in the form

$$R + R^* = 1, \tag{6.22}$$

$$R\alpha^* + R^*(\tau\alpha)^* = 0, \ \alpha \in \Gamma_1. \tag{6.23}$$

In view of (6.22), (6.23) can be rewritten in the form

$$R\alpha^* + R^*(\tau\alpha)^* = 0, \ \alpha \in \Gamma_1. \tag{6.24}$$

Hence from Lemmas 6.7 and 6.8 it follows that to prove the consistency of the system of equations (6.6), (6.7), it suffices to verify for any $\alpha, \beta \in \Gamma_1$ the equation $(\tau\alpha - \alpha, \tau\beta) + (\tau\beta - \beta, \tau\alpha) = (\tau\alpha - \alpha, \tau\beta - \beta)$. This equation is equivalent with condition a) in the definition of admissible triple. The homogeneous system corresponding to (6.6) and (6.7) is equivalent with the following system:

$$r^{12} + r^{21} = 0, \quad (\tau\alpha \otimes 1)(r) = (\alpha \otimes 1)(r), \quad \alpha \in \Gamma_1.$$

Its solutions are skew-symmetric tensors from $\mathfrak{h}_0 \otimes \mathfrak{h}_0$. The lemma is proved.

We choose nonzero vectors $e_\alpha^+ \in \mathfrak{g}_1^\alpha, \alpha \in \Gamma_1$. For any $\alpha \in \Gamma$ we denote by e_α^- the element of $\mathfrak{g}_{-1}$ such that $(e_\alpha^-, e_\beta^+) = \delta_{\alpha\beta}$ for all $\beta \in \Gamma$. For any $\alpha, \beta \in \Gamma$ we set $A_{\alpha\beta} = \beta(h_\alpha)$, where $h_\alpha \overset{\text{def}}{=} 2\alpha^*/(\alpha,\alpha)$. It is known [3] that $A_{\alpha\beta} \in \mathbb{Z}$, $A_{\alpha\beta} < 0$ for $\alpha \neq \beta$. Moreover, it is known (cf. [3], point 4, and also Lemma 9 of [4]) that

$$[e_\alpha^+, e_\beta^-] = \delta_{\alpha\beta} h_\alpha, \; [h_\alpha, e_\beta^+] = A_{\alpha\beta} e_\beta^+, \quad [h_\alpha, e_\beta^-] = -A_{\alpha\beta} e_\beta^-, \tag{6.25}$$

$(ad e_\alpha^+)^{1-A_{\alpha\beta}} e_\beta^+ = (ad e_\alpha^-)^{1-A_{\alpha\beta}} e_\beta^- = 0$ for $\alpha \neq \beta$. Let G and G_j denote the same things as in point 6.5, in the situation when A = C, m = h. We set $G^+ = \overset{\infty}{\underset{j=1}{\oplus}} G_j$, $G^- = \overset{\infty}{\underset{j=1}{\oplus}} G_{-j}$. It is known [3] that the algebra G^+ is generated by the elements $e_\alpha^+ z$, $\alpha \in \Gamma$, G^- by the elements $e_\alpha^- z^{-1}$, and $G_0 = \mathfrak{h}$ by the elements h_α. For any $S \subset \Gamma$ we denote by G_S (respectively, G_S^+) the subalgebra in G, generated by the elements $e_\alpha^+ z, h_\alpha, e_\alpha^- z^{-1}, \alpha \in S$ (respectively, by the elements $e_\alpha^+ z$, $\alpha \in S$).

LEMMA 6.9. Let $S \subset \Gamma$, $S \neq \Gamma$. Then

a) G_S is a semisimple finite-dimensional Lie algebra with Weyl generators $e_\alpha^+ z$, $e_\alpha^- z^{-1}$, k_α;

b) G_S^+ is the sum of some of the subspaces $\mathfrak{g}_j^\gamma z^j$, $\gamma \in \mathfrak{h}^* \setminus \{0\}$;

c) $G_S^+ \subset \overset{n-1}{\underset{j=1}{\oplus}} G_j$.

Proof. From the results of [3] it is easy to see that the matrix $(A_{\alpha\beta})$, $\alpha, \beta \in S$ is the Cartan matrix of a semisimple finite-dimensional Lie algebra. Whence and from (6.25) follows a). Since $\dim \mathfrak{g}_i^\gamma = 1$ for $\gamma \neq 0$, to prove b) it suffices to show that if $\gamma \in \mathfrak{h}^*$, $G_S^+ \cap \mathfrak{g}_j^\gamma z^j \neq 0$, then $\gamma \neq 0$. In fact, from a) it follows that in this case $\gamma(h_\alpha) \neq 0$ for some $\alpha \in S$. If $G_S^+ \not\subset \overset{h-1}{\underset{j=1}{\oplus}} G_j$, then $G_S^+ \cap G_h \neq 0$, which contradicts b), since $[\mathfrak{h}, G_h] = [\mathfrak{h}, \mathfrak{h} z^h] = 0$. The lemma is proved.

From Lemma 6.9 there quickly follows Lemma 6.2 ($\mathfrak{a}_i$ is the image of $G_{\Gamma_i}^+$ under the canonical homomorphism $G \to \mathfrak{g}$).

Suppose it is given an admissible triple $(\Gamma_1, \Gamma_2, \tau)$ and an isomorphism $\varphi_\alpha : \mathfrak{g}_1^\alpha \rightrightarrows \mathfrak{g}_1^{\tau(\alpha)}$. It will be assumed that the elements e_α^+, $\alpha \in \Gamma$ are chosen so that $\varphi_\alpha(e_\alpha^+) = e_{\tau(\alpha)}^+$ for $\alpha \in \Gamma_1$ (such a choice is possible in view of condition b) of the definition of admissible triple). From Lemma 6.9 it follows that there exists an isomorphism $T: G_{\Gamma_1} \rightrightarrows G_{\Gamma_2}$ such that $T(e_\alpha^+ z) = e_{\tau(\alpha)}^+ z$, $T(e_\alpha^- z^{-1}) = e_{\tau(\alpha)}^- z^{-1}$, $T(h_\alpha) = h_{\tau(\alpha)}$.

Whence follows Lemma 6.3.

We define a linear opeartor $\tilde{T}: G^+ \to G^+$ in the following way: if $w \in G_{\Gamma_1}^+$, then $\tilde{T}(w) = T(w)$; now if $\gamma \in \mathfrak{h}^*$, $j \in \mathbb{N}$, $\mathfrak{g}_j^\gamma z^j \not\subset G_{\Gamma_1}^+$, $\tilde{T}(\mathfrak{g}_j^\gamma z^j) = 0$.

LEMMA 6.10. The operator $\tilde{T}$ is nilpotent.

Proof. Let us assume that $w \in \mathfrak{g}_j^\gamma$, $z^j \subset G_{\Gamma_1}^+$, $w \neq 0$, where for any $k > 0$ $T^k(w) \in G_{\Gamma_1}^+$. Let $T(w) \in \mathfrak{g}_j^{\gamma'} z^j$. It is clear that $\gamma = \sum_{\alpha \in S} n_\alpha \alpha$, where $n_\alpha > 0$, $S \subset \Gamma_1$, $\gamma' = \sum_{\alpha \in S} n_\alpha \cdot \tau(\alpha) = \sum_{\alpha \in S'} n'_\alpha \cdot \alpha$, where $n'_\alpha > 0$, $S' \subset \Gamma_1$. Here $\sum_{\alpha \in S} n_\alpha = \sum_{\alpha \in S'} n'_\alpha = j$. Just as in the proof of Lemma 6.7, we deduce from this that $S' = \tau(S)$. Thus, $\tau^k(S) \subset \Gamma_1$ for any k, which contradicts the admissibility of $(\Gamma_1, \Gamma_2, \tau)$. The lemma is proved.

From Lemma 6.10 follows Lemma 6.4.

We start the proof of assertion 1) of Theorem 6.1. Let $r \in \mathfrak{h} \otimes \mathfrak{h}$ satisfy (6.6) and (6.7). Then $r = (R \otimes 1)t_0$, where $R: \mathfrak{h} \to \mathfrak{h}$ satisfies (6.22) and (6.24). We define $f_+ : G^+ \to G^+$ by the formula $f_+ = \tilde{T}/(\tilde{T} - 1)$. We define $f_- : G^- \to G^-$ by the formula $f_- = 1 - f_+^*$. Let $f: G \to G$ be a linear operator whose restrictions to G^+, G^-, $\mathfrak{h}$ are equal to f_+, f_-, R. It is clear that f satisfies (6.14)-(6.16). It remains to show that f also satisfies (6.17) [it is easy to see that the solution of (1.4) constructed from f is given by (6.8)].

We consider the triple (C_1, C_2, θ) corresponding to f (cf. Lemma 6.6).

LEMMA 6.11. 1) If $\alpha \in \Gamma_i$, then $h_\alpha \in C_i$.

2) Let $\alpha \in \Gamma_1$, $\bar{h}_\alpha$ and $\bar{\bar{h}}_{\tau(\alpha)}$ be the images of h_α and $h_{\tau(\alpha)}$ in $C_1 | C_1^\perp$ and $C_2 | C_2^\perp$. Then $\theta(\bar{h}_\alpha) = \bar{\bar{h}}_{\tau(\alpha)}$.

Proof. From (6.24) and the equation $(\tau(\alpha), \tau(\alpha)) = (\alpha, \alpha)$ it follows that $R(h_{\tau(\alpha)} - h_\alpha) = h_{\tau(\alpha)}$ for $\alpha \in \Gamma_1$. Whence it is easy to derive the lemma. The lemma is proved.

It is easy to see that $C_1 = G_{\Gamma_1} + G^+ + V_1$, $C_2 = G_{\Gamma_2} + G^- + V_2$, where V_1, V_2 are vector subspaces in $\mathfrak{h}$. Whence it follows that C_1 and C_2 are Lie subalgebras in G, while $C_i | C_i^\perp = G_{\Gamma_i} \otimes \mathfrak{v}_i$, where $\mathfrak{v}_i$ is an Abelian algebra consisting of elements of degree 0. We shall show that θ is an isomorphism of Lie algebras. Since the operator θ is orthogonal, it suffices to show that $\theta(w) = T(w)$ for $w \in G_{\Gamma_1}$. The operators θ and T are orthogonal and preserve the graduation (the orthogonality of T follows from condition a) in the definition of admissible triple). Hence it suffices to prove the equation $\theta(w) = T(w)$ in the case when $w \in G_{\Gamma_1}$ is a homogeneous element of nonnegative degree. For $w \in G_{\Gamma_1}^+$ this equation can be verified directly, and for $\deg w = 0$ it follows from Lemma 6.11.

6.7. In this point we use the system of notation of point 6.5 (in particular, the automorphism A is not assumed to be Coxeter). Let $f: G \to G$ satisfy (6.14)-(6.17). We have: $G = \bigoplus_{\lambda \in C} G^\lambda$, where $G^\lambda \overset{\text{def}}{=} \bigcup_n \operatorname{Ker}(f - \lambda)^n$.

LEMMA 6.12. If $\lambda + \mu \neq 1$, then $[G^\lambda, G^\mu] \subset G^\nu$, where $\nu = \lambda\mu/(\lambda + \mu - 1)$. If $\lambda + \mu = 1$, $\lambda\mu \neq 0$, then $[G^\lambda, G^\mu] = 0$.

Proof. If $\lambda + \mu \neq 1$, then we set $V = G^\nu$, $\nu = \lambda\mu/(\lambda + \mu - 1)$. If $\lambda + \mu = 1$, $\lambda\mu \neq 0$, then we set $V = 0$. It is necessary to show that if $(f - \lambda)^k x = 0$, $(f - \mu)^l y = 0$, then $[x, y] \in V$. This assertion is proved by induction on $k + l$ with the help of the identity

$$[(f - \lambda) x, (f - \mu) y] = (f - \lambda) [x, (f - \mu) y] + (f - \mu) [(f - \lambda) x, y] + ((\lambda + \mu - 1) f - \lambda\mu) [x, y],$$

which follows from (6.17).

LEMMA 6.13. If ψ is an automorphism of a nonsolvable finite-dimensional Lie algebra, then $\det(\psi - 1) = 0$.

The proof reduces to the case of a semisimple algebra, and then to the case of a simple algebra, considered in Proposition 2.3. The lemma is proved.

We set $G' = \bigoplus_{\lambda \neq 0,1} G^\lambda$.

LEMMA 6.14. G' is a finite-dimensional solvable subalgebra in G.

Proof. From Lemma 6.15 it follows that G' is a subalgebra. From (6.15) and (6.16) it follows that $G_i \subset G^0$ for $i \gg 0$, $G_i \subset G^1$ for $i \ll 0$. Hence $\dim G' < \infty$. We define $\psi: G' \to G'$ by the formula $\psi = f/(f - 1)$. Then $\det \psi \neq 0$, $\det(\psi - 1) \neq 0$. From (6.17) and (6.18) it follows that ψ is an automorphism of G' as a Lie algebra. It remains to use Lemma 6.13.

LEMMA 6.15. 1) $(G^0)^\perp = G^0 \oplus G'$, $(G^1)^\perp = G^1 \oplus G'$. 2) $G^0 \oplus G'$ is a subalgebra in G, and G^0 is an ideal in $G^0 \oplus G^1$. 3) $G^1 \oplus G'$ which is a subalgebra in G, and G^1 is an ideal in $G^1 \oplus G^1$.

Proof. Assertion 1) follows from (6.16). Assertions 2) and 3) follow from Lemma 6.12. The lemma is proved.

We set $\mathfrak{a} \overset{\text{def}}{=} \det G_0 = \mathfrak{g}_0$, $\mathfrak{a}^\lambda \overset{\text{def}}{=} \mathfrak{a} \cap G^\lambda$, $\mathfrak{a}' \overset{\text{def}}{=} \mathfrak{a} \cap G'$, $\mathfrak{n}^\lambda \overset{\text{def}}{=} \{x \in \mathfrak{a} | \times [x, \mathfrak{a}_\lambda] \subset \mathfrak{a}_\lambda\}$. According to Lemma 1 of [3], the algebra $\mathfrak{a}$ is reductive in $\mathfrak{g}$.

LEMMA 6.16. $f(\mathfrak{n}^\lambda) \subset \mathfrak{n}^\lambda$.

Proof. By induction on k we show that if $x \in \mathfrak{n}^\lambda$, $y \in \mathfrak{a}$, $(f - \lambda)^k y = 0$, then $[f(x), y] \in \mathfrak{a}^\lambda$. From (6.17) it follows that $[f(x), f(y)] - f([f(x), y]) \in \mathfrak{a}^\lambda$, and by the inductive hypothesis $[f(x), (f - \lambda)y] \in \mathfrak{a}^\lambda$. Hence $(f - \lambda)[f(x), y] \in \mathfrak{a}^\lambda$, and consequently, $[f(x), y] \in \mathfrak{a}^\lambda$.

Proposition 6.3. There exist complementary Borel subalgebras $\mathfrak{b}_+, \mathfrak{b}_- \subset \mathfrak{a}$ such that a) $f(\mathfrak{b}_+) \subset \mathfrak{b}_+$, $f(\mathfrak{b}_-) \subset \mathfrak{b}_-$; b) $\subset \mathfrak{b}_+ \cap \mathfrak{b}_-$, $\mathfrak{b}_+ \supset \mathfrak{a}^0 \supset [\mathfrak{b}_+, \mathfrak{b}_+]$, $\mathfrak{b}_- \supset \mathfrak{a}^1 \supset [\mathfrak{b}_-, \mathfrak{b}_-]$.

Proof. $\mathfrak{a}^0 \subset (\mathfrak{a}^0)^\perp$, so from the Cartan criterion there follows the solvability of $\mathfrak{a}^0$. Since $\mathfrak{a}^0$ is an ideal in $\mathfrak{a}^0 \oplus \mathfrak{a}'$, from Lemma 6.14 follows the solvability of $\mathfrak{a}^0 \oplus \mathfrak{a}'$. Hence $\mathfrak{a}^0 \oplus \mathfrak{a}'$ is contained in some Borel subalgebra b_+. In exactly the same way one proves that $\mathfrak{a}^1 \oplus \mathfrak{a}'$ is contained in some Borel subalgebra b_-. Since

$\mathfrak{a}^0 \oplus \mathfrak{a}' \oplus \mathfrak{a}' = \mathfrak{a}$, one has $\mathfrak{b}_+ + \mathfrak{b}_- = \mathfrak{a}$, i.e., b+ and b- are complementary. Since $(\mathfrak{a}^0)^\perp = \mathfrak{a}^0 \oplus \mathfrak{a}' \subset \mathfrak{b}_+$, one has $\mathfrak{a}^0 \supset \mathfrak{b}_+^\perp = [\mathfrak{b}_+, \mathfrak{b}_+]$ (here $\perp$ denotes the orthogonal complement of $\mathfrak{a}$). Analogously, $\mathfrak{a}^1 \supset [\mathfrak{b}_-, \mathfrak{b}_-]$. Since $\mathfrak{b}_+ \supset \mathfrak{a}^0 \supset [b_+, b_+]$, one has $n^0 = b_+$. Hence from Lemma 6.16 it follows that $f(\mathfrak{b}_+) \subset \mathfrak{b}_+$. Analogously, $f(\mathfrak{b}_-) \subset \mathfrak{b}_-$. The lemma is proved.

We set $\mathfrak{h} = \mathfrak{b}_+ \cap \mathfrak{b}_-$. $\mathfrak{h}$ is a Cartan subalgebra in $\mathfrak{a}$, while $f(\mathfrak{h}) \subset \mathfrak{h}$. It is clear that $\mathfrak{h} = \mathfrak{a}' \oplus \mathfrak{h}^0 \oplus \mathfrak{h}^1$, where $\mathfrak{h}^0 = \mathfrak{a}^0 \cap \mathfrak{h}$, $\mathfrak{h}^1 = \mathfrak{a}^1 \cap \mathfrak{h}$. Here $\mathfrak{a}' \perp (\mathfrak{h}^0 \oplus \mathfrak{h}')$, and $\mathfrak{h}^0$ and $\mathfrak{h}^1$ are isotropic.

LEMMA 6.17. $[\mathfrak{h}, G^0] \subset G^0$, $[\mathfrak{h}, G^1] \in G^1$, $[\mathfrak{h}, G^1] \subset G'$.

Proof. Since $[G^0 + G', G^0] \subset G^0$, one has $[\mathfrak{h}^0 \oplus \mathfrak{a}', G^0] \subset G^0$. Whence we deduce that $[\mathfrak{h}, G^0]' \subset G^0$. It is known (cf. [3]) that $G = \bigoplus_{\alpha \in \mathfrak{h}^*} G_\alpha$, where $G_\alpha = \{w \in G \mid \forall a \in \mathfrak{h}\ [a, w] = \alpha(a) w\}$, so it suffices to show that if α, $\beta \in \mathfrak{h}^*$, $\alpha \neq \beta$, $G_\alpha \neq 0$, $G_\cdot \neq 0$, then the restriction of $\alpha - \beta$ to $\mathfrak{h}^0 \otimes \mathfrak{a}'$ is not equal to zero. In fact, if $(\alpha - \beta) \times (\mathfrak{h}^0 \oplus \mathfrak{a}') = 0$, then $(\alpha - \beta, \alpha - \beta) = 0$, and this is impossible by virtue of point 5 of [3].

In exactly the same way one proves that $[\mathfrak{h}, G^1] \subset G^1$. Since $G' = (G^0 + G^1)^\perp$, one has $[\mathfrak{h}, G'] \subset G'$. The lemma is proved.

For any $i \in \mathbb{Z}$, $\alpha \in \mathfrak{h}^*$ we set $G_i^\alpha = \{w \in G_i \mid \forall a \in \mathfrak{h}, [a, w] = \alpha(a) w\}$. The elements of the set $\Sigma \stackrel{\text{def}}{=} \{(\alpha, i) \mid G^* \neq 0\}$ are called weights. We have: $G = \bigoplus_{(\alpha,i) \in \Sigma} G_i^\alpha$. We set $\Sigma' = \{(\alpha, i) \in \Sigma \mid a \neq 0\}$. With each weight $(\alpha, i) \in \Sigma'$ we associate a functional λ_i^α: $\mathfrak{h} \to C$ by the formula $\lambda_i(a) = \alpha(a) + i$. In [3] it is shown that the functionals λ_i^α, $(\alpha, i) \in \Sigma'$ form an affine system of roots in the sense of [7].

LEMMA 6.18. 1) There exists a Weyl chamber K such that $G^0 = \mathfrak{h}^0 \oplus G^+$, $G^1 = \mathfrak{h}^1 \oplus G^-$, $G' = \mathfrak{a}'$, where $G^+ \stackrel{\text{def}}{=} \bigoplus_{(\alpha,i) \in \Sigma_+} G_i^\alpha$, $G^- \stackrel{\text{def}}{=} \bigoplus_{(\alpha,i) \in \Sigma_-} G_i^\alpha$, Σ_+ (respectively, Σ_-) is the set of weights, positive (negative) with respect to K.

2) $f(G^+) \subset G^+$, $f(G^-) \subset G^-$.

Proof. Let $(\alpha, i) \in \Sigma'$. Then $\dim G_i^\alpha = 1$ (cf. [3]). Hence from Lemma 6.17 it follows that either $G_i^\alpha \subset G^0$, or $G_i^\alpha \subset G^1$, or $G_i^\alpha \subset G'$. From (6.16) it follows that the scalar product on G' is nondegenerate. Hence if $G_i^\alpha \subset G'$, then $G_{-i}^{-\alpha} \subset G'$, and consequently, G' would contain a subalgebra, isomorphic with $sl(2)$ (cf. Lemma 2 of [3]), and this contradicts Lemma 6.14. Thus, $G_i^\alpha \subset G^0$ or $G_i^\alpha \subset G^1$. From the fact that G^0 and G' are isotropic it follows that $G_i^\alpha \subset G^0 \Leftrightarrow G_{-i}^{-\alpha} \subset G^1$. We set $S = \{(\alpha, i) \in \Sigma' \mid G_i^\alpha \subset G^0\}$. We have shown that $S \cup (-S) = \Sigma'$. Moreover, from (6.15) it follows that if $(\alpha, i) \in \Sigma'$, $i > 0$, then $(\alpha, i) \in S$. Whence follows the existence of a Weyl chamber K such that all weights, simple with respect to K, belong to S. Then $G_i^\alpha \subset G^0$ for $(\alpha, i) \in \Sigma_+$, $G_i^\alpha \subset G^1$ for $(\alpha, i) \in \Sigma_-$, whence follows assertion 1). To prove 2) it suffices to note that $G^+ = (G^1 + \mathfrak{h})^\perp$, $G^- = (G^0 + \mathfrak{h})^\perp$. The lemma is proved.

We denote by Γ the set of simple weights corresponding to K. For any $S \subset \Gamma$ we denote by G_S the subalgebra in G generated by the subspaces G_i^α and $G_{-i}^{-\alpha}$, $(\alpha, i) \in S$. We set $G_S^+ = G_S \cap G^+$. We consider the triple (C_1, C_2, θ) corresponding to f (cf. Lemma 6.6).

LEMMA 6.19. There exist subsets $\Gamma_i \subset \Gamma$ and vector subspaces $V_i \subset \mathfrak{h}$ $(i = 1, 2)$ such that $C_1 = G_{\Gamma_1} + G^+ + V_1$, $C_2 = G_{\Gamma_2} + G^- + V_2$.

Proof. Since $C_1 = I_m(f - 1) \supset \mathfrak{h}^0 \oplus \mathfrak{a}'$, one has $[\mathfrak{h}^0 \otimes \mathfrak{a}', C_1] = C_1$. Whence it follows that $[\mathfrak{h}, C_1] \subset C_1$ (cf. the proof of Lemma 6.17). Further, $C_1 \supset G^0 \supset G^+$. Whence, and from the results of [3] it follows that C_1 has the form required. One proves the assertion about C_2 analogously. The lemma is proved.

From Lemma 6.19 it follows that $C_i/C_i^+ = G_{\Gamma_i} \otimes \mathfrak{v}_i$, where $\mathfrak{v}_i$ is an Abelian subalgebra. The isomorphism θ maps G_{Γ_1} onto G_{Γ_2}, preserving the grading, so θ induces a bijection $\tau: \Gamma_1 \to \Gamma_2$.

LEMMA 6.20. The triple $(\Gamma_1, \Gamma_2, \tau)$ is admissible.

Proof. We define $\psi: G^+ \to G^+$ by the formula $\psi = f/(f - 1)$. Since $G^+ \subset G^0$, ψ is well-defined and ψ is nilpotent. It is easy to verify that if $(\alpha, i) \in \Gamma_1$, $\tau(\alpha, i) = (\beta, j)$, then $\psi(G_i^\alpha) = G_j^\beta$; now if $(\alpha, i) \in \Gamma \setminus \Gamma_1$, then $\psi \times (G_\alpha^i) = 0$. Whence follows condition b) of the definition of admissible triple. Condition a) follows from the orthogonality of θ. The lemma is proved.

For any $(\alpha, i) \in \Sigma$ we denote by $n(\alpha, i)$ the sum of the coefficients of the expansion of (α, i) with respect to the elements of Γ. We introduce in G a new grading (we call it the K-grading), setting $\deg G_i^\alpha = n(\alpha, i)$.

LEMMA 6.21. f preserves the K-grading.

Proof. It suffices to verify that the operator ψ from the proof of Lemma 6.20 preserves the K-grading, and this fact follows from the analogous property of θ. The lemma is proved.

Let $\Gamma = \{(\alpha_0, i_0), \ldots, (\alpha_r, i_r)\}$. According to [3], the functionals $\alpha_0, \ldots, \alpha_r$ generate $\mathfrak{h}^*$ and among them is exactly one linear relation $\sum_{s=0}^{r} k_s\alpha_s = 0$. Here the coefficients k_s can be normalized by the condition $\sum_{s=0}^{r} k_s i_s = m$ and then $k_s \in \mathbb{N}$, and $\sum_{s=0}^{r} k_s$ is equal to the Coxeter number h of the pair $(\mathfrak{g}, \sigma)$, where σ is the automorphism of the Dynkin diagram of $\mathfrak{g}$ corresponding to A. Hence there exists exactly one element $a_0 \in \mathfrak{h}$ such that $\alpha_s(a_0) = 1/h - i_s/m$, $s = 0, 1, \ldots, r$. We set $C = A \cdot \exp(2\pi i \cdot ad\, a_0)$. It is easy to see that C is a Coxeter automorphism. We set $\omega = e^{2\pi i/h}$, $\mathfrak{g}_i^C = \{x \in \mathfrak{g} \mid Cx = \omega^i x\}$, $G^C = \bigoplus_{j \in \mathbb{Z}} \mathfrak{g}_j^C z^j$. We define a linear operator $\varphi: G \to G^C$ as follows: if $x \in \mathfrak{g}$, $xz^j \in G_j^\alpha$, then $\varphi(xz^j) \overset{\text{def}}{=} xz^{h(\alpha, j)}$. It is easy to see that this is well-defined and φ is an isomorphism of Lie algebras, preserving the scalar product, and carrying the K-grading of G into the usual grading of G^C. Hence the operator $f: G^C \to C^C$, defined by the formula $\tilde{f} = \varphi f \varphi^{-1}$, satisfies (6.14)-(6.17). We denote by X(u) and $\tilde{X}$(u) the solutions of (1.4) corresponding to f and $\tilde{f}$ (cf. point 6.5).

LEMMA 6.22. The solutions X(u) and $\tilde{X}$(u) are equivalent.

Proof. It is easy to verify that $\tilde{X}(u) = (e^{u\, ad\, a_0} \otimes 1) X(u)$, $[a_0 \otimes 1 + 1 \otimes a_0, X(u)] = 0$. The lemma is proved.

We proceed directly to the proof of assertion 2) of Theorem 6.1. Let X(u) be a normalized trigonometric solution of (1.4), satisfying (6.12), $f: G \to G$ be the corresponding operator. Lemma 6.22 shows that, replacing X(u) by an equivalent solution, one can make $i_0 = \ldots = i_r = 1$, and thus A = C. It is easy to see that in this case the operator ψ from the proof of Lemma 6.20 coincides with the operator $\tilde{T}$, with which we were concerned in Lemma 6.10. Hence the restriction of f to G^+ is equal to $\tilde{T}(T-1)$. Turning to the proof of Lemma 6.11, we get that the restriction of f to $\mathfrak{h}$ satisfies (6.22) and (6.24). Thus, our operator f coincides with the operator f from point 6.6, and consequently, X(u) has the form (6.8).

7. Rational Solutions, not Having a Pole at Infinity

7.1. Let X(u) be a rational solution of (1.4), not having a pole at infinity, and with residue t at zero. Then

$$X(u) = \frac{t}{u} + r, \quad r \to \mathfrak{g} \otimes \mathfrak{g}. \tag{7.1}$$

It is easy to verify that the function X(u), defined by (7.1), is a solution of (1.4) if and only if r satisfies the system of equations (1.2), (1.3). The problem of the complete classification of solutions of this system seems hopeless to us, since it contains the subproblem of the classification of commutative subalgebras in $\mathfrak{g}$ [in fact, if $\mathfrak{a}$ is a commutative subalgebra in $\mathfrak{g}$ and $r \in \mathfrak{a} \otimes \mathfrak{a}$, then r satisfies (1.2)]. Hence we restrict ourselves to giving some methods of construction of solutions of the system of equations (1.2), (1.3).

7.2. 1) Let $a, b \in \mathfrak{g}$, $[a, b] = b$. Then $r = a \otimes b - b \otimes a$ is a solution of the system (1.2), (1.3). It is easy to verify that for $\mathfrak{g} = sl(2)$ this construction gives all nonzero solutions. We note that if $a, b \in sl(2)$, $[a, b] = b$, then there exists a matrix $T \in SL(2)$, such that

$$a = T\begin{pmatrix} \frac{1}{2} & 0 \\ 0 & -\frac{1}{2} \end{pmatrix} T^{-1}, \quad b = T\begin{pmatrix} 0 & 1 \\ 0 & 0 \end{pmatrix} T^{-1}.$$

2) Let A be a finite-dimensional commutative associative algebra over C, $\mathfrak{a} \overset{\text{def}}{=} A \otimes A^*$. We introduce on $\mathfrak{a}$ a Lie algebra structure in the following way: $[e_i, e_j] = 0$, $[e^i, e^j] = 0$, $[e_i, e^j] = \overset{*}{\alpha}{}^j_{ik} e^k$, where $\{e_i\}$ is a basis A, $\{e^i\}$ is the dual basis in A*, $e_i e_j = \alpha_{ij}^k e_k$. It is easy to verify that the tensor $r = e_i \otimes e^i - e^i \otimes e_i \in \mathfrak{a} \otimes \mathfrak{a}$ is a solution of the system (1.2), (1.3). If in addition there is given a homomorphism $f: \mathfrak{a} \to \mathfrak{g}$, then $(f \otimes f)(r)$ is a solution belonging to $\mathfrak{g} \otimes \mathfrak{g}$. For A = C this method of construction of solutions becomes method 1).

7.3. Let $r \in \mathfrak{g} \otimes \mathfrak{g}$ be a solution of (1.2), (1.3). We denote by $\mathfrak{a}$ the smallest vector subspace in $\mathfrak{g}$ such that $r \in \mathfrak{a} \otimes \mathfrak{a}$. Then $\mathfrak{a}$ is a Lie subalgebra and the tensor r is nondegenerate as an element of $\mathfrak{a} \otimes \mathfrak{a}$. Let B be the bilinear form on $\mathfrak{a}$, which is the inverse with respect to r [i.e., if $\{e_\mu\}$ is a basis in $\mathfrak{a}$, $r = r^{\mu\nu} e_\mu \otimes e_\nu$, $(S_{\mu\nu})$ is the matrix, inverse to $(r^{\mu\nu})$, then $B(e_\mu, e_\nu) = S_{\mu\nu}$]. According to Proposition 2.4, the form B is a 2-cocycle [i.e., is skew-symmetric and satisfies (2.10)]. Conversely, to each pair $(\mathfrak{a}, B)$, where $\mathfrak{a}$ is a subalgebra in $\mathfrak{g}$, B is a nondegenerate 2-cocycle on $\mathfrak{a}$, there corresponds a solution of (1.2), (1.3).

We recall that among the 2-cocycles there are, in particular, 2-coboundaries, i.e., forms of the form $B(x, y) = l([x, y])$, where $l \in \mathfrak{a}^*$. We call the functional $l \in \mathfrak{a}^*$ nondegenerate, if the form $l([x, y])$ is nondegenerate. Lie algebras $\mathfrak{a}$, on which there exist nondegenerate linear functionals, were investigated, for example, in [13-15]. Such algebras are called Frobenius algebras. Thus, from a Frobenius Lie algebra $\mathfrak{a}$, a nondegenerate functional $l \in \mathfrak{a}^*$, and an imbedding $\mathfrak{a} \subset \mathfrak{g}$ one constructs a solution of (1.2), (1.3). This solution essentially does not depend on l, since according to [15], all nondegenerate functionals on $\mathfrak{a}$ are obtained by application of inner automorphisms of $\mathfrak{a}$ to a fixed functional.

Example. $\mathfrak{a}$ is the set of matrices of size $n \times n$, for which the lower k rows are equal to zero. As A. G. Élashvili informed us, the algebra $\mathfrak{a}$ is Frobenius if and only if n is divisible by k. Let this condition hold. Then the functional $l: \mathfrak{a} \to C$, defined by $l(A) = \sum_{i=1}^{n-k} a_{i,i+k}$, where $A = (a_{ij})$, is nondegenerate. The corresponding tensor $r \in \mathfrak{a} \otimes \mathfrak{a}$, satisfying (1.2) and (1.3), has the form

$$r = \sum_{i=1}^{k} \sum_{j=1}^{k} \sum_{(a,b,c,d) \in S} (e_{i+ka,\, j+kb} \otimes e_{j+kc,\, i+kd} - e_{j+kc,\, i+kd} \otimes e_{i+ka,\, j+kb}),$$

where $S = \{(a, b, c, d) \mid a, b, c, d \in Z,\ b + d - a - c = 1,\ 0 \leqslant b \leqslant a < m - 1,\ b \leqslant c < m - 1,\ 0 \leq d < m\}$, $m = n/k$, e_{rs} is the matrix whose element at the intersection of the r-th row and s-th column is equal to 1, and all other elements are equal to zero. In order to get a solution of (1.2), (1.3), lying in $sl(n) \otimes sl(n)$, it suffices to apply to r the map $f \otimes f$, where $f: \mathfrak{a} \subset sl(n)$ is defined by $f(A) = A - (1/N)(\mathrm{Tr}\, A) \cdot E$.

LITERATURE CITED

1. A. A. Belavin, "Discrete groups and integrability of quantum systems," Funkts. Anal., 14, No. 4, 18-26 (1980).
2. A. V. Zhiber and A. B. Shabat, "Klein–Gordon equations with nontrivial group," Dokl. Akad. Nauk SSSR, 247, No. 5, 1103-1107 (1979).
3. V. G. Kats, "Automorphisms of finite order of semisimple Lie algebras," Funkts. Anal. Anal., 3, No. 3, 94-96 (1969).
4. V. G. Kats, "Simple irreducible graded Lie algebras of finite height," Izv. Akad. Nauk SSSR, Ser. Mat., 32, No. 6, 1323-1367 (1968).
5. P. P. Kulish and E. K. Sklyanin, "Solutions of the Yang–Baxter equation," in: Differential Geometry of Lie Groups and Mechanics [in Russian], Zap. Nauchn. Sem. Leningr. Otd. Mat. Inst., 95, 129-160 (1980).
6. A. N. Leznov, M. V. Savel'ev, and V. G. Smirnov, "Theory of representations of groups and integration of nonlinear dynamical systems," Preprint IFVE, 80-51, Serpukhov: IFVE (1980).
7. I. G. MacDonald, "Affine systems of roots and the Dedekind η-function," Matematika, 16, No. 4, 3-49 (1972).
8. J.-P. Serre, Lie Algebras and Lie Groups, W. A. Benjamin (1965).
9. I. V. Cherednik, "Method of construction of factorized S-matrices in elementary functions," Teor. Mat. Fiz., 43, No. 1, 117-119 (1980).
10. O. I. Bogoyavlensky, "On perturbations of the periodic Toda lattice," Commun. Math. Phys., 51, 201-209 (1976).
11. V. G. Kac, "Infinite-dimensional algebras, Dedekind's η-function, classical Möbius function and the very strange formula," Adv. Math., 30, No. 2, 85-136 (1978).
12. A. V. Michailov, M. A. Olshanetsky, and A. M. Perelomov, Preprint ITEP-64, Moscow: ITEP (1980).

12a. S. A. Bulgadaev, "Two-dimensional integrable field theories connected with simple Lie algebras," P. L., 96B, 151-153 (1980).

13. A. I. Ooms, "On Lie algebras having a primitive universal enveloping algebra," J. Algebra, 32, No. 3, 488-500 (1975).
14. A. I. Ooms, "On Lie algebras with primitive envelopes – Supplements," Proc. Am. Math. Soc., 58, 67-72 (1976).
15. A. I. Ooms, "On Frobenius Lie algebras," Commun. Algebra, 8(1), 13-52 (1980).
16. A. Weil, Varietes Abeliennes et Courbes Algebriques, Hermann, Paris (1948).
17. A. Weil, "On algebraic groups of transformations," Am. J. Math., 77, 355-391 (1955).

Докл. Акад. Наук СССР
Том 268 (1983), № 2

Soviet Math. Dokl.
Vol. 27 (1983), No. 1

HAMILTONIAN STRUCTURES ON LIE GROUPS, LIE BIALGEBRAS AND THE GEOMETRIC MEANING OF THE CLASSICAL YANG-BAXTER EQUATIONS

UDC 512.81

V. G. DRINFEL'D

1. Recall that a *Poisson bracket* (or a Hamiltonian structure) on a smooth manifold M is an operation $(\varphi, \psi) \to \{\varphi, \psi\}$ on $C^\infty(M)$ which makes $C^\infty(M)$ a Lie algebra and which is such that $\{\varphi, \psi\chi\} = \{\varphi, \psi\}\chi + \{\varphi, \chi\}\psi$ for any $\varphi, \psi, \chi \in C^\infty(M)$. A *Hamiltonian manifold* is a manifold together with a Hamiltonian structure. A map $f\colon M_1 \to M_2$ of Hamiltonian manifolds is said to be *Hamiltonian* if it is smooth and if $f^*\{\varphi, \psi\} = \{f^*\varphi, f^*\psi\}$ for any $\varphi, \psi \in C^\infty(M_2)$. Here $(f^*\varphi)(x) \overset{\text{def}}{=} \varphi(f(x))$.

If M_1 and M_2 are Hamiltonian manifolds, then $M_1 \times M_2$ will always be considered to have the Hamiltonian structure characterized by the following properties:

1) The projections $\pi_i\colon M_1 \times M_2 \to M_i$, $i = 1, 2$, are Hamiltonian maps.
2) $\{\pi_1^*\varphi, \pi_2^*\psi\} = 0$ for any $\varphi \in C^\infty(M_1)$ and $\psi \in C^\infty(M_2)$.

2. Let G be a Lie group and let $\mathfrak{g}$ be its Lie algebra. Choose a basis $\{I_\mu\}$ of $\mathfrak{g}$. We let ∂_μ (∂'_μ) be the right-invariant (left-invariant) vector field on G defined by $I_\mu \in \mathfrak{g}$.

THEOREM 1. *Let $R \in \mathfrak{g} \otimes \mathfrak{g}$ be such that $R = r^{\mu\nu} I_\mu \otimes I_\nu$, where $r^{\mu\nu} = -r^{\nu\mu}$. If $\varphi, \psi \in C^\infty(G)$, set $\{\varphi, \psi\} \overset{\text{def}}{=} r^{\mu\nu}\partial_\mu\varphi \cdot \partial_\nu\psi$ and $\{\varphi, \psi\}' \overset{\text{def}}{=} r^{\mu\nu}\partial'_\mu\varphi \cdot \partial'_\nu\psi$. Then the following conditions are equivalent*:

1) *The operation $(\varphi, \psi) \to \{\varphi, \psi\}$ is a Poisson bracket.*
2) *The operation $(\varphi, \psi) \to \{\varphi, \psi\}'$ is a Poisson bracket.*
3) *R satisfies the classical Yang-Baxter equation* [3]; *that is,*

$$[R^{12}, R^{13}] + [R^{12}, R^{23}] + [R^{13}, R^{23}] = 0 \tag{1}$$

where

$$[R^{12}, R^{13}] \overset{\text{def}}{=} r^{\alpha\beta}r^{\mu\nu}[I_\alpha, I_\mu] \otimes I_\beta \otimes I_\nu, \qquad [R^{12}, R^{13}] \overset{\text{def}}{=} r^{\alpha\beta}r^{\mu\nu}I_\alpha \otimes [I_\beta, I_\mu] \otimes I_\nu,$$

$$[R^{13}, R^{23}] \overset{\text{def}}{=} r^{\alpha\beta}r^{\mu\nu}I_\alpha \otimes I_\mu \otimes [I_\beta, I_\nu].$$

Specifying a nondegenerate Hamiltonian structure is equivalent to specifying a nondegenerate 2-form ω such that $d\omega = 0$. Thus Theorem 1 implies the following corollary.

COROLLARY. *Suppose that $r^{\mu\nu} = -^{\nu\mu}$ is such that* $\det(r^{\mu\nu}) \neq 0$. *Let (b_{kl}) be the matrix which is the inverse of $(r^{\mu\nu})$ and let B be the bilinear form on $\mathfrak{g}$ with matrix (b_{kl}). Then R satisfies equation* (1) *if and only if B is a 2-cocycle; that is, if and only if*

$$B([x, y], z) + B([y, z], x) + B([z, x], y) = 0.$$

1980 *Mathematics Subject Classification.* Primary 58F05, 22E60; Secondary 22E70, 35Q20, 70H99.

Another geometric interpretation of (1), which is very close to the one cited above, is proposed in [4]. Namely, if we set $\Omega^0 = \mathbf{R}$ and $\Omega^1 = \mathfrak{g}^*$ (where the notation is as in [4]), then the Hamiltonian operators will be in one-to-one correspondence with the skew-symmetric solutions of (1). Thus, the classical Yang-Baxter equations can be considered as a special case of the Gel'fand-Dorfman scheme.

3. DEFINITION. A Hamiltonian structure on G is said to be *grouped* if the map μ: $G \times G \to G$ defined by $\mu(x, y) = xy$ is Hamiltonian. A Lie group together with a grouped Hamiltonian structure on it will be called a *Hamilton-Lie group*. A homomorphism of Hamilton-Lie groups is a group homomorphism which is a Hamiltonian map.

Let G be a Hamilton-Lie group. Then there is a Lie algebra structure on $\mathfrak{g}^*$; if l_1, $l_2 \in \mathfrak{g}^*$ and φ_1, $\varphi_2 \in C^\infty(G)$ are such that the differential of φ_i at the identity $e \in G$ (henceforth denoted by $d_e\varphi_i$) is equal to l_i, then $[l_1, l_2] \overset{\text{def}}{=} d_e\{\varphi_1, \varphi_2\}$. If the Poisson bracket on G is given by the formula $\{\varphi, \psi\} = \eta^{\mu\nu}\partial_\mu\varphi \cdot \partial_\nu\psi$, then the structure constants $f_\lambda^{\mu\nu}$ of $\mathfrak{g}^*$ are equal to $(\partial_\lambda\eta^{\mu\nu})(e)$.

4. DEFINITION. Let $\mathfrak{g}$ be a finite-dimensional vector space and suppose that both $\mathfrak{g}$ and $\mathfrak{g}^*$ have Lie algebra structures. The Lie algebra structures are called *compatible* if

(2) $$c_{rs}^k f_k^{ij} = c_{\alpha r}^i f_s^{j\alpha} - c_{\alpha r}^j f_s^{i\alpha} - c_{\alpha s}^i f_r^{j\alpha} + c_{\alpha s}^j f_r^{i\alpha},$$

where c_{rs}^k and f_k^{ij} are the structure constants of $\mathfrak{g}$ and $\mathfrak{g}^*$ with respect to bases of $\mathfrak{g}$ and $\mathfrak{g}^*$ which are dual to one another. We will say that $\mathfrak{g}$ *is given a Lie bialgebra structure* if $\mathfrak{g}$ and $\mathfrak{g}^*$ are given Lie algebra structures which are compatible with one another.

THEOREM 2. *Suppose that $\mathfrak{g}$ and $\mathfrak{g}^*$ have fixed Lie algebra structures. Define the linear map* φ: $\mathfrak{g}^* \otimes \mathfrak{g}^* \to \mathfrak{g}^*$ *by setting* $\varphi(l_1 \otimes l_2) = [l_1, l_2]$. *Then the following conditions are equivalent:*

1) *The Lie algebra structures on $\mathfrak{g}$ and $\mathfrak{g}^*$ are equivalent.*

2) *The map* φ^*: $\mathfrak{g} \to \mathfrak{g} \otimes \mathfrak{g}$ *is a 1-cocycle (it being understood that $\mathfrak{g}$ acts on $\mathfrak{g} \otimes \mathfrak{g}$ by means of the adjoint representation).*

3) *There is a Lie algebra structure on $\mathfrak{g} \oplus \mathfrak{g}^*$ inducing the given Lie algebra structures on $\mathfrak{g}$ and $\mathfrak{g}^*$ which is such that the bilinear form* Q: $(\mathfrak{g} \oplus \mathfrak{g}^*) \times (\mathfrak{g} \oplus \mathfrak{g}^*) \to \mathbf{R}$, *given by the formula*

$$Q((x_1, l_1), (x_2, l_2)) = l_1(x_2) + l_2(x_1),$$

is invariant with respect to the adjoint representation of $\mathfrak{g} \oplus \mathfrak{g}^*$.

We remark that the Lie algebra structure on $\mathfrak{g} \oplus \mathfrak{g}^*$ referred to in condition 3) is unique if it exists.

5. THEOREM 3. 1) *Let G be a Hamilton-Lie group with Lie algebra $\mathfrak{g}$. Then the Lie algebra structures on $\mathfrak{g}$ and $\mathfrak{g}^*$ (see §3) are compatible.*

2) *Let $\mathfrak{g}$ be a Lie bialgebra over $\mathbf{R}$ and let G be a connected, simply-connected Lie group with the Lie algebra $\mathfrak{g}$. Then there exists exactly one grouped Hamiltonian structure on G inducing the given Lie algebra structure on $\mathfrak{g}^*$.*

The proof is based on the following fact: the Poisson bracket on G given by the formula $\{\varphi, \psi\} = \eta^{\mu\nu}\partial_\mu\varphi \cdot \partial_\nu\psi$ is grouped if and only if the function η: $G \to \mathfrak{g} \otimes \mathfrak{g}$ defined by the formula $\eta(g) = \eta^{\mu\nu}(g)I_\mu \otimes I_\nu$ is a 1-cocycle; that is, if and only if $\eta(gh) = \mathrm{Ad}_g \cdot \eta(h) + \eta(g)$.

REMARK. Theorem 3 asserts that connected, simply-connected Hamilton-Lie groups are in one-to-one correspondence with Lie bialgebras over **R**. It is not difficult to show that the homomorphisms of connected, simply-connected Hamilton-Lie groups are in one-to-one correspondence with bialgebra homomorphisms.

6. Let $R \in \mathfrak{g} \otimes \mathfrak{g}$ be such that $R = r^{\mu\nu} I_\mu \otimes I_\nu$, where $r^{\mu\nu} = -r^{\nu\mu}$. Relation (2) will be satisfied if we set $f_\lambda^{\mu\nu} = c_{i\lambda}^{\mu} r^{i\nu} + c_{i\lambda}^{\nu} r^{\mu i}$ (in this case the cocycle φ^* of Theorem 2 is a coboundary). It is easy to verify that the $f_\lambda^{\mu\nu}$ are structure constants of a Lie algebra if and only if the tensor $\langle R, R \rangle \overset{\text{def}}{=} [R^{12}, R^{13}] + [R^{12}, R^{23}] + [R^{13}, R^{23}]$ is invariant with respect to the adjoint representation of $\mathfrak{g}$. In this case the grouped Poisson bracket on G corresponding to the $f_\lambda^{\mu\nu}$ is given by the formula

$$\{\varphi, \psi\}_R = r^{\mu\nu}(\partial_\mu \varphi \cdot \partial_\nu \psi - \partial'_\mu \varphi \cdot \partial'_\nu \psi).$$

Now suppose that $G \subset \mathrm{GL}(n_1, \mathbf{R}) \times \cdots \times \mathrm{GL}(n_k, \mathbf{R})$ and let $S_i\colon G \to \mathrm{Mat}(n_i, \mathbf{R})$ be the projection onto the ith component. Let $I_\mu^{(i)} \in \mathrm{Mat}(n_i, \mathbf{R})$ denote the projection of I_μ on the ith component and set $R_{ij} \overset{\text{def}}{=} r^{\mu\nu} I_\mu^{(i)} \otimes I_\nu^{(j)}$. Then $\{S_i, S_j\}_R = [R_{ij}, S_i \otimes S_j]$ (here, $\{S_i, S_j\}_R$ and $S_i \otimes S_j$ are functions $G \to \mathrm{Mat}(n_i, \mathbf{R}) \otimes \mathrm{Mat}(n_j, \mathbf{R})$). The Poisson bracket between elements of the transition matrix for many Hamiltonian systems which are integrable by the inverse scattering method has this form (see formula (I.2.11) of [2] and also §2.3 of [3]).

7. EXAMPLES OF LIE BIALGEBRAS. a) If $\mathfrak{g}$ is a Lie algebra, we obtain Lie bialgebra structures on $\mathfrak{g}$ and $\mathfrak{g}^*$ by setting $[l_1, l_2] = 0$ for any $l_1, l_2 \in \mathfrak{g}^*$. The Hamilton-Lie group with the zero Poisson bracket corresponds to the bialgebra $\mathfrak{g}$. The additive group $\mathfrak{g}^*$ endowed with the Berezin-Kirillov bracket [1] corresponds to the bialgebra $\mathfrak{g}^*$.

b) If $\dim \mathfrak{g} = 2$, then any Lie algebra structures on $\mathfrak{g}$ and $\mathfrak{g}^*$ are compatible.

c) Let $\mathfrak{g}$ be a Lie algebra. In §6, we have shown that each skew-symmetric tensor $R \in \mathfrak{g} \otimes \mathfrak{g}$ satisfying (1) corresponds to a Lie bialgebra structure on $\mathfrak{g}$. Such tensors can be constructed with the help of the corollary to Theorem 1.

d) Given a Lie algebra structure on $\mathfrak{g}$, it is natural to try to find all Lie algebra structures on $\mathfrak{g}^*$ compatible with it. We discuss this question in the case when $\mathfrak{g}$ is a simple Lie algebra over **C**. In this case any 1-cocycle $\mathfrak{g} \to \mathfrak{g} \otimes \mathfrak{g}$ is a coboundary. Therefore, it is sufficient to find all skew-symmetric $R \in \mathfrak{g} \otimes \mathfrak{g}$ such that the tensor $\langle R, R \rangle \in \mathfrak{g} \otimes \mathfrak{g} \otimes \mathfrak{g}$ is invariant. It is easy to see that if R is skew-symmetric, so is $\langle R, R \rangle$. On the other hand, since $\mathfrak{g}$ is simple, any invariant skew-symmetric tensor in $\mathfrak{g} \otimes \mathfrak{g} \otimes \mathfrak{g}$ has the form $a \cdot c_{\lambda\mu\nu} I_\lambda \otimes I_\mu \otimes i_\nu$. Here $\{I_\lambda\}$ is a basis of $\mathfrak{g}$ which is orthonormal with respect to the Killing form, and the $c_{\lambda\mu\nu}$ are the structure constants. A. A. Belavin and the author have succeeded in finding all skew-symmetric solutions of the equation $\langle R, R \rangle = a \cdot c_{\lambda\mu\nu} I_\lambda \otimes I_\mu \otimes I_\nu$ when $a \neq 0$. This was accomplished by passing from the tensor R to an operator $A\colon \mathfrak{g} \to \mathfrak{g}$ by the formula $R = A(I_\mu) \otimes I_\mu$ and then applying a Cayley transformation to the skew-symmetric operator A.

The author is grateful to A. A. Belavin, A. A. Kirillov and Yu. I. Manin for their constant interest in this work. Moreover, the author is indebted to Yu. I. Manin, who predicted that conditions 1) and 3) of Theorem 3 were equivalent (oral communication).

Physical-Technical Institute of Low Temperatures
Academy of Sciences of the Ukrainian SSR

Received 4/JUNE/82

BIBLIOGRAPHY

1. A. G. Reĭman, Zap. Nauchn. Sem. Leningrad. Otdel. Mat. Inst. Steklov. (LOMI) **95** (1980), 3; English transl. in J. Soviet Math. **19** (1982), no. 5.

2. E. K. Sklyanin, Zap. Nauchn. Sem. Leningrad. Otdel. Mat. Inst. Steklov. (LOMI) **95** (1980), 55; English transl. in J. Soviet Math. **19** (1982), no. 5.

3. P. P. Kulish and E. K. Sklyanin, Zap. Nauchn. Sem. Leningrad. Otdel. Mat. Inst. Steklov. (LOMI) **95** (1980), 129; English transl. in J. Soviet Math. **19** (1982), no. 5.

4. I. M. Gel'fand and I. Ya. Dorfman, Funktsional. Anal. i Prilozhen. **14** (1980), no. 3, 71; English transl. in Functional Anal. Appl. **14** (1980).

Translated by D. B. O'SHEA

WHAT IS A CLASSICAL R-MATRIX?

M. A. Semenov-Tyan-Shanskii

UDC 517.43+519.46

The method of the classical r-matrix appeared in the papers of Sklyanin [1] and [2] as a "by-product" of the quantum inverse-scattering method and has already acquired considerable popularity. The classical Yang–Baxter equation, which is connected with the classical r-matrices, was discussed in detail in [3-5]. In this paper we establish a connection between the r-matrix method and the Riemann-problem method, which is the most effective tool presently available for integrating nonlinear equations. This leads us to a new approach to the Yang–Baxter equation. Our study is the outcome of a careful study of the papers [3-5]. The results of N. Yu. Reshetikhin and L. D. Faddeev* played an important role in clarifying the algebraic meaning of the r-matrix. The advice of E. K. Sklyanin and A. G. Reiman was helpful to me in the preparation of this paper. To them, I express my deep gratitude. Our approach to the various topics is mainly suitable for the mathematician who is familiar with Hamiltonian mechanics on groups and with the standard methods of handling integrable equations of Lax form. A number of preliminary results were published in the author's note [7].

A few words on the structure of the paper. Sections 1-3 are introductory. They deal mainly with the translation of the tensor notations, which are standard in the r-matrix method, into the more general mathematical language of the Lie algebras. Also, the connections with the definitions of paper [5] are discussed in detail. A somewhat paradoxical observation we make in Sec. 3 is that the Yang–Baxter equation, which lies at the basis of the r-matrix method, is never actually fulfilled in the rigorous sense (at least in those cases where a well-posed Riemann problem exists). The modified Yang–Baxter equation, introduced in Sec. 3, permits the construction of an abstract version of the Riemann-problem

*See Faddeev's lectures at Les Houches [6]. A detailed discussion can be found in the paper "Hamiltonian structures of integrable models of field theory," Theor. Math. Phys., 56, No. 3 (1983).

V. A. Steklov Mathematics Institute, Leningrad Branch, Academy of Sciences of the USSR.
Translated from Funktsional'nyi Analiz i Ego Prilozheniya, Vol. 17, No. 4, pp. 17-33, October-December, 1983. Original article submitted April 15, 1983; revision submitted May 16, 1983.

method (Sec. 4). The proof of this theorem does not depend upon the classification of the solutions to the Yang–Baxter equation, and covers numerous concrete applications. In Sec. 5 these results are extended to systems with quadratic Poisson brackets. Further, in Sec. 6 the r-matrix method is applied to two-dimensional systems which admit a zero-curvature representation. The fundamental idea of the two-dimensionalization method – the shift to a central extension of the current algebra – was proposed by A. G. Reiman and the author. Here, we extend the r-matrix method to nonultralocal Poisson brackets and give a simple criterion for ultralocality. In the concluding section, we discuss a classification of the r-matrices over the finite-dimensional semisimple Lie algebras, which extends the classification given in [4]. Due to the already considerable length of this paper, it was not possible to include concrete examples (even as illustrations). It is useful for the reader to keep in mind, say, the family of examples covered by the Adler–Kostant procedure [9-12]. At the present time, we have no examples of nonlinear equations related to r-matrices, other than those included in the Adler–Kostant scheme. The quantization problem is not considered in this paper. The main task is to include the quantum inverse-scattering method in the algebraic scheme. Until now, only the first steps in this direction have been made.

The author dedicates this paper to M. I. Gel'fand on the occasion of his birthday.

1. Classical r-Matrices and Double Lie Algebras

The "correct" algebraic version of the notion r-matrix is that of double Lie algebra. The relationship with "Lie bialgebras" [5] will be discussed in Sec. 2.

Let $\mathfrak{g}$ be a Lie algebra. A linear operator $R \in \operatorname{End} \mathfrak{g}$ will be referred to as a classical r-matrix if the bracket on $\mathfrak{g}$, given by

$$[X, Y]_R = [RX, Y] + [X, RY], \tag{1}$$

is a Lie bracket, i.e., satisfies the Jacobi identity. (The skew-symmetry of (1) for any R is obvious!). We call the pair $(\mathfrak{g}, R)$ a double Lie algebra, and use the symbol $\mathfrak{g}_R$ to denote the algebra $\mathfrak{g}$ equipped with the bracket (1). Along with the two Lie brackets on $\mathfrak{g}$ we shall consider on the space $C^\infty(\mathfrak{g}^*)$ the corresponding Lie–Poisson brackets:†

$$\{h_1, h_2\}_0 (L) = L ([X_1, X_2]), \tag{2}$$

and

$$\{h_1, h_2\}_R (L) = L ([X_1, X_2]_R), \quad L \in \mathfrak{g}^*, \quad X_i = dh_i (L) \in \mathfrak{g}. \tag{3}$$

The symbols ad* and ad_R^* stand, respectively, for the coadjoint representations of the algebras $\mathfrak{g}$ and $\mathfrak{g}_R$. From (1) it readily follows that

$$\mathrm{ad}_R^* X\cdot L = \mathrm{ad}^* RX\cdot L + R^* (\mathrm{ad}^* X\cdot L). \tag{4}$$

The following simple theorem lies at the basis of all applications.

THEOREM 1. Let $(\mathfrak{g}, R)$ be a double Lie algebra. Then:

(i) The ad*-invariant functions on $\mathfrak{g}^*$ are in involution with respect to both Lie–Poisson brackets on $\mathfrak{g}^*$;

(ii) The equation of motion on $\mathfrak{g}^*$ given by an invariant Hamiltonian h with respect to the bracket (3) can be written in the equivalent forms

$$\frac{dL}{dt} = \mathrm{ad}_R^* \, dh(L) \cdot L; \tag{5}$$

and

$$\frac{dL}{dt} = \mathrm{ad}^* M_h \cdot L; \quad M_h = R(dh(L)). \tag{6}$$

Proof. (i) It is well known that the ad*$\mathfrak{g}$-invariant functions are central with respect to the bracket (2). Let us verify that they also commute with respect to the bracket (3).

†Let $\mathfrak{g}$ and $\mathfrak{g}^*$ be, respectively, an arbitrary Lie algebra and its dual. The Lie–Poisson bracket of two functions on $\mathfrak{g}^*$, f and g, is defined as $\{f, g\}(L) = L([df, dg])$. This definition is due to Lie ([13], Vol. 3, Ch. 25, Sec. 115, formula (75)). Later, this bracket was rediscovered by F. A. Berezin and used in the method of orbits of Kirillov–Kostant. It is also known as the Berezin–Kirillov bracket.

In fact, suppose that h_1 and h_2 are two such functions, with differentials X_1 and X_2. The $ad^* \mathfrak{g}$ -invariance is expressed by the equality $ad^* dh(L)\cdot L = 0$. Then $\{h_1, h_2\}_R(L) = L([X_1, X_2]_R) = L([RX_1, X_2] + [X_1, RX_2]) = ad^*X_2\cdot L(RX_1) - ad^* X_1\cdot L(RX_2) = 0$.

(ii) Formula (5) is the standard form in which one writes the equations that are Hamiltonian with respect to a Lie–Poisson bracket. The equivalence of (5) and (6) is a straightforward consequence of (4), because $ad^* X\cdot L = 0$.

A related theorem is given in [9] and [11] in connection with what is known as the Adler–Kostant scheme [14] and [15]. We shall see below that the method of the r-matrix generalizes the latter.

2. Relation with the Classical Formalism

In our approach, formula (1) is taken as the definition of the classical r-matrix. We may legitimately ask whether the Poisson bracket associated with such an r-matrix can be written in standard tensor form and whether the r-matrix satisfies the Yang–Baxter equation. Generally speaking, the answers to these two questions are negative.

Let us try first to attach a rigorous meaning to the tensor formalism. Thus, recall that the Lie–Poisson bracket of two linear functions in $\mathfrak{g}^*$ is still linear. This allows us to extend the Lie–Poisson bracket to arbitrary linear maps, so that if $A \in \mathrm{Hom}(\mathfrak{g}^*, V)$ and $B \in \mathrm{Hom}(\mathfrak{g}^*, W)$, then $\{A \otimes B\} \in \mathrm{Hom}(\mathfrak{g}^*, V \otimes W)$. In particular, let $A = B = 1$ be the identity map of $\mathfrak{g}^*$ (distinct copies of $\mathfrak{g}^*$ are indexed by the numbers 1, 2, 3, ...); set, by definition,

$$\{L_1 \otimes L_2\}_R = \{1 \otimes 1\}_R (L), \quad L \in \mathfrak{g}^*. \tag{7}$$

We shall assume that: a) there is a nondegenerate invariant bilinear form on $\mathfrak{g}$, and b) the operator $R \in \mathrm{End}\, \mathfrak{g}$ is skew-symmetric ("the "unitarity condition").

Proposition 2. (i) Identify $\mathfrak{g}$ and $\mathfrak{g}^*$ via the bilinear form on $\mathfrak{g}$, and let $r \in \mathfrak{g} \otimes \mathfrak{g}$ be the element corresponding to R under the canonical isomorphism $\mathrm{End}\, \mathfrak{g} \simeq \mathfrak{g} \otimes \mathfrak{g}$. Then

$$\{L_1 \otimes L_2\} = [r, L \otimes 1 + 1 \otimes L]. \tag{8}$$

(ii) Conditions a) and b) are necessary for (8) to hold.

Formula (8) is the starting point in the standard formulation of the r-matrix method (see [2 and 6]).

According to our definition, for general Lie algebras the r-matrix belongs to $\mathrm{End}\, \mathfrak{g} \simeq \mathfrak{g} \otimes \mathfrak{g}^*$ rather than $\mathfrak{g} \otimes \mathfrak{g}$, as is customary. V. G. Drinfel'd has proposed in [5] an alternative axiomatization of the notion of r-matrix, interpreting it (for arbitrary Lie algebras $\mathfrak{g}$) as an element of $\mathfrak{g} \otimes \mathfrak{g}$. Such an r-matrix defines a structure of Lie algebra on the space $\mathfrak{g}^*$.

Definition ([5]). The Lie brackets on $\mathfrak{g}$ and $\mathfrak{g}^*$ are said to be compatible if the mapping $\mathfrak{g} \to \mathfrak{g} \otimes \mathfrak{g}$, dual to the Lie bracket $\mathfrak{g}^* \otimes \mathfrak{g}^* \to \mathfrak{g}^*$, is a 1-cocyle on $\mathfrak{g}$. A pair $(\mathfrak{g}, \mathfrak{g}^*)$ with compatible Lie brackets is called a Lie bialgebra.

Let $r \in \mathrm{Hom}\, (\mathfrak{g}^*, \mathfrak{g})$ be a skew-symmetric operator. Define a bracket on $\mathfrak{g}^*$ by $\{f, g\} = ad^*r(f)g - ad^*r(g)\cdot f$, $f, g \in \mathfrak{g}^*$. If this bracket fulfills the Jacobi identity, then $(\mathfrak{g}, \mathfrak{g}^*)$ is a Lie algebra. For Lie algebras Theorem 1 is no longer true: the two relevant Lie–Poisson brackets live, generally speaking, on distinct spaces. The connection between Lie algebras and integrable systems may be established only if the spaces $\mathfrak{g}^*$ and $\mathfrak{g}$ can be identified.

Proposition 3. The double Lie algebra $(\mathfrak{g}, R)$ is a Lie bialgebra in the sense of [5] if and only if conditions a) and b) of Proposition 2 are fulfilled.

Condition a) arises naturally in the investigation of finite-dimensional dynamical systems: it guarantees that Eqs. (6) are in the Lax form. For systems with linear Poisson brackets, however, there is no reason why we should restrict ourselves to unitary r-matrices. (Compare with the examples of integrable systems given in [9]). The Lie algebras connected with infinite-dimensional systems are central extensions of algebras of currents and for them a) is not satisfied. It turns out that for such algebras the unitary r-matrices are isolated by the constraint that the Poisson bracket be ultralocal (see Sec. 7 below). Another class of problems which lead to unitary r-matrices is provided by the finite-difference systems with quadratic Poisson brackets (Sec. 6).

3. Yang–Baxter Equations

We next discuss the Jacobi identity for bracket (1) and its connection with the Yang–Baxter equation.

Proposition 4. Bracket (1) satisfies the Jacobi identity if and only if

$$[X, [RY, RZ] - R([Y, Z]_R)] + \text{c. p.} = 0 \tag{9}$$

for all $X, Y, Z \in \mathfrak{g}$.

In order to compare formula (9) with the Yang–Baxter equation, we rewrite the Jacobi identity in tensor notation as $\{L_1 \otimes \{L_2 \otimes L_3\}_R\}_R + \{L_2 \otimes \{L_3 \otimes L_1\}_R\}_R + \{L_3 \otimes \{L_1 \otimes L_2\}_R\}_R = 0$. Here, as above $\{L_1 \otimes \{L_2 \otimes L_3\}\}$ is a Lie–Poisson bracket of identity mappings computed at the point $L \in \mathfrak{g}^*$: $\{L_1 \otimes \{L_2 \otimes L_3\}\} = \{1 \otimes \{1 \otimes 1\}\}(L)$. Now suppose that algebra $(\mathfrak{g}, R)$ satisfies the conditions of Proposition 2, so that the bracket is given by formula (8). Identifying the operator R with a skew-symmetric tensor $r_{ik} \in \mathfrak{g}_i \otimes \mathfrak{g}_k \subset \mathfrak{g} \otimes \mathfrak{g} \otimes \mathfrak{g}$ (i, k = 1, 2, 3), we obtain $\{L_1 \otimes \{L_2 \otimes L_3\}_R\}_R + \text{c. p.} = [L \otimes 1 \otimes 1 + 1 \otimes L \otimes 1 + 1 \otimes 1 \otimes L, B]$, $B = [r_{13}, r_{12}] + [r_{23}, r_{13}] + [r_{23}, r_{12}]$. The tensor B is an element of $\mathfrak{g} \otimes \mathfrak{g} \otimes \mathfrak{g}$ and we shall regard it as a mapping $\mathfrak{g} \otimes \mathfrak{g} \to \mathfrak{g}$. It is readily verified that

$$B(Y, Z) = [RY, RZ] - R([Y, Z]_R).$$

Therefore, one can reexpress identity (9) as

$$[X, B(Y, Z)] + [Y, B(Z, X)] + [Z, B(X, Y)] = 0. \tag{10}$$

To obtain the Yang–Baxter equation, one simply sets the individual terms in (10) equal to zero:

$$[r_{13}, r_{12}] + [r_{23}, r_{13}] + [r_{23}, r_{12}] = 0$$

or

$$[RY, RZ] - R([Y, Z]_R) = 0. \tag{11}$$

In this form, the Yang–Baxter equation makes sense for arbitrary Lie algebras $\mathfrak{g}$ and arbitrary operators $R \in \operatorname{End} \mathfrak{g}$.

Exercise 1. Rewrite Eq. (11) in tensor form for nonunitary r-matrices. 2. Show that there are nonunitary r-matrices which satisfy (11), and explain the connection with Theorem 1.1 of [3].

Our goal is to reduce the Lax equations (6) to a factorization problem (Riemann problems) following the pattern of [9] or [11]. It turns out that this cannot be achieved for r-matrices satisfying Eq. (11): this equation is, in a certain sense, degenerate. This circumstance justifies the following.

Definition. The modified equation Yang–Baxter is

$$[RX, RY] - R([X, Y]_R) = -[X, Y]. \tag{12}$$

Obviously, (12) implies (9), and thus with every solution of (12) there is associated a double Lie algebra.

Remarks: 1. One may consider, instead of (12), the equation $[RX, RY] - R([X, Y]_R) = -\alpha^2[X, Y]$, $\alpha \neq 0$, which is reduced to (12) by the dilatation $R \to \alpha R$. Equation (12) is normalized so that the operator R = 1 is one of its solutions.

2. Equation (12) was considered, in an equivalent form, in [3-4], but it seems that these authors have not noticed the fundamental role that it plays in connection with the Riemann problem.

3. The singular integral operators associated with the standard r-matrices [3] satisfy identity (12) (see discussion below). Thus, the right-hand side of the Yang–Baxter identity for such r-matrices, interpreted as a distribution, does not vanish.

An important class of solutions of Eq. (12) is produced as follows. Let $\mathfrak{g}$ be an arbitrary Lie algebra, and let $\mathfrak{g}_+$ and $\mathfrak{g}_-$ be subalgebras of such that $\mathfrak{g} = \mathfrak{g}_+ \dot{+} \mathfrak{g}_-$. We let $P_\pm$ denote the projection on $\mathfrak{g}_\pm$ parallel to $\mathfrak{g}_\mp$ and set $R = P_+ - P_-$.

Proposition 5. (i) The operator R satisfies identity (12).

(ii) $(\mathfrak{g}, R)$ is a double Lie algebra, with

$$[X,Y]_R=2[X_+,Y_+]-2[X_-,Y_-],\quad X_\pm=P_\pm X,\quad Y_\pm=P_\pm Y.$$

The proof is obvious. For this case, Theorem 1 is nothing else but the Kostant–Adler theorem (in the form given in [9]). The global version of Theorem 1 amounts to a reduction of Eqs. (6) to a factorization problem. It turns out that this can be also achieved in the general case, proceeding merely from Eq. (12), which leads to a generalization of the Kostant–Adler procedure to arbitrary r-matrices.

Example. Let $\mathfrak{g}=\mathrm{Mat}(n,\mathbf{C})\otimes\mathbf{C}[\lambda,\lambda^{-1}]$ be the associative matrix algebra over the ring $\mathbf{C}[\lambda,\lambda^{-1}]$. We equip $\mathfrak{g}$ with the inner product $(X, Y) = \mathrm{Res}_{\lambda=0}\,\mathrm{tr}\,X(\lambda)Y(\lambda)$. The operator $R\in\mathrm{End}\,\mathfrak{g}$ is defined as

$$RX=\begin{cases} X, & \text{if } X=\sum_{n\geq 0}X_n\lambda^n,\\ -X, & \text{if } X=\sum_{n<0}X_n\lambda^n.\end{cases} \tag{13}$$

Proposition 6. (i) Operator (13) is skew-symmetric and satisfies the identity

$$RX\cdot RY=R(RX\cdot Y+X\cdot RY)-XY. \tag{14}$$

(ii). The multiplication defined in $\mathfrak{g}$ by the formula $X\times_R Y = RX\cdot Y + X\cdot RY$ is associative.

In tensor notation, the Yang r-matrix corresponding to the operator (13) takes the simplest form $r(\lambda,\mu) = P/(\lambda-\mu)$ (see [2]). Here the distribution is understood as principal value. The integral operator associated with it is the usual Hilbert transform, and formula (14) is essentially the classical Poincaré–Bertrand formula [16].

Remark. The example we have just considered leads us to the notion of double associative algebra. The classification of the operators satisfying the Poincaré–Bertrand identity (14) is similar to the classification of the solutions of the Yang–Baxter equation. We note also that $X\times_R Y = 2X_+Y_+ - 2X_-Y_-$, and hence that the operator $Y\mapsto X\times_R Y$ is Toeplitz.

4. Factorization Theorem

Let $\mathfrak{g}$ be a Lie algebra and let $R\in\mathrm{End}\,\mathfrak{g}$ satisfy identity (14). Set $\mathfrak{g}_\pm = \mathrm{Im}(R\pm 1)$, $\mathfrak{k}_\pm=\mathrm{Ker}(R\mp 1)$.

Proposition 7. (i) $R\pm 1\colon \mathfrak{g}_R\to\mathfrak{g}$ are Lie-algebra homomorphisms. (ii) $\mathfrak{g}_\pm\subset\mathfrak{g}$ are Lie subalgebras. (iii) $\mathfrak{k}_\pm\subset\mathfrak{g}_R$ are ideals, and $\mathfrak{g}_\pm=\mathfrak{g}_R/\mathfrak{k}_\mp$.

Proof. Formula

$$(R\pm 1)([X,Y]_R)=[(R\pm 1)X,\ (R\pm 1)(Y)] \tag{15}$$

is a reformulation of (12). Assertions (ii) and (iii) are straightforward consequences of (i).

Proposition 8 ([4]). (i) $\mathfrak{k}_\pm\subset\mathfrak{g}_\pm$ are ideals. (ii). The map $\theta\colon\mathfrak{g}_+/\mathfrak{k}_+\to\mathfrak{g}_-/\mathfrak{k}_-$ defined by the rule $\theta\colon (R+1)X\mapsto(R-1)X$ is a Lie-algebra isomorphism.

Following [3 and 4], the operator θ will be referred to as the Cayley transform of the operator R.

Proposition 9 (Infinitesimal Form of the Factorization Theorem). Consider the sequence of maps

$$\mathfrak{g}_R\xrightarrow{R+1,\,R-1}\mathfrak{g}_+\oplus\mathfrak{g}_-\xrightarrow{1,\,-1}\mathfrak{g}.$$

(i) The image of $\mathfrak{g}_R$ in $\mathfrak{g}_+\oplus\mathfrak{g}_-$ is the subalgebra

$$\tilde{\mathfrak{g}}_R=\{(X_+,X_-)\in\mathfrak{g}_+\oplus\mathfrak{g}_-;\ \theta(\overline{X}_+)=\overline{X}_-\}.$$

(Here $\overline{X}_\pm$ stands for the residues class $X_\pm(\mathrm{mod}\ \mathfrak{k}_\pm)$.)

(ii) The map $\mathfrak{g}_R\to\tilde{\mathfrak{g}}_R$ is a Lie-algebra isomorphism.

(iii) Every element $X\in\mathfrak{g}$ can be uniquely expressed as $X = X_+ - X_-$, $(X_+, X_-)\in\tilde{\mathfrak{g}}_R$.

Conversely, suppose that there is given a Lie algebra $\mathfrak{g}$ together with subalgebras $\tilde{\mathfrak{g}}_+$, and $\tilde{\mathfrak{g}}_-$.

Proposition 10. Let $\tilde{\mathfrak{g}}_R \subset \tilde{\mathfrak{g}}_+ \oplus \tilde{\mathfrak{g}}_-$ be a Lie subalgebra.

(i) Suppose that every element $X \in \mathfrak{g}$ can be uniquely expressed in the form

$$X = X_+ - X_-, \quad (X_+, X_-) \in \tilde{\mathfrak{g}}_R. \tag{16}$$

Then the operator R, acting according to the rule $RX = X_+ + X_-$, satisfies identity (12).

(ii) Suppose, in addition, that there are given ideals $\tilde{\mathfrak{k}}_\pm \subset \tilde{\mathfrak{g}}_\pm$ and an isomorphism θ: $\tilde{\mathfrak{g}}_+/\tilde{\mathfrak{k}}_+ \to \tilde{\mathfrak{g}}_-/\tilde{\mathfrak{k}}_-$, such that $\tilde{\mathfrak{g}}_R = \{(X_+, X_-) \in \tilde{\mathfrak{g}}_+ \oplus \tilde{\mathfrak{g}}_-;\ \theta(\overline{X}_+) = \overline{X}_-\}$. Then $\tilde{\mathfrak{g}}_\pm = \mathrm{Im}\,(R \pm 1)$, $\tilde{\mathfrak{k}}_\pm = \mathrm{Ker}(R \mp 1)$, and θ is the Cayley transform of the operator R. Thus, in order that an r-matrix exist it is necessary and sufficient that the factorization problem (16) be uniquely solvable.

Remark. The Yang–Baxter equation (11) is clearly a degenerate case of Eq. (12). That is, $R: \mathfrak{g}_R \to \mathfrak{g}$ is a Lie-algebra homomorphism. Therefore, in this case, instead of the two subalgebras $\mathfrak{g}_\pm = \mathrm{Im}\,(R \pm 1)$ there is only one, and the factorization problem (16) cannot be formulated.

We now turn to the global version of the factorization theorem. Let $(\mathfrak{g}, R)$ be a double Lie algebra, and assume that R satisfies the identity (14). Let G be a Lie group with Lie algebra $\mathfrak{g}$, and let $G_\pm$ and $K_\pm$ be analytic subgroups of G corresponding to the Lie subalgebras $\mathfrak{g}_\pm = \mathrm{Im}\,(R \pm 1)$, and $\mathfrak{k}_\pm = \mathrm{Ker}\,(R \mp 1)$. A homomorphism of Lie groups, corresponding to a homomorphism of their Lie algebras, will be denoted by the same letter as the latter. Also, we let $\overline{x}$ denote the image of an element $x \in G_\pm$ under the canonical projection $G_\pm \to G_\pm/K_\pm$.

THEOREM 11. Let h be an $\mathrm{ad}^*\mathfrak{g}$-invariant function on $\mathfrak{g}^*$. The solution of Eq. (6) with Hamiltonian h is given by the formula

$$L(t) = \mathrm{Ad}_G^*\, g_+(t) \cdot L_0 = \mathrm{Ad}_G^*\, g_-(t) \cdot L_0, \tag{17}$$

where g_+, g_- is the solution of the factorization problem

$$\exp 2tX_0 = g_+(t)\, g_-(t)^{-1}, \quad g_\pm \in G_\pm, \quad \theta(\overline{g}_+) = \overline{g}_-, \quad X_0 = dh(L_0). \tag{18}$$

This problem is solvable for sufficiently small t.

Proof. The solvability of problem (18) for small t follows from Proposition 9. Now, if L(t) is defined by (17), then (6) is readily verified by means of differentiation.

In the above proof, the solution of Eq. (6) was guessed in advance. In the particular case of the Adler–Kostant scheme two geometric proofs are available [9, 11]. We leave the reader to verify that these two proofs can be also carried out, with almost no modifications, in the general case.

5. Quadratic Poisson Brackets

If the given r-matrix is unitary, then one can associate with it quadratic Poisson brackets, written usually in the tensor form

$$\{L_1 \overset{\otimes}{,} L_2\} = [r, L_1 \otimes L_2]. \tag{19}$$

One is lead to formula (19) when one computes Poisson brackets for elements of the monodromy matrix (see Sec. 7 below). For finite-difference equation, (19) gives the "fundamental Poisson brackets" at a node of the lattice [6]. Brackets (19) and their quantum analogs are studied in Sklyanin's paper [17].

The invariant object connected with brackets (19) is a Poisson structure on the underlying group [5]. Let G be a Lie group with Lie algebra $\mathfrak{g}$. Consider in $\mathfrak{g}$ a basis $\{e_\mu\}$ and the left- (right-) invariant vector fields ∂_μ (respectively, ∂'_μ) on G, defined by the elemen e_μ. Suppose that the tensor $r = \Sigma r_{\mu\nu} e_\mu \otimes e_\nu \in \mathfrak{g} \otimes \mathfrak{g}$ is skew symmetric and such that the bracket

$$\{\varphi, \psi\} = \sum r_{\mu\nu} (\partial_\mu \varphi \partial_\nu \psi - \partial'_\mu \varphi \partial'_\nu \psi), \quad \varphi, \psi \in C^\infty(G), \tag{20}$$

satisfies the Jacobi identity. The characteristic property of (20) is that the multiplication in G is a Hamiltonian map relative to it. For this reason, V. G. Drinfel'd has proposed the term "Hamilton–Lie groups" for the Lie groups endowed with a bracket (20). [the rigorous definition of a Hamilton–Lie group is this. Let A(G) be the ring of functions on G. A Poisson structure on G turns A(G) into a Lie algebra. The group G is valled a Hamilton–Lie group if: (i) the comultiplication in A(G) is a Lie algebra homomorphism; (ii) the map $x \to x^{-1}$ induces a antiautomorphism of A(G). This definition represents a reformulation of a well-known property of the monodromy matrix for finite-difference systems; see Proposition 15 below.]

Here, we are mainly interested in the connection between the brackets (19), (20) and integrable systems. It turns out (in full analogy with Proposition 2) that such a connection is established only if there is a nondegenerate invariant inner product on $\mathfrak{g}$. Under this assumption, analogs of Theorems I and II hold true for Hamilton–Lie groups.

For infinite-dimensional Lie algebras (which are usually encountered in applications) the correct object is an operator $R \in \mathrm{Hom}(\mathfrak{g}^*, \mathfrak{g})$ rather than a tensor $r \in \mathfrak{g} \otimes \mathfrak{g}$, (the kernel of R can be a distribution). Let us rewrite (20) in operator notation. To this end, we identify the tangent space to the group at the identity with the Lie algebra of right-invariant vector fields. For $L \in G$, we let ρ_L, λ_L: $\mathfrak{g} \to T_LG$ denote the differentials of the right and left translations. If ρ_L^*, $\lambda_L^*: T_L^*G \to \mathfrak{g}^*$ are the dual maps, then

$$\{\varphi_1, \varphi_2\}(L) = \rho_L^*(X_2)(R(\rho_L^*(X_1)) - \lambda_L^*(X_2)(R(\lambda_L^*(X_1)),\quad X_i = d\varphi_i(L). \tag{20'}$$

Formula (20) acquires a simpler form provided G is a matrix group. In this case, $\rho_L^*(X) = LX$, $\lambda_L^*(X) = XL$, and hence

$$\{\varphi_1, \varphi_2\} = LX_2(R(LX_1)) - X_2L(R(X_1L)). \tag{20''}$$

To put (20) back in the tensor form (19), we shall assume that G is an affine matrix group. Then the matrix elements L_{ik} can be thought of as generators of the affine ring of G, and computing their Poisson brackets we recover (19).

Below it will be assumed that there is a nondegenerate invariant form on $\mathfrak{g}$. We use it to identify $\mathfrak{g}$ and $\mathfrak{g}^*$ and view R as a linear operator acting in $\mathfrak{g}$.

Proposition 12. Suppose that the operator $R \in \mathrm{End}\,\mathfrak{g}$ is skew-symmetric and satisfies the Yang–Baxter equation (12). Then, bracket (20) satisfies the Jacobi identity.

Remark. If R satisfies the Yang–Baxter equation (11), then two other Poisson brackets arise on G:

$$\{\varphi_1, \varphi_2\}'(L) = (\rho_L^* X_2, R(\rho_L^* X_1)); \tag{21'}$$

and

$$\{\varphi_1, \varphi_2\}''(L) = (\lambda_L^* X_2, R(\lambda_L^* X_1)),\quad X_i = \mathrm{grad}\,\varphi_i(L), \tag{21''}$$

or, for matrix groups,

$$\{\varphi_1, \varphi_2\}'(L) = (LX_2, R(LX_1)),\quad \{\varphi_1, \varphi_2\}''(L) = (X_2L, R(X_1L)).$$

The Jacobi identity for these brackets is discussed in [5] and in the paper of Gel'fand and Dorfman [18]. Brackets (21) are not connected with integrable systems, because the analog of Theorem I does not hold for them.

The proof of the Jacobi identity for the bracket (20) is rather tedious and we omit it. A suitable proof technique was proposed by Gel'fand and Dorfman [19]. In their method one takes the de Rham complex of the group G to be the basic complex. A slightly more general theorem was announced in [5]; its proof may be carried out using analogous tools, but we do not need this result.

Bracket (20) can be (under certain supplementary constraints on the class of Lie algebras) endowed with an additional Poisson structure for $C^\infty(\mathfrak{g}^*)$. As we shall show, this structure is an abstract analog of the "second Hamiltonian structure" of Gel'fand–Dikii [20]. We shall assume that $\mathfrak{g}$ is the Lie algebra of an associative algebra (for which we use the same letter) and the multiplication in $\mathfrak{g}$ is symmetric with respect to an inner product (fixed once for all). Recall that in Sec. 1 we have associated with bracket (1) the Poisson bracket (3) on $C^\infty(\mathfrak{g}^*)$. Now define a quadratic bracket on $C^\infty(\mathfrak{g})$ by

$$\{h_1, h_2\}(L) = (L, [X_1, R(LX_2)] + [R(X_1L), X_2]),\quad X_i = \mathrm{grad}\,h_i(L). \tag{22}$$

This formula is clearly formally equivalent to (20), but the two have distinct domains of applicability: for an infinite-dimensional Lie algebra $\mathfrak{g}$ there is not necessarily a corresponding group, whereas formula (22) remains valid for such algebras.

THEOREM 13. Let $R \in \operatorname{End} \mathfrak{g}$ be a skew-symmetric operator satisfying the modified Yang–Baxter equation (12). Then:

(i) Bracket (22) is skew-symmetric and satisfies the Jacobi identity.

(ii) Brackets (22) and (3) form a Hamiltonian pair, i.e., their linear combinations are still Poisson brackets.

(iii) The invariant functions on $\mathfrak{g}$ are in involution with respect to (22). Brackets (22) and (3) generate a hierarchy of Hamiltonians in the sense of [19].

Statement (i) is a variant of the previous proposition. However, since it is not assumed that a group corresponding to the Lie algebra $\mathfrak{g}$ exists, an independent proof is needed. One may again use the Gel'fand–Dorfman technique, but with another choice of the underlying complex (for example, one can work with the cochain complex of the Lie algebra Vect $\mathfrak{g}$ of vector fields on $\mathfrak{g}$, with values in $C^\infty(\mathfrak{g})$). To prove (ii), assume first that $\mathfrak{g}$ is an algebra with identity. Pick $\lambda \in C$ and $h \in C^\infty(\mathfrak{g})$, and define the function h^λ by $h^\lambda(L + \lambda I) := h(L)$. Then $\{h_1^\lambda, h_2^\lambda\}(\lambda I + L) = \{h_1, h_2\}(L) + \lambda\{h_1, h_2\}_R(L)$, which shows that the right-hand side is a Poisson bracket. In the general situation we adjoin an identity to $\mathfrak{g}$ and note that the same formula (22) defines a Poisson bracket on the extended algebra $\mathfrak{g} + CI$, which is uniquely determined by its restriction to a hyperplane λ = const. To prove the involutiveness of the invariant functions on $\mathfrak{g}$ with respect to the bracket (22) one proceeds as in Theorem 1. The Hamiltonian equation defined by the Hamiltonian h with respect to (22) has the Lax form

$$\frac{dL}{dt} = [L, M], \quad M = R(XL), \quad X = \operatorname{grad} h(L). \tag{23}$$

Note that the invariance of h implies $XL = LX$. Now (iii) follows by comparing (6) and (23).

Theorem 13 shows that the quadratic bracket (22) is similar in its properties to the "second Hamiltonian structure" of Gel'fand–Dikii. We propose to the reader to verify that the latter can be obtained from (22) as a particular case, if one takes for $\mathfrak{g}$ the algebra of formal pseudodifferential operators [15] and defines $R \subset \operatorname{End} \mathfrak{g}$ by the rule:

$$RX = \begin{cases} X, & \text{if } X \text{ is a differential operator,} \\ -X, & \text{if } X \text{ is a Volterra operator.} \end{cases}$$

We return now to the group bracket (20) and discuss it in connection with integrable systems.

THEOREM 14. Let G be a Lie group with Lie algebra $\mathfrak{g}$. Suppose that there is a nondegenerate invariant inner product on $\mathfrak{g}$. Let $R \in \operatorname{End} \mathfrak{g}$ be a skew-symmetric operator satisfying the modified Yang–Baxter equation (12). Then:

(i) Central functions on *G* are in involution with respect to the Poisson bracket (20).

(ii) If G is a matrix group, then the equation of motion on G defined by a central function h is in the Lax form

$$\frac{dL}{dt} = [L, M], \quad M = R(L \operatorname{grad} h(L)). \tag{24}$$

In the general case, the equation of motion takes the form

$$\frac{dL}{dt} = \lambda_L M - \rho_L M, \quad M = R(\rho_L^* \operatorname{grad} h(L)). \tag{24'}$$

(iii) Let $G_\pm$, $K_\pm$ be the subgroups of G defined in the formulation of Theorem 11, and let $\theta: G_+/K_+ \to G_-/K_-$ be the Cayley transform of the operator R. The solution of Eq. (24) is given by the formula

$$L(t) = g_\pm(t) L_0 g_\pm(t)^{-1}, \tag{25}$$

where g_+, g_- is the solution of the factorization problem

$$\exp(2tL_0 \operatorname{grad} h(L_0)) = g_+(t) g_-(t)^{-1}, \quad g_\pm \in G_\pm, \quad \theta(g_+) = g_-. \tag{26}$$

(In the general case the one-parameter group is given by the formula $\exp 2t\rho_L^* (\operatorname{grad} h(L_0))$.)

Assertions (i) and (ii) are proved in Theorem 1. A simple differentiation shows that the curve (26) satisfies the differential equation (24). Unfortunately I was not able to carry over the proofs of [9, 11] to this case.

Remark. Theorem 14 does not hold for general Hamilton–Lie groups. In fact, in the absence of a canonical identification of the spaces $\mathfrak{g}$ and $\mathfrak{g}^*$, one does not even succeed in defining the one-parameter subgroup (26).

Let us give another variant of Theorem 14, fitted for finite-difference systems. Set $\mathbf{G} = G \times \ldots \times G$ (N copies), and define a Poisson structure on $\mathbf{G}$ as the product of the Poisson structures (20) on the factors. In tensor notation, $\{L_1^{(n)} \otimes L_2^{(m)}\} = [r, L_1^{(n)} \otimes L_2^{(m)}]\,\delta_{mn}$ (see [6]). Let $L = (L_1, \ldots, L_N) \in \mathbf{G}$ and set

$$\psi_m(L) = \widehat{\prod_{1\leqslant k<m}} L_k, \quad T(L) = \widehat{\prod_{1\leqslant k\leqslant N}} L_k.$$

Function ψ_m satisfies the difference equation $\psi_{m+1} = L_m\psi_m$, and $T(L)$ is the monodromy matrix of this system.

Proposition 15. The map $T: \mathbf{G} \to G: L \to T(L)$ preserves Poisson brackets.

This proposition is the starting point in the application of the orbit method to difference systems (see [6]). It asserts that the multiplication in G is a Hamiltonian map with respect to bracket (20) and serves as the main motivation for the term "Hamilton–Lie group" introduced in [5].

Let φ be a central function on G. Define a Hamiltonian $\mathbf{G}$ by $h(L) = \varphi(T(L))$.

THEOREM 16. Let G be a matrix group.

(i) The Hamiltonian equation on $\mathbf{G}$ with Hamiltonian h has the form

$$\frac{dL_m}{dt} = L_m M_m - M_{m+1} L_m, \quad M_m = R(\psi_m T(L)\,\mathrm{grad}\,\varphi(T(L))\,\psi_m^{-1}). \tag{27}$$

The monodromy matrix $T(L)$ satisfies Novikov's equation

$$\frac{dT}{dt} = [T, M], \quad M = R(T\,\mathrm{grad}\,\varphi(T)).$$

(ii) Let g_m^+, g_m^- be the solution of the factorization problem

$$\psi_m^0 \exp(2tT(L^0)\,\mathrm{grad}\,\varphi(T(L^0)))(\psi_m^0)^{-1} = g_m^+(t)(g_m^-(t))^{-1},$$
$$\psi_m^0 = \psi_m(L^0), \quad g_m^{\pm} \in G_{\pm}, \quad \theta(\overline{g_m^+}) = \overline{g_m^-}. \tag{28}$$

Then the solution of Eq. (27) with initial data $L^0 = (\overset{0}{L_1}, \ldots, \overset{0}{L_N})$ is given by

$$L_m(t) = g_{m+1}^{\pm}(t)\, L_m^0\, (g_m^{\pm}(t))^{-1}. \tag{29}$$

The case of general Lie groups differs from the one just considered only in notation: instead of (27) we must write

$$\frac{dL_m}{dt} = \lambda_{L_m} M_m - \rho_{L_m} M_{m+1}, \quad M_m = R(\mathrm{Ad}\,\psi_m\,\rho_{T(L)}^*\,\mathrm{grad}\,\varphi);$$

the other formulas undergo similar modifications.

Comment: the Hamiltonian h is invariant under the "gauge action" of the group $\mathbf{G}$ on itself, $g(L_1, \ldots, L_N) = (g_1 L_1 g_2^{-1}, \ldots, g_{k+1}^{-1}, {}_{+1}, \ldots, g_N L_N g_1^{-1})$. The right-hand side of the Lax equation (27) is an infinitesimal gauge transformation. The gradient of h, transported to the identity of the group, is the vector of $\overset{N}{\oplus}\,\mathfrak{g}$ with components $\psi_m T(L)\,\mathrm{grad}\,\varphi(T(L))\psi_m^{-1} \in \mathfrak{g}$, $m = 1, \ldots, N$.

Formulas (28) and (29) show that the dynamics of our system is given by dressing transformations in the terminology of Zakharov–Shabat [21].†

In order to apply Theorem 16 to concrete differential equations, one must first select a suitable Lie group and an r-matrix. The simplest case with interesting applications can

†For a group-theoretical treatment of "dressing" transformations, see [7].

be described as follows. Let $\overline{G}$ = GL(n, C) and put G = $\Omega_1\overline{G}$ — the loop group of $\overline{G}$, i.e., the group of $\overline{G}$-valued functions on the circle. The Lie algebra of G can be identified with $\mathfrak{g}$ = $gl(n) \otimes \mathbf{C}[\lambda, \lambda^{-1}]$. The simplest unitary r-matrix on $\mathfrak{g}$ is the already considered Hilbert transform (13). The factorization problem associated with the r-matrix (13) is the usual matrix Riemann problem.

The next topic we wish to discuss is the description of those submanifolds of the group G that are invariant under the Poisson bracket. Generally speaking, bracket (20), like the simpler Lie–Poisson bracket (3), is degenerate. The description of the minimal invariant submanifolds for the Lie–Poisson bracket is well known: they are all orbits of the coadjoint representation of G. No analogous result is known for the group bracket (20). However, let us mention the following simple fact.

Proposition 17. Let $\mathcal{O}_1, \mathcal{O}_2 \subset G$ be submanifolds invariant under the bracket (20). Then their product, $\mathcal{O}_1 \cdot \mathcal{O}_2 \subset G$ is also invariant.

Thus, the aforementioned classification problem reduces to the classification of the indecomposable invariant manifolds. The loop group G = $\Omega_1\overline{G}$ contains the subgroup G_{alg} = GL(n, C[λ, λ^{-1}]) (the group of points of GL(n) over the ring C[λ, λ^{-1}]). It is readily verified that the submanifold $G_{alg} \subset G$ is invariant under bracket (20). The minimal invariant submanifolds in G_{alg} correspond to difference systems whose Lax operators depend rationally on the spectral parameter. (Interesting examples can be found in [6, 17]). The factorization problem for such invariant submanifolds can be effectively solved [11].

6. Two-Dimensionalization

The integrable nonlinear equations for functions of two variables admit the zero-curvature representation, i.e., can be written as the condition that the curvature of some connection vanishes. An algebraic mechanism which produces nonlinear Hamiltonian equations admitting the zero-curvature representation was proposed in [8]. Below we discuss a generalization of this mechanism in the framework of the r-matrix method. The basic idea of [8] is easily accessible if one examines formula (6), which represents the generalization of the Lax equation to an arbitrary Lie algebra. In order that Eq. (6) express the vanishing of the curvature of a connection in a two-dimensional space with coordinates (x, t), it is necessary first of all that our Lie algebra be an algebra of functions of x with values in a suitable "smaller" Lie algebra $\bar{\mathfrak{g}}$ (which parametrizes the inner states of our system "at one point"). Furthermore, it is necessary that

$$\mathrm{ad}^* M \cdot L = \frac{dM}{dx} + [L, M]. \tag{30}$$

It is not hard to find a Lie algebra whose coadjoint action is given by gauge transformations: take a central extension of the algebra $\mathfrak{g} = C^\infty(S^1; \bar{\mathfrak{g}})$. According to the general scheme described in Sec. 1, we must find the invariants of the coadjoint action (30) and introduce on $\mathfrak{g}$ a structure of double Lie algebra by means of an appropriate r-matrix.

We now turn to a more formal discussion. Thus, let $\bar{\mathfrak{g}}$ be a Lie algebra equipped with a nondegenerate inner product. Let $\mathfrak{g} = C^\infty(S^1; \bar{\mathfrak{g}})$ denote the Lie algebra of $\bar{\mathfrak{g}}$-valued functions on the circle (0, 2π), with the pointwise commutator. The following formula defines an invariant inner product on $\mathfrak{g}$:

$$(X, Y) = \int_0^{2\pi} dx\,(X(x), Y(x)). \tag{31}$$

Next, ω(X, Y) = (X, dY/dx) is a nontrivial 2-cocycle on g (the "Maurer–Cartan cocycle"). Let $\hat{\mathfrak{g}} = \mathfrak{g} + \mathbf{C}$ be the central extension of $\mathfrak{g}$, associated with cocycle ω.

Proposition 18. Identify the Lie algebra $\mathfrak{g}$ with its dual space via the inner product (31).

(i) The coadjoint representation of the Lie algebra $\mathfrak{g}$ in the space $\hat{\mathfrak{g}}^* = \mathfrak{g} \oplus \mathbf{C}$ is given by the rule

$$\mathrm{ad}^* M \cdot (L, e) = \left([L, M] + e\frac{dM}{dx}, 0\right). \tag{32}$$

Consequently, the coadjoint representation leaves invariant the hyperplanes e = const and for e ≠ 0 it is given by infinitesimal gauge transformations.

(ii) Suppose $\bar{\mathfrak{g}}$ is a matrix Lie algebra with corresponding matrix Lie group $\bar{G}$, and set $G = C^\infty(S^1; \bar{G})$. Then the coadjoint representation of G in the space $\hat{\mathfrak{g}}^*$ is given by the rule

$$\mathrm{Ad}^* g \cdot (L, e) = \left(gLg^{-1} + e\frac{dg}{dx}g^{-1}, e\right). \tag{33}$$

Remark. The group $\hat{G}$ corresponding to algebra $\hat{\mathfrak{g}}$, is a rather intricate object: the bundle $\hat{G} \to G$ is not trivial. Fortunately, due to the fact that under the coadjoint representation the center acts trivially, it suffices to consider only the group G.

(iii) Consider the auxiliary linear equation

$$e\frac{d\psi}{dx} = L\psi, \quad L \in \mathfrak{g}. \tag{34}$$

Let $T(L) \in \bar{G}$ be the monodromy matrix of Eq. (34). Two pairs, (L, e) and (L', e'), (with $e \neq 0$), lie on the same orbit in $\hat{\mathfrak{g}}^*$ if and only if $e = e'$ and the matrices T(L) and T(L') are conjugate in $\bar{G}$.

COROLLARY. Let φ be a central function on $\bar{G}$. The functionals of the form $L \mapsto \varphi(T(L))$ on $\hat{\mathfrak{g}}^*$ are gauge-invariant and generate the ring of invariant functionals on the hyperplane $e = \mathrm{const} \neq 0$.

Now suppose that $\bar{\mathfrak{g}}$ is endowed with a structure of double-Lie algebra. Extend by linearity the operator $R \in \mathrm{End}\, \bar{\mathfrak{g}}$ to $\mathfrak{g} = C^\infty(S^1; \bar{\mathfrak{g}})$. This yields a structure of double Lie algebra on $\mathfrak{g}$. We have the following simple but general result.

Proposition 19. Let $(\mathfrak{g}, R)$ be a double Lie algebra, and let ω be a 2-cocycle on $\mathfrak{g}$. Suppose that R satisfies the modified Yang–Baxter equation (12). Then

$$\omega_R(X, Y) = \omega(RX, Y) + \omega(X, RY) \tag{35}$$

is a cocylce on $\mathfrak{g}_R$.

Therefore, the central extension of $\mathfrak{g}$ associated with ω is also equipped with a structure of double Lie algebra. In the case in which we are interested, the cocycle is given by an unbounded skew-adjoint operator; moreover, it is required that its domain be invariant under R. Theorem 1 yields a collection of Hamiltonians in involution on $\hat{\mathfrak{g}}_R^*$ and the equation of motion (6) takes the zero-curvature form.

In applications one is interested in local Hamiltonians and integrals of motions. Generally speaking, the functionals $\varphi(T(L))$ defined in Proposition 18 are not local. Suppose now that $\bar{\mathfrak{g}}$ is an algebra of functions of the spectral parameter λ. Under this assumption, the usual technique of "asymptotic expansion of the Lax operator at a pole" may be used to produce local functionals (see [8, 12] for the exact assertions).

The Lie–Poisson bracket of algebra $\mathfrak{g}_R = C^\infty(S^1; \bar{\mathfrak{g}}_R)$ can be expressed as $\{\varphi_1, \varphi_2\}(L) = \int_0^{2\pi} (L(x), [X_1(x), X_2(x)])\, dx$, $X_i = \mathrm{grad}\, \varphi_i(L)$, and hence it is generally speaking ultralocal. (More formally, the Hamiltonian operator defining this bracket involves no differentiations with respect to x.) Clearly, the Lie–Poisson bracket on algebra $\hat{\mathfrak{g}}_R$ is ultralocal if and only if $\omega_R = 0$. A simple criterion is given by the following proposition.

Proposition 20. The Lie–Poisson bracket of algebra $\hat{\mathfrak{g}}_R$ is ultralocal if and only if $R = -R^*$.

In fact, $R(dY/dx) = (d/dx)RY$, and so the equalities $\omega_R = 0$ and $R + R^* = 0$ are equivalent.

When the r-matrix $R \in \mathrm{End}\, \bar{\mathfrak{g}}$ is unitary, there is an associated Poisson bracket on the group $\bar{G}$.

Proposition 21. Equip the group $\bar{G}$ with the Poisson bracket (20) constructed from the operator R. The mapping $\hat{\mathfrak{g}}_R^* \to \bar{G}\colon (L, e) \mapsto T(L)$ is Hamiltonian.

This statement represents a "continuous version" of Proposition 15 (cf. [2, 6]).

Theorems 1 and 11 apply in the nonultralocal case too. However, in this case the Poisson brackets of the elements of the monodromy are no longer defined. In fact, consider the functionals of the form $h_\Phi(L) = \mathrm{tr}\, \Phi T(L)$, $\Phi \in \bar{G}$. By the definition of the Lie–Poisson bracket

$$\{h_1, h_2\}_R (L, e) = \int dx \operatorname{tr} L ([X_1, X_2]_R) + e\omega_R (x_1, x_2), \quad X_i = \operatorname{grad} h_i.$$

It is readily verified that the gradient of h_Φ equals $X_\Phi(x) = \psi(x)\Phi T(N)\psi(x)^{-1}$, where ψ is the fundamental solution of the differential equation $e(d\psi/dx) = L\psi$, $\psi(0) = I$. Obviously,

$$X_\Phi (x + 2\pi) = T (L) X_\Phi (x) T (L)^{-1}. \tag{36}$$

The operator d/dx, defined on smooth functions satisfying the boundary condition (36), is essentially self-adjoint. However, its domain is not invariant under the operator R, and for this reason the bilinear form (35) does not satisfy the Jacobi identity and is not a cocycle on the Lie algebra of those functions which satisfy (36).

Proposition 22. The elements of the monodromy matrix of the linear equation (34) are smooth functions on $\hat{\mathfrak{g}}_R^*$ if and only if either $\omega_R = 0$ or $R = 1$ (the second case is obviously of no interest).

Note that the functionals of the eigenvalues of the monodromy matrix are always smooth. Indeed, their gradients are periodic functions on the circle; the space of periodic functions is clearly invariant under R.

7. Classification of r-Matrices for Finite-Dimensional Semisimple Lie Algebras

The classification of unitary r-matrices satisfying the modified Yang–Baxter equations for finite-dimensional semisimple Lie algebras is given in [4]. If one drops the unitarity assumption, the class of solutions becomes considerably larger (for example, the solutions described in Proposition 5 are, as a rule, nonunitary). The closest to the Belavin–Drinfel'd classification is the class of graded r-matrices that we describe in the present section. Our main purpose is to emphasize the simple, geometric ideas of the construction, in particular, the connection with von Neumann's theory of extensions of linear operators. For this reason, we do not strive for high generality. The qualified reader should extend our results to affine Lie algebras with no difficulty.

Let $\mathfrak{g}$ be an arbitrary complex Lie algebra, and let $R \in \operatorname{End} \mathfrak{g}$ satisfy identity (12). Set $\mathfrak{g}_\pm \neq \operatorname{Im}(R \pm 1)$, and $\mathfrak{k}_\pm = \operatorname{Ker}(R \mp 1)$. Let R_0 denote the partial linear operator in $\mathfrak{g}$ with domain $D_0 = \mathfrak{k}_+ \oplus \mathfrak{k}_-$ and acting as $R_0(X_+ - X_-) = X_+ + X_-$. Clearly, operator R is an extension of R_0. The problem of the description of all r-matrices that satisfy (12) and are such that $\operatorname{Im}(R \pm 1)$ and $\operatorname{Ker}(R \pm 1)$ are given subalgebras, is conveniently divided into two subproblems:

a) Describe all the extensions of the operator R_0 such that $\operatorname{Im}(R \pm 1) = \mathfrak{g}_\pm$, and $\operatorname{Ker}(R \pm 1) = \mathfrak{k}_\mp$.

b) Isolate among these extensions the operators which satisfy identity (12).

Problem a) is standard in the theory of extensions (see, for example, [22]).

Let us fix splittings $\mathfrak{g}_\pm = \mathfrak{m}_\pm \dotplus \mathfrak{k}_\pm$. An operator $\theta \in \operatorname{Hom}(\mathfrak{m}_+, \mathfrak{m}_-)$ will be said to be regular if the subspace $(1 - \theta)\mathfrak{m}_+ \subset \mathfrak{g}$ complements $\mathfrak{k}_+ \oplus \mathfrak{k}_-$ (note that $\dim \mathfrak{m}_+ = \operatorname{codim}(\mathfrak{k}_+ \oplus \mathfrak{k}_-)$).

Proposition 23. (i) The linear extensions of the operator R_0 for which $\operatorname{Im}(R + 1) = \mathfrak{g}_+$ and $\operatorname{Im}(R - 1) \subseteq \mathfrak{g}_-$ are given by the von Neumann formulas

$$\begin{gathered} X = X_+ - X_- + (1 - \theta) X_0; \quad X_\pm \in \mathfrak{k}_\pm; \quad X_0 \in \mathfrak{m}_+; \\ R_0 X = X_+ + X_- + (1 + \theta) X_0, \end{gathered} \tag{37}$$

where $\theta \in \operatorname{Hom}(\mathfrak{m}_+, \mathfrak{m}_-)$ is regular.

(ii) Suppose that θ is a linear isomorphism; then $\operatorname{Im}(R_\theta - 1) = \mathfrak{g}_-$, $\operatorname{Ker}(R_\theta - 1) = \mathfrak{k}_+$.

(iii) The operator R_θ satisfies identity (12) if and only if $[\theta X, \theta Y] \bmod \mathfrak{k}_- = \theta([X, Y] \bmod \mathfrak{k}_+)$.

If $\mathfrak{m}_+ = \mathfrak{m}_-$, (37) becomes simpler:

$$X = X_+ - X_- + X_0; X_\pm \in \mathfrak{k}_\pm, X_0 \in \mathfrak{m}, R_\theta X = X_+ + X_- + (1 + \theta)(1 - \theta)^{-1} X_0, \ \det \theta \neq 0, \ \det (1 - \theta) \neq 0.$$

Now let $\mathfrak{g}$ be a finite-dimensional semisimple Lie algebra. Fix a Cartan subalgebra $\mathfrak{h} \subset \mathfrak{g}$

and a system of simple roots $P \in \mathfrak{h}^*$. Define $x_0 \in \mathfrak{h}$ by the conditions $\alpha(x_0) = 1$, $\alpha \in P$, and let $\mathfrak{g}_j = \{X \in \mathfrak{g};\ \mathrm{ad}\, x_0 \cdot X = jX\}$. Then $\mathfrak{g} = \bigoplus_{j\in\mathbb{Z}} \mathfrak{g}_j$ and $[\mathfrak{g}_i, \mathfrak{g}_j] \subset \mathfrak{g}_{i+j}$.

Definition. An operator $R \in \mathrm{End}\,\mathfrak{g}$ is graded if $[R, \mathrm{ad}\, x_0] = 0$. A subspace $\mathfrak{v} \subset \mathfrak{g}$ is graded if $\mathfrak{v} = \bigoplus_j \mathfrak{v} \cap \mathfrak{g}_j$.

A subalgebra $\mathfrak{p} \subset \mathfrak{g}$ is said to be parabolic if it contains a Borel subalgebra. A parabolic subalgebra $\mathfrak{p}$ is graded if $\mathfrak{p} \supset \mathfrak{h}$.

Choose a graded parabolic subalgebra $\mathfrak{p}$; let $\mathfrak{b} \subset \mathfrak{p}$ be the corresponding Borel subalgebra with opposite Borel subalgebra $\bar{\mathfrak{b}}$ and let τ be the Cartan involution in $\mathfrak{g}$, which interchanges $\mathfrak{b}$ and $\bar{\mathfrak{b}}$. Further, let $\mathfrak{n} \subset \mathfrak{p}$ be the nilradical, and $\mathfrak{m} = \mathfrak{p} \cap \tau\mathfrak{p}$. Then the subalgebra $\mathfrak{m} \subset \mathfrak{p}$ is reductive, $\mathfrak{p} = \mathfrak{m} \dotplus \mathfrak{n}$, and the subalgebras $\mathfrak{m}$ and $\mathfrak{n}$ are graded. Let $\mathfrak{m}'$ be the semisimple component of $\mathfrak{m}$ and let $\mathfrak{h}_{\mathfrak{p}} \subset \mathfrak{h}$ be the center of $\mathfrak{m}$.

Two parabolic algebras, $\mathfrak{p}$ and $\mathfrak{q}$ are said to be associated if subalgebras $\mathfrak{h}_{\mathfrak{p}}$ and $\mathfrak{h}_{\mathfrak{q}}$ are conjugate in $\mathfrak{g}$.

Let $R \in \mathrm{End}\,\mathfrak{g}$ be a graded r-matrix satisfying identity (12). Obviously, in this case the subalgebras $\mathfrak{g}_\pm$ and $\mathfrak{k}_\pm$ are graded, and one can also choose graded splitting subspaces $\mathfrak{m}_\pm \subset \mathfrak{g}_\pm$. An operator $R_\theta \supset R_0$ is graded only if $\theta \in \mathrm{Hom}(\mathfrak{m}_+, \mathfrak{m}_-)$ is a graded map of degree zero.

We shall say that a given r-matrix is subordinate to the pair $(\mathfrak{p}_+, \mathfrak{p}_-)$ of parabolic subalgebras if $\mathfrak{g}_\pm \subset \mathfrak{p}_\pm$, $\mathfrak{k}_\pm \supset \mathfrak{n}_\pm$.

Proposition 24. Every graded r-matrix satisfying identity (12) is subordinated to some pair $(\mathfrak{p}_+, \mathfrak{p}_-)$ of parabolic subalgebras.

Modifying, if necessary, the order in the root system, we may assume that $\mathfrak{p}_+$ contains a positive Borel subalgebra, whereas the opposite Borel subalgebra lies in $\mathfrak{p}_-$.

Proposition 25. Subalgebras $\mathfrak{p}_+$ and $\mathfrak{p}_-$ are associated. Subalgebras $\mathfrak{h}_{\mathfrak{p}_+}, \mathfrak{h}_{\mathfrak{p}_-} \subset \mathfrak{h}$ are conjugate with respect to the Weyl group.

We distinguish two possible cases.

1) Subalgebras $\mathfrak{h}_{\mathfrak{p}_+}$ and $\mathfrak{h}_{\mathfrak{p}_-}$ coincide. In this case $\mathfrak{p}_+ = \tau\mathfrak{p}_-$, i.e., $\mathfrak{p}_+$ and $\mathfrak{p}_-$ are opposite.

2) Subalgebras $\mathfrak{h}_{\mathfrak{p}_+}$ and $\mathfrak{h}_{\mathfrak{p}_-}$ do not coincide.

Suppose that $\mathfrak{h}_{\mathfrak{p}_+} = \mathfrak{h}_{\mathfrak{p}_-}$.

Proposition 26. For a generic r-matrix subordinate to the pair $(\mathfrak{p}_+, \mathfrak{p}_-)$. either $\mathrm{Im}\,(R + 1) = \mathfrak{p}_+$ and $\mathrm{Im}\,(R - 1) = \mathfrak{h}_{\mathfrak{p}} + \mathfrak{n}_-$, or $\mathrm{Im}\,(R + 1) = \mathfrak{h}_{\mathfrak{p}} + \mathfrak{n}_+$ and $\mathrm{Im}\,(R - 1) = \mathfrak{p}_-$. (Here "generic" means that the set of all r-matrices for which these identities do not hold has positive codimension in the set of all r-matrices subordinate to the pair $(\mathfrak{p}_+, \mathfrak{p}_-)$.) In the first case, the r-matrices are described by the formulas $X = X_+ - X_- + X_0$; $X_+ \in \mathfrak{m}' + \mathfrak{n}_+$; $X_- \in \mathfrak{n}_-$; $X_0 \in \mathfrak{h}_{\mathfrak{p}}$; $R_\theta X = X_+ + X_- + (1+\theta)(1-\theta)^{-1}X_0$; $\theta \in \mathrm{End}\,\mathfrak{h}_{\mathfrak{p}}$; $\det\theta \neq 0$; $\det(1-\theta) \neq 0$.

The second case is reduced to the first by inverting the order in the root system.

If the subalgebras $\mathfrak{h}_{\mathfrak{p}_+}$ and $\mathfrak{h}_{\mathfrak{p}_-}$ do not coincide, then, for a generic r-matrix subordinate to a pair $(\mathfrak{p}_+, \mathfrak{p}_-)$, one has $\mathfrak{g}_\pm = \mathfrak{p}_\pm$ and $\mathfrak{k}_\pm = \mathfrak{n}_\pm$.

Let W be the Weyl group of algebra $\mathfrak{g}$. Let G and $H \subset G$ be respectively the adjoint Lie group of $\mathfrak{g}$ and the Cartan subgroup corresponding to the subalgebra $\mathfrak{h} \subset \mathfrak{g}$. We shall say that an element $x \in G$ represents an element $s \in W$ if $\mathrm{Ad}\, x|_{\mathfrak{h}} = s$.

Proposition 27. Let $\mathfrak{m}_\pm = \mathfrak{p}_\pm \cap \tau\mathfrak{p}_\pm$ be the reductive component of $\mathfrak{p}_\pm$ and let $\theta \in \mathrm{Hom}(\mathfrak{m}_+, \mathfrak{m}_-)$ be a graded isomorphism of Lie algebras. Then, $\theta = \theta_0 \dotplus \theta'$. $\theta_0 \in \mathrm{Hom}(\mathfrak{h}_{\mathfrak{p}_+}, \mathfrak{h}_{\mathfrak{p}_-})$, $\theta' \in \mathrm{Hom}(\mathfrak{m}'_+, \mathfrak{m}'_-)$. Let $s \in W$ be an element of the Weyl group such that $s\mathfrak{h}_{\mathfrak{p}_+} = \mathfrak{h}_{\mathfrak{p}_-}$ The isomorphism θ' is the restriction to $\mathfrak{m}'_+$ of an inner automorphism $\mathrm{Ad}\, x$, where x represents $s \in W$. Therefore, θ' is uniquely defined up to an inner automorphism $\mathrm{Ad}\, h$, $h \in H$.

We let $P_\pm$ denote the projection on $\mathfrak{m}_\pm$ parallel to a graded complement in $\mathfrak{g}$.

LEMMA. An operator $\theta \in \mathrm{Hom}(\mathfrak{m}_+, \mathfrak{m}_-)$ is regular if and only if the operators $1 - P_+\theta \in \mathrm{End}\,\mathfrak{m}_+$, $1 - P_-\theta^{-1} \in \mathrm{End}\,\mathfrak{m}_-$ are nondegenerate.

COROLLARY. If $\theta_0 \in \mathrm{Hom}(\mathfrak{h}_{\mathfrak{p}_+}, \mathfrak{h}_{\mathfrak{p}_-})$ is a linear isomorphism, then the operator θ described in Proposition 27 is regular.

Proposition 28. In case 2), a generic operator subordinate to the pair $(\mathfrak{p}_+, \mathfrak{p}_-)$, is given by the formulas $X = X_+ - X_- + (1 - \theta')X_0' + (1 - \theta_0)X_0$, $X_\pm \in \mathfrak{n}_\pm$, $X_0' \in \mathfrak{m}_+'$, $X_0 \in \mathfrak{h}_{\mathfrak{r}_+}$, $R_\theta X = X_+ + X_- + (1 + \theta')X_0' + (1 + \theta_0)X_0$.

The formulas for singular operators subordinate to $(\mathfrak{p}_+, \mathfrak{p}_-)$, are analogous, but more complicated, and we omit them. We leave to the reader the useful exercise of clarifying the connection between the results discussed here and the approach in [4].

LITERATURE CITED

1. E. K. Sklyanin, "On complete integrability of the Landau–Lifshitz equation," Preprint LOMI, E-3-79, Leningrad, LOMI (1980).
2. E. K. Sklyanin, "The quantum inverse scattering method," Zap. Nauchn. Sem. LOMI, 95, 55-128 (1980).
3. A. A. Belavin and V. G. Drinfel'd, "On the solutions to the classical Yang–Baxter equation for simple Lie algebras," Funkts. Anal. 16, No. 3, 1-29 (1982).
4. A. A. Belavin and V. G. Drinfel'd, "Triangle equations and simple Lie algebras," Preprint ITF 1982-18, Chernogolovka, ITF.
5. V. G. Drinfel'd, "Hamiltonian structures on Lie groups, Lie bialgebras, and the geometrical meaning of the Yang–Baxter equations," Dokl. Akad. Nauk SSSR, 268, No. 2, 285-287 (1983).
6. L. D. Faddeev, "Integrable models in 1 + 1 dimensional quantum field theory," Preprint S.Ph.T. 82/76, CEN Saclay (1982).
7. M. A. Semenov-Tyan-Shanskii, "Classical r-matrices and the orbit method," Zap. Nauchn. Sem. LOMI, 123, 77-91 (1983).
8. A. G. Reyman, and M. A. Semenov-Tyan-Shanskii, "Algebras of currents and nonlinear partial differential equations," Dokl. Akad. Nauk SSSR, 251, No. 6, 1310-1313 (1980).
9. A. G. Reyman, "Integrable Hamiltonian systems connected with graded Lie algebras," Zap. Nauchn. Sem. LOMI, 95, 3-54 (1980).
10. M. Adler and P. van Moerbeke, "Completely integrable systems, Euclidean Lie algebras, and curves," Adv. Math., 38, No. 2, 267-317 (1980).
11. A. G. Reyman and M. A. Semenov-Tyan-Shanskii [Semenov-Trian-Shansky], "Reduction of Hamiltonian systems, affine Lie algebras, and Lax equations, I, II," Invent. Math., 54, No. 1, 81-100 (1979), and 63, No. 3, 423-432 (1981).
12. P. P. Kulish and A. G. Reyman, "Hamiltonian structure of polynomial bundles," Zap. Nauchn. Sem. LOMI, 123, 67-76 (1983).
13. S. Lie (Unter Mitwirkung von F. Engel), Theorie der Transformationsgruppen, Bd. 1-3, Teubner, Leiptzig (1888, 1890, 1893).
14. B. Kostant, "Quantization and representation theory," in: Proc. Research Symp. on Representations of Lie groups, Oxford 1977, London Math. Soc. Lecture Notes Series, Vol. 34 (1979).
15. M. Adler, "On a trace functional for formal pseudodifferential operators and the symplectic structure for the KdV type equations," Invent. Math., 50, No. 2, 219-248 (1979).
16. N. I. Muskhelishvili, Singular Integral Equations [in Russian], Nauka, Moscow (1968).
17. E. K. Sklyanin, "Some algebraic structures connected with the Yang–Baxter equation," Funkts. Anal., 16, No. 4, 27-34 (1982).
18. I. M. Gel'fand and I. Ya. Dorfman, "Hamiltonian operators and the classical Yang–Baxter equation," Funkts. Anal., 16, No. 4, 1-9 (1982).1-9 (1982).
19. I. M. Gel'fand and I. Ya. Dorfman, "Schouten brackets and Hamiltonian operators," Funkts. Anal., 14, No. 3, 71-74 (1980).
20. I. M. Gel'fand and L. A. Dikii, "A family of Hamiltonian structures connected with nonlinear integrable equations," Preprint IPM Akad. Nauk SSSR, No. 136, IPM, Moscow (1978).
21. V. E. Zakharov and A. B. Shabat, "Integration of nonlinear equations of mathematical physics by the method of the inverse scattering problem II," Funkts. Anal., 13, No. 3, 13-22 (1979).
22. I. M. Glazman and Yu. I. Lyubich, Finite-Dimensional Linear Analysis in Problems [in Russian], Nauka, Moscow (1969).

QUANTUM GROUPS

3. Quantum Groups

Through the development of QISM it became apparent that upon quantization of classical systems some of the structures undergo quantum deformation [108–109),145)]. In the course of constructing trigonometric solutions of YBE, Kulish and Reshetikhin [141)] introduced a deformation of the universal enveloping algebra of $\mathfrak{sl}(2)$. In the reprint [11] Sklyanin defines an elliptic version in connection with the eight-vertex model, and found three series of representations of this algebra. These are the first occurrences of new algebraic objects, now called 'quantum groups'.

The example of Kulish-Reshetikhin [141)] has subsequently been generalized to include arbitrary simple or affine Lie algebras in the reprints [12],[14]. As for the generalization of the Sklyanin algebra to $\mathfrak{sl}(n)$, see [60–64)]. Drinfel'd [12] observed that Hopf algebras afford an appropriate language for the description of quantum groups as well as for the whole QISM. The full scope of the theory of quantum groups is expounded in Drinfel'd's report [13]. In fact quantum groups have also been discovered quite independently by Woronowicz [260–263)] from the viewpoint of operator algebras. Faddeev, Reshetikhin and Takhtajan [15],[86),247)] proposed another approach to quantum groups in the spirit closer to the original QISM.

The discovery of quantum groups has stimulated many works concerning the related algebraic structures[10),50–51),76),105–107),169–170),195–196),219–220),241–244,264)], the super-analog[140)], representation theory[47),52–54),68),95),104),133),156),239),248–252)] (including the features arising from the special values of q[74),157–160)]), non-commutative geometry [171–172),178–180)], harmonic analysis and special functions[102),132),134–137),174–177),184–186),204),257)]. One of the reasons for the widespread interest in quantum groups is their role in the theory of link invariants, and also in conformal field theory. As for these topics the reader is referred to the reprint volume by Kohno mentioned in the introduction (see also [82),200),236)]). At present there is no survey article available that covers the latest developments. For this reason we have also included in the list of references some quantum-group papers that may not be directly connected to YBE.

SOME ALGEBRAIC STRUCTURES CONNECTED WITH THE YANG–BAXTER EQUATION

E. K. Sklyanin UDC 517.43+519.46

One of the strongest methods of investigating the exactly solvable models of quantum and statistical physics is the quantum inverse problem method (QIPM; see the review papers [1-3]). The problem of enumerating the discrete quantum systems that can be solved by the QIPM reduces to the problem of enumerating the operator-valued functions L(u) that satisfy the relation

$$R(u-v)L'(u)L''(v) = L''(v)L'(u)R(u-v) \tag{1}$$

for a fixed solution R(u) of the so-called quantum Yang–Baxter equation

$$R_{12}(u-v)R_{13}(u)R_{23}(v) = R_{23}(v)R_{13}(u)R_{12}(u-v). \tag{2}$$

Here we use the notation $L' = L \otimes 1$, $L'' = 1 \otimes L$ (see [1, 3]). More detailed information concerning equations (1) and (2) and the notation used here can be found in the review papers to which we have already referred.

In the classical case, (1) is replaced by the equation

$$\{L'(u), L''(v)\} = [r(u-v), L'(u)L''(v)]_-, \tag{3}$$

while (2) becomes the classical Yang–Baxter equation

$$[r_{12}(u-v), r_{13}(u)]_- + [r_{12}(u-v), r_{23}(v)]_- + [r_{13}(u), r_{23}(v)]_- = 0. \tag{4}$$

Here we use {,} to denote the Poisson bracket, and $[A, B]_- = AB - BA$ stands for the commutator of the matrices A and B. We shall also make use of the notation $[A, B]_+ = AB + BA$ for the anticommutator.

The problem of enumerating the solutions to equations (1) and (3) has received little attention. This contrasts with the intense study of both the quantum and classical Yang–Baxter equations, which has led to a number of successes. These have revealed, in particular, the deep relationship between the Yang–Baxter equation, the theory of Lie groups [4, 5], and algebraic geometry [6, 7]. However, important results were obtained in [8, 9], where solutions to (1) and (3) corresponding to lattice versions of the nonlinear Schrödinger and sine–Gordon equations were found.

The present paper is devoted to a study of equations (1) and (3) in the case when R(u) and r(u) are, respectively, the simplest solution to equation (1), found by R. Baxter [10], and its classical analog [11]. During our investigation it turned out that it is necessary to bring into the picture new algebraic structures, namely, the quadratic algebras of Poisson brackets and the quadratic generalization of the universal enveloping algebra of a Lie algebra. The theory of these mathematical objects is surprisingly reminiscent of the theory of

---

V. A. Steklov Mathematics Institute, Leningrad Branch, Academy of Sciences of the USSR. Translated from Funktsional'nyi Analiz i Ego Prilozheniya, Vol. 16, No. 4, pp. 27-34, October-December, 1982. Original article submitted May 24, 1982.

Lie algebras, the difference being that it is more complicated. In our opinion, it deserves the greatest attention of mathematicians.

In the basic text we shall use the following convention on indices: Latin indices a, b, c run over the values 0, 1, 2, 3, while the Greek indices α, β, γ run over the values 1, 2, 3. In formulas (11)-(13), (22), (26), and (30)-(32), the triple of indices α, β, γ denotes a cyclic permutation of (1 2 3).

1. Classical Case

Let r(u) be the simplest solution to equation (4), obtained in [11]:

$$r(u)=\sum_{\alpha=1}^{3} w_\alpha(u)\sigma_\alpha\otimes\sigma_\alpha, \tag{5}$$

where σ_α are the standard Pauli matrices

$$\sigma_1=\begin{pmatrix}0&1\\1&0\end{pmatrix},\quad \sigma_2=\begin{pmatrix}0&-i\\i&0\end{pmatrix},\quad \sigma_3=\begin{pmatrix}1&0\\0&-1\end{pmatrix},$$

and the coefficients $w_\alpha(u)$ can be expressed in terms of the Jacobi elliptic functions

$$w_1(u)=\rho\frac{1}{sn\,(u,k)},\quad w_2(u)=\rho\cdot\frac{dn\,(u,k)}{sn\,(u,k)},\quad w_3(u)=\rho\frac{cn\,(u,k)}{sn\,(u,k)} \tag{6}$$

($\rho > 0$ and $k\in[0,\ 1]$ are fixed).

Notice that the coefficients w_α lie on a quadratic

$$w_\alpha^2-w_\beta^2=J_{\alpha\beta}, \tag{7}$$

which is uniformized by the parameter u. The constants $J_{\alpha\beta}$, which can be easily expressed through ρ and k using (6), satisfy the obvious equality

$$J_{12}+J_{23}+J_{31}=0. \tag{8}$$

In the sequel it will be convenient to represent $J_{\alpha\beta}$ in the form

$$J_{\alpha\beta}=J_\alpha-J_\beta, \tag{9}$$

where the constants J_α are determined modulo the transformation $J_\alpha \mapsto J_\alpha + c$.

We shall seek a solution L(u) to equation (3) in the form

$$L(u)=S_0+i\sum_{\alpha=1}^{3} w_\alpha(u)S_\alpha\sigma_\alpha, \tag{10}$$

where the S_a are, for the moment, unknown. Now substitute (5) and (10) into (3) and take advantage of the easily verifiable identities

$$w_\alpha(u-v)w_\gamma(v)-w_\beta(u-v)w_\gamma(u)=w_\alpha(u)w_\beta(v), \tag{11}$$

$$w_\alpha(u-v)w_\beta(u)w_\alpha(v)-w_\beta(u-v)w_\alpha(u)w_\beta(v)=-J_{\alpha\beta}w_\gamma(u). \tag{12}$$

Observe that, in fact, one needs to verify only (12), because (11) is equivalent to the Yang-Baxter equation (4). In the end, we obtain the following quadratic algebra of Poisson brackets for the variables S_a:

$$\{S_\alpha,S_0\}=2J_{\beta\gamma}S_\beta S_\gamma,\quad \{S_\alpha,S_\beta\}=-2S_0S_\gamma. \tag{13}$$

Here is the place to give a number of general definitions concerning algebras of Poisson brackets.

Definition 1. A *Poisson brackets algebra* (PBA) is a set endowed with two structures: one of a commutative ring over the field C with multiplication f, g $\mapsto$ fg = gf, and one of a Lie algebra with binary operation (Poisson bracket) f, g $\mapsto$ {f, g} = −{g, f}, the two of them being related by the Leibniz relation {fg, h} = f{g, h} + g{f, h}.

Definition 2. The *center* Z of a PBA is its center as a Lie algebra relative to the operation {,}.

Definition 3. A PBA is said to be *nondegenerate* if its center is one-dimensional.

Definition 4. A *homogeneous Poisson brackets Lie algebra* (HPBA) is a polynomial algebra with the generators $\{x_j\}_1^N$ and with the Poisson bracket

$$\{x_j, x_k\} = C_{jk}(x), \tag{14}$$

where the $C_{jk}(x)$ are homogeneous polynomials of degree n in the generators x_j satisfying $C_{jk}(x) = -C_{kj}(x)$ and the Jacobi identity

$$\{\{x_j, x_k\}, x_l\} + \{\{x_k, x_l\}, x_j\} + \{\{x_l, x_j\}, x_k\} = 0. \tag{15}$$

In particular, a quadratic HPBA is described by the relations

$$\{x_j, x_k\} = \sum_{l, m=1}^{N} c_{jk}^{lm} x_l x_m, \tag{16}$$

where the tensor of the structure constants c_{jk}^{lm} must have the symmetries

$$c_{jk}^{lm} = -c_{kj}^{lm} = c_{jk}^{ml} = -c_{kj}^{ml} \tag{17}$$

and satisfy the system of quadratic equations

$$\sum_{m=1}^{N} c_{jk}^{mn} c_{ml}^{pq} + c_{kl}^{mn} c_{mj}^{pq} + c_{lj}^{mn} c_{mk}^{pq} = 0, \tag{18}$$

which ensures that the Jacobi identity (15) is fulfilled.

Let us give several examples of HPBAs.

1. When n = 0, any HPBA reduces to a linear symplectic structure [12]

$$\{x_j, x_k\} = c_{jk}, \quad c_{jk} = -c_{kj}. \tag{19}$$

2. A linear HPBA is given by relations

$$\{x_j, x_k\} = \sum_{l=1}^{N} c_{jk}^{l} x_l, \tag{20}$$

where c_{jk}^{l} is the tensor of the structure constants of some Lie algebra (the Berezin–Kirillov–Kostant symplectic structure [12]).

Let us return to the Poisson brackets (13).

Proposition 1. Relations (13) define a quadratic HPBA (which we shall denote by $\mathcal{P}$).

To carry out the proof, it suffices to verify that the identities (17) and (18) are satisfied. A straightforward computation shows that (17) always holds, while (18) is valid provided (8) is satisfied, and this is true because of the definition (7) of the $J_{\alpha\beta}$. On the other hand, one can show that the Jacobi identity for (13) is a consequence of the classical Yang–Baxter equation (4). Indeed, take $L'(u_1) = L(u_1) \otimes 1 \otimes 1$, $L''(u_2) = 1 \otimes L(u_2) \otimes 1$, and $L'''(u_3) = 1 \otimes 1 \otimes L(u_3)$, and consider the expression $\{\{L'(u_1), L''(u_2)\}, L'''(u_3)\}$. Applying (3) and (4), we obtain the Jacobi identity for the operators $L'(u_1)$, $L''(u_2)$, and $L'''(u_3)$. Now, taking the residues at the poles of the coefficients $w_\alpha(u)$, we are led to the required Jacobi identity for the variables S_α.

Let $\mathcal{P}^*$ be the extension of algebra $\mathcal{P}$ consisting of analytic functions in the arguments S_α.

THEOREM 1. The center $Z_{\mathcal{P}^*}$ of algebra $\mathcal{P}^*$ consists of functions of the two quadratic central functions K_0 and K_1:

$$K_0 = \sum_{\alpha=1}^{3} S_\alpha^2, \quad K_1 = S_0^2 + \sum_{\alpha=1}^{3} J_\alpha S_\alpha^2 \tag{21}$$

(notice that under a translation $J_\alpha \mapsto J_\alpha + c$ of all J_α's, K_1 changes to $K_1 + cK_0$).

Proof. Let $f \in Z_{\mathcal{P}}$. This means that $\{f, S_\alpha\} = 0$, or, by virtue of (13), that

$$\{f, S_0\} = \sum_{\alpha=1}^{3} \frac{\partial f}{\partial S_\alpha} J_{\beta\gamma} S_\beta S_\gamma = 0,$$
$$\{f, S_\alpha\} = -\frac{\partial f}{\partial S_0} J_{\beta\gamma} S_\beta S_\gamma + \frac{\partial f}{\partial S_\beta} S_0 S_\gamma - \frac{\partial f}{\partial S_\gamma} S_0 S_\beta = 0. \quad (22)$$

The general solution of the system of linear homogeneous equations (22) is of the form $df = f_0 dK_0 + f_1 dK_1$ (provided (8) holds), which immediately proves the theorem.

Problem. Show that the center $Z_{\mathcal{P}}$ of the algebra $\mathcal{P}$ consists of polynomials in K_0 and K_1.

COROLLARY. Let $\Gamma(K_0, K_1)$ be the two-dimensional algebraic variety defined by equations (21) for fixed K_0 and K_1. Then the PBA $\mathcal{P}_\Gamma$, defined as the restriction of algebra $\mathcal{P}$ to Γ, is nondegenerate in the sense of Definition 2.

The topology of Γ depends, in an essential way, upon the choice of the parameters K_0 and K_1. Notice that for the parametrization (6) that we selected, one has $J_1 \geqslant J_2 \geqslant J_3$, whence S_0^2 lies on the segment $[K_1 - K_0J_1, K_1 - K_0J_3]$.

a) $K_1 \in (K_0J_1, \infty)$, b) $K_1 \in (K_0J_3, K_0J_1)$, в) $K_1 \in (-\infty, K_0J_3)$.

In cases a) and b), respectively, the variety Γ is homeomorphic to the sphere S^2 and to the disjoint union of two spheres (which merge into a torus for $J_2 = J_3$). In case c) there are no real solutions.

To conclude this section, let us describe the completely integrable dynamical system related to the L operator (10). As the phase space, we take the product of N copies of the variety $\Gamma(K_0, K_1)$. In other words, we consider a ring with N nodes, and associate to each node a quartet of dynamical variables $S_\alpha^{(n)}$ that satisfy constraints (21). Let the Poisson brackets between the variables $S_\alpha^{(n)}$ corresponding to a fixed node be given by relation (13), and let the Poisson brackets between variables corresponding to distinct nodes be equal to zero. We introduce the L operator in the n-th node by the formula $L_n(u) \equiv L(u, S_\alpha^{(n)})$, where $L(u, S_\alpha)$ is given by (10). Construct the monodromy matrix [1, 2] $T(u) = L_N(u)L_{N-1}(u) \ldots L_1(u)$. By virtue of (3), the traces $t(u) \equiv \operatorname{tr} T(u)$ of the monodromy matrix are in involution: $\{t(u), t(v)\} = 0$, and this allows us to take $\ln t(u)$ as a generating function for the integrals of motion. In a manner analogous to [9], the local integrals of motion $H^{(j)}$ are derived from the expansion of $\ln |t(u)|^2$ in powers of $(u - u_0)$,

$$H^{(j)} = \frac{\partial^j}{\partial u^j} \ln |t(u)|^2 \Big|_{u=u_0}$$

at the point u_0 determined from the nondegeneracy condition for the L operator

$$\det L(u_0) \equiv K_1 + K_0 (w_1^2(u_0) - J_1) = 0.$$

In particular, the simplest two-point Hamiltonian $H^{(0)}$ has the form

$$H^{(0)} = \sum_n H_{n+1, n},$$
$$H_{n+1, n} = \ln \operatorname{tr} L_{n+1}(u_0) L_n(u_0) = \ln \left(S_0^{(n+1)} S_0^{(n)} + \sum_{\alpha=1}^{3} \left(\frac{K_1}{K_0} - J_\alpha \right) S_\alpha^{(n+1)} S_\alpha^{(n)} \right). \quad (23)$$

The Hamiltonian (23) is nothing but the discrete variant of the Landau–Lifshits Hamiltonian from ferromagnetism theory [11]. As in the continuous case, the discrete nonlinear Schrödinger equation [8, 9] and the sine–Gordon equation [2, 8] are degenerate cases of (23). A more detailed investigation of the model described by the Hamiltonian (23) in the framework of the inverse problem method will be published separately.

2. Quantum Case

Let R(u) be the solution to Eq. (2) discovered by R. Baxter [10]:

$$R(u)=1+\sum_{\alpha=1}^{3} W_\alpha(u)\sigma_\alpha\otimes\sigma_\alpha, \tag{24}$$

Here the coefficients $W_\alpha(u)$

$$W_1(u)=\frac{sn\,(i\eta,k)}{sn\,(u+i\eta,k)},\quad W_2(u)=\frac{dn}{sn}(u+i\eta,k)\frac{sn}{dn}(i\eta,k), \\ W_3(u)=\frac{cn}{sn}(u+i\eta,k)\frac{sn}{cn}(i\eta,k) \tag{25}$$

lie on the algebraic curve

$$\frac{W_\alpha^2-W_\beta^2}{W_\gamma^2-1}=J_{\alpha\beta}. \tag{26}$$

We shall assume that the constants η and k are real. Then it is easily seen that the constants $J_{\alpha\beta}$ are real, too. Notice that among the equations (26) only two are independent, because the $J_{\alpha\beta}$ are related via

$$J_{12}+J_{23}+J_{31}+J_{12}J_{23}J_{31}=0, \tag{27}$$

This represents the quantum analogue of the classical relation (8). As in the classical case, it is convenient to express $J_{\alpha\beta}$ in the form

$$J_{\alpha\beta}=-\frac{J_\alpha-J_\beta}{J_\gamma}, \tag{28}$$

where the constants J_α are uniquely defined modulo the transformation $J_\alpha \mapsto cJ_\alpha$.

In full analogy with the classical case, we shall seek the solution L(u) to Eq. (1) in the form

$$L(u)=S_0+\sum_{\alpha=1}^{3} W_\alpha(u)S_\alpha, \tag{29}$$

where the S_α are temporarily unknown quantities. Insert (24) and (29) in (1), and apply the identities

$$W_\beta(u-v)W_\gamma(u)-W_\alpha(u-v)W_\gamma(v)+W_\alpha(u)W_\beta(v)- \\ -W_\gamma(u-v)W_\beta(u)W_\alpha(v)=0, \tag{30}$$

$$\frac{W_\beta(u-v)W_\gamma(u)W_\beta(v)-W_\gamma(u-v)W_\beta(u)W_\gamma(v)}{W_\alpha(u)-W_\alpha(u-v)W_\alpha(v)}=J_{\beta\gamma}. \tag{31}$$

Recall that in (30) and (31), as well as in the sequel, we use the convention on the indices $\alpha\beta\gamma$ mentioned in the introduction. As in the classical case, identity (30) is equivalent to the Yang–Baxter equation.

As a result, we obtain the following commutation relations for the variables S_α:

$$[S_\alpha,S_0]_-=-iJ_{\beta\gamma}[S_\beta,S_\gamma]_+, \\ [S_\alpha,S_\beta]_-=i[S_0,S_\gamma]_+. \tag{32}$$

Relations (32) generate a two-sided ideal I in a standard manner in the free associative algebra $\mathcal{A}$ with the four generators S_α. The quotient algebra $\mathcal{F}=\mathcal{A}/I$ corresponding to the ideal I is the basic object of investigation in the present section.

First, let us clarify the relationship between the algebra $\mathcal{F}$ and the quadratic Poisson brackets algebra $\mathcal{P}$ described in the previous section. To this end, we introduce Planck's constant h and apply the well-known principle of correspondence [12] between classical and quantum mechanics. According to the latter, the commutator of two observables becomes the Poisson bracket when $h\to 0$:

$$[,]_-\sim -ih\{,\}. \tag{33}$$

Setting $\eta = \rho h$ in (25) and passing to the limit $h \to 0$, we obtain the following relations:

$$W_\alpha(u) = ihw_\alpha(u) + O(h^2), \quad R(u) = 1 + ihr(u) + O(h^2),$$
$$J_{\alpha\beta} = h^2 J_{\alpha\beta} + O(h^4), \quad J_\alpha = 1 - h^2 J_\alpha + O(h^4). \tag{34}$$

Now, using expansion (34) and assuming that the quantum quantities S_a become the corresponding classical quantities S_a when $h \to 0$, according to the rule $S_0 \sim hS_0$, $S_\alpha \sim S_\alpha$, it is not hard to convince ourselves that the quantum equalities (1), (2), and (32) are transformed into the classical ones (3), (4), and (13), respectively. At the same time one has the relation $L(u) \sim hL(u)$. Thus, the quadratic PBA $\mathcal{P}$ is the classical limit of the algebra $\mathcal{F}$.

It is instructive to compare the quantization of the quadratic HPBA $\mathcal{P}$ with that of an HPBA of degree $n < 2$. For $n = 0$ and $n = 1$, the classical relations (19) and (20) become, when quantized, the relations: $[x_j, x_k]_- = -ihc_{jk}$ and $[x_j, x_k]_- = -ih \sum_{l=1}^{N} c^l_{jk} x_l$, respectively (see [12]); i.e., the tensor of structure constants of the HPBA is simply multiplied by $(-ih)$. This allows us to obtain strong results concerning the relation between a given classical HPBA and the corresponding quantum algebra [13].

When $n = 2$ the picture is significantly more intricate, because the simple correspondence rule $c \mapsto -ihc$ is no longer available. Indeed, Planck's constant enters in an essentially nonhomogeneous way in the quantum structure constants $J_{\alpha\beta}$. This can be seen at least from the relation (27), which becomes the classical relation (8) only at the limit $h \to 0$. This circumstance, as well as the absence (in the generic case) of a continuous symmetry, strongly encumbers the investigation of algebra $\mathcal{F}$.

Since for $J_{\alpha\beta} = 0$ the algebra $\mathcal{F}$ degenerates into the trivial deformation of the universal Lie enveloping algebra $\mathfrak{U}(so(3))$, it is natural to expect a certain similarity between the properties of the algebras $\mathcal{F}$ and $\mathfrak{U}(so(3))$. In particular, it would be desirable to obtain for $\mathcal{F}$ some sort of analogue of the well-known Poincaré–Birkhoff–Witt theorem [12] for universal enveloping algebras.

Problem. Prove (or disprove) that the dimension of the subspace of homogeneous polynomials of degree p in the variables S_a in the algebra $\mathcal{F}$ is $(p + 3)(p + 2)(p + 1)/6$, i.e., coincides with the dimension of the space of polynomials of degree p in four commuting variables S_a.

The author succeeded in proving this statement for $p = 3$ by direct calculations. In so doing, it turned out, as expected, that in order to count the linearly independent monomials one must make essential use of relation (27), which plays, for the algebra $\mathcal{F}$, the same role as the Jacobi identity plays for the structure constants of a Lie algebra.

The next important problem is that of analyzing the structure of the center $Z_\mathcal{F}$ of the algebra $\mathcal{F}$.

THEOREM 2. The quadratic operators K_0 and K_1

$$K_0 = \sum_{a=0}^{3} S_a^2, \quad K_1 = S_0^2 + \sum_{\alpha=1}^{3} (1 - J_\alpha) S_\alpha^2 \tag{35}$$

belong to $Z_\mathcal{F}$.

The fact that K_0 and K_1 commute with S_a is straightforward, choosing some basis in the subspace of cubic polynomials in the variables S_a. It also follows from the following lemma, given in [1]:

LEMMA. The quantum determinant $D(u)$, defined by the formula

$$D(u) = \operatorname{tr} P_- L(u) \otimes L(u - 2i\eta) = S_0^2 - \sum_{\alpha=1}^{3} W_\alpha(u) W_\alpha(u - 2i\eta) S_\alpha^2, \tag{36}$$

where $P_- = (1 - \sigma_\alpha \otimes \sigma_\alpha)/4$ is the antisymmetrization operator, belongs to $Z_\mathcal{F}$.

The operators K_0 and K_1 are derived from $D(u)$ as the coefficients of its expansion into linearly independent elliptic functions of argument u.

In the limit $h \to 0$, K_0 and K_1 become, respectively, K_0 and K_1 in (21). Notice that here, in contrast to the case of Lie algebras [13], the centers of the classical and quantum algebras do not coincide.

Problem. Prove (or disprove) that the center of the algebra $\mathcal{F}$ consists of polynomials K_0 and K_1.

In conclusion, let us approach the problem of finding the representations of the algebra $\mathcal{F}$. By a representation φ of $\mathcal{F}$ in a linear space V we mean any homomorphism $\varphi: \mathcal{F} \to \mathrm{End}\, V$, as is natural. A linear representation will be called self-adjoint if V is endowed with a φ Hermitian bilinear form relative to which the operators $\varphi(S_a)$ are self-adjoint. The reducible and decomposable representations are defined in the standard way [12]. Obviously, every reducible self-adjoint representation of $\mathcal{F}$ is decomposable.

The following are examples of irreducible finite-dimensional self-adjoint representations of algebra $\mathcal{F}$:

1. The two-dimensional representation by Pauli matrices:

$$S_0 = 1,\ S_\alpha = \sigma_\alpha.$$

2. The three-dimensional representation

$$S_0 = \begin{pmatrix} J_3 & 0 & J_1 - J_2 \\ 0 & J_1 + J_2 - J_3 & 0 \\ J_1 - J_2 & 0 & J_3 \end{pmatrix}, \quad S_1 = \sqrt{2J_2J_3}\begin{pmatrix} 0 & 1 & 0 \\ 1 & 0 & 1 \\ 0 & 1 & 0 \end{pmatrix},$$

$$S_2 = \sqrt{2J_3J_1}\begin{pmatrix} 0 & -i & 0 \\ i & 0 & -i \\ 0 & i & 0 \end{pmatrix}, \qquad S_3 = 2\sqrt{J_1J_2}\begin{pmatrix} 1 & 0 & 0 \\ 0 & 0 & 0 \\ 0 & 0 & -1 \end{pmatrix}.$$

Notice that this representation is self-adjoint only when $J_\alpha > 0$.

Presently, there are no other examples of representations of the algebra $\mathcal{F}$ in the generic case. It is highly probable that one could succeed in finding new representations of $\mathcal{F}$ using the construction of "multiplication" of R matrices proposed in [1, 4]. Some series of representations of algebra $\mathcal{F}$ for the degenerate cases $k = 0$ and $k = 1$ are given in [2, 8, 9].

3. Possible Applications and Generalizations

The main field to which the algebraic structures described in the present work find themselves applicable is the theory of classical and quantum completely integrable systems. Relations (1) and (3) provide a systematic procedure for obtaining completely integrable lattice approximations to various continuous completely integrable systems.

In the derivation of relations (13) and (32), a great role was played by the successful choice of the substitutions (10) and (29) for the L operator. The problem of enumerating all L operators (possibly with a more complicated functional dependence upon the spectral parameter for a given R matrix is still open. A possible approach to the solution of this problem was suggested in [14].

One of the possible directions in which the examples given in the present paper could be further generalized is to search for quadratic PBAs and the corresponding quantum algebras for other known examples of solutions to the Yang–Baxter equations (2) and (4).

In addition, one can consider other real forms of the same algebra $\mathcal{F}$ that we studied here, and look for their (possibly infinite-dimensional) representations.

The author is grateful to L. D. Faddeev and A. G. Reiman for the interest manifested in the work and for useful remarks.

LITERATURE CITED

1. P. P. Kulish and E. K. Sklyanin, "Quantum spectral transform method. Recent developments," Lect. Notes Phys., 151, 61-119 (1982).
2. A. G. Izergin and V. E. Korepin, "Quantum inverse problem method," Fiz. Elem. Chastits At. Yad., 13, No. 3, 501-541 (1982).
3. P. P. Kulish and E. K. Sklyanin, "On the solutions to the Yang–Baxter equation," Zap. Nauchn. Sem. Leningr. Otd. Mat. Inst., 95, 129-160 (1980).

4. P. P. Kulish, N. Yu. Reshetikhin, and E. K. Sklyanin, "Yang—Baxter equation and representation theory," Lett. Math. Phys., 5, No. 5, 393-403 (1981).
5. A. A. Belavin and V. G. Drinfel'd, "On the solutions to the classical Yang—Baxter equation for simple Lie algebras," Funkts. Anal. Prilozhen., 16, No. 3, 1-29 (1982).
6. A. A. Belavin, "Discrete groups and integrability of quantum systems," Funkts. Anal. Prilozhen., 14, No. 4, 18-26 (1980).
7. I. M. Krichever, "Baxter's equations and algebraic geometry," Funkts. Anal. Prilozhen., 15, No. 2, 22-35 (1981).
8. A. G. Izergin and V. E. Korepin, "Lattice versions of quantum field theory models in two dimensions," Nucl. Phys. B., 205, No. 3, 401-413 (1982).
9. A. G. Izergin and V. E. Korepin, "A lattice model connected with the nonlinear Schrödinger equation," Dokl. Akad. Nauk SSSR, 259, No. 1, 76-79 (1981).
10. R. J. Baxter, "Partition function for the eight-vertex lattice model," Ann. Phys., 70, No. 1, 193-228 (1972).
11. E. K. Sklyanin, "On complete integrability of the Landau—Lifshitz equation," Preprint LOMI E-3-1979, Leningrad (1979).
12. A. A. Kirillov, Elements of Representation Theory [in Russian], Nauka, Moscow (1978).
13. I. M. Gel'fand, "The center of the infinitesimal group ring," Mat. Sb., 26, No. 1, 103-112 (1950).
14. V. E. Korepin, "An analysis of the bilinear relation of the six-vertex model," Dokl. Akad. Nauk SSSR, 265, No. 6, 1361-1364 (1982).

SOME ALGEBRAIC STRUCTURES CONNECTED WITH THE YANG–BAXTER EQUATION. REPRESENTATIONS OF QUANTUM ALGEBRAS

E. K. Sklyanin

UDC 517.43+519.46

The present paper is the continuation of [1]. The object of our analysis is the algebra $\mathcal{F}$, (introduced in [1], formula (32)†) generated by commutation relations of the form

$$[S_0, S_\alpha]_- = iJ_{\beta\gamma}[S_\beta, S_\gamma]_+, \quad [S_\alpha, S_\beta]_- = i[S_0, S_\gamma]_+. \tag{1}$$

As in [1], here and from now on $[A, B]_\pm = AB \pm BA$, and the Latin indices a, b, c assume the values 0, 1, 2, 3, whereas the Greek ones, α, β, γ, assume the values 1, 2, 3; also, in formulas (1), (3), and (25), the triplet of indices (α, β, γ) stands for any cyclic permutation of (1, 2, 3). We select a real form of algebra $\mathcal{F}$, requiring that the structure constants $J_{\alpha\beta}$ be real, and defining an involutive anti-automorphism by $(aAB)^* = \bar{a}B^*A^*$, where A, B $\in \mathcal{F}$, $a \in \mathbb{C}$.

We shall assume that the structure constants $J_{\alpha\beta}$ obey constraint (1-27)

$$J_{12} + J_{23} + J_{31} + J_{12}J_{23}J_{31} = 0, \tag{2}$$

which can be suitably solved by parametrizing $J_{\alpha\beta}$ in the form (1-28):

$$J_{\alpha\beta} = -\frac{J_\alpha - J_\beta}{J_\gamma}. \tag{3}$$

As shown in [1], algebra $\mathcal{F}$ contains two linearly independent central elements ("Casimir operators") that are quadratics in S_a and which we find convenient to write here in the following form rather than (1-35):

$$K_0 = \sum_{a=0}^{3} S_a^2, \quad K_2 = \sum_{\alpha=1}^{3} J_\alpha S_\alpha^2. \tag{4}$$

(In [1] operator $K_1 = K_0 - K_2$ was used instead of K_2).

In [1] we raised the problem of constructing representations of relations (1) by self-adjoint operators S_a in a Hilbert space. (Such representations will be referred to as self-adjoint.) We also gave the simplest examples of self-adjoint representations.

In this paper we construct (under certain restrictions on the constants $J_{\alpha\beta}$) three infinite series of finite-dimensional self-adjoint representations of algebra $\mathcal{F}$. In the classical limit (1-34) these series yield three cases (a, b, c), of the topological classification of phase spaces Γ (see Sec. 2).

The paper is organized as follows: in Sec. 1 we describe the construction of the representations which constitute series a) (the principal analytic series). These representations are parametrized by a discrete parameter l which runs over the nonnegative half-integers: $l = 0, 1/2, 1, 3/2, \ldots$; they are finite-dimensional (of dimension $2l + 1$) and act in spaces of θ-functions on a torus. We then show that in the limit $J_{\alpha\beta} \to 0$ the representations of series a) become the standard finite-dimensional representations of the Lie algebra $su(2)$.

Section 2 is devoted to the construction of the representations forming series b) (the nonanalytic series) and c) (the dual nonanalytic series). The representations in series b) are parametrized by three continuous, real parameters m, γ, δ, and are realized in spaces of functions defined on discrete subsets of the real axis and which are square integrable with

†Henceforth references to our first paper [1] will be given like this: (1-32).

V. A. Steklov Mathematics Institute, Leningrad Branch, Academy of Sciences of the USSR.
Translated from Funktsional'nyi Analiz i Ego Prilozheniya, Vol. 17, No. 4, pp. 34-48, October-December, 1983. Original article submitted May 20, 1983.

respect to a certain weight. The representations in series c) act in the same Hilbert spaces of analytic functions as the representations in series a), from which they are obtained via automorphisms of algebra $\mathcal{F}$. The parameter n of the representations in series c) runs, as in case a), over the (nonnegative) half-integers.

In Sec. 3 we investigate degenerate cases of algebra $\mathcal{F}$, which have important applications in the theory of integrable quantum systems. The corresponding representations act in spaces of trigonometric polynomials. Finally, in Sec. 4 we draw the conclusions and discuss a number of open problems.

The author dedicates this paper to I. M. Gel'fand on the occasion of his birthday.

1. Representations of Series a)

We are first concerned with the reduction of the commutation relations (1) to canonical form. The algebra $\mathcal{F}$ given by relations (1) is characterized by structure constants $J_{\alpha\beta}$ which satisfy Eq. (2).

Proposition 1. Two algebras $\mathcal{F}$. whose collections of structure constants differ by a cyclic permutation, for example, $(J_{12}, J_{23}, J_{31}) \mapsto (J_{23}, J_{31}, J_{12})$, or by an inversion, i.e., by a transposition accompanied by a simultaneous change of signs, for example, $(J_{12}, J_{23}, J_{31}) \mapsto (-J_{23}, -J_{12}, -J_{31})$, are isomorphic.

For a proof, it suffices to guess the corresponding isomorphisms. For the examples above, these are $(S_0, S_1, S_2, S_3) \mapsto (S_0, S_2, S_3, S_1)$ and, respectively, $(S_0, S_1, S_2, S_3) \mapsto (S_0, S_3, -S_2, S_1)$.

THEOREM 1. Every collection of structure constants $J_{\alpha\beta}$ can be reduced by means of cyclic permutations and inversions to one of the following canonical forms:

$$
\begin{array}{lll}
1a) \ J_{12} > 0, & 1 > J_{23} > 0, & J_{31} < 0; \\
1b) \ J_{12} > 0, & J_{23} > 1, & J_{31} < 0; \\
1c) \ J_{12} > 0, & J_{23} = 1, & J_{31} = -1; \\
2a) \ J_{12} > 0, & J_{23} = 0. & J_{31} = -J_{12}; \\
2a') \ J_{12} = 0, & 1 > J_{23} > 0, & J_{31} = -J_{23}; \\
2b) \ J_{12} = 0, & J_{23} > 1. & J_{31} = -J_{23}; \\
2c) \ J_{12} = 0, & J_{23} = 1, & J_{31} = -1; \\
3) \ J_{12} = 0, & J_{23} = 0, & J_{31} = 0.
\end{array}
$$

Proof. Equation (2) shows that only three cases are possible: 1) $J_{\alpha\beta} \neq 0$; 2) only one of the three constants $J_{\alpha\beta}$ vanishes; and 3) $J_{\alpha\beta} = 0$. Using once more Eq. (2), we see that in case 1) two of the three constants $J_{\alpha\beta}$ have the same sign, whereas the third has opposite sign. Performing, if necessary, an inversion, we can always achieve that two constants be positive and one negative. Further, by a cyclic permutation, we can replace J_{31} by a negative constant. Then, depending upon the values of J_{23}, we distinguish three subvariants: 1a), 1b), and 1c). Moreover, in case 1c) we may use the equality $J_{23} = 1$ to rewrite Eq. (2) in the form $(J_{12} + 1)(J_{31} + 1) = 0$, which yields $J_{31} = -1$, since, by hypothesis, $J_{12} > 0$. Case 2) is analysed similarly.

In this paper we restrict ourselves to representations of algebra $\mathcal{F}$ for the cases 1a) (Secs, 1, 2) and 2a)-2a') (Sec. 3). In Sec. 3 we make also several observations concerning the remaining cases.

For the moment we merely note that the generic cases are 1a) and 1b), whereas all the other cases may be regarded as degenerate. Observe also that case 3) corresponds to the situation when algebra $\mathcal{F}$ degenerates in a trivial extension of the universal enveloping algebra of the Lie algebra $su(2)$, and hence is of no interest.

Before turning to the representations of $\mathcal{F}$, let us introduce some functional spaces. Let $k \in (0, 1)$ be the modulus of Jacobi's elliptic functions, and let K and K' be the corresponding total elliptic integrals [2]. We let $\Theta^p_{g_1g_2}$ denote the space of Jacobi θ-functions of modulus k and order $p \in Z+$, and with characteristics $g_1, g_2 \in \{0, 1\}$, i.e., the space of functions f(u) holomorphic on C such that

$$f(u + 2K) = e^{-\pi i g_1} f(u), \quad f(u + 2iK') = e^{-\pi i g_2} e^{-p\frac{\pi i}{K}(u+iK')} f(u). \tag{5}$$

We distinguish $\Theta^p_{g_1g_2}$ the subspaces $\Theta^{p+}_{g_1g_2}$ and $\Theta^{p-}_{g_1g_2}$ of even and, respectively, odd functions.

It is known [3] that dim $\Theta^{p}_{g_1g_2} = p$, while dim $\Theta^{p+}_{00} = n + 1$, if $p = 2n$ or $p = 2n + 1$. For the four standard θ-functions of the first order we shall adhere to the notations of [3]: $\theta_{g_1g_2}(u) \in \Theta^{1}_{g_1g_2}$. For readers' convenience, we indicate the correspondence between our notations and the standard ones (see, for example, [2]): $\theta_{00}(u) = \theta_3(u)$, $\theta_{01}(u) = \theta_4(u)$, $\theta_{10}(u) = \theta_2(u)$, $\theta_{11}(u) = \theta_1(u)$. Note that $\theta_{g_1g_2}(2u) \in \Theta^{4}_{00}$, whereas $\theta_{00}(2u)$, $\theta_{01}(2u)$, $\theta_{10}(2u) \in \Theta^{4+}_{00}$, and $\theta_{11}(2u) \in \Theta^{4-}_{00}$. For θ-functions of zero argument we shall use the notation $\theta_{g_1g_2} = \theta_{g_1g_2}(0)$.

The key to the theory of representations of algebra $\mathcal{F}$ is the following result:

<u>THEOREM 2.</u> Let k, η, and l be arbitrary. Then the operators S_α ($\alpha = 0, 1, 2, 3$), which act on functions f(u) meromorphic on C according to the rule:

$$(S_\alpha f)(u) = \frac{s_\alpha(u - l\eta) f(u + \eta) - s_\alpha(-u - l\eta) f(u - \eta)}{\theta_{11}(2u)}, \tag{6}$$

where $s_0(u) = \theta_{11}(\eta)\theta_{11}(2u)$, $s_1(u) = \theta_{10}(\eta)\theta_{10}(2u)$, $s_2(u) = i\theta_{00}(\eta)\theta_{00}(2u)$, and $s_3(u) = \theta_{01}(\eta)\theta_{01}(2u)$, definte a representation of the algebra $\mathcal{F}$ with structure constants

$$\begin{aligned} J_{12} &= \frac{\theta_{01}^2(\eta)\,\theta_{11}^2(\eta)}{\theta_{00}^2(\eta)\,\theta_{10}^2(\eta)} = k'^2\,\frac{\mathrm{sn}^2}{\mathrm{cn}^2\,\mathrm{dn}^2}(\eta, k), \\ J_{23} &= \frac{\theta_{10}^2(\eta)\,\theta_{11}^2(\eta)}{\theta_{00}^2(\eta)\,\theta_{01}^2(\eta)} = k^2\,\frac{\mathrm{sn}^2\,\mathrm{cn}^2}{\mathrm{dn}^2}(\eta, k), \\ J_{31} &= -\frac{\theta_{00}^2(\eta)\,\theta_{11}^2(\eta)}{\theta_{01}^2(\eta)\,\theta_{10}^2(\eta)} = -\frac{\mathrm{sn}^2\,\mathrm{dn}^2}{\mathrm{cn}^2}(\eta, k). \end{aligned} \tag{7}$$

The author's proof of Theorem 2 uses the standard addition theorems for θ-functions [3] and is a straightforward and very tedious verification of relations (1), for which reason we omit it. It would be interesting to find a more conceptual proof, based on geometric or algebraic ideas. To this end, the representation of the Zamolodchikov algebra for the XYZ-model found in [4] might turn to be useful.

Straightforward computations yield also the following two results.

<u>Proposition 2.</u> The structure constants $J_{\alpha\beta}$ given by formula (7) admit a representation (3) with

$$J_1 = \frac{\theta_{10}(2\eta)\,\theta_{10}}{\theta_{10}^2(\eta)}, \quad J_2 = \frac{\theta_{00}(2\eta)\,\theta_{00}}{\theta_{00}^2(\eta)}, \quad J_3 = \frac{\theta_{01}(2\eta)\,\theta_{01}}{\theta_{01}^2(\eta)}, \tag{8}$$

and hence satisfy Eq. (2).

<u>Proposition 3.</u> The Casimir operators (4) for representation (6) are scalar and given by

$$K_0 = 4\theta_{11}^2((2l + 1)\eta), \quad K_2 = 4\theta_{11}(2(l + 1)\eta)\theta_{11}(2l\eta). \tag{9}$$

Therefore, k and η parametrize the structure constants, while l is the representation parameter. Henceforth, we identify any two representations differing by a trivial automorphism $S_\alpha \mapsto cS_\alpha$ of algebra $\mathcal{F}$.

There is an intimate connection between parametrization (7) of the structure constants $J_{\alpha\beta}$ and the quantum L-operator (1-29). Namely, inserting (1-25) into (1-26) we obtain a parametrization of $J_{\alpha\beta}$ which turns into (7) following the substitution $k \leftrightarrow k'$. Thus, the spectral parameter u of the quantum L-operator and the functional variable u in the space of representation (6) can be in some sense identified (to emphasize this we denote them by the same letter). Unfortunately, at this moment, we cannot comment upon this remarkable fact, which undoubtedly holds for profound reasons.

As the next theorem shows, for real η, parametrization (7) corresponds to case 1a) in the classification of Theorem 1.

<u>THEOREM 3.</u> Parametrization (7) defines a one-to-one mapping of the domain $\{0 < k < 1,\ 0 < \eta < K(k)\}$ onto that part of the manifold (2) defined by the inequality of point 1a) of Theorem 1.

<u>Proof.</u> By Definition 2, the quantities $J_{\alpha\beta}(\eta, k)$ satisfy Eq. (2). The required inequalities are readily verified. The inverse mapping is given by the equations

$$k=\frac{J_{12}+1}{J_{12}+J_{23}}\sqrt{-J_{23}J_{31}}, \quad \operatorname{sn}^2(\eta,k)=\frac{J_{12}+J_{23}}{J_{12}+1},$$

which, as one can easily see, are uniquely solvable in the indicated domain of variation of the parameters k and η, provided that $J_{\alpha\beta}$ satisfy Eq. (2) and the inequalities of 1a), Theorem 1.

Until now we have assumed, for the sake of simplicity, that representation (6) acts in the infinite-dimensional space of functions meromorphic on C. The next step is to identify the finite-dimensional invariant subspaces.

THEOREM 4. The space $\Theta^{p+}_{g_1g_2}$ is invariant under the action of the operators S_α given by formula (6) if and only if the following conditions are satisfied:

1) $(p-4l)\eta = 0 \pmod{2K}$; 2) p is even; 3) $g_1 = g_2 = 0$.

Proof. Let $f(u) \in \Theta^{p+}_{g_1g_2}$. Then the evenness of the function $(S_\alpha f)(u)$ follows from the evenness of f(u) and the oddness of $\theta_{11}(2u) \in \Theta^{4-}_{00}$. Further, using the fact that $s_\alpha(u) \in \Theta^4_{00}$, and the characteristic properties (5) of θ-functions, we get $s_\alpha(\pm u - l\eta)f(u \pm \eta) \in \Theta^{p+4}_{g_1, g_2\pm(p-4l)\eta/K}$ and hence, since the characteristics g_1, g_2 of θ-functions are defined only mod 2, function $(S_\alpha f)(u)$ has the desired automorphy properties if and only if condition 1) is satisfied. Now suppose that 1) holds. Then it remains to verify that $(S_\alpha f)(u)$ is holomorphic if and only if 2) and 3) are fulfilled. To this end, note that the numerator of expression (6) for $(S_\alpha f)(u)$ belongs to $\Theta^{(p+4)-}_{g_1g_2}$ and apply the following simple lemma (see [3, p. 75]):

LEMMA 1. The following two statements are equivalent:

a) For every function $h(u) \in \Theta^{q-}_{g_1g_2}$: $h(0) = h(K) = h(iK') = h(K + iK') = 0$.

b) q is even and $g_1 = g_2 = 0$.

Therefore, the zeros of the numerator and denominator in (6) mutually reduce if and only if conditions 2) and 3) of the theorem are satisfied, and this completes the proof.

According to Theorem 4, to every pair (l, p) satisfying conditions 1) and 2) there corresponds a (p/2 + 1)-dimensional representation of algebra $\mathcal{F}$ in the space Θ^{p+}_{00}. In this paper we study only the representations corresponding to the case p = 4l, and these will be referred to as representations of the principal analytic series or, briefly, representations of series a). By condition 2), the representations of series a) are indexed by a parameter l which runs over the nonnegative half-integers: l = 0, 1/2, 1, 3/2, ..., and have dimensions 2l + 1.

Let us separate the real and imaginary parts of the complex variable u: u = x + iy, $\bar{u} = x - iy$, and, assuming that $(2l+1)\eta < K$, introduce in the space Θ^{4l+}_{00} of a representation of series a) the following positive-definite inner product:

$$(f,g)=\int_0^{2K}dx\int_0^{2K'}dy\,\overline{f(u)}\,g(u)\,\mu(u,\bar{u}), \tag{10}$$

where

$$\mu(u,\bar{u})=\frac{\theta_{11}(2u)\,\theta_{11}(2\bar{u})}{\prod\limits_{j=0}^{2l-1}\theta_{00}(u+\bar{u}+(2j-2l-1)\eta)\,\theta_{00}(u-\bar{u}+(2j-2l-1)\eta)}. \tag{11}$$

Note that by virtue of the condition $f, g \in \Theta^{4l+}_{00}$ and formula (11) for $\mu(u, \bar{u})$, the integrand in (10) is a periodic function of x and y (with respective periods 2K and 2K'), and hence the integration domain in (10) can be regarded as a torus T^2. The positive-definiteness of the form (10) follows from the fact that $\mu(u, \bar{u})$ is positive throughout the complement of a set of measure zero in T^2, as may be readily verified. The condition $(2l + 1)\eta < K$ guarantees that function $\mu(u, \bar{u})$ has no singularities for real x and y.

One can rewrite (10) as

$$(f,g)=\int_C \omega, \tag{12}$$

where the meromorphic 2-form ω(u, v) on the two-dimensional complex manifold $T^2 \times T^2$ is given by the formula

$$\omega(u,v)=\overline{f(\bar v)}\,g(u)\,\mu(u,v)\frac{dv\wedge du}{2i},$$

while the 2-cycle C is defined by the condition $v=\bar u$. The singularities of $\omega(u,v)$ lie on the manifolds defined by the equations $u\pm v+(2j-2l-1)\eta=K-iK'$ (modulo the periods $2K$, $2iK'$), $j=0,1,\ldots,2l+1$.

THEOREM 5. Operators S_α (6) are self-adjoint with respect to the metric (10): $S_\alpha^*=S_\alpha$.

Proof. It will be enough to verify the equality $(f, S_\alpha g)=(S_\alpha f, g)$ for arbitrary f, $g\in\Theta_{00}^{4l+}$ Using (6) and (12), one can reexpress $(f, S_\alpha g)$ as

$$(f,S_\alpha g)=\int_C\overline{f(\bar v)}\,\mu(u,v)\,\theta_{11}^{-1}(2u)[s_\alpha(u-l\eta)\,g(u+\eta)-s_\alpha(-u-l\eta)\,g(u-\eta)]\frac{dv\wedge du}{2i}. \tag{13}$$

Since the pole of $\theta_{11}^{-1}(2u)$ is cancelled by the numerator of $\mu(u,v)$, one can examine the terms of (13) separately. Performing translations of variables, $u\mapsto u-\eta$, in the first terms, and $u\mapsto u+\eta$, in the second, we obtain the equality

$$(f,S_\alpha g)=\int_{C_1}\overline{f(\bar v)}\,g(u)\,[\theta_{11}^{-1}(2u-2\eta)\,s_\alpha(u-(l+1)\eta)\,\mu(u-\eta,v)-\theta_{11}^{-1}(2u+2\eta)\,s_\alpha(-u-(l+1)\eta)\,\mu(u+\eta,v)]\frac{dv\wedge du}{2i}, \tag{14}$$

where the cycle C_1 is, generally speaking, different from C, because, for $2(l+1)\eta>K$, the singularities of $\mu(u,v)$ cross the cycle C as a result of the shifts $u\mapsto u\pm\eta$. An entirely similar computation yields for $(S_\alpha f, g)$ the expression

$$(S_\alpha f,g)=\int_{C_1}\overline{f(\bar v)}\,g(u)[\theta_{11}^{-1}(2v-2\eta)\,\overline{s_\alpha(\bar v-(l+1)\eta)}\,\mu(u,v-\eta)-\theta_{11}^{-1}(2v+2\eta)\,\overline{s_\alpha(-\bar v-(l+1)\eta)}\,\mu(u,v+\eta)]\frac{dv\wedge du}{2i}. \tag{15}$$

A detailed investigation further reveals that the cycles C_1 and C_2 can be selected to be homologous on the manifold obtained from $T^2\times T^2$ by removing the union of singularities of $\mu(u\pm\eta, v)$ and $\mu(u, v\pm\eta)$. This allows us to identify C_1 and C_2 and write in the following form a condition necessary in order that expressions (14) and (15) coincide for any f and g:

$$\theta_{11}^{-1}(2u-2\eta)\,s_\alpha(u-(l+1)\eta)\,\mu(u-\eta,v)-\theta_{11}^{-1}(2u+2\eta)\,s_\alpha(-u-(l+1)\eta)\mu(u+\eta,v)=$$
$$=\theta_{11}^{-1}(2v-2\eta)\overline{s_\alpha(\bar v-(l+1)\eta)}\,\mu(u,v-\eta)-\theta_{11}^{-1}(2v+2\eta)\,\overline{s^\alpha(-\bar v-(l+1)\eta)}\qquad\mu(u,v-\eta).$$

The last equality is verified by direct computation using the standard properties of θ-functions [3].

With the proof of Theorem 5 completed, this is a good place to discuss the possibility of generalizing our result to the case $(2l+1)\eta\geqslant K$. A careful analysis of the proof given above shows that it remains valid for $(2l+1)\eta\geqslant K$ provided that we replace the cycle C in the definition of the inner product (12) by a cycle $C_l(\eta)$, which depends continuously on η in such a manner that it does not intersect the singularities of $\mu(u,v)$ as η varies and is homologous to C for η small enough (more precisely, for $(2l+1)\eta<K$).

Generally speaking, it may happen that the inner product (12) is no longer positive-definite as a result of such a deformation of cycle C. An examination of the degenerate case (see Sec. 3) shows, however, that we can preserve positive definiteness for some interval of variation of η, and allows us to make the following conjecture.

Conjecture. The inner product redefined in the way indicated above is positive definite for $2l\eta<K$.

The existence of a metric of the form (12) in the space Θ_{00}^{4l+} of a representation of series a) permits us (at least for $(2l+1)\eta<K$) to use the well-developed apparatus of the theory of Hilbert spaces of analytic functions (see, for example, [5, Chap. 3]). In particular, we can introduce a reproducing kernel $E_l(u,\bar u)$ such that for every function $f\in\Theta_{00}^{4l+}$

$$\int_C E_l(u,\bar v)\,f(v)\,\mu(v,\bar v)\,\frac{d\bar v\wedge dv}{2i}=f(u).$$

With the aid of the reproducing kernel one defines the coherent state $e_{\bar u}\in\Theta_{00}^{4l+}$, $e_{\bar u}(u)=E_l(u,\bar u)$ as well as the kernel $S(u,\bar u)=(Se_{\bar u})(u)$ and the symbol $S'(u,\bar u)=S(u,\bar u)/E_l(u,\bar u)$ of an arbitrary operator S.

Unfortunately, the standard recipe for the construction of the reproducing kernel

$$E_l(u,\bar{u}) = \sum_{j=0}^{2l} \frac{f_j(u)\overline{f_j(u)}}{(f_j, f_j)}, \tag{16}$$

where f_j constitute an orthogonal basis in Θ_{00}^{4l+}, is seemingly of little use in the determination of the explicit form of $E_l(u, \bar{u})$, because the calculation of integrals of the form (12) is very difficult.

Conjecture. The reproducing kernel $E_l(u, \bar{u})$ is given, up to a constant factor, by the expression

$$E_l(u,\bar{u}) = \mathrm{const} \prod_{j=0}^{2l-1} \theta_{00}(u + \bar{u} + (2j - 2l + 1)\eta)\,\theta_{00}(u - \bar{u} + (2j - 2l + 1)\eta). \tag{17}$$

If this conjecture (which arose as a result of the consideration of various degenerate cases; see Sec. 3) turns out to be true, then, using formulas (6) and (17), one can readily calculate the kernels $S_\alpha(u, \bar{u})$ of the operators S_α:

$$S_0(u,\bar{u}) = \frac{\theta_{11}(\eta)}{\theta_{11}(2l\eta)}\theta_{00}(u + \bar{u})\,\theta_{00}(u - \bar{u}) \cdot \sigma(u,\bar{u});$$

$$S_1(u,\bar{u}) = \frac{\theta_{10}(\eta)}{\theta_{10}(2l\eta)}\theta_{01}(u + \bar{u})\,\theta_{01}(u - \bar{u}) \cdot \sigma(u,\bar{u});$$

$$S_2(u,\bar{u}) = -i\frac{\theta_{00}(\eta)}{\theta_{00}(2l\eta)}\theta_{11}(u + \bar{u})\,\theta_{11}(u - \bar{u}) \cdot \sigma(u,\bar{u}),$$

$$S_3(u,\bar{u}) = -\frac{\theta_{01}(\eta)}{\theta_{01}(2l\eta)}\theta_{10}(u + \bar{u})\,\theta_{10}(u - \bar{u}) \cdot \sigma(u,\bar{u}),$$

where

$$\sigma(u,\bar{u}) = \mathrm{const}\,\theta_{11}(4l\eta)\,E_{l-\frac{1}{2}}(u,\bar{u}),$$

and const is the same as in (17).

To conclude this section, let us discuss the connection between the representations of algebra $\mathcal{F}$ that we have constructed and the representations of the Lie algebra $su(2)$.

Note that for $\eta \to 0$ the structure constant $J_{\alpha\beta} \to 0$. Moreover, $S_0/\eta \to 2\theta_{01}\theta_{10}/\theta_{00}$, and the operators

$$\mathfrak{S}_\alpha = \lim_{\eta\to 0} \frac{\theta_{00}}{4\eta\theta_{01}\theta_{10}} S_\alpha$$

form the Lie algebra $su(2)$: $[\mathfrak{S}_\alpha, \mathfrak{S}_\beta]_- = i\mathfrak{S}_\gamma$. From (6) we derive the following expressions for the operators $\mathfrak{S}_\alpha$

$$\mathfrak{S}_\alpha = \frac{\theta_{00}}{2\theta_{01}\theta_{10}}\theta_{11}^{-1}(2u)\left[\mathfrak{s}_\alpha(u)\frac{d}{du} - l\frac{d\mathfrak{s}_\alpha(u)}{du}\right], \tag{18}$$

where $\mathfrak{s}_1(u) = \theta_{10}\theta_{10}(2u)$, $\mathfrak{s}_2(u) = i\,\theta_{00}\theta_{00}(2u)$, $\mathfrak{s}_3(u) = \theta_{01}\theta_{01}(2u)$. The operators $\mathfrak{S}_\alpha$ act in the space Θ_{00}^{4l+} and are self-adjoint with respect to the inner product given by the formula (10), where $\mu(u, \bar{u})$ is replaced by

$$\mu(u,\bar{u}) = \frac{\theta_{11}(2u)\,\theta_{11}(2\bar{u})}{\theta_{00}^{2(l+1)}(u+\bar{u})\,\theta_{00}^{2(l+1)}(u-\bar{u})}. \tag{19}$$

THEOREM 6. Representation (18) of the Lie algebra $su(2)$ is an irreducible representation of spin l.

Proof. It suffices to verify that representation (18) is isomorphic to the following standard realization of the irreducible representation of spin l of $su(2)$ in the space of polynomials $\varphi(z)$ of degree $2l$ (see, for example, [6, Vol. 1, p. 289]):

$$\left.\begin{aligned} \mathfrak{S}_1 &= \frac{1-z^2}{2}\frac{d}{dz} + lz; \\ \mathfrak{S}_2 &= i\,\frac{1+z^2}{2}\frac{d}{dz} - ilz; \\ \mathfrak{S}_3 &= z\frac{d}{dz} - l \end{aligned}\right\} \tag{20}$$

with the inner product

$$(\varphi, \chi) = \text{const} \int_C \frac{\overline{\varphi(z)}\,\chi(z)}{(1+z\bar z)^{2(l+1)}} \frac{d\bar z \wedge dz}{2i}. \tag{21}$$

The desired unitary equivalence of the representations (18)-(19) and (20)-(21) is described by the formulas

$$z(u) = \frac{k'}{\mathrm{dn}\,(2u, k) + k\,\mathrm{cn}\,(2u, k)}, \quad \varphi(z(u)) = \text{const}\, \frac{z^l(u)}{\theta_{01}^l(2u)} f(u).$$

2. Representations of Series b) and c)

As shown by Theorem 6, the representations of $\mathcal{F}$ in series a) correspond to the finite-dimensional irreducible representations of Lie algebra $su(2)$. However, in contrast to the Lie algebra, algebra $\mathcal{F}$ admits two other series of representations which no longer have analogs among the representations of $su(2)$.

The simplest way to see why algebra $\mathcal{F}$ must admit two other series of representations is to examine its classical limit (1-13, 34). We must correct here a mistake which slipped into [1] in the topological classification of phase spaces Γ which depends upon the values of the classical Casimir operators K_0, K_1 (1-21): on p. 266 in [1] rows 15-18 from the top should read:

"Four cases are possible:

a) $K_1 \in (K_0J_1, \infty)$, b) $K_1 \in (K_0J_2, K_0J_1)$,
c) $K_1 \in (K_0J_3, K_0J_2)$, d) $K_1 \in (-\infty, K_0J_3)$.

In cases a) and c), the variety Γ is homeomorphic to the disjoint union of two spheres S^2 and in case b) — to a torus T^2; finally, in case d) there are no linear solutions."

To prove this assertion, consider the function $\xi(S_1, S_2, S_3) = K_1 - J_1S_1^2 - J_2S_2^2 - J_3S_3^2$ on the sphere $S_1^2 + S_2^2 + S_3^2 = K_0$. From (1-21) it follows that variety Γ is given by the equation $S_0^2 = \xi(S_1, S_2, S_3)$, and hence it is topologically equivalent to two copies of the variety defined by the inequality $\xi \geqslant 0$ on the above sphere, glued along the boundary $\xi = 0$. The above topological classification is now easily obtained by examining the topology of the level surfaces of function ξ, which has two maxima, two minima, and two saddle points.

Thus, in the classical case there are three distinct domains of variation of the parameter K_1/K_0, such that when one passes from one domain to another the topological type of the phase space undergoes drastic changes. It is natural therefore to expect that to these three domains will correspond in the quantum case three series of representations with distinct qualitative properties.

Returning to the quantum case, we describe first the representations of series b). Thus, we subject the representations (6) of algebra $\mathcal{F}$ in the space of functions $f(u)$ meromorphic on C, to the similarity transformation $S_\alpha \mapsto U^{-1}S_\alpha U$, with $(Uf)(u) = \exp(-(\pi i/k)lu)f(u - i(K'/2))$. Following the substitution $2l\eta = 2im\eta - \eta + K$, the operator S_α takes the form

$$(S_a f)(u) = (\theta_{01}^{-1}(2u)\,[s_a'(2u + \eta - 2im\eta) f(u+\eta) + \varepsilon_a s_a'(2u - \eta + 2im\eta) f(u - \eta)], \tag{22}$$

where

$$s_0'(u) = \theta_{11}(\eta)\,\theta_{00}(u), \quad s_1'(u) = -i\theta_{10}(\eta)\theta_{01}(u),$$
$$s_2'(u) = \theta_{00}(\eta)\,\theta_{11}(u), \quad s_3'(u) = -\theta_{01}(\eta)\,\theta_{10}(u),$$
$$\varepsilon_0 = -\varepsilon_1 = \varepsilon_2 = \varepsilon_3 = 1.$$

Clearly, operators (22) satisfy the original relations (1) and hence define a representation of algebra $\mathcal{F}$ in the space of functions meromorphic on C. A straightforward computation yields the following result.

Proposition 4. The Casimir operators (4) of representation (22) are scalar and are given by

$$K_0 = 4\theta_{10}^2(2im\eta), \quad K_2 = 4\theta_{10}((2im+1)\eta)\theta_{10}((2im-1)\eta). \tag{23}$$

As in the construction of representation of series a), the next step is to find the invariant subspaces. We restrict the functions $f(u)$ to the lattice $u_j = \gamma + j\eta$, where γ is, for the moment, an arbitrary real parameter, while j runs over the integers. Suppose that constant η is commensurable with the period 2K, i.e., $\eta = 2Kp/q$, where p and q are coprime positive integers. Then it is readily checked that the functions $f(uj)$ of the form $f(uj) = \exp(2\pi i(\delta/\eta)uj)\varphi_j$, where $\varphi_{j+q} = \varphi_j$, form a q-dimensional space invariant under the action of the operators S_α (22). These operators act on the functions φ_j according to the rule

$$(S_a\varphi)_j = \theta_{01}^{-1}(2\gamma + 2j\eta)[e^{2\pi i\delta} s_a'(2\gamma + (2j+1)\eta - 2im\eta)\varphi_{j+1} + \varepsilon_a e^{-2\pi i\delta} s_a'(2\gamma + (2j-1)\eta + 2im\eta)\varphi_{j-1}]. \quad (24)$$

THEOREM 7. Operators S_α are self-adjoint with respect to the inner product

$$(\varphi, \chi) = \sum_{j=0}^{q-1} \varphi_j \bar{\chi}_j \theta_{01}(2u_j).$$

The proof consists, as in the case of Theorem 5, of a direct verification of the equality $(\varphi, S_a\chi) = (S_a\varphi, \chi)$.

The resulting finite-dimensional self-adjoint representation of algebra $\mathcal{F}$ (which exists only for η commensurable with 2K) will be referred to as a representation of the nonanalytic series, or a representation of series b). By construction, the representations of series b) are characterized by a triplet of real parameters (m, γ, δ).

Proposition 5. The representations of series b) corresponding to two parameter triplets, (m, γ, δ) and $(m', \gamma', \delta') = (m + 2K'r/\eta, \gamma + 2Ks/q, \delta - 2\gamma r/K + t/q)$, with r, s, t arbitrary integers, are unitary equivalent.

Proof. The unitary operator which establishes the asserted equivalence

$$S_a(m', \gamma', \delta') = \text{const}\, U^{-1} S_a(m, \gamma, \delta)\, U,$$

is given by the rule

$$(U\varphi)_j = \exp\left[2\pi i\left(\frac{t}{q} j + 2r\frac{p}{q} j^2\right)\right]\varphi_{j-p'},$$

where p' is an integer such that $pp' \equiv s \pmod q$.

COROLLARY. It suffices to let the parameters (m, γ, δ) vary in the intervals: $m \in [0, 2K'/\eta)$, $\gamma \in [0, 2K/q)$, and $\delta \in [0, 1/q)$.

We turn next to a description of the representations forming series c). We need some preliminary facts concerning the automorphisms of algebra $\mathcal{F}$.

Recall that we have agreed to identify representations which differ by the trivial automorphism $S_\alpha \mapsto cS_\alpha$.

It is readily seen that the operators X_α (α = 1, 2, 3) of the form X_α: $(S_0, S_\alpha, S_\beta, S_\gamma) \mapsto (S_0, S_\alpha, -S_\beta, -S_\gamma)$ are automorphisms of algebra $\mathcal{F}$ (recall that α, β, γ is a permutation of 1, 2, 3).

Proposition 6. A transformation Y_α (α = 1, 2, 3) of the form Y_α: $(S_0, S_\alpha, S_\beta, S_\gamma) \to (y_0 S_\alpha, y_\alpha S_0, y_\beta S_\gamma, y_\gamma S_\beta)$ is an automorphism of $\mathcal{F}$ if and only if the following equalities are satisfied

$$\frac{y_0 y_\alpha}{y_\beta y_\gamma} = -1, \quad \frac{y_0 y_\beta}{y_\alpha y_\gamma} = -J_{\gamma\alpha}, \quad \frac{y_0 y_\gamma}{y_\alpha y_\beta} = J_{\alpha\beta}. \quad (25)$$

The proof reduces to a straightforward verification of the commutation relations (1).

Remark 1. Equations (25) determine the automorphisms Y_α only modulo automorphisms of the forms $S_\alpha \mapsto cS_\alpha$ and X_α.

Remark 2. If the structure constants $J_{\alpha\beta}$ satisfy condition 1a) of Theorem 1 (as we have always assumed), then, from the three automorphisms Y_1, Y_2, Y_3, only Y_3 preserves the Hermitianness of the operators S_α.

We choose the following coefficients y_α for the automorphism Y_3: $y_0 = y_3^{-1} = -(\theta_{11}(\eta)/\theta_{01}(\eta))$, $y_1 = -y_2^{-1} = -(\theta_{10}(\eta)/\theta_{00}(\eta))$.

Now everything is ready for the presentation of series c). Namely, we take the operators S_α (6) which give the representation of series a) in space Θ_{00}^{4l+}, and subject them to the automorphism Y_3. Replacing the representation parameter l by n, we get

$$(S_\alpha f)(u) = \frac{s''_\alpha(u-n\eta) f(u+\eta) - s''_\alpha(-u-n\eta) f(u-\eta)}{\theta_{11}(2u)}, \tag{26}$$

where

$$s''_0(u) = -\theta_{11}(\eta)\theta_{01}(2u), \quad s''_1(u) = -i\theta_{10}(\eta)\theta_{00}(2u),$$
$$s''_2(u) = \theta_{00}(\eta)\theta_{10}(2u), \quad s''_3(u) = -\theta_{01}(\eta)\theta_{11}(2u).$$

The resulting representation will be said to belong to the dual nonanalytic series, or to series c). Clearly, these representations act in the same spaces Θ_{00}^{4n+} equipped with the same metric, as the representations of series a), and the parameter n runs over the nonnegative half-integers.

Proposition 7. The Casimir operators (4) of a representation (26) from series c) are scalar and given by

$$K_0 = 4\theta_{01}^2((2n+1)\eta), \quad K_2 = 4\theta_{01}(2(n+1)\eta)\theta_{01}(2n\eta). \tag{27}$$

THEOREM 8. The classical limits of the Casimir operators K_0 and $K_1 = K_0 - K_2$, $K_0 = \lim_{\eta\to 0} K_0$, $K_1 = \lim_{\eta\to 0} \eta^{-2}K_1$, of series a), b), and c) fall respectively in cases a), b), and c) of the topological classification of phase spaces Γ (see the beginning of this section).

Proof. To simplify the calculations, it is convenient to use the fact that, by virtue of (3), the quantities J_α (8) are defined up to a constant factor, and redefine J_α by $J_\alpha \mapsto J_\alpha\theta_{01}^2/\theta_{01}^2(\eta)$. Following this substitution, the operator K_2 becomes $K_2\theta_{01}^2/\theta_{01}^2(\eta)$. The redefined J_α's admit a quasiclassical expansion of the form (1-34): $J_\alpha = 1 - \eta^2 J_\alpha + O(\eta^4)_2$, $\eta \to 0$, with $J_1 = 1$, $J_2 = k^2$, and $J_3 = 0$. Now take the new operator K, form the ratio $K_1/K_0 = (K_0 - K_2)/K_0$, and compute it for the representations of series a), b), and c), using (9), (23), and (27): we thus get

$$\begin{aligned} &\text{a)} \quad \frac{K_1}{K_0} = \mathrm{sn}^2(\eta, k)\frac{1}{\mathrm{sn}^2((2l+1)\eta, k)}; \\ &\text{b)} \quad \frac{K_1}{K_0} = \mathrm{sn}^2(\eta, k)\frac{\mathrm{dn}^2}{\mathrm{cn}^2}(2im\eta, k); \\ &\text{c)} \quad \frac{K_1}{K_0} = \mathrm{sn}^2(\eta, k)\cdot k^2\,\mathrm{sn}^2((2n+1)\eta, k). \end{aligned} \tag{28}$$

Passing here to the classical limit $\eta \to 0$, $l\eta \to \lambda$, $m\eta \to \mu$, $n\eta \to \nu$, $K_1/K_0\eta^2 \to K_1/K_0$, we get

$$\text{a)} \quad \frac{K_1}{K_0} = \frac{1}{\mathrm{sn}^2(2\lambda, k)} \in [1, \infty) = [J_1, \infty),$$

$$\text{b)} \quad \frac{K_1}{K_0} = \frac{\mathrm{dn}^2}{\mathrm{cn}^2}(2i\mu, k) \in [k^2, 1] = [J_2, J_1],$$

$$\text{c)} \quad \frac{K_1}{K_0} = k^2\,\mathrm{sn}^2(2\nu, k) \in [0, k^2] = [J_3, J_2],$$

and the conclusion of our theorem is now plain.

Remark. As expressions (28) show, the quantities K_1/K_0 for the representations of series a), b), and c) lie in disjoint intervals, and hence the representations of these three series cannot be mutually equivalent (except, maybe, for values of K_1/K_0 on the boundaries of the intervals of variation).

An interesting question is whether it is possible to produce new series of representations by using combinations of the already known automorphisms and representations. A negative answer is given by the next two propositions.

Proposition 8. The action of the operators X_α on a representation of series a) yields an equivalent representation.

Proof. We may easily guess the form of the unitary operators U_α which establish the stated isomorphisms: $X_\alpha[S_\alpha] = \text{const}\ U^{-1}S_\alpha U_\alpha$. In fact, $(U_1 f)(u) = e^{\pi i l} f(u + K)$; $(U_3 f)(u) = e^{\pi i l} e^{(\pi i/K)l(2u+iK')} f(u + iK')$, $U_2 = U_3U_1$.

Note that the following relations hold: $U_\alpha^2 = (-1)^{2l}$, $U_\alpha U_\beta = (-1)^{2l} U_\beta U_\gamma = U_\gamma$. Therefore, for integer l, the group generated by the operators U_α is commutative and consists of four elements: 1 and U_α, whereas for half-integer l the same group is not commutative and consists of eight elements: ± 1, $\pm U_\alpha$.

Proposition 9. The actions of the operators X_1, X_3, and Y_3 on a representation of series b) give the following results:

$$X_1[S_a(m,\gamma,\delta)] = S_a(m,\gamma+K,\delta);$$
$$X_3[S_a(m,\gamma,\delta)] = V_3^{-1}S_a(m+K'/\eta,\gamma,\delta-\gamma/K)V_3;$$
$$Y_3[S_a(m,\gamma,\delta)] = W_3^{-1}S_a(m+K'/2\eta,\gamma,\delta-\gamma/2K-p/2)W_3,$$

where

$$(V_3\varphi)_j = \exp\left(-2\pi i\frac{p}{q}j^2\right)\varphi_j;$$
$$(W_3\varphi)_j = \exp\left(-\pi i\frac{p}{q}j(j-q)\right)\varphi_j.$$

3. Degenerate Cases

In the previous two sections we have discussed the foundations of representation theory for algebra $\mathcal{F}$ in case 1a) according to the classification of Theorem 1. A detailed analysis of the second generic case, 1b), and of the case 1c), which occupies and intermediate position between 1a) and 1b), can form the subject of a separate investigation and goes beyond the purpose of the present paper. Here we merely note that in case 1b) the representation theory for algebra $\mathcal{F}$ can be developed, as in case 1a), proceeding from the model representation (6), where one must now replace η by $\eta + iK'$. As for the case 1c), it is obtained from case 1a) in the limit $k \to 1$, $\eta = 2K\zeta$, $\zeta = \text{const}$.

In this section we present a number of results concerning the cases 2a) and 2a'), and which appear as degenerate cases of 1a) when $J_{23} \to 0$ ($k \to 0$) and, respectively, $J_{12} \to 0$ ($k \to 1$). This cases have important physical applications, because to them correspond integrable systems such as the XXZ-magnet of the type "light plane" (2a) or "light axis" (2a'), as well as the sine-Gordon lattice model (2a). On the other hand, the investigation of these degenerate cases may serve as a source of conjectures concerning the general case. In particular, L. A. Takhtadzhyan has informed the author that he succeeded in proving for cases 2a) and 2a') the conjectures formulated in [1] concerning the dimension of the space of homogeneous polynomials in algebra $\mathcal{F}$ (the analog of the Poincaré–Birkhoff–Witt theorem) and the structure of the center of $\mathcal{F}$ (in the general case these conjectures are still open). The conjectures formulated in Section 1 of the present paper concerning the positive-definiteness of the inner product and the form of the reproducing kernel (17) for representations of series a) are based on the consideration of precisely these cases.

We treat first case 2a). The appropriate parametrization of the structure constants $J_{\alpha\beta}$: $J_{12} = \tan^2\eta$, $J_{23} = 0$, $J_{31} = -\tan^2\eta$, $\eta \in (0, \pi/2)$, is obtained from (7) in the limit $k \to 0$. The representations of algebra $\mathcal{F}$ in case 2a) also result by passing to the limit in representations of series 1a). Thus, representation (6) of series a) yields the following representation (up to an inessential constant factor), following the similarity transformation $S_a \mapsto U^{-1}S_aU$, with $(Uf)(u) = \exp(-(\pi i/K)lu)\, f(u - k/2 - i(K'/2))$, and in the limit $k \to 0$:

$$(S_0f)(u) = s_0[f(u+\eta) + f(u-\eta)],$$
$$(S_af)(u) = s_a(u-l\eta)f(u+\eta) - s_a(u+l\eta)f(u-\eta), \tag{29}$$

where

$$s_0 = \sin\eta,\quad s_1(u) = -i\cos\eta,\quad s_2(u) = -\sin 2u,\quad s_3(u) = \cos 2u.$$

Moreover, the Casimir operators (9) become

$$K_0 = S_0^2 + S_1^2 + S_2^2 + S_3^2 = 4\sin^2(2l+1)\eta;$$
$$K_2 = \frac{\cos 2\eta}{\cos^2\eta}S_1^2 + S_2^2 + S_3^2 = 4\sin 2(l+1)\eta\sin 2l\eta. \tag{30}$$

This representation acts in the $(2l+1)$-dimensional space of trigonometric polynomials, spanned by the basis $f_j(u) = \exp(2i(j-l)u)$, $j = 0, 1, \ldots, 2l$. This space is obtained from the image of Θ_{00}^{4l+} under U by letting $k \to 0$. Formula (10) for the inner product takes the form $(f, g) = \int_0^\pi dx \int_{-\infty}^{\infty} dy\, \overline{f(u)}\, g(u)\mu(u,\bar u)$, where $\mu(u,\bar u) = \text{const}\Big/\prod_{j=0}^{2l+1}\cos(u-\bar u+(2j-2l-1)\eta)$. With the aid of basis f_j the reproducing kernel $E_l(u, \bar{u})$, given by formula (16) is readily seen to be

$$E_l(u, \bar{u}) = \text{const} \prod_{j=0}^{2l-1} \cos(u - \bar{u} + (2j - 2l + 1)\eta).$$

Note that f_j are eigenvectors of the commuting operators S_0 and S_1. The presence of such a pair of commuting operators, which may be thought of as the analog of a Cartan subalgebra in a Lie algebra, is a characteristic feature of the degenerate cases 2a) and 2a') (as well as 2b), 2c), and 3)) of algebra $\mathcal{F}$, which is absent in the general case. The similarity with a Lie algebra is further strengthened by the fact that the operators $S_\pm = (S_2 \pm iS_3)/2$ play the same role as the raising and lowering operators in the Lie algebra $su(2)$, as shown by the following formulas:

$$S_0 f_j = 2\sin\eta\cos 2\eta\,(j - l)\cdot f_j;\quad S_1 f_j = 2\cos\eta\sin 2\eta\,(j - l)\cdot f_j;$$
$$S_+ f_j = \sin 2\eta\,(2l - j)\cdot f_{j+1};\qquad S_- f_j = \sin 2\eta j\cdot f_{j-1}. \tag{31}$$

Using (30) and (31) we can easily derive the relation between the norms of the vectors f_j and f_{j+1}:

$$(f_{j+1}, f_{j+1}) = \frac{(S_+ f_j, S_+ f_j)}{\sin^2 2\eta\,(2l - j)} = \frac{(f_j, S_- S_+ f_j)}{\sin^2 2\eta\,(2l - j)} = \frac{(f_j, [K_0 - (S_0 + S_1)^2] f_j)}{4\sin^2 2\eta\,(2l - j)} = \frac{\sin 2\eta\,(j + 1)}{\sin 2\eta\,(2l - j)}(f_j, f_j). \tag{32}$$

An analysis of (32) should convince the reader that the condition $4l\eta < \pi$ is sufficient to guarantee that $(f_j, f_j) > 0$ for $j = 1, \ldots, 2l$, if $(f_0, f_0) > 0$. This is precisely the result which motivated the conjecture advanced in Sec. 1 concerning the positive-definiteness of the inner product for a representation of series a), in the nondegenerate case 1a) of algebra $\mathcal{F}$.

Finally, we remark that representations of the type just described were first considered in [7], where explicit formulas are given for the matrix-elements of the generators with respect to a basis of eigenfunctions of the "Cartan subalgebra."

Now consider representations of series b) under the restrictive hypothesis $\eta = \pi p/q$. Letting $k \to 0$ in formulas (24) we obtain the following q-dimensional representation

$$(S_a\varphi)_j = e^{2\pi i\delta} s'_a\,(2\gamma + (2j + 1)\eta - 2im\eta)\,\varphi_{j+1} + \varepsilon_a e_2^{\,\pi i\delta} s'_a\,(2\gamma + (2j - 1)\eta + 2im\eta)\,\varphi_{j-1},$$

where

$$\varphi_{j+q} = \varphi_j,\ \varepsilon_0 = -\varepsilon_1 = \varepsilon_2 = \varepsilon_3 = 1,\ s'_0(u) = \sin\eta,\ s'_1(u) = -i\cos\eta,$$
$$s'_2(u) = \sin u,\ s'_3(u) = -\cos u,\ m \in (-\infty, \infty),\ \gamma \in [0, \pi/q),\ \delta \in [0, 1/q).$$

The appropriate inner product is $(\varphi, \chi) = \sum_{j=0}^{q-1} \bar{\varphi}_j \chi_j$.

The Casimir operators become $K_0 = 4\cosh^2 2m\eta$ and $K_2 = 4\cos(2im + 1)\eta\cos(2im - 1)\eta$. The above representation is equivalent to the representation used in paper [8] to solve the sine-Gordon lattice model.

Here the representations of series c) from case 2a) have no analogs, because, as it is readily seen, Eqs. (25), which define the automorphism Y_3, have no solution for $J_{23} = 0$ ($\alpha = 3$, $\beta = 1$ and $\gamma = 2$).

The case 2a') is analysed similarly. The parametrization of the structure constants $J_{12} = 0$, $J_{23} = \tanh^2\eta$, $J_{31} = -\tanh^2\eta > 0$, is obtained from (7) in the limit $k \to 1$. Subjecting representation (6) of series a) to the similarity transformation $S_\alpha \mapsto U^{-1}S_\alpha U$ with $(Uf)(u) = f(u - K/2)$, and letting $k \to 1$, we obtain a representation of the form (29), where $s_0(u) = \sinh\eta$, $s_1(u) = -\sinh 2u$, $s_2(u) = i\cosh u$, and $s_3(u) = \cosh\eta$.

This representation can be realized in the $(2l + 1)$-dimensional space of trigonometric polynomials spanned by the basis $f_j(u) = \exp(2(j - l)u)$, $j = 0, 1, \ldots, 2l$, and equipped with the inner product $(f, g) = \int_{-\infty}^{\infty} dx \int_0^{\pi} dy\, \overline{f(u)}\, g(u)\, \mu(u, \bar{u})$, where $\mu(u, \bar{u}) = \text{const} \Big/ \prod_{j=0}^{2l+1} \cosh(u + \bar{u} + (2j - 2l - 1)\eta)$.

The corresponding reproducing kernel is $E_l(u,\bar{u}) = \text{const}\prod_{j=0}^{2l-1}\cosh(u+\bar{u}+(2j-2l+1)\eta)$. and the Casimir operators become

$$K_0 = S_0^2 + S_1^2 + S_2^2 + S_3^2 = 4\sinh^2(2l+1)\eta;$$

$$K_2 = S_1^2 + S_2^2 + \frac{\text{ch}\, 2\eta}{\text{ch}^2\eta} S_3^2 = 4\sinh 2(l+1)\eta\, \text{sh}\, 2l\eta.$$

As in case 2a), the operators S_0 and S_3 are diagonal in the basis f_j, whereas the operators $S_\pm = (S_1 \pm iS_2)/2$ shift the index j. The relevant formulas result from (31) and (32) if we replace trigonometric polynomials by hyperbolic ones and S_1 by S_3. The representations of series c) are obtained by the action of the automorphism Y_3 (25) described above, where the parameters are now $y_0 = -y_3^{-1} = \tan\eta$, $y_1 = -y_2^{-1} = -1$.

The representations of series b) have no finite limits as $k \to 1$.

4. Conclusions

The author would consider his task accomplished if he has succeeded in convincing the reader that the subject of [1] and of the present paper, that is, algebra $\mathcal{F}$, provides an interesting and promising area of investigation. Besides being a natural generalization of the universal enveloping algebra of the Lie algebra $su(2)$, algebra $\mathcal{F}$ displays a number of new features, such as, for example, the nontrivial relationship between the classical and quantum cases, and the existence of the series b) and c), which contribute to the diversity and complexity of its theory.

Paper [1] and the present work do not exhaust in the least the circle of problems connected with algebra $\mathcal{F}$. Many important questions are still to be settled. Among them one can mention, for example, all those related to irreducibility of the constructed representations. Also, it is not known whether the constructed representations include all the finite-dimensional irreducible representations of algebra $\mathcal{F}$ (the author hopes that this is the case). The author expresses his deep gratitude to A. M. Vershik, I. M. Gel'fand, A. G. Reiman, M. A. Semenov-Tyan-Shanskii, L. A. Takhtadzhyan, and L. D. Faddeev for the constant interest in this work and for useful observations.

LITERATURE CITED

1. E. K. Sklyanin, "Some algebraic structures connected with the Yang–Baxter equation," Funct. Anal. Appl., 16, No. 4, 27-34 (1982).
2. H. Bateman and A. Erdelyi, Higher Transcendental Functions, Vol. 3. Elliptic and Automorphic Functions. Lamé and Mathieu Functions. [Russian translation], Nauka, Moscow (1967).
3. H. Weber, Lehrbuch der Algebra, B. III. Elliptische Funktionen und Algebraische Zahlen, F. Vieweg und Sohn, Braunschweig (1908).
4. I. V. Cherednik, "On some S-matrices connected with Abelian varieties," Dokl. Akad. Nauk SSSR, 249, No. 5, 1095-1098 (1979).
5. P. Halmos, A Hilbert Space Problem Book, Van Nostrand, Princeton, New Jersey (1967).
6. A. Barut and R. Ronchka, Theory of Group Representations and Its Applications [Russian translation], Nauka, Moscow (1980).
7. P. P. Kulish and N. Yu. Reshetikhin, "The quantum linear problem for the sine-Gordon equation, and higher representations," in: Problems in Quantum Field Theory and Statistical Physics, Vol. 2 (Zap. Nauchn. Sem. LOMI 101), Nauka, Leningrad (1980), pp. 101-110.
8. A. G. Izergin and V. E. Korepin, "Lattice versions of quantum field theory models in two dimensions," Nucl. Phys., B, B205 [FS5], No. 3, 401-413 (1982).

Докл. Акад. Наук СССР
Том 283 (1985), № 5

Soviet Math. Dokl.
Vol. 32 (1985), No. 1

HOPF ALGEBRAS AND THE QUANTUM YANG-BAXTER EQUATION

UDC 512.554.3+512.667.7

V. G. DRINFEL'D

Let $\mathfrak{a}$ be a given finite-dimensional simple Lie algebra over $\mathbf{C}$ with a fixed invariant inner product. According to [1], the function $r(u) = u^{-1}I_\mu \otimes I_\mu$, where $\{I_\mu\}$ is an orthonormal basis in $\mathfrak{a}$ (summation over repeated indices is always assumed to be carried out), satisfies the classical Yang-Baxter equation (CYBE). If, in addition, a representation $\rho\colon \mathfrak{a} \to \operatorname{End} V$ is given, the question arises (see [4]) whether the quantum Yang-Baxter equation (QYBE) has a solution which can be written as a formal series

$$R(u,h) = 1 + h\cdot(\rho\otimes\rho)(r(u)) + \sum_{k=2}^{\infty} A_k(u)h^k, \qquad A_k(u) \in \operatorname{End}(V\otimes V).$$

Here h is Planck's constant. Since $r(u)$ is homogeneous, it is natural to require that $R(\lambda u, \lambda h) = R(u,h)$. We thus need to find out whether a solution of the QYBE exists in the form of a formal series

$$(1) \qquad R(u) = 1 + u^{-1}\rho(I_\mu)\otimes\rho(I_\mu) + \sum_{k=2}^{\infty} R_k u^{-k}, \qquad R_k \in \operatorname{End}(V\otimes V).$$

It is well known (see [5] and [7]) that such solutions exist for classical $\mathfrak{a}$ and for certain ρ. Using the concepts of Lie bialgebras [3] and quantization of Lie bialgebras (see below), we shall construct the Hopf algebra $Y(\mathfrak{a})$ and use it to prove that solutions of the QYBE in the form (1) exist for all ρ if $\mathfrak{a} = \mathfrak{sl}(n)$ and, for other $\mathfrak{a}$, the QYBE has solutions in the form (1) for some but not all ρ. Moreover, we shall indicate (see Theorem 4) an efficient method of constructing solutions of QYBE in the form (1) which generalizes the method devised for $\mathfrak{sl}(n)$ in [7]. In addition to $Y(\mathfrak{a})$, we shall also introduce another class of Hopf algebras (the quantized Kac-Moody algebras) which is apparently related to trigonometric solutions of QYBE. It is the author's opinion that Hopf algebras should be regarded not only as a technical tool for the construction of solutions of QYBE but also as a natural language for the quantum method of the inverse problem.

1. The following heuristic concepts underlie our approach to the quantization of solutions of CYBE. From the solution $r(u)$ of CYBE for an algebra $\mathfrak{a}$ we construct in the standard way (see (2.13) of [8]) the Poisson bracket on the group of currents G corresponding to $\mathfrak{a}$. The group G thereby becomes a Hamilton-Lie group [3] and the algebra of functions on G (denoted by B_0) is a Poisson Hopf algebra, i.e., a Poisson algebra (see [6], §1) equipped with a Poisson homomorphism $\Delta\colon B_0 \to B_0 \otimes B_0$ with the coassociative property. The quantization of $r(u)$ must lead to quantization of B_0, i.e., construction of a Hopf algebra B over $\mathbf{C}[[h]]$ such that $B/hb = B_0$ and $[b_1, b_2] \equiv h\{b_1, b_2\} \bmod h^2$ for $b_1, b_2 \in B$. It is convenient to consider instead of B_0 its dual object, i.e., the algebra $U\mathfrak{g}$, where $\mathfrak{g} = \operatorname{Lie}(G)$. The algebra $U\mathfrak{g}$ is a co-Poisson Hopf algebra, i.e., a vector space equipped with multiplication, comultiplication $\Delta\colon U\mathfrak{g} \to U\mathfrak{g}\otimes U\mathfrak{g}$, and a Poisson cobracket $\delta\colon U\mathfrak{g} \to U\mathfrak{g}\otimes U\mathfrak{g}$ which satisfy axioms dual to the axioms defining a Poisson

1980 *Mathematics Subject Classification* (1985 *Revision*). Primary 81E13; Secondary 17B67, 16A24.

Hopf algebra. Hence, quantization of solutions of CYBE is closely related to quantization of co-Poisson Hopf algebras. Let us now turn to the exact formulation.

THEOREM 1. *Let $\mathfrak{g}$ be a Lie algebra and let $\delta\colon U\mathfrak{g} \to U\mathfrak{g} \otimes U\mathfrak{g}$ be a map that turns $U\mathfrak{g}$ into a co-Poisson Hopf algebra with the standard multiplication and comultiplication. Then $\delta(\mathfrak{g}) \subset \mathfrak{g}\otimes\mathfrak{g}$, and the map $\varphi\colon \mathfrak{g} \to \mathfrak{g}\otimes\mathfrak{g}$ induced by δ turns $\mathfrak{g}$ into a Lie bialgebra. Conversely, if $(\mathfrak{g}, \varphi)$ is a Lie bialgebra, then there is a unique map $\delta\colon U\mathfrak{g} \to U\mathfrak{g} \otimes U\mathfrak{g}$ which is an extension of φ and which turns $U\mathfrak{g}$ into a co-Poisson Hopf algebra.*

DEFINITION. Let $(\mathfrak{g}, \varphi)$ be a Lie algebra and $\delta\colon U\mathfrak{g} \to U\mathfrak{g} \otimes U\mathfrak{g}$ the corresponding Poisson cobracket. *Quantization* of $(\mathfrak{g}, \varphi)$ is a topological Hopf algebra (A, Δ) over $\mathbf{C}[[h]]$ which is a topologically free $\mathbf{C}[[h]]$-module and satisfies the following conditions: 1) A/hA is identical with $U\mathfrak{g}$ as a Hopf algebra, and 2) $h^{-1}(\Delta(a) - \sigma(\Delta(a))) \bmod h = \delta(a \bmod h)$ for $a \in A$, where $\sigma\colon A \otimes A \to A \otimes A$ and $\sigma(x \otimes y) = y \otimes x$.

To a solution $r(u) = u^{-1} I_\mu \otimes I_\mu$ of a CYBE corresponds a Lie bialgebra $(\mathfrak{g}, \varphi)$, where $\mathfrak{g} = \mathfrak{a}[u]$ and $\varphi(a(u)) = [a(u) \otimes 1 + 1 \times a((v), r(u-v)]$ (we identify $\mathfrak{a}[u] \otimes \mathfrak{a}[u]$ with $\mathfrak{a} \otimes \mathfrak{a}[u]$). Since $\mathfrak{g}$ is a graded algebra ($\mathfrak{g} = \bigoplus_k \mathfrak{a}u^k$) and φ is homogeneous of degree -1, it is natural to impose the following condition of homogeneity on the quantization (A, Δ) of the bialgebra $(\mathfrak{g}, \varphi)$: (A, Δ) should be a graded Hopf algebra over the graded algebra $\mathbf{C}[[h]]$ (we assume that $\deg h = 1$) and the grading of $\mathfrak{g}$ and grading of A induce the same grading in $U\mathfrak{g}$.

THEOREM 2. *The bialgebra $(\mathfrak{a}[u], \varphi)$ admits a unique homogeneous quantization (A, Δ). The algebra A regarded as an associative topological algebra with unity is generated by elements I_λ and J_λ with defining relations*

$$(2)\qquad [I_\lambda, I+\mu] = c_{\lambda\mu\nu} I_\nu, \qquad [I_\lambda, J_\mu] = c_{\lambda\mu\nu} J_\nu;$$

$$(3)\qquad [J_\lambda, [J_\mu, I_\nu]] - [I_\lambda, [J_\mu, J_\nu]] = h^2 a_{\lambda\mu\nu\alpha\beta\gamma}\{I_\alpha, I_\beta, I_\gamma\};$$

$$(4)\qquad \begin{aligned}&[[J_\lambda, J_\mu], [I_r, J_s]] + [[J_r, J_s], [I_\lambda, J_\mu]]\\ &\quad = h^2 (a_{\lambda\mu\nu\alpha\beta\gamma} c_{rs\nu} + a_{rs\nu\alpha\beta\gamma} c_{\lambda\mu\nu})\{I_\alpha, I_\beta, J_\gamma\},\end{aligned}$$

where the $c_{\lambda\mu\nu}$ are the structure constants of $\mathfrak{a}$, and

$$a_{\lambda\mu\nu\alpha\beta\gamma} = \tfrac{1}{24} c_{\lambda\alpha i} c_{\mu\beta j} c_{\nu\gamma k} c_{ijk}, \qquad \{x_1, x_2, x_3\} = \sum_{i\neq j\neq k} x_i x_j x_k.$$

Here, $\deg I_\lambda = 0$ *and* $\det J_\lambda = 1$. *Moreover,*

$$(5)\qquad \Delta(I_\lambda) = I_\lambda \otimes 1 + 1 \otimes I_\lambda, \qquad \Delta(J_\lambda) = J_\lambda \otimes 1 + 1 \otimes J_\lambda + \tfrac{1}{2} h c_{\lambda\mu\nu} I_\nu \otimes I_\mu.$$

REMARK. For $\mathfrak{a} = \mathfrak{sl}(2)$, we find that (2) implies (3), and for $\mathfrak{a} = \mathfrak{sl}(2)$, (4) follows from (2) and (3).

We shall call the *Yangian* of $\mathfrak{a}$, denoted by $Y(\mathfrak{a})$ (in honor of C. N. Yang, who found the first solution of the QYBE in the form (1); see [9]), the Hopf algebra over $\mathbf{C}$ obtained by setting $h = 1$ in (2)–(5). In other words, $Y(\mathfrak{a}) = A'/(h-1)A'$, where A' is the algebraic direct sum of homogeneous components of A.

2. For any $a, b \in \mathbf{C}$, we define $T_{a,b}\colon Y(\mathfrak{a}) \otimes Y(\mathfrak{a}) \to Y(\mathfrak{a}) \otimes Y(\mathfrak{a})$ by $T_{a,b} = T_a \otimes T_b$, where T_a is an automorphism of $Y(\mathfrak{a})$ such that $T_a(J_\mu) = J_\mu + aI_\mu$ and $T_a(I_\mu) = I_\mu$. We denote by Δ' the composition of the comultiplication $\Delta\colon Y(\mathfrak{a}) \to Y(\mathfrak{a}) \otimes Y(\mathfrak{a})$ and the operator of transposition of factors $Y(\mathfrak{a}) \otimes Y(\mathfrak{a}) \to Y(\mathfrak{a}) \otimes Y(\mathfrak{a})$.

THEOREM 3. *There is a unique formal series*

$$\mathcal{R}(u) = 1 + \sum_{k=1}^{\infty} \mathcal{R}_k u^{-k}, \qquad \mathcal{R}_k \in Y(\mathfrak{a}) \otimes Y(\mathfrak{a})$$

such that $(\mathrm{id}\otimes\Delta)\mathcal{R}(u)=\mathcal{R}^{12}(u)\mathcal{R}^{13}(u)$ *and*

$$T_{0,\mu}\Delta'(a)=\mathcal{R}(u)^{-1}(T_{0,\mu}\Delta(a))\mathcal{R}(u)\quad \text{for } a\in Y(\mathfrak{a});$$

this $\mathcal{R}(u)$ *satisfies the QYBE. Moreover,*

$$\mathcal{R}^{12}(u)\mathcal{R}^{21}(-u)=1,\qquad T_{v,w}\mathcal{R}(u)=\mathcal{R}(u+w-v),$$
$$\ln\mathcal{R}(u)=u^{-1}I_\mu\otimes I_\mu+u^{-2}(J_\mu\otimes I_\mu-I_\mu\otimes J_\mu)+O(u^{-3}).$$

Let $\rho\colon Y(\mathfrak{a})\to \mathrm{Mat}(n,\mathbf{C})$ be an irreducible representation $R_\rho(u)=(\rho\otimes\rho)(\mathcal{R}(u))$. Then $R_\rho(u)$ is a solution of the QYBE in the form (1).

THEOREM 4. *$R_\rho(u)$ is, up to a scalar factor, a unique solution of the equation*

$$P_\lambda^+(u,v)R(v-u)=R(v-u)P_\lambda^-(u,v),$$

where $P_\lambda^\pm(u,v)=(\rho(J_\lambda)+u\rho(I_\lambda))\otimes 1+1\otimes(\rho(J_\lambda)+v\rho(I_\lambda))\pm\frac{1}{2}c_{\lambda\mu\nu}I_\nu\otimes I_\mu$. *In particular, $R_\rho(u)$ is, up to a scalar factor, a rational function of u.*

EXAMPLE 1. Let $\mathfrak{a}=\mathfrak{sl}(n)$, $(X,Y)=\mathrm{Tr}(XY)$ for $X,Y\in\mathfrak{a}$, $\rho\colon Y(\mathfrak{a})\to\mathrm{Mat}(n,\mathbf{C})$ and $\rho(I_\mu)=I_\mu$, $\rho(J_\mu)=0$. Then $R_\rho(u)=f(u)(1+u^{-1}\sigma)$, where $\sigma\colon \mathbf{C}^n\otimes\mathbf{C}^n\to\mathbf{C}^n\otimes\mathbf{C}^n$ is the transposition of factors and $f(u)\in 1+u^{-1}\mathbf{C}[[u^{-1}]]$ is determined by $\prod_{k=1}^n f(u-k)=1-u^{-1}$.

EXAMPLE 2. Let $\mathfrak{a}=\mathfrak{o}(n)$, $(X,Y)=\frac{1}{2}\mathrm{Tr}(XY)$, $\rho\colon Y(\mathfrak{a})\to\mathrm{Mat}(n,\mathbf{C})$, $\rho(I_\mu)=I_\mu$ and $\rho(J_\mu)=0$. Then $R_\rho(u)=(1+u^{-1})\varphi(u)R(u)$, where $R(u)$ is defined by formula (A.5) in the Appendix to [4] for $\eta=1$ and $\varphi(u)\in 1+u^{-1}\mathbf{C}[[u^{-1}]]$ can be obtained from $\varphi(u)\varphi(u+1-n/2)=(1-u^{-2})^{-1}$.

THEOREM 5. *Let $\rho\colon\mathfrak{a}\to\mathrm{Mat}(n,\mathbf{C})$ be an irreducible representation and $R(u)$ the solution of the QYBE in the form (1). It is then possible to extend ρ to a representation $\pi\colon Y(\mathfrak{a})\to\mathrm{Mat}(n,\mathbf{C})$ so that $R(u)=f(u)R_\pi(u)$, where $f(u)$ is a scalar. The representation π is determined uniquely by $R(u)$ and ρ up to replacement of π by $\pi\circ T_a$, $a\in\mathbf{C}$.*

Assuming that the hypotheses of Theorem 4 hold, we denote by A_ρ the associative algebra with unity over $\mathbf{C}$ with generators $t_{ij}^{(k)}$, $1\le i,j\le n$, $k\in\mathbf{N}$, and with the defining relation

$$R_\rho(u-v)\cdot(T(u)\otimes E_n)\cdot(E_n\otimes T(v))=(E_n\otimes T(v))\cdot(T(u)\otimes E_n)\cdot R_\rho(u-v),$$

where E_n is an identity matrix of order n, $T(u)=(t_{ij}(u))$ and $t_{ij}(u)=\delta_{ij}+\sum_{k=1}^\infty t_{ij}^{(k)}u^{-k}$ (this relation is central to the theory of quantum integrable systems [8]). There is a homomorphism $\Delta\colon A_\rho\to A_\rho\otimes A_\rho$ such that $\Delta(t_{ij}(u))=\sum_k t_{ik}(u)\otimes t_{kj}(u)$ and (A_ρ,Δ) is a Hopf algebra.

THEOREM 6. *There is an epimorphism of Hopf algebras $\varphi\colon A_\rho\to Y(\mathfrak{a})$ such that $\varphi(T(u))=(\rho\otimes\mathrm{id})\mathcal{R}(u)$. There are elements $c_1,c_2,\ldots$ in the center of A_ρ generating $\mathrm{Ker}\,\varphi$ as an ideal and such that $\Delta(c(u))=c(u)\otimes c(u)$, where $c(u)=1+\sum_i c_iu^{-i}$. In particular, in the situation of Example 1, $c(u)$ can be taken as the quantum determinant $T(u)$ (see [8]) and, in Example 2, we can set $c(u)=a(u)$, where $a(u)$ is determined by*

$$\sum_i t_{ij}(u)t_{ik}\left(u+\frac{n}{2}-1\right)=a(u)\cdot\delta_{jk}.$$

3. One way to construct the representation $Y(\mathfrak{a})$ is to choose a representation ρ of the algebra $\mathfrak{a}$ such that $\rho(L_{\lambda\mu\nu})=0$, where $L_{\lambda\mu\nu}=a_{\lambda\mu\nu\alpha\beta\gamma}\{I_\alpha,I_\beta,I_\gamma\}$, and setting $\rho(J_\mu)=0$. We choose a simple root α of the algebra $\mathfrak{a}$ and denote by ρ_t the irreducible representation of $\mathfrak{a}$ with highest weight $t\omega_\alpha$, where ω_α is the fundamental weight. Let n_α be the coefficient of α in the expansion of the maximum root $\alpha_{\max}$. It can easily be seen that $k_\alpha|n_\alpha$, where $k_\alpha=(\alpha_{\max},\alpha_{\max})/(\alpha,\alpha)$.

THEOREM 7. *If $n_\alpha = 1$, then $\rho_t(L_{\lambda\mu\nu}) = 0$ for all t. If $n_\alpha = k_\alpha$, then $\rho_1(L_{\lambda\mu\nu}) = 0$.*

REMARK. Solutions of the QYBE corresponding to representations of $\mathfrak{a}$ with the highest weight $t\omega_\alpha$, $n_\alpha = 1$, were obtained in [5] for classical $\mathfrak{a}$.

Theorem 7 yields examples of nontrivial finite-dimensional representations of $Y(\mathfrak{a})$ for all $\mathfrak{a}$ except for E_8. The following theorem implies the existence of such representations for all $\mathfrak{a}$.

THEOREM 8. *Let V be a one-dimensional space, and let $v_0 \in V$, $v_0 \neq 0$. If $\mathfrak{a} = \mathfrak{sl}(2)$, then for any $b \in \mathbf{C}$ there is a representation $\rho\colon Y(\mathfrak{a}) \to \mathrm{End}(\mathfrak{a} \otimes V)$ such that $\rho(I_\mu)v_0 = 0$, $\rho(J_\mu)v_0 = bI_\mu$, $\rho(I_\mu)x = [I_\mu, x]$ and $\rho(J_\mu)x = (x, I_\mu)v_0$ for $x \in \mathfrak{a}$. If $\mathfrak{a} \neq \mathfrak{sl}(2)$, then such ρ exists for a unique b, namely for $b = s^3 c(\mathfrak{a})$, where s is the ratio of the Killing form to a given inner product on $\mathfrak{a}$, $c(\mathfrak{sl}(n)) = -(32n^2)^{-1}$, $c(\mathfrak{o}(n)) = -(n-4)/16(n-2)^3$, $c(\mathfrak{sl}(n)) = -(n+4)/16(n+2)^3$ and $c(\mathfrak{a}) = -5/144(\dim \mathfrak{a} + 2)$ for exceptional $\mathfrak{a}$.*

THEOREM 9. *Let $\mathfrak{a} = \mathfrak{sl}(n)$, and let $\mathrm{Tr}(XY)$ be taken as the inner product of $\mathfrak{a}$. Then there is a homomorphism of algebras (but not a Hopf-algebra homomorphism) $\varphi\colon Y(\mathfrak{a}) \to U\mathfrak{a}$ such that $\varphi(I_\lambda) = I_\lambda$ and $\varphi(J_\lambda) = \frac{1}{4} d_{\lambda\alpha\beta} I_\alpha I_\beta$, where $d_{\lambda\alpha\beta} = \mathrm{Tr}(I_\lambda I_\alpha I_\beta + I_\lambda I_\beta I_\alpha)$.*

For $\mathfrak{a} = \mathfrak{sl}(n)$ it follows from Theorems 3 and 9 that there is a solution of the QYBE in the form

$$R(u) = 1 + u^{-1}(I_\mu \otimes I_\mu) + \sum_{k=2}^{\infty} R_k u^{-k}, \qquad R_k \in U\mathfrak{a} \otimes U\mathfrak{a}, \tag{6}$$

and hence a solution in the form (1) for any ρ. If $\mathfrak{a}$ does not belong to the A series, then there are no solutions of the QYBE in the form (6). In fact, it can be shown that in that case there are representations $\rho\colon \mathfrak{a} \to \mathrm{End}\, V$ (for example, the adjoint representation) which cannot be extended to a representation $Y(\mathfrak{a}) \to \mathrm{End}\, V$, and hence there are no solutions of the QYBE in the form (1) for such ρ.

4. Assume that a matrix $A = (A_{ij})$ satisfies the conditions $A_{ij} \in \mathbf{Z}$, $A_{ii} = 2$ and $A_{ij} < 0$ for $i \neq j$, and let there be $d_i \neq 0$ such that $d_i A_{ij} = d_j A_{ji}$. We choose a fixed d_i and denote by $\mathfrak{g}_A$ the Lie algebra with generators X_i, Y_i, H_i and the Kac-Moody relations

$$[X_i, Y_j] = 0 \quad \text{for } i \neq j, \quad [H_i, H_j] = 0, \quad [H_i, X_j] = A_{ij} X_j, \quad [H_i, Y_j] = -A_{ij} Y_j; \tag{7}$$

$$[X_i, Y_i] = H_i, \quad (\mathrm{ad}\, X_i)^{1-A_{ij}} X_j = (\mathrm{ad}\, Y_i)^{1-A_{ij}} Y_j = 0 \quad \text{for } i \neq j. \tag{8}$$

Then there is a 1-cocycle $\varphi\colon \mathfrak{g}_A \to \Lambda^2 \mathfrak{g}_A$ such that $\varphi(H_i) = 0$, $\varphi(X_i) = d_i X_i \wedge H_i$ and $\varphi(Y_i) = d_i Y_i \wedge H_i$; $(\mathfrak{g}_A, \varphi)$ is then a Lie bialgebra. It turns out that this bialgebra admits quantization. Let us denote by Q_A the associative algebra with unity over $\mathbf{C}[[h]]$ which is complete in the h-adic topology and whose generators X_i, Y_i, H_i satisfy the defining relations (7) and also

$$[X_i, Y_i] = (d_i h)^{-1} \cdot \sinh d_i h H_i,$$

$$\sum_{r=0}^{N} (-1)^r \check{C}_N^r(d_i h) X_i^r X_j X_i^{N-r} = \sum_{r=0}^{N} (-1)^r \check{C}_N^r(d_i h) Y_i^r Y_j Y_i^{N-r} = 0 \quad \text{for } i \neq j, \tag{8'}$$

$$N = 1 - A_{ij}.$$

Here,

$$\check{C}_N^r(\alpha) = \prod_{k=N-r+1}^{N} \sinh k\alpha \Big/ \prod_{k=1}^{r} \sinh k\alpha.$$

We can show that there is a homomorphism $\Delta\colon Q_A \to Q_A \hat{\otimes} Q_A$ such that

$$\Delta(H_i) = H_i \otimes 1 + 1 \otimes H_i,$$
$$\Delta(X_i) = X_i \otimes \exp\left(\tfrac{1}{2}d_i hH_i\right) + \exp\left(-\tfrac{1}{2}d_i hH_i\right) \otimes X_i,$$
$$\Delta(Y_i) = Y_i \otimes \exp\left(\tfrac{1}{2}d_i hH_i\right) + \exp\left(-\tfrac{1}{2}d_i hH_i\right) \otimes Y_i$$

and that (Q_A, Δ) is a quantization of $(\mathfrak{g}_A, \varphi)$. If A is the Cartan matrix of an affine system of roots, then the quotient of the bialgebra $\mathfrak{g}_A$ over its center corresponds to the trigonometric solution of the CYBE given by (6.1) of [2]. It thus seems likely that there is a relation between Q_A and trigonometric solutions of QYBE similar to the relation between $Y(\mathfrak{a})$ and rational solutions of QYBE. Moreover, as Leznov first pointed out, the algebras Q_A are closely related to quantization of the Liouville equation and of its analogs corresponding to simple Lie algebras; formula (30) of [10] is based on the relations (7) and (8′).

We also note that the algebra Q_A for a matrix A of order 1 was introduced already in [11] and [12] (in [11] as an algebra and in [12] as a Hopf algebra).

The author is grateful to A. A. Belavin, I. M. Gel′fand, A. V. Zelevinskiĭ, Yu. I. Manin, B. L. Feĭgin, and to the staff of the Laboratory of Mathematical Methods in Physics of the Leningrad Branch of the Steklov Institute of Mathematics for their valuable comments.

Physico-Technical Institute of Low Temperatures
Academy of Sciences of the Ukrainian SSR
Kharkov

Received 22/APR/85

BIBLIOGRAPHY

1. A. A. Belavin, Funktsional. Anal. i Prilozhen. **14** (1980), no. 4, 18 26; English transl. in Functional Anal. Appl. **14** (1980).

2. A. A. Belavin and V. G. Drinfel′d, Funktsional. Anal. i Prilozhen. **16** (1982), no. 3, 1-29; English transl. in Functional Anal. Appl. **16** (1982).

3. V. G. Drinfel′d, Dokl. Akad. Nauk SSSR **268** (1983), 285-287; English transl. in Soviet Math. Dokl. **27** (1983).

4. P. P. Kulush and E. K. Sklyanin, Zap. Nauchn. Sem. Leningrad. Otdel. Mat. Inst. Steklov. (LOMI) **95** (1980), 129–160; English transl. in J. Soviet Math. **19** (1962), no. 5.

5. N. Yu. Reshetikhin, Teoret. Mat. Fiz. **63** (1985), 347–366; English transl. in Theoret. Math. Phys. **63** (1985).

6. E. K. Sklyanin, Funktsional. Anal. i Prilozhen. **17** (1983), no. 4, 34–48; English transl. in Functional Anal. Appl. **17** (1983).

7. P. P. Kulish, N. Yu. Reshetikhin and E. K. Sklyanin, Lett. Math. Phys. **5** (1981), 393–403.

8. P. P. Kulish and E. K. Sklyanin, Integrable Quantum Field Theories (Tvärminne, 1981), Lecture Notes in Phys., vol. 151, Springer-Verlag, 1982, pp. 61–119.

9. C. N. Yang, Phys. Rev. Lett. **19** (1967), 1312–1315.

10. A. N. Leznov and I. A. Fedoseev, Teoret. Mat. Fiz. **53** (1982), 358–373; English transl. in Theoret. Math. Phys. **53** (1982).

11. P. P. Kulish and N. Yu. Reshetikhin, Zap. Nauchn. Sem. Leningrad. Otdel. Mat. Inst. Steklov (LOMI) **101** (1981), 101–110; English transl. in J. Soviet Math. **23** (1983), no. 4.

12. E. K. Sklyanin, Uspekhi Mat. Nauk **40** (1985), no. 2(242), 214. (Russian)

Translated by D. MATHON and J. MATHON

Proceedings of the International Congress of Mathematicians
Berkeley, California, USA, 1986

Reprinted with permission from American Mathematical Society.

Quantum Groups

V. G. DRINFEL'D

This is a report on recent works on Hopf algebras (or quantum groups, which is more or less the same) motivated by the quantum inverse scattering method (QISM), a method for constructing and studying integrable quantum systems, which was developed mostly by L. D. Faddeev and his collaborators. Most of the definitions, constructions, examples, and theorems in this paper are inspired by the QISM. Nevertheless I will begin with these definitions, constructions, etc. and then explain their relation to the QISM. Thus I reverse the history of the subject, hoping to make its logic clearer.

1. What is a quantum group? Recall that both in classical and in quantum mechanics there are two basic concepts: state and observable. In classical mechanics states are points of a manifold M and observables are functions on M. In the quantum case states are 1-dimensional subspaces of a Hilbert space H and observables are operators in H (we forget the self-adjointness condition). The relation between classical and quantum mechanics is easier to understand in terms of observables. Both in classical and in quantum mechanics observables form an associative algebra which is commutative in the classical case and noncommutative in the quantum case. So quantization is something like replacing commutative algebras by noncommutative ones.

Now let us consider elements of a group G as states and functions on G as observables. The notion of group is usually defined in terms of states. To quantize it one has to translate it first into the language of observables. This translation is well known, but let us recall it nevertheless. Consider the algebra $A = \text{Fun}(G)$ consisting of functions on G which are supposed to be smooth if G is a Lie group, regular if G is an algebraic group, and so on. A is a commutative associative unital algebra. Clearly $\text{Fun}(G \times G) = A \otimes A$ if one understands the sign "$\otimes$" in the appropriate sense (e.g. if G is a Lie group then $\otimes$ should be understood as the topological tensor product). So the group operation considered as a mapping $f: G \times G \to G$ induces an algebra homomorphism $\Delta: A \to A \otimes A$ called "comultiplication." To formulate the associativity property of the group operation in terms of Δ one expresses it as the commutativity of the diagram

$$\begin{array}{ccccc} & \stackrel{f\times \mathrm{id}}{\nearrow} & G\times G & \stackrel{f}{\searrow} & \\ G\times G\times G & & & & G \\ & \stackrel{\mathrm{id}\times f}{\searrow} & G\times G & \stackrel{f}{\nearrow} & \end{array}$$

and then the functor $X \mapsto \mathrm{Fun}(X)$ is applied. The result is the coassociativity property of Δ, i.e., the commutativity of the diagram

$$\begin{array}{ccccc} & \stackrel{\Delta}{\nearrow} & A\otimes A & \stackrel{\mathrm{id}\otimes\Delta}{\searrow} & \\ A & & & & A\otimes A\otimes A \\ & \stackrel{\Delta}{\searrow} & A\otimes A & \stackrel{\Delta\otimes\mathrm{id}}{\nearrow} & \end{array}$$

The translation of the notions of the group unit e and the inversion mapping $g \mapsto g^{-1}$ is quite similar. We consider the mappings $\varepsilon\colon A \to k$, $S\colon A \to A$ given by $\varphi \mapsto \varphi(e)$ and $\varphi(g) \mapsto \varphi(g^{-1})$ respectively (here k is the ground field, i.e. $k = \mathbf{R}$ if G is a real Lie group, $k = \mathbf{C}$ if G is a complex Lie group, etc). Then we translate the identities $g \cdot e = e \cdot g = g$, $g \cdot g^{-1} = g^{-1} \cdot g = e$ into the language of commutative diagrams and apply the functor $X \mapsto \mathrm{Fun}(X)$. As a result we obtain the commutative diagrams

$$\begin{array}{ccc} A & \xrightarrow{\mathrm{id}} & A \\ {\scriptstyle\Delta}\downarrow & & \| \\ A\otimes A & \xrightarrow{\mathrm{id}\otimes\varepsilon} & A\otimes k \end{array} \qquad \begin{array}{ccc} A & \xrightarrow{\mathrm{id}} & A \\ {\scriptstyle\Delta}\downarrow & & \| \\ A\otimes A & \xrightarrow{\varepsilon\otimes\mathrm{id}} & k\otimes A \end{array} \qquad (1)$$

$$\left.\begin{array}{c} \begin{array}{ccccccc} A & \xrightarrow{\Delta} & A\otimes A & \xrightarrow{\mathrm{id}\otimes S} & A\otimes A & \xrightarrow{m} & A \\ & \stackrel{\varepsilon}{\searrow} & & k & & \stackrel{i}{\nearrow} & \end{array} \\ \\ \begin{array}{ccccccc} A & \xrightarrow{\Delta} & A\otimes A & \xrightarrow{S\otimes\mathrm{id}} & A\otimes A & \xrightarrow{m} & A \\ & \stackrel{\varepsilon}{\searrow} & & k & & \stackrel{i}{\nearrow} & \end{array} \end{array}\right\} \qquad (2)$$

Here m is the multiplication (i.e. $m(a \otimes b) = ab$) and $i(c) = c \cdot 1_A$. The commutativity of (1) (resp. (2)) is expressed by the words "ε is the counit" (resp. "S is the antipode"). The properties of $(A, m, \Delta, i, \varepsilon, S)$ listed above mean that A is a commutative Hopf algebra.

Now there is a general principle: the functor $X \mapsto \mathrm{Fun}(X)$ from the category of "spaces" to the category of commutative associative unital algebras (perhaps, with some additional structures or properties) is an antiequivalence. This principle becomes a theorem if "space" is understood as "affine scheme," or if "space" is understood as "compact topological space" and "algebra" is understood as "C^*-algebra." From the above principle it follows that the category of groups is antiequivalent to the category of commutative Hopf algebras.

Now let us define the category of quantum spaces to be dual to the category of (not necessarily commutative) associative unital algebras. Denote by Spec A (spectrum of A) the quantum space corresponding to an algebra A. Finally define a quantum group to be the spectrum of a (not necessarily commutative) Hopf algebra. So the notions of Hopf algebra and quantum group are in fact equivalent, but the second one has some geometric flavor.

Let me make some general remarks on the definition of Hopf algebra. First of all, it is known that for given A, m, Δ the counit $\varepsilon : A \to k$ is unique and is a homomorphism. The antipode $S : A \to A$ is also unique and it is an antihomomorphism (with respect to both multiplication and comultiplication). Secondly, in the noncommutative case one also requires the existence of the "skew antipode" $S' : A \to A$ which is in fact the antipode for the opposite multiplication and the same comultiplication. S' is also the antipode for the opposite comultiplication and the same multiplication. It is known that $SS' = S'S = \mathrm{id}$. In the commutative or cocommutative case $S' = S$, but in general $S' \neq S$ and $S^2 \neq \mathrm{id}$. The proofs of these statements can be found in standard texts on Hopf algebras [**1–3**]. An informal discussion of some aspects of the notion of Hopf algebra can be found in [**4**]. The reader should keep in mind that our usage of the term "Hopf algebra" is not generally accepted: some people do not require the existence of the unit, the counit and the antipode (so their Hopf algebras correspond to quantum semigroups).

It is important that a quantum group is not a group, nor even a group object in the category of quantum spaces. This is because for noncommutative algebras the tensor product is not a coproduct in the sense of category theory.

Now the question arises whether there exist natural examples of noncommutative Hopf algebras. An easy way of constructing such an algebra is to consider A^*, A being a commutative but not cocommutative algebra (recall that the dual space of a Hopf algebra A has a Hopf algebra structure, the multiplication mapping $A^* \otimes A^* \to A^*$ being induced by the comultiplication of A and the comultiplication of A^* being induced by the multiplication of A). In this way more or less all cocommutative noncommutative Hopf algebras are obtained. The words "more or less" are due to the fact that $V^{**} \neq V$ for a general vector space V. But at the heuristic level $V^{**} = V$ and therefore any cocommutative Hopf algebra is of the form $(\mathrm{Fun}(G))^*$ for some group G. Clearly $(\mathrm{Fun}(G))^*$ is commutative iff G is commutative. Note that $(\mathrm{Fun}(G))^*$ is nothing but the group algebra of G. An important class of cocommutative Hopf algebras is formed by universal enveloping algebras (the comultiplication $U\mathfrak{g} \to U\mathfrak{g} \otimes U\mathfrak{g}$ is defined by the formula $\Delta(x) = x \otimes 1 + 1 \otimes x$, $x \in \mathfrak{g}$). If $\mathfrak{g}$ is the Lie algebra of a Lie group G then $U\mathfrak{g}$ may be considered as the subalgebra of $(C^\infty(G))^*$ consisting of distributions $\varphi \in C_0^{-\infty}(G)$ such that $\mathrm{Supp}\,\varphi \subset \{e\}$. One can identify $(U\mathfrak{g})^*$ with the completion of the local ring of $e \in G$ or with $\mathrm{Fun}(\hat{G})$ where $\hat{G}$ is the formal group corresponding to $\mathfrak{g}$ (the second realization of $(U\mathfrak{g})^*$ makes sense even if G does not exist, which may well happen if $\dim \mathfrak{g} = \infty$).

The most interesting and mysterious Hopf algebras are those which are neither commutative nor cocommutative. Though Hopf algebras were intensively studied both by "pure" algebraists [**2–12**] and by specialists in von Neumann algebras [**13–26**], I believe that most of the examples of noncommutative noncocommutative Hopf algebras invented independently of integrable quantum system theory are counterexamples rather than "natural" examples (however there are remarkable exceptions, e.g. [**12**], [**16**], and [**2**, pp. 89-90]). We are going to discuss a general method for constructing noncommutative noncocommutative Hopf algebras, which was proposed in [**27, 28**] under the influence of the QISM. This method is based on the concept of quantization. It can be considered as a realization of the ideas of G. I. Kac and V. G. Palyutkin (see the end of [**16**]).

2. Quantization. Roughly speaking, a quantization of a commutative associative algebra A_0 over k is a (not necessarily commutative) deformation of A_0 depending on a parameter h (Planck's constant), i.e. an associative algebra A over $k[[h]]$ such that $A/hA = A_0$ and A is a topologically free $k[[h]]$-module. Given A, we can define a new operation on A_0 (the Poisson bracket) by the formula

$$\{a \bmod h, b \bmod h\} = \frac{[a,b]}{h} \bmod h. \tag{3}$$

Thus A_0 becomes a Poisson algebra (i.e. a Lie algebra with respect to $\{\ ,\ \}$ and a commutative associative algebra with respect to multiplication, these two structures being compatible in the following sense:

$$\{a, bc\} = \{a,b\}c + b\{a,c\}).$$

Now we shall slightly change our point of view on quantization.

DEFINITION. A *quantization* of a Poisson algebra A_0 is an associative algebra deformation A of A_0 over $k[[h]]$ such that the Poisson bracket on A_0 defined by (3) is equal to the bracket given a priori.

Of course, this approach to quantization is as old as quantum mechanics. It was explained to mathematicians by F. A. Berezin, J. Vey, A. Lichnerowicz, M. Flato, D. Sternheimer, and others.

We shall need a Hopf algebra version of the above definition. In this case A_0 is a Poisson-Hopf algebra (i.e. a Hopf algebra structure and a Poisson algebra structure on A_0 are given such that the multiplication is the same for both structures and the comultiplication $A_0 \to A_0 \otimes A_0$ is a Poisson algebra homomorphism, the Poisson bracket on $A_0 \otimes A_0$ being defined by $\{a\otimes b, c\otimes d\} = ac\otimes\{b,d\} + \{a,c\}\otimes bd$) and A is a Hopf algebra deformation of A_0. We shall also use the dual notion of quantization of co-Poisson-Hopf algebras (a co-Poisson-Hopf algebra is a cocommutative Hopf algebra B with a Poisson cobracket $B \to B \otimes B$ compatible with the Hopf algebra structure).

We discuss the structure of Poisson-Hopf algebras and co-Poisson-Hopf algebras in §§3 and 4. Then we consider the quantization problem.

3. Poisson groups and Lie bialgebras. A *Poisson group* is a group G with a Poisson bracket on Fun(G) which makes Fun(G) a Poisson-Hopf algebra. In other words the Poisson bracket must be compatible with the group operation, which means that the mapping $\mu\colon G \times G \to G$, $\mu(g_1, g_2) = g_1 g_2$, must be a Poisson mapping in the sense of [**33**], i.e. $\mu^*\colon \mathrm{Fun}(G) \to \mathrm{Fun}(G \times G)$ must be a Lie algebra homomorphism. Specifying the meaning of the word "group" and the symbol Fun(G), we obtain the notions of Poisson-Lie group, Poisson formal group, Poisson algebraic group, etc. According to our general principles the notions of Poisson group and Poisson-Hopf algebra are equivalent.

There exists a very simple description of Poisson-Lie groups in terms of *Lie bialgebras.*

DEFINITION. A *Lie bialgebra* is a vector space $\mathfrak{g}$ with a Lie algebra structure and a Lie coalgebra structure, these structures being compatible in the following sense: the cocommutator mapping $\mathfrak{g} \to \mathfrak{g} \otimes \mathfrak{g}$ must be a 1-cocycle ($\mathfrak{g}$ acts on $\mathfrak{g} \otimes \mathfrak{g}$ by means of the adjoint representation).

If G is a Poisson-Lie group then $\mathfrak{g} = \mathrm{Lie}\,(G)$ has a Lie bialgebra structure. To define it write the Poisson bracket on $C^\infty(G)$ as

$$\{\varphi, \psi\} = \eta^{\mu\nu} \partial_\mu \varphi \cdot \partial_\nu \psi, \qquad \varphi, \psi \in C^\infty(G), \tag{4}$$

where $\{\partial_\mu\}$ is a basis of right-invariant vector fields on G. The compatibility of the bracket with the group operation means that the function $\eta\colon G \to \mathfrak{g} \otimes \mathfrak{g}$ corresponding to $\eta^{\mu\nu}$ is a 1-cocycle. The 1-cocycle $f\colon \mathfrak{g} \to \mathfrak{g} \otimes \mathfrak{g}$ corresponding to η defines a Lie bialgebra structure on $\mathfrak{g}$ (the Jacobi identity for $f^*\colon \mathfrak{g}^* \otimes \mathfrak{g}^* \to \mathfrak{g}^*$ holds because f^* is the infinitesimal part of the bracket (4)).

THEOREM 1. *The category of connected and simply-connected Poisson-Lie groups is equivalent to the category of finite dimensional Lie bialgebras.*

The analogue of Theorem 1 for Poisson formal groups over a field of characteristic 0 can be proved in the following way. The algebra of functions on the formal group corresponding to $\mathfrak{g}$ is nothing but $(U\mathfrak{g})^*$. A Poisson-Hopf structure on $(U\mathfrak{g})^*$ is equivalent to a co-Poisson-Hopf structure on $U\mathfrak{g}$. So it suffices to prove the following easy theorem.

THEOREM 2. *Let $\delta\colon U\mathfrak{g} \to U\mathfrak{g} \otimes U\mathfrak{g}$ be a Poisson cobracket which makes $U\mathfrak{g}$ a co-Poisson-Hopf algebra* (*the Hopf structure on $U\mathfrak{g}$ is usual*). *Then $\delta(\mathfrak{g}) \subset \mathfrak{g} \otimes \mathfrak{g}$ and* ($\mathfrak{g}, \delta|\mathfrak{g}$) *is a Lie bialgebra. Thus we obtain a one-to-one correspondence between co-Poisson-Hopf structures on $U\mathfrak{g}$ inducing the usual Hopf structure and Lie bialgebra structures on $\mathfrak{g}$ inducing the given Lie algebra structure.*

Now let us discuss the notion of Lie bialgebra. First of all there is a one-to-one correspondence between Lie bialgebras and *Manin triples.* A Manin triple $(\mathfrak{p}, \mathfrak{p}_1, \mathfrak{p}_2)$ consists of a Lie algebra $\mathfrak{p}$ with a nondegenerate invariant scalar product on it and isotropic Lie subalgebras $\mathfrak{p}_1, \mathfrak{p}_2$ such that $\mathfrak{p}$ is the direct sum of $\mathfrak{p}_1$ and $\mathfrak{p}_2$ as a vector space. The correspondence mentioned above is constructed in the following way: if $(\mathfrak{p}, \mathfrak{p}_1, \mathfrak{p}_2)$ is a Manin triple then we put $\mathfrak{g} = \mathfrak{p}_1$ and define the cocommutator $\mathfrak{g} \to \mathfrak{g} \otimes \mathfrak{g}$ to be dual to the commutator mapping $\mathfrak{p}_2 \otimes \mathfrak{p}_2 \to \mathfrak{p}_2$ (note that $\mathfrak{p}_2$ is naturally isomorphic to $\mathfrak{g}^*$). Conversely, if a Lie bialgebra $\mathfrak{g}$ is given we put $\mathfrak{p} = \mathfrak{g} \oplus \mathfrak{g}^*$, $\mathfrak{p}_1 = \mathfrak{g}$, $\mathfrak{p}_2 = \mathfrak{g}^*$ and define the commutator $[x, l]$ for $x \in \mathfrak{g}$, $l \in \mathfrak{g}^*$ so that the natural scalar product on $\mathfrak{p}$ should be invariant. Note that if $(\mathfrak{p}, \mathfrak{p}_1, \mathfrak{p}_2)$ is a Manin triple then so is $(\mathfrak{p}, \mathfrak{p}_2, \mathfrak{p}_1)$. Therefore the notion of Lie bialgebra is self-dual.

Here are some examples of Lie bialgebras. Examples 3.2–3.4 are important for the inverse scattering method.

EXAMPLE 3.1. If $\dim \mathfrak{g} = 2$ then any linear mappings $\bigwedge^2 \mathfrak{g} \to \mathfrak{g}$ and $\mathfrak{g} \to \bigwedge^2 \mathfrak{g}$ define a Lie bialgebra structure on $\mathfrak{g}$. A 2-dimensional Lie bialgebra is called nondegenerate if the composition $\bigwedge^2 \mathfrak{g} \to \mathfrak{g} \to \bigwedge^2 \mathfrak{g}$ is nonzero. In this case there exists a basis $\{x_1, x_2\}$ of $\mathfrak{g}$ such that $[x_1, x_2] = \alpha x_2$ and the cocommutator is given by $x_1 \mapsto 0, x_2 \mapsto \beta x_2 \wedge x_1$. Here $\alpha\beta \neq 0$ and $\alpha\beta$ does not depend on the choice of x_1, x_2.

EXAMPLE 3.2. Let $\mathfrak{g}$ be a Kac-Moody algebra (in the sense of [**55**]) with a fixed invariant scalar product $\langle\ ,\ \rangle$, $\mathfrak{h}$ the Cartan subalgebra, $\mathfrak{b}_\pm \supset \mathfrak{h}$ the Borel subalgebras. Put $\mathfrak{p} = \mathfrak{g} \times \mathfrak{g}$, $\mathfrak{p}_1 = \{(x, y) \in \mathfrak{g} \times \mathfrak{g} | x = y\} \approx \mathfrak{g}$, $\mathfrak{p}_2 = \{(x, y) \in \mathfrak{b}_- \times \mathfrak{b}_+ | x_\mathfrak{h} + y_\mathfrak{h} = 0\}$. Define the scalar product of $(x_1, y_1) \in \mathfrak{p}$ and $(x_2, y_2) \in \mathfrak{p}$ to equal $\langle x_1, x_2\rangle - \langle y_1, y_2\rangle$. Since $(\mathfrak{p}, \mathfrak{p}_1, \mathfrak{p}_2)$ is a Manin triple, $\mathfrak{g}$ has a Lie bialgebra structure. The cocommutator $\varphi \colon \mathfrak{g} \to \bigwedge^2 \mathfrak{g}$ can be described explicitly in terms of the canonical generators X_i^+, X_i^-, H_i (here $X_i^\pm \in [\mathfrak{b}_\pm, \mathfrak{b}_\pm]$ and H_i is the image of the simple root $\alpha_i \in \mathfrak{h}^*$ under the isomorphism $\mathfrak{h}^* \to \mathfrak{h}$): $\varphi(H_i) = 0$, $\varphi(X_i^\pm) = \frac{1}{2} X_i^\pm \wedge H_i$. Note that $\mathfrak{b}_+$ and $\mathfrak{b}_-$ are subbialgebras of $\mathfrak{g}$. If $\mathfrak{g} = \mathfrak{sl}(2)$ then $\mathfrak{b}_+$ and $\mathfrak{b}_-$ are of the type described in Example 3.1. The Manin triple corresponding to $\mathfrak{b}_+$ is $(\mathfrak{g} \times \mathfrak{h}, \operatorname{Im}\psi_+, \operatorname{Im}\psi_-)$ where $\psi_\pm \colon \mathfrak{b}_\pm \hookrightarrow \mathfrak{g} \times \mathfrak{h}$ is defined by $\psi_\pm(a) = (a, \pm a_\mathfrak{h})$ and the scalar product on $\mathfrak{g} \times \mathfrak{h}$ equals $\langle\ ,\ \rangle_\mathfrak{g} - \langle\ ,\ \rangle_\mathfrak{h}$.

EXAMPLE 3.3. Fix a simple Lie algebra $\mathfrak{a}$ ($\dim \mathfrak{a} < \infty$) and an invariant scalar product on it. Set $\mathfrak{p} = \mathfrak{a}((u^{-1}))$, $\mathfrak{p}_1 = \mathfrak{a}[u]$, $\mathfrak{p}_2 = u^{-1}\mathfrak{a}[[u^{-1}]]$ and define the scalar product on $\mathfrak{p}$ by $(f, g) = \operatorname{res}_{u=\infty}(f(u), g(u))\, du$. The Manin triple $(\mathfrak{p}, \mathfrak{p}_1, \mathfrak{p}_2)$ defines a Lie bialgebra structure on $\mathfrak{g} = \mathfrak{a}[u]$. The cocommutator in $\mathfrak{g}$ is given by the formula

$$a(u) \mapsto [a(u) \otimes 1 + 1 \otimes a(v), r(u, v)]. \tag{5}$$

Here $a \in \mathfrak{a}[u]$, $\mathfrak{g} \otimes \mathfrak{g}$ is identified with $(\mathfrak{a} \otimes \mathfrak{a})[u, v]$ and $r(u, v) = t/(u - v)$ where t is the element of $\mathfrak{a} \otimes \mathfrak{a}$ corresponding to our scalar product. The right-hand side of (5) has no pole at $u = v$ because t is invariant.

EXAMPLE 3.4 (cf. [**29–31**]). Fix a nonsingular irreducible projective algebraic curve X over $\mathbf{C}$. Denote by E (resp. A, O_x) the field of rational functions on X (resp. the adèle ring, the completion of the local ring of X at x). Fix an absolutely simple Lie algebra G over E ($\dim G < \infty$) and a rational differential ω on X, $\omega \neq 0$. Define an invariant $\mathbf{C}$-valued scalar product on $G \otimes_E A$ by

$$(u, v) = \sum_{x \in X} \operatorname{res}_x \{\omega \cdot \operatorname{Tr} \rho(u_x) \rho(v_x)\}$$

where $\rho\colon G \to \mathfrak{gl}(n, E)$ is an exact representation, $u = (u_x)_{x\in X}$, $v = (v_x)_{x\in X}$. Then G is a maximal isotropic subspace of $G \otimes_E A$. If there exists an open isotropic $\mathbf{C}$-subalgebra $\Lambda \subset G \otimes_E A$ such that $G \otimes_E A = G \oplus \Lambda$ then we obtain a Lie bialgebra structure on G. To every subset $S \subset X$, $S \neq \varnothing$, there corresponds a subbialgebra $G_S = \{a \in G \mid$ the image of a in $G \otimes A^S$ belongs to the image of Λ in $G \otimes A^S\}$, where A^S is the ring of adèles without x-components, $x \in S$. The bialgebra of Example 3.3 is, in fact, G_S, where

$$X = \mathbf{P}^1, \quad \omega = d\lambda, \quad S = \{\infty\}, \quad \mathfrak{g} = \mathfrak{a} \otimes E, \quad \Lambda = \mathfrak{a} \otimes \left(m_\infty \times \prod_{x \in X\setminus S} O_x \right),$$

and m_∞ is the maximal ideal of O_∞. If $\mathfrak{g}$ is an affine Kac-Moody algebra with the bialgebra structure of Example 3.2 then $\mathfrak{g}' \stackrel{\text{def}}{=} [\mathfrak{g}, \mathfrak{g}]/(\text{the center of } \mathfrak{g})$ is, in fact, G_S for $X = \mathbf{P}^1$, $\omega = \lambda^{-1} d\lambda$, $S = \{0, \infty\}$, $G = \mathfrak{g}' \otimes_{\mathbf{C}[\lambda, \lambda^{-1}]} \mathbf{C}(\lambda)$ ($\mathfrak{g}'$ has a natural structure of a $\mathbf{C}[\lambda, \lambda^{-1}]$-algebra),

$$\Lambda = \mathfrak{c} \times \prod_{x \in X\setminus S} (\mathfrak{g}' \otimes_{\mathbf{C}[\lambda,\lambda^{-1}]} O_x),$$

where $\mathfrak{c} \subset (\mathfrak{g}' \otimes_{\mathbf{C}[\lambda,\lambda^{-1}]} \mathbf{C}((\lambda^{-1}))) \times (\mathfrak{g}' \otimes_{\mathbf{C}[\lambda,\lambda^{-1}]} \mathbf{C}((\lambda)))$ is the closure of

$$\mathfrak{p}_2' = \{(x, y)\} \in \mathfrak{g}' \times \mathfrak{g}' \mid x \in \mathfrak{b}'_-,\ y \in \mathfrak{b}'_+,\ x_{\mathfrak{h}} + y_{\mathfrak{h}} = 0\}.$$

4. Classical Yang-Baxter equation. Let $\mathfrak{g}$ be a Lie algebra and suppose that $\varphi\colon \mathfrak{g} \to \bigwedge^2 \mathfrak{g}$ is the coboundary of $r \in \bigwedge^2 \mathfrak{g}$. It can be shown that $(\mathfrak{g}, \varphi)$ is a Lie bialgebra iff

$$[r^{12}, r^{13}] + [r^{12}, r^{23}] + [r^{13}, r^{23}] \text{ is } \mathfrak{g}\text{-invariant.} \tag{6}$$

Here, for instance,

$$[r^{12}, r^{13}] \stackrel{\text{def}}{=} \sum_{i,j} [a_i, a_j] \otimes b_i \otimes b_j$$

where $r = \sum_i a_i \otimes b_i$ (one may imagine that $r^{12} = \sum_i a_i \otimes b_i \otimes 1 \in (U\mathfrak{g})^{\otimes 3}$, $r^{13} = \sum_i a_i \otimes 1 \otimes b_i \in (U\mathfrak{g})^{\otimes 3}$, $r^{23} = \sum_i 1 \otimes a_i \otimes b_i \in (U\mathfrak{g})^{\otimes 3}$). In particular, if

$$[r^{12}, r^{13}] + [r^{12}, r^{23}] + [r^{13}, r^{23}] = 0, \tag{7}$$

then $(\mathfrak{g}, \varphi)$ is a Lie bialgebra. Equation (7) is just the classical Yang-Baxter equation (CYBE), or the classical triangle equation. There is an important case which is intermediate between (6) and (7). Suppose that $r \in \mathfrak{g} \otimes \mathfrak{g}$ satisfies (7) but the skew-symmetry condition $r^{12} + r^{21} = 0$ is replaced by "$r^{12} + r^{21}$ is $\mathfrak{g}$-invariant" (then the coboundary $\varphi = \partial r$ is skew-symmetric). In this situation $(\mathfrak{g}, \varphi)$ is also a Lie bialgebra. If $r^{12} + r^{21} = P$, $\rho = r - \frac{1}{2} P$ then $\rho \in \bigwedge^2 \mathfrak{g}$, $\varphi = \partial\rho$ and $[\rho^{12}, \rho^{13}] + [\rho^{12}, \rho^{23}] + [\rho^{13}, \rho^{23}] = \frac{1}{4}[P^{12}, P^{23}]$.

DEFINITION. A *coboundary Lie bialgebra* is a pair $(\mathfrak{g}, r)$, where $\mathfrak{g}$ is a Lie bialgebra, $r \in \bigwedge^2 \mathfrak{g}$, $\partial r =$ the cocommutator of $\mathfrak{g}$. A coboundary Lie bialgebra $(\mathfrak{g}, r)$ is said to be *triangular* if r satisfies (7). A *quasitriangular Lie bialgebra* is a pair $(\mathfrak{g}, r)$, where $\mathfrak{g}$ is a Lie bialgebra, $r \in \mathfrak{g} \otimes \mathfrak{g}$, $\partial r =$ the cocommutator of $\mathfrak{g}$, and r satisfies (7).

It was noticed in [**32, 27**] that the expression $[r^{12}, r^{13}] + [r^{12}, r^{23}] + [r^{13}, r^{23}]$, $r \in \bigwedge^2 \mathfrak{g}$, is equal to $\frac{1}{2}\{r, r\}$ where $\{\ ,\ \}$ is the bilinear operation on $\bigwedge^* \mathfrak{g}$ such that $\{a, b\} = [a, b]$ for $a, b \in \mathfrak{g}$ and

$$\{a, b\} = -(-1)^{(k+1)(l+1)}\{b, a\},$$
$$\{a, b \wedge c\} = \{a, b\} \wedge c + (-1)^{(k+1)l} b \wedge \{a, c\}$$

for $a \in \bigwedge^k \mathfrak{g}$, $b \in \bigwedge^l \mathfrak{g}$, $c \in \bigwedge^* \mathfrak{g}$ ($\bigwedge^* \mathfrak{g}$ is a Poisson superalgebra with an odd Poisson bracket; if elements of $\bigwedge^* \mathfrak{g}$ are considered as left-invariant polyvector fields on the Lie group corresponding to $\mathfrak{g}$ then the operation $\{\ ,\ \}$ in $\bigwedge^* \mathfrak{g}$ is the Schouten bracket [**34**]).

The Poisson bracket on the Poisson-Lie group G corresponding to a coboundary bialgebra $(\mathfrak{g}, r)$ is of the form

$$\{\varphi, \psi\} = r^{\mu\nu}(\partial'_\mu \varphi \cdot \partial'_\nu \psi - \partial_\mu \varphi \cdot \partial_\nu \psi), \tag{8}$$

where ∂_μ (respectively ∂'_μ) are the right-invariant (respectively left-invariant) vector fields on G corresponding to a basis of $\mathfrak{g}$ and $r^{\mu\nu}$ are the coordinates of r. The following property of the bracket (8) is crucial for the Hamiltonian version of the classical inverse scattering method (a systematic exposition of this method is given in [**35**]; see also [**36–38**]):

$$\varphi, \psi \in C^\infty(G), \quad \varphi(gxg^{-1}) = \varphi(x), \quad \psi(gyg^{-1}) = \psi(y) \Rightarrow \{\varphi, \psi\} = 0. \tag{9}$$

Let us reconsider Examples 3.1–3.4. The cocommutator of a nondegenerate 2-dimensional bialgebra is not a coboundary. Now let us consider the cocommutator (5) on $\mathfrak{g} = \mathfrak{a}[u]$. First suppose that $r(u, v)$ is an $\mathfrak{a} \otimes \mathfrak{a}$-valued polynomial. Then we can consider r as an element of $\mathfrak{g} \otimes \mathfrak{g}$ and the cocommutator (5) as the coboundary of r. The CYBE and the condition $r \in \bigwedge^2 \mathfrak{g}$ take the form

$$\begin{aligned}[r^{12}(u_1, u_2), r^{13}(u_1, u_3)] + [r^{12}(u_1, u_2), r^{23}(u_2, u_3)]& \\ + [r^{13}(u_1, u_3), r^{23}(u_2, u_3)] = 0,&\end{aligned} \tag{10}$$

$$r^{12}(u_1, u_2) + r^{21}(u_2, u_1) = 0. \tag{11}$$

So every polynomial solution of (10), (11) defines a triangular bialgebra structure on $\mathfrak{g}$. In Example 3.3, $r(u, v)$ is a nonpolynomial solution of (10), (11). Therefore the corresponding bialgebra is "pseudotriangular" (nevertheless (9) holds).

The bialgebra $\mathfrak{g}$ of Example 3.2 has a quasitriangular structure if $\dim \mathfrak{g} < \infty$: consider the element $t \in \mathfrak{g} \otimes \mathfrak{g}$ corresponding to the scalar product on $\mathfrak{g}$, represent t as $t_{+-} + t_0 + t_{-+}$, $t_{+-} \in [\mathfrak{b}_+, \mathfrak{b}_+] \otimes [\mathfrak{b}_-, \mathfrak{b}_-]$, $t_{-+} \in [\mathfrak{b}_-, \mathfrak{b}_-] \otimes [\mathfrak{b}_+, \mathfrak{b}_+]$, $t_0 \in \mathfrak{h} \otimes \mathfrak{h}$. and set $r = t_{+-} + \frac{1}{2} t_0$. If $\dim \mathfrak{g} = \infty$ then r does not belong to the algebraic tensor product $\mathfrak{g} \otimes \mathfrak{g}$. Therefore $(\mathfrak{g}, r)$ is "pseudoquasitriangular." Let us examine more attentively the non-twisted affine case (the twisted affine case is similar). In this case $\mathfrak{g}' = [\mathfrak{g}, \mathfrak{g}]/(\text{the center of } \mathfrak{g}) = \mathfrak{a}[\lambda, \lambda^{-1}]$ for some simple algebra $\mathfrak{a}$, $\dim \mathfrak{a} < \infty$. Denote by $\bar{r}$ the analogue of r for $\mathfrak{g}'$. Consider $\bar{r}$ as a power series in $\lambda = \lambda \otimes 1$ and

$\mu = 1 \otimes \lambda$ with coefficients in $\mathfrak{a} \otimes \mathfrak{a}$. Then $\bar{r}(\lambda, \mu)$ is the Laurent decomposition of a rational function of λ/μ having a single pole at $\lambda/\mu = 1$ with residue $t_\mathfrak{a}$ ($t_\mathfrak{a} \in \mathfrak{a} \otimes \mathfrak{a}$ corresponds to the scalar product on $\mathfrak{a}$). This rational function satisfies (10) and (11). Moral [**39**, §3]: if a solution $r(u, v)$ of (10), (11) has a pole at $u = v$ then the corresponding Lie bialgebra is "pseudoquasitriangular" rather than "pseudotriangular"; in other words $r^{12}(u_1, u_2) + r^{21}(u_2, u_1)$ is $t_\mathfrak{a} \cdot \delta(u_1 - u_2)$ rather than 0.

I. V. Cherednik proved [**31**] that the bialgebras of Example 4 also correspond to solutions $r(u, v)$ of (10), (11) having a pole at $u = v$. Moreover, it follows from [**40–42**] that there is a one-to-one correspondence between "nondegenerate" solutions of (10), (11) up to an equivalence relation (such solutions always have a pole at $u = v$) and quadruples (X, ω, G, Λ) of the type described in Example 3.4. Besides, a classification of nondegenerate solutions of (10), (11) is given in [**40, 41**]. In terms of (X, ω, G, Λ) the results of [**40, 41**] can be stated in the following way.

(1) For Λ to exist ω should have no zeros and therefore there are three possibilities: (a) X is an elliptic curve and ω is regular, (b) $X = \mathbf{P}^1$, $\omega = \lambda^{-1} d\lambda$, (c) $X = \mathbf{P}^1$, $\omega = d\lambda$. The corresponding solutions $r(u, v)$ are elliptic, trigonometric or rational functions of $u - v$.

(2) In case (a) for Λ to exist G has to be isomorphic to $\mathfrak{sl}(n)$, and for every n there are finitely many choices of Λ all of which are listed. In case (b) G is of the type described at the end of Example 3.4, and all the choices of Λ are listed (they are similar to the Λ described at the end of Example 3.4). In case (c) little is known.

Some important ideas and results of M. A. Semenov-Tian-Shansky concerning Poisson-Lie groups and the CYBE are in [**39, 43**].

5. Some historical remarks. The Poisson bracket (8) and the CYBE were introduced by E. K. Sklyanin [**44, 37**] as the classical limits of the corresponding quantum objects. The compatibility of the bracket (8) with the group operation was formulated by him almost explicitly. The abstract notion of Poisson-Lie group appeared later [**27**]. The classification of the solutions of the CYBE was based on vast experimental material due to many authors (see the Appendix of [**45**]). In particular, the solution $r(u, v) = t/(u - v)$ was found in [**45, 46**].

6. Quantization of Lie bialgebras (examples). By definition, a quantization of a Lie bialgebra $\mathfrak{g}$ is a quantization of $U\mathfrak{g}$ in the sense of §2, where $U\mathfrak{g}$ is considered as a co-Poisson-Hopf algebra according to Theorem 2. If A is a quantization of $\mathfrak{g}$ we call $\mathfrak{g}$ the *classical limit* of A.

EXAMPLE 6.1 (cf. [**12**], [**2**, pp. 89-90]). Consider the $\mathbf{C}[[h]]$-algebra A generated (in the h-adic sense, i.e. as an algebra complete in the h-adic topology) by x_1, x_2 with the defining relation $[x_1, x_2] = \alpha x_2$, $\alpha \in \mathbf{C}^*$. Define $\Delta\colon A \to A \otimes A$ by $\Delta(x_1) = x_1 \otimes 1 + 1 \otimes x_1$, $\Delta(x_2) = x_2 \otimes \exp(\frac{1}{2}h\beta x_1) + \exp(-\frac{1}{2}h\beta x_1) \otimes x_2$, $\beta \in \mathbf{C}^*$. Then A is a quantization of the bialgebra of Example 3.1. From the results of §9 it follows that the quantization is unique up to substitutions $h \mapsto h + \sum_{i=2}^{\infty} c_i h^i, c_i \in \mathbf{C}$.

EXAMPLE 6.2. Let $\mathfrak{g}, \mathfrak{h}, \alpha_i, \langle\ ,\ \rangle, H_i$ denote the same objects as in Example 3.2. Consider the $\mathbf{C}[[h]]$-algebra $U_h\mathfrak{g}$ generated (in the h-adic sense) by $\mathfrak{h}$ and X_i^+, X_i^- with the defining relations

$$[a_1, a_2] = 0 \quad \text{for } a_1, a_2 \in \mathfrak{h}, \qquad [a, X_i^{\pm}] = \pm\alpha_i(a) X_i^{\pm} \quad \text{for } a \in \mathfrak{h},$$

and also

$$[X_i^+, X_j^-] = 2\delta_{ij} h^{-1} \mathrm{sh}(hH_i/2),$$

$$i \neq j,\ n = 1 - A_{ij}, q = \exp\frac{h}{2}\langle H_i, H_i\rangle \Rightarrow$$

$$\Rightarrow \sum_{k=0}^{n} (-1)^k \binom{n}{k}_q q^{-k(n-k)/2} (X_i^{\pm})^k X_j^{\pm} (X_i^{\pm})^{n-k} = 0.$$

Here (A_{ij}) is the Cartan matrix and $\binom{n}{k}_q$ is the Gauss polynomial [**47**, §3.3], i.e., $\binom{n}{k}_q = (q^n - 1)(q^{n-1} - 1) \cdots (q^{n-k+1} - 1)/(q^k - 1)/(q^{k-1} - 1) \cdots (q - 1)$. It can be shown that $U_h\mathfrak{g}$ is a topologically free $\mathbf{C}[[h]]$-module and that there exists a homomorphism $\Delta\colon U_h\mathfrak{g} \to U_h\mathfrak{g} \otimes U_h\mathfrak{g}$ such that

$$\Delta(X_i^{\pm}) = X_i^{\pm} \otimes \exp(hH_i/4) + \exp(-hH_i/4) \otimes X_i^{\pm},$$

$\Delta(a) = a \otimes 1 + 1 \otimes a$ for $a \in \mathfrak{h}$. $(U_h\mathfrak{g}, \Delta)$ is a quantization of $\mathfrak{g}$. It can be shown using the results of §9 that $U_h\mathfrak{g}$ is the unique (up to substitutions $h \mapsto h + \sum_{i=2}^{\infty} c_i h^i$, $c_i \in \mathbf{C}$) quantization A of $\mathfrak{g}$ for which there exist a cocommutative Hopf subalgebra $C \subset A$ and a mapping $\theta\colon A \to A$ such that the mapping $C/hC \to U\mathfrak{g}$ is injective and its image is equal to $U\mathfrak{h}$, $\theta^2 = \mathrm{id}$, $\theta(C) = C$, θ is an algebra automorphism and a coalgebra antiautomorphism, $\theta \bmod h$ is the Cartan involution (if $A = U_h\mathfrak{g}$ put $C = U\mathfrak{h}[[h]]$, $\theta(X_i^{\pm}) = -X_i^{\mp}$, $\theta|_{\mathfrak{h}} = -\mathrm{id}$). So it is natural to call $U_h\mathfrak{g}$ *the* quantized Kac-Moody algebra corresponding to $\mathfrak{g}$. $U_h\mathfrak{sl}(2)$ was introduced by Kulish-Reshetikhin [**48**] and E. K. Sklyanin [**49**]. For general $\mathfrak{g}$ the algebra $U_h\mathfrak{g}$ was introduced by M. Jimbo [**50**, **51**] and the author [**28**]. All these works were motivated by the QISM. For $\mathfrak{g}$ affine or finite dimensional $U_h\mathfrak{g}$ is closely related to trigonometric solutions of the quantum Yang-Baxter equation (QYBE); see [**50–52**] and §13.

EXAMPLE 6.3 [**28**, **54**]. Let $\mathfrak{a}$ and $\mathfrak{g}$ denote the same objects as in Example 3.3. Since $\mathfrak{g}$ is graded ($\deg au^k = k$ for $a \in \mathfrak{a}$) and the cocommutator has degree -1, it is natural to look for a quantization Q which is a graded Hopf algebra over $\mathbf{C}[[h]]$, where $\mathbf{C}[[h]]$ is graded so that $\deg h = 1$. It can be shown using the results of §9 that Q exists and is unique. As an h-adic algebra, Q is generated by the Lie algebra $\mathfrak{a}$ ($\deg a = 0$ for $a \in \mathfrak{a}$) and elements $J(a)$ of degree 1, $a \in \mathfrak{a}$, with the defining relations

$$J(\lambda a + \mu b) = \lambda J(a) + \mu J(b), \quad J([a, b]) = [a, J(b)] \quad \text{for } a, b \in \mathfrak{a},\ \lambda, \mu \in \mathbf{C}, \tag{12}$$

$$\sum_i [a_i, b_i] = 0 \Rightarrow \sum_i [J(a_i), J(b_i)] \tag{13}$$
$$= \frac{h^2}{12} \sum_i ([[a_i, I_\alpha], [b_i, I_\beta]], I_\gamma) \cdot \{I_\alpha, I_\beta, I_\gamma\}, \quad a_i, b_i \in \mathfrak{a},$$

$$\sum_i [[a_i, b_i], c_i] = 0 \Rightarrow \sum_i [[J(a_i), J(b_i)], J(c_i)] \tag{14}$$
$$= \frac{h^2}{4} \sum_i \sum_{\alpha,\beta,\gamma} f(a_i, b_i, c_i, I_\alpha, I_\beta, I_\gamma)\{I_\alpha, I_\beta, J(I_\gamma)\}, \quad a_i, b_i, c_i \in \mathfrak{a},$$

where $\{I_\alpha\}$ is an orthonormal basis of $\mathfrak{a}$,

$$f(a, b, c, x, y, z) = \operatorname*{Alt}_{a,b} \operatorname*{Sym}_{a,c} ([a, [b, x]], [[c, y], z]),$$
$$\{x_1, x_2, x_3\} = \frac{1}{6} \sum_{i \neq j \neq k} x_i x_j x_k.$$

$\Delta : Q \to Q \otimes Q$ is defined by the formulas

$$\begin{aligned} &\Delta(a) = a \otimes 1 + 1 \otimes a, \\ &\Delta(J(a)) = J(a) \otimes 1 + 1 \otimes J(a) + \tfrac{h}{2}[a \otimes 1, t] \qquad a \in \mathfrak{a}, \end{aligned} \tag{15}$$

where $t = \sum_\alpha I_\alpha \otimes I_\alpha$. Note that the classical limit of $J(a)$ is $au \in \mathfrak{g}$. So the formula (12) and the left-hand sides of (13), (14) are quite natural. (15) is also natural because $[a \otimes 1, t]$ is the image of au under the cocommutator mapping. The terrific right-hand sides of (13), (14) are derived from (12), (15). Note also that if $\mathfrak{a} = \mathfrak{sl}(2)$ then (13) is superfluous and if $\mathfrak{a} \neq \mathfrak{sl}(2)$ then (14) is superfluous. The Hopf algebra over $\mathbf{C}$ obtained from Q by setting $h = 1$ is denoted by $Y(\mathfrak{a})$ and called the Yangian of $\mathfrak{a}$. $Y(\mathfrak{a})$ is related to rational solutions of the QYBE (see §12), the simplest of which was found by C. N. Yang [**53**].

Fortunately Q has another system of generators $\xi^+_{ik}, \xi^-_{ik}, \varphi_{ik}$, $i \in \Gamma, k = 0, 1, 2, \ldots$, $\deg \xi^\pm_{ik} = \deg \varphi_{ik} = k$, where Γ is the set of simple roots of $\mathfrak{a}$. The classical limits of $\xi^\pm_{ik}$ and φ_{ik} are equal to $X^\pm_i u^k$ and $H_i u^k$, where $X^\pm_i$ and H_i are the generators of $\mathfrak{a}$ used in Example 3.2. Here is the complete set of relations between $\xi^\pm_{ik}$ and φ_{ik}:

$$[\varphi_{ik}, \varphi_{jl}] = 0, \qquad [\varphi_{i0}, \xi^\pm_{jl}] = \pm 2B_{ij}\xi^\pm_{jl}, \qquad [\xi^+_{ik}, \xi^-_{jl}] = \delta_{ij}\varphi_{i,k+l},$$
$$[\varphi_{i,k+1}, \xi^\pm_{jl}] - [\varphi_{ik}, \xi^\pm_{j,l+1}] = \pm hB_{ij}(\varphi_{ik}\xi^\pm_{jl} + \xi^\pm_{jl}\varphi_{ik})$$
$$[\xi^\pm_{i,k+1}, \xi^\pm_{jl}] - [\xi^\pm_{ik}, \xi^\pm_{j,l+1}] = \pm hB_{ij}(\xi^\pm_{ik}\xi^\pm_{jl} + \xi^\pm_{jl}\xi^\pm_{ik}),$$
$$i \neq j, n = 1 - A_{ij} \Rightarrow \operatorname{Sym}_{k_1,\ldots,k_n}[\xi^\pm_{ik_1}, [\xi^\pm_{ik_2}, \ldots [\xi^\pm_{ik_n}, \xi^\pm_{jl}]], \ldots] = 0,$$

where (A_{ij}) is the Cartan matrix of $\mathfrak{a}$, $B_{ij} = \frac{1}{2}(H_i, H_j)$. The connection between the two presentations of Q is described in [**54**].

A similar realization exists for quantized affine Kac-Moody algebras [**54**]. For instance, if $\mathfrak{g}$ is untwisted, i.e. $[\mathfrak{g}, \mathfrak{g}]/(\text{the center of } \mathfrak{g}) = \mathfrak{a}[\lambda, \lambda^{-1}]$, then $U_h\mathfrak{g}$ has the following presentation. Generators: $c, D, \xi^{\pm}_{ik}, \kappa_{ik}$, where $i \in \Gamma$, $k \in \mathbf{Z}$. Defining relations:

$$[c, \kappa_{ik}] = [c, \xi^{\pm}_{ik}] = [c, D] = 0, \qquad [D, \kappa_{ik}] = k\kappa_{ik}, \qquad [D, \xi^{\pm}_{ik}] = k\xi^{\pm}_{ik},$$
$$[\kappa_{ik}, \kappa_{jl}] = 4\delta_{k,-l}k^{-1}h^{-2}\mathrm{sh}\,(khB_{ij})\mathrm{sh}(khc/2),$$
$$[\kappa_{ik}, \xi^{\pm}_{jl}] = \pm 2(kh)^{-1}\mathrm{sh}(khB_{ij})\exp(\mp|k| \cdot hc/4)\xi^{\pm}_{j,k+l},$$
$$\xi^{\pm}_{i,k+1}\xi^{\pm}_{jl} - e^{\pm hB_{ij}}\xi^{\pm}_{jl}\xi^{\pm}_{i,k+1} = e^{\pm hB_{ij}}\xi^{\pm}_{ik}\xi^{\pm}_{j,l+1} - \xi^{\pm}_{j,l+1}\xi^{\pm}_{ik},$$
$$[\xi^{+}_{ik}, \xi^{-}_{jl}] = \delta_{ij}h^{-1}\{\psi_{i,k+l}e^{hc(k-l)/4} - \varphi_{i,k+l}e^{hc(l-k)/4}\},$$
$$i \neq j,\ n = 1 - A_{ij},\ q = \exp \tfrac{h}{2}(H_i, H_i) \Rightarrow$$
$$\Rightarrow \mathrm{Sym}_{k_1,\ldots,k_n} \sum_{r=0}^{n} (-1)^r \binom{n}{r}_q q^{-r(n-r)/2} \xi^{\pm}_{ik_1} \cdots \xi^{\pm}_{ik_r} \xi^{\pm}_{jl} \xi^{\pm}_{ik_{r+1}} \cdots \xi^{\pm}_{ik_n} = 0.$$

Here φ_{ip}, ψ_{ip} are defined from the relations

$$\sum_p \varphi_{ip}u^p = \exp h\left(\frac{1}{2}\kappa_{i0} + \sum_{p<0}\kappa_{ip}u^p\right),$$
$$\sum_p \psi_{ip}u^p = \exp h\left(\frac{1}{2}\kappa_{i0} + \sum_{p>0}\kappa_{ip}u^p\right).$$

The connection between the two presentations of $U_h\mathfrak{g}$ is described in [**54**]. Note that $U_h\mathfrak{g}'$ (i.e. $U_h\mathfrak{g}$ "without D" and with the auxiliary relation $c = 0$) degenerates into $Y(\mathfrak{a})$: if A is the subalgebra of $U_h\mathfrak{g}' \otimes_{\mathbf{C}[[h]]} \mathbf{C}((h))$ generated by $U_h\mathfrak{g}'$ and $h^{-1}\cdot\mathrm{Ker}\, f$, where f is the composition $U_h\mathfrak{g}' \to U\mathfrak{g}' = U(\mathfrak{a}[\lambda, \lambda^{-1}]) \xrightarrow{\lambda=1} U\mathfrak{a}$, then $A/hA = Y(\mathfrak{a})$.

The problem of describing Δ in terms of $\xi^{\pm}_{ik}$ is not solved for Yangians nor for quantized affine algebras.

7. Quantized universal enveloping algebras and h-formal groups. A *quantized universal enveloping algebra* (QUE-algebra) is a Hopf algebra A over $\mathbf{C}[[h]]$ such that A is a topologically free $\mathbf{C}[[h]]$-module and A/hA is a universal enveloping algebra. In other words, a QUE-algebra is a quantization of some Lie bialgebra. It is easy to show that a cocommutative QUE-algebra A is the h-adic completion of $U\mathfrak{g}$ for some Lie algebra $\mathfrak{g}$ over $\mathbf{C}[[h]]$ which is a topologically free $\mathbf{C}[[h]]$-module (in fact, $\mathfrak{g} = \{a \in A |\ \Delta(a) = a \otimes 1 + 1 \otimes a\}$).

Another important class of Hopf algebras over $\mathbf{C}[[h]]$ consists of *quantized formal series Hopf algebras* (QFSH-algebras). A QFSH-algebra is a topological Hopf algebra B over $\mathbf{C}[[h]]$ such that (1) as a topological $\mathbf{C}[[h]]$-module B is isomorphic to $\mathbf{C}[[h]]^I$ for some set I, (2) B/hB is isomorphic to $\mathbf{C}[[u_1, u_2, \ldots]]$ as a topological algebra.

An *h-formal group* is the spectrum of a QFSH-algebra. To understand h-formal groups in down-to-earth terms let us fix a "coordinate system," i.e. elements $x_i \in B$ such that (1) B/hB is the ring of formal power seris in $\bar{x}_i = x_i$ mod h, and (2) $\varepsilon(x_i) = 0$, where $\varepsilon\colon B \to \mathbf{C}[[h]]$ is the counit. $\Delta(x_i)$ can be expressed as $F_i(x_1, x_2, \ldots; y_1, y_2, \ldots; h)$, where $y_i = 1 \otimes x_i \in B \otimes B$ and x_i is identified with $x_i \otimes 1 \in B \otimes B$. So an h-formal group is defined by a "quantum group law" $F(x, y, h)$ and "commutation relations," i.e., formulas expressing $h^{-1}[x_i, x_j]$ as a formal series in x_i and h.

Now let us discuss the relation between QUE-algebras and QFSH-algebras. First of all, the dual of a QUE-algebra is a QFSH-algebra and vice versa (in both cases the dual should be understood in an appropriate sense). For instance, the dual of the QUE algebra of Example 6.1 is generated as an algebra by ξ_1 and ξ_2 where

$$\xi_1(f(x_1, x_2)) = \frac{\partial}{\partial x_1} f(x_1, 0)|_{x_1=0}, \qquad \xi_2\left(\sum_i \varphi_i(x_1)x_2^i\right) = \varphi_1\left(\frac{\alpha}{2}\right).$$

The commutation relation between ξ_1 and ξ_2 is of the form $[\xi_1, \xi_2] = -h\beta\xi_2$ and the comultiplication is given by $\Delta(\xi_1) = \xi_1 \otimes 1 + 1 \otimes \xi_1$, $\Delta(\xi_2) = \xi_2 \otimes \exp(-\frac{\alpha}{2}\xi_1) + \exp(\frac{\alpha}{2}\xi_1) \otimes \xi_2$. The dual of $U_h\mathfrak{sl}(n)$ also has an explicit description. Consider the representation ρ: $U_h\mathfrak{sl}(n) \to \mathrm{Mat}\,(n, \mathbf{C}[[h]])$ such that $\rho(H_i) = e_{ii} - e_{i+1,i+1}$, $\rho(X_i^+) = e_{i,i+1}$, $\rho(X_i^-) = 2h^{-1}\mathrm{sh}(h/2)e_{i+1,i}$ where e_{ij} are the matrix units (the scalar product on $\mathfrak{sl}(n)$ is supposed to equal $\mathrm{Tr}(XY)$). The matrix elements ρ_{ij}, $1 \leq i, j \leq n$, belong to $(U_h\mathfrak{sl}(n))^*$. It is easy to show that

$$\rho_{ij}\rho_{il} = e^{-h/2}\rho_{il}\rho_{ij} \quad \text{if } j < l, \tag{16}$$

$$\rho_{ij}\rho_{kj} = e^{-h/2}\rho_{kj}\rho_{ij} \quad \text{if } i < k, \tag{17}$$

$$\rho_{il}\rho_{kj} = \rho_{kj}\rho_{il} \quad \text{if } i < k,\ l > j, \tag{18}$$

$$[\rho_{il}, \rho_{kj}] = (e^{h/2} - e^{-h/2})\rho_{ij}\rho_{kl} \quad \text{if } i > k, l > j, \tag{19}$$

$$\sum_{i_1,\ldots,i_n} \rho_{1i_1}\rho_{2i_2}\cdots\rho_{ni_n} \cdot (-e^{-h/2})^{l(i_1,\ldots,i_n)} = 1,$$

where $l(i_1, \ldots, i_n)$ is the number of inversions in the permutation $(i_1, \ldots, i_n)$. In fact $(U_h\mathfrak{sl}(n))^*$ is the quotient of the ring of noncommutative power series in $\tilde{\rho}_{ij} = \rho_{ij} - \delta_{ij}$ by the ideal generated by the above relations. The comultiplication is usual: $\Delta(\rho_{ij}) = \sum_k \rho_{ik} \otimes \rho_{kj}$. It is natural to call $\mathrm{Spec}(U_h\mathfrak{sl}(n))^*$ the quantized formal $\mathrm{SL}(n)$. One can also consider the subalgebra of $(U_h\mathfrak{sl}(n))^*$ consisting of the matrix elements of finite-dimensional representations of $U_h\mathfrak{sl}(n)$, i.e. of the polynomials in ρ_{ij}. The spectrum of this subalgebra can be considered as a quantization of the algebraic group $\mathrm{SL}(n)$.

Somewhat unexpectedly the category of QFSH-algebras finitely generated as topological algebras is *equivalent* (not only antiequivalent) to the category of QUE-algebras whose classical limit is a finite dimensional Lie bialgebra (the finiteness conditions are not very essential here). The following example can help to understand this equivalence. Consider the QUE-algebra of Example 6.1, i.e. $[x_1, x_2] = \alpha x_2$, $\Delta(x_1) = x_1 \otimes 1 + 1 \otimes x_1$, $\Delta(x_2) = x_2 \otimes \exp(\frac{1}{2}h\beta x_1) + \exp(-\frac{1}{2}h\beta x_1) \otimes x_2$. Put $y_i = hx_i$. Then $[y_1, y_2] = \alpha h y_2$, $\Delta(y_1) = y_1 \otimes 1 + 1 \otimes y_1$, $\Delta(y_2) = y_2 \otimes \exp(\frac{1}{2}\beta y_1) + \exp(-\frac{1}{2}\beta y_1) \otimes y_2$, i.e. we have obtained a QFSH-algebra. Here is the general construction. If A is a QUE-algebra we put $\delta_1(a) = a - \varepsilon(a) \cdot 1$, $\delta_2(a) = \Delta(a) - a \otimes 1 - 1 \otimes a + \varepsilon(a) \cdot (1 \otimes 1)$, etc. ($\varepsilon$ is the counit).

Then $\{a \in A \mid \delta_n(a) \in h^n A^{\otimes n}$ for any $n\}$ is a QFSH-algebra. Conversely, if B is a QFSH-algebra and m is its maximal ideal then the h-adic completion of $\sum_n h^{-n} m^n$ is a QUE-algebra.

DEFINITION. The QUE-dual of a QUE-algebra A is the QUE-algebra corresponding to A^* in the above sense (or, which is the same, the dual of the QFSH-algebra corresponding to A).

It can be shown that the classical limit of the QUE-dual of A is a Lie bialgebra dual to the classical limit of A.

8. The square of the antipode. Let A be a quantization of a Lie bialgebra $\mathfrak{g}$. Denote by S the antipode of A. In general $S^2 \neq \mathrm{id}$, but $S^2 \equiv \mathrm{id} \bmod h$. Here is a description of $h^{-1}(S^2 - \mathrm{id}) \bmod h$ in terms of $\mathfrak{g}$. Denote by D the composition of the cocommutator $\mathfrak{g} \to \mathfrak{g} \otimes \mathfrak{g}$ and the commutator $\mathfrak{g} \otimes \mathfrak{g} \to \mathfrak{g}$. It is easy to show that $D\colon \mathfrak{g} \to \mathfrak{g}$ is a bialgebra derivation (this means that e^{tD} is a bialgebra automorphism) and $h^{-1}(S^2 - \mathrm{id}) \bmod h$ is the unique derivation of $U\mathfrak{g}$ whose restriction to $\mathfrak{g}$ is equal to $-D/2$.

If $\mathfrak{g} = \mathfrak{a}[u]$ is the bialgebra of Example 3.3 and the scalar product on $\mathfrak{a}$ is the Killing form then $D = d/du$. For bialgebras corresponding to nondegenerate solutions of (10), (11), D becomes equal to d/du after a suitable normalization.

9. Obstruction theory. Let $\mathfrak{g}$ be a Lie bialgebra. Denote by $\mathrm{Der}^i\,\mathfrak{g}$ the $(i+2)$th cohomology group of $\mathrm{Ker}(C^*(\mathfrak{g} \oplus \mathfrak{g}^*) \to C^*(\mathfrak{g}) \oplus C^*(\mathfrak{g}^*))$ where $\mathfrak{g} \oplus \mathfrak{g}^*$ is equipped with a Lie algebra structure defined in §3 and $C^*(\mathfrak{a})$ is the standard complex $0 \to \mathfrak{a}^* \to \bigwedge^2 \mathfrak{a}^* \to \bigwedge^3 \mathfrak{a}^* \to \cdots$. Obviously $\mathrm{Der}^i\,\mathfrak{g} = 0$ for $i < 0$. $\mathrm{Der}^0\,\mathfrak{g}$ is the set of bialgebra derivations of $\mathfrak{g}$. It can be shown that (1) if two quantizations of $\mathfrak{g}$ coincide modulo h^n then their difference $\bmod h^{n+1}$ "belongs" to $\mathrm{Der}^1\,\mathfrak{g}$, and (2) if $\mathrm{Der}^2\,\mathfrak{g} = 0$ then $\mathfrak{g}$ has a quantization. I do not know whether every Lie bialgebra can be quantized.

10. Coboundary, triangular, and quasitriangular Hopf algebras. A pair (A, R) consisting of a Hopf algebra A and an invertible element $R \in A \otimes A$ will be called a *coboundary Hopf algebra* if

$$\Delta'(a) = R\Delta(a)R^{-1}, \qquad a \in A, \tag{20}$$

and $R^{12}R^{21} = 1, R^{12} \cdot (\Delta \otimes \mathrm{id})(R) = R^{23} \cdot (\mathrm{id} \otimes \Delta)(R)$, $(\varepsilon \otimes \varepsilon)(R) = 1$. Here $\Delta' = \sigma \circ \Delta$, $\sigma\colon A \otimes A \to A \otimes A$, $\sigma(x \otimes y) = y \otimes x$ and the symbols R^{12}, R^{21}, R^{23} have the following meaning: if $R = \sum_i a_i \otimes b_i$ then

$$R^{21} = \sum_i b_i \otimes a_i, \qquad R^{23} = \sum_i 1 \otimes a_i \otimes b_i, \qquad R^{12} = \sum_i a_i \otimes b_i \otimes 1$$

(and sometimes $R^{12} = R$). (A, R) is called a *quasitriangular Hopf algebra* if R satisfies (20) and the equations

$$(\Delta \otimes \mathrm{id})(R) = R^{13}R^{23}, \qquad (\mathrm{id} \otimes \Delta)(R) = R^{13}R^{12}. \tag{21}$$

A quasitriangular Hopf algebra is said to be *triangular* if $R^{12}R^{21} = 1$. By definition, a coboundary (resp. triangular, quasitriangular) QUE-algebra is a coboundary (triangular, quasitriangular) Hopf algebra (A, R) such that A is a QUE-algebra and $R \equiv 1 \bmod h$.

It can be shown that if (A, R) is a coboundary (triangular, quasitriangular) QUE-algebra, $r = h^{-1}(R-1) \bmod h$ and $\mathfrak{g}$ is the classical limit of A then $r \in \mathfrak{g}\otimes\mathfrak{g}$ (a priori $r \in U\mathfrak{g} \otimes U\mathfrak{g}$) and $(\mathfrak{g}, r)$ is a coboundary (triangular, quasitriangular) Lie bialgebra. There are also other arguments in favor of the above definitions.

It is easy to show that (1) a triangular Hopf algebra is a coboundary Hopf algebra, (2) if (A, R) is a quasitriangular QUE-algebra and $\overline{R} = (R^{12}R^{21})^{-1/2}R$ then $(A, \overline{R})$ is a coboundary QUE-algebra, (3) if (A, R) is quasitriangular then R satisfies the QYBE, i.e.

$$R^{12}R^{13}R^{23} = R^{23}R^{13}R^{12},$$

(4) if (A, R) is a coboundary QUE-algebra and R satisfies the QYBE then (A, R) is triangular, (5) equations (21) mean that the mapping $A^* \to A$ given by $l \mapsto (l \otimes \mathrm{id})(R)$ is an algebra homomorphism and a coalgebra antihomomorphism.

To understand the various types of Hopf algebras one may use the Tannaka-Krein approach [**56**], i.e., instead of considering a Hopf algebra A one may consider the category Rep_A of its representations together with the forgetful functor $F\colon \mathrm{Rep}_A \to \{\text{vector spaces}\}$. Recall that a representation of a Hopf algebra A is a representation of A as an algebra, and the tensor product of representations $\rho_1\colon A \to \mathrm{End}\, V_1$ and $\rho_2\colon A \to \mathrm{End}\, V_2$ is equal to the composition $A \xrightarrow{\Delta} A \otimes A \xrightarrow{\rho_1\otimes\rho_2} \mathrm{End}(V_1 \otimes V_2)$. For any A the functor $\otimes\colon \mathrm{Rep}_A \times \mathrm{Rep}_A \to \mathrm{Rep}_A$ is associative. More precisely, for this functor there is an associativity constraint [**56**] compatible with F. In general the A-modules $V_1 \otimes V_2$ and $V_2 \otimes V_1$ are not isomorphic, but if (20) holds then R defines an isomorphism between them. Moreover, if $R^{12}R^{21} = 1$ we obtain a commutativity constraint [**56**] for $\otimes\colon \mathrm{Rep}_A \times \mathrm{Rep}_A \to \mathrm{Rep}_A$. It is not compatible with F unless $R = 1$. The condition (21) is the compatibility of the commutativity and the associativity constraints [**56**]. So if (A, R) is triangular then Rep_A is a tensor category [**56**], but F is not a tensor functor unless $R = 1$. V. V. Lyubashenko's result [**57**] means that a tensor category C with a functor $F\colon C \to \{\text{vector spaces}\}$ having the above properties is more or less Rep_A for a unique triangular Hopf algebra A.

If (A, Δ_0) is a cocommutative Hopf algebra and ϕ is an invertible element of $A\otimes A$ such that $\phi^{12}\cdot(\Delta_0\otimes\mathrm{id})(\phi) = \phi^{23}\cdot(\mathrm{id}\otimes\Delta_0)(\phi)$ then we can obtain a triangular Hopf algebra (A, Δ, R) by setting $\Delta(a) = \phi\Delta_0(a)\phi^{-1}$, $R = \phi^{21}(\phi^{12})^{-1}$. It can be shown that all triangular QUE-algebras can be obtained in this way (this is natural from the Tannaka-Krein point of view). In particular, a triangular QUE-algebra is isomorphic (as an algebra) to a universal enveloping algebra.

Finally I am going to mention a version of formula (20) which plays a crucial role in the QISM (e.g., see [**71**]) and which was in fact the origin of (20). Given a representation $\rho\colon A \to \mathrm{Mat}(n, \mathbf{C})$, its matrix elements $t_{ij} \in A^*$ satisfy the commutation relations $R_\rho \tilde{T}\overset{\approx}{T} = \overset{\approx}{T}\tilde{T}R_\rho$, where $R_\rho = (\rho \otimes \rho)(R)$,

$T = (t_{ij})$, $\tilde{T} = T \otimes E_n$, $\tilde{\tilde{T}} = E_n \otimes T$ (E_n is the unit matrix). Moreover, for a family of representations $\rho_\lambda \colon A \to \mathrm{Mat}(n, \mathbf{C})$, $\lambda \in \mathbf{C}$, we have

$$R(\lambda, \mu)\tilde{T}(\lambda)\tilde{\tilde{T}}(\mu) = \tilde{\tilde{T}}(\mu)\tilde{T}(\lambda)R(\lambda, \mu), \tag{22}$$

where $R(\lambda, \mu) = (\rho_\lambda \otimes \rho_\mu)(R)$, $T(\lambda) = (t_{ij}(\lambda))$, $t_{ij}(\lambda)$ are the matrix elements of ρ_λ. The more traditional view on (22) is to start from a function $R(\lambda, \mu) \in \mathrm{End}(\mathbf{C}^n \otimes \mathbf{C}^n)$ and consider the algebra B with generators $t_{ij}(\lambda), \lambda \in \mathbf{C}$, $1 \leq i, j \leq n$ and defining relations (22). Note that the formula

$$\Delta(t_{ij}(\lambda)) = \sum_k t_{ik}(\lambda) \otimes t_{kj}(\lambda)$$

defines a Hopf algebra structure on B (in the weak sense, i.e. without antipode), and if $R(\lambda, \mu) = (\rho_\lambda \otimes \rho_\mu)(R)$ for some $A, \{\rho_\lambda\}$ and $R \in A \otimes A$ such that (20) holds, then there is a natural homomorphism $B \to A$. So Hopf algebras have already appeared implicitly in [**71**]. Precise statements concerning the connection between triangular Hopf algebras and the QYBE can be found in [**57, 79**].

11. QISM from the Hopf algebra point of view. The QISM is a method for constructing and studying integrable quantum systems, which was created in 1978–1979 by L. D. Faddeev, E. K. Sklyanin, and L. A. Takhtajan [**69–71, 37**]. It is closely related to the classical inverse scattering method, Baxter's works on exactly solvable models of classical statistical mechanics (e.g. see [**58**]), and one-dimensional relativistic factorized scattering theory (C. N. Yang, A. B. Zamolodchikov, ...). Here is an exposition of the basic construction of the QISM, which slightly differs from the standard expositions [**70–72, 36, 37**], where Hopf algebras occur implicitly and (22) is used instead of (20).

Let A be a Hopf algebra, $C = \{l \in A^* \mid l(ab) = l(ba)$ for any $a, b \in A\}$. Then C is a subalgebra of A^*. Now suppose that (20) holds. Then it is easy to show that C is commutative (this is the quantum analogue of (9)). Fix a representation $\pi \colon A^* \to \mathrm{End}\, V$ ("the quantum L-operator") and a number N ("the number of sites of a 1-dimensional lattice"). Elements of $V^{\otimes N}$ are considered as quantum states. The image of C under the representation $\pi^{\otimes N} \colon A^* \to \mathrm{End}\, V^{\otimes N}$ is a commutative algebra of observables. Physicists usually fix an element $H \in A^*$ and consider $\pi^{\otimes N}(H)$ as a Hamiltonian.

If (A, R) is quasitriangular then R induces a homomorphism $A^* \to A$ and therefore a representation of A induces that of A^*. Usually only those representations of A^* are considered which come from representations of A. This means that instead of $\pi^{\otimes N}$ and C one works with $\rho^{\otimes N}$ and C', where ρ is a representation of A and C' is the image of C in A, i.e. $C' = \{(l \otimes \mathrm{id})(R) \mid l \in C\}$. To use this scheme one has to construct a quasitriangular (A, R) and as many representations of A as possible (these representations can be taken as ρ, and their characters belong to C). The Hopf algebras A used in the QISM are quantizations of Lie bialgebras G_S of Example 3.4 (i.e. of Lie bialgebras corresponding to nondegenerate solutions of the CYBE). Yangians and quantized affine algebras are typical examples. On these algebras there exists not a quasitriangular but a "pseudotriangular" structure (see §§12,13), but this suffices to use the above scheme.

Deeper aspects of the QISM, such as the description of the spectrum of $\rho^{\otimes N}(C')$ and the limit $N \to \infty$ (see [**65**, **36**, **71**]), are not yet understood from the Hopf algebra point of view.

12. The pseudotriangular structure on Yangians and rational solutions of the QYBE. Recall that $Y(\mathfrak{a})$ is obtained by quantizing the bialgebra $\mathfrak{g} = \mathfrak{a}[u]$ of Example 3.3 and setting $h = 1$. Since $\mathfrak{g}$ is pseudotriangular it is natural to expect $Y(\mathfrak{a})$ to be pseudotriangular too. To be more precise let us consider the operators $T_\lambda\colon \mathfrak{g} \to \mathfrak{g}$ given by $T_\lambda(f(u)) = f(u+\lambda)$. Note that though $r = t/(u-v)$ does not belong to $\mathfrak{g}\otimes\mathfrak{g}$, the expression

$$(T_\lambda \otimes \mathrm{id})r = t/(u+\lambda-v) = \sum_{k=0}^{\infty} t(v-u)^k \lambda^{-k-1}$$

is a power series in λ^{-1} whose coefficients belong to $\mathfrak{g}\otimes\mathfrak{g}$. Now define the automorphism $T_\lambda\colon Y(\mathfrak{a}) \to Y(\mathfrak{a})$ by the formulas $T_\lambda\,|_{\mathfrak{a}} = \mathrm{id}$, $T_\lambda(J(a)) = J(a)+\lambda a$ for $a \in \mathfrak{a}$. It is natural to expect that there exists a formal series

$$\mathcal{R}(\lambda) = 1 + \sum_{k=1}^{\infty} \mathcal{R}_k \lambda^{-1}, \qquad \mathcal{R}_k \in Y(\mathfrak{a}) \otimes Y(\mathfrak{a}),$$

which behaves as if $\mathcal{R}(\lambda)$ were equal to $(T_\lambda \otimes \mathrm{id})R$ for some R satisfying (20), (21) and the equation $R^{12}R^{21} = 1$. This is really so [**28**]. More precisely, there is a unique $\mathcal{R}(\lambda)$ such that $(\Delta \otimes \mathrm{id})\mathcal{R}(\lambda) = \mathcal{R}^{13}(\lambda)\mathcal{R}^{23}(\lambda)$ and $(T_\lambda \otimes \mathrm{id})\Delta'(a) = \mathcal{R}(\lambda)((T_\lambda \otimes \mathrm{id})\Delta(a))\mathcal{R}(\lambda)^{-1}$ for $a \in Y(\mathfrak{a})$. Besides, $\mathcal{R}(\lambda_1 - \lambda_2)$ satisfies the QYBE, i.e.

$$\begin{aligned}\mathcal{R}^{12}(\lambda_1-\lambda_2)\mathcal{R}^{13}(\lambda_1-\lambda_3)\mathcal{R}^{23}(\lambda_2-\lambda_3)\\ = \mathcal{R}^{23}(\lambda_2-\lambda_3)\mathcal{R}^{13}(\lambda_1-\lambda_3)\mathcal{R}^{12}(\lambda_1-\lambda_2).\end{aligned}$$

Moreover, $\mathcal{R}^{12}(\lambda)\mathcal{R}^{21}(-\lambda) = 1$, $(T_\mu \otimes T_\nu)\mathcal{R}(\lambda) = \mathcal{R}(\lambda+\mu-\nu)$, and the coefficient $\mathcal{R}_1$ is equal to t.

By the way, we are lucky that $Y(\mathfrak{a})$ is pseudotriangular and not triangular: otherwise $Y(\mathfrak{a})$ would be isomorphic (as an algebra) to a universal enveloping algebra and life would be dull.

Given a representation $\rho\colon Y(\mathfrak{a}) \to \mathrm{Mat}(n, \mathbf{C})$ we obtain a matrix solution $R(\lambda_1 - \lambda_2) = (\rho\otimes\rho)(\mathcal{R}(\lambda_1-\lambda_2))$ of the QYBE. It is of the form

$$R(\lambda_1-\lambda_2) = 1 + (\rho\otimes\rho)(t)\cdot(\lambda_1-\lambda_2)^{-1} + \sum_{k=2}^{\infty} A_k(\lambda_1-\lambda_2)^{-k}. \tag{23}$$

If ρ is irreducible then $R(\lambda)$ is a rational function up to a scalar factor [**28**].

Now let us temporarily forget about $Y(\mathfrak{a})$ and adopt the point of view of a person who wants to construct matrix solutions of the QYBE. According to [**45**] it is natural to look for solutions of the form

$$R(\lambda_1 - \lambda_2, h) = 1 + h(\rho \otimes \rho)(r(\lambda_1 - \lambda_2)) + \sum_{k=2}^{\infty} h^k a_k(\lambda_1 - \lambda_2),$$

where ρ is a representation of $\mathfrak{a}$ and r is an $\mathfrak{a} \otimes \mathfrak{a}$-valued solution of the CYBE. The simplest r is given by $r(\lambda) = t/\lambda$. Since this r is homogeneous it is natural to require that $R(\alpha\lambda, \alpha h) = R(\lambda, h)$. In this case $R(\lambda, h)$ is uniquely determined by $R(\lambda, 1)$, which is of the form (23). So the natural problem is to describe for a given $\rho\colon \mathfrak{a} \to \mathrm{Mat}(n, \mathbf{C})$ all the solutions of the QYBE of the form (23). Theorem [**28**]: *if ρ is irreducible then all such solutions are of the form $(\tilde{\rho} \otimes \tilde{\rho})(\mathcal{R}(\lambda_1 - \lambda_2))$ for some representation $\tilde{\rho}\colon Y(\mathfrak{a}) \to \mathrm{Mat}(n, \mathbf{C})$ such that $\tilde{\rho}|_{\mathfrak{a}} = \rho$.*

So it is clear that the problem of describing all the irreducible finite-dimensional representations of $Y(\mathfrak{a})$ is very important. It was solved for $\mathfrak{a} = \mathfrak{sl}(2)$ by V. E. Korepin and V. O Tarasov [**59–61**]. For classical $\mathfrak{a}$ specialists in the QISM have constructed by more or less elementary means a lot of R-matrices of the form (23) (see, for instance, [**62**, **63**] and the Appendix of [**45**]) and therefore a lot of representations of $Y(\mathfrak{a})$. Some results concerning this problem were obtained in [**28**, **64**]. In [**54**] a parametrization of the set of irreducible finite-dimensional representations of $Y(\mathfrak{a})$ (in the spirit of Cartan's highest weight theorem) is obtained using the second presentation of $Y(\mathfrak{a})$ (see §6) and the results for $Y(\mathfrak{sl}(2))$.

$Y(\mathfrak{a})$ has two "large" commutative subalgebras: the "Cartan" subalgebra generated by φ_{ik} (see §6) and the subalgebra $\mathcal{A}$ generated by the coefficients of $(l \otimes \mathrm{id})\mathcal{R}(\lambda)$, $l \in C$, where $C = \{l \in (Y(\mathfrak{a}))^* | \; l(ab) = l(ba)$ for any $a, b \in Y(\mathfrak{a})\}$. It is desirable to describe the spectra of these subalgebras in irreducible representations of $Y(\mathfrak{a})$. For $\mathcal{A}$ this problem is especially important (see §11). For classical $\mathfrak{a}$ and at least some representations of $Y(\mathfrak{a})$ the spectrum of $\mathcal{A}$ is known [**65–68**]. The analogy between the structure of the Bethe equations and the corresponding Cartan matrices (N. Yu. Reshetikhin [**68**]) suggests that the spectrum of $\mathcal{A}$ can be described for all $\mathfrak{a}$.

Finally, I am going to mention a realization of $Y(\mathfrak{a})$ which is often useful and which appeared much earlier than the general definition of $Y(\mathfrak{a})$. Fix a nontrivial irreducible representation $\rho\colon Y(\mathfrak{a}) \to \mathrm{Mat}(n, \mathbf{C})$. Set $R(\lambda) = (\rho \otimes \rho)(\mathcal{R}(\lambda))$. Then $Y(\mathfrak{a})$ is almost isomorphic to the Hopf algebra A_ρ with generators $t_{ij}^{(k)}$, $1 \leq i, j \leq n$, $k = 1, 2, \ldots$, defining relations

$$\tilde{T}(\lambda)\tilde{\tilde{T}}(\mu)R(\mu - \lambda) = R(\mu - \lambda)\tilde{\tilde{T}}(\mu)\tilde{T}(\lambda)$$

and comultiplication

$$\Delta(t_{ij}(\lambda)) = \sum_k t_{ik}(\lambda) \otimes t_{kj}(\lambda),$$

where $t_{ij}(\lambda) = \delta_{ij} + \sum_{k=1}^{\infty} t_{ij}^{(k)} \lambda^{-k}$, $T(\lambda) = (t_{ij}(\lambda))$, $\tilde{T}(\lambda) = T(\lambda) \otimes E_n$, $\tilde{\tilde{T}}(\lambda) = E_n \otimes T(\lambda)$. More precisely, there is an epimorphism $A_\rho \to Y(\mathfrak{a})$ given by $T(\lambda) \mapsto (\mathrm{id} \otimes \rho)(\mathcal{R}(\lambda))$, and to obtain $Y(\mathfrak{a})$ from A_ρ one has to add an auxiliary relation of the form $c(\lambda) = 1$, where $c(\lambda) \in A_\rho[[\lambda^{-1}]]$, $\Delta(c(\lambda)) = c(\lambda) \otimes c(\lambda)$, and $[a, c(\lambda)] = 0$ for any $a \in A_\rho$. For instance, if $\mathfrak{a} = \mathfrak{sl}(n)$ and $\rho: Y(\mathfrak{a}) \to \mathrm{Mat}(n, \mathbf{C})$ is the "vector" representation then $c(\lambda)$ is the "quantum determinant" of $T(\lambda)$ in the sense of [**73**] and $R(\lambda)$ is proportional to the Yang R-matrix $1 + \lambda^{-1}\sigma$, where $\sigma: \mathbf{C}^n \otimes \mathbf{C}^n \to \mathbf{C}^n \otimes \mathbf{C}^n$, $\sigma(x \otimes y) = y \otimes x$.

13. The double. Trigonometric solutions of the QYBE. If $\mathfrak{g}$ is a Lie bialgebra denote by $D(\mathfrak{g})$ the space $\mathfrak{g} \oplus \mathfrak{g}^*$ with the Lie algebra structure defined in §3. It can be shown that (1) the image r of the canonical element of $\mathfrak{g} \otimes \mathfrak{g}^*$ under the embedding $\mathfrak{g} \otimes \mathfrak{g}^* \hookrightarrow D(\mathfrak{g}) \otimes D(\mathfrak{g})$ satisfies (7) and therefore $(D(\mathfrak{g}), r)$ is a quasitriangular Lie bialgebra, and (2) the embedding $\mathfrak{g} \hookrightarrow D(\mathfrak{g})$ is a Lie bialgebra homomorphism and the embedding $\mathfrak{g}^* \hookrightarrow D(\mathfrak{g})$ is an algebra homomorphism and coalgebra antihomomorphism. $D(\mathfrak{g})$ is called the *double* of $\mathfrak{g}$. This construction has important applications [**43**].

Now let us define the *quantum double*. Let A be a Hopf algebra. Denote by A° the algebra A^* with the opposite comultiplication. It can be shown that there is a unique quasitriangular Hopf algebra $(D(A), R)$ such that (1) $D(A)$ contains A and A° as Hopf subalgebras, (2) R is the image of the canonical element of $A \otimes A^\circ$ under the embedding $A \otimes A^\circ \hookrightarrow D(A) \otimes D(A)$, and (3) the linear mapping $A \otimes A^\circ \to D(A)$ given by $a \otimes b \mapsto ab$ is bijective. As a vector space, $D(A)$ can be identified with $A \otimes A^\circ$, and the Hopf algebra structure on $A \otimes A^\circ$ can be found from the commutation relations

$$e_s e^t = \mu_s^{kjn} m_{plk}^t \sigma_n^p e^l e_j \quad \text{and} \quad e^t e_s = \mu_s^{njk} m_{klp}^t \sigma_n^p e_j e^l,$$

which follow from (1)–(3). Here $\{e_s\}$ and $\{e^t\}$ are dual bases of A and A°, while σ, m, and μ are the matrices of the skew antipode $A \to A$, the multiplication map $A \otimes A \otimes A \to A$, and the comultiplication map $A \to A \otimes A \otimes A$. If A is a QUE-algebra we will understand A° in the QUE-sense. Then $D(A)$ is a QUE-algebra, too. If $\mathfrak{g}$ is the classical limit of A then $(D(\mathfrak{g}), r)$ is the classical limit of $(D(A), R)$

Let $\mathfrak{g}, \mathfrak{b}_+, \mathfrak{b}_-, \mathfrak{h}$ denote the same objects as in Example 3.2. Then $D(\mathfrak{b}_+) = \mathfrak{g} \times \mathfrak{h}$ with the trivial Lie bialgebra structure on $\mathfrak{h}$. The quasitriangular (or pseudoquasitriangular) structure on $\mathfrak{g}$ (see §4) is defined by the image of $r \in D(\mathfrak{b}_+) \otimes D(\mathfrak{b}_+)$ under the natural mapping $D(\mathfrak{b}_+) \otimes D(\mathfrak{b}_+) \to \mathfrak{g} \otimes \mathfrak{g}$. It can be shown that in the quantum case the situation is quite similar. Denote by $U_h\mathfrak{b}_+$ the subalgebra of $U_h\mathfrak{g}$ generated by $\mathfrak{h}$ and X_i^+. Then $U_h\mathfrak{b}_- = (U_h\mathfrak{b}_+)^\circ$ and $D(U_h\mathfrak{b}_+) = U_h\mathfrak{g} \otimes U\mathfrak{h}$. Denote by $\overline{R}$ the image of $R \in D(U_h\mathfrak{b}_+) \otimes D(U_h\mathfrak{b}_+)$ under the natural mapping $D(U_h\mathfrak{b}_+) \otimes D(U_h\mathfrak{b}_+) \to U_h\mathfrak{g} \otimes U_h\mathfrak{g}$. Then $\overline{R}$ defines a quasitriangular structure on $U_h\mathfrak{g}$ if $\dim \mathfrak{g} < \infty$ and a "pseudoquasitriangular" structure if $\dim \mathfrak{g} = \infty$.

If $\mathfrak{g} = \mathfrak{sl}(2)$ with the scalar product $\mathrm{Tr}(XY)$ then

$$\overline{R} = \sum_{k=0}^{\infty} h^k Q_k(h) \{\exp \frac{h}{4}[H \otimes H + k(H \otimes 1 - 1 \otimes H)]\} (X^+)^k \otimes (X^-)^k,$$

where $Q_k(h) = e^{-kh/2}\prod_{r=1}^{k}(e^h-1)/(e^{rh}-1)$. In general,

$$\overline{R} = \sum_{\beta\in\mathbf{Z}_+^r}\left\{\exp h\left[\frac{1}{2}t_0 + \frac{1}{4}(H_\beta\otimes 1 - 1\otimes H_\beta)\right]\right\}P_\beta,$$

where t_0 is the element of $\mathfrak{h}\otimes\mathfrak{h}$ corresponding to the canonical scalar product on $\mathfrak{h}$, $\mathbf{Z}_+ = \{n\in\mathbf{Z}\mid n\geq 0\}$, r is the rank of $\mathfrak{g}$, for every $\beta = (\beta_1,\beta_2,\dots)\in\mathbf{Z}^r$ the symbol H_β denotes $\sum_i\beta_i H_i$, and $P_\beta\in U_h\mathfrak{b}_+\otimes U_h\mathfrak{b}_-$ is a polynomial in $u_i = X_i^+\otimes 1$ and $v_i = 1\otimes X_i^-$ homogeneous with respect to each variable and such that $\deg_{u_i}P_\beta = \det_{v_i}P_\beta = \beta_i$. In addition: (1) $P_0 = 1$, (2) $P_\beta\equiv 0 \bmod h^{k_\beta}$, $P_\beta\not\equiv 0 \bmod h^{k_\beta-1}$, where $k_\beta = \inf\{n\mid H_\beta$ is representable as a sum of n positive roots of $\mathfrak{g}\}$, and (3) the coefficients of $h^{-|\beta|}P_\beta$ are rational functions of e^{ch} for $c\in\mathbf{C}$, where $|\beta| = \sum_i\beta_i$.

If $\dim\mathfrak{g}<\infty$ then $R_\rho \stackrel{\text{def}}{=} (\rho\otimes\rho)(\overline{R})$ makes sense and satisfies the QYBE for any representation $\rho\colon U_h\mathfrak{g}\to \mathrm{Mat}(n,\mathbf{C}[[h]])$. For instance, if $\mathfrak{g} = \mathfrak{sl}(n)$ and ρ is the representation mentioned in §7 then

$$R_\rho = e^{-h/2n}\left\{\sum_{i\neq j}e_{ii}\otimes e_{jj} + e^{h/2}\sum_i e_{ii}\otimes e_{ii} + (e^{h/2}-e^{-h/2})\sum_{i<j}e_{ij}\otimes e_{ji}\right\},$$

where e_{ij} are the matrix units (this formula is implicit in [**52**]). Note that the relations (16)–(19) simply mean that $R_\rho\tilde{T}\tilde{\tilde{T}} = \tilde{\tilde{T}}\tilde{T}R_\rho$, where $T = (\rho_{ij})$.

Now let $\mathfrak{g}$ be affine and set $\mathfrak{g}' = [\mathfrak{g},\mathfrak{g}]/(\text{the center of }\mathfrak{g})$. Denote by R' the analogue of $\overline{R}$ for $U_h\mathfrak{g}'$. Suppose that a representation $\rho\colon U_h\mathfrak{g}'\to\mathrm{Mat}(n,\mathbf{C}[[h]])$ is given. Strictly speaking, the expression $(\rho\otimes\rho)(R')$ is meaningless. But if we define $T_\lambda\in\mathrm{Aut}\,U_h\mathfrak{g}$ setting $T_\lambda(X_i^\pm) = \lambda^{\pm 1}X_i^\pm$ and put $R'(\lambda) = (T_\lambda\otimes\mathrm{id})R'$ then $R_\rho(\lambda)\stackrel{\text{def}}{=}(\rho\otimes\rho)(R'(\lambda))$ is a well-defined formal series in λ such that

$$R_\rho^{12}(\lambda_1/\lambda_2)R_\rho^{13}(\lambda_1/\lambda_3)R_\rho^{23}(\lambda_2/\lambda_3) = R_\rho^{23}(\lambda_2/\lambda_3)R_\rho^{13}(\lambda_1/\lambda_3)R_\rho^{12}(\lambda_1/\lambda_2).$$

Using an idea of M. Jimbo [**50**] it can be proved that if ρ is irreducible then $R_\rho(\lambda)$ is a rational function of λ up to a scalar factor. Moreover, for a given ρ this rational function can be found by solving certain linear equations [**50**]. Substituting $\lambda = e^u$ we obtain trigonometric solutions of the QYBE. Explicit formulas for the solutions corresponding to the classical $\mathfrak{g}$ and some ρ were found by elementary methods [**48, 74–77**].

14. Final remarks. Because of the lack of space I cannot discuss here the relation between $Y(\mathfrak{sl}(n)), U_h\mathfrak{sl}(n)$ and Hecke algebras (see [**52, 64**]), between Yangians and Hermitian symmetric spaces (see [**62**] and [**28**, Theorem 7]), between $Y(\mathfrak{sl}(n))$ and representations of infinite-dimensional classical groups (see [**80**]). By the same reason I cannot discuss compact quantum groups [**81–83**] as well as the attempts to generalize the notion of superalgebra, which are based on the QYBE [**78**] or triangular Hopf algebras [**79**].

REFERENCES

1. J. W. Milnor and J. C. Moore, *On the structure of Hopf algebras*, Ann. of Math. (2) **81** (1965), 211–264.

2. M. E. Sweedler, *Hopf algebras*, Mathematical Lecture Note Series, Benjamin, N.Y., 1969.

3. E. Abe, *Hopf algebras*, Cambridge Tracts in Math., No. 74, Cambridge Univ. Press, Cambridge-New York, 1980.

4. G. M. Bergman, *Everybody knows what a Hopf algebra is*, Contemp. Math. **43** (1985), 25–48.

5. M. E. Sweedler, *Integrals for Hopf algebras*, Ann. of Math. (2) **89** (1969), 323–335.

6. R. G. Larson, *Characters of Hopf algebras*, J. Algebra **17** (1971), 352–368.

7. W. C. Waterhouse, *Antipodes and group-likes in finite Hopf algebras*, J. Algebra **37** (1975), 290–295.

8. D. E. Radford, *The order of the antipode of a finite dimensional Hopf algebra is finite*, Amer. J. Math. **98** (1976), 333–355.

9. H. Shigano, *A correspondence between observable Hopf ideals and left coideal subalgebras*, Tsukuba J. Math. **1** (1977), 149–156.

10. E. J. Taft and R. L. Wilson, *There exist finite dimensional Hopf algebras with antipodes of arbitrary even order*, J. Algebra **62** (1980), 283–291.

11. H. Shigano, *On observable and strongly observable Hopf ideals*, Tsukuba J. Math. **6** (1982), 127–150.

12. B. Pareigis, *A noncommutative noncocommutative Hopf algebra in "nature"*, J. Algebra **70** (1981), 356–374.

13. G. I. Kac, *Generalizations of the group duality principle*, Dokl. Akad. Nauk SSSR **138** (1961), 275–278. (Russian)

14. ——, *Compact and discrete ring groups*, Ukrainian Math. J. **14** (1962), 260–270. (Russian)

15. ——, *Ring groups and the duality principle*, Proc. Moscow Math. Soc. **12** (1963), 259–303; **13** (1965), 84–113. (Russian)

16. G. I. Kac and V. G. Palyutkin, *An example of a ring group generated by Lie groups*, Ukrainian Math. J. **16** (1964), 99–105. (Russian)

17. ——, *Finite ring groups*, Proc. Moscow Math. Soc. **15** (1966), 224–261. (Russian)

18. G. I. Kac, *Group extensions which are ring groups*, Mat. Sb. **76** (1968), 473–496. (Russian)

19. ——, *Some arithmetical properties of ring groups*, Funktsional Anal. i Prilozhen. **6** (1972), no. 2, 88–90. (Russian)

20. G. I. Kac and L. I. Vainerman, *Non-unimodular ring groups and Hopf-von Neumann algebras*, Mat. Sb. **94** (1974), 194–225. (Russiam)

21. M. Enock and J.-M. Schwartz, *Une dualité dans les algèbres de von Neumann*, Bull. Soc. Math. France, mém. No 44, 1975.

22. J.-M. Schwartz, *Relations entre "ring groups" et algèbres de Kac*, Bull. Sci. Math. (2) **100** (1976), 289–300.

23. M. Enock, *Produit croisé d'une algèbre de von Neumann par une algèbre de Kac*, J. Funct. Anal. **26** (1977), 16–47.

24. M. Enock and J.-M. Schwartz, *Produit croisé d'une algèbre de von Neumann par une algèbre de Kac*. II, Publ. Res. Inst. Math. Sci. **16** (1980), 189–232.

25. J. De Cannière, M. Enock, and J-M. Schwartz, *Algèbres de Fourier associées à une algèbre de Kac*, Math. Ann. **245** (1979), 1–22.

26. ——, *Sur deux résultats d'analyse harmonique non-commutative: une application de la théorie des algèbres de Kac*, J. Operator Theory **5** (1981), 171–194.

27. V. G. Drinfeld, *Hamiltonian structures on Lie groups, Lie bialgebras and the geometric meaning of the classical Yang-Baxter equations*, Dokl. Akad. Nauk SSSR **268** (1983), 285–287. (Russian)

28. ——, *Hopf algebras and the quantum Yang-Baxter equation*, Dokl. Akad. Nauk SSSR **283** (1985), 1060–1064. (Russian)

29. I. M. Gelfand and I. V. Cherednik, *Abstract Hamiltonian formalism for classical Yang-Baxter bundles*, Uspekhi Mat. Nauk **38** (1983), no. 3, 3–21. (Russian)

30. I. V. Cherednik, *Bäcklund-Darboux transformations for classical Yang-Baxter bundles*, Funktsional Anal. i Prilozhen. **17** (1983), no. 2, 88–89. (Russian)

31. ——, *On the definition of τ-functions for generalized affine Lie algebras*, Funktsional Anal. i Prilozhen. **17** (1983), no. 3, 93–95. (Russian)

32. I. M. Gelfand and I. Ya. Dorfman, *Hamiltonian operators and the classical Yang-Baxter equation*, Funktsional Anal. i Prilozhen. **16** (1982), no. 4, 1–9. (Russian)

33. A. Weinstein, *The local structure of Poisson manifolds*, J. Diff. Geometry **18** (1983), 523–557.

34. J. A. Schouten, *Über differentialkomitanten zweier kontravarianter Grössen*, Nederl. Akad. Wetensch. Proc. Ser. A **43** (1940), 449–452.

35. L. D. Faddeev and L. A. Takhtajan, *Hamiltonian approach to solitons theory*, Springer-Verlag, Berlin-New York, 1987.

36. L. D. Faddeev, *Integrable models in* $(1+1)$*-dimensional quantum field theory* (Lectures in Les Houches, 1982), Elsevier Science Publishers B. V., 1984.

37. E. K. Sklyanin, *The quantum version of the inverse scattering method*, Zap. Nauchn. Sem. LOMI **95** (1980), 55–128.

38. L. D. Faddeev and N. Yu. Reshetikhin, *Hamiltonian structures for integrable models of field theory*, Theoret. Math. Phys. **56** (1983), 323–343. (Russian)

39. M. A. Semenov-Tian-Shansky, *What is the classical r-matrix*, Funktsional. Anal. i Prilozhen. **17** (1983), no. 4, 17–33. (Russian)

40. A. A. Belavin and V. G. Drinfeld, *On the solutions of the classical Yang-Baxter equations for simple Lie algebras*, Funktsional. Anal. i Prilozhen. **16** (1982), no. 3, 1–29. (Russian)

41. ——, *Triangle equations and simple Lie algebras*, Soviet Scientific Reviews Section C **4** (1984), 93–165. Harwood Academic Publishers, Chur (Switzerland)-New York.

42. ——, *On the classical Yang-Baxter equation for simple Lie algebras*, Funktsional. Anal. i Prilozhen. **17** (1983), no. 3, 69–70. (Russian)

43. M. A. Semenov-Tian-Shansky, *Dressing transformations and Poisson group actions*, Publ. Res. Inst. Math. Sci. **21** (1986).

44. E. K. Sklyanin, *On complete integrability of the Landau-Lifshitz equation*, LOMI preprint E-3-1979, Leningrad, 1979.

45. P. P. Kulish and E. K. Sklyanin, *On the solutions of the Yang-Baxter equations*, Zap. Nauchn. Sem. LOMI **95** (1980), 129–160. (Russian)

46. A. A. Belavin, *Discrete groups and integrability of quantum systems*, Funktsional. Anal. i Prilozhen. **14** (1980), 18–26. (Russian)

47. G. E. Andrews, *The theory of partitions*, Addison-Wesley, London, 1976.

48. P. P. Kulish and N. Yu. Reshetikhin, *The quantum linear problem for the sine-Gordon equation and higher representations*, Zap. Nauchn. Sem. LOMI **101** (1981), 101–110. (Russian)

49. E. K. Sklyanin, *On an algebra generated by quadratic relations*, Uspekhi Mat. Nauk **40** (1985), no. 2, 214.

50. M. Jimbo, *Quantum R-matrix for the generalized Toda system*, Commun. Math. Phys. **102** (1986), 537–547.

51. ——, *A q-difference analogue of $U(\mathfrak{g})$ and the Yang-Baxter equation*, Lett. Math. Phys. **10** (1985), 63–69.

52. ——, *A q-analogue of $U(\mathfrak{gl}(N+1))$, Hecke algebra and the Yang-Baxter equation*, Lett. Math. Phys. **11** (1986), 247–252.

53. C. N. Yang, *Some exact results for the many-body problem in one dimension with repulsive delta-function interaction*, Phys. Rev. Letters **19** (1967), 1312–1314.

54. V. G. Drinfeld, *A new realization of Yangians and quantized affine algebras*, FTINT Preprint No. 30–86, Kharkov, 1986 (to appear in Dokl. Akad. Nauk SSSR). (Russian)

55. V. G. Kac, *Infinite dimensional Lie algebras*, Birkhäuser, Boston-Basel-Stuttgart, 1983.

56. P. Deligne and J. Milne, *Tannakian categories*, Lecture Notes in Math., vol. 900, Springer-Verlag, Berlin-Heidelberg-New York, 1982, pp. 101–228.

57. V. V. Lyubashenko, *Hopf algebras and vectorsymmetries*, Uspekhi Mat. Nauk **41** (1986), no. 5, 185–186.

58. R. J. Baxter, *Partition function of the eight-vertex lattice model*, Ann. Phys. (N.Y.) **70** (1972), 193–228.

59. V. E. Korepin, *The analysis of the bilinear relation of the six-vertex model*, Dokl. Akad. Nauk SSSR **265** (1982), 1361-1364. (Russian)

60. V. O. Tarasov, *On the structure of quantum L-operators for the R-matrix of the XXZ-model*, Theoret. Math. Phys. **61** (1984), 163–173. (Russian)

61. ——, *Irreducible monodromy matrices for the R-matrix of the XXZ-model and lattice local quantum Hamiltonians*, Theoret. Math. Phys. **63** (1985), 175–196. (Russian)

62. N. Yu. Reshetikhin, *Integrable models of quantum 1-dimensional magnetics with* O(n) *and* Sp$(2k)$*-symmetry*, Theoret. Math. Phys. **63** (1985), 347–366. (Russian)

63. P. P. Kulish, N. Yu. Reshetikhin, and E. K. Sklyanin, *Yang-Baxter equation and representation theory*. I, Lett. Math. Phys. **5** (1985), 393–403.

64. V. G. Drinfeld, *Degenerate affine Hecke algebras and Yangians*, Funktsional. Anal. i Prilozhen. **20** (1986), no. 1, 69–70. (Russian)

65. L. A. Takhtajan, *Quantum inverse scattering method and the algebrized matrix Bethe ansatz*, Zap. Nauchn. Sem. LOMI **101** (1981), 158–183. (Russian)

66. P. P. Kulish and N. Yu. Reshetikhin, *The generalized Heisenberg ferromagnetic and the Gross-Neveu Model*, J. Experim. Theoret. Phys. **80** (1981), 214–227. (Russian)

67. N. Yu. Reshetikhin, *Functional equation method in the theory of exactly solvable quantum systems*, J. Experim. Theoret. Phys. **84** (1983), 1190–1201. (Russian)

68. ——, *Exactly solvable quantum systems on the lattice, which are connected with classical Lie algebras*, Zap. Nauchn. Sem. LOMI **123** (1983), 112–125. (Russian)

69. L. D. Faddeev and E. K. Sklyanin, *The quantum-mechanical approach to completely integrable models of field theory*, Dokl. Akad. Nauk SSSR **243** (1978), 1430–1433. (Russian)

70. L. D. Faddeev, E. K. Sklyanin, and L. A. Takhtajan, *The quantum inverse problem* I, Theoret. Math. Phys. **40** (1979), 194–220. (Russian)

71. L. D. Faddeev and L. A. Takhtajan, *The quantum inverse problem method and the XYZ Heisenberg model*, Uspekhi Mat. Nauk **34** (1979), no. 5, 13–63. (Russian)

72. L. D. Faddeev, *Quantum completely integrable models in field theory*, Soviet Scientific Reviews Section C **1** (1980), 107–155; Harwood Academic Publishers, Chur (Switzerland)-New York.

73. P. P. Kulish and E. K. Sklyanin, *Quantum spectral transform method. Recent developments*, Lecture Notes in Phys., vol. 151, Springer-Verlag, Berlin-Heidelberg-New York, 1982, pp. 61–119.

74. I. V. Cherednik, *On the method of constructing factorized S-matrices in elementary functions*, Theoret. Math. Phys. **43** (1980), 117–119. (Russian)

75. A. V. Zamolodchikov and V. A. Fateev, *A model factorized S-matrix and the integrable Heisenberg lattice with spin* 1, Yadernaya Fiz. **32** (1980), 581–590. (Russian)

76. A. G. Izergin and V. E. Korepin, *The inverse scattering approach to the quantum Shabat-Mikhailov model*, Comm. Math. Phys. **79** (1981), 303–316.

77. V. V. Bazhanov, *Trigonometric solutions of triangle equations and classical Lie algebras*, Phys. Lett. **159B** (1985), 321–324.

78. D. I. Gurevich, *Quantum Yang-Baxter equation and generalization of the formal Lie theory*, Seminar on Supermanifolds (D. A. Leĭtes, ed.), D. Reidel (to appear).

79. V. V. Lyubashenko, *Vectorsymmetries*, ibid.

80. G. I. Olshansky, *An extension of* $U(\mathfrak{g})$ *for infinite dimensional Lie algebras* $\mathfrak{g}$ *and Yangians*, Dokl. Akad. Nauk SSSR (to appear).

81. S. L. Woronowicz, *Twisted* SU(2) *group. An example of a non-commutative differential calculus*, Publications RIMS Kyoto University (to appear).

82. ——, *Compact matrix pseudogroups*, Preprint No. 1/87, Warsaw University, 1987.

83. L. L. Vaksman and Ya. S. Soibelman, *The algebra of functions on quantized* SU(2), Funktsional. Anal. i Prilozhen. (to appear).

PHYSICO-ENGINEERING INSTITUTE OF LOW TEMPERATURES, UKR. SSR ACADEMY OF SCIENCES, KHARKOV 310164, USSR

Letters in Mathematical Physics **10** (1985) 63–69. 0377–9017/85.15.

A q-Difference Analogue of U($\mathfrak{g}$) and the Yang–Baxter Equation

MICHIO JIMBO
Research Institute for Mathematical Sciences, Kyoto University, Kyoto, 606, Japan

(Received: 24 April 1985)

Abstract. A q-difference analogue of the universal enveloping algebra U($\mathfrak{g}$) of a simple Lie algebra $\mathfrak{g}$ is introduced. Its structure and representations are studied in the simplest case $\mathfrak{g} = \mathfrak{sl}(2)$. It is then applied to determine the eigenvalues of the trigonometric solution of the Yang–Baxter equation related to $\mathfrak{sl}(2)$ in an arbitrary finite-dimensional irreducible representation.

1. Consider the following commutation relations among the generators e, f, h:

$$[h, e] = 2e\,, \qquad [h, f] = -2f\,, \qquad [e, f] = \sinh(\eta h)/\sinh \eta \tag{1}$$

where η is a parameter. When $\eta = 0$, these are precisely the familiar commutator rules for $\mathfrak{sl}(2)$; for general η, (1) no longer defines a Lie algebra but an associative algebra, which may be regarded as a deformation of the universal enveloping algebra of $\mathfrak{sl}(2)$.

The relations (1) first appeared in the work of Kulish–Reshetikhin [1] on the Yang–Baxter (YB) equation

$$(\check{R}(u)\otimes 1)(1\otimes \check{R}(u+v))(\check{R}(v)\otimes 1) = (1\otimes \check{R}(v))(\check{R}(u+v)\otimes 1)(1\otimes \check{R}(u)). \tag{2}$$

Here $\check{R}(u) \in \mathrm{End}(V\otimes V)$ is a matrix valued function of $u \in \mathbb{C}$, and (2) is to be understood as an equation in $\mathrm{End}(V\otimes V\otimes V)$. As for the formulation of the YB equation and its significance in the theory of completely integrable systems, the reader is referred to [2]. In [1] is treated the case where V is an irreducible representation space of $\mathfrak{g} = \mathfrak{sl}(2)$ and $\check{R}(u) = \check{R}(u, \eta)$ is trigonometric (i.e., a rational function of the variable $x = e^u$). With the aid of (1), these authors were able to derive recurrence relations for the matrix elements of $\check{R}(u)$.

If we set $u \to u\eta$ and let $\eta \to 0$, then $\check{R}(u, \eta)$ above reduces to a rotational function of u.

The limit $\check{R}_0(u)$ gains the invariance

$$[\check{R}_0(u), X\otimes 1 + 1\otimes X] = 0\,, \quad X \in \mathfrak{g}\,.$$

An independent construction of such solutions with group invariance was given in [3]. There a general formula for $\check{R}_0(u)$ is obtained which is valid for any representation space of $\mathfrak{sl}(2)$.

The purpose of this Letter is to show that the trigonometric case of [1] can be put on the same footing as the rational case by employing the representation theory of the algebra (1). In Section 2 we formulate this type of algebra in general, which makes sense for an arbitrary symmetrizable generalized Cartan matrix. In Section 3 the simplest case of $\mathfrak{sl}(2)$ is considered in detail. Construction of the trigonometric $\check{R}(n, \eta)$ related to $\mathfrak{sl}(2)$ is given in Section 4. The general formula (13) is quite analogous to the rational case [3] and, in fact, reduces to it as $\eta \to 0$.

2. Let $A = (a_{ij})$ be a symmetrizable generalized Cartan matrix in the sense of [4], and let $\{\alpha_i\}_{1 \leq i \leq N}$, $\{h_i\}_{1 \leq i \leq N}$ be the simple roots and co-roots such that $\langle h_i, \alpha_j \rangle = a_{ij}$. For a nonzero parameter t we write $t_i = t^{(\alpha_i | \alpha_i)/2}$ so that $t_i^{a_{ij}} = t^{(\alpha_i | \alpha_j)} = t_j^{a_{ji}}$, where $(|)$ denotes the invariant inner product in $\mathfrak{h}^* = \oplus \mathbb{C}\alpha_i$.

Consider the following relations among the generators $\{k_i^{\pm 1}, e_i, f_i\}_{1 \leq i \leq N}$:

$$k_i \cdot k_i^{-1} = k_i^{-1} \cdot k_i = 1, \qquad [k_i, k_j] = 0, \tag{3A}$$

$$k_i e_j k_i^{-1} = t_i^{a_{ij}} e_j, \qquad k_i f_j k_i^{-1} = t_i^{-a_{ij}} f_j, \tag{3B}$$

$$[e_i, f_j] = \delta_{ij}(k_i^2 - k_i^{-2})/(t_i^2 - t_i^{-2}), \tag{3C}$$

$$\sum_{\nu=0}^{1-a_{ij}} (-)^{\nu} \begin{bmatrix} 1 - a_{ij} \\ \nu \end{bmatrix}_{t_i^2} e_i^{1-a_{ij}-\nu} e_j e_i^{\nu} = 0 \quad (i \neq j), \tag{3D}$$

$$\sum_{\nu=0}^{1-a_{ij}} (-)^{\nu} \begin{bmatrix} 1 - a_{ij} \\ \nu \end{bmatrix}_{t_i^2} f_i^{1-a_{ij}-\nu} f_j f_i^{\nu} = 0 \quad (i \neq j). \tag{3E}$$

Here t ($= e^{\eta/2}$ in (1)) is an arbitrary parameter, and the symbol

$$\begin{bmatrix} m \\ n \end{bmatrix}_t = \begin{bmatrix} m \\ m - n \end{bmatrix}_t$$

is defined by

$$\begin{bmatrix} m \\ n \end{bmatrix}_t = \begin{cases} \dfrac{(t^m - t^{-m})(t^{m-1} - t^{-m+1}) \dots (t^{m-n-1} - t^{-m+n-1})}{(t - t^{-1})(t^2 - t^{-2}) \dots (t^n - t^{-n})} & (m > n > 0) \\ 1 & (n = 0, m) \end{cases}$$

and $[{}^m_n]_t = 0$ otherwise. It is related to the more standard Gauss polynomial

$$\begin{bmatrix} m \\ n \end{bmatrix}_q = \frac{(1 - q^m)(1 - q^{m-1}) \dots (1 - q^{m-n+1})}{(1 - q)(1 - q^2) \dots (1 - q^n)}$$

via

$$\begin{bmatrix} m \\ n \end{bmatrix}_t = \begin{bmatrix} m \\ n \end{bmatrix}_{t^2} \cdot t^{-n(m-n)}.$$

We denote by U_t the algebra generated by $\{k_i^{\pm 1}, e_i, f_i\}$ under the defining relations (3A–E). If we put $k_i = t_i^{h_i}$ and let $t \to 1$, then (3A–E) reduce to the usual defining relations of the Kac–Moody Lie algebra $\mathfrak{g}(A)$. Thus, U_t may be thought of as a q-difference analogue of the universal enveloping algebra of $\mathfrak{g}(A)$. Such an algebra was first introduced in [1] in the simplest case $\mathfrak{g}(A) = \mathfrak{sl}(2)$, for which (3D–E) are void.

The structure and representation theory of U_t is under investigation. Here we only note the following fact which will be of use later.

PROPOSITION 1. *There is a unique algebra homomorphism* $\delta^{(N)}: U_t \to U_t \otimes \cdots \otimes U_t$ *(N fold tensor product) such that*

$$\delta^{(N)}(k_i) = k_i \otimes \cdots \otimes k_i,$$

$$\delta^{(N)}(e_i) = \sum_{\nu=1}^{N} k_i \otimes \cdots \otimes k_i \otimes \overset{\nu}{\check{e}_i} \otimes k_i^{-1} \otimes \cdots \otimes k_i^{-1}, \tag{4}$$

$$\delta^{(N)}(f_i) = \sum_{\nu=1}^{N} k_i \otimes \cdots \otimes k_i \otimes \overset{\nu}{\check{f}_i} \otimes k_i^{-1} \otimes \cdots \otimes k_i^{-1}.$$

The proof uses the identity ([5], Equation (3.3.6))

$$\prod_{\nu=0}^{n-1} (1 - zq^{\nu}) = \sum_{\nu=0}^{n} (-)^{\nu} \begin{bmatrix} n \\ \nu \end{bmatrix}_q z^{\nu} q^{\nu(\nu-1)/2}.$$

3. Hereafter we restrict our discussion to the simplest case $\mathfrak{g}(A) = \mathfrak{sl}(2)$. In this case the following hold:

(i) An element of U_t is uniquely representable as a linear combination of monomials $f^m k^l e^n$, where $m, n \in \mathbb{Z}_+ = \{0, 1, 2, \ldots\}$ and $l \in \mathbb{Z}$.

(ii) The 'Casimir element'

$$C = \frac{(kt - k^{-1}t^{-1})^2}{(t^2 - t^{-2})^2} + fe \tag{5}$$

belongs to the center of U_t.

(iii) For $\lambda \in \mathbb{C}$, the Verma module $M_\lambda = U_t v_\lambda$, defined via the relations $kv_\lambda = t^\lambda v_\lambda$, $ev_\lambda = 0$, is reducible if $\lambda \in \mathbb{Z}_+$. When $\lambda \in \mathbb{Z}_+$, $L_\lambda = M_\lambda / U_t f^{\lambda+1} v_\lambda$ is irreducible and $\dim L_\lambda = \lambda + 1$.

(iv) A nondegenerate symmetric bilinear form (,) is introduced in L_λ by setting

$$(f^m v_\lambda, f^{m'} v_\lambda) = \delta_{mm'} [m]_m [\lambda]_m \tag{6}$$

where

$$[\lambda]_m = \prod_{j=1}^{m} \left(\frac{(t^{2(\lambda-j+1)} - t^{-2(\lambda-j+1)})}{(t^2 - t^{-2})} \right).$$

It is contravariant in the sense $(Xv, v') = (v, \theta(X)v')$ $(X \in U_t)$, where θ signifies the anti-automorphism such that $\theta(e) = f$, $\theta(f) = e$ and $\theta(k) = k$.

(v) For $\lambda, \mu \in \mathbb{Z}_+$ the tensor product $L_\lambda \otimes L_\mu$, viewed as a U_t-module by Proposition 1, is completely reducible:

$$L_\lambda \otimes L_\mu \simeq \bigoplus_{\nu=0}^{\min(\lambda,\mu)} L_{\lambda+\mu-2\nu}.$$

A highest weight vector $v^{(\nu)}$ in the component $L_{\lambda+\mu-2\nu}$ is given by

$$v^{(\nu)} = \sum_{r=0}^{\nu} c_{\nu r} f^{\nu-r} v_\lambda \otimes f^r v_\mu, \qquad (7)$$

$$\delta^{(2)}(e) v^{(\nu)} = 0$$

where

$$c_{\nu r} = (-t^{-\lambda-\mu+2(\nu-1)})^r \frac{[\nu]_r [\lambda-\nu+r]_r}{[r]_r [\mu]_r}.$$

The proof of (i)–(v) is done in an analogous way as in the case of the Lie algebra $\mathfrak{sl}(2)$, using the formula

$$[e, f^m] = f^{m-1} \frac{t^{2m}-t^{-2m}}{t^2-t^{-2}} \cdot \frac{k^2 t^{-2(m-1)} - k^{-2} t^{2(m-1)}}{t^2-t^{-2}}.$$

In fact we have the following realization of U_t.

PROPOSITION 2. *Let $\bar{e}, \bar{f}, \bar{h}$ be the standard Chevalley basis of the Lie algebra $\mathfrak{sl}(2)$, and let $\bar{M}_\lambda$ be the usual Verma module with highest weight λ. Then there is an algebra homomorphism ρ: $U_t \to \mathrm{End}(\bar{M}_\lambda)$ such that*

$$\rho(k) = t^{\bar{h}}, \qquad \rho(e) = \Phi_{\lambda+1}\bar{e}, \qquad \rho(f) = \bar{f}, \qquad (8)$$

where

$$\Phi_{\lambda+1} = \Phi_{-\lambda-1} = \frac{4(t^{2(\bar{h}-1)} + t^{-2(\bar{h}-1)} - t^{2(\lambda+1)} - t^{-2(\lambda+1)})}{((\bar{h}-1)^2 - (\lambda+1)^2)(t^2-t^{-2})^2}.$$

In view of the relation between λ and the usual Casimir operator $\bar{C} = \bar{h}(\bar{h}+2)/4 + \bar{f}\bar{e}$,

$$4\bar{C} + 1 = (\lambda+1)^2 \quad \text{in } \mathrm{End}(\bar{M}_\lambda),$$

(8) can be regarded as an embedding of U_t in the (properly completed) universal enveloping algebra of $\mathfrak{sl}(2)$.

Note that in the general case the simple choice $\rho(f) = \bar{f}$ is impossible because of (3D–E).

4. Let us proceed to the construction of the solution $\check{R}$ of the YB equation. We fix a representation space L_λ ($\lambda \in \mathbb{Z}_+$) and regard $\check{R}$ as an element of $\mathrm{End}(L_\lambda \otimes L_\lambda)$. Put $x = e^u$ and write $\check{R}(x)$ for $\check{R}(u)$.

It has been shown in [6] that a solution to the linear equations

$$[\check{R}(x), e\otimes k^{-1} + k\otimes e] = 0\,, \tag{9}$$

$$\check{R}(x)(xf\otimes k + k^{-1}\otimes f) = (f\otimes k + xk^{-1}\otimes f)\check{R}(x) \tag{10}$$

satisfies the YB equation (2). It was shown also that (9) and (10) imply

$$[\check{R}(x), f\otimes k^{-1} + k\otimes f] = 0\,, \tag{11}$$

$$[\check{R}(x), k\otimes k] = 0\,. \tag{12}$$

Here $\check{R}(x)$ is related to $R(x)$ in [6] through $\check{R}(x) = PR(x)$, where $P\in \mathrm{End}(L_\lambda\otimes L_\lambda)$ signifies the transposition $P(u\otimes v) = v\otimes u$. The point is, it is $\check{R}(x)$ and not $R(x)$ which can be diagonalized independently of x. In view of Proposition 1, (9), (11) and (12) mean that $\check{R}(x)$ commutes with the action of $\delta^{(2)}(U_t)$. Hence, it must have the form

$$\check{R}(x) = \sum_{\nu=0}^{\lambda} \rho_\nu(x) P_\nu$$

where P_ν is the projector onto the irreducible component $L_{2\lambda-2\nu}$ of $L_\lambda\otimes L_\lambda$. To determine the eigenvalues $\rho_\nu(x)$, we need:

LEMMA.

(i) $P_\mu f\otimes kP_\nu = 0\,, \qquad P_\mu k^{-1}\otimes fP_\nu = 0 \quad (\mu\neq \nu, \nu\pm 1)\,,$

(ii) $P_{\nu+1} f\otimes kP_\nu = -t^{4(\lambda-\nu)}P_{\nu+1}k^{-1}\otimes fP_\nu\,,$

$P_{\nu-1} f\otimes kP_\nu = -t^{-4(\lambda-\nu+1)}P_{\nu-1}k^{-1}\otimes fP_\nu\,,$

(iii) $P_\nu f\otimes kP_\nu = P_\nu k^{-1}\otimes fP_\nu\,.$

Proof. First note that $f\otimes k$, $k^{-1}\otimes f$ commute with $\delta^{(2)}(f)$. For the highest weight vector $v^{(\nu)}$ of (7), put

$$f\otimes kv^{(\nu)} = \sum_{j\geqslant 0} d_j\,\delta^{(2)}(f)^j v^{(\nu+1-j)}\,,$$

$$k^{-1}\otimes fv^{(\nu)} = \sum_{j\geqslant 0} d_j'\,\delta^{(2)}(f)^j v^{(\nu+1-j)}\,.$$

Applying $\delta^{(2)}(f)^{2(\lambda-\nu)+1}$ to these and using $\delta^{(2)}(f)^n v^{(\nu)} = 0$ iff $n\geqslant 2(\lambda-\nu)+1$, we see that the coefficients d_j, d_j' vanish except for $j = 0, 1, 2$. This shows (i). To show (ii), extend the contravariant form $(\,,)$ on L_λ to the one on $L_\lambda\otimes L_\lambda$ by $(u\otimes v, u'\otimes v') = (u, u')(v, v')$. We have then

$$\begin{aligned}
&((f\otimes k + t^{4(\lambda-\nu)}k^{-1}\otimes f)v^{(\nu)}, v^{\nu+1)})\\
&= ((f\otimes k + k^{-1}\otimes f\cdot k^2\otimes k^2)v^{(\nu)}, v^{(\nu+1)})\\
&= (\delta^{(2)}(f)\cdot 1\otimes k^2 v^{(\nu)}, v^{(\nu+1)})\\
&= (1\otimes k^2 v^{(\nu)}, \delta^{(2)}(e)v^{(\nu+1)})\\
&= 0\,,
\end{aligned}$$

from which follows $d_0 = -t^{4(\lambda-\nu)}d_0'$. Likewise, we compute

$$((f\otimes k + t^{-4(\lambda-\nu+1)}k^{-1}\otimes f)\,\delta^{(2)}(f)^{2(\lambda-\nu)}v^{(\nu)},\ \delta^{(2)}(f)^{2(\lambda-\nu+1)}v^{(\nu-1)})$$

to obtain $d_2 = -t^{-4(\lambda-\nu+1)}d_2'$. For the proof of (iii), it suffices to show that $\delta^{(2)}(e)(f\otimes k - k^{-1}\otimes f)v^{(\nu)}$ is proportional to $v^{(\nu-1)}$ or, equivalently, that it is an eigenvector of $\delta^{(2)}(C)$ with the eigenvalue $(t^{2(\lambda-\nu)+3} - t^{-2(\lambda-\nu)-3})^2/(t^2 - t^{-2})^2$. Using

$$e\otimes k^{-1}v^{(\nu)} = -k\otimes ev^{(\nu)}, \qquad e\otimes fv^{(\nu)} = -t^2(1\otimes fe)(k\otimes k)v^{(\nu)},$$
$$f\otimes ev^{(\nu)} = -t^{-2}(fe\otimes 1)(k^{-1}\otimes k^{-1})v^{(\nu)}$$

and that

$$(t^2 - t^{2})^2 fe = (t^{\lambda+1} - t^{-\lambda-1})^2 - (tk - t^{-1}k^{-1})^2 \quad \text{on } L_\lambda$$

(cf. (5)) we get

$$(t^2 - t^{-2})\,\delta^{(2)}(e)(f\otimes k - k^{-1}\otimes f)v^{(\nu)}$$
$$= -(t^2 + t^{-2})((t^{2(\lambda-\nu+1)} + t^{-2(\lambda-\nu+1)})k^{-1}\otimes$$
$$\otimes k - (t^{2(\lambda+1)} + t^{-2(\lambda+1)}))v^{(\nu)}$$
$$(t^2 - t^{-2})^2\,\delta^{(2)}(C)k^{-1}\otimes kv^{(\nu)}$$
$$= (t^{2(\lambda-\nu)+3} - t^{-2(\lambda-\nu)-3})^2 k^{-1}\otimes$$
$$\otimes kv^{(\nu)} - (t^2 - t^{-2})(t^{2(\lambda+1)} + t^{-2(\lambda+1)})(t^{2(\lambda-\nu+1)} - t^{-2(\lambda-\nu+1)})v^{(\nu)}.$$

Our assertion follows from these equalities. □

By virtue of the lemma, (10) now reduces to the single recursion relation for $\rho_\nu(x)$

$$\frac{\rho_{\nu+1}(x)}{\rho_\nu(x)} = \frac{x - t^{4(\lambda-\nu)}}{1 - xt^{4(\lambda-\nu)}},$$

whose solution is

$$\rho_\nu(x) = \rho_\lambda(x)\prod_{j=1}^{\lambda-\nu}\frac{1 - xt^{4j}}{x - t^{4j}}.$$

Assuming $|t| < 1$, let us introduce the function

$$F(x) = \prod_{n=1}^{\infty}(1 - xt^{4n}).$$

Then the $\check{R}$ matrix up to scalar multiple can be written as

$$\check{R}(x) = x^{-\Lambda/2}\frac{F(x^{-1}t^{2\Lambda})}{F(xt^{2\Lambda})}, \tag{13}$$

where Λ is related to (5) via

$$\delta^{(2)}(C) = \left(\frac{t^{\Lambda+1} - t^{-\Lambda-1}}{t^2 - t^{-2}}\right)^2 .$$

Formula (13) is the q-difference analogue of the group invariant solution (Equation (34) in [3]). The latter is recovered in the limit $x = t^u$, $t \to 1$.

Note added in Proof. After completing the manuscript, the author learned that the same algebra (3A–E) has also been introduced in the recent work of V. G. Drinfel'd (*Doklady Akad. Nauk. SSSR*, 1985). He would like to thank Prof. L. D. Faddeev for drawing his attention to the papers of E. K. Sklyanin (*Funct. Anal. Appl.* **16**, 263 (1982); **17**, 273 (1983)), where the representations of U_t in the case $\mathfrak{g} = \mathfrak{sl}(2)$ is discussed as a degenerate case of its elliptic-function version (Sklyanin algebra).

References

1. Kulish, P. P. and Reshetikhin, N. Yu., *J. Soviet Math.* **23**, 2435 (1983). Russian original *Zapiski nauch. semin. LOMI* **101**, 112 (1980).
2. Kulish, P. P. and Sklyanin, E. K., *J. Soviet Math.* **19**, 1596 (1982). Russian original *Zapiski nauch. semin. LOMI* **95**, 129 (1980).
3. Kulish, P. P., Reshetikhin, N. Yu., and Sklyanin, E. K., *Lett. Math. Phys.* **5**, 393 (1981).
4. Kac, V. G., *Infinite Dimensional Lie Algebras*, Birkhäuser, Boston, 1983.
5. Andrews, G. E., *The Theory of Partitions*, Addison-Wesley, 1976.
6. Jimbo, M., RIMS preprint **506** (1985). to appear in *Commun. Math. Phys.*

Quantization of Lie Groups and Lie Algebras

L. D. Faddeev, N. Yu. Reshetikhin, and L. A. Takhtajan
Steklov Mathematical Institute
Leningrad Branch
Leningrad
USSR

The Algebraic Bethe Ansatz—the quantum inverse scattering method—emerges as a natural development of the following directions in mathematical physics: the inverse scattering method for solving nonlinear equations of evolution [1], the quantum theory of magnets [2], the method of commuting transfer-matrices in classical statistical mechanics [3] and factorizable scattering theory [4, 26]. It was formulated in our papers [5-7]. Two simple algebraic formulae lie at the foundation of the method:

$$RT_1T_2 = T_2T_1R \qquad (*)$$

and

$$R_{12}R_{13}R_{23} = R_{23}R_{13}R_{12}. \qquad (**)$$

Their exact meaning will be explained in the next section. In the original context of the Algebraic Bethe Ansatz, T plays the role of quantum monodromy matrix of the auxiliary linear problem and is a matrix with operator-valued entries, whereas R is the usual "c-number" matrix. The second formula can be considered as a compatibility condition for the first one.

Realizations of the formulae (*) and (**) for the particular models naturally led to new algebraic objects which can be viewed as deformations of Lie-algebraic structures [8–11]. V. Drinfeld [12, 13] has shown that in the description of these constructions the language of Hopf algebras [14] is quite useful. In this way he obtained a deep generalization of the results of [9–11]. Part of these results were also obtained by M. Jimbo [15, 16].

However, from our point of view, these authors did not use formula (*) to the full strength. We decided, using the experience gained in the analysis of concrete models, to look again at the basic constructions of deformations. Our aim is to show that one can naturally define the quantization (q-deformation) of simple Lie groups and Lie algebras using exclusively the main formulae (*) and (**). Following the spirit of the non-commutative geometry [17] we will quantize instead of the Lie group G the algebra of functions Fun(G) on it. The quantization of the universal enveloping algebra $U(\mathfrak{G})$ of the Lie algebra $\mathfrak{G}$ will be based on a generalization of the relation

$$U(\mathfrak{G}) = C_e^{-\infty}(G),$$

where $C_e^{-\infty}(G)$ is a subalgebra in $C^{-\infty}(G)$ of distributions with support in the unit element e of G.

We begin with some general definitions. After that we treat two important examples and finally we discuss our constructions from the point of view of deformation theory. In this paper we use a formal algebraic language and do not consider the problems connected with topology and analysis. A detailed presentation of our results will be given elsewhere.

1. Quantum Formal Groups

Let V be an n-dimensional complex vector space (the reader can replace the field $\mathbf{C}$ by any field of characteristic zero). Consider a non-degenerate matrix $R \in \mathrm{Mat}(V^{\otimes 2}, \mathbf{C})$, satisfying the equation

$$R_{12}R_{13}R_{23} = R_{23}R_{13}R_{12}, \tag{1}$$

where the lower indices describe the imbedding of the matrix R into $\mathrm{Mat}(V^{\otimes 3}, \mathbf{C})$.

Definition 1. Let $A = A(R)$ be the associative algebra over $\mathbf{C}$ with the generators 1, t_{ij}, $i, j = 1, \ldots, n$, satisfying the relations

$$RT_1T_2 = T_2T_1R, \tag{2}$$

where $T_1 = T \otimes I$, $T_2 = I \otimes T \in \mathrm{Mat}(V^{\otimes 2}, A)$, $T = (t_{ij})_{i,j=1}^n \in \mathrm{Mat}(V, A)$ and I is a unit matrix in $\mathrm{Mat}(V, \mathbf{C})$. The algebra $A(R)$ is called the *algebra of functions on the quantum formal group corresponding to the matrix R.*

In the case $R = I^{\otimes 2}$ the algebra $A(R)$ is generated by the matrix elements of the group $GL(n, \mathbf{C})$ and is commutative.

Theorem 1. *The algebra A is a bialgebra (a Hopf algebra) with comultiplication* $\Delta: A \to A \otimes A$

$$\Delta(1) = 1 \otimes 1$$

$$\Delta(t_{ij}) = \sum_{k=1}^{n} t_{ik} \otimes t_{kj}, \qquad i, j = 1, \ldots, n.$$

Let $A' = \mathrm{Hom}(A, \mathbf{C})$ be the dual space to the algebra A. Comultiplication in A induces multiplication in A':

$$(l_1 l_2, a) \equiv (l_1 l_2)(a) = (l_1 \otimes l_2)(\Delta(a))$$

where $l_1, l_2 \in A'$ and $a \in A$. Thus A' has a structure of an associative algebra with unit $1'$, where $1'(t_{ij}) = \delta_{ij}$, $i, j = 1, \ldots, n$.

Definition 2. Let $U(R)$ be the subalgebra in $A(R)'$, generated by elements $1'$, and $l_{ij}^{(\pm)}$, $i, j = 1, \ldots, n$, where

$$(1', T_1 \cdots T_k) = I^{\otimes k},$$

$$(L^{(+)}, T_1 \cdots T_k) = R_1^{(+)} \cdots R_k^{(+)}, \tag{3}$$

$$(L^{(-)}, T_1 \cdots T_k) = R_1^{(-)} \cdots R_k^{(-)}.$$

Here $L^{(\pm)} = (l_{ij}^{(\pm)})_{i,j=1}^n \in \mathrm{Mat}(V, U)$,

$$T_i = I \otimes \cdots \underbrace{\otimes T \otimes}_{i} \cdots \otimes I \in \mathrm{Mat}(V^{\otimes k}, A), \qquad i = 1, \ldots, k,$$

the matrices $R_i^{(\pm)} \in \mathrm{Mat}(V^{\otimes(k+1)}, \mathbf{C})$ act nontrivially on factors number 0 and i in the tensor product $V^{\otimes(k+1)}$ and coincide there with the matrices $R^{(\pm)}$, where

$$R^{(+)} = PRP, \qquad R^{(-)} = R^{-1}; \tag{4}$$

here P is the permutation matrix in $V^{\otimes 2}$: $P(v\otimes w) = w\otimes v$ for $v, w \in V$. The left hand side of the formula (3) denotes the values of $1'$ and of the matrices-functionals $L^{(\pm)}$ on the homogeneous elements of the algebra A of degree k. When $k=0$ the right hand side of formula (3) is equal to I. The algebra $U(R)$ is called the *algebra of regular functionals on* $A(R)$.

Due to the equation (1) and

$$R_{12}R_{23}^{(-)}R_{13}^{(-)} = R_{13}^{(-)}R_{23}^{(-)}R_{12},$$

Definition 2 is consistent with the relations (2) in the algebra A.

Remark 1. The apparent doubling of the number of generators of the algebra $U(R)$ in comparison with the algebra $A(R)$ is explained as follows: due to the formula (3) some of the matrix elements of the matrices $L^{(\pm)}$ are identical or equal to zero. In the interesting examples (see below) matrices $L^{(\pm)}$ are of Borel type.

Theorem 2. (1) *In the algebra* $U(R)$ *the following relations take place*:

$$R^{(+)}L_1^{(\pm)}L_2^{(\pm)} = L_2^{(\pm)}L_1^{(\pm)}R^{(+)}, \qquad R^{(+)}L_1^{(+)}L_2^{(-)} = L_2^{(-)}L_1^{(+)}R^{(+)}, \tag{5}$$

where $R^{(+)} = PRP$ and $L_1^{(\pm)} = L^{(\pm)}\otimes I$, $L_2^{(\pm)} = I\otimes L^{(\pm)} \in \mathrm{Mat}(V^{\otimes 2}, A')$.

(2) *Multiplication in the algebra* $A(R)$ *induces comultiplication* δ *in* $U(R)$,

$$\delta(1') = 1'\otimes 1'$$

$$\delta(l_{ij}^{(\pm)}) = \sum_{k=1}^{n} l_{ik}^{(\pm)}\otimes l_{kj}^{(\pm)}, \qquad i, j = 1, \ldots, n,$$

so $U(R)$ *acquires a structure of a bialgebra.*

The algebra $U(R)$ can be considered as a quantization of the universal enveloping algebra, which is defined by the matrix R.

Let us also remark that in the framework of the scheme presented one can easily formulate the notion of quantum homogeneous spaces.

Definition 3. A subalgebra $B\subset A = A(R)$ which is a left coideal, $\Delta(B)\subset A\otimes B$, is called the *algebra of functions on a quantum homogeneous space corresponding to the matrix* R.

Now we shall discuss concrete examples of the general construction presented above.

2. Finite-Dimensional Example

Let $V=\mathbf{C}^n$; a matrix R of the form

$$R=\sum_{\substack{i\neq j\\ i,j=1}}^{n} e_{ii}\otimes e_{jj}+q\sum_{i=1}^{n} e_{ii}\otimes e_{ii}+(q-q^{-1})\sum_{1\leq j<i\leq n} e_{ij}\otimes e_{ji}, \tag{6}$$

where $e_{ij}\in \mathrm{Mat}(\mathbf{C}^n)$ are matrix units and $q\in\mathbf{C}$, satisfies equation (1). It is natural to call the corresponding algebra $A(R)$ the algebra of functions on the q-deformation of the group $GL(n,\mathbf{C})$ and denote it by $\mathrm{Fun}_q(GL(n,\mathbf{C}))$.

Theorem 3. *The element*

$$\det_q T=\sum_{s\in S_n}(-q)^{l(s)}t_{1s_1}\cdots t_{ns_n},$$

where summation goes over all elements s of the symmetric group S_n and $l(s)$ is the length of the element s, generates the center of the algebra $\mathrm{Fun}_q(GL(n,\mathbf{C}))$.

Definition 4. The quotient algebra of the algebra $\mathrm{Fun}_q(GL(n,\mathbf{C}))$ defined by an additional relation $\det_q T=1$ is called the *algebra of functions on the q-deformation of the group $SL(n,\mathbf{C})$* and is denoted by $\mathrm{Fun}_q(SL(n,\mathbf{C}))$.

Theorem 4. *The algebra $\mathrm{Fun}_q(SL(n,\mathbf{C}))$ has an antipode γ, which on the generators t_{ij} is given by:*

$$\gamma(t_{ij})=(-q)^{i-j}\tilde{t}_{ji},\qquad i,j=1,\ldots,n,$$

where

$$\tilde{t}_{ij}=\sum_{s\in S_{n-1}}(-q)^{l(s)}t_{1s_1}\cdots t_{i-1s_{i-1}}t_{i+1s_{i+1}}\cdots t_{ns_n}$$

and $s=(s_1,\ldots,s_{i-1},s_{i+1},\ldots,s_n)=s(1,\ldots,j-1,j+1,\ldots,n)$. *The antipode γ has the properties $T\gamma(T)=\gamma(T)T=I$ and $\gamma^2(T)=DTD^{-1}$, where $D=\mathrm{diag}(1,q^2,\ldots,q^{2(n-1)})\in\mathrm{Mat}(\mathbf{C}^n)$.*

In the case $n=2$, the matrix R in explicit form is given by

$$R=\begin{pmatrix} q & 0 & 0 & 0\\ 0 & 1 & 0 & 0\\ 0 & q-q^{-1} & 1 & 0\\ 0 & 0 & 0 & q\end{pmatrix}, \tag{7}$$

and the relations (2) reduce to the following simple formulae:

$$t_{11}t_{12}=qt_{12}t_{11}, \qquad t_{11}t_{21}=qt_{21}t_{11},$$

$$t_{12}t_{21}=t_{21}t_{12}, \qquad t_{12}t_{22}=qt_{22}t_{12},$$

$$t_{21}t_{22}=qt_{22}t_{21}, \qquad t_{11}t_{22}-t_{22}t_{11}=(q-q^{-1})t_{12}t_{21},$$

and

$$\det_q T=t_{11}t_{22}-qt_{12}t_{21}.$$

In this case

$$\gamma(T)=\begin{pmatrix} t_{22} & -q^{-1}t_{12} \\ -qt_{21} & t_{11} \end{pmatrix}.$$

Remark 2. When $|q|=1$ relations (2) admit the following $*$-anti-involution: $t_{ij}^{*}=t_{ij}$, $i,j=1,\ldots,n$. The algebra $A(R)$ with this anti-involution is nothing but the algebra $\mathrm{Fun}_q(SL(n,\mathbf{R}))$. In the case $n=2$ this algebra and the matrix R of the form (7) appeared for the first time in the paper [18]. The subalgebra $B\subset \mathrm{Fun}_q(SL(n,\mathbf{R}))$, generated by the elements 1 and $\sum_{k=1}^{n} t_{ik}t_{jk}$, $i,j=1,\ldots,n$ is the left coideal and may be called the algebra of functions on the q-deformation of the symmetric homogeneous space of rank $n-1$ for the group $SL(n,\mathbf{R})$. In the case $n=2$ we obtain the q-deformation of the Lobachevsky plane.

Remark 3. When $q\in\mathbf{R}$ the algebra $\mathrm{Fun}_q(SL(n,\mathbf{C}))$ admits the following $*$-anti-involution: $\gamma(t_{ij})=t_{ji}^{*}$, $i,j=1,\ldots,n$. The algebra $\mathrm{Fun}_q(SL(n,\mathbf{C}))$ with this anti-involution is nothing but the algebra $\mathrm{Fun}_q(SU(n))$. In the case $n=2$ this algebra was introduced in the papers [19-20].

Remark 4. The algebras $\mathrm{Fun}_q(G)$, where G is a simple Lie group, can be defined in the following way. For any simple group G there exists a corresponding matrix R_G satisfying equation (1) which generalize the matrix R of the form (6) for the case $G=SL(n,\mathbf{C})$. This matrix R_G depends on the parameter q and as $q\to 1$

$$R_G=I+(q-1)r_G+O((q-1)^2),$$

where

$$r_G=\sum_i \frac{\rho(H^i)\otimes\rho(H^i)}{2}+\sum_{\alpha\in\Delta_+}\rho(X_\alpha)\otimes\rho(X_{-\alpha}).$$

Here ρ is the vector representation of Lie algebra $\mathfrak{G}$, H^i, X_α its Cartan–Weyl basis, and Δ_+ the set of positive roots. The explicit form of the matrices R_G can be extracted from [16], [21]. The corresponding algebra $A(R)$ is defined by the relations (2) and an appropriate anti-involution compatible with them. It may be called the algebra of functions on the q-deformation of the Lie group G.

Let us discuss now the properties of the algebra $U(R)$. It follows from the explicit form (6) of the matrix R and Definition 2 that the matrices-functionals $L^{(+)}$ and $L^{(-)}$ are, correspondingly, the upper- and lower-triangular matrices. Their diagonal parts satisfy the simple relation:

$$\mathrm{diag}(L^{(+)})\,\mathrm{diag}(L^{(-)}) = I.$$

Theorem 5. *The following equality holds*:

$$U(R) = U_q(\mathfrak{sl}(n, \mathbf{C})),$$

where $U_q(\mathfrak{sl}(n, \mathbf{C}))$ *is the* q*-deformation of the universal enveloping algebra* $U(\mathfrak{sl}(n, \mathbf{C}))$ *of the Lie algebra* $\mathfrak{sl}(n, \mathbf{C})$, *introduced in the papers* [12] *and* [15]. *The center of* $U(R)$ *is generated by the elements*

$$C_k = \sum_{s,s' \in S_n} (-q)^{l(s)+l(s')} l^{(+)}_{s_1 s'_1} \cdots l^{(+)}_{s_k s'_k} l^{(-)}_{s_{k+1} s'_{k+1}} \cdots l^{(-)}_{s_n s'_n}, \qquad k = 0, \ldots, n-1.$$

Remark 5. It is instructive to compare the relations for the elements $l^{(\pm)}_{ij}$, $i, j = 1, \ldots, n$, which follow from (5), with those given in the papers [12] and [15]. The elements $l^{(\pm)}_{ij}$ can be considered as the q-deformation of the Cartan–Weyl basis, whereas the elements $l^{(+)}_{i\,i+1}$, $l^{(-)}_{i-1\,i}$, $l^{(+)}_{ii}$ are the q-deformation of the Chevalley basis. It was this basis that was used in [12] and [15]; the complicated relations between the elements of the q-deformation of the Chevalley basis, presented in these papers, follow from the simple formulae (3), (5) and (6).

Remark 6. It follows from the definition of the algebra $U_q(\mathfrak{sl}(n, \mathbf{C}))$ that it can be considered as the algebra $\mathrm{Fun}_q((G_+ \times G_-)/H)$, where $G_\pm$ and H are, correspondingly, Borel and Cartan subgroups of Lie group $SL(n, \mathbf{C})$. Moreover in the general case the q-deformation $U_q(\mathfrak{G})$ of the universal enveloping algebra of the simple Lie algebra $\mathfrak{G}$ can be considered as the quantization of the group $(G_+ \times G_-)/H$. For the infinite-dimensional case (see the next section) this observation provides a key to the formulation of the quantum Riemann problem.

In the case $n=2$ we have the following explicit formulae:

$$L^{(+)}=\sqrt{q}\begin{pmatrix} e^{-hH/2} & hX \\ 0 & e^{hH/2} \end{pmatrix},$$

$$L^{(-)}=\frac{1}{\sqrt{q}}\begin{pmatrix} e^{hH/2} & 0 \\ hY & e^{-hH/2} \end{pmatrix}, \qquad q=e^{-h},$$

where the generators 1 and $e^{\pm hH/2}$, X, Y of the algebra $U_q(\mathfrak{sl}(2,\mathbf{C}))$ satisfy the relations

$$e^{\pm hH/2}X=q^{\pm 1}Xe^{\pm hH/2}, \qquad e^{\pm hH/2}Y=q^{\mp 1}Ye^{\pm hH/2},$$

$$XY-YX=-\frac{(q-q^{-1})}{h^2}(e^{hH}-e^{-hH}),$$

which appear for the first time in the paper [8].

3. Infinite-Dimensional Example

Replace in the general construction of Section 1 the finite-dimensional vector space V by the infinite-dimensional $\mathbf{Z}$-graded vector space $\tilde{V}=\bigoplus_{n\in\mathbf{Z}}\lambda^n V=\bigoplus_{n\in\mathbf{Z}}V_n$, where λ is a formal variable (spectral parameter). Denote by $\mathbf{S}$ the shift operator (multiplication by λ) and as matrix R consider an element $\tilde{R}\in \mathrm{Mat}(\tilde{V}^{\otimes 2},\mathbf{C})$, satisfying equation (1) and commuting with the operator $\mathbf{S}\otimes\mathbf{S}$. Infinite-dimensional analogs of the algebras $A(R)$ and $U(R)$—the algebras $A(\tilde{R})$ and $U(\tilde{R})$—are introduced as before by Definitions 1 and 2; in addition elements $\tilde{T}\in\mathrm{Mat}(\tilde{V},A(\tilde{R}))$ and $\tilde{L}^{(\pm)}\in\mathrm{Mat}(\tilde{V},U(\tilde{R}))$ commute with $\mathbf{S}$. Theorems 1 and 2 are valid for this case.

Let us discuss a meaningful example of this construction. Choose the matrix $\tilde{R}$ to be a matrix-valued function $R(\lambda,\mu)$, defined by the formula

$$R(\lambda,\mu)=\frac{\lambda q^{-1}R^{(+)}-\mu qR^{(-)}}{\lambda q^{-1}-\mu q},$$

where the matrices $R^{(\pm)}$ are given by (4). The role of the elements $\tilde{T}$ now played by the infinite formal Laurent series

$$T(\lambda)=\sum_{m\in\mathbf{Z}}T_m\lambda^m,$$

with the relations

$$R(\lambda, \mu)T_1(\lambda)T_2(\mu) = T_2(\mu)T_1(\lambda)R(\lambda, \mu).$$

The comultiplication in the algebra $A(\tilde{R})$ is given by the formula

$$\Delta(t_{ij}(\lambda)) = \sum_{k=1}^{n} t_{ik}(\lambda)\otimes t_{kj}(\mu), \qquad i, j = 1, \ldots, n.$$

The following result is the analog of Theorem 3.

Theorem 6. *The element*

$$\det_q T(\lambda) = \sum_{s\in S_n} (-q)^{l(s)} t_{1s_1}(\lambda)\cdots t_{ns_n}(\lambda q^{n-1})$$

generates the center of the algebra $A(\tilde{R})$.

Let us now briefly discuss the properties of the algebra $U(\tilde{R})$. It is generated by the formal Taylor series

$$L^{(\pm)}(\lambda) = \sum_{m\in \mathbf{Z}_+} L_m^{(\pm)}\lambda^{\pm m},$$

which act on the elements of the algebra $A(\tilde{R})$ by the formulae (3). The relations in $U(\tilde{R})$ have the form

$$R^{(+)}(\lambda, \mu)L_1^{(\pm)}(\lambda)L_2^{(\pm)}(\mu) = L_2^{(\pm)}(\mu)L_1^{(\pm)}(\lambda)R^{(+)}(\lambda, \mu),$$

$$R^{(+)}(\lambda, \mu)L_1^{(+)}(\lambda)L_2^{(-)}(\mu) = L_2^{(-)}(\mu)L_1^{(+)}(\lambda)R^{(+)}(\lambda, \mu).$$

Due to the shortage of space we shall not discuss here an interesting question about the connection of the algebra $U(\tilde{R})$ with the q-deformation of the Kac–Moody algebras, introduced in the paper [13]. We shall only point out that the algebra $U(\tilde{R})$ has a natural limit when $q\to 1$. In this case the subalgebra, generated by elements $L^{(+)}(\lambda)$, coincides with the Yangian $Y(\mathfrak{sl}(n, \mathbf{C}))$, introduced in the papers [12, 22].

4. Deformation Theory and Quantum Groups

Consider the contraction of the algebras $A(R)$ and $U(R)$ when $q\to 1$. For definiteness we have in mind a finite-dimensional example. The algebra $A(R) = \mathrm{Fun}_q(G)$ when $q\to 1$ goes into the commutative algebra $\mathrm{Fun}(G)$ with the Poisson structure given by the formula

$$\{g \underset{,}{\otimes} g\} = [r_G, g\otimes g]. \qquad (8)$$

Here g_{ij}, $i, j = 1, \ldots, n$ are the coordinate functions on the Lie group G. Passing from the Lie group G to its Lie algebra $\mathfrak{G}$, from (8) we obtain the Poisson structure on the Lie algebra

$$\{h \underset{,}{\otimes} h\} = [r_G, h \otimes I + I \otimes h]. \tag{9}$$

Here $h = \sum_{i=1}^{\dim \mathfrak{G}} h_i X^i$, where X^i, $i = 1, \ldots, \dim \mathfrak{G}$, form a basis of $\mathfrak{G}$. If we define $h^{(\pm)} = h_\pm + h_{\mathfrak{H}}/2$, where $h_\pm$ and $h_{\mathfrak{H}}$ are respectively the nilpotent and Cartan components of h, we can rewrite the formula (9) in the form

$$\{h^{(\pm)} \underset{,}{\otimes} h^{(\pm)}\} = [r_G, h^{(\pm)} \otimes I + I \otimes h^{(\pm)}],$$
$$\{h^{(\pm)} \underset{,}{\otimes} h^{(\mp)}\} = 0. \tag{10}$$

This Poisson structure and its infinite-dimensional analogs were studied in [23] (see also [24-25]). Thus the Lie algebra $\mathfrak{G}$ has a structure of a Lie bialgebra, where the cobracket $\mathfrak{G} \to \mathfrak{G} \wedge \mathfrak{G}$ is defined by the Poisson structure (10).

Analogously the contraction $q \to 1$ of the algebra $U(R)$ leads to the Lie bialgebra structure on the dual vector space $\mathfrak{G}^*$ to the Lie algebra $\mathfrak{G}$. The Lie bracket on $\mathfrak{G}^*$ is dual to the Poisson structure (10) and the cobracket $\mathfrak{G}^* \to \mathfrak{G}^* \wedge \mathfrak{G}^*$ is defined by the canonical Lie-Poisson structure on $\mathfrak{G}^*$.

This consideration explains in more detail in what senses the algebras $A(R)$ and $U(R)$ define deformations of the corresponding Lie group and Lie algebra. Moreover it shows how an additional structure is defined on these "classical" objects. Returning to the relations (*) and (**) we can say now that (*) constitutes a deformation of the Lie-algebraic relations with t_{ij} playing the role of generators, R being the array of "quantum" structure constants, and (**) generalizes the Jacobi identity.

References

[1] C. Gardner, J. Green, M. Kruskal, and R. Miura, *Phys. Rev. Lett.* **19** (1967) 19, 1095-1097.
[2] H. Bethe, *Z. Phys.* **71** (1931) 205-226.
[3] R. Baxter, *Exactly Solved Models in Statistical Mechanics.* Academic Press, London, 1982.
[4] C. N. Yang, *Phys. Rev. Lett.* **19** (1967) 23, 1312-1314.
[5] E. Sklyanin, L. Takhtajan, and L. Faddeev, *TMF* **40** (1979) 2, 194-220 (in Russian).
[6] L. Takhtajan and L. Faddeev, *Usp. Math. Nauk* **34** (1979) 5, 13-63 (in Russian).
[7] L. Faddeev, Integrable Models in 1+1-Dimensional Quantum Field Theory, in *Les Houches Lectures 1982*, Elsevier, Amsterdam, 1984.
[8] P. Kulish and N. Reshetikhin, *Zap. Nauch. Semin. LOMI* **101** (1981) 101-110 (in Russian).
[9] E. Sklyanin, *Func. Anal. and Appl.* **16** (1982) 4, 27-34 (in Russian).
[10] E. Sklyanin, *ibid*, **17**, 4, 34-48 (in Russian).

[11] E. Sklyanin, *Usp. Mat. Nauk* **40** (1985) 2, 214 (in Russian).
[12] V. Drinfeld, *DAN SSSR* **283** (1985) 5, 1060-1064 (in Russian).
[13] V. Drinfeld, Talk at IMC-86, Berkeley, 1986.
[14] E. Abe, *Hopf Algebras*, Cambridge Tracts in Math., 74, Cambridge University Press, Cambridge-New York, 1980.
[15] M. Jimbo, *Lett. Math. Phys.* **11** (1986) 247-252.
[16] M. Jimbo, *Commun. Math. Phys.* **102** (1986) 4, 537-548.
[17] A. Connes, *Noncommutative Differential Geometry*. Extract Publ. Math., I.H.E.S., **62**, 1986.
[18] L. Faddeev and L. Takhtajan, Liouville model on the lattice. Preprint, Université Paris VI, June 1985; *Lect. Notes in Physics* **246** (1986) 166-179.
[19] L. Varksman and J. Soibelman, *Funct. Anal. and Appl.* (to appear).
[20] S. Woronowicz, *Publ. RIMS, Kyoto Univ.* **23** (1987) 117-181.
[21] V. Bazhanov, *Phys. Lett.* **159B** (1985) 4-5-6, 321-324.
[22] A. Kirillov and N. Reshetikhin, *Lett. Math. Phys.* **12** (1986) 199-208.
[23] N. Reshetikhin and L. Faddeev, *TMF* **56** (1983) 323-343 (in Russian).
[24] M. Semenov-Tian-Shansky, *Publ. RIMS Kyoto Univ.* **21** (1985) 6, 1237-1260.
[25] L. Faddeev and L. Takhtajan, *Hamiltonian Methods in the Theory of Solitons*. Springer-Verlag, 1987.
[26] A. Zamolodchikov and A. Zamolodchikov, *Annals of Physics* **120** (1979) 2, 253-291.

VARIETY OF SOLUTIONS I — QUANTIZATION

4. Variety of solutions I — quantization

The articles assembled in this section present explicit solutions of YBE, which are 'quantization' of the classical r matrices due to Belavin and Drinfel'd [8].

The group-invariant R matrices known at an earlier stage [182),207),265),274–276)] correspond to the simplest rational solutions in [8].

According to the Belavin-Drinfel'd classification CYBE admit non-degenerate elliptic solutions only for the Lie algebras of type A_n. Belavin's reprint [16],[40)] gives the corresponding quantum R matrix, generalizing the eight-vertex model to an $n+1$ state model (see [48),214)] for a proof of YBE).

In the trigonometric case, there exist a series of solutions for each pair of a simple Lie algebra $\mathfrak{g}$ and its Dynkin diagram automorphism σ. The reprint by Perk and Schultz [17] describes the solutions for the A_n type Lie algebra (with σ being the identity). These solutions have been found by several authors [11),56),69),202–203),221)]. The first example of 'quantization' other than these was the Izergin-Korepin model [110)], which corresponds to $\mathfrak{g} = \mathfrak{sl}(3)$ together with its order 2 automorphism. When $\mathfrak{g}$ is of type A, B, C or D the quantum R matrices have been found in the reprints by Bazhanov [18],[26)] and Jimbo [19]. A similar work for Lie superalgebras is also available [29)].

All these solutions are in fact constructed in the vector representation of the underlying Lie algebra, or rather of the quantum group. For the Lie algebra $\mathfrak{sl}(2)$ solutions in the higher spin representations are treated in [114),141),143),231),237–238)]. R matrices corresponding to the spin representations of $O(n)$ [209–211)] and the lowest dimensional representations of G_2[147),209–210)] are also known explicitly. A description of the constant solutions (i.e. those that do not depend on the spectral parameters) is given in [78)].

In the reprint [13], Drinfel'd showed the existence of a universal R matrix living in the tensor product of quantum groups. This allows one in principle to get matrix solutions of YBE in each representation space. The explicit form of the universal R for $\mathfrak{g} = \mathfrak{sl}(n)$ can be found in [50),81),219)]. The fusion procedure described in the reprints in Section 6 affords a practical way of constructing solutions out of fundamental representations.

Nuclear Physics B180[FS2] (1981) 189–200

DYNAMICAL SYMMETRY OF INTEGRABLE QUANTUM SYSTEMS

A.A. BELAVIN

The Landau Institute for Theoretical Physics,
The Academy of Sciences of the USSR,
142432 Chernogolovka, Moscow Region, USSR

Received 17 September 1980

We consider the equations of triangles (alias Yang-Baxter equations), which a factorized two-particle S-matrix obeys. These equations are shown to possess a symmetry which consists of discrete Lorentz transformations acting independently on states of particles with different momenta. It is this symmetry which ensures compatibility of the overconstrained equations of triangles. The use of it enables one to construct the factorized two-particle S-matrix requiring invariance (automorphity) with respect to discrete Lorentz transformations.

1. Introduction

Sect. 2 of this paper outlines equations of triangles (or Yang-Baxter equations), which are coded expressions of hidden symmetry for one-dimensional quantum and classical integrable systems [1–12] as well as for two-dimensional lattice statistical models of the Baxter type [4, 8, 10]. In sect. 3 the invariance principle is introduced for such systems, which requires symmetry under the discrete subgroup of the Lorentz group, acting independently on particles with different momenta. It is shown that this principle is applicable for the determination of all elements of the S-matrix satisfying the equations of triangles. Thus it is argued that the hidden symmetry is closely connected with invariance under the action of the above-mentioned group. In sect. 4 we present explicit examples of two-particle S-matrices. In sect. 5 we discuss possible generalizations of our approach, including the multi-dimensional case.

2. Equations of triangles and integrable systems

Equations of triangles were first derived by Yang [3] who has studied problems of distinguishable non-relativistic particles with the two-body δ-interaction. Yang discovered that for Bethe's ansatz to be self-consistent, two-particle scattering amplitudes should satisfy certain functional relations (the equations of triangles). Later on, the same equations were constructed in relativistic scattering theory [9–11] as factorization conditions of a multiparticle S-matrix. Factorization of the S-matrix ensures conservation of a number of particles and their momenta. Factorization equations in relativistic theory involving N different kinds of particles, are of the form

$$S_{i_1 i_2}^{k_1 k_2}(u_1 - u_2) S_{k_1 i_3}^{i_1 k_3}(u_1 - u_3) S_{k_2 k_3}^{i_2 j_3}(u_2 - u_3) = S_{i_2 i_3}^{k_2 k_3}(u_2 - u_3) S_{i_1 k_3}^{k_1 j_3}(u_1 - u_3) S_{k_1 k_2}^{j_1 j_2}(u_1 - u_2), \quad (1)$$

where S is a two-particle S-matrix; i_1, i_2, i_3 (j_1, j_2, j_3) denote the kinds of initial (final) particles; they take values from 1 up to N. Summation from 1 to N is implied in (1) over all the repeated indices k_1, k_2, k_3. The quantities u_1, u_2, u_3 denote the rapidities of three colliding particles related to the energy (momentum) as $E = m\mathrm{ch}u$ ($p = m\mathrm{sh}u$). Henceforth it will be convenient to employ a more compact notation for eq. (1). Consider the linear spaces H_a, $a = 1, 2, 3$ (each being of dimensionality N) of states of the ath particle. Let $\mathrm{H}_{ab} = \mathrm{H}_a \cdot \mathrm{H}_b$ and $\mathrm{H} = \mathrm{H}_1 \cdot \mathrm{H}_2 \cdot \mathrm{H}_3$. Further, let $\mathcal{Q}(\mathrm{H}_a)$, $\mathcal{Q}(\mathrm{H}_{ab})$, and $\mathcal{Q}$ be the algebras of matrices over H_a, H_{ab} and H, respectively. Then, the natural embeddings $\mathcal{Q}(\mathrm{H}_a) \subseteq \mathcal{Q}(\mathrm{H}_{ab}) \subseteq \mathcal{Q}$ exist as the embeddings of C^* algebras. In what follows the elements of the algebras $\mathcal{Q}(\mathrm{H}_a)$, $\mathcal{Q}(\mathrm{H}_{ab})$ will be denoted by the corresponding upper suffix, for example, $g^{(1)} \in \mathcal{Q}(\mathrm{H}_1)$, $h^{(2)} \in \mathcal{Q}(\mathrm{H}_2)$, $S^{(12)} \in \mathcal{Q}(\mathrm{H}_{12})$, etc. Using this notation allows one to represent (1) as follows:

$$S^{(12)}(u_1 - u_2) S^{(13)}(u_1 - u_3) S^{(23)}(u_2 - u_3) \\ = S^{(23)}(u_2 - u_3) S^{(13)}(u_1 - u_3) S^{(12)}(u_1 - u_2) . \tag{2}$$

In a number of papers [4–6] Baxter has investigated statistical lattice two-dimensional models. In these models fluctuating variables are positioned on the lattice edges and run over N values. The statistical weight of the given configuration is defined as the product of the vertex weights $S_{i_1 i_2}^{j_1 j_2}$, where, i_1, i_2, j_1, j_2 are values of the variables on the edges, belonging to some vertex, over all lattice vertices. The transfer matrix of the model is defined as

$$\hat{T}_{(i)}^{(j)}(u) = \hat{T}_{i_1 \cdots i_M}^{j_1 \cdots j_M}(u) = S_{i_1 k}^{j_1 k_1}(u) S_{i_2 k_1}^{j_2 k_2}(u) \cdots S_{i_M k_{M-1}}^{j_M k}(u) . \tag{3}$$

Baxter has shown that if (1) is fulfilled, then

$$\left[\hat{T}(u), \hat{T}(v) \right] = 0 . \tag{4}$$

This has enabled Baxter to work out the free energy of the eight-vertex model exactly. A set of commuting operators $\hat{T}(u)$ is related with quantum one-dimensional spin chains of the Heisenberg model type. For the hamiltonian of the Heisenberg model, the following relation with $\hat{T}(u)$ is valid:

$$\hat{H}_{\mathrm{Heis}} = \mathrm{d} \ln \hat{T}(u) / \mathrm{d}u |_{u=0} . \tag{5}$$

Thanks to eq. (4) the model has a series of motion integrals

$$\left[\hat{H}_{\mathrm{Heis}}, \hat{T}(u) \right] = 0 \tag{6}$$

and is completely integrable on the quantum level. In refs. [7, 8] Faddeev, Sklyanin and Takhtadjyan have developed a quantum inverse scattering method making it possible to find eigenvectors and eigenvalues of $\hat{T}(u)$, using (1). The relation between factorized scattering theory and these lattice theories has been established by Zamolodchikov in [10].

Let us finally mention the relation between the solutions of (1) and the classical systems integrable by the inverse scattering method. This relation has first been found by Sklyanin in [13] for the Baxter model. Let us assume that $S^{(12)}(u) = 1 + \hbar X^{(12)}(u)$, where $\hbar \ll 1$. Then the first non-vanishing term in (1) or (2) is of the form

$$[X^{(12)}(u), X^{(13)}(u+v)] + [X^{(12)}(u), X^{(23)}(v)] + [X^{(13)}(u+v), X^{(23)}(v)] = 0. \tag{7}$$

Let us introduce the explicit parametrization of the quantities $X^{(12)}(u)$:

$$X^{(12)}(u) = v_{\mu\nu}(u) I_\mu^{(1)} I_\nu^{(2)}. \tag{8}$$

Here I_μ is the basis of a certain Lie subalgebra with the commutation relations

$$[I_\mu, I_\nu] = F_{\mu\nu}^\lambda I_\lambda, \tag{9}$$

where $F_{\mu\nu}^\lambda$ are the structural constants. Eqs. (7) are functional equations for the quantities $v_{\mu\nu}(u)$. Let us assume that eq. (7) has a non-trivial solution. Consider the symplectic manifold, in which there are M functions S_μ, $\mu = 1, 2, \ldots, M$ (where M is the number of generators of the Lie subalgebra under consideration). Let the Poisson brackets for the functions S_μ satisfy the relation

$$\{S_\mu S_\nu\} = F_{\mu\nu}^\lambda S_\lambda. \tag{10}$$

Due to eqs. (7), (10), the matrix valued functions

$$Q^{(1)}(u) = v_{\mu\nu}(u) S_\mu I_\nu^{(1)},$$

$$Q^{(2)}(u) = v_{\mu\nu}(u) S_\mu I_\nu^{(2)},$$

which depend on u, satisfy the relation

$$\{Q^{(1)}(u) \otimes Q^{(2)}(u+v)\} + [Q^{(1)}(u), X^{(12)}(v)] + [Q^{(2)}(u+v), X^{(12)}(v)] = 0. \tag{11}$$

It follows from (11) that for any integers n and l

$$\{\operatorname{Tr} Q^n(u) \operatorname{Tr} Q^l(v)\} = 0. \tag{12}$$

Let us introduce a hamiltonian system, considering the trace of the kth power of $Q(u_0)$ as its hamiltonian functional:

$$\mathcal{H} = \operatorname{Tr} Q^k(u_0).$$

It follows from (12) that this hamiltonian system possesses a number of motion integrals for the equations of motion

$$\dot{S}_\mu = \{S_\mu \mathcal{H}\}. \tag{13}$$

This approach allows one to obtain a classical field theory possessing an infinite set of motion integrals. To obtain one-dimensional systems, it is necessary to introduce the quantities $S_\mu(x)$ depending not only on time but also on the coordinate x, defining the Poisson brackets as

$$\{S_\mu(x) S_\nu(y)\} = F^\lambda_{\mu\nu} S_\lambda(x) \delta(x-y). \tag{14}$$

Then the operator $\hat{L}(u)$ can be defined as

$$\hat{L}(u) = \frac{\partial}{\partial x} + v_{\mu\nu}(u) S_\mu(x) I_\nu. \tag{15}$$

Introducing the monodromy matrix $M(x, x_0; u)(M(x_0, x_0; u) = 1)$ as a solution of the equation

$$\hat{L}(u) M(x, x_0; u) = 0, \tag{16}$$

we shall obtain [from the equation analogous to (11)]

$$\{\operatorname{Tr} M(x, x_0; u) \operatorname{Tr} M(x, x_0; v)\} = 0. \tag{17}$$

Thus, any non-trivial solution of eqs. (1) leads to the existence of integrable quantum, classical or statistical systems. The problem is how to find solutions of these equations. At first glance it seems amazing that such solutions actually exist, since the number of equations in (1) greatly exceeds the number of unknown functions. For example, if no symmetries are required, there are N^4 components of $S^{j_1 j_2}_{i_1 i_2}$, while the number of equations (1) equals N^6.

3. Principle of invariance under the discrete subgroup of the Lorentz group

Eqs. (1) are invariant under the global isotopic transformations

$$S^{(12)}(u) \to \tilde{S}^{(12)}(u) = (G^{(1)})^{-1} (G^{(2)})^{-1} S^{(12)}(u) G^{(1)} G^{(2)}. \tag{18}$$

Here G is an arbitrary $N\times N$ matrix of GL(N). It is clear that eqs. (1) or (2) at $u_1=u_2=u_3$ have, apart from the trivial solution $S^{(12)}(0)=1$, another solution $S^{j_1j_2}_{i_1i_2}(0)=p^{j_1j_2}_{i_1i_2}=\delta^{j_2}_{i_1}\delta^{j_1}_{i_2}$. The first solution corresponds to the absence of interaction $S^{(12)}(0)=1$, the second, $S^{(12)}(0)=P^{(12)}$, to the total reflection of particles moving with zero relative velocity. From now on we shall consider u to be complex and look for solutions of eq. (1) in the class of meromorphic functions of u. Let us consider an integer lattice in the complex rapidity plane

$$u_k=k_1+\tau k_2. \tag{19}$$

Here $k(k_1,k_2)$ is an integer vector on the lattice and τ is a complex number ($\operatorname{Im}\tau>0$).

Let a certain matrix G_k be assigned to each point k of the lattice. Let the matrices have the properties

$$G_kG_l=s(k,l)G_{k+l},\qquad G_k^{-1}=G_{-k},\qquad G_0=1. \tag{20}$$

Here $s(k,l)$ is a scalar multiplier. This means that the matrices G_k form the representation (projective due to $s(k,l)$, of the commutative group of lattice shifts. Then we can easily verify that by means of the matrices G_k we can construct a solution of (1) for all k lattice points:

$$S^{(12)}(u_k)=(G_k^{(1)})^{-1}P^{(12)}G_k^{(1)}=G_k^{(2)}P^{(12)}(G^{(2)})^{-1}. \tag{21}$$

Here the number 1 on the matrix G_k symbolizes that $G_k^{(1)}$ acts only on the indices i_1,j_1 of the particles with rapidity equal to u_1. The same statement is true for 2. The second equality in (21) follows from the invariance of $P^{(12)}$ under the transformations (18). Due to the group properties (20), G_k has the form

$$G_k=G_{k_1,k_2}=g^{k_1}h^{k_2}. \tag{22}$$

Here $g=G_{1,0}$, $h=G_{0,1}$. Eqs. (20) will be satisfied if

$$gh=\omega hg, \tag{23}$$

where ω is a scalar multiplier. The last thing we are to do is to restore the analytic meromorphic matrix function $S^{(12)}(u)$ with the known values (21) in all k lattice points. By this we mean the meromorphism of all its matrix elements. Since eq. (1) is homogeneous, one can look for the solution in the class of entire functions:

$$S^{(12)}(u_k+1)=(g^{(1)})^{-1}S^{(12)}(u_k)g^{(1)}=g^{(2)}S^{(12)}(u_k)(g^{(2)})^{-1},$$

$$S^{(12)}(u_k+\tau)=\lambda(h^{(1)})^{-1}S^{(12)}(u_k)h^{(1)}=\lambda h^{(2)}S^{(12)}(u_k)(h^{(2)})^{-1}. \tag{24}$$

Now let us require that at arbitrary u the matrix possess the same properties:

$$S^{(12)}(u+1)=(g^{(1)})^{-1}S^{(12)}(u)g^{(1)}=g^{(2)}S^{(12)}(u)(g^{(2)})^{-1}, \tag{25'}$$

$$S^{(12)}(u+\tau)=\lambda\exp(2\pi iu)(h^{(1)})^{-1}S^{(12)}(u)h^{(1)}=\lambda\exp(2\pi iu)$$

$$\times h^{(2)}S^{(12)}(u)(h^{(2)})^{-1}. \tag{25''}$$

Here for compatibility of (25) the multiplier in (25″) should be chosen as a periodic function of u; the automorphity requirements (25) together with the initial condition $S^{(12)}(0)=P^{(12)}$ uniquely determine the two-particle S-matrix. The solution of (25) may be derived, for instance, by means of a Fourier series expansion.

For the $S^{(12)}(u)$ thus constructed to be the solution of eq. (1), the matrices g and h should satisfy certain conditions which we explain below.

Note now that the argument of $S^{(12)}(u)$ is equal to the rapidity difference u_1-u_2 of the colliding particles. Therefore, (25) is essentially the requirement of invariance of the two-particle S-matrix under the discrete Lorentz transformation of the momenta of one of the two particles before and after the collision (corresponding, for example, to the rapidity u_1), this Lorentz transformation being accompanied by the simultaneous "isotopic" transformation of the particles of rapidity u_1 in the in and out states.

4. $\mathbf{Z}_N\times\mathbf{Z}_N$ symmetric *S*-matrix

Let us now consider the example which will enable us to realise the above construction explicitly. Let us take g and h as the matrices for which $g^N=h^N=1$, $\omega=\exp(2\pi i/N)$. They may be chosen in the form

$$g=\begin{bmatrix}1 & 0 & 0 & \cdots & 0\\ 0 & \omega & 0 & \cdots & 0\\ \vdots & & & & \\ 0 & 0 & 0 & \cdots & \omega^{N-1}\end{bmatrix},\qquad h=\begin{bmatrix}0 & 1 & 0 & \cdots & 0\\ 0 & 0 & 1 & \cdots & 0\\ \vdots & & & & \\ 1 & 0 & 0 & \cdots & 0\end{bmatrix} \tag{26}$$

Then it is convenient to introduce a basis of matrices

$$I_\alpha=I_{\alpha_1\alpha_2}=g^{\alpha_1}h^{\alpha_2},$$

$$\alpha=(\alpha_1,\alpha_2),\qquad \alpha_1,\alpha_2=0,1,\ldots,N-1. \tag{27}$$

For $N=2$, $g=\sigma_z$, $h=\sigma_x$; therefore

$$I_{0,0}=1;\ I_{1,0}=\sigma_z;\ I_{0,1}=\sigma_x;\ I_{1,1}=i\sigma_y.$$

Any matrix $S^{(12)}(u)$ can be written in the form

$$S^{(12)}(u) = w_{\alpha,\beta}(u) I_\alpha^{(1)} I_\beta^{(2)} . \tag{28}$$

The second of equations (25) gives rise to the restriction for $S^{(12)}(u)$, which is $Z_N \times Z_N$ invariance:

$$S^{(12)}(u) = w_\alpha(u) I_\alpha^{(1)} \bar{I}_\alpha^{(2)} . \tag{29}$$

The bar above $I_\alpha^{(2)}$ denotes hermitian conjugation. We can easily verify that

$$S^{(12)}(0) = P^{(12)} = I_\alpha^{(1)} \bar{I}_\alpha^{(2)} . \tag{30}$$

Substituting (29) and (26) into (25) and (30) we get

$$w_\alpha(u+1) = \omega^{\alpha_2} w_\alpha(u),$$
$$w_\alpha(u+\tau) = \lambda \exp(2\pi i u)\omega^{\alpha_1} w_\alpha(u), \qquad w_\alpha(0) = 1 . \tag{31}$$

The solution of (31) can be obtained by Fourier expansion and is of the form

$$w_\alpha(u) = \Theta_\alpha(u+\eta)/\Theta_\alpha(\eta) . \tag{32}$$

Here

$$\Theta_\alpha(u) = \sum_{m=-\infty}^{\infty} \exp\left[i\pi\left(m + \frac{\alpha_2}{N}\right)^2 \tau + 2\pi i\left(m + \frac{\alpha_2}{N}\right)\left(u + \frac{\alpha_1}{N}\right)\right] . \tag{33}$$

For $N = 2$ this solution coincides with one investigated by Baxter [5, 6].

Let us show in this example how (25) reduces the number of eqs. (1) to the number of unknown functions. It is evident from (29) that the number of functions $w_\alpha(u)$ is equal to N^2. By means of shifts in τ and (25″) any of eqs. (1) can be turned into the equation for which $i_1 = i_2 = i_3 = 1$. Besides, eq. (25) requires that

$$j_1 + j_2 + j_3 = i_1 + i_2 + i_3, \qquad (\mathrm{mod}\, N) .$$

Thus the number of independent equations also equals N^2. By a similar method one can apparently construct solutions of eqs. (1) in terms of Θ-functions of many variables; it suffices to replace the quantity u by the n-dimensional vector $\boldsymbol{u}$ and to treat the indices i, j as vector indices: $\boldsymbol{i} = (i_1, \ldots, i_n)$. Instead of (25) the requirement of automorphism with respect to $2n$ shifts should be introduced:

$$u_\alpha \to u_\alpha + \delta_{\alpha\beta},$$
$$u_\alpha \to u_\alpha + \tau_{\alpha\beta},$$
$$\alpha, \beta = 1, \ldots, n,$$
$$\tau_{\alpha\beta} = \tau_{\beta\alpha}, \qquad \mathrm{Im}|\tau_{\alpha\beta}| > 0, \tag{34}$$

the matrices g and h being replaced by $2n$ matrices g_α and h_α. For instance, $g_\alpha = 1 \otimes \cdots \otimes g \otimes \cdots \otimes 1$ (and a similar expression for the matrices h_α). Here the index means that the matrix g_α acts on the symbols i_α, j_α as g, whereas its action on the remaining symbols is as the unitary matrix. The proof that the thus constructed $S^{(12)}(u)$ satisfies (1) at all u, should be based on the fact that at $v = 0$ eqs. (1) are fulfilled. Since up to now no direct proof, and accordingly no restrictions on the form of g and h have been found, we give below indirect arguments in favour of the fact that (32) satisfies eq. (1).

ANALYSIS OF EQ. (7)

It follows from (33) that $\Theta_{0,0}(1+\tau) = 0$. Put the quantity η in (32) close to $1+\tau$:

$$\eta = 1 + \tau + \hbar, \qquad \hbar \ll 1. \tag{35}$$

Then $S^{(12)}(u)$ up to a multiplicative factor is of the form

$$S^{(12)}(u) \simeq 1 + \hbar X^{(12)}(u). \tag{36}$$

The quantity $X^{(12)}(u)$ is the meromorphic double-periodic function, since the multiplier $\Theta_{0,0}(u)$ has been omitted in obtaining (36). Below we shall prove that $X^{(12)}(u)$ satisfies eq. (7). This testifies to the fact that the overall S-matrix provided by eq. (32) satisfies (1). Another argument in favour of this is the verification performed by Babudjhan and Tetelman in the degenerate case resulting from (32) at $|\tau| \to \infty$. Now consider eq. (7); this stands independently, because of the above-mentioned relation of these equations to classical integrable systems. Note, that eqs. (7) have first been studied by V.A. Fateev who has obtained a number of interesting results. Let us assume that the solution of (7) has only one pole at u_0:

$$X^{(12)}(u) = \frac{t_0^{(12)}}{u - u_0}. \tag{37}$$

Substitution of (37) into (7) shows that this is possible only in two cases. Either

$$u_0 \neq 0, \qquad \left[t_0^{(12)}, t_0^{(13)}\right] = 0, \tag{38}$$

or

$$u_0 = 0, \qquad [t_0^{(12)}, t_0^{(13)}] + [t_0^{(12)}, t_0^{(23)}] = 0. \tag{39}$$

First take the second possibility, (39). It is easy to verify that (39) is satisfied if

$$t_0^{(12)} = I_\mu^{(1)} I_\mu^{(2)}, \tag{40}$$

where I_μ form the basis of a Lie subalgebra of gl(N) and

$$\operatorname{Tr} I_\mu I_\nu = \delta_{\mu\nu}.$$

It can also be shown that (40) is the general solution of (39). Let us now make the assumption that $X^{(12)}(u)$ has several poles. Then eqs. (7) show that the sum of the positions of two poles should be the position of another pole, if the possibility (38) is again neglected. This gives rise to the conclusion that the poles of $X^{(12)}(u)$ form a lattice:

$$X^{(12)}(u) \to \frac{t_k^{(12)}}{u - u_k}, \qquad u_k = k_1 + \tau k_2, \tag{41}$$

at

$$u \to u_k, \qquad k = (k_1, k_2), \qquad \operatorname{Im} \tau > 0.$$

We have the following relations on residues t_k:

$$[t_k^{(12)}, t_{l+k}^{(13)}] + [t_k^{(12)}, t_l^{(23)}] = 0,$$

$$[t_k^{(21)}, t_{l+k}^{(31)}] + [t_k^{(21)}, t_l^{(32)}] = 0,$$

$$[t_k^{(12)}, t_{l+k}^{(13)}] + [t_k^{(12)}, t_{-l}^{(32)}] = 0. \tag{42}$$

At $k = l = 0$, eqs. (42) coincide with (39), the complete solution of which, (40), is known. All eqs. (42) will be satisfied if there exist matrices G_k, forming a projective representation for a group of lattice shifts, i.e.,

$$G_k G_l = s(k,l) G_{k+l}, \qquad G_k^{-1} = G_{-k}, \qquad G_0 = 1. \tag{43}$$

Then the solution of (42) may be written as

$$t_k^{(12)} = (G_k^{(1)})^{-1} \cdot I_\mu^{(1)} \cdot G_k^{(1)} \cdot I_\mu^{(2)} = G_k^{(2)} I_\mu^{(2)} (G_k^{(2)})^{-1} I_\mu^{(1)}. \tag{44}$$

The second equality in (44) requires that the matrix belongs to the same Lie group for which I_μ form the Lie algebra. It is clear from (44) that the group properties (43) should be satisfied up to a scalar multiplier (i.e., G_k form a projective representation). Then we can again introduce the generators

$$g = G_{1,0}, \qquad h = G_{0,1}, \tag{45}$$

and write G_k in the form

$$G_k = g^{k_1} h^{k_2}, \tag{46}$$

$$gh = \omega hg. \tag{46'}$$

To extrapolate the solution to arbitrary u let us require that g and h be elements of finite order [S.P. Novikov has explained to me, that this follows from (46')]. For instance

$$g^N = h^N = 1, \qquad \omega = \exp(2\pi i/N). \tag{47}$$

Requiring that the matrix $X^{(12)}(u)$ satisfy the automorphic properties consistent with (44) we get

$$X^{(12)}(u+1) = (g^{(1)})^{-1} X^{(12)}(u) g^{(1)}, \; X^{(12)}(u+\tau) = (h^{(1)})^{-1} X^{(12)}(u) h^{(1)}. \tag{48}$$

It is clear from (48) that $X^{(12)}(u)$ is a double-periodic function with the real period N and the complex period $N\tau$. Such a function exists and is defined up to the addition of a constant (if the values of the residues and the positions of the poles are given), if the sum of the residues in the parallelogram of periods equals zero. Then with (44) taken into account we have the following conditions for g and h:

$$\sum_{k_1=0}^{N-1} \sum_{k_2=0}^{N-1} G_k^{-1} I_\mu G_k = 0, \tag{49}$$

for all μ. This condition also seems to be the true restriction for g and h for the case of $S^{(12)}(u)$ satisfying (1).

Finally, the constant is excluded due to the automorphity condition, (48). Then it is clear that by virtue of the Louiville theorem, the function thus constructed actually satisfies (7). If g and h are taken as matrices (26) and I_α (27) as generators, all the conditions including (49) are satisfied. On the other hand, this solution is a limit transition (35) from (32). Thus we have partly proved that (32) satisfies (1).

5. Conclusion

Generalizations of this approach are possible along various lines.

First of all, the tensor product of three N-dimensional vector spaces may be replaced by three different linear spaces. For instance, the symbols i, j may assume continuous values. Then $S^{(12)}(u)$ from the tensor product of matrices turns into the tensor product of linear integral operators. Secondly, the two vectors p_1, p_2 may be treated as arguments, which are affected by a certain group of transformations different from those of the Lorentz group. Then it is required that $S^{(12)}(p_1,p_2)$ be invariant with respect to the group. Eq. (2) is now of the form

$$S^{(12)}(p_1,p_2)S^{(13)}(p_1,p_3)S^{(23)}(p_2,p_3) = S^{(23)}(p_2,p_3)S^{(13)}(p_1,p_3)S^{(12)}(p_1,p_2) \tag{50}$$

Then a certain discrete subgroup of the given group of transformations separately acting on p_1, p_2 and p_3 should play the role of shifts with respect to the lattice.

Finally, the most interesting problem is the problem of many dimensional integrable systems (Yang-Mills). In a remarkable work by Zamolodchikov [12] an important step has been taken in this direction. Zamolodchikov has studied 3-dimensional lattice systems possessing properties similar to (1). In terms of the scattering theory, infinite 1-dimensional strings, experiencing a three string collision in 2 + 1 dimensional space-time, play the role of particles. The "tetrahedron equations", relating the amplitudes of the 3-string collisions $S^{123}(n_1, n_2, n_3)$, play the role of the triangle equations. The quantity S^{123} depends on unit vectors n_i normal to the "world-plane" of the string propagation in a Lorentz-invariant way, i.e., S^{123} is dependent on the scalar product of the vector $n_i n_j$. It is possible that the quantity $S^{123}(n_1, n_2, n_3)$ may be defined by the automorphity requirement with respect to the action of a certain discrete subgroup of the Lorentz-group.

Note added in proof

We now have a proof that eqs. (29), (32), (33) indeed satisfy eq. (2) in the general case. The principal idea is as follows.

Denote the difference of the left- and right-hand sides of eq. (2) as $\phi^{(123)}(u,v)$. Owing to eq. (30), $\phi^{(123)}(0,v) = \phi^{(123)}(u,0) = 0$, whereas from eqs. (25′), (25″)

$$\phi^{(123)}(u+1,v) = (g^{(1)})^{-1}\phi^{(123)}(u,v)g^{(1)},$$

$$\phi^{(123)}(u+\tau,v) = \lambda^2 \exp(4\pi iu + 2\pi iv)(h^{(1)})^{-1}\phi^{(123)}(u,v)h^{(1)},$$

$$\phi^{(123)}(u,v+1) = (g^{(2)})^{-1}\phi^{(123)}(u,v)g^{(2)},$$

$$\phi^{(123)}(u,v+\tau) = \lambda^2 \exp(2\pi iu + 4\pi iv)(h^{(2)})^{-1}\phi^{(123)}(u,v)h^{(2)}.$$

Now perform a Fourier series expansion. Then it is easy to prove that the entire function $\phi^{(123)}(u,v)$ satisfying these equations vanishes identically. This completes the proof.

References

[1] H. Bethe, Z. Phys. 205 (1931)
[2] L. Onsager, Phys. Rev. 65 (1944) 117
[3] C.N. Yang, Phys. Rev. 168 (1968) 1920
[4] R.J. Baxter, Ann. of Phys. 70 (1972) 193
[5] R.J. Baxter, Ann. of Phys. 76 (1973) 1, 25, 48
[6] R.J. Baxter, Phil. Trans. Roy. Soc. 289 (1978) 315
[7] F.K. Sklyanin, L.A. Takhtadjyan and L.D. Faddeev, Theor. Mat. Fiz. 40 (1979) 194
[8] L.D. Faddeev, preprint LOMI P-2-79, Leningrad (1979)
[9] A.B. Zamolodchikov and Al.B. Zamolodchikov, Ann. of Phys. 120 (1979) 253
[10] A.B. Zamolodchikov, Phys. Rev. D, to be published
[11] M. Karowski, H. Thun, T. Truong and P. Weisz, Phys. Lett., 67B (1977) 321
[12] A.B. Zamolodchikov. ZhETF79 (1980) 641
[13] E.K. Sklyanin, preprint LOMI, E-3-1979, Leningrad (1979)
[14] A.V. Mikhailov, ZhETF Pis'ma 40 (1979) 443
[15] A.V. Mikhailov, Kiev Conf., September 1979 (North-Holland, Amsterdam)

FAMILIES OF COMMUTING TRANSFER MATRICES IN q-STATE VERTEX MODELS

Jacques H. H. Perk and Cherie L. Schultz
Institute for Theoretical Physics
State University of New York at Stony Brook
Stony Brook, New York 11794, U.S.A.

in

Proceedings of RIMS Symposium on Non-Linear Integrable Systems-Classical Theory and Quantum Theory, Kyoto, Japan May 13 - May 16, 1981, ed. by M. Jimbo and T. Miwa (World Science Publishing Co., Singapore, 1983).

Contents

Abstract: Using the method of commuting transfer matrices new exactly-solvable q-state vertex models and their associated spin-chain Hamiltonians are found. A partial review of related work is given.

INTRODUCTION

In recent years great progress is made in developing new techniques and deriving exact results in many branches of physics and mathematics. However, it is becoming more and more clear that these developments are not independent of each other. In fact, this conference deals with such seemingly different topics as soliton equations and other integrable systems in classical mechanics, exactly solvable models of classical and quantum statistical mechanics, lattice gauge theory and quantum field theory. But intimate connections between these fields have been displayed by many of the speakers. In the present contribution we want to address the problem of the existence of new classes of exactly solvable lattice models with short-range interactions in statistical mechanics. We shall restrict ourselves to models on a two-dimensional spatial lattice and their associated quantum problems in one space and one time dimension, the most famous example being the two-dimensional Ising model[1] and the one-dimensional XY-model (Ising model in transverse field)[2]. In spite of the fact that nature apparently has three space and one time dimensions, such exact results of nonperturbative nature had and still have a large impact in many fields. In the first place they triggered an entirely different view on phase transitions and critical exponents[3]. They have led to a vast amount of experiments on systems of low dimensionality[4]. They also have been used to study nonperturbative phenomena in quantum field theory defined by lattice renormalization[5]. Finally, it is generally believed that (lattice) gauge theories in four dimensions share many features with their two-dimensional counterparts[6] and some albeit modest advance has been made in higher dimension[7,8]. Therefore, the construction and classification of integrable lattice models seems a highly useful activity.

It all started in 1931 when Bethe[9], inspired by the ideas of Bloch[10], solved exactly the eigenvalue problem of the isotropic Heisenberg antiferromagnetic chain. His results have been further elaborated and generalized by many authors[11,12]. Then in 1967 it was realized that the same techniques could be utilized to solve the six-vertex model[13], with as first example square ice. Subsequently, in a pioneering piece of work, McCoy and Wu[14] initiated interest in commutation relations between transfer matrices and hamiltonians and much work has been done on this since[15-17]. The existence of these commutation relations implies the existence of an infinite number of conservation laws which seem to imply complete integrability. It is a cornerstone in Baxter's solutions of the symmetric 8-vertex model[18], the related XYZ-model[19], and (very recently) the hard-hexagon model[20]. It is also deeply connected with Yang's consistency equation[21] for applying Bethe's hypothesis to a system of particles moving on a line and interacting via a delta-function potential[22], which is at the basis of the theory of factorizable S-matrices[23]. Furthermore, the commutation relations supplemented with certain functional relations and analyticity assumptions[24,25] lead to much simpler derivations of free energies and ground state energies. However, because of certain ambiguities in the method, such as the CDD zeros and poles[26] in the theory of factorizable S-matrices, it is always necessary to check the results by other means such as Bethe's Ansatz. Also, the most outstanding problem of determining the Green's functions of these models, which are essential in linking theory and experiment or in obtaining a renormalized quantum field theory, has not been solved at all. So far real progress has been made only for the Ising model[27], the XY-model[28], the (continuum) hard boson system[29] and low order perturbation expansion about it[30]. But in these papers the full structure of the Bethe-Ansatz has not been encountered; for that case only the normalization of the wave function has been found[31]. Finally it has become clear that the Bethe-Ansatz is the quantization[32] of the inverse scattering method, which became very explicit in second quantized language[33,34].

Most lattice models mentioned above have two degrees of freedom per bond (or site) of the lattice. Recently there is a great interest in generalizations with a higher number of degrees of freedom both on the lattice[24,35-37] and in the continuum[38,39]. This is also important if one attempts to understand unambiguously in terms of lattice renormalization the cut-off procedure needed in the Bethe-Ansatz solution of the SU(N) generalizations[40] of the Thirring model[41]. This had been carried out for the (one-component) Thirring model by Luther[42] and Lüscher[43] on the basis of results of Johnson, Krinsky, and McCoy[44] for the XYZ-model. But work of Sutherland[45] indicates that the straightforward generalizations of the Heisenberg-Ising model to more components are not solvable on the lattice in general.

Let us now define the q-state vertex model on a quadratic lattice with M rows and N columns and with periodic boundary conditions. The states are defined by assigning a color to each bond between two neighboring lattice sites, q being the number of different colors. We associate a Boltzmann weight factor W with each vertex depending on the q^4 color configurations of the four bonds joining in a given site. We can write these weights W symbolically as the transfermatrix at that one site

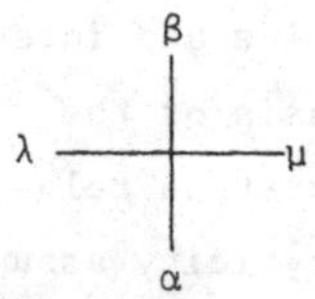

$$\equiv R^{\lambda\mu}(\alpha,\beta) \equiv R_{\lambda\alpha|\mu\beta}\,,\qquad (\alpha,\beta,\lambda,\mu = 1,\cdots,q), \tag{1}$$

and the transfer matrix of one row is

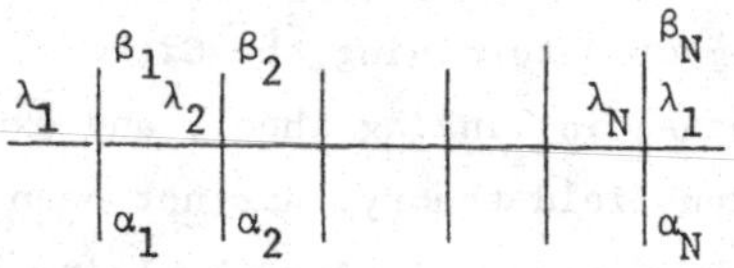

$$= T_{\alpha_1\cdots\alpha_N|\beta_1\cdots\beta_N} =$$

$$= \mathrm{Tr}_{(q)}\,\{R(\alpha_1,\beta_1)\,\cdots\,R(\alpha_N,\beta_N)\} =$$

$$= \sum_{\lambda_1=1}^{q}\cdots\sum_{\lambda_N=1}^{q} R^{\lambda_1\lambda_2}(\alpha_1,\beta_1)R^{\lambda_2\lambda_3}(\alpha_2,\beta_2)\cdots R^{\lambda_N\lambda_1}(\alpha_N,\beta_N)\;. \tag{2}$$

Then the partition function is given by

$$Z \equiv \sum_{\text{all states}} \prod_{\text{all sites}} W = \operatorname{Tr}_{(q^N)} T^M, \tag{3}$$

where the last trace is taken over a q^N-dimensional vector space

In general this q^4-vertex model will not be exactly solvable although it covers all periodic lattice systems with short-range interactions and with a finite number of states per unit cell. At this moment the best known criterium for solvability of a model of this type is the existence of a family of commuting transfer matrices $\{T(u)\}$, where u is a continuous parameter, i.e. $[T(u),T(u')] = 0$. A sufficient and presumably to large extent necessary[17] condition is what is now known as the star-triangle relation of Baxter[25] or the factorization equation. More precisely, for the transfermatrices T and $\bar{T}$, defined by the vertex weights $R^{\lambda\mu}(\alpha,\beta)$ and $\bar{R}^{\lambda\mu}(\alpha,\beta)$ respectively, to commute we require the existence of a nonsingular $q^2 \times q^2$ matrix X such that all q^6 equations

$$\sum_{i,j,k} X^{\lambda j}_{\mu k}\, R^{j\lambda'}(\alpha,i)\bar{R}^{k\mu'}(i,\beta) = \sum_{i,j,k} \bar{R}^{\mu k}(\alpha,i)R^{\lambda j}(i,\beta)X^{j\lambda'}_{k\mu'} \tag{4}$$

are satisfied simultaneously, Indeed, using invariance of trace under the transformation X it then follows that

$$T\bar{T} = \operatorname{Tr}_{(q^2)} \prod_{j=1}^{N} \sum_{\gamma_j} R(\alpha_j,\gamma_j) \otimes \bar{R}(\gamma_j,\beta_j)$$
$$= \operatorname{Tr}_{(q^2)} \prod_{j=1}^{N} \sum_{\gamma_j} \bar{R}(\alpha_j,\gamma_j) \otimes R(\gamma_j,\beta_j) = \bar{T}\, T. \tag{5}$$

The equation (4) has been expressed most elegantly in terms of diagrams by Baxter[18,25] and Zamolodchikov[8,23]

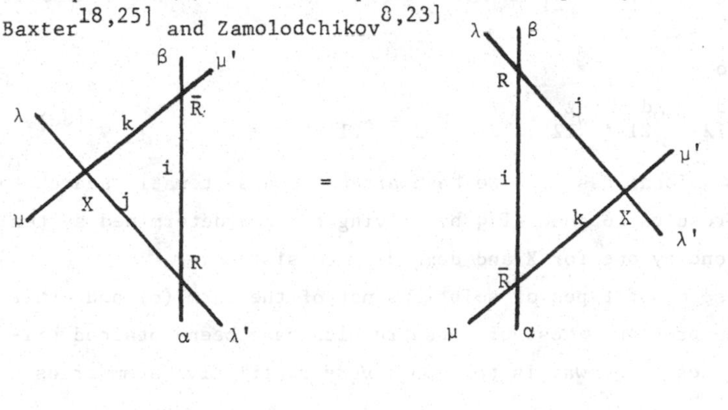

(4a)

and is also known in a more restricted context as the condition for factorization of the S-matrix provided we consider R, $\bar{R}$, and X to be the same matrix-valued function of the rapidities (in the nonrelativistic case momenta) θ, $\bar{\theta}$, and $\theta-\bar{\theta}$ respectively.

We shall restrict ourselves to solutions of condition (4) of the form

$$R^{\lambda\mu}(\alpha,\beta) = W^d_{\alpha\lambda}\delta_{\alpha\beta}\delta_{\lambda\mu} + W^r_{\alpha\beta}\delta_{\alpha\lambda}\delta_{\beta\mu} + W^\ell_{\alpha\beta}\delta_{\alpha\mu}\delta_{\beta\lambda}, (W^r_{\alpha\alpha} \equiv W^\ell_{\alpha\alpha} \equiv 0), \tag{6}$$

and similarly for $\bar{R}$ and X. So the allowed (nonvanishing) vertex weights are

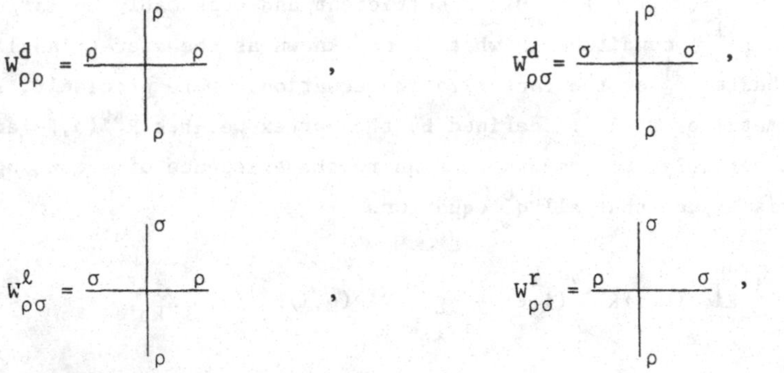

(7)

with conservation of color at each vertex. This is a natural generalization of the 8-vertex model for q = 2, for which the commuting families are[17]

a) the free-fermion model

$$W^d_{11}W^d_{22} + W^d_{12}W^d_{21} = W^\ell_{12}W^\ell_{21} + W^r_{12}W^r_{21} , \tag{8a}$$

b) the 6-vertex model

$$W^r_{12} = W^r_{21} = 0 \text{ or } W^\ell_{12} = W^\ell_{21} = 0 , \tag{8b}$$

c) the Baxter model

$$W^d_{11} = W^d_{22}, W^d_{12} = W^d_{21} , W^\ell_{12} = W^\ell_{21}, W^r_{12} = W^r_{21} , \tag{8c}$$

and some "one-dimensional cases." We have attempted a systematic search to generalize these results to general q by solving the overdetermined system (4) of equations one by one for X and demanding consistency.

There are also other types of solutions not of the form (6) and other ways to attack the problem. Most of these results have been obtained primarily for S-matrices. One way is to impose very restrictive symmetries like O(n)[46], U(n)[47], Z_4[48], possibly combined with supersymmetry[49], which reduce the number of parameters to very few. One can also solve differential equations with respect to the rapidity [36,50]. An important condition seems to be complete X-symmetry, X being an additive group with as elements the color indices: the model is X-symmetric if $S^{\beta\mu}_{\lambda\alpha} \equiv R^{\lambda\mu}(\alpha,\beta) = 0$ if $\lambda+\alpha \neq \mu+\beta$ within the group X, and completely X-symmetric if in addition $R^{\lambda+\nu,\mu+\nu}(\alpha+\nu,\beta+\nu) = R^{\lambda\mu}(\alpha,\beta)$. It has been argued[51-53] that in that case all S-matrix solutions are given in terms of Jacobian-or more general θ-functions. There are also constructions in terms of elementary functions, with or without such symmetries[37,54,55]. Finally, S-matrices have been inferred from known integrable systems in classical mechanics, such as the Toda lattice of period 3 or higher[56,57], both directly or by explicit quantization[58-60].

Before giving our results we should make one more remark. We are only interested in inequivalent, irreducible solutions of families of commuting transfer matrices. It can be seen that the solution

$$R'^{\lambda\mu}(\alpha,\beta) \equiv \sum_{\alpha',\beta',\lambda',\mu'} R^{\lambda'\mu'}(\alpha',\beta')S_{\alpha\alpha'}S^{-1}_{\beta'\beta}T_{\lambda\lambda'}T^{-1}_{\mu'\mu} \quad ,$$
$$\bar{R}'^{\lambda\mu}(\alpha,\beta) \equiv \sum_{\alpha',\beta',\lambda',\mu} \bar{R}^{\lambda'\mu'}(\alpha',\beta')\, S_{\alpha\alpha'}S^{-1}_{\beta'\beta}U_{\lambda\lambda'}U^{-1}_{\mu'\mu} \quad , \tag{9}$$

is equivalent to the original one R, $\bar{R}$ by changing X to $X' = \sum XUU^{-1}TT^{-1}$ correspondingly. This is related to a well-known symmetry property of vertex models[61]. Restricted to our case (6) we have the following symmetry. If the transfermatrices specified by the weights W and $\bar{W}$ commute, then transfermatices with

$$W^{d'}_{\rho\rho} = C\omega^2_\rho W^d_{\rho\rho} \;, \qquad W^{d'}_{\rho\sigma} = C\omega_\rho{}^2 \varepsilon_{\rho\sigma} W^d_{\rho\sigma} \;,$$

$$W^{\ell'}_{\rho\sigma} = C\omega_\rho\omega_\sigma \exp(g^\ell_\sigma - g^\ell_\rho)W^\ell_{\rho\sigma} \;,$$

$$W^{r'}_{\rho\sigma} = C\omega^{-1}_\rho\omega^{-1}_\sigma \exp(g^r_\sigma - g^r_\rho)W^r_{\rho\sigma} \;, \tag{10a}$$

$$\omega^4_\rho = 1, \qquad \varepsilon_{\rho\sigma} = \varepsilon_{\sigma\rho} = \pm 1, \qquad \text{and } \bar{W}' \text{ similar with}$$

$$\bar{\varepsilon}_{\rho\sigma} = \varepsilon_{\rho\sigma} \;, \quad \bar{g}^\ell_\rho + \bar{g}^r_\rho = g^\ell_\rho + g^r_\rho \;, \tag{10b}$$

also commute.

There are two obvious ways for the model to be reducible. The first is direct-product weights

$$R^{\lambda\mu}(\alpha,\beta) = R'^{\lambda'\mu'}(\alpha',\beta')R''^{\lambda''\mu''}(\alpha'',\beta''), \tag{11}$$

$\alpha = (\alpha',\alpha'')$, etc. each index being a pair of independent indices. In this case the condition (4) is satisfied for the third one if it is satisfied for any two of R', R'', R. The partition function and the correlation function factorize $Z = Z'\,Z''$ and the free energy satisfies $f = f' + f''$. This kind of factorization is apparently the case for the S-matrix for supersymmetric models[49], so that their stat. mech. analogues are less interesting. It preserves complete X-symmetry but not the color conservation condition (6). The second way is direct-sum weights, i.e. $R^{\lambda\mu}(\alpha,\beta) = 0$, unless all indices α,β,λ,μ in subset A' or all in subset A''. In this case the partition function is additive, $Z = Z' + Z''$, so that the free energy satisfies $f = \min(f',f'')$. We may have first-order phase transitions in this case as is illustrated in the one-dimensional KDP-model of Nagle[62].

RESULTS

Let us now give our solutions of the conditions (4), (6), which are more explicitly given by, for $q \geq 3$,

$$X^{\ell}_{\beta\gamma} W^{d}_{\alpha\beta} \bar{W}^{d}_{\alpha\gamma} = X^{\ell}_{\beta\gamma} W^{d}_{\alpha\gamma} \bar{W}^{d}_{\alpha\beta} , \tag{12a}$$

$$X^{d}_{\gamma\alpha} W^{d}_{\beta\alpha} \bar{W}^{\ell}_{\beta\gamma} = X^{d}_{\beta\alpha} W^{d}_{\gamma\alpha} \bar{W}^{\ell}_{\beta\gamma} , \tag{12b}$$

$$X^{d}_{\alpha\beta} W^{r}_{\beta\gamma} \bar{W}^{d}_{\gamma\alpha} = X^{d}_{\alpha\gamma} W^{r}_{\beta\gamma} \bar{W}^{d}_{\beta\alpha} , \tag{12c}$$

$$X^{\ell}_{\alpha\beta} W^{\ell}_{\beta\alpha} \bar{W}^{\ell}_{\alpha\beta} = X^{\ell}_{\beta\alpha} W^{\ell}_{\alpha\beta} \bar{W}^{\ell}_{\beta\alpha} , \tag{13a}$$

$$X^{\ell}_{\alpha\beta} W^{r}_{\alpha\beta} \bar{W}^{r}_{\beta\alpha} = X^{\ell}_{\beta\alpha} W^{r}_{\beta\alpha} \bar{W}^{r}_{\alpha\beta} , \tag{13b}$$

$$X^{r}_{\alpha\beta} W^{r}_{\beta\alpha} \bar{W}^{\ell}_{\alpha\beta} = X^{r}_{\beta\alpha} W^{r}_{\alpha\beta} \bar{W}^{\ell}_{\beta\alpha} , \tag{13c}$$

$$\sum_{\beta} X^{r}_{\alpha\beta} W^{\ell}_{\alpha\beta} \bar{W}^{r}_{\beta\alpha} = \sum_{\beta} X^{r}_{\beta\alpha} W^{\ell}_{\beta\alpha} \bar{W}^{r}_{\alpha\beta} , \tag{14}$$

$$X^{d}_{\beta\gamma} W^{\ell}_{\alpha\gamma} \bar{W}^{\ell}_{\gamma\beta} + X^{\ell}_{\beta\gamma} W^{\ell}_{\alpha\beta} \bar{W}^{d}_{\beta\gamma} = X^{\ell}_{\alpha\gamma} W^{d}_{\beta\gamma} \bar{W}^{\ell}_{\alpha\beta} , \tag{15a}$$

$$X^{d}_{\gamma\beta} W^{r}_{\gamma\beta} \bar{W}^{r}_{\alpha\gamma} + X^{\ell}_{\beta\gamma} W^{d}_{\beta\gamma} \bar{W}^{r}_{\alpha\beta} = X^{\ell}_{\alpha\gamma} W^{r}_{\alpha\beta} \bar{W}^{d}_{\beta\gamma} , \tag{15b}$$

$$X^{r}_{\alpha\gamma} W^{d}_{\beta\gamma} \bar{W}^{\ell}_{\beta\gamma} + X^{r}_{\alpha\beta} W^{r}_{\beta\gamma} \bar{W}^{d}_{\gamma\beta} = X^{d}_{\beta\gamma} W^{r}_{\alpha\gamma} \bar{W}^{\ell}_{\beta\alpha} , \tag{15c}$$

$$X^{d}_{\gamma\beta}W^{\ell}_{\alpha\beta}\bar{W}^{d}_{\beta\gamma}+X^{\ell}_{\gamma\beta}W^{\ell}_{\alpha\gamma}\bar{W}^{\ell}_{\gamma\beta} = X^{d}_{\gamma\alpha}W^{\ell}_{\alpha\beta}\bar{W}^{d}_{\alpha\gamma}+X^{\ell}_{\alpha\gamma}W^{\ell}_{\gamma\beta}\bar{W}^{\ell}_{\alpha\gamma}, \tag{16a}$$

$$X^{d}_{\beta\gamma}W^{d}_{\beta\gamma}\bar{W}^{r}_{\alpha\beta}+X^{\ell}_{\gamma\beta}W^{r}_{\gamma\beta}\bar{W}^{r}_{\alpha\gamma} = X^{d}_{\alpha\gamma}W^{d}_{\alpha\gamma}\bar{W}^{r}_{\alpha\beta}+X^{\ell}_{\alpha\gamma}W^{r}_{\alpha\gamma}\bar{W}^{r}_{\gamma\beta}, \tag{16b}$$

$$X^{r}_{\alpha\beta}W^{d}_{\gamma\beta}\bar{W}^{d}_{\gamma\beta}+X^{r}_{\alpha\gamma}W^{r}_{\gamma\beta}\bar{W}^{\ell}_{\beta\gamma} = X^{r}_{\alpha\beta}W^{d}_{\gamma\alpha}\bar{W}^{d}_{\gamma\alpha}+X^{r}_{\gamma\beta}W^{r}_{\alpha\gamma}\bar{W}^{\ell}_{\gamma\alpha}, \tag{16c}$$

$$X^{d}_{\alpha\alpha}W^{r}_{\alpha\beta}\bar{W}^{d}_{\beta\alpha}+X^{r}_{\alpha\beta}W^{d}_{\alpha\beta}\bar{W}^{\ell}_{\alpha\beta} = X^{d}_{\alpha\beta}W^{r}_{\alpha\beta}\bar{W}^{d}_{\alpha\alpha}+X^{\ell}_{\beta\alpha}W^{d}_{\beta\alpha}\bar{W}^{r}_{\alpha\beta}, \tag{17a}$$

$$X^{d}_{\alpha\alpha}W^{d}_{\beta\alpha}\bar{W}^{\ell}_{\alpha\beta}+X^{r}_{\beta\alpha}W^{r}_{\alpha\beta}\bar{W}^{d}_{\alpha\beta} = X^{d}_{\beta\alpha}W^{d}_{\alpha\alpha}\bar{W}^{\ell}_{\alpha\beta}+X^{\ell}_{\beta\alpha}W^{\ell}_{\alpha\beta}\bar{W}^{d}_{\beta\alpha}, \tag{17b}$$

$$X^{\ell}_{\alpha\beta}W^{d}_{\alpha\alpha}\bar{W}^{d}_{\alpha\beta}+X^{d}_{\alpha\beta}W^{\ell}_{\alpha\beta}\bar{W}^{\ell}_{\beta\alpha} = X^{\ell}_{\alpha\beta}W^{d}_{\alpha\beta}\bar{W}^{d}_{\alpha\alpha}+X^{d}_{\beta\alpha}W^{r}_{\beta\alpha}\bar{W}^{r}_{\alpha\beta}, \tag{17c}$$

$$X^{d}_{\alpha\alpha}W^{\ell}_{\alpha\beta}\bar{W}^{d}_{\alpha\alpha}+X^{r}_{\beta\alpha}W^{d}_{\beta\beta}\bar{W}^{r}_{\alpha\beta}+\sum_{\gamma}X^{r}_{\gamma\alpha}W^{\ell}_{\gamma\beta}\bar{W}^{r}_{\alpha\gamma} = X^{d}_{\alpha\beta}W^{\ell}_{\alpha\beta}\bar{W}^{d}_{\beta\alpha}+X^{\ell}_{\alpha\beta}W^{d}_{\alpha\alpha}\bar{W}^{\ell}_{\alpha\beta}, \tag{18a}$$

$$X^{d}_{\alpha\alpha}W^{d}_{\alpha\alpha}\bar{W}^{r}_{\alpha\beta}+X^{r}_{\alpha\beta}W^{\ell}_{\alpha\beta}\bar{W}^{d}_{\beta\beta}+\sum_{\gamma}X^{r}_{\alpha\gamma}W^{\ell}_{\alpha\gamma}\bar{W}^{r}_{\gamma\beta} = X^{d}_{\beta\alpha}W^{d}_{\beta\alpha}\bar{W}^{r}_{\alpha\beta}+X^{\ell}_{\alpha\beta}W^{r}_{\alpha\beta}\bar{W}^{d}_{\alpha\alpha}, \tag{18b}$$

$$X^{r}_{\alpha\beta}W^{d}_{\alpha\alpha}\bar{W}^{d}_{\alpha\alpha}+X^{d}_{\beta\beta}W^{\ell}_{\beta\alpha}\bar{W}^{r}_{\alpha\beta}+\sum_{\gamma}X^{r}_{\gamma\beta}W^{\ell}_{\gamma\alpha}\bar{W}^{r}_{\alpha\gamma} = X^{r}_{\alpha\beta}W^{d}_{\alpha\beta}\bar{W}^{d}_{\alpha\beta}+X^{d}_{\alpha\alpha}W^{r}_{\alpha\beta}\bar{W}^{\ell}_{\beta\alpha}, \tag{18c}$$

$$X^{r}_{\gamma\alpha}W^{d}_{\alpha\alpha}\bar{W}^{r}_{\alpha\beta}+X^{r}_{\gamma\beta}W^{\ell}_{\alpha\beta}\bar{W}^{d}_{\beta\beta}+X^{d}_{\gamma\gamma}W^{\ell}_{\alpha\gamma}\bar{W}^{r}_{\gamma\beta}+\sum_{\delta}X^{r}_{\gamma\delta}W^{\ell}_{\alpha\delta}\bar{W}^{r}_{\delta\beta} = X^{\ell}_{\alpha\beta}W^{r}_{\gamma\beta}\bar{W}^{\ell}_{\alpha\gamma}, \tag{19}$$

and the equations which follow from (15), (17), (18), (19) under the simultaneous transposition of $W^{\ell}_{\rho\sigma}$, $W^{r}_{\rho\sigma}$, $\bar{W}^{\ell}_{\rho\sigma}$, $\bar{W}^{r}_{\rho\sigma}$, $X^{\ell}_{\rho\sigma}$, $X^{r}_{\rho\sigma}$, i.e. $W^{\ell}_{\rho\sigma} \rightarrow W^{\ell}_{\sigma\rho}$, etc. (Note that eqs. (12), (13), (14,) and (16) are invariant under this change.)

The first family of commuting transfermatrices follows if one assumes that all (or sufficiently many) of the vertex weights W and $\bar{W}$ are nonvanishing, and solves for a (nonsingular) similarity transformation X. Then it follows from the consistency of (12), (13), (15), (16),(and (18) if too many elements of X vanish), that all asymmetry in W and $\bar{W}$ can be described by the equivalence transformations (10); it is therefore no restriction to assume that $W^{d}_{\rho\sigma}$, $W^{\ell}_{\rho\sigma}$, $W^{r}_{\rho\sigma}$, and similarly $\bar{W}$ and X, are symmetric in ρσ. The further consistency requirements, implied by (15) and (16), can be expressed as

$$W^{d}_{\rho\sigma} = W^{d}_{\lambda\mu}, \quad W^{\ell}_{\rho\sigma} = W^{\ell}_{\lambda\mu}, \quad W^{r}_{\rho\sigma} = W^{r}_{\lambda\mu}, \tag{20a}$$

if ρ,σ,λ,μ all different, and

$${W^{d}_{\rho\sigma}}^{2} - {W^{d}_{\rho\tau}}^{2} = {W^{\ell}_{\rho\sigma}}^{2} - {W^{\ell}_{\rho\tau}}^{2} = {W^{r}_{\rho\sigma}}^{2} - {W^{r}_{\rho\tau}}^{2}, \tag{20b}$$

proportional to the same expressions in terms of $\bar{W}$ and X. Hence, for q=3 and q=4 we can solve the conditions (12) - (19) after parametrizing in terms of Jacobian elliptic functions, and for $q \geq 5$ all index dependence has to drop out. After a lot of algebra one arrives at the same solution as found by Fateev[36] within the frame work of factorizable S-matrices and under the additional assumptions that W^d, W^ℓ, and W^r are symmetric. To be explicit, up to the symmetries (10), the complete solution is given by , for q=3,

$$W^\ell_{\rho\sigma} = ps(u), \quad W^r_{\rho\sigma} = ps(\eta-u),$$

$$W^d_{\rho\sigma} = ps(2\eta), \quad W^d_{\tau\tau} = F(u)+ps(2\eta), \tag{21a}$$

where

$$\begin{aligned} p &= c, \quad \text{for } \rho\sigma = 12 \text{ or } 21 \quad \text{and } \tau\tau = 33, \\ p &= n, \quad \text{for } \rho\sigma = 13 \text{ or } 31 \quad \text{and } \tau\tau = 22, \\ p &= d, \quad \text{for } \rho\sigma = 23 \text{ or } 32 \quad \text{and } \tau\tau = 11, \end{aligned} \tag{21b}$$

and

$$F(u) \equiv - cs(2\eta) - ds(2\eta) + ns(u)ns(\eta-u)sn(\eta). \tag{22}$$

Here we have used Glaisher's notation[63] for elliptic functions of modulus k. There are two fixed parameters η and k and one running parameter u for this family of commuting transfermatrices. This is also the case for q=4, where

$$W^\ell_{\rho\sigma} = ps(u) \;, \quad W^r_{\rho\sigma} = ps(\eta-u),$$

$$W^d_{\rho\sigma} = ps(\eta) \;, \quad W^d_{\tau\tau} = F(u) = \text{const.} \;, \tag{23a}$$

with

$$\begin{aligned} p &= c \quad \text{for } \rho\sigma = 12,\ 21,\ 34,\ 43 \;, \\ p &= n \quad \text{for } \rho\sigma = 13,\ 31,\ 24,\ 42 \;, \\ p &= d \quad \text{for } \rho\sigma = 23,\ 32,\ 14,\ 41 \;. \end{aligned} \tag{23b}$$

Finally for $q \geqq 5$ the solution is[35, 46] , $(\rho \neq \sigma)$,

$$\begin{aligned} W^\ell_{\rho\sigma} &\equiv W^\ell = u^{-1} \;, & W^r_{\rho\sigma} &\equiv W^r = (1-u)^{-1}, \\ W^d_{\rho\sigma} &\equiv W^o = -\tfrac{1}{2}(q-2) \;, & W^d_{\tau\tau} &\equiv W^d = W^\ell + W^r + W^o \;, \end{aligned} \tag{24}$$

with no other parameters besides u. Here again the solution is unique upto the symmetries (10). Note that for q=3 or 4 solution (24) is a limiting case of solutions (21) or (23) respectively, as can be seen by taking $u \to \eta u$, $\eta \to 0$, and using (10).

To find other families of commuting transfer matrices some of the vertex weights have to be chosen zero. There seems to be only two possibilities for irreducible families although this has not been proven yet. The second class of models we have consists of generalizations of the six-vertex model for q=2. These are models with all $W^{\ell}_{\rho\sigma} = 0$, or equivalently by rotation (or reflection) symmetry all $W^{r}_{\rho\sigma} = 0$. Two commuting families are given by

$$W^{d}_{\rho\rho} = X^2_\rho \frac{\mathrm{sh}(\eta+\varepsilon_\rho u)}{\mathrm{sh}\eta}, \quad W^{d}_{\rho\sigma} = X^2_\rho \frac{\mathrm{sh}u}{\mathrm{sh}\eta} \exp(g_{\rho\sigma}), \quad (\rho\neq\sigma)$$
$$W^{r}_{\rho\sigma} = X_\rho X_\sigma \exp\{\omega_\sigma - \omega_\rho + u \,\mathrm{sign}\,(\sigma-\rho)\}, \quad W^{\ell}_{\rho\sigma} = 0, \tag{25}$$

and

$$W^{d}_{\rho\rho} = X^2_\rho \frac{\mathrm{sh}(\eta+\varepsilon_\rho u)}{\mathrm{sh}\eta}, \quad W^{d}_{\rho\sigma} = X^2_\rho \frac{\mathrm{sh}u}{\mathrm{sh}\eta} \exp(-g_{\rho\sigma}),$$
$$W^{\ell}_{\rho\sigma} = X_\rho X_\sigma \exp\{\omega_\sigma - \omega_\rho - u\,\mathrm{sign}(\sigma-\rho)\}, \quad W^{r}_{\rho\sigma} = 0, \tag{26}$$

where $g_{\rho\sigma} = -g_{\sigma\rho}$, $\varepsilon_\rho = \pm 1$, and η are constants, u is the nontrivial variable, ω_ρ are the gauge parameters g_ρ of (10), and X_ρ are variables related to the conservation of total number of bonds of given color ρ in each row. The two families (25) and (26) also commute mutually. The symmetric case $W^{r}_{\rho\sigma} = W^{r}_{\sigma\rho}$ had been given before[35], and the special case $X_\rho \equiv 1$, $\varepsilon_\rho \equiv 1$, $g_{\rho\sigma} \equiv 0$, $\omega_\rho = \lambda\rho$ is very similar to the completely X-symmetric solution[37,57] of (4),(6).

The last family is, up to the symmetries (10), given by

$$W^{d}_{\rho\rho} = \left[\frac{(K_\rho s+1)(s+K_\rho)}{K_\rho}\right]^{1/2}, \qquad W^{d}_{\rho\sigma} = 0, \quad \rho\neq\sigma,$$
$$W^{r}_{\rho\sigma} = s\left[\frac{(K_\rho s+1)(K_\sigma s+1)}{K_\rho K_\sigma (s+K_\rho)(s+K_\sigma)}\right]^{1/4},$$
$$W^{\ell}_{\rho\sigma} = \left[\frac{(s+K_\rho)(s+K_\sigma)}{K_\rho K_\sigma (K_\rho s+1)(K_\sigma s+1)}\right]^{1/4}, \tag{27}$$

where

$$K_\rho = 1, \quad \rho = 1,\cdots,n ,$$

$$K_\rho = \frac{n+\sqrt{n^2+4(q-n-1)}}{2} \equiv K, \quad \rho = n+1,\cdots,q, \tag{28}$$

and

$$W = W(s), \bar{W} = W(\bar{s}), \quad X = W(t), \tag{29}$$

with

$$(Ks+1)(K\bar{s}+1)(Kt+1) = K(s+K)(\bar{s}+K)(t+K). \tag{30}$$

For given q there are q+1 families, apart from q=2 where all three choices for n coincide. The special choice n=q, $W^d_{\rho\rho} = s+1$, $W^r_{\rho\sigma} = s$, $W^\ell_{\rho\sigma} = 1$, was first introduced by Stroganov[24] for q=3, and in ref. 35 for general q; for even q it can be related to case V of ref. 47. In this case n=q the partition function counts in how many ways one can cover the bonds of the lattice with closed colored polygons, without intersections or common bonds. The special case n=0, all $K_\rho \equiv K$, was given in ref. 24 for q=3 and in ref. 35.

For all families we can construct a one-dimensional hamiltonian commuting with the family according to standard procedure[19]. We take first minus the logarithmic derivative of the transfermatrix and then the limit to the right-shift point

$$W^\ell_{\rho\sigma} = 0, \; W^r_{\rho\sigma} = 1-\delta_{\rho\sigma}, \; W^d_{\rho\sigma} = \delta_{\rho\sigma} \tag{31a}$$

or the left-shift point

$$W^r_{\rho\sigma} = 0, \; W^\ell_{\rho\sigma} = 1-\delta_{\rho\sigma}, \; W^d_{\rho\sigma} = \delta_{\rho\sigma}. \tag{31b}$$

The transfermatrices corresponding to (31a) and (31b) shift the configuration of a given row by one step to the right (or left) in going to the next row. We shall use the following basis of q × q matrices

$$(\Lambda^{\rho d})_{ij} = \delta_{i\rho}\delta_{j\rho} ,$$

$$(\Lambda^{\rho\sigma x})_{ij} = \delta_{i\rho}\delta_{j\sigma} + \delta_{j\rho}\delta_{i\sigma} ,$$

$$(\Lambda^{\rho\sigma y})_{ij} = -i\delta_{i\rho}\delta_{j\sigma} + i\delta_{j\rho}\delta_{i\sigma} . \tag{32}$$

for $\rho < \sigma$.

We have the following hamiltonians. Corresponding to (21) we have (for q=3),

$$H = -\tfrac{1}{2} \sum_J \{ (cs(2\eta)+cs(\eta))\Lambda_J^{12x}\Lambda_{J+1}^{12x}+(cs(2\eta)-cs(\eta))\Lambda_J^{12y}\Lambda_{J+1}^{12y}$$

$$+ (ns(2\eta)+ns(\eta))\Lambda_J^{13x}\Lambda_{J+1}^{13x}+(ns(2\eta)-ns(\eta))\Lambda_J^{13y}\Lambda_{J+1}^{13y}$$

$$+ (ds(2\eta)+ds(\eta))\Lambda_J^{23x}\Lambda_{J+1}^{23x}+(ds(2\eta)-ds(\eta))\Lambda_J^{23y}\Lambda_{J+1}^{23y}$$

$$+ 2cs(2\eta)\Lambda_J^{3d}\Lambda_{J+1}^{3d} + 2ns(2\eta)\Lambda_J^{2d}\Lambda_{J+1}^{2d}$$

$$+ 2ds(2\eta)\Lambda_J^{1d}\Lambda_{J+1}^{1d} \}, \qquad (33)$$

corresponding to (23) we have, (for q=4),

$$H = -\sum_J \{ cs(\eta)(\Lambda_J^{12x}\Lambda_{J+1}^{12x} + \Lambda_J^{34x}\Lambda_{J+1}^{34x})$$

$$+ ns(\eta)(\Lambda_J^{13x}\Lambda_{J+1}^{13x} + \Lambda_J^{24x}\Lambda_{J+1}^{24x})$$

$$+ ds(\eta)(\Lambda_J^{23x}\Lambda_{J+1}^{23x} +\Lambda_J^{14x}\Lambda_{J+1}^{14x})\}, \qquad (34)$$

and corresponding to (24) we have, (for $q \geq 5$),

$$H = -\tfrac{1}{4}\sum_J \{(q-4)\sum_{\rho<\sigma} \Lambda_J^{\rho\sigma x}\Lambda_{J+1}^{\rho\sigma x}+ q\sum_{\rho<\sigma}\Lambda_J^{\rho\sigma y}\Lambda_{J+1}^{\rho\sigma y} + 2(q-4)\sum_\rho \Lambda_J^{\rho d}\Lambda_{J+1}^{\rho d}\}. \qquad (35)$$

In order to go to the left-shift operator in the limit $u\to 0$, we first have added a normalization factor sn(u) to (21a) and (23a) and a factor u to (24). For case (27), (28) we find, at the shift point s = 0,

$$H = -\sum_J \{\sum_\rho \tfrac{1}{2}(K_\rho + K_\rho^{-1})\Lambda_J^{\rho d}\Lambda_{J+1}^{\rho d}+ \sum_{\rho<\sigma}\tfrac{1}{4}(K_\rho^{-1}+K_\sigma^{-1}-K_\rho-K_\sigma)(\Lambda_J^{\rho d}\Lambda_{J+1}^{\sigma d}+\Lambda_J^{\sigma d}\Lambda_{J+1}^{\rho d})$$

$$+ \sum_{\rho<\sigma} \tfrac{1}{2}(K_\rho K_\sigma)^{-\frac{1}{2}}(\Lambda_J^{\rho\sigma x}\Lambda_{J+1}^{\rho\sigma x} - \Lambda_J^{\rho\sigma y}\Lambda_{J+1}^{\rho\sigma y}) \}. \qquad (36)$$

Finally for the six-vertex cases (25), (26) the associated hamiltonian is a linear combination of $-(\partial/\partial u)\ell nT$ and $-(\partial/\partial x_\rho)\ell nT$ at u=0, $x_\rho = 1$. We find

$$H = - \sum_J \{ \sum_\rho \varepsilon_\rho \coth\eta \, \Lambda_J^{\rho d} \Lambda_{J+1}^{\rho d} + \sum_{\rho<\sigma} (\Lambda_J^{\rho d} \Lambda_{J+1}^{\sigma d} - \Lambda_J^{\sigma d} \Lambda_{J+1}^{\rho d})$$

$$+ (2\mathrm{sh}\eta)^{-1} \sum_{\rho<\sigma} [\mathrm{ch}\, g_{\rho\sigma} (\Lambda_J^{\rho\sigma x} \Lambda_{J+1}^{\rho\sigma x} + \Lambda_J^{\rho\sigma y} \Lambda_{J+1}^{\rho\sigma y})$$

$$- i\, \mathrm{sh}\, g_{\rho\sigma} (\Lambda_J^{\rho\sigma x} \Lambda_{J+1}^{\rho\sigma y} - \Lambda_J^{\rho\sigma y} \Lambda_{J+1}^{\rho\sigma x})] + \sum_\rho a_\rho \Lambda_J^{\rho d} \}. \tag{37}$$

As mentioned before, in this case the number n_ρ of bonds of each color cannot change in going from one row of vertical bonds to another. Therefore the counting operators $\sum_J \Lambda_J^{\rho d} = n_\rho$ are conserved. The conjugate variables are the "magnetic fields" a_ρ of the spin chain and the field weight factors X_ρ of the vertex model. The signs ε_ρ play a significant role. For real η, $a_\rho = 0$ and all $\varepsilon_\rho = +1$, one can show by simple variational arguments that the ground state energy per spin of the hamiltonian (37) is

$$E_0/N = - \mathrm{ch}\eta , \tag{38}$$

for arbitrary $n_1, n_2, \cdots n_q$, the trial state being given by $111\cdots122\cdots23\cdots qq\cdots q$. With applied field we have

$$E_0/N = - \mathrm{ch}\eta - a_i , \tag{39}$$

where $a_i = \max a_\rho$ and the ground state is $iii\cdots i$. If $\varepsilon_\rho = +1$, $(\rho=1,\cdots,\ell)$, $\varepsilon_\rho = -1$, $(\varepsilon_\rho = \ell+1,\cdots,q)$, the same arguments apply as long as $a_{\ell+1} = \cdots = a_q = 0$. In the limit $\eta \to \infty$ and with $\varepsilon_\rho = -1$, $a_\rho = 0$, the absolute ground state is $123\cdots q\ 123\cdots q \cdots 123\cdots$, and it is straightforward to do perturbation expansion about this point, giving

$$E_0/N = \frac{2}{q} -1 -2\, e^{-2\eta} -2e^{-4\eta} + a_6 e^{-6\eta} + \cdots , \tag{40}$$

with $a_6 = -2$, $(q>3)$, $a_6 = 2$, $(q=3)$. For $\eta = 0$, $\varepsilon_\rho \equiv -1$, $g_{\rho\sigma} \equiv 0$ the groundstate energy has been determined by Sutherland[45]. Similar arguments can be given for the vertex models.

Following the method of Stroganov[24,25] we can derive an inversion relation for the maximal eigenvalue of the transfermatrix, which upon iteration gives the partition function per site in the thermodynamic limit. For the family (27), (28) we find the functional recursion relation

$$z(s) z(\phi(s)) = \phi(s), \ \phi(s) = - \frac{1}{s} - K - \frac{1}{K} , \tag{41}$$

giving

$$z = \lim_{M,N\to\infty} Z^{1/MN} = \frac{K^{3/2}\, x^{1/2}}{K-x}\, R(x)R(1/x),$$

$$x \equiv \frac{Ks+1}{s+K},$$

$$R(x) \equiv \prod_{p=1}^{\infty} \frac{1-x/K^{4p-1}}{1-x/K^{4p+1}}, \qquad (42)$$

in agreement with previous results for subcases n=0, q[24,35], which were calculated with normalization $W^{d}_{\rho\rho} \equiv 1$. Note that the recursion relation (41) is also satisfied by solutions of the form z(s)p(s), where p(s) is an arbitrary function[64] satisfying $p(s)p(\phi(s)) = 1$. This ambiguity, however, is almost fully suppressed by taking into account the reflection (rotation) symmetry of the lattice, which gives z(1/s) = z(s)/s following from the interchange of W^{r}'s and W^{ℓ}'s. In cases (25),(26), with $g_{\rho\sigma} \equiv 0, x_{\rho} \equiv 1$, we have

$$z(u)z(-u) = \mathrm{sh}(\eta+u)\,\mathrm{sh}(\eta-u)/\mathrm{sh}^2\eta. \qquad (43)$$

For q=2 this determines together with the discrete lattice symmetry z(u) = z(η-u) the free energy.

ACKNOWLEDGEMENT

The authors are indebted to Professor B.M.McCoy for helpful discussions and to Professor R. J. Baxter for valuable remarks. This work is supported in part by the National Science Foundation under Grants No. DMR-79-08556 and No. 79-06376A01.

REFERENCES

1. L. Onsager, Phys. Rev. 65 (1944) 117; B. Kaufman, Phys. Rev. 76 (1949) 1232; C. N. Yang, Phys. Rev. 85(1952) 808.
2. E. Lieb, T. Schultz, and D. Mattis, Ann. Phys. 16 (1961) 407; S. Katsura, Phys. Rev. 127 (1962) 1508; Y. Nambu, Progr. Theor. Phys. 5 (1950) 1.
3. M. E. Fisher, Rep. Progr. Phys. 30 (1967) 615; P. Heller, ibid. p. 731; Various contributions in "Phase Transitions and Critical Phenomena", C. Domb and M. S. Green, eds., (Academic Press, London-New York 1972ff.).
4. L. J. de Jongh and A. R. Miedema, Adv. Phys. 23 (1974) 1; M. Steiner, J. Villain, and C. G. Windsor, Adv. Phys. 25 (1976) 87.
5. B. M. McCoy and T. T. Wu, in "Bifurcation Phenomena in Mathematical Physics and Related Topics", C. Bardos and D. Bessis, eds., (D. Reidel Publ. Comp, Dordrecht, 1980), p. 69.
6. A. M. Polyakov, Phys. Lett. 72B (1977) 224; A. A. Migdal, Sov. Phys. JETP 42 (1975) 743.
7. B.M. McCoy, Lattice gauge theory in four dimensions: Decoupling points and dual symmetric perturbation theory, this conference; B.M.McCoy and T.T.Wu, to be published.

8. A. B. Zamolodchikov, Sov. Phys. JETP 52 (1980) 325.
9. H. Bethe, Z. Physik 71 (1931) 205.
10. F. Bloch, Z. Physik 61 (1930) 206.
11. C. N. Yang and C. P. Yang, Phys. Rev. 150 (1966) 321,327; 151 (1966) 258.
12. J. des Cloizeaux and M. Gaudin, J. Math. Phys. 7 (1966) 1384.
13. E. H. Lieb, Phys. Rev. 162 (1967) 162; Phys. Rev. Lett. 18 (1967) 692, 1046; 19 (1967) 108; B. Sutherland, ibid. 19 (1967) 103; C. P. Yang, ibid. 19 (1967) 586; B. Sutherland, C. N. Yang, and C. P. Yang, ibid. 19 (1967) 588.
14. B. M. McCoy and T. T. Wu, Nuovo Cimento 56B (1968) 311.
15. B. Sutherland, J. Math. Phys. 11 (1970) 3183; M. Suzuki, Progr. Theor. Phys. 46 (1971) 1337; Phys. Lett. 34A (1971) 94, 338; S. Krinsky, Phys. Lett. 39A (1972) 169.
16. E. H. Lieb and F. Y. Wu, Phase Transitions and Critical Phenomena I, C. Domb and M. S. Green, eds., (Ac. Press, London, 1972), p. 331, in particular the section written by E. Barouch, p. 366.
17. P. W. Kasteleyn, Fundamental Problems in Statistical Mechanics III, E. G. D. Cohen, ed., (North-Holland/American Elsevier, Amsterdam, 1975), p.103.
18. R. J. Baxter, Ann. Phys. 70 (1972) 193; 76 (1973) 1, 25, 48; Philos. Trans. R. Soc. Lond. A289 (1978) 315.
19. R. J. Baxter, Ann. Phys. 70 (1972) 323.
20. R. J. Baxter, J. Phys. A13 (1980) L61.
21. C. N. Yang, Phys. Rev. Lett. 19 (1967) 1312; Phys. Rev. 168 (1968) 1920.
22. E. H. Lieb and W. Liniger, Phys. Rev. 130 (1963) 1605; E. H. Lieb, Phys. Rev. 130 (1963) 1616; J. B. McGuire, J. Math. Phys. 6 (1965) 432; 7 (1966) 123; M. Gaudin, Phys. Lett. 24A (1967) 55; C. N. Yang and C. P. Yang, J. Math. Phys. 10 (1969) 1115.
23. A. B. and Al. B. Zamolodchikov, Ann. Phys. 120 (1979) 253; M. Karowski, Phys. Reports 49C (1979) 229; M. Karowski, in "Field Theoretical Methods in Particle Physics", W. Rühl, ed., (Plenum Press, New York-London, 1980), p. 307; J. B. McGuire, J. Math. Phys. 5 (1964) 622; E. Brezin and J. Zinn-Justin, C. R. Acad. Sc. Paris 263 (1966) B670; F. A. Berezin and V. N. Sushko, Sov. Phys. JETP 21 (1965) 865.
24. Yu. G. Stroganov, Phys. Lett. 74A (1979) 116.
25. R. J. Baxter, Fundamental Problems in Statistical Mechanics V, E. G. D. Cohen, ed., (North-Holland Publ. Co., Amsterdam, 1980), p. 109.
26. L. Castillejo, R. H. Dalitz, and F. J. Dyson, Phys. Rev. 101 (1956) 453.
27. T. T. Wu, B. M. McCoy, C. A. Tracy, and E. Barouch, Phys. Rev. B13 (1976) 316; R. Z. Bariev, Physica 83A (1976) 388, 93A (1978) 354; M. Sato, T. Miwa, and M. Jimbo, Proc. Japan Acad. 53A (1977) 147, 153, 183; Publ. RIMS 14 (1978) 223; 15 (1979) 201, 577, 871; 16 (1980) 531; 17 (1981) 137; B. M. McCoy, J. H. H. Perk, and T. T. Wu, Phys. Rev. Lett. 46 (1981) 757; M. Jimbo and T. Miwa, Proc. Japan Acad. 56A (1980) 405; 57A (1981).
28. H. G. Vaidya and C. A. Tracy, Physica 92A (1978) 1; J. H. H. Perk and H. W. Capel, Physica 89A (1977) 265; 92A (1978) 163; 100A (1980) 1; J. H. H. Perk, Phys. Lett. 79A (1980) 1; R. Z. Bariev, Phys. Lett. 68A (1978) 175.
29. H. G. Vaidya and C. A. Tracy, J. Math. Phys. 20 (1979) 2291; M. Jimbo, T. Miwa, Y. Môri, and M. Sato, Physica 1D (1980) 80; M. Girardeau, J. Math. Phys. 1 (1960) 516; T. D. Schultz, J. Math. Phys. 4 (1963) 666; A. Lenard, J. Math. Phys. 5 (1964) 930; 7 (1966) 1268.
30. D. B. Creamer, H. B. Thacker, and D. Wilkinson, Phys. Rev. D23 (1981) 3081; M. Jimbo and T. Miwa, preprint RIMS-370.
31. M. Gaudin, B. M. McCoy, and T. T. Wu, Phys. Rev. D23 (1981) 417.
32. B. Sutherland, Rocky Mountain J. Math. 8 (1978) 413.
33. L. A. Takhtadzhan and L. D. Faddeev, Russian Math. Surveys (Uspekhi Mat. Nauk) 34:5 (1979) 11.

34. H. B. Thacker, Rev. Mod. Phys. 53 (1981) 253.
35. C. L. Schultz, Phys. Rev. Lett. 46 (1981) 629.
36. V. A. Fateev, CERN preprint Ref. TH. 2963.
37. D. V. and G. V. Chudnovsky, Phys. Lett. 79A (1980) 36.
38. B. Sutherland, Phys. Rev. Lett. 20 (1968) 98; C. K. Lai and C. N. Yang, Phys. Rev. A3 (1971) 393.
39. D. J. Gross and A. Neveu, Phys. Rev. D10 (1974) 3235.
40. N. Andrei and J. H. Lowenstein, Phys. Rev. Lett. 43 (1979) 1698; Phys. Lett. 90B (1980) 106, 91B (1980) 401; A. E. Arinstein, Phys. Lett. 95B (1980) 280; A. A. Belavin, Phys. Lett. 87B (1979) 117.
41. W. E. Thirring, Ann. Phys. 3 (1958) 91.
42. A. Luther, Phys. Rev. B14 (1976) 2153.
43. M. Lüscher, Nucl. Phys. B117 (1976) 475.
44. J. D. Johnson, S. Krinsky, and B.M.McCoy, Phys. Rev. A8(1973) 2526.
45. B. Sutherland, Phys. Rev. B12 (1975) 3795.
46. A. B. and Al.B. Zamolodchikov, JETP Lett. 26 (1977) 457; Phys. Lett. 72B (1978) 481; Nucl. Phys. B133(1978) 525; Ann. Phys. 120 (1979) 253; M. Karowski and H. J. Thun, to be published in Nucl. Phys. B[FS].
47. B. Berg, M. Karowski, P. Weisz, and V. Kurak, Nucl. Phys. B134(1978) 125; B. Berg and P. Weisz, Nucl. Phys. B146 (1978) 205; V. Kurak and J.A. Swieca, Phys. Lett. 82B (1979) 289; E. Abdalla, B. Berg, and P. Weisz, Nucl. Phys. B157 (1979) 387; B. Berg and P. Weisz, Commun. Math. Phys. 67 (1979) 241; R. Köberle, V. Kurak,and J.A.Swieca, Phys.Rev.D20(1979)897.
48. A.B.Zamolodchikov. Commun. Math.Phys.69(1979)165; K.Sogo, M.Uchinami, A. Nakamura, and M.Wadati, preprint Univ. of Tokyo (1981). Z(N) invariant models have been treated also: R.Köberle and J.A.Swieca, Phys.Lett.86B (1979)209; A.E.Arinshtein, V.A.Fateyev, and A.B.Zamolodchikov, Phys.Lett. 87B(1979)389; M.Karowski, Springer Lecture Notes in Physics 126(1980)344.
49. R. Shankar and E. Witten, Phys. Rev. D17 (1978)2134; R.Köberle and V. Kurak, São Paulo preprints IFUSP/P-200, P-201.
50. A.B.Zamolodchikov and V.A.Fateev, Sov. J. Nucl. Phys. 32 (1980) 298.
51. D.V. and G.V.Chudnovsky, C.R. Acad. Sc. Paris 291 (1980) A619; Phys. Lett. 81A (1981) 105; Lett. Math. Phys. 5 (1981) 43; Saclay preprint DPh-T/80/131.
52. A.A.Belavin, JETP Lett. 32 (1980) 169; Funct. Anal. Appl. 14 (1980) 260; Nucl. Phys. B180 [FS2] (1981) 189.
53. I. V. Cherednik, Sov. Phys. Doklady 24 (1979) 974.
54. I. V. Cherednik, Theor. Math. Phys. 43 (1980) 356, (note that his X-symmetric solution does not fully agree with ref. 37).
55. D. V. and G. V. Chudnovsky, Phys. Lett. 98B (1981) 83.
56. M. Toda. Phys. Reports 18C (1975) 1; Suppl. Progr. Theor. Phys.59(1976)1.
57. M.A. Olshanetsky and A.M.Perelomov, Phys. Reports 71C (1981) 313.
58. O. Babelon, H. J. deVega,and C. M. Viallet, Preprint Paris LPTHE 81/5, in agreement with ref. 37.
59. D. V. and G. V. Chudnovsky, Saclay preprint DPh-T/80/131.
60. A.G.Izergin and V.E.Korepin. LOMI preprint E-3-80, in agreement with example 7.4 of ref. 59.
61. F.J.Wegner, Physica 68A (1973) 570; A.Gaaff and J. Hijmans, Physica 80A (1975) 149, 97A (1979) 244.
62. J.F.Nagle, Am. J. Phys. 36 (1968) 1114.
63. M. Abramowitz and I.A.Stegun, Handbook of Mathematical Functions, (Dover Publ., New York 1965) Ch. 16.
64. The ambiguity of the periodic factor was not taken into account in the results (37) and (41) of ref. 35.

(Received in August, 1981)

Commun. Math. Phys. 113, 471–503 (1987)

Communications in Mathematical Physics

Integrable Quantum Systems and Classical Lie Algebras

V. V. Bazhanov

Institute for High Energy Physics, Serpukhov, Moscow Region, USSR

Abstract. We have obtained six new infinite series of trigonometric solutions to triangle equations (quantum R-matrices) associated with the nonexceptional simple Lie algebras: $sl(N)$, $sp(N)$, $o(N)$. The R-matrices are given in two equivalent representations: in an additive one (as a sum of poles with matrix coefficients) and in a multiplicative one (as a ratio of entire matrix functions). These R-matrices provide an exact integrability of anisotropic generalizations of $sl(N)$, $sp(N)$, $o(N)$ invariant one-dimensional lattice magnetics and two-dimensional periodic Toda lattices associated with the above algebras.

Table of Contents

1. Introduction

In the theory of two-dimensional integrable systems of quantum field theory and statistical physics a specific importance is attached to the special system of algebraic functional equations called triangle equations (or Yang-Baxter equations)

$$R_{i_1i_2}^{j_1j_2}(\theta)R_{j_1i_3}^{k_1j_3}(\theta+\theta')R_{j_2j_3}^{k_2k_3}(\theta')=R_{i_2i_3}^{j_2j_3}(\theta')R_{i_1j_3}^{j_1k_3}(\theta+\theta')R_{j_1j_2}^{k_1k_2}(\theta). \tag{1.1}$$

Here θ, θ' are complex variables, the indices run over N various values. The summation over the repeated indices is assumed.

Each solution of the triangle Eqs. (1.1) can be treated, on the one hand, as an exact factorized S-matrix in some $1+1$-dimensional field theory [1, 2] and, on the other hand, as a vertex weight matrix of an exactly soluble statistical model on plane lattice [3, 4]. Besides, the triangle equation arises as the consistency condition for the Bethe anzatz solution of quantum-field models [5–7] and those of one-dimensional magnetics [7]. Triangle equations also make a part of the quantum inverse problem technique [8, 9]. Finally, studies of triangle equations resulted in a new mathematical object called quantum groups [10].

Thus, each solution of the triangle equations is associated with quite a number of exactly soluble models from field theory and lattice statistics.

Usually Eq. (1.1) is written in a more compact form using matrix notations. We shall consider $R_{i_1i_2}^{j_1j_2}(\theta)$ as matrix elements of some matrix $R(\theta)$ acting in the tensor product of two vector spaces $\mathbb{C}^N\otimes\mathbb{C}^N$. The matrix $R(\theta)$ is called a (quantum) R-matrix. Introduce the matrices $R_{12}(\theta)$, $R_{13}(\theta)$, $R_{23}(\theta)$, acting in the tensor product of three vector spaces $\mathbb{C}^N\otimes\mathbb{C}^N\otimes\mathbb{C}^N$ according to the rule

$$(R_{12})_{i_1i_2i_3}^{j_1j_2j_3}=R_{i_1i_2}^{j_1j_2}(\theta)\delta_{i_3}^{j_3}$$

$[R_{13}(\theta)$ and $R_{23}(\theta)$ are defined similarly, they act identically in the second and first spaces, respectively]. In new notations, Eqs. (1.1) become

$$R_{12}(\theta)R_{13}(\theta+\theta')R_{23}(\theta')=R_{23}(\theta')R_{13}(\theta+\theta')R_{12}(\theta). \tag{1.1'}$$

The quantum R-matrix is called quasi-classical if it depends on the additional parameter φ (playing the role of the Planck constant) so that for small φ,

$$R(\theta,\varphi)=1+2\varphi r(\theta)+O(\varphi^2). \tag{1.2}$$

Substituting (1.2) into (1.1′), one obtains the classical triangle equations, quite essential for the theory of integrable classical systems [11]:

$$[r_{12}(\theta), r_{13}(\theta+\theta')+r_{23}(\theta')]+[r_{13}(\theta+\theta'), r_{23}(\theta')]=0 \tag{1.3}$$

[,] denoting a commutator. The quantity $r(\theta)$ is called a classical r-matrix.

Note, that Eq. (3) is written only with the help of commutators. Therefore, one may assume that $r\in\mathcal{G}\otimes\mathcal{G}$, $\mathcal{G}$ being a Lie algebra. It becomes clear then that the solutions of (3) can be written in an invariant form, i.e. independent of the representation of $\mathcal{G}$. However, the corresponding solutions to the quantum Eqs. (1.1) depend essentially on the representation of $\mathcal{G}$.

Intensive studies [12–24] of Eqs. (1.1) and (1.3) over the last years led to the discovery of a great number of new integrable models and provided extensive "experimental" material that has clarified essentially the general structure of R-matrices. Let us enumerate here some characteristic properties of R-matrices.

(i) All known R-matrices are meromorphic functions of θ expressed via rational, trigonometric, or elliptic functions only.

(ii) Trigonometric and elliptic R-matrices possess an automorphicity property. In the trigonometric case, this property is described as

$$R(\theta+\pi)=(U\otimes 1)R(\theta)(U\otimes 1)^{-1}=(1\otimes U)^{-1}R(\theta)(1\otimes U), \quad (1.4)$$

U being the matrix in $\mathbb{C}^N$ of finite order g, i.e. $U^g=1$.

(iii) Most R-matrices (excluding R-matrices [18, 19], as well as some others) possess a crossing symmetry

$$R(\theta)=(V\otimes 1)(PR(-\theta-\varrho)P)^{t_1}(V\otimes 1)^{-1}, \quad (1.5)$$

where V is a matrix in $\mathbb{C}^N$, ϱ is some constant, t_1 means taking the transpose in the first space $\mathbb{C}^N$.

(iv) The relation $R(0)\simeq P$ takes place, P being a permutation matrix in $\mathbb{C}^N\otimes\mathbb{C}^N$,

$$P^{i_2 j_2}_{i_1 j_1}=\delta_{i_1 j_2}\delta_{j_1 i_2}, \quad P(x\otimes y)=y\otimes x; \quad x, y\in\mathbb{C}^N. \quad (1.6)$$

It is convenient to choose a normalization in which $R(\theta)$ has a pole at $\theta=0$, i.e.

$$\operatorname{Res} R(\theta)|_{\theta=0}\simeq P. \quad (1.7)$$

It follows from (1.4)–(1.7) that $R(\theta)$ has poles for $\theta=k\pi$, $\theta=-\varrho+k\pi$, $k\in\mathbb{Z}$. Equations (1.7) and (1.1) imply

(v) Unitarity

$$R(\theta)PR(-\theta)P=\Phi(\theta)\cdot E. \quad (1.8)$$

E being a unit matrix in $\mathbb{C}^N\otimes\mathbb{C}^N$, $\Phi(\theta)$ some (scalar) function of θ.

Note, that the properties (i)–(v) alone [i.e. without Eq. (1.1)], are rather hard restrictions on the form of $R(\theta)$. In [22], matrix functions of the form

$$R(\theta)=A+P\operatorname{ctg}\theta-P^{t_1}\operatorname{ctg}(\theta+\varrho), \quad (1.9)$$

where A is independent of θ, satisfying the requirements (i)–(v), were considered. [Relation (1.6) is fulfilled trivially because $U=1$, $g=1$.] The great bulk of the solutions obtained (though not all of them) proved to automatically satisfy the triangle Eq. (1.1). In this way more than 30 new solutions to triangle equations were constructed in [22]. Comparing this observation with the method [20] for proving Eqs. (1.1) for elliptic R-matrices [19] by the Liouville theorem and with the analogous method used in [24] for Eq. (1.3), we found out that the properties (i)–(v) complemented by conditions

(vi) $R(\theta)$ has no poles at $\theta\neq k\pi$, $\theta\neq-\varrho'+k\pi$, $k\in\mathbb{Z}$,

$$R^{\pm}=\lim_{\theta\to\pm i\infty}R(\theta); \quad |R^{\pm}|<\infty, \quad R^{+}_{12}R^{+}_{13}R^{+}_{23}=R^{+}_{23}R^{+}_{13}R^{+}_{12} \quad (1.10)$$

result in the fulfillment of (1.1). For the proof see [25] and Sect. 4 of this paper.

The concept of automorphicity (ii) for classical and quantum R-matrices was introduced in [19] where its connection with automorphisms of the Lie algebra was also established for the case $\mathcal{G}=sl(N)$. In [24] a rather complete classification of nondegenerate[1] solutions (1.3) for all simple Lie algebras was made using this idea. In particular, all elliptic and trigonometric r-matrices were found. Elliptic r-matrices turned out to be connected with the $sl(N)$ algebra and be exhausted by those constructed in [19] where the corresponding quantum R-matrix (in the fundamental representation) was also found.

[1] An r-matrix is called nondegenerate provided $\det\|r_{\mu\nu}\|\neq 0$, where $r=r_{\mu\nu}E^{\mu}\otimes E^{\nu}$, E^{μ} is the basis in $\mathcal{G}$

As for the trigonometric r-matrices, constructed in [24], which are not degenerate elliptic ones, their corresponding quantum R-matrices were known only in a few particular cases [15, 17].

In the present paper, six new infinite series of trigonometric quantum R-matrices corresponding to the fundamental representations of the nonexceptional simple Lie algebras $sl(N)$, $sp(N)$, $o(N)$ have been constructed. The technique we applied consisted of constructing R-matrices obeying the above-mentioned properties (i)–(vi) and possessing the quasi-classical limit (1.2) with the classical r-matrices found in [24] [2]. The R-matrices constructed are given in two equivalent representations: in an additive one (as a sum of poles with matrix coefficients) and in a multiplicative one (as a ratio of entire matrix functions). From the point of view of the above properties (i)–(iv) these representations complement each other: the additive representation possesses the explicit crossing-symmetry while the multiplicative one possesses the explicit unitarity. The main results of the paper were published briefly in [25–27]. Note that our methods were also generalized to the case of R-matrices connected with Lie superalgebras [28, 29].

An alternative approach to the construction of quasiclassical trigonometric R-matrices were introduced in [30]. The point of this approach is an interwinding relation for the quantum L-operators for generalized periodic Toda lattices. In this way the connection between quasiclassical trigonometric R-matrices and quantum Kac-Moody algebras [10, 30] is established.

We shall proceed as follows. In Sect. 2 some necessary information from the theory of Lie algebras is presented. In Sect. 3, trigonometric classical r-matrices of [24] are considered. In Sect. 4 the corresponding quantum R-matrices are constructed. Section 5 contains a discussion of the integrable quantum systems associated with these R-matrices: 1) two-dimensional generalized Toda lattices; 2) anisotropic generalizations of the $sl(N)$, $sp(N)$, and $o(N)$ invariant models of one-dimensional magnetics. In Sect. 6, factorized representations for trigonometric R-matrices are obtained. A representation of this type was first obtained in [31] for an elliptic R-matrix [19]. It is expressed with the help of the ordered exponent

$$R(\theta)=P\exp\left\{\tfrac{1}{2}\int\limits_{\theta-2\varphi}^{\theta+2\varphi} r(s)ds\right\}, \tag{1.12}$$

where $r(\theta)$ is the corresponding classical r-matrix, normalized at $\theta=0$ by the condition

$$r(\theta)=\frac{P-1}{\theta}+\text{const}+O(\theta); \tag{1.13}$$

P is the permutation operator (1.6). As it was recognized in [32], Eq. (1.12) may be rewritten in the form

$$R(\theta)=\sigma(\theta-2\varphi)\sigma^{-1}(\theta+2\varphi), \tag{1.14}$$

where the function

$$\sigma(\theta)=P\exp\left\{\tfrac{1}{2}\int\limits_{\theta}^{\theta_0} r(s)ds\right\}\sigma_0 \tag{1.15}$$

[2] This technique, however, is not applicable to the R-matrices for the non-fundamental representations of $\mathscr{G}$, because these R-matrices do not possess the property (vi)

can be interpreted as a matrix generalization of the Weierstrasse elliptic σ-function [$\frac{1}{2}r(s)$ is treated, accordingly, as a matrix generalization of the elliptic ζ-function].

We make use of this analogy to introduce the matrix generalizations of the trigonometric functions, corresponding to the obtained trigonometric R-matrices.

2. Some Information from the Theory of Simple Lie Algebras

This section presents briefly some necessary information from the theory of simple Lie algebras. For details and proof see [24].

Let $\mathcal{G}=\mathcal{G}(N)$ be a Lie algebra of N by N matrices. In this paper we restrict ourselves to the consideration of the classical matrix Lie algebras $sl(N)$, $sp(N)$, $o(N)$ in the fundamental representations:

$$\begin{aligned} sl(n)&\stackrel{\text{def}}{=}\{X\in \mathrm{Mat}(n,\mathbb{C})|\mathrm{Sp}\,X=0\}\,,\\ sp(2n)&\stackrel{\text{def}}{=}\{X\in \mathrm{Mat}(2n,\mathbb{C})|X^t=-\tilde{S}X\tilde{S}^{-1}\}\,,\\ o(n)&\stackrel{\text{def}}{=}\{X\in \mathrm{Mat}(n,\mathbb{C})|X^t=-SXS^{-1}\}\,, \end{aligned} \tag{2.1}$$

where t denotes taking the transpose, while the n by n matrix S and the $2n$ by $2n$ matrix $\tilde{S}$ are of the form

$$S=\begin{bmatrix} 0 & & & 1\\ & & \cdot^{\cdot^{\cdot}} & \\ & & 1 & \\ & 1 & & \\ 1 & & & 0 \end{bmatrix};\quad \tilde{S}=\left[\begin{array}{c|c} 0 & S\\ \hline -S & 0 \end{array}\right] \tag{2.2}$$

Let $\{E_i\}$ be the basis of generators in $\mathcal{G}(N)$. The scalar product in $\mathcal{G}$ is specified by the Killing form [note, that the definitions (2.1) entail $\mathrm{Sp}\,X=0$]:

$$(X,Y)=\mathrm{Sp}(XY);\qquad X,Y\in\mathcal{G}.$$

The Killing form has the invariance property

$$(X,[Y,Z])=([X,Y],Z), \tag{2.3}$$

[,] denoting a commutator. The tensor of the Killing form is

$$g_{ij}=(E_i,E_j)=\mathrm{Sp}(E_iE_j). \tag{2.4}$$

In order to describe the trigonometric solutions of Eqs. (1.3), one has to use the concept of the Coxeter automorphism of algebra $\mathcal{G}$ (see, e.g., [24]).

An automorphism of a Lie algebra is a one-to-one linear transformation that preserves the commutation operation. An inner automorphism is a product of a finite number of automorphisms of the form $e^{\mathrm{ad}\,x}$, $x\in\mathcal{G}$, with $\mathrm{ad}\,x$ being a linear operator acting by the rule

$$\mathrm{ad}\,x(y)=[x,y];\qquad x,y\in\mathcal{G}. \tag{2.5}$$

Let $\mathcal{G}$ be a simple Lie algebra. Any automorphism of $\mathcal{G}$ can be represented as $\varphi_{\mathrm{int}}\cdot\varphi_\tau$, where φ_{int} is an inner automorphism and φ_τ is induced by the automorphism τ of the Dynkin diagram of algebra $\mathcal{G}$.

Let us fix the automorphism τ of the Dynkin diagram. The automorphism $A_c = \varphi_{\text{int}} \cdot \varphi_\tau$ is said to be a Coxeter one provided

1) algebra $\mathcal{G}_0 \stackrel{\text{def}}{=} \{X \in \mathcal{G} | A_c[X] = X\}$ is Abelian,

2) A_c has minimum order among the automorphisms $A' = \varphi'_{\text{int}} \cdot \varphi_\tau$ such that the algebra $\mathcal{G}'_0$ is abelian.

In the following the pair of the algebra and its Coxeter automorphisms will be denoted by the symbol $\tilde{\mathcal{G}} = (\mathcal{G}, A_c)$. For classical Lie algebras $A_{n-1} = sl(n)$, $B_n = o(2n+1)$, $C_n = sp(2n)$, $D_n = o(2n)$ there are seven infinite series of pairs: $\tilde{\mathcal{G}} = A^{(1)}_{n-1}$, $A^{(2)}_{2n}$, $A^{(2)}_{2n-1}$, $B^{(1)}_n$, $C^{(1)}_n$, $D^{(1)}_n$, $D^{(2)}_n$. In these notations, the numbers in parentheses show the order of the automorphism τ of the Dynkin diagram.

The explicit form of the Coxeter automorphisms for the algebras (2.1) under consideration is given in Table 1 borrowed from the paper [24]. The notations: h is the order of an automorphism, $A_c[X]$ is the image of an element $X \in \mathcal{G}$ under the action of an automorphism, $\omega = \exp(2\pi i/h)$, σ_x in the definition of the matrix T for the series, $D^{(2)}_n$ denotes a diagonal 2 by 2 block with the elements $\sigma_{11} = \sigma_{22} = 0$; $\sigma_{12} = \sigma_{21} = 1$.

The eigenvalues of A_c equal ω^j, $j \in \mathbb{Z}$. Therefore, $\mathcal{G}$ may be represented in the form

$$\mathcal{G} = \sum_{j=0}^{h-1} \mathcal{G}_j; \qquad \mathcal{G}_j \stackrel{\text{def}}{=} \{X \in \mathcal{G} | A_c[X] = \omega^j X\}. \tag{2.6}$$

Note that by the definition of a Coxeter automorphism the algebra $\mathcal{G}_0$ is abelian.

Let $r = \dim \mathcal{G}_0$. There exist elements $e_0, \ldots, e_r \in \mathcal{G}_1$; $f_0, \ldots, f_r \in \mathcal{G}_{-1}$; $\bar{h}_0, \ldots, \bar{h}_{r-1} \in \mathcal{G}_0$, such that

1) $e_0, \ldots, e_r$ form the basis in $\mathcal{G}_1$; $f_0, \ldots, f_r$ form one in $\mathcal{G}_{-1}$; $\bar{h}_0, \ldots, \bar{h}_{r-1}$ form the basis in $\mathcal{G}_0$, normalized by the condition

$$(\bar{h}_i, \bar{h}_j) = 2\delta_{ij}; \tag{2.7}$$

2) The following relations are satisfied:

$$[\bar{h}_a, \bar{h}_b] = 0, \tag{2.8}$$

$$[\bar{h}_a, e_i] = \alpha^a_i e_i, \qquad [\bar{h}_a, f_i] = -\alpha^a_i f_i, \tag{2.9}$$

$$[e_i, f_j] = 2\delta_{ij} \frac{(\alpha_i, \bar{h})}{(\alpha_i, \alpha_i)}, \tag{2.10}$$

Table 1. Coxeter automorphisms of the classical Lie algebras

$\tilde{\mathcal{G}}$	$\mathcal{G}(N)$	h	$A[X]$	T
$A^{(1)}_{n-1}$	$sl(n)$	n	TXT^{-1}	$\omega^{(n-1)/2}\,\text{diag}(1, \omega^{-1}, \omega^{-2}, \ldots, \omega^{1-n})$
$A^{(2)}_{2n}$	$sl(2n+1)$	$4n+2$	$-TX^tT^{-1}$	$S\,\text{diag}(1, \xi, \ldots, \xi^{2n})$; $\xi = -\omega$
$A^{(2)}_{2n-1}$	$sl(2n)$	$4n-2$	$-TX^tT^{-1}$	$S\,\text{diag}(1, \xi, \ldots, \xi^{n-2}, \xi^{n-1}, \xi^{n-1}, \xi^n, \ldots, \xi^{2n-2})$; $\xi = -\omega$
$C^{(1)}_n$	$sp(2n)$	$2n$	TXT^{-1}	$\omega^{(2n-1)/2}\,\text{diag}(1, \omega^{-1}, \ldots, \omega^{1-2n})$
$B^{(1)}_n$	$o(2n+1)$	$2n$	TXT^{-1}	$\text{diag}(1, \omega^{-1}, \ldots, \omega^{1-2n}, 1)$
$D^{(1)}_n$	$o(2n)$	$2n-2$	TXT^{-1}	$\text{diag}(1, \omega^{-1}, \ldots, \omega^{2-n}, \omega^{1-n}, \omega^{1-n}, \omega^{-n}, \ldots, \omega^{3-2n}, 1)$
$D^{(2)}_n$	$o(2n)$	$2n$	TXT^{-1}	$\text{diag}(\omega^{-1}, \omega^{-2}, \ldots, \omega^{1-n}, \sigma_x, \omega^{-n-1}, \ldots, \omega^{1-2n})$

with

$$(\alpha_i, \alpha_i) = \sum_{a=1}^{r-1} \alpha_i^a \alpha_i^a, \quad (\alpha_i, \bar{h}) = \sum_{a=1}^{r-1} \alpha_i^a \bar{h}_a. \tag{2.11}$$

The r-dimensional vectors $\alpha_j = (\alpha_j^0, \ldots, \alpha_j^{r-1})$ are called simple weights for the $(\mathcal{G}, A_c)$ pair. The explicit realizations for the generators $\bar{h}_a$, e_i, f_i and for the systems of simple weights are presented in Appendix A.

The matrix

$$t = \lambda \sum g^{ij} E_i \otimes E_j, \tag{2.12}$$

plays in important role in describing the solutions of the triangle equations. Here we choose $\lambda = 1$ for $A_{n-1}^{(1)}$ and $\lambda = 2$ for other series; tensor g^{ij} is inverse to tensor of the Killing form (2.4). It follows from the invariance of the Killing form (2.3) that for any $X \in \mathcal{G}$,

$$[t, X \otimes 1 + 1 \otimes X] = 0. \tag{2.13}$$

As may be easily checked, $(A \otimes A)[t] = t$. Hence, t may be represented as

$$t = \sum_{j=0}^{h-1} t_j; \quad t_j \in \mathcal{G}_j \otimes \mathcal{G}_{-j},$$
$$t_j = h^{-1} \sum_{n=0}^{h-1} \omega^{-nj} (A_c \otimes 1)^n [t], \tag{2.14}$$

h being the order of A_c. Note, (2.7), (2.12), (2.14) entail that

$$t_0 = \frac{\lambda}{2} \sum_a \bar{h}_a \otimes \bar{h}_a. \tag{2.15}$$

3. Classical Triangle Equations

The most complete results for the solutions of the classical triangle Eq. (1.3) were obtained in [24]. Below we present some necessary information from this work.

3.1. Classical r-Matrices

As has been noted in the Introduction, Eq. (1.3) is written in terms of commutators. Therefore, one can assume $r(\theta) \in \mathcal{G}(N) \otimes \mathcal{G}(N)$, $\mathcal{G} = \mathcal{G}(N)$ being a Lie algebra of N by N matrices.

For any Lie algebra, there is a simplest rational solution to Eq. (1.3) of the form [19, 23],

$$r(\theta) = \frac{t}{\theta}. \tag{3.1}$$

For these solutions, Eq. (1.3) reduces to relation (2.13). The trigonometric solutions for Eq. (1.3) are obtained [24] via "averaging" (the term introduced in [33]) the elementary pole (3.1) over the one-dimensional lattice $\theta = k\pi/\mu$, $k \in \mathbb{Z}$, using the Coxeter automorphism A_c of the algebra $\mathcal{G}$

$$r(\theta) = h^{-1} \sum_{j=0}^{h-1} A_c^j[t] \operatorname{ctg}((\theta\mu - k\pi)/h) = it_0 + 2i \sum_{j=0}^{h-1} \frac{\exp(2i\theta\mu j/h) t_j}{\exp(2i\mu\theta) - 1}, \tag{3.2}$$

where t_j are defined in (2.14), μ is a constant. Thus, the trigonometric solutions are determined by the choice of the pair $\tilde{\mathscr{G}}=(\mathscr{G}, A_c)$. Remember, that for classical Lie algebras there are six infinite series of pairs listed in Table 1. Formula (3.2) was obtained in [24] for the generic case, and in [34, 35] for the series $A_{n-1}^{(1)}$, $B_n^{(1)}$, $C_n^{(1)}$, $D_n^{(1)}$. Note that the function $r(\theta)$ is unambiguously determined by the following three properties:

1) $r(\theta)$ is meromorphic in θ with the set of poles $\theta=k\pi/\mu$, $k\in\mathbb{Z}$, and at zero it has the residue t,

$$\operatorname{Res} r(\theta)|_{\theta=0}\simeq t\,. \tag{3.3}$$

2) Automorphicity (quasi-periodicity)

$$r(\theta+\pi)=(A_c\otimes 1)[r(\theta)]=(1\otimes A_c)^{-1}[r(\theta)]\,. \tag{3.4}$$

3) Asymptotic behavior

$$r(\theta)|_{\theta\to\pm i\infty}=\mp it_0\,, \tag{3.5}$$

where t_0 is given in (2.14). With the help of Eqs. (2.14) and (3.2) one can easily derive an additional property of $r(\theta)$:

4) Classical unitarity

$$r(\theta)=-Pr(-\theta)P\,, \tag{3.6}$$

P being the permutation matrix (1.6).

Reference [24] contained a highly simple and elegant proof of the fact that $r(\theta)$ satisfies the classical triangle Eq. (1.3). This proof will be given below. It is based on the abovesaid properties 1)–4) of the function $r(\theta)$. Let us denote by $\psi(\theta,\theta')$ the left-hand side of Eq. (1.3) and consider it as a function of θ, keeping θ' fixed. By virtue of (3.3), (3.4) it is quasi-periodic,

$$\Psi(\theta+\pi g/\mu,\theta')=(A_c\otimes 1\otimes 1)[\Psi(\theta,\theta')]\,, \tag{3.7}$$

and has simple poles at $\theta=k\pi$, $\theta=-\theta'+k\pi$, $k\in\mathbb{Z}$. Using relations (2.13), (3.3), (3.6), (3.7), one can readily exhibit that the residues in these poles vanish. It follows from (3.5) that $\psi(\theta,\theta')$ is finite when $\theta\to\pm i\infty$. Together with the periodicity of $\psi(\theta,\theta')$ in θ (with the period $g\pi/\mu$) this means that $\psi(\theta,\theta')$ is independent of θ. Similarly, we can show $\psi(\theta,\theta')$ to be independent of θ'. Now, tending $\theta\to i\infty$, and then $\theta'\to i\infty$, using (3.5) and remembering that $t_0\in\mathscr{G}_0\otimes\mathscr{G}_0$, where $\mathscr{G}_0$ is an Abelian subalgebra of $\mathscr{G}$, one finds $\psi(\theta,\theta')\equiv 0$.

In our next section we shall construct the quantum R-matrices, corresponding to the classical r-matrices (3.2), for the fundamental representations of the algebras (2.1) in question. In this connection, it is useful to rewrite relation (3.2) taking into account the explicit form of t for the representations (2.1). Choosing some particular basis in $\mathscr{G}$, it is easy to show that[3]

$$\begin{aligned} t_{sl(N)}&=\lambda P\,,\\ t_{sp(N)}&=P-\tilde{S}P^{t_1}\tilde{S}^{-1}\,,\\ t_{o(N)}&=P-SP^{t_1}S^{-1}\,, \end{aligned} \tag{3.8}$$

[3] In the case of $sl(N)$ algebra we drop in (3.8) a term, proportional to unity in $\mathbb{C}^N\otimes\mathbb{C}^N$, which does not affect Eq. (1.3)

where λ was defined after Eq. (2.12), P is the permutation matrix (1.6), t_1 means taking the transpose in the first space $\mathbb{C}^N$. Here and after we use boldface letters to denote products of the form $A\otimes 1$, e.g.

$$\mathbf{S}=S\otimes 1\,,\quad \tilde{\mathbf{S}}=\tilde{S}\otimes 1\,,\quad \mathbf{U}=U\otimes 1\,, \tag{3.9}$$

where $\tilde{S}$, S were defined in (2.2).

Let U be a matrix in $\mathbb{C}^N$ of a finite order g, $U^g=1$. Introduce the function

$$Z(\theta)=Z(\theta,P)=g^{-1}\sum_{k=0}^{g-1}\mathbf{U}^k P\mathbf{U}^{-k}\operatorname{ctg}((\theta-k\pi)/g)\,. \tag{3.10}$$

It is easy to check that $Z(\theta)$ obeys the relations

$$\operatorname{Res}Z(\theta)|_{\theta=0}=P\,, \tag{3.11}$$

$$Z(\theta+\pi)=\mathbf{U}Z(\theta)\mathbf{U}^{-1}\,,\qquad Z(-\theta)=-PZ(\theta)P\,. \tag{3.12}$$

Let us now choose in Eq. (3.2) $\mu=2$ for the series $A^{(2)}_{2n}$, $A^{(2)}_{2n-1}$ and $\mu=1$ for the other series. Using (3.8)–(3.12) and the expression for $A[X]$ from Table 1, we can represent the function $r(\theta)$ in the form

$$r(\theta)=Z(\theta)\,,\qquad \mathscr{G}=A^{(1)}_{n-1}\,, \tag{3.13}$$

$$r(\theta)=Z(\theta)-(B\otimes 1)Z^{t_1}(\theta-\Delta)(B\otimes 1)^{-1}, \mathscr{G}\neq A^{(1)}_{n-1}\,. \tag{3.14}$$

The values of g and the matrices U and B for each series are presented in Table 2, t_1 denotes taking the transpose in the first space $\mathbb{C}^N$

$$\Delta=\begin{cases}0\,, & \mathscr{G}=B^{(1)}_n, C^{(1)}_n, D^{(1)}_n, D^{(2)}_n\,,\\ \pi/2 & \mathscr{G}=A^{(2)}_{2n}, A^{(2)}_{2n-1}\,.\end{cases} \tag{3.15}$$

Note, that the matrix U possesses the property

$$BUB^{-1}=U^{-1}\,. \tag{3.16}$$

Let us give one more useful formula resulting from (2.14) and (3.8):

$$t_G=\begin{cases}\tilde{P}\,, & \mathscr{G}=A^{(1)}_{n-1}\\ \tilde{P}-\tilde{K}\,, & \mathscr{G}\neq A^{(1)}_{n-1}\end{cases}\,, \tag{3.17}$$

Table 2. The parameters of the solutions of triangle equations. The quantities h, T for each series are given in Table 1. S, $\tilde{S}$ are defined by Eqs. (2.2)

Ser.	g	B	U	ϱ	$\{c_a\}$, $a=0,\ldots,n-1$
$A^{(1)}_{n-1}$	h	-	T	--	-
$A^{(2)}_{2n}$	$h/2$	T	$T(T^t)^{-1}$	$N\varphi-\frac{\pi}{2}$	$c_a=-1$
$A^{(2)}_{2n-1}$	$h/2$	T	$T(T^t)^{-1}$	$N\varphi-\frac{\pi}{2}$	$c_a=2(a-n+1)/(2n-1)$
$C^{(1)}_n$	h	$\tilde{S}$	T	$(N+2)\varphi$	$c_a=(2a-n+1)/n$
$B^{(1)}_n$	h	S	T	$(N-2)\varphi$	$c_a=-a/n$
$D^{(1)}_n$	h	S	T	$(N-2)\varphi$	$c_a=0$
$D^{(2)}_n$	h	S	T	$(N-2)\varphi$	$c_{n-1}=0$; $c_a=-(2a-n+2)/n$, $a<n-1$

where

$$\tilde{P}=g^{-1}\sum_{k=0}^{g-1}\mathbf{U}^kP\mathbf{U}^{-k}, \tag{3.18}$$

$$\tilde{K}=g^{-1}\sum_{k=0}^{g-1}\mathbf{U}^kK\mathbf{U}^{-k}, \tag{3.19}$$

$$K=\mathbf{B}P^{t_1}\mathbf{B}^{-1}, \quad K^2=NK, \tag{3.20}$$

$$(P^{t_1})_{i_1i_2}^{j_1j_2}=\delta_{i_1i_2}\delta_{j_1j_2}. \tag{3.21}$$

Here N is the matrix dimension of $\mathscr{G}$.

3.2. *Classical Toda Lattices*

In this subsection we consider a class of integrable classical systems related to the r-matrices (3.2), called by the two-dimensional generalized periodic Toda lattices [36, 37].

Consider the pair $\tilde{\mathscr{G}}=(\mathscr{G},A_c)$ where $\mathscr{G}$ is a simple Lie algebra and A_c is its Coxeter automorphism.

The system with the Lagrangian

$$\mathscr{L}=\gamma^{-1}\int dx[\tfrac{1}{2}(\partial_\mu u)^2-V(u)],$$

$$V(u)=m^2\sum_{i=0}^{r}\frac{2}{(\alpha_i,\alpha_i)}e^{2(\alpha_i,u)}, \tag{3.22}$$

$$(\alpha_i,u)=\sum_{a=0}^{r-1}\alpha_i^a u_a,$$

is called a two-dimensional Toda lattice associated with $\tilde{\mathscr{G}}$. Here $u_a=u_a(x,t)$, $a=0,\dots,r-1$, is a set of scalar fields in two-dimensional space-time, $r=\dim\mathscr{G}_0$, $\{\alpha_j\}$ is a system of simple weights $\tilde{\mathscr{G}}$ [see Eqs. (2.7)–(2.11)]. The equations of motion

$$(\partial_t^2-\partial_x^2)u_a=-\frac{\partial}{\partial u_a}V(u) \tag{3.23}$$

can be represented in Zakharov-Shabat's form

$$\partial_t L-\partial_x M+[L,M]=0, \tag{3.24}$$

$$L(\theta)=\partial_t u+\lambda m e^{\mathrm{ad}u}I_+ +\lambda^{-1}m e^{-\mathrm{ad}u}I_-, \tag{3.25}$$

$$M(\theta)=\partial_x u-\lambda m e^{\mathrm{ad}u}I_+ +\lambda^{-1}m e^{-\mathrm{ad}u}I_-, \tag{3.26}$$

$$u=\sum_{a=0}^{r-1}u^a\tilde{h}_a, \quad \lambda=\exp(-2i\theta/g), \tag{3.27}$$

$$I_+=\sum_{i=0}^{r}e_i, \quad I_-=\sum_{i=0}^{r}f_i. \tag{3.28}$$

The operator ad$u$ was defined by (2.5), the values of g can be found in Table 2.

The canonical Poisson bracket has the form

$$\{\pi_a(x),u_b(y)\}=\delta(x-y)\delta_{ab}, \tag{3.29}$$

with $\pi=\gamma^{-1}\partial_t u$.

The most efficient approach to integrable systems with an ultralocal (i.e. having no derivatives of δ-function) Poisson bracket is the r-matrix technique [11]. It is based on the fact that a Poisson bracket of two L-operators can be written in the so-called r-matrix form:

$$\{L(\theta,x)\otimes L(\theta,x')\}=i\gamma\delta(x-x')[L(\theta)\otimes 1+1\otimes L(\theta'),r(\theta'-\theta)], \tag{3.30}$$

where $r(\theta)$ is a classical r-matrix.

For the L-operator (3.25) and r-matrix (3.13), (3.14) this relation was proved in [24]. We will not present the proof here since relation (3.30) can be treated as a quasi-classical limit of the relation (5.4) for the corresponding quantum L-operator, discussed in Sect. 5.

4. Quantum Triangle Equations

In the present section we shall construct quantum R-matrices, corresponding to the classical r-matrices (3.14), for the series $A_{2n}^{(2)}$, $A_{2n-1}^{(2)}$, $B_n^{(1)}$, $C_n^{(1)}$, $D_n^{(1)}$, $B_n^{(2)}$. The quantum R-matrix for the series $A_n^{(1)}$ was already known; it will be given in the end of this section. As has been pointed out in the Introduction, quantum R-matrices depend essentially on the representation of algebra $\mathcal{G}$. In the present paper we limit ourselves to the consideration of fundamental representations of $\mathcal{G}$, defined in (2.1).

4.1. A Simple Theorem

The basic idea of our approach is to employ the following theorem [25].

Let $R(\theta)$ be a meromorphic function of θ with the following properties

(i) Automorphicity (quasi-periodicity) and invariance:

$$R(\theta+\pi)=(U\otimes 1)R(\theta)(U\otimes 1)^{-1}=(1\otimes U)^{-1}R(\theta)(1\otimes U), \tag{4.1}$$

where U is the matrix in $\mathbb{C}^N$ of finite order g, i.e. $U^g=1$. It follows from (4.1) that $R(\theta)$ is periodic with the period $g\pi$.

(ii) Crossing symmetry:

$$R(\theta)=(V\otimes 1)(PR(-\theta-\varrho)P)^{t_1}(V\otimes 1)^{-1}. \tag{4.2}$$

Here ϱ is a constant, $\varrho\neq 0$; V is a matrix in $\mathbb{C}^N$, t_1 denotes taking the transpose with respect to the first space $\mathbb{C}^N$, P is the permutation matrix (1.6).

(iii) Unitarity:

$$R(\theta)PR(-\theta)P=E\Phi(\theta), \tag{4.3}$$

with E being a unit matrix in $\mathbb{C}^N\otimes\mathbb{C}^N$, while $\Phi(\theta)$ is some scalar function of θ.

(iv) Asymptotic behavior:

$$R(\theta) \text{ is finite when } \theta\to\pm i\infty, \tag{4.4}$$

$$R_{12}^+R_{13}^+R_{23}^+=R_{23}^+R_{13}^+R_{12}^+;\quad R^+=\lim_{\theta\to\pm i\infty}R(\theta). \tag{4.5}$$

(v) Pole structure: $R(\theta)$ has simple poles, which are located at $\theta=k\pi$, $\theta=-\varrho+k\pi$, $k\in\mathbb{Z}$, only. The residue of $R(\theta)$ for $\theta=0$ is proportional to P:

$$\operatorname{Res}R(\theta)|_{\theta=0}\simeq P. \tag{4.6}$$

The remaining residues are fixed by the properties (i)–(ii).

Theorem. *It follows from* (i)–(v) *that* $R(\theta)$ *satisfies Eq.* (1.1).

Proof. Let us denote by $\psi(\theta,\theta')$ the difference between the right- and left-hand sides of Eq. (1.1) and consider $\psi(\theta,\theta')$ as a function of θ keeping θ' fixed. It follows from (i), (v), that $\psi(\theta,\theta')$, similarly to $R(\theta)$, possesses the automorphicity property, is periodic with the period $g\pi$ and has poles at $\theta=k\pi$, $\theta=-\varrho+k\pi$, $\theta=-\theta'+k\pi$, $\theta=-\theta'-\varrho+k\pi$, $k\in\mathbb{Z}$. Using (ii), (iii), and (v), one can show that the residues at these poles vanish. (Note that due to automorphicity it is sufficient to consider only four poles.) Next, Eq. (4.4) entails that $\psi(\theta,\theta')$ is finite for $\theta\to\pm i\infty$. This means [provided periodicity of $\psi(\theta,\theta')$ is taken into account] that $\psi(\theta,\theta')$ is a constant, i.e. independent of θ. Similarly $\psi(\theta,\theta')$ may be proved to be independent of θ'. Letting now $\theta\to i\infty$ and then $\theta'\to i\infty$ and using Eq. (4.5), we obtain $\Psi(\theta,\theta')\equiv 0$. Q.E.D.

Remark 1. It is not difficult to obtain a generalization of the above theorem to the elliptic case. For R-matrices with one series of poles [19] [i.e. having no property (ii)] the theorem was used in [20].

Remark 2. The property of R-matrix to be quasi-classical has not been used in the proof. Therefore, the theorem can be applied to nonquasi-classical R-matrices as well, in particular to those of [22] (the corresponding calculations were performed by O. Vasiliev).

It would also be interesting to investigate new R-matrices, found in [38], from the point of view of the above theorem.

4.2. Quantization of Trigonometric r-Matrices

Below we construct a quantum R-matrix $R(\theta,\varphi)$ which obeys the conditions of the theorem of the previous subsection and has the quasi-classical limit (1.2) with the classical r-matrix (3.14).

An analysis of known R-matrices for the cases $A_2^{(2)}$ [17] and $C_1^{(1)}$ [15] shows that they may be represented in the form

$$R(\theta,\varphi)=R_0(\varphi)+\sin 2\varphi\left[Z(\theta)-\mathbf{C}(\varphi)\mathbf{B}Z^{t_1}(\theta+\varrho)(\mathbf{C}(\varphi)\mathbf{B})^{-1}\right], \tag{4.7}$$

$$R_0(\varphi)=1+(\cos 2\varphi-1)t_0^2, \tag{4.8}$$

$$\mathbf{C}(\varphi)=C(\varphi)\otimes 1, \quad \mathbf{B}=B\otimes 1, \quad \mathbf{U}=U\otimes 1, \tag{4.9}$$

where t_0 is given in (3.17), the other notations are defined by Eqs. (2.14), (3.10), (3.14) and Table 1, $C(\varphi)$ is a diagonal matrix in $\mathbb{C}^N$, such that

$$[C(\varphi),U]=[C(\varphi),\tilde{K}]=0, \tag{4.10}$$

$$C(\varphi)B=BC^{-1}(\varphi), \tag{4.11}$$

where $\tilde{K}$ is defined in (3.19)[4]. It is convenient to write $C(\varphi)$ in the form

$$C(\varphi)=\exp(i\varphi(c,\tilde{h})); \qquad (x,y)=\sum_{a=0}^{r-1}x_a y_a, \tag{4.12}$$

[4] Condition (4.11) eliminates gauge freedom in choosing $C(\varphi)$ connected with the transformations $R\to(D\otimes D)R(D\otimes D)$, where D is a diagonal matrix commuting with the matrix B

where $\{\bar{h}_a\}$ is the basis in $\mathscr{G}_0$, $r=\dim\mathscr{G}_0$. Note that in our realization of $\mathscr{G}$,

$$\bar{h}_a=e_{a,a}-e_{N-a-1,N-a-1}; \qquad (e_{ij})_{\alpha\beta}=\delta_{i\alpha}\delta_{j\beta}, \tag{4.13}$$

N being the matrix dimension of $\mathscr{G}$ (see Sect. 2 and Appendix A).

It seems very reasonable to use Ansatz (4.7)–(4.12) in a general case. Indeed, with the help of relations (3.11), (3.17), (4.8) it is easy to show that $R(\theta,\varphi)$, so defined, satisfies conditions (i), (ii), (v) of the previous subsection. Note that in (4.2) $V=C(\varphi)B$. Next, it follows from the definition (2.14) and the explicit form of Coxeter automorphisms (Table 1), that t_0 is a diagonal matrix and

$$t_0^{2+k}=t_0^k; \qquad k>1. \tag{4.14}$$

Exploiting (3.10), (3.17), (4.7)–(4.9), (4.14) we get

$$R(\theta,\varphi)|_{\theta\to\pm i\infty}=R^\pm+O(e^{\pm 2i\theta/g}), \qquad R^\pm=e^{\mp 2i\varphi t_0}. \tag{4.15}$$

Since $t_0\in\mathscr{G}_0\otimes\mathscr{G}_0$, $\mathscr{G}_0$ being an abelian algebra, relation (4.5) is satisfied trivially.

Thus, conditions (i), (ii), (iv), (v) are checked and it is left to check the unitarity condition (4.3). The substitution of (4.7) into (4.3) leads to a system of equations for $C(\varphi)$ and ϱ, which allows us to define $C(\varphi)$ and ϱ unambiguously provided (4.11) is taken into account. The required calculations are simple, but tedious. The result is given in the author's work [25] [5] and we reproduce it in Table 2. Fortunately there exists a more simple method to prove unitarity for R-matrix (4.7), using the quantum L-operator for generalized periodic Toda lattices. This proof will be given in Sect. 5. The most essential point of the proof is that ϱ and $C(\varphi)$ from Table 2 are determined unambiguously by the system of equations

$$-(\alpha_j,\alpha_j)+(\alpha_j,c)+\frac{2(\varrho+\Delta\varphi)}{g\varphi}=0; \qquad j=0,1,\dots,r, \tag{4.16}$$

where α_j^a, $j=0,\dots,r$, is the system of simple weights of $(\mathscr{G},A_c)$ (see, Sect. 2), Δ is defined by (3.15).

Assuming now that the unitarity condition is proved, let us calculate the function $\Phi(\theta)$ from the left-hand side of Eq. (4.3). Denote by $\hat{\Phi}(\theta)$ the left-hand side of (4.3). $\hat{\Phi}(\theta)$ is a meromorphic function of θ and may have the poles at $\theta=k\pi$, $\theta=\pm\varrho+k\pi$, $k\in\mathbb{Z}$. By means of (4.1), (4.6), (4.15) it is easy to show that

$$\text{a)}\ \hat{\Phi}(\theta+\pi)=(U\otimes 1)\hat{\Phi}(\theta)(U\otimes 1)^{-1}, \qquad \lim_{\theta\to\pm i\infty}\hat{\Phi}(\theta)=E, \tag{4.17}$$

where E is a unit matrix in $\mathbb{C}^N\otimes\mathbb{C}^N$.

b) When $\theta=k\pi$, $k\in\mathbb{Z}$, $\hat{\Phi}(\theta)$ has the second order poles

$$\hat{\Phi}(\theta)=-\frac{\sin^2 2\varphi\cdot E}{(\theta-k\pi)^2}+O(1)+O((\theta-k\pi)). \tag{4.18}$$

Next, at $\theta=-\varrho+k\pi$, $k\in\mathbb{Z}$, the residues of $R(\theta)$ are degenerate matrices

$$R_k=\operatorname{Res}R(\theta)|_{\theta=-\varrho+k\pi}=\mathrm{U}^k\mathrm{C}(\varphi)K\mathrm{C}^{-1}(\varphi)\mathrm{U}^{-k}, \tag{4.19}$$

[5] In the present paper we use slightly modified (but, of course, equivalent) definitions of Coxeter automorphisms (cf. Table 1 and Table 2 form [25] and present paper)

where the notations (3.20) and (4.9) are used. Since $\det K=0$,

$$\det R_k=0 \tag{4.20}$$

and consequently,

$$\det(\mathrm{Res}\,\hat{\Phi}(\theta)|_{\theta=\mp\varrho+k\pi})=0. \tag{4.21}$$

The only way to make (4.21) agree with the matrix structure of the left-hand side of (4.3) is to require

$$\text{c) } \mathrm{Res}\,\hat{\Phi}(\theta)|_{\theta=\mp\varrho+k\pi}=0. \tag{4.22}$$

Now, using the Liouville theorem, we can recover the function $\Phi(\theta)$ unambiguously by the properties (a)–(c) given above,

$$\hat{\Phi}(\theta)=R(\theta)PR(-\theta)P=\left(1-\frac{\sin 2\varphi}{\sin^2\theta}\right)E. \tag{4.23}$$

Thus, the quantum R-matrix for the series $A_{2n}^{(2)}$, $A_{2n-1}^{(2)}$, $B_n^{(1)}$, $C_n^{(1)}$, $D_n^{(1)}$, $B_n^{(2)}$, defined by (4.7) and by the values of $C(\varphi)$ and ϱ from Table 2, satisfies all the conditions of the theorem from the previous subsection and, hence, satisfy Eq. (1.1) as well. In the quasi-classical limit (1.2), Eq. (4.7) yields the corresponding classical r-matrix (3.14).

For $\tilde{\mathscr{G}}=A_{n-1}^{(1)}$ the quantum R-matrix has been known before [18]. In our notations it looks like

$$R(\theta,\varphi)=R_0+\sin 2\varphi Z(\theta), \tag{4.24}$$

where $Z(\theta)$, R_0 was defined by (3.10), (4.8) respectively and the matrix U entering the definition of $Z(\theta)$ was given in Table 2. Clearly, (4.30) has a correct quasi-classical limit with the classical r-matrix (3.13).

To conclude this section, we note that the R-matrices (4.7), (4.24) have the **PT**-symmetry and invariance properties

$$R^{t_1t_2}(\theta,\varphi)=PR(\theta,\varphi)P, \tag{4.25}$$

$$(X\otimes X)R(\theta)(X\otimes X)^{-1}=R(\theta), \tag{4.26}$$

where t_1t_2 denotes taking the transpose in $\mathbb{C}^N\otimes\mathbb{C}^N$, and X is any diagonal matrix obeying the relation

$$BXB^{-1}=X^{-1}. \tag{4.27}$$

5. Integrable Quantum Systems

5.1. Quantum Toda Lattices

In this section the quantum variant of generalized periodic Toda lattices will be discussed. For these models we construct the quantum L-operators [26, 30, 39] satisfying relation (5.5). Apparently, this means that they can be integrated by the quantum inverse scattering technique [8, 9].

For the series $\tilde{\mathscr{G}}=A_{n-1}^{(1)}$ this problem was solved in [40] (see also [41, 42]). We shall handle the other series from Tables 1, 2.

In the quantum case the Poisson bracket (3.29) is replaced by the commutator

$$[\pi_a(x), u_b(y)] = -i\delta_{ab}\delta(x-y), \tag{5.1}$$

where $\pi_a^{-1} = \gamma\partial_t u_a$, $\gamma = 2\varphi$.

The ultraviolet regularization is achieved by introducing a spatial lattice with a (small) spacing δ. Let us define the variables

$$u_a(n) = \delta^{-1}\int_{x_n}^{n_n+\delta} u(y)dy, \quad p_a(n) = \tfrac{1}{2}\int_{x_n}^{x_n+\delta} \partial_t u(y)dy, \tag{5.2}$$

$$[p_a(n), u_b(n')] = -i\varphi\delta_{ab}\delta_{nn'}. \tag{5.3}$$

To apply the quantum inverse problem method [8, 9], it is necessary to construct a quantum L-operator $\hat{L}(\theta, p, u)$, satisfying the relation

$$L_n^1(\theta)L_n^2(\theta')R^{12}(\theta'-\theta) = R^{12}(\theta'-\theta)L_n^2(\theta')L_n^1(\theta) + O(\delta^2), \tag{5.4}$$

$$L_n(\theta) = \hat{L}(\theta, p(n), u(n)),$$

$$L^1 = L\otimes 1, \quad L^2 = 1\otimes L.$$

Studying the known L-operators for the cases of $A_1^{(1)}$ [8] and $A_2^{(2)}$ [17], we observed that they could be written in the form [26, 30],

$$\hat{L}(\theta, p, u) = e^p(1 - i\delta m[\lambda e^{\mathrm{ad}u}I_+ + \lambda e^{-\mathrm{ad}u}I_-])e^p, \tag{5.5}$$

where $p = (p, \bar{h})$; $u = (u, \bar{h})$ and the rest of the notations were defined in (2.5), (3.27), (3.28). Note, that in the quasi-classical limit we have $\varphi\to 0$, $p\to -i\varphi\delta\pi(x)$ and

$$\hat{L}(\theta, p, u) \simeq 1 + \delta L_c(\theta, \pi, u) + O(\varphi^2), \tag{5.6}$$

where $L_c(\theta, \pi, n)$ is the classical L-operator defined by relation (3.25).

It turns out that Eq. (5.5) works for all the algebras under consideration. We shall show below that the L-operator (5.5) and R-matrix (4.7) obey relation (5.4).

Let us enumerate some features of the L-operator (5.5),

$$\hat{L}(\theta + \pi) = U^{-1}\hat{L}(\theta)U, \tag{5.7}$$

where the explicit form of U is presented in Tables 1 and 2,

$$\hat{L}(\theta)(C(\varphi)B)\hat{L}^t(\theta - \varrho)(C(\varphi)B)^{-1} = 1 + O(\delta^2). \tag{5.8}$$

Here t denotes taking the transpose of L as a matrix in $\mathbb{C}^N$; the matrices B, $C(\varphi)$ and the constant ϱ are presented in Table 2. For $\theta\to\pm i\infty$,

$$\hat{L}(\theta)|_{\theta\to\pm i\infty} = J_\pm e^{\mp 2i\theta/g} + O(1), \tag{5.9}$$

$$J_\pm = im\delta e^p e^{\pm\mathrm{ad}u} I_\pm e^p. \tag{5.10}$$

Equations (5.7) and (5.9) follow directly from the definition (5.5). Equation (5.8) is proved in Appendix B, using a new representation for ϱ and the matrix $C(\varphi)$ in internal terms of the pair $(\mathcal{G}, A_c)$,

$$C(\varphi) = \exp\left(i\varphi\sum_a c_a\bar{h}_a\right), \tag{5.11}$$

where $\bar{h}_a$, $a = 0, \ldots, r-1$, is a basis in $\mathcal{G}_0$, c_a and ϱ are determined by the equations

$$-(\alpha_j, \alpha_j) + (\alpha_j, c) + \frac{2(\varrho + \Delta\varphi)}{g\varphi} = 0; \quad j = 0, 1, \ldots, r, \tag{5.12}$$

where α_j^a, $j=0,\ldots,r$, is the system of simple weights of $(\mathscr{G}, A_c)$ (see Sect. 2), Δ is defined in (3.15). The solution to system (5.12) always exists and is unique.

Now turn to the proof of Eq. (5.4). As in Sect. 3, the main idea will be to apply the Liouville theorem. Consider the difference $\Psi(\theta,\theta')$ between the right-hand side and left-hand side of Eq. (5.4) as a function of θ for fixed θ'. Since $L(\theta)$ is an entire function of θ, then $\Psi(\theta,\theta')$ as well as $R(\theta-\theta')$ has simple poles at $\theta=-\theta'+k\pi$; $\theta=-\theta'-\varrho+k\pi$; $k\in\mathbb{Z}$. As follows from (4.1) and (5.7), $\Psi(\theta,\theta')$ possesses automorphicity:

$$\Psi(\theta+\pi,\theta')=(U\otimes 1)^{-1}\Psi(\theta,\theta')(U\otimes 1), \tag{5.13}$$

and is therefore periodic in θ with the period $g\pi$. The residue of $R(\theta)$ at $\theta=0$ is of the form

$$\operatorname{Res} R(\theta)|_{\theta=0}=\sin 2\varphi\cdot P, \tag{5.14}$$

and at $\theta=-\varrho$ it is defined by (4.19). With the help of (5.8), (5.13), (5.14) one can show that the residues of $\Psi(\theta,\theta')$ are of the order of $O(\delta^2)$. Thus,

$$\Psi(\theta,\theta')=\text{entire function of } \theta+O(\delta^2). \tag{5.15}$$

Consider next the limits $\theta\to\pm i\infty$. Using (4.15) and (5.9), we obtain

$$\Psi(\theta,\theta')|_{\theta\to\pm i\infty}=\Psi_1^\pm(\theta')e^{\mp 2i\theta/g}+\Psi_0^\pm(\theta')+O(e^{\pm 2i\theta/g}), \tag{5.16}$$

$$\Psi_1^\pm(\theta')=[J_\pm\otimes 1, R^\mp(1\otimes e^{2P})]+O(\delta^2), \tag{5.17}$$

$$\Psi_0^\pm(\theta')=(e^{2P}\otimes\hat{L}(\theta'))R^\mp-R^\mp(1\otimes\hat{L}(\theta'))(e^{2P}\otimes 1)+\text{const}\, e^{\mp 2i\theta'/g}+O(\delta^2), \tag{5.18}$$

where $J_\pm$ and $R^\pm$ are defined in (5.10) and (4.15), respectively. Using (2.9), (2.15), (5.1), one can easily exhibit that the commutator in (5.17) vanishes. Hence, up to the terms of the order of $O(\delta^2)$, $\Psi(\theta,\theta')$ is independent of θ:

$$\Psi(\theta,\theta')=\Psi_0^+(\theta')+O(\delta^2)=\Psi_0^-(\theta')+O(\delta^2). \tag{5.19}$$

In the same fashion it is proved that $\Psi(\theta,\theta')$ does not depend on θ', i.e. that $\Psi^+(\theta')$ is a constant. The simplest way to calculate it is to let $\theta'\to-i\infty$ in (5.18). Using (5.12) one obtains

$$\Psi_0^+(\theta')=O(\delta^2). \tag{5.20}$$

It follows from (5.19), (5.20) that $\Psi(\theta,\theta')\equiv O(\delta^2)$, which does prove (5.4).

Note that in the quasi-classical limit (1.2), (5.6) relation (5.4) reduces to Eq. (3.30).

One more important observation is as follows. When proving (5.4), we used 1) automorphicity of the R-matrix (4.7); 2) the expression for the residues (4.19), (5.14) and Eqs. (5.12), which define $C(\varphi)$ and ϱ; 3) the asymptotics (4.15). These properties define the R-matrix unambiguously. Thus, relations (5.4) and (5.5) may be taken as the basis to calculate the R-matrices (4.7)[6]. This program was recently realized in [30, 39].

Using Eq. (5.4) it is not difficult to prove the unitarity of R-matrix (4.7). Applying Eq. (4.5) twice, we have

$$[L^1(\theta)L^2(\theta'), \hat{\Phi}(\theta-\theta')]=O(\delta^2),$$

[6] This possibility was pointed out to the author by V. A. Fateev

where $\hat{\Phi}(\theta)$ denotes left-hand side of Eq. (4.3). Expanding the product of two L-operators in a series in δ, we obtain that $\hat{\Phi}(\theta)$ commutes with any matrix of the form $1\otimes X$ and $X\otimes 1$, $X\in G$. Hence, $\hat{\Phi}(\vartheta)$ is proportional to the unity in $\mathbb{C}^N\otimes\mathbb{C}^N$.

5.2. Integrable Models of Magnetics

The R-matrices (4.7), (4.24) allow us to construct integrable models for magnetics in a standard way. The Hamiltonian is of the form

$$\mathbb{H}=\sum_n H_{n,n+1}, \tag{5.21}$$

$$H_{1,2}=i\partial/\partial\theta(PR_{12}(\theta,\varphi))|_{\theta=0}\,. \tag{5.22}$$

When φ is purely imaginary, $\mathbb{H}$ is real, and with account of (4.25), hermitian. The magnetics (5.21) are anisotropic generalizations of $sl(n)$, $o(n)$, $sp(n)$-invariant magnetics considered in [7].

6. Factorized Representations for Quantum *R*-Matrices

6.1. Preliminary Remarks

Let $f(z)$ be a meromorphic periodic function of z, $f(z+\pi)=f(z)$, bounded when $z\to\pm i\infty$, and having in the strip $0\leqq \operatorname{Re} z<\pi$ simple poles at $z=a_k$, with the residues c_k and zeros at $z=b_k$, $k=1,\dots,r$. It is well known that such a function can be represented both as a sum of cotangents,

$$f(z)=\sum_{k=1}^{r} c_k \operatorname{ctg}(z-a_k)+\text{const}, \tag{6.1}$$

and as a product of sines

$$f(z)=\text{const}\prod_{k=1}^{r}\frac{\sin(z-b_k)}{\sin(z-a_k)}. \tag{6.2}$$

The trigonometric R-matrices, considered in Sect. 4, are meromorphic *quasi-periodic matrix functions* of θ. They are bounded at $\theta\to\pm i\infty$ and have simple poles only. The representations (4.7), (4.24) for these R-matrices may be treated as a matrix analogue of the representation (6.1) for ordinary periodic functions. Note that the representation (4.7) allows one to check without difficulty the crossing-symmetry (4.2) of the corresponding R-matrices. However, checking the unitarity property (4.3) is not trivial.

In this section we shall construct new multiplicative representations for the R-matrices (4.7), (4.24), playing the same role as the representations (6.2) for ordinary functions. For that, we introduce elementary matrix multipliers (in general, noncommuting) which should be naturally considered as matrix generalizations of the functions $\sin z$ and $\cos z$. For these new representations, the unitarity of R-matrix becomes an easily checked feature.

6.2. Matrix Generalizations of Trigonometric Functions

As is known (see [43], Sect. 22.4), the function

$$\operatorname{ctg} z=z^{-1}+\sum_{m=-\infty}^{\infty}{}' [(z-m\pi)^{-1}+(m\pi)^{-1}], \tag{6.3}$$

may be chosen as the basis for the theory of trigonometric functions. The prime at the sum sign means that the value $m=0$ is omitted. Then the function $\sin z$ is defined as the solution of the equation

$$\frac{d}{dz}\ln f(z)=\operatorname{ctg} z \tag{6.4}$$

with the initial condition,

$$f(z)=z+O(z^2); \quad z\simeq 0. \tag{6.5}$$

Let us introduce the matrix analogues for $\operatorname{ctg}(z-\Delta)$ and $\sin(z-\Delta)$, where Δ is a constant. Let $\zeta(\theta)$ be a meromorphic matrix function in $\mathbb{C}^N\otimes\mathbb{C}^N$, with the following properties:

(i) Quasi-periodicity:

$$\zeta(\theta+\pi)=(U\otimes 1)\zeta(\theta)(U\otimes 1)^{-1}=(1\otimes U)^{-1}\zeta(\theta)(1\otimes U), \tag{6.6}$$

where U is the matrix of finite order g, i.e. $U^g=1$.

(ii) $\zeta(\theta)$ has only simple poles when $\theta=k\pi+\Delta$, $k\in\mathbb{Z}$. For $\theta\simeq\Delta$

$$\zeta(\theta)=\frac{M}{\theta-\Delta}+N+O(\theta-\Delta), \tag{6.7}$$

where

$$M^2=M, \tag{6.8}$$

$$MN(1-M)=0. \tag{6.9}$$

(iii) $\zeta(\theta)$ is finite when $\theta\to\pm i\infty$,

$$|\zeta(\theta)|<\infty, \quad \theta\to\pm i\infty. \tag{6.10}$$

It is natural to consider the function $\zeta(\theta)$ as a matrix generalization of the function $\operatorname{ctg}(z-\Delta)$. The meaning of the requirements (6.8), (6.9) will clear up later on.

Define a matrix generalization of the sine function, $\sigma(\theta)$, as the solution of equation

$$\frac{d\sigma}{d\theta}=\zeta(\theta)\sigma(\theta) \tag{6.11}$$

with the initial condition

$$\sigma(\theta_0)=\sigma_0; \quad \theta_0\neq\Delta+k\pi, \quad k\in\mathbb{Z}, \tag{6.11'}$$

σ_0 being a matrix in $\mathbb{C}^N\otimes\mathbb{C}^N$.

Let us show $\sigma(\theta)$ to be an entire function of θ. It is sufficient to prove that $\sigma(\theta)$ is entire in the vicinity of singular points of Eq. (6.11), at $\theta=\Delta+k\pi$, $k\in\mathbb{Z}$. Let us consider Eq. (6.11) in the vicinity of the point $\theta=\Delta$.

Represent $\sigma(\theta)$ as

$$\sigma(\tilde\theta)=e^{M\ln\tilde\theta}\Sigma(\tilde\theta), \tag{6.12}$$

$$e^{M\ln\tilde\theta}=1-M+M\tilde\theta; \quad \tilde\theta=\theta-\Delta. \tag{6.13}$$

The second equality has been obtained taking into account Eq. (6.8). The substitution of (6.7), (6.12) into (6.11) yields

$$\frac{d\Sigma(\tilde\theta)}{d\theta}=e^{-M\ln\tilde\theta}(\zeta(\theta)e^{M\ln\tilde\theta}-M)\Sigma(\tilde\theta)=\left(\frac{MN(1-M)}{\tilde\theta}+a_0+O(\tilde\theta)\right)\Sigma(\tilde\theta), \tag{6.14}$$

Therefore $\Sigma(\theta)$ and, hence, $\sigma(\theta)$ are regular at $\tilde{\theta}=0$. Due to the quasi-periodicity of $\zeta(\theta)$, Eq. (6.6), the above reasoning hold for other singular points of Eq. (6.11), i.e. $\theta=\Delta+k\pi$, $k\in\mathbb{Z}$. The scheme of the above proving that $\sigma(\theta)$ is single-valued was borrowed from [32].

The function $\sigma(\theta)$ may be written as an ordered exponent,

$$\sigma(\theta)=\sigma(\theta,\theta_0)\sigma_0\,, \tag{6.15}$$

$$\sigma(\theta_1,\theta_2)=P\exp\left\{-\int_{\theta_1}^{\theta_2}\zeta(s)ds\right\}=\sigma(\theta_1)\sigma^{-1}(\theta_2)\,. \tag{6.16}$$

The integration contour here is arbitrary because of the already proved uniqueness of $\sigma(\theta)$.

Note, it follows from the conditions (i)–(iii) on $\zeta(\theta)$ [without taking into account (6.9)] that

$$\zeta(\theta)=Z(\theta-\Delta,M)+\delta\,. \tag{6.17}$$

Here δ is a constant matrix in $\mathbb{C}^N\otimes\mathbb{C}^N$, commuting with $U\otimes 1$, whereas the function $Z(\theta,M)$ is defined by relation (3.10), with P replaced by M,

$$Z(\theta,M)=g^{-1}\sum_{k=0}^{g-1}\mathrm{U}^kM\mathrm{U}^{-k}\operatorname{ctg}((\theta-k\pi)/g)\,. \tag{6.18}$$

6.3. Two Types of ζ- and σ-Functions

The above rather general definitions of the functions $\sigma(\theta)$ and $\zeta(\theta)$ were given without the connection with the structure of the R-matrices (4.7), (4.24). Now we are going to concretize these definitions.

Let $\tilde{\mathcal{G}}=(\mathcal{G},A_c)$ be one of the pairs from Tables 1 and 2 and g, B, U, $C(\varphi)$ take the corresponding values from Table 2. [We identify the matrix U in (6.6) and (4.1).] Let us define two types of the functions $\zeta(\theta)$ with the same matrix U but different residues (the function ζ_2 is not defined for $A_{n-1}^{(1)}$):

$$\zeta_1(\theta)=Z(\theta,P_-)-i\delta_1/2\,, \tag{6.19}$$

$$\zeta_2(\theta)=(Z(\theta-\Delta,K)-i\delta_2)/N\,. \tag{6.20}$$

Here $Z(\theta,M)$ is defined in (6.18), N is the matrix dimension of $\mathcal{G}$, Δ is given in (3.15), matrix K is defined in (3.20),

$$P_\pm=\tfrac{1}{2}(1\pm P);\qquad P_\pm^t=P_\pm\,, \tag{6.21}$$

$$\delta_1=ig^{-1}\sum_{k=1}^{g-1}(U^{-k}\otimes U^k)\operatorname{ctg}(k\pi/g)\,, \tag{6.22}$$

$$\delta_2=ig^{-1}G_0\sum_{k=1}^{g-1}(U\otimes 1)^{-k}\operatorname{ctg}(k\pi/g)\operatorname{Tr}(U^k)\,, \tag{6.23}$$

$$\tilde{\delta}_1=\delta_1G_0\,. \tag{6.24}$$

The matrices P and P^t were defined by (1.6) and (3.21). The projection matrix in $\mathbb{C}^N\otimes\mathbb{C}^N$, G_0, entering into (6.24) is defined by

$$(G_\alpha)_{ik}^{jl}=\delta_{i-\alpha,j}\delta_{k+\alpha,l}\delta_{j+l,N-1}\,;\qquad \alpha=0,1,\ldots,N-1\,. \tag{6.25}$$

Remember, the matrix indices run over the values 0, 1, ..., $N-1$, the indices i, j refer to the first space $\mathbb{C}^N$, and the indices k, l refer to the second one. Henceforth the indices in δ-symbols are considered modulo N, i.e.

$$\delta_{ij}=\delta_{i(\mathrm{mod}\,N),\,j(\mathrm{mod}\,N)}\,. \tag{6.26}$$

The matrices δ_1, δ_2 may be calculated using the explicit form of U from Table 1. Table 3, given on page 498, represents the vectors $\{d^a_{1,2}\}$, $a=0,\ldots,r-1$, related to δ_1, δ_2 as

$$\delta_1=G_0(\Delta_1\otimes 1); \quad \delta_2=G_0(\Delta_2\otimes 1), \quad \Delta_i=(d_i, \bar{h}), \tag{6.27}$$

where (,) and h_a are defined by (4.12), (4.13).

Immediately from the definitions (6.19) and (6.20), it is easy to show that the functions $\zeta_1(\theta)$ possess classical unitarity

$$\zeta_{1,2}(-\theta)=-P\zeta_{1,2}(\theta)P\,, \quad \delta_{1,2}=-P\delta_{1,2}P\,, \tag{6.28}$$

and have the following asymptotic expansions at $\theta\to\pm i\infty$,

$$\zeta_1(\theta)=\frac{i}{2}[\mp(1-\tilde{P})-\delta_1]+O(e^{\pm 2i\theta/g}), \quad \zeta_2(\theta)=\frac{i}{N}[\mp\tilde{K}-\delta_2]+O(e^{\pm 2i\theta/g}), \tag{6.29}$$

where $\tilde{P}$ and $\tilde{K}$ are defined by (3.18) and (3.19).

Applying (6.19)–(6.23), one can easily verify that $\zeta_{1,2}(\theta)$ satisfies conditions (6.6)–(6.10). Hence, the functions $\sigma_{1,2}(\theta)$, defined via $\zeta_{1,2}(\theta)$ with the help of (6.11), are the entire functions of θ.

Note, the function $\sigma_1(\theta)$ for the series $A^{(1)}_{n-1}$ (in a more general, elliptic case) was introduced in [32].

Using the definitions of the functions $\sigma_{1,2}(\theta)$ and their analytical properties, one can obtain their explicit expressions via the function (6.18). These calculations are performed in Appendix C.

Let us enumerate some properties of $\sigma_{1,2}(\theta)$ following from (6.6), (6.7), (6.11), (6.15), and (6.19)–(6.23):

$$\sigma_{1,2}(\theta+\pi)=U\sigma_{1,2}(\theta)\Lambda_{1,2}\,, \tag{6.30}$$

where $\Lambda_{1,2}$ is independent of θ,

$$\sigma_1(0)=P_+X_1\,, \quad \mathrm{Res}\,\sigma_1^{-1}(\theta)|_{\theta=0}=X_1'P_-\,, \tag{6.31}$$

$$\sigma_2(\Delta)=(N-K)X_2\,, \quad \mathrm{Res}\,\sigma_2^{-1}(\theta)_{\theta=\Delta}=X_2'K\,, \tag{6.32}$$

where X_1, X_2', X_1', X_2 are some matrices.

Next, by virtue of (6.28), for the ratio of two functions $\sigma(\theta)$ defined in (6.16), one has

$$\sigma_{1,2}(\theta_1,\theta_2)=P\sigma_{1,2}(-\theta_1,-\theta_2)P\,. \tag{6.33}$$

6.4. *Factorization of Quantum R-Matrices*

Here we present the factorized representations for the quantum R-matrices (4.7) and (4.24). The derivation of these representations is contained in Appendix D. The definition of the functions $\sigma_{1,2}$ are given in Sects. 6.2 and 6.3; the ratio of two functions $\sigma_{1,2}$, $\sigma_{1,2}(\theta_1,\theta_2)$, is defined in Eq. (6.16); the matrices δ_1, δ_2 are given in (6.22), (6.23). The matrix $C(\varphi)$ and the constant ϱ entering into (4.7) are determined from (5.11) and (5.12) and also presented in Table 2; N is the matrix dimension of algebra $\mathcal{G}$.

Let us define the matrix

$$\mathscr{C}=1-G_0+(C(\varphi)\otimes 1)G_0, \tag{6.34}$$

where G_0 is given by (6.25). As follows from (4.12), (4.13),

$$P\mathscr{C}P=\mathscr{C}^{-1}. \tag{6.35}$$

Below it will be suitable to use the R-matrices $\tilde{R}(\theta)$ differing from (4.7) and (4.24) by the normalization

$$\tilde{R}(\theta)=\sin\theta \quad R(\theta)/\sin(\theta+2\varphi). \tag{6.36}$$

6.4.1. $\tilde{\mathcal{G}}=A_{n-1}^{(1)}; \quad \mathcal{G}=sl(n)$.

$$\tilde{R}(\theta)=\mathscr{L}\sigma_1(\theta-2\varphi,\theta+2\varphi)\mathscr{L}, \tag{6.37}$$

$$\mathscr{L}=\exp(-i\varphi\delta_1). \tag{6.38}$$

This formula is a particular case of a more general result of [31] where an elliptic R-matrix [19] connected with algebra $sl(N)$ was considered. Note that in the trigonometric limit the elliptic R-matrix of [31] goes not into $\tilde{R}(\theta)$, but into the equivalent R-matrix $\mathscr{L}^{-1}\tilde{R}\mathscr{L}^{-1}$. That is why Eq. (6.37) contains an additional factor $\mathscr{L}$ comparing with formula (4) from [31].

6.4.2. $\tilde{\mathcal{G}}=B_n^{(1)},\ D_n^{(1)},\ D_n^{(2)}; \quad \mathcal{G}=o(2n+1), o(2n), o(2n)$.

$$\tilde{R}(\theta)=\mathscr{C}^{-1}\sigma_2(\theta-\varrho,\theta+2\varphi)\mathscr{L}\sigma_1(\theta-2\varphi,\theta+2\varphi)\mathscr{L}\sigma_2(\theta-2\varphi,\theta+\varrho)\mathscr{C}^{-1}, \tag{6.39}$$

$$\mathscr{L}=\exp(-i\varphi(\delta_1+\bar{\delta}_1)); \quad \varrho=(N-2)\varphi. \tag{6.40}$$

6.4.3. $\tilde{\mathcal{G}}=C_n^{(1)}; \quad \mathcal{G}=sp(2n)$.

$$\tilde{R}(\theta)=\mathscr{C}^{-1}\sigma_2(\theta-\varrho,\theta-2\varphi)\mathscr{L}\sigma_1(\theta-2\varphi,\theta+2\varphi)\mathscr{L}\sigma_2(\theta+2\varphi,\theta+\varrho)\mathscr{C}^{-1}, \tag{6.41}$$

$$\mathscr{L}=\exp(i\varphi(\bar{\delta}_1-\delta_1)); \quad \varrho=(N+2)\varphi. \tag{6.42}$$

Note the difference of the signs in Eqs. (6.39), (6.40), and (6.41), (6.42).

6.4.4. $\tilde{\mathcal{G}}=A_{2n}^{(2)}; \quad \mathcal{G}=sl(2n+1)$.

$$\tilde{R}(\theta)=\mathscr{C}^{-1}\sigma_2(\theta-N\varphi)\mathscr{L}\sigma_1(\theta-2\varphi,\theta+2\varphi)\mathscr{L}\sigma_2^{-1}(\theta+N\varphi)\mathscr{C}^{-1}, \tag{6.43}$$

$$\mathscr{L}=\exp(-i\varphi\delta_1)\mathscr{C}^{-1}. \tag{6.44}$$

The initial conditions for $\sigma_2(\theta)$ in (6.11) are chosen in the form

$$\sigma_2(0)=\sigma_0=1-G_0+(-1)^n(G_n+G_{n+1})/2. \tag{6.45}$$

Here G_α are defined in (6.25). Equation (6.43) was reported in [25].

6.4.5. $\tilde{\mathcal{G}}=A_{2n-1}^{(2)}; \quad \mathcal{G}=sl(2n)$.

$$\tilde{R}(\theta)=\mathscr{C}^{-1}\sigma_2(\theta-N\varphi)\mathscr{L}F(\theta)\mathscr{L}\sigma_2^{-1}(\theta+N\varphi)\mathscr{C}^{-1}. \tag{6.46}$$

The initial conditions for $\sigma_2(\theta)$ are chosen in the form

$$\sigma_2(0)=\sigma_0=1-G_0+\hat{\sigma}_0\,, \tag{6.47}$$

$$(\hat{\sigma}_0)^{jl}_{ik}=\delta_{i+k,N-1}\delta_{j+l,N-1}\Sigma_{kl}\,, \tag{6.48}$$

where the $2n$ by $2n$ matrix $\|\Sigma_{kl}\|$ is (the omitted matrix elements equal zero)

$$
\|\Sigma_{kl}\| =
\begin{array}{r}
(n-1)\left\{\begin{array}{c} \\ \\ \\ \\ \\ \end{array}\right. \\
\\ \\ \\ \\ \\ \\ \\ \\ \\
\end{array}
\begin{array}{l}
\overbrace{\qquad\qquad\qquad\quad}^{n-2} \\
\left[\begin{array}{cccc|cccc|cccc}
1 & & & & 0 & 1 & 0 & 0 & & & & \\ \cline{5-8}
1 & 1 & & & & & & & & & & \\
 & 1 & \ddots & & & & & & & & & \\
 & & \ddots & 1 & & & & & & & & \\
 & & & 1 & 2 & & & & & & & \\ \hline
 & & & & 1 & 0 & 1 & 1 & & & & \\
 & & & & 1 & 0 & -1 & 1 & & & & \\ \hline
 & & & & & & & 2 & 1 & & & \\
 & & & & & & & & 1 & \ddots & & \\
 & & & & & & & & & \ddots & 1 & \\
 & & & & & & & & & & 1 & 1 \\
 & & & & 0 & 1 & 0 & 0 & & & & 1 \\ \cline{5-8}
\end{array}\right]
\end{array}
$$

The matrix $\mathscr{L}$ is of the form

$$\mathscr{L}=\mathscr{C}\exp[i\varphi(-\delta_2-\delta_1+\tilde{\bar{\delta}}_1-\bar{\delta}_1)]\,, \tag{6.49}$$

where

$$\bar{\delta}_1=G_0 iN^{-1}\sum_{k=1}^{N-1}(V^{-k}\otimes V^k)\operatorname{ctg}\frac{k\pi}{N}\,, \tag{6.50}$$

$$\begin{aligned} &V=\operatorname{diag}(\omega^{n-1},\omega^{n-2},\ldots,\omega,1,-1,\omega^{-1},\ldots,\omega^{1-n}),\\ &\omega=\exp(2\pi i/N)\,. \end{aligned} \tag{6.51}$$

Next,

$$F(\theta)=(1-G_0)\sigma_1(\theta-2\varphi,\theta+2\varphi)+G_0\bar{\sigma}_1(\theta-2\varphi,\theta+2\varphi)\,, \tag{6.52}$$

where $\bar{\sigma}_1(\theta_1,\theta_2)$ is defined by relations (6.11), (6.19) where U, g, δ_1, and P are replaced by V, N, $\bar{\delta}_1$, and $\bar{P}$ respectively, where

$$\bar{P}=\sigma_0^{-1}P\sigma_0=1-G_0+G_0\hat{P}\,.$$

In the same basis as Eq. (6.48), we have

$$\hat{P}=\left[\begin{array}{c|c|c} \overbrace{\qquad\qquad}^{n-1} & & \begin{matrix} & & 1\\ & \iddots & \\ 1 & & \end{matrix} \\ \hline & \begin{matrix} 1 & \\ & -1 \end{matrix} & \\ \hline \begin{matrix} & & 1\\ & \iddots & \\ 1 & & \end{matrix} & & \end{array}\right] \quad \text{(first block row: } (n-1) \text{ rows)}$$

Appendix A

This appendix contains an explicit realization of relations (2.7)–(2.10). Remember that the discussed realizations of the algebras $sl(N)$, $o(N)$, $sp(N)$ are defined by (2.1), the Coxeter automorphisms are given in Table 1. Let be the matrix dimension of $\mathscr{G}$, $\mathscr{G}_0$ be the subalgebra of $\mathscr{G}$, defined by (2.6), $r=\dim\mathscr{G}_0$. Define the N by N matrices

$$\tilde{h}_a=-e_{a,a}+e_{N-a-1,N-a-1}, \qquad a=0,\ldots,r-1, \tag{A.1}$$

$$(e_{\alpha\beta})_{ij}=\delta_{\alpha i}\delta_{\beta j}, \tag{A.2}$$

and the r-dimensional vectors

$$(\varepsilon_\alpha)_i=\delta_{\alpha i}; \qquad i=1,\ldots,r. \tag{A.3}$$

Equations (2.7)–(2.10) contain the elements e_i, f_i, and $\tilde{h}_i$ and the simple weights α_i. We choose a normalization so that $f_i=e_i^t$. Below we list the sets $\{e_i,\alpha_i\}$.

1. Series $A_{2n}^{(2)}$; $\mathscr{G}=sl(2n+1)$, $n\geqq 1$, $r=n$.

Generators:

$$e_0=e_{2n,0}, \quad e_i=e_{i-1,i}+e_{2n-i,2n+1-i}, \quad i=1,\ldots,n-1; \quad e_n=e_{n-1,n}+e_{n,n+1}.$$

Simple weights:

$$\alpha_0=2\varepsilon_1; \quad \alpha_i=-\varepsilon_i+\varepsilon_{i+1}, \quad i=1,\ldots,n-1; \quad \alpha_n=-\varepsilon_n.$$

2. Series $A_{2n-1}^{(2)}$, $\mathscr{G}=sl(2n)$, $n\geqq 2$, $r=n$.

Generators:

$$e_0=e_{2n-1,0}, \quad e_i=e_{i-1,i}+e_{2n-1-i,2n-i}, \quad i=1,\ldots,n-1;$$
$$e_n=(e_{i-1,n}+e_{n-1,n+1}).$$

Simple weights:

$$\alpha_0=2\varepsilon_1; \quad \alpha_i=-\varepsilon_i+\varepsilon_{i+1}, \quad i=1,\ldots,n-1, \quad \alpha_n=-\varepsilon_{n-1}-\varepsilon_n.$$

3. Series $B_n^{(1)}$, $\mathscr{G}=o(2n+1)$, $n\geqq 1$, $r=n$.

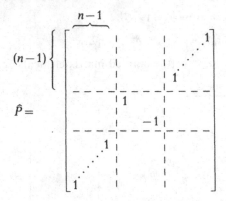

Generators:

$$e_0=(e_{2n-1,0}-e_{2n,1}),\quad e_i=e_{i-1,i}-e_{2n-i,2n+1-i},\quad i=1,\dots,n-1;$$
$$e_n=e_{n-1,n}-e_{n,n+1}.$$

Simple weights:

$$\alpha_0=\varepsilon_1+\varepsilon_2;\quad \alpha_i=-\varepsilon_i+\varepsilon_{i+1},\quad i=1,\dots,n-1;\quad \alpha_n=-\varepsilon_n.$$

4. Series $C_n^{(1)}$, $\mathcal{G}=sp(2n)$, $n\geqq 1$, $r=n$.

Generators:

$$e_0=e_{2n-1,0},\quad e_i=e_{i-1,i}-e_{2n-i-1,2n-i},\quad i=1,\dots,n-1;\quad e_n=e_{n-1,n}.$$

Simple weights:

$$\alpha_0=2\varepsilon_1,\quad \alpha_i=-\varepsilon_i+\varepsilon_{i+1},\quad \alpha_n=-2\varepsilon_n.$$

5. Series $D_n^{(1)}$, $\mathcal{G}=o(2n)$, $n\geqq 3$, $r=n$.

Generators:

$$e_0=(e_{2n-2,0}-e_{2n-1,1}),\quad e_i=e_{i-1,i}-e_{2n-1-i,2n-i},\quad i=1,\dots,n-1;$$
$$e_n=(e_{n-2,n}-e_{n-1,n+1}).$$

Simple weights:

$$\alpha_0=\varepsilon_1+\varepsilon_2;\quad \alpha_i=-\varepsilon_i+\varepsilon_{i+1},\quad \alpha_n=-\varepsilon_{n-1}-\varepsilon_n.$$

6. Series $D_{n+1}^{(2)}$, $\mathcal{G}=o(2n+2)$, $n\geqq 2$, $r=n$.

Generators:

$$e_0=e_{n,0}+e_{n+1,0}-e_{2n+1,n}-e_{2n+1,n+1},\quad e_i=e_{i-1,i}-e_{2n+1-i,2n+2-i},$$
$$i=1,\dots,n-1;\quad e_n=e_{n-1,n}-e_{n-1,n+1}+e_{n,n+2}-e_{n+1,n+2}.$$

Simple weights:

$$\alpha=\varepsilon_1,\quad \alpha_i=-\varepsilon_i+\varepsilon_{i+1},\quad \alpha_n=-\varepsilon_n.$$

Appendix B

Here we shall obtain relation (5.8) for the quantum *L*-operator (5.5). List first some properties of the matrices $\bar{h}_a$, e_k, f_k, entering into relations (2.7)–(2.10)

$$Be_k^tB^{-1}=-\exp(+2i\Delta/g)e_k,\quad Bf_k^tB^{-1}=-\exp(-2i\Delta/g)f_k,\qquad \text{(B.1)}$$

where B and g are defined in Table 2, Δ is defined in (3.15). For $G\neq sl(N)$, when $\Delta=0$, the Eq. (C.1) are trivial consequences of the definitions (2.1). In the case of $G=sl(N)$, $\tilde{G}=A_{2n}^{(2)}, A_{2n-1}^{(2)}$, when $\Delta=\pi/2$, the Eqs. (C.1) follow from the definition (2.6) of the subspaces G_j (remember that $e_k\in G_1$, $f_k\in G_{-1}$) and from the relation $\exp(2i\Delta/g)=\omega=\exp(2\pi i/h)$, h being the order for the Coxeter automorphism A_c. Next, for our realizations of G and A_c the matrices $\bar{h}_a\in G_0$ are diagonal (see Appendix A). Therefore, Eq. (2.9) may be rewritten in the form

$$(e_k)_{ij}(\bar{h}_{ii}^a-\bar{h}_{jj}^a)=\alpha_k^a(e_k)_{ij},\quad (f_k)_{ij}(\bar{h}_{ii}^a-\bar{h}_{jj}^a)=-\alpha_k^a(f_k)_{ij}.\qquad \text{(B.2)}$$

Besides, it follows from (A.1), (2.1), (5.3) that

$$\bar{h}_a B + B\bar{h}_a = 0; \qquad e^P e^{(\alpha, u)} e^{-P} = e^{(\alpha, u - i\varphi\bar{h})}, \tag{B.3}$$

where the notations are the same as in (3.22), (5.3), (5.5).

Let us come to the proof of (5.8). With the help of the definitions (3.28), (5.5) and the properties (B.1)–(B.3) one can easily show that

$$BL'(\theta - \varrho')B^{-1} = e^{-P}[1 + \mathrm{im}\,\delta(e^{\mathrm{ad}\,u}\tilde{I}_+\lambda + e^{-\mathrm{ad}\,u}\tilde{I}_-\lambda^{-1})]e^{-P}, \tag{B.4}$$

where ϱ' is some constant, and the matrices $\tilde{I}_\pm$ are defined by relations (3.28), with e_k, f_k, changed by $\tilde{e}_k$, $\tilde{f}_k$,

$$\begin{aligned} \tilde{e}_k &= e_k \exp(-i\varphi(\alpha_k, \alpha_k) + 2i(\varrho' + \Delta)/g), \\ \tilde{f}_k &= f_k \exp(+i\varphi(\alpha_k, \alpha_k) - 2i(\varrho' + \Delta)/g). \end{aligned} \tag{B.5}$$

The last two equalities may be written down as similarity transformations,

$$\tilde{e}_k = C'(\varphi) e_k C'^{-1}(\varphi); \qquad f_k = C'(\varphi) f_k C'^{-1}(\varphi), \tag{B.6}$$

$$C'(\varphi) = \exp(c, \bar{h}), \tag{B.7}$$

provided there exists an r-dimensional vector c satisfying the system of equations

$$-(\alpha_k, \alpha_k) + (c, \alpha_k) + 2i(\varrho' + \Delta)/\varphi g = 0, k = 0, \ldots, r, \qquad r = \dim \mathscr{G}_0, \tag{B.8}$$

where α_K are simple weights of $\hat{\mathscr{G}}$. Obviously, the vector $\frac{1}{2}c$ defines a point, equidistant with respect to $(r+1)$ points with the coordinates α_k, $k = 0, \ldots, r$, in an r-dimensional Euclidean space. Since $\alpha_k \neq \alpha_l$ when $K \neq l$, the vector c and the constant ϱ' are determined by (B.8) unambiguously.

Using the explicit form of the system of simple weights $\{\alpha_k\}$, given in Appendix A, one can easily get convinced that the quantities $C'(\varphi)$ and ϱ', defined via (B.7) and (B.8), coincide exactly with the result of [25] for $C(\varphi)$ and ϱ given in Table 2,

$$C'(\varphi) = C(\varphi), \qquad \varrho' = \varrho. \tag{B.9}$$

Taking into account (B.6), (B.9), we may rewrite (B.4) in the form

$$C(\varphi)BL'(\theta - \varrho)B^{-1}C^{-1}(\varphi) = e^{-P}[1 + im\delta(e^{\mathrm{ad}\,u}I_+\lambda + e^{-\mathrm{ad}\,u}I_-\lambda^{-1})]e^{-P}, \tag{B.10}$$

whereof one can immediately obtain Eq. (5.8).

Appendix C

In this appendix we shall discuss the main properties of the functions $\sigma_1(\theta)$ and $\sigma_2(\theta)$, defined by the relations (6.11), (6.19), (6.20).

1. The Function $\sigma_1(\theta)$. A particular choice of the initial condition (6.11′) is inessential for the representations (6.37), (6.39), (6.41), (6.43), since they contain only the ratio $\sigma_1(\theta_1; \theta_2)$. We can use this freedom to simplify the calculations. It is convenient to replace (6.11′) by the initial condition of the form

$$\sigma_1(\theta) = P_+ + \theta P_- + O(\theta^2), \tag{C.1}$$

where $P_\pm$ is defined in (6.21). This is possible because of the special structure of the expansion coefficients in (6.7) for $\zeta_1(\theta)$ (remember that $\Delta=0$ in this case). It follows from the definitions (6.7), (6.22) that

$$M=P_-\,,\qquad N=-iP_+\delta_1=-i\delta_1 P_-\,, \tag{C.2}$$

whereof it can be readily shown that in Eq. (6.14) $a_0\simeq P_- a_0'$, and hence

$$\Sigma(\theta)=(1+\theta P_- a_0')\Sigma(0)+O(\theta^2)\,.$$

Substituting the last equation and Eq. (C.2) into (6.12), we get convinced that (C.1) is equivalent to the initial condition $\Sigma(0)=1$ for Eq. (6.14).

The function $\sigma_1(\theta)$, normalized by the condition (C.1), possesses both the general properties (6.30), (6.31) with $M=P_-$, and the additional property [32],

$$P\sigma_1(\theta)=\sigma_1(-\theta)\,. \tag{C.3}$$

The latter is a consequence of (6.11) and the invariance of (C.1) under the transformations (C.3). Consider next the ratio of the form (6.16),

$$\Phi_1(\theta)=\sigma_1(\theta-2\varphi,\theta+2\varphi)=\sigma_1(\theta-2\varphi)\sigma_1^{-1}(\theta+2\varphi)\,. \tag{C.4}$$

The function $\Phi_1(\theta)$ is a meromorphic function of θ and possesses the following quasi-periodicity property

$$\Phi_1(\theta+\pi)=(U\otimes 1)\Phi_1(\theta)(U\otimes 1)^{-1}\,. \tag{C.5}$$

As follows from (6.31), it has poles at $\theta=-2\varphi+k\pi$, $k\in\mathbb{Z}$, and

$$\operatorname{Res}\Phi_1(\theta)|_{\theta=-2\varphi}=A_1(\varphi)P_-\,. \tag{C.6}$$

Some analysis of the form of ζ_1 shows that it is sufficient to choose the matrix $A_1(\varphi)$ to be diagonal. Then it follows from (C.5), (C.6), the definition (6.16) and the finiteness of ζ_1 at $\theta\to\pm i\infty$, that

$$\Phi_1(\theta)=A_1(\varphi)Z(\theta+2\varphi,P_-)+B_1(\varphi)\,. \tag{C.7}$$

Here Z is defined in (6.18), and the matrices $A_1(\varphi)$ and $B_1(\varphi)$ are calculated by considering the limits of (C.7) when $\theta\to\pm i\infty$. Using (6.29), one obtains

$$A_1(\varphi)=-2e^{2i\varphi\delta_1}\sin 2\varphi\cos(2\varphi\tilde{P})\,,\qquad B_1(\varphi)=e^{2i\varphi\delta_1}\cos(2\varphi(\tilde{P}-1))\,, \tag{C.8}$$

with δ_1 defined by (6.22) and $\tilde{P}$ defined by (3.18). In the derivation of (C.8) we exploited the property

$$\tilde{P}=\tilde{P}^3\,. \tag{C.9}$$

Similarly one may consider the function $\sin(\theta+2\varphi)\Phi_1(\theta)/\sin(\theta-2\varphi)$. Writing for it the representation of the type (C.7) one has

$$\begin{aligned}\Phi_1(\theta)&=\frac{\sin(\theta-2\varphi)}{\sin(\theta+2\varphi)}[-Z(\theta-2\varphi,P_+)A_1(\varphi)+B'(\varphi)]\,,\\ B'(\varphi)&=e^{2i\varphi\delta_1}\cos[2\varphi(\tilde{P}+1)]\,,\end{aligned} \tag{C.10}$$

where $A_1(\varphi)$ is given by (C.8). This representation, in particular, gives

$$\Phi_1(2\varphi)=-\sin^{-1}4\varphi\quad P_+A_1(\varphi)\,. \tag{C.11}$$

Let us obtain one more expression for $\Phi_1(\theta)$. It follows from (C.3) and (C.4) that

$$\Phi_1(k\pi) = \mathrm{U}^k P \mathrm{U}^{-k}, \qquad k \in \mathbb{Z}. \tag{C.12}$$

Consider the function $\sin(\theta+2\varphi)\Phi_1(\theta)/\sin\theta$. It is meromorphic and has poles at $\theta = k\pi$, $k \in \mathbb{Z}$, with the residues given by (C.12). At $\theta \to \pm i\infty$ its asymptotics is easily calculated by means of (C.7), (C.8). Therefore, it is easy to write for it an expression like (C.7). Thus, one comes to

$$\Phi_1(\theta) = \frac{\sin\theta}{\sin(\theta+2\varphi)}\{e^{2i\delta_1\varphi} + (\cos 2\varphi - 1)\tilde{P}^2 + \sin 2\varphi Z(\theta, P)\}. \tag{C.13}$$

2. *The Function* $\sigma_2(\theta)$. The initial conditions become important for the function $\sigma_2(\theta)$, when one considers the series $A_{2n}^{(2)}$ and $A_{2n-1}^{(2)}$. They will be specified at the appropriate place (see Appendix D). Here we shall examine the following ratio of two functions $\sigma_2(\theta)$ for the case of $\mathcal{G} = o(N)$ and $sp(N)$

$$\Phi_2(\theta, \varphi) = \sigma_2(\theta - N\varphi, \theta) = \sigma_2(\theta - N\varphi)\sigma_2^{-1}(\theta), \tag{C.14}$$

where N is the matrix dimension of algebra $\mathcal{G}$. Repeating the analysis that has led to relations (C.7), (C.8), we obtain

$$\Phi_2(\theta) = A_2(\varphi)Z(\theta, K) + B_2(\varphi), \tag{C.15}$$

K being defined in (3.20). The matrices $A_2(\varphi)$ and $B_2(\varphi)$ are of the form

$$A_2(\varphi) = -\sin\varphi\, e^{i\varphi\delta_2}\cos(\varphi Y), \qquad B_2(\varphi) = e^{i\varphi\delta_2}\cos\varphi \tilde{K}, \tag{C.16}$$

where $\tilde{K}$, δ_2, and G_0 are defined by (3.19), (6.23), and (6.25)

$$\tilde{K} = G_0 + Y, \quad G_0^2 = G_0, \quad G_0 Y = Y, \quad Y^3 = Y. \tag{C.17}$$

It follows from (C.14) that

$$\Phi_2(\theta, \varphi)\Phi_2(\theta - N\varphi, -\varphi) = 1.$$

Putting here $\theta = N\varphi$ and substituting (C.15), we obtain

$$Z(N\varphi, K)A_2(-\varphi)K = -A_2^{-1}(\varphi)B_2(\varphi)A_2(-\varphi)K. \tag{C.18}$$

This relation will be used in Appendix D.

Appendix D

In this appendix we will prove Eqs. (6.37), (6.39), (6.44), (6.46).

1. $\tilde{\mathcal{G}} = A_{n-1}^{(1)}$, $\mathcal{G} = sl(n)$.

Equation (6.37) is a simple consequence of relations (4.24), (C.4), (C.13).

2. $\tilde{\mathcal{G}} = B_n^{(1)}$, $D_n^{(1)}$, $D_n^{(2)}$, $\mathcal{G} = o(2n+1)$, $o(2n)$, $o(2n)$.

Let us calculate the function

$$\hat{R}(\theta) = \sigma_2(\theta + 2\varphi, \theta - \varrho)\mathscr{C}R(\theta)\mathscr{C}\sigma_2(\theta + \varrho, \theta - 2\varphi), \tag{D.1}$$

where $R(\theta)$ was given in (4.7), while all other notations were defined at the beginning of our Sect. 6. $\hat{R}(\theta)$ may have poles when $\theta=\pm\varrho+k\pi$, $\theta=2\varphi+k\pi$, $\theta=k\pi$. Let us show that the residues of $\hat{R}(\theta)$ turn out to be nonzero only when $\theta=k\pi$, $k\in\mathbb{Z}$. It follows from (4.10) and (6.30) that

$$\hat{R}(\theta+\pi)=\mathrm{U}\hat{R}(\theta)\mathrm{U}^{-1}. \tag{D.2}$$

Therefore, it is sufficient to consider only one pole of each series, e.g. $\pm\varrho, 2\varphi, 0$. Assuming in (4.23) $\theta=-\varrho$ and applying (3.20) and (6.35), one obtains

$$R(\varrho)=\mathscr{C}^{-1}(1-K/N)X, \tag{D.3}$$

where X is some matrix, $\mathscr{C}$ is defined by (6.34). Together with Eqs. (3.21) and (6.32), the last equality yields

$$\operatorname{Res}\hat{R}(\theta)|_{\theta=\varrho}=0. \tag{D.4}$$

By analogy, one can prove that

$$\operatorname{Res}\hat{R}(\theta)|_{\theta=-\varrho}=0. \tag{D.5}$$

For the consideration of the case $\theta=2\varphi$ we shall employ the formula [which is valid for $G=o(N)$]

$$\mathscr{C}=\exp(-i\varphi(\tilde{\delta}_1-\delta_2)). \tag{D.6}$$

It is proved by an immediate substitution the values from Tables 2 and 3, into (D.6). Using (3.11) and (C.15) we get

$$\operatorname{Res}\hat{R}(\theta)|_{\theta=2\varphi}=\sigma_2(4\varphi, 4\varphi-N\varphi)\mathscr{C}F, \tag{D.7}$$

$$F=R(2\varphi)\mathscr{C}A_2(-\varphi)K, \tag{D.8}$$

with K and A defined by (3.19) and (C.16), respectively. Substituting (4.7) into (D.8) and using (C.18), one has

$$F=(R_0+\sin 2\varphi Z(2\varphi,P))\mathscr{C}A_2(-\varphi)K+\sin 2\varphi\mathscr{C}A_2^{-1}(\varphi)B_2(\varphi)A_2(-\varphi)K, \tag{D.9}$$

Table 3. The values of d_1 and d_2 entering into Eqs. (6.27)

$\tilde{\mathscr{G}}$		$\{d_1^a\}, \{d_2^a\}$	
$A_{n-1}^{(1)}$		$d_1^a=(4a+2-n)/n$,	$0\leqq a\leqq n/2$
$A_{2n}^{(2)}$		$d_1^a=(4a+1-2n)/(2n+1)$,	$0\leqq a\leqq n-1$
		$d_2^a=0$,	
$A_{2n-1}^{(2)}$	$d_1^{(n-1)}=0$;	$d_1^a=(4a+3-2n)/(2n-1)$,	$1\leqq a\leqq n-2$
	$d_2^{(n-1)}=0$;	$d_2^a=(2a+1)/(2n-1)$,	
$B_n^{(1)}$	$d_1^0=0$;	$d_1^a=(2a-n)/n$,	$1\leqq a\leqq n-1$
	$d_2^0=0$;	$d_2^a=(a-n)/n$,	
$C_n^{(1)}$		$d_1^a=(2a-n+1)/n$,	$1\leqq a\leqq n-1$
		$d_2^a=0$,	
$D_n^{(1)}$	$d_1^0=d_1^{(n-1)}=0$;	$d_1^a=(2a+1-n)/(n-1)$,	$1\leqq a\leqq n-1$
	$d_2^0=d_2^{(n-1)}=0$;	$d_2^a=(2a+1-n)/(n-1)$,	
$D_n^{(2)}$	$d_1^{(n-1)}=0$;	$d_1^a=(2a+2-n)/n$,	$0\leqq a\leqq n-2$
	$d_2^{(n-1)}=0$;	$d_2^a=0$,	

where R_0 is given in (4.8). After some calculations involving the relations

$$P_+K=K, \quad P_-K=0, \tag{D.10}$$

where $P_\pm$ are defined in (6.21), we obtain that

$$F=\exp(-i\tilde{\delta}_1\varphi)\{[\cos(2\varphi t_0)+\cos(2\varphi\tilde{P})]\sin\varphi\cos(\varphi Y) \\ -\sin 2\varphi\cos[\varphi(G_0+Y)]\}K. \tag{D.11}$$

G_0 and Y are specified in (6.25) and (C.17). Now it is not difficult to calculate the explicit form of the matrices t_0, G_0, Y proceeding from the definitions (3.17), (3.18), (3.19), (C.17) and verify that for all the cases we are considering

$$F\equiv 0. \tag{D.12}$$

Substituting the last equality into (C.7), one comes to

$$\operatorname{Res}\hat{R}(\theta)|_{\theta=2\varphi}=0. \tag{D.13}$$

Thus, we have demonstrated that $\hat{R}(\theta)$ has no poles at $\theta\neq k\pi$, $k\in\mathbb{Z}$. The residues of $\hat{R}$ at $\theta=k\pi$, $k\in\mathbb{Z}$, are calculated with the help of (5.14), (6.35), (6.33),

$$\operatorname{Res}\hat{R}(\theta)|_{\theta=k\pi}=\sin 2\varphi\mathbf{U}^kP\mathbf{U}^{-k}, \tag{D.14}$$

whereas the asymptotics for $\theta\to\pm i\infty$ is found applying (4.15), (6.29), (D.6). Restoring $\hat{R}(\theta)$ by (D.14), (D.15) with the help of the Liouville theorem and using (C.4), (C.13), we obtain that

$$\sin\theta\hat{R}(\theta)/\sin(\theta+2\varphi)=\mathscr{L}\sigma_1(\theta-2\varphi,\theta+2\varphi)\mathscr{L}, \tag{D.16}$$

where $\mathscr{L}$ was defined in (6.40). Equations (D.1) and (D.16) lead immediately to the representation (6.39).

3. $\tilde{\mathscr{G}}=C_n^{(1)}$, $\mathscr{G}=sp(2n)$.

The proof of (6.41) is in a complete analogy with the above proof of Eq. (6.39). Note that in this case Eqs. (D.6) and (D.10) are replaced by the relations

$$\mathscr{C}=\exp(i\varphi\tilde{\delta}_1), \tag{D.17}$$

$$P_+K=0, \quad P_-K=K. \tag{D.18}$$

4. $\tilde{\mathscr{G}}=A_{2n}^{(2)}$, $A_{2n-1}^{(2)}$, $\mathscr{G}=sl(2n+1)$, $sl(2n)$.

Remember, that the initial conditions for $\sigma_2(\theta)$ in Eq. (6.11) for the series $A_{2n}^{(2)}$ and $A_{2n-1}^{(2)}$ are given by the relations (6.45) and (6.47) respectively. Thus, we have

$$\sigma_2(0)=\sigma_2(\theta,0)\sigma_0. \tag{D.19}$$

From (6.6) one has

$$\sigma_2(\theta+\pi)=\mathbf{U}\sigma_2(\theta)\mathbf{V}^{-1}, \tag{D.20}$$

$$\mathbf{U}=1-G_0+G_0(U\times 1), \tag{D.21}$$

$$\mathbf{V}=1-G_0+G_0(V\times 1), \tag{D.22}$$

where

$$V^{-k}=\sigma_0^{-1}U^{-k}\sigma_2(k\pi,0)\sigma_0. \tag{D.23}$$

Using (C.15), it can be shown

$$\mathrm{Sp}(G_0V^k)=\begin{cases}0, & k\neq 0(\mathrm{mod}\,N)\\ N, & k=0(\mathrm{mod}\,N)\end{cases}. \tag{D.24}$$

It follows then that the eigenvalues of the matrix (G_0V) are

$$\lambda_\alpha=\exp(2\pi i\alpha/N), \quad \alpha=0,\ldots,N-1. \tag{D.25}$$

For σ_0 in the form of (6.45) and (6.47) the matrix V is diagonal and defined by the relation

$$V=-U; \quad \mathscr{G}=A_{2n}^{(2)}, \tag{D.26}$$

and by the relation (6.41) for $\tilde{\mathscr{G}}=A_{2n-1}^{(2)}$.

Turn now to the proof of (6.43). The matrices (6.20), (6.21) in this case become

$$K=\sum_{\alpha=0}^{N-1}(-\omega)^{-\alpha}G_\alpha, \quad \tilde{K}=G_0, \tag{D.27}$$

with the matrices G_α defined by (6.25). Substituting (D.27) into (6.20) and summing over k, one has

$$\zeta_2(\theta)=N^{-1}\left\{G_0\,\mathrm{tg}\theta+i\cos^{-1}\theta\sum_{\alpha=1}^{N-1}(-1)^\alpha G_\alpha\exp[i\theta(N-2\alpha)/N]\right\}. \tag{D.28}$$

This entails, in particular, that

$$\zeta_2(0)=iN^{-1}\sum_{\alpha=1}^{N-1}(-1)^\alpha G_\alpha. \tag{D.29}$$

The relation (C.15) in this case takes the form

$$\sigma_2(\theta-N\varphi,\theta)=-N\sin\varphi \quad \zeta_2(\theta)+\cos(\varphi G_0). \tag{D.30}$$

Then it follows from (D.29), (D.30), (6.45) and the relation $G_\alpha G_\beta=G_{\alpha+\beta}$ that

$$\sigma_2(\theta)=\frac{(-1)^n}{2}(e^{i\theta/N}G_n+e^{-i\theta/N}G_{n+1})+1-G_0. \tag{D.31}$$

Consider now the function

$$\hat{R}(\theta)=\sigma_2^{-1}(\theta-N\varphi)\mathscr{C}R(\theta)\mathscr{C}\sigma_2(\theta+N\varphi). \tag{D.32}$$

By virtue of the properties of $\sigma_2(\theta)$, $R(\theta)$ has no poles at $\theta=\pi/2\pm N\varphi+k\pi$, $k\in\mathbb{Z}$. The residues at poles at $\theta=k\pi$ are easy to calculate with the help of (5.14), (6.33), (D.20), (D.26),

$$\mathrm{Res}\,\hat{R}(\theta)|_{\theta=k\pi}=\sin 2\varphi\,\mathrm{U}^kP\mathrm{U}^{-k}. \tag{D.33}$$

Exploiting the asymptotic values following from (D.31)

$$\sigma_2(\theta)|_{\theta\to\pm i\infty}=\frac{(-1)^n}{2}G_{(N\pm1)/2}\,e^{\mp i\theta/N}+1-G_0+\ldots, \tag{D.34}$$

as well as Eqs. (3.17), (4.15) and

$$G_{(N\pm 1)/2}\mathscr{C}\, e^{\pm 2i\varphi\tilde{P}}\mathscr{C}G_{(N\mp 1)/2}=\mathscr{C}^{-1}\, e^{\pm 2i\varphi\tilde{P}}\mathscr{C}^{-1}G_0\,, \tag{D.35}$$

we obtain

$$\hat{R}(\theta)|_{\theta\to\pm i\infty}=\mathscr{C}^{-1}\, e^{\mp 2i\varphi\tilde{P}}\mathscr{C}^{-1}\,, \tag{D.36}$$

where $\tilde{P}$ and $\mathscr{C}$ are defined by (3.18), (6.34). As follows from (D.33), (D.36), (C.13),

$$\mathscr{C}\tilde{R}(\theta)\mathscr{C}=\sin(\theta+2\varphi)\sin^{-1}\theta\, e^{-i\varphi\delta_1}\sigma_1(\theta-2\varphi,\theta+2\varphi)\, e^{-i\delta\varphi_1}\,. \tag{D.37}$$

Substituting (D.37) in (D.32), we obtain the representation (6.43). The proof of Eq. (6.46) is similar (though more cumbersome) and is therefore omitted here.

Acknowledgement. The author thanks A. G. Shadrikov for reading the manuscript and very useful comments.

References

1. Karowski, M., Thun, H., Truong, T., Weisz, P.: On the uniqueness of purely elastic S-matrix in $(1+1)$ dimension. Phys. Lett. B **67**, 321–323 (1967);
   Zamolodchikov, A.B.: Exact two-particle S-matrix of quantum solitons in sine-Gordon model. Pis'ma v ZhETF **25**, 499–502 (1977)
2. Zamolodchikov, A.B., Zamolodchikov, A.B.: Factorized S-matrix in two dimension and exact solution of some relativistic field theory models. Ann. Phys. **120**, 253–291 (1979)
3. Baxter, R.J.: Partition function of the eight-vertex lattice model. Ann. Phys. **70**, 193–228 (1972)
4. Baxter, R.J.: Exactly solved models in statistical mechanics. London: Academic Press 1982
5. Yang, S.N.: S-matrix for the one-dimensional N-body problem with repulsive or attractive δ-function interaction. Phys. Rev. **168**, 1920–1923 (1968)
6. Belavin, A.A.: Exact solution of the two-dimensional model with asymptotic freedom. Phys. Lett. B **87**, 117–121 (1979)
7. Reshetikhin, N.Yu.: $O(N)$ invariant quantum field theoretical models: Exact solution. Nucl. Phys. B **251** [FS13], 565–580 (1985)
8. Sklyanin, E.K., Takhtadjan, L.A., Faddeev, L.D.: Quantum inverse problem method I. Teor. Mat. Fiz. **40**, 194–220 (1979)
9. Faddeev, L.D.: Quantum completely integrable models in field theory. Sov. Sci. Rev. C **1**, 107–155 (1980)
10. Drinfeld, V.G.: Hopf algebras and quantum triangle equations. Dokl. Akad. Nauk SSSR. **283**, 1060–1064 (1985)
11. Sklyanin, E.K.: Quantum variant of the inverse scattering method. Zap. Nauchn. Semin. LOMI. **95**, 55–128 (1980)
12. Perk, J.H., Schultz, C.L.: Diagonalization of the transfer matrix of a nonintersecting string model. Physica **122** A, 50–70 (1983)
13. Zamolodchikov, A.B., Zamolodchikov, Al.B.: Relativistic factorized S-matrix in two-dimensions having $O(N)$ isotopic symmetry. Nucl. Phys. B **133**, 525–535 (1978)
14. Berg, B., Karowski, M., Kurak, V., Weisz, P.: Factorized symmetric S-matrices in two dimensions. Nucl. Phys. B **134**, 125–132 (1977)
15. Zamolodchikov, A.B., Fateev, V.A.: Model factorized S-matrix and an integrable Heisenberg chain with spin 1. Yad. Fiz. **32**, 581–590 (1980) [English transl.: Sov. J. Nucl. Phys. **32**, 298 (1980)]
16. Fateev, V.A.: Factorized S matrix for particles with different parities and integrable 21-vertex statistical model. Yad. Fiz. **33**, 1419–1430 (1981) [English transl.: Sov. J. Nucl. Phys. **33**, 761 (1980)]
17. Izergin, A.G., Korepin, V.E.: The inverse scattering method approach to the Shabat-Mikhailov model. Commun. Math. Phys. **79**, 303–316 (1981)

18. Cherednik, I.V.: On method of constructing factorized S matrices in elementary functions. Teor. Mat. Fiz. **43**, 117–119 (1980) [English transl.: Theor. Math. Phys. **43**, 356 (1980)]
19. Belavin, A.A.: Discrete groups and the integrability of quantum systems. Funkts. Anal. Prilozh. **14**, 18–26 (1980);
Belavin, A.A.: Dynamical symmetry of integrable quantum systems. Nucl. Phys. B **180** [FS2], 189–200 (1981)
20. Cherednik, I.V.: On properties of factorized S-matrices in elliptic functions. Yad. Fiz. **32**, 549–557 (1982)
21. Bazhanov, V.V., Stroganov, Yu.G.: A new class of factorized S-matrices and triangle equations. Phys. Lett. B **105**, 278–280 (1981)
22. Bazhanov, V.V., Stroganov, Yu.G.: Trigonometric and S_n symmetric solution of triangle equations with variable on the faces. Nucl. Phys. B **205** [FS5], 505–526 (1982)
23. Kulish, P.P., Sklyanin, E.K.: Solutions of the Yang-Baxter equations. Zap. Nauchn. Semin. LOMI. **95**, 129–160 (1982) [English transl.: J. Soviet Math. **19**, 1596 (1982)]
24. Belavin, A.A., Drinfeld, V.G.: Solutions of the classical Yang-Baxter equation for simple Lie algebras. Funkts. Anal. Prilozh. **16**(3), 1–39 (1982) [English transl.: Funct. Anal. Appl. **16**, 159 (1982)];
Belavin, A.A., Drinfeld, V.G.: Triangle equations and simple Lie algebras. Sov. Sci. Rev. **4**C, 93–165 (1984)
25. Bazhanov, V.V.: Trigonometric solutions of the triangle equations and classical Lie algebras. Phys. Lett. **159** B, 321–324 (1985)
26. Bazhanov, V.V.: Exactly soluble models connected with simple Lie algebras. In: Special topics in gauge theories. Proc. XIX Int. Simp. Arenshoop, November, 1985, pp. 181–199
27. Bazhanov, V.V.: Quantum R-matrices and matrix generalizations of trigonometric functions. Preprint IHEP 86–39, Serpukhov: 1986 (in Russian) (submitted to Teor. Mat. Fiz.)
28. Leites, D.A., Serganova, V.V.: On solutions of the classical Yang-Baxter equations for simple Lie superalgebras. Teor. Mat. Fiz. **58**, 26–37 (1984)
29. Bazhanov, V.V., Shadrikov, A.G.: Quantum triangle equations and Lie superalgebras. Preprint IHEP 86–43. Serpukhov: 1986 (submitted to Teor. Mat. Fiz.)
30. Jimbo, M.: Quantum R-matrix for the generalized Toda system. Commun. Math. Phys. **102**, 537–547 (1986)
31. Bazhanov, V.V., Stroganov, Yu.G.: On connection between the solutions of the quantum and classical triangle equations. In: Proc. 6th. Int. Sem. on high energy physics and field theory. Protvino **1**, 51–53 (1983)
32. Takhtadjan, L.A.: Solutions of triangle equations with $Z_n \times Z_n$ symmetry and matrix analogs of Weierstrass ζ- and σ-functions. Zap. Nauchn. Semin. LOMI. **133**, 258–276 (1984)
33. Reshetikhin, N.Yu., Faddeev, L.D.: Hamiltonian structures for integrable models of field theory. Teor. Mat. Fiz. **56**, 311–321 (1983) [English transl.: Theor. Math. Phys. **56**, 847 (1984)]
34. Kulish, P.P.: Quantum difference nonlinear Schrödinger equations. Lett. Math. Phys. **5**, 191–197 (1981)
35. Olive, D.I., Turok, N.: Algebraic structure of Toda systems. Nucl. Phys. B **220** [FS8], 491 (1983)
36. Leznov, A.N., Saveliev, M.V., Smirnov, V.G.: Group representation theory and integrable nonlinear systems. Teor. Mat. Fiz. **48**, 3–12 (1981);
Bogoyavlensky, O.I.: On perturbations of the periodic Toda latice. Commun. Math. Phys. **51**, 201–209 (1976)
37. Mikhailov, A.V., Olshanetsky, M.A., Perelomov, A.M.: Two-dimensional generalized Toda lattice. Commun. Math. Phys. **79**, 473–488 (1981)
38. Jimbo, M., Miwa, T.: Classification of solutions to the star-triangle relations for a class of 3- and 4-state IRF models. Nucl. Phys. B **257** [FS14], 1–18 (1985)
39. Jimbo, M.: Quantum R matrix related to the generalized Toda system: an algebraic approach. Preprint RIMS-521, Kyoto: 1985
40. Babelon, O., de Vega, H.J., Viallet, C.M.: Solutions of the factorization equations from Toda field Theory. Nucl. Phys. B **190**, 542 (1981)

41. Izergin, A.G., Korepin, V.E.: Lattice version of quantum field theory models in two dimensions. Nucl. Phys. B **205** [FS5], 401–413 (1982);
Izergin, A.G., Korepin, V.E.: The most general *L*-operator for the *R*-matrix of the XXX model. Lett. Math. Phys. **8**, 259–265 (1984)
42. Babelon, O.: Representation of the Yang-Baxter algebra associated to Toda field theory. Nucl. Phys. B **230** [FS10], 241–249 (1983)
43. Whittaker, E.T., Watson, G.N.: A course of modern analysis. Cambridge: Cambridge University Press 1927

Communicated by Ya. G. Sinai

Received March 15, 1987

Commun. Math. Phys. 102, 537–547 (1986)

Communications in Mathematical Physics

Quantum R Matrix for the Generalized Toda System

Michio Jimbo

RIMS, Kyoto University, Kyoto, 606 Japan

Abstract. We report the explicit form of the quantum R matrix in the fundamental representation for the generalized Toda system associated with non-exceptional affine Lie algebras.

1. Introduction

It has been known for some time that the Yang–Baxter (YB) equations play a crucial rôle in classical and quantum integrable systems (see e.g. [1]). The structure of the classical YB equation is now fairly well understood [2–3]. In ref. [3] a classification of non-degenerate solutions related to simple Lie algebras is given, subject to the unitarity condition. Unfortunately such classification is yet unavailable in the quantum case. One of the consequences of [3] is that the trigonometric solutions, up to certain equivalence, are finite in number, and that they allow a neat description in terms of Dynkin diagrams. An immediate question would be whether it is possible to quantize all these solutions. The most typical ones among them are the classical solutions associated with the generalized Toda system (GTS). In this paper we report on the corresponding quantum solutions for the case of non-exceptional affine Lie algebras.

To be more specific, we consider the solutions $r(x)$ of the classical YB equation

$$[r^{12}(x), r^{13}(xy)] + [r^{12}(x), r^{23}(y)] + [r^{13}(xy), r^{23}(y)] = 0 \tag{1.1}$$

for the GTS of type $A_n^{(1)}$, $B_n^{(1)}$, $C_n^{(1)}$, $D_n^{(1)}$, $A_{2n}^{(2)}$, $A_{2n-1}^{(2)}$ and $D_{n+1}^{(2)}$, as given in Eq. (2.3), (3.1–4). Here the notations are standard: $r(x)$ is a $\mathfrak{G} \otimes \mathfrak{G}$-valued rational function, $\mathfrak{G}$ being a finite dimensional simple Lie algebra, and $r^{12}(x) = r(x) \otimes I$, etc. The problem is to find an $R(x) = R(x, \hbar)$ containing an arbitrary parameter $\hbar$, such that (i) it satisfies the quantum YB equation

$$R^{12}(x)R^{13}(xy)R^{23}(y) = R^{23}(y)R^{13}(xy)R^{12}(x), \tag{1.2}$$

and (ii) as $\hbar \to 0$,

$$R(x, \hbar) = \kappa(x, \hbar)(I + \hbar r(x) + \cdots) \tag{1.3}$$

holds with some scalar $\kappa(x, \hbar)$. In contrast to the classical case (1.1), the quantum

Eq. (1.2) is formulated for a function $R(x)$ with values in $\mathfrak{U}(\mathfrak{G})\otimes\mathfrak{U}(\mathfrak{G})$, where $\mathfrak{U}(\mathfrak{G})$ denotes the universal enveloping algebra of $\mathfrak{G}$. Existence of such a solution would imply that for any finite dimensional representation V_i $(i=1,2,3)$ of $\mathfrak{G}$ there correspond matrices $R^{ij}(x)\in\mathrm{End}(V_i\otimes V_j)$ satisfying (1.2)[4]. The main result of the present article is the explicit construction of $R(x)\in\mathrm{End}(V\otimes V)$, taking $V_1=V_2=V_3=V$ to be the fundamental representation. Construction of the "universal" $(=\mathfrak{U}(\mathfrak{G})\otimes\mathfrak{U}(\mathfrak{G})$-valued) solution is an interesting future problem (cf. [4, 5]).

The method of construction is described in Sect. 2. The line of arguments essentially follows that of ref. [5] (except for the examination of the sufficiency part). In Sect. 3 explicit forms of solutions are presented. The solutions for the type $A_n^{(1)}$[6] and $A_2^{(2)}$[7] have been known. The quantum "spin" Hamiltonians obtained as the first log derivative of the transfer matrix are also given.

2. The GTS and the YB Equation

First let us recall the formulation of the GTS and the corresponding classical r-matrix. Let $\hat{\mathfrak{G}}$ be an affine Lie algebra, and $\mathfrak{h}$ be a Cartan subalgebra thereof. The GTS associated with $\hat{\mathfrak{G}}$ is the following equation for a $\mathfrak{h}$-valued function $q=q(t)$ [8]:

$$q_{tt}=-\nabla_q U,\quad U=\sum_{\alpha\in\pi} e^{2\alpha(q)}.$$

Here π denotes the set of simple roots of $\hat{\mathfrak{G}}$. It is known to be representable in the Lax form $L_t=[A,L]$. In terms of the standard Chevalley basis $\{e_\alpha, f_\alpha, h_\alpha\}$, L and A are given by [8]

$$\begin{aligned} L&=p+e^{\mathrm{ad}q}e+e^{-\mathrm{ad}q}f,\\ A&=-e^{\mathrm{ad}q}e+e^{-\mathrm{ad}q}f, \end{aligned}\tag{2.1}$$

where $p=q_t\in\mathfrak{h}$, $e=\sum_{\alpha\in\pi}e_\alpha$ and $f=\sum_{\alpha\in\pi}f_\alpha$.

In order to describe the corresponding classical r-matrix, we employ the homogeneous picture of $\hat{\mathfrak{G}}$ (cf. [9]). We find it simpler than the principal picture adopted in [3], for then the degree of the rational function $r(x)$ will become independent of the rank of $\hat{\mathfrak{G}}$. Thus let $\mathfrak{G}$ be a complex finite-dimensional simple Lie algebra, and let σ be its diagram automorphism of order $k(=1,2,3)$. Put $\mathfrak{G}_j=\{X\in\mathfrak{G}/\sigma(X)=\omega^j X\}$, where ω is a primitive k^{th} root of unity. Let $\mathfrak{G}_j=\bigoplus_{\alpha\in\Delta_j}\mathfrak{G}_{j,\alpha}$ be its root space decomposition with respect to a Cartin subalgebra $\not{h}_0$ of $\mathfrak{G}_0$. Fixing an invariant bilinear form (,) on $\mathfrak{G}$, we choose $X_{j,\alpha}\in\mathfrak{G}_{j,\alpha}$, and normalize them as $(X_{j,\alpha},X_{-j,-\alpha})=1$. We write $E_\alpha=X_{0,\alpha}$, $F_\alpha=X_{0,-\alpha}$ $(\alpha\in\pi_0)$, $E_0=X_{1,-\theta}$ and $F_0=X_{-1,\theta}$, where π_0 is the set of simple roots of $\mathfrak{G}_0$ and θ denotes the highest weight of $\mathfrak{G}_0$ in $\mathfrak{G}_{-1}$. As is well known [10], if $\mathfrak{G}$ is of type X_N, then the loop algebra $\mathfrak{G}^{(k)}[\lambda,\lambda^{-1}]=\bigoplus_{j\in\mathbb{Z}}\lambda^j\mathfrak{G}_{j\,\mathrm{mod}\,k}$ gives a realization of the affine Lie algebra of type $X_N^{(k)}$ modulo the center. In this picture the Chevalley basis is given by

$$e_0=\lambda E_0,\ f_0=\lambda^{-1}F_0,\ e_\alpha=E_\alpha,\ f_\alpha=F_\alpha\quad(\alpha\in\pi_0).\tag{2.2}$$

(With the above normalization the diagonal of the Cartan matrix is (α,α).) For an orthonormal basis $\{I_\mu\}$ of $\mathfrak{G}$, we set

$$t=\sum_\mu I_\mu\otimes I_\mu=\sum_{j=0}^{k-1} t_j,\ (\sigma\otimes 1)t_j=\omega^j t_j.$$

Set further

$$r_0=\sum_{\alpha\in\Delta_0}\operatorname{sgn}\alpha X_{0,\alpha}\otimes X_{0,-\alpha}.$$

The classical r-matrix for the GTS of type $X_N^{(k)}$ is then given by the formula

$$r(x)=r_0-t_0+\frac{2}{1-x^k}\sum_{j=0}^{k-1}x^j t_j. \tag{2.3}$$

This r-matrix is related to the L-operator of (2.1) through the fundamental Poisson bracket relation

$$\{L(\lambda)\otimes L(\mu)\}=[r(\lambda/\mu),L(\lambda)\otimes 1+1\otimes L(\mu)]. \tag{2.4}$$

Here the λ-dependence of L is explicitly exhibited, regarding $\mathfrak{G}\otimes 1$ and $1\otimes\mathfrak{G}$ as realized in $\mathfrak{G}^{(k)}[\lambda,\lambda^{-1}]\otimes 1$ and $1\otimes\mathfrak{G}^{(k)}[\mu,\mu^{-1}]$, respectively. In the left-hand side of (2.4) the Poisson bracket is introduced by letting p and q be canonically conjugate; namely, writing $p=\sum p_iH_i$, $q=\sum q_iH_i$ for an orthonormal basis $\{H_i\}$ of $\mathfrak{h}_0$, one has $\{p_i,q_j\}=\delta_{ij}$.

To find the corresponding quantum R matrix, we quantize the relation (2.4) following the line of ref. [5]. Let now p and q denote $\mathfrak{h}_0$-valued operators acting on some Hilbert space satisfying the Heisenberg commutation relations $[p_i,q_j]=\hbar\delta_{ij}$, where $\hbar$ is an arbitrary parameter. We introduce further the elements $\hat{E}_\alpha,\hat{F}_\alpha$ of $\mathfrak{U}(\mathfrak{G})$ (or more precisely its completion) with the properties

$$[H,\hat{E}_\alpha]=\alpha(H)\hat{E}_\alpha,\quad [H,\hat{F}_\alpha]=-\alpha(H)\hat{F}_\alpha\quad (H\in\mathfrak{h}_0), \tag{2.5}$$

$$[\hat{E}_\alpha,\hat{F}_\beta]=\delta_{\alpha\beta}\sinh(2\hbar H_\alpha)/\sinh(2\hbar), \tag{2.6}$$

$$\hat{E}_\alpha\to E_\alpha,\quad \hat{F}_\alpha\to F_\alpha\quad\text{as}\quad \hbar\to 0. \tag{2.7}$$

Here H_α denotes the image of $\alpha\in\mathfrak{h}_0^*$ under the identification $\mathfrak{h}_0^*\simeq\mathfrak{h}$ via the bilinear form $(\ ,\ )$. Eventually we shall restrict to the fundamental representation of $\mathfrak{G}$ and identify $\hat{E}_\alpha,\hat{F}_\alpha$ with E_α,F_α. However the following arguments go through under (2.5–7). Define $\hat{e}_\alpha,\hat{f}_\alpha$ as in (2.2) and put $\hat{e}=\sum\hat{e}_\alpha$, $\hat{f}=\sum\hat{f}_\alpha$. In place of the classical L-operator (2.1) we use (cf. [5])

$$\begin{aligned} L(\lambda)&=e^p(1+\varepsilon(e^{\mathrm{ad}\,q}\hat{e}+e^{-\mathrm{ad}\,q}\hat{f}))e^p\\ &=\left(1+\varepsilon\sum_{\alpha\in\pi}e^{\alpha(q)}(e^{\alpha(p)}K_\alpha\hat{e}_\alpha+e^{-\alpha(p)}K_\alpha\hat{f}_\alpha)\right)e^{2p},\end{aligned}$$

where $K_\alpha=\exp(\hbar H_\alpha)$. In the second line the operators are normal-ordered (q to the left, p to the right). For the quantum R matrix we require the relation

$$\begin{aligned} &R(\lambda/\mu)L_1(\lambda)L_2(\mu)\equiv L_2(\mu)L_1(\lambda)R(\lambda/\mu)\mod\varepsilon^2,\\ &L_1(\lambda)=L(\lambda)\otimes 1,\quad L_2(\mu)=1\otimes L(\mu).\end{aligned} \tag{2.8}$$

Reducing the expressions $L_1(\lambda)L_2(\mu)$, $L_2(\mu)L_1(\lambda)$ into the normal-ordered form and comparing the coefficients of $e^{\alpha(q)}e^{\pm\alpha(p)}$, we find that (2.8) is equivalent to

$$[R(x), H\otimes 1 + 1\otimes H] = 0 \quad (H\in\mathfrak{h}_0), \tag{2.9}$$

$$R(x)(\hat{e}_\alpha\otimes K_\alpha^{-1} + K_\alpha\otimes\hat{e}_\alpha) = (\hat{e}_\alpha\otimes K_\alpha + K_\alpha^{-1}\otimes\hat{e}_\alpha)R(x), \tag{2.10}$$

$$R(x)(\hat{f}_\alpha\otimes K_\alpha^{-1} + K_\alpha\otimes\hat{f}_\alpha) = (\hat{f}_\alpha\otimes K_\alpha + K_\alpha^{-1}\otimes\hat{f}_\alpha)R(x). \tag{2.11}$$

Here $x = \lambda/\mu$ (recall that the λ or μ dependence enters through $\hat{e}_0\otimes 1 = \lambda\hat{E}_0\otimes 1$, $1\otimes\hat{e}_0 = 1\otimes\mu\hat{E}_0$, etc.). Below we shall discuss the uniqueness of solutions of the system (2.9–11) and its sufficiency for the validity of the YB Eq. (1.2). In the sequel we fix finite-dimensional irreducible representation spaces $V_i (i = 1, 2)$ of $\mathfrak{G}$ and consider (2.9–11) in $\mathrm{End}(V_1\otimes V_2)$.

Proposition 1. *For a general value of* $\hbar$, *the dimension of the solution space of the linear system* (2.10) *is at most 1.*

Proof. It suffices to show that the dimension is 1 for the special value $\hbar = 0$. In this case the proof reduces to the following lemma, which we formulate in a slightly more general way. Let $V_i (i = 1,\ldots, N)$ be finite dimensional irreducible $\mathfrak{G}$-modules. For $X\in\mathfrak{G}$ we write $X^{(i)} = 1\otimes\cdots 1\otimes \overset{i}{X}\otimes 1\cdots\otimes 1$. Consider the linear equations for $R\in\mathrm{End}(V_1\otimes\cdots\otimes V_N)$,

$$[R, E_\alpha^{(1)} + \cdots + E_\alpha^{(N)}] = 0 \quad (\alpha\in\pi), \quad [R, \lambda_1 E_0^{(1)} + \cdots + \lambda_N E_0^{(N)}] = 0. \tag{2.12}$$

Lemma. *For general values of* λ_i, *the only solution of* (2.12) *is* $R = \mathrm{const}\ I$.

Proof of Lemma. First we note that $[\sum\lambda_i^m X^{(i)}, \sum\lambda_i^n Y^{(i)}] = \sum\lambda_i^{m+n}[X, Y]^{(i)}$. Since E_0 is the lowest weight vector of the ad irreducible $\mathfrak{G}_0$-module $\mathfrak{G}_1$ [10], (2.12) implies that $[R, \lambda_1 X^{(1)} + \cdots + \lambda_N X^{(N)}] = 0$ for any $X\in\mathfrak{G}_1$. Hence we have

$$[R, \lambda_1^j X^{(1)} + \cdots + \lambda_N^j X^{(N)}] = 0 \tag{2.13}$$

for any $X\in[\mathscr{L}_1, [\mathscr{L}_2, \ldots, [\mathscr{L}_{r-1}, \mathscr{L}_r]\ldots]]$, where $\mathscr{L}_s$ denotes either $\mathfrak{G}_1$ or $\bigoplus_{\alpha\in\Pi_0}\mathbb{C}E_\alpha$, and $\mathfrak{G}_1$ appears j times in the sequence $\{\mathscr{L}_s\}$. It can be checked that such elements generate $\mathfrak{G}_{j\,\mathrm{mod}\,k}$. Taking j to be $j, j+k, j+2k, \ldots$ in (2.13), we conclude that $[R, X^{(i)}] = 0$ holds for any i and $X\in\mathfrak{G}_j$. In other words R commutes with $\mathfrak{U}(\mathfrak{G})\otimes\cdots\otimes\mathfrak{U}(\mathfrak{G})$. The lemma now follows from the fact that, for an irreducible V_i, $\mathfrak{U}(\mathfrak{G})$ spans $\mathrm{End}(V_i)$.

Corollary. *If* (2.10) *admits a non-trivial solution, it has the form* $R = I + \hbar R_1 + \cdots$ *up to constant multiple. In particular* $\det R \not\equiv 0$, $\mathrm{tr}\, R \not\equiv 0$.

Proposition 2. *A solution of* (2.10) *satisfies both* (2.9) *and* (2.11).

Proof. It is enough to consider the case of a non-trivial solution $R(x)$. Using (2.5) and (2.6), one checks that $R_1 = [R, H\otimes 1 + 1\otimes H]$ $(H\in\mathfrak{h}_0)$, $R_2 = R(\hat{f}_\alpha\otimes K_\alpha^{-1} + K_\alpha\otimes\hat{f}_\alpha) - (\hat{f}_\alpha\otimes K_\alpha + K_\alpha^{-1}\otimes\hat{f}_\alpha)R$ both solve (2.10). It follows that $R_i = \kappa_i R$ with some scalar κ_i. Taking the trace of $R_i R^{-1}$, we find $\kappa_i = 0$.

Proposition 3. *Assume that* (2.10) *admits non-trivial solutions* $R^{ij}(x)\in\mathrm{End}(V_i\otimes V_j)$ *for* $(i,j) = (1,2), (1,3), (2,3)$. *Then the* YB *Eq.* (1.2) *is satisfied.*

Proof. Put $Q_1 = R^{12}(x)R^{13}(xy)R^{23}(y)$, $Q_2 = R^{23}(y)R^{13}(xy)R^{12}(x)$, where $x = \lambda/\mu$ and $y = \mu/\nu$. The relation (2.8) implies that both $Q_i (i = 1,2)$ have the intertwining property,

$$Q_i L_1(\lambda)L_2(\mu)L_3(\nu) \equiv L_3(\nu)L_2(\mu)L_1(\lambda)Q_i \mod \varepsilon^2.$$

Hence their ratio $Q = Q_1^{-1}Q_2$ should satisfy

$$\begin{aligned}&[Q, H^{(1)} + H^{(2)} + H^{(3)}] = 0 \quad (H \in \mathfrak{h}_0),\\ &[Q, K_\alpha^{\pm 1} \otimes K_\alpha^{\pm 1} \otimes \hat{e}_\alpha + K_\alpha^{\pm 1} \otimes \hat{e}_\alpha \otimes K_\alpha^{\mp 1} + \hat{e}_\alpha \otimes K_\alpha^{\mp 1} \otimes K_\alpha^{\mp 1}] = 0,\end{aligned} \tag{2.14}$$

and those obtained by replacing $\hat{e}_\alpha \leftrightarrow \hat{f}_\alpha$. Arguing similarly as above, one can show that (2.14) has the only solution $Q = \text{const}\ I$ for a general $\hbar$.

Comparing the determinant we have

$$R^{12}(x)R^{13}(xy)R^{23}(y) = \zeta R^{23}(y)R^{13}(xy)R^{12}(x),$$

where ζ is a root of unity. Letting $\hbar \to 0$ we find that $\zeta = 1$.

Thus the YB equation is reduced to solving the homogeneous linear Eqs. (2.10) for $R(x)$. In the next section we give the result by taking $V_1 = V_2 = V$ to be the fundamental representation of $\mathfrak{G}$ and $\hat{e}_\alpha = e_\alpha$, $\hat{f}_\alpha = f_\alpha$.

3. Quantum *R* Matrix (Main Results)

In the sequel we adopt the following realization of classical Lie algebras: $\mathrm{sl}(n) = \{X \in \mathrm{Mat}(n) | \mathrm{tr}\, X = 0\}$, $o(n) = \{X \in \mathrm{sl}(n) | X = -S^t X S\}$, $\mathrm{sp}(2n) = \{X \in \mathrm{sl}(2n) | X = -\tilde{S}^{-1} {}^t X \tilde{S}\}$, where $S = (\delta_{\alpha, n+1-\beta})_{1 \leq \alpha, \beta \leq n}$ and $\tilde{S} = \begin{pmatrix} 0 & S \\ -S & 0 \end{pmatrix}$. Diagram automorphisms of order 2 are given by $\sigma(X) = -S^t X S$ for $\mathrm{sl}(n)$ and $\sigma(X) = T X T^{-1}$ for $o(2n)$ with

$$T = \left[\begin{array}{ccc|ccc} 1 & & & & & \\ & \ddots & & & & \\ & & 1 & & & \\ & & 0 & 1 & & \\ \hline & & 1 & 0 & & \\ & & & & 1 & \\ & & & & & \ddots \\ & & & & & & 1 \end{array}\right].$$

By convention the indices α, β run over $1, 2, \ldots, N$, where N is the size of the matrix: $N = n + 1, 2n + 1, 2n, 2n, 2n + 1, 2n, 2n + 2$ for $\mathfrak{G} = A_n^{(1)}, B_n^{(1)}, C_n^{(1)}, D_n^{(1)}, A_{2n}^{(2)}, A_{2n-1}^{(2)}, D_{n+1}^{(2)}$. We put $\alpha' = N + 1 - \alpha$. $E_{\alpha\beta}$ will denote the matrix $(\delta_{i\alpha}\delta_{j\beta})$. Let further $\varepsilon_\alpha = 1\ (1 \leq \alpha \leq n)$, $= -1\ (n + 1 \leq \alpha \leq 2n)$ for $\mathfrak{G} = C_n^{(1)}$ and $\varepsilon_\alpha = 1$ in the remaining cases. Under these notations the classical r-matrix (2.3) reads as follows:

$\mathfrak{G} = A_n^{(1)}$:

$$(1 - x)r(x) = (1 + x)\left(\sum E_{\alpha\alpha} \otimes E_{\alpha\alpha} - \frac{1}{N} I\right) + 2\left(\sum_{\alpha < \beta} + x \sum_{\alpha > \beta}\right) E_{\alpha\beta} \otimes E_{\beta\alpha}, \tag{3.1}$$

$\mathfrak{G} = B_n^{(1)}, C_n^{(1)}, D_n^{(1)}$:

$$(1-x)r(x) = (1+x)\sum (E_{\alpha\alpha}\otimes E_{\alpha\alpha} - E_{\alpha\alpha}\otimes E_{\alpha'\alpha'})$$
$$+2\Big(\sum_{\alpha<\beta} + x\sum_{\alpha>\beta}\Big)(E_{\alpha\beta}\otimes E_{\beta\alpha} - \varepsilon_\alpha\varepsilon_\beta E_{\alpha\beta}\otimes E_{\alpha'\beta'}), \tag{3.2}$$

$\mathfrak{G} = A_{2n}^{(2)}, A_{2n-1}^{(2)}$:

$$(1-x^2)r(x) = (1+x)^2\sum E_{\alpha\alpha}\otimes E_{\alpha\alpha} - (1-x)^2\sum E_{\alpha\alpha}\otimes E_{\alpha'\alpha'} - \frac{4x}{N}I$$
$$+2(1+x)\Big(\sum_{\alpha<\beta} + x\sum_{\alpha>\beta}\Big)E_{\alpha\beta}\otimes E_{\beta\alpha}$$
$$+2(1-x)\Big(-\sum_{\alpha<\beta} + x\sum_{\alpha>\beta}\Big)E_{\alpha\beta}\otimes E_{\alpha'\beta'}, \tag{3.3}$$

$\mathfrak{G} = D_{n+1}^{(2)}$:

$$(1-x^2)r(x) = (1+x^2)\sum_{\alpha\neq n+1,n+2}(E_{\alpha\alpha}\otimes E_{\alpha\alpha} - E_{\alpha\alpha}\otimes E_{\alpha'\alpha'})$$
$$+2x(E_{n+1,n+1} - E_{n+2,n+2})\otimes(E_{n+1,n+1} - E_{n+2,n+2})$$
$$+2\Big(\sum_{\alpha<\beta,\alpha,\beta\neq n+1,n+2} + x^2\sum_{\alpha>\beta,\alpha,\beta\neq n+1,n+2}\Big)(E_{\alpha\beta}\otimes E_{\beta\alpha} - E_{\alpha\beta}\otimes E_{\alpha'\beta'})$$
$$+(1+x)\Big(\sum_{\alpha<n+1,\beta=n+1,n+2} + x\sum_{\alpha>n+2,\beta=n+1,n+2}\Big)(E_{\alpha\beta}\otimes E_{\beta\alpha}$$
$$-E_{\alpha\beta}\otimes E_{\alpha'\beta'} + E_{\beta'\alpha'}\otimes E_{\alpha'\beta'} - E_{\beta'\alpha'}\otimes E_{\beta\alpha})$$
$$+(1-x)\Big(\sum_{\alpha<n+1,\beta=n+1,n+2} - x\sum_{\alpha>n+2,\beta=n+1,n+2}\Big)(E_{\alpha\beta}\otimes E_{\beta'\alpha}$$
$$-E_{\alpha\beta}\otimes E_{\alpha'\beta} - E_{\beta'\alpha'}\otimes E_{\beta'\alpha} + E_{\beta'\alpha'}\otimes E_{\alpha'\beta}). \tag{3.4}$$

Corresponding quantum R-matrices are given by the following formulas ($k = e^{-2h}$ denotes an arbitrary parameter).

$\mathfrak{G} = A_n^{(1)}$:

$$R(x) = (x-k^2)\sum E_{\alpha\alpha}\otimes E_{\alpha\alpha} + k(x-1)\sum_{\alpha\neq\beta} E_{\alpha\alpha}\otimes E_{\beta\beta}$$
$$-(k^2-1)\Big(\sum_{\alpha<\beta} + x\sum_{\alpha>\beta}\Big)E_{\alpha\beta}\otimes E_{\beta\alpha}. \tag{3.5}$$

$\mathfrak{G} = B_n^{(1)}, C_n^{(1)}, D_n^{(1)}, A_{2n}^{(2)}, A_{2n-1}^{(2)}$:

$$R(x) = (x-k^2)(x-\xi)\sum_{\alpha\neq\alpha'} E_{\alpha\alpha}\otimes E_{\alpha\alpha} + k(x-1)(x-\xi)\sum_{\alpha\neq\beta,\beta'} E_{\alpha\alpha}\otimes E_{\beta\beta}$$
$$-(k^2-1)(x-\xi)\Big(\sum_{\alpha<\beta,\alpha\neq\beta'} + x\sum_{\alpha>\beta,\alpha\neq\beta'}\Big)E_{\alpha\beta}\otimes E_{\beta\alpha}$$
$$+\sum a_{\alpha\beta}(x)E_{\alpha\beta}\otimes E_{\alpha'\beta'}, \tag{3.6}$$

where

$$a_{\alpha\beta}(x)=\begin{cases}(k^2x-\xi)(x-1) & (\alpha=\beta,\alpha\neq\alpha')\\ k(x-\xi)(x-1)+(\xi-1)(k^2-1)x & (\alpha=\beta,\alpha=\alpha')\\ (k^2-1)(\varepsilon_\alpha\varepsilon_\beta\xi k^{\bar\alpha-\bar\beta}(x-1)-\delta_{\alpha\beta'}(x-\xi)) & (\alpha<\beta)\\ (k^2-1)x(\varepsilon_\alpha\varepsilon_\beta k^{\bar\alpha-\bar\beta}(x-1)-\delta_{\alpha\beta'}(x-\xi)) & (\alpha>\beta).\end{cases}$$

Here ξ and $\bar\alpha$ are given respectively by

$$\xi=k^{2n-1},\quad k^{2n+2},\quad k^{2n-2},\quad -k^{2n+1},\quad -k^{2n},$$

for $\mathfrak{G}=B_n^{(1)}, C_n^{(1)}, D_n^{(1)}, A_{2n}^{(2)}, A_{2n-1}^{(2)}$;

$$\bar\alpha=\begin{cases}\alpha-\frac12 & (1\leqq\alpha\leqq n)\\ \alpha+\frac12 & (n+1\leqq\alpha\leqq 2n)\end{cases}$$

for $\mathfrak{G}=C_n^{(1)}$, and

$$\bar\alpha=\begin{cases}\alpha+\frac12 & \left(1\leqq\alpha<\dfrac{N+1}{2}\right)\\ \alpha & \left(\alpha=\dfrac{N+1}{2}\right)\\ \alpha-\frac12 & \left(\dfrac{N+1}{2}<\alpha\leqq N\right)\end{cases}$$

in the remaining cases.

$\mathfrak{G}=D_{n+1}^{(2)}$:

$$\begin{aligned}R(x)&=(x^2-k^2)(x^2-\xi^2)\sum_{\alpha\neq n+1,n+2}E_{\alpha\alpha}\otimes E_{\alpha\alpha}+k(x^2-1)(x^2-\xi^2)\sum_{\substack{\alpha\neq\beta,\beta'\\ \alpha\,\mathrm{or}\,\beta\neq n+1,n+2}}\\ &\cdot E_{\alpha\alpha}\otimes E_{\beta\beta}-(k^2-1)(x^2-\xi^2)\Bigg(\sum_{\substack{\alpha<\beta,\alpha\neq\beta'\\ \alpha,\beta\neq n+1,n+2}}+x^2\sum_{\substack{\alpha>\beta,\alpha\neq\beta'\\ \alpha,\beta\neq n+1,n+2}}\Bigg)E_{\alpha\beta}\otimes E_{\beta\alpha}\\ &-\tfrac12(k^2-1)(x^2-\xi^2)\Bigg((x+1)\Bigg(\sum_{\alpha<n+1,\beta=n+1,n+2}+x\sum_{\alpha>n+2,\beta=n+1,n+2}\Bigg)\\ &\cdot(E_{\alpha\beta}\otimes E_{\beta\alpha}+E_{\beta'\alpha'}\otimes E_{\alpha'\beta'})+(x-1)\Bigg(-\sum_{\alpha<n+1,\beta=n+1,n+2}+x\sum_{\alpha>n+2,\beta=n+1,n+2}\Bigg)\\ &\cdot(E_{\alpha\beta}\otimes E_{\beta'\alpha}+E_{\beta'\alpha'}\otimes E_{\alpha'\beta})\Bigg)+\sum_{\alpha,\beta\neq n+1,n+2}a_{\alpha\beta}(x)E_{\alpha\beta}\otimes E_{\alpha'\beta'}+\tfrac12\sum_{\alpha\neq n+1,n+2,\beta=n+1,n+2}\\ &\cdot(b_\alpha^+(x)(E_{\alpha\beta}\otimes E_{\alpha'\beta'}+E_{\beta'\alpha'}\otimes E_{\beta\alpha})+b_\alpha^-(x)(E_{\alpha\beta}\otimes E_{\alpha'\beta}+E_{\beta\alpha'}\otimes E_{\beta\alpha}))\\ &+\sum_{\alpha=n+1,n+2}(c^+(x)E_{\alpha\alpha}\otimes E_{\alpha'\alpha'}+c^-(x)E_{\alpha\alpha}\otimes E_{\alpha\alpha}\\ &+d^+(x)E_{\alpha\alpha'}\otimes E_{\alpha'\alpha}+d^-(x)E_{\alpha\alpha'}\otimes E_{\alpha\alpha'}),\end{aligned}\tag{3.7}$$

where for $\alpha,\beta \neq n+1, n+2$

$$a_{\alpha\beta}(x)=\begin{cases}(k^2x^2-\xi^2)(x^2-1) & (\alpha=\beta)\\ (k^2-1)(\xi^2k^{\bar\alpha-\bar\beta}(x^2-1)-\delta_{\alpha\beta'}(x^2-\xi^2)) & (\alpha<\beta)\\ (k^2-1)x^2(k^{\bar\alpha-\bar\beta}(x^2-1)-\delta_{\alpha\beta'}(x^2-\xi^2)) & (\alpha>\beta),\end{cases}$$

$$b^{\pm}_{\alpha}(x)=\begin{cases}\pm k^{\alpha-1/2}(k^2-1)(x^2-1)(x\pm\xi) & (\alpha<n+1)\\ k^{\alpha-n-5/2}(k^2-1)(x^2-1)x(x\pm\xi) & (\alpha>n+2)\end{cases},$$

$$c^{\pm}(x)=\pm\tfrac{1}{2}(k^2-1)(\xi+1)x(x\mp 1)(x\pm\xi)+k(x^2-1)(x^2-\xi^2),$$
$$d^{\pm}(x)=\pm\tfrac{1}{2}(k^2-1)(\xi-1)x(x\pm 1)(x\pm\xi),$$

and $\xi=k^n, \bar\alpha=\alpha+1(\alpha<n+1), =n+3/2(\alpha=n+1,n+2), =\alpha-1(\alpha>n+2)$.

Among these, the solutions for $A^{(1)}_n$ [6] and $A^{(2)}_2$ [7] have been known. In the case $A^{(1)}_n, R(x)$ splits into a direct sum of copies of 1×1 and 2×2 elementary blocks,

$$(x-k^2),\quad \begin{pmatrix} k(x-1) & -(k^2-1)\\ -(k^2-1)x & k(x-1)\end{pmatrix}.$$

Likewise (3.6) consists of blocks

$$((x-k^2)(x-\xi)),\quad \begin{pmatrix} k(x-1) & -(k^2-1)\\ -(k^2-1)x & k(x-1)\end{pmatrix}\times(x-\xi) \tag{3.8}$$

and an $N\times N$ piece $(a_{\alpha\beta}(x))$. In the case $D^{(2)}_{n+1}$ the elementary blocks are (3.8) (with x,ξ replaced by x^2 and ξ^2), 4×4 pieces

$$\begin{bmatrix} k(x^2-1) & 0 & \tfrac{1}{2}(k^2-1)(x-1) & -\tfrac{1}{2}(k^2-1)(x+1)\\ 0 & k(x^2-1) & -\tfrac{1}{2}(k^2-1)(x+1) & \tfrac{1}{2}(k^2-1)(x-1)\\ -\tfrac{1}{2}(k^2-1)x(x-1) & -\tfrac{1}{2}(k^2-1)x(x+1) & k(x^2-1) & 0\\ -\tfrac{1}{2}(k^2-1)x(x+1) & -\tfrac{1}{2}(k^2-1)x(x-1) & 0 & k(x^2-1)\end{bmatrix}\times(x^2-\xi^2)$$

and an $(N+2)\times(N+2)$ piece. They are all subject to the symmetry

$$\begin{aligned}&[R(x),H\otimes 1+1\otimes H]=0\quad (H\in\mathfrak{h}_0),\\ &PR(x)P=(S\otimes S)R(x)(S\otimes S)={}^tR(x),\\ &R(x^{-1},k^{-1})=\gamma(x,k)^{-1}\,{}^tR(x,k),\end{aligned} \tag{3.9}$$

with $\gamma(x,k)=-k^2x(\mathfrak{G}=A^{(1)}_n), =k^2\xi x^2(\mathfrak{G}=B^{(1)}_n,\ C^{(1)}_n,\ D^{(1)}_n,\ A^{(2)}_{2n},\ A^{(2)}_{2n-1}), = k^2\xi^2x^4(\mathfrak{G}=D^{(2)}_{n+1})$. Aside from these symmetries, they have the following properties.

(i) inversion relation

$$\check R(x)\check R(x^{-1})=\rho(x)I, \tag{3.10}$$

$$\rho(x)=\begin{cases}(x-k^2)(x^{-1}-k^2) & (\mathfrak{G}=A^{(1)}_n)\\ (x-k^2)(x-\xi)(x^{-1}-k^2)(x^{-1}-\xi) & (\mathfrak{G}=B^{(1)}_n,\ldots,A^{(2)}_{2n-1}),\\ (x^2-k^2)(x^2-\xi^2)(x^{-2}-k^2)(x^{-2}-\xi^2) & (\mathfrak{G}=D^{(2)}_{n+1})\end{cases}$$

where we have set $\check R(x)=PR(x)$.

(ii) As $k \to 1$,

$$R(x,k) = \kappa_1(x)(I + (k-1)(r(x) + \kappa_2(x)I) + \cdots)$$

for appropriate scalars $\kappa_i(x)$.
(iii) As $x \to 1$,

$$\check{R}(x,k) = \check{\kappa}_1(k)(I + (x-1)(s(k) + \check{\kappa}_2(k)I) + \cdots),$$

with some scalars $\check{\kappa}_i(k)$. Here the tensor $s(k)$ is given by

$\mathfrak{G} = A_n^{(1)}$:

$$s(k) = \tfrac{1}{2}(1-k^2)\sum_{\alpha\neq\beta} \operatorname{sgn}(\beta-\alpha)E_{\alpha\alpha}\otimes E_{\beta\beta} + \tfrac{1}{2}(1+k^2)\sum_{\alpha} E_{\alpha\alpha}\otimes E_{\alpha\alpha} + k\sum_{\alpha\neq\beta} E_{\alpha\beta}\otimes E_{\beta\alpha}, \tag{3.11}$$

$\mathfrak{G} = B_n^{(1)}, \ldots, A_{2n-1}^{(2)}$:

$$\begin{aligned} s(k) = {}& \tfrac{1}{2}(1-k^2)(1-\xi)\sum_{\alpha\neq\beta} \operatorname{sgn}(\beta-\alpha)E_{\alpha\alpha}\otimes E_{\beta\beta} \\ &+ \tfrac{1}{2}(1+k^2)(1-\xi)\sum_{\alpha\neq\alpha'} E_{\alpha\alpha}\otimes E_{\alpha\alpha} \\ &+ (k(1-\xi) - \tfrac{1}{2}(1-k^2)(1+\xi))\sum \delta_{\alpha\alpha'}E_{\alpha\alpha}\otimes E_{\alpha\alpha} \\ &+ k(1-\xi)\sum_{\alpha\neq\beta,\beta'} E_{\alpha\beta}\otimes E_{\beta\alpha} + (k^2-\xi)\sum_{\alpha\neq\alpha'} E_{\alpha\alpha'}\otimes E_{\alpha'\alpha} \\ &- (1-k^2)\left(\sum_{\alpha<\beta} \varepsilon_\alpha\varepsilon_\beta k^{\bar{\alpha}-\bar{\beta}}\xi + \sum_{\alpha>\beta} \varepsilon_\alpha\varepsilon_\beta k^{\bar{\alpha}-\bar{\beta}}\right)E_{\alpha'\beta}\otimes E_{\alpha\beta'}, \end{aligned} \tag{3.12}$$

$\mathfrak{G} = D_{n+1}^{(2)}$:

$$\begin{aligned} s(k) = {}& (1-k^2)(1-\xi^2)\sum_{\alpha\neq\beta,\alpha,\beta\neq n+1,n+2} \operatorname{sgn}(\beta-\alpha)E_{\alpha\alpha}\otimes E_{\beta\beta} \\ &+ \tfrac{1}{2}(1-k^2)(1-\xi^2)\sum_{\alpha=n+1,n+2,\beta\neq n+1,n+2} \operatorname{sgn}(\beta-\alpha)((E_{\alpha\alpha}+E_{\alpha'\alpha})\otimes E_{\beta\beta} \\ &+ E_{\beta'\beta'}\otimes(E_{\alpha'\alpha'}+E_{\alpha\alpha'})) \\ &+ (1+k^2)(1-\xi^2)\sum_{\alpha\neq n+1,n+2} E_{\alpha\alpha}\otimes E_{\alpha\alpha} \\ &+ 2k(1-\xi^2)\sum_{\alpha=n+1,n+2} (E_{\alpha\alpha}\otimes E_{\alpha\alpha} + E_{\alpha\alpha'}\otimes E_{\alpha'\alpha}) \\ &- \tfrac{1}{2}(1-k^2)\sum_{\alpha=n+1,n+2} ((1-\xi)^2(E_{\alpha\alpha}\otimes E_{\alpha\alpha} + E_{\alpha\alpha'}\otimes E_{\alpha\alpha'}) \\ &+ (1+\xi)^2(E_{\alpha\alpha'}\otimes E_{\alpha'\alpha} + E_{\alpha\alpha}\otimes E_{\alpha'\alpha'})) \\ &+ 2k(1-\xi^2)\sum_{\substack{\alpha\neq\beta,\beta' \\ \alpha \text{ or } \beta\neq n+1,n+2}} E_{\alpha\beta}\otimes E_{\beta\alpha} + 2(k^2-\xi^2)\sum_{\alpha\neq n+1,n+2} E_{\alpha\alpha'}\otimes E_{\alpha'\alpha} \\ &- 2(1-k^2)\left(\sum_{\alpha<\beta,\alpha,\beta\neq n+1,n+2} \xi^2 k^{\bar{\alpha}-\bar{\beta}} + \sum_{\alpha>\beta,\alpha,\beta\neq n+1,n+2} k^{\bar{\alpha}-\bar{\beta}}\right) \end{aligned}$$

$$\cdot E_{\alpha'\beta} \otimes E_{\alpha\beta'} - (1-k^2)\Bigg((1+\xi)\Bigg(\sum_{\alpha<n+1,\beta=n+1,n+2} \xi k^{\bar\alpha-\bar\beta}$$

$$+ \sum_{\alpha>n+2,\beta=n+1,n+2} k^{\bar\alpha-\bar\beta}\Bigg)(E_{\alpha'\beta}\otimes E_{\alpha\beta'} + E_{\beta\alpha'}\otimes E_{\beta'\alpha})$$

$$+(1-\xi)\Bigg(- \sum_{\alpha<n+1,\beta=n+1,n+2} \xi k^{\bar\alpha-\bar\beta} + \sum_{\alpha>n+2,\beta=n+1,n+2} k^{\bar\alpha-\bar\beta}\Bigg)$$

$$\cdot(E_{\alpha'\beta}\otimes E_{\alpha\beta} + E_{\beta\alpha'}\otimes E_{\beta\alpha})\Bigg).$$

Remark 1. It can be shown that under the first condition of (3.9), $R'(x) = (D(x)\otimes I)$ $R(x)(D(x)\otimes I)^{-1}$, where $D(x)\in \exp \mathfrak{h}_0$, $D(xy)=D(x)D(y)$, is again a solution to the YB equation. Choosing an appropriate $D(x)$ one obtains the R matrix in the principal picture.

Remark 2. Except for the case $\mathfrak{G} = D^{(2)}_{n+1}$, our R matrix satisfies $[\check{R}(x), \check{R}(y)] = 0$ so that $\check{R}(x)$ is diagonalizable independently of x. This is a consequence of (3.10), $\check{R}(1) \propto I$ and that $\deg R(x) \leqq 2$.

It is well known that a quantum R matrix gives rise to an integrable vertex model in statistical mechanics, whose transfer matrices

$$T(x) = \mathrm{tr}_{V_0}(R^{01}(x)R^{02}(x)\dots R^{0N}(x)) \in \mathrm{End}(V_1\otimes\cdots\otimes V_N)$$

commute among themselves: $[T(x), T(y)] = 0$. Hence their log derivatives provide a mutually commuting family of "spin" Hamiltonians. The first one is given by

$$T(1)^{-1}\frac{dT}{dx}(1) = \sum_{j=1}^{N} \frac{d\check{R}^{ii+1}}{dx}(1) = \sum_{j=1}^{N} s^{jj+1}(k) + \text{const } I.$$

In our case the tensors $s(k) \in \mathrm{End}(V\otimes V)$ are given by (3.11–13).

Acknowledgement. The author is indebted to stimulating discussions with Prof. T. Miwa and Prof. M. Hashizume.

References

1. Kulish, P. P., Sklyanin, E. K.: Solutions of the Yang–Baxter equation, J. Sov. Math. **19**, 1596 (1982)
2. Semenov-Tyan-Shanskii, M. A.: What is a classical *R*-matrix? Funct. Anal. Appl. **17**, 259 (1983)
   Drinfel'd, V. G.: Hamiltonian structures on Lie groups, Lie bi-algebras and the geometric meaning of the classical Yang–Baxter equations. Sov. Math. Dokl. **27**, 68 (1983)
   Reshetikhin, N. Yu., Faddeev, L. D.: Hamiltonian structures for integrable models of field theory Theor. Math. Phys. **56**, 847 (1984)
3. Belavin, A. A., Drinfel'd, V. G.: Solutions of the classical Yang–Baxter equation for simple Lie algebras. Funct. Anal. Appl. **16**, 159 (1982)
4. Kulish, P. P., Reshetikhin, N. Yu., Sklyanin, E. K.: Yang–Baxter equation and representation theory I. Lett. Math. Phys. **5**, 393 (1981)
5. Kulish, P. P., Reshetikhin, N. Yu.: Quantum linear problem for the sine-Gordon equation and higher representations. J. Sov. Math. **23**, 2435 (1983)
6. Babelon, O., de Vega, H. J., Viallet, C. M.: Solutions of the factorization equations from Toda field theory. Nucl. Phys. **B190**, 542 (1981)

Cherednik, I. V.: On a method of constructing factorized S matrices in elementary functions. Theor. Math. Phys. **43**, 356 (1980)
Chudnovsky, D. V., Chudnovsky, G. V.: Characterization of completely X-symmetric factorized S-matrices for a special type of interaction. Phys. Lett. **79A**, 36 (1980)
Schultz, C. L.: Solvable q-state models in lattice statistics and quantum field theory. Phys. Rev. Lett. **46**, 629 (1981)
Perk, J. H. H., Schultz, C. L.: New families of commuting transfer matrices in q-state vertex models. Phys. Lett. **84A**, 407 (1981)
7. Izergin, A. G., Korepin, V. E.: The inverse scattering method approach to the quantum Shabat–Mikhailov model. Commun. Math. Phys. **79**, 303 (1981)
8. Bogoyavlensky, O. I.: On perturbations of the periodic Toda lattice. Commun. Math. Phys. **51**, 201 (1976)
9. Olive, D. I., Turok, N.: Algebraic structure of Toda systems. Nucl. Phys. **B220** [FS8], 491 (1983)
10. Kac, V. G.: Infinite dimensional Lie algebras. Boston, MA: Birkhäuser 1983

Communicated by H. Araki

Received March 6, 1985

Note added in proof. After submitting the manuscript, the author learned that similar results have been obtained independently by Bazhanov (IHEP preprint 85-18, Serpukhov 1985)

VARIETY OF SOLUTIONS II —
IRF MODELS AND OTHERS

5. Variety of solutions II — IRF models and others —

In statistical mechanics, there are usually three ways of formulating two dimensional lattice models. These are called vertex models, interaction-round-the-face (IRF) medels and (Ising-type) spin models, depending on the form of the interaction. The YBE is written in different forms accordingly. Mathematically the three formulations are mutually equivalent[8)], but in each particular case one is often more natural and convenient than the others. The solutions of Section 4 all belong to the category of vertex models.

Baxter's series of reprints [20] concern the construction of the Bethe eigenvectors for the transfer matrix of the eight-vertex model. Here it is shown that one can associate to the eight-vertex model a family of solvable IRF models, called the eight-vertex solid-on-solid (8VSOS) models. As it turned out, a rich class of interesting models are included in the 8VSOS models, the Ising and the hard-hexagon models[18)] being the first two in the series. Detailed analysis of them was carried through by Andrews, Baxter and Forrester[5)] (In fact there is a subtle distinction between the original 8VSOS models and those treated by Andrews, Baxter and Forrester, the latter being called the restricted SOS models.)

The 8VSOS models (both restricted and unrestricted) are related to the Lie algebra $\mathfrak{sl}(2)$ in the sense that the eight-vertex model is. The reprint of Jimbo, Miwa and Okado [21] presents IRF models related to the other classical simple Lie algebras (see also [72-73),75),117),121-124)]). These models are parametrized by elliptic functions. Pasquier's reprint [22] revealed their true nature in the trigonometric limit; they can be generated from the vertex models built on quantum groups via a Wigner calculus.

There are also other IRF models discovered by direct methods[30-31),118),120),183)]. Pasquier [197-198)] extended the Andrews-Baxter-Forrester models to a direction different from the Jimbo-Miwa-Okado models, taking the fluctuation variables to range over the Dynkin diagrams of ADE. For further generalizations see [93-94),146),148),201)].

The $\mathbf{Z}_N$ model of Zamolodchikov and Fateev in the reprint [23] is an N state spin model generalizing the Ising model for $N = 2$. An elliptic extension of it has been found in [131)].

The free fermion model [35-37),90-92)] is a vertex model isolated from the family of section 4. It has been noted[35),111)] that the set of spectral parameters for this model can be regarded as an element in the group $SL(2)$, where the usual addition of the parameters in YBE is replaced by the group multiplication law.

The issue of full classification of solutions to YBE seems still an open problem (cf. [120),162-163)]). Krichever's work[138)] is an approach to this from algebro-geometric viewpoint.

Eight-Vertex Model in Lattice Statistics and One-Dimensional Anisotropic Heisenberg Chain. I. Some Fundamental Eigenvectors*

RODNEY BAXTER†

Institute for Theoretical Physics, State University of New York, Stony Brook, New York 11790

Received March 29, 1972

We obtain some simple eigenvectors of the transfer matrix of the zero-field eight-vertex model. These are also eigenvectors of the Hamiltonian of the one-dimensional anisotropic Heisenberg chain. We also obtain new equations for the matrix **Q**(v) introduced in earlier papers.

1. INTRODUCTION AND SUMMARY

In two previous papers [1, 2] (the results of which were anounced earlier [3, 4]), we obtained equations for the eigenvalues of the transfer matrix **T** of the two-dimensional zero-field eight-vertex lattice model, and of the Hamiltonian $\mathscr{H}$ of the one-dimensional anisotropic Heisenberg ring (the "XYZ model"). We were thus able to calculate the partition function of the eight-vertex model, and the ground-state energy of $\mathscr{H}$, for infinitely large systems.

In this and a subsequent paper we shall further obtain expressions for the eigenvectors of **T**. Since $\mathscr{H}$ and **T** commute (for appropriate values of their parameters), these are also the eigenvectors of $\mathscr{H}$.

In this present paper, we obtain some eigenvectors which have a particularly simple form. In a later paper we shall show that we can generalize these vectors to give a basis in which **T** breaks up into diagonal blocks. The general eigenvectors can then be obtained by a Bethe ansatz similar to that used for the "ice-type" models [5]. The eigenvectors we shall now obtain can be regarded as akin to the $n = 0$ case of these ice models (but see the end of this section for a fuller discussion of this point).

* Work supported in part by National Science Foundation No. P2P2847.

† On leave of absence from Research School of Physical Sciences, The Australian National University, Canberra A.C.T. 2600, Australia.

As a by-product of the working, in Section 6 we obtain new equations for the matrix $\mathbf{Q}(v)$ which was introduced in [1]. In particular, for the ice-type models these enable us to write down explicit expressions for the elements of $\mathbf{Q}(v)$ in the $n = \frac{1}{2}N$ subspace.

We define the model and state our main result straight away. Consider a square N by N lattice, wound on a torus, and place an arrow on each bond of the lattice so that at each vertex there are an even number of arrows pointing in (or out). Then there are eight possible configurations of arrows at each vertex, as shown in Fig. 1.

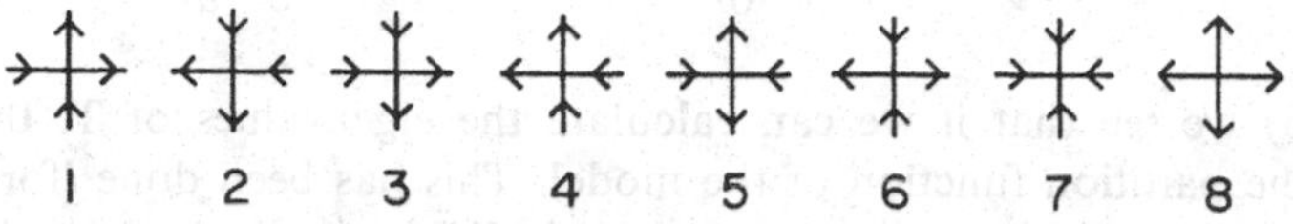

FIG. 1. The eight arrow configurations allowed at a vertex.

Associating energies $\epsilon_1, \ldots, \epsilon_8$ with these vertex configurations, the partition function is

$$Z = \sum \exp\left(-\beta \sum_{j=1}^{8} N_j \epsilon_j\right), \tag{1}$$

where the summation is over all allowed configurations of arrows on the lattice and N_j is the number of vertices of type j.

We impose the "zero field" condition

$$\begin{aligned} \epsilon_1 &= \epsilon_2, & \epsilon_3 &= \epsilon_4, \\ \epsilon_5 &= \epsilon_6, & \epsilon_7 &= \epsilon_8. \end{aligned} \tag{2}$$

We can then write the vertex weights

$$\omega_j = \exp(-\beta\epsilon_j) \tag{3}$$

as

$$\begin{aligned} \omega_1 &= \omega_2 = a, & \omega_3 &= \omega_4 = b, \\ \omega_5 &= \omega_6 = c, & \omega_7 &= \omega_8 = d. \end{aligned} \tag{4}$$

Look at some particular row of vertical bonds in the lattice and let $\alpha_J = +$ or $-$ according to whether there is an up or down arrow in column J ($J = 1, \ldots, N$). Let α denote the set $\{\alpha_1, \ldots, \alpha_N\}$, so that α defines the configuration of arrows on the whole row of bonds and has 2^N possible values. Then in Section 3 of [1] we show that

$$Z = \mathrm{Tr}\, \mathbf{T}^N \tag{5}$$

where $\mathbf{T}$ is the 2^N by 2^N transfer matrix. Its elements are

$$T_{\alpha,\alpha'} = \mathrm{Tr}\{\mathbf{R}(\alpha_1, \alpha_1')\, \mathbf{R}(\alpha_2, \alpha_2') \cdots \mathbf{R}(\alpha_N, \alpha_N')\}, \tag{6}$$

where the $\mathbf{R}(\alpha_J, \alpha_J')$ are 2 by 2 matrices:

$$\begin{aligned} \mathbf{R}(+,+) &= \begin{pmatrix} a & 0 \\ 0 & b \end{pmatrix}, & \mathbf{R}(+,-) &= \begin{pmatrix} 0 & d \\ c & 0 \end{pmatrix}, \\ \mathbf{R}(-,+) &= \begin{pmatrix} 0 & c \\ d & 0 \end{pmatrix}, & \mathbf{R}(-,-) &= \begin{pmatrix} b & 0 \\ 0 & a \end{pmatrix}, \end{aligned} \tag{7}$$

From (5) we see that if we can calculate the eigenvalues of $\mathbf{T}$, then we can calculate the partition function of the model. This has been done (for an infinite lattice) in [1]. A related problem (not solved in [1]) is to calculate the eigenvectors, and this is the problem to which we address ourselves. A knowledge of the eigenvectors is a first step towards calculating correlations and spontaneous polarizations of the model, but we hasten to add that these further calculations are very complicated and have not yet yielded tractable results.

We show that it is convenient to introduce parameters ρ, k, η, v (in general complex numbers), related to a, b, c, d by the equations

$$\begin{aligned} a &= \rho\Theta(-2\eta)\, \Theta(\eta - v)\, H(\eta + v), \\ b &= -\rho\Theta(-2\eta)\, H(\eta - v)\, \Theta(\eta + v), \\ c &= -\rho H(-2\eta)\, \Theta(\eta - v)\, \Theta(\eta + v), \\ d &= \rho H(-2\eta)\, H(\eta - v)\, H(\eta + v). \end{aligned} \tag{8}$$

For the purpose of discussing the matrix $\mathbf{Q}(v)$ (Section 6), it is sufficient to take $H(u)$ and $\Theta(u)$ to be the elliptic theta functions of argument u and modulus k (Section 8.192 of [6], hereinafter referred to as GR). The parametrization (8) is then the same as that of Ref. [1].

However, for our main purpose of discussing eigenvectors, it is necessary to suppose that there exist integers L, m_1, m_2 such that

$$L\eta = 2m_1K + im_2K', \tag{9}$$

where K, K' are the complete elliptic integrals of moduli k, $k' = (1 - k^2)^{1/2}$, respectively. It is then convenient to redefine $H(u)$, $\Theta(u)$ as

$$\begin{aligned} H(u) &= H_{Jb}(u) \exp[i\pi m_2(u - K)^2/(4KL\eta)], \\ \Theta(u) &= \Theta_{Jb}(u) \exp[i\pi m_2(u - K)^2/(4KL\eta) , \end{aligned} \tag{10}$$

where $H_{Jb}(u)$, $\Theta_{Jb}(u)$ are the normal Jacobian elliptic theta functions given in GR. The effect of this is to ensure that

$$H(u + 2L\eta) = H(u), \qquad \Theta(u + 2L\eta) = \Theta(u), \tag{11}$$

and simply to renormalize a, b, c, d.

Note that (9) is quite a mild restriction, since we can approach arbitrarily close to any desired value of η by taking sufficient large integer values of L, m_1 and m_2.

Using the definitions (10), we further define two functions:

$$\xi(+ \mid u) = H(u), \qquad \xi(- \mid u) = \Theta(u). \tag{12}$$

The set of numbers $\alpha = \{\alpha_1, ..., \alpha_N\}$ labels the 2^N possible states of the arrows on the N vertical bonds in a row of the lattice (we can think of these as N spins, each of which can be either up or down). We use the notation Ψ to denote a vector in this 2^N-dimensional space, and $[\Psi]_\alpha$ to denote the element of Ψ corresponding to the state α. Define a vector Ψ and a function $g(v)$ by

$$[\Psi]_\alpha = \sum_{j=1}^{L} \omega^j \prod_{J=1}^{N} \xi[\alpha_J \mid s + 2(J - j)\eta], \tag{13}$$

$$g(v) = \rho\Theta(0)\, H(v)\, \Theta(-v), \tag{14}$$

where ω is an L-th root of unity. Provided

$$N = L \times \text{integer} \tag{15}$$

we find that

$$\mathbf{T}\Psi = [\omega^{-1} g^N(v - \eta) + \omega g^N(v + \eta)]\Psi. \tag{16}$$

Thus Ψ is an eigenvector of the transfer matrix $\mathbf{T}$. There are more than one such eigenvectors, since s is arbitrary and there are L choices of ω. However, we show in Section 7 that there are at most $2N$ linearly independent eigenvectors that can be formed by taking linear combinations of Ψ's.

It is instructive to look at the case when $k \to 0$, η and v being held finite. In this case $d \to 0$ and we regain the ice-type models. The transfer matrix then breaks up into diagonal blocks connecting states with the same number of n of down arrows (spins). Since $K \to \frac{1}{2}\pi$ and $K' \to \infty$, to satisfy the restriction (9) we must set $m_2 = 0$, when it becomes

$$L\eta = m_1\pi. \tag{17}$$

The most general limiting form of (13) is obtained by choosing $s \pm \frac{1}{2}iK'$ to remain finite as $k \to 0$. From Section 8.192 of GR we then have

$$\xi(\alpha_J \mid s) \sim \exp\{\tfrac{1}{2}i(1 + \alpha_J)[\tfrac{1}{2}iK' \pm (s - K)]\} \tag{18}$$

and by substituting this expression into (13) we find that any Ψ is a sum of $2N$ linearly independent eigenvectors $\Psi_0, \Psi_{1,\pm}, \ldots, \Psi_{N-1,\pm}, \Psi_N$, given by

$$[\Psi_{n,\pm}]_\alpha = 0, \text{ if } \alpha_1 + \cdots + \alpha_N \neq N - 2n; = \exp\left\{\mp i\eta \sum_{J=1}^{N} J(1 - \alpha_J)\right\} \tag{19}$$

if $\alpha_1 + \cdots + \alpha_N = N - 2n$. In the latter case there must be n negative α_J's (down arrows). Writing $x_1, \ldots, x_n$ for the positions (J-values) of these negative α_J's, we see from (19) that

$$[\Psi_{n,\pm}]_\alpha = \exp[\mp 2i\eta(x_1 + \cdots + x_n)]. \tag{20}$$

Each such $\Psi_{n,\pm}$ satisfies the eigenvalue Eq. (16), the corresponding value of ω being

$$\omega = \exp[\mp 2inm_1\pi/L] \tag{21}$$

[using the restrictions (15) and (17)]. Thus in the ice-type limit there are precisely $2N$ linearly independent eigenvectors of the form (13). We surmise that the same is true for general values of k.

At first sight an eigenvector of the form (20) appears to contradict the results of Lieb [5] for the ice-type models, and Yang [7] for the related Heisenberg chain. However, this is not of course so. Both authors introduce a parameter Δ, which in our notation is given by

$$\begin{aligned}\Delta &= (a^2 + b^2 - c^2)/(2ab) \\ &= \cos(2\eta).\end{aligned} \tag{21}$$

Our form (20) of Ψ is equivalent to choosing $k_1 = \cdots = k_n = \pm 2\eta$ in [5] and [7]. In this case we see that

$$2\Delta e^{ik_j} - 1 - e^{i[k_j+k_l]} = 0 \tag{22a}$$

and hence the equations relating the coefficients A_P are automatically satisfied. Also, from (15) and (17) we see that

$$\exp(iNk_j) = 1, \tag{22b}$$

so the cyclic boundary conditions are satisfied.

We point out that values of η satisfying (17) for quite small values of L do explicitly occur in certain models. For instance, the XY model [8] is the Heisenberg chain with $\Delta = 0$, so from (21) we see that $\eta = \pi/4$. Also, the ice model [5] has vertex weights $a = b = c = 1$, hence $\Delta = \frac{1}{2}$ and $\eta = \pi/6$.

2. Simplification of $\mathbf{T}\psi$

We wish to find the eigenvectors of the matrix $\mathbf{T}$ defined by (6). This expression suggests first trying vectors whose elements are products of single-spin functions, i.e., a vector ψ with elements

$$[\psi]_\alpha = \phi_1(\alpha_1)\, \phi_2(\alpha_2) \cdots \phi_N(\alpha_N) \tag{23}$$

for all 2^N choices of $\alpha = \{\alpha_1, ..., \alpha_N\}$.

Such a vector is not translation invariant, i.e., replacing each α_J by α_{J+1} does not leave it unchanged. However, we can if we wish form a translation-invariant vector by performing all N cyclic permutations of the functions $\phi_1, ..., \phi_N$ in (23) and summing the N vectors thereby formed.

The product $\mathbf{T}\psi$ is of course a vector with elements

$$[\mathbf{T}\psi]_\alpha = \sum_{\alpha'} T_{\alpha,\alpha'} [\psi]_{\alpha'}, \tag{24}$$

where the summation is over all 2^N values of $\alpha' = \{\alpha_1', ..., \alpha_N'\}$, i.e., over $\alpha_1' = \pm, ..., \alpha_N' = \pm$. The point in looking at vectors of the type (23) is that (24) then has a fairly simple form. From (6) we find that

$$[\mathbf{T}\psi]_\alpha = \mathrm{Tr}\{\mathbf{U}_1(\alpha_1)\, \mathbf{U}_2(\alpha_2) \cdots \mathbf{U}_N(\alpha_N)\}, \tag{25}$$

where

$$\mathbf{U}_J(\alpha_J) = \sum_{\alpha_J'} \mathbf{R}(\alpha_J, \alpha_J')\, \phi_J(\alpha_J'). \tag{26}$$

Thus $\mathbf{U}_J(+)$ and $\mathbf{U}_J(-)$ are 2 by 2 matrices. From (7) we see that

$$\mathbf{U}_J(+) = \begin{pmatrix} a\phi_J(+) & d\phi_J(-) \\ c\phi_J(-) & b\phi_J(+) \end{pmatrix}, \qquad \mathbf{U}_J(-) = \begin{pmatrix} b\phi_J(-) & c\phi_J(+) \\ d\phi_J(+) & a\phi_J(-) \end{pmatrix}. \tag{27}$$

The next few steps of our argument closely parallel those of Appendix C of [1]. Note that (25) is unaffected if we replace each $\mathbf{U}_J(\pm)$ by

$$\mathbf{U}_J^*(\pm) = \mathbf{M}_J^{-1} \mathbf{U}_J(\pm)\, \mathbf{M}_{J+1} \tag{28}$$

for $J = 1,\ldots, N$, provided

$$\mathbf{M}_{N+1} = \mathbf{M}_1 . \tag{29}$$

The $\mathbf{M}_J$ are of course 2 by 2 matrices. The product (28) is most conveniently handled if we introduce p_J, p_J', r_J, r_J', related to the elements of $\mathbf{M}_J$ in such a way that

$$\mathbf{M}_J = \begin{pmatrix} r_J' & r_J p_J \\ r_J' p_J' & r_J \end{pmatrix}. \tag{30}$$

The trace in (25) will simplify to the sum of two products if we can choose the $\phi_J(\pm)$ and $\mathbf{M}_J$ so that all $\mathbf{U}_J{}^*(+)$ and $\mathbf{U}_J{}^*(-)$ are lower left-triangular matrices, i.e., their top right elements vanish. Evaluating the matrix product in (28), we find that this is so if

$$\begin{aligned} (ap_{J+1} - bp_J)\, \phi_J(+) + (d - cp_J p_{J+1})\, \phi_J(-) &= 0, \\ (c - dp_J p_{J+1})\, \phi_J(+) + (bp_{J+1} - ap_J)\, \phi_J(-) &= 0, \end{aligned} \tag{31}$$

for $J = 1,\ldots, N$.

Clearly (31) is a pair of homogeneous linear equations for $\phi_J(+)$ and $\phi_J(-)$, so the determinant of the coefficients must vanish, i.e.,

$$(a^2 + b^2 - c^2 - d^2)\, p_J p_{J+1} = ab(p_J{}^2 + p_{J+1}^2) - cd[1 + p_J{}^2 p_{J+1}^2] \tag{32}$$

for $J = 1,\ldots, N$.

Given p_J, (32) is a quadratic equation for p_{J+1}. Thus if we take p_1 to be given, we can construct the entire sequence $p_1, \ldots, p_{N+1}$, at each stage having a choice of two alternatives. To satisfy (29) we require that $p_{N+1} = p_1$, i.e., the sequence closes cyclically. It is obviously important to find the simplest possible parametrization of this sequence. In the next section we find that this leads us to introduce elliptic functions.

3. Parametrization in Terms of sn Functions

We seek to parametrize $a, b, c, d, p_1, \ldots, p_{N+1}$ so as to be readily able to handle the Eqs. (32).

Let

$$S = (a^2 + b^2 - c^2 - d^2)/(2ab), \tag{33}$$

$$X = cd/(ab). \tag{34}$$

Solving (32) for p_{J+1}, we find

$$p_{J+1} = (Sp_J \pm X^{1/2} D_J)/(1 - Xp_J{}^2), \tag{35}$$

where

$$D_J = \{1 - Yp_J^2 + p_J^4\}^{1/2} \tag{36}$$

and

$$Y = X + X^{-1} - X^{-1}S^2. \tag{37}$$

Note that the discrimant D_J is the square root of a quartic polynomial in p_J. Such expressions can be parametrized in terms of the Jacobian elliptic sn(u), cn(u), dn(u) functions, of argument u and modulus k, by using the properties

$$\begin{aligned} \mathrm{cn}^2(u) &= 1 - \mathrm{sn}^2(u), \\ \mathrm{dn}^2(u) &= 1 - k^2\,\mathrm{sn}^2(u) \end{aligned} \tag{38}$$

(Sections 8.141–8.144 of GR). In our case it is sufficient to define the modulus k so that

$$k + k^{-1} = Y \tag{39}$$

and a parameter u_J such that

$$p_J = k^{1/2}\,\mathrm{sn}(u_J). \tag{40}$$

From (36) and (38) it is then apparent that

$$D_J = \mathrm{cn}(u_J)\,\mathrm{dn}(u_J). \tag{41}$$

The elliptic functions satisfy a number of addition theorems. In particular, from Sections 8.156.1 of GR we have

$$\mathrm{sn}(u \pm v) = \frac{\mathrm{sn}\,u\,\mathrm{cn}\,v\,\mathrm{dn}\,v \pm \mathrm{sn}\,v\,\mathrm{cn}\,u\,\mathrm{dn}\,u}{1 - k^2\,\mathrm{sn}^2\,u\,\mathrm{sn}^2\,v}. \tag{42}$$

Using (40) and (41), we see that the r.h.s. of (35) is of the same form as (42). This suggests defining a further parameter η such that

$$X = k\,\mathrm{sn}^2(2\eta). \tag{43}$$

From (37) and (39) we can then verify that

$$S = \{1 - YX + X^2\}^{1/2} = \mathrm{cn}(2\eta)\,\mathrm{dn}(2\eta). \tag{44}$$

The complete elliptic integrals K, K' of moduli k, $k' = (1 - k^2)^{1/2}$ are defined in Sections 8.112 of GR. Incrementing η by K leaves (43) unchanged but negates the r.h.s. of (44) (Sections 8.151.2 of GR). Thus whatever the sign of S, there exists an η such that (43) and (44) are simultaneously satisfied.

Using (40), (41), (43), (44) in (35), we see from (42) that

$$p_{J+1} = k^{1/2}\,\mathrm{sn}(u_J \pm 2\eta). \tag{45}$$

Comparing (40) and (45), we see that the parametrization of the sequence $p_1 ,..., p_{N+1}$ is now simple. We define $\sigma_1 ,..., \sigma_N$ such that

$$\sigma_J = \pm 1 \tag{46}$$

for $J = 1,..., N$ (some may be positive, some negative), and set

$$u_J = t + 2(\sigma_1 + \cdots + \sigma_{J-1})\eta \tag{47}$$

for $J = 1,..., N + 1$ (t is arbitrary). Then p_J is given by (40) for $J = 1,..., N + 1$ and the Eqs. (32) are satisfied identically.

The requirement that $p_{N+1} = p_1$ leads to some complications that we wish to postpone. For the moment note that it is certainly satisfied if

$$\sigma_1 + \cdots + \sigma_N = 0. \tag{48}$$

The Eqs. (32) followed from (31). In order to handle these we further need to parametrize a, b, c, d so as to satisfy (43) and (44) [and hence (39)]. To do this, note from (34) and (43) that we can introduce a parameter z such that

$$\begin{aligned} c &= az^{-1}k^{1/2}\,\mathrm{sn}(2\eta) \\ d &= bzk^{1/2}\,\mathrm{sn}(2\eta). \end{aligned} \tag{49}$$

Equation (43) is then identically satisfied. Substituting these forms for c and d into (33) and (44), we get

$$[1 - z^2k\,\mathrm{sn}^2(2\eta)]b^2 - 2\,\mathrm{cn}(2\eta)\,\mathrm{dn}(2\eta)ab + [1 - z^{-2}k\,\mathrm{sn}^2(2\eta)]a^2 = 0. \tag{50}$$

This is a quadratic equation for b/a, with a discriminant that is a quartic in z. As with the equation relating p_{J+1} and p_J, we find that it is natural to introduce a further parameter y such that

$$z = k^{1/2}\,\mathrm{sn}(y). \tag{51}$$

Using the formulas (38) and (42), it then follows that we can choose

$$b/a = \mathrm{sn}(y - 2\eta)/\mathrm{sn}(y). \tag{52}$$

We can now obtain all the ratios $a : b : c : d$ from (49), (51) and (52). We find it convenient to set

$$y = v + \eta \tag{53}$$

and we then find that

$$a : b : c : d = \mathrm{sn}(v + \eta) : \mathrm{sn}(v - \eta) : \mathrm{sn}(2\eta) : k\, \mathrm{sn}(2\eta)\, \mathrm{sn}(v - \eta)\, \mathrm{sn}(v + \eta). \tag{54}$$

With this parametrization, Eqs. (39), (43) and (44) are satisfied identically.

4. Parametrization in Terms of Theta Functions

Up to now we have considered only the requirements (31) that the topright elements of $\mathbf{U}_J{}^*(\pm)$ vanish. We shall of course need to calculate the diagonal elements of these matrices. To do this it turns out to be convenient to normalize a, b, c, d and $\phi_J(\pm)$ in such a way as to introduce the Jacobi elliptic functions $H(u)$, $\Theta(u)$ defined in Section 8.192 of GR. These are related to $\mathrm{sn}(u)$ by

$$k^{1/2}\, \mathrm{sn}(u) = H(u)/\Theta(u) \tag{55}$$

(Sections 8.191.1 of GR).

Noting that $\mathrm{sn}(u)$ is an odd function, it follows from (54) and (55) that we can choose the normalization of a, b, c, d so that they are given by (8), ρ being regarded as a normalization constant. The theta functions are entire functions in the complex plane and satisfy the quasi-periodic conditions

$$\begin{aligned} H(u + 2K) &= -H(u), \qquad \Theta(u + 2K) = \Theta(u), \\ H(u + 2iK') &= -H(u) \exp[\pi(K' - iu)/K], \\ \Theta(u + 2iK') &= -\Theta(u) \exp[\pi(K' - iu)/K] \end{aligned} \tag{56}$$

(Sections 8.182.1–4 of GR).

Also, $H(u)$ has only simple zeros, occurring at

$$u = 2mK + 2inK', \quad m, n \text{ integers} \tag{57}$$

[Section 8.151.1 of GR, using (55)].

Since $\mathrm{sn}(u)$ is an odd function, we can easily deduce from (55) the identity

$$H(u)\, \Theta(-u) + H(-u)\, \Theta(u) = 0. \tag{58}$$

These properties (56)–(58) are sufficient for us to be able to prove two addition theorems which are all we need to determine $\phi_J(\pm)$ and the diagonal elements of $U_J{}^*(\pm)$.

They are

$$\Theta(u)\,\Theta(v)\,H(w)\,H(u+v+w)+H(u)\,H(v)\,\Theta(w)\,\Theta(u+v+w)$$
$$=\Theta(0)\,\Theta(u+v)\,H(v+w)\,H(w+u), \tag{59}$$
$$\Theta(u)\,\Theta(v)\,\Theta(w)\,\Theta(u+v+w)+H(u)\,H(v)\,H(w)\,H(u+v+w)$$
$$=\Theta(0)\,\Theta(u+v)\,\Theta(v+w)\,\Theta(w+u) \tag{60}$$

for all values of u, v, w.

To prove (59), regard u and v as constants. We can then think of the ratio of the l.h.s. to the r.h.s. as a function $F(w)$. From (56) we can verify that this function is doubly periodic, with periods $2K$ and $2iK'$. Within a given period rectangle the r.h.s. has zeros (simple) only at $w=-u$ and $w=-v$, but by using (58) we see that these are also zeros of the l.h.s. Thus $F(w)$ is entire and doubly periodic. It is therefore bounded, and by the Cauchy–Liouville theorem it must be a constant. Setting $w=0$, we find that this constant is unity. Hence $F(w)\equiv 1$, which proves (59).

The proof of (60) is similar, except that is convenient to first rearrange the equation so that all Θ functions occur on the l.h.s., all H functions on the r.h.s.

5. Evaluation of $\mathbf{T}\psi$

We are now in a position to calculate $\phi_J(\pm)$, $U_J{}^*(\pm)$, ψ and $\mathbf{T}\psi$. From (40) and (55) we see that

$$p_J=H(u_J)/\Theta(u_J) \tag{61}$$

where $u_1,\ldots,u_{N+1}$ are given by (47), in particular,

$$u_{J+1}=u_J+2\sigma_J\eta, \tag{62}$$

where $\sigma_J=\pm 1$. The vertex weights a, b, c, d we take to be given by (8).

Using these formulations together with the formulas (58)–(60), we can evaluate all the expressions occurring in (31). For instance,

$$ap_{J+1}-bp_J=\rho\Theta(-2\eta)\{\Theta(\eta-v)\,\Theta(u_J)\,H(\eta+v)\,H(u_{J+1})$$
$$+H(\eta-v)\,H(u_J)\,\Theta(\eta+v)\,\Theta(u_{J+1})\}/[\Theta(u_J)\,\Theta(u_{J+1})]. \tag{63}$$

We can simplify this expression by using the identity (59). If $\sigma_J=+1$ we replace u, v, w in (59) by $\eta-v$, u_J, $\eta+v$; if $\sigma_J=-1$ we replace them by $\eta+v$, u_{J+1}, $\eta-v$. In either case we obtain

$$ap_{J+1}-bp_J=\rho\Theta(0)\,H(2\eta)\,\Theta(-2\eta)\,H[u_J+\sigma_J(\eta+v)]$$
$$\times\,\Theta[u_J+\sigma_J(\eta-v)]/[\Theta(u_J)\,\Theta(u_{J+1})]. \tag{64}$$

Similarly,

$$d - cp_J p_{J+1} = \rho\Theta(0)\, H(-2\eta)\, H[u_J + \sigma_J(\eta + v)]\, \Theta(2\eta) \times H[u_J + \sigma_J(\eta - v)]/[\Theta(u_J)\, \Theta(u_{J+1})], \tag{65}$$

so from the first of the Eqs. (31), using (58), we see that we can choose

$$\begin{aligned} \phi_J(+) &= H[u_J + \sigma_J(\eta - v)], \\ \phi_J(-) &= \Theta[u_J + \sigma_J(\eta - v)]. \end{aligned} \tag{66}$$

It is convenient to introduce the two functions $\xi(\pm \mid u)$ defined by (12). Equation (66) can then be written as

$$\phi_J(\pm) = \xi[\pm \mid u_J + \sigma_J(\eta - v)]. \tag{67}$$

Equations (31) ensure that $\mathbf{U}_J{}^*(\pm)$ is a 2 by 2 matrix of the form

$$\mathbf{U}_J{}^*(\pm) = \begin{pmatrix} A_J(\pm) & 0 \\ C_J(\pm) & B_J(\pm) \end{pmatrix}. \tag{68}$$

The diagonal elements can be simplified by using (31). For instance, from Eqs. (27)–(30) we find that

$$A_J(+) = r'_{J+1}(F_J + p'_{J+1}G_J)/[r_J'(1 - p_J p_J')], \tag{69}$$

where

$$\begin{aligned} F_J &= a\phi_J(+) - cp_J\phi_J(-), \\ G_J &= d\phi_J(-) - bp_J\phi_J(+). \end{aligned} \tag{70}$$

The first of equations (31) can be written as

$$p_{J+1}F_J + G_J = 0; \tag{71}$$

so (69) is equivalent to

$$A_J(+) = r'_{J+1}(1 - p_{J+1}p'_{J+1})F_J/[r_J'(1 - p_J p_J')]. \tag{72}$$

Using (8), (61) and (66), F_J can also be evaluated by (59), giving

$$F_J = \rho\Theta(0)\, H(v - \eta)\, \Theta(\eta - v)\, \Theta(u_{J+1})\, H[u_J - \sigma_J(\eta + v)]/\Theta(u_J). \tag{73}$$

Similarly, we can calculate the other diagonal elements. We find that if we choose r_J, r_J' so that

$$\begin{aligned} r_J &= \Theta(u_J), \\ \det \mathbf{M}_J &= r_J r_J'(1 - p_J p_J') = 1, \end{aligned} \tag{74}$$

then

$$\begin{aligned} A_J(\pm) &= g(v - \eta)\, \xi[\pm \mid u_J - \sigma_J(\eta + v)], \\ B_J(\pm) &= g(v + \eta)\, \xi[\pm \mid u_J + \sigma_J(3\eta - v)], \end{aligned} \tag{75}$$

where the function $g(v)$ is defined by

$$g(v) = \rho\Theta(0)\, H(v)\, \Theta(-v). \tag{76}$$

We shall not need $C_J(\pm)$ in this paper, but for future reference record that, using (27)–(30) and (74),

$$\begin{aligned} C_J(+) &= r_J' r_{J+1}' \{(b p_{J+1}' - a p_J')\, \phi_J(+) + (c - d p_J' p_{J+1}')\, \phi_J(-)\}, \\ C_J(-) &= r_J' r_{J+1}' \{(d - c p_J' p_{J+1}')\, \phi_J(+) + (a p_{J+1}' - b p_J')\, \phi_J(-)\}. \end{aligned} \tag{77}$$

From (23), (47) and (67) we see that the vector ψ we have constructed depends on v, t and $\sigma_1, ..., \sigma_N$. Exhibiting this dependence explicitly, we can write ψ as a vector function $\psi(v \mid t, \sigma)$, where σ denotes the set $\{\sigma_1, ..., \sigma_N\}$. The element of this vector corresponding to the state $\alpha = \{\alpha_1, ..., \alpha_N\}$ is

$$[\psi(v \mid t, \sigma)]_\alpha = \prod_{J=1}^{N} \xi[\alpha_J \mid t + 2(\sigma_1 + \cdots + \sigma_{J-1})\eta + \sigma_J(\eta - v)]. \tag{78}$$

Note that the arguments of the functions ξ in (75) differ from those in (67) only by replacing v by $v \pm 2\eta$. Replacing each $\mathbf{U}_J(\pm)$ in (25) by $\mathbf{U}_J^*(\pm)$, and expanding the trace by using (68), it follows that

$$\mathbf{T}\psi(v \mid t, \sigma) = g^N(v - \eta)\, \psi(v + 2\eta \mid t, \sigma) + g^N(v + \eta)\, \psi(v - 2\eta \mid t, \sigma). \tag{79}$$

These Eqs. (78) and (79) are true for all complex numbers v, t, and all allowed values ± 1 of $\sigma_1, ..., \sigma_N$, provided the boundary condition (29) is satisfied.

6. The Matrix $\mathbf{Q}(v)$

This section is not strictly relevant to our main aim of discussing the eigenvectors of $\mathbf{T}$, but sheds some new light on the method previously used for obtaining the eigenvalues.

We suppose that N is even and

$$\sigma_1 + \cdots + \sigma_N = 0. \tag{80}$$

Thus $\frac{1}{2}N$ of the σ_j are $+1$, the other $\frac{1}{2}N$ are -1. From (47) we have $u_{N+1} = u_1$, and the boundary condition (29) is automatically satisfied (for *all* values of η).

Let $\mathbf{Q}_R(v)$ be a 2^N by 2^N matrix whose columns are all vectors $\psi(v \mid t, \sigma)$, using different values of t or σ for different columns. From (79) we then have

$$\mathbf{T}\mathbf{Q}_R(v) = g^N(v - \eta)\, \mathbf{Q}_R(v + 2\eta) + g^N(v + \eta)\, \mathbf{Q}_R(v - 2\eta). \tag{81}$$

From (7) and (8) we can verify that replacing v by $2K - v$ is equivalent to transposing $\mathbf{T}$. Replacing v by $2K - v$ in (81) and transposing, we obtain

$$\mathbf{Q}_L(v)\mathbf{T} = g^N(v - \eta)\, \mathbf{Q}_L(v + 2\eta) + g^N(v + \eta)\, \mathbf{Q}_L(v - 2\eta), \tag{82}$$

where

$$\mathbf{Q}_L(v) = \mathbf{Q}'(2K - v), \tag{83}$$

the prime denoting transposition. We have used (76) and the relations

$$H(2K - v) = H(v), \qquad \Theta(2K - v) = \Theta(v).$$

By using methods similar to those used to derive (59) and (60), we can show that there exist single-valued functions $f_1(u), f_2(u)$, periodic of period $4K$, such that

$$H(u)\, H(v) + \Theta(u)\, \Theta(v) = f_1(u + v) f_2(u - v). \tag{84}$$

From (78) we can then deduce that the scalar product

$$\psi'(2K - u \mid t', \sigma')\, \psi(v \mid t, \sigma)$$

is a symmetric function of u and v, for any values of t', σ', t, σ. (The proof is not completely trivial; we note that all allowed values of $\{\sigma_1', \ldots, \sigma_N'\}$ are permutations of one another, consider the effect of interchanging σ_j' and σ_{j+1}', and obtain a proof by recursion).

From this symmetry and the definitions of $\mathbf{Q}_R(v)$, $\mathbf{Q}_L(v)$, it follows that

$$\mathbf{Q}_L(u)\, \mathbf{Q}_R(v) = \mathbf{Q}_L(v)\, \mathbf{Q}_R(u). \tag{85}$$

Suppose we can construct $\mathbf{Q}_R(v)$, $\mathbf{Q}_L(v)$ so that they are nonsingular at some value v_0 of v. Then from (85) we can further define a matrix $\mathbf{Q}(v)$ such that

$$\mathbf{Q}(v) = \mathbf{Q}_R(v)\, \mathbf{Q}_R^{-1}(v_0) = \mathbf{Q}_L^{-1}(v_0)\, \mathbf{Q}_L(v). \tag{86}$$

Post-multiplying (81) by $\mathbf{Q}_R^{-1}(v_0)$, and pre-multiplying (82) by $\mathbf{Q}_L^{-1}(v_0)$, we therefore find

$$\mathbf{TQ}(v) = g^N(v-\eta)\,\mathbf{Q}(v+2\eta) + g^N(v+\eta)\,\mathbf{Q}(v-2\eta) \tag{87}$$

$$= \mathbf{Q}(v)\mathbf{T}. \tag{88}$$

Similarly, from (85) we obtain

$$\mathbf{Q}(u)\,\mathbf{Q}(v) = \mathbf{Q}(v)\,\mathbf{Q}(u). \tag{89}$$

These functional matrix relations (87)–(89) were derived in [1]. Together with the analyticity and quasi-periodic properties of $\mathbf{Q}(v)$ (which can be obtained from our construction), they enable one to obtain equations for the eigenvalues of the transfer matrix $\mathbf{T}$.

The above methods provide a different (though obviously related) definition of $\mathbf{Q}(v)$ to that of [1], which may help us to understand $\mathbf{Q}(v)$ a little better. From (86) we have

$$\mathbf{Q}(v)\,\mathbf{Q}_R(v_0) = \mathbf{Q}_R(v). \tag{90}$$

Each column of $\mathbf{Q}_R(v)$ is a ψ-vector. Thus a typical column of the matrix Eq. (90) is

$$\mathbf{Q}(v)\,\psi(v_0 \mid t, \sigma) = \psi(v \mid t, \sigma). \tag{91}$$

This result must be true for all values of t and σ used in the column vectors of $\mathbf{Q}_R(v)$. However, from the reasoning of Section 6 of [1], we see that (87)–(89) define the eigenvectors of $\mathbf{Q}(v)$ (the same as of $\mathbf{T}$), and each eigenvalue to within a multiplicative constant. The definition (86) fixes these constants so as to ensure that $\mathbf{Q}(v)$ is the identity matrix when $v = v_0$. Thus $\mathbf{Q}(v)$ is then uniquely defined and (91) must be true whatever values of t and σ we select.

This is quite a startling result, since (91) gives an infinite number of equations as t, σ are varied, and $\mathbf{Q}(v)$ has to be chosen so as to satisfy all of them. Nevertheless we conjecture that this can be done.

It is instructive to consider the case when $k \to 0$, η and v being held finite during this limiting procedure. In this case $d \to 0$ and we regain the ice-type "six-vertex" models [5].

The most general form for ψ is obtained if we hold not t, but $t - \frac{1}{2}iK'$ finite. In this limit we can deduce from Sections 8.112, 8.192 of GR that

$$K \to \tfrac{1}{2}\pi, \qquad K' \to \infty, \tag{92}$$

$$\begin{aligned} H(t) &\sim \exp[i(K + \tfrac{1}{2}iK' - t)], \\ \Theta(t) &\sim 1. \end{aligned} \tag{93}$$

Using (12), we can combine these last two formulas as

$$\xi(\alpha_J \mid t) \sim \exp\{\tfrac{1}{2}i(1+\alpha_J)(K+\tfrac{1}{2}iK'-t)\}. \tag{94}$$

Substituting this into (78), we then obtain

$$\psi(v \mid t, \sigma) = \sum_{n=0}^{N} \Phi_n(v \mid \sigma) \exp\{i(N-n)(K+\tfrac{1}{2}iK'-t)\}, \tag{95}$$

where $\Phi_n(v \mid \sigma)$ is a vector whose elements are zero except for states with n negative α_J (down arrows) and $N-n$ positive α_J (up arrows). For such states its elements are

$$[\Phi_n(v \mid \sigma)]_\alpha = \exp\left\{-\tfrac{1}{2}i \sum_{J=1}^{N} (1+\alpha_J)[2(\sigma_1+\cdots+\sigma_{J-1})\eta + \sigma_J(\eta - v)]\right\}. \tag{96}$$

Thus for any value of t, $\psi(v \mid t, \sigma)$ lies in a subspace spanned by $\Phi_0(v \mid \sigma),\ldots, \Phi_N(v, \sigma)$. Further, any $\Phi_n(v \mid \sigma)$ can be formed by Fourier analyzing $\psi(v \mid t, \sigma)$. Thus (91) is equivalent to

$$\mathbf{Q}(v)\, \Phi_n(v_0 \mid \sigma) = \Phi_n(v \mid \sigma) \tag{97}$$

for $n = 0,\ldots, N$ and all σ satisfying (80). Hence $\mathbf{Q}(v)$ breaks up into $N+1$ diagonal blocks connecting states with the same number n of down arrows. This is to be expected, since we know the transfer matrix $\mathbf{T}$ has this property for the ice-type models [5].

Look at some particular value of n in (97). The corresponding diagonal block of $\mathbf{Q}(v)$ is of dimension $\binom{N}{n}$ by $\binom{N}{n}$, while there are $\binom{N}{\frac{1}{2}N}$ values of $\sigma_1,\ldots, \sigma_N$ that satisfy (80). Thus in general there are more equations obtainable from (97) than there are elements of $\mathbf{Q}(v)$. However, we expect that for general values of v_0 they can be solved (they can for $n = 1$).

When $n = \frac{1}{2}N$ we have just the right number of equations for the number of unknowns. In this case (which is interesting since this subspace contains the maximum eigenvalue of $\mathbf{T}$) we can obtain an explicit expression for $\mathbf{Q}(v)$ by letting $v_0 \to -i\infty$. From (96) we then find after dividing by a constant factor $\exp(\frac{1}{2}iNv_0)$ that

$$[\Phi_n(v_0 \mid \sigma)]_\alpha = 0 \tag{98}$$

unless $\sigma_1 = \alpha_1,\ \sigma_2 = \alpha_2,\ldots, \sigma_N = \alpha_N$. From (97) it follows that the element of $\mathbf{Q}(v)$ between states α and β is given by

$$[\mathbf{Q}(v)]_{\alpha,\beta} = [\Phi_n(v) \mid \beta)]_\alpha / [\Phi_n(v_0 \mid \beta)]_\beta\,. \tag{99}$$

In the $n = \frac{1}{2}N$ subspace α, β satisfy

$$\alpha_1 + \cdots + \alpha_N = \beta_1 + \cdots + \beta_N = 0. \tag{100}$$

Using this restriction on β, it follows from (96) and (99) that

$$[\mathbf{Q}(v)]_{\alpha,\beta} = \exp\left\{\tfrac{1}{2}i\eta \sum\sum_{1\leqslant J<K\leqslant N} (\alpha_J\beta_K - \alpha_K\beta_J) + \tfrac{1}{2}iv\sum_{J=1}^{N}\alpha_J\beta_J\right\} \tag{101}$$

[removing the constant factor $\exp(-\frac{1}{2}iNv_0)$].

This is a comparitively simple expression for the elements of $\mathbf{Q}(v)$. Since $\mathbf{T}$ commutes with $\mathbf{Q}(v)$ and hence has the same eigenvectors, it is possible that some new insights into the structure of the transfer matrix could be obtained from (101).

It seems appropriate to give a word or warning regarding the choice of v_0. For symmetry reasons it is tempting to try $v_0 = \pm\eta$, 0 or K. However, for $n = 2$ and $N = 4$ we can verify from (101) that at these values $\mathbf{Q}(v_0)$ is singular and (86) cannot be satisfied. This seems likely to be true for arbitrary n, N. Thus either such values of v_0 should be avoided or the above definitions should be modified.

7. Eigenvectors of T

We return to our prime aim, which is to find vectors of the type $\psi(v \mid t, \sigma)$, or simple linear combinations of these, which are eigenvectors if $\mathbf{T}$. To do this we see from (79) that we should like $\mathbf{T}\psi(v \pm 2\eta \mid t, \sigma)$ to simplify into the sum of two terms in the same way as $\mathbf{T}\psi(v \mid t, \sigma)$ does, i.e. given t, we should like there to exist t', σ' such that

$$\psi(v + 2\eta \mid t, \sigma) = \psi(v \mid t', \sigma') \tag{102}$$

[Similarly for $\psi(v - 2\eta \mid t, \sigma)$]. From (78) we can see that this will in general only be so if the σ_J are all equal.

Consider the case when

$$\sigma_1 = \sigma_2 = \cdots = \sigma_N = 1. \tag{103}$$

From (47) we see that

$$u_{N+1} = u_1 + 2N\eta. \tag{104}$$

We can satisfy the boundary condition (29) by using the periodic or quasi-periodic properties of the elliptic functions. Suppose there exist intergers L, m_1, m_2 such that

$$L\eta = 2m_1K + im_2K' \tag{105}$$

(as we pointed out in the Introduction, this is quite a weak condition); then since $\mathrm{sn}(u)$ is periodic of periods $4K$, $2iK'$, we have

$$\mathrm{sn}(u + 2L\eta) = \mathrm{sn}(u). \tag{106}$$

Thus if

$$N = L \times \text{integer}, \tag{107}$$

we see from (104) and (40) that

$$p_{N+1} = p_1 . \tag{108}$$

This goes part of the way toward satisfying the boundary conditions. In particular it ensures that $\mathbf{M}_{N+1}^{-1}\mathbf{M}_1$ is diagonal. In fact this is sufficient to ensure that the trace in (25) decomposes into the sum of two products, but we get some irritating extra factors due to our normalization in terms of theta functions, which in general are only quasi-periodic over a period $2L$.

The easiest way out of these problems seems to be to change our definitions of the theta functions, renormalizing them to ensure the required periodicity. From now on we denote the normal Jacobi theta functions defined in GR as $H_{Jb}(u)$, $\Theta_{Jb}(u)$. We use $H(u)$, $\Theta(u)$ to denote the renormalized theta functions defined by (10). Since we can establish from (56) that

$$\begin{aligned} H_{Jb}(u + 4mK) &= H_{Jb}(u), \\ H_{Jb}(u + 2inK') &= (-1)^n H_{Jb}(u) \exp[\pi(n^2K' - inu)/K] \end{aligned} \tag{109}$$

for all u and all integers m, n, it follows that the function $H(u)$ defined by (10) is periodic of period $2L\eta$. Similarly, so is $\Theta(u)$.

We can verify that this renormalization of the theta functions leaves the identities (58)–(60) unchanged. Thus the working of Section 4, notably Eqs. (78) and (79), remains true. The condition (107) is now sufficient to ensure that the boundary condition (29) is completely satisfied.

From (78) and (103) we see that $\psi(v \mid t, \sigma)$ involves t and v only via their difference $t - v$. It follows that if we define a vector

$$\psi_j = \psi(v \mid s + v + (1 - 2j)\eta, \sigma), \tag{110}$$

with elements

$$[\psi_j]_\alpha = \prod_{J=1}^{N} \xi[\alpha_J \mid s + 2(J - j)\eta], \tag{111}$$

then (79) implies that

$$\mathbf{T}\psi_j = g^N(v - \eta)\, \psi_{j+1} + g^N(v + \eta)\, \psi_{j-1} . \tag{112}$$

From the periodicity of our modified theta functions we see that

$$\psi_{j+L} = \psi_j ; \tag{113}$$

so if ω is an L-th root of unity and

$$\Psi = \sum_{j=1}^{L} \omega^j \psi_j , \tag{114}$$

then

$$\mathbf{T}\Psi = [\omega^{-1} g^N(v - \eta) + \omega g^N(v + \eta)]\Psi. \tag{115}$$

Thus Ψ is an eigenvector of the transfer matrix $\mathbf{T}$, as stated in the Introduction.

By varying s and making different choices of ω, we can construct an infinite number of eigenvectors. Clearly only a finite number of these can be linearly independent; we now show that there at most $2N$ such linearly independent eigenvectors.

To do this we need the quasiperiodic properties of our renormalized theta functions. From the definitions (9), (10) and the properties (56) of the Jacobi functions, we find that

$$\begin{aligned} &H(u + 4pK + 2iqK') \\ &\quad = H(u) \exp\{2\pi i(pm_2 - qm_1)[u + (2p - 1)K + iqK']/(L\eta)\} \end{aligned} \tag{116}$$

for all integers p, q. The same relation holds if we replace the function H by Θ.

From (111) it is apparent that there exists a vector function $\Phi(s)$ such that

$$\psi_j = \Phi(s - 2j\eta) \tag{117}$$

for all values of s and j. The elements of $\Phi(s)$ are products of our renormalized theta functions and differ only in whether we use an H or Θ function in a given position. Since (116) applies to both these functions, we can verify, using (107), that the vector function $\Phi(s)$ satisfies the quasi-periodic relation

$$\begin{aligned} &\Phi(s + 4pK + 2iqK') \\ &\quad = \Phi(s) \exp\{2\pi i N(pm_2 - qm_1)[s + (2p - 1)K + iqK']/(L\eta)\} \end{aligned} \tag{118}$$

for all integers p and q.

Setting $p = m_1$, $q = m_2$ and using (9), it is apparent that $\Phi(s)$ is periodic of period $2L\eta$. More generally, if l is the highest common factor of m_1 and m_2, then $\Phi(s)$ is periodic of period $2L\eta/l$. Hence there exists a Fourier expansion

$$\Phi(s) = \sum_{n=-\infty}^{\infty} \mathbf{x}_n \exp[i\pi nls/(L\eta)], \tag{119}$$

where the $\mathbf{x}_n$ are fixed vectors, independent of s.

Without loss of generality we can suppose that the integers L, m_1 and m_2 in (9) have no common factors, and hence L and l have no common factor. Substituting the form (119) of $\Phi(s)$ into (117) and (114), it follows that Ψ is an eigenvector of $\mathbf{T}$ for all values of s if and only if each $\mathbf{x}_n$ is an eigenvector. The eigenvalue corresponding to $\mathbf{x}_n$ is, using (115),

$$\omega_n^{-1} g^N(v-\eta) + \omega_n g^N(v+\eta), \tag{120}$$

where ω_n is the L-th root of unity given by

$$\omega_n = \exp(2\pi i n l/L). \tag{121}$$

We appear to have constructed an infinite set of eigenvectors $\mathbf{x}_n$. However, substituting the form (119) of $\Phi(s)$ into (118) and equating Fourier coefficients, we find that

$$\mathbf{x}_{n'} = \mathbf{x}_n \exp\{\pi i l[(n+n')(2pK+iqK') + (n-n')K]/(L\eta)\} \tag{122}$$

for

$$n' = n + 2N(pm_2 - qm_1)/l \tag{123}$$

and all integer values of n, p and q.

Since l is a factor of m_1 and m_2, we see from (123) that n' is an integer, as it should be. Also, as it is the highest such factor, for any integer r we can choose p and q so that

$$(pm_2 - qm_1)/l = r. \tag{124}$$

It therefore follows from (122) that the vectors $\mathbf{x}_n$, $\mathbf{x}_{n\pm 2n}$, $\mathbf{x}_{n\pm 4N}$, etc. are simply scalar multiples of one another. With appropriate normalization they therefore give the same eigenvector.

Thus there are at most $2N$ distinct eigenvectors $\mathbf{x}_n$, namely, $\mathbf{x}_1, \mathbf{x}_2, \ldots, \mathbf{x}_{2N}$. From (120) and (121) we see that they form L sets of $2N/L$ eigenvectors, all members of a set having the same eigenvalue.

We have discussed these vectors in the "ice" limit $k \to 0$ in the Introduction and shown that they are then linearly independent. It seems reasonable to suppose that this will also be so for nonzero k.

8. Heisenberg Chain

We can regard the transfer matrix $\mathbf{T}$ defined by Eqs. (6)–(8) as a function $\mathbf{T}(v)$ of v (regarding ρ, k, η as constants). Then in [1] we showed that two matrices $\mathbf{T}(u)$, $\mathbf{T}(v)$ commute. (Note that the eigenvectors we have constructed are independent of v, in agreement with this property.)

In [2] we further showed that the logarithmic derivative of $\mathbf{T}(v)$ at $v = \eta$ was related to the Hamiltonian of the fully anisotropic one-dimensional Ising-Heisenberg ring (the "XYZ model"). Using our present notation and renormalized theta functions, we can verify from (6)–(8) that

$$[\mathbf{T}(\eta)]^{-1}\,\mathbf{T}'(\eta) = \tfrac{1}{2}\mathscr{H}/\mathrm{sn}(2\eta) + Np\mathbf{E}, \tag{125}$$

where

$$p = \frac{1}{2}\left\{\frac{H'(2\eta)}{H(2\eta)} + \frac{\Theta'(2\eta)}{\Theta(2\eta)}\right\} - \frac{\Theta'(0)}{\Theta(0)}, \tag{126}$$

$$\begin{aligned}\mathscr{H} = \sum_{J=1}^{N} \{[1 + k\,\mathrm{sn}^2(2\eta)]\,\sigma_J{}^x\sigma_{J+1}^x + [1 - k\,\mathrm{sn}^2(2\eta)]\,\sigma_J{}^y\sigma_{J+1}^y \\ + \mathrm{cn}(2\eta)\,\mathrm{dn}(2\eta)\,\sigma_J{}^z\sigma_{J+1}^z\},\end{aligned} \tag{127}$$

$\sigma_J{}^x$, $\sigma_J{}^y$, $\sigma_J{}^z$ being the 2 by 2 Pauli matrices acting on the spin (arrow) at site J. [We have used (55), (58) and Sections 8.158.1 of GR.]

From (127) it is apparent that $\mathscr{H}$ is the Hamiltonian of the anisotropic Heisenberg ring, while from (125) we see that it must commute with $\mathbf{T}(v)$ and hence have the same eigenvectors. Thus the vectors $\mathbf{\Psi}$ defined by (13) are also eigenvectors of $\mathscr{H}$. Substituting the form given by (16) for the eigenvalue of $\mathbf{T}(v)$ into (125), we find that the corresponding eigenvalue of $\mathscr{H}$ is

$$\Lambda = N\,\mathrm{sn}(2\eta)\left\{\frac{2\Theta'(0)}{\Theta(0)} - \frac{H'(-2\eta)}{H(-2\eta)} - \frac{\Theta'(-2\eta)}{\Theta(-2\eta)}\right\}. \tag{128}$$

Note that this is independent of the choice of ω in (16). Thus all the $2N$ linearly independent eigenvectors that can be formed by varying Ψ are degenerate for the Heisenberg chain, all having the same eigenvalue.

We remark that we can deduce directly that each ψ_j given by (111) (and hence each Ψ) is an eigenvector of $\mathscr{H}$. To do this we write, in obvious notation,

$$\mathscr{H} = \sum_{J=1}^{N} e_1 \otimes \cdots \otimes e_{J-1} \otimes H_{J,J+1} \otimes e_{J+2} \otimes \cdots \otimes e_N, \tag{129}$$

$$\psi_j = \phi_1 \otimes \phi_2 \otimes \cdots \otimes \phi_N, \tag{130}$$

(e_J being the identity 2 by 2 matrix operating on the spin J), and show that there exists a scalar λ_J and a 2-dimensional vector s_J such that

$$(H_{J,J+1} - \lambda_J)\,\phi_J \otimes \phi_{J+1} = s_J \otimes \phi_{J+1} - \phi_J \otimes s_{J+1}. \tag{131}$$

It then follows that, provided $s_{N+1} = s_1$,

$$\mathscr{H}\psi_j = (\lambda_1 + \cdots + \lambda_N)\,\psi_j\,. \tag{132}$$

Using this method, we find an alternative expression for the eigenvalue Λ, namely,

$$\Lambda = N\{\mathrm{cn}(2\eta)\,\mathrm{dn}(2\eta) + 2k^2\,\mathrm{sn}^2(2\eta)F\}, \tag{133}$$

where

$$F = N^{-1}\sum_{J=1}^{N} \mathrm{sn}[t + 2J\eta]\,\mathrm{sn}[t + 2(J+1)\eta]. \tag{134}$$

Remembering that we require N and η to be chosen so that $\mathrm{sn}(u)$ is periodic of period $2N\eta$, we can verify that F is an entire function of the parameter t that occurs in its definition. Since it is doubly periodic it is therefore bounded, and hence a constant, by the Cauchy–Liouville theorem. Thus in fact F does not depend on t, despite appearances.

This makes the mathematical identity implied by the equivalence of the two formulas (128) and (133) more reasonable, though still not obvious. Equating the two formulas, the identity can be written as

$$[\Theta'(0)/\Theta(0)] - [\Theta'(-2\eta)/\Theta(-2\eta)] - k^2\,\mathrm{sn}(2\eta)F = 0. \tag{135}$$

We have managed to prove this identity when $m_2 = 0$. In this case the fact that F is independent of t means that we can replace (134) by its average over all real t, namely,

$$F = (2K)^{-1}\int_0^{2K} dt\,\mathrm{sn}(t)\,\mathrm{sn}(t + 2\eta). \tag{136}$$

The identity (135) must then be true for all real η. We prove it by analytically continuing the l.h.s. of (135) to the whole complex η-plane and showing that the resulting function is doubly periodic and analytic in a domain containing the period parallelogram. By the Cauchy–Liouville theorem it is therefore a constant, and since it vanishes when $\eta = 0$, it is zero. This proves the identity (135).

When $m_2 \neq 0$ the situation is more complicated, but presumably a proof of (135) can be constructed in a similar manner.

9. Inhomogeneous Lattice Model

Note that the recurrence relation (32) involves the vertex weights a, b, c, d only via the expressions S and X defined in (33) and (34). From (43) and (44) we see that these are functions of k and η, but not of v.

It follows that all the above working (except for the Heisenberg chain) can be performed if v has a different value v_J on each column J of the lattice (but k and η are the same for all columns). In particular, we can still construct the eigenvectors Ψ satisfying an equation of the form (16), only now

$$[\Psi]_\alpha = \sum_{j=1}^{L} \omega^j \prod_{J=1}^{N} \xi[\alpha_J \mid s + 2(J - j)\eta - v_J], \tag{137}$$

and

$$\mathbf{T}\Psi = [\omega^{-1}\tau_1 + \omega\tau_2]\Psi, \tag{138}$$

where

$$\begin{aligned} \tau_1 &= \sum_{J=1}^{N} g(v_J - \eta), \\ \tau_2 &= \sum_{J=1}^{N} g(v_J + \eta). \end{aligned} \tag{139}$$

From (137) we see that adding the same constant to each $v_1, \ldots, v_N$ is equivalent to subtracting that constant from the arbitrary parameter s, and therefore does not affect the set of eigenvectors that can be obtained by varying s. This agrees with the observation in Section 10 of [1] that two transfer matrices commute if their v_J's for a given column differ by a constant which is the same for all columns.

We do not claim that varying the v_J in this way from column to column is of any great physical significance, but we do feel it aids in understanding the mathematics of the solution. In a subsequent paper we intend to show how by starting with Ψ as a "zero-particle" eigenvector we can introduce dislocations or "particles" into its definition and obtain further eigenvectors by a Bethe-type ansatz. Any mathematical insight we can gain into such eigenvectors is welcome.

Acknowledgments

The author thanks Dr. R. Gibberd for bringing the form (133) of the Heisenberg chain eigenvalue to his attention.

References

1. R. J. Baxter, *Ann. Phys.* (*N.Y.*) **70** (1972), 193–228.
2. R. J. Baxter, *Ann. Phys.* (*N.Y.*) **70** (1972), 229–241.
3. R. J. Baxter, *Phys. Rev. Lett.* **26** (1971), 832–834.

4. R. J. Baxter, *Phys. Rev. Lett.* **26** (1971), 834–835.
5. E. H. Lieb, *Phys. Rev.* **162** (1967), 162–172 *Phys. Rev. Lett.* **18** (1967), 1046–1048 **19** (1967), 108–110.
6. I. S. Gradshteyn and I. M. Ryzhik, "Tables of Integrals, Series and Products," pp. 909–925, Academic Press, New York, 1965.
7. C. N. Yang and C. P. Yang, *Phys. Rev.* **150** (1966), 327–339.
8. S. Katsura, *Phys. Rev.* **127** (1962), 1508–1518.

Eight-Vertex Model in Lattice Statistics and One-Dimensional Anisotropic Heisenberg Chain. II. Equivalence to a Generalized Ice-type Lattice Model*

RODNEY BAXTER†

Institute for Theoretical Physics, State University of New York, Stony Brook, New York 11790

Received May 24, 1972

We establish an equivalence between the zero-field eight-vertex model and an Ising model (with four-spin interaction) in which each spin has L possible values, labeled 1,..., L, and two adjacent spins must differ by one (to modulus L). Such an Ising model can also be thought of as a generalized ice-type model and we will later show that the eigenvectors of the transfer matrix can be obtained by a Bethe-type ansatz.

1. INTRODUCTION AND SUMMARY

This is the second paper of a series in which we intend to obtain the eigenvectors of the transfer matrix **T** of the zero-field eight-vertex model. Since **T** commutes with the Hamilton $\mathscr{H}$ of the one-dimensional anisotropic Heisenberg ring (for appropriate values of their parameters), these are also the eigenvectors of $\mathscr{H}$.

In the previous paper [1], Paper I, we found some special eigenvectors. Equations for all the eigenvalues have already been obtained from a functional matrix relation [2].

In this paper we consider the effect of slightly altering the special eigenvectors found in Paper I. We show that this leads us to construct a family of vectors such that if ψ is a vector of the family, then $\mathbf{T}\psi$ is a linear combination of vectors of the same family. We then observe that with respect to this basis **T** is the transfer matrix of an Ising model (with four-spin interactions) in which each spin can have L values, labeled 1,..., L, where L is a positive integer. Two adjacent spins must differ by one, and we show from this property that we can regard the Ising model as a generalized ice-type model [3]. In the next paper of the series we shall obtain the general eigenvectors of **T** by a Bethe-type ansatz.

* Work supported in part by National Science Foundation Grant No. GP-32998X.

† On leave of absence from the Research School of Physical Sciences, Australian National University, Canberra A.C.T. 2600, Australia.

2. The Transfer Matrix

In this section we recapitulate some of our notation and define the transfer matrix $\mathbf{T}$ of the eight vertex model.

Consider a vertex of the lattice and the surrounding four bonds. We place arrows on the bonds so that there are an even number of arrows pointing into (and out of) the vertex. There are then eight allowed configurations of arrows round the vertex, as shown in Fig. 1 of Paper I. We associate Boltzmann weights a, b, c, d with these configurations as in (I.4).

We associate parameters $\alpha, \beta, \lambda, \mu$ with the bonds as indicated in Fig. 1 of this paper, the parameters having value $+(-)$ if the arrow on the corresponding bond points up or to the right (down or to the left). We define a function R such that

$$R(\alpha, \beta \mid \lambda, \mu) = a, b, c, d \text{ or } 0, \tag{2.1}$$

according to whether the arrow configuration specified by $\alpha, \beta, \lambda, \mu$ is allowed and has Boltzmann weight a, b, c, d, or is not allowed (weight zero).

We can regard λ, μ as indices and define 2 by 2 matrices $\mathbf{R}(\alpha, \beta)$ such that

$$\mathbf{R}(\alpha, \beta) = \begin{pmatrix} R(\alpha, \beta \mid +, +) & R(\alpha, \beta \mid +, -) \\ R(\alpha, \beta \mid -, +) & R(\alpha, \beta \mid -, -) \end{pmatrix}. \tag{2.2}$$

Then

$$\mathbf{R}(+, +) = \begin{pmatrix} a & 0 \\ 0 & b \end{pmatrix}, \quad \mathbf{R}(+, -) = \begin{pmatrix} 0 & d \\ c & 0 \end{pmatrix},$$
$$\mathbf{R}(-, +) = \begin{pmatrix} 0 & c \\ d & 0 \end{pmatrix}, \quad \mathbf{R}(-, -) = \begin{pmatrix} b & 0 \\ 0 & a \end{pmatrix}. \tag{2.3}$$

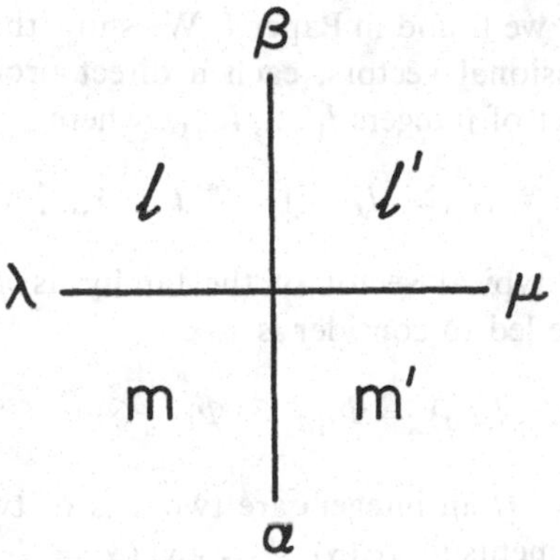

Fig. 1. Arrangement of bond parameters $\alpha, \beta, \lambda, \mu$ and spins l, l', m, m' round a vertex. The weight of this configuration in the Ising-like model is $W(m, m' \mid l, l')$.

We consider a rectangular lattice of N columns, would on a cylinder. The configuration of arrows on a row of vertical bonds is then specified by the set $\boldsymbol{\alpha} = \{\alpha_1, ..., \alpha_N\}$, where α_J is a parameter associated with the bond in column J, having value $+$ for an up arrow, $-$ for a down arrow. The transfer matrix **T** is then defined as a 2^N by 2^N matrix with elements

$$T_{\boldsymbol{\alpha},\boldsymbol{\beta}} = \mathrm{Tr}\{\mathbf{R}(\alpha_1, \beta_1)\, \mathbf{R}(\alpha_2, \beta_2) \cdots \mathbf{R}(\alpha_N, \beta_N)\}. \tag{2.4}$$

We use the notation ψ to denote a 2^N-dimensional vector in the space operated on by the matrix **T**, and $[\psi]_{\boldsymbol{\alpha}}$ to denote the element of ψ corresponding to the state $\boldsymbol{\alpha}$ ($\boldsymbol{\alpha}$ has 2^N values). We look in particular at vectors whose elements are products of functions of $\alpha_1, ..., \alpha_N$, respectively. For instance, if $N = 3$ we might consider a vector ψ whose elements are of the form

$$[\psi]_{\boldsymbol{\alpha}} = f(\alpha_1)\, g(\alpha_2)\, h(\alpha_3) \tag{2.5}$$

for $\alpha_1 = \pm$, $\alpha_2 = \pm$, $\alpha_3 = \pm$.

We can regard $f(+)$ and $f(-)$ as the elements of a two-dimensional vector f, and similarly for g and h. We can then think of the 2^N-dimensional vector ψ defined by (2.5) as a direct product of N two-dimensional vectors and write (2.5) as

$$\psi = f \otimes g \otimes h. \tag{2.6}$$

We shall use this notation.

3. Reduction to an Ising-Like Model

In Appendix B we consider the effect of making small changes in the special eigenvectors of **T** that we found in Paper I. We show that this leads us to consider a family of 2^N-dimensional vectors, each a direct product like (2.6), a member being specified by a set of integers $l_1, ..., l_{N+1}$, where

$$l_{J+1} = l_J \pm 1, \qquad J = 1, ..., N. \tag{3.1}$$

We therefore write a typical vector of the family as $\psi(l_1, ..., l_{N+1})$. The form of this vector that we are led to consider is

$$\psi(l_1, ..., l_{N+1}) = \Phi_{l_1, l_2} \otimes \Phi_{l_2, l_3} \otimes \cdots \otimes \Phi_{l_N, l_{N+1}}, \tag{3.2}$$

where $\Phi_{l, l+1}$ and $\Phi_{l+1, l}$ (l an integer) are two sets of two-dimensional vectors, as yet arbitrary, with elements $\Phi_{l, l+1}(\alpha)$, $\Phi_{l+1, l}(\alpha)$ ($\alpha = +$ or $-$).

In Appendix B we also consider whether we can choose the $\Phi_{l,m}$ so that $\mathbf{T}\psi(l_1, ..., l_{N+1})$ is a linear combination of vectors of the family (3.2). We find that we can, provided there exist two other sets of two-dimensional vectors $z_{l,l+1}$, $z_{l+1,l}$, with elements $z_{l,l+1}(\lambda)$, $z_{l+1,l}(\lambda)$, and a set of coefficients $W(m, m' \mid l, l')$ such that

$$\sum_{\beta,\mu} R(\alpha, \beta \mid \lambda, \mu)\, \Phi_{l,l'}(\beta)\, z_{m',l'}(\mu) = \sum_{m} W(m, m' \mid l, l')\, \Phi_{m,m'}(\alpha)\, z_{m,l}(\lambda) \quad (3.3)$$

for $\alpha, \lambda = \pm$ and all integer values of l, l', m' such that $\mid l - l' \mid = \mid m' - l' \mid = 1$. The summation on the r.h.s. of (1.9) is over integer values of m such that $\mid m - m' \mid = 1$, $\mid m - l \mid = 1$. Since m' and l are either equal or differ by 2, there are only 2 or 1 allowed values of m, respectively, in this summation.

Note that (3.3) is a local property of the lattice, involving only the allowed arrangements of arrows round a single vertex. We emphasize this because our working in terms of the transfer matrix tends to obscure this property.

The $\Phi_{l,l\pm1}(\alpha)$, $z_{l,l\pm1}(\lambda)$ and $W(m, m' \mid l, l')$ are now unknowns which must be chosen to satisfy (3.3).

Note that there are more equations than unknowns in (3.3). Taking $l' = 1, 2, \ldots, p$ (p large) and all allowed values of α, λ, l and m', we see that we obtain $16p$ equations from (3.3). On the other hand, since any normalization factors of the 4 sets of vectors $\Phi_{l,m}$, $z_{l,m}$ can be incorporated into the 6 sets of coefficients W, there are only $10p$ independent unknowns. Thus we have no right to expect to be able to solve (3.3). Indeed for the eight-vertex model in the presence of ferroelectric fields [i.e., $\omega_1 \neq \omega_2$ or $\omega_3 \neq \omega_4$ in (1.4)], we cannot in general do so. However, for the zero-field eight-vertex model we can. We give the necessary formulas in Appendix C.

Equations (3.3) are the essential step in the working of this paper. Two side conditions that we also need are that the two vectors $z_{l-1,l}$ and $z_{l+1,l}$ be linearly independent of one another for each value of l, and that

$$z_{l_{N+1}\pm1,l_{N+1}} = z_{l_1\pm1,l_1} \quad (3.4)$$

for either choice of sign (the same on both sides) in (3.4). This latter condition comes from the fact that the lattice is wound cyclically onto a cylinder. Notice that it is certainly satisfied if $l_{N+1} = l_1$.

Using (3.3) and these conditions, we show in Appendix B that

$$\mathbf{T}\psi(l_1, \ldots, l_{N+1}) = \sum \left\{ \prod_{J=1}^{N} W(m_J, m_{J+1} \mid l_J, l_{J+1}) \right\} \psi(m_1, \ldots, m_{N+1}), \quad (3.5)$$

the summation being over all integers $m_1, \ldots, m_{N+1}$ such that

$$m_{J+1} = m_J \pm 1, \qquad J = 1, \ldots, N, \quad (3.6)$$

and

$$m_J = l_J \pm 1, \qquad J = 1, \ldots, N+1. \quad (3.7)$$

The choice of sign in (3.7) must be the same for $J = 1$ and $J = N + 1$. Thus if $l_{N+1} = l_1$, it follows that $m_{N+1} = m_1$.

It is therefore clear that we have constructed a family of vectors $\psi(l_1, ..., l_{N+1})$ such that $\mathbf{T}\psi(l_1, ..., l_{N+1})$ is a linear combination of vectors of the same family. Using (3.5), we can give a graphical interpretation of $\mathbf{T}$ with respect to this basis. Associate a "spin" $l_{I,J}$ with each face (I, J) of the lattice, where $l_{I,J}$ can take any integer value, and allow only configurations in which adjacent spins differ by unity. With each allowed configuration of the four spins l, l', m, m' round a vertex (as shown in Fig. 1) associate a "Boltzmann weight" $W(m, m' \mid l, l')$. Then with respect to the basis set of vectors $\psi(l_1, ..., l_{N+1})$, $\mathbf{T}$ is the row-to-row transfer matrix of this Ising-like spin problem. In this sense the zero-field eight-vertex problem and the spin problem are equivalent.

One rather unphysical feature of this equivalence is that the spins have an infinite number of possible values. However, we show in Section 6 that for certain values of a, b, c, d the vectors $\Phi_{l,l\pm1}$, $z_{l,l\pm1}$ are unchanged by incrementing l by some integer L. In fact we can find such special values of a, b, c, d arbitrarily close to any given values if we take L sufficiently large. In this case we can restrict the spins in the Ising-like problem to have values $1, ..., L$, and require that adjacent spins differ by one to modulus L.

The simplest such special case is when $ab = cd$, in which case $L = 4$. The eight-vertex model is then simply the superposition of two independent normal Ising models [4]. On the other hand, since L is even and adjacent spins must differ by one, we can restrict our Ising-like model to one in which only even spins (2 or 4) occur on one sublattice and odd spins (1 or 3) on the other. Thus each spin really has only two possible values. Thus the normal spin-$\frac{1}{2}$ (two values per spin) Ising model is equivalent to a spin-$\frac{1}{2}$ Ising model in which all four spins round a vertex interact, and not all configurations of these four spins are allowed. As we shall show in a later paper for the general case, this new Ising model can be solved by a Bethe-type ansatz.

4. Formulation as an Ice-Type Model

Returning to the general case, an important feature of the Ising-like problem we have obtained is that it can be thought of as a generalized "ice-type" problem. To see this, draw arrows on all vertical bonds, pointing up (down) if the spin to the right of the bond is one greater (less) than the spin to the left. Also draw arrows on all horizontal bonds, pointing to the right (left) if the spin below the bond is one greater (less) than the spin above.

For a given value of l (say), there are six possible sets of values of the spins l, l', m, m' round a vertex. We show these in Fig. 2, together with the corresponding arrow configurations on the four bonds. We see that in each case there are two arrows pointing into the vertex and two out, as in the ice model [3]. Thus the number n of down arrows in a row of vertical bonds is the same for each row of the lattice.

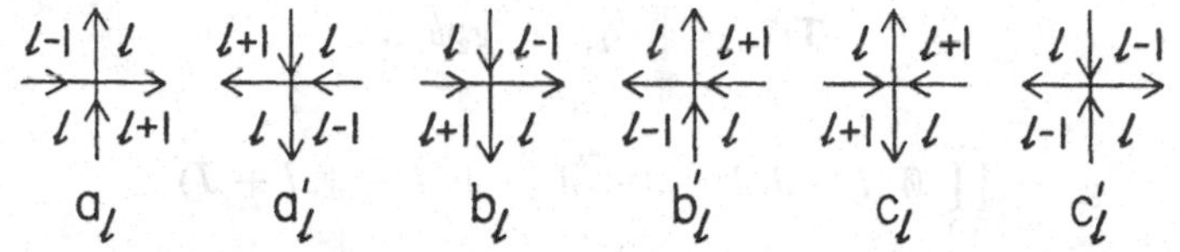

FIG. 2. The six types of vertex in the Ising-like model with their corresponding weights, e.g., $a_l = W(l, l+1 \mid l-1, l)$.

This means that we can split our family of vectors ψ of the form (3.2) into $N+1$ subfamilies $\mathscr{F}_0, \mathscr{F}_1, \ldots, \mathscr{F}_N$, such that if $\psi \in \mathscr{F}_n$, then in the definition (3.2) of ψ, n Φ's are of the type $\Phi_{l,l-1}$, while $N-n$ are of the type $\Phi_{l,l+1}$. Providing the boundary condition (3.4) is satisfied, it follows that $\mathbf{T}\psi$ is a linear combination of vectors belonging to $\mathscr{F}_n$.

We must study this boundary condition. Since there are n decreasing steps and $N-n$ increasing steps in the sequence $l_1, \ldots, l_{N+1}$, we see that if $\psi \in \mathscr{F}_n$, then

$$l_{N+1} = l_1 + N - 2n. \tag{4.1}$$

Thus in general (3.4) is satisfied if $N = 2n$. However, we can allow other values of n if we consider the special cases discussed above in which the vectors $z_{l\pm 1,l}$ are unchanged by incrementing l by L. The boundary condition (3.4) is then satisfied if

$$N - 2n = L \times \text{integer}. \tag{4.2}$$

5. The Case of $n = 0$

To fix our ideas, consider the case when $n = 0$, i.e., all arrows are up. The sequence $l_1, \ldots, l_{N+1}$ is then increasing and $l_J = l + J - 1$ for $J = 1, \ldots, N+1$ (writing l for l_1). The boundary condition (3.4) is satisfied if

$$N = L \times \text{integer}. \tag{5.1}$$

(We are forced to restrict attention to the special values of a, b, c, d for which L exists.) From (3.2) we see that $\mathscr{F}_0$ is the family of vectors

$$\psi(l, l+1, \ldots, l+N) = \phi_l \otimes \phi_{l+1} \otimes \cdots \otimes \phi_{l+N-1}, \tag{5.2}$$

where

$$\phi_l = \Phi_{l,l+1} . \tag{5.3}$$

Incrementing l by L leaves the two-dimensional vector ϕ_l unchanged, so there are only L vectors in $\mathscr{F}_0$, corresponding to taking $l = 1,..., L$ in (5.2).

Let us write $\psi(l, l+1,..., l+N)$ simply as ψ_l. Then from (3.5) we can verify that

$$\mathbf{T}\psi_l = g_1\psi_{l+1} + g_2\psi_{l-1} , \tag{5.4}$$

where

$$g_1 = \prod_{J=1}^{N} W(l+J, l+J+1 \mid l+J-1, l+J),$$
$$g_2 = \prod_{J=1}^{N} W(l+J-2, l+J-1 \mid l+J-1, +J). \tag{5.5}$$

When we calculate the weights W that occur in (5.5) we find that g_1 and g_2 are constants, independent of l. Also, $\psi_{l+L} = \psi_l$ for any integer l. Thus, if ω is the Lth root of unity and

$$\Psi = \sum_{l=1}^{L} \omega^l \psi_l , \tag{5.6}$$

we see from (5.4) that

$$\mathbf{T}\Psi = [\omega^{-1}g_1 + \omega g_2]\Psi, \tag{5.7}$$

i.e., Ψ is an eigenvector of the transfer matrix. These are the special eigenvectors that we found in Paper I. In our next paper we shall consider the cases when $n > 0$ and show that we can then obtain the eigenvectors of $\mathbf{T}$ by a generalized Bethe ansatz.

6. The Vectors $\Phi_{l,m}$, $z_{l,m}$

So far we have discussed the general nature of our results, deliberately avoiding the mathematical details of the solution of (3.3). We now state what we find the two-dimensional vectors $\Phi_{l,l\pm1}$, $z_{l,l\pm1}$ and the coefficients $W(m, m' \mid l, l')$ to be. (The derivation is explained in Appendix C.)

We find that to solve (3.3) it is first convenient to introduce parameters ρ, k, η, v which are related to the Boltzmann weights a, b, c, d of the eight-vertex model by

$$\begin{aligned} a &= \rho\Theta(-2\eta)\,\Theta(\eta - v)\,H(\eta + v), \\ b &= -\rho\Theta(-2\eta)\,H(\eta - v)\,\Theta(\eta + v), \\ c &= -\rho H(-2\eta)\,\Theta(\eta - v)\,\Theta(\eta + v), \\ d &= \rho H(-2\eta)\,H(\eta - v)\,H(\eta + v), \end{aligned} \tag{6.1}$$

where $H(u)$, $\Theta(u)$ are related to the elliptic theta functions of argument u and modulus k (Section 8.192 of [5], hereinafter referred to as GR). We leave their precise definition for the moment and note only that we require that their ratio be the same as that of the theta functions, i.e.,

$$H(u)/\Theta(u) = k^{1/2}\,\mathrm{sn}(u), \tag{6.2}$$

where $\mathrm{sn}(u)$ is the elliptic sine-amplitude function of argument u and modulus k (Sections 8.14, 8.191.1 of GR).

Eliminating ρ between any two of Eqs. (6.1), we see that only such ratios of $H(u)$ and $\Theta(u)$ occur, so (6.1) and (6.2) are sufficient to define k, η, v.

Remember that the $\Phi_{l,m}$ $(m = l \pm 1)$ are two-dimensional vectors with elements $\Phi_{l,m}(\alpha)$, i.e.,

$$\Phi_{l,m} = \begin{pmatrix} \Phi_{l,m}(+) \\ \Phi_{l,m}(-) \end{pmatrix} \tag{6.3}$$

(and similarly for the $z_{l,m}$). To define these vectors it is convenient to introduce a notation borrowed from quantum mechanics and write $|\,u\rangle$ for the two-dimensional vector

$$|\,u\rangle = \begin{pmatrix} H(u) \\ \Theta(u) \end{pmatrix} \tag{6.4}$$

(u any complex number). Then we find that

$$\begin{aligned} \Phi_{l,l+1} &= |\, s + 2l\eta + \eta - v\rangle, \\ \Phi_{l+1,l} &= |\, t + 2l\eta + \eta + v\rangle, \\ z_{l+1,l} &= |\, s + 2l\eta\rangle, \\ z_{l-1,l} &= h(w + 2l\eta)\,|\, t + 2l\eta\rangle, \end{aligned} \tag{6.5}$$

for all integers l, where s and t are some constants which can be chosen arbitrarily, the function $h(u)$ is defined by

$$h(u) = H(u)\,\Theta(-u), \tag{6.6}$$

and

$$w = \tfrac{1}{2}(s + t - 2K), \tag{6.7}$$

K and K' being the complete elliptic integrals of moduli k and $k' = (1 - k^2)^{1/2}$ (Sections 8.110–112 of GR).

Note that (6.2) and (6.4) define $|\,u\rangle$ to within a normalization factor. The function $\mathrm{sn}(u)$ is doubly periodic, with periodic, with periods $4K$, $2iK'$.

Thus if there exist integers L, m_1, m_2 such that

$$L\eta = 2m_1K + im_2K', \tag{6.8}$$

then $\mathrm{sn}(u)$ is periodic of period $2L\eta$. It follows that to within a normalization factor the vectors $z_{l,l\pm1}$ defined in (6.5) are periodic functions of l, with period L. As we remarked above, this periodicity is convenient to our working (though not strictly essential) since it enables us to consider values of n other than $\frac{1}{2}N$.

We write the usual Jacobian elliptic theta functions defined in GR as $H_{Jb}(u)$ and $\Theta_{Jb}(u)$. Like $\mathrm{sn}(u)$, they are periodic of period $4K$, but they are only quasi-periodic of period $2iK'$. If we wish to discuss the special values of η given by (6.8) (with $m_2 \neq 0$), it is therefore convenient to define $H(u)$, $\Theta(u)$ in the above equations to be given by

$$\begin{aligned} H(u) &= H_{Jb}(u) \exp[i\pi m_2(u-K)^2/(4KL\eta)], \\ \Theta(u) &= \Theta_{Jb}(u) \exp[i\pi m_2(u-K)^2/(4KL\eta)]. \end{aligned} \tag{6.9}$$

This ensures that $H(u)$, $\Theta(u)$ are completely periodic of period $2L\eta$. Hence the vectors $z_{l,l\pm1}$ are periodic of period l.

We emphasize that this renormalization of the theta functions is purely a matter of mathematical convenience. The reader who finds it confusing can focus attention on the case $m_2 = 0$, when the theta functions used here are the same as those of Jacobi. Alternatively, he can consider the case $n = \frac{1}{2}N$, when the restriction (6.8) on η is unnecessary and the re-normalization (6.9) is irrelevant, affecting none of our equations.

7. The Coefficients W

It remains only to quote the results for the coefficients W in (3.3). The working is outlined in Appendix C. We find that for all integers l:

$$\begin{aligned} W(l, l+1 \mid l-1, l) &= a_l = \rho' h(v+\eta), \\ W(l, l-1 \mid l+1, l) &= a_l' = \rho' h(v+\eta)\, h(w_{l+1})/h(w_l), \\ W(l+1, l \mid l, l-1) &= b_l = \rho' h(v-\eta)\, h(w_{l-1})/h(w_l), \\ W(l-1, l \mid l, l+1) &= b_l' = \rho' h(v-\eta), \\ W(l+1, l \mid l, l+1) &= c_l = \rho' h(2\eta)\, h(w_l+\eta-v)/[h(w_l)\, h(w_{l+1})], \\ W(l-1, l \mid l, l-1) &= c_l' = \rho' h(2\eta)\, h(w_l-\eta+v), \end{aligned} \tag{7.1}$$

where

$$\rho' = \rho\Theta(0), \tag{7.2}$$

$$w_l = \tfrac{1}{2}(s + t) - K + 2l\eta, \tag{7.3}$$

and the function $h(u)$ is defined by (6.6).

We have introduced the coefficients a_l, a_l', b_l, b_l', c_l, c_l' into (7.1) simply as a shorthand way of writing the coefficients W. They can be thought of as the Boltzmann weights: of the ice-type vertices 1,..., 6, respectively, shown in Fig. 2. Note that they are not to be confused with the Boltzmann weights a, b, c, d of the original eight-vertex model.

We have remarked above that changing the normalizations of the vectors $\Phi_{l,l\pm 1}$, $z_{l,l\pm 1}$ simply changes the coefficients W. The normalizations we use in (6.5) are chosen so as to simplify the coefficients W as far as possible. In particular, the extra normalization factor $h(w + 2l\eta)$ in $z_{l-1,l}$ is introduced so as to ensure that both a_l and b_l' are independent of l. This turns out to be a help in handling the Bethe ansatz equations for the eigenvectors of **T**.

8. An Inhomogeneous Lattice

Equation (3.5) can be regarded as a similarity transformation of **T** which converts the eight-vertex model transfer matrix to that of the Ising-like model. This equation is derived in Appendix B from the local property (3.3) and the boundary conditions (3.4). In this derivation the only interaction between columns of the lattice occurs through the two-dimensional vectors $z_{l\pm 1,l}$. From (6.5) we note that these depend on k, η and the arbitrary parameters s, t, but they do not depend on v. Thus the working still goes through if we suppose the Boltzmann weights a, b, c, d to be given by (6.1), where k and η are the same for all sites of the lattice but v can vary from column to column. The only change is that in evaluating the coefficients W and the vectors $\Phi_{l,l\pm 1}$ from (7.1) and (6.5), one must first consider to which column of the lattice they belong and insert the appropriate value of v.

This generalization can be helpful in studying the structure of the eigenvectors of the transfer matrix that we shall obtain.

9. Summary

We have shown that the zero-field eight-vertex model can be translated into an Ising-like model with four spin interactions. This model can in turn be thought of as a generalized ice-type problem. In a subsequent paper we shall show that we can obtain the eigenvectors of the transfer matrix by a method similar to that used for the ice models [3], namely, by a generalized Bethe ansatz.

The essential step in our working is the fact that Eqs. (3.3) can be solved for the coefficients $W(m, m' \mid l, l')$ and the two-dimensional vectors $\Phi_{l,l\pm1}$, $z_{l,l\pm1}$. This is a purely local property of the lattice. We can in fact establish the equivalence of the eight-vertex model and the Ising-like model directly from (3.3) by imagining the lattice to be built up successively from top to bottom and right to left, associating as we go appropriate combinations of vectors $\Phi_{l,l\pm1}$, $z_{l,l\pm1}$ with the bonds of the lattice.

We have left the detailed working of this paper to the Appendices. In Appendix A we summarize the results of Paper I, in which we obtained some special eigenvectors of **T**. In Appendix B we show how perturbing these leads us to consider vectors of the form (3.2) and to look for the relations (3.3). We go on to show that if these relations are satisfied, together with the boundary conditions (3.4), then Eq. (3.5) (which is the main result of this paper) follows. In Appendix C we show that (3.3) can indeed be satisfied, and end by showing that it is a corollary of two mathematical identities that involve elliptic theta functions.

Appendix A

We summarize here some results of Paper I concerning certain special eigenvectors of the transfer matrix **T**. We make a few notational changes to suit our present purpose.

Replace j and s in (I.111) by $-l$ and $s - \eta - v$. Using the notation outlined above in Eqs. (2.5) and (2.6), the 2^N-dimensional vector ψ_j defined by (I.111) can then be written as (dropping the suffix j):

$$\psi = \phi_l \otimes \phi_{l+1} \otimes \cdots \otimes \phi_{l+N-1}, \tag{A.1}$$

where for any integer l, ϕ_l is a two-dimensional vector with elements

$$\begin{aligned} \phi_l(+) &= H(s + 2l\eta + \eta - v), \\ \phi_l(-) &= \Theta(s + 2l\eta + \eta - v). \end{aligned} \tag{A.2}$$

From the form (2.4) of the elements of the transfer matrix **T**, it then follows that

$$[\mathbf{T}\psi]_{\alpha} = \mathrm{Tr}\{\mathbf{U}_l(\alpha_1)\, \mathbf{U}_{l+1}(\alpha_2) \cdots \mathbf{U}_{l+N-1}(\alpha_N)\}, \tag{A.3}$$

where

$$\mathbf{U}_l(\alpha) = \sum_{\beta} \mathbf{R}(\alpha, \beta)\, \phi_l(\beta) \tag{A.4}$$

for $\alpha = +$ and $-$, and for all integers l. The summation in (A.4) is over the two values $+$ and $-$ of β.

Note that our suffix l plays a different role to the suffix J used in (I.26)–(I.32), denoting the vector ϕ_l used in (A.4) rather than the position of a particular $\mathbf{U}$ in the product on the r.h.s. of (A.3).

The essence of the working of Paper I is that the r.h.s. of (A.3) is unchanged if we replace each $\mathbf{U}_l(\alpha)$ by

$$\mathbf{U}_l^*(\alpha) = \mathbf{M}_l^{-1}\mathbf{U}(\alpha)\,\mathbf{M}_{l+1} \tag{A.5}$$

(all integers l and $\alpha = +$ and $-$), provided that the boundary condition

$$\mathbf{M}_{l+N} = \mathbf{M}_l \tag{A.6}$$

is satisfied. Further, we can choose the 2 by 2 matrices $\mathbf{M}_l$ so as to make each matrix $\mathbf{U}_l^*(\alpha)$ lower-left triangular.

The working is formally equivalent to setting $\sigma_J = 1$ in (I.67)–(I.75), replacing u_J by $s + 2l\eta$, and suffixes J by l. The only result that we need here is that

$$\mathbf{U}_l^*(\alpha) = \begin{pmatrix} b_l'\phi_{l-1}(\alpha) & 0 \\ c_l\tau_l(\alpha) & a_{l+1}\phi_{l+1}(\alpha) \end{pmatrix} \tag{A.7}$$

for $\alpha = \pm$ and all integers l. We leave the coefficients a_l, b_l', c_l and the two-dimensional vectors τ_l undefined. We shall derive them in Appendix C, using a more direct method than that of Paper I.

Note that the coefficients a_l, b_l', c_l are not to be confused with the Boltzmann weights a, b, c, d of the eight-vertex model.

Replacing each $\mathbf{U}_l(\alpha)$ in (A.3) by $\mathbf{U}_l^*(\alpha)$, given by (A.7), we see that the r.h.s. of (A.3) becomes the trace of a product of lower-left triangular matrices. It is therefore easy to evaluate and by doing this we can construct the special eigenvectors discussed in Paper I, and in Section 5 of this paper.

Appendix B

Here we consider the effect of slightly altering the special eigenvectors found in Paper I. We show that this leads us to consider a family of vectors of the type (3.2), and to see if we can choose the two-dimensional vectors $\Phi_{l,l\pm1}$, $z_{l,l\pm1}$ so that (3.3) is satisfied. We show that if this can be done and the boundary condition (3.4) satisfied, then Eq. (3.5) is valid.

We have summarized the results of Paper I that we need in Appendix A. Suppose we alter the last vector, ϕ_{l+N-1}, in (A.1). Most of the reasoning still goes through, but now when we replace each $\mathbf{U}_l(\alpha)$ in (A.3) by $\mathbf{U}_l^*(\alpha)$, we find that the first $N-1$ matrices on the r.h.s. are lower-left triangular, while the last is not. We can still expand the trace explicitly, only now we get $N+1$ terms, a typical one being proportional to

$$\phi_{l+1} \otimes \phi_{l+2} \otimes \phi_{l+3} \otimes \tau_{l+3} \otimes \phi_{l+3} \otimes \phi_{l+4} \otimes \phi_{l+5} \otimes \cdots. \tag{B.1}$$

Typically we get a break in the sequence, characterized by a ϕ_{l+J} followed by a τ_{l+J} followed by a ϕ_{l+J}. For the moment we ignore what happens at the end of the sequence.

Now consider the effect of premultiplying the vector (B.1) by the transfer matrix $\mathbf{T}$. From (2.4) we see that the product is a 2^N-dimensional vector with elements

$$\mathrm{Tr}\{\mathbf{U}_{l+1}(\alpha_1)\, \mathbf{U}_{l+2}(\alpha_2)\, \mathbf{U}_{l+3}(\alpha_3)\, \mathbf{V}_{l+3}(\alpha_4)\, \mathbf{U}_{l+3}(\alpha_5)\, \mathbf{U}_{l+4}(\alpha_6) \cdots\}, \tag{B.2}$$

where $\mathbf{U}_l(\alpha)$ is given by (A.4) and

$$\mathbf{V}_l(\alpha) = \sum_\beta \mathbf{R}(\alpha, \beta)\, \tau_l(\beta) \tag{B.3}$$

for any integer l and $\alpha = +$ or $-$.

Again we can apply the transformation (A.5) so as to replace each $\mathbf{U}_l(\alpha)$ by $\mathbf{U}_l^*(\alpha)$ in (B.2). We see that we must then replace $\mathbf{V}_l(\alpha)$ (for any integer l) by

$$\mathbf{V}_l^*(\alpha) = \mathbf{M}_{l+1}^{-1} \mathbf{V}_l(\alpha)\, \mathbf{M}_l\,. \tag{B.4}$$

Let us write the 2 by 2 matrices $\mathbf{V}_l^*(\alpha)$ explicitly as

$$\mathbf{V}_l^*(\alpha) = \begin{pmatrix} f_l(\alpha) & g_l(\alpha) \\ h_l(\alpha) & r_l(\alpha) \end{pmatrix} \tag{B.5}$$

for all integers l and $\alpha = \pm$. Replacing each $\mathbf{U}_l(\alpha)$, $\mathbf{V}_l(\alpha)$ in (B.2) by $\mathbf{U}_l^*(\alpha)$, $\mathbf{V}_l^*(\alpha)$ and evaluating the trace, using (A.7) and (B.5), we get a number of terms, some typical ones being proportional to

$$\begin{aligned} &\phi_l \otimes \phi_{l+1} \otimes \phi_{l+2} \otimes f_{l+3} \otimes \phi_{l+2} \otimes \phi_{l+3} \otimes \cdots, \\ &\phi_{l+2} \otimes \tau_{l+2} \otimes \phi_{l+2} \otimes g_{l+3} \otimes \phi_{l+4} \otimes \phi_{l+5} \otimes \cdots, \\ &\phi_{l+2} \otimes \phi_{l+3} \otimes \phi_{l+4} \otimes h_{l+3} \otimes \phi_{l+2} \otimes \phi_{l+3} \otimes \cdots, \\ &\phi_{l+2} \otimes \phi_{l+3} \otimes \phi_{l+4} \otimes r_{l+3} \otimes \phi_{l+4} \otimes \phi_{l+5} \otimes \cdots. \end{aligned} \tag{B.6}$$

[We are using the notation outlined in (2.5) and (2.6).]

We should like the vectors (B.6) to be of the same general type as (B.1), since then **T** would map a vector of this family onto the space spanned by the family and we might hope to use this closure to calculate the eigenvectors of **T**. Comparing (B.1) and (B.6), we see that necessary conditions for this to be so are that

$$\begin{aligned} f_l &\propto \tau_{l-1}, \qquad g_l \propto \phi_l, \\ h_l &= 0, \qquad r_l \propto \tau_{l+1}. \end{aligned} \tag{B.7}$$

Thus we should like to be able to choose the two-dimensional vectors ϕ_l, τ_l and the 2 by 2 matrices **M** so that (A.7) is satisfied and

$$\mathbf{V}_l^*(\alpha) = \begin{pmatrix} a_l'\tau_{l-1}(\alpha) & c_{l+1}'\phi_l(\alpha) \\ 0 & b_{l+1}\tau_{l+1}(\alpha) \end{pmatrix} \tag{B.8}$$

for all integers l and $\alpha = \pm$, where a_l', b_l, c_l' are some coefficients.

These requirements can be written more explicitly if we define two 2-dimensional vectors x_l, y_l, with elements $x_l(\pm)$, $y_l(\pm)$, such that

$$\mathbf{M}_l = \begin{pmatrix} y_l(+) & x_l(+) \\ y_l(-) & x_l(-) \end{pmatrix}. \tag{B.9}$$

Premultiplying Eqs. (A.5), (B.4) by $\mathbf{M}_l$, $\mathbf{M}_{l+1}$, respectively, and performing the matrix multiplications explicitly, using (2.2), (A.4), (A.7), (B.3), (B.8) and (B.9), we find that these requirements are equivalent to four sets of equations, one of them being

$$\sum_{\beta,\mu} R(\alpha, \beta \mid \lambda, \mu)\, \phi_l(\beta)\, x_{l+1}(\mu) = a_{l+1}\phi_{l+1}(\alpha)\, x_l(\lambda) \tag{B.10}$$

for all integers l and $\alpha = \pm$, $\lambda = \pm$. The summation on the r.h.s. is over $\beta = \pm$ and $\mu = \pm$.

The structure of these equations becomes clearer if we use an abbreviated notation which utilizes the fact that we already think of $\phi_l(+)$, $\phi_l(-)$ as the elements of a two-dimensional vector ϕ_l (similarly for x_l). We write (B.10) as

$$\mathbf{R}\{\phi_l \otimes x_{l+1}\} = a_{l+1}\phi_{l+1} \otimes x_l\,. \tag{B.11a}$$

Using this notation, the other three sets of equations that we obtain can be written as

$$\begin{aligned} \mathbf{R}\{\phi_l \otimes y_{l+1}\} &= b_l'\phi_{l-1} \otimes y_l + c_l\tau_l \otimes x_l\,, \\ \mathbf{R}\{\tau_l \otimes x_l\} &= c_{l+1}'\phi \otimes y_{l+1} + b_{l+1}\tau_{l+1} \otimes x_{l+1}\,, \\ \mathbf{R}\{\tau_l \otimes y_l\} &= a_l'\tau_{l-1} \otimes y_{l+1} \end{aligned} \tag{B.11b}$$

for all integers l.

The first two of these sets of equations (B.11) are equivalent to requiring that $\mathbf{U}_l^*(\alpha)$ be given by (A.7), the second two to requiring that $\mathbf{V}_l^*(\alpha)$ be given by (B.8). We know from Paper I that we can satisfy (A.7), and hence the first two equations (B.11). It is far from obvious that we can also satisfy the second two. Nevertheless it turns out that we can. The working is given in Appendix C.

It remains to see whether Eqs. (B.11) are sufficient to obtain the required closure, namely, that premultiplying a 2^N-dimensional vector of the form (B.1) by $\mathbf{T}$ gives a linear combination of vectors of the same general form.

To do this we must characterize such vectors more precisely. Allowing the possibility of more than one break in the sequence in (B.1), we see that the rules for constructing such vectors are, for $J = 1,..., N-1$:

(i) a two-dimensional vector ϕ_l in position J must be followed by either ϕ_{l+1} or τ_l in position $J+1$,

(B.12)

(ii) a vector τ_l in position J must be followed by either ϕ_l or τ_{l-1} in position $J+1$.

(The last possibility follows from considering the effect of having two τ-breaks in the increasing ϕ-sequence, and progressively moving them to adjacent positions.)

These rules can be stated more simply if we adopt a notation that is already suggested by (A.5) and (B.4), and define two sets of two-dimensional vectors $\Phi_{l,l+1}$, $\Phi_{l+1,l}$ such that

$$\Phi_{l,l+1} = \phi_l\,, \qquad \Phi_{l+1,l} = \tau_l \tag{B.13}$$

for all integers l. The rules (B.12) then become

a vector $\Phi_{l\pm1,l}$ in position J must be followed by either $\Phi_{l,l+1}$ or $\Phi_{l,l-1}$ in position $J+1$ $(J = 1,..., N-1)$. (B.14)

It follows that any vector of the form (B.1) can be specified by a set of integers $l_1,..., l_{N+1}$ such that $l_{J+1} = l_J \pm 1$ for $J = 1,..., N$, and is given explicitly by (3.2).

Now premultiply (3.2) by the transfer matrix $\mathbf{T}$. Using (2.4), (A.4), (B.3) and (B.13), we see that

$$[\mathbf{T}\psi(l_1,..., l_{N+1})]_\alpha = \mathrm{Tr}\{\mathbf{S}_{l_1,l_2}(\alpha_1)\,\mathbf{S}_{l_2,l_3}(\alpha_2)\cdots\mathbf{S}_{l_N,l_{N+1}}(\alpha_N)\}, \tag{B.15}$$

where

$$\mathbf{S}_{l,l+1}(\alpha) = \mathbf{U}_l(\alpha), \qquad \mathbf{S}_{l+1,l}(\alpha) = \mathbf{V}_l(\alpha) \tag{B.16}$$

for all integers l and $\alpha = \pm$.

We now make the tsansformations (A.5), (B.4), i.e., we replace each $\mathbf{S}_{l,m}(\alpha)$ in (B.15) by

$$\mathbf{S}^*_{l,m}(\alpha) = \mathbf{M}_l^{-1}\mathbf{S}_{l,m}(\alpha)\,\mathbf{M}_m\,, \qquad m = l \pm 1. \tag{B.17}$$

From (A.5) and (B.4) we see that

$$\mathbf{S}^*_{l,l+1}(\alpha) = \mathbf{U}_l{}^*(\alpha), \qquad \mathbf{S}^*_{l+1,l} = \mathbf{V}_l{}^*(\alpha). \tag{B.18}$$

We now substitute the explicit forms (A.7), (B.8) of $\mathbf{U}_l{}^*(\alpha)$, $\mathbf{V}_l{}^*(\alpha)$ into (B.18) and use (B.13) to replace ϕ_l, τ_l by $\Phi_{l,l+1}$, $\Phi_{l+1,l}$ (for any integer l). A little inspection then shows that the element (λ, μ) of the matrix $\mathbf{S}^*_{l,l'}(\alpha)$ can be written as

$$[\mathbf{S}^*_{l,l'}(\alpha)]_{\lambda,\mu} = W(l-\lambda, l'-\mu \mid l, l')\, \Phi_{l-\lambda,l'-\mu}(\alpha) \tag{B.19}$$

for $\lambda, \mu, \alpha = \pm 1$, and $l' = l-1$ or $l+1$. The coefficients $W(m, m' \mid l, l')$ are related to the a_l, a_l', b_l, b_l', c_l, c_l' as in (7.1), and

$$W(l-1, l+2 \mid l, l+1) = W(l+2, l-1 \mid l+1, l) = 0. \tag{B.20}$$

(This last equation follows from the vanishing elements of $\mathbf{U}_l{}^*(\alpha)$, $\mathbf{V}_l{}^*(\alpha)$.)

Using (B.17), we see that we can replace each $\mathbf{S}_{l,m}(\alpha)$ in (B.15) by $\mathbf{S}^*_{l,m}(\alpha)$ provided that

$$\mathbf{M}_{l_{N+1}} = \mathbf{M}_{l_1}. \tag{B.21}$$

Doing this, writing the trace explicitly and using (B.19), we find that

$$[\mathbf{T}\psi(l_1, ..., l_{N+1})]_\alpha = \sum_{\lambda_1,...,\lambda_N} \left\{ \prod_{J=1}^{N} W(l_J - \lambda_J, l_{J+1} - \lambda_{J+1} \mid l_J, l_{J+1}) \right\} \times \psi(l_1 - \lambda_1, ..., l_{N+1} - \lambda_{N+1}), \tag{B.22}$$

where $\lambda_{N+1} \equiv \lambda_1$ and the summation is over $\lambda_1 = \pm 1, ..., \lambda_N = \pm 1$.

Setting

$$m_J = l_J - \lambda_J, \qquad J = 1, ..., N+1, \tag{B.23}$$

we see that we can replace the summation in (B.22) by a summation over the integers $m_1, ..., m_{N+1}$, subject to the restrictions

$$\begin{gathered} m_J = l_J \pm 1, \qquad J = 1, ..., N+1, \\ m_1 - l_1 = m_{N+1} - l_{N+1}. \end{gathered} \tag{B.24}$$

Since l_J and l_{J+1} differ by one, it follows that m_J and m_{J+1} differ by one or three. However, the latter case gives a zero contribution since from (B.20) the summand then vanishes. Thus we can impose the further restriction

$$m_{J+1} = m_J \pm 1, \qquad J = 1, ..., N. \tag{B.25}$$

Equations (B.22)–(B.25) then become the result (3.5)–(3.7) quoted in the text. Thus Eqs. (B.11), together with the boundary condition (B.21) and the requirement that each matrix $\mathbf{M}_l$ be non-singular, are sufficient to establish the closure property (3.5).

We can use (B.13) to combine the four equations (B.11) into one. Define two sets of two-dimensional vectors $z_{l+1,l}$ and $z_{l-1,l}$ by

$$z_{l+1,l} = x_l\,, \qquad z_{l-1,l} = y_l \tag{B.26}$$

for all integers l. Using (B.13) and (B.26), we then find that all four equations (B.11) can be written as

$$\mathbf{R}\{\Phi_{l,l'} \otimes z_{m',l'}\} = \sum_m W(m, m' \mid l, l')\, \Phi_{m,m'} \otimes z_{m,l} \tag{B.27}$$

for all integer values of l, l', m' such that $| l - l' | = | m' - l' | = 1$. The summation on the r.h.s. is over integer values of m such that $| m - m' | = 1$, $| m - l | = 1$. The coefficients $W(m, m' \mid l, l')$ are again related to $a_l ,..., c_l'$ by (7.1).

We note that (B.27) is simply Eq. (3.3) written in abbreviated notation. Thus our conditions (B.11) are the conditions (3.3) quoted in the text.

From (B.9) and (B.26) we see that the two column vectors of the matrix $\mathbf{M}_l$ are $z_{l-1,l}$ and $z_{l+1,l}$. Thus $\mathbf{M}_l$ is nonsingular if these are linearly independent. Also, the boundary conditions (B.21) and (3.4) are equivalent. This complete the proof that the conditions (3.3)–(3.4) are sufficient to ensure the closure property (3.5)–(3.7).

Appendix C

Here we show that we can solve (3.3), or equivalently (B.11). The solution leads us to introduce elliptic functions.

As we remarked in Appendix B, the first two sets of Eqs. (B.11) have already been solved in Paper I. There we attempted to show in some detail how the parametrization in terms of elliptic functions arises. We shall therefore not stress this aspect, referring the reader to Paper I for complete details.

First look at the first of the Eqs. (B.11), which is written explicitly in (B.10). Taking $\alpha = \pm$, $\lambda = \pm$, we obtain four equations (for a given value of l) which are homogeneous and linear in the unknowns $\phi_l(+)$, $\phi_l(-)$, $\phi_{l+1}(+)$, $\phi_{l+1}(-)$. The determinant of the coefficients must therefore vanish, giving

$$(a^2 + b^2 - c^2 - d^2)\, p_l p_{l+1} = ab(p_l^2 + p_{l+1}^2) - cd[1 + p_l^2 p_{l+1}^2], \tag{C.1}$$

where

$$p_l = x_l(+)/x_l(-) \tag{C.2}$$

for all integers l.

This symmetric quadratic recursion relation between p_l and p_{l+1} was discussed in Paper I [Eqs. (I.32)–(I.54)]. We showed that it led us to introduce parameters k, η, v, defined in terms of the Boltzmann weights a, b, c, d by

$$a : b : c : d = \mathrm{sn}(v + \eta) : \mathrm{sn}(v - \eta) : \mathrm{sn}(2\eta) : k\,\mathrm{sn}(2\eta)\,\mathrm{sn}(v - \eta)\,\mathrm{sn}(v + \eta), \tag{C.3}$$

where $\mathrm{sn}(u)$ is the elliptic sine-amplitude function of argument u and modulus k (8.14 of GR). It followed that if

$$p_l = k^{1/2}\,\mathrm{sn}(u_l) \tag{C.4}$$

for some value of l, then

$$p_{l+1} = k^{1/2}\,\mathrm{sn}(u_l \pm 2\eta). \tag{C.5}$$

Equations (B.10) can now be solved for $\phi_l(\pm)$, $\phi_{l+1}(\pm)$. Using the relation (6.2) and the addition formulae (I.59), (I.60), we find that

$$\begin{aligned} \phi_l(+)/\phi_l(-) &= k^{1/2}\,\mathrm{sn}[u_l \pm (\eta - v)], \\ \phi_{l+1}(+)/\phi_{l+1}(-) &= k^{1/2}\,\mathrm{sn}[u_l \pm (3\eta - v)]. \end{aligned} \tag{C.6}$$

Equations (C.4)–(C.6) apply to some particular value of l and the same choice of sign must be made in each equation. To satisfy them for each value of l it appears that we must make the same choice of sign for all values of l, say positive. We then see that (C.4)–(C.6) are satisfied by setting

$$p_l = k^{1/2}\,\mathrm{sn}(s + 2l\eta), \tag{C.7}$$

$$\phi_l(+)/\phi_l(-) = k^{1/2}\,\mathrm{sn}(s + 2l\eta + \eta - v) \tag{C.8}$$

for all integers l. The parameter s is arbitrary, but the same for all l.

In an exactly similar way we can solve the last of Eqs. (B.11) and obtain

$$y_l(+)/y_l(-) = k^{1/2}\,\mathrm{sn}(t + 2l\eta), \tag{C.9}$$

$$\tau_l(+)/\tau_l(-) = k^{1/2}\,\mathrm{sn}(t + 2l\eta + \eta + v), \tag{C.10}$$

for all integers l; t is arbitrary.

We are free to choose any convenient normalization of the vectors x_l, ϕ_l, y_l, τ_l and to write a, b, c, d in any way that satisfies (C.3). We find that the coefficients a_l,

a_l' in (B.11) are greatly simplified if we define a, b, c, d by (6.1) and for the moment choose

$$\begin{aligned} x_l(-) &= \Theta(s + 2l\eta), \\ \phi_l(-) &= \Theta(s + 2l\eta + \eta - v), \\ y_l(-) &= \Theta(t + 2l\eta), \\ \tau_l(-) &= \Theta(t + 2l\eta + \eta + v). \end{aligned} \tag{C.11}$$

Using (6.2) and the fact that sn(u) is an odd function, we see that

$$H(u)\,\Theta(-u) + H(-u)\,\Theta(u) = 0 \tag{C.12}$$

for all complex numbers u. It follows that (6.1) implies (C.3). Also, from (6.2), (C.2), (C.4), (C.5), (C.9)–(C.11), we see that the elements $x_l(+)$, $\phi_l(+)$, $y_l(+)$, $\tau_l(+)$ are given by replacing the minus signs on the left of (C.11) by plus signs, and the functions $\Theta(u)$ on the right by $H(u)$.

Using the notation (6.4), it follows that for the moment we are choosing the vectors x_l, ϕ_l, y_l, τ_l to be given by

$$\begin{aligned} x_l &= | s + 2l\eta\rangle, \qquad & \phi_l &= | s + 2l\eta + \eta - v\rangle, \\ y_l &= | t + 2l\eta\rangle, \qquad & \tau_l &= | t + 2l\eta + \eta + v\rangle. \end{aligned} \tag{C.13}$$

With these normalizations, the coefficients a_l, a_l' in the first and last of Eqs. (B.11) are found to be

$$a_l = a_l' = \rho' h(v + \eta) \tag{C.14}$$

for all integers l, where the function $h(u)$ is defined by (6.6), i.e.,

$$h(u) = H(u)\,\Theta(-u) \tag{C.15}$$

and, as in (7.2),

$$\rho' = \rho\Theta(0). \tag{C.16}$$

To summarize our results so far: Given the parametrization (6.1) of a, b, c, d in terms of ρ, k, η, v, the first and last of Eqs. (B.11) are satisfied by (C.13) and (C.14). The parameters s and t are arbitrary.

It is quite simple to verify this directly, using only the properties (I.59), (I.60), (C.12) of the functions $H(u)$, $\Theta(u)$. These properties are certainly satisfied by elliptic theta functions of Jacobi (Section 8.192 of GR), which we write as $H_{Jb}(u)$, $\Theta_{Jb}(u)$. More generally, they are also satisfied by the modified theta functions:

$$\begin{aligned} H(u) &= H_{Jb}(u) \exp\{A(u - K)^2\}, \\ \Theta(u) &= \Theta_{Jb}(u) \exp\{A(u - K)^2\}, \end{aligned} \tag{C.17}$$

for any values of the constant A. Throughout this appendix we take $H(u)$, $\Theta(u)$ to be defined by (C.17) for some value of A, which is not yet defined. We use this extra degree of freedom in Section 6 to ensure that under certain circumstances [i.e., when (6.8) is satisfied] the functions $H(u)$, $\Theta(u)$ can be chosen to have a convenient periodicity property.

All the mathematical formulas written in this appendix, as well as (I.59) and (I.60), apply to these generalized theta functions, for any value of A.

It remains to check if we can satisfy the middle two Eqs. (B.11), using the forms (C.13) of the vectors. Remember that these equations are written in the same short-hand notation that abbreviates (B.10) to (B.11a). Thus for a given value of l each is actually four equations, corresponding to taking $\alpha = \pm$, $\lambda = \pm$.

One way to check the second equation is to hold l fixed, regard the vectors ϕ_l, y_{l+1}, y_l, x_l as known, given by (C.13), and to solve for the four unknowns $b_l'\phi_{l-1}(\pm)$, $c_l\tau_l(\pm)$. This involves only solving two independent pairs of linear homogeneous equations in two unknowns. For instance, using (2.2) and (2.3) to write the l.h.s. explicitly and solving for $c_l\tau_l(+)$, we find that

$$[x_l(+)\, y_l(-) - x_l(-)\, y_l(+)]\, c_l\tau_l(+) = F_l\phi_l(+) + G_l\phi_l(-), \tag{C.18}$$

where

$$\begin{aligned} F_l &= a y_{l+1}(+)\, y_l(-) - b y_{l+1}(-)\, y_l(+), \\ G_l &= d y_{l+1}(-)\, y_l(-) - c y_{l+1}(+)\, y_l(+). \end{aligned} \tag{C.19}$$

All the terms in this equation are regarded as known, given by (6.1) and (C.13), except for $c_l\tau_l(+)$, which we are to evaluate.

Our choice of grouping the terms on the r.h.s. of (C.18) is dictated by the fact that F and G can be simplified by using the addition theorem (I.59). Substituting the forms for a, b, c, d, $y_l(\pm)$, $y_{l+1}(\pm)$ given by (6.1) and (C.13) [or (C.11) and its analogue], we find from (I.59) that

$$\begin{aligned} F_l &= \rho\Theta(0)\, H(2\eta)\, \Theta(-2\eta)\, H(t + 2l\eta + \eta + v)\, \Theta(t + 2l\eta + \eta - v), \\ G_l &= \rho\Theta(0)\, H(-2\eta)\, \Theta(2\eta)\, H(t + 2l\eta + \eta + v)\, H(t + 2l\eta + \eta - v). \end{aligned} \tag{C.20}$$

Using (C.12)–(C.16), it follows that the r.h.s. of (C.18) is

$$\begin{aligned} \rho' h(2\eta)\, H(t + 2l\eta + \eta + v)[&\Theta(t + 2l\eta + \eta - v)\, H(s + 2l\eta + \eta - v) \\ &- H(t + 2l\eta + \eta - v)\, \Theta(s + 2l\eta + \eta - v)]. \end{aligned} \tag{C.21}$$

Also, the bracketted factor on the l.h.s. of (C.18) is [using (C.13)]

$$H(s + 2l\eta)\, \Theta(t + 2l\eta) - \Theta(s + 2l\eta)\, H(t + 2l\eta). \tag{C.22}$$

To proceed further it is clearly necessary to obtain some mathematical identity which simplifies the expression $H(u)\,\Theta(v) - \Theta(u)\,H(v)$. Such an identity can be obtained by noting that the zeros of this expression occur at $u - v = 4mK + 2inK'$, $u + v = (4m+2)K + 2inK'$, for any integers m, n. From the analyticity and quasi-periodicity of the theta functions, it follows that there must exist two entire functions $f(u)$, $g(u)$ such that

$$H(u-v)\,\Theta(u+v) - \Theta(u-v)\,H(u+v) = f(u)\,g(v) \tag{C.23}$$

for all complex numbers u, v.

Setting $u = 0$ and using (C.5), it follows that we can choose

$$g(v) = H(v)\,\Theta(-v) = h(v), \tag{C.24}$$

where $h(u)$ is the function defined by (6.6) and (C.15). To obtain $f(u)$ we set $v = K$ in (C.23) and use the symmetry relations

$$H(2K-u) = H(u), \qquad \Theta(2K-u) = \Theta(u) \tag{C.25}$$

together with (C.5) and (C.24). This gives

$$f(u) = 2h(u-K)/h(K). \tag{C.26}$$

Thus we have established the identity

$$H(u-v)\,\Theta(u+v) - \Theta(u-v)\,H(u+v) = 2h(u-K)\,h(v)/h(K), \tag{C.27}$$

where $h(u)$ is the function defined by (6.6) and (C.15). This identity applies for any value of the parameter A in the definition (C.17) of our generalized theta functions.

We now use this identity to simplify the expressions (C.21) and (C.22). We find that they both have a factor $2h[(s-t)/2]/h(K)$. Substituting the resulting expressions into (C.18), these common factors cancel, leaving

$$c_l\tau_l(+) = \rho' h(2\eta)\,h(w_l+\eta-v)\,H(t+2l\eta+\eta+v)/h(w_l), \tag{C.28}$$

where

$$w_l = \tfrac{1}{2}(s+t) - K + 2l\eta. \tag{C.29}$$

We now go through similar working to solve the second of the equations (B.11) for $c_l\tau_l(-)$. We find that the result is the same as the r.h.s. of (C.28), except that $H(t+2l\eta+\eta+v)$ is replaced by $\Theta(t+2l\eta+\eta+v)$. Taking the ratios of these results and using (6.2), we see that the vector τ_l we have obtained satisfied (C.10), and hence this first test of the consistency of the equations is satisfied. (Note that s

and t can still be chosen arbitrarily.) Using the given normalization (C.13) of the vector τ_l, we see from (C.28) that

$$c_l = \rho' h(2\eta)\, h(w_l + \eta - v)/h(w_l). \tag{C.30}$$

The next step is to solve for $b_l' \phi_{l-1}(\pm)$. The same techniques work, we find that the result is consistent with (C.13) and that

$$b_l' = \rho' h(v - \eta)\, h(w_{l+1})/h(w_l). \tag{C.31}$$

This verifies the second of the Eqs. (B.11).

The verification of the third can be done in exactly the same way, solving now for $c_{l+1}' \phi_l$ and $b_{l+1} \tau_{l+1}$. We find that it is satisfied provided that

$$c_{l+1}' = \rho' h(2\eta)\, h(w_l + \eta + v)/h(w_{l+1}), \tag{C.32}$$

$$b_{l+1} = \rho' h(v - \eta)\, h(w_l)/h(w_{l+1}). \tag{C.33}$$

These equations are valid for all integers l.

Having gone through all this working, we find that we have actually proved two mathematical identities. Combining the notations used in (B.11) and (6.4), these can be written as

$$\mathbf{R}\{| s \pm (\eta - v)\rangle \otimes | s \pm 2\eta\rangle\} = r_1 \,| s \pm (3\eta - v)\rangle \otimes | s\rangle, \tag{C.34a}$$

$$\begin{aligned} \mathbf{R}\{| s \pm (\eta - v)\rangle \otimes | t \pm 2\eta\rangle\} = r_2 \,| s \mp (\eta + v)\rangle \otimes | t\rangle \\ + r_3 \,| t \pm (\eta + v)\rangle \otimes | s\rangle, \end{aligned} \tag{C.34b}$$

where a, b, c, d are given by (6.1) and the coefficients r_1, r_2, r_3 by

$$\begin{aligned} r_1 &= \rho' h(v + \eta), \\ r_2 &= \rho' h(v - \eta)\, h(w \pm 2\eta)/h(w), \qquad (C.35) \\ r_3 &= \rho' h(2\eta)\, h[w \pm (\eta - v)]/h(w), \\ w &= \tfrac{1}{2}(s + t - 2K). \qquad (C.36) \end{aligned}$$

These equations are mathematical identities, satisfied for all values, real or complex, of s and t. The choice of sign (upper or lower) must be made consistently throughout either of Eqs. (C.34) and the subsidiary definitions (C.35).

These identities are sufficient to establish Eqs. (B.11), and hence (3.3). The first and last of the Eqs. (B.11) are obtained by choosing the upper and lower signs in (C.34a) and replacing s by $s + 2l\eta$, $t + 2(l + 1)$, respectively. The second (third) of Eqs. (B.11) is obtained by choosing the upper (lower) signs in (C.34b) and replacing s, t by $s + 2l\eta$, $t + 2l\eta$ ($t + 2l\eta + 2\eta$, $s + 2l\eta + 2\eta$).

We end this Appendix by noting from (B.11) and (C.31) that we can arrange that b_l' becomes independent of l by renormalizing the vector y_l to be given by

$$y_l = h(w_l) \mid t + 2l\eta\rangle \tag{C.37}$$

for all integers l. The vectors x_l, ϕ_l, τ_l remain given by (C.13). Using (B.11) to make the appropriate adjustments in a_l, a_l', b_l, b_l', c_l, c_l', we find from (C.14) and (C.30)–(C.33) that these coefficients are given by (7.1). This renormalization turns out to be a help when we come to handling the Bethe ansatz equations for the eigenvectors of the transfer matrix.

References

1. R. J. Baxter, *Ann. Phys.*, **76** (1973), 1–24.
2. R. J. Baxter, *Ann. Phys. N.Y.* **70** (1972), 193–228.
3. E. H. Lieb, *Phys. Rev.* **162** (1967), 162–172 *Phys. Rev. Lett.* **18** (1967), 1046–1048 **19** (1967), 108–110.
4. F. Y. Wu, *Phys. Rev. B* **4** (1971), 2312–2314.

Eight-Vertex Model in Lattice Statistics and One-Dimensional Anisotropic Heisenberg Chain. III. Eigenvectors of the Transfer Matrix and Hamiltonian

RODNEY BAXTER*,†

Institute for Theoretical Physics, State University of New York, Stony Brook, New York 11790

Received September 5, 1972

We obtain the eigenvectors of the transfer matrix of the zero-field eight vertex model. These are also the eigenvectors of the Hamiltonian of the corresponding one-dimensional anistropic Heisenberg chain.

1. INTRODUCTION AND SUMMARY

This is the third and final paper of a series in which we obtain the eigenvectors of the transfer matrix **T** of the zero-field eight-vertex model.

In Paper I (referred to as I) [1], we found some special eigenvectors. In Paper II (referred to as II) [2] we generalized these to form a basis set of vectors, with respect to which **T** becomes the transfer matrix of an Ising-like problem. In this problem each spin can have L values and the four spins round a square interact. Most importantly, two adjacent spins must differ by unity. From this last property it follows that the problem can be thought of as a generalized ice-type problem and we may hope to obtain the eigenvectors of **T** by an appropriate extension of the Bethe ansatz technique that works for the ice models [3]. We show here that this is so.

In Section 8 of I we showed that **T** commutes with the Hamiltonian $\mathscr{H}$ of an anisotropic one-dimensional Heisenberg chain. Thus the eigenvectors we construct here are also those of $\mathscr{H}$.

In this section we present our results for the eigenvectors and eigenvalues of **T**, and show that the equations for the eigenvalues are the same as those we obtained in our original solution of the eight-vertex model [4]. This previous solution gave no information regarding the eigenvectors.

* Work supported in part by National Science Foundation Grant No. P2P2847-000.

† On leave of absence from the Research School of Physical Sciences, The Australian National University, Canberra A. C. T. 2600, Australia.

The motivation of this series of papers, namely to obtain the eigenvectors, is something of a long-range one. In principle a knowledge of the eigenvectors should enable one to calculate correlations and spontaneous polarizations. However, our eigenvectors are more complicated than the eigenvectors found by Lieb [3] for the ice models (they contain them as special cases). As yet it has not been found possible to calculate the correlations and staggered polarizations of the ice models (e.g., the F model), due to the mathematical complexity of evaluating the required matrix elements. Thus our results must wait in hope for this to be done. The calculation of the correlations and spontaneous polarizations of the eight-vertex model should not then be far off.

Basic Vectors

We first quote the principal results of II. There we used the notation $|\, u\rangle$ to denote the two-dimensional vector

$$|\, u\rangle = \begin{pmatrix} H(u) \\ \Theta(u) \end{pmatrix}, \tag{1.1}$$

where $H(u)$, $\Theta(u)$ are the modified elliptic theta functions of argument u and modulus k, defined by (II.6.9). The modulus k, together with three other parameters ρ, η, v, are defined from the Boltzmann weights a, b, c, d of the 8 vertex model by the relations (II.6.1).

We then defined two sets of two-dimensional vectors $\Phi_{l,l+1}$ and $\Phi_{l+1,l}$ (l any integer) by

$$\Phi_{l,l+1} = |\, s + 2l\eta + \eta - v\rangle \tag{1.2a}$$

$$\Phi_{l+1,l} = |\, t + 2l\eta + \eta + v\rangle. \tag{1.2b}$$

The parameters s and t are arbitrary. (Except that the vectors $z_{l+1,l}$, and $z_{l-1,l}$ in (II.6.5) must be linearly independent for each value of l. In particular we exclude the case $s = t$.)

The next step is to define a family $\mathscr{F}$ of 2^N-dimensional vectors, a typical member being

$$\psi(l_1, ..., l_{N+1}) = \Phi_{l_1,l_2} \otimes \Phi_{l_2,l_3} \otimes \cdots \otimes \Phi_{l_N,l_{N+1}}, \tag{1.3}$$

where $l_1, ..., l_{N+1}$ are any integers such that

$$l_{J+1} = l_J \pm 1, \qquad J = 1, ..., N. \tag{1.4}$$

Pre-multiplying (1, 3) by the transfer matrix $\mathbf{T}$ of the eight-vertex model, as defined in (II.2.4), we found [Eq. (II.3.5)] that, providing a boundary condition was satisfied,

$$\mathbf{T}\psi(l_1, ..., l_{N+1}) = \sum \left\{ \prod_{J=1}^{N} W(m_J, m_{J+1} \mid l_J, l_{J+1}) \right\} \psi(m_1, ..., m_{N+1}), \tag{1.5}$$

where the coefficients W are defined by (II.7.1) and the summation is over all integers $m_1, ..., m_{N+1}$ such that

$$m_J = l_J \pm 1, \qquad J = 1, ..., N+1, \tag{1.6}$$

$$m_{J+1} = m_J \pm 1, \qquad J = 1, ..., N, \tag{1.7}$$

$$m_1 - l_1 = m_{N+1} - l_{N+1} . \tag{1.8}$$

The signs in (1.4), (1.6), (1.7) may be chosen arbitrarily for each J, except that for $J = 1$ and $N + 1$ in (1.6) they must be the same. This ensures (1.8).

The boundary condition needed in the derivation of (1.5) is certainly satisfied if $l_{N+1} = l_1$. More generally, let K, K' be the complete elliptic integrals of moduli k, $k' = (1 - k^2)^{1/2}$, respectively, (Section 8.112 of Gradshteyn and Ryzhik [5], hereinafter referred to as GR). Then if there exist integers L, m_1, m_2 such that

$$L\eta = 2m_1 K + im_2 K', \tag{1.9}$$

then the modified theta functions $H(u)$, $\Theta(u)$ defined by (II.6.9) that we use in the working are periodic of period $2L\eta$. In this case it follows that the boundary condition is satisfied if

$$l_{N+1} - l_1 = L \times \text{integer}. \tag{1.10}$$

Note from (1.8) that if l_1 and l_{N+1} satisfy (1.10), then so do m_1 and m_{N+1}. Thus we see that we have defined a family $\mathscr{F}$ of 2^N-dimensional vectors such that pre-multiplying any member of the family by $\mathbf{T}$ gives a linear combination of vectors of the family.

We further point out in Section 4 of II that $\mathscr{F}$ breaks up into $N + 1$ subfamilies $\mathscr{F}_0, ..., \mathscr{F}_N$, such that $\mathscr{F}_n$ consists of vectors of the form (1.3) with $N - n$ increasing steps in the sequence $l_1, ..., l_{N+1}$, and n decreasing steps [i.e. $N - n$ of the two-dimensional vectors on the r.h.s. of (1.3) are of the type (1.2a), the other n of the type (1.2b)]. Pre-multiplying a vector of $\mathscr{F}_n$ by $\mathbf{T}$ gives a linear combination of vectors of $\mathscr{F}_n$, provided the boundary condition (1.10) is satisfied. Since there are $N - n$ increasing steps in the sequence $l_1, ..., N + 1$, and n decreasing steps, it follows that $l_{N+1} = l_1 + N - 2n$. Hence (1.10) is satisfied if

$$N - 2n = L \times \text{integer}. \tag{1.11}$$

We can therefore attempt to construct eigenvectors of $\mathbf{T}$ which are linear combinations of vectors of $\mathscr{F}_n$, for some fixed value of n. The case $n = 0$ is trivial and is discussed in I, and in Section 5 of II. In this paper we consider the general case when $0 < n < N$.

Bethe Ansatz

We can think of the decreasing steps (i.e. l_J followed by $l_J - 1$) in the sequence $l_1, ..., l_{N+1}$ as "down arrows" or "particles" (the arrow formulation is given in Section 4 of II). A basis vector of the family $\mathscr{F}_n$ can be specified by prescribing the value l of l_1 and the positions $x_1, ..., x_n$ of these decreasing steps. We then have

$$\begin{aligned} &l_1 = l, \\ &l_{J+1} = l_J + 1 \quad \text{for} \quad J = 1, ..., N, \\ &\qquad\qquad \text{provided} \quad J \neq x_1, x_2, ..., \text{ or } x_n, \\ &l_{J+1} = l_J - 1 \quad \text{if} \quad J = x_1, x_2, ..., \text{ or } x_n. \end{aligned} \tag{1.12}$$

Taking $x_1, ..., x_n$ to be arranged in increasing order and successively evaluating $l_1, l_2, ..., l_{N+1}$, it follows that

$$l_J = l + J - 1 - 2m \qquad \text{if} \quad x_m < J \leqslant x_{m+1} \tag{1.13}$$

(defining $x_0 = 0$, $x_{n+1} = N + 1$).

We attempt to construct eigenvectors Ψ of $\mathbf{T}$ which are linear combinations of vectors of $\mathscr{F}_n$. Thus we define

$$\Psi = \sum_{l=1}^{L} \sum_{X} f(l \mid x_1, ..., x_n)\, \psi(l_1, ..., l_{N+1}), \tag{1.14}$$

where the X summation is over all positions $X = \{x_1, ..., x_n\}$ such that $1 \leqslant x_1 < x_2 < \cdots < x_n \leqslant N$, and $l_1, ..., l_{N+1}$ are given in terms of $l, x_1, ..., x_n$ by (1.12) and (1.13).

Using (1.5), we wish to choose the coefficients $f((l \mid X)$ so that

$$\mathbf{T}\Psi = \Lambda\Psi, \tag{1.15}$$

i.e., Ψ is an eigenvector of $\mathbf{T}$ with corresponding eigenvalue Λ.

The resulting equations for the $f(l \mid X)$ are given in Section 2. We see that they are very similar to the equations for the eigenvectors of the ice models [3]. By generalizing the technique that Lieb used for these models, we are led to try a generalized Bethe ansatz for $f(l \mid X)$, namely

$$f(l \mid X) = \sum_{P} A(P)\, g_{P1}(l, x_1)\, g_{P2}(l - 2, x_2) \cdots g_{Pn}(l - 2n + 2, x_n), \tag{1.16}$$

where the summation is over all $n!$ permutations $P = \{P1, ..., Pn\}$ of the integers $\{1, ..., n\}$, the $A(P)$ are $n!$ coefficients, and $g_1(l, x), ..., g_n(l, x)$ are n "single particle" functions of l and x.

These coefficients A and functions g are unknown. We attempt to choose them to satisfy the equations for $f(l \mid X)$ and Λ that follow from (1.14) and (1.15).

Results of the Bethe Ansatz

The working is given in Sections 2–6. After a series of mathematical flukes we find that the eigenvalue equations can indeed be satisfied. We are led to introduce n numbers (in general complex) $u_1, \ldots, u_n$ such that u_j is associated with the single particle function $g_j(l, x)$. We can also introduce n "wave numbers" $k_1, \ldots, k_n$ such that k_j is defined in terms of u_j by

$$\exp(ik_j) = h(u_j + \eta)/h(u_j - \eta) \tag{1.17}$$

for $j = 1, \ldots, n$. The function $h(u)$ is given by

$$h(u) = H(u)\,\Theta(-u). \tag{1.18}$$

It is an entire function, odd, periodic of period $2L\eta$.

The single-particle function $g_j(l, x)$ is then given by

$$g_j(l, x) = e^{ik_j x} h(w_{l+x-1} - \eta - u_j)/[h(w_{l+x-2})\, h(w_{l+x-1})] \tag{1.19a}$$

for $j = 1, \ldots, n$ and all integers l and x, where

$$w_l = \tfrac{1}{2}(s + t) - K + 2l\eta. \tag{1.19b}$$

The coefficients $A(P)$ are found to be given by

$$A(P) = \epsilon_P \prod_{1 \leqslant j < m \leqslant n} h(u_{Pj} - u_{Pm} + 2\eta), \tag{1.20}$$

where ϵ_P is the signature ($\pm$) of the permutation $P = \{P1, \ldots, Pn\}$. The eigenvalue Λ is given by

$$\Lambda = \phi(v - \eta) \prod_{j=1}^{n} \frac{h(v - u_j + 2\eta)}{h(v - u_j)} + \phi(v + \eta) \prod_{j=1}^{n} \frac{h(v - u_j - 2\eta)}{h(v - u_j)}, \tag{1.21}$$

where the function $\phi(v)$ is defined to be

$$\phi(v) = [\rho\Theta(0)\, h(v)]^N. \tag{1.22}$$

We also get n equations which come from the cyclic boundary conditions we impose on the lattice model. They are

$$\exp(iNk_j) = -\prod_{m=1}^{n} [h(u_j - u_m + 2\eta)/h(u_j - u_m - 2\eta)] \tag{1.23}$$

for $j = 1,..., n$. These equations define $u_1 ,..., u_n$ (there will be many solutions, corresponding to different eigenvectors). Once these are known, the eigenvalue Λ of the transfer matrix can be evaluated from (1.21). The eigenvector Ψ can be obtained by evaluating the coefficients $A(P)$ and the single-particle functions $g_j(l, x)$ from (1.20) and (1.19), substituting these into (1.16) to give $f(l \mid X)$, and in turn substituting this result into (1.14) to give Ψ.

We remark that these results are completely analogous to those of the normal Bethe ansatz used by Lieb [3] and Yang and Yang [6]. The relation (1.17) between the wave numbers k_j and the parameters u_j is analogous to the difference-kernel transformation that takes k to α. The main difference is that in our working it is $u_1 ,..., u_n$, rather than $k_1 ,..., k_n$, that occur naturally.

Functional Equation for the Eigenvalues

We point out here that the equations (1.21), (1.23) are the same as those we originally obtained for the eigenvalues [4]. To see this, regard $u_1 ,..., u_n$ as known and define a function $Q(v)$ by

$$Q(v) = \prod_{j=1}^{n} h(v - u_j). \tag{1.24}$$

Using (1.17), the equations (1.21), (1.23) can be written as

$$\Lambda = [\phi(v - \eta)\, Q(v + 2\eta) + \phi(v + \eta)\, Q(v - 2\eta)]/Q(v), \tag{1.25}$$

and

$$\phi(u_j - \eta)\, Q(u_j + 2\eta) + \phi(u_j + \eta)\, Q(u_j - 2\eta) = 0 \tag{1.26}$$

for $j = 1,..., n$.

Note from (1.17) and (1.23) that $u_1 ,..., u_n$ depend on k and η, but not on v. From (1.19), (1.20) it follows that the coefficients $f(l \mid X)$ are independent of v. Also, from (1.2) and (1.3) we see that the basis vectors $\psi(l_1 ,..., l_{N+1})$ depend on s, t, v only through the linear combinations $s - v$, $t + v$. Since s and t are arbitrary, we can replace them by $s + v$, $t - v$. [This has no effect on (1.19).] The basis vectors are then independent of v.

It follows that the eigenvectors of $\mathbf{T}$ that we have constructed are independent of v. Assuming that these vectors form a complete set (i.e. span all 2^N-dimensional space), we can construct a 2^N by 2^N non-singular matrix $\mathbf{P}$ whose columns are the eigenvectors of $\mathbf{T}$. Exhibiting the dependence of $\mathbf{T}$ on v explicitly by writing it as $\mathbf{T}(v)$, it follows that

$$\mathbf{P}^{-1}\mathbf{T}(v)\mathbf{P} = \mathbf{T}_d(v), \tag{1.27}$$

where $\mathbf{T}_d(v)$ is the diagonal matrix whose diagonal elements are the 2^N eigenvalues $\Lambda(v)$ of $\mathbf{T}(v)$. The matrix $\mathbf{P}$ is independent of v.

Note that (1.27) implies that two transfer matrices $\mathbf{T}(u)$, $\mathbf{T}(v)$ commute (k and η are regarded here as constants, so must be the same for both). This can be proved directly, and was the first step in our original solution of the eight-vertex model [4].

Let $u_1, \ldots, u_n$ be a solution of (1.23), or equivalently (1.26). Hopefully there will be 2^N independent solutions (allowing all possible values of n), corresponding to the eigenvectors of $\mathbf{T}(v)$. For each solution (1.24) and (1.25) define an eigenvalue $\Lambda(v)$ of $\mathbf{T}(v)$ and a function $Q(v)$. Let $\mathbf{Q}_d(v)$ be a diagonal matrix with these functions $Q(v)$ as diagonal elements, arranged in the same order as the $\Lambda(v)$ in $\mathbf{T}_d(v)$. From (1.25) it is then apparent that the 2^N by 2^N diagonal matrices $\mathbf{T}_d(v)$, $\mathbf{Q}_d(v)$ satisfy the relation

$$\mathbf{T}_d(v)\,\mathbf{Q}_d(v) = \phi(v-\eta)\,\mathbf{Q}_d(v+2\eta) + \phi(v+\eta)\,\mathbf{Q}_d(v-2\eta). \tag{1.28}$$

Now define a matrix $\mathbf{Q}(v)$ by

$$\mathbf{Q}(v) = \mathbf{P}\mathbf{Q}_d(v)\,\mathbf{P}^{-1}. \tag{1.29}$$

From (1.27)–(1.29) we see that the matrices $\mathbf{T}(v)$, $\mathbf{Q}(u)$, $\mathbf{Q}(v)$ all commute for any values of u and v, and that

$$\mathbf{T}(v)\,\mathbf{Q}(v) = \phi(v-\eta)\,\mathbf{Q}(v+2\eta) + \phi(v+\eta)\,\mathbf{Q}(v-2\eta). \tag{1.30}$$

This is the functional matrix relation we derived directly in [4]. Given this relation and the commutation properties one can reason backwards to (1.25). The equations (1.26) for $u_1, \ldots, u_n$ then follow simply as a consequence of the fact that the elements, and from (1.27) the eigenvalues $\Lambda(v)$, of $\mathbf{T}(v)$ are entire functions of v. Thus when $v = u_1, \ldots, u_n$ the numerator of the r.h.s. of (1.25) must vanish, giving (1.26).

Our present results agree precisely with those of [4] provided we take $n = \frac{1}{2}N$ in (1.24). Remember that this is the case when the boundary condition is automatically satisfied. In this case the re-normalization (II.6.9) of the theta functions does not affect our final equations.

For the cases when η satisfies (1.9) we see from (1.11) that we also allow n in (1.24) to have the values (taking N to be even):

$$n = \tfrac{1}{2}N + L \times \text{integer}. \tag{1.31}$$

It seems likely that these solutions can be regarded as aberrations of the $n = \frac{1}{2}N$ case, in which some of the u_j form complete "strings," each string being made up of L u_j's of the form $u_j = \text{constant} + 2j\eta$ for $j = 1, \ldots, L$. Such strings will cancel out of the equations (1.25), (1.26).

Dependence of Eigenvectors on s and t

We are still free to choose the parameters s and t arbitrarily. From (1.17) and (1.21)–(1.23) we see that these do not enter into the equations for $u_1, ..., u_n$ and the eigenvalue Λ. They do however occur in the expressions (1.2), (1.19), (1.20) that are used in constructing the eigenvectors.

Let choose some solution $u_1, ..., u_n$ of (1.23) and then vary s and t. The eigenvectors that we form will all have the same eigenvalue and must lie in some sub-space of 2^N-dimensional space. In particular, from the Frobenius theorem the eigenvector corresponding to the maximum eigenvalue should be unique. Thus in this case varying s and t can only alter the eigenvector by a multiplicative normalization factor.

That this should be so is by no means obvious from the above equations. It is possible that we can obtain further insight into the structure of the eigenvector by using this requirement.

Summary of Remaining Sections

We have now stated our results and discussed them. In the following Sections 2 to 6 we go through the mathematical working required to obtain the coefficients $f(l \mid X)$ in the expression (1.14) for the eigenvector Ψ. In Section 7 we summarize this working and in Section 8 we indicate an allowed extension to a certain class of inhomogeneous lattices.

2. Transfer Matrix Equations

From (1.5), (1.14), (1.15) the transfer matrix equations for Λ and $f(l \mid X)$ are:

$$\Lambda f(l \mid X) = \sum \left\{ \prod_{J=1}^{N} W(l_J, l_{J+1} \mid m_J, m_{J+1}) \right\} f(m \mid Y), \tag{2.1}$$

where the $l_1, ..., l_{N+1}$ are given in terms of l and $x_1, ..., x_n$ by (1.13); $m_1, ..., m_{N+1}$ are similarly expressed in terms of m and $y_1, ..., y_n$. The summation is over all $m_1, ..., m_{N+1}$ such that $m_J = l_J \pm 1$, $m_{J+1} = m_J \pm 1$.

We can think of this summation graphically. Remember that $W(l, l' \mid m, m')$ is the weight of the configuration of spins round a vertex shown in Fig. 1. Since adjacent spins must differ by unity, there are six sets of possible spin configurations at a vertex. We show these in Fig. 2. In this figure we have drawn broken lines on vertical bonds (horizontal bonds) across which the spins increase from left to right (down to up), and we have drawn solid lines on bonds for the other cases. In the arrow notation of Paper II, broken lines are arrows pointing up or to the right, solid lines are arrows pointing down or to the left.

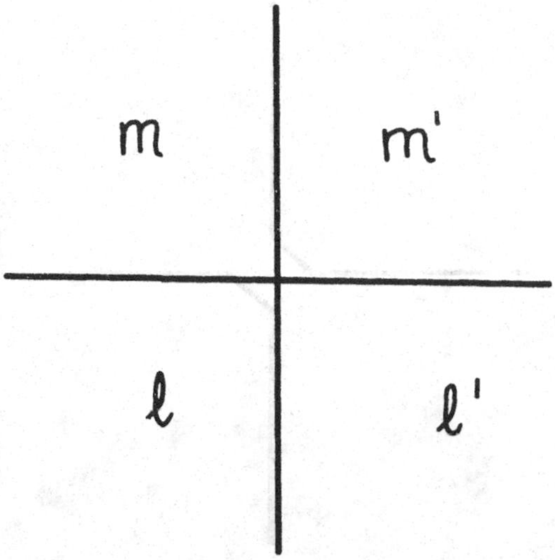

FIG. 1. Arrangements of spins round a vertex.

The weights a_l ,..., c_l' shown in Fig. 2 are the appropriate weights $W(l, l' \mid m, m')$. They are given explicitly in (II.7.2). Note in particular that a_l , b_l' are independent of l.

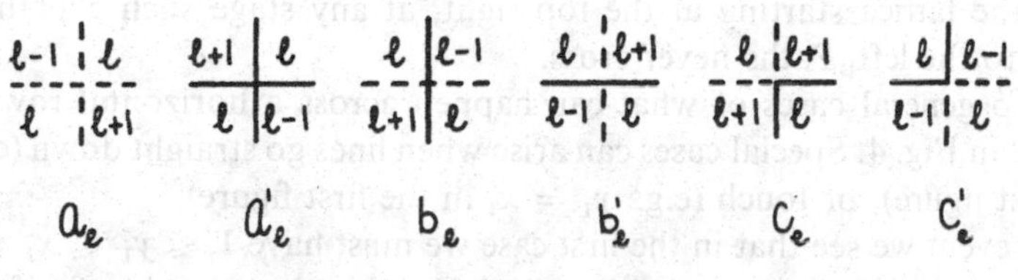

FIG. 2. The six types of allowed configurations at a vertex.

Now look at (2.1). We see that this is a transfer matrix equation relating two rows of the lattice. Note that l and $X = \{x_1 ,..., x_n\}$ specify the state of the lower row, while m and Y specify the state of the upper. Since $x_1 ,..., x_n$ are the positions of the decreasing steps in the sequence $l_1 ,..., l_{N+1}$, they are the positions of the solid lines in the lower row, $l = l_1$ is the value of the first spin (the one to the left of bond 1) in the lower row.

Similarly, $y_1 ,..., y_n$ are the positions of the solid lines in the upper row, m is the value of the left-hand upper spin.

Let us draw the second vertex in Fig. 2 as in Fig. 3.

We see then that a sequence of solid lines follows a continuous and unique path

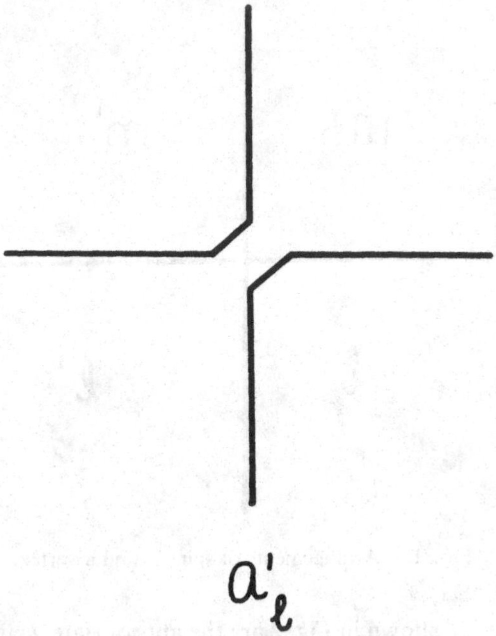

FIG. 3. The second configuration of Fig. 2.

through the lattice starting at the top right, at any stage such a path is moving down or to the left. Paths never cross.

The two general cases of what can happen across a horizontal row are shown for $n = 2$ in Fig. 4. Special cases can arise when lines go straight down (e.g., $y_2 = x_1$ in the first figure), or touch (e.g., $y_1 = x_1$ in the first figure).

In any event we see that in the first case we must have $1 \leqslant y_1 \leqslant x_1 \leqslant y_2 \leqslant x_2$, in the second case $x_1 \leqslant y_1 \leqslant x_2 \leqslant y_2 \leqslant N$. ($N$ is the number of columns of the lattice.)

It follows that for $n = 2$ we can write (2.1) more explicitly as

$$\begin{aligned} \Lambda f(l \mid x_1, x_2) = {} & \sum_{y_1=1}^{x_1} \sum_{y_2=x_1}^{x_2}{}^{*} D_L(l, X, Y) f(l+1 \mid y_1, y_2) \\ & + \sum_{y_1=x_1}^{x_2} \sum_{y_2=x_2}^{N}{}^{*} D_R(l, X, Y) f(l-1 \mid y_1, y_2), \end{aligned} \tag{2.2}$$

where the * on the summation indicates that the cases $y_1 = x_1 = y_2$, $y_1 = x_2 = y_2$ are excluded. The factors D_L, D_R are the products of the weights of the vertices

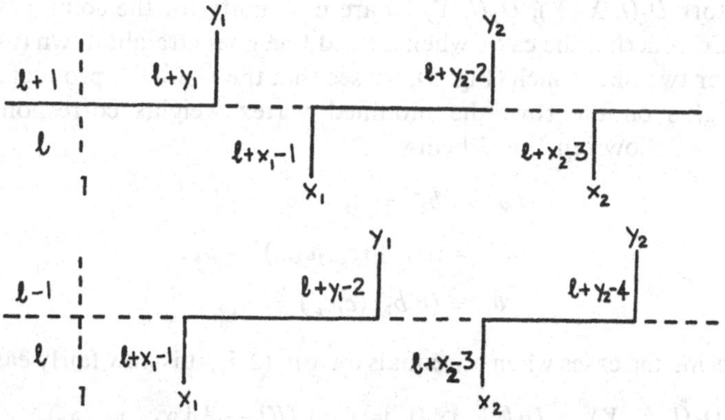

FIG. 4. General arrangements of solid lines between two rows for $n = 2$. Most of the broken lines on vertical bonds are omitted. The spins shown are those of the face immediately to the left of the vertical bond drawn.

in the row. For instance, for the configurations shown in Fig. 4 we see by using Fig. 2 that

$$D_L(l, X, Y) = b'^{y_1-1}c'_{l+y_1}a^{x_1-y_1-1}c_{l+x_1-2}b'^{y_2-x_1-1}c'_{l+y_2-2}a^{x_2-y_2-1}c_{l+x_2-4}b'^{N-x_2},$$
$$D_R(l, X, Y) = a^{x_1-1}c_{l+x_1-2}b'^{y_1-x_1-1}c'_{l+y_1-2}a^{x_2-y_1-1}c_{l+x_2-4}b'^{y_2-x_2-1}c'_{l+y_2-4}a^{N-y_2}. \tag{2.3}$$

(Since a_l, b_l' are independent of l we have written them simply as a, b': they are not to be confused with the vertex weights of the original eight-vertex model.)

Substituting these expressions into (2.2), we see that this equation simplifies if we define

$$\tilde{f}(l \mid x_1, x_2) = \left(\frac{b'}{a}\right)^{x_1+x_2} \frac{c'_{l+x_1-1}c'_{l+x_2-3}}{a^2} f(l \mid x_1, x_2). \tag{2.4}$$

The equation (2.2) then becomes

$$\Lambda\tilde{f}(l \mid x_1, x_2) = r_{l+x_1}r_{l+x_2-2}\Big\{b'^N \sum_{y_1=1}^{x_1} \sum_{y_2=x_1}^{x_2}{}^{*} \tilde{D}_L(l, X, Y)\tilde{f}(l+1 \mid y_1, y_2)$$
$$+ a^N \sum_{y_1=x_1}^{x_2} \sum_{y_2=x_2}^{N}{}^{*} \tilde{D}_R(l, X, Y)\tilde{f}(l-1 \mid y_1, y_2)\Big\}. \tag{2.5}$$

where

$$r_l = (c_{l-2}c'_{l-1})/(ab') \tag{2.6}$$

The factors $\tilde{D}_L(l, X, Y)$, $\tilde{D}_R(l, X, Y)$ are now unity for the configuratons shown in Fig. 4. Considering the cases when a solid line goes straight down (third diagram in Fig. 2), or two lines touch (Fig. 3), we see that the $\tilde{D}$ are the products of modified vertex weights on the row, the modified vertex weights corresponding to the configurations shown in Fig. 2 being

$$\begin{aligned} \tilde{a}_l &= \tilde{b}_l' = \tilde{c}_l = \tilde{c}_l' = 1, \\ \tilde{a}_l' &= (aa_l')/(c_{l-1}c_{l+1}') = p_l, \\ \tilde{b}_l &= (b'b_l)/(c_lc_l') = q_l. \end{aligned} \tag{2.7}$$

Considering the cases when a y equals an x in (2.5), it is now fairly easy to see that

$$\begin{aligned} \tilde{D}_L(l, X, Y) &= U(l+1 \mid 0, y_1, x_1)\, U(l-1 \mid x_1, y_2, x_2), \\ \tilde{D}_R(l, X, Y) &= U(l-1 \mid x_1, y_1, x_2)\, U(l-3 \mid x_2, y_2, N+1), \end{aligned} \tag{2.8}$$

where the function $U(l \mid x, y, x')$ is defined for $x < x'$, $x \leqslant y \leqslant x'$ by:

$$\begin{aligned} U(l \mid x, y, x') &= q_{l+x-1} && \text{if } y = x \\ &= 1 && \text{if } x < y < x' \\ &= p_{l+x'-2} && \text{if } y = x'. \end{aligned} \tag{2.9}$$

For clarity we have considered the case $n = 2$. The extension to arbitrary n is straightforward: $\tilde{f}$ becomes $\tilde{f}(l \mid x_1, \dots, x_n)$, the r-factor in (2.5) becomes $r_{l+x_1} r_{l+x_2-2} r_{l+x_3-4} \cdots r_{l+x_n-2n+2}$, the two summations are over $y_1, \dots, y_n$ such that $1 \leqslant y_1 \leqslant x_1$, $x_1 \leqslant y_2 \leqslant x_2, \dots, x_{n-1} \leqslant y_n \leqslant x_n$ and $x_1 \leqslant y_1 \leqslant x_2, \dots, x_{n-1} \leqslant y_{n-1} \leqslant x_n$, $x_n \leqslant y_n \leqslant N$ (but no two y's can be equal), and the D are given by

$$\begin{aligned} \tilde{D}_L(l, X, Y) &= \prod_{j=1}^{n} U(l+1-2j \mid x_{j-1}, y_j, x_j), \\ \tilde{D}_R(l, X, Y) &= \prod_{j=1}^{n} U(l-1-2j \mid x_j, y_j, x_{j+1}), \end{aligned} \tag{2.10}$$

where we adopt the conventions $x_0 = 0$, $x_{n+1} = N + 1$.

3. Bethe Ansatz and Resulting Equations

It is apparent that the transfer matrix equation (2.5) is of the same type (though considerably more complicated) as the equations obtained by Lieb for the ice-models (cf. [3], in particular Eq. (2.9) of Lieb's first paper). We might hope to solve it by the same general techniques, and indeed it turns out that we can.

We therefore need to find an appropriately generalized Bethe ansatz for f. Some inspection suggests that we try

$$\tilde{f}(l \mid X) = \sum_P A(P)\, \tilde{g}_{P1}(l, x_1)\, \tilde{g}_{P2}(l-2, x_2) \cdots \tilde{g}_{Pn}(l-2n+2, x_n), \tag{3.1}$$

where the summation is over all permutations $P = \{P1,..., Pn\}$ of the integers $1,..., n$. The $A(P)$ are $n!$ coefficients, to be chosen appropriately, and the $g_j(l, x)$ $(j = 1,..., n)$ can be thought of as n "single-particle" functions associated with a solid line in our graph. They are also at our disposal.

Again we focus attention on the case $n = 2$. As a first step we set

$$\tilde{f}(l \mid x_1, x_2) = \tilde{g}_1(l, x_1)\, \tilde{g}_2(l-2, x_2) \tag{3.2}$$

on the r.h.s. of (2.5). We can perform the summations explicitly by first allowing $y_1 = y_2$ and then subtracting off the spurious terms thereby introduced (Lieb calls these "diagonal" terms). Leaving out the r, b'^N and a^N factors, the first summation in (2.5) gives

$$\begin{aligned}[F_1(l+1, x_1) - C_1(l+1)][F_2(l-1, x_2) - G_2(l-1, x_1)]\\ - p_{l+x_1-1}q_{l+x_1-2}\,\tilde{g}_1(l+1, x_1)\,\tilde{g}_2(l-1, x_1),\end{aligned} \tag{3.3}$$

while the second gives

$$\begin{aligned}[F_1(l-1, x_2) - G_1(l-1, x_1)][C_2'(l-3) - G_2(l-3, x_2)]\\ - p_{l+x_2-3}q_{l+x_2-4}\,\tilde{g}_1(l-1, x_2)\,\tilde{g}_2(l-3, x_2).\end{aligned} \tag{3.4}$$

The functions $F_j(l, x)$, $G_j(l, x)$, $C_j'(l)$ are defined by

$$F_j(l, x) = C_j(l) + \sum_{y=1}^{x-1} \tilde{g}_j(l, y) + p_{l+x-2}\,\tilde{g}_j(l, x), \tag{3.5}$$

$$G_j(l, x) = C_j(l) + \sum_{y=1}^{x} \tilde{g}_j(l, y) - q_{l+x-1}\,\tilde{g}_j(l, x),$$

$$C_j'(l) = C_j(l) + \sum_{y=1}^{N} \tilde{g}_j(l, y). \tag{3.6}$$

The parameters $C_j(l)$ are at our disposal.

Expanding the products in (3.3) and (3.4), we can classify the resulting terms into:

(i) *Wanted terms*: These are products of n F's, or n G's. Each x_j occurs once and only once in the product.

(ii) *Unwanted internal terms*: These are products in which there are n F, G or $\tilde{g}$ functions, but one of the x_j occurs as an argument in two of the functions simultaneously. Note that the "diagonal" terms are included in this category.

(iii) *Unwanted boundary terms*: These are terms which contain a C or a C'.

We endeavour to choose the *wanted terms* so as to cancel with the l.h.s. of the transfer matrix equation (2.5). For instance, the wanted term in (3.3) is $F_1(l+1, x_1)\, F_2(l-1, x_2)$. Multiplying this by the factor $r_{l+x_1} r_{l+x_2-2}$ in (2.5), we see that it is of the same form as the r.h.s. if there exist $\lambda_1, ..., \lambda_n$ such that

$$\lambda_j \tilde{g}_j(l, x) = r_{l+x} F_j(l+1, x) \tag{3.7}$$

for $j = 1, ..., n$ and all integers l, x. Similarly, the wanted term $G_1(l-1, x_1)$ $G_2(l-3, x_2)$ in (3.4) will be of the same form as the l.h.s. if there exist $\mu_1, ..., \mu_n$ such that

$$-\mu_j \tilde{g}_j(l, x) = r_{l+x} G_j(l-1, x). \tag{3.8}$$

Taken together, the two wanted terms on the r.h.s. of (2.5) will then cancel with the l.h.s. if

$$\Lambda = b'^N \lambda_1, ..., \lambda_n + a^N \mu_1, ..., \mu_n\,. \tag{3.9}$$

Note that these equations (3.7)–(3.9), together with the definitions (3.5), are unaffected if we permute the n single-particle functions $\tilde{g}_1(l, x), ..., \tilde{g}_n(l, x)$ in (3.2). Thus if we can find solutions of (3.7) and (3.8) we are then free to use the general ansatz (3.1) and to attempt to use the coefficients $A(P)$, together with any remaining freedom in the single-particle functions, to cancel the unwanted terms on the r.h.s. of (2.5).

The *unwanted internal terms* in (3.3) can be written as $-B_{1,2}(l, x_1)$, where

$$B_{i,j}(l, x) = F_i(l+1, x)\, G_j(l-1, x) + p_{l+x-1} q_{l+x-2} \tilde{g}_i(l+1, x)\, \tilde{g}_j(l-1, x) \tag{3.10}$$

for $i, j = 1, ..., n$ and any integers l, x. Interchanging the suffixes 1 and 2 on the single-particle functions in (3.3) and using the general ansatz (3.1), we see that these will cancel if

$$B_{1,2}(l, x)\, A(1, 2) + B_{2,1}(l, x)\, A(2, 1) = 0 \tag{3.11}$$

for all integers l, x.

The unwanted internal terms in (3.4) will also cancel if (3.11) is satisfied. More generally, for arbitrary n we find the unwanted internal terms cancel if

$$B_{i,j}(l, x)\, A(P) + B_{j,i}(l, x)\, A(Q) = 0 \tag{3.12}$$

for all integers l, x and all permutations $P = \{..., i, j, ...\}$, $Q = \{..., j, i, ...\}$. (P and Q differ only in interchanging adjacent elements i and j.)

Note that these requirements imply that the ratio $B_{i,j}(l, x)/B_{j,i}(l, x)$ must be independent of l and x for $i = 1,\ldots, n$ and $j = 1,\ldots, n$. This is a very strong condition.

Interchanging the function suffixes 1 and 2 in (3.4), we see that the *unwanted boundary terms* in (3.3) and (3.4) are of similar type. They can be made to cancel by requiring that

$$\left(\frac{b'}{a}\right)^N C_1(l+1)\, A(1, 2) = C_1'(l-3)\, A(2, 1). \tag{3.13}$$

For arbitrary n we find that they cancel if

$$\left(\frac{b'}{a}\right)^N C_{P1}(l+1)\, A(P1,\ldots, Pn) = C'_{P1}(l+1-2n)\, A(P2, P3,\ldots, Pn, P1) \tag{3.14}$$

for all integers l and all permutations $P = \{P1,\ldots, Pn\}$.

To summarize the working so far: The Bethe ansatz (3.1) will work if we can satisfy (3.7), (3.8), (3.12) and (3.14). The functions $\tilde{g}_j(l, x)$, the coefficients $A(P)$, and the parameters $C_j(l)$, λ_j, μ_j are at our disposal. Nevertheless we have many more equations than unknowns, and these equations cannot be satisfied for arbitrary weights a_l, a_l', b_l, b_l', c_l, c_l'. However, for the weights we are interested in, namely those given by (II.7.1), it turns out that they are soluble.

4. Single Particle Equations

We look first at the equations (3.7) and (3.8), together with the subsidiary definitions (3.5). To handle these we define one more function:

$$\Gamma_j(l, x) = C_j(l) + \sum_{y=1}^{x} \tilde{g}_j(l, y), \tag{4.1}$$

$j = 1,\ldots, n$, all integers l, x.

We can express $\tilde{g}_j(l, x)$, $F_j(l, x)$, $G_j(l, x)$ as linear combinations of $\Gamma_j(l, x)$ and $\Gamma_j(l, x-1)$. Doing this, the equations (3.7), (3.8) become:

$$\begin{aligned} &\lambda_j[\Gamma_j(l, x) - \Gamma_j(l, x-1)] \\ &\quad = r_{l+x}[p_{l+x-1}\Gamma_j(l+1, x) + p'_{l+x-1}\Gamma_j(l+1, x-1)], \\ &-\mu_j[\Gamma_j(l, x) - \Gamma_j(l, x-1)] \\ &\quad = r_{l+x}[q'_{l+x-2}\Gamma_j(l-1, x) + q_{l+x-2}\Gamma_j(l-1, x-1)], \end{aligned} \tag{4.2}$$

where

$$p_x' = 1 - p_x, \qquad q_x' = 1 - q_x. \tag{4.3}$$

Note that we can consider some particular value of j. Also, the coefficients r, p, p', q, q' depend on l and x only via $l + x$. Under quite general conditions it follows that the solution of (4.2) must be of the form

$$\Gamma_j(l, x) = \exp(ik_j'x)\, \gamma_j(l + x), \tag{4.4}$$

where i is now the square root of -1, k_j' is an unknown wave-number, and $\gamma_j(x)$ is some function of the integer variable x.

Substituting (4.4) into (4.2), replacing $l + x$ in the first equation by y, in the second by $y + 1$, we get two equations, homogeneous and linear in the three unknowns $\gamma_j(y - 1), \gamma_j(y), \gamma_j(y + 1)$. Solving them we get, dropping the suffixes j,

$$\begin{aligned} &e^{-ik'}\gamma(y-1) : \gamma(y) : e^{ik'}\gamma(y+1) \\ &\quad = \lambda'\mu' - \mu' r_y + R_y : \lambda'\mu' - S_y : \lambda'\mu' - \lambda' r_{y+1} + R_y', \end{aligned} \tag{4.5}$$

where

$$\lambda' = e^{ik'}\lambda, \qquad \mu' = e^{-ik'}\mu, \tag{4.6}$$

$$\begin{aligned} R_y &= r_y r_{y+1} p_{y-1} q_{y-1}', \\ S_y &= r_y r_{y+1} p_{y-1} q_{y-1}, \\ R_y' &= r_y r_{y+1} p_{y-1}' q_{y-1}. \end{aligned} \tag{4.7}$$

The ratio of the first two terms in (4.5) must be equal to the ratio of the second two with y replaced by $y - 1$. This gives the equation

$$\begin{aligned} &\lambda'\mu'(\lambda' + \mu') - \lambda'\mu'(R_{y-1}' + R_y + S_{y-1} + S_y + r_y^2)/r_y \\ &\quad + \lambda' R_y + \mu' R_{y-1}' + (S_{y-1}S_y - R_{y-1}'R_y)/r_y = 0. \end{aligned} \tag{4.8}$$

This equation must be satisfied for all integers y and for n values of λ', μ', corresponding to $j = 1, \ldots, n$. We should therefore like the coefficients in this equation (regarded as a biquadratic equation relating λ' and μ') to be independent of y. Using (II.7.1) we find that this is indeed the case. The main tool that we need in the working is an addition formula that is satisfied by our elliptic functions $h(u)$, defined by (1.18), namely:

Formula

$$\begin{aligned} &h(x+y)\,h(x-y)\,h(z+t)\,h(z-t) - h(x+t)\,h(x-t)\,h(z+y)\,h(z-y) \\ &\quad = h(y-t)\,h(y+t)\,h(x+z)\,h(x-z) \end{aligned} \tag{4.9}$$

for all complex numbers x, y, z, t.

The proof is simple: $h(u)$ is an entire function, odd, and for any integers m, n there exist coefficients $d_{m,n}$, $e_{m,n}$ such that

$$h(u + 2mK + inK') = h(u) \exp[d_{m,n}u + e_{m,n}]. \tag{4.10}$$

for all complex numbers u. [This is a very weak way of writing the quasi-periodic properties of $h(u)$: (4.10) is satisfied whether we use the Jacobi theta functions in (1.18), or our modified theta functions defined by (II.6.9).]

We regard x as a complex variable and y, z, t as constants. Define a function $f(x)$ as the ratio of the l.h.s. of (4.9) to the r.h.s. Then $f(x)$ is the ratio of two entire functions. From (4.10) it is doubly periodic, with periods $2K$, iK'. The r.h.s. has zeros, which are simple, only at $x = \pm z + 2mK + inK'$. When $x = \pm z$ it is clear that the l.h.s. also vanishes, so it must do so at all zeros of the r.h.s. Thus $f(x)$ is entire and doubly periodic. It is therefore bounded, and by the Cauchy–Liouville theorem must be a constant. Setting $x = t$ we find that this constant is unity, so $f(x) \equiv 1$. This proves the formula.

We return to looking at (4.8). From (II.7.1), (2.6) and (2.7) we have

$$r_l = \frac{h^2(2\eta)\, h(w_l - 3\eta + v)\, h(w_l - 3\eta - v)}{h(v - \eta)\, h(v + \eta)\, h(w_l - 4\eta)\, h(w_l - 2\eta)}, \tag{4.11}$$

$$p_l = \frac{h^2(v + \eta)\, h(w_l - 2\eta)\, h(w_l + 2\eta)}{h^2(2\eta)\, h(w_l - \eta - v)\, h(w_l + \eta + v)}, \tag{4.12}$$

$$q_l = \frac{h^2(v - \eta)\, h(w_l - 2\eta)\, h(w_l + 2\eta)}{h^2(2\eta)\, h(w_l + \eta - v)\, h(w_l - \eta + v)}, \tag{4.13}$$

where

$$w_l = \tfrac{1}{2}(s + t) - K + 2l\eta. \tag{4.14}$$

(Remember that s, t, K, η are constants, so $w_{l+1} = w_l + 2\eta$.)

From (4.3), (4.12), (4.13) we find, using our formula (4.9):

$$p_l' = -\frac{h(v - \eta)\, h(v + 3\eta)\, h^2(w_l)}{h^2(2\eta)\, h(w_l - \eta - v)\, h(w_l + \eta + v)} \tag{4.15}$$

$$q_l' = -\frac{h(v + \eta)\, h(v - 3\eta)\, h^2(w_l)}{h^2(2\eta)\, h(w_l - \eta + v)\, h(w_l + \eta - v)}. \tag{4.16}$$

Substituting these expressions into (4.7) gives

$$\begin{aligned} R_y &= -h(v - 3\eta)\, h(v + \eta)/h^2(v - \eta), \\ S_y &= h(w_y - 4\eta)\, h(w_y)/h^2(w_y - 2\eta), \\ R_y' &= -h(v + 3\eta)\, h(v - \eta)/h^2(v + \eta). \end{aligned} \tag{4.17}$$

We see immediately that the coefficients R_y, R'_{y-1} in (4.8) are independent of y. By using (4.9) once more we can also show that the last coefficient is independent of y. The coefficient of $\lambda'\mu'$ is more difficult, but we can show that it is an entire, doubly-periodic function of w_y, and hence a constant.

Having established that (4.5) is internally consistent, we can go on to solve it. The equation that led to (4.8) is

$$(\lambda'\mu' - S_y)(\lambda'\mu' - S_{y-1}) = (\lambda'\mu' - \lambda' r_y + R')(\lambda'\mu' - \mu' r_y + R) \tag{4.18}$$

(dropping the redundant suffixes on R, R'). Guided by the form (4.17) of S_y, we introduce a parameter u such that

$$\lambda'\mu' = h(u - 2\eta)\, h(u + 2\eta)/h^2(u). \tag{4.19}$$

The integer y enters (4.18) only through the complex number w_y that occurs in (4.17). Thus we expect (4.18) to be identically satisfied by any complex number w_y. Comparing (4.17) and (4.19) we see that $\lambda'\mu' - S_y$ vanishes if $w_y = 2\eta + u$. Thus one of the factors on the r.h.s., say the first, must then vanish. (Choosing the second is simply equivalent to replacing u by $-u$.) This gives us an equation for λ', namely

$$\lambda' = [(\lambda'\mu' + R')/r_y]_{w_y = 2\eta + u}\,. \tag{4.20}$$

We can evaluate this by using (4.19), (4.17), (4.11) and the formula (4.9). We find that

$$\lambda' = zh(u - 2\eta)/h(u), \tag{4.21}$$

where

$$z = [h(v - \eta)\, h(v + \eta + u)]/[h(v + \eta)\, h(v - \eta + u)]. \tag{4.22}$$

From (4.19) we therefore have

$$\mu' = z^{-1} h(u + 2\eta)/h(u). \tag{4.23}$$

Substituting these results into (4.5) and making repeated use of the formula (4.9), we find that (4.5) becomes

$$\begin{aligned} &e^{-ik'}\gamma(y-1) : \gamma(y) : e^{ik'}\gamma(y+1) \\ &\quad = z^{-1}[h(w_{y-2} - u)/h(w_{y-2})] : h(w_{y-1} - u)/h(w_{y-1}) : z[h(w_y - u)/h(w_y)] \end{aligned} \tag{4.24}$$

It is apparent that these equations are internally consistent.

We still have some freedom in our choice of the single-particle functions (in addition to the choice of the parameter u), since we can multiply each function $\Gamma_j(l, x)$ by a factor $\exp(\chi_j l)$, where $\chi_1, \ldots, \chi_n$ are some constants. This re-defines λ_j,

μ_j in (4.2), and k_j', $\gamma_j(y)$ in (4.4). Put another way, the choice of k_j' is at our disposal.

To remove this arbitrariness, we note that for those values of η such that the (modified) elliptic functions are periodic of period $2L\eta$, we expect $f(l \mid X)$ to be a periodic function of l, of period L. From (2.4), (3.1), (4.1) and (4.4) we therefore expect $\tilde{g}_j(l, x)$, $\Gamma_j(l, x)$ and $\gamma_j(l)$ to be periodic functions of l, with period L. From (4.24) and (4.14) it follows that

$$e^{iLk'} = z^L. \tag{4.25}$$

Incrementing u by $2K$ or iK' leaves the definition (4.19) unchanged, but multiplies the definition (4.22) of z by an Lth root of unity. It follows that we can always choose

$$e^{ik_j'} = z_j\,. \tag{4.26}$$

(We now restore the suffixes j to λ, μ, $\gamma(y)$, and to the related parameters u, z.)

From (4.4) and (4.24) we can now see that

$$\Gamma_j(l, x) = \tau_j e^{ik_j'x} h(w_{l+x-1} - u_j)/h(w_{l+x-1}), \tag{4.27}$$

where $\tau_1, \ldots, \tau_n$ are arbitrary constants, independent of l and x. From (4.1)

$$\tilde{g}_j(l, x) = \Gamma_j(l, x) - \Gamma_j(l, x-1). \tag{4.28}$$

Using (4.27), (4.26), (4.22) and the formula (4.9), this gives

$$\tilde{g}_j(l, x) = -\tau_j e^{ik_j'x} \frac{h(2\eta)\, h(u_j)\, h(w_{l+x-1} + v - \eta)\, h(w_{l+x-1} - \eta - u_j - v)}{h(v - \eta)\, h(v + u_j + \eta)\, h(w_{l+x-2})\, h(w_{l+x-1})}. \tag{4.29}$$

From the transformation (2.4) we see that the single-particle functions $g_j(l, x)$ in (1.16) are related to the functions $\tilde{g}_j(l, x)$ in (3.1) by

$$g_j(l, x) = (a/b')^x\, [a/c'_{l+x-1}]\, \tilde{g}_j(l, x). \tag{4.30}$$

Using (4.29), the definitions (II.7.1) of a, b', c_x', and the definitions (4.26), (4.22) of $\exp(ik_j')$, it follows that

$$g_j(l, x) = \left\{ \frac{h(u_j + v + \eta)}{h(u_j + v - \eta)} \right\}^x \frac{h(w_{l+x-1} - u_j - v - \eta)}{h(w_{l+x-2})\, h(w_{l+x-1})}, \tag{4.31}$$

where we have chosen

$$\tau_j = -[h(v - \eta)\, h(v + u_j + \eta)]/[h(v + \eta)\, h(u_j)]. \tag{4.32}$$

For each $j = 1,..., n$, we have now obtained the single-particle functions in terms of one unknown parameter u_j. The corresponding single-particle eigenvalues λ_j, μ_j are given by (4.6), (4.21), (4.23), (4.22) and (4.26) to be

$$\lambda_j = h(u_j - 2\eta)/h(u_j), \qquad \mu_j = h(u_j + 2\eta)/h(u_j). \tag{4.33}$$

Hence from (3.9) the corresponding eigenvalue Λ of the transfer matrix I is

$$\Lambda = [\rho' h(v - \eta)]^N \prod_{j=1}^{n} \frac{h(u_j - 2\eta)}{h(u_j)} + [\rho' h(v + \eta)]^N \prod_{j=1}^{n} \frac{h(u_j + 2\eta)}{h(u_j)}. \tag{4.34}$$

5. Equations Arising from Unwanted Internal Terms

We now look at the equations (3.12), together with the definition (3.10). These determine the coefficients $A(P)$.

From (3.7), (3.8), (4.11), (4.29) and (4.33) we find that

$$F_j(l + 1, x = -\tau_j e^{ik_j' x} \frac{h(v + \eta)\, h(u_j - 2\eta)\, h(w_{l+x-1} - \eta - u_j - v)}{h(2\eta)\, h(u_j + v + \eta)\, h(w_{l+x-1} - \eta - v)}, \tag{5.1}$$

$$G_j(l - 1, x) = \tau_j e^{ik_j' x} \frac{h(v + \eta)\, h(u_j + 2\eta)\, h(w_{l+x-1} - \eta - u_j - v)}{h(2\eta)\, h(u_j + v + \eta)\, h(w_{l+x-1} - \eta - v)}. \tag{5.2}$$

Substituting these expressions into (3.10), using also (4.12), (4.13) and (4.29), we find that we can again use the formula (4.9) and obtain

$$B_{j,m}(l, x) = \tau_j \tau_m e^{i(k_j' + k_m')x} \times \frac{h^2(v + \eta)\, h(w_{l+x-1} - \eta - u_j - u_m - v)\, h(u_m - u_j + 2\eta)}{h(2\eta)\, h(w_{l+x-1} - \eta - v)\, h(u_j + v + \eta)\, h(u_m + v + \eta)}. \tag{5.3}$$

We note that interchanging u_j and u_m affects only the last term in the numerator of (5.3). Thus

$$B_{j,m}(l, x)/B_{m,j}(l, x) = h(u_m - u_j + 2\eta)/h(u_j - u_m + 2\eta). \tag{5.4}$$

This ratio is therefore independent of l and x and the equations (3.12) are consistent. They determine $A(P)$ to within a normalization constant and are satisfied by

$$A(P) = \epsilon_P \prod_{1 \leqslant j < m \leqslant n} h(u_{P_j} - u_{P_m} + 2\eta), \tag{5.5}$$

where ϵ_P is the signature of the permutation P (+ for an even number of interchanges—for an odd number).

6. Equations Arising from Unwanted Boundary Terms

The final step in the solution of the Bethe ansatz is to look at the equations (3.14). From (4.1) and (3.6) we see that

$$C_j(l) = \Gamma_j(l, 0), \tag{6.1}$$

$$C_j'(l) = \Gamma_j(l, N). \tag{6.2}$$

From (4.27) it follows that

$$\frac{C_j'(l+1-2n)}{C_j(l+1)} = e^{iNk_j'} \frac{h(w_l)\, h[w_l - u_j + (N-2n)\,\eta]}{h[w_l + (N-2n)\,\eta]\, h[w_l - u_j]}. \tag{6.3}$$

However, we have imposed the boundary condition that either $n = \frac{1}{2}N$, or if there exists an integer L such that $2L\eta$ is a period of the elliptic functions, then $N - 2n = L \times$ integer. In either case we see that the functions h on the r.h.s. of (6.3) cancel, so the result is independent of l. This is necessary for the equations (3.14) to be consistent.

Using the definitions (4.26), (4.22) of $\exp(ik_j)$, together with the definitions (II.7.1) of a, b', we see that

$$\left(\frac{a}{b'}\right)^N \frac{C_j'(l+1-2n)}{C_j(l+1)} = \left\{\frac{h(u_j + v + \eta)}{h(u_j + v - \eta)}\right\}^N. \tag{6.4}$$

Using this result, together with the result (5.5) for $A(P)$, in (3.14), we get the n equations

$$\left\{\frac{h(u_j + v + \eta)}{h(u_j + v - \eta)}\right\}^N = -\prod_{m=1}^{n} \frac{h(u_j - u_m + 2\eta)}{h(u_j - u_m - 2\eta)} \tag{6.5}$$

for $j = 1, \ldots, n$. (We have used the fact that $h(u)$ is an odd function.)

These equations determine $u_1, \ldots, u_n$, which up to now have been at our disposal.

7. Summary of Sections 2 to 6

In Section 2 we set up the eigenvalue equations for Λ and $f(l \mid X)$, finding it convenient to transform from $f(l \mid X)$ to $\tilde{f}(l \mid X)$ as in Eq. (2.4). In Section 3 we made the Bethe ansatz (3.1) for $\tilde{f}(l \mid X)$. This is equivalent to the Bethe ansatz (1.16) for $f(l \mid X)$ if we relate the single-particle functions g and $\tilde{g}$ by (4.30). We found that this worked provided we could satisfy (3.7)–(3.9), (3.12) and (3.14). In Sections 4–6 we have shown that we can indeed do so, despite the fact that at each stage we have more equations than unknowns.

The results are contained in equations (6.5), (5.5), (4.34) and (4.31). The equations (6.5) determine $u_1, \ldots, u_n$. The coefficients $A(P)$, the eigenvalue Λ and the single-particle functions $g_j(l, x)$ are then given by (5.5), (4.34) and (4.31) respectively.

Using the definition (4.14) and $\rho' = \rho\Theta(0)$ [Eq. (II.7.2)], we see that these results are the same as those quoted in Eqs. (1.17)–(1.23) of Section 1, provided we replace each u_j by $u_j - v$.

8. Inhomogeneous System

We end by remarking that all the above working still goes through if we consider an eight-vertex model with weights a, b, c, d given by (II.6.1), in which k and η (and trivially ρ) are the same for each site of the lattice, but v can vary from column to column.

Let v_x be the value of v on column x, where $x = 1, 2, \ldots, N$. The equations (1.2), (1.3) for our basis vectors $(l_1, \ldots, l_{N+1})$ become

$$\psi(l_1, \ldots, l_{N+1}) = \Phi^{(1)}_{l_1, l_2} \otimes \Phi^{(2)}_{l_2, l_3} \otimes \cdots \otimes \Phi^{(N)}_{l_N, l_{N+1}}, \tag{8.1}$$

where

$$\begin{aligned} \Phi^{(x)}_{l, l+1} &= | s + 2l\eta + \eta - v_x \rangle, \\ \Phi^{(x)}_{l+1, l} &= | t + 2l\eta + \eta + v_x \rangle, \end{aligned} \tag{8.2}$$

for $x = 1, \ldots, N$ and all integers l.

The transfer matrix equation (1.5) in this basis becomes

$$\mathbf{T}\psi(l_1, \ldots, l_{N+1}) = \sum \left\{ \prod_{J=1}^{N} W^{(J)}(m_J, m_{J+1} \mid l_J, l_{J+1}) \right\} \psi(m_1, \ldots, m_{N+1}), \tag{8.3}$$

where $W^{(J)}(m, m' \mid l, l')$ is given by (II.7.1) with v replaced by v_J ($J = 1, \ldots, N$).

The eight-vertex model therefore transforms to an Ising-like model whose weights $a_l, a_l', b_l, b_l', c_l, c_l'$ (shown in Fig. 2) depend on which column (J or x) is being considered. We therefore write them as $a_l(J), \ldots, c_l'(J)$, where $J = 1, \ldots, N$. They are given by (II.7.1) with v replaced by v_J.

The transformation (2.4) becomes

$$\tilde{f}(l \mid x_1, x_2) = s_l(x_1)\, s_{l-2}(x_2)\, f(l \mid x_1, x_2), \tag{8.4}$$

where

$$s_l(x) = \left\{ \prod_{y=1}^{x-1} b'(y) \right\} c'_{l+x-1}(x) \left\{ \prod_{y=x+1}^{N} a(y) \right\} \tag{8.5}$$

(b' and a are still independent of l). We make the same ansatz (1.16), which is equivalent to (3.1) if

$$\tilde{g}_j(l, x) = s_l(x)\, g_j(l, x). \tag{8.6}$$

The definitions (2.6), (2.7) apply for each column separately of the lattice, so $r_l(x)$, $p_l(x)$, $q_l(x)$ are given by (4.11) and (4.12) with v replaced by v_x. In the Bethe ansatz equations (3.5)–(3.14) each p_l, q_l, r_l is replaced by $p_l(x)$, $q_l(x)$, $r_l(x)$, and a^N, b'^N by

$$\prod_{y=1}^{N} a(y), \qquad \prod_{y=1}^{N} b'(y). \tag{8.7}$$

Similarly, the coefficients r_{l+x}, p_{l+x-1}, p'_{l+x-1}, q_{l+x-2}, q'_{l+x-2} in (4.2) all have argument x (i.e., they are evaluated from (4.11) and (4.12) with v replaced by v_x). We can no longer use (4.4) to simplify (4.2), but by considering a particular value of x and using the previous results of Section 4 it is not difficult to see that (4.2) is satisfied by (4.27), (4.29) and (4.33), with $\exp(ik_j'x)$ replaced by

$$\prod_{y=1}^{x} \frac{h(v_y - \eta)\, h(v_y + \eta + u_j)}{h(v_y + \eta)\, h(v_y - \eta + u_j)}, \tag{8.8}$$

and v in (4.29) by v_x.

From (8.5), (8.6) and (4.29) it follows that (with an appropriate choice of τ_j):

$$g_j(l, x) = \left\{\prod_{y=1}^{x-1} h(u_j + v_y + \eta)\right\} \frac{h(w_{l+x-1} - \eta - u_j - v_x)}{h(w_{l+x-2})\, h(w_{l+x-1})} \prod_{y=x+1}^{N} h(u_j + v_y - \eta). \tag{8.9}$$

This defines the single-particle functions. Using the rule (8.7), the equation (4.34) for the eigenvalue Λ becomes:

$$\Lambda = \left\{\prod_{J=1}^{N} \rho' h(v_J - \eta)\right\}\left\{\prod_{j=1}^{n} \frac{h(u_j - 2\eta)}{h(u_j)}\right\} + \left\{\prod_{J=1}^{N} \rho' h(v_J + \eta)\right\}\left\{\prod_{j=1}^{n} \frac{h(u_j - 2\eta)}{h(u_j)}\right\}. \tag{8.10}$$

The equation (5.5) for the coefficients $A(P)$ is the same for the inhomogeneous system as for the homogeneous, i.e.,

$$A(P) = \epsilon_P \prod_{1 \leqslant j < m \leqslant n} h(u_{P_j} - u_{P_m} + 2\eta). \tag{8.11}$$

The equations (6.5) for $u_1, \ldots, u_n$ become

$$\prod_{y=1}^{N} \frac{h(u_j + v_y + \eta)}{h(u_j + v_y - \eta)} = -\prod_{m=1}^{n} \frac{h(u_j - u_m + 2\eta)}{h(u_j - u_m - 2\eta)}, \qquad j = 1, \ldots, n. \tag{8.12}$$

These results (8.9)–(8.12) are formally not much more complicated than those for the homogeneous lattice. In some ways they make the properties of the solution clearer: for instance it is apparent from (8.9) that each single-particle function is a product of N factors, one from each column of the lattice.

References

1. R. J. Baxter, Some Fundamental Eigenvectors, *Annals of Physics,* **76** (1973), 1–24.
2. R. J. Baxter, Equivalence to a generalized ice-type lattice model, *Annals of Physics,* **76** (1973), 25–47.
3. E. H. Lieb, *Phys. Rev.* **162** (1967), 162–172; *Phys. Rev. Lett.* **18** (1967), 1046–1048 and **19** (1967), 108–110.
4. R. J. Baxter, *Annals of Physics,* **70** (1972), 193–228.
5. I. S. Gradshteyn and I. M. Ryzhik, "Tables of Integrals, Series and Products," pp. 909–925, Academic Press, New York, 1965.
6. C. N. Yang and C. P. Yang, *Phys. Rev.* **150** (1966), 321–327.

Commun. Math. Phys. 116, 507–525 (1988)

Communications in Mathematical Physics

Solvable Lattice Models Related to the Vector Representation of Classical Simple Lie Algebras

Michio Jimbo, Tetsuji Miwa, and Masato Okado

Research Institute for Mathematical Sciences, Kyoto University, Kyoto 606, Japan

Abstract. A series of solvable lattice models with face interaction are introduced on the basis of the affine Lie algebra $X_n^{(1)} = A_n^{(1)}, B_n^{(1)}, C_n^{(1)}, D_n^{(1)}$. The local states taken on by the fluctuation variables are the dominant integral weights of $X_n^{(1)}$ of a fixed level. Adjacent local states are subject to a condition related to the vector representation of X_n. The Boltzmann weights are parametrized by elliptic theta functions and solve the star-triangle relation.

1. Introduction

Through the last decade of investigation the significance of the Yang-Baxter equation (YBE) in integrable systems has been commonly acknowledged. A number of two dimensional models have been solved in statistical mechanics and in quantum field theory on this basis. Very recently there is also renewed interest for the YBE because of its connection with other branches of mathematics, such as the braid group, link invariants and operator algebras [1–4].

In the study of the YBE, the idea of Lie algebras and representation theory has turned out to be particularly fruitful. Such a viewpoint has been developed in the framework of the quantum inverse method. Motivated by the connection with soliton theory, a quasi-classical version of the YBE was formulated (the classical YBE). The classification of its solutions associated with simple Lie algebras was accomplished by Belavin and Drinfeld [5]. Kulish et al. [6] initiated the representation theoretical construction of quantum R matrices (= solutions to the YBE) corresponding to the classical ones. These works led several authors [7–11] to the discovery of a novel algebraic structure underlying the problem, the quantum group as formulated by Drinfeld [9].

In the statistical mechanics language, the works mentioned above are concerned with the vertex model. Here the fluctuation variables are placed on the edges of a two dimensional lattice, and each element of the R matrix provides the statistical weight (the Boltzmann weight) for a configuration round a lattice site, or a vertex. There is also a dual object, the face model, in which the variables live on the sites, and the Boltzmann weight is attached to a configuration round a face.

(This was introduced by Baxter [12] and was called the interaction-round-a-face model. Here we prefer a simpler terminology, the face model.) The YBE is formulated accordingly and is more often called the star-triangle relation (STR).

Baxter's solution of the hard hexagon model [13] showed that the face model has interest of its own right. For face models one can apply Baxter's corner transfer matrix method to evaluate the local state probability (LSP), the probability that the variable at a site takes a particular state. Andrews et al. [14] presented a series of face models generalizing the hard hexagon model, and obtained their LSPs in terms of modular functions. Subsequently it was recognized [15] that their models are naturally associated with the affine Lie algebra $A_1^{(1)}$, and that the modular functions in the LSP result are the branching coefficients for the pair $(A_1^{(1)} \oplus A_1^{(1)}, A_1^{(1)})$. This sort of result was first obtained in the $\mathbf{Z}_n$ Ising type models [16], and has been further extended to other cases; those related to symmetric tensors of $A_1^{(1)}$ [17, 18] or to the vector representation of $A_n^{(1)}$ [19, 20] (in the sense to be described below).

This paper is directly motivated by these latest works. Before giving a detailed introduction of the theme let us mention our motivation in relation to the conformal field theory. The enumeration of all the solvable conformal field theories is a goal still far from our sight. Recently, the classification of the unitary discrete series has been completed, in which the conformal anomaly satisfies $0<c<1$. Aiming at theories with $c \geqq 1$ three different approaches are being pursued. The first is to exploit larger symmetries than the Virasoro algebra such as affine Lie algebras, super Virasoro algebras, parafermion algebras, W_n algebras and so on. The second is to approach from the modular invariance. The third is to work with solvable lattice models. One advantage in the third approach is that the master equation is at hand; the YBE or the STR. Recent studies [21–30] have revealed that the known solutions are just the tip of the iceberg. The aim of this paper is to manifest its body by showing the role of affine Lie algebras in this game.

Consider an affine Lie algebra of type $X_n^{(1)}$ ($X_n = A_n, B_n, C_n$ or D_n) and an irreducible representation (π, V_π) of its classical part X_n. We put forward the construction of solvable face models that have the following features: Firstly, the local states, i.e. the values of the fluctuation variables, are taken to be the dominant integral weights of $X_n^{(1)}$ of a fixed level l. We choose l to be not too small so that there exists the level l highest weight representation of $X_n^{(1)}$ whose highest weight vector generates an irreducible X_n module isomorphic to V_π.

Secondly, we require that the states (a, b) of neighboring variables should be admissible in the following sense. Let $V(\bar{b})$ denote the irreducible X_n module having the classical part $\bar{b}$ of b as the highest weight. We say that

(*) (a, b) is *admissible* if and only if $V(\overline{\sigma(b)})$ appears in the irreducible decomposition of $V(\overline{\sigma(a)}) \otimes V_\pi$ for any Dynkin diagram automorphism σ.

The condition (*) was introduced in [25, 26] to rule out the divergent restricted weights, and was shown to fit the description of the LSP result in terms of the affine Lie algebra characters [17, 18, 20]. It was also utilized by Wenzl [31] in his construction of the irreducible representations of the Hecke algebra, and by Tsuchiya and Kanie [32] in the fusion rule of vertex operators in the conformal field theory. Recently, Pasquier [33] touched upon the possibility of constructing face models of this sort.

In general, the solution to the star-triangle relation satisfying these two requirements is not unique (see Sect. 5). Therefore we need a further specification of the model. In the $A_n^{(1)}$ family [19] it was noticed that the factors G_a appearing in the second inversion relation [see (2.13b)] are equal to the principally specialized characters. We exploit this remarkable coincidence as the third Merkmal in finding solutions of type $X_n^{(1)}$.

The cases $X_n^{(1)}=A_n^{(1)}$ and $\pi=$ the vector representation [19, 20] or symmetric tensors [26, 27] have been treated previously. Here we shall work out the solutions in the case $X_n^{(1)}=B_n^{(1)}$, $C_n^{(1)}$, $D_n^{(1)}$ and $\pi=$ the vector representation. In fact, as was done by Andrews et al. [14] in the original 8VSOS model, we shall first construct models such that the local states take *generic* complex weights in the dual space of the Cartan subalgebra. The admissibility (*) is replaced by a weaker condition that $b-a$ is a weight appearing in V_π. The solutions of the STR are parametrized by degree 2 elliptic theta functions, in contrast to the degree 1 parametrization in the $A_n^{(1)}$ family. The resulting models (which we call unrestricted models) admit infinitely many local states, differing mutually by integral weights. We then restrict them to *dominant* integral weights of level l and show that the STR closes among the finitely many Boltzmann weights thus obtained. These are the main results of the present article.

Let us include one remark on the above formulation of the face model. The representations (π, V_π) of X_n dealt with in this paper are limited in that the weight multiplicities are merely *one*. For general π we need presumably to introduce extra fluctuation variables on edges corresponding to the weight multiplicities.

The text is organized as follows. In Sect. 2 we give the formulas for the Boltzmann weights for the unrestricted models of type $A_n^{(1)}$, $B_n^{(1)}$, $C_n^{(1)}$, and $D_n^{(1)}$. Proof of the STR is given in Sect. 3. Section 4 deals with the restriction process described above. Discussions are included in Sect. 5.

2. Unrestricted Models

The models we consider will be built upon the affine Lie algebra $X_n^{(1)}$, where X_n denotes one of the finite dimensional simple Lie algebras A_n, B_n, C_n, or D_n. Our basic reference on the affine Lie algebras is Kac-Peterson [34]. We denote by Λ_j $(0\leqq j\leqq n)$ the fundamental weights, and set $\varrho=\Lambda_0+\ldots+\Lambda_n$, $\mathfrak{h}_{\mathbb{C}}^*=\sum_{j=0}^{n}\mathbb{C}\Lambda_j$. For an element $a\in\mathfrak{h}_{\mathbb{C}}^*$, $\bar{a}$ signifies its calssical part. Following Bourbaki [35] we introduce orthonormal vectors ε_i to express $\bar{\Lambda}_j$ as in Table 1. We list also the set $\mathscr{A}$ of weights that belong to the vector representation of X_n.

We write an elcment of $\mathscr{A}$ as

$$
\begin{aligned}
\hat{\mu}&=\varepsilon_\mu-\frac{1}{n+1}(\varepsilon_1+\ldots+\varepsilon_{n+1}) \quad (1\leqq\mu\leqq n+1) \quad \text{for} \quad A_n,\\
&=\pm\varepsilon_i \quad \text{or} \quad 0 \quad (\mu=\pm i, 1\leqq i\leqq n, \quad \text{or} \quad \mu=0) \quad \text{for} \quad B_n,\\
&=\pm\varepsilon_i \quad (\mu=\pm i, 1\leqq i\leqq n) \quad \text{for} \quad C_n, D_n.
\end{aligned}
$$

Table 1

A_n	$\mathscr{A}=\{\varepsilon_1-\varepsilon,\dots,\varepsilon_{n+1}-\varepsilon\}$,
	$\bar{\Lambda}_i=\varepsilon_1+\dots+\varepsilon_i-i\varepsilon \quad (1\leqq i\leqq n), \quad \varepsilon=\frac{1}{n+1}\sum_{j=1}^{n+1}\varepsilon_j$.
B_n	$\mathscr{A}=\{\pm\varepsilon_1,\dots,\pm\varepsilon_n,0\}$,
	$\bar{\Lambda}_i=\varepsilon_1+\dots+\varepsilon_i \quad (1\leqq i\leqq n-1)$,
	$=\frac{1}{2}(\varepsilon_1+\dots+\varepsilon_n) \quad (i=n)$.
C_n	$\mathscr{A}=\{\pm\varepsilon_1,\dots,\pm\varepsilon_n\}$,
	$\bar{\Lambda}_i=\varepsilon_1+\dots+\varepsilon_i \quad (1\leqq i\leqq n)$.
D_n	$\mathscr{A}=\{\pm\varepsilon_1,\dots,\pm\varepsilon_n\}$,
	$\bar{\Lambda}_i=\varepsilon_1+\dots+\varepsilon_i \quad (1\leqq i\leqq n-2)$,
	$=\frac{1}{2}(\varepsilon_1+\dots+\varepsilon_{n-2}+\varepsilon_{n-1}-\varepsilon_n) \quad (i=n-1)$,
	$=\frac{1}{2}(\varepsilon_1+\dots+\varepsilon_{n-2}+\varepsilon_{n-1}+\varepsilon_n) \quad (i=n)$.

For $a\in\mathfrak{h}_{\mathbb{C}}^*$ we put

$$a_\mu=\langle a+\varrho,\hat{\mu}\rangle \quad (\mu\neq 0),$$
$$=-\tfrac{1}{2} \quad (\mu=0),$$

so that we have

$$\bar{a}+\bar{\varrho}=\sum_{i=1}^{n+1}a_i\varepsilon_i,\ \sum_{i=1}^{n+1}a_i=0 \quad \text{for} \quad A_n^{(1)},$$
$$=\sum_{i=1}^{n}a_i\varepsilon_i \quad \text{for} \quad B_n^{(1)},C_n^{(1)},D_n^{(1)}.$$

Consider now a two dimensional square lattice $\mathscr{L}$. We shall introduce face models on $\mathscr{L}$ that have the following basic features:

(1) The fluctuation variable placed on each lattice site assumes its values in $\mathfrak{h}_{\mathbb{C}}^*$. We call these values local states.

(2) Adjacent local states differ by an element in $\mathscr{A}$, i.e. by a weight in the vector representation of X_n. More precisely, this means that the Boltzmann weights $W\begin{pmatrix} a & b \\ d & c\end{pmatrix}$ describing the interaction of four fluctuation variables round a face [12] satisfy the condition

$$W\begin{pmatrix} a & b \\ d & c\end{pmatrix}=0 \quad \text{unless} \quad b-a,c-b,d-a,c-d\in\mathscr{A}. \tag{2.1}$$

Because of (2.1) the local states appearing in a possible configuration on $\mathscr{L}$ are actually confined to the set $a^0+\sum_{\hat{\mu}\in\mathscr{A}}\mathbb{Z}\hat{\mu}$, where a^0 is a fixed element of $\mathfrak{h}_{\mathbb{C}}^*$.

Under the setting above we have found a system of Boltzmann weights $W\left(\begin{matrix} a & b \\ d & c\end{matrix}\middle| u\right)$ that depend on the *spectral parameter* $u\in\mathbb{C}$ and solve the *star-triangle relation* (STR)

$$\sum_g W\begin{pmatrix} f & g \\ e & d \end{pmatrix}\!\Big|u\Big) W\begin{pmatrix} b & c \\ g & d \end{pmatrix}\!\Big|v\Big) W\begin{pmatrix} a & b \\ f & g \end{pmatrix}\!\Big|u+v\Big)$$

$$=\sum_g W\begin{pmatrix} a & b \\ g & c \end{pmatrix}\!\Big|u\Big) W\begin{pmatrix} a & g \\ f & e \end{pmatrix}\!\Big|v\Big) W\begin{pmatrix} g & c \\ e & d \end{pmatrix}\!\Big|u+v\Big). \tag{2.2}$$

The solutions are parametrized in terms of the elliptic theta function

$$[u]=\theta_1\left(\frac{\pi u}{L}, p\right),$$
$$\theta_1(u,p)=2p^{1/8}\sin u \prod_{k=1}^{\infty}(1-2p^k\cos 2u+p^{2k})(1-p^k), \tag{2.3}$$

where $L \neq 0$ is an arbitrary complex parameter. Explicitly they are given by the following formulas. We shall write

$$\mu \overset{\kappa}{\underset{\nu}{\square}} \sigma = W\begin{pmatrix} a & a+\hat{\kappa} \\ a+\hat{\mu} & a+\hat{\mu}+\hat{\nu} \end{pmatrix}\!\Big|u\Big) \qquad (\hat{\mu}+\hat{\nu}=\hat{\kappa}+\hat{\sigma}).$$

$A_n^{(1)}$:

$$\mu \overset{\mu}{\underset{\mu}{\square}} \mu = \frac{[1+u]}{[1]}, \tag{2.4a}$$

$$\mu \overset{\mu}{\underset{\nu}{\square}} \nu = \frac{[a_\mu - a_\nu - u]}{[a_\mu - a_\nu]} \qquad (\mu \neq \nu), \tag{2.4b}$$

$$\mu \overset{\nu}{\underset{\nu}{\square}} \mu = \frac{[u]}{[1]}\left(\frac{[a_\mu - a_\nu + 1][a_\mu - a_\nu - 1]}{[a_\mu - a_\nu]^2}\right)^{1/2} \qquad (\mu \neq \nu). \tag{2.4c}$$

$B_n^{(1)}$, $C_n^{(1)}$, $D_n^{(1)}$:

$$\mu \overset{\mu}{\underset{\mu}{\square}} \mu = \frac{[\lambda - u][1+u]}{[\lambda][1]} \qquad (\mu \neq 0), \tag{2.5a}$$

$$\mu \overset{\mu}{\underset{\nu}{\square}} \nu = \frac{[\lambda - u][a_\mu - a_\nu - u]}{[\lambda][a_\mu - a_\nu]} \qquad (\mu \neq \pm\nu), \tag{2.5b}$$

$$\mu \overset{\nu}{\underset{\nu}{\square}} \mu = \frac{[\lambda - u][u]}{[\lambda][1]}\left(\frac{[a_\mu - a_\nu + 1][a_\mu - a_\nu - 1]}{[a_\mu - a_\nu]^2}\right)^{1/2} \qquad (\mu \neq \pm\nu), \tag{2.5c}$$

$$\mu \overset{\nu}{\underset{-\mu}{\square}} -\nu = \frac{[u][a_\mu + a_\nu + 1 + \lambda - u]}{[\lambda][a_\mu + a_\nu + 1]}(G_{a\mu}G_{a\nu})^{1/2} \qquad (\mu \neq \nu), \tag{2.5d}$$

$$\mu \underset{-\mu}{\overset{\mu}{\square}} -\mu = \frac{[\lambda+u][2a_\mu+1+2\lambda-u]}{[\lambda][2a_\mu+1+2\lambda]} - \frac{[u][2a_\mu+1+\lambda-u]}{[\lambda][2a_\mu+1+2\lambda]} H_{a\mu}, \tag{2.5e}$$

$$= \frac{[\lambda-u][2a_\mu+1-u]}{[\lambda][2a_\mu+1]} + \frac{[u][2a_\mu+1+\lambda-u]}{[\lambda][2a_\mu+1]} G_{a\mu}(\mu \neq 0), \tag{2.5f}$$

$$H_{a\mu} = \sum_{\kappa(\neq\mu)} \frac{[a_\mu+a_\kappa+1+2\lambda]}{[a_\mu+a_\kappa+1]} G_{a\kappa}. \tag{2.5g}$$

Here $\mu, \nu = 1, \ldots, n+1$ (for $A_n^{(1)}$), $0, \pm 1, \ldots, \pm n$ (for $B_n^{(1)}$) or $\pm 1, \ldots, \pm n$ (for $C_n^{(1)}$, $D_n^{(1)}$). In the above we used the following notations. The *crossing parameter* λ in (2.5) is fixed to be

$$\lambda = -tg/2 \tag{2.6}$$

with $t = (\text{long root})^2/2$ and g = the dual Coxeter number of $X_n^{(1)}$ (see Table 2). The factor $G_{a\mu}$ is given by

$$G_{a\mu} = G_{a+\hat{\mu}}/G_a = \sigma \frac{h(a_\mu+1)}{h(a_\mu)} \prod_{\kappa(\neq\pm\mu,0)} \frac{[a_\mu-a_\kappa+1]}{[a_\mu-a_\kappa]} \quad (\mu \neq 0), \quad G_{a0} = 1, \tag{2.7}$$

$$G_a = \prod_{1 \leq i < j \leq n+1} [a_i - a_j] \quad \text{for} \quad A_n^{(1)}, \tag{2.8a}$$

$$= \varepsilon(a) \prod_{i=1}^{n} h(a_i) \prod_{1 \leq i < j \leq n} [a_i - a_j][a_i + a_j] \quad \text{for} \quad B_n^{(1)}, C_n^{(1)}, D_n^{(1)}. \tag{2.8b}$$

Here $\sigma = -1$ for $C_n^{(1)}$ and $= 1$ otherwise, and $\varepsilon(a)$ is a sign factor such that $\varepsilon(a+\hat{\mu})/\varepsilon(a) = \sigma$. The function $h(a)$ is given in Table 2. Up to a common factor independent of a, the expression for $\pm G_a$ coincides with the denominator formula of $X_n^{(1)}$ evaluated at $a+\varrho$, and hence with the principally specialized character for the *dual* affine Lie algebra (see [34] for the definition of the dual affine Lie algebra).

Table 2

type	$A_n^{(1)}$	$B_n^{(1)}$	$C_n^{(1)}$	$D_n^{(1)}$
g	$n+1$	$2n-1$	$n+1$	$2n-2$
t	1	1	2	1
$h(a)$	1	$[a]$	$[2a]$	1

Proof of the STR (2.2) [including that of the equality of (2.5e) and (2.5f) for $\mu \neq 0$] will be given in Sect. 3. Besides the spectral parameter u, these weights contain two arbitrary parameters L, p entering in (2.3). We call the models defined by (2.1, 4–5) *unrestricted* $X_n^{(1)}$ *models* (as opposed to *restricted models* to be discussed in Sect. 4). These terminologies, unrestricted and restricted, go back to Andrews–Baxter–Forrester [14].

We have verified that the following equivalences between the representations of classical simple Lie algebras extend to those between the corresponding unrestricted face models:

(B_1, the vector representation)
$\sim$(A_1, the symmetric tensor representation of degree 2),

(D_3, the vector representation)
$\sim$(A_3, the skew-symmetric tensor representation of degree 2).

The symmetric tensor representations of the $A_n^{(1)}$ face model have been constructed in [26, 27]. We hope to discuss a similar construction for the skew-symmetric tensors in a future publication (see [36] for the vertex models).

The Boltzmann weights enjoy the following properties.

Initial condition

$$W\left(\begin{matrix} a & b \\ d & c \end{matrix}\middle| 0\right) = \delta_{bd}. \tag{2.9}$$

Reflection symmetry

$$W\left(\begin{matrix} a & b \\ d & c \end{matrix}\middle| u\right) = W\left(\begin{matrix} a & d \\ b & c \end{matrix}\middle| u\right). \tag{2.10}$$

Dynkin diagram symmetry

$$W\left(\begin{matrix} \sigma(a) & \sigma(b) \\ \sigma(d) & \sigma(c) \end{matrix}\middle| u\right) = W\left(\begin{matrix} a & b \\ d & c \end{matrix}\middle| u\right), \tag{2.11}$$

where σ is any Dynkin diagram automorphism acting on $\mathfrak{h}_{\mathbb{C}}^*$ and the parameter L is set to $t(l+g)$ (l: the level of a).

Rotational symmetry (valid except for $A_n^{(1)}$)

$$W\left(\begin{matrix} a & b \\ d & c \end{matrix}\middle| u\right) = \left(\frac{G_b G_d}{G_a G_c}\right)^{1/2} W\left(\begin{matrix} d & a \\ c & b \end{matrix}\middle| \lambda - u\right). \tag{2.12}$$

The following *inversion relations* will play a role in the evaluation of the local state probabilities, though we do not discuss it in this paper.

$$\sum_g W\left(\begin{matrix} a & g \\ d & c \end{matrix}\middle| u\right) W\left(\begin{matrix} a & b \\ g & c \end{matrix}\middle| -u\right) = \delta_{bd}\varrho_1(u), \tag{2.13a}$$

$$\sum_g \bar{W}\left(\begin{matrix} a & b \\ d & g \end{matrix}\middle| \lambda - u\right) \bar{W}\left(\begin{matrix} c & d \\ b & g \end{matrix}\middle| \lambda + u\right) = \delta_{ac}\varrho_2(u). \tag{2.13b}$$

Here we have set

$$\bar{W}\left(\begin{matrix} a & b \\ d & c \end{matrix}\middle| u\right) = \left(\frac{G_a G_c}{G_b G_d}\right)^{1/2} W\left(\begin{matrix} a & b \\ d & c \end{matrix}\middle| u\right),$$

$$\varrho_1(u) = \frac{[1+u][1-u]}{[1]^2}, \qquad \varrho_2(u) = \frac{[\lambda+u][\lambda-u]}{[1]^2} \quad \text{for} \quad A_n^{(1)},$$

$$\varrho_1(u) = \varrho_2(u) = \frac{[\lambda+u][\lambda-u][1+u][1-u]}{[\lambda]^2[1]^2} \quad \text{for} \quad B_n^{(1)}, C_n^{(1)}, D_n^{(1)}.$$

In fact the first inversion relation (2.13a) is a direct consequence of the STR (2.2) and the initial condition (2.9). For types other than $A_n^{(1)}$ (see [20] for $A_n^{(1)}$), the second inversion relation (2.13b) follows from (2.12) and (2.13a).

3. Proof of the STR

This section is devoted to the proof of the STR (2.2). Since a proof in the case of $A_n^{(1)}$ was given in a more general situation [27], we only consider the remaining cases $B_n^{(1)}$, $C_n^{(1)}$, and $D_n^{(1)}$. Throughout this section L, τ are complex numbers satisfying $L \neq 0$, $\operatorname{Im}\tau > 0$ [$p = e^{2\pi i\tau}$ in (2.3)]. We shall frequently use the quasi-periodicity property of the symbol $[u]$

$$[u+L] = -[u], \qquad [u+L\tau] = -e^{-\pi i\tau - 2\pi i u/L}[u], \tag{3.1}$$

along with the following standard lemma (cf. [37]).

Lemma 1. *If $f(u)$ is entire, not identically zero and satisfies*

$$f(u+L) = e^{-2\pi i B} f(u), \qquad f(u+L\tau) = e^{-2\pi i(A_1 + A_2 u/L)} f(u),$$

then A_2 is a non-negative integer, $f(u)$ has A_2 zeros $\operatorname{mod} L\mathbb{Z} + L\mathbb{Z}\tau$ *and* $\sum$ *zeros* $\equiv L(B\tau + A_2/2 - A_1)$.

First we prove the inversion relation (2.13a). It is divided into the following three types:

Proposition 2.

$$\text{(i)}\quad \begin{array}{c}\mu\\ \mu\,\boxed{u}\,\mu\\ \mu\end{array}\ \begin{array}{c}\mu\\ \mu\,\boxed{-u}\,\mu\\ \mu\end{array} = \varrho(u) \qquad (\mu \neq 0),$$

$$\text{(ii)}\quad \begin{array}{c}\mu\\ \mu\,\boxed{u}\,\nu\\ \nu\end{array}\ \begin{array}{c}\mu\\ \mu\,\boxed{-u}\,\nu\\ \nu\end{array} + \begin{array}{c}\nu\\ \mu\,\boxed{u}\,\mu\\ \nu\end{array}\ \begin{array}{c}\mu\\ \nu\,\boxed{-u}\,\nu\\ \mu\end{array} = \varrho(u) \qquad (\mu \neq \pm\nu),$$

$$\begin{array}{c}\mu\\ \mu\,\boxed{u}\,\nu\\ \nu\end{array}\ \begin{array}{c}\nu\\ \mu\,\boxed{-u}\,\mu\\ \nu\end{array} + \begin{array}{c}\nu\\ \mu\,\boxed{u}\,\mu\\ \nu\end{array}\ \begin{array}{c}\nu\\ \nu\,\boxed{-u}\,\mu\\ \mu\end{array} = 0 \qquad (\mu \neq \pm\nu),$$

$$\text{(iii)}\quad \sum_{k} \begin{array}{c}\kappa\\ \mu\,\boxed{u}\,-\kappa\\ -\mu\end{array}\ \begin{array}{c}\nu\\ \kappa\,\boxed{-u}\,-\nu\\ -\kappa\end{array} = \varrho(u)\,\delta_{\mu\nu},$$

where

$$\varrho(u) = \frac{[\lambda+u][\lambda-u][1+u][1-u]}{[\lambda]^2[1]^2}.$$

In (iii) *the sum is taken over* $\kappa = \pm 1, \ldots, \pm n, 0$ *for* $B_n^{(1)}$, *and* $= \pm 1, \ldots, \pm n$ *for* $C_n^{(1)}$, $D_n^{(1)}$.

Before going to the proof we need to prepare two lemmas.

Lemma 3. *For any a, b, c, u, v, w, λ (u+v+w=λ), we have the identity*

$$0=-[\lambda-u][\lambda-v][\lambda-w]\frac{[b+c+u][c+a+v][a+b+w]}{[b+c][c+a][a+b]}$$
$$+[\lambda-u][v][w]\frac{[2a-u][a+b+\lambda-v][a+c+\lambda-w]}{[2a][a+b][a+c]}$$
$$+[u][\lambda-v][w]\frac{[b+a+\lambda-u][2b-v][b+c+\lambda-w]}{[b+a][2b][b+c]}$$
$$+[u][v][\lambda-w]\frac{[c+a+\lambda-u][c+b+\lambda-v][2c-w]}{[c+a][c+b][2c]}$$
$$+[u][v][w]\sum_{\omega}\tfrac{1}{2}\frac{[a+\lambda-u+\omega][b+\lambda-v+\omega][c+\lambda-w+\omega]}{[a+\omega][b+\omega][c+\omega]}e^{2\pi i\theta(\omega)}.$$

Here the summation $\sum_{\omega}$ *is over the half periods* $\omega=0, L/2, L\tau/2, L(1+\tau)/2$, *and*

$$\begin{aligned}\theta(\omega)&=0 \quad \omega=0, L/2,\\ &=\lambda/L \quad \omega=L\tau/2, L(1+\tau)/2.\end{aligned} \tag{3.2}$$

Proof. Regarding the right-hand side as a function of *a*, one can verify that it does not have any pole. Next we apply Lemma 1, taking $B=A_2=0$, $A_1=(v+w)/L$. □

Lemma 4. *Set*

$$\varphi(u)=\frac{1}{C}\frac{d}{du}\log[u], \quad C=\lim_{u\to 0}\frac{[u]}{u}. \tag{3.3}$$

We have then

$$\varphi(u+1)+\varphi(u-1)-2\varphi(u)=\frac{[1]^2[2u]}{[u]^2[u-1][u+1]}. \tag{3.4}$$

Proof. By the definition, the function $\varphi(u)$ satisfies

$$\varphi(u+L)=\varphi(u), \quad \varphi(u+L\tau)=\varphi(u)-2\pi i/LC.$$

From this one can verify that (the left-hand side)–(the right-hand side) of (3.4) is a doubly periodic odd function and has no poles. Hence it must be zero. □

Proof of Proposition 2. Equation (i) is trivial. Equation (ii) is easily checked by using the addition formula,

$$[x+z][x-z][y+w][y-w]-[x+w][x-w][y+z][y-z]$$
$$=[x+y][x-y][z+w][z-w].$$

Let us prove (iii) with μ, $\nu\neq 0$. In this case we use the expression (2.5f). Set

$$F_{\mu\nu}(z)=\frac{[a_\mu+z+1+\lambda-u][a_\nu+z+1+\lambda+u]}{[a_\mu+z+1][a_\nu+z+1]}\frac{\sigma h(z+1)[2z]}{[1]h(z)[2z+1]}\prod_{\kappa(\neq 0)}\frac{[z-a_\kappa+1]}{[z-a_\kappa]}.$$

From (2.6) and Table 2, one finds that $F_{\mu\nu}(z)$ is a doubly periodic function. If $\mu\neq\nu$, its poles are located at $z=a_\kappa$ $(\kappa\neq 0)$, $-1/2+\omega$ $(\omega=0, L/2, L\tau/2, L(1+\tau)/2)$. If $\mu=\nu$, it has an additional pole at $z=-a_\mu-1$. The relation $\sum \mathrm{Res}\, F_{\mu\nu}(z)=0$ gives rise to

$$\sum_k \frac{[a_\mu+a_\kappa+1+\lambda-u][a_\nu+a_\kappa+1+\lambda+u]}{[a_\mu+a_\kappa+1][a_\nu+a_\kappa+1]} G_{a\kappa}$$
$$-\tfrac{1}{2}\sum_\omega \frac{[a_\mu+1/2+\lambda-u+\omega][a_\nu+1/2+\lambda+u+\omega]}{[a_\mu+1/2+\omega][a_\nu+1/2+\omega]} e^{2\pi i\theta(\omega)}$$
$$-\delta_{\mu\nu}\frac{[\lambda-u][\lambda+u][2a_\mu][2a_\mu+2]}{[1]^2[2a_\mu+1]^2} G_{a\mu}^{-1}=0. \tag{3.5}$$

Here $\theta(\omega)$ is given in (3.2), and the summation $\sum\limits_\omega$ is over $\omega=0, L/2, L\tau/2, L(1+\tau)/2$. (Note again that the sum $\sum\limits_\kappa$ includes $\kappa=0$ for $B_n^{(1)}$.) Combining (3.5) with Lemma 3 specialized as $a=b=a_\mu+1/2$, $u=-v$, $w=\lambda$, we obtain (iii) with $\mu, \nu \neq 0$.

Next let us prove (iii) in the case of $\mu=\nu=0$. Let $f(u)$ be (the left-hand side)–(the right-hand side) of (iii). In view of Lemma 1, it suffices to check $f(0)=f(\pm\lambda)=f(\pm 1)=0$. (Note that $f(u)$ is an even function.) The only nontrivial step is to show $f(1)=0$. Pick any index $\mu\neq 0$. Regarding $f(1)$ as a function of a_μ we denote it by $g(a_\mu)$, namely

$$g(a_\mu)=\left(\frac{[\lambda+1][2\lambda-1]}{[\lambda][2\lambda]}-\frac{[1][\lambda-1]}{[\lambda][2\lambda]}H_{a0}\right)$$
$$\times\left(\frac{[\lambda-1][2\lambda+1]}{[\lambda][2\lambda]}+\frac{[1][\lambda+1]}{[\lambda][2\lambda]}H_{a0}\right)$$
$$-\frac{[1]^2}{[\lambda]^2}\sum_{\kappa(\neq 0)}\frac{[a_\kappa+3/2+\lambda][a_\kappa-1/2+\lambda]}{[a_\kappa+1/2]^2}G_{a\kappa}.$$

This is doubly periodic in a_μ. Let us show that it is holomorphic everywhere. It is easy to see that the apparent poles $a_\mu=0$ or $a_\mu=a_\kappa(\mu\neq\pm\kappa)$ are regular points, and that the coefficient of $(a_\mu+1/2)^{-2}$ at $a_\mu=-1/2$ vanishes. The vanishing of the coefficient of $(a_\mu+1/2)^{-1}$ means

$$0=-\frac{[1][\lambda+1]^2[2\lambda-1]}{[\lambda]^2[2\lambda]}+\frac{[1][\lambda-1]^2[2\lambda+1]}{[\lambda]^2[2\lambda]}-\frac{[1]^2[\lambda-1][\lambda+1]}{[\lambda]^2}$$
$$\times\left\{2\varphi(2\lambda)-\varphi(\lambda-1)-\varphi(\lambda+1)+2\varphi(1/2)+2\sum_{\kappa(\neq\pm\mu,0)}\varphi(1/2-a_\kappa)\right\}$$
$$+\frac{2[1]^2[\lambda-1][\lambda+1]}{[\lambda]^2[2\lambda]}\sum_{\kappa(\neq\mu,0)}\frac{[a_\kappa+1/2+2\lambda]}{[a_\kappa+1/2]}G_{a\kappa}. \tag{3.6}$$

Consider a doubly periodic function

$$F(z)=\frac{[z+1/2+2\lambda][2z][z+1][z+3/2]}{[z+1/2][1][2z+1][z][z-1/2]}\prod_{\kappa(\neq\pm\mu,0)}\frac{[z-a_\kappa+1]}{[z-a_\kappa]}.$$

$\sum \mathrm{Res}\, F(z)=0$ this time gives rise to

$$\sum_{\kappa(\neq\mu,0)}\frac{[a_\kappa+1/2+2\lambda]}{[a_\kappa+1/2]}G_{a\kappa}=\tfrac{1}{2}\sum_{\omega(\neq 0)}\frac{[2\lambda+\omega]}{[\omega]}e^{2\pi i\theta(\omega)}$$
$$+\frac{[2\lambda]}{2}\left\{\varphi(2\lambda)+2\varphi(1/2)+2\sum_{\kappa(\neq\pm\mu,0)}\varphi(1/2-a_\kappa)\right\}. \tag{3.7}$$

Furthermore, by specializing $u=-\lambda=-v=-w$, $a\to 0$ in Lemma 3, we have

$$\sum_{\omega(\neq 0)} \frac{[2\lambda+\omega]}{[\omega]} e^{2\pi i\theta(\omega)} = [2\lambda](2\varphi(\lambda)-\varphi(2\lambda)). \tag{3.8}$$

Thanks to (3.7–8) and Lemma 4, (3.6) is reduced to

$$\frac{[\lambda-1][2\lambda+1]}{[1][\lambda+1][2\lambda]} - \frac{[\lambda+1][2\lambda-1]}{[1][\lambda-1][2\lambda]} + \frac{[1]^2[2\lambda]}{[\lambda]^2[\lambda-1][\lambda+1]} + 4\varphi(\lambda) - 2\varphi(2\lambda) = 0.$$

This can be checked by a similar method as in the proof of Lemma 4.

Now we have proved that $f(1)$ is independent of $a_\mu(\mu\neq 0)$. Letting $a_\mu=\mu$ for all μ and using the specialized values

$$G_{a\kappa} = \frac{[1-2\lambda]}{[1]}\delta_{\kappa n}, \quad H_{a0} = \frac{[\lambda+1][2\lambda-1]}{[1][\lambda-1]},$$

we find $f(1)=0$. This completes the proof of (iii) with $\mu=v=0$.

The remaining case is (iii) with $\mu=0$, $v\neq 0$. It reads as

$$\begin{aligned} 0 = &\left(\frac{[\lambda+u][2\lambda-u]}{[\lambda][2\lambda]} - \frac{[u][\lambda-u]}{[\lambda][2\lambda]} H_{a0}\right) \frac{[a_v+1/2+\lambda+u]}{[a_v+1/2]} \\ &+ \sum_{\kappa(\neq v,0)} \frac{[u]}{[\lambda]} \frac{[a_\kappa+1/2+\lambda-u][a_\kappa+a_v+1+\lambda+u]}{[a_\kappa+1/2][a_\kappa+a_v+1]} G_{a\kappa} \\ &- \frac{[a_v+1/2+\lambda-u]}{[a_v+1/2]} \left(\frac{[\lambda+u][2a_v+1+u]}{[\lambda][2a_v+1]} - \frac{[u][2a_v+1+\lambda+u]}{[\lambda][2a_v+1]} G_{av}\right). \end{aligned}$$

From Lemma 1, it suffices to check that the right-hand side vanishes at $u=0, \pm\lambda$. The nontrivial case $u=\lambda$ reduces to the identity

$$\sum_\kappa \frac{[a_v+a_\kappa+1+2\lambda]}{[a_v+a_\kappa+1]} G_{a\kappa} = \frac{[2\lambda][2a_v+1+\lambda]}{[\lambda][2a_v+1]} (v\neq 0). \tag{3.9}$$

This is obtained by specializing $u=\lambda$ in (iii) with $\mu=v\neq 0$. □

Remark. From (3.9) it follows also that (2.5e) is equal to (2.5f) in the case $\mu\neq 0$.

Now let us proceed to the proof of the STR. First put

$$\begin{aligned} &X(a|\kappa, \mu, v; \alpha, \beta, \gamma|u, v) \\ &= \sum_g W\left(\begin{matrix} a+\hat\kappa & g \\ a+\hat\kappa+\hat\mu & a+\hat\kappa+\hat\mu+\hat v \end{matrix}\middle| u\right) W\left(\begin{matrix} a+\hat\alpha & a+\hat\alpha+\hat\beta \\ g & a+\hat\alpha+\hat\beta+\hat\gamma \end{matrix}\middle| v\right) \\ &\times W\left(\begin{matrix} a & a+\hat\alpha \\ a+\hat\kappa & g \end{matrix}\middle| u+v\right) - \sum_g W\left(\begin{matrix} a & a+\hat\alpha \\ g & a+\hat\alpha+\hat\beta \end{matrix}\middle| u\right) \\ &\times W\left(\begin{matrix} a & g \\ a+\hat\kappa & a+\hat\kappa+\hat\mu \end{matrix}\middle| v\right) W\left(\begin{matrix} g & a+\hat\alpha+\hat\beta \\ a+\hat\kappa+\hat\mu & a+\hat\kappa+\hat\mu+\hat v \end{matrix}\middle| u+v\right), \end{aligned}$$

where $\hat\kappa+\hat\mu+\hat v=\hat\alpha+\hat\beta+\hat\gamma$ is assumed.

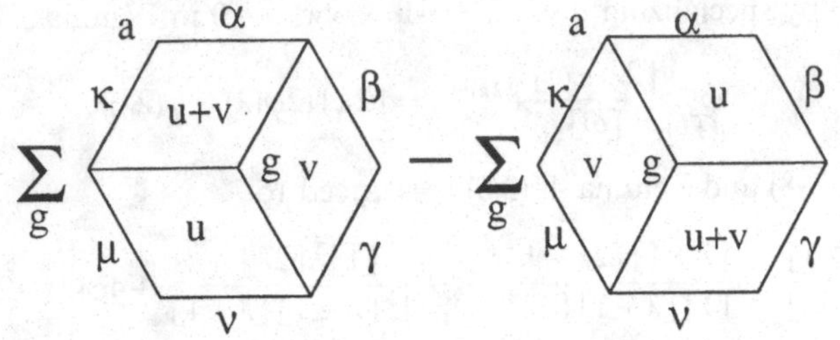

Fig. 1. The type of the STR denoted by $X(a|\kappa, \mu, \nu; \alpha, \beta, \gamma|u, v)$

Proposition 5. *Set* $f(u)=X(a|\kappa, \mu, \nu; \alpha, \beta, \gamma|u, v)$. *Then* $f(u)$ *has zeros at* $0, -v, \lambda-v$ *and* λ.

Proof. That $f(0)=f(-v)=0$ is a direct consequence of the initial condition (2.9) and the inversion relation (2.13a) proved in Proposition 2.

Thanks to the rotational symmetry (2.12), we also have

$$X(a|\kappa, \mu, \nu; \alpha, \beta, \gamma|u, v)$$
$$=-\left(\frac{G_{a+\hat{\alpha}}G_{a+\hat{\kappa}+\hat{\mu}}}{G_a G_{a+\hat{\alpha}+\hat{\beta}+\hat{\gamma}}}\right)^{1/2} X(a+\hat{\alpha}|-\alpha, \kappa, \mu; \beta, \gamma, -\nu|v, \lambda-u-v).$$

This shows $f(\lambda-v)=f(\lambda)=0$. □

From (3.1) the Boltzmann weights enjoy the following quasi-periodicity property:

$$W\left(\begin{matrix} a & b \\ d & c \end{matrix}\middle| u+L\right)=W\left(\begin{matrix} a & b \\ d & c \end{matrix}\middle| u\right),$$

$$W\left(\begin{matrix} a & b \\ d & c \end{matrix}\middle| u+L\tau\right)=e^{-2\pi i\tau-2\pi i(2u-\lambda+\xi)/L}\,W\left(\begin{matrix} a & b \\ d & c \end{matrix}\middle| u\right),$$

where $\xi=1, -a_\mu+a_\nu, 0, -a_\mu-a_\nu-1, -2a_\mu-1$ for the weights (2.5a–e) respectively. From these we have for $f(u)=X(a|\kappa, \mu, \nu; \alpha, \beta, \gamma|u, v)$,

$$f(u+L)=f(u),$$
$$f(u+L\tau)=e^{-2\pi i(2\tau+(4u+2v-2\lambda+\zeta)/L)}f(u),$$

where ζ depends only on $a_i (i=1, \ldots, n)$.

Now let us assume that $\zeta\not\equiv 0 \bmod L\mathbb{Z}+L\mathbb{Z}\tau$. Then we can show that $f(u)\equiv 0$ as follows. We have found zeros at $u=0, -v, \lambda-v$ and λ. Lemma 1 states that $f(u)$ has exactly four zeros $\bmod L\mathbb{Z}+L\mathbb{Z}\tau$, and that

$$2\lambda-2v\equiv L\left(4-\left(2\tau+\frac{2v-2\lambda+\zeta}{L}\right)\right) \bmod L\mathbb{Z}+L\mathbb{Z}\tau.$$

This contradicts the assumption $\zeta\not\equiv 0$.

From the symmetry (2.10) and (2.12), the following are equivalent:

$$X(a|\kappa, \mu, \nu; \alpha, \beta, \gamma|u, v)=0,$$
$$X(a|\alpha, \beta, \gamma; \kappa, \mu, \nu|u, v)=0,$$
$$X(a|\mu, \nu, -\gamma; -\kappa, \alpha, \beta|u, v)=0.$$

Using this, we can reduce the proof of the STR to the case $\zeta \not\equiv 0 \bmod L\mathbb{Z}+L\mathbb{Z}\tau$ except for

$$(1) \qquad X(a|\kappa, \mu, \nu; \nu, \mu, \kappa|u, v)=0 \quad (\kappa \neq \pm\mu, \pm\nu, \mu \neq \pm\nu),$$

$$(2) \qquad X(a|0, 0, 0; 0, 0, 0|u, v)=0,$$

$$(3) \qquad X(a|\mu, -\mu, \mu; \mu, -\mu, \mu|u, v)=0 \quad (\mu \neq 0),$$

$$(4) \qquad X(a|0, \mu, 0; 0, \mu, 0|u, v)=0 \quad (\mu \neq 0).$$

We now prove these exceptional cases. Equations (1) and (2) follow straightforwardly. Let us consider (3). It has the form $S(a,\mu)-S(a+\hat{\mu}, -\mu)=0$, where

$$\begin{aligned} S(a,\mu)= &\frac{[u][1+\lambda-u][v][1+\lambda-v][w][1+\lambda-w]}{[\lambda]^3[1]^3} G_{a\mu}^{-1} G_{a+\hat{\mu}-\mu} \\ &+\left(\frac{[u][2a_\mu+1-\lambda+u]}{[\lambda][2a_\mu+1]} G_{a+\hat{\mu}-\mu} + \frac{[\lambda-u][2a_\mu+1+u]}{[\lambda][2a_\mu+1]}\right) \\ &\times\left(\frac{[v][2a_\mu+1-\lambda+v]}{[\lambda][2a_\mu+1]} G_{a+\hat{\mu}-\mu} + \frac{[\lambda-v][2a_\mu+1+v]}{[\lambda][2a_\mu+1]}\right) \\ &\times\left(\frac{[\lambda-w][2a_\mu+1+w]}{[\lambda][2a_\mu+1]} G_{a\mu} + \frac{[w][2a_\mu+1-\lambda+w]}{[\lambda][2a_\mu+1]}\right) \\ &+\sum_{\kappa(\neq\pm\mu)} \frac{[u][-a_\mu+a_\kappa+\lambda-u][v][-a_\mu+a_\kappa+\lambda-v][w][-a_\mu+a_\kappa+\lambda-w]}{[\lambda]^3[-a_\mu+a_\kappa]^3} \\ &\times G_{a\mu}^{-1} G_{a+\hat{\mu}-\kappa}. \end{aligned}$$

Here $w=\lambda-u-v$.

Now consider a function

$$\begin{aligned} F(z)= &\frac{C[-a_\mu+z+\lambda-u][-a_\mu+z+\lambda-v][-a_\mu+z+\lambda-w]}{[-a_\mu+z]^3} \\ &\times \frac{\sigma h(z+1)[2z]}{[1]h(z)[2z+1]} \prod_{\kappa(\neq\pm\mu,0)} \frac{[z-a_\kappa+1]}{[z-a_\kappa]}. \end{aligned}$$

This is doubly periodic. A similar calculation as before shows ($A=2a_\mu+1$)

$$\begin{aligned} &(S(a,\mu)-S(a+\hat{\mu},-\mu))\frac{[\lambda]^3}{3} \\ &=-\frac{[2A]}{[A]^4}+\frac{[A-1][A+1]}{[1]^2[A]^2}(\varphi(A+1)+\varphi(A-1)-2\varphi(A)). \end{aligned}$$

This vanishes by virtue of Lemma 4.

The proof of (4) reduces to the identity

$$H_{a0}-H_{a+\hat{\mu}0}=\frac{[1]^2[2\lambda][2a_\mu+1]}{[a_\mu-1/2][a_\mu+1/2]^2[a_\mu+3/2]}, \tag{3.10}$$

where H_{a0} is given in (2.5g). Without loss of generality we can assume that $\mu=n$. The function H_{a0} is doubly periodic in each variable $a_1,\ldots,a_n$. The poles in a_i

within the period rectangle are $a_i = \pm 1/2$ and their residues are independent of $(a_1, \ldots, a_n)$. Therefore $H_{a0} - H_{a+\hat{n}0}$ is independent of $(a_1, \ldots, a_{n-1})$. By setting $a_i = i (i = 1, \ldots, n-1)$, (3.10) reduces to $(A = a_n)$,

$$\frac{[1]^2[1-2n][1+2A]}{[A-1/2][A+1/2]^2[A+3/2]}$$
$$= \frac{[2n-2]}{[1]}\left(\frac{[n+1+A][n-1-A]}{[n+A][n-2-A]} - \frac{[n+A][n-A]}{[n-1+A][n-1-A]}\right)$$
$$+ \sum_{B=A,A+1} (-)^{B-A}\left(-\frac{[2n-2][n+B][n-B]}{[1][n-1+B][n-1-B]}\right.$$
$$\left. + \frac{[B+3/2-2n][B+n]}{[B+1/2][B-n+1]} + \frac{[B-3/2+2n][B-n]}{[B-1/2][B+n-1]}\right).$$

It is easy to see that (the left-hand side)–(the right-hand side) is independent of A. Setting $A = n$ we can check the difference is vanishing.

Thus we have completed the proof of the STR.

4. Restricted Models

Let X_n denote A_n $(n \geqq 1)$, B_n $(n \geqq 2)$, C_n $(n \geqq 1)$ or D_n $(n \geqq 3)$. In this section we construct solvable face models whose local states are the dominant integral weights of the affine Lie algebra $X_n^{(1)}$ of a fixed level l. Our procedure is to restrict the models discussed in the previous sections as follows. We set

$$L = t(l+g),$$

where t is given in Table 2. By definition a local state a of level l is a level l dominant integral weight of $X_n^{(1)}$. It reads as follows:

$A_n^{(1)}$: $$a = (L - a_1 + a_{n+1} - 1)\Lambda_0 + \sum_{i=1}^{n} (a_i - a_{i+1} - 1)\Lambda_i,$$

where $a_i - a_j \in \mathbb{Z}$, $\sum_{i=1}^{n+1} a_i = 0$, and $L + a_{n+1} > a_1 > a_2 > \ldots > a_{n+1}$.

$B_n^{(1)}$: $$a = (L - a_1 - a_2 - 1)\Lambda_0 + \sum_{i=1}^{n-1} (a_i - a_{i+1} - 1)\Lambda_i + (2a_n - 1)\Lambda_n,$$

where either $a_i \in \mathbb{Z}$ (all i) or $a_i \in \mathbb{Z} + 1/2$ (all i), and $L > a_1 + a_2$, $a_1 > a_2 > \ldots > a_n > 0$.

$C_n^{(1)}$: $$a = (L/2 - a_1 - 1)\Lambda_0 + \sum_{i=1}^{n-1} (a_i - a_{i+1} - 1)\Lambda_i + (a_n - 1)\Lambda_n,$$

where $a_i \in \mathbb{Z}$ and $L/2 > a_1 > a_2 > \ldots > a_n > 0$.

$D_n^{(1)}$: $$a = (L - a_1 - a_2 - 1)\Lambda_0 + \sum_{i=1}^{n-1} (a_i - a_{i+1} - 1)\Lambda_i + (a_{n-1} + a_n - 1)\Lambda_n,$$

where either $a_i \in \mathbb{Z}$ (all i) or $a_i \in \mathbb{Z} + 1/2$ (all i), and $L > a_1 + a_2$, $a_1 > a_2 > \ldots > a_n$, $a_{n-1} + a_n > 0$.

Let $\bar{a}$ denote the classical part of a, and let $V(\bar{a})$ be the irreducible X_n module with the highest weight $\bar{a}$. A pair of weights (a, b) is called *admissible* if and only if a, b are local states of level l and $V(\sigma(\bar{b}))$ appears in the tensor module $V(\sigma(\bar{a}))\otimes V(\bar{\Lambda}_1)$ for any Dynkin diagram automorphism. (The representation $V(\bar{\Lambda}_1)$ is called the vector representation of X_n.) This condition is equivalent to $b-a\in\mathscr{A}$ (see Table 1) except for the case $X_n=B_n$ with $a=b$ and $a_n=1/2$; in this case (a, b) is non-admissible because $V(\bar{a})$ is not contained in $V(\bar{a})\otimes V(\bar{\Lambda}_1)$. Note also that if (a, b) is admissible the multiplicity of $V(\bar{b})$ in $V(\bar{a})\otimes V(\bar{\Lambda}_1)$ is exactly one.

We define the restricted weight of a configuration $\begin{pmatrix} a & b \\ d & c \end{pmatrix}$ round a face to be $W\left(\begin{matrix} a & b \\ d & c \end{matrix}\middle| u\right)$ as given in (2.4) and (2.5) if $(a, b), (b, c), (a, d), (d, c)$ are admissible, and to be 0 otherwise.

Theorem 7. *The restricted weights are finite and satisfy the* STR *among themselves.*

Proof. We consider the case $X_n^{(1)}=B_n^{(1)}, C_n^{(1)}, D_n^{(1)}$. The $A_n^{(1)}$ case is similar. First we show that $W\left(\begin{matrix} a & b \\ d & c \end{matrix}\middle| u\right)$ in (2.5) is finite if a is a local state and not both (a, b) and (a, d) are non-admissible. The factor $G_{a\mu}$ of (2.7) is finite because $0<|a_\mu+a_\nu|<L$ if $\mu\neq\pm\nu$ and $\mu, \nu\neq 0$. (Note also that $0<|2a_\mu|<L$ if $X_n=C_n$.)

The weight of type $\mu\overset{\mu}{\underset{\nu}{\square}}\nu$ or $\mu\overset{\nu}{\underset{\nu}{\square}}\mu$ with $\mu\neq\pm\nu$ is finite if $[a_\mu-a_\nu]\neq 0$. So it is except for the case $a_\mu=1/2$ and $(\mu, \nu)=(0, -n), (-n, 0)$, in which both (a, b) and (a, d) are non-admissible. The weight of type $0\overset{0}{\underset{0}{\square}}0$ is finite because $[a_\mu+1/2]=0$ means that $\mu=-n$, $a_n=1/2$ and that both (a, b) and (a, d) are non-admissible. The weight of type $\mu\overset{\nu}{\underset{-\mu}{\square}}-\nu\,(\mu\neq\nu)$ is finite because the factor $[a_\mu+a_\nu+1]$ is cancelled by the same factor in $\sqrt{G_{a\mu}G_{a\nu}}$. Finally the weight of type $\mu\overset{\mu}{\underset{-\mu}{\square}}-\mu\;(\mu\neq 0)$ is finite because we have two different expressions (2.5e) and (2.5f); the former is relevant if $[2a_\mu+1+2\lambda]\neq 0$ and so is the latter if $[2a_\mu+1]\neq 0$. (Note that $[a_\mu+a_\kappa+1]$ $(\kappa\neq\pm\mu)$ in (2.5g) is cancelled by $G_{a\kappa}$.)

Next we prove that if $(a, d), (d, c)$ are admissible and (a, b) is non-admissible then the weight $W\left(\begin{matrix} a & b \\ d & c \end{matrix}\middle| u\right)$ is vanishing with the exceptions mentioned below. As for the weight of type $\mu\overset{\nu}{\underset{\nu}{\square}}\mu\,(\mu\neq\pm\nu)$ this is because $[a_\mu-a_\nu-1]=0$. As for the weight of type $\mu\overset{\nu}{\underset{-\mu}{\square}}-\nu\,(\mu\neq\nu)$ this is because $G_{a\nu}=0$ unless $X_n=B_n$, $a_n=1/2$ and $\nu=0, -$n.

Since the original weights $W\left(\begin{matrix} a & b \\ d & c \end{matrix}\middle| u\right)$ satisfy the STR, the proof of the theorem is completed if it is shown that the contribution from the terms with non-

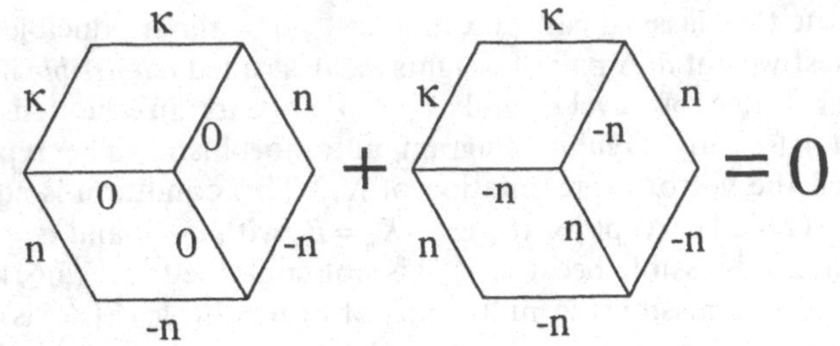

Fig. 2. The cancellation of the two unwanted terms in the STR

admissible pairs is zero. With the above consideration the last step is to check the cancellation of Fig. 2 in the case $X_n = B_n$ and $b_n = f_n = 1/2$. This is straightforward. □

5. Discussion

In this paper we have presented further elliptic solutions to the STR. They give solvable models with face interactions on a two dimensional square lattice $\mathscr{L}$. The local states of the model belong to the dual space of the Cartan subalgebra of the affine Lie algebra $X_n^{(1)} = A_n^{(1)}, B_n^{(1)}, C_n^{(1)}$ or $D_n^{(1)}$. Two local states a, b are admitted to occupy adjacent sites of $\mathscr{L}$, a being located at the left or the upper neighbor of b, if and only if $b-a$ is a weight of the vector representation of X_n. We propose to call the model the $X_n^{(1)}$ face model (or the unrestricted face model of type $X_n^{(1)}$).

The naming comes from the fact that a hierarchy of restricted models is obtained from the unrestricted one in such a way that the local states are the level l dominant integral weights of $X_n^{(1)}$.

We note that a solvable face model is not uniquely determined by the specification of the local states and the selection rule of adjacent states. In fact, the unrestricted models of type $C_n^{(1)}$ and $D_n^{(1)}$ are not distinguishable in this sense. There are three different restricted models with the same local states and the selection rule of Fig. 3; the Akutsu-Kuniba-Wadati model [24], the level 2 restricted model of type $B_n^{(1)}$ with $a_i \in \mathbb{Z} + 1/2$ and the level 2 restricted model of type $D_{2n+1}^{(1)}$ with $a_i \in \mathbb{Z}$. From our unrestricted models the whole series of restricted ones are produced simultaneously by restricting the local states a and the parameter L. We think this fact justifies our naming.

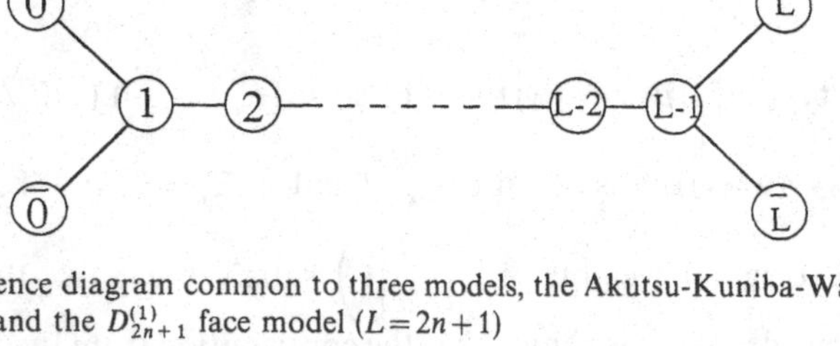

Fig. 3. The incidence diagram common to three models, the Akutsu-Kuniba-Wadati model, the $B_n^{(1)}$ face model and the $D_{2n+1}^{(1)}$ face model ($L = 2n+1$)

It will be fully supported if the following programs can be executed:

(i) The construction of unrestricted models of type $X_n^{(1)}$ (including $E_6^{(1)}$, $E_7^{(1)}$, $E_8^{(1)}$, $G_2^{(1)}$, and $F_4^{(1)}$) corresponding to arbitrary irreducible representations of X_n, and the restriction to level l. In general, the weight multiplicities of an irreducible representation of X_n may exceed one. For such a case a modification of the STR will be necessary.

(ii) The computation of the local state probabilities: The symmetric tensor of the $A_1^{(1)}$ face model [17, 18] and the vector representation of the $A_n^{(1)}$ face model [20] have been treated. The results were mysteriously related to the irreducible decomposition of the tensor product representations of $A_n^{(1)}$. A key to this link was the identification of the factors G_a (2.8) in the second inversion relation with the specialized characters of $A_n^{(1)}$. It is tantalizing to note that the G_a of the $X_n^{(1)}$ face models coincide with the specialized characters of the *dual* affine Lie algebra of $X_n^{(1)}$, i.e. $A_{2n-1}^{(2)}$ for $B_n^{(1)}$ and $D_{n+1}^{(2)}$ for $C_n^{(1)}$. ($D_n^{(1)}$ is self dual.) We hope to discuss on this point in a future publication.

Before ending, a remark is in order about the relation with solvable vertex models. In [19, 20] we have found the $A_n^{(1)}$ face models via the vertex-face correspondence [38] starting from Belavin's elliptic solution [39]. Is there a similar correspondence in the $B_n^{(1)}$, $C_n^{(1)}$, $D_n^{(1)}$ cases? In [5] a classification scheme was given of the solutions to the classical Yang-Baxter equation in terms of simple Lie algebras. Belavin's solution mentioned above corresponds to the elliptic solution of type A_n in their table. Under the non-degeneracy condition assumed by Belavin-Drinfeld no other elliptic solutions exist. Therefore, if possibly the vertex-face correspondence can be extended to the general cases the vertex counterpart must not have the classical limit in Belavin-Drinfeld's sense. We also note that the trigonometric limit of the vertex-face correspondence, between Belavin's vertex model and the $A_n^{(1)}$ face model, becomes trivial and does not give a proper correspondence between the models in the limit. For the moment we do not know any internal link between the trigonometric vertex models of [39–41] and the trigonometric limit of the face models given in this paper.

Acknowledgements. We would like to thank A. Tsuchiya, Y. Kanie, A. Kuniba, and T. Yajima for useful discussions.

References

1. Kuniba, A., Akutsu, Y., Wadati, M.: Virasoro algebra, von Neumann algebra and critical eight-vertex SOS models. J. Phys. Soc. Jpn. **55** 3285–3288 (1986)
2. Akutsu, Y., Wadati, M.: Knot invariants and critical statistical systems, J. Phys. Soc. Jpn. **56** 839–842 (1987)
3. Kohno, T.: Monodromy representations of braid groups and Yang-Baxter equations, preprint 1987. Ann. Inst. Fourier (to appear)
4. Turaev, V.G.: The Yang-Baxter equation and invariants of links, LOMI preprint E-3-87, 1987
5. Belavin, A.A., Drinfeld, V.G.: Solutions of the classical Yang-Baxter equation for simple Lie algebras. Funct. Anal. Appl. **16** 159–180 (1982)
6. Kulish, P.P., Reshetikhin, N.Yu., Sklyanin, E.K.: Yang-Baxter equation and representation theory. I. Lett. Math. Phys. **5**, 393 (1981)
7. Sklyanin, E.K.: Some algebraic structure connected with the Yang-Baxter equation. Funct. Anal. Appl. **16**, 263–270 (1982); **17**, 273–284 (1983)

8. Kulish, P.P., Reshetikhin, N.Yu.: Quantum linear problem for the sine-Gordon equation and higher representations. J. Sov. Math. **23**, 2435–2441 (1983)
9. Drinfeld, V.G.: Quantum groups, ICM 86 report
10. Jimbo, M.: A q-difference analogue of $U(\mathfrak{g})$ and the Yang-Baxter equation. Lett. Math. Phys. **10**, 63–69 (1985)
11. Jimbo, M.: A q-analogue of $U(\mathfrak{gl}(N+1))$, Hecke algebra, and the Yang-Baxter equation. Lett. Math. Phys. **11**, 247–252 (1986)
12. Baxter, R.J.: Exactly solved models in statistical mechanics. London: Academic 1982
13. Baxter, R.J.: Hard hexagons: Exact solution. J. Phys. A: Math. Gen. **13**, L61–L70 (1980)
14. Andrews, G.E., Baxter, R.J., Forrester, P.J.: Eight-vertex SOS model and generalized Rogers-Ramanujan-type identities. J. Stat. Phys. **35**, 193–266 (1984)
15. Date, E., Jimbo, M., Miwa, T., Okado, M.: Automorphic properties of local height probabilities for integrable solid-on-solid models. Phys. Rev. B **35**, 2105–2107 (1987)
16. Jimbo, M., Miwa, T., Okado, M.: Solvable lattice models with broken $\mathbb{Z}_n$ symmetry and Hecke's indefinite modular forms. Nucl. Phys. B **275** [FS 17], 517–545 (1986)
17. Date, E., Jimbo, M., Kuniba, A., Miwa, T., Okado, M.: Exactly solvable SOS models: Local height probabilities and theta function identies, Nucl. Phys. B **290** [FS 20], 231–273 (1987)
18. Date, E., Jimbo, M., Kuniba, A., Miwa, T., Okado, M.: Exactly solvable SOS models. II. Proof of the star-triangle relation and combinatorial identities. Adv. Stud. Pure Math. **16** (to appear)
19. Jimbo, M., Miwa, T., Okado, M.: Solvable lattice models whose states are dominant integral weights of $A^{(1)}_{n-1}$, Lett. Math. Phys. **14**, 123–131 (1987)
20. Jimbo, M., Miwa, T., Okado, M.: Local state probabilities of solvable lattice models: An $A^{(1)}_{n-1}$ family, preprint RIMS-594, Kyoto Univ. (1987). Nucl. Phys. (to appear)
21. Baxter, R.J., Andrews, G.E.: Lattice gas generalization of the hard hexagon model. I. Star-triangle relation and the local densities. J. Stat. Phys. **44**, 249–271 (1986)
22. Kuniba, A., Akutsu, Y., Wadati, M.: Exactly solvable IRF models. I. A three state model, J. Phys. Soc. Jpn. **55**, 1092–1101 (1986)
23. Akutsu, Y., Kuniba, A., Wadati, M.: Exactly solvable IRF models. II. S_N generalizations. J. Phys. Soc. Jpn. **55**, 1466–1474 (1986)
24. Akutsu, Y., Kuniba, A., Wadati, M.: Exactly solvable IRF models. III. A new hierarchy of solvable models. J. Phys. Soc. Jpn. **55**, 1880–1886 (1986)
25. Date, E., Jimbo, M., Miwa, T., Okado, M.: Fusion of the eight-vertex SOS model, Lett. Math. Phys. **12**, 209–215 (1986)
26. Jimbo, M., Miwa, T., Okado, M.: An $A^{(1)}_{n-1}$ family of solvable lattice models. Mod. Phys. Lett. B **1**, 73–79 (1987)
27. Jimbo, M., Miwa, T., Okado, M.: Symmetric tensors of the $A^{(1)}_{n-1}$ family, preprint RIMS 592, Kyoto University (1987). Algebraic Analysis (Festschrift for M. Sato's 60th birthday) Academic Press 1988
28. Pasquier, V.: Exact solubility of the D_n series. J. Phys. A: Math. Gen. **20**, L217–L220 (1987)
29. Kashiwara, M., Miwa, T.: A class of elliptic solutions to the star-triangle relation. Nucl. Phys. B **275** [FS17], 121–134 (1986)
30. Kuniba, A., Yajima, T.: Local state probabilities for solvable RSOS models: A_n, D_n, $D^{(1)}_n$, and $A^{(1)}_n$, preprint. Tokyo University 1987
31. Wenzl, H.: Representations of Hecke algebras and subfactors. Thesis, University of Pennsylvania 1985
32. Tsuchiya, A., Kanie, Y.: Vertex operators on conformal field theory on $\mathbb{P}^1$ and monodromy representations of braid groups. Adv. Stud. Pure Math. **16** (to appear)
33. Pasquier, V.: Continuum limit of lattice models built on quantum groups. Saclay preprint SPHT/87-125
34. Kac, V.G., Peterson, D.H.: Infinite-dimensional Lie algebras, theta functions and modular forms. Adv. Math. **53**, 125–264 (1984)
35. Bourbaki, N.: Groupes et Algebres de Lie, Chaps. 4–6. Paris: Hermann 1968
36. Cherednik, I.V.: Some finite dimensional representations of generalized Sklyanin algebra. Funct. Anal. Appl. **19**, 77–79 (1985)

37. Richey, M.P., Tracy, C.A.: $\mathbb{Z}_n$ Baxter model: symmetries and the Belavin parametrization. J. Stat. Phys. **42**, 311–348 (1986)
38. Baxter, R.J.: Eight-vertex model in lattice statistics and one-dimensional anisotropic Heisenberg chain. II. Equivalence to a generalized ice-type lattice model, ibid. **76**, 25–47 (1973)
39. Belavin, A.A.: Dynamical symmetry of integrable quantum systems. Nucl. Phys. B **180** [FS 2], 189–200 (1981)
40. Bazhanov, V.V.: Trigonometric solutions of the triangle equation and classical Lie algebras. Phys. Lett. **159** B, 321–324 (1985)
41. Jimbo, M.: Quantum R matrix for the generalized Toda system. Commun. Math. Phys. **102**, 537–547 (1986)
42. Jimbo, M.: Quantum R matrix related to the generalized Toda system: an algebraic approach. In: Proceedings of the Symposium on field theory, quantum gravity and strings. Meudon and Paris VI, 1984/85. Lecture Notes in Physics, Vol. 246, pp. 335–361. Berlin, Heidelberg, New York: Springer 1986

Communicated by H. Araki

Received December 8, 1987

Commun. Math. Phys. 118, 355–364 (1988)

Communications in Mathematical Physics

Etiology of IRF Models

V. Pasquier

Service de Physique Théorique, Institut de Recherche Fondamentale, CEA, CEN-Saclay, F-91191 Gif-sur-Yvette Cedex, France

Abstract. We show that a class of $2D$ statistical mechanics models known as IRF models can be viewed as a subalgebra of the operator algebra of vertex models. Extending the Wigner calculus to quantum groups, we obtain an explicit intertwiner between two representations of this subalgebra.

Two major progresses have recently been made in the understanding of two dimensional lattice models. The first is the classification of rational and trigonometric solutions of the Yang-Baxter equations for a class of models known as vertex models [1–5]. The second is the discovery of many representatives of another class known as IRF (interacting round a face) models and their study in connection with conformal field theories [6–11]. In the appendix of [11] it was shown that both classes correspond to different representations of the same algebra, and it is the aim of this letter to complete the identification by building an explicit intertwiner between them. The method we use is an application to the quantum group case of Ocneanu's cell technique [12].

Quantum Wigner Calculus. Consider the associative algebra $\hat{U}(SU(2))$ [4, 5] generated by the symbols $q^{h/2}$, J_+, J_- under the following relations:

$$q^{h/2}q^{-h/2}=q^{-h/2}q^{h/2}=1\ , \quad q^{h/2}J_+q^{-h/2}=qJ_+\ , \quad q^{h/2}J_-q^{-h/2}=q^{-1}J_-\ ,$$
$$[J_+J_-]=\frac{q^h-q^{-h}}{q-q^{-1}}\ . \tag{1}$$

We denote by $\Delta^{(N)}$ the coproduct homomorphism $\Delta^{(N)}\hat{U}\to\hat{U}\otimes^N$ (N fold tensor product)

$$\Delta^{(N)}(q^{h/2})=q^{h/2}\otimes q^{h/2}\ldots\otimes q^{h/2}\ ,$$
$$\Delta^{(N)}(J_\pm)=\sum_{v=1}^{N}q^{h/2}\otimes\ldots\otimes q^{h/2}\otimes \overset{v}{J_\pm}\otimes q^{-h/2}\ldots\otimes q^{-h/2}\ . \tag{2}$$

In what follows, unless we specify it, q is not a root of unity. For $j \in \frac{1}{2}\mathcal{N}$, V_j denotes an irreducible $\hat{U}$ module of spin j and $|jm\rangle$ its canonical bases. Let $V_{j_1}\, V_{j_2}\, V_J$ be three such modules, an intertwiner between $V_{j_1} \otimes V_{j_2}$ and V_J is given by the (q) Wigner coefficients

$$\langle (j_1 j_2) JM | j_1 m_1 j_2 m_2 \rangle =$$ 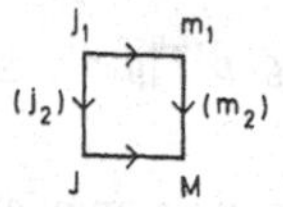

. (3)

Following the arrows from the upper left to the down right corner, we describe the two states $|(j_1 j_2)JM\rangle$ and $|j_1 m_1\rangle \otimes |j_2 m_2\rangle$. The Wigner coefficients can be deduced from the recursion relations:

$$\begin{aligned} &((J-M+1)(J+M))^{1/2}\langle JM-1|j_1 m_1 j_2 m_2\rangle \\ &\quad = q^{-m_2}((j_1-m_1)(j_1+m_1+1))^{1/2}\langle JM|j_1 m_1+1 j_2 m_2\rangle \\ &\quad + q^{m_1}((j_2-m_2)(j_2+m_2+1))^{1/2}\langle JM|j_1 m_1 j_2 m_2+1\rangle , \end{aligned}$$

where

$$(n) = \frac{q^n - q^{-n}}{q - q^{-1}} . \tag{4}$$

We list them for $j_2 = \frac{1}{2}, 1$ Table 1.

Table 1. Wigner Coefficients[a]

$(jm\|j_1 m_1 \frac{1}{2} \mu)$		
$j=$	$\mu=\frac{1}{2}$	$\mu=-\frac{1}{2}$
$j_1+\frac{1}{2}$	$q^{(m-j_1-\frac{1}{2})/2}\left(\frac{(j_1+m+\frac{1}{2})}{(2j_1+1)}\right)^{1/2}$	$q^{(m+j_1+\frac{1}{2})/2}\left(\frac{(j_1-m+\frac{1}{2})}{(2j_1+1)}\right)^{1/2}$
$j_1-\frac{1}{2}$	$-q^{(m+j_1+\frac{1}{2})/2}\left(\frac{(j_1-m+\frac{1}{2})}{(2j_1+1)}\right)^{1/2}$	$q^{(m-j_1-\frac{1}{2})/2}\left(\frac{(j_1+m+\frac{1}{2})}{(2j_1+1)}\right)^{1/2}$

$(jm\|j_1 m_1 1 \mu)$			
$j=$	$\mu=1$	$\mu=0$	$\mu=-1$
j_1+1	$q^{m-j_1-1}\left(\frac{(j_1+m)(j_1+m+1)}{(2j_1+1)(2j_1+2)}\right)^{1/2}$	$q^m\left(\frac{(2)(j_1-m+1)(j_1+m+1)}{(2j_1+1)(2j_1+2)}\right)^{1/2}$	$q^{m+j_1+1}\left(\frac{(j_1-m)(j_1-m+1)}{(2j_1+1)(2j_1+2)}\right)^{1/2}$
j_1	$-q^m\left(\frac{(2)(j_1+m)(j_1-m+1)}{(2j_1)(2j_1+2)}\right)^{1/2}$	$\frac{q^{m-j_1-1}(j_1+m)-q^{j_1+m+1}(j_1-m)}{((2j_1)(2j_1+2))^{1/2}}$	$q^m\left(\frac{(2)(j_1-m)(j_1+m+1)}{(2j_1)(2j_1+2)}\right)^{1/2}$
j_1-1	$q^{m+j_1}\left(\frac{(j_1-m)(j_1-m+1)}{(2j_1)(2j_1+1)}\right)^{1/2}$	$-q^m\left(\frac{(2)(j_1-m)(j_1+m)}{(2j_1)(2j_1+1)}\right)^{1/2}$	$q^{m-j_1}\left(\frac{(j_1+m+1)(j_1+m)}{(2j_1)(2j_1+1)}\right)^{1/2}$

[a] The exponents of q are ordinary [not (q)] numbers

Let $V_j V_{g_1} V_{g_2}$ be $\hat{U}$-modules and T a homomorphism of $V_{g_1} \otimes V_{g_2}$ in $V_{g_2} \otimes V_{g_1}$ such that $\Delta^{(2)}(X)T = T\Delta^{(2)}(X)$ for all $X \in \hat{U}$. Matrix elements of T can be described in the bases $\langle g_2\alpha| \otimes \langle g_1\beta|, |g_1\alpha'\rangle \otimes |g_2\beta'\rangle$:

$$\langle g_2\alpha| \otimes \langle g_1\beta|T|g_1\alpha'\rangle \otimes |g_2\beta'\rangle =$$ 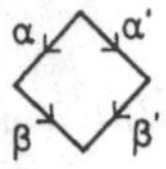 $$= \sigma_{\alpha\beta,\alpha'\beta'} \ . \tag{5}$$

Alternatively, we can recouple angular momentums:

$$V_j \otimes V_{g_2} = \overset{\oplus}{j_1} V_{j_1} \ , \quad V_j \otimes V_{g_1} = \overset{\oplus}{j_1'} V_{j_1'} \ ,$$
$$V_{j_1} \otimes V_{g_1} = \overset{\oplus}{J} V_J \ , \quad V_{j_1'} \otimes V_{g_2} = \overset{\oplus}{J'} V_{J'} \ ,$$

and describe matrix elements of T as:

$$\langle (jg_2)j_1 g_1 JM|T|(jg_1)j_1' g_2 J'M'\rangle = \delta_{JJ'}\delta_{MM'}\langle (jg_2)j_1 g_1 J\|T\|(jg_1)j_1' g_2 J\rangle \ . \tag{6}$$

The reduced matrix element being denoted

$$\sigma^{(jJ)}_{j_1 j_1'} =$$ 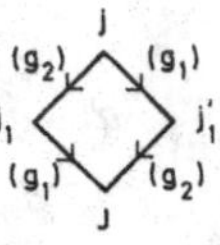 .

The two representations are related by Wigner coefficients through the following equations:

$$\sum_{j_1' m_1'}$$ 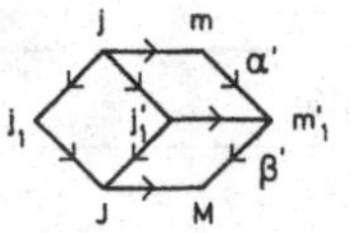 $$= \sum_{\substack{\alpha\beta \\ m_1}}$$ 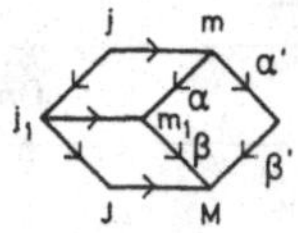 .

$$\sum_{j_1' m_1'} \sigma^{(jJ)}_{j_1 j_1'} \langle j_1' m_1' | jmg_1\alpha'\rangle \langle JM|j_1' m_1' g_2\beta'\rangle$$
$$= \sum_{\alpha\beta m_1} \sigma_{\alpha\beta,\alpha'\beta'} \langle j_1 m_1 | jmg_2\alpha\rangle \langle JM|j_1 m_1 g_1\beta\rangle \ . \tag{7}$$

We call them respectively vertex and path representations of T.

Path Representation of the R Matrix. Let V_g be a $\hat{U}$ module. In End$(V_g \otimes V_g)$, the vertex R matrix obeying the Yang Baxter equation [3–5] is characterized by the following equations:

$$[R(x), \Delta^{(2)}(X)] = 0 \quad \forall X \in \hat{U} \ ,$$
$$R(x)(xj_- \otimes q^{h/2} + q^{-h/2} \otimes j_-) = (j_- \otimes q^{h/2} + xq^{-h/2} \otimes j_-)R(x) \ . \tag{8}$$

The first equation expresses that $R(x)$ is in the commutant of $\Delta^{(2)}(\hat{U})$. Hence it can be written

$$R(x) = \sum_j \varrho_j(x) P^{(j)} \ . \tag{9}$$

With $P^{(j)}$ projectors onto irreducible components V_j of $V_g \otimes V_g$. The second equation determines $\varrho_j(x)$ and we quote the result from [4]:

$$\frac{\varrho_{j-1}(x)}{\varrho_j(x)} = \frac{x-q^{2j}}{1-xq^{2j}} \,. \tag{10}$$

The vertex matrix elements of $P^{(j)}$ are by definition

$$\sigma_{\alpha\beta,\alpha'\beta'} = \quad$$ 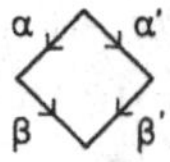 $$= \sum_m \langle g\alpha g\beta | jm\rangle \langle jm | g\alpha' g\beta'\rangle \tag{11}$$

due to the charge conservations, σ breaks into block matrices according to the total charge $Q=\alpha+\beta=\alpha'+\beta'$. The path representation of $P^{(j)}$ follows from (7).

Examples. (All matrices are symmetric)

$g=\frac{1}{2}$: (6 vertex model)

There are two projectors $P^{(0)}, P^{(1)}=1-P^{(0)}$, the matrix elements of $P^{(0)}$ are:

a) Vertex:

$$Q=1\,, \quad \sigma=0\,,$$

$$Q=0\,, \quad \begin{pmatrix} \sigma_{1/2-1/2,\,1/2-1/2} & \sigma_{1/2-1/2,\,-1/21/2} \\ \sigma_{-1/21/2,\,1/2-1/2} & \sigma_{-1/21/2,\,-1/21/2} \end{pmatrix} = \frac{1}{(2)} \begin{pmatrix} q & -1 \\ -1 & q^{-1} \end{pmatrix} .$$

b) Path:

$$J=j\pm 1\,, \quad \sigma=0$$

$$J=j\,, \quad \begin{pmatrix} \sigma_{j-\frac{1}{2},j-\frac{1}{2}} & \sigma_{j-\frac{1}{2},j+\frac{1}{2}} \\ \sigma_{j+\frac{1}{2},j-\frac{1}{2}} & \sigma_{j+\frac{1}{2},j+\frac{1}{2}} \end{pmatrix} = \frac{1}{(2j+1)(2)} \begin{pmatrix} (2j) & * \\ \sqrt{(2j)(2j+2)} & (2j+2) \end{pmatrix},$$

$$R(x)=(1-xq^2)P^{(1)}+(x-q^2)P^{(0)}\,. \tag{12}$$

Both expressions of $P^{(0)}$ are known representations of the Temperley and Lieb-Jones algebra [13–14].

$g=1$.

There are 3 projectors $P^{(0)}$, $P^{(1)}$, $P^{(2)}=1-P^{(0)}-P^{(1)}$. The matrix elements of $P^{(0)}$ and $P^{(1)}$ are:

a) Vertex:

$P^{(0)}$:

$$Q=1,2\,, \quad \sigma=0\,,$$

$$Q=0\,, \quad \begin{pmatrix} \sigma_{1-1,1-1} & \sigma_{1-1,00} & \sigma_{1-1,-11} \\ \sigma_{00,1-1} & \sigma_{00,00} & \sigma_{00,-11} \\ \sigma_{-11,1-1} & \sigma_{-11,00} & \sigma_{-11,-11} \end{pmatrix} = \frac{1}{(3)} \begin{pmatrix} q^{-2} & -q^{-1} & 1 \\ -q^{-1} & 1 & -q \\ 1 & -q & q^2 \end{pmatrix} .$$

$P^{(1)}$:

$$Q=1\ ,\quad \begin{pmatrix}\sigma_{10,10} & \sigma_{10,01}\\ \sigma_{01,10} & \sigma_{01,01}\end{pmatrix}=\frac{(2)}{(4)}\begin{pmatrix}q^{-2} & -1\\ -1 & q^{2}\end{pmatrix},$$

$$Q=-1\ ,\quad \begin{pmatrix}\sigma_{0-1,0-1} & \sigma_{0-1,-10}\\ \sigma_{-10,0-1} & \sigma_{-10,-10}\end{pmatrix}=\frac{(2)}{(4)}\begin{pmatrix}q^{-2} & -1\\ -1 & q^{2}\end{pmatrix},$$

$$Q=0\ ,\quad \begin{pmatrix}\sigma_{1-1,1-1} & \sigma_{1-1,00} & \sigma_{1-1,-11}\\ \sigma_{00,1-1} & \sigma_{00,00} & \varrho_{00,-11}\\ \sigma_{-11,1-1} & \sigma_{-11,00} & \sigma_{-11,-11}\end{pmatrix}$$

$$=\frac{(2)}{(4)}\begin{pmatrix}1 & q^{-1}-q & -1\\ q^{-1}-q & (q^{-1}-q)^2 & q-q^{-1}\\ -1 & q-q^{-1} & 1\end{pmatrix}.$$

b) Path:

$P^{(0)}$:

$$J=j\pm 1\ ,\quad \sigma=0\ ,$$

$$J=j\ ,\quad \sigma=\begin{pmatrix}\sigma_{j-1,j-1} & \sigma_{j-1,j} & \sigma_{j-1,j+1}\\ \sigma_{j,j-1} & \sigma_{j,j} & \sigma_{j,j+1}\\ \sigma_{j+1,j-1} & \sigma_{j+1,j} & \sigma_{j+1,j+1}\end{pmatrix}$$

$$=\frac{1}{(3)(2j+1)}\begin{pmatrix}(2j-1) & * & *\\ \sqrt{(2j+1)(2j-1)} & (2j+1) & *\\ \sqrt{(2j+3)(2j-1)} & \sqrt{(2j+3)(2j+1)} & (2j+3)\end{pmatrix}.$$

$P^{(1)}$:

$$J=j+1\begin{pmatrix}\sigma_{j,j} & \sigma_{j,j+1}\\ \sigma_{j+1,j} & \sigma_{j+1,\sigma+1}\end{pmatrix}=\frac{(2)}{(4)(2j+2)}\begin{pmatrix}(2j) & *\\ \sqrt{(2j)(2j+4)} & (2j+4)\end{pmatrix},$$

$$J=j-1\begin{pmatrix}\sigma_{j,j} & \sigma_{j,j-1}\\ \sigma_{j-1,j} & \sigma_{j-1,j-1}\end{pmatrix}=\frac{(2)}{(4)(2j)}\begin{pmatrix}(2j+2) & *\\ -\sqrt{(2j+2)(2j-2)} & (2j-2)\end{pmatrix},$$

$$J=j\begin{pmatrix}\sigma_{j-1,j-1} & \sigma_{j-1,j} & \sigma_{j-1,j+1}\\ \sigma_{j,j-1} & \sigma_{j,j} & \sigma_{j,j+1}\\ \sigma_{j+1,j-1} & \sigma_{j+1,j} & \sigma_{j+1,j+1}\end{pmatrix}$$

$$=\frac{(2)}{(4)}\begin{pmatrix}\left(1-\frac{(2)}{(2j)(2j+1)}\right) & * & *\\ -\left(\frac{(2j-1)}{(2j+1)}\right)^{1/2}\frac{q^{2j+1}+q^{-2j-1}}{(2j)}, & \frac{2+q^{4j+2}+q^{-4j-2}}{(2j)(2j+2)} & *\\ -\frac{((2j-1)(2j+3))^{1/2}}{(2j+1)}, & \left(\frac{(2j+3)}{(2j+1)}\right)^{1/2}\frac{q^{2j+1}+q^{-2j-1}}{(2j+2)}, & \left(1-\frac{(2)}{(2j+1)(2j+2)}\right)\end{pmatrix},$$

$$R(x)=(1-xq^4)(1-xq^2)P^{(2)}+(x-q^4)(1-xq^2)P^{(1)}+(x-q^2)(x-q^4)P^{(0)} \tag{13}$$

Path Algebra. Let us fix V_g an irreducible $\hat{U}$ module. We define paths $(j)_{(g)}=(j_0=0, j_1, \ldots, j_N)$ by a sequence $j_k \in \frac{1}{2}\mathcal{N}$ such that $V_{j_{k+1}} \subset V_{j_k} \otimes V_g$. We consider the matrix algebra generated by matrix units $((j)_{(g)}, (j')_{(g)})$, $j_N = j'_N$ under the following multiplication law:

$$((j),(j'))((k),(k'))=\delta_{(j'),(k)}((j),(k')) \ . \tag{14}$$

In a similar way, we define a base of matrix units of $\mathrm{End}(V_g \otimes^N)$ by $((\mu),(\mu'))$; $(\mu)=(\mu_1, \mu_2, \ldots, \mu_N)$ $\mu_k \in \{-g, -g+1, \ldots, g\}$. The first algebra is isomorphic to the commutant $\Delta^{(N)}(\hat{U})'$ of $\Delta^{(N)}(\hat{U})$ in $\mathrm{End}(V_g \otimes^N)$ and the inclusion (i) in $\mathrm{End}(V_g \otimes^N)$ can be described with Wigner coefficients:

$$i((j),(j'))=\sum_{((\mu),(\mu'))} C^{((\mu),(\mu'))}_{((j),(j'))}((\mu),(\mu')) \ ,$$

with

$$C^{((\mu),(\mu'))}_{((j),(j'))}=\sum_{(m)(m')}$$ 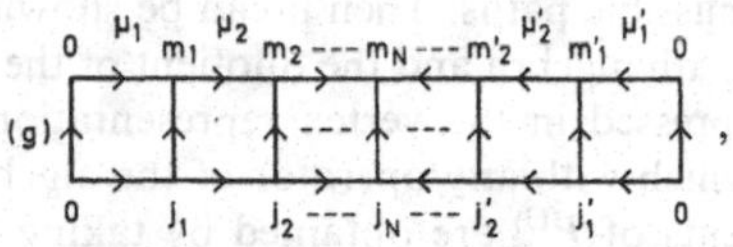 ,

where

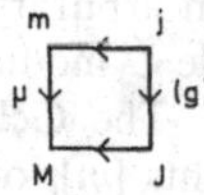

$$= (jmg\mu|(jg)JM) \ . \tag{15}$$

The fact that (i) defines a homomorphism requires the orthogonality relations:

$$\sum_{m,\mu}(JM|jmg\mu)(jmg\mu|J'M')=\delta_{JJ'}\delta_{MM'}=\sum_{m,\mu}$$ 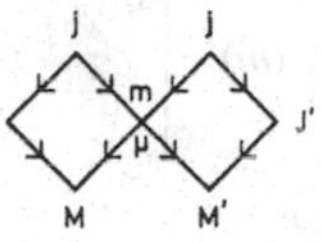 . (16)

A trace on $\Delta^{(N)}(U)'$ is defined [14, 15] by

$$\begin{aligned} \mathrm{tr}((\mu),(\mu')) &= \delta_{(\mu),(\mu')} \prod_{k=1}^{N} q^{-2\mu_k} \ , \\ \mathrm{tr}((j),(j')) &= \delta_{(j),(j')}(2j_N+1) \ . \end{aligned} \tag{17}$$

Both expressions define the same trace due to the identity

$$(2j+1)\sum_{M,\mu}(JM|jmg\mu)(j'm'g\mu|JM)q^{-2\mu}=(2j+1)\sum_{M,\mu}$$ 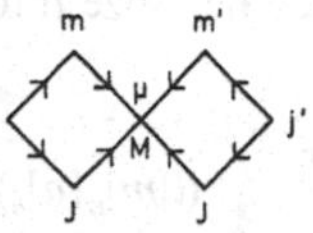

$$=(2J+1)\delta_{jj'}\delta_{mm'} \ . \tag{18}$$

The proofs can easily be done using the graphical representations.

Restricted Algebras. We define the (q) dimension of V_j by $\dim V_j = (2j+1)$. When $q = e^{\pi i/L}$, $L \in \mathcal{N}$ we restrict to values of j such that $(\dim V_j) > 0$. This imposes $j \leqq j_{\max} = \frac{L}{2} - 1$. In this case we define paths by the connection matrices $A^{(g)}_{jj'}$, $0 \leqq j, j', g \leqq j_{\max}$ determined by the recursion relations:

$$\begin{aligned} &A^{(0)} = 1\,, \\ &A^{(1/2)} = 0 \quad \text{if} \quad |j - j'| \neq \tfrac{1}{2}\,, \\ &1 \quad \text{if} \quad |j - j'| = \tfrac{1}{2}\,, \\ &A^{(g-1)} + A^{(g)} = A^{(g-\frac{1}{2})} A^{(\frac{1}{2})}\,. \end{aligned} \tag{19}$$

A path $(j)_g$ is admissible if $0 \leqq j_k \leqq j_{\max}$ and $A^{(g)}_{j_k, j_{k+1}} = 1$. The path algebra is a in (14) with admissible paths. Then it can be shown that (i) defines an isomorphism between the path algebra and the quotient of the algebra generated by the projectors $P^{(j)}$ (expressed in the vertex representation) by the ideal of operators the product of which with any operator of the algebra is traceless. In particular, the matrix elements of $P^{(j)}$ are obtained by taking the generic expression (12), (13) (setting $q = e^{i\pi/L}$) and restricting them to admissible paths.

GL(n). What precedes can be extended to an arbitrary Lie group [9, 11]; let us consider $\hat{U}(gl(n))$ defined in [3, 5]. Irreducible $\hat{U}$ modules are characterised by a Young pattern $[m] = (m_{1n} \geqq m_{2n} \cdots \geqq m_{nn} \geqq 0)$. The Gelfand Zetlin bases of $[m]$ consists of states (m) which are highest weights $[m]_k$ of $\hat{U}(gl(k))$ for the natural inclusion $gl(1) \subset \ldots \subset gl(n)$. They are denoted by the symbol [16]

$$(m) = \begin{bmatrix} m_{1n} & & m_{2n} & \ldots & & m_{nn} \\ & m_{1,n-1} & & \ldots & m_{n-1,n-1} & \\ & & & m_{11} & & \end{bmatrix} = \begin{bmatrix} [m] \\ [m]_{n-1} \\ [m]_1 \end{bmatrix},$$

$$m_{ij} \geqq m_{i,j-1} \geqq m_{i+1,j}\,. \tag{20}$$

Using similar notations as for $\hat{U}(SU(2))$, we denote the Wigner coefficients for $V_{[M]} \subset V_{[m]} \otimes V_{[g]}$ by

$$(([m][g])[M][M]_k | [m][m]_k [g][g]_k) =$$ 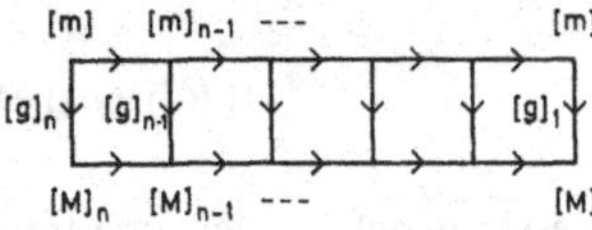 . (21)

They factorize into the product of reduced coefficients

$$(([m]_k [g]_k)[M]_k [M]_{k-1} | [m]_k [m]_{k-1} [g]_k [g]_{k-1}) =$$ 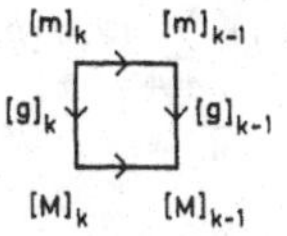 . (22)

If $[g] = (1, 0, 0, \ldots)$ is the fundamental representation, $[g]_k = (\mu_k, 0, \ldots) = \mu_k$ with $\mu_n = 1 \geqq \mu_{n-1} \cdots \geqq \mu_1 \geqq 0$. The corresponding reduced coefficients are listed in

Table 2. Reduced Wigner Coefficients of $\hat{U}(gl(n))$[a]

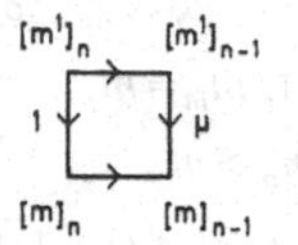

$$m_{j,n} = m'_{j,n} + \delta_{i,j}$$
$$m_{j,n-1} = m'_{j,n-1} \quad \text{if} \quad \mu = 0$$
$$= m'_{j,n-1} + \delta_{j,k} \quad \text{if} \quad \mu = 1$$

μ	
0	$q^{1/2\left(\sum_{j=1, j\neq i}^{n} l'_{jn} - \sum_{j=1}^{n-1} l_{j,n-1} + n - 1\right)} \left[\dfrac{\prod_{j=1}^{n-1} (l_{j,n-1} - l'_{in} - 1)}{\prod_{j=1, j\neq 1}^{n} (l'_{jn} - l'_{in})} \right]^{1/2}$
1	$q^{1/2(l'_{in} - l_{k,n-1} + 1)} S(k-i) \left[\prod_{j=1, j\neq k}^{n-1} \dfrac{(l'_{in} - l_{j,n-1} + 1)}{(l_{k,n-1} - l_{j,n-1})} \prod_{j=1, j\neq i}^{n} \dfrac{(l'_{jn} - l_{k,n-1} + 1)}{(l'_{jn} - l'_{in})} \right]^{1/2}$

$$l_{ij} = m_{ij} - i$$
$$S(i) = \textit{sign of } i$$

[a] The exponents of q are ordinary [not (q)] numbers

Table 2. Since $V^{(1,0,\ldots)} \otimes V^{(1,0,\ldots)} = V^{(2,0,\ldots)} \oplus V^{(1,1,0,\ldots)}$. We can build two projectors $P^{(1,1,0,\ldots)}$, $P^{(2,0,\ldots)} = 1 - P^{(1,1,0,\ldots)}$. Let us denote by α the state $\mu_k = 1$ $k \geq n - \alpha$, $\mu_k = 0$ $k < n - \alpha$. Then, the vertex matrix elements of $P^{(1,1,\ldots)}$ are denoted by $\sigma_{\alpha\beta,\alpha'\beta'}$. Due to the charge conservations, $\sigma_{\alpha\beta,\alpha'\beta'} = 0$ unless $\{\alpha, \beta\} = \{\alpha', \beta'\}$, and we have for $\alpha > \beta$:

$$\begin{pmatrix} \sigma_{\alpha\beta,\alpha\beta} & \sigma_{\alpha\beta,\beta\alpha} \\ \sigma_{\beta\alpha,\alpha\beta} & \sigma_{\beta\alpha,\beta\alpha} \end{pmatrix} = \frac{1}{(2)} \begin{pmatrix} q & -1 \\ -1 & q^{-1} \end{pmatrix} . \tag{23}$$

From the $gl(n)$ generalisation of (7) we deduce the path matrix elements

$$\sigma_{[m^1][m^{1'}]} =$$ 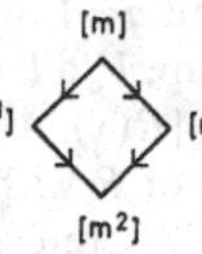 $$= \sigma^{([m][m^2])}_{[m^1],[m^{1'}]} . \tag{24}$$

We set

$$i_1 = [m^1] = (m_{kn} + \delta_{ki_1}) ,$$

$$i'_1 = [m^{1'}] = (m_{kn} + \delta_{ki'_1}) ,$$

$$(i,j) = [m^2] = (m_{kn} + \delta_{ki} + \delta_{kj}) , \tag{25}$$

then $i_1, i_1' \in \{i,j\}$ and:

$$i=j, \qquad \sigma_{i,i}=0$$
$$j=i+1,\ m_{in}=m_{jn}:\ \sigma_{i,i}=1$$
$$i<j,\ m_{in}>m_{jn}:$$
$$\begin{pmatrix}\sigma_{ii} & \sigma_{ij}\\ \sigma_{ji} & \sigma_{jj}\end{pmatrix}=\frac{1}{(2)(l_i-l_j)}\begin{pmatrix}(l_i-l_j+1) & *\\ \sqrt{(l_i-l_j+1)(l_i-l_j-1)} & (l_i-l_j-1)\end{pmatrix} \tag{26}$$

with $l_i=m_{in}-i$.

Both expressions correspond to known representations of the Hecke algebra [5, 17]. The R matrix:

$$R(x)=(1-xq^2)P^{(2,0,\dots)}+(x-q^2)P^{(1,1,0,\dots)} \tag{27}$$

is obtained in [18] in the vertex representation and studied in [8, 11] in the path representation. For $V^{[g]}$ an irreducible $\hat{U}$ module, the path algebra is defined as for $SU(2)$. The Wigner coefficients intertwine the path algebra and $\Delta^{(N)}(U)'$ in $\mathrm{End}\,(V_g)\otimes^N$. The expression of the trace in the path algebra is:

$$\mathrm{tr}\,(([m]),([m'])) = \delta_{[m],[m']}\prod_{i,j}\frac{(l_i-l_{i+j})}{(j)}\ ,$$

where

$$([m])=([m]_1,\dots,[m]_N) \quad \text{denotes a path}\ . \tag{28}$$

When $q=e^{i\pi/L}$, we must restrict $[m]$ to values such that

$$\dim V_{[m]}=\prod_{i,j}\frac{(l_i-l_{i+j})}{(j)}>0\ . \tag{29}$$

This imposes $m_{1n}-m_{nn}\leq L-n$ and the connection matrices defining admissible paths are determined by the recursion relations:

$$A^{(0)}=1\ ,$$
$$A^{(1,0,\dots)}_{[m][m']}=\begin{array}{l}\text{the matrix of the generic model restricted to } [m][m']\\ \text{such that } \dim V_{[m]},\ \dim V_{[m']}>0\ ,\end{array} \tag{30}$$
$$A^{[g]}A^{[1,0,\dots,0]}=\sum_{[g']}A^{[1,0,\dots]}_{[g][g']}A^{[g']}\ .$$

Discrete Groups. We consider the limit $q=1$, where the R matrix degenerates to its rational limit. Let Γ be a discrete group, V_g a Γ module and $\Delta^{(2)}$ the homomorphism of Γ in $\mathrm{End}\,(V_g\otimes V_g)$. Assume that $[R(x),\Delta^2(\gamma)]=0$ for all $\gamma\in\Gamma$. The preceding method yields a path representation of R with intertwiner given by the Γ Wigner coefficients. The case where Γ is a discrete subgroup of $SU(2)$ and V_g its two dimensional module is considered in [19–21]. The connection matrix of the path algebra: A^g is the incidence matrix of an extended Coxeter diagram and the components of its Perron-Frobenius vector are the dimensions of the representations of Γ [22]. Other models associated with ordinary Coxeter diagrams have been discussed [10]. It is natural to ask whether they correspond to a Hopf algebra [3].

Then each vertex of the Coxeter diagram should be associated to a $\hat{U}$ module V^a of (q) dimension S^a with S^a the component of the Perron-Frobenius vector normalised so that its smallest component $S^0=1$. One must therefore define commuting matrices A^a with positive integer coefficients characterising the paths built on V^a and determined by the identities (expressing the associativity of the tensor product):

$$A^{(0)}=1\,,$$

$$A^{(1)}=\text{the incidence matrix of the Coxeter diagram}\,,$$

$$A^{(a)}A^{(1)}=\sum_{(b)} A^{(1)}_{ab}A^{(b)}\,. \tag{31}$$

Remarkably, this is only possible for A_n, D_{2n}, E_6, E_8 as in Ocneanu's classification of subfactors [12]. We hope to be able to discuss this point further separately.

Acknowledgements. I wish to thank A. Ocneanu for explaining me his cell technique, G. Elliot and D. Evans for helpful discussions, P. Ginsparg for bringing the discrete group case to my attention.

References

1. Kulish, P.P., Reshetikhin, N.Yu., Sklyanin, E.: Lett. Math. Phys. **5**, 393 (1981)
2. Sklyanin, E.K.: Funct. Anal. Appl. **16**, 263 (1982). **17**, 273 (1983)
3. Drinfeld, V.G.: ICM86 report
4. Jimbo, M.: Lett. Math. Phys. **10**, 63 (1985)
5. Jimbo, M.: Lett. Math. Phys. **11**, 247 (1986)
6. Andrews, G.E., Baxter, R.J., Forrester, P.J.: J. Stat. Phys. **35**, 193 (1984)
7. Huse, D.A.: Phys. Rev. B**30**, 3908 (1984)
8. Date, E., Jimbo, M., Miwa, T., Okado, M.: Advanced Studies in Pure Mathematics **16**, 17 (1988) and references therein
9. Jimbo, M., Miwa, T., Okado, M.: RIMS-600
10. Pasquier, V.: Nucl. Phys. B**285**, 162 (FS19)
11. Pasquier, V.: Nucl. Phys. B**295**, [FS21], 491 (1988)
12. Ocneanu, A.: Pennsylvania University Preprint
13. Temperley, H.N.V., Lieb, E.: Proc. R. Soc. Lond. A**322**, 251 (1971)
14. Jones, V.R.F.: Invent. Math. **72**, 1 (1983)
15. Turaev, V.G.: LOMI Preprint E-3-1987
16. Baird, G.E., Bidenharn, L.C.: J. Math. Phys. **12**, 1847 (1965)
17. Wenzl, H.: Thesis Pennsylvania University (1985)
18. Babelon, O., de Vega, H.J., Viallet, J.M.: Nucl. Phys. B**190**, 542 (1981)
19. Kuniba, A., Yajima, T.: J. Phys. A**21**, 519 (1988)
20. Pasquier, V.: J. Phys. A**20**, 1229 (1987)
21. Ginsparg, P.: Nucl. Phys..B**295** [FS21], 153 (1988)
22. McKay, J.: Am. Math. Soc. **81**, 153 (1981)

Communicated by A. Connes

Received March 1, 1988

Volume 92A, number 1 PHYSICS LETTERS 18 October 1982

SELF-DUAL SOLUTIONS OF THE STAR-TRIANGLE RELATIONS IN Z_N-MODELS

V.A. FATEEV and A.B. ZAMOLODCHIKOV
L.D. Landau Institute for Theoretical Physics, The Academy of Sciences of the USSR, Moscow, USSR

Received 19 July 1982

The self-dual solution of the star-triangle relations in Z_N models is presented. The corresponding partition functions are calculated.

In the past years much attention in the study of two-dimensional lattice statistics has been paid to the investigation of Z_N models [1–5]. The general case of a Z_N model can be defined as follows. We associate a spin variable $z(x)$ which takes its values in the group Z_N $[z^N(x) = 1]$ to every site x, where x is a two-dimensional integer-valued vector labeling the sites of a two-dimensional rectangular lattice. Then the partition function of the statistical Z_N models with nearest-neighbour interaction can be represented in the form:

$$Z = \sum_{\{z\}} \prod_{x} \prod_{\sigma=\pm 1} w^{(\sigma)}(z(x), z(x+e_\delta)) . \tag{1}$$

Here the sum runs over all values of the variable z in every site of the lattice. The functions $w^{(\sigma)}$ $(\sigma = \pm 1)$ are the weight functions determining the interaction between spins on the neighbouring sites of the lattice in horizontal $(\sigma = 1)$ and vertical $(\sigma = -1)$ directions. The vectors $e_1 = (1,0)$ and $e_{-1} = (0,1)$ are the basic vectors of the lattice.

If external fields are absent the most general interaction between two neighbouring spins, after proper normalization, has the form:

$$w^{(\sigma)}(z_1, z_2) = 1 + \sum_{i=1}^{N-1} x_i^{(\sigma)}(z_1\bar{z}_2)^i . \tag{2}$$

We suppose that the functions $w^{(\sigma)}(z_1, z_2)$ are real. It imposes on the parameters $x_i^{(\sigma)}$ (the statistical weights of the Z_N models) the condition

$$x_i^{(\sigma)} = x_{N-i}^{(\sigma)} . \tag{3}$$

An isotropic system corresponds to the case $x_i^{(1)} = x_i^{(-1)}$.

The Z_N models possess the important property of the Kramers–Wannier symmetry which helps us to treat the phase structure of these models [2–5]. Namely, the partition function (1) of these models acquires only an irrelevant multiplier under the following "dual" transformation of the statistical weights $x_i^{(\sigma)}$

$$\tilde{x}_i^{(\sigma)} = \left(1 + \sum_{k=1}^{N-1} x_k^{(-\sigma)}\omega^{ki}\right)\left(1 + \sum_{k=1}^{N-1} x_k^{(-\sigma)}\right)^{-1} , \tag{4}$$

where $\omega = \exp(2\pi i/N)$. The equations

$$\tilde{x}_i^{(\sigma)} = x_i^{(\sigma)} \tag{5}$$

determine the region of self-duality in the space of parameters $x_i^{(\sigma)}$.

The exact solution of Z_N models in the general case is unknown. A complete solution was constructed only for the case $N = 2$ (Ising model). For other values of N there are particular values of the parameters x_i for which the Z_N models admit an exact solution. A nontrivial example is so called critical Potts model [6–10] which corresponds to the case $x_1 = x_2 = \ldots = x_{N-1} = x_c$ where x_c is a certain critical value ‡1. In this note we introduce other particular values of the parameters x_i for which Z_n Models can be solved exactly. The exact solution can be obtained if there exists a family of functions $x_i(\alpha)$ which depends on the auxiliary parameter α, such that

‡1 For the case $N = 4$ a more general exactly solvable case is known, the so called self-dual (critical) Ashkin–Teller model [8,10,11].

$$x_i^{(1)} = x_i(\alpha)\,, \quad x_i^{(-1)} = x_i(\pi - \alpha)$$

and the functions $x_i(\alpha)$ satisfy the star-triangle relations [8–10,12]:

$$\sum_{k=0}^{N-1} x_{n_1-k}(\alpha)x_{n_2-k}(\pi-\alpha-\alpha')x_{n_3-k}(\alpha')$$
$$= c(\alpha,\alpha')x_{n_2-n_3}(\pi-\alpha)x_{n_1-n_3}(\alpha+\alpha')$$
$$\times x_{n_1-n_2}(\pi-\alpha')\,. \tag{6}$$

Here

$$x_0 = 1\,, \quad x_{-n}(\alpha) = x_{N-n}(\alpha) = x_n(\alpha)\,, \tag{7}$$

and $c(\alpha, \alpha')$ is an arbitrary symmetric function of α and α'. Eq. (6) possesses the following property of self-duality: if functions $x_n(\alpha)$ are the solution of eq. (6) then the functions

$$\tilde{x}_n(\alpha) = \left(\sum_{k=0}^{N-1} x_k(\pi-\alpha)\omega^{kn}\right)\left(\sum_{k=0}^{N-1} x_k(\pi-\alpha)\right)^{-1} \tag{8}$$

also satisfy eq. (6) [possibly with another function $c(\alpha, \alpha')$]. This fact can easily be verified using the properties of Fourier transformation and condition (7).

It follows from this property that eq. (6) admits self-dual solutions, i.e. the solutions satisfying the relations

$$\tilde{x}_n(\alpha) \equiv \left(\sum_{k=0}^{N-1} x_k(\pi-\alpha)\omega^{kn}\right)\left(\sum_{k=0}^{N-1} x_k(\pi-\alpha)\right)^{-1}$$
$$= x_n(\alpha)\,. \tag{9}$$

The general solution of eq. (6) seems to be of great interest for the analytical study of Z_N models. Here we present the particular solution of eq. (6) which possesses the property of self-duality (9).

Using relations (9) one can change the star-triangle relation (6) to

$$\sum_{l=0}^{N-1} x_{n_1-l}(\alpha)x_{n_2}(\alpha+\alpha')x_{n_3-l}(\alpha')\omega^{-n_2 l}$$
$$= \sum_{l=0}^{N-1} x_l(\alpha)x_{n_1-n_3}(\alpha+\alpha')x_{l-n_2}(\alpha')\omega^{-l(n_1-n_3)-n_2n_3}\,, \tag{10}$$

which is appropriate for the self-dual case. Besides the known solutions (corresponding to the critical Potts and critical Ashkin–Teller models, see refs. [6–10]), eqs. (10) possess the following solution:

$$x_0 = 1\,, \quad x_n(\alpha) = \prod_{k=0}^{n-1} \frac{\sin(\pi k/N + \alpha/2N)}{\sin[\pi(k+1)/N - \alpha/2N]}\,, \tag{11}$$

as one can verify by means of a direct calculation. For $N = 2,3$ the solution (11) coincides with the known one corresponding to the critical Potts model, for $N = 4$ (11) is a particular case of the critical Ashkin–Teller model. We suppose that for $N = 5,7$ the solution (11) describes the critical bifurcation points in the phase diagram [5].

The partition functions corresponding to solutions (11) can be calculated easily by means of the standard trick described in refs. [8,13,14]; the result has the form

$$f(\alpha) = -\int_0^\infty \frac{\operatorname{sh}\frac{1}{2}\alpha x \operatorname{sh}\frac{1}{2}(\pi-\alpha)x \operatorname{sh}\frac{1}{2}\pi x(N-1)}{\operatorname{ch}^2\frac{1}{2}\pi x \operatorname{ch}\frac{1}{2}\pi N x}\frac{dx}{x}\,, \tag{12}$$

where $f(\alpha)$ denotes the "specific" (per lattice vertex) free energy of the infinite lattice.

Taking the limit $N \to \infty$ in solution (11) one obtains the U(1) invariant lattice model – the XY model with a specific interaction. In this model the plane rotator $\boldsymbol{n}(x) = \{\cos\varphi(x), \sin\varphi(x)\}$ is associated with each lattice vertex and the weight function of the nearest neighbour interaction has the form:

$$w(\varphi,\varphi'|\alpha) = 1$$
$$+ 2\sum_{k=1}^{\infty} \frac{\Gamma(k+\alpha/2\pi)\Gamma(1-\alpha/2\pi)}{\Gamma(k+1-\alpha/2\pi)\Gamma(\alpha/2\pi)} \cos k(\varphi-\varphi')$$
$$= \frac{[\Gamma(1-\alpha/2\pi)\Gamma(1/2)}{\Gamma(1/2-\alpha/2\pi)]}\,|\sin[(\varphi-\varphi')/2]|^{-\alpha/\pi}\,.$$

Note, that solutions (12) of the star-triangle relations could be obtained alternatively from that presented in ref. [12] ($D = 1$) through a conformal transformation.

References

[1] F.Y. Wu and Y.K. Wang, J. Math. Phys. 17 (1976) 439.
[2] A.B. Zamolodchikov, Zh. Eksp. Teor. Fiz. 75 (1978) 341.
[3] V.S. Dotzenko, Zh. Eksp. Teor. Fiz. 75 (1978) 1083.
[4] S. Elitzur, R.B. Pearson and J. Shigemitsu, Phys. Rev. D19 (1979) 3698.
[5] F.C. Alcaraz and R. Koberle, J. Phys. A13 (1980) L153.

[6] R.J. Baxter and I.G. Enting, J. Phys. A11 (1978) 2463.
[7] R.J. Baxter, H.N.V. Temperly and S.E. Ashley, Proc. Roy. Soc. London A358 (1978) 535.
[8] R.J. Baxter, Exactly solved models, in: Fundamental problems in statistical mechanics, Vol. V, ed. E.G.D. Cohen (North-Holland, Amsterdam, 1980).
[9] R.J. Baxter, Australian National University Preprint (1981), J. Stat. Phys., to be published.
[10] Yu. Bashilov and S. Pokrovsky, Commun. Math. Phys., to be published.
[11] J. Ashkin and E. Teller, Phys. Rev. 64 (1943) 178.
[12] A.B. Zamolodchikov, Phys. Lett. 97B (1980) 63.
[13] Yu.G. Stroganov, Phys. Lett 74A (1979) 116.
[14] A.B. Zamolodchikov, Sov. Sci. Rev. 2 (1980).

MISCELLANEOUS TOPICS

6. Miscellaneous Topics

In classical representation theory it is often useful to consider the decomposition of tensor products of fundamental representations. This affords in particular a way of realizing 'higher' representations explicitly. The fusion procedure is an analog of this technique which enables one to generate new solutions to YBE out of elementary ones. Systematic study of this sort has begun in the reprint [24] by Kulish, Reshetikhin and Sklyanin for the rational R matrix associated with $\mathfrak{gl}(n)$. They showed that the products of R matrices, with the spectral parameters suitably shifted and followed by projecton operators, yield new R matrices corresponding to symmetric or anti-symmetric tensor representations. (Such an idea is also implicit in the paper[240].) As Cherednik shows in the reprint [26], this method can be generalized to include arbitrary irreducible representations. It is applicable further to the elliptic solutions of Baxter and Belavin.

The reprint of Date et al. [25] (the first half of the paper [73]) applies this idea to the 8VSOS model, thereby generating further series of IRF models corresponding to the higher spin represntations of $\mathfrak{sl}(2)$. Some of these have been found by direct methods in [1-2),24),149-154]. Extension to the case of $\mathfrak{sl}(n)$ generalizatons of IRF models in an arbitrary irreducible representation is described in [117].

Let us mention here some other topics discussed in the literature. Ogievetsky, Reshetikhin and Wiegmann [187-188),208),213] have shown that the spectrum of the R matrices associated with simple Lie algebras in arbitrary representations can be described by using the Bethe Ansatz technique (cf.[3]). See the reprint [27]. Bazhanov, Stroganov[27-28),34] and Takhtajan[245] give a multiplicative representation for certain R matrices that can be viewed as matrix generalizations of the classical trigonometric functions or the Weierstrass functions. The direct connection appearing here between the classical and quantum R matrices is yet to be understood. There are also variants of the ordinary R matrices; Cherednik[59] studies factorizable S matrices in the presence of reflecting walls related to root systems, and Gaudin[97] gives explicit R matrices of infinite rank.

YANG–BAXTER EQUATION AND REPRESENTATION THEORY: I

P.P. KULISH, N.Yu. RESHETIKHIN and E.K. SKLYANIN
Steklov Mathematical Institute, Leningrad Branch, Fontanka 27, 191011, Leningrad, U.S.S.R.

ABSTRACT. The problem of constructing the GL(N, $\mathbb{C}$)-invariant solutions to the Yang–Baxter equation (factorized S-matrices) is considered. In case $N = 2$ all the solutions for arbitrarily finite-dimensional irreducible representations of GL(2, $\mathbb{C}$) are obtained and their eigenvalues are calculated. Some results for the case $N > 2$ are also presented.

1. INTRODUCTION

In recent papers [1] devoted to the generalization of the inverse spectral transform method to the quantum case, a new mathematical object, to so-called R-matrix, has arisen (for review of the quantum inverse spectral transform method see [2]). An important property of the R-matrix is that it satisfies the so-called Yang–Baxter equation (YBE):

$$R_{12}(u)R_{13}(u+v)R_{23}(v) = R_{23}(v)R_{13}(u+v)R_{12}(u). \tag{1}$$

Equation (1), which arose in papers of C.N. Yang [3] and R.J. Baxter [4], plays a fundamental role in the theory of completely integrable quantum systems [5]. Equation (1) has also a nice interpretation in terms of scattering theory. Namely, the R-matrix may be regarded as a two-particle scattering matrix.

Let us explain the notation in (1). The objects R_{ab} are linear operators in the tensor product of three linear spaces $V_{123} = V_1 \otimes V_2 \otimes V_3$. Furthermore, the operator R_{ab} is the canonical embedding of an operator $\widetilde{R}_{ab}$ acting in $V_a \otimes V_b$ into V_{123} (i.e., for example $R_{12} = \widetilde{R}_{12} \otimes I_3$, I_3 being the identity operator in V_3). The parameter u, which we shall call the spectral parameter, is assumed to run over the complex plane $\mathbb{C}$.

In the present article we shall construct several families of solutions to the YBE. The idea of our approach is most easily explained by considering first a simpler equation [5, 7, 8]:

$$[r_{12}(u), r_{13}(u+v)] + [r_{12}(u), r_{23}(v)] + [r_{12}(u+v), r_{23}(v)] = 0, \tag{2}$$

which may be regarded as the classical analogue of (1). To get (2) from (1) assume that $R_{ab}(u)$ depends on an additional parameter η in such a way that

Letters in Mathematical Physics 5 (1981) 393-403.

$$R_{ab}(u,\eta)|_{\eta=0} = I. \tag{3}$$

Differentiating (1) twice with respect to η at $\eta = 0$ and denoting

$$\frac{\partial}{\partial\eta} R_{ab}(u,\eta)|_{\eta=0} = r_{ab}(u) \tag{4}$$

one gets (2).

Since (2) is written in terms of commutators only, it may be regarded as an abstract Lie algebra equation. Given a solution $r_{ab}^{(0)}$ of (2) in a certain Lie algebra $\mathfrak{G}$, i.e., $r_{ab}^{(0)} \in \mathfrak{G} \otimes \mathfrak{G}$* we may obtain new solutions by considering linear representations of $\mathfrak{G}$. More precisely, let T_a be representations of $\mathfrak{G}$, V_a the corresponding $\mathfrak{G}$-moduli. Then $r_{ab}(u) = T_a \otimes T_b r_{ab}^{(0)}(u)$ is a solution to (2) with values in End $(V_a \otimes V_b)$.

A particular solution of (2) for a semisimple Lie algebra $\mathfrak{G}$ is constructed as follows.

STATEMENT 1. [5, 8]. Let $\{e_i\}$ be a basis of $\mathfrak{G}$, let k be its Killing form. Put $k_{ij} = k(e_i, e_j)$, $k^{ij} = (k^{-1})_{ij}$. Then $r_{ab}(u)$ given by

$$r_{ab}^{(k)}(u) = \frac{1}{u} \sum_{i,j} k^{ij} e_i \otimes e_j \tag{5}$$

satisfies (2).

By analogy with (2) it is natural to look for a 'universal' solution of (1) taking values in the universal enveloping algebra $\mathfrak{U}(\mathfrak{G} \times \mathfrak{G})$ for a given Lie algebra $\mathfrak{G}$. (In contrast with (2) Equation (1) cannot be written down in terms of commutators only, so clearly one has to pass to the enveloping algebra.)

The following conjecture seems plausible.

CONJECTURE 1. Fix a Lie algebra $\mathfrak{G}$ and a solution $r_{ab}^{(0)}(u) \in \mathfrak{G} \otimes \mathfrak{G}$ of Equation (2). Then:

(a) There exists a smooth family $R_{ab}(u, \eta) \in \mathfrak{U}(\mathfrak{G} \times \mathfrak{G})$ which satisfies (1) for all η sufficiently close to zero and gives $r_{ab}^{(0)}(u)$ in the classical limit in the sense of (3, 4).

(b) The solution $R_{ab}(u, \eta)$ with the properties listed above is unique up to multiplication by a central element $f(u, \eta) \in Z(\mathfrak{G}) \otimes Z(\mathfrak{G})$.

Our approach to constructing the universal R-matrix consists in building it of 'elementary' R-matrices corresponding to irreducible representations of $\mathfrak{G}$. In other words, instead of dealing directly with Conjecture 1 we are going to examine another conjecture.

CONJECTURE 2. Let the conditions of Conjecture 1 be satisfied. Then for any two irreducible $\mathfrak{G}$-moduli V_{Λ_a}, V_{Λ_b} there exists a family of operators $R_{ab}^{\Lambda_a\Lambda_b}(u, \eta)$ with values in End $(V_{\Lambda_a} \otimes V_{\Lambda_b})$ such that:

(a) $R_{ab}^{\Lambda_a\Lambda_b}(u, \eta)$ gives $r_{ab}^{(0)}$ in the classical limit (3, 4).

(b) For each triplet V_{Λ_a}, V_{Λ_b}, V_{Λ_c} the YBE

*We consider $\mathfrak{G} \otimes \mathfrak{G}$ as a subspace of $\mathfrak{U}(\mathfrak{G} \times \mathfrak{G})$.

$$R_{ab}^{\Lambda_a\Lambda_b}(u,\eta)R_{ac}^{\Lambda_a\Lambda_c}(u+v,\eta)R_{bc}^{\Lambda_b\Lambda_c}(v,\eta) = R_{bc}^{\Lambda_b\Lambda_c}(v,\eta)R_{ac}^{\Lambda_a\Lambda_c}(u+v,\eta)R_{ab}^{\Lambda_a\Lambda_b}(u,\eta) \tag{6}$$

is valid.

(c) For given V_{Λ_a}, V_{Λ_b} the solution $R_{ab}^{\Lambda_a\Lambda_b}(u,\eta)$ is unique up to a scalar factor $f(u,\eta)$.

Clearly, Conjecture 1 implies Conjecture 2 (by specifying the representation). The converse of part (a) of Conjecture 1 is

THEOREM 1. *Assume Conjecture 2 to be true. Let $I = \Sigma_\alpha P_{\Lambda_\alpha}$ be a decomposition of $\mathfrak{U}(\mathfrak{G} \times \mathfrak{G})$ into irreducible $\mathfrak{G}$-moduli, P_{Λ_α} being the corresponding projectors. Then the R-matrix given by*

$$R(u,\eta) = \sum_{\alpha,\beta} R_{\alpha\beta}^{\Lambda_\alpha\Lambda_\beta} P_{\Lambda_\alpha} \otimes P_{\Lambda_\beta} \tag{7}$$

satisfies (1)*.

In the present article we shall examine Conjecture 2 in case $\mathfrak{G} = \mathrm{gl}(N, \mathbb{C})$. We consider only finite-dimensional irreducible representations T_Λ of $\mathrm{gl}(N,\mathbb{C})$ with the highest weight $\Lambda = (m_1, \ldots, m_N)$, $m_1 \geq m_2 \geq \ldots \geq m_N$. Furthermore, we shall consider only the $\mathrm{GL}(N,\mathbb{C})$ invariant solutions of (1)**. The group invariance is understood here in the following sense [5]:

$$T_{\Lambda_a}(g) \otimes T_{\Lambda_b}(g) R_{ab}(u,\eta) T_{\Lambda_a}(g^{-1}) \otimes T_{\Lambda_b}(g^{-1}) = R_{ab}(u,\eta), \quad g \in \mathrm{GL}(N,\mathbb{C}). \tag{8}$$

We shall prove Conjecture 2 in case $N = 2, 3$ and for a lot of representations in case $N > 3$. The plan of our paper is as follows. In Section 2, the construction of a family of R-matrices will be described, which gives all the solutions of (6) in case $N = 2$ and some solutions in case $N > 2$. In Section 3, the case $N = 2$ is considered in detail and the eigenvalues of the R-matrices are calculated. The case $N > 2$ is discussed in Section 4. In the Conclusion, a list of unsolved problems is given.

2. CONSTRUCTION OF R-MATRICES

In the present section we shall construct solutions of (6) for finite-dimensional irreducible representations of $\mathrm{gl}(N,\mathbb{C})$ corresponding to the highest weights of the form $\Lambda = (m, 0, \ldots, 0) \equiv m+$, $m > 0$ and $\Lambda = (1, \ldots, 1, 0, \ldots, 0) = (1^m) \equiv m-$. These representations are specified by the condition

* Cf. the definition of a reducible R-matrix in [5].

** Strong evidence in favour of Conjecture 2 for non-group-invariant R-matrices is presented by the recently-found generalizations of Baxter's solution [4] (two-dimensional representation of $\mathrm{gl}(2,\mathbb{C})$ in our classification) to the three-dimensional representation [9] and for arbitrary representations [10] (the latter in case of the so-called XXZ-model.

that the corresponding Young diagrams consist of only one row (resp. column). The projectors from the space $\otimes_{k=1}^{m} \mathbb{C}^N$ onto the corresponding representation spaces are given by symmetrizers and antisymmetrizers

$$P^{\pm}_{1\dots m} = \frac{1}{m!} \sum_{\sigma} (\pm 1)^{\sigma} \mathscr{P}_{\sigma}. \tag{9}$$

The sum in (9) is taken over all permutations $\sigma = (\sigma_1, \dots, \sigma_m)$ of $(1, \dots, m)$, $\mathscr{P}_\sigma$ being the permutation operator in the space $\otimes_{k=1}^{m} \mathbb{C}^N$.

The family of operators $R_{ab}^{\Lambda_a \Lambda_b}(u, \eta)$ shall be built up from simple blocks $R_{ab}^{11}(u, \eta)$ correspondir to the weights $\Lambda_a = \Lambda_b = 1+ = 1- \equiv 1$:

$$R_{ab}^{11}(u, \eta) = uI + \eta \mathscr{P} \tag{10}$$

where I is the identity operator and $\mathscr{P}$ is a transposition operator in the space $\mathbb{C}^N \otimes \mathbb{C}^N$ (i.e., $\mathscr{P}(v \otimes w) = w \otimes v$). Using the standard e_{ij} in the space End $(\mathbb{C}^N)$ the operator $\mathscr{P}$ can be written as follows:

$$\mathscr{P} = \sum_{i,j=1}^{N} e_{ij} \otimes e_{ji}, \quad (e_{ij})_{kl} = \delta_{ki}\delta_{lj}. \tag{11}$$

The solution (10) has been obtained first in [3]. Dividing (10) by u and differentiating it with respect to η at $\eta = 0$ one gets easily with help of (11) that the classical r-matrix corresponding to (10) in the sense of (3) and (4) has the form (5).

The following property of $R^{11}(u)$ is basic for our construction.

STATEMENT 2. There exist two values of the spectral parameter $u(u = \pm\eta)$ for which $R^{11}(u)$ is a projector

$$R_{ab}^{11}(\pm\eta) = \pm 2\eta P_{ab}^{\pm}. \tag{12}$$

The solutions $R^{m\epsilon, n\delta}(u)(\epsilon, \delta = \pm)$ are constructed by tensoring of the fundamental solutions $R^{11}(u)$ with the subsequent (anti-)symmetrization. To describe the construction let us introduce the following notation:

$$R_{\{a\}b}(u, \eta, m) \equiv R_{a_1 b}^{11}\left(u + \frac{m-1}{2}\eta\right) \dots R_{a_j b}^{11}\left(u + \frac{m-1-2j}{2}\eta\right) \dots R_{a_m b}^{11}\left(u - \frac{m-1}{2}\eta\right). \tag{13}$$

The quantity $R_{\{a\}b}$ is an operator in the space $V_{a_1} \otimes \dots \otimes V_{a_m} \otimes V_b$ $(V_{a_1} = \dots = V_{a_m} = V_b = \mathbb{C}^N)$. Using (13), the solution $R^{m\epsilon, n\delta}(u)$ is given by*

*For the simplest case $R^{1,2+}$ the construction can be found in [11].

$$R^{m\pm,1}_{\{a\}b}(u) = P^{\pm}_{\{a\}}R_{\{a\}b}(u, \eta, m)P^{\pm}_{\{a\}}, \tag{14}$$

$$R^{m\epsilon, n\delta}_{ab}(u) = P^{\delta}_{\{b\}}R^{m\epsilon,1}_{ab_1}\left(u + \frac{n-1}{2}\eta\right) \dots R^{m\epsilon,1}_{ab_j}\left(u + \frac{n-1-2j}{2}\eta\right) \dots$$
$$\dots R^{m\epsilon,1}_{ab_n}\left(u - \frac{n-1}{2}\eta\right)P^{\delta}_{\{b\}}. \tag{15}$$

THEOREM 2. *The R-matrices $R^{m\epsilon, n\delta}(u)$ defined by* (10), (14) *and* (15) *satisfy the YBE* (6) *for arbitrary choice of weights $\Lambda_a = m_a\pm$. The classical limit* (4) *of $R^{m\epsilon, n\delta}(u, \eta)$ after a proper normalization is* (5).

The proof of Theorem 2 is lengthy and we hope to publish it elsewhere. It uses the following identities:

$$\begin{aligned} P^{\pm}_{\{a\}}R^{11}_{a_1 b}\left(u \pm \frac{\eta}{2}\right)R^{11}_{a_2 b}\left(u \mp \frac{\eta}{2}\right) &= R^{11}_{a_2 b}\left(u \mp \frac{\eta}{2}\right)R^{11}_{a_1 b}\left(u \pm \frac{\eta}{2}\right)P^{\pm}_{\{a\}} \\ &= P^{\pm}_{\{a\}}R^{11}_{a_1 b}\left(u \pm \frac{\eta}{2}\right)R^{11}_{a_2 b}\left(u \mp \frac{\eta}{2}\right)P^{\pm}_{\{a\}} \\ &= P^{\pm}_{\{a\}}R^{11}_{a_2 b}\left(u \mp \frac{\eta}{2}\right)R^{11}_{a_1 b}\left(u \pm \frac{\eta}{2}\right)P^{\pm}_{\{a\}}, \end{aligned} \tag{16}$$

$$P^{\pm}_{12\dots n}R^{11}_{1n+1}(\pm n\eta)P^{\pm}_{23\dots n+1} = \pm(n+1)\eta P^{\pm}_{12\dots n+1}, \tag{17}$$

$$\begin{aligned} P^{\pm}_{\{a\}}R_{\{a\}b}(u, \eta, n) &= R_{\{a\}b}(u, -\eta, u)P^{\pm}_{\{a\}} \\ &= P^{\pm}_{\{a\}}R_{\{a\}b}(u, \eta, n)P^{\pm}_{\{a\}} = P^{\pm}_{\{a\}}R_{\{a\}b}(u, -\eta, n)P^{\pm}_{\{a\}}. \end{aligned} \tag{18}$$

The identity (16) follows from (12), YBE (1) and the characteristic property of the projector: $P^2 = P$. The identity (17) generalizes (12) and gives an opportunity to express the projector onto any irreducible representation of the type considered in terms of R^{11}. The identity (18) following from (16) and (17) is the direct generalization of (16) for an arbitrary projector.

Let us sketch the proof of the Theorem 2 for the case $\Lambda_a = \Lambda_b = 1, \Lambda_c = 2+$. In other words, we need to prove the equality

$$R^{11}_{ab}(u-v)R^{12+}_{ac}(u)R^{12+}_{bc}(v) = R^{12+}_{bc}(v)R^{12+}_{ac}(u)R^{11}_{ab}(u-v). \tag{19}$$

To prove (19), it is sufficient to get the left-hand side of (19), represent it using (16) in the form

$$R^{11}_{ab}(u-v)P^{+}_{\{c\}}R^{11}_{ac_1}\left(u+\frac{\eta}{2}\right)R^{11}_{ac_2}\left(u-\frac{\eta}{2}\right)P^{+}_{\{c\}}R^{11}_{bc_1}\left(v+\frac{\eta}{2}\right)R^{11}_{bc_2}\left(v-\frac{\eta}{2}\right) \equiv (*)$$

or by virtue of (16) and $P^2 = P$.

$$(*) = P^+_{\{c\}} R^{11}_{ab}(u-v) R^{11}_{ac_2}\left(u-\frac{\eta}{2}\right) R^{11}_{ac_1}\left(u+\frac{\eta}{2}\right) R^{11}_{bc_2}\left(v-\frac{\eta}{2}\right) R^{11}_{bc_1}\left(v+\frac{\eta}{2}\right) P^+_{\{c\}}$$

and to apply the YBE subsequently to the triplets $R_{ab}R_{ac_2}R_{bc_2}$ and $R_{ab}R_{ac_1}R_{bc_1}$ obtaining the right-hand side of (19).

REMARK. The proof given above uses only the Statement 2 and the identity $P^2 = P$. Therefore it is applicable in any case when R-matrix $R_{ab}(u)$ is a projector $R_{ab}(u_0) \sim P$ at some value u_0 of the spectral parameter u. We shall use this remark in Section 4.

3. EIGENVALUES OF R-MATRICES. CASE $N = 2$

Theorem 2 checks parts (a) and (b) of Conjecture 2 for representations considered. To prove the uniqueness of constructed R-matrices (part (c)) of Conjecture 2) we shall need more information on their structure. In particular, we shall obtain some formulas for eigenvalues of R-matrices which are of independent interest.

STATEMENT 3. Let e_{ij} be the generators (11) of GL(N, $\mathbb{C}$) in the fundamental representation and p the same generators in an arbitrary representation T_Λ. Then the function $\mathcal{R}^{\Lambda,1}(u, \eta)$ defined by

$$\mathcal{R}^{\Lambda,1}(u,\eta) = u \cdot I + \eta \sum_{i,j=1}^{N} p_{ij} e_{ji} \tag{20}$$

satisfies the YBE (6) for $\Lambda_a = \Lambda$ $\Lambda_b = \Lambda_c = 1$ and $R^{11}_{bc}(v)$ of the form (10).

The representation T_Λ in (20) can be arbitrary because the YBE in this case is equivalent to the classical one (2). The question of how the solutions (20) and (14) are connected is answered by the following proposition.

STATEMENT 4. If T_Λ is an irreducible representation with the highest weight Λ of the form $m\pm$ then the solution (20) coincides with (14) up to a scalar factor

$$\mathcal{R}^{m\pm,1}(u) = \prod_{k=1}^{m-1} (u \pm k\eta)^{-1} R^{m\pm,1}\left(u \pm \frac{m-1}{2}\eta\right). \tag{21}$$

Let us consider now the YBE (6) for Λ_a, Λ_b corresponding to arbitrary irreducible finite-dimensional representations of gl(N, $\mathbb{C}$), $\Lambda_c = 1$ assuming existence of the corresponding R-matrices. Using the explicit form (20) for $R^{\Lambda,1}(u)$ we rewrite (6) in the form

$$R_{ab}^{\Lambda_a\Lambda_b}(u)\left(u+v+\eta\sum_{i,j=1}^{N}p_{ij}^{a}e_{ji}\right)\left(v+\eta\sum_{i,j=1}^{N}p_{ij}^{b}e_{ji}\right)$$
$$=\left(v+\eta\sum_{i,j=1}^{N}p_{ij}^{b}e_{ji}\right)\left(u+v+\eta\sum_{i,j=1}^{N}p_{ij}^{a}e_{ji}\right)R_{ab}^{\Lambda_a\Lambda_b}(u). \tag{22}$$

Using the notation

$$d_{ij}=p_{ij}^{a}-p_{ij}^{b},\qquad d_{ij}^{2}=\sum_{k=1}^{N}(p_{ik}^{a}-p_{ik}^{b})(p_{kj}^{a}-p_{kj}^{b}),$$
$$C_2(p)=\sum_{i,j=1}^{N}p_{ij}p_{ji} \tag{23}$$

and relation

$$\sum_{k=1}^{N}(p_{ik}^{a}p_{kj}^{b}-p_{ik}^{b}p_{kj}^{a})=\tfrac{1}{4}[C_2(p^a+p^b),d_{ij}^{2}], \tag{24}$$

and excluding matrices e_{ij} from (22) we get the equation

$$\left[R_{ab}^{\Lambda_a\Lambda_b}(u),ud_{ij}-\frac{\eta}{2}d_{ij}^{2}\right]=\frac{\eta}{4}\{[C_2(p^a+p^b),d_{ij}],R_{ab}^{\Lambda_a\Lambda_b}(u)\} \tag{25}$$

where [,] stands for commutator and { , } for anticommutator.

The group invariance (8) of $R_{ab}^{\Lambda_a\Lambda_b}(u)$ implies that its spectral decomposition has the form

$$R_{ab}^{\Lambda_a\Lambda_b}(u)=\sum_{k}\rho_k(u)P_{\Lambda_k} \tag{26}$$

where P_{Λ_k} is the projector onto the irreducible space V_{Λ_k} in the Klebsch–Gordan series $V_{\Lambda_a}\otimes V_{\Lambda_b}=\Sigma_k V_{\Lambda_k}$. The Equations (25) – (26) allow, in principle, to determine the eigenvalues $\rho_k(u)$ up to a scalar factor.

We postpone the discussion of the general case $N>2$ to the next section and examine first the case $N=2$. In the case of gl(2, $\mathbb{C}$) the irreducible representation T_Λ of the highest weight $\Lambda=(m_1,m_2)$ has the dimension $m_1-m_2+1=2L+1$. The decomposition (26) takes now the form*

$$R_{ab}^{L_aL_b}(u)=\sum_{J=|L_a-L_b|}^{L_a+L_b}\rho_J(u)P_J. \tag{27}$$

*For the sake of simplicity in this section we assume $m_1+m_2=0$.

Using the standard basis $(S_3, S_\pm)$ of sl(2, $\mathbb{C}$) $[S_3, S_\pm] = \pm S_\pm$, $[S_+, S_-] = 2S_3$ one of the Equations (25) can be rewritten as

$$[R_{ab}^{L_a L_b}(u), u(S_3^a - S_3^b)]$$
$$= \eta \{R_{ab}^{L_a L_b}(u), S_-^a S_+^b - S_+^a S_-^b\}. \tag{28}$$

To determine the eigenvalues $\rho_J(u)$, we shall use the ladder operators [12]

$$X_\pm = \tfrac{1}{2}(S_+^a S_-^b - S_-^a S_+^b) \mp (S_3^a A + S_3^b B)(\mathbb{J} + \tfrac{1}{2} \mp \tfrac{1}{2}),$$
$$A = \frac{\mathbb{J}(\mathbb{J}+1) - L^a(L^a+1) - L^b(L^b+1)}{2\mathbb{J}(\mathbb{J}+1)}, \tag{29}$$
$$B = A - 1$$

which change by 1 the total angular momentum J; $\mathbb{J}(\mathbb{J}+1) = \Sigma_{j=1}^3 (S_j^a + S_j^b)^2$

$$[\mathbb{J}, X_\pm] = \pm X_\pm. \tag{30}$$

Using (29), one can rewrite (28) in the form:

$$R_{ab}^{L_a L_b}(u)\left[X_+ \frac{-u + \eta(\mathbb{J}+1)}{2\mathbb{J}+1} + X_- \frac{u + \eta\mathbb{J}}{2\mathbb{J}+1}\right]$$
$$= \left[X_+ \frac{-u - \eta(\mathbb{J}+1)}{2\mathbb{J}+1} + X_- \frac{u - \eta\mathbb{J}}{2\mathbb{J}+1}\right] R_{ab}^{L_a L_b}(u). \tag{31}$$

Combining (31) with (27) and (30) we arrive at the recurrence relations

$$\rho_{J+1}(u) = \frac{u + \eta(J+1)}{u - \eta(J+1)} \rho_J(u) \tag{32}$$

which determine the eigenvalues $\rho_J(u)$ up to a common factor, thus proving part (c) of Conjecture 2. The calculation of the factor gives the following result.

THEOREM 3. *In case $N = 2$ the R-matrix $R_{ab}^{m+, n+}(u)$ $(L_a = m/2, L_b = n/2)$ constructed in Section 2 has the spectral decomposition* (27) *with the eigenvalues $\rho_J(u)$ given by the following formula**

*As M.A. Semenov-Tian-Shansky pointed out to us, there is a striking resemblance between (33) and some formulas familiar in the theory of intertwining operators for semisimple Lie groups. Indeed, the factorization property of the intertwining operators [13] essentially coincides with YBE. No explanation is known so far of these parallels.

$$\rho_J(u) = g_{mn}(u) \prod_{k=1}^{J} \frac{u+\eta k}{u-\eta k}$$

$$= g_{mn}(u)(-1)^J \frac{\Gamma(u/\eta) + J + 1)\Gamma(-(u/\eta) + 1)}{\Gamma(-(u/\eta) + J + 1)\Gamma(u/\eta) + 1)}, \tag{33}$$

$$g_{mn}(u) = \prod_{k=-1}^{m-2} \prod_{j=0}^{n-1} \left(u + \eta \left(\frac{m+n}{2} - 1 \right) - \eta(k+j) \right) \prod_{l=1}^{m+n/2} \frac{u-\eta l}{u+\eta l}, \quad (m \geqslant n).$$

COROLLARY 1. *The recurrence relation* (32) *proves part* (c) *of Conjecture* 2 *for finite-dimensional representations of* gl(2, $\mathbb{C}$).

COROLLARY 2. *Since the finite-dimensional representations of* gl(2, $\mathbb{C}$) *exhaust all irreducible representations of the Lie algebra* su(2) *being a real form of* sl(2, $\mathbb{C}$) *and by virtue of Theorem 1 the part* (a) *of Conjecture* 1 *is proven for* $\mathfrak{G}$ = su(2). *The corresponding universal R-matrix has the form*

$$R(u, \eta) = (-1)^{\mathbb{J}} \frac{\Gamma((u/\eta) + \mathbb{J} + 1)}{\Gamma(-(u/\eta) + \mathbb{J} + 1)} f(u, \eta; (S^a)^2, (S^b)^2), \tag{34}$$

where f is an arbitrary function of Casimir operators.

4. GENERAL CASE $N > 2$

To obtain relations between the eigenvalue $\rho_k(u)$ in (26) in the general case $N > 2$ we shall use Equation (25) and the Wigner–Eckart theorem.

The tensor operators d, d^2 (25) act in $V_{\Lambda_a} \otimes V_{\Lambda_b}$ and due to the Wigner–Eckart (WE) theorem are defined by their reduced matrix elements and Klebsch–Gordan coefficients.

For any two weights Λ_k, Λ_l from the Klebsch–Gordan (KG) series (26) such that the reduced matrix element $\langle \Lambda_k \| d \| \Lambda_l \rangle$ of d does not vanish we get the following relations between the eigenvalues

$$\rho_k(u) = \frac{u + (\eta/4)(C_2(\Lambda_k) - C_2(\Lambda_l)) - \eta D_{kl}}{u - (\eta/4)(C_2(\Lambda_k) - C_2(\Lambda_l)) - \eta D_{kl}} \rho_l(u),$$

$$c_2(\Lambda) = \sum_{i=1}^{N} m_i^2 + \sum_{i<j} (m_i - m_j), \quad \Lambda = (m_1, \ldots, m_N), \tag{35}$$

$$D_{kl} = \tfrac{1}{2} \langle \Lambda_k \| d^2 \| \Lambda_l \rangle / \langle \Lambda_k \| d \| \Lambda_l \rangle.$$

Here $C_2(\Lambda)$ is the eigenvalue of the Casimir operator (23) in representation with the highest weight Λ.

The set of relations (35) imposes constraints on the eigenvalues of $GL(N, \mathbb{C})$-invariant R-matrix. For general Λ_a, Λ_b in (26) this system is overdetermined and the question of its consistency is under consideration.

We have succeeded in calculation of the eigenvalues ρ_k for the representations Λ_a, Λ_b considered in Section 2 for which the KG series is multiplicity free.

THEOREM 4. *The spectral decomposition* (26) *for R-matrices* (15), (20) *has the form* ($m \geqslant n$)

(1)
$$R^{m\pm, n\pm}(u) = \prod_{k=0}^{n-1} f_m^{\pm}\left(u \pm \eta\left(\frac{m+n}{2} - 1 - k\right)\right) f_{n+1}^{\pm}\left(u \pm \eta\frac{m+n}{2}\right) \sum_{k=0}^{n} \rho_k^{\pm}(u) P_{\Lambda_k^{\pm}} \quad (36)$$

$$f_m^{\pm}(u) = \prod_{k=0}^{m-2} (u \mp k\eta), \qquad \rho_k^{\pm}(u) = \prod_{l=0}^{k-1} \frac{u \mp \eta((m+n/2) - l)}{u \pm \eta((m+n/2) - l)},$$

$$\Lambda_k^{+} = (m+n-k, k, 0, \ldots, 0), \qquad \Lambda_k^{-} = (2, \ldots, 2, 1, \ldots, 1, 0, \ldots, 0) = (2^k, 1^{m+n-2k}) \quad (37)$$

(2)
$$R^{m+, n+}(u) = \prod_{k=0}^{n-1} f_m^{+}\left(u + \eta\left(\frac{m-n}{2} + k\right)\right) f_n^{-}\left(u - \eta\left(\frac{m+n}{2} - 1\right)\right) \times$$

$$\times \left(\left(u + \eta\frac{m+n}{2}\right) P_{\Lambda_1} + \left(u - \eta\frac{m+n}{2}\right) P_{\Lambda_2}\right), \quad (38)$$

$$\Lambda_1 = (m+1, 1, \ldots, 1, 0, \ldots, 0) = (m+1, 1^{n-1}), \quad \Lambda_2 = (m, 1^n).$$

(3)
$$\mathcal{R}^{\Lambda,1}(u) = \sum_{k=1}^{N} (u + \eta(m_k + 1 - k)) P_{\Lambda_k},$$

$$\Lambda = (m_1, \ldots, m_N), m_k > m_{k+1}, \Lambda_k = (\ldots, m_k + 1, \ldots). \quad (39)$$

As we pointed out, the construction of Section 2 may be applied under the single condition that the R-matrix is a projector at some point $u = u_0$. We see that R-matrices (36) and (38) can be used for constructing new R-matrices following the same pattern. In particular, for $N = 3$ R-matrices can be constructed for all finite-dimensional irreducible representations of $GL(3, \mathbb{C})$.

However this method cannot be directly applied to all representations in the general case $N > 3$, because a multiplicity appears in KG-series and not all R-matrices satisfy the above condition (e.g., (39)).

5. CONCLUSION

In the present paper we have obtained all the group-invariant solutions of the YBE for arbitrary finite-dimensional irreducible representations of $GL(2, \mathbb{C})$ and a number of solutions for

GL(N, $\mathbb{C}$), $N > 2$. It seems quite interesting to extend these results in the following directions:

(a) To study arbitrary finite-dimensional and infinite-dimensional representations of GL(N, $\mathbb{C}$).

(b) To investigate R-matrices without group symmetry (e.g., [4, 8]).

(c) To consider arbitrary Lie algebras.

(d) It is known [5] that each solution of the YBE generates a completely integrable quantum chain model. Of great interest is the problem of determining the ground state energy and excitations of the models generated by the R-matrices obtained in the present paper.

ACKNOWLEDGEMENTS

We gratefully acknowledge useful conversations with L.D. Faddeev and M.A. Semenov-Tian-Shansky. We are also grateful to M.A. Semenov-Tian-Shansky for a careful reading of the manuscript and suggesting several improvements in the text.

REFERENCES

1. Sklyanin, E.K., *Doklady AN SSSR* **244**, 1337 (1979).
   Faddeev, L.D., Sklyanin, E.K., and Takhtajan, L.A., *Teor. Mat. Fiz.* **40**, 194 (1979).
   Kulish, P.P. and Sklyanin, E.K., *Phys. Lett. A* **70A**, 461 (1979).
2. Faddeev, L.D., in: *Soviet Scientific Reviews*, Harvard Academic, London, **C1**, 1980, p. 107.
3. Yang, C.N., *Phys. Rev. Lett.* **19**, 1312 (1967).
4. Baxter, R.J., *Ann. Phys.* **70**, 193 (1972).
5. Kulish, P.P. and Sklyanin, E.K., *Zapiski nauch. semin. LOMI* **95**, 129 (1980).
6. Zamolodchikov, A.B. and Zamolodchikov, Al. B., *Ann. Phys.* **120**, 253 (1979).
7. Sklyanin, E.K., preprint LOMI-E-3-79, Leningrad (1979).
8. Belavin, A.A., *Funkt. Analiz i ego pril* **14**, 18 (1980).
9. Fateev, V.A. and Zamolodchikov, A.B., *Yadernaya Fizika* **32**, 581 (1980);
   Fateev, V.A., preprint TH 2950-CERN (1980).
10. Kulish, P.P. and Reshetikhin, N.Yu., *Zapiski nauch. semin. LOMI* **101**, 112 (1980).
11. Karowski, M., *Nucl. Phys. B.* **153**, 244 (1979).
12. Fedorov, F.I., *Gruppa Lorentza,* Nauka, Moskwa, 1979, p. 117.
13. Shiffman, G., *Bull. Math. Soc. France* **99**, 3 (1971).

(Received March 20, 1981)

Advanced Studies in Pure Mathematics 16, 1988
Conformal Field Theory and Solvable Lattice Models
pp. 17–122

Exactly Solvable SOS Models II:

Proof of the star-triangle relation and combinatorial identities

E. Date, M. Jimbo, A. Kuniba, T. Miwa and M. Okado

Dedicated to Professor Nagayoshi Iwahori on his 60th birthday

§1 Introduction

This is a continuation of the paper [1], hereafter referred to as Part I. As announced therein, we give detailed proofs of (i) the star-triangle relation (STR) for the restricted solid-on-solid (SOS) models [2], and (ii) the combinatorial identities used in the evaluation of the local height probabilities (LHPs) [1, 3]. We try to keep this paper self-contained so that the mathematical content is comprehensible without reading Part I. Below we shall outline the setting and the content of each section.

1.1. The fusion models

The STR is the following system of equations for functions $W(a, b, c, d|u)$ $(a, b, c, d \in \boldsymbol{Z}, u \in \boldsymbol{C})$, to be called Boltzmann weights:

$$\sum_g W(a, b, g, f|u)\, W(f, g, d, e|u+v)\, W(g, b, c, d|v)$$
$$=\sum_g W(f, a, g, e|v)\, W(a, b, c, g|u+v)\, W(g, c, d, e|u).$$

Section 2 deals with the construction of solutions to the STR by the fusion procedure. Using the Boltzmann weights of the eight vertex SOS (8VSOS) model [4] as an elementary block, we construct "composed blocks" satisfying the STR. As in the 8VSOS case the resulting weights depend on the elliptic "nome" p as well as the parameter u.

This construction given in section 2.1 is known as the block spin transformation in the renormalization group theory. Namely, we sum up the freedoms $\ell_i \in \boldsymbol{Z}$ associated with sites i of a given lattice $\mathscr{L}_1$, leaving free the ones in $\mathscr{L}_N = N\mathscr{L}_1$. (See Fig. 1.1.) The ℓ_i is called a height in the sequel. In general, the locality of the Hamiltonian is not preserved by this

Received March 27, 1987.
Revised June 2, 1987.

transformation. However, by starting from a specially chosen inhomogeneous 8VSOS Hamiltonian we can retain the locality. (See Fig. 1.2.) Here it is crucial that the 8VSOS weights satisfy the STR.

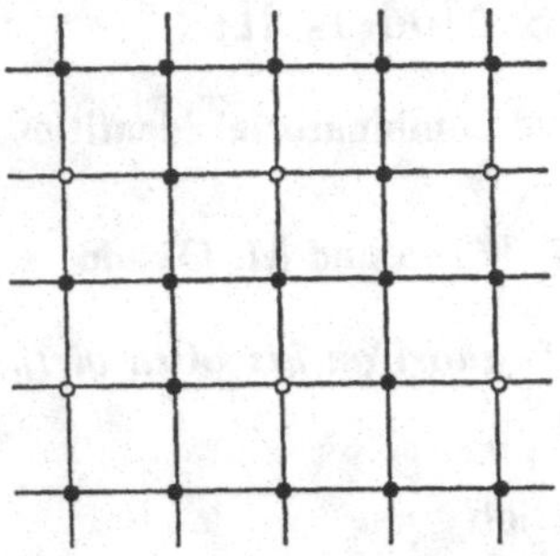

Fig. 1.1

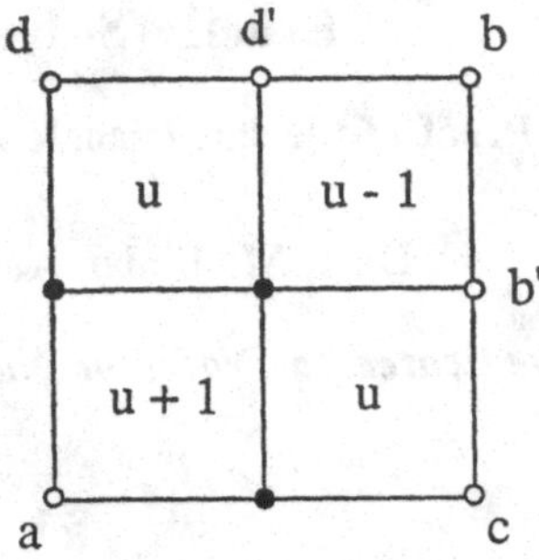

Fig. 1.2

Fig. 1.1. The block spin transformation with $N=2$. The solid circles are summed up.

Fig. 1.2. The locality of the renormalized weight. Upon summing over solid circles, the weight is independent of b' and d'.

In the 8VSOS model the adjacent heights ℓ_i, ℓ_j are restricted to $|\ell_i-\ell_j|=1$, while the fusion procedure gives rise to the constraint

$$\ell_i-\ell_j=-N,\ -N+2,\ \cdots,\ N. \tag{1.1.1}$$

A pair of integers (ℓ_i, ℓ_j) satisfying (1.1.1) is called weakly admissible.

At this stage the models are not quite realistic (any more than the 8VSOS model is), for the ℓ_i ranges over all integers. Following Andrews, Baxter and Forrester (ABF) [5], we next pick up (section 2.2) for each positive integer L a finite subset of the weights such that the STR is satisfied among themselves and ℓ_i is restricted to

$$\ell_i=1,\ \cdots,\ L-1.$$

In fact, this restriction process fetches us another constraint for adjacent heights

$$\ell_i+\ell_j=N+2,\ \cdots,\ 2L-N-2. \tag{1.1.2}$$

A weakly admissible pair (ℓ_i, ℓ_j) is called admissible if it satisfies (1.1.2). Although the block spin transformation alone produces models equivalent to the original one, the restricted models labelled by (L, N) are all inequivalent because of (1.1.2).

Section 2.3 is a side remark on the vertex-SOS correspondence first established by Baxter [4] for $N=1$. It is straightforward to generalize it to the correspondence between our restricted SOS models and the fusion vertex models [6]. Actually, it was this path that led us to the construction of section 2.1.

1.2. The corner transfer matrix (CTM) method

The CTM is a powerful tool in evaluating the local height probability $P(a)$

$$P(a)=\frac{1}{Z}\sum_{\text{the central height}=a}\prod W(\ell_1,\ell_2,\ell_3,\ell_4), \tag{1.2.1a}$$

$$Z=\sum\prod W(\ell_1,\ell_2,\ell_3,\ell_4). \tag{1.2.1b}$$

Here the product is taken over all faces, ℓ_1, ℓ_2, ℓ_3, ℓ_4 are four surrounding heights of a face, and the sum extends over all possible two dimensional configurations of heights in (1.2.1b) (with the restriction that the central height is fixed to a in (1.2.1a)). The method has recourse to the existence of limiting values of the weights such that the heights along diagonals lying northeast and southwest are frozen to be equal (see Fig. 1.3). In fact, such a limit is realized at $p=\pm 1$. To put it in a different way, the model shrinks to a sort of one dimensional system in this limit. The eigenvalues of the CTM therein are cast into one dimensional configuration sums. On the other hand, the STR implies a simple dependence of these eigenvalues with respect to the nome p. Therefore the evaluation of the LHPs for general p is reduced to that of the 1D configuration sums.

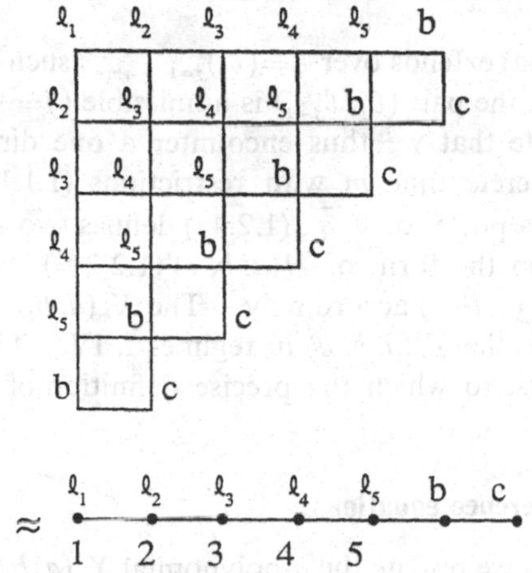

Fig. 1.3 The equivalence to a one dimensional system in the frozen limit. The integer under ℓ_i signifies its multiplicities in the CTM.

Section 3 is devoted to the derivation of the 1D configuration sums. Since the matrix elements of the CTM are products of face weights, the task is to compute the weight $W(a, b, c, d|u)$ in the above mentioned limit. The result is written in terms of a parameter w (related to the parameter u) which survives the limit as

$$\lim W(a, b, c, d|u) \sim \delta_{ac} w^{\varepsilon H(d, a, b)}.$$

Here $\sim$ means that some scaling factor has been dropped. The weight function $H(a, b, c)$ reads as

(1.2.2) Regimes II, III ($\varepsilon = -$): $H(a, b, c) = \frac{|a-c|}{4}$,

Regimes I, IV ($\varepsilon = +$):

(1.2.3a) $H(a, b, c) = \min\left(n-b, \frac{\min(a, c) - b + N}{2}\right)$ if $n \geq b$,

(1.2.3b) $= \min\left(b-n-1, \frac{b - \max(a, c) + N}{2}\right)$ if $n+1 \leq b$.

Here $n = [L/2]$. (If L is even the argument in regimes I, IV is a little bit more complicated than this. See section 3.3 for details.) With the definition of $H(a, b, c)$ above, the 1D configuration sums assume the form

(1.2.4a) $$\sum_{\ell_1 = a, \ell_{m+1} = b, \ell_{m+2} = c} q^{\phi_m(\ell_1, \cdots, \ell_{m+2})},$$

(1.2.4b) $$\phi_m(\ell_1, \cdots, \ell_{m+2}) = \sum_{j=1}^{m} jH(\ell_j, \ell_{j+1}, \ell_{j+2}).$$

The sum in (1.2.4a) extends over $\ell = (\ell_j)_{j=1,\cdots,m+2}$ such that $\ell_1 = a$, $\ell_{m+1} = b$, $\ell_{m+2} = c$ and that the pair (ℓ_j, ℓ_{j+1}) is admissible ($j = 1, \cdots, m+1$). It is intriguing to note that we thus encounter a one dimensional Brownian motion in a discrete time m with restrictions (1.1.1–2) and the weight (1.2.4b). Up to a power of $\sqrt{q}$, (1.2.4a) defines two kinds of polynomials in q depending on the form of $H(a, b, c)$ (1.2.2–3). We denote them by $X_m(a, b, c)$ or $Y_m(a, b, c)$ accordingly. The $X_m(a, b, c)$ is used in regimes II, III, while so is the $Y_m(a, b, c)$ in regimes I, IV. The b, c represent the boundary heights, to which the precise definition of the LHPs refers (see Part I).

1.3. Linear difference equations

In section 4, we rewrite the q-polynomial $X_m(a, b, c)$ or $Y_m(a, b, c)$ into series involving the Gaussian polynomials. This is done by a systematic use of linear difference equations.

Let us first consider the case where $H(a,b,c)$ is given by (1.2.2). The first step deals with the solution $f_m^{(N)}(b,c)$ of the linear difference equation

$$f_m^{(N)}(b,c)=\sum_d{}' f_{m-1}^{(N)}(d,b)q^{m|d-c|/4}. \tag{1.3.1}$$

Here the sum $\sum_d'$ extends over d such that (d,b) is weakly admissible. We fix the initial condition

$$f_0^{(N)}(b,c)=\delta_{b0}.$$

The $f_m^{(N)}(b,c)$ is called the fundamental solution. In the asymptotic region $1\ll a,b,c\ll L-1$ $X_m(a,b,c)$ coincides with $f_m^{(N)}(b-a,c-a)$. An explicit formula of $f_m^{(N)}(b,c)$ is given in Theorem 4.1.1. Using this we show that $f_m^{(N)}(b,c)$ also satisfies an equation at equal m.

$$\sum_{d=b-N,b-N+2,\dots,N-b} q^{(a+md)/4} f_{m-1}^{(N)}(a+d,a-b)=(a\to -a). \tag{1.3.2}$$

This additional identity characteristic to the fundamental solution plays a key role as explained below. The second step is to represent $X_m(a,b,c)$ as a linear superposition of $f_m^{(N)}(b,c)$ (Theorem 4.4.1). The $X_m(a,b,c)$ is characterized by the linear difference equation

$$\begin{aligned} &X_m(a,b,c)=\sum_d{}'' X_{m-1}(a,d,b)q^{m|d-c|/4}, \\ &X_0(a,b,c)=\delta_{ab}. \end{aligned} \tag{1.3.3}$$

Here the sum $\sum_d''$ extends over d such that (d,b) is admissible. The equation (1.3.3) can be viewed as (1.3.1) supplemented by the "boundary condition"

$$(\sum_d{}'-\sum_d{}'')X_{m-1}(a,d,b)q^{m|d-c|/4}=0.$$

Note that the d appearing in $\sum_d'-\sum_d''$ are "close" to the boundaries $d=1$ or $d=L-1$ in the sense that $d+b\le N$ or $d+b\ge 2L-N$. The extra equation (1.3.2) is responsible for this boundary condition.

The case when $H(a,b,c)$ is given by (1.2.3) follows a similar line (section 4.3). We denote the corresponding fundamental solution by $g_m^{(N)}(b,c)$. The expression for the $f_m^{(N)}(b,c)$ is piecewise analytic in (b,c), which reflects the piecewise analyticity of the function $|a-c|/4$. If $n\gg b$ or $n\ll b$, the behavior of $H(a,b,c)$ is essentially the same as (1.2.2). In fact,

$$H(a,b,c)=-\frac{|a-c|}{4}+\frac{a+c-2b+2N}{4}$$

for $n \gg b$, and $g_m^{(N)}(b, c)$ is expressed simply in terms of $f_m^{(N)}(b, c)$ with q replaced by q^{-1}. So is the case $n \ll b$. The expression of $g_m^{(N)}(b, c)$ is obtained throughout the intermediate regions of (b, c) as well by patchwork (see Fig. 4.1).

Section 4.2 is devoted to deriving several different expressions for $f_m^{(N)}(b, c)$. This is necessary in section 5 when we consider the $m \to \infty$ limit of $X_m(a, b, c)$ with q replaced by q^{-1}. (This is the 1D configuration sum relevant to Regime II.) The various identities proved in this section have emerged from computer experiments by using MACSYMA and FORTRAN.

1.4. The 1D configuration sums as modular functions

With the preliminaries in section 4, we proceed to the proofs of the main result of Part I; namely that the 1D configuration sums in the $m \to \infty$ limit give rise to modular functions. Except for the rather simple case of regime I, the modular functions are grasped as the branching coefficients appearing in theta function identities. (See also Appendix C.) Since they are fully exposed in Part I (Appendices A–B), we merely recall the basic definitions necessary in the proofs.

The result in regime III (resp. IV) is given in Theorem 5.1.1 (resp. Theorem 5.1.2). The branching coefficients $c_{j_1 j_2 j_3}^{(\varepsilon)}(\tau)$ ($\varepsilon = \pm$) used therein are defined through the theta function identity

$$\Theta_{j_1, m_1}^{(-,\varepsilon)}(z, q)\Theta_{j_2, m_2}^{(-,+)}(z, q)/\Theta_{1,2}^{(-,+)}(z, q) = \sum_{j_3} c_{j_1 j_2 j_3}^{(\varepsilon)}(\tau)\Theta_{j_3 m_3}^{(-,\varepsilon)}(z, q).$$

Here the sum extends over $j_3 \in \boldsymbol{Z}$ such that $0 < j_3 < m_3$ (resp. $0 < j_3 \leq m_3$) if $\varepsilon = +$ (resp. $\varepsilon = -$). The theta function $\Theta_{j,m}^{(-,\varepsilon)}(z, q)$ is defined by

$$\Theta_{j,m}^{(-,\varepsilon)}(z, q) = \sum_{n \in \boldsymbol{Z}, \gamma = n + j/2m} \varepsilon^n q^{m\gamma^2}(z^{-m\gamma} - z^{m\gamma}).$$

By residue calculus we obtain an expression of $c_{j_1 j_2 j_3}^{(\pm)}(\tau)$ as a threefold sum (see (5.1.1)). It is rather straightforward to identify $X_m(a, b, c)$ (or $Y_m(a, b, c)$) with this form of $c_{j_1 j_2 j_3}^{(\pm)}(\tau)$.

The difficulty arises when we consider the $m \to \infty$ limit of $X_m(a, b, c; q^{-1})$. For simplicity sake we explain it in the case $a = \langle b - mN \rangle$ (see eq. (B.1) for the definition of $\langle\ \rangle$) and $c = b + N$. The dominant contribution to this quantity comes from the configuration $(\bar{\ell}_j)_{j=1,\cdots,m+2}$, where $\bar{\ell}_j = \langle b + (j - m - 1)N \rangle$. Setting $k = -[(b - mN - 1)/(L-2)]$ we have

$$\phi_m(\bar{\ell}_1, \cdots, \bar{\ell}_{m+2}) = \frac{m(m+1)N + k(k-1)(L-2) - 2k(1 - b + mN)}{4}.$$

The trouble is that the term proportional to m^2 is fractional (note the implicit dependence on m through k). This fact suggests the occurrence of subtle cancellation in the expression of $X_m\ (a, b, c)$ as a series of the Gaussian polynomials. To overcome this difficulty is the highlight of our combinatorial analysis.

Our goal is Theorem 5.2.1, by which the $m\to\infty$ limit of $X_m(a, \bar{\ell}_{m+1}, \bar{\ell}_{m+2}; q^{-1})$ is identified with the branching coefficient $e^{\ell}_{jk}(\tau)$. The $e^{\ell}_{jk}(\tau)$ are characterized by theta function identities of ℓ variables (see (B.5) of Part I). We refer to Appendix B of Part I as for the Lie algebra theoretic interpretation of $e^{\ell}_{jk}(\tau)$. (See also [3] as for $c^{(+)}_{j_1j_2j_3}(\tau)$.) In [7] it was found that the matrix inverse to $(e^{\ell}_{jk}(\tau))$ is given simply by theta series $(\tilde{p}_{kj}(\tau)/\eta(\tau))$ (see (5.2.4)). Upon adjusting the precise power of q, our task is reduced to showing

$$\lim_{\substack{m\to\infty \\ nN\equiv\rho \bmod 2(L-2)}} \sum_{0<a<L} q^{M(m,a,b)} X_m(a, b, b+N; q^{-1})\tilde{p}^{L-2}_{a-1,b'-1}(\tau)$$

$$=\eta(\tau) \qquad \text{if } \langle b'\rangle=\langle b-\rho\rangle,$$

$$=0 \qquad \text{otherwise,}$$

where

$$M(m, a, b)=\frac{m(m+1)N}{4}-\frac{1}{4(L-2)}\left(mN+\frac{L}{2}-b\right)^2 +\frac{1}{4L}\left(\frac{L}{2}-a\right)^2+\frac{1}{24}.$$

This is proved in section 5.2 exploiting the expressions of $f^{(N)}_m(b, c)$ given in section 4.2.

The text is followed by four Appendices A-D. Appendix A determines the lowest order of the branching coefficients $c^{(\pm)}_{j_1j_2j_3}(\tau)$. Appendix B is the proof of the fact that $(\bar{\ell}_j)=(\langle b+jN\rangle)$ is a ground state configuration. Appendix C expresses $c^{(\pm)}_{j_1j_2j_3}(\tau)$ with $m_2=3$, 4 in terms of theta zero values. Appendix D gives the free energy per site for the fusion models.

1.5. Notations

In sections 4,5, we retain the following notations in Part I.

$$\text{(1.5.1)} \qquad \begin{bmatrix} m \\ j \end{bmatrix}=(q)_m/(q)_{m-j}(q)_j \qquad \text{if } 0\leq j\leq m,$$

$$=0 \qquad \text{otherwise,}$$

where

(1.5.2) $$(z)_m=(1-z)(1-zq)\cdots(1-zq^{m-1}).$$

(1.5.3) $$E(z,q)=\prod_{k=1}^{\infty}(1-zq^{k-1})(1-z^{-1}q^k)(1-q^k).$$

(1.5.4) $$\theta_1(u,p)=2|p|^{1/8}\sin u\prod_{k=1}^{\infty}(1-2p^k\cos 2u+p^{2k})(1-p^k).$$

(1.5.5) $$\varphi(q)=\prod_{k=1}^{\infty}(1-q^k).$$

(1.5.6) $$\eta(\tau)=q^{1/24}\varphi(q),\qquad q=e^{2\pi i\tau}.$$

Throughout this paper we shall fix this relation between q and τ.

(1.5.7) $$\begin{aligned}\varepsilon_j^{\ell}&=1/2 \qquad \text{if } j\equiv 0 \bmod \ell,\\ &=1 \qquad \text{otherwise.}\end{aligned}$$

In sections 2, 3, we use the following notations of [3]. Let $H(u)$ and $\Theta(u)$ denote the Jacobian elliptic theta functions with the half periods K and iK' (see [8]).

(1.5.8) $$[u]=\theta_1(\pi\lambda u/2K,p)=\zeta H(\lambda u)\Theta(\lambda u),$$

where $p=e^{-\pi K'/K}$, $\zeta=p^{-1/8}\varphi(p)/\varphi(p^2)^2$ and λ is a free parameter. (We have changed the definition of $[u]$ in [3] by the factor ζ.)

(1.5.9) $$[u]_m=[u][u-1]\cdots[u-m+1],$$

(1.5.10) $$\begin{bmatrix}u\\m\end{bmatrix}=\frac{[u]_m}{[m]_m}.$$

The symbol (1.5.10) is used only in sections 2, 3 and in Appendix D. It is not to be confused with the Gaussian polynomial (1.5.1).

§ 2. The Fusion Models

2.1. Fusion of SOS models

Let $\mathscr{L}$ be a two dimensional square lattice. An SOS model on $\mathscr{L}$ consists of (i) an integer variable ℓ_i on each site (=lattice point) i of $\mathscr{L}$, and (ii) a function $W(a,b,c,d)$ of a quadruple of integers (a,b,c,d). We call ℓ_i a height variable and $W(a,b,c,d)$ a Boltzmann weight (or simply a weight). The integers a,b,c,d represent a configuration of heights round a face (=an elementary square), ordered *anticlockwise* from the southwest corner.

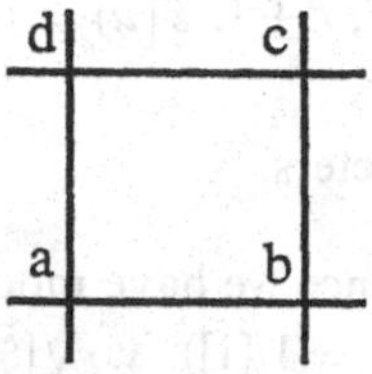

Fig. 2. 1 A height configuration round a face.

Take three systems of Boltzmann weights W, W' and W'', all depending on a complex variable u. The STR for W, W' and W'' is the following set of functional equations.

$$\begin{aligned}&\sum_g W(a, b, g, f\,|\,u)W'(f, g, d, e\,|\,u+v)W''(g, b, c, d\,|\,v)\\ &=\sum_g W''(f, a, g, e\,|\,v)W'(a, b, c, g\,|\,u+v)W(g, c, d; e\,|\,u)\\ &\qquad \text{for any } a, b, c, d, e, f.\end{aligned} \tag{2.1.1}$$

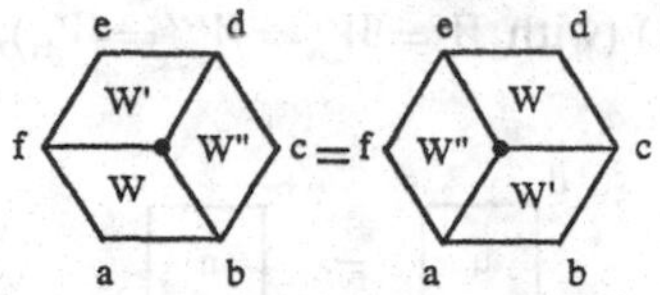

Fig. 2. 2 The STR for the system of weights W, W', W''.

As for the significance of the STR in the theory of solvable lattice models, see [8]. Our aim here is to construct a class of solutions to the STR on the basis of a known one by the *fusion procedure*.

As a seed solution, we take the 8-vertex SOS model of Baxter [4]. By definition, its Boltzmann weight $W_{11}(a, b, c, d\,|\,u)$ is set to 0 unless

$$|a-b|=|b-c|=|c-d|=|d-a|=1. \tag{2.1.2}$$

The nonzero weights are parametrized in terms of the elliptic theta function (1.5.4)

$$[u]=\theta_1(\pi\lambda u/2K, p) \tag{2.1.3}$$

as follows.

$$W_{11}(\ell\pm1,\ \ell\pm2,\ \ell\pm1,\ \ell\,|\,u)=\frac{[u+1]}{[1]}, \tag{2.1.4a}$$

$$W_{11}(\ell\mp1,\ \ell,\ \ell\pm1,\ \ell\,|\,u)=\frac{[\xi+\ell\pm1][u]}{[\xi+\ell][1]}, \tag{2.1.4b}$$

(2.1.4c) $$W_{11}(\ell\pm1, \ell, \ell\pm1, \ell\,|\,u) = \frac{[\xi+\ell\mp u]}{[\xi+\ell]},$$

Here λ and ξ are free parameters.

Remark. For convenience we have modified the weights $W(a, b\,|\,d, c)$ of [5], eq. (1.2.12b) (with $\rho' = 1/[1]$) as $\sqrt{[\xi+a]/[\xi+c]}\,W_{11}(a, b, c, d\,|\,u) = \sqrt{-1}^{\,c-a}\,W(a, b\,|\,d, c)$. The l.h.s. will be the symmetrized weight $S_{11}(a, b, c, d\,|\,u)$ defined below (2.1.24). Our variables are related to those in [5] through

(2.1.5) $$u = (v-\eta)/2\eta, \quad \lambda = 2\eta, \quad \xi = w_0/2\eta.$$

We shall often write the weights (2.1.4) graphically as

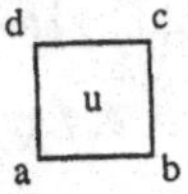

$$\begin{array}{c} d \quad\quad c \\ \boxed{\quad u \quad} \\ a \quad\quad b \end{array} = W_{11}(a, b, c, d\,|\,u).$$

Besides the STR (2.1.1) (with $W = W' = W'' = W_{11}$), they satisfy the symmetry

(2.1.6) $$\begin{array}{c} d \quad\quad c \\ \boxed{\quad u \quad} \\ a \quad\quad b \end{array} = \begin{array}{c} b \quad\quad c \\ \boxed{\quad u \quad} \\ a \quad\quad d \end{array}$$

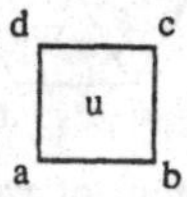

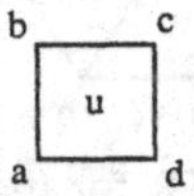

For $u=0$ and $u=-1$ they simplify to:

(2.1.7) $$\begin{array}{c} d \quad\quad c \\ \boxed{\;u=0\;} \\ a \quad\quad b \end{array} = \delta_{ac},$$

where (2.1.2) is implied,

(2.1.8a) $$\begin{array}{c} d \quad\quad c \\ \boxed{\;u=-1\;} \\ a \quad\quad b \end{array} = 0 \qquad \text{if } |b-d|=2,$$

(2.1.8b) $$\begin{array}{c} \ell \quad\quad \ell\pm1 \\ \boxed{\;u=-1\;} \\ \ell\pm1 \quad\quad \ell \end{array} = - \begin{array}{c} \ell \quad\quad \ell\pm1 \\ \boxed{\;u=-1\;} \\ \ell\mp1 \quad\quad \ell \end{array} \left(= \frac{[\xi+\ell\pm1]}{[\xi+\ell]} \right).$$

These properties will play a role in the following construction.

An elementary step of the fusion procedure is provided by

Lemma 2.1.1. (i) *Put*

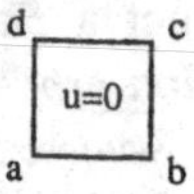

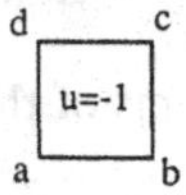

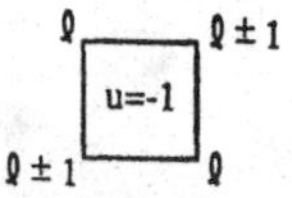

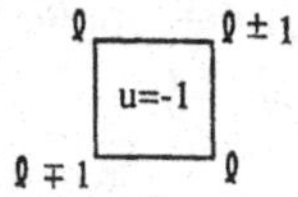

(2.1.9) $$W'_{21}(a, b, c, d\,|\,u)=\sum_{a'} W_{11}(a, a', c', d\,|\,u+1)W_{11}(a', b, c, c'\,|\,u).$$

Then the r.h.s. is independent of the choice of c' provided that $|c-c'|=|c'-d|=1$.

(ii) *For all a, b, c, d we have $W'_{21}(a, b, c, d\,|-1)=0$.*

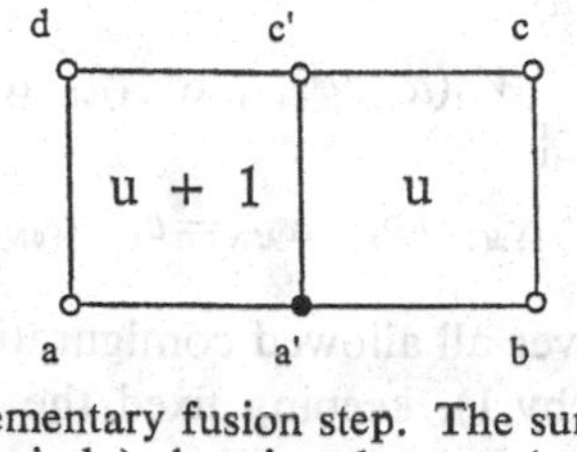

Fig. 2. 3 An elementary fusion step. The sum is taken over a' (solid circle), keeping the rest (open circles) fixed. It is independent of c'.

Proof. To see (i), it suffices to check the case $c=d$ (otherwise the possible choice of c' is unique). We are to show

(2.1.10) $0=\sum_{a'}$ 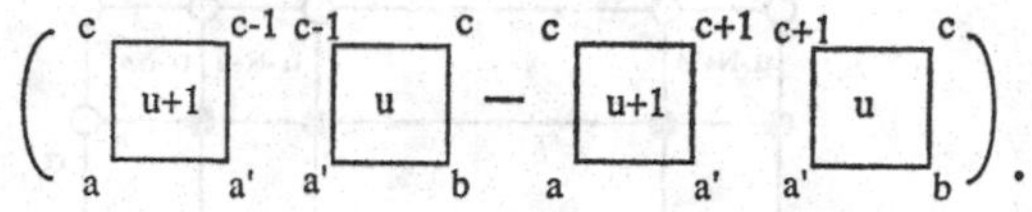

Let $c''=c+1$ (or $c-1$), and multiply each summand of (2.1.10) by 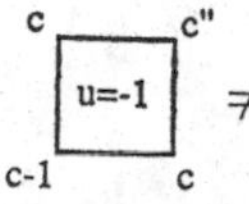

$\neq 0$. Thanks to (2.1.8b) the result can be put into the form

(2.1.11) $$\sum_{c'=c\pm 1} W_{11}(a, a', c', c\,|\,u+1)W_{11}(a', b, c, c'\,|\,u)W_{11}(c', c, c'', c\,|-1)$$
$$=\sum_{c'=c\pm 1} W_{11}(a', b, c', a\,|-1)W_{11}(a, c', c'', c\,|\,u)W_{11}(c', b, c, c''\,|\,u+1).$$

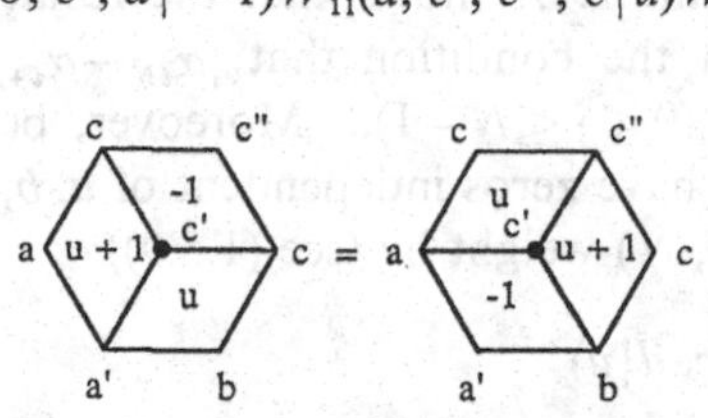

Fig. 2. 4 The STR used in the proof of the c'-independence.

To get the second line we have used the STR (2.1.1) with $u\to -1$ and $v\to u+1$. If $|a-b|=2$, then $W_{11}(a', b, c', a\,|-1)=0$ by (2.1.8a). If $a=b$, then take the sum over a' in (2.1.11). Owing to (2.1.8b), the summands contribute with opposite signs for each fixed c'. This proves (2.1.10).

To show (ii), choose $c'=b\pm 2$ in (2.1.9) and use (2.1.8a). This choice is not allowed if $c=b\mp 1$ and $d=b\pm 1$, in which case apply (2.1.7) or (2.1.8b). □

Now let M and N be positive integers. Define

$$
\begin{aligned}
&W'_{MN}(a,b,c,d\,|\,u)\\
(2.1.12)\qquad &=\sum \prod_{\substack{0\le i\le M-1\\ 0\le j\le N-1}} W_{11}(\alpha_{ij},\alpha_{i+1j},\alpha_{i+1j+1},\alpha_{ij+1}\,|\,u-i-j+M-1),\\
&\alpha_{00}=a,\quad \alpha_{M0}=b,\quad \alpha_{MN}=c,\quad \alpha_{0N}=d,
\end{aligned}
$$

where the sum is taken over all allowed configurations $\{\alpha_{ij}\}$ (*i.e.* the neighboring pairs must differ by 1), keeping fixed the corner heights a, b, c, d and the right/top boundary heights

$$
(2.1.13)\qquad \alpha_{M1},\cdots,\alpha_{MN-1},\quad \alpha_{1N},\cdots,\alpha_{M-1N}.
$$

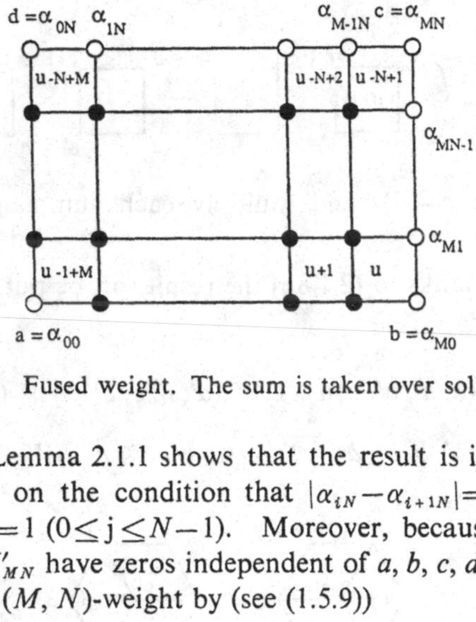

Fig. 2.5 Fused weight. The sum is taken over solid circles.

Repeated use of Lemma 2.1.1 shows that the result is independent of the choice of (2.1.13) on the condition that $|\alpha_{iN}-\alpha_{i+1N}|=1$ $(0\le i\le M-1)$ and $|\alpha_{Mj}-\alpha_{Mj+1}|=1$ $(0\le j\le N-1)$. Moreover, because of Lemma 2.1.1 (ii) the weights W'_{MN} have zeros independent of a, b, c, d. Factoring them out we define the (M, N)-weight by (see (1.5.9))

$$
\begin{aligned}
&W_{MN}(a,b,c,d\,|\,u)\\
(2.1.14)\qquad &=W'_{MN}(a,b,c,d\,|\,u)[1]^{MN}/([N]_N\prod_{j=1}^{N}[u+M-j]_{M-1}),\quad (N\le M),\\
&=W'_{MN}(a,b,c,d\,|\,u)[1]^{MN}/([M]_M\prod_{j=1}^{M}[u+M-j]_{N-1}),\quad (M\le N).
\end{aligned}
$$

By the construction it is obvious that $W_{MN}(a,b,c,d\,|\,u)$ vanishes unless

$$a-b,\quad c-d=-M,\ -M+2,\ \cdots,\ M,$$
(2.1.15)
$$a-d,\quad b-c=-N,\ -N+2,\ \cdots,\ N.$$

They also inherit the symmetry (2.1.6) for W_{11} in the form

(2.1.16) $$W_{MN}(a,b,c,d\,|\,u)=W_{NM}(a,d,c,b\,|\,u+M-N).$$

Theorem 2.1.2 (Theorem 1 of [2]). *For a triple of positive integers M, N, P, we have the following STR*

(2.1.17)
$$\sum_g W_{MN}(a,b,g,f\,|\,u)W_{MP}(f,g,d,e\,|\,u+v)W_{NP}(g,b,c,d\,|\,v)$$
$$=\sum_g W_{NP}(f,a,g,e\,|\,v)W_{MP}(a,b,c,g\,|\,u+v)W_{MN}(g,c,d,e\,|\,u).$$

Proof. It is sufficient to prove the STR for the W'_{MN}, since the normalization factors in (2.1.14) cancel out. We will show the STR for the case $M=2$, $N=P=1$. The general cases are proved similarly. From the definition of W'_{21} (2.1.9), the l.h.s. of (2.1.17) becomes

(2.1.18)
$$\sum_g \sum_{a',f'} (W_{11}(a,a',f'',f\,|\,u+1)W_{11}(a',b,g,f''\,|\,u))$$
$$\times(W_{11}(f,f',e',e\,|\,u+v+1)W_{11}(f',g,d,e'\,|\,u+v))$$
$$\times W_{11}(g,b,c,d\,|\,v).$$

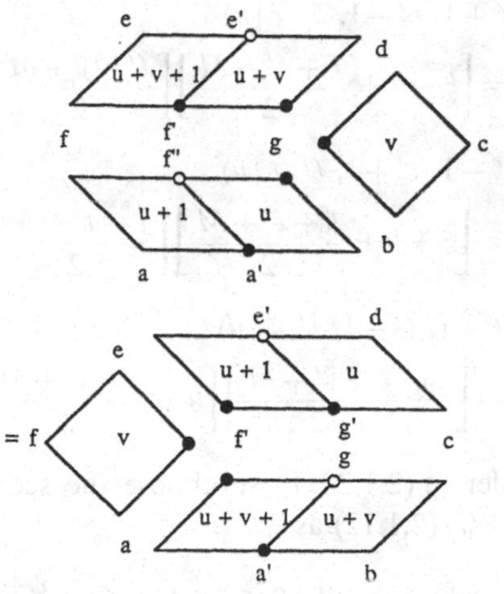

Fig. 2.6 Proof of the STR of type (2, 1, 1).

Here f'' and e' are arbitrary provided that $|e-e'|=|e'-d|=|f-f''|=|f''-g|=1$. Performing first the summation over a' and using Lemma

2.1.1 we can set $f''=f'$ in (2.1.18). Then by applying twice the STR for the weight W_{11}, (2.1.18) is transformed into

$$\sum_{f'}\sum_{a',g} W_{11}(f,a,f',e|v)(W_{11}(a,a',g,f'|u+v+1)W_{11}(a',b,c,g|u+v))$$
$$\times(W_{11}(f',g,e',e|u+1)W_{11}(g,c,d,e'|u)).$$

Again by Lemma 2.1.1 and the definition of W'_{21} (2.1.9), this is the r.h.s. of (2.1.17). □

We shall refer to (2.1.17) as the STR of type (M,N,P).

Below we list explicit formulas of the W_{MN}. In the course of the derivation we will use the identity

$$[x+z][x-z][y+w][y-w]-[x+w][x-w][y+z][y-z] = [x+y][x-y][z+w][z-w]. \tag{2.1.19}$$

First consider the case $N=1$.

Lemma 2.1.3 (eq. (5) of [2]). *The* $(M,1)$-*weight is given by*

$$W_{M1}(\ell+1,\ell'+1,\ell',\ell|u) = \left[\xi+\frac{\ell+\ell'-M}{2}\right]\left[u+\frac{\ell'-\ell+M}{2}\right]\Big/[1][\xi+\ell], \tag{2.1.20a}$$

$$W_{M1}(\ell+1,\ell'-1,\ell',\ell|u) = \left[\xi-u+\frac{\ell+\ell'-M}{2}\right]\left[\frac{\ell'-\ell+M}{2}\right]\Big/[1][\xi+\ell], \tag{2.1.20b}$$

$$W_{M1}(\ell-1,\ell'+1,\ell',\ell|u) = \left[\xi+u+\frac{\ell+\ell'+M}{2}\right]\left[\frac{\ell-\ell'+M}{2}\right]\Big/[1][\xi+\ell], \tag{2.1.20c}$$

$$W_{M1}(\ell-1,\ell'-1,\ell',\ell|u) = \left[\xi+\frac{\ell+\ell'+M}{2}\right]\left[u+\frac{\ell-\ell'+M}{2}\right]\Big/[1][\xi+\ell]. \tag{2.1.20d}$$

Proof. To derive (2.1.20a), we choose the sequence $\{\alpha_{i1}\}$ (2.1.13) in the definition of W'_{M1} (2.1.12) as

$$\alpha_{i1}=\ell-i \quad \text{if } 0\leq i\leq M-\frac{\ell'-\ell+M}{2},$$
$$=\ell'-M+i \quad \text{if } M-\frac{\ell'-\ell+M}{2}\leq i\leq M.$$

For such a choice of $\{\alpha_{i1}\}$, the sequence $\{\alpha_{i0}\}$ in (2.1.13) is uniquely determined to be $\alpha_{i0}=\alpha_{i1}+1$. Then (2.1.20a) follows immediately. We show the formula (2.1.20b) by an induction on M. Assume that it is true for M. By the definition and the induction hypothesis we have

$$\begin{aligned}
&W_{M+1,1}(\ell+1,\ell'-1,\ell',\ell\,|\,u)\\
&=(W_{M1}(\ell+1,\ell'-2,\ell'-1,\ell\,|\,u+1)W_{11}(\ell'-2,\ell'-1,\ell',\ell'-1\,|\,u)\\
&\quad+W_{M1}(\ell+1,\ell',\ell'-1,\ell\,|\,u+1)W_{11}(\ell',\ell'-1,\ell',\ell'-1\,|\,u))[1]/[u+1]\\
&=\left(\left[\xi-u-1+\frac{\ell+\ell'-1-M}{2}\right]\left[\frac{\ell'-1-\ell+M}{2}\right]\Big/[1][\xi+\ell]\right.\\
&\quad\times\frac{[\xi+\ell'][u]}{[1][\xi+\ell'-1]}+\left[\xi+\frac{\ell+\ell'-1-M}{2}\right]\left[u+1+\frac{\ell'-1-\ell+M}{2}\right]\\
&\quad\left.\Big/[1][\xi+\ell]\times\frac{[\xi+\ell'-1-u]}{[\xi+\ell'-1]}\right)[1]/[u+1].
\end{aligned}$$

Using the identity (2.1.19) with

$$2x=\xi+\frac{\ell+\ell'-3-M}{2},\qquad 2y=\xi+\frac{3\ell'-\ell-1+M}{2},$$

$$2z=\xi-2u-1+\frac{\ell+\ell'-1-M}{2},\qquad 2w=\xi+\frac{\ell+\ell'+1-M}{2},$$

we obtain (2.1.20b) for $M+1$. The equalities (2.1.20c) and (2.1.20d) are shown similarly. □

Now we turn to the general (M,N)-weights. The definition (2.1.12) can be viewed as defining the W'_{MN} in terms of the W'_{M1}:

$$W'_{MN}(a,b,c,d\,|\,u)=\sum\prod_{i=1}^{N}W'_{M1}(a_{i-1},b_{i-1},b_i,a_i\,|\,u-i+1),\tag{2.1.21}$$
$$a_0=a,\quad a_N=d,\quad b_0=b,\quad b_N=c.$$

Here the sum extends over $a_1,\cdots,a_{N-1}$ such that $|a_i-a_{i+1}|=1$ for $0\le i\le N-1$. Using (2.1.21) we show the

Lemma 2.1.4 (Appendix of [2]). *Assuming* $N\le M$, *we have (see* (1.5.10))

$$\begin{aligned}
&W_{MN}(\ell+2j-N,\ell+2i-M+N,\ell+2i-M,\ell\,|\,u)\\
&=\begin{bmatrix}M-i\\N-j\end{bmatrix}\begin{bmatrix}\xi+\ell+i+j-M-1\\j\end{bmatrix}\begin{bmatrix}i+u\\j\end{bmatrix}\begin{bmatrix}\xi+\ell+i+u\\N-j\end{bmatrix}\\
&\quad\Big/\begin{bmatrix}\xi+\ell+j\\N-j\end{bmatrix}\begin{bmatrix}\xi+\ell+2j-N-1\\j\end{bmatrix},
\end{aligned}\tag{2.1.22a}$$

$$W_{MN}(\ell+2j-N,\ \ell+2i-M-N,\ \ell+2i-M,\ \ell\,|\,u)$$

(2.1.22b) $$=\begin{bmatrix} i \\ j \end{bmatrix}\begin{bmatrix} \xi+\ell+i+j-M-1-u \\ j \end{bmatrix}\begin{bmatrix} M-i+u \\ N-j \end{bmatrix}\begin{bmatrix} \xi+\ell+i \\ N-j \end{bmatrix}$$
$$\Big/\begin{bmatrix} \xi+\ell+j \\ N-j \end{bmatrix}\begin{bmatrix} \xi+\ell+2j-N-1 \\ j \end{bmatrix}.$$

Proof. These can be checked by an induction on N. The case $j=0$ or N is straightforward, since in (2.1.21) the choice of $\{a_j\}$ is unique. The induction proceeds in a similar way as in the derivation of (2.1.20b). □

Choosing the sequence in $\{\alpha_{Mj}\}$ (2.1.13) suitably, the general (M, N)-weight is expressed as a sum of products of weights of type (2.1.22). In fact, we get

$$W_{MN}(\ell+2j-N,\ \ell'+2i-N,\ \ell',\ \ell\,|\,u)\begin{bmatrix} N \\ i \end{bmatrix}$$

(2.1.23a) $$=\sum_{k=\max(0,i+j}^{\min(i,j)} W_{Mi}(l+2k-i,\ \ell'+i,\ \ell',\ \ell\,|\,u-N+i)$$
$$\times W_{M,N-i}(\ell+2j-N,\ \ell'+2i-N,\ \ell'+i,\ \ell+2k-i\,|\,u),$$

(2.1.23b) $$=\sum_{k=\max(0,j-i)}^{\min(N-i,j)} W_{M,N-i}(\ell+2k-N+i,\ \ell'-N+i,\ \ell',\ \ell\,|\,u-i)$$
$$\times W_{Mi}(\ell+2j-N,\ l'+2i-N,\ \ell'-N+i,\ \ell+2k-N+i\,|\,u).$$

Let us modify the weight W_{MN} as

(2.1.24a) $$S_{MN}(a,b,c,d\,|\,u)=\left(\frac{(a,b)_M(d,a)_N}{(d,c)_M(c,b)_N}\right)^{1/2} W_{MN}(a,b,c,d\,|\,u),$$

$$(\ell,\ell')_M=(\ell',\ell)_M$$

(2.1.24b) $$=\begin{bmatrix} M \\ \dfrac{\ell-\ell'+M}{2} \end{bmatrix}^{-1}\frac{\left[\xi+\dfrac{\ell+\ell'-M}{2},\ \xi+\dfrac{\ell+\ell'+M}{2}\right]}{\sqrt{[\xi+\ell][\xi+\ell']}},$$

where we set

(2.1.24c) $$[A,B]=[A][A+1]\cdots[B],\qquad [A,A-1]=1.$$

(We have changed the definition of $(\ell,\ell')_M$ from that of ref. 2 by the common factor $[M]_M$). The STR remains valid for S_{MN}, because the factor in front of the r.h.s. of (2.1.24a) cancels out in the STR.

As a result of this modification S_{MN} aquires the following symmetry.

Theorem 2.1.5 (Theorem 2 of [2]).

(2.1.25a) $S_{MN}(a, b, c, d|u)=S_{NM}(a, d, c, b|M-N+u)$

(2.1.25b) $=S_{NM}(c, b, a, d|M-N+u)$

(2.1.25c) $=S_{MN}(c, d, a, b|u)$

(2.1.25d) $=(g_a g_c/g_b g_d)S_{MN}(b, a, d, c|-M+N-1-u),$

where $g_\ell=\varepsilon_\ell\sqrt{[\xi+\ell]}$, $\varepsilon_\ell=\pm 1$ *and* $\varepsilon_\ell\varepsilon_{\ell+1}=(-)^\ell$.

For the proof we prepare

Lemma 2.1.6.

$$S_{MN}(a, b, c, d|u)=(g_a g_c/g_b g_d)S_{MN}(d, c, b, a|-M+N-1-u). \tag{2.1.26}$$

Proof. Direct calculations using explicit formulas (2.1.22) give

$$W_{MN}(a, b, c, d|u) = \frac{(d, c)_M}{(a, b)_M}\frac{g_a g_c}{g_b g_d} W_{MN}(d, c, b, a|-M+N-1-u) \tag{2.1.27}$$

for the case $|b-c|=N$. Next using (2.1.23a) and the definition (2.1.24) of S_{MN}, we have

$$\begin{aligned}
&S_{MN}(\ell+2j-N, \ell'+2i-N, \ell', \ell|u)\begin{bmatrix} N \\ i \end{bmatrix}\\
&=\left(\frac{(\ell+2j-N, \ell'+2i-N)_M(\ell, \ell+2j-N)_N}{(\ell, \ell')_M(\ell', \ell'+2i-N)_N}\right)^{1/2}\\
&\times\sum_{k=\max(0,i+j-N)}^{\min(i,j)} W_{Mi}(\ell+2k-i, \ell'+i, \ell', \ell|u-N+i)\\
&\qquad\times W_{M,N-i}(\ell+2j-N, \ell'+2i-N, \ell'+i, \ell+2k-i|u).
\end{aligned}$$

By applying the equality (2.1.27) for the extreme case ($|b-c|=N$), the r.h.s. becomes

$$\begin{aligned}
&\frac{g_{\ell'}g_{\ell+2j-N}}{g_\ell g_{\ell'+2i-N}}\times\left(\frac{(\ell,\ell')_M(\ell, \ell+2j-N)_N}{(\ell+2j-N, \ell'+2i-N)_M(\ell', \ell'+2i-N)_N}\right)^{1/2}\\
&\times\sum_{k=\max(0,i+j-N)}^{\min(i,j)} W_{M,N-i}(\ell+2k-i, \ell'+i, \ell'+2i-N, \ell+2j-N|\\
&\qquad\qquad -M+N-i-1-u)\\
&\qquad\times W_{Mi}(\ell, \ell', \ell'+i, \ell+2k-i|-M+N-1-u).
\end{aligned}$$

This is just $\begin{bmatrix} N \\ i \end{bmatrix}$ times the r.h.s. of (2.1.26). □

Proof of Theorem 2.1.5. We know already (2.1.25a) by (2.1.16). Using this and Lemma 2.1.6 alternately, we have

$$\begin{aligned} S_{MN}(a, b, c, d\,|\,u) &= S_{NM}(a, d, c, b\,|\,M-N+u) \\ &= \frac{g_a g_c}{g_b g_d} S_{NM}(b, c, d, a\,|\,-u-1) \\ &= \frac{g_a g_c}{g_o g_d} S_{MN}(b, a, d, c\,|\,-M+N-1-u) \\ &= S_{MN}(c, d, a, b\,|\,u), \end{aligned}$$

which proves (2.1.25c-d). Eq. (2.1.25b) follows from (2.1.25a, c). □

2.2. Restricted SOS models

Hereafter we consider exclusively the SOS models with $M=N$. We shall also specialize the parameters in (2.1.3–4) to

$$\lambda = 2K/L, \tag{2.2.1a}$$

$$\xi = 0, \tag{2.2.1b}$$

where L is a positive integer satisfying

$$L \geq N+3. \tag{2.2.2}$$

The condition (2.2.1a) gives rise to the symmetry

$$[L-u] = [u]. \tag{2.2.3}$$

Let ℓ_i, ℓ_j be adjacent heights. In addition to the restriction (2.1.15) (with $M=N$)

$$\ell_i - \ell_j = -N, -N+2, \cdots, N, \tag{2.2.4}$$

we impose further the constraint

$$\ell_i + \ell_j = N+2, N+4, \cdots, 2L-N-2. \tag{2.2.5}$$

These two conditions imply in particular that each height variable can assume at most $L-1$ states

$$\ell_i = 1, 2, \cdots, L-1. \tag{2.2.6}$$

We remark that if $N=1$, then conversely (2.2.5) follows from (2.2.4) and (2.2.6).

The condition (2.2.5) naturally enters for the following reason. As an illustration, take the following (2, 2)-weight

$$S_{22}(\ell, \ell, \ell, \ell \mid u) = \frac{[\ell-1-u][\ell+u]}{[\ell-1][\ell]} + \frac{[\ell-1][\ell+2]}{[\ell][\ell+1]} \frac{[u][1+u]}{[1][2]}.$$

Because of the specialization (2.2.1), this now has poles at $\ell=1$ or $L-1$. We forbid such configurations to occur by requiring (2.2.5).

In Part I, we called a pair of heights (a, b) *admissible* if it satisfies (2.2.4–5). By abuse of language, we call a weight $S_{NN}(a, b, c, d \mid u)$ admissible if the pairs (a, b), (b, c), (c, d) and (d, a) are all admissible.

In this paragraph we shall show that admissible weights are well defined (Theorem 2.2.1), and that they satisfy the STR among themselves (Theorem 2.2.4). (The latter statement does not follow directly from the STR (2.1.17) for the unspecialized weights, since non-admissible configurations can occur in the summand even if $a, b, \cdots, f$ are all admissible.) Following ABF [5], we call the resulting models *restricted SOS models.*

In order to prove their existence we make use of the explicit formulas in the previous paragraph for the symmetrized weights $S_{NN}(a, b, c, d \mid u)$. We shall frequently use the parametrization

$$\text{(2.2.7)} \qquad a=\ell+N-2r, \quad b=\ell+2(N-k), \quad c=\ell+N-2s, \quad d=\ell.$$

Thanks to the 180° rotational symmetry (2.1.24c), we can assume without loss of generality

$$\text{(2.2.8a)} \qquad 0 \leq k \leq N.$$

Then the weight $S_{NN}(\ell+N-2r, \ell+2(N-k), \ell+N-2s, \ell \mid u)$ is admissible if and only if

$$\text{(2.2.8b)} \qquad \max(0, \ell+2N-L-k+1) \leq r, s \leq \min(\ell-1, k).$$

In terms of (2.2.7) the formulas (2.1.22–24) read as follows.

$$S_{NN}(\ell+N-2r, \ell+2(N-k), \ell+N-2s, \ell \mid u)$$

$$\text{(2.2.9a)} \qquad = \sqrt{S} \sum_{i=\max(0, s-r)}^{\min(k-r, s)} U(i)$$

$$\text{(2.2.9b)} \qquad = \sqrt{S} \sum_{i=\max(k-N, s-r)}^{\min(k-r, s)} D(i),$$

where

$$\text{(2.2.10)} \quad S = \left(\frac{(\ell, \ell+N-2r)_N (\ell+N-2r, \ell+2(N-k))_N}{(\ell, \ell+N-2s)_N (\ell+N-2s, \ell+2(N-k))_N} \right) \Big/ \begin{bmatrix} N \\ k-s \end{bmatrix}^2$$

$$=\frac{1}{\begin{bmatrix}N\\k-r\end{bmatrix}\begin{bmatrix}N\\k-s\end{bmatrix}}$$

$$\times\frac{\begin{bmatrix}N\\s\end{bmatrix}\begin{bmatrix}\ell+N-r\\N+1\end{bmatrix}\begin{bmatrix}\ell+2N-k-r\\N+1\end{bmatrix}\begin{bmatrix}\ell+N-2s\\1\end{bmatrix}}{\begin{bmatrix}N\\r\end{bmatrix}\begin{bmatrix}\ell+N-s\\N+1\end{bmatrix}\begin{bmatrix}\ell+2N-k-s\\N+1\end{bmatrix}\begin{bmatrix}\ell+N-2r\\1\end{bmatrix}},$$

(2.2.11)
$$\begin{aligned}U(i)=&\,W_{N,N-k+s}(\ell+N-k-s+2i,\ \ell+2N-k-s,\\&\qquad\ell+N-2s,\ \ell\,|\,u-k+s)\\&\times W_{N,k-s}(\ell+N-2r,\ \ell+2(N-k),\ \ell+2N-k-s,\\&\qquad\ell+N-k-s+2i\,|\,u)\\=&\,\frac{\begin{bmatrix}s\\s-i\end{bmatrix}\begin{bmatrix}N-i\\k-r-i\end{bmatrix}\begin{bmatrix}\ell+N-k-s+i-1\\N-k+i\end{bmatrix}\begin{bmatrix}\ell+2N-k-s+i\\r-s+i\end{bmatrix}}{\begin{bmatrix}\ell+N-k+i\\s-i\end{bmatrix}\begin{bmatrix}\ell+N-2r-1\\k-r-i\end{bmatrix}\begin{bmatrix}\ell+N-k-s+2i-1\\N-k+i\end{bmatrix}}\\&\times\begin{bmatrix}\ell+N-r-s+i\\r-s+i\end{bmatrix}^{-1}\\&\times\begin{bmatrix}\ell+N-k+u\\s-i\end{bmatrix}\begin{bmatrix}\ell+N-r-s-1-u\\k-r-i\end{bmatrix}\begin{bmatrix}N-k+u\\N-k+i\end{bmatrix}\begin{bmatrix}i+u\\r-s+i\end{bmatrix},\end{aligned}$$

(2.2.12)
$$\begin{aligned}D(i)=&\,W_{N,k-s}(\ell+k-2r+s-2i,\ \ell+N-k-s,\ \ell+N-2s,\ \ell\\&\qquad|\,u-N+k-s)\\&\times W_{N,N-k+s}(\ell+N-2r,\ \ell+2(N-k),\ \ell+N-k-s,\\&\qquad\ell+k-2r+s-2i\,|\,u)\\=&\,\frac{\begin{bmatrix}k-r+s-i\\s-i\end{bmatrix}\begin{bmatrix}N-s\\k-r-i\end{bmatrix}\begin{bmatrix}\ell+N-k-r-1\\N-k+i\end{bmatrix}\begin{bmatrix}\ell+N-s\\r-s+i\end{bmatrix}}{\begin{bmatrix}\ell+N-2r+s-i\\s-i\end{bmatrix}\begin{bmatrix}\ell+k-2r+s-2i-1\\k-r-i\end{bmatrix}\begin{bmatrix}\ell+N-2r-1\\N-k+i\end{bmatrix}}\\&\times\begin{bmatrix}\ell+k-r-i\\r-s+i\end{bmatrix}^{-1}\\&\times\begin{bmatrix}\ell+N-r-i+u\\s-i\end{bmatrix}\begin{bmatrix}\ell+N-r-i-1-u\\k-r-i\end{bmatrix}\\&\times\begin{bmatrix}N-k+r-s+i+u\\N-k+i\end{bmatrix}\begin{bmatrix}-N+k+u\\r-s+i\end{bmatrix}.\end{aligned}$$

Theorem 2.2.1. *An admissible weight is well defined.*

Proof. Let A be a real number and B a non-negative integer. Call a term $\begin{bmatrix} A \\ B \end{bmatrix}$ of type I if $B \leq A < L$, of type II if $B < L$. We can easily check the following:

$$0 < \begin{bmatrix} A \\ B \end{bmatrix} < \infty \qquad \text{if } \begin{bmatrix} A \\ B \end{bmatrix} \text{ is of type I,}$$

$$\begin{bmatrix} A \\ B \end{bmatrix} < \infty \qquad \text{if } \begin{bmatrix} A \\ B \end{bmatrix} \text{ is of type II.}$$

Now consider the expressions (2.2.9). Using (2.2.7–8) and $\max(k-N, s-r) \leq i \leq \min(k-r, s)$, we can verify that (i) all the factors appearing in (2.2.10) and in the denominator of (2.2.11–12) are of type I, and that (ii) those in the numerator of (2.2.11–12) are of type II. The theorem follows from these facts. □

In what follows a weight $S_{NN}(a, b, c, d \mid u)$ is always understood to be admissible.

Lemma 2.2.2.

$$S_{NN}(L-a, L-b, L-c, L-d \mid u) = S_{NN}(a, b, c, d \mid u). \tag{2.2.13}$$

Proof. We consider first the specialization (2.2.1a), regarding ξ as yet arbitrary. From (2.1.4) and (2.2.3), we see that each unsymmetrized (1, 1)-weight $W_{11}(a, b, c, d \mid u)$ is invariant under the transformation

$$\xi \longrightarrow -\xi, \; a \longrightarrow L-a, \; b \longrightarrow L-b, \; c \longrightarrow L-c, \; d \longrightarrow L-d. \tag{2.2.14}$$

Therefore the unsymmetrized (N, N)-weights have the same nature by the definition. Moreover we can see from (2.1.23b) that the symmetrizing factor $(a, b)_N (d, a)_N / (d, c)_N (c, b)_N$ is also invariant under (2.2.14). Letting ξ tend to 0 we get (2.2.13). □

Lemma 2.2.3. *Put*

$$\bar{N} = N + \ell - k - 1, \quad \bar{\ell} = k + 1, \quad \bar{k} = \ell - 1. \tag{2.2.15}$$

If $N \geq \bar{N}$, then we have

$$\begin{aligned} &S_{NN}(\ell + N - 2r, \ell + 2(N-k), \ell + N - 2s, \ell \mid u) \\ &= \frac{\begin{bmatrix} N-u \\ N-\bar{N} \end{bmatrix}}{\begin{bmatrix} N \\ \bar{N} \end{bmatrix}} S_{\bar{N}\bar{N}}(\bar{\ell} + \bar{N} - 2r, \bar{\ell} + 2(\bar{N} - \bar{k}), \bar{\ell} + \bar{N} - 2s, \bar{\ell} \mid u). \end{aligned} \tag{2.2.16}$$

Proof. First we show that $D(i+1)/D(i)$ is invariant under the change

(2.2.17) $$\ell \longrightarrow \bar{\ell}, \quad N \longrightarrow \bar{N}, \quad k \longrightarrow \bar{k}.$$

A direct calculation by using (2.2.12) shows that

$$D(i+1)/D(i) = D_1(i)D_2(i)$$

where

$$D_1(i) = \frac{[\ell+N-2r+s-i][\ell+N-r-i][\ell+k-2r+s-2i-2] \cdot [\ell+k-r-i][s-i]}{[\ell+k-2r+s-2i][\ell+k-2r-i-1][N-k+r-s+i+1] \cdot [N-k+i+1][r-s+i+1]}$$
$$\times \frac{[N-k+r-s+i+1+u][-N+k-r+s-i+u]}{[\ell+N-r-i-1-u][\ell+N-r-i+u]},$$

$$D_2(i) = \frac{[\ell-r-i-1][k-r-i]}{[\ell-r+s-i-1][k-r+s-i]}.$$

Note that $\ell+N=\bar{\ell}+\bar{N}$, $\ell+k=\bar{\ell}+\bar{k}$, $N-k=\bar{N}-\bar{k}$. Therefore under (2.2.17) all factors in $D_1(i)$ are invariant, while in $D_2(i)$ the two factors get interchanged both in the numerator and in the denominator.

Because of the symmetry (2.1.25b), we may assume $r \leq s$. Since $D_2(\ell-r-1)=0$, it suffices to consider $D(i)$ with $i \leq l-r-1$. We thus have

$$s-r \leq i \leq \min(\ell-r-1, k-r, s).$$

The transformation (2.2.17) does not change both ends of this interval. Therefore we get

(2.2.18) $$\frac{S_{NN}(\ell+N-2r, \ell+2(N-k), \ell+N-2s, \ell \,|\, u)}{S_{\bar{N}\bar{N}}(\bar{\ell}+\bar{N}-2r, \bar{\ell}+2(\bar{N}-\bar{k}), \bar{\ell}+\bar{N}-2s, \bar{\ell} \,|\, u)} = \left(\frac{S}{\bar{S}}\right)^{1/2} \frac{D(s-r)}{\bar{D}(s-r)}$$

where $\bar{S}$, $\bar{D}$ are obtained from S, D by applying (2.2.17). Using (2.2.10, 12) we find

$$\frac{S}{\bar{S}} = \left(\frac{[\ell-r, k-r][N-s+1, N-r]}{[\bar{N}+1, N][\bar{N}-s+1, \bar{N}-r]}\right)^2,$$

$$\frac{D(s-r)}{\bar{D}(s-r)} = \frac{[\bar{N}-s+1, \bar{N}-r][\ell+N-k-u \; N-u]}{[\ell-r, k-r][N-s+1, N-r]},$$

where $[A, B]$ is defined in (2.1.24c). So

$$\text{the r.h.s. of (2.2.18)} = \frac{[\bar{N}+1-u,\ N-u]}{[\bar{N}+1,\ N]} = \frac{\begin{bmatrix} N-u \\ N-\bar{N} \end{bmatrix}}{\begin{bmatrix} N \\ \bar{N} \end{bmatrix}}.$$

This completes the proof of (2.2.16). □

Now let us proceed to the proof of the STR. We prepare several lemmas. We call (a, b) lower (resp. upper) non-admissible if $a+b \leq N$ (resp. $a+b \geq 2L-N$). A weakly admissible pair (a, b) cannot be both lower and upper non-admissible.

Lemma 2.2.4. *Assume that the pairs (d, c) and (c, b) are admissible and that (d, a), (a, b) are weakly admissible but not both admissible. The symmetrized weight $S_{NN}(a, b, c, d \mid u)$ is then finite-valued. It is vanishing if one of the following occurs:*

(i) $a=0$,

(ii) *either (d, a) or (a, b) is admissible,*

(iii) *(d, a) is lower non-admissible and (a, b) is upper non-admissible,*

(iv) *(d, a) is upper non-admissible and (a, b) is lower non-admissible.*

Proof. By virtue of the symmetry (2.1.25a), we can assume without loss of generality that

$$b \geq d. \tag{2.2.19}$$

We consider the following three cases for (d, a)

(1) $N+2 \leq d+a \leq 2L-N-2$,

(2) $d+a \leq N$,

(3) $d+a \geq 2L-N$,

and for (a, b)

(1′) $N+2 \leq a+b \leq 2L-N-2$,

(2′) $a+b \leq N$,

(3′) $a+b \geq 2L-N$.

The case (1)–(1′) is excluded by the assumption of the lemma. The cases (1)–(2′), (3)–(1′) and (3)–(2′) do not occur because, under the assumption (2.2.19), (2′) implies (2) and (3) implies (3′).

The symmetries (2.1.25a) and (2.2.13) allows us to reduce (3)–(3′) to (2)–(2′) and (1)–(3′) to (2)–(1′). So we have only three cases to check:

Case 1: (2)–(1′). *Case* 2: (2)–(2′). *Case* 3: (2)–(3′).

Let us use the parametrization (2.2.7). From the assumptions we have

(2.2.20a) $$0 \leq r \leq k \leq N,$$

(2.2.20b) $$\max(0, \ell+2N-L-k+1) \leq s \leq \min(\ell-1, k).$$

The condition (2) is nothing but

(2.2.21) $$\ell - r \leq 0,$$

while (1′), (2′), (3′) can be written respectively as follows:

(2.2.22a) $$1 \leq \ell+N-k-r \leq L-N-1,$$

(2.2.22b) $$\ell+N-k-r \leq 0,$$

(2.2.22c) $$\ell+N-k-r \geq L-N.$$

Use the formula (2.2.9a) for $S_{NN}(a, b, c, d|u)$, in which i is restricted to

(2.2.23) $$\max(0, s-r) \leq i \leq \min(k-r, s).$$

From (2.2.20–21) and (2.2.23) we can write

(2.2.24) $$\sqrt{S}\,U(i) = \frac{\sqrt{AB[a]}}{C} \times (\text{a non-zero finite-valued factor}),$$

where

$$A = [\ell-r, \ell+N-r], \qquad B = [\ell+N-k-r, \ell+2N-k-r],$$
$$C = [\ell+N-k-r+i, \ell+N-r-s+i].$$

For $A = [I, J]$ let A_{ini} and A_{fin} signify respectively I and J. We write $A \subset B$ if $A_{ini} \geq B_{ini}$ and $A_{fin} \leq B_{fin}$. We find from (2.2.20–21,23) that

(2.2.25a) $$C \subset B,$$

(2.2.25b) $$[0] \subset A,$$

(2.2.25c) $$C_{fin} < L.$$

In Case 1 and Case 3, $C_{ini} > 0$ follows respectively from (2.2.22a) and (2.2.22c). Therefore (2.2.24) is equal to 0 from (2.2.25b-c). In Case 2 the finiteness is a direct consequence of (2.2.24–25) because $-L < C_{ini}$.

Finally we note that $S_{NN}(0, b, c, d|u) = 0$ follows immediately from (2.2.24). □

Lemma 2.2.5. *Assume that the pairs* (a, d) *and* (d, c) *are admissible and that* (a, b) *and* (b, c) *are weakly admissible, then* $S_{NN}(a, b, c, d|u)$ *is finite-valued.*

Proof. From Theorem 2.1.5, we have

$$S_{NN}(a, b, c, d\,|\,u)=\frac{g_a g_c}{g_b g_d} S_{NN}(b, c, d, a\,|\,-1-u).$$

Here $S_{NN}(b, c, d, a\,|\,-1-u)$ is finite-valued from Lemma 2.2.4. We have $g_d \neq 0$ because of the admissibility of the pair (a, d). As for g_b the factor $\sqrt{[b]}$ in the numerator of (2.2.24) cancels it out. This proves the lemma. □

Lemma 2.2.6. *Consider Case* 2 *in the proof of Lemma* 2.2.4. *We assume* (2.2.19) *and that* $[a]\neq 0$. *Setting*

(2.2.26) $$\bar{r}=\ell+N-r$$

we have

(2.2.27a) $$\sqrt{S}\, W_{NN}(a, b, c, d\,|\,u)=\begin{bmatrix} N \\ \frac{c-b+N}{2} \end{bmatrix}^{-1} \sum_i \sqrt{S}\, U(i),$$

where

(2.2.27b) $$\max(0, s-r, s-\bar{r})\leq i\leq \min(k-r, k-\bar{r}, s)$$

Each summand is then non-vanishing.

Proof. From (2.2.24–25), we find that

$$\sqrt{S}\, U(i)\neq 0 \qquad \text{if } C_{ini}\leq 0\leq C_{fin}.$$
$$=0 \qquad \text{otherwise.}$$

The first condition is rewritten as

$$s-\bar{r}\leq i\leq k-\bar{r}.$$

The lemma follows immediately from this. □

Lemma 2.2.7. *Let* (d, c), (c, b) *be admissible. Furthermore we suppose that* $a\neq 0$ *and the pairs* (d, a), (a, b), $(d, -a)$ $(-a, b)$ *are all weakly admissible. Then*

(2.2.28a) $$W_{NN}(a, b, c, d\,|\,u)/W_{NN}(-a, b, c, d\,|\,u)=-1.$$

(2.2.28b) $$W_{NN}(d, a, b, c\,|\,u)/W_{NN}(d, -a, b, c\,|\,u)=1,$$

(2.2.28c) $$W_{NN}(c, d, a, b\,|\,u)/W_{NN}(c, d, -a, b\,|\,u)=1,$$

(2.2.28d) $$W_{NN}(b, c, d, a\,|\,u)/W_{NN}(c, b, d, -a\,|\,u)=1.$$

Proof. First we show (2.2.28a). Note that the weak admissibility of $(d, -a)$, $(-a, b)$ implies that (d, a), (a, b) are lower non-admissible. We may assume (2.2.19). Retaining the parametrization (2.2.7), we have $-a = \ell+N-2\bar{r}$ where $\bar{r}$ is given by (2.2.26). In Lemma 2.2.6, the range of i (2.2.27b) is invariant under the change $r \to \bar{r}$. So it is sufficient to show that

$$U(i)/\bar{U}(i) = -1$$

for all i satisfying (2.2.27b), where $\bar{U}(i)$ is given by (2.2.11) with $\bar{r}$ in place of r. For the calculation we note the simple facts:

$$(2.2.29) \qquad \begin{aligned} &[A_1, B_1][A_2, B_2]/[A_1, B_2][A_2, B_1] = 1 \\ &\qquad \text{if } \max(A_1, A_2) \leq \min(B_1, B_2)+1. \end{aligned}$$

$$(2.2.30) \qquad \begin{aligned} [A, B] &= (-)^{B-A}[-B, -A] && \text{if } A, B \in \mathbf{Z} \text{ and } A \leq 0 \leq B, \\ &= (-)^{B-A+1}[-B, -A] && \text{otherwise.} \end{aligned}$$

Writing down the ratio we have

$$\begin{aligned} \frac{U(i)}{\bar{U}(i)} = {} & \frac{[N-k+\bar{r}+1, \ell+2N-k-s+i]_1 [N-k+r+1, N-i]_1 \cdot [-k+r+i, r-\bar{r}-1]_{\bar{2}}}{[-k+\bar{r}+i, \bar{r}-r-1]_{\bar{3}} [\bar{r}-r+1, \bar{r}-s+i]_2 \cdot [N-k+r+1, \ell+2N-k-s+i]_1} \\ & \times \frac{[r-\bar{r}+1, r-s+i]_3 [\ell+N-k-s+i-u, \bar{r}-s-1-u]_{\bar{4}} \cdot [-r+s+1+u, i+u]_4}{[N-k+\bar{r}+1, N-i]_1 [\ell+N-k-s+i-u, r-s-1-u]_{\bar{4}} \cdot [-\bar{r}+s+1+u, i+u]_4} \\ & \times \frac{[1, k-\bar{r}-i]_3 [1, \bar{r}-s+i]_2}{[1, k-r-i]_2 [1, r-s+i]_3}. \end{aligned}$$

We grouped together the members to which we apply (2.2.29) by putting the suffix $j=1, 2, 3, 4$. For those with barred suffix, we apply also (2.2.30). We thus find

$$U(i)/\bar{U}(i) = (-)^{(k-\bar{r}-i)+(k-r-i)+(\bar{r}-r)-1} = -1$$

as desired. This completes the proof of (2.2.28a).

Next we proceed to (2.2.28b). The rest are shown similarly. First note the following formula:

$$(2.2.31) \qquad \frac{(d, a)_N}{(d, -a)_N} = (-)^a \frac{\sqrt{[-a]}}{\sqrt{[a]}} = \frac{g_{-a}}{g_a}.$$

This is a direct consequence of (2.1.24b). From (2.1.24a), Theorem 2.1.5, (2.2.31) and (2.2.30a), we have

$$\frac{W_{NN}(a,b,c,d|u)}{W_{NN}(a,-b,c,d|u)}=\frac{\sqrt{(c,b)_N(a,-b)_N}}{\sqrt{(a,b)_N(c,-b)_N}}\frac{S_{NN}(a,b,c,d|u)}{S_{NN}(a,-b,c,d|u)}$$

$$=\frac{\sqrt{(c,b)_N(a,-b)_N}}{\sqrt{(a,b)_N(c,-b)_N}}\frac{g_{-b}}{g_b}\frac{S_{NN}(b,c,d,a|-1-u)}{S_{NN}(-b,c,d,a|-1-u)}$$

$$=\frac{(c,b)_N}{(c,-b)_N}\frac{g_{-b}}{g_b}(-)=1. \qquad \square$$

So much for the preparation. Now it is rather straightforward to show

Theorem 2.2.8 (Theorem 3 of [2]). *The set of the admissible weights $S_{NN}(a,b,c,d|u)$ satisfy the STR among themselves.*

Proof. The symmetrized weights for the unrestricted models satisfy the STR. We are to show that if the exterior pairs (a,b), (b,c), (c,d), (d,e), (e,f), (f,a) are all admissible, the terms with non-admissible inner heights cancel out among themselves in each side of the STR.

Let us consider the l.h.s. of Fig. 2.2. Set

$$R(g)=S_{NN}(a,b,g,f|u)S_{NN}(f,g,d,e|u+v)S_{NN}(g,b,c,d|v).$$

If admissible pairs and non-admissible pairs coexist among the inner pairs (g,b), (g,d) and (g,f), then $R(g)$ vanishes because of Lemma 2.2.4. The same is true if lower and upper non-admissible pairs coexist.

From Lemma 2.2.2, it is enough to consider the case that the inner pairs are lower non-admissible.

If a pair (a,b) is admissible, it is clear from (2.1.24b) that the factor $(a,b)_N$ is strictly positive. So Lemma 2.2.4–5 and Theorem 2.1.5 allow us to prove the cancellation of the unsymmetrized weights

$$R'(g)=W_{NN}(a,b,g,f|u)W_{NN}(f,g,d,e|u+v)W_{NN}(g,b,c,d|v)$$

instead of the symmetrized ones.

Under the condition that $a>0$, if the pair (a,g) is weakly admissible and lower non-admissible then so is $(a,-g)$. Therefore we have

$$\text{the summation of the lower non-admissible terms}$$
$$=R'(0)+\sum_{g>0}(R'(g)+R'(-g)),$$

where

$$R'(g)=W_{NN}(a,b,g,f|u)W_{NN}(f,g,d,e|u+v)W(g,b,c,d|v).$$

$R'(0)$ vanishes from Lemma 2.2.4 while $R'(g)+R'(-g)=0$ follows from Lemma 2.2.7. We have now proved that non-admissible terms in the l.h.s. cancel out.

The proof is the same for the r.h.s. □

2.3. Vertex-SOS correspondence

In [4] through an attempt to obtain the eigenvectors of its row-to-row transfer matrix, Baxter found an equivalence of the eight vertex model to an SOS model. We shall extend this equivalence to the fusion models. The result of this paragraph is not used in the rest of this paper.

First we recall the fusion of the eight vertex model [6]. We denote by $R^{\alpha\beta}_{\gamma\delta}$ the Boltzmann weight of the eight vertex model associated with a vertex as indicated in Fig. 2.7.

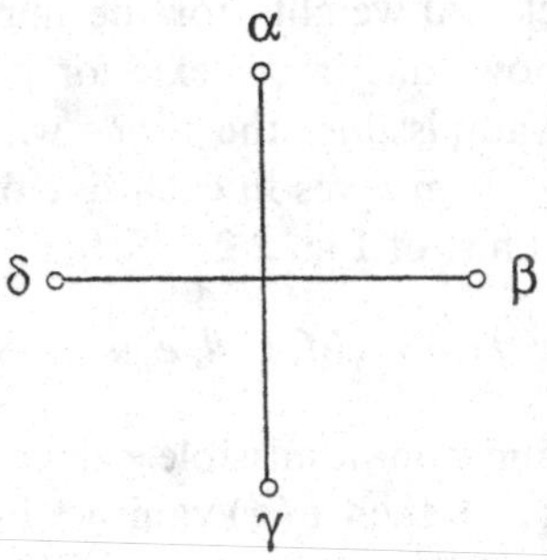

Fig. 2.7 A vertex configuration.

We use the following parametrization ($\lambda\neq 0$):

$$(2.3.1)\qquad \begin{aligned} R^{\alpha\alpha}_{\alpha\alpha}(u)&=\rho_0\Theta(\lambda)\Theta(\lambda u)H(\lambda(u+1)),\\ R^{\alpha\beta}_{\alpha\beta}(u)&=\rho_0\Theta(\lambda)H(\lambda u)\Theta(\lambda(u+1)),\\ R^{\alpha\beta}_{\beta\alpha}(u)&=\rho_0 H(\lambda)\Theta(\lambda u)\Theta(\lambda(u+1)),\\ R^{\alpha\alpha}_{\beta\beta}(u)&=\rho_0 H(\lambda)H(\lambda u)H(\lambda(u+1)),\\ \rho_0&=1/\Theta(0)H(\lambda)\Theta(\lambda). \end{aligned}$$

Here $\alpha,\beta=\pm 1$ and $\alpha\neq\beta$.

Let $V=Cv_+\oplus Cv_-\simeq C^2$. We define $E_{\alpha\beta}\in\operatorname{End}(V)$ $(\alpha,\beta=\pm 1)$ by $E_{\alpha\beta}v_\gamma=v_\alpha\delta_{\beta\gamma}$, and set $R(u)=\sum R^{\alpha\beta}_{\gamma\delta}(u)E_{\gamma\alpha}\otimes E_{\delta\beta}\in\operatorname{End}(V\otimes V)$. Let $V_1,\cdots,V_M$ be copies of V. Given $T\in\operatorname{End}(V\otimes V)$ we define $T^{jk}\in\operatorname{End}(V_1\otimes\cdots\otimes V_M)$ by $T^{jk}=\iota_{jk}T\pi_{jk}$, where ι_{jk} is the natural injection $\iota_{jk}\colon V_j\otimes V_k\to$

$V_1 \otimes \cdots \otimes V_M$ and π_{jk} is the natural projection π_{jk}: $V_1 \otimes \cdots \otimes V_M \to V_j \otimes V_k$. In this notation the Yang-Baxter equation reads as

(2.3.2) $$R^{12}(u)R^{13}(u+v)R^{23}(v)=R^{23}(v)R^{13}(u+v)R^{12}(u).$$

We denote by $I \in \mathrm{End}(V \otimes V)$ the identity, and by $C \in \mathrm{End}(V \otimes V)$ the transposition: $Cv_1 \otimes v_2 = v_2 \otimes v_1$. We set $P=(C+I)/2$. We have

(2.3.3a) $$R(0)=C,$$

(2.3.3b) $$R(-1)=C-I.$$

Set $u=-1$ in (2.3.1) and multiply it by P^{12}. Then, because of (2.3.3b) and $P(C-I)=0$, the l.h.s. is zero. Therefore we have

(2.3.4) $$P^{12}R^{23}(u)R^{13}(u-1)C^{12}=P^{12}R^{23}(u)R^{13}(u-1).$$

Now we denote by $P_{1\ldots M}$ the projection on the space of the symmetric tensors in $V_1 \otimes \cdots \otimes V_M$:

$$P_{1\ldots M}=\frac{1}{M!}(C^{1M}+\cdots+C^{M-1M}+I)\cdots(C^{12}+I).$$

We call $P_{1\ldots M}$ a symmetrizer for short. We prepare further copies $V_{\bar{1}}, \cdots, V_{\bar{N}}$ of V and use $T^{j\bar{k}}$ in the sense similar to T^{jk} with V_k replaced by $V_{\bar{k}}$. We define an operator $R'_{1\ldots M\bar{1}}(u) \in \mathrm{End}(V_1 \otimes \cdots \otimes V_M \otimes V_{\bar{1}} \otimes \cdots \otimes V_{\bar{N}})$ by

$$R'_{1\ldots M\bar{1}}(u)=P_{1\ldots M}R^{1\bar{1}}(u+M-1)\cdots R^{M\bar{1}}(u).$$

Lemma 2.3.1.

(2.3.5a) (i) $R'_{1\ldots M\bar{1}}(u)P_{1\ldots M}=R'_{1\ldots M\bar{1}}(u).$

(ii) $R'_{1\ldots M\bar{1}}(u)=0$

(2.3.5b) *for* $u=-1, \cdots, -M+1, -1+iK'/\lambda, \cdots,$
$-M+1+iK'/\lambda.$

Proof. (i) This follows immediately from (2.3.4).

(ii) For $u=-1, \cdots, -M+1$ $R'_{1\ldots M\bar{1}}(u)=0$ because of (2.3.3) and the identity $P^{12}C^{1\bar{1}}(C^{2\bar{1}}-I)=0$. The latter half is proved similarly. □

This lemma tells that

$$R_{1\ldots M\bar{1}}(u)=R'_{1\ldots M\bar{1}}(u)[1]^{M-1}/[u+M-1]_{M-1}$$

is holomorphic. For $M \geq N$ we define the (M, N)-weight

$$R_{MN}(u)\in \mathrm{End}(\underbrace{V\otimes\cdots\otimes V}_{M}\otimes\underbrace{V\otimes\cdots\otimes V}_{N})$$
$$\simeq \mathrm{End}(V_1\otimes\cdots\otimes V_M\otimes V_{\bar 1}\otimes\cdots\otimes V_{\bar N})$$

by

$$R_{MN}(u)=P_{\bar 1\ldots\bar N}R_{1\ldots M\bar N}(u)\cdots R_{1\ldots M\bar 1}(u-N+1)[1]^N/[N]_N.$$

For $M<N$ we define it through the following commutative diagram.

$$\begin{array}{ccc}
\underbrace{V\otimes\cdots\otimes V}_{M}\otimes\underbrace{V\otimes\cdots\otimes V}_{N} & \xrightarrow{C_{MN}} & \underbrace{V\otimes\cdots\otimes V}_{N}\otimes\underbrace{V\otimes\cdots\otimes V}_{M} \\
R_{MN}(u)\Big\downarrow & & \Big\downarrow R_{NM}(u+M-N) \\
\underbrace{V\otimes\cdots\otimes V}_{M}\otimes\underbrace{V\otimes\cdots\otimes V}_{N} & \xrightarrow[C_{MN}]{} & \underbrace{V\otimes\cdots\otimes V}_{N}\otimes\underbrace{V\otimes\cdots\otimes V}_{M}
\end{array}$$

The C_{MN} is the transposition of $\underbrace{V\otimes\cdots\otimes V}_{M}$ and $\underbrace{V\otimes\cdots\otimes V}_{N}$. As (2.3.5a) we have from (2.3.4)

$$R_{MN}(u)=R_{MN}(u)P_{1\ldots M}=R_{MN}(u)P_{\bar 1\ldots\bar N}. \tag{2.3.6}$$

Theorem 2.3.2 ([6]). *Fix a triple of integers* (M, N, P), *and set* $V_1=\underbrace{V\otimes\cdots\otimes V}_{M}$, $V_2=\underbrace{V\otimes\cdots\otimes V}_{N}$ *and* $V_3=\underbrace{V\otimes\cdots\otimes V}_{P}$. *We define* R^{12}_{MN}, R^{13}_{MP}, $R^{23}_{NP}\in \mathrm{End}(V_1\otimes V_2\otimes V_3)$ *as in* (2.3.1). *Then they satisfy the Yang-Baxter equation*:

$$R^{12}_{MN}(u)R^{13}_{MP}(u+v)R^{23}_{NP}(v)=R^{23}_{NP}(v)R^{13}_{MP}(u+v)R^{12}_{MN}(u). \tag{2.3.7}$$

Proof. If we discard all the symmetrizers appearing in (2.3.7), the equality follows by a repeated use of (2.3.1). Multiplying this identity by the symmetrizers from the left and using (2.3.6) we obtain (2.3.7). □

Now we establish an equivalence between R_{MN} and W_{MN}. Choose arbitrary constants s^+ and s^-, and fix ξ in (2.1.4) by

$$\xi=\frac{s^+ + s^-}{2}-\frac{K}{\lambda}.$$

We set

$$\phi_{ab}(u)=\begin{pmatrix} H(\lambda(s^\varepsilon+a-\varepsilon u)) \\ \Theta(\lambda(s^\varepsilon+a-\varepsilon u))\end{pmatrix} \qquad \text{if } \varepsilon=b-a=\pm 1,$$
$$=0 \qquad \text{otherwise.}$$

We define a vector $\phi_{M,ab}(u)$ in $\underbrace{V\otimes\cdots\otimes V}_{M}$ by

$$(2.3.8)\qquad \phi_{M,ab}(u)=P_{1\ldots M}(\phi_{a_0a_1}(u+M-1)\otimes\cdots\otimes\phi_{a_{M-1}a_M}(u)) \quad (a_0=a,\ a_M=b).$$

The a_i are integers satisfying $|a_i-a_{i+1}|=1$, The definition (2.3.8) is independent of the choice of these integers. We note that $W_{MN}(a, b, c, d|u)$ are invariant under the change of $(a, b, c, d, \xi, s^+, s^-)$ to $(-a, -b, -c, -d, -\xi, -s^-+2K/\lambda, -s^++2K/\lambda)$. This is useful when we check the following identity due to Baxter [4] (see Fig. (2.8)):

$$(2.3.9)\qquad R(u-v)(\phi_{dc}(u)\otimes\phi_{cb}(v))=\sum_a W_{11}(a, b, c, d|u-v)\phi_{ab}(u)\otimes\phi_{da}(v).$$

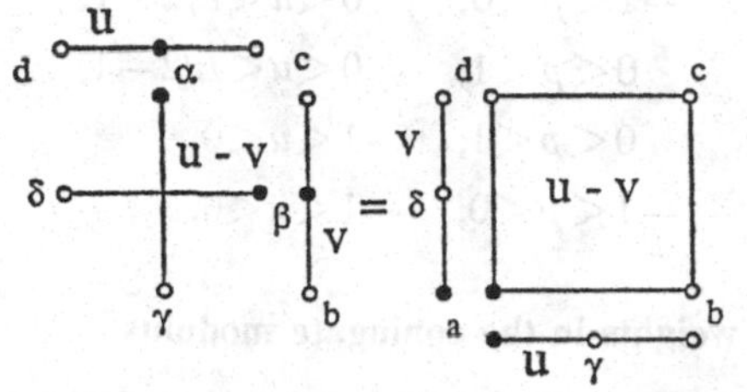

Fig. 2.8 The vertex-SOS correspondence.

By a similar argument as in Theorem 2.3.2 the identity (2.3.9) is generalized to

Theorem 2.3.3 ([2]).

$$R_{MN}(u-v)(\phi_{M,dc}(u)\otimes\phi_{N,cb}(v))=\sum_a W_{MN}(a, b, c, d|u-v)\phi_{M,ab}(u)\otimes\phi_{N,da}(v).$$

We omit the proof, which is similar to that of Theorem 2.3.2. A simple case is schematically shown in Fig. 2.9.

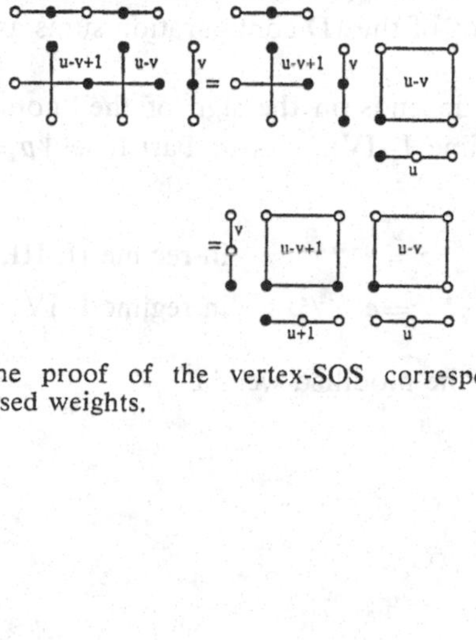

Fig. 2.9 The proof of the vertex-SOS correspondence for fused weights.

Remark. The first formula in section 4 of [2] should be coorrected as

$$R(u)=\frac{1}{\Theta(0)}\sum_{a=0}^{3} w_a(u)\sigma_a\otimes\sigma_a.$$

§ 3. One Dimensional Configuration Sums

In this section we transform the LHPs into 1D configuration sums by means of the CTM method. As in the case $N=1$ [5], we consider the following four regimes.

Regime I:	$-1<p<0,$	$0<u<L/2-1,$
Regime II:	$0<p<1,$	$0<u<L/2-1,$
Regime III:	$0<p<1,$	$-1<u<0,$
Regime IV:	$-1<p<0,$	$-1<u<0.$

3.1. Boltzmann weights in the conjugate modulus

In order to evaluate the LHPs we appeal to Baxter's corner transfer matrix (CTM) trick [8]. The method is summarized in Appendix A of [5] for $N=1$. Apart from the change of notation (2.1.5), the reasoning therein applies equally well for general N. As a result, the LHP $P(a)$ is given in terms of a 1D configuration sum (in an appropriate limit $m\to\infty$)

$$X_m(a,b,c)=\sum q^{\phi_m(\ell_1,\cdots,\ell_{m+2})},\qquad \ell_1=a,\quad \ell_{m+1}=b,\quad \ell_{m+2}=c, \tag{3.1.1}$$

where the sum is taken over sequences $\ell_2,\cdots,\ell_m$ which are admissible in the sense that (ℓ_j,ℓ_{j+1}) satisfies (2.2.4–5) for $1\le j\le m+1$. (For the definition of q, see Table 1 in Part I). The goal of section 3 is to determine the function $\phi_m(\ell_1,\cdots,\ell_{m+2})$. The precise definition of the LHPs and the expressions in terms of the 1D configuration sums are given in section 2 of Part I.

The working depends on the sign of the "nome" p: $p>0$ (regime II, III), or $p<0$ (regime I, IV). As in Part I, let $|p|=e^{-\varepsilon/L}$ and define the variable x by

$$\begin{aligned} x&=e^{-4\pi^2/\varepsilon} && \text{in regime II, III,}\\ &=e^{-2\pi^2/\varepsilon} && \text{in regime I, IV.}\end{aligned}$$

We shall consider the modified weight

(3.1.2) $$\mathcal{S}_N(a, b, c, d) = S_{NN}(a, b, c, d)\frac{G_a G_c}{G_b G_d} F^{-N},$$

where G_a and F will be specified below (3.1.6). The LHPs are unaffected by such a modification.

The line of argument to obtain $\phi_m(\ell_1, \cdots, \ell_{m+2})$ goes as follows [5, 8]. Let $A(u)$ denote the CTM corresponding to the southeast quadrant of the lattice. It is an operator acting on the subspace $\mathcal{H}_{b,c}$ of $C^{L-1}\otimes\cdots\otimes C^{L-1}$ ($m+2$ fold tensor product) spanned by the vectors $e_{\ell_1}\otimes\cdots\otimes e_{\ell_{m+2}}$ such that $\ell_{m+1}=b$, $\ell_{m+2}=c$ and $\{\ell_j\}_{j=1}^{m+2}$ is admissible. Here e_i signifies the standard basis of C^{L-1}, and (b, c) is a fixed admissible pair. Define the face operators $\mathcal{U}_i$ $(2\le i\le m+1)$ by ([5], eq. (A2))

(3.1.3) $$\mathcal{U}_i e_{\ell_1}\otimes\cdots\otimes e_{\ell_{m+2}} = \sum_{\ell'_i} \mathcal{S}_N(\ell_i, \ell_{i+1}, \ell'_i, \ell_{i-1}) e_{\ell_1}\otimes\cdots\otimes e_{\ell'_i}\otimes\cdots\otimes e_{\ell_{m+2}}.$$

(For $i=m+1$ the sum is confined to only one term $\ell'_i=\ell_i=b$.) The CTM is defined to be $A(u)=\mathcal{F}_2\mathcal{F}_3\cdots\mathcal{F}_{m+1}$, $\mathcal{F}_j=\mathcal{U}_{m+1}\mathcal{U}_m\cdots\mathcal{U}_j$ ([5], eq. (A14)). Then, in the large lattice limit, (i) $P_a=\mu_a/\sum_{a'}\mu_{a'}$ with $\mu_a= g_a^2\,\mathrm{trace}_{\ell_1=a}(A(-t))$, where $t=2$ or $2-L$ and g_a is given in Theorem 2.1.5 (cf. [5], eqs (A26, 28)); (ii) the eigenvalues of $A(u)$ have the form $x^{n_i u}$ with $2n_i \in Z$ (up to a common factor). Finally (iii) the n_i's are calculated by considering the limit

(3.1.4) $$x\longrightarrow 0, \quad u\longrightarrow 0, \quad w=x^u \text{ fixed.}$$

In this limit the face operators $\mathcal{U}_i$, and hence $A(u)$ also, become diagonal. (For even L the last step requires a slight modification. See the discussion at the end of section 3.3.) In what follows we mean by lim the limit (3.1.4).

To study the limit of (3.1.2) in this sense, we exploit the conjugate modulus transformation $p\to x$:

(3.1.5) $$[u]=\theta_1\left(\frac{\pi u}{L}, p\right)=\kappa\, x^{\mu}(u).$$

The quantities κ, μ, (u), $G_a=G_{L-a}$ and F are given as follows.

(3.1.6)

	Regime II, III	Regime I, IV
κ	$\sqrt{2\pi L/\varepsilon}\, x^{L/8}$	$\sqrt{\pi L/\varepsilon}\, x^{L/16}$
μ	$u(u-L)/2L$	$u(2u-L)/2L$
(u)	$E(x^u, x^L)$	$E(x^u, -x^{L/2})$
G_a	$w^{a(a-L)/4L}$	$w^{a(a-L)/2L}$
F	$x^{u(u+1)/2L}$	$x^{u(2u+2-L)/2L}$

The symmetries (2.1.25a-d) (with $M=N$) and (2.2.13) remain valid for (3.1.2). As in section 2 we find it convenient to use the graphical notation

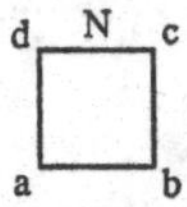

$$=\mathscr{S}_N(a, b, c, d).$$

Unless otherwise stated, we shall always assume $a\leq c$, $d\leq b$ and use the parametrization (2.2.7)

$$\text{(3.1.7)}\qquad \begin{array}{l} a=\ell+N-2r, \quad b=\ell+2(N-k), \quad c=\ell+N-2s, \quad d=\ell, \\ 0\leq k\leq N, \quad \max(0, \ell+2N-L-k+1)\leq s\leq r\leq \min(\ell-1, k). \end{array}$$

Rewriting (2.2.11) we get the expression for the modified weight

(3.1.8)

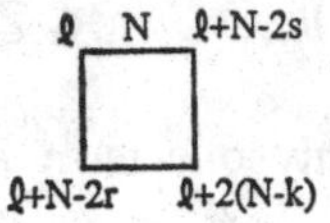

$$=\sum_{i=0}^{\min(s,k-r)} w^{\alpha}x^{\beta(i)}\mathscr{V}(i),$$

where

$$\text{(3.1.9a)}\qquad \begin{array}{ll} \alpha=\dfrac{r-s-N+k}{2} & \text{in regime II, III,} \\ \quad =r & \text{in regime I, IV,} \end{array}$$

$$\text{(3.1.9b)}\qquad \beta(i)=\frac{i(i+1)}{2}+i(r-s+N-k)+\frac{(r-s)(N-k+1)}{2}.$$

The $\mathscr{V}(i)$ has the same form as $\sqrt{S}\,U(i)$ (2.2.9–11) with the symbol $[\ell]$ therein replaced by (ℓ) (3.1.6). We shall show that

$$\text{(3.1.10)}\qquad \phi_m(\ell_1, \cdots, \ell_{m+2})=\sum_{j=1}^{m} jH(\ell_j, \ell_{j+1}, \ell_{j+2})$$

where in regimes II and III

$$\text{(3.1.11)}\qquad H(a, b, c)=|a-c|/4,$$

and in regimes I, IV

$$\text{(3.1.12)}\qquad \begin{array}{ll} H(a, b, c)=\min\left(n-b, \dfrac{\min(a, c)-b+N}{2}\right) & \text{if } b\leq n, \\ \quad =\min\left(b-n-1, \dfrac{b-\max(a, c)+N}{2}\right) & \text{if } b\geq n+1, \\ & n=[L/2]. \end{array}$$

Note that $H(a, b, c)$ (3.1.12) has the symmetry:

$$H(2n+1-a,\ 2n+1-b,\ 2n+1-c)=H(a, b, c). \tag{3.1.13}$$

3.2. Regime II, III

In this case we have by the definition (3.1.6)

$$(\ell \pm u)=1+O(x) \qquad \text{if} \quad 0<\ell<L. \tag{3.2.1}$$

Theorem 3.2.1

(3.2.2) $\lim$ 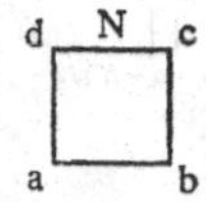 $=\delta_{ac} w^{-|b-d|/4}$.

Proof. From the representation (3.1.8), (2.2.9–11) together with (3.2.1) it follows that $\mathscr{V}(i)=O(1)$ for all i in the sum. The power $\beta(i)$ (3.1.9b) is non-negative, and vanishes if and only if $i=0$ and $r=s$. In this case one can check that $\lim \mathscr{V}(0)=1$, so the r.h.s. of (3.1.8) tends to $w^{-(N-k)/2}=w^{-|b-d|/4}$. This completes the proof. □

From Theorem 3.2.1 follows the expression (3.1.10–11) of $\phi_m(\ell_1, \cdots, \ell_{m+2})$.

3.3. Regime I, IV

In this case, the factor $(\ell)=E(x^\ell, -x^{L/2})$ shows a behavior different from that of regime II, III:

$$\begin{aligned}(\ell)&=1+O(\sqrt{x}) && \text{if } 1\leq \ell<L/2,\\ &=2(1+O(x)) && \text{if } \ell=L/2 \text{ for } L \text{ even,}\end{aligned} \tag{3.3.1a}$$

$$(\ell)=x^{L/2-\ell}(L-\ell)=-x^\ell(-\ell) \tag{3.3.1b}$$

Note that if $\ell \geq L/2$, then (ℓ) gives rise to an extra power $L/2-\ell$. To avoid technical complexity we assume throughout regimes I, IV that

$$L\geq 2N+1. \tag{3.3.2}$$

This implies

$$(\ell),\ (\ell \pm u)=1+O(\sqrt{x}) \qquad \text{provided } 1\leq \ell \leq N. \tag{3.3.3}$$

We shall carry out the computation by following the four steps.

$$\text{(3.3.4)} \quad \textit{Step 1:} \quad \lim \begin{array}{ccc} d & N & c \\ & \square & \\ a & & b \end{array} = \lim \begin{array}{ccc} d+1 & N-1 & c \\ & \square & \\ a & & b-1 \end{array} \qquad \text{if } d<b.$$

$$\text{(3.3.5)} \quad \textit{Step 2:} \quad \lim \begin{array}{ccc} b & N & c \\ & \square & \\ a & & b \end{array} = 0 \quad \text{unless } a=c,$$

$$\text{or} \quad a+c=L \quad \text{with} \quad L \text{ even.}$$

$$\text{(3.3.6)} \quad \textit{Step 3:} \quad \lim \begin{array}{ccc} b & N & a \\ & \square & \\ a & & b \end{array} = \prod_{j=(a+b-N)/2}^{a-1} \mathscr{W}(j) \times \prod_{j=a+1}^{(a+b+N)/2} \mathscr{W}(L-j),$$

where

$$\text{(3.3.7)} \quad \begin{aligned} \mathscr{W}(j) &= 1 && \text{if } 0<j<L/2, \\ &= \frac{1+w}{2} && \text{if } j=L/2, \\ &= w && \text{if } L/2<j<L. \end{aligned}$$

$$\text{(3.3.8)} \quad \textit{Step 4:} \quad \lim \begin{array}{ccc} b & N & L\text{-}a \\ & \square & \\ a & & b \end{array} = (-1)^{N+b+1-L/2} \frac{1-w}{1+w} \lim \begin{array}{ccc} b & N & a \\ & \square & \\ a & & b \end{array}$$

$$\text{if } 2a \neq L \quad \text{and} \quad L \text{ even.}$$

For $N=1$ these assertions (3.3.4–8) can be verified directly by using the representation (3.1.7–9) of the modified weight and the fact that

$$\text{(3.3.9)} \qquad \lim \frac{(\ell-u)}{(\ell)} = \mathscr{W}(\ell), \qquad \lim \frac{(\ell+u)}{(\ell)} = w^{-1}\mathscr{W}(L-\ell).$$

(The weight $\mathscr{S}_0(a, b, c, d)$ in (3.3.4) is understood as 1 if $a=b=c=d$, 0 otherwise.) In what follows we assume (3.3.4–8) to be true for $N-1$ in place of N.

Step 1.

First note that in the representations (3.1.8–9) of $\mathscr{S}_N(a, b, c, d)$ and $\mathscr{S}_{N-1}(a, b-1, c, d+1)$ in (3.3.4) the suffix i ranges over the common interval $0 \leq i \leq \min(s, k-r)$. Let $\mathscr{A}(i) = w^r x^{\beta(i)} \mathscr{V}(i)$, $\mathscr{A}'(i) = w^r x^{\beta'(i)} \mathscr{V}'(i)$ stand for the corresponding summands. In the following Lemmas 3.3.1–2 we show that for all i either $\lim \mathscr{A}(i)/\mathscr{A}'(i) = 1$ or else $\lim \mathscr{A}(i) = \lim \mathscr{A}'(i)$

$=0$ holds, thereby proving (3.3.4). Without loss of generality we shall assume $\ell \leq L/2$.

Lemma 3.3.1. *We have*

$$x^{\beta(\ell)-\beta'(\ell)}\mathscr{V}(i)/\mathscr{V}'(i)=O(1) \qquad \textit{for all } 0\leq i\leq \min(s, k-r). \tag{3.3.10}$$

Moreover the l.h.s. tends to 1 *if* $2N+\ell-k-r>L/2$, *or* $i=0$ *and* $r=s$.

Proof. Dropping the factors of the form (3.3.3) we find that the l.h.s. of (3.3.10) behaves like $A(1+O(\sqrt{x}))$ with

$$A=x^{i+(r-s)/2}\left(\frac{(\ell-r)(\ell-s)}{(\ell+2N-k-r)(\ell+2N-k-s)}\right)^{1/2}\frac{(\ell+2N-s-k+i)}{(\ell-s+i)}.$$

By taking into account the contributions from the factors (ℓ) (cf. (3.3.1)) it can be shown that $A=O(x^{\nu})$, where $0\leq\nu\leq i+(r-s)/2$, and that $A=1+O(\sqrt{x})$ if $2N+\ell-k-r>L/2$ (see Fig. 3.1).

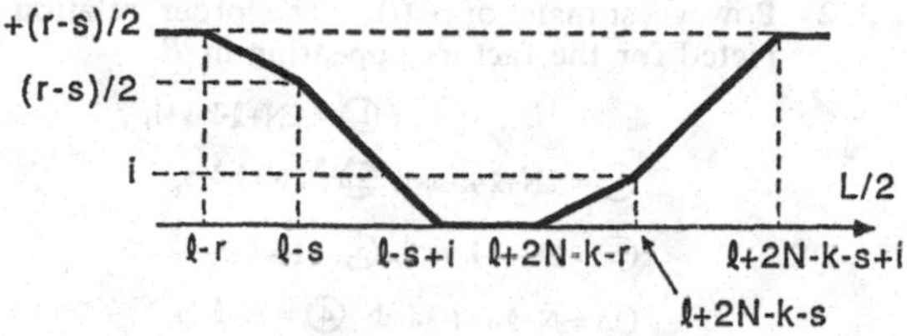

Fig. 3.1 Power estimate of the l.h.s. of (3.3.9).

Finally when $i=0$ and $r=s$ the l.h.s. of (3.3.10) becomes $(N-k+u)/(N-k)$, which tends to 1 in the limit by (3.3.3). This completes the proof of Lemma 3.3.1. □

Lemma 3.3.2. *Assume* $L/2\geq\ell, 2N+\ell-k-r+1$. *Then we have*

$$\lim x^{\beta(\ell)}\mathscr{V}(i)=0$$

except when $i=0$ *and* $r=s$.

Proof. Proceeding in the same way as in Lemma 3.3.1, we have $\mathscr{V}(i)=B\times O(1)$ where

$$B=\left(\frac{(2N+\ell-k-r+1,\ 2N+\ell-k-s)(N+\ell-2s)}{(N+\ell-r+1,\ N+\ell-s)}\right)^{1/2}$$
$$\times(2N+\ell-k-s+1,\ 2N+\ell-k-s+i).$$

Here we have set $(a, b)=(a)(a+1)\cdots(b)$ for $a\leq b$ and $(a, a-1)=1$.

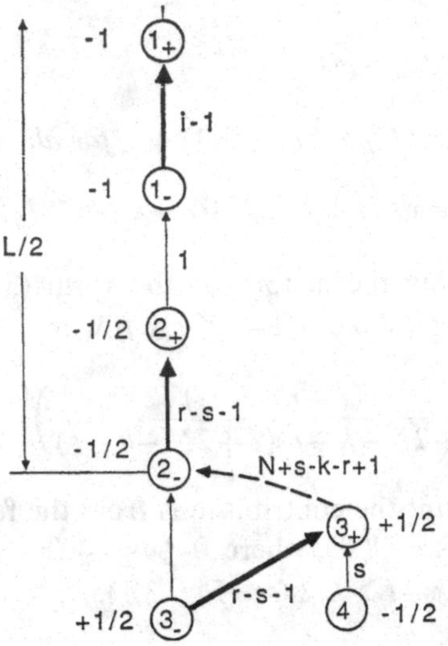

Fig. 3.2 Power estimate of $\mathscr{V}(i)$. The order relation is depicted for the factors appearing in B.

$(1_+) = 2N+\ell-k-s+i$,

$(1_-) = 2N+\ell-k-s+1$, $(2_+) = 2N+\ell-k-s$,

$(2_-) = 2N+\ell-k-r+1$, $(3_+) = N+\ell-s$,

$(3_-) = N+\ell-r+1$ and $(4) = N+\ell-2s$.

Counting the total power of x (see Fig. 3.2), we obtain the estimate $x^{\beta(i)}B = O(x^{\gamma(i)})$ with

$$\gamma(i)=i(N-k+1)+\frac{1}{4}(r-s)(N+s+1-k-r+N-k+2) \qquad \text{if } N+s+1\geq k+r,$$

$$=i(N-k+1)+\frac{1}{2}\min(r-s,\ s+N-k+1) \qquad \text{otherwise.}$$

Under the assumption of the lemma, $\gamma(i)>0$ in either case as desired. □

Step 2.

Here we make use of the STR of type $(N, 1, N)$ as in Fig. 3.3: The variables are so chosen that one of the summands in the r.h.s. vanishes. This provides us with the relation

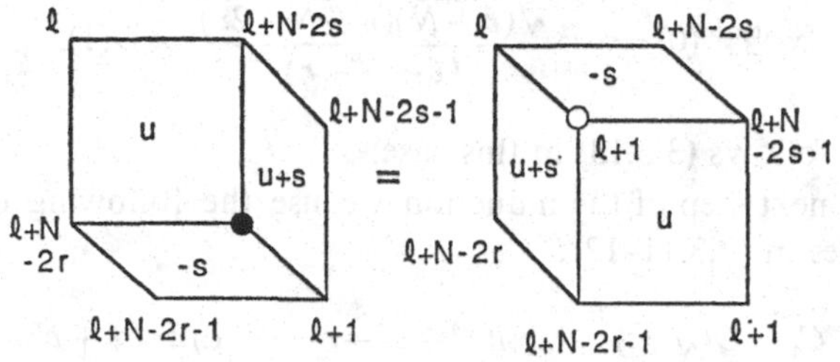

Fig. 3.3 The STR of type $(N, 1, N)$ used in deriving (3.3.10). In the r.h.s. the term with the center spin $\ell - 1$ vanishes.

(3.3.11)

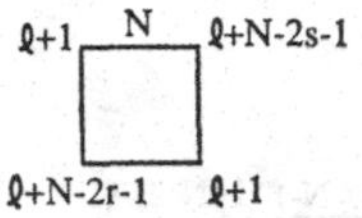

$$= w\,C_1 \;\begin{array}{c}\ell \quad N \quad \ell+N-2s \\ \boxed{} \\ \ell+N-2r \qquad \ell\end{array}\; + \; x^{(r+s)/2+1}\,C_2 \;\begin{array}{c}\ell \quad N \quad \ell+N-2s \\ \boxed{} \\ \ell+N-2r \qquad \ell+2\end{array}\;,$$

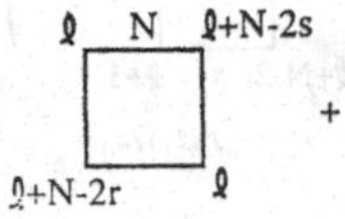

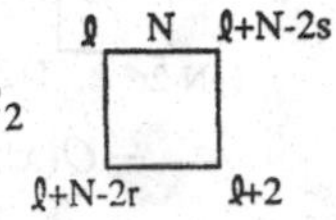

(3.3.12)

$$C_1 = \left(\frac{(\ell+N-2r-1)(\ell+N-2s-1)}{(\ell+N-2r)(\ell+N-2s)}\right)^{1/2}$$
$$\times \frac{(\ell+N-r-s)(\ell+1+u)}{(\ell+N-r-s-1-u)(\ell+1)},$$

$$C_2/C_1 = \left(\frac{(\ell-r)(\ell-s)(\ell+N-r+1)(\ell+N-s+1)}{(N-r)(N-s)(r+1)(s+1)}\right)^{1/2}$$
$$\times \frac{(N-r-s-1)(u)}{(\ell+N-r-s)(\ell+1+u)}.$$

Eq. (3.3.5) is an immediate consequence of the following lemma.

Lemma 3.3.3. *Assuming $r > s$, we have*

(3.3.13)
$$\begin{array}{c}\ell \quad N \quad \ell+N-2s \\ \boxed{} \\ \ell+N-2r \qquad \ell\end{array} = O(x^{\min\,(r-s,\,|\ell+N-r-s-L/2|)/2}).$$

Proof. Using (3.3.2) and the symmetry (2.2.13) we may assume further $\ell + N < L$, so that $0 \le s < r \le \min(\ell-1, N)$. We prove (3.3.13) by an induction on s.

Suppose $s = 0$. Then there is only one term in the sum (3.1.8), and we have

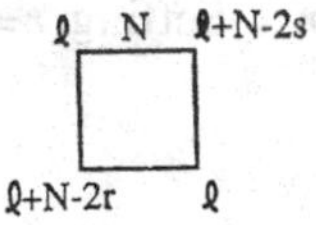

$$x^{\beta(0)}\mathscr{V}(0)=x^{r/2}\frac{\sqrt{(\ell+N)(\ell+N-2r)}}{(\ell+N-r)}\times O(1),$$

from which follows (3.3.13) in this case.

To the next step of the induction we use the following estimates for the quantities in (3.3.11–12):

$$\text{(3.3.14)}\qquad \begin{aligned} C_1&=O(\sqrt{x}) && \text{if } \ell+N-r-s\le L/2<\ell+N-2s,\\ &=O(\sqrt{x}^{\,-1}) && \text{if } \ell+N-2r\le L/2<\ell+N-r-s,\\ &=O(1) && \text{otherwise.}\end{aligned}$$

$$\text{(3.3.15)}\qquad x^{(r+s)/2+1}C_2/C_1=O(1).$$

(3.3.16)

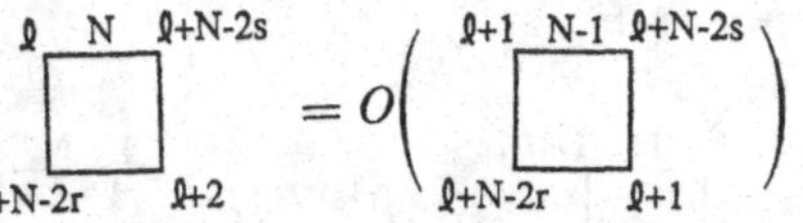

$$=O(x^{\min(r-s,\,|\ell+N-r-s-L/2|)/2}).$$

In (3.3.16) we used (3.3.4) and the induction hypothesis. The assertion (3.3.13) follows from (3.3.14–16) by virtue of (3.3.11). □

Step 3.

To prove (3.3.6) we may assume that $b\le L/2$, for it is invariant under the change $a\to L-a$, $b\to L-b$.

In the case $a=b+N$ (*i.e.* $r=0$), we have

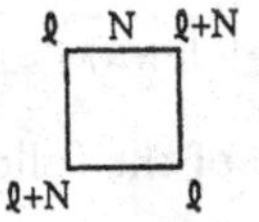

$$=\prod_{j=0}^{N-1}\frac{(\ell+j-u)}{(\ell+j)},$$

so (3.3.6) is obvious from (3.3.9).

Next suppose that (3.3.6) is true for r. Setting $r=s$ in (3.3.12) and noting $\ell+1\le L/2$, we have

$$\lim wC_1=\mathscr{W}(L-\ell-1)/\mathscr{W}(\ell+N-2r-1),$$

$$\lim x^{r+1}C_2=\varepsilon_{\ell+1}^{L/2}\frac{1-w}{\mathscr{W}(\ell+N-2r-1)}\lim x^{r+1}\frac{(\ell+N-r+1)}{(\ell+N-2r)}(N-2r-1).$$

Substitute these equations into (3.3.11), apply (3.3.4) and use the induction hypothesis for

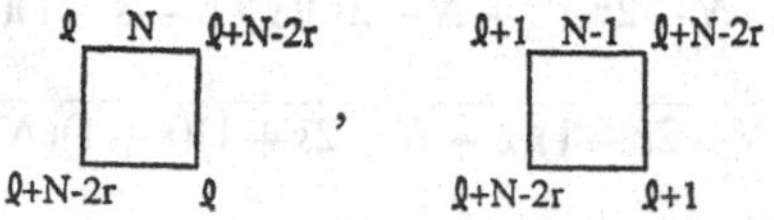

Upon simplifying the result, we are left with the proof of

Lemma 3.3.4. *Assuming* $0 \leq r \leq N-1$ *and* $\ell+1 \leq L/2$ *we have*

$$\mathscr{W}(L-N-\ell+2r)$$
$$=\mathscr{W}(L-\ell-1)+\varepsilon_{\ell+1}^{L/2}(1-w)\lim x^{r+1}\frac{(\ell+N-r+1)}{(\ell+N-2r)}(N-2r-1),$$

where $\varepsilon_{\ell}^{L/2}$ *is defined by* (1.5.7).

The proof can be done by case checking.

Step 4.

Here we utilize another STR of type $(N, 1, N)$ as shown below.

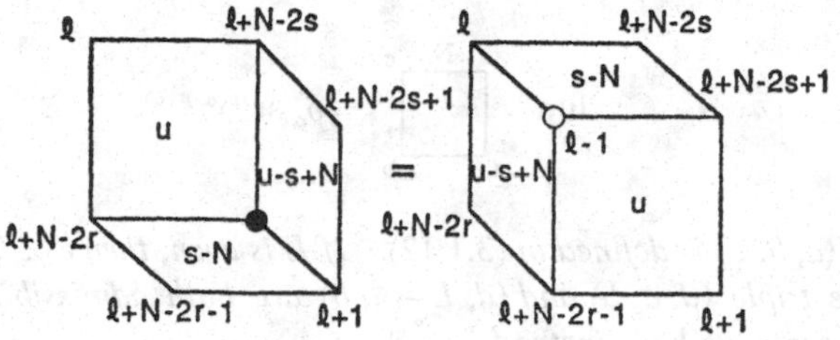

Fig. 3.4 The STR of type $(N, 1, N)$ used in deriving (3.3.17).

Explicitly we have

(3.3.17) $x^{(r-s)/2}$ 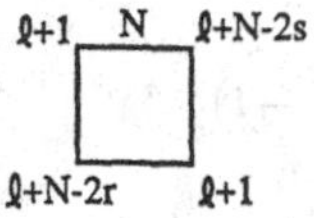

$= C_3$ 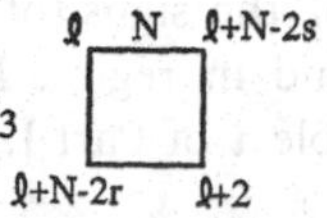

$+ C_4$ 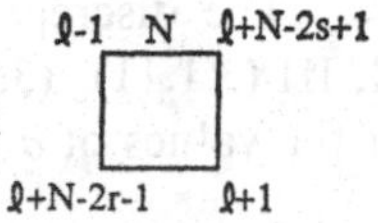

(3.3.18) $$C_3=\left(\frac{(s+1)(N-s)(\ell-r)(\ell+N-r+1)}{(r+1)(N-r)(\ell-s)(\ell+N-s+1)}\right)^{1/2}\frac{(r-s+1)(\ell-u)}{(\ell-r+s)(1+u)},$$

$$C_4=\left(\frac{(\ell+N-2r)(\ell+N-2s)(s)(N-s+1)(\ell-r-1)\cdot(\ell+N-r)}{(\ell+N-2r-1)(\ell+N-2s+1)(r+1)(N-r)(\ell-s)\cdot(\ell+N-s+1)}\right)^{1/2}$$

$$\times\frac{(\ell+1)(r-s+1+u)}{(\ell-r+s)(1+u)}.$$

Because of (3.3.4), we can apply (3.3.8) to the weights in the r.h.s. of (3.3.17). Doing this and proceeding in the same way as in step 3, we find that the proof is reduced to the following.

Lemma 3.3.5. *Let* $C_\nu^0=\lim x^{(s-r)/2}C_\nu$, $\nu=3, 4$. *Assuming* $s<r$ *we have*

$$\mathscr{W}(\ell-r)=C_3^0-C_4^0\frac{\mathscr{W}(\ell-r)\mathscr{W}(L-\ell-N+2r)}{\mathscr{W}(\ell+N-2r-1)\mathscr{W}(L-\ell-N+r)}. \tag{3.3.19}$$

This can also be verified by case checking, so we omit the proof.

Summarizing (3.3.4–8) we obtain the following result.

Theorem 3.3.6. *If L is odd, then*

(3.3.20) $\lim$ 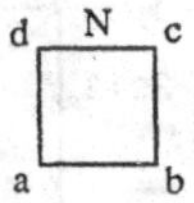 $=\delta_{ac}w^{H(d,a,b)}$,

where $H(a, b, c)$ *is defined in* (3.1.12). *If L is even, then* (3.3.20) *holds except when the triple* (d, a, b) *and* $(d, L-a, b)$ *are both admissible (with* $2a\neq L$*). In these cases we have instead*

(3.3.21) $\lim$ 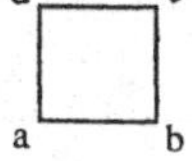 $=\delta_{ac}w^{|a-n|-1}\frac{1+w}{2}$

$$-\delta_{a,L-c}(-1)^{b-n+N}w^{|a-n|-1}\frac{1-w}{2}.$$

Remark. The discrepancy of the signs (of H in the exponent of w) in regime II, III (3.1.11), (3.2.2) and in regime I, IV (3.3.20) is due to the difference of the values of σ in Table 1 of Part I.

We thus find that (3.1.10, 12) is true for odd L. For even L the face operator $\mathscr{U}_i^0$ (3.1.3) in the limit is non-diagonal and contains 2 by 2 blocks of the form

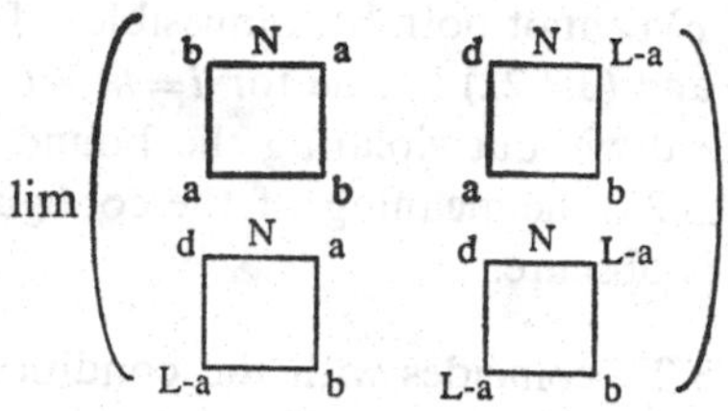

Lemma 3.3.7. *The face operators $\mathcal{U}_i^0$ are mutually commutative.*

Proof. It is sufficient to prove $\mathcal{U}_i^0\mathcal{U}_{i+1}^0=\mathcal{U}_{i+1}^0\mathcal{U}_i^0$. This amounts to showing that

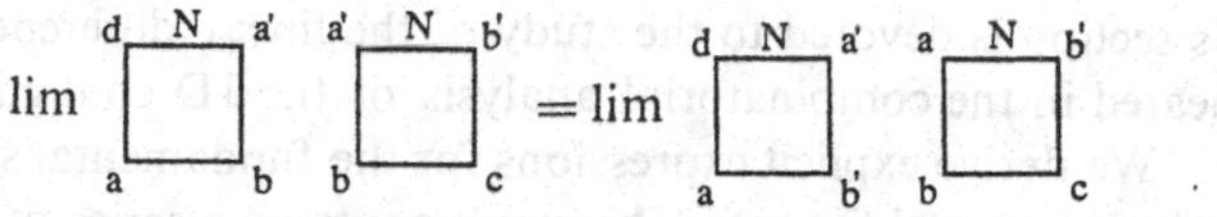

for all a, b, c, d, a', b'.

We may assume that (a, b) and $(a, L-b)$ are both admissible, for otherwise we must have $a=a'$, $b=b'$. Suppose for instance $a'=L-a\neq a$ and $b'=L-b\neq b$. From (3.3.21), both sides are then equal to

$$w^{|a-n|+|b-n|-2}((1-w)/2)^2.$$

The remaining cases can be verified similarly. □

Thus the face operators can be diagonalized simultaneously. For an admissible sequence (ℓ_j), define the vectors $|\ell_1, \cdots, \ell_{m+2}\rangle$ inductively by

$$|\ell_1, \ell_2\rangle=e_{\ell_1}\otimes e_{\ell_2},$$
$$|\ell_1, \cdots, \ell_{k+2}\rangle=|\ell_1, \cdots, \ell_{k+1}\rangle\otimes e_{\ell_{k+2}}+\varepsilon|\ell_1, \cdots, L-\ell_{k+1}\rangle\otimes e_{\ell_{k+2}}$$
$$\text{if } (\ell_k, \ell_{k+1}, \ell_{k+2}),\quad (\ell_k, L-\ell_{k+1}, \ell_{k+2}) \text{ are admissible,}$$
$$=|\ell_1, \cdots, \ell_{k+1}\rangle\otimes e_{\ell_{k+2}} \quad \text{otherwise.}$$

Here $\varepsilon=\pm(-1)^{\ell_k-n+N}$ or 0 according as $\ell_k<n$, $>n$ or $=n$. Then we have

$$\text{(3.3.22)}\qquad \mathcal{U}_i|\ell_1, \cdots, \ell_{m+2}\rangle=w^{H(\ell_{i-1},\ell_i,\ell_{i+1})}|\ell_1, \cdots, \ell_{m+2}\rangle$$

for $2\leq i\leq m$, where $H(a, b, c)$ is given by the same equation (3.1.12). In general, the vectors $|\ell_1, \cdots, \ell_{m+2}\rangle$ may not belong to the space $\mathcal{H}_{b,c}$ (see 3.1) where the boundary heights $\ell_{m+1}=b$, $\ell_{m+2}=c$ are specified. Suppose further that they satisfy the additional conditions

$$\text{(3.3.23)}\qquad (b+c-N)/2<n-N \quad \text{or} \quad (b+c-N)/2\geq n+1.$$

Then (b, c) and $(L-b, c)$ cannot both be admissible. This guarantees that $|\ell_1, \cdots, \ell_{m+2}\rangle \in \mathcal{H}_{b,c}$, and (3.3.22) is true for $i=m+1$ as well. Thus the CTM can be diagonalized without violating the boundary conditions. In the cases other than (3.3.23) the meaning of the configuration sum (3.1.1), (3.1.10, 12) for even L is obscure.

Remark. Eq. (3.3.23) coincides with the condition for the sequence $\cdots bcbc \cdots$ to be a ground state configuration in regime IV, with the exception $(b+c-N)/2=n-N$. See eq. (2.9b) of Part I.

§ 4. Combinatorial Identities

This section is devoted to the study of the linear difference equations that appeared in the combinatorial analysis of the 1D configuration sums in Part I. We derive explicit expressions for the fundamental solutions and rewrite them in several forms. The result contains a series of new combinatorial identities.

As in sections 2–3, we call a pair of integers (a, b) *weakly admissible* if the following relation holds

(4.0.1) $$a-b=-N, \ -N+2, \cdots, N.$$

A weakly admissible pair (a, b) is called *admissible* if it further satisfies

(4.0.2) $$a+b=N+2, \ N+4, \cdots, 2L-N-2.$$

4.1. Fundamental solution for the linear difference equation

For a weakly admissible pair (b, c) and integers $m, N \geq 0$, let $f_m^{(N)}(b, c)$ denote the solution to the following linear difference equation.

(4.1.1a) $$f_m^{(N)}(b, c)=\sum_d{}' f_{m-1}^{(N)}(d, b) q^{m|d-c|/4},$$

(4.1.1b) $$f_0^{(N)}(b, c)=\delta_{b0}.$$

Here the sum $\sum'$ is taken over d such that the pair (d, b) is weakly admissible. We set $f_m^{(N)}(b, c)=0$ if (b, c) is not weakly admissible.

One may consider m and b as discrete time and space variables, respectively. There are $N+1$ possible values of c for a given b. In this sense the equation (4.1.1) is a system of $N+1$ simultaneous equations in $1+1$ dimensions. It is of order 1 with respect to m.

Remark. By the definition $f_m^{(N)}(b, c)$ enjoys the following.
Reflection symmetry:

(4.1.2) $$f_m^{(N)}(b, c)=f_m^{(N)}(-b, -c), \qquad \text{for } m\geq 0.$$

Support property:

(4.1.3) $$f_m^{(N)}(b, c)=0 \quad \text{unless } |b|\leq mN, \quad |c|\leq(m+1)N \quad \text{and } b\equiv mN \bmod 2.$$

In particular, for $N=0$, (4.1.3) asserts that

(4.1.4) $$f_m^{(0)}(b, c)=\delta_{b0}\delta_{c0}.$$

We seek for the solution to (4.1.1) in the form of a double sum

$$\sum_{j,k} A(m, j, k)q^{B(m,j,k)N+jb+kc}.$$

The coefficients A and B are polynomials in m, j, k. The sum extends over different regions of (j, k) according as the regions of (b, c) specified below.

Given an integer μ set

(4.1.5) $$R_\mu=\{b \in Z\,|\,(\mu-1)N\leq b\leq(\mu+1)N\},$$

where the left (resp. right) equality sign is taken if $\mu-1\leq 0$ (resp. $\mu+1\geq 0$). We also set $R_{\mu,\nu}=R_\mu\times R_\nu$. For a weakly admissible pair (b, c) and integer $m(\geq 1)$, let μ, ν $(=\mu\pm 1)$ be integers such that

(4.1.6a) $$(b, c) \in R_{\mu,\nu},$$

(4.1.6b) $$\mu\equiv m+1, \quad \nu\equiv m+2 \bmod 2.$$

Equation (4.1.6) uniquely determines μ and ν except when $b=0$ $(\mu=\pm 1)$ of $c=0$ $(\nu=\pm 1)$. In these cases either choice is allowed.

Theorem 4.1.1 ((3.4) of Part I). *For all weakly admissible pairs $(b, c) \in R_{\mu,\nu}$ and integers $m, N\geq 1$,*

(4.1.7a) $$(q)_{m-1}f_m^{(N)}(b, c) =\Big(\sum_{\substack{k\leq(m+\mu-1)/2\\ j\geq(m+\nu)/2}}-\sum_{\substack{j\leq(m+\nu)/2-1\\ k\geq(m+\mu+1)/2}}\Big)(-1)^{j-k}q^{Q_{j,k}^{(m)}(b,c)}\begin{bmatrix} m-1\\ j\end{bmatrix}\begin{bmatrix} m\\ k\end{bmatrix},$$

(4.1.7b) $$Q_{j,k}^{(m)}(b, c) =\frac{1}{2}(j-k)(j-k+1)-\Big(j-\frac{m-1}{2}\Big)\Big(k-\frac{m}{2}\Big)N +\frac{b}{2}\Big(j-\frac{m-1}{2}\Big)+\frac{c}{2}\Big(k-\frac{m}{2}\Big).$$

Proof. First we check the properties (4.1.2) and (4.1.3) for $f_m^{(N)}(b, c)$ given by (4.1.7). The former immediately follows from $Q_{j,k}^{(m)}(-b, -c)=Q_{m-1-j,m-k}^{(m)}(b, c)$. To show the latter suppose for instance $b>mN$. Then we have $\mu\geq m+1$ and $\nu\geq m$, so that $j\geq m$ or $k\geq m+1$. Therefore the product of the Gaussian polynomials identically vanishes. The other cases are similar.

For $m=1$ (4.1.7) can be directly verified by using (4.1.1). In the following we assume $m\geq 2$. In view of (4.1.2–3) we may assume

$$1\leq\mu\leq m-1 \quad \text{if } m \text{ is even } (m\geq 2), \tag{4.1.8a}$$

$$0\leq\mu\leq m-1 \quad \text{if } m \text{ is odd } (m\geq 3). \tag{4.1.8b}$$

There are two cases to consider: (i) $\nu=\mu+1$, (ii) $\nu=\mu-1$. Here we prove the case (i). The case (ii) can be verified similarly. Since we have $c>(\nu-1)N=\mu N$, the r.h.s. of (4.1.1a) reads as

$$\begin{aligned}
&\sum_d{}' f_{m-1}^{(N)}(d, b)q^{m|d-c|/4}=K_1+K_2+K_3,\\
&K_1=\sum_{d=b-N,b-N+2,\cdots,\mu N} f_{m-1}^{(N)}(d, b)q^{m(c-d)/4},\\
&K_2=\sum_{d=\mu N+2,\mu N+4,\cdots,c} f_{m-1}^{(N)}(d, b)q^{m(c-d)/4},\\
&K_3=\sum_{d=c+2,c+4,\cdots,b+N} f_{m-1}^{(N)}(d, b)q^{m(d-c)/4}.
\end{aligned} \tag{4.1.9}$$

In K_1, K_2 and K_3 the pair (d, b) belongs to $R_{\mu-1,\mu}$, $R_{\mu+1,\mu}$ and $R_{\mu+1,\mu}$, respectively. We substitute the expression (4.1.7) into $K_1\sim K_3$ and perform the d-summations. By using the identities

$$\begin{bmatrix} m-2 \\ j \end{bmatrix}\frac{1}{1-q^{m-j-1}}=\begin{bmatrix} m-1 \\ j \end{bmatrix}\frac{1}{1-q^{m-1}}, \tag{4.1.10a}$$

$$\begin{bmatrix} m-2 \\ j \end{bmatrix}\frac{1}{1-q^{j+1}}=\begin{bmatrix} m-1 \\ j+1 \end{bmatrix}\frac{1}{1-q^{m-1}}, \tag{4.1.10b}$$

we obtain the following:

$$\begin{aligned}
&K_i=K_i^{(+)}+K_i^{(-)}, \qquad i=1, 2, 3,\\
&K_i^{(\pm)}=\pm\Big(\sum_{\substack{j\geq(m+\mu-1)/2\\ k\leq(m+\mu-3)/2+\varepsilon_i}}-\sum_{\substack{j\leq(m+\mu-3)/2\\ k\geq(m+\mu-1)/2+\varepsilon_i}}\Big)(-1)^{j-k}q^{Q_{j,k}^{(m-1)}(0,b)+Q_i^{(\pm)}(j)}\\
&\qquad\times\begin{bmatrix} m-1 \\ j+1-\varepsilon_{4-i} \end{bmatrix}\begin{bmatrix} m-1 \\ k \end{bmatrix}\frac{1}{(q)_{m-1}},
\end{aligned} \tag{4.1.11}$$

where

$$\varepsilon_i = [i/2],$$

(4.1.12a) $$Q_1^{(+)}(j) = Q_2^{(-)}(j) = \frac{m}{4}(c - 2\mu N) + \frac{1}{2}\mu N(j+1),$$

(4.1.12b) $$Q_1^{(-)}(j) = \frac{m}{4}(c - 2b + 2N + 4) + \frac{1}{2}(b - N - 2)(j+1),$$

(4.1.12c) $$Q_2^{(+)}(j) = -\frac{m}{4}c + \frac{c}{2}(j+1),$$

(4.1.12d) $$Q_3^{(+)}(j) = -\frac{m}{4}c + \frac{1}{2}(c+2)(j+1),$$

(4.1.12e) $$Q_3^{(-)}(j) = -\frac{m}{4}c + \frac{1}{2}(b + N + 2)(j+1).$$

First consider $K_1^{(-)}$ (resp. $K_3^{(-)}$). Under the replacement $(j, k) \to (k, j)$ (resp. $(j, k) \to (k-1, j+1)$) in (4.1.11), the sums $(\sum - \sum)$ exchange the sign while the summands are invariant due to the property

$$Q_{k,j}^{(m-1)}(0, b) + Q_1^{(-)}(k) = Q_{j,k}^{(m-1)}(0, b) + Q_1^{(-)}(j),$$
$$Q_{k-1,j+1}^{(m-1)}(0, b) + Q_3^{(-)}(k-1) = Q_{j,k}^{(m-1)}(0, b) + Q_3^{(-)}(j).$$

Thus we have $K_1^{(-)} = K_3^{(-)} = 0$. Next we observe from (4.1.12a) that

(4.1.13)
$$\begin{aligned} &K_1^{(+)} + K_2^{(-)} \\ &= -\sum_{\substack{k=(m+\mu-1)/2 \\ j \in \mathbf{Z}}} (-1)^{j-k} q^{Q_{j,k}^{(m-1)}(0,b) + Q_1^{(+)}(j)} \begin{bmatrix} m-1 \\ j \end{bmatrix} \begin{bmatrix} m-1 \\ k \end{bmatrix} \frac{1}{(q)_{m-1}} \\ &= \frac{(-1)^{k-1}}{(q)_{m-1}} (q^{(3-\mu-m)/2})_{m-1} \begin{bmatrix} m-1 \\ k \end{bmatrix} \\ &\qquad \times q^{m^2/8 - \{((\mu-1)N - c + 2)/4\} m + (\mu^2 + (2b-4)\mu + 3)/8}. \end{aligned}$$

Here we have used the formula ([9], Theorem 3.3)

(4.1.14) $$\sum_{i \in \mathbf{Z}} (-z)^i q^{i(i-1)/2} \begin{bmatrix} M \\ i \end{bmatrix} = (z)_M.$$

In general the product $(q^\rho)_{m-1}$ vanishes for $\rho = 0, -1, \cdots, -(m-2)$. In view of this and (4.1.8, 13) it turns out that $K_1^{(+)} + K_2^{(-)} = 0$. Finally we combine $K_2^{(+)}$ and $K_3^{(+)}$. Simplifying the sum of Gaussian polynomials by the formula

(4.1.15)
$$\begin{bmatrix} m-1 \\ j \end{bmatrix} + \begin{bmatrix} m-1 \\ j+1 \end{bmatrix} q^{j+1} = \begin{bmatrix} m \\ j+1 \end{bmatrix},$$

we have

(4.1.16)
$$K_2^{(+)} + K_3^{(+)} = \Big(\sum_{\substack{j \ge (m+\mu-1)/2 \\ k \le (m+\mu-1)/2}} - \sum_{\substack{j \le (m+\mu-3)/2 \\ k \ge (m+\mu+1)/2}} \Big) (-1)^{j-k} q^{Q_{j,k}^{(m-1)}(0,b) + Q_2^{(+)}(j)} \times \begin{bmatrix} m-1 \\ k \end{bmatrix} \begin{bmatrix} m \\ j+1 \end{bmatrix} \frac{1}{(q)_{m-1}}.$$

Replace (j, k) in (4.1.16) by $(k-1, j)$ and recall the assumption $\nu=\mu+1$. Then the resulting expression coincides with $f_m^{(N)}(b, c)$ (4.1.7) because of the identity

$$Q_{k-1,j}^{(m-1)}(0, b) + Q_2^{(+)}(k-1) = Q_{j,k}^{(m)}(b, c). \qquad \square$$

We call $f_m^{(N)}(b-a, c-a)$ the *fundamental solution* of the equation (4.1.1). It follows from (4.1.1) (though not obvious in (4.1.7)) that the function $f_m^{(N)}(b, c)$ or $\sqrt{q}\, f_m^{(N)}(b, c)$ is a *polynomial* in q with *positive* coefficients.

The fundamental solution satisfies an extra set of linear difference equations *at equal m* (Lemma 4.1.2 below). This is crucial when we consider the linear difference equation in the bounded domain of (b, c) with the restriction (4.0.2) (see section 4.4.)

Lemma 4.1.2 ((3.6) of Part I). *For* $1 \le b \le N$ *and* $m \ge 1$,

(4.1.17)
$$\sum_{d=b-N, b-N+2, \cdots, N-b} q^{(a+md)/4} f_{m-1}^{(N)}(a+d, a-b) = (a \to -a).$$

Proof. There are two cases to consider:

(4.1.18a) (i) $(a+b-N, a-b+N) \subset R_\rho$ $(\rho \equiv m \bmod 2)$,

(4.1.18b) (ii) $(a+b-N, a-b+N) \subset R_{\rho-1} \cup R_{\rho+1}$ $(\rho \pm 1 \equiv m \bmod 2)$.

We prove here the case (i). The case (ii) is similar. From (4.1.18a) we see $(a+d, a-b) \in R_{\rho,\rho-1}$, $(a+d, a+b) \in R_{\rho,\rho+1}$. Substitute (4.1.7) into (4.1.17). After performing the d-summations therein the formula (4.1.10) leads us to the following expressions.

(4.1.19)
$$\text{the l.h.s. of (4.1.17)} = (A(+,+) + A(+,-))/(q)_{m-1},$$
$$\text{the r.h.s. of (4.1.17)} = (A(-,+) + A(-,-))/(q)_{m-1},$$

where $A(\varepsilon_1, \varepsilon_2)$ is given by $(\varepsilon_1, \varepsilon_2 = \pm 1)$

$$A(\varepsilon_1, \varepsilon_2) = \varepsilon_2 \Big(\sum_{\substack{k \le (m+\rho-2)/2 \\ j \ge (m+\rho-1-\varepsilon_1)/2}} - \sum_{\substack{k \ge (m+\rho)/2 \\ j \le (m+\rho-3-\varepsilon_1)/2}} \Big) \times \begin{bmatrix} m-1 \\ k \end{bmatrix} \begin{bmatrix} m-1 \\ j+(1+\varepsilon_1)/2 \end{bmatrix} (-1)^{j-k} \times q^{Q_{j,k}^{(m-1)}(a, a-\varepsilon_1 b) + \varepsilon_1 a/4 + \varepsilon_1 \varepsilon_2 (b-N-1+\varepsilon_2)(j+1+(\varepsilon_1 - 1)m/2)/2}. \tag{4.1.20}$$

It turns out that $A(\pm, -) = 0$. This can be seen from the invariance of the summand in (4.1.20) under the transformation $(j, k) \to (k-(1\pm 1)/2, j+(1\pm 1)/2)$. Now we calculate the difference of the remaining terms in (4.1.19).

$$\begin{aligned} & A(+, +) - A(-, +) \\ & = -\Big(\sum_{\substack{j \ge (m+\rho)/2 \\ k \le (m+\rho-2)/2}} - \sum_{\substack{j \le (m+\rho-2)/2 \\ k \ge (m+\rho)/2}} \Big) (-1)^{j-k} \begin{bmatrix} m-1 \\ j \end{bmatrix} \begin{bmatrix} m-1 \\ k \end{bmatrix} \\ & \times \big(q^{Q_{j-1,k}^{(m-1)}(a, a-b) + a/4 + (1/2)(b-N)j} + q^{Q_{j,k}^{(m-1)}(a, a+b) - a/4 - (1/2)(N-b)(m-j-1)} \big). \end{aligned}$$

Again, this vanishes identically thanks to the invariance of the summand under the change $(j, k) \to (k, j)$. □

4.2. Various representations for $f_m^{(N)}(b, b+N)$

Here we establish various representations for the function

$$\hat{f}_m^{(N)}(v) = q^{m(m+1)N/4 - v/2} f_m^{(N)}(b, b+N; q^{-1}), \tag{4.2.1a}$$

$$v = \frac{mN-b}{2}, \tag{4.2.1b}$$

other than the one obtained directly from (4.1.7). They are utilized in section 5 in order to examine the $m \to \infty$ behavior of the 1D configuration sum for regime II (see section 3.4 of Part I). Note from (4.1.3) that $\hat{f}_m^{(N)}(v) = 0$ unless $0 \le v \le mN$.

First we rewrite the double sum (4.1.7) into a single sum.

Theorem 4.2.1 ((3.27) of Part I). *For* $m \ge 1$, $N \ge 1$ *and* $0 \le v \le mN$,

$$\begin{aligned} \hat{f}_m^{(N)}(v) = & \sum_{j \in Z} (-1)^j \begin{bmatrix} v+m-(N+1)j-1 \\ m-1 \end{bmatrix} \begin{bmatrix} m-1 \\ j \end{bmatrix} q^{\mathscr{P}} \\ & + \sum_{j \in Z} (-1)^j \begin{bmatrix} v+m-(N+1)j-1 \\ m-1 \end{bmatrix} \begin{bmatrix} m-1 \\ j-1 \end{bmatrix} q^{\mathscr{P} - (N+2)j + v + m}, \end{aligned} \tag{4,2.2a}$$

$$\mathscr{P} = j(j-1)/2 + j(mN - v + 1). \tag{4.2.2b}$$

Proof. Set $[v/N]=\lambda$. We have

$$0\leq\lambda\leq m, \tag{4.2.3a}$$

$$(b, b+N)=(mN-2v, (m+1)N-2v)\in R_{\mu,\nu}, \quad \mu=m-2\lambda-1, \quad \nu=m-2\lambda. \tag{4.2.3b}$$

Thus the formula (4.1.7) with (j, k, q) replaced by $(k-1, m-j, q^{-1})$ yields

$$\begin{aligned}&(q)_{m-1}f_m^{(N)}(b, b+N; q^{-1})\\ &=\Big(\sum_{\substack{j\geq\lambda+1\\ k\geq m-\lambda+1}}-\sum_{\substack{j\leq\lambda\\ k\leq m-\lambda}}\Big)(-1)^{j+k}\begin{bmatrix}m\\ j\end{bmatrix}\begin{bmatrix}m-1\\ k-1\end{bmatrix}\\ &\quad\times q^{(j-k)(j-k+1)/2-(j-m/2)(k-(m+1)/2)N+((b+N)/2)(j-m/2)-(b/2)(k-(m+1)/2)}.\end{aligned} \tag{4.2.4}$$

We rewrite the sum

$$\Big(\sum_{\substack{j\geq\lambda+1\\ k\geq m-\lambda+1}}-\sum_{\substack{j\leq\lambda\\ k\leq m-\lambda}}\Big)$$

in (4.2.4) as

$$\Big(\sum_{\substack{j\in\mathbf{Z}\\ k\geq m-\lambda+1}}-\sum_{\substack{j\leq\lambda\\ k\in\mathbf{Z}}}\Big).$$

Applying (4.1.14) for the sum over $j\in\mathbf{Z}$ or $k\in\mathbf{Z}$, we get

$$\begin{aligned}\hat{f}_m^{(N)}(v)=&\sum_{k\geq m-\lambda+1}(-1)^k q^{(k/2)(k-1)+(k-1)v}\begin{bmatrix}m-1\\ k-1\end{bmatrix}\frac{(q^{1-v-(N+1)k+(m+1)N})_m}{(q)_{m-1}}\\ &+\sum_{j\leq\lambda}(-1)^j q^{(j/2)(j-1)+j(mN-v)}\begin{bmatrix}m\\ j\end{bmatrix}\frac{(q^{1+v-(N+1)j})_{m-1}}{(q)_{m-1}}\end{aligned} \tag{4.2.5}$$

$$\begin{aligned}=&-\sum_{j\leq\lambda}(-1)^j q^{(j/2)(j-1)+j(mN-v)}\begin{bmatrix}m-1\\ j-1\end{bmatrix}\frac{(q^{1+v-(N+1)j})_{m-1}}{(q)_{m-1}}\\ &\times(1-q^{v+m-(N+1)j})+\sum_{j\leq\lambda}(-1)^j q^{(j/2)(j-1)+j(mN-v)}\\ &\times\Big(q^j\begin{bmatrix}m-1\\ j\end{bmatrix}+\begin{bmatrix}m-1\\ j-1\end{bmatrix}\Big)\frac{(q^{1+v-(N+1)j})_{m-1}}{(q)_{m-1}}.\end{aligned} \tag{4.2.6}$$

We replaced k by $m+1-j$ in (4.2.5) and used the relation

$$(z)_M=(-z)^M q^{M(M-1)/2}(z^{-1}q^{1-M})_M, \qquad (z)_M=(z)_{M-1}(1-q^{M-1}z).$$

Notice that among the four terms in (4.2.6) the first and the last cancel each other. By virtue of (4.2.3) we can extend the sum $\sum_{j\leq\lambda}$ to $\sum_{j\in\mathbf{Z}}$ in the remaining terms and rewrite

$$\frac{(q^{1+v-(N+1)j})_{m-1}}{(q)_{m-1}} \quad \text{as} \quad \begin{bmatrix} v+m-(N+1)j-1 \\ m-1 \end{bmatrix}.$$

Thus we arrive at (4.2.2). □

Theorem 4.2.2. *For $m \geq 0$ and $N \geq 0$,*

$$\hat{f}_m^{(N)}(v) = \hat{f}_m^{(N)}(mN-v). \tag{4.2.7}$$

Proof. Since the case $mN=0$ is trivial we assume $mN \geq 1$. Consider the first term in (4.2.2a)

$$\sum_{j \in Z} (-1)^j q^{(j/2)(j-1)+j(mN-v+1)} \begin{bmatrix} m-1 \\ j \end{bmatrix} \frac{(q^{1+v-(N+1)j})_{m-1}}{(q)_{m-1}}. \tag{4.2.8}$$

Expand the product $(q^{1+v-(N+1)j})_{m-1}$ by (4.1.14) and then take the j-sum by the same formula. The expression (4.2.8) is cast into the form with v replaced by $mN-v$. The second term can be handled similarly. □

The following is a slight modification of (4.2.2).

Lemma 4.2.3. *For $m \geq 1$ and $N \geq 1$,*

$$\begin{aligned} \hat{f}_m^{(N)}(v) = &\sum_{j \in Z} (-1)^j \begin{bmatrix} v+m-(N+1)j \\ v-(N+1)j \end{bmatrix} \begin{bmatrix} m \\ j \end{bmatrix} q^{\mathscr{P}} \\ &-\sum_{j \in Z} (-1)^j \begin{bmatrix} v+m-(N+1)j-1 \\ v-(N+1)j-1 \end{bmatrix} \begin{bmatrix} m \\ j \end{bmatrix} q^{\mathscr{P}+m-j}, \end{aligned} \tag{4.2.9}$$

where $\mathscr{P}$ has been defined in (4.2.2b).

Proof. We write down (4.2.2a) slightly modifying the second term

$$\begin{aligned} \hat{f}_m^{(N)}(v) = &\sum_{j \in Z} (-1)^j \begin{bmatrix} v+m-(N+1)j-1 \\ v-(N+1)j \end{bmatrix} \begin{bmatrix} m-1 \\ j \end{bmatrix} q^{\mathscr{P}} \\ &+\sum_{j \in Z} (-1)^j \begin{bmatrix} v+m-(N+1)j-1 \\ v-(N+1)j \end{bmatrix} \begin{bmatrix} m-1 \\ j-1 \end{bmatrix} q^{\mathscr{P}+m-j} \\ &-\sum_{j \in Z} (-1)^j \begin{bmatrix} v+m-(N+1)j-1 \\ v-(N+1)j \end{bmatrix} \begin{bmatrix} m-1 \\ j-1 \end{bmatrix} q^{\mathscr{P}+m-j} \\ &\times (1-q^{v-(N+1)j}). \end{aligned} \tag{4.2.10}$$

In (4.2.10) simplify the first two terms via the identity

$$\begin{bmatrix} m-1 \\ j \end{bmatrix} + \begin{bmatrix} m-1 \\ j-1 \end{bmatrix} q^{m-j} = \begin{bmatrix} m \\ j \end{bmatrix} \tag{4.2.11}$$

and rewrite the last term using

$$\begin{aligned}(4.2.12)\qquad &\begin{bmatrix} v+m-(N+1)j-1 \\ v-(N+1)j \end{bmatrix}\begin{bmatrix} m-1 \\ j-1 \end{bmatrix}(1-q^{v-(N+1)j}) \\ &=\begin{bmatrix} v+m-(N+1)j-1 \\ v-(N+1)j-1 \end{bmatrix}\begin{bmatrix} m \\ j \end{bmatrix}(1-q^{j}).\end{aligned}$$

We have thus

$$\begin{aligned}(4.2.13)\qquad \hat{f}_m^{(N)}(v)=&\sum(-1)^j\begin{bmatrix} v+m-(N+1)j-1 \\ v-(N+1)j \end{bmatrix}\begin{bmatrix} m \\ j \end{bmatrix}q^{\varphi} \\ &-\sum_{j\in Z}(-1)^j\begin{bmatrix} v+m-(N+1)j-1 \\ v-(N+1)j-1 \end{bmatrix}\begin{bmatrix} m \\ j \end{bmatrix}(q^{\varphi+m-j}-q^{\varphi+m}).\end{aligned}$$

Applying the formula (4.2.11) (with (m, j) replaced by $(v+m-(N+1)j, v-(N+1)j)$) for the first and the third terms in (4.2.13) we obtain (4.2.9). □

The $\hat{f}_m^{(N)}(v)$ is also characterized by a recurrence relation as given below. It is necessary to change both m and N therein.

Lemma 4.2.4. *For $m\geq 0$ and $N\geq 1$,*

$$(4.2.14)\qquad \hat{f}_m^{(N)}(v)=\sum_{0\leq i\leq m}\begin{bmatrix} m \\ i \end{bmatrix}q^{i(i+(N-1)m-v)}\hat{f}_{m-i}^{(N-1)}(v-Ni).$$

Proof. We show that each term in (4.2.9) satisfies the equation (4.2.14). Explicitly, we are to show

$$\begin{aligned}(4.2.15)\qquad &\sum_{j\in Z}(-1)^j\begin{bmatrix} v+m-(N+1)j-\delta \\ v-(N+1)j-\delta \end{bmatrix}\begin{bmatrix} m \\ j \end{bmatrix}q^{(j/2)(j-1)+j(mN-v+1)+\delta m} \\ &=\sum_{\substack{0\leq i\leq m \\ k\in Z}}(-1)^k\begin{bmatrix} m \\ i \end{bmatrix}\begin{bmatrix} v+m-i-(i+k)N-\delta \\ v-(i+k)N-\delta \end{bmatrix}\begin{bmatrix} m-i \\ k \end{bmatrix} \\ &\quad\times q^{(k/2)(k-1)+(i+k)(m(N-1)+i-v)+k+\delta(m-i-k)},\end{aligned}$$

where $\delta=0$ or 1 corresponding to the first and the second term in (4.2.9), respectively. In fact, each j-summand in the l.h.s. is equal to the sum in the r.h.s. with the restriction $i+k=j$:

$$\begin{aligned}(4.2.16)\qquad &\begin{bmatrix} v+m-(N+1)j \\ v-(N+1)j \end{bmatrix}\begin{bmatrix} m \\ j \end{bmatrix} \\ &=\sum_{\substack{0\leq i\leq j \\ k\in Z}}(-1)^i\begin{bmatrix} m \\ i \end{bmatrix}\begin{bmatrix} v+m-i-jN \\ v-jN \end{bmatrix}\begin{bmatrix} m-i \\ j-i \end{bmatrix}q^{(i/2)(i-1)-jm}.\end{aligned}$$

Let us prove (4.2.16). Noting the identity $\begin{bmatrix} m \\ i \end{bmatrix}\begin{bmatrix} m-i \\ j-i \end{bmatrix}=\begin{bmatrix} m \\ j \end{bmatrix}\begin{bmatrix} j \\ i \end{bmatrix}$, we cancel the factor $\begin{bmatrix} m \\ j \end{bmatrix}$. Replacing v by $v+(N+1)j$, we get ($m\geq 0$, $v\geq 0$, $j\geq 0$)

$$(4.2.17)\qquad \begin{bmatrix} v+m \\ m \end{bmatrix}=\sum_{0\leq i\leq j}(-1)^i\begin{bmatrix} j \\ i \end{bmatrix}\begin{bmatrix} v+m+j-i \\ m-i \end{bmatrix}q^{(i/2)(i-1)-jm}.$$

Multiply (4.2.17) by z^m and sum it over $m\geq 0$. On account of the formula (4.1.14) and

$$(4.2.18)\qquad \frac{1}{(z)_M}=\sum_{n\geq 0}\begin{bmatrix} M+n-1 \\ n \end{bmatrix}z^n,$$

([9], Theorem 3.3) (4.2.17) is equivalent to the obvious identity

$$(4.2.19)\qquad \frac{1}{(z)_{v+1}}=\frac{(zq^{-j})_j}{(zq^{-j})_{v+j+1}}. \qquad \square$$

The recurrence relation leads us to an expression of $\hat{f}_m^{(N)}(v)$ in terms of the q-multinomial coefficient $\begin{bmatrix} m \\ x_0, \cdots, x_N \end{bmatrix}$. It is defined by [9]

$$\begin{bmatrix} m \\ x_0, \cdots, x_N \end{bmatrix}=(q)_m\Big/\prod_{j=0}^{m}(q)_{x_j}$$

$$\text{if } x_0+\cdots+x_N=m \quad\text{and}\quad x_j\geq 0 \quad\text{for all } j,$$
$$=0 \qquad \text{otherwise.}$$

Theorem 4.2.5 ((3.29) of Part I). *For $m\geq 0$ and $N\geq 1$,*

$$(4.2.20)\qquad \hat{f}_m^{(N)}(v)=\sum\begin{bmatrix} m \\ x_0, \cdots, x_N \end{bmatrix}q^{\sum_{j<k}(k-j-1)x_jx_k},$$

where the outer sum is taken over all non-negative integers $x_0, \cdots, x_N$ such that

$$(4.2.21a)\qquad \sum_{j=0}^{N}x_j=m,$$

$$(4.2.21b)\qquad \sum_{j=0}^{N}jx_j=v.$$

Proof. Repeated use of Lemma 4.2.4 yields

(4.2.22a)
$$\hat{f}_m^{(N)}(v)=\sum\begin{bmatrix} m \\ x_N \end{bmatrix}\begin{bmatrix} m-x_N \\ x_{N-1} \end{bmatrix}\cdots\begin{bmatrix} m-x_2-\cdots-x_N \\ x_1 \end{bmatrix} q^{\mathscr{X}} \times \hat{f}_{\tilde{m}}^{(0)}\Big(v-\sum_{j=1}^{N} jx_j\Big),$$

(4.2.22b)
$$\tilde{m}=m-\sum_{j=2}^{N} x_j,$$

(4.2.22c)
$$\mathscr{X}=\sum_{j=1}^{N} x_j\Big(x_j+(j-1)\Big(m-\sum_{k=j+1}^{N} x_k\Big)-v+\sum_{k=j+1}^{N} kx_k\Big)$$

where the sum in (4.2.22a) extends over non-negative integers $x_1, \cdots, x_N$ satisfying $x_1+\cdots+x_N\leq m$. Because of (4.1.4) the sum can be restricted to (4.2.21b). Eliminating m in (4.2.22c) by introducing an extra variable x_0 as in (4.2.21a), we have $\mathscr{X}=\sum_{j<k}(k-j-1)x_j x_k$. The product of the Gaussian polynomials in (4.2.22a) is nothing but the q-multinomial coefficient $\begin{bmatrix} m \\ x_0, \cdots, x_N \end{bmatrix}$. □

Remark. For $N=1$, Theorem 4.2.5 together with (4.1.3) and (4.2.1) provides us with a simple expression (cf. (2.3.6) of [5])

(4.2.23)
$$f_m^{(1)}(b, c; q)=q^{bc/4}\begin{bmatrix} m \\ (m+b)/2 \end{bmatrix} \qquad (c=b\pm 1).$$

Although the multinomial expression (4.2.20) looks neat, it is not quite adequate for examining the large m behavior of $H_m^{(N)}(\beta)$ defined in section 5 (see (5.2.6)). The rest of this section is devoted to introducing $h_m^{(N)}(w)$ and its limit $h_\infty^{(N)}(w)=\lim_{m\to\infty} h_m^{(N)}(w)$ and deriving various representations for them (see (4.2.28) and (4.2.41)).

Lemma 4.2.6. *For $m\geq 1$ and $N\geq 1$,*

(4.2.24a)
$$\hat{f}_m^{(N)}(v)=\sum(-1)^{y_0}\Big(\begin{bmatrix} v+m-2s_0-s_1+y_N-1 \\ v-s_1-y_0 \end{bmatrix}\frac{1}{(q)_{y_0}\cdots(q)_{y_N}} - \begin{bmatrix} v+m-2s_0-s_1+y_N \\ v-s_1-y_0 \end{bmatrix}\frac{1}{(q)_{y_0}\cdots(q)_{y_{N-1}}}\Big)q^{\mathscr{Y}},$$

(4.2.24b)
$$s_0=\sum_{j=0}^{N} y_j,$$

(4.2.24c)
$$s_1=\sum_{j=0}^{N} jy_j$$

$$\mathscr{Q}=\sum_{0\le j<k\le N}(j-k)y_j y_k+\frac{y_0(y_0-1)}{2}-y_0(v-s_1)-y_N \tag{4.2.24d}$$
$$+(mN+1-s_1)s_0.$$

The sum in (4.2.24a) *extends over all non-negative integers* $y_0, \cdots, y_N$.

Proof. Using (4.2.11) we rewrite the r.h.s. of (4.2.24a) as

$$\sum_{s_0\ge 0}\Bigg(\sum_{y_0+\cdots+y_N=s_0}(-1)^{y_0}\begin{bmatrix} v+m-2s_0-s_1+y_N-1 \\ v-s_1-y_0\end{bmatrix}\frac{q^{\mathscr{Q}+y_N}}{(q)_{y_0}\cdots(q)_{y_N}}$$
$$-\sum_{y_0+\cdots+y_N=s_0}(-1)^{y_0}\begin{bmatrix} v+m-2s_0-s_1+y_N-1 \\ v-s_1-y_0-1\end{bmatrix} \tag{4.2.25}$$
$$\times\frac{q^{\mathscr{Q}+m-2s_0+y_0+y_N}}{(q)_{y_0}\cdots(q)_{y_{N-1}}}\Bigg).$$

We compare the first and the second terms in the sum with those in (4.2.2) through the identification of s_0 with j. A little calculation shows that it is sufficient to show

$$\sum_{y_0+\cdots+y_K=s_0}(-1)^{y_1+\cdots+y_N}\begin{bmatrix} v+m-2s_0-s_1+y_N-1 \\ v-s_1-y_0\end{bmatrix}$$
$$\times\frac{q^{\tilde{\mathscr{Q}}}}{(q)_{y_0}\cdots(q)_{y_N}} \tag{4.2.26a}$$
$$=\begin{bmatrix} v+m-(N+1)s_0-1 \\ v-(N+1)s_0,\ m-s_0-1,\ s_0\end{bmatrix},$$

$$\tilde{\mathscr{Q}}=\frac{1}{2}\sum_{j=1}^{N}y_j^2+\sum_{1\le j<k\le N}(j-k+1)y_j y_k+\left(v+\frac{1}{2}\right)\sum_{j=1}^{N}y_j$$
$$-s_0\left(\sum_{j=1}^{N}(j+1)y_j\right). \tag{4.2.26b}$$

Here we have eliminated y_0 by (4.2.24b) and written the product of Gaussian polynomials in (4.2.2a) as the q-trinomial coefficient. We are going to show (4.2.26) by an induction with respect to v and $m(\ge 1)$. Let us denote the identity (4.2.26) (to be proved) by $[v, m, s_0]$. Clearly $[0, 1, s_0]$ holds as an equality $\delta_{0s_0}=\delta_{0s_0}$; so is $[v, m, s_0]$ for $v<0$ as $0=0$. Assume $[\hat{v}, \hat{m}, s_0]$ for $-\infty<\hat{v}\le v$, $1\le\hat{m}\le m$, $\hat{v}+\hat{m}<v+m$, $-\infty<s_0<\infty$. We split the both sides of $[v, m, s_0]$ into three terms using the standard formulas for the q-multinomial:

(4.2.27a)
$$\begin{bmatrix} C \\ C_1,\ C_2,\ C_3 \end{bmatrix} = \begin{bmatrix} C-1 \\ C_1,\ C_2-1,\ C_3 \end{bmatrix} + \begin{bmatrix} C-1 \\ C_1-1,\ C_2,\ C_3 \end{bmatrix} q^{C_2} + \begin{bmatrix} C-1 \\ C_1,\ C_2,\ C_3-1 \end{bmatrix} q^{C_1+C_2},$$

(4.2.27b)
$$\begin{bmatrix} D \\ D_1 \end{bmatrix} = \begin{bmatrix} D-1 \\ D_1 \end{bmatrix} + \begin{bmatrix} D-1 \\ D_1-1 \end{bmatrix} q^{D-D_1-y_N} - \begin{bmatrix} D-1 \\ D_1-1 \end{bmatrix} q^{D-D_1-y_N}(1-q^{y_N})$$

where $C=v+m-(N+1)s_0-1$, $C_1=v-(N+1)s_0$, $C_2=m-s_0-1$, $C_3=s_0$, $D=v+m-2s_0-s_1+y_N-1$ and $D_1=v-s_1-y_0$. The contributions to (4.2.26a) from the first, the second and the third terms in (4.2.27a) are respectively equal to those from (4.2.27b) due to $[v, m-1, s_0]$, $[v-1, m, s_0]$ and $[v-N-1, m-1, s_0-1]$. Thus we have proved $[v, m, s_0]$. □

Theorem 4.2.7. *For $m \geq 1$ and $N \geq 1$,*

(4.2.28a)
$$\hat{f}_m^{(N)}(v) = \sum_{w\in \mathbf{Z}} (-1)^w q^{-w(w-1)/2+wm} \begin{bmatrix} Nm-2w \\ v-w \end{bmatrix} h_m^{(N)}(w),$$

(4.2.28b)
$$h_m^{(N)}(w) = \sum{}' (-1)^{\bar{s}_1} \frac{q^{\mathscr{R}}}{(q)_{y_1}\cdots(q)_{y_{N-1}}} h_\infty^{(N)}(w-\bar{s}_1),$$

(4.2.28c)
$$\bar{s}_0 = \sum_{j=1}^{N-1} y_j, \qquad \bar{s}_1 = \sum_{j=1}^{N-1} j y_j,$$

(4.2.28d)
$$\mathscr{R} = -\sum_{j=1}^{N-1} j y_j^2 - 2 \sum_{1\leq j<k\leq N-1} k y_j y_k + (1+mN-2(w-\bar{s}_1))\bar{s}_0 + (w-m-(\bar{s}_1+1)/2)\bar{s}_1,$$

(4.2.28e)
$$h_\infty^{(N)}(w) = \delta_{w0} \qquad \text{if } N=1,$$
$$= q^{w(w-1)} \sum_j (-1)^{Nj} \frac{q^{N^2j^2/2-(N+1)wj}}{(q)_{w-Nj}} \left(\frac{q^{Nj/2}}{(q)_j} - \frac{q^{-Nj/2+w}}{(q)_{j-1}} \right), \qquad \textit{otherwise}.$$

Here the sum $\sum'$ in (4.2.28b) is taken over non-negative integers $y_1, \cdots, y_{N-1}$ under the restriction $0 \leq \bar{s}_1 \leq w$. The j-sum in (4.2.28e) extends over $0 \leq j \leq [w/N]$ or $1 \leq j \leq [w/N]$ for the first or second term, respectively.

Proof. The case $N=1$ is clear from (4.2.1) and (4.2.23) (See also (4.2.33) below.) For $N \geq 2$ we reduce (4.2.28) to (4.2.24). Expanding $\begin{bmatrix} Nm-2w \\ v-w \end{bmatrix}$ using the formula (4.1.14), we get

$$\text{(4.2.29)}\quad \begin{aligned}\begin{bmatrix} Nm-2w \\ v-w \end{bmatrix} &= (q^{Nm-v-w+1})_{v-w}/(q)_{v-w} \\ &= \sum_{y_0\geq 0}(-1)^{y_0} q^{y_0(y_0-1)/2+(Nm-v-w+1)y_0}\frac{1}{(q)_{y_0}(q)_{v-w-y_0}}.\end{aligned}$$

Combine the factor $1/(q)_{v-w-y_0}$ in (4.2.29) with $1/(q)_{w-\bar{s}_1-Nj}$ coming from $h_\infty^{(N)}(w-\bar{s}_1)$ as

$$\text{(4.2.30)}\quad \frac{1}{(q)_{v-w-y_0}(q)_{w-\bar{s}_1-Nj}} = \frac{1}{(q)_{v-\bar{s}_1-y_0-Nj}}\begin{bmatrix} v-\bar{s}_1-y_0-Nj \\ w-\bar{s}_1-Nj \end{bmatrix}.$$

After replacing w by $w+\bar{s}_1+Nj$, we perform the w-summation by applying (4.1.14). Then we have

$$\text{(4.2.31a)}\quad \begin{aligned}\hat{f}_m^{(N)}(v) &= \sum(-1)^{y_0}\frac{q^{\mathscr{R}'}}{(q)_{y_0}(q)_{y_1}\cdots(q)_{y_{N-1}}(q)_{v-\bar{s}_1-y_0-Nj}} \\ &\times\left(\frac{(q^{m-2\bar{s}_0-y_0-j})_{v-\bar{s}_1-y_0-Nj}}{(q)_j}-\frac{(q^{m-2\bar{s}_0-y_0-j+1})_{v-\bar{s}_1-y_0-Nj}}{(q)_{j-1}}\right).\end{aligned}$$

$$\text{(4.2.31b)}\quad \begin{aligned}\mathscr{R}' &= \mathscr{R}|_{w=0}+y_0(y_0-1)/2+(Nm-v+1)y_0 \\ &+\bar{s}_1(\bar{s}_1/2+m-2\bar{s}_0-y_0+1/2)-Nj(j-m+2\bar{s}_0+y_0).\end{aligned}$$

Here the sum $\sum$ in (4.2.31a) extends over all non-negative integers j, y_0, $y_1, \cdots, y_{N-1}$. Setting

$$\text{(4.2.32)}\quad j=y_N,\quad \bar{s}_0=s_0-y_0-y_N,\quad \bar{s}_1=s_1-Ny_N,$$

we obtain the expression identical with (4.2.24). □

Remark. From (4.2.28b–e) the following formula holds:

$$\text{(4.2.33)}\quad h_m^{(1)}(w)=\delta_{w0},\qquad h_m^{(N)}(0)=1.$$

The $h_\infty^{(N)}(w)$ is characterized by the following recurrence relation along with the initial condition $h_m^{(1)}(w)=\delta_{w0}$.

Lemma 4.2.8. *For* $N\geq 2$,

$$\text{(4.2.32)}\quad \begin{aligned}h_\infty^{(N)}(w) &= \sum_{k\geq 0}\frac{(-1)^{Nk}}{(q)_k}q^{-(N^2/2-2N+1)k^2+((N-3)w-N/2+1)k} \\ &\times h_\infty^{(N-1)}(w-(N-1)k).\end{aligned}$$

Proof. Substitute (4.2.28e) into the r.h.s. of (4.2.34). After replacing j by $j-k$, we combine the products as

(4.2.35)
$$\frac{1}{(q)_k(q)_{j-k-\delta}}=\frac{1}{(q)_{j-\delta}}\begin{bmatrix} j-\delta \\ k \end{bmatrix},$$

where $\delta=0$ or 1 corresponding to the first and the second terms in (4.2.28e). Performing the k-summations in the resulting expression by applying the formula (4.1.14), we get

(4.2.36)
$$\begin{aligned}&(\text{the r.h.s. fo (4.2.34)})\\ &=q^{w(w-1)}\sum_{j\geq 0}(-1)^{(N-1)j}\frac{q^{(N-1)^2j^2/2-Nwj}}{(q)_{w-(N-1)j}}\\ &\quad\times\left(\frac{(q^{(N-1)j-w+1})_j q^{(N-1)j/2}}{(q)_j}-\frac{(q^{(N-1)j-w+1})_{j-1}q^{-(N-1)j/2+w}}{(q)_{j-1}}\right)\\ &=q^{w(w-1)}\sum_{j\geq 0}(-1)^{Nj}\frac{q^{N^2j^2/2-(N+1)wj}}{(q)_{w-Nj}}\\ &\quad\times\left(\frac{(1-q^{w-Nj})q^{Nj/2}}{(q)_j(1-q^{w-(N-1)j})}+\frac{q^{-(3N/2-1)j+2w}}{(q)_{j-1}(1-q^{w-(N-1)j})}\right).\end{aligned}$$

where we used the formulas such as

$$(q)_{w-(N-1)j}=(q)_{w-Nj}(1-q^{w-Nj+1})\cdots(1-q^{w-(N-1)j}),$$
$$(q^{(N-1)j-w+1})_j=(-1)^j q^{(N-1/2)j^2-(w-1/2)j}(q^{w-Nj})_j,\quad etc.$$

The first and the second terms in (4.2.36) decompose into two terms

(4.2.37a) $$(1-q^{w-Nj})q^{Nj/2}=(1-q^{w-(N-1)j})q^{Nj/2}-(1-q^j)q^{-Nj/2+w},$$

(4.2.37b) $$q^{-(3N/2-1)j+2w}=-(1-q^{w-(N-1)j})q^{-Nj/2+w}+q^{-Nj/2+w},$$

where the first terms give rise to $h_\infty^{(N)}(w)$ while the second terms cancel each other. □

Similarly as Lemma 4.2.4 leads to Theorem 4.2.5, Lemma 4.2.8 leads to the following.

Lemma 4.2.9. *For* $N\geq 1$,

(4.2.38a) $$h_\infty^{(N)}(w)=\sum{}'(-1)^{w+\sum_{1\leq j\leq N-1}x_j}\frac{q^{\mathcal{T}_N(w)}}{(q)_{x_1}\cdots(q)_{x_{N-1}}},$$

(4.2.38b)
$$\begin{aligned}\mathcal{T}_N(w)=&-\sum_{1\leq j\leq N-1}\left(j-\frac{1}{2}\right)x_j^2-2\sum_{1\leq j<k\leq N-1}jx_jx_k\\ &+\frac{1}{2}\sum_{1\leq j\leq N-1}x_j+\frac{1}{2}w(w-1),\end{aligned}$$

where the sum $\sum'$ *in* (4.2.38a) *is taken over non-negative integers* $x_1, \cdots, x_{N-1}$ *under the restriction*

(4.2.38c) $$\sum_{1\le j\le N-1} jx_j = w.$$

Proof. The case $N=1$ is clearly true. (See (4.2.28e).) For $N\ge 2$, we are going to show that the r.h.s. of (4.2.38a) satisfies the recurrence relation (4.2.34). Let us denote x_{N-1} by i. Direct calculation shows that

(4.2.39a) $$\mathcal{T}_N(w)=\mathcal{T}_{N-1}(w-(N-1)i)-(N^2/2-2N+1)i^2 +((N-3)w-N/2+1)i,$$

(4.2.39b) $$w+\sum_{1\le j\le N-1} x_j = Ni+w-(N-1)i+\sum_{1\le j\le N-2} x_j.$$

Thus the r.h.s. of (4.2.38a) can be written as

(4.2.40a) $$\sum_{i\ge 0}\frac{(-1)^{Ni}}{(q)_i} q^{-(N^2/2-2N+1)i^2+((N-3)w-N/2+1)i} \times \sum{}''(-1)^{w-(N-1)i+\sum_{1\le j\le N-2} x_j} \frac{q^{\mathcal{T}_{N-1}(w-(N-1)i)}}{(q)_{x_1}\cdots(q)_{x_{N-2}}}.$$

Here the sum $\sum''$ is taken over non-negative integers $x_1, \cdots, x_{N-2}$ such that

(4.2.40b) $$\sum_{1\le j\le N-2} jx_j = w-(N-1)i.$$

This is nothing but the r.h.s. of (4.2.38a) with N and w replaced by $N-1$ and $w-(N-1)i$, respectively. Thus we conclude that the r.h.s. of (4.2.38a) satisfies (4.2.34). □

Lemma 4.2.9 gives another expression for $h_m^{(N)}(w)$ as given below. It is remarkable that as a consequence we have so many different expressions for one and the same quantity $\hat{f}_m^{(N)}(v)$.

Theorem 4.2.10. *For* $N\ge 1$,

(4.2.41a) $$h_m^{(N)}(w)=\sum{}'(-1)^{w+\sum_{1\le j\le N-1} x_j} q^{\mathcal{T}_N(w)} \times\begin{bmatrix} m \\ x_{N-1}\end{bmatrix}\begin{bmatrix} 2m-2x_{N-1} \\ x_{N-2}\end{bmatrix}\cdots \times\begin{bmatrix}(N-1)m-2(x_2+2x_3+\cdots+(N-2)x_{N-1}) \\ x_1\end{bmatrix}.$$

Here $\mathcal{T}_N(w)$ *is given by* (4.2.38b). *The sum* $\sum'$ *is taken over non-negative integers* $x_1, \cdots, x_{N-1}$ *subject to*

$$\sum_{1\le j\le N-1} jx_j = w. \tag{4.2.41b}$$

Proof. Expand all the Gaussian polynomials in a way like (see (4.1.14))

$$\begin{aligned}\begin{bmatrix} m \\ x_{N-1} \end{bmatrix} &= \frac{(q^{m-x_{N-1}+1})_{x_{N-1}}}{(q)_{x_{N-1}}} \\ &= \sum_{y_{N-1}} (-1)^{y_{N-1}} \frac{q^{y_{N-1}(y_{N-1}-1)/2+y_{N-1}(m-x_{N-1}+1)}}{(q)_{y_{N-1}}(q)_{x_{N-1}-y_{N-1}}}, \quad etc.\end{aligned} \tag{4.2.42}$$

Then the r.h.s. of (4.2.41a) is cast into the form

$$\begin{aligned}&\sum_{0\le \bar{s}_1\le w} \sum{}' \frac{q^{\sum_{1\le j\le N-1} y_j((y_j+1)/2+m(N-j))}}{(q)_{y_1}\cdots(q)_{y_{N-1}}} \\ &\quad\times \sum{}'' (-1)^{w+\sum_{1\le j\le N-1}(x_j-y_j)} \\ &\qquad\times \frac{q^{\mathcal{T}_N(w)-\sum_{1\le j\le N-1} x_j y_j - 2\sum_{1\le j<k\le N-1}(k-j)x_k y_j}}{(q)_{x_1-y_1}\cdots(q)_{x_{N-1}-y_{N-1}}},\end{aligned} \tag{4.2.43}$$

where the sum $\sum'$ (resp. $\sum''$) is taken over non-negative integers $y_1, \cdots, y_{N-1}$ (resp. $x_1, \cdots, x_{N-1}$) satisfying $\sum_{1\le j\le N-1} jy_j = \bar{s}_1$ (resp. $\sum_{1\le j\le N-1} jx_j = w$). If we replace x_j by $x_j + y_j$, the condition $\sum_{1\le j\le N-1} jx_j = w$ changes to

$$\sum_{1\le j\le N-1} jx_j = w - \bar{s}_1. \tag{4.2.44}$$

After a little calculation (4.2.43) turns out to be written as follows

$$\begin{aligned}&\sum_{0\le \bar{s}_1\le w} (-1)^{\bar{s}_1} \sum{}' \frac{q^{\mathcal{R}}}{(q)_{y_1}\cdots(q)_{y_{N-1}}} \\ &\quad\times \sum{}''' (-1)^{w-\bar{s}_1+\sum_{1\le j\le N-1} x_j} \frac{q^{\mathcal{T}_N(w-\bar{s}_1)}}{(q)_{x_1}\cdots(q)_{x_{N-1}}}.\end{aligned} \tag{4.2.45}$$

Here the sum $\sum'$ is taken in the same manner as (4.2.43) and the sum $\sum''$ extends over all non-negative integers $x_1, \cdots, x_{N-1}$ with the condition (4.2.44). The power $\mathcal{R}$ is given by (4.2.28c-d). Now we apply Lemma 4.2.9 to identify the sum $\sum'''$ in (4.2.45) with $h_\infty^{(N)}(w-\bar{s}_1)$. Comparing the resulting expression with (4.2.28b-d) we arrive at (4.2.41). □

Remark. For $N=2$, (4.2.41) yields the simple formula

(4.2.46) $$h_m^{(2)}(w) = \begin{bmatrix} m \\ w \end{bmatrix}.$$

4.3. Fundamental solution for the modified linear difference equation

In this paragraph we assume that $L \geq 2N+3$. Let $H(n, a, b, c)$ denote the function $H(a, b, c)$ (3.1.12) with n regarded as an integer parameter (not necessarily $[L/2]$). Note that $H(n, a, b, c)$ now acquires the translational invariance

(4.3.1) $$H(n, a, b, c) = H(n+1, a+1, b+1, c+1).$$

In this paragraph we study the function $g_m^{(N)}(n; b, c)$ ((b, c): weakly admissible, $m \geq 0$, $N \geq 1$) characterized by the following linear difference equation.

(4.3.2a) $$g_m^{(N)}(n; b, c) = \sum_d{}' g_{m-1}^{(N)}(n; d, b) q^{mH(n, d, b, c)},$$

(4.3.2b) $$g_0^{(N)}(n; b, c) = \delta_{b0}.$$

Here the sum $\sum'$ is taken over d such that the pair (d, b) is weakly admissible. We set $g_m^{(N)}(n; b, c) = 0$ if (b, c) is not weakly admissible.

We recall that the $H(a, b, c)$ is akin to the negative of $|a-c|/4$ employed in the linear difference equation (4.1.1). The effect of this modification is fully absorbed by patchwork construction of $g(n; b, c)$ (see (4.3.12)). By the definition (4.3.2) together with the symmetries of $H(n, a, b, c)$ (3.1.13), (4.3.1) the following formulas are valid.

(4.3.3) $$g_m^{(N)}(n; b, c) = g_m^{(N)}(-n-1; -b, -c) \qquad \text{for } m \geq 0,$$

(4.3.4) $$g_m^{(N)}(n; b, c) = 0 \quad \text{unless } |b| \leq mN, \quad |c| \leq (m+1)N, \quad \text{and } b \equiv mN \bmod 2.$$

In order to describe $g_m^{(N)}(n; b, c)$ we prepare some notations. For a weakly admissible pair of integers (b, c) define the four regions $S_1(n), \cdots, S_4(n)$ as follows.

(4.3.5)
$$\begin{aligned} S_1(n) &= \{(b, c) \mid b+c < 2n+1-N\}, \\ S_2(n) &= \{(b, c) \mid b+c > 2n+1-N,\ b < n+1/2\}, \\ S_3(n) &= \{(b, c) \mid b+c < 2n+1+N,\ b > n+1/2\}, \\ S_4(n) &= \{(b, c) \mid b+c > 2n+1+N\}. \end{aligned}$$

We set

(4.3.6) $$\kappa=[(2n+1)/2N],$$

and further subdivide the regions $S_{2,3}(n)$.

(4.3.7)
$$\begin{aligned} S_2^{(\pm)}(n)&=\{(b, c)\,|\,(b, c)\in S_2(n),\ b\in R_{\kappa\pm1}\}, &&\text{if } m+\kappa \text{ is even,}\\ &=\{(b, c)\,|\,(b, c)\in S_2(n),\ b\gtrless 2n+1-(\kappa+1)N\}, &&\text{otherwise.}\\ S_3^{(\pm)}(n)&=\{(b, c)\,|\,(b, c)\in S_3(n),\ b\lesseqgtr 2n+1-\kappa N\}, &&\text{if } m+\kappa \text{ is even,}\\ &=\{(b, c)\,|\,(b, c)\in S_3(n),\ b\in R_{\kappa+1\mp1}\}. &&\text{otherwise.}\end{aligned}$$

These regions are schematically shown in Fig. 4.1.

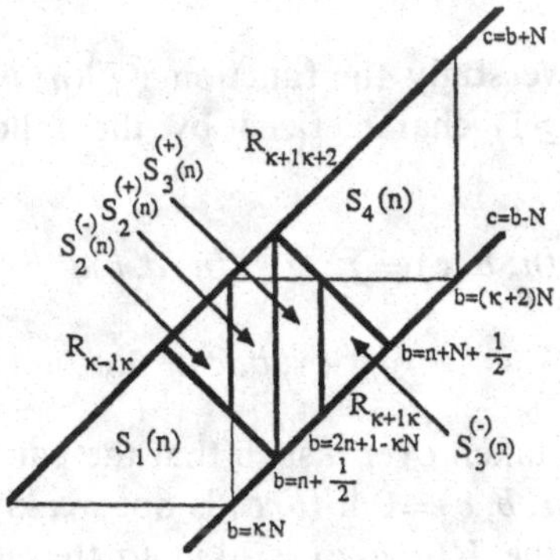

Fig. 4.1 a Regions of (b, c) when $m+\kappa$ is even. $S_1(n)$ and $S_3(n)$ are the infinite domains bounded by the bold lines.

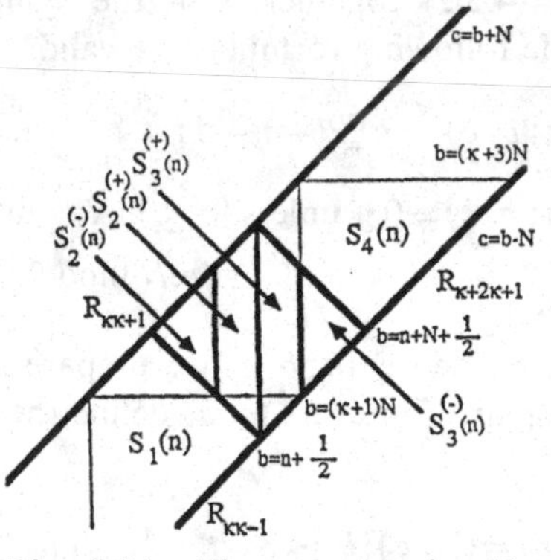

Fig. 4.1 b The case $m+\kappa$ is odd.

The following relations are readily derivable from (4.3.5–7)

(4.3.8a) $(b, c)\in S_1(n)\longleftrightarrow(-b, -c)\in S_4(-n-1)$,

(4.3.8b) $(b, c)\in S_2^{(\pm)}(n)\longleftrightarrow(-b, -c)\in S_3^{(\pm)}(-n-1)$.

For a weakly admissible pair $(b, c) \in R_{\mu,\nu}$ with μ, ν satisfying (4.1.6) we define a function $\bar{f}_m^{(N)}(b, c; \mu, \nu)$ by

$$(q)_{m-1}\bar{f}_m^{(N)}(b, c; \mu, \nu)$$

(4.3.9a)
$$\Big(\sum_{\substack{j\geq 1+(m+\nu)/2\\ k\geq (m-\mu+1)/2}} - \sum_{\substack{j\leq (m+\nu)/2\\ k\leq (m-\mu-1)/2}}\Big)(-1)^{j+k}q^{P_{j,k}^{(m)}(b,c)}\begin{bmatrix} m-1\\ j-1\end{bmatrix}\begin{bmatrix} m\\ k\end{bmatrix},$$

(4.3.9b)
$$P_{j,k}^{(m)}(b, c) = -\frac{j}{2}b+\frac{k}{2}c+\Big(\frac{m}{2}j+\frac{m+1}{2}k-jk\Big)N + \frac{1}{2}(j-k)(j-k-1), \quad \text{if } m\geq 1,$$

(4.3.9c)
$$\bar{f}_0^{(N)}(b, c; \mu, \nu) = \delta_{b0}.$$

The $\bar{f}_m^{(N)}$ satisfies the following linear difference equation.

Lemma 4.3.1. *For* $m\geq 1$,

(4.3.10)
$$\bar{f}_m^{(N)}(b, c; \mu, \nu) = \sum_d{}' \bar{f}_{m-1}^{(N)}(d, b: \lambda, \mu)q^{m(\min(d,c)-b+N)/2},$$

where the sum $\sum'$ *is taken over d such that the pair* (d, b) *is weakly admissible and* λ *is specified by* $d \in R_\lambda$, $\lambda\equiv m \bmod 2$ *(see (4. 1.5–8)).*

Proof. Comparing (4.1.7) and (4.3.9) it is straightforward to see $(m\geq 0)$

(4.3.11)
$$\bar{f}_m^{(N)}(b, c; \mu, \nu) = f_m^{(N)}(b, c; q^{-1})q^{Nm^2/4-(m/4)(b-c-N)-b/4}.$$

We substitute this into (4.3.10). The resulting equation for $f_m^{(N)}(b, c; q^{-1})$ turns out to be equal to (4.1.1a) with q replaced by q^{-1} (note that $\min(d, c) = (d+c-|d-c|)/2$). Thus we have (4.3.10) by Theorem 4.1.1. □

Now we construct $g_m^{(N)}(n; b, c)$ from $\bar{f}_m^{(N)}$ by patchwork.

Theorem 4.3.2 ((3.14) of Part I). *For all weakly admissible pairs* $(b, c) \in R_{\mu,\nu}$ *and integers* m, $N\geq 1$,

$$g_m^{(N)}(n; b, c)$$

(4.3.12a)
$$= \bar{f}_m^{(N)}(b, c; \mu, \nu), \qquad \text{if } (b, c) \in S_1(n)\cap R_{\mu,\nu},$$

(4.3.12b)
$$= \bar{f}_m^{(N)}(-b, -c; -\mu, -\nu), \qquad \text{if } (b, c) \in S_4(n)\cap R_{\mu,\nu},$$

(4.3.12c)
$$= \bar{f}_m^{(N)}(b, 2n-b-N; \kappa\pm(1-\rho), \kappa\mp\rho), \qquad \text{if } (b, c) \in S_2^{(\pm)}(n),$$

(4.3.12d) $\qquad = \bar{f}^{(N)}_m(-b, -2n-2+b-N; -\kappa-1\pm\rho, -\kappa-1\pm(\rho-1)),$

$\qquad\qquad$ *if* $(b, c) \in S_3^{(\pm)}(n)$,

where $\rho=0$ *or* 1 *according as* $m+\kappa$ *is even or odd, respectively.*

Proof. The properties (4.3.3) and (4.3.4) can be directly checked by using (4.1.3) and (4.3.8, 11, 12). In view of (4.3.3, 8) we restrict ourselves to the case $(b, c) \in S_1(n)$ or $S_2^{(\pm)}(n)$. If $(b, c) \in S_1(n)$ and $b<n-N+1/2$, then for any d such that (d, b) is weakly admissible,

$$H(n, d, b, c)=(\min(d, c)-b+N)/2 \quad \text{and} \quad (d, b) \in S_1(n).$$

By virtue of (4.3.12a) equation (4.3.2a) in this case reduces to (4.3.10), which has been already proved. We list up the non-trivial cases (see Fig. 4.1).

If $m+\kappa$ is odd:

(4.3.13a) $\qquad (b, c) \in S_1(n) \cap R_{\kappa,\kappa\pm1},$

(4.3.13b) $\qquad\qquad \in S_2^{(\pm)}(n).$

If $m+\kappa$ is even:

(4.3.13c) $\qquad (b, c) \in S_1(n) \cap R_{\kappa-1,\kappa-1\pm1},$

(4.3.13d) $\qquad\qquad \in S_1(n) \cap R_{\kappa+1,\kappa},$

(4.3.12e) $\qquad\qquad \in S_2^{(\pm)}(n).$

We prove here the case (4.3.13a). The other cases are similar. Let us write down the r.h.s. of (4.3.2a) taking (4.3.12) and (4.3.13a) into account.

$$\begin{aligned}
&\sum_{b-N\le d\le \kappa N}{}' \bar{f}^{(N)}_{m-1}(d, b; \kappa-1, \kappa)q^{m((\min(d,c)-b+N)/2)} \\
&+ \sum_{\kappa N\le d<2n+1-N-b}{}' \bar{f}^{(N)}_{m-1}(d, b; \kappa+1, \kappa)q^{m((\min(d,c)-b+N)/2)} \\
&+ \sum_{2n+1-N-b<d<n+1/2}{}' \bar{f}^{(N)}_{m-1}(d, 2n-d-N; \kappa+1, \kappa)q^{m((c-b+N)/2)} \\
&+ \sum_{n+1/2<d\le b+N}{}' \bar{f}^{(N)}_{m-1}(-d, -2n-2+d-N; -\kappa-1, -\kappa-2) \\
&\qquad\qquad \times q^{m((c-b+N)/2)}.
\end{aligned} \tag{4.3.14}$$

Here d in the sums $\sum'$ runs over the series $d=b-N, b-N+2, \cdots, b+N$ divided into the four regions as above. The term $d=\kappa N$ is contained in the first (resp. second) sum if $(d, b) \in R_{\kappa-1,\kappa}$ (resp. $R_{\kappa+1,\kappa}$). On the other hand by using Lemma 4.3.1 the l.h.s. of (4.3.2a) is expressed as

$$\begin{aligned}(4.3.15)\qquad &\sum_{b-N\le d\le \kappa N}{}' \bar{f}^{(N)}_{m-1}(d, b; \kappa-1, \kappa)q^{m((\min(d,c)-b+N)/2)}\\ &+\sum_{\kappa N\le d<2n+1-N-b}{}' \bar{f}^{(N)}_{m-1}(d, b; \kappa+1, \kappa)q^{m((\min(d,c)-b+N)/2)}\\ &+\sum_{2n+1-N-b<d\le b+N}{}' \bar{f}^{(N)}_{m-1}(d, b; \kappa+1, \kappa)q^{m((c-b+N)2)},\end{aligned}$$

where the d-sum is taken in the same manner as (4.3.14). Comparing (4.3.14) with (4.3.15) we are led to the following equality to show

$$(4.3.16)\qquad T_1=T_2+T_3,$$

$$(4.3.17a)\qquad T_1=\sum_{2n+2-N-b\le d\le b+N}{}' \bar{f}^{(N)}_{m-1}(d, b; \kappa+1, \kappa),$$

$$(4.3.17b)\qquad T_2=\sum_{2n+2-N-b\le d\le \tilde{n}}{}' \bar{f}^{(N)}_{m-1}(d, 2n-d-N; \kappa+1, \kappa),$$

$$(4.3.17c)\qquad T_3=\sum_{\tilde{n}+2\le d\le b+N}{}' \bar{f}^{(N)}_{m-1}(-d, -2n-2+d-N; -\kappa-1, -\kappa-2),$$

with d taking the values in $\{b-N, b-N+2, \cdots, b+N\}$ and $\tilde{n}=n$ or $n-1$ according as $b+N\equiv n$ or $n-1$ mod 2. Substituting (4.3.9) into (4.3.17) and performing the d-summations, we get

$$(4.3.18)\qquad T_i=T_i^{(+)}+T_i^{(-)},\qquad i=1, 2, 3,$$

$$(4.3.19a)\qquad \begin{aligned}T_i^{(\pm)}=&\pm\Big(\sum_{\substack{j\ge(m+\kappa+1)/2\\ k\ge(m-\kappa-1)/2}}-\sum_{\substack{j\le(m+\kappa-1)/2\\ k\le(m-\kappa-3)/2}}\Big)(-1)^{j+k}\\ &\times\begin{bmatrix}m-2\\ \alpha_i-1\end{bmatrix}\begin{bmatrix}m-1\\ \beta_i\end{bmatrix}\frac{q^{P_i(\pm)}}{1-q^{\gamma_i}},\end{aligned}$$

$$(4.3.19b)\qquad \begin{aligned}&(\alpha_1, \alpha_2, \alpha_3)=(j, j, k)\quad (\beta_1, \beta_2, \beta_3)=(k, k, j),\\ &(\gamma_1, \gamma_2, \gamma_3)=(j, j+k, j+k),\end{aligned}$$

$$(4.3.19c)\qquad \begin{aligned}&P_1(+)=P^{(m-1)}_{j,k}(b+N, b),\\ &P_1(-)=P_2(-)=P^{(m-1)}_{j,k}(2n-N-b, b),\\ &P_2(+)=P^{(m-1)}_{j,k}(\tilde{n}, 2n-\tilde{n}-N),\\ &P_3(+)=P^{(m-1)}_{k,j}(-\tilde{n}-2, \tilde{n}-2n-N),\\ &P_3(-)=P^{(m-1)}_{k,j}(-b-N-2, b-2n).\end{aligned}$$

First consider $T_1^{(+)}$ and rewrite the factor $\dfrac{1}{1-q^j}\begin{bmatrix}m-1\\ j-1\end{bmatrix}$ therein by the formula (4.1.10b). Under the replacement (j, k) by $(m-1-k, m-1-j)$ the summand in (4.3.19a) for $T_1^{(+)}$ is invariant because of

$$P^{(m-1)}_{m-1-k,m-1-j}(b+N, b)=P^{(m-1)}_{j,k}(b+N, b),$$

while the sum $(\sum - \sum)$ changes the sign. Hence we have $T_1^{(+)}=0$. Next we combine $T_2^{(+)}$ and $T_3^{(+)}$. Using the formula

(4.3.20)
$$\frac{1}{1-q^{j+k}}\left(\begin{bmatrix} m-2 \\ j-1 \end{bmatrix}\begin{bmatrix} m-1 \\ k \end{bmatrix}+\begin{bmatrix} m-2 \\ k-1 \end{bmatrix}\begin{bmatrix} m-1 \\ j \end{bmatrix}q^j\right) = (j\leftrightarrow k) = \frac{1}{1-q^{m-1}}\begin{bmatrix} m-1 \\ j \end{bmatrix}\begin{bmatrix} m-1 \\ k \end{bmatrix},$$

we have

(4.3.21)
$$T_2^{(+)}+T_3^{(+)} = \Big(\sum_{\substack{j\geq(m+\varepsilon+1)/2 \\ k\geq(m-\varepsilon-1)/2}} - \sum_{\substack{j\leq(m+\varepsilon-1)/2 \\ k\leq(m-\varepsilon-3)/2}}\Big)(-1)^{j+k} \times\begin{bmatrix} m-1 \\ j \end{bmatrix}\begin{bmatrix} m-1 \\ k \end{bmatrix}\frac{q^P}{1-q^{m-1}},$$

where $P=P_{j,k}^{(m-1)}(n, n-N)$ or $P_{k,j}^{(m-1)}(-n-1, -n-1-N)$ according as $\tilde{n}=n$ or $n-1$. The same argument exploiting the transformation $(j, k)\rightarrow(m-1-k, m-1-j)$ leads us to $T_2^{(+)}+T_3^{(+)}=0$. Finally we verify the equality that holds among the remaining terms in (4.3.16):

(4.3.22)
$$T_1^{(-)}-T_2^{(-)}=T_3^{(-)}.$$

Using the identity

$$\left(\frac{1}{1-q^j}-\frac{1}{1-q^{j+k}}\right)\begin{bmatrix} m-2 \\ j-1 \end{bmatrix}\begin{bmatrix} m-1 \\ k \end{bmatrix} = \frac{1}{1-q^{j+k}}\begin{bmatrix} m-1 \\ j \end{bmatrix}\begin{bmatrix} m-2 \\ k-1 \end{bmatrix}q^j,$$

the l.h.s. of (4.3.22) becomes

(4,3.23)
$$-\Big(\sum_{\substack{j\geq(m+\varepsilon+1)/2 \\ k\geq(m-\varepsilon-1)/2}} - \sum_{\substack{j\leq(m+\varepsilon-1)/2 \\ k\leq(m-\varepsilon-3)/2}}\Big)(-1)^{j+k} \times\begin{bmatrix} m-1 \\ j \end{bmatrix}\begin{bmatrix} m-2 \\ k-1 \end{bmatrix}\frac{q^{P_{j,k}^{(m-1)}(2n-2N-b,\, b)}}{1-q^{j+k}}.$$

This is identical with $T_3^{(-)}$ since we have

$$P_{j,k}^{(m-1)}(2n-2-N-b, b)=P_{k,j}^{(m-1)}(-b-N-2, b-2n).$$

This establishes (4.3.16). □

4.4. 1D configuration sums as superpositions of the fundamental solutions

Here we prove that the 1D configuration sums defined in section 3 (see (3.1.1), (3.1.10–12)) are expressed as linear superpositions of the fundamental solutions discussed in section 4.1–3. We begin by redefining them in terms of linear difference equations and initial conditions.

Regime III

(4.4.1a) $$X_m(a,b,c)=\sum_d{}'' X_{m-1}(a,d,b)q^{m|d-c|/4},$$

(4.4.1b) $$X_0(a,b,c)=\delta_{a,b}.$$

Regime I ($L\geq 2N+3$)

(4.4.2a) $$Y_m(a,b,c)=\sum_d{}'' Y_{m-1}(a,d,b)q^{mH(d,b,c)},$$

(4.4.2b) $$Y_0(a,b,c)=\delta_{a,b}.$$

Here the sum $\sum''$ is taken over d such that the pair (d,b) is admissible and the function $H(a,b,c)$ has been given in (3.1.12). Note that we retain the original definition of n: $n=[L/2]$. We set $X_m(a,b,c)=Y_m(a,b,c)=0$ unless (b,c) is admissible and $0<a<L$.

Remark. By the definition (4.4.1, 2) the 1D configuration sums have the following properties.

(4.4.3a) $X_m(a,b,c)=0$ if $a\not\equiv b+mN \bmod 2$,

(4.4.3b) $X_m(a,b,c)=X_m(L-a,L-b,L-c)$,

(4.4.4a) $Y_m(a,b,c)=0$ if $a\not\equiv b+mN \bmod 2$,

(4.4.4b) $Y_m(a,b,c)=Y_m(L-a,L-b,L-c)$ if L is odd.

Theorem 4.4.1 ((3.5) of Part I). *For* $m\geq 0$,

(4.4.5a) $X_m(a,b,c)=q^{-a/4}(F_m(a,b,c)-F_m(-a,b,c))$,

(4.4.5b) $F_m(a,b,c)=\sum_{\lambda\in Z} q^{-L\lambda^2+(L/2-a)\lambda+a/4}f_m^{(N)}(b-a-2L\lambda,\ c-a-2L\lambda)$.

Proof. The property (4.4.3a) is clear from (4.4.5) and (4.1.3). So is (4.4.3b) by the invariance of the summand in $q^{-a/4}F_m(\pm a,b,c)$ under the change

$$(a,b,c,\lambda)\longrightarrow\left(L-a,L-b,L-c,-\lambda+\frac{1\mp 1}{2}\right).$$

Since the function $X_m(a,b,c)$ in (4.4.5) is a linear superposition of $f_m^{(N)}(b,c)$, it also satisfies the same equation (4.1.1). In order to prove

(4.4.1a) it is enough to show the cancellation of non-admissible summands corresponding to the following values of d (see (4.0.1, 2)).

$$d=b-N,\ b-N+2,\cdots,N-b,\qquad \text{if } b\le N,$$
(4.4.6)
$$d=2L-N-b,\ 2L-N-b+2,\cdots,b+N,\qquad \text{if } L-b\le N.$$

The latter case can be reduced to the former thanks to the symmetry (4.4.3b). Thus we are to show that the following is equal to zero for $1\le b\le N$.

(4.4.7)
$$\begin{aligned}&\sum_{d=b-N,b-N+2,\cdots,N-b}\sum_{\lambda\in Z}\\ &\quad\times(q^{-L\lambda^2+(L/2-a)\lambda+a/4}f^{(N)}_{m-1}(d-a-2L\lambda,b-a-2L\lambda)\\ &\quad-q^{-L\lambda^2+(L/2+a)\lambda-a/4}f^{(N)}_{m-1}(d+a-2L\lambda,b+a-2L\lambda))q^{m|c-d|/4}\\ &=\sum_{\lambda\in Z}q^{-L\lambda^2-a\lambda+mc/4}\sum_{d}\\ &\quad\times(q^{(a+2L\lambda-md)/4}f^{(N)}_{m-1}(d-a-2L\lambda,b-a-2L\lambda)\\ &\quad-q^{(-a-2L\lambda-md)/4}f^{(N)}_{m-1}(d+a+2L\lambda,b+a+2L\lambda)).\end{aligned}$$

Here we have used the fact $c>N-b\ge d$ (see (4.0.2)) in reducing $|c-d|$. The vanishing of (4.4.7) follows directly from Lemma 4.1.2. □

Theorem 4.4.2 ((3.15 of Part I). *For $m\ge 0$,*

(4.4.8a)
$$Y_m(a,b,c)=q^{a/2}(G_m(a,b,c)-G_m(-a,b,c)),$$

(4.4.8b)
$$\begin{aligned}G_m(a,b,c)=&\sum_{\lambda\in Z}q^{2L\lambda^2+(2a-L)\lambda-a/2}\\ &\times g^{(N)}_m(n-a-2L\lambda;b-a-2L\lambda,c-a-2L\lambda).\end{aligned}$$

Proof. We assume that

(4.4.9)
$$(b,c)\in S_1(n)\quad\text{or}\quad S_2^{(\pm)}(n).$$

The other cases are similar. As in the proof of Theorem 4.4.1, we show the cancellation of the non-admissible summands. From (4.4.9) together with $n=[L/2]$ and $L\ge 2N+3$ we have

(4.4.10a)
$$L-b>N,$$

(4.4.10b)
$$\begin{aligned}&(d,b)\in S_1(n),\ H(n,d,b,c)=\frac{d-b+N}{2},\\ &\text{if } b\le N \text{ and } d\in\{b-N,b-N+2,\cdots,N-b\}.\end{aligned}$$

In view of (4.4.6) and (4.4.10a) we assume $b\le N$ without loss of generality. Then the cancellation identity reads as

$$
\begin{aligned}
0= & \sum_{d=b-N,b-N+2,\cdots,N-b}\ \sum_{\lambda\in Z} \\
& \times\big(q^{2L\lambda^2+(2a-L)\lambda-a/2}g^{(N)}_{m-1}(n-a-2L\lambda;\, d-a-2L\lambda,\, b-a-2L\lambda) \\
& -q^{2L\lambda^2+(2a+L)\lambda+a/2}g^{(N)}_{m-1}(n+a+2L\lambda;\, d+a+2L\lambda,\, b+a+2L\lambda)\big) \\
(4.4.11)\quad & \times q^{mH(n,d,b,c)} \\
= & \sum_{\lambda\in Z} q^{2L\lambda^2+2a\lambda+m(N-b)/2}\sum_{d} \\
& \times\big(q^{(-a-2L\lambda+md)/2}\bar{f}^{(N)}_{m-1}(d-a-2L\lambda,\, b-a-2L\lambda;\, \mu_-,\nu_-) \\
& -q^{(a+2L\lambda+md)/2}\bar{f}^{(N)}_{m-1}(d+a+2L\lambda,\, b+a+2L\lambda;\, \mu_+,\nu_+)\big),
\end{aligned}
$$

where $\mu_\pm, \nu_\pm$ are integers defined by $(d\pm(a+2L\lambda), b\pm(a+2L\lambda))\in R_{\mu_\pm,\nu_\pm}$. We used (4.4.10b) and (4.3.12a) in deriving (4.4.11). Substitute (4.3.11) into (4.4.11) and replace $(a+2L\lambda, d)$ by $(a, -d)$. The resulting sum $\sum_d$ is equal to (the l.h.s.—the r.h.s.) of (4.1.17) with q replaced by q^{-1}. Thus (4.4.11) follows from Lemma 4.1.2. □

Remark. From (4.4.8) and (4.3.12) $Y_m(a, b, c)$ enjoys the following symmetries

(4.4.12a) $Y_m(a, b, c)=Y_m(L-a, L-b, L-c)$, if $(b, c)\in S_4(n)$,

(4.4.12b) $\quad = Y_m(a, b, 2n-b-N)$, if $(b, c)\in S_2(n)$,

(4.4.12c) $\quad = Y_m(a, b, 2n+2-b+N)$, if $(b, c)\in S_3(n)$.

In this way the evaluation of $Y_m(a, b, c)$ for $(b, c)\in S_4(n)$, $S_2(n)$ and $S_3(n)$ reduces to that for $S_1(n)$, $S_1(n)$ and $S_4(n)$, respectively.

§ 5. One Dimensional Configuration Sums as Modular Functions

In this section we identify the limit $m\to\infty$ of the 1D configuration sums $X_m(a, b, c; q^{\pm1})$ and $Y_m(a, b, c; q^{\pm1})$ (up to some power corrections in q) with modular functions appearing in theta function identities. Regime I was fully treated in Part I. In what follows we shall deal with the other regimes III, IV and II.

5.1. Regimes III and IV

The modular functions $c^{(\pm)}_{j_1j_2j_3}(\tau)$ (which we called the branching coefficients) have been defined in Appendix A of Part I. For our present purpose let us quote the expression (A.6) therein.

Assume $j_1, m_1\in Z/2$, $j_2, m_2\in Z$, $0<j_i\le m_i\ne 0$ $(i=1, 2)$ and $m_3=m_1+m_2-2>0$. We choose $j_3\in Z+j_1$ with the restriction $0<j_3\le m_3$ for

$c^{(-)}_{j_1j_2j_3}(\tau)$ and $0<j_3<m_3$ for $c^{(+)}_{j_1j_2j_3}(\tau)$. The branching coefficient $c^{(\pm)}_{j_1j_2j_3}(\tau)$ is given by

(5.1.1a) $$c^{(\pm)}_{j_1j_2j_3}(\tau)=\varepsilon^{m_3}_{j_3}\frac{q^{\gamma(j_1j_2j_3)}}{\varphi(q)^3}\Big(\sum_{\xi+1\le t,\eta\le\mathscr{L}_1\in Z}-\sum_{\xi\ge t,\eta-1\ge\mathscr{L}_1\in Z}\Big)\text{ (summand)},$$

(5.1.1b) $$\text{(summand)}=(-1)^{t+(\varepsilon_1+\varepsilon_2)/2}(\pm1)^{n_1}q^{t(t-1)/2+t\mathscr{L}_1+\mathscr{L}_0},$$

(5.1.1c) $$\mathscr{L}_0=\sum_{i=1}^{2}(m_in_i^2+j_in_i),$$

(5.1.1d) $$\mathscr{L}_1=\frac{j_3+1}{2}+\sum_{i=1}^{2}\varepsilon_i(m_in_i+j_i/2),$$

(5.1.1e) $$\gamma(j_1,j_2,j_3)=\frac{j_1^2}{4m_1}+\frac{j_2^2}{4m_2}-\frac{1}{8}-\frac{j_3^2}{4m_3}.$$

Here the sum is over integers t, n_1, n_2 and $\varepsilon_1, \varepsilon_2=\pm1$ with the restriction written as above. The integers $\xi=\xi(\varepsilon_1, n_1)$ and $\eta=\eta(\varepsilon_1, n_1)$ can be chosen arbitrarily for fixed ε_1 and n_1. As for further properties of these branching coefficients (the modular invariance, the small q behavior etc, $\cdots$) see Appendix A in Part I. The modular functions $c^{(+)}_{j_1j_2j_3}(\tau)$ $(j_1, m_1\in Z)$ appear as the branching coefficinets in the irreducible decompositions of tensor products of $A_1^{(1)}$ modules. The 1D configuration sums in regime III coincide with the $c^{(+)}_{j_1j_2j_3}(\tau)$:

Theorem 5.1.1.

(5.1.2a) $$\lim_{m\text{ even}\to\infty}X_m(a,b,c)=\lim_{m\text{ odd}\to\infty}X_m(a,c,b)=q^{\rho}c^{(+)}_{rsa}(\tau),$$

(5.1.2b) $$r=\frac{b+c-N}{2},\qquad s=\frac{b-c+N}{2}+1,$$

(5.1.2c) $$\rho=\frac{b-a}{4}-\gamma(r,s,a),$$

where $c^{(+)}_{rsa}(\tau)$ *is given by* (5.1.1) *with*

(5.1.2d) $$m_1=L-N,\quad m_2=N+2,\quad m_3=L.$$

Proof. Substitute (4.1.7) into (4.4.5) and replace (j, k) by

$$\Big(j+\frac{m}{2}, k+\frac{m}{2}\Big)\quad\text{or}\quad\Big(k+\frac{m-1}{2}, j+\frac{m+1}{2}\Big)$$

according as m is even or odd, respectively. Using that

$$\lim \begin{bmatrix} M \\ \frac{M+j}{2} \end{bmatrix} = \frac{1}{\varphi(q)}$$

(j fixed, $M \to \infty$, with $M \equiv j \bmod 2$), we obtain

(5.1.3a)
$$\varphi(q)^3 \lim_{m \text{ even} \to \infty} X_m(a, b, c) = \varphi(q)^3 \lim_{m \text{ odd} \to \infty} X_m(a, c, b)$$
$$= \varphi(q)^3 q^{-a/4}(F(a, b, c) - F(-a, b, c)),$$

(5.1.3b)
$$\varphi(q)^3 q^{-a/4} F(\pm a, b, c)$$
$$= \sum_{\lambda \in Z} \Big(\sum_{\substack{j \geq \nu^{\pm}/2 \\ k \leq (\mu^{\pm}-1)/2}} - \sum_{\substack{j \leq \nu^{\pm}/2-1 \\ k \geq (\mu^{\pm}+1)/2}} \Big) (-1)^{j+k} q^{A^{(\pm)}(\lambda, j, k)},$$

(5.1.3c)
$$A^{(\pm)}(\lambda, j, k) = -L\lambda^2 + \left(\frac{L}{2} \mp a\right)\lambda - \frac{1 \mp 1}{4} a$$
$$+ Q^{(0)}_{j,k}(b \mp a - 2L\lambda, c \mp a - 2L\lambda),$$

where $Q^{(m)}_{j,k}(b, c)$ is defined in (4.1.7) and the integers $\mu^{\pm}$, $\nu^{\pm}$ are determined by

(5.1.3d) $(b \mp a - 2L\lambda, c \mp a - 2L\lambda) \in R_{\mu^{\pm}, \nu^{\pm}}$, $\quad \mu^{\pm}$: odd, $\nu^{\pm}$: even.

On the other hand we have from (5.1.1) (note that $\varepsilon_a^{L/2} = 1$ for $0 < a < L/2$)

(5.1.4a)
$$\varphi(q)^3 q^{\rho} c^{(+)}_{rsa}(\tau)$$
$$= \Big(\sum_{\xi+1 \leq t, \eta \leq \mathscr{L}_1} - \sum_{\xi \geq t, \eta - 1 \geq \mathscr{L}_1} \Big) (-1)^{t+(\varepsilon_1+\varepsilon_2)/2} q^{B(\varepsilon_1, \varepsilon_2, t, n_1, n_2)},$$

(5.1.4b)
$$B(\varepsilon_1, \varepsilon_2, t, n_1, n_2) = t(t-1)/2 + t\mathscr{L}_1(\varepsilon_1, \varepsilon_2) + (b-a)/4$$
$$+ (L-N)n_1^2 + rn_1 + (N+2)n_2^2 + sn_2,$$

(5.1.4c) $\mathscr{L}_1(\varepsilon_1, \varepsilon_2) = (a+1)/2 + \varepsilon_1((L-N)n_1 + r/2) + \varepsilon_2((N+2)n_2 + s/2),$

where the sum in (5.1.4a) is taken in the same manner as in (5.1.1a). We have explicitly exhibited the $(\varepsilon_1, \varepsilon_2)$-dependence of the summands. Note that the condition $\mathscr{L}_1 \in Z$ is fulfilled because of the restriction $a - b \equiv mN \equiv 0 \bmod 2$. As the first step to identify (5.1.4) with (5.1.3), we seek a transformation from the variables (t, n_1, n_2) to (λ, j, k) such that the summands in (5.1.4a) are transformed into those in (5.1.3a, b). This is achieved in the following way. We have

(5.1.5) $$B(\mp, \varepsilon_2, t, n_1, n_2) = A^{(\pm)}(\lambda, j, k),$$

if (t, n_1, n_2) is chosen as follows:

(5.1.6)

$(\varepsilon_1, \varepsilon_2)$	$(+,+)$	$(+,-)$	$(-,+)$	$(-,-)$
$t=t(\varepsilon_1, \varepsilon_2)$	$2\lambda+j+k$	$2\lambda+j+k$	$-2\lambda-j-k$	$-2\lambda-j-k$
$n_1=n_1(\varepsilon_1, \varepsilon_2)$	$-\lambda$	$-\lambda$	$-\lambda$	$-\lambda$
$n_2=n_2(\varepsilon_1, \varepsilon_2)$	$-\lambda-k$	$\lambda+j$	$\lambda+j$	$-\lambda-k$

From (5.1.4c) and (5.1.6) we deduce

(5.1.7a) $\mathscr{L}_1(-\varepsilon_1, -\varepsilon_2)=a+1-\mathscr{L}_1(\varepsilon_1, \varepsilon_2),$

(5.1.7b) $\mathscr{L}_1(+,+)=(a+b)/2+1-(L+2)\lambda-(N+2)k,$

(5.1.7c) $\mathscr{L}_1(+,-)=(a+c-N)/2-(L+2)\lambda-(N+2)j.$

Our next step is to partch up the summation domains in (5.1.4a) so as to make them coincide with those in (5.1.3b). To describe the constraints on the summation variables in (5.1.4a) we introduce the domain $D_\lambda(\kappa; \varepsilon_1, \varepsilon_2)$ $(\kappa, \varepsilon_1, \varepsilon_2=\pm 1)$ by

$$D_\lambda(\kappa; \varepsilon_1, \varepsilon_2)=\{(j,k)\in \boldsymbol{Z}^2 \,|\, \kappa t(\varepsilon_1, \varepsilon_2)>\kappa(\xi(\varepsilon_1, n_1(\varepsilon_1, \varepsilon_2))+1/2), \quad \kappa\mathscr{L}_1(\varepsilon_1, \varepsilon_2)>\kappa(\eta(\varepsilon_1, n_1(\varepsilon_1, \varepsilon_2))-1/2)\}. \tag{5.1.8}$$

For example $D_\lambda(+:+,+)$ reads as (see (5.1.6, 7))

$$D_\lambda(+;+,+)=\{(j,k)\in \boldsymbol{Z}^2 \,|\, j+k\geq \xi(+,-\lambda)-2\lambda+1, \quad (N+2)k\leq -\eta(+,-\lambda)-(L+2)\lambda+(a+b)/2+1\}.$$

Using the relation (5.1.5) and the fact $(-1)^{j+k}=(-1)^t$ (see (5.1.6)) we recombine (5.1.4a) as follows.

$$\begin{aligned}\varphi(q)^3 q^\rho c^{(+)}_{rsa}(\tau) &= \sum_{\lambda\in \boldsymbol{Z}}\Big(\sum_{(j,k)\in D_\lambda(+;-,+)\cup D_\lambda(-;-,-)} - \sum_{(j,k)\in D_\lambda(-;-,+)\cup D_\lambda(+;-,-)}\Big) \\ &\quad\times(-1)^{j+k} q^{A^{(+)}(\lambda,j,k)} \\ &\quad -\sum_{\lambda\in \boldsymbol{Z}}\Big(\sum_{(j,k)\in D_\lambda(+;+,+)\cup D_\lambda(-;+,-)} - \sum_{(j,k)\in D_\lambda(-;+,+)\cup D_\lambda(+;+,-)}\Big) \\ &\quad\times(-1)^{j+k} q^{A^{(-)}(\lambda,j,k)}.\end{aligned} \tag{5.1.9}$$

Compare this expression with the one given in (5.1.3). They are identical if there exist $\xi(\mp, -\lambda)$ and $\eta(\mp, -\lambda)$ such that

(5.1.10a) $$D_\lambda(\pm;\mp,\pm)\cup D_\lambda(\mp,\mp,\mp)$$
$$=\{(j,k)\in \boldsymbol{Z}^2\,|\,j\geq \nu^{\pm}/2,\ k\leq(\mu^{\pm}-1)/2\},$$

(5.1.10b) $$D_\lambda(\mp;\mp,\pm)\cup D_\lambda(\pm,\mp,\mp)$$
$$=\{(j,k)\in \boldsymbol{Z}^2\,|\,j\leq \nu^{\pm}/2-1,\ k\geq(\mu^{\pm}+1)/2\}.$$

Schematic explanation of the patch-up procedure (5.1.10a) is given in Fig. 5.1.

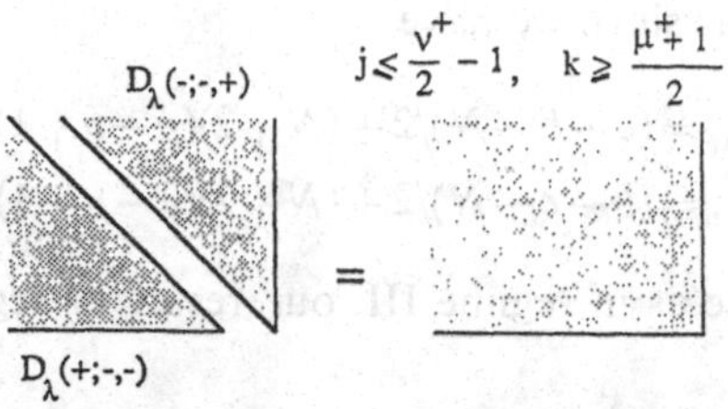

Fig. 5.1 An example of the patch up procedure of the summation domains in the (j,k) plane.

Thus the remaining task is to check (5.1.10). In the following we denote $\xi(-,-\lambda)$, $\eta(-,-\lambda)$, μ^+, ν^+ simply by ξ, η, μ, ν and verify the upper case in (5.1.10a). The other cases are similar. Now that we have

(5.1.11a) $$D_\lambda(+;-,+)=\{(j,k)\in \boldsymbol{Z}^2\,|\,-2\lambda-j-k\geq\xi+1,$$
$$(a-c+N)/2+1+(L+2)\lambda+(N+2)j\geq\eta\},$$

(5.1.11b) $$D_\lambda(-;-,-)=\{(j,k)\in \boldsymbol{Z}^2\,|\,-2\lambda-j-k\leq\xi,$$
$$(a-b)/2+(L+2)\lambda+(N+2)k\leq\eta-1\},$$

the condition that assures the existence of η is stated as follows.

(5.1.12) $$E_1\cap E_2\neq\varnothing,$$

(5.1.13a) $$E_1=\left\{\eta\in \boldsymbol{Z}\,\middle|\,\left\{j\in \boldsymbol{Z}\,|\,j\geq\frac{1}{N+2}\left(\eta-(L+2)\lambda-\frac{a-c+N}{2}-1\right)\right\}\right.$$
$$\left.=\{j\in \boldsymbol{Z}\,|\,j\geq\nu/2\}\right\},$$

(5.1.13b) $$E_2=\left\{\eta\in \boldsymbol{Z}\,\middle|\,\left\{k\in \boldsymbol{Z}\,|\,k\leq\frac{1}{N+2}\left(\eta-(L+2)\lambda-\frac{a-b}{2}-1\right)\right\}\right.$$
$$\left.=\{k\in \boldsymbol{Z}\,|\,k\leq(\mu-1)/2\}\right\}.$$

It is easy to see

$$E_1=\{\eta \in \mathbf{Z} \mid \eta_{\min}<\eta\leq\eta_{\max}\},$$

(5.1.14a)
$$\eta_{\substack{\max\\ \min}}=(L+2)\lambda+(a-c+N)/2+1+(N+2)\times\begin{cases}\nu/2\\ \nu/2-1\end{cases},$$

$$E_2=\{\eta \in \mathbf{Z} \mid \eta'_{\min}\leq\eta<\eta'_{\max}\},$$

(5.1.14b)
$$\eta'_{\substack{\max\\ \min}}=(L+2)\lambda+(a-b)/2+1+(N+2)\times\begin{cases}(\mu+1)/2\\ (\mu-1)/2\end{cases}.$$

Now (5.1.12) is clear since we have

$$\eta'_{\max}-\eta_{\min}=(c-b+N)/2+(N+2)(\mu-\nu+1)/2+2>0,$$
$$\eta_{\max}-\eta'_{\min}=(b-c+N)/2+(N+2)(\nu-\mu+1)/2\geq 0. \qquad \square$$

Much the same as in regime III, our result in regime IV is stated as follows.

Theorem 5.1.2. *For* $(b, c) \in S_1(n)$,

(5.1.15a)
$$\begin{aligned}&\lim_{m\ \mathrm{even}\to\infty} Y_m(a, b, c; q^{-1})q^{(m/4)(c-b)+m(m+1)N/4}\\ &=\lim_{m\ \mathrm{odd}\to\infty} Y_m(a, c, b; q^{-1})q^{((m+1)/4)(b-c)+m(m+1)N/4}\\ &=q^{\sigma}c^{(-)}_{rsa}(\tau), \qquad \textit{if } L \textit{ is odd},\end{aligned}$$

(5.1.15b)
$$\begin{aligned}&\lim_{m\ \mathrm{even}\to\infty} \varepsilon_a^{L/2}(Y_m(a, b, c; q^{-1})\\ &\qquad \mp Y_m(L-a, b, c; q^{-1}))q^{(m/4)(c-b)+(m(m+1)N)/4}\\ &=\lim_{m\ \mathrm{odd}\to\infty} \varepsilon_a^{L/2}(Y_m(a, c, b; q^{-1})\\ &\qquad \mp Y_m(L-a, c, b; q^{-1}))q^{((m+1)/4)(b-c)+(m(m+1)N)/4}\\ &=q^{\sigma}c^{(\pm)}_{rsa}(\tau), \qquad \textit{if } L \textit{ is even}.\end{aligned}$$

Here

(5.1.15c)
$$\sigma=\frac{b-a}{2}-\gamma(r, s, a),$$

where the integers r, s *are defined in* (5.1.2b) *and the* $c^{(\pm)}_{rsa}(\tau)$ *is given by* (5.1.1) *with*

(5.1.15d)
$$m_1=L/2-N, \quad m_2=N+2, \quad m_3=L/2.$$

Proof. As was done for Theorem 5.1.1 the proof consists of two steps:

(i) Write down the l.h.s. (resp. r.h.s.) as a sum over (λ, j, k) (resp. (t, n_1, n_2)) and find the transformations between (λ, j, k) and (t, n_1, n_2) such that the summands are mapped to each other.

(ii) Utilizing the parameters ξ and η, patch up the summation domains in the r.h.s. to make them coincide with those in the l.h.s.

Here we shall only demonstrate (i). The step (ii) is almost the same as the one for Theorem 5.1.1.

We substitute (4.3.9) and (4.3.12a) into (4.4.8) and replace q by q^{-1} and (j, k) by

$$\left(j+\frac{m}{2}+1, \frac{m}{2}-k\right) \quad \text{or} \quad \left(k+\frac{m+1}{2}, \frac{m-1}{2}-j\right)$$

according as m is even or odd. Then in the limit $m\to\infty$ we get

$$\begin{aligned}&\varphi(q)^3 \lim_{m\ \mathrm{even}\to\infty} Y_m(a, b, c; q^{-1})q^{(m/4)(c-b)+(m(m+1)N)/4}\\ &=\varphi(q)^3 \lim_{m\ \mathrm{odd}\to\infty} Y_m(a, c, b; q^{-1})q^{((m+1)/4)(b-c)+(m(m+1)N)/4}\\ &=\varphi(q)^3 q^{-a/2}(G(a, b, c)-G(-a, b, c)),\end{aligned} \tag{5.1.16a}$$

$$\begin{aligned}&\varphi(q)^3 q^{-a/2} G(\pm a, b, c)\\ &=\sum_{\lambda\in Z}\Big(\sum_{\substack{j\geq\nu^\pm/2\\ k\leq(\mu^\pm-1)/2}} - \sum_{\substack{j\leq\nu^\pm/2-1\\ k\geq(\mu^\pm+1)/2}}\Big)(-1)^{j+k} q^{\bar{A}^{(\pm)}(\lambda, j, k; a)},\end{aligned} \tag{5.1.16b}$$

$$\begin{aligned}\bar{A}^{(\pm)}(\lambda, j, k; a) = &-2L\lambda^2-(\pm 2a-L)\lambda\\ &-\frac{1\mp 1}{2}a+\frac{j+1}{2}(b\mp a-2L\lambda)\\ &+\frac{k}{2}(c\mp a-2L\lambda)-\left(j+\frac{1}{2}\right)kN\\ &+\frac{1}{2}(j-k)(j-k+1),\end{aligned} \tag{5.1.16c}$$

where the integers $\mu^\pm, \nu^\pm$ is given by (5.1.3d). Next we write down the branching coefficients $c_{rsa}^{(\pm)}(\tau)$ in the r.h.s. of (5.1.15a, b)

$$\begin{aligned}\varphi(q)^3 q^\sigma c_{rsa}^{(\pm)}(\tau) = \varepsilon_a^{L/2}\Big(&\sum_{\xi+1\leq t, \eta\leq\bar{\mathscr{L}}_1\in Z} - \sum_{\xi\geq t, \eta-1\geq\bar{\mathscr{L}}_1\in Z}\Big)\\ &\times(-1)^{t+(\varepsilon_1+\varepsilon_2)/2}(\pm 1)^{n_1} q^{\bar{B}(\varepsilon_1, \varepsilon_2, t, n_1, n_2)},\end{aligned} \tag{5.1.17a}$$

$$\begin{aligned}\bar{B}(\varepsilon_1, \varepsilon_2, t, n_1, n_2) = &\frac{t(t-1)}{2}+t\bar{\mathscr{L}}_1(\varepsilon_1, \varepsilon_2)+\frac{b-a}{2}\\ &+(L/2-N)n_1^2+rn_1+(N+2)n_2^2+sn_2,\end{aligned} \tag{5.1.17b}$$

(5.1.17c) $\mathscr{L}_1(\varepsilon_1,\varepsilon_2)=\dfrac{a+1}{2}+\varepsilon_1((L/2-N)n_1+r/2)+\varepsilon_2((N+2)n_2+s/2).$

Consider the case L is odd. The condition $\mathscr{L}_1\in \boldsymbol{Z}$ asserts that $n_1\in 2\boldsymbol{Z}$. In this case we can transform (5.1.17) into (5.1.16) through the following rule:

(5.1.18) $$\bar{B}(\mp,\varepsilon_2,t,n_1,n_2)=\bar{A}^{(\pm)}(\lambda,j,k;a),$$

where (t,n_1,n_2) is given in terms of (λ,j,k) as

(5.1.19)

$(\varepsilon_1,\varepsilon_2)$	$(+,+)$	$(+,-)$	$(-,+)$	$(-,-)$
t	$4\lambda+j+k$	$4\lambda+j+k$	$-4\lambda-j-k$	$-4\lambda-j-k$
n_1	-2λ	-2λ	-2λ	-2λ
n_2	$-2\lambda-k$	$2\lambda+j$	$2\lambda+j$	$-2\lambda-k$

From (5.1.18) and (5.1.19) together with the patch-up argument we conclude (5.1.15a). (Note that $\varepsilon_a^{L/2}=1$ in this case.) In the case L is even, the condition $\mathscr{L}_1\in\boldsymbol{Z}$ is satisfied irrespective of the parity of n_1. Thus $\varphi(q)^3q^{\sigma}c_{rsa}^{(\pm)}(\tau)$ in (5.1.17) consists of two kinds of the summands corresponding to $n_1\in 2\boldsymbol{Z}$ or $n_1\in 2\boldsymbol{Z}+1$. The former can be treated in the same way as the case L is odd and is identified with

(5.1.20) $$\begin{aligned}&\varphi(q)^3\lim_{m\ \text{even}\to\infty}\varepsilon_a^{L/2}Y_m(a,b,c;q^{-1})q^{(m/4)(c-b)+(m(m+1)N)/4}\\ &=\varphi(q)^3\lim_{m\ \text{odd}\to\infty}\varepsilon_a^{L/2}Y_m(a,c,b;q^{-1})q^{((m+1)/4)(b-c)+(m(m+1)N)/4}.\end{aligned}$$

Transform the latter summands ($n_1\in 2\boldsymbol{Z}+1$) via

(5.1.21)

$(\varepsilon_1,\varepsilon_2)$	$(+,+)$	$(+,-)$	$(-,+)$	$(-,-)$
t	$4\lambda+j+k+2$	$4\lambda+j+k+2$	$-4\lambda-j-k-2$	$-4\lambda-j-k-2$
n_1	$-2\lambda-1$	$-2\lambda-1$	$-2\lambda-1$	$-2\lambda-1$
n_2	$-2\lambda-k-1$	$2\lambda+j+1$	$2\lambda+j+1$	$-2\lambda-k-1$

Under the rule (5.1.21) we have

(5.1.22) $$\bar{B}(\pm,\varepsilon_2,t,n_1,n_2)=\bar{A}^{(\pm)}(\lambda,j,k;L-a).$$

Thus their contribution to $\varphi(q)^3q^{\sigma}c_{rsa}^{(\pm)}(\tau)$ is equal to

$$\begin{aligned}\mp\varphi(q)^3 & \lim_{m\ \text{even}\to\infty} \varepsilon_a^{L/2} Y_m(L-a, b, c; q^{-1}) q^{(m/4)(c-b)+(m(m+1)N/4} \\ (5.1.23) \qquad & = \mp\varphi(q)^3 \lim_{m\ \text{odd}\to\infty} \varepsilon_a^{L/2} Y_m(L-a, c, b; q^{-1}) \\ & \times q^{((m+1)/4)(b-c)+(m(m+1)N)/4}.\end{aligned}$$

From (5.1.20) and (5.1.23) we arrive at (5.1.15b). □

By virtue of the symmetries (4.4.12) the evaluation of the limit $m\to\infty$ of $Y_m(a, b, c; q^{-1})$ for $(b, c) \in S_2(n) \sim S_4(n)$ reduces to the case $(b, c) \in S_1(n)$ described above.

Remark. It is not easy to find the lowest power in q of the $c^{(\pm)}_{j_1 j_2 j_3}(\tau)$ by using (5.1.1). In Appendix A this is done by manipulating the 1D configuration sums.

5.2. Regime II

In this paragraph we prove

Theorem 5.2.1. *Let* $1\leq a,\ b\leq L-1$. *We set* $\bar{\ell}_j=\langle b+(j-1)N\rangle$. *The* $(\bar{\ell}_j)$ *is a ground state configuration, and we have*

$$\begin{aligned}(5.2.1) \qquad & \lim_{m\to\infty} q^{\phi_m(\bar{\ell}_1,\cdots,\bar{\ell}_{m+2})+\nu(a,b)} X_m(a, \bar{\ell}_{m+1}, \bar{\ell}_{m+2}; q^{-1}) \\ & = e^{L-2}_{b-1,a-1}(\tau),\end{aligned}$$

where

$$\nu(a, b)=-\frac{1}{4(L-2)}\left(\frac{L}{2}-b\right)^2+\frac{1}{4L}\left(\frac{L}{2}-a\right)^2+\frac{1}{24}.$$

As for the symbol $\langle\ \rangle$ see Appendix B. There we prove that $(\bar{\ell}_j)$ is a ground state configuration in the sense of Part I, section 2.4. An explicit form of $\phi_m(\bar{\ell}_1, \cdots, \bar{\ell}_{m+2})$ is given in (B.1). In fact, the following relation holds (see Lemma B.3 and the last remark in Appendix B)

$$\begin{aligned}(5.2.2) \qquad & X_m(a, \bar{\ell}_{m+1}, \bar{\ell}_{m+2}; q^{-1}) \\ & = X_{m-1}(a, \bar{\ell}_m, \bar{\ell}_{m+1}; q^{-1}) q^{-m|\bar{\ell}_m-\bar{\ell}_{m+2}|/4}(1+O(q^{\alpha m})),\end{aligned}$$

where α is some positive constant. Thus the proof of (5.2.1) is reduced to the case $1\leq\bar{\ell}_{m+1}, \bar{\ell}_{m+2}\leq L-1$, $\bar{\ell}_{m+2}=\bar{\ell}_{m+1}+N$ by the repeated use of (5.2.2).

The modular functions $e^{\ell}_{jk}(\tau)$ have been described in Appendix B of Part I. They are the branching coefficients appearing in the irreducible decompositions of level 1 highest weight modules of the affine Lie algebra $A^{(1)}_{2\ell-1}$ with respect to the embedded $C^{(1)}_{\ell}$. Here we shall exploit the characterization of $(e^{\ell}_{jk}(\tau))_{0\leq j,k\leq\ell}$ in terms of its inverse matrix.

Lemma 5.2.2 ((B. 11, 12) of Part I). *Set*

$$\tilde{p}^{\ell}_{kj}(\tau)=\varepsilon^{\ell}_{j}(p^{\ell}_{kj}(\tau)+p^{\ell}_{k,-j}(\tau)), \tag{5.2.3a}$$

$$p^{\ell}_{kj}(\tau)=\sum_{k'\equiv k \bmod 2(\ell+2)}(-)^{(k'-j)/2}q^{(k'-j)(k'-j+2)/8-(k'+1)^2/4(\ell+2)+j^2/4\ell+1/8},$$

$$\text{if } j,k\in \boldsymbol{Z} \text{ and } j\equiv k \bmod 2, \tag{5.2.3b}$$

$$=0, \quad \text{otherwise}.$$

Then we have

$$\sum_{0\leq k\leq\ell,k\equiv j \bmod 2} e^{\ell}_{jk}(\tau)\tilde{p}^{\ell}_{kj'}(\tau)=\eta(\tau)\varepsilon^{\ell}_{j'}(\tilde{\delta}^{2\ell}_{jj'}+\tilde{\delta}^{2\ell}_{j-j'}), \tag{5.2.4}$$

where $\tilde{\delta}^{2\ell}_{jj'}=1$ *if* $j'\equiv j \bmod 2\ell$ *and* $=0$ *otherwise.*

Remark. By the definition, $p^{\ell}_{kj}(\tau)$ has the following symmetries

$$p^{\ell}_{kj}(\tau)=-p^{\ell}_{-k-2,-j}(\tau)=p^{\ell}_{k+2(\ell+2),j}(\tau)=p^{\ell}_{k,j+2\ell}(\tau). \tag{5.2.5}$$

For $\beta\in \boldsymbol{Z}$, define the function $H^{(N)}_{m}(\beta)$ by

$$H^{(N)}_{m}(\beta)=\sum_{j\in \boldsymbol{Z}}(-1)^{j}q^{j^2/2}f^{(N)}_{m}(\beta-2j,\beta+N-2j;q^{-1}), \tag{5.2.6}$$

where $f^{(N)}_{m}(b,c;q)$ is the fundamental solution described in sections 4.1–2. By (4.2.1) $H^{(N)}_{m}(\beta)$ can also be expressed as

$$H^{(N)}_{m}(\beta)=(-1)^{\omega}q^{-m(m+1)N/4+\omega^2/2}\sum_{v\in \boldsymbol{Z}}(-1)^{v}q^{v(v-2\omega+1)/2}\hat{f}^{(N)}_{m}(v), \tag{5.2.7a}$$

$$\omega=\frac{Nm-\beta}{2}. \tag{5.2.7b}$$

Note that the summand in (5.2.7a) is supported in the interval $0\leq v\leq mN$. In the case $N=1$ (4.2.23), (5.2.6) together with the formula (4.1.14) give

$$H^{(1)}_{m}(\beta)=(-1)^{\omega}q^{-m(m+1)/4+\omega^2/2}(q^{1-\omega})_{m}. \tag{5.2.8}$$

From (5.2.7) and Theorem 4.2.2, we deduce the symmetry:

$$H^{(N)}_{m}(\beta)=-q^{(\beta+1)/2}H^{(N)}_{m}(-\beta-2). \tag{5.2.9}$$

In what follows we make use of several representations of $\hat{f}^{(N)}_{m}(v)$ derived in section 4.2, and determine the large m behavior of $H_m^{(N)}(\beta)$ in various regions of β. The results are given in Lemma 5.2.3–6.

Lemma 5.2.3 ((3.28) of Part I). *For* $N\geq 1$, $\beta\geq mN$,

$$H_m^{(N)}(\beta)=O(q^{-m(m+1)N/4+\omega^2/2}). \tag{5.2.10}$$

Proof. First we show that

$$\hat{f}_m^{(N)}(v)=1+\cdots \qquad \text{if } 0\leq v\leq mN \ (m\geq 1), \tag{5.2.11a}$$

$$=0 \qquad \text{otherwise.} \tag{5.3.11b}$$

By the definition (4.2.1), (5.2.11b) is clear. Recall the expression (4.2.2) for $\hat{f}_m^{(N)}(v)$. Note that the Gaussian polynomial is of the form $1+O(q)$. Under the assumption $0\leq v\leq mN$, we see ($\mathscr{P}$ defined in (4.2.2b))

$$\min_{j\geq 0}\mathscr{P}=\mathscr{P}|_{j=0}=0.$$

Thus the lowest order of the first term in (4.2.2a) is unity. Because of the symmetry (4.2.7) we can assume that $v\leq mN/2$. Then it is shown that the second term in (4.2.2a) does not contain the power lower than q^1. Thus we have (5.2.11a). Since the minimum of $v(v-2\omega+1)/2$ in the interval $0\leq v\leq mN$ is attained at $v=0$ (note that $\omega\leq 0$), we obtain (5.2.10). □

The expression (4.2.2) yields the following estimate of $\hat{f}_m^{(N)}(v)$ as m tends to ∞ (when v is fixed):

$$\hat{f}_m^{(N)}(v)=\begin{bmatrix} v+m-1 \\ m-1 \end{bmatrix}+O(q^{mN-v+1}), \tag{5.2.12}$$

Applying this to (5.2.7) with $\beta=mN$ ($\omega=0$) we obtain

$$\lim_{m\to\infty} q^{m(m+1)N/4}H_m^{(N)}(mN)=\sum_{v\in Z}(-1)^v\frac{q^{v(v+1)/2}}{(q)_v}=\varphi(q), \tag{5.2.13}$$

where we have used the formula (4.1.14) in the limit $M\to\infty$.

Lemma 5.2.4 ((3.30) of Part I). *Assume that* $N\geq 2$. *Then for* $mN>\beta>m(N-C)$, $C=\min(2, 4(L-N-2)/(L-4))$,

$$H_m^{(N)}(\beta)=O(q^{-m(m+1)N/4+\omega(m+1-\omega/2)}). \tag{5.2.14}$$

Proof. Substitute the expression (4.2.20) into (5.2.7). After eliminating x_0 by (4.2.21a), we perform the x_1-summation utilizing the formula (4.1.14). The result takes the form:

(5.2.15a)
$$H_m^{(N)}(\beta)=(-1)^{\omega}q^{-m(m+1)N/4+\omega^2/2}$$
$$\times\sum(-1)^{\sum_{2\le j\le N} jx_j}\begin{bmatrix} m \\ x_2,\cdots,x_N,\tilde{m}\end{bmatrix} q^{\mathscr{S}_1}(q^{\mathscr{S}_0})_{\tilde{m}},$$

(5.2.15b)
$$\tilde{m}=m-\sum_{2\le j\le N} x_j,$$

(5.2.15c)
$$\mathscr{S}_0=1-\omega+\sum_{2\le j\le N}(j-1)x_j,$$

(5.2.15d)
$$\mathscr{S}_1=\sum_{2\le j\le N}(j^2-2j+2)x_j^2/2+\sum_{2\le k<j\le N}(jk-2k+1)x_jx_k$$
$$+\sum_{2\le j\le N}((j-1)m-(\omega-1/2)j)x_j,$$

where the sum $\sum$ in (5.2.15a) is taken over all non-negative integers $x_2, \cdots, x_N$. In (5.2.15a) the product $(q^{\mathscr{S}_0})_{\tilde{m}}$ is non-zero if

(5.2.16)
$$\mathscr{S}_0\ge 1 \quad\text{or}\quad \mathscr{S}_0\le -\tilde{m}.$$

Under the assumption $Nm>\beta>m(N-C)$ $(0<\omega<mC/2\le m)$, we can discard the latter possibility, for

(5.2.17)
$$\mathscr{S}_0+\tilde{m}=1+m-\omega+\sum_{2\le j\le N}(j-2)x_j>0.$$

In the former case in (5.2.16), the lowest order term of the product $(q^{\mathscr{S}_0})_{\tilde{m}}$ as well as the q-multinomial coefficient in (5.2.15a) is unity. Thus apart from the obvious overall factor in (5.2.15a), the lowest power of $H_m^{(N)}(\beta)$ is given by

(5.2.17a)
$$\min_D \mathscr{S}_1,$$

where D is the domain of $(x_2, \cdots, x_N)$ specified by

(5.2.17b)
$$D=\{(x_2,\cdots,x_N)\in \boldsymbol{Z}^{N-1}|x_2,\cdots,x_N\ge 0,\ \tilde{m}\ge 0,\ \mathscr{S}_0\ge 1\}.$$

Let us evaluate (5.2.17). For $N=2$ it is easily seen by using the condition $m\ge x_2\ge\omega$ that the minimum is attained by $x_2=\omega$. This gives the value $\mathscr{S}_1=\omega(m+1-\omega)$ leading to (5.2.14). In the sequel we prove the case $N\ge 3$. First we seek for the point that attains the minimum in $\{(x_2,\cdots,x_N)\in \boldsymbol{R}^{N-1}|x_2,\cdots,x_N\ge 0,\ \tilde{m}\ge 0,\ \mathscr{S}_0\ge 1\}$. Consider the derivative of $\mathscr{S}_1$ with respect to x_i $(2\le i\le N)$. Then there appear the terms $(i-1)m-\omega i$ (see (5.2.15d)). If we rewrite this as $(i-1)(m-\omega)-\omega$ and use the condition $\omega\le\sum_{2\le j\le N}(j-1)x_j$, we have for $2\le i\le N$

(5.2.18)
$$\frac{\partial\mathscr{S}_1}{\partial x_i}\ge(i^2-3i+3)x_i+\sum_{2\le k<i}((i-3)k+2)x_k$$
$$+\sum_{i<k\le N}(k-2)(i-1)x_k+(i-1)(m-\omega)+i/2>0.$$

This implies that the points in question are located on the hyperplane $\sum_{2\leq j\leq N}(j-1)x_j=\omega$. If $N\geq 4$ we derive further by using $\omega=\sum_{2\leq j\leq N}(j-1)x_j$ that

$$(N-2)\frac{\partial \mathscr{S}_1}{\partial x_N}-(N-1)\frac{\partial \mathscr{S}_1}{\partial x_{N-1}} = \frac{N-3}{2}x_N+\frac{N-4}{2}x_{N-1}+\sum_{2\leq k\leq N-2}(k-1)x_k+\frac{\omega-1}{2}\geq 0. \tag{5.2.19}$$

Thus we must set $x_N=0$ to attain the minimum. Repeated use of this argument reduces the problem to the case $x_N=\cdots=x_4=0$. Then, $x_2+2x_3=\omega$ and (5.2.19) reads as

$$\frac{\partial \mathscr{S}_1}{\partial x_3}-2\frac{\partial \mathscr{S}_1}{\partial x_2}=x_3-1/2. \tag{5.2.20}$$

From this we see that the minimum is attained at $x_2=\omega-1$, $x_3=1/2$, $x_4=\cdots=x_N=0$. Actually x_2 and x_3 range within $\boldsymbol{Z}$. Therefore (5.2.20) leads us to the conclusion for all $N(\geq 3)$ that

$$\min_D \mathscr{S}_1=\mathscr{S}_1|_{N=3,x_2=\omega,x_3=0}=\mathscr{S}_1|_{N=3,x_2=\omega-2,x_3=1}=\omega(m+1-\omega). \tag{5.2.21}$$

From this and (5.2.15a) we establish (5.2.14). □

Remark. In the case $N=1$ it follows from the expression (5.2.8) that

$$H_m^{(1)}(\beta)=0, \qquad \text{for } m>\beta\geq -m. \tag{5.2.22}$$

Lemma 5.2.5. *Assume that* $N\geq 2$. *For* $0\leq w\leq mN/2$, *set* $u=w/m$ $(0\leq u\leq N/2)$. *Then we have as* $m\to\infty$

$$h_m^{(N)}(um)=O(q^{(N-2)u^2m^2/2N+\text{linear terms in } m}), \tag{5.2.23}$$

where $h_m^{(N)}(w)$ *has been defined in* (4.2.28b-d).

Proof. The case $N=2$ is obvious from (4.2.46) and so is the case $u=0$ from $h_m^{(N)}(0)=1$ (see (4.2.33)). In the following we assume $N\geq 3$ and $u\neq 0$. We employ the expression (4.2.28b-e) for the function $h_m^{(N)}(w)$. Our proof here consists of three steps:

(i) Under the conditions $\sum_{1\leq j\leq N-1} jy_j=\bar{s}_1$ and $y_j\geq 0$, extract the minimum of the power $\mathscr{R}$ (4.2.28d) as a function of $\bar{s}_1$.

(ii) Supply (i) with the contributions from the j-summand in $h_\infty^{(N)}(w-\bar{s}_1)$.

(iii) Under the conditions $0\leq \bar{s}_1\leq w$ and $0\leq j\leq (w-\bar{s}_1)/N$ (or $1\leq j\leq (w-\bar{s}_1)/N$, see (4.2.28e)), minimize the total power obtained in (ii).

In the working (i) ~ (iii) we shall regard $y_j \in \boldsymbol{R}$ rather than $y_j \in \boldsymbol{Z}$. This does not effect the estimations to the order of m^2.

(i) Let $\bar{s}_1$ be a non-negative constant. Eliminate y_1 by using the latter of (4.2.28c). Then the power $\mathscr{R}$ in (4.2.28d) is written as

$$\mathscr{R}=\sum_{2\leq j,k\leq N-1}\mathscr{R}_{j,k}y_jy_k-(mN-2(w-\bar{s}_1)+1)\sum_{2\leq j\leq N-1}(j-1)y_j \tag{5.2.24a}$$
$$+((N-1)m-w+(\bar{s}_1+1)/2)\bar{s}_1,$$

$$\mathscr{R}_{j,k}=\mathscr{R}_{k,j}=(\min(j,k)-1)\max(j,k). \tag{5.2.24b}$$

Our aim here is to evaluate the quantity

$$\min_{\bar{D}} \mathscr{R}, \tag{5.2.25a}$$

$$\bar{D}=\{(y_2,\cdots,y_{N-1})\in \boldsymbol{R}^{N-1}\mid \sum_{2\leq j\leq N-1} jy_j\leq \bar{s}_1,\ y_j\geq 0\}. \tag{5.2.25b}$$

The quadratic form $\sum_{2\leq j,k\leq N-1}\mathscr{R}_{j,k}y_jy_k$ is positive definite. This follows from the fact that the inverse of $(\mathscr{R}_{jk})_{j,k=2,\dots,N-1}$ is given by

$$\begin{pmatrix} 2 & -1 & & & & \\ -1 & 2 & -1 & & & \\ & -1 & 2 & \cdot & & \\ & & \cdot & \cdot & & \\ & & & \cdot & \cdot & \\ & & & \cdot & 2 & -1 \\ & & & & -1 & \dfrac{N-2}{N-1} \end{pmatrix} \tag{5.2.26}$$

Thus the minimum of $\mathscr{R}$ is attained by the point $\bar{M}$ where its inward derivatives with respect to $\bar{D}$ are non-negative. It is given by

$$\begin{aligned} \bar{M}:\ & y_2=\cdots=y_{N-2}=0,\quad y_{N-1}=\frac{\bar{s}_1}{N-1} && \text{for } N\geq 4, \\ :\ & y_2=\bar{s}_1/2 && \text{for } N=3. \end{aligned} \tag{5.2.27}$$

The condition is checked as follows (recall the assumption $0\leq w\leq mN/2$):

$$-\frac{\partial\mathscr{R}}{\partial y_{N-1}}\bigg|_{\bar{M}}=(N-2)(mN-2w+1)>0 \qquad \text{for } N\geq 3,$$

$$\left((N-1)\frac{\partial\mathscr{R}}{\partial y_j}-j\frac{\partial\mathscr{R}}{\partial y_{N-1}}\right)\bigg|_{\bar{M}}=(N-1-j)(mN-2w+1)>0$$
for $N\geq 4$ and $2\leq j\leq N-2$.

In this way we obtain

(5.2.28) $$\min_{\bar D} \mathscr{R}=\mathscr{R}|_{\bar M}=\bar s_1((N-3)(w-(\bar s_1+1)/2)+m)/(N-1).$$

(ii) Define the functions $\mathscr{U}_1(\bar s_1, j; w)$, $\mathscr{U}_2(\bar s_1, j; w)$ by

(5.2.29a) $$\begin{aligned}\mathscr{U}_1(\bar s_1, j; w)=&\bar s_1((N-3)(w-(\bar s_1+1)/2)+m)/(N-1)\\ &+(w-\bar s_1)(w-\bar s_1-1)+N^2j^2/2\\ &-(N+1)(w-\bar s_1)j+Nj/2,\end{aligned}$$

(5.2.29b) $$\begin{aligned}\mathscr{U}_2(\bar s_1, j; w)=&\bar s_1((N-3)(w-(\bar s_1+1)/2)+m)/(N-1)\\ &+(w-\bar s_1)^2+N^2j^2/2\\ &-(N+1)(w-\bar s_1)j-Nj/2.\end{aligned}$$

In addition to (5.2.28), $\mathscr{U}_1$ and $\mathscr{U}_2$ respectively count the power of q coming from the first and the second terms in (4.2.28e) with w replaced by $w-\bar s_1$. These are positive definite quadratic forms of $\bar s_1$ and j.

(iii) From (i), (ii) and (4.2.28b-e) we now have the estimation

(5.2.30a) $$h_m^{(N)}(w)=O(q^{\mathscr{U}}),$$

(5.2.30b) $$\mathscr{U}=\min(\mathscr{U}_1^*, \mathscr{U}_2^*),$$

(5.2.30c) $$\mathscr{U}_i^*=\min_{D_i^*} \mathscr{U}_i(\bar s_1, j; w), \qquad (i=1, 2),$$

(5.2.30d) $$D_1^*=\{(\bar s_1, j)\in \boldsymbol{R}^2 | 0\le \bar s_1\le w, \quad 0\le j\le (w-\bar s_1)/N\},$$

(5.2.30e) $$D_2^*=\{(\bar s_1, j)\in \boldsymbol{R}^2 | 0\le \bar s_1\le w, \quad 1\le j\le (w-\bar s_1)/N\}.$$

(we put $\mathscr{U}_2^*=\infty$ if $w<N$.) Consider the point $M^*=(0, w/N)$ on the boundaries of D_1^* and D_2^*, and the derivatives of $\mathscr{U}_i$ at M^*

(5.2.31a) $$\left.\frac{\partial \mathscr{U}_1}{\partial j}\right|_{M^*}=\frac{N}{2}-w, \qquad \left.\frac{\partial \mathscr{U}_2}{\partial j}\right|_{M^*}=-\frac{N}{2}-w,$$

(5.2.31b) $$\left.\left(N\frac{\partial \mathscr{U}_1}{\partial \bar s_1}-\frac{\partial \mathscr{U}_1}{\partial j}\right)\right|_{M^*}=\left.\left(N\frac{\partial \mathscr{U}_2}{\partial \bar s_1}-\frac{\partial \mathscr{U}_2}{\partial j}\right)\right|_{M^*}=\frac{mN-2w+N}{N-1}.$$

If we set $w=um$ $(0<u\le N/2)$ and let $m\to\infty$, (5.2.31a) tends to $-\infty$ while (5.2.31b) remains positive because of the assumption $0\le w\le mN$. Thus we can employ the same argument as in (i) to find

(5.2.32) $$\mathscr{U}_i^* \xrightarrow[m\to\infty]{} \mathscr{U}_i\left(0, \frac{um}{N}; um\right)=\frac{N-2}{2N}m^2+\text{linear terms in } m.$$

From this and (5.2.30) we establish (5.2.23). □

Lemma 5.2.6 ((3.31) of Part I). *Assume that* $N \geq 2$. *For* $m(N-C) \geq |\beta|$ (*C defined in Lemma* 5.2.4), *set* $t=\beta/m$ ($|t| \leq N-C$). *Then we have*

(5.2.33) $$H_m^{(N)}(mt)=O(q^{-t^2m^2/4N+\text{linear terms in } m}).$$

Proof. In view of the symmetry (5.2.9) it is sufficient to verify the case $m(N-C) \geq \beta \geq 0$ ($N-C \geq t \geq 0$). Substitute (4.2.28a) into (5.2.7) and replace v by $v+w$. The v-summation using the formula (4.1.14) yields ω defined in (5.2.7b))

(5.2.34) $$H_m^{(N)}(\beta)=(-1)^\omega q^{-m(m+1)N/4+\omega^2/2} \times \sum_{0 \leq w \leq mN/2} (q^{w-\omega+1})_{mN-2w} q^{w(m-\omega+1)} h_m^{(N)}(w).$$

The product $(q^{w-\omega+1})_{mN-2w}$ is non-zero only if

(5.2.35) $$w \geq \omega \quad \text{or} \quad w \geq mN-\omega+1.$$

By virtue of the assumption $m(N-C) \geq \beta \geq 0$ (i.e. $mC/2 \leq \omega \leq mN/2$), the latter can be discarded, which leads to

(5.2.36) $$(q^{w-\omega+1})_{mN-2w}=1+O(q).$$

From (5.2.34–36) and Lemma 5.2.5, we obtain

$$H_m^{(N)}(mt)=O(q^{\sigma m^2+\text{linear terms in } w}),$$
$$\sigma=-\frac{N}{4}+\frac{1}{2}\left(\frac{N-t}{2}\right)^2+\min_{(N-t)/2 \leq u \leq N/2}\left(u\left(1-\frac{N-t}{2}\right)+\frac{N-2}{2N}u^2\right)$$
$$=-t^2/4N.$$

where we used the assumption $0 \leq t (\leq N-C)$. □

Proof of Theorem 5.2.1. It is sufficient to show that

(5.2.37) $$\lim_{\substack{m \to \infty \\ mN \equiv \rho \bmod 2(L-2)}} x_m(a, b, b+N)=e_{b-\rho-1,a-1}^{L-2}(\tau),$$

where $0<b$, $b+N<L$ and

(5.2.38a) $$x_m(a, b, b+N)=q^{M(m,a,b)}X_m(a, b, b+N; q^{-1}),$$

(5.2.38b) $$M(m, a, b)=\frac{m(m+1)N}{4}-\frac{1}{4(L-2)}\left(mN+\frac{L}{2}-b\right)^2 +\frac{1}{4L}\left(\frac{L}{2}-a\right)^2+\frac{1}{24}.$$

Let us introduce the function $\tilde{y}_m(b, b')$ by

$$\tilde{y}_m(b, b') = \sum_{0<a<L} x_m(a, b, b+N)\, \tilde{p}^{L-2}_{a-1, b'-1}(\tau), \tag{5.2.39}$$

where $\tilde{p}^l_{kj}(\tau)$ is given in (5.2.3). Thanks to Lemma 5.2.2, (5.2.37) is equivalent to the following statement

$$\begin{aligned} \lim_{\substack{m\to\infty \\ mN\equiv\rho \bmod 2(L-2)}} \tilde{y}_m(b, b') &= \eta(\tau) \qquad \text{if } b'-1 \equiv \pm(b-\rho-1) \bmod 2(L-2), \\ &= 0 \qquad \text{otherwise.} \end{aligned} \tag{5.2.40}$$

Using Theorem 4.4.1 we express $x_m(a, b, b+N)$ in terms of $f^{(N)}_m(*, *+N; q^{-1})$. Substitute the explicit form (5.2.3) of $\tilde{p}^l_{a-1, b'-1}(\tau)$ into (5.2.39). After a little calculation we find

$$\tilde{y}_m(b, b') = q^{1/24} \varepsilon^{L-2}_{b'-1}(y_m(b, b') + y_m(b, 2-b')), \tag{5.2.41a}$$

$$y_m(b, b') = \sum_{\beta\in Z,\ \beta\equiv b-b' \bmod 2(L-2)} z_m(\beta, b), \tag{5.2.41b}$$

$$z_m(\beta, b) = q^{m(m+1)N/4 - \omega(mN-\omega+L/2-b)/(L-2)} H^{(N)}_m(\beta), \tag{5.2.41c}$$

where the function $H^{(N)}_m(\beta)$ has been defined by (5.2.6) and ω by (5.2.7b). First consider the case $N=1$. Then $y_m(b, b')$ is written by using (5.2.8) as

$$\begin{aligned} y_m(b, b') = &\sum_{\substack{\nu\ge 0 \\ 2\nu\equiv b-b'-m \bmod 2(L-2)}} (-1)^\nu q^{\nu(2m+L(\nu+1)-2b)/(2L-4)} (q^{\nu+1})_m \\ + &\sum_{\substack{\nu\ge 1 \\ 2\nu\equiv b-b'-m \bmod 2(L-2)}} (-1)^\nu q^{(L\nu^2+(2m-L+2b)\nu+2m(b-1))/(2L-4)} (q^\nu)_m. \end{aligned}$$

In this form it is straightforward to take the limit $m\to\infty$. We obtain (recall that $b>0$)

$$\begin{aligned} \lim_{m\to\infty,\ m\equiv\rho \bmod 2(L-2)} y_m(b, b') &= \varphi(q) \qquad \text{if } b'\equiv b-\rho \bmod 2(L-2), \\ &= 0 \qquad \text{otherwise.} \end{aligned} \tag{5.2.42}$$

This proves (5.2.40) for $N=1$. Next we treat the case $N\ge 2$. Much the same as in (5.2.42), we find for general N using (5.2.13) and (5.2.41c) that

$$\lim_{m\to\infty} z_m(mN, b) = \varphi(q). \tag{5.2.43}$$

Thus the remaining task is to show that the contributions from all other values of β vanish in the limit $m\to\infty$. By Lemma 5.2.3, 5.2.4 and 5.2.6 we deduce the following estimates:

$$z_m(\beta, b) = O(q^{-\omega(2mN-L\omega+L-2b)/(2L-4)})$$
$$\text{if } \beta > mN \ (\omega < 0),$$
$$= O(q^{\omega(2(L-N-2)m-(L-4)\omega+L-4+2b)/(2L-4)})$$
$$\text{if } mN > \beta > m(N-C) \ (0 < \omega < mC/2),$$
$$= O(q^{Am^2 + \text{linear terms in } m}),$$
$$\text{if } \beta = mt, \ |t| \leq N-C \ (mC/2 \leq \omega \leq m(N-C/2)),$$

where $A=(L-N-2)(N^2-t^2)/(4N(L-2))>0$ (see (2.2.2)). Now it is clear that $z_m(\beta, b)$ in these regions converges to zero as m tends to ∞. This is also the case for the remaining region $\beta < -m(N-C)$ due to the symmetry (5.2.9). □

Appendix A. Minimum/Maximum Configurations

In section 5 we identified the 1D configuration sums with modular forms. It is important to know the lowest power in the q-expansion of these modular forms, for they are related to the critical exponents of the models. (See section 4 in Part I.) In regime I the modular forms are given in the form of infinite product (Part I (A. 14)), and the lowest power can be easily read off. In regime II the modular forms $e^l_{jk}(\tau)$ have been encountered previously and the lowest power is known (see (5.1.6) in [10] or (2.4) in [7]). In regimes III and IV, where the modular forms are characterized by theta function identities (Part I (A.3)), it is not straightforward to pinpoint the lowest power (the fractional power mod Z follows immediately). Here we do it by singling out the sequence that attains the minimum (regime III)/maximum (regime IV) of

$$\phi_m(l_1, \cdots, l_{m+2}) = \sum_{j=1}^{m} jH(l_j, l_{j+1}, l_{j+2}), \tag{A.1}$$

under the condition that

$$l_1 = a, \ l_{m+1} = b, \ l_{m+2} = c. \tag{A.2}$$

A.1. Regime III

$$H(a, b, c) = \frac{|a-c|}{4}$$

We define $2m+1$ integers a_j $(-m \leq j \leq m)$ by

$$a_j = b + jN \quad \text{if } j \equiv m \bmod 2,$$
$$= c + jN \quad \text{otherwise.}$$

Note that if $a > a_m$ or $a < a_{-m}$ there is no sequence satisfying (A.2). Without loss of generality we assume that $a \leq a_0$. We also assume that $a_{-1} \geq 1$. Because of the symmetry (Part I (A. 5)) of the branching coefficient this does not restrict our aim.

Lemma A.1. *Under these assumptions let* μ *be a positive integer such that* $a_{-\mu} \leq a \leq a_{1-\mu}$. *The minimum of* (A.1) *among admissible sequence* $(l_j)_{j=1,\cdots,m+2}$ *satisfying* (A.2) *is attained by the following sequence*:

$$
\begin{aligned}
\bar{\ell}_j &= a+(j-1)N && \text{if } 1 \leq j \leq \mu, \\
&= c && \text{else if } j \equiv m \bmod 2, \\
&= b && \text{otherwise.}
\end{aligned} \tag{A.3}
$$

The minimum is

$$
\begin{aligned}
&\phi_m(\bar{\ell}_1, \cdots, \bar{\ell}_{m+2}) \\
&\quad = \frac{a-b-\mu(2a-b-c)-\mu(\mu-1)N}{4} && \text{if } \mu \equiv m \bmod 2, \\
&\quad = \frac{a-c-\mu(2a-b-c)-\mu(\mu-1)N}{4} && \text{otherwise.}
\end{aligned} \tag{A.4}
$$

Proof. Because of the assumption $a_{-1} \geq 1$ the $(\bar{\ell}_j)$ of (A.3) is admissible. Therefore it is sufficient to show that this attains the minimum of (A.1) among weakly admissible sequences satisfying (A.2). Assume that (ℓ_j) attains the minimum. Take any successive four $\ell_j, \ell_{j+1}, \ell_{j+2}, \ell_{j+3}$ $(1 \leq j \leq m-1)$. They are subject to the following: If $\ell_{j+1} > \ell_{j+3}$ then the weight (A.1) strictly decreases if we replace ℓ_{j+1} by $\ell_{j+1}-2$. Therefore this replacement should violate the weak admissibility of the sequence. This implies that $\ell_j = \ell_{j+1}+N$. Similarly, if $\ell_{j+1} < \ell_{j+3}$ then we have $\ell_j = \ell_{j+1}-N$. The $(\bar{\ell}_j)$ is the unique one which satisfies these restrictions as well as (A.2). Therefore the minimum of (A.1) is attained solely by the $(\bar{\ell}_j)$. It is straightforward to compute (A.4). □

A.2. Regime IV

$$
\begin{aligned}
H(a, b, c) &= \min\left(n-b, \frac{\min(a, c)-b+N}{2}\right) && \text{if } b \leq n, \\
&= \min\left(b-n-1, \frac{b-\max(a, c)+N}{2}\right) && \text{if } b \geq n+1.
\end{aligned} \tag{A.5}
$$

We seek for the sequence $(\bar{\ell}_j)$ that attains the maximum of (A.1) where the weight function $H(a, b, c)$ is given by (A.5). As noted in Part I (eqs.

(3.9–11)) this resembles the negative of the weight in regime III. Therefore it may be expected that the maximum is attained again by (A.3). This is partially true as we shall see below.

We assume that

$$\frac{b+c-N}{2} \leq n-N. \tag{A.6}$$

This is not restrictive, for the branching coefficients relevant to regime IV are all obtained as the $m \to \infty$ limit of $Y_m(a, b, c; q^{-1})$ with b, c satisfying (A.6).

Because of the restriction (A.6) we cannot assume that $a \leq a_0$ as we did in Lemma A.1. We need two more candidates maximizing (A.1) (see Fig. A.1): Let μ be a positive integer such that $a_{\mu-1} \leq a \leq a_\mu$. We set

$$\begin{aligned} \bar{\ell}_j^{(1)} &= a-(j-1)N && \text{if } 1 \leq j \leq \mu, \\ &= c && \text{else if } j \equiv m \bmod 2, \\ &= b && \text{otherwise.} \end{aligned} \tag{A.7a}$$

With $\mu \geq 2$ as above we set

$$\begin{aligned} \bar{\ell}_j^{(2)} &= a-(j-1)N && \text{if } 1 \leq j \leq \mu-1, \\ &= c+N && \text{if } j=\mu \equiv m \bmod 2, \\ &= b+N && \text{if } j=\mu \not\equiv m \bmod 2, \\ &= c && \text{else if } j \equiv m \bmod 2, \\ &= b && \text{otherwise.} \end{aligned} \tag{A.8a}$$

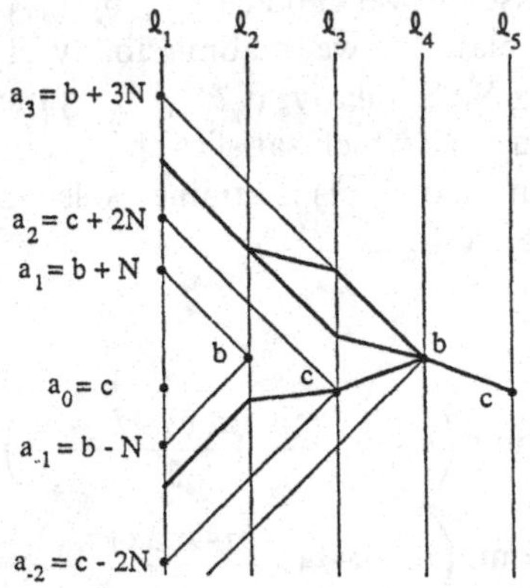

Fig. A. 1 The configurations maximizing the weight $\sum_{j=1}^{m} jH(\ell_j, \ell_{j+1}, \ell_{j+2})$.

The $(\bar{\ell}_j^{(1)})$ differs from the $(\bar{\ell}_j^{(2)})$ only at $j=\mu$. We shall use these sequences under the condition

$$\text{(A.7b)} \qquad \frac{\bar{\ell}_{m-1}^{(1)}+b}{2} \leq n,$$

$$\text{(A.8b)} \qquad \frac{\bar{\ell}_{m-1}^{(2)}+b}{2} \geq n+1,$$

respectively. The sequences $(\bar{\ell}_j)$, $(\bar{\ell}_j^{(1)})$ and $(\bar{\ell}_j^{(2)})$ are weakly admissible, but they are not necessarily admissible. For a given a there exists a unique path connecting a to b, c among these three.

Lemma A.2. *The maximum of* (A.1), *among weakly admissible sequences satisfying* (A.2), *is uniquely attained by one of* (A.3), (A.7–8). *Let* $a_{-\mu} \leq a \leq a_{1-\mu}$ $(1-m \leq \mu \leq m)$. *The maximum value* $\phi_{\max}$ *is*

$$\text{(A.9)} \qquad \begin{aligned} \phi_{\max} &= \frac{m(c-b)+m(m+1)N+\mu(2a-b-c)+\mu(\mu-1)N}{4} \\ &\qquad\qquad \textit{if } \mu \equiv m \text{ mod } 2, \\ &= \frac{(m+1)(c-b)+m(m+1)N+\mu(2a-b-c)+\mu(\mu-1)N}{4} \\ &\qquad\qquad \textit{otherwise.} \end{aligned}$$

The proof of this lemma will be given in Lemma A.3–8. The sequences used in the proof are weakly admissible unless otherwise stated. For $(\ell_j)_{j=j_0,\cdots,j_1}$ and $(\ell'_j)_{j=j_0,\cdots,j_1}$ such that

$$\text{(A.10a)} \qquad \ell_j = \ell'_j \quad \text{if } j=j_0, j_0+1, j_1-1, j_1,$$

or when $j_0=1$

$$\text{(A.10b)} \qquad \ell_j = \ell'_j \quad \text{if } j=1, j_1-1, j_1,$$

we denote by $(\ell_j)<(\ell'_j)$ and say the latter *dominates* the former if

$$\phi_m(\ell_1, \cdots, \ell_{m+2}) < \phi_m(\ell'_1, \cdots, \ell'_{m+2}),$$

where $\ell_j = \ell'_j$ $(j<j_0$ or $j>j_1)$ are supplemented arbitrarily. We abbreviate $H(\ell_j, \ell_{j+1}, \ell_{j+2})$ (resp. $H(\ell'_j, \ell'_{j+1}, \ell'_{j+2})$) to H_j (resp. H'_j). If we replace ϕ_m of (A.1) with $\sum_{j=1}^m \alpha_j H(\ell_j, \ell_{j+1}, \ell_{j+2})$ where $\alpha_j < \alpha_{j+1}$ $(j=1, \cdots, m)$ the proof goes well without change. In fact we need this generalization later in the proof of Lemma A.9.

Lemma A.3. *Consider an (ℓ_j) such that*

$$\ell_j = \ell_m - (m-j)N \qquad \text{if } i+1 \leq j \leq m-1$$

where $1 \leq i \leq m-1$. We also assume that $\ell_m < c$ and $\ell_i > \ell_{i+1} - N$. Then the (ℓ'_j) defined by

$$\begin{aligned} \ell'_j &= \ell_j + 2 && \text{if } i+1 \leq j \leq m, \\ &= \ell_j && \text{otherwise,} \end{aligned}$$

dominates the (ℓ_j).

Proof. We have $H'_{i-1} \geq H_{i-1}$, $H'_i = H_i - 1$, $H'_j = H_j = 0$ $(i+1 \leq j \leq m-1)$, and $H'_m = H_m + 1$. This implies $(\ell_j) < (\ell'_j)$. □

Lemma A.3 tells that if (ℓ_j) attains the maximum and $\ell_m < c$, then (ℓ_j) must be of the form (A.3). Now we consider the case $\ell_m > c$.

Lemma A.4. *Assume that $(\ell_j)_{j=i,\dots,i+4}$ and $(\ell'_j)_{j=i,\dots,i+4}$ satisfy* (A.10), *and that*

$$\frac{\ell_{i+3} + \ell_{i+4} - N}{2} \leq n - N, \qquad c \leq \ell_{i+2} = \ell'_{i+2} - 2.$$

If $\ell_{i+2} \geq n+1$ then $(\ell_j) < (\ell'_j)$, and if $\ell'_{i+2} \leq n$ then $(\ell_j) > (\ell'_j)$.

Proof. Assume that $\ell_{i+2} \geq n+1$. Then we have $H'_i \geq H_i - 1$, $H'_{i+1} \geq H_{i+1} + 1$ and $H'_{i+2} = H_{i+2}$. From this follows $(\ell_j) < (\ell'_j)$. The other case is proved similarly. □

Lemma A.5. *Assume that $(\ell_j)_{j=i,\dots,i+4}$ and $(\ell'_j)_{j=i,\dots,i+4}$ satisfy* (A.10), *and that*

$$\frac{\ell_{i+3} + \ell_{i+4} - N}{2} \leq n - N, \quad \ell_{i+2} = \ell_{i+1} + N, \quad \ell'_{i+2} = \ell_{i+4}.$$

Then we have $(\ell_j) < (\ell'_j)$.

Proof. We have $H'_i \geq H_i - n + (\ell_{i+2} + \ell_{i+4})/2$, $H'_{i+1} \geq H_{i+1} + n + 1 - (\ell_{i+2} + \ell_{i+4})/2$ *and* $H'_{i+2} = H_{i+2}$. Noting that

$$n + 1 - \frac{\ell_{i+2} + \ell_{i+4}}{2} \geq n + 1 - \frac{\ell_{i+3} + N + \ell_{i+4}}{2} \geq 1,$$

we have $(\ell_j) < (\ell'_j)$. □

From Lemma A.4–5 we can conclude that if (ℓ_j) satisfying (A.2) attains the maximum of ϕ_m and if $\ell_m > c$ then one of the following is valid:
(i) $\ell_m = b + N$,
(ii) $\ell_m = \ell_{m-1} - N$.
Note that $(\bar{\ell}_j^{(1)})$ and $(\bar{\ell}_j^{(2)})$ satisfies (i) and (ii), respectively. Now we distinguish these two as in (A.7b) and (A.8b).

Lemma A.6. *Assume that* $(\ell_j)_{j=i,\cdots,i+4}$ *and* $(\ell'_j)_{j=i\cdots,i+4}$ *satisfy* (A.10), *and that*

$$\frac{\ell_{i+3}+\ell_{i+4}-N}{2} \leq n-N, \quad \ell_{i+2}=\ell_{i+1}-N, \quad \ell'_{i+2}=\ell_{i+3}+N.$$

We set

$$\omega = \frac{\ell_{i+1}+\ell_{i+3}}{2} - n.$$

Then we have

$$\text{(A.11a)} \qquad (\ell_j) > (\ell'_j) \qquad \textit{if } \omega \leq 0,$$

$$\text{(A.11b)} \qquad (\ell_j) < (\ell'_j) \qquad \textit{if } \omega \geq 1.$$

In either case we have

$$\max(H_{i+1}, H'_{i+1}) = \frac{\ell_{i+3}-\ell_{i+1}}{2} + N.$$

Proof. We consider (A.11a) first. We have $H_i \geq H'_i + \omega$, $H_{i+1} \geq H'_i + 1 - \omega$ and $H_{i+2} = H'_{i+2}$. Therefore (A.11a) is valid. If $\omega \geq 1$, we have $H'_j \geq H_j - \omega + 1$, $H'_{i+1} \geq H_{i+1} + \omega$ and $H_{i+2} = H'_{i+2}$. This implies (A.11b). □

Thus we have proved the last four $\ell_{m-1}, \ell_m, \ell_{m+1}, \ell_{m+2}$ are as expected in (A.7–8) if (ℓ_j) attains the maximum. (Remember that we are assuming that $\ell_m > c$). We now prove

$$\text{(A.12)} \qquad \ell_j = \ell_{j+1} + N \qquad \text{if } j = 1, \cdots, m-2.$$

Assume that

$$\text{(A.13a)} \qquad \ell_j - \ell_{j+1} = N \qquad \text{if } j = i+1, \cdots, m-1,$$

$$\text{(A.13b)} \qquad < N \qquad \text{if } j = i.$$

If $(\ell_i+\ell_{i+2})/2\leq n$ then $(\ell_{i+2}-\ell_{i+3}-N)/2\leq n-N$ and $\ell_{i+1}>\ell_{i+3}$. This is a contradiction, for ℓ_i must be equal to $\ell_{i+1}+N$ from Lemmas A. 4–6. Therefore it is sufficient to show (A.12) when $(\ell_i+\ell_{i+2})/2\geq n+1$.

Lemma A.7. *Let* (ℓ_j) *satisfy* (A.2), (A.13). *We also assume*

$$\frac{\ell_i+\ell_{i+2}}{2}\geq n+2.$$

Define (ℓ'_j) *by* $\ell'_j=\ell_j$ $(j=1,\cdots,i,m+1,m+2)$ *and* $\ell'_j=\ell_j-2$ $(j=i+1,\cdots,m)$, *then we have* $(\ell'_j)>(\ell_j)$.

Proof. We have $H'_j=H_j$ $(j=1,\cdots,i-2)$, $H'_{i-1}\geq H_{i-1}$, $H'_i\geq H_i-1$, $H'_j=H_j=0$ $(j=i+1,\cdots,m-2)$ and $H'_{m-1}\geq H_{m-1}+1$. Therefore (ℓ'_j) dominates (ℓ_j). □

Lemma A.8. *Let* (ℓ_j) *satisfy* (A.2), (A.13). *We also assume*

$$\frac{\ell_i+\ell_{i+2}}{2}=n+1. \tag{A.15}$$

There are three cases to consider:

(*Case* 1) $\ell_i-(m-i-1)N\leq b$

Define (ℓ'_j) *by*

$$\begin{aligned}\ell'_j&=\ell_i-(j-i)N && \text{if } j=1,\cdots,m-3,\\ &=b+N && \text{if } j=m-2 \text{ and } \frac{\ell'_{m-3}+b}{2}\geq n+1,\\ &=\ell'_{m-3}-N && \text{if } j=m-2 \text{ and } \frac{\ell'_{m-3}+b}{2}\leq n,\\ &=b && \text{if } j=m-1,m+1,\\ &=c && \text{if } j=m,m+2.\end{aligned}$$

(*Case* 2) $b<l_i-(m-i-1)N\leq c+N$

Define (ℓ'_j) *by*

$$\begin{aligned}\ell'_j&=\ell_i-(j-i)N && \text{if } j=1,\cdots,m-2,\\ &=b+N && \text{if } j=m-1 \text{ and } \frac{\ell'_{m-2}+b}{2}\geq n+1,\\ &=\ell'_{m-3}-N && \text{if } j=m-1 \text{ and } \frac{\ell'_{m-2}+b}{2}\leq n,\\ &=b && \text{if } j=m+1,\\ &=c && \text{if } j=m,m+2.\end{aligned}$$

(*Case* 3) $c+N<\ell_i-(m-i-1)N\leq b+2N$

Define (ℓ'_j) *by*

$$\begin{aligned}\ell'_j&=\ell_i-(j-i)N && \textit{if } j=1,\cdots,m,\\ &=b && \textit{if } j=m+1,\\ &=c && \textit{if } j=m+2.\end{aligned}$$

In each case the (ℓ'_j) *dominates the* (ℓ_j).

Proof. From (A.14) and $\ell_{i+1}=\ell_{i+2}+N$, we have

$$\ell_{i+1}-n-1=\frac{\ell_{i+1}-\ell_i+N}{2}>0.$$

We denote this quantity by ρ. Relevant values of H_j and H'_j are:

j	i	$i+1$	$\cdots$	$m-4$	$m-3$	$m-2$	$m-1$
H_j	ρ	0	$\cdots$	0	0	0	$\frac{b-\ell_{m-1}}{2}+N$
H'_j							
(*Case* 1)	0	0	$\cdots$	0	$\frac{b-\ell_{m-1}}{2}+\rho$	$\frac{c-b+N}{2}$	$\frac{b-c+N}{2}$
(*Case* 2)	0	0	$\cdots$	0	0	$\frac{c-\ell_{m-1}+N}{2}+\rho$	$\frac{b-c+N}{2}$
(*Case* 3)	0	0	$\cdots$	0	0	0	$\frac{b-\ell_{m-1}}{2}+N+\rho$

Noting that $(\ell_{m-1}-c-N)/2\geq 1$ we have $(\ell'_j)>(\ell_j)$. □

We have proved that the maximum of ϕ_m, among weakly admissible sequences satisfying (A.2), is attained by one of $(\bar{\ell}_j)$, $(\bar{\ell}_j^{(1)})$, $(\bar{\ell}_j^{(2)})$.

The maximum value (A.9) is obtained by a straightforward computation.

Lemma A.9. *The maximum of* ϕ_m *among admissible sequences is attained by one of* $(\bar{\ell}_j)$, $(\bar{\ell}_j^{(1)})$, $(\bar{\ell}_j^{(2)})$, *if it is admissible. Otherwise we need the following modification*: (*We assume that m is even. If m is odd, we must interchange b and c.*)

(i) $\mu=1$ *and* $a+c<N+2$

(A.15a) $$\bar{\ell}_2=N+2-a,$$

(ii) $\mu=-1$ *and* $a+b>2(L-N-1)$

(A.15b) $$\bar{\ell}_2^{(2)}=2L-N-2-a$$

The maximum value is modified to

(A.16) $$\text{the r.h.s. of (A.9)}+\min\left(0, \frac{a+c-N-2}{2}, L-N-1-\frac{a+b}{2}\right).$$

Proof. Using the assumption that $L\geq 2N+3$, we can verify that the sequences $(\bar{\ell}_j)$, $(\bar{\ell}_j^{(1)})$, $(\bar{\ell}_j^{(2)})$ are admissible except for (i) and (ii). In these cases we have $H(a, \ell_2, \ell_3)\leq H(a, \ell_2, b)$. Therefore ℓ_3 must be equal to b. Since $H(\ell_2 b, c)$ is independent of ℓ_2 for the admissible values of ℓ_2, our task is to pick up ℓ_2 that maximizes $H(a, \ell_2, b)$. Thus we obtain (A.15), and then (A.16) follows immediately. □

Appendix B. Ground State Configuration in Regime II

Let (b, c) be any admissible pair. In this appendix we shall determine the admissible sequences (ℓ_j) that maximize the weight

$$\phi_m(\ell_1, \cdots, \ell_{m+2})=\sum_{j=1}^{m} j\frac{|\ell_j-\ell_{j+2}|}{4},$$

under the restriction

$$\ell_{m+1}=b, \qquad \ell_{m+2}=c.$$

For $\ell\in \boldsymbol{Z}$ we denote by $\langle\ell\rangle$ the unique integer satisfying

(B.1) $$1\leq\langle\ell\rangle\leq L-1, \qquad \langle\ell\rangle-1\equiv\pm(\ell-1) \bmod 2(L-2).$$

Lemma B.1. *Define* $(\bar{\ell}_j^{(\pm)})$ *by*

$$\begin{aligned}\bar{\ell}_j^{(\pm)}&=\langle b\pm(m-j+1)N\rangle && \text{if } 1\leq j\leq m,\\ &=b && \text{if } j=m+1,\\ &=c && \text{if } j=m+2.\end{aligned}$$

We denote by $\mu_m(b)$ *a positive integer determined by*

$$1-\mu_m(b)(L-2)\leq b-mN<L-1-\mu_m(b)(L-2).$$

Then we have

$$\phi_m(\bar{\ell}_1^{(-)}, \cdots, \bar{\ell}_{m+2}^{(-)})=\mathscr{G}_m(b, c), \qquad \phi_m(\bar{\ell}_1^{(+)}, \cdots, \bar{\ell}_{m+2}^{(+)})=\mathscr{G}_m(L-b, L-c),$$

where

(B.2) $\mathcal{G}_m(b, c)$

$$=\frac{m^2N+m(c-b)+\mu_m(b)(\mu_m(b)-1)(L-2)+2\mu_m(b)(b-mN-1)}{4}.$$

Proof. The integer $\mu_m(b)$ gives the number of reflections along the $(\bar{\ell}_j^{(-)})$ (see Fig. B.1). If $\mu_m(b)=0$ the weight is given by $(m^2N+m(c-b))/4$. The deficiency in the weight at the i-th reflection (counting from the right) is $(1+mN-(i-1)(L-2)-b)/2$. Therefore the exact value of the weight is given by (B.2). □

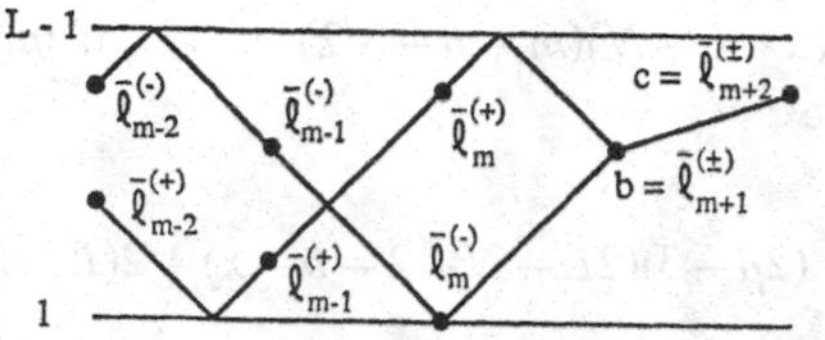

Fig. B.1 The configuration $(\bar{\ell}_j^{(\pm)})$.

Lemma B.2. $m>\mu_m(b)+(\mu_m(b)-1)/N$.

The proof is easy. From this we know, in particular, that $m-\mu_m(b)$ tends to ∞ when $m\to\infty$. Therefore if m is sufficiently large the sequence $\bar{\ell}_j=\langle b+Nj\rangle$ must contain a pair $(\bar{\ell}_i, \bar{\ell}_{i+1})$ such that $\bar{\ell}_{i+1}=\bar{\ell}_i+N$.

Lemma B.3. *Let x be an integer such that $\langle b-N\rangle<x<\langle b+N\rangle$, and set*

$$\mathcal{D}_m(x, b, c)=\max(\mathcal{G}_m(b, c), \mathcal{G}_m(L-b, L-c))-\left(\mathcal{G}_{m-1}(x, b)+\frac{m|x-c|}{4}\right).$$

Then $\mathcal{D}_m(x, b, c)>0$ *and* $\liminf_{m\to\infty}\mathcal{D}_m(x, b, c)/m>0.$

Proof. We set $\mu^{(-)}=\mu_m(b)$, $\mu^{(+)}=\mu_m(L-b)$ and $\mu=\mu_{m-1}(x)$. There are five cases:

	(i)	(ii)	(iii)	(iv)	(v)
$\mu^{(+)}$	μ	$\mu+1$	$\mu+1$	$\mu-1$	μ
$\mu^{(-)}$	μ	$\mu+1$	μ	μ	$\mu+1$

We set $\mathcal{D}^{(-)}=4(\mathcal{G}_m(b, c)-\mathcal{G}_{m-1}(x, b)-m|x-c|/4)$ and $\mathcal{D}^{(+)}=4(\mathcal{G}_m(L-b, L-c)-\mathcal{G}_{m-1}(x, b)-m|x-c|/4)$. The following prove the claim of this lemma:

Case (i), (v), $x\geq c$.

$$\mathcal{D}^{(+)}=(2m-1)N+b-x \qquad \text{if } \mu=0,$$
$$=(2\mu-1)(2L-N-2-b-x) \qquad \text{if } \mu\geq 1.$$

Case (i), (iii), (iv), $x \leq c$.

$$\mathscr{D}^{(-)} = (2m-2\mu-1)(x+N-b).$$

Case (ii), (iii), $x \geq c$.

$$\mathscr{D}^{(+)} = (2\mu+1)(2L-2-N-b-x).$$

Case (ii), (v), $x \leq c$.

$$\begin{aligned} \mathscr{D}^{(-)} &= 2\mu(L-N-2)+x+b-N-2 && \text{if } m=\mu+1, \\ &= 2(x-b+N)(m-\mu-3/2) && \text{if } m \geq \mu+2. \end{aligned}$$

Case (iv), $x \geq c$.

$$\mathscr{D}^{(+)} > (2\mu-3)(2L-N-2-b-x)+2(L-2). \qquad \square$$

From Lemma B.3 follows that the sequence $\bar{\ell}_j = \langle b+jN \rangle$ is a ground state configuration in the sense of section 2 of Part I. In fact, we have $\mathscr{G}_m(\bar{\ell}_{m+1}, \bar{\ell}_{m+2}) > \mathscr{G}_m(L-\bar{\ell}_{m+1}, L-\bar{\ell}_{m+2})$. Therefore $(\bar{\ell}_j)$ maximizes the weight $\phi_m(\ell_1, \cdots, \ell_{m+2})$.

Appendix C. Branching Coefficients and String Functions

Here we relate the branching coefficients $c^{(\varepsilon)}_{j_1 j_2 j_3}(\tau)$ and the string functions of $A_1^{(1)}$ (see Part I, Appendix B). This observation is due to V. G. Kac. Our LHP results in regime III, IV for $N=2$ are shown to coincide with the particular cases studied in [11] by using the expressions in C.3.

C.1. Products of theta functions

For a positive integer m and a real number μ, define [10]

(C.1) $$\Theta^{(\pm)}_{\mu,m}(u, \tau) = \sum_{\gamma = n+\mu/m, n \in \boldsymbol{Z}} (\pm)^n q^{m\gamma^2/2} z^{-m\gamma}, \quad q=e^{2\pi i\tau}, \quad z=e^{2\pi i u},$$

The *Theta Null Werte* have the product representation (see (1.5.3))

(C.2) $$\Theta^{(\varepsilon)}_{\mu,m}(0, \tau) = q^{\mu^2/2m} E(-\varepsilon q^{\mu+m/2}, q^m), \qquad \varepsilon = \pm.$$

The theta function (C.1) obey the standard multiplication formula

(C.3) $$\begin{aligned} &\Theta^{(\varepsilon_1)}_{\mu_1,m_1}(u, \tau)\Theta^{(\varepsilon_2)}_{\mu_2,m_2}(u, \tau) \\ &= \sum_{\nu \in \boldsymbol{Z}/(m_1+m_2)\boldsymbol{Z}} \varepsilon_1^{\nu} \Theta^{(\varepsilon)}_{\alpha,\beta}(0, \tau) \Theta^{(\varepsilon_1\varepsilon_2)}_{m_1\nu+\mu_1+\mu_2, m_1+m_2}(u, \tau), \end{aligned}$$

where $\varepsilon = \varepsilon_1^{m_2}\varepsilon_2^{m_1}$, $\alpha = m_1 m_2(\nu+\mu_1/m_1-\mu_2/m_2)$, $\beta = m_1 m_2(m_1+m_2)$. The theta functions introduced in Part I, Appendix A and those here are connected by

(C.4) $$\Theta^{(\varepsilon_1,\varepsilon_2)}_{j,m}(z,q)=\Theta^{(\varepsilon_2)}_{j,2m}(u/2,\tau)+\varepsilon_1\Theta^{(\varepsilon_2)}_{-j,2m}(u/2,\tau).$$

C.2. Branching coefficients and string functions

Recall the definition of the branching coefficients $c^{(\varepsilon)}_{j_1j_2j_3}(\tau)$:

(C.5) $$\Theta^{(-,\varepsilon)}_{j_1,m_1}(z,q)\Theta^{(-,+)}_{j_2,m_2}(z,q)/\Theta^{(-,+)}_{1,2}(z,q)=\sum_{j_3} c^{(\varepsilon)}_{j_1j_2j_3}(q)\Theta^{(-,\varepsilon)}_{j_3,m_3}(z,q).$$

Here the sum ranges over $j_3\in \boldsymbol{Z}+j_1$ such that $0<j_3\leq m_3$ (if $\varepsilon=+$, then $j_3<m_3$). On the other hand, the level m string functions $c^m_{jk}(\tau)$ $(=c^{m-k\,k}_{m-j\,j}(\tau)$ in the notation of [10]) for $A^{(1)}_1$ are characterized by the identity

(C.6) $$\Theta^{(-,+)}_{k-1,m+2}(z,q)/\Theta^{(-,+)}_{1,2}(z,q)=\sum_{0\leq j\leq m,\, j\equiv k \bmod 2} c^m_{jk}(\tau)\varepsilon^m_j\Theta^{(+,+)}_{j,m}(z,q).$$

In the l.h.s. of (C.5), replace the part $\Theta^{(-,+)}_{j_2,m_2}(z,q)/\Theta^{(-,+)}_{1,2}(z,q)$ by the r.h.s. of (C.6), use (C.4) and apply (C.3). The result reads

$$\sum c^{m_2-2}_{j,j_2-1}(\tau)\Theta'^{(+)}_{2m_1j_3-2m_3j_1,8m_1(m_2-2)m_3}(0,\tau)\Theta^{(-,\varepsilon)}_{j_3,m_3}(z,q)$$

where the sum is taken over $j\in \boldsymbol{Z}/2(m_2-2)\boldsymbol{Z}$ and $j_3\in \boldsymbol{Z}/4(m_2-2)m_3\boldsymbol{Z}$ under the conditions $j\equiv j_2-1 \bmod 2$, $j_3\equiv j+j_1 \bmod 2(m_2-2)$. Equating the coefficients of linearly independent $\Theta^{(-,\varepsilon)}_{j_3,m_3}(z,q)$'s, we get an expression of $c'^{(\varepsilon)}_{j_1j_2j_3}(\tau)$ in terms of string functions $c^m_{jk}(\tau)$ and the Theta Null Werte (C.2). Explicit formula for $c^m_{jk}(\tau)$ for small m can be found in [10], pp. 219–220.

C.3. Branching coefficients for $m_2=3,4$

We give below the resulting formulas for $m_2=3,4$. In the case $m_1\in \boldsymbol{Z}+1/2$, we find it convenient to replace the j_3 sum in the r.h.s. of (C.5) by $0<j_3<2m_3$ with the restriction $j_3+1\equiv j_1+j_2 \bmod 2$ (recall that by the definition in Appendix A.1 of Part I $c'^{(\varepsilon)}_{j_1j_2j_3}(\tau)=-\varepsilon c^{(\varepsilon)}_{j_1j_2\,2m_3-j_3}(\tau)$). Recall also that if $j_1, m_1\in \boldsymbol{Z}$ and $j_3+1\equiv j_1+j_2 \bmod 2$, then $c'^{(\varepsilon)}_{j_1j_2j_3}(\tau)=0$. In the sequel we set

$$k=m_1j_3-m_3j_1,\quad \ell=m_1j_3+m_3j_1,\quad n=m_1m_3,$$

and assume that $j_1\in \boldsymbol{Z}$, $j_3+1\equiv j_1+j_2 \bmod 2$.

The case $m_2=3$.

Here we use the formula $c^1_{00}(\tau)=\eta(\tau)^{-1}$.

(i) $m_1\in \boldsymbol{Z}+1/2$.

$$c'^{(\varepsilon)}_{j_1j_2j_3}(\tau)=\eta(\tau)^{-1}(\Theta'^{(+)}_{2k,8n}(0,\tau)-\Theta'^{(+)}_{2\ell,8n}(0,\tau)).$$

(ii) $m_1 \in Z$.

$$c^{(\varepsilon)}_{j_1j_2j_3}(\tau)=\eta(\tau)^{-1}\varepsilon^{m_3}_{j_3}(\Theta^{(\varepsilon)}_{k,2n}(0,\tau)-\Theta^{(\varepsilon)}_{\ell,2n}(0,\tau)).$$

The case $m_2=4$.

Put $\gamma_\pm(\tau)=c^2_{00}(\tau)\pm c^2_{20}(\tau)$, $\gamma_0(\tau)=c^2_{11}(\tau)$. Then we have

$$\gamma_+(\tau)=e^{-\pi i/24}\eta\left(\frac{\tau+1}{2}\right)\eta(\tau)^{-2},\ \gamma_-(\tau)=\eta\left(\frac{\tau}{2}\right)\eta(\tau)^{-2},$$

$$\gamma_0(\tau)=\eta(2\tau)\eta(\tau)^{-2}.$$

We give the results in the case $j_3+1\equiv j_1+j_2 \bmod 4$. The other case $j_3+1\equiv j_1+j_2+2 \bmod 4$ can be obtained by negating $\gamma_-(\tau)$.

(i) $m_1 \in Z+1/2$, $j_2=1, 3$.

$$c^{(\varepsilon)}_{j_1j_2j_3}(\tau)=\frac{1}{2}\gamma_+(\tau)(\Theta^{(+)}_{k,4n}(0,\tau)-\Theta^{(+)}_{\ell,4n}(0,\tau))$$

$$+\frac{1}{2}\gamma_-(\tau)(\Theta^{(-)}_{k,4n}(0,\tau)-(-)^{j_1}\Theta^{(-)}_{\ell,4n}(0,\tau)).$$

(ii) $m_1 \in Z+1/2$, $j_2=2$.

$$c^{(\varepsilon)}_{j_1j_2j_3}(\tau)=\gamma_0(\tau)(\Theta^{(+)}_{k,4n}(0,\tau)-\Theta^{(+)}_{\ell,4n}(0,\tau)).$$

(iii) $m_1 \in Z$, $j_2=1, 3$.

$$c^{(\varepsilon)}_{j_1j_2j_3}(\tau)=\frac{1}{2}\gamma_+(\tau)\varepsilon^{m_3}_{j_3}(\Theta^{(\varepsilon)}_{k/2,n}(0,\tau)-\Theta^{(\varepsilon)}_{\ell/2,n}(0,\tau))$$

$$+\frac{1}{2}\gamma_-(\tau)\varepsilon^{m_3}_{j_3}(\Theta^{((-)^{m_1}\varepsilon)}_{k/2,n}(0,\tau)-(-)^{j_1}\Theta^{((-)^{m_1}\varepsilon)}_{\ell/2,n}(0,\tau)).$$

(iv) $m_1 \in Z$, $j_2=2$.

$$c^{(\varepsilon)}_{j_1j_2j_3}(\tau)=\gamma_0(\tau)\varepsilon^{m_3}_{j_3}(\Theta^{(\varepsilon)}_{k/2,n}(0,\tau)-\Theta^{(\varepsilon)}_{\ell/2,n}(0,\tau)).$$

Appendix D. Free Energy

In this appendix we shall give the free energy for the fusion vertex models (section 2.3) and the restricted SOS models (section 2.2), and discuss its critical behavior. The calculation is based on the inversion relation method [8, 12]. As it turns out, the inversion relations for the restricted SOS models are formally identical with those for the vertex models (with the parameter λ replaced by $2K/L$); consequently the free energy itself has the same form under this correspondence.

D.1. Unitarity and crossing symmetry

First let us recall the unitarity relation for vertex models. As in section 2.3 we denote by $C \in \mathrm{End}(V\otimes V)$ the transposition operator.

Lemma D.1. *Let $R(u)$ be a solution to the Yang-Baxter equation* (2.3.2) *(with $V_1=V_2=V_3$) satisfying the initial condition* (2.3.3a). *Then we have*

(D.1) $$R^{12}(u)R^{21}(-u)=\rho(u)I,$$

where $R^{12}(u)=R(u)$, $R^{21}(u)=CR(u)C$ and $\rho(u)$ is a scalar function (Fig. D.1).

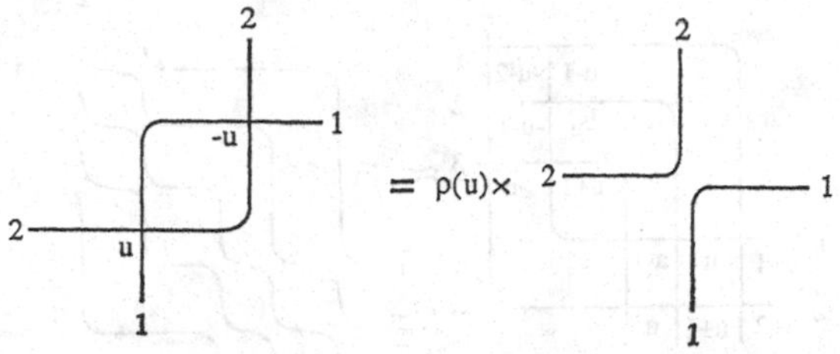

Fig. D. 1 The unitarity relation for vertex models. The r.h.s. is proportional to the scalar operator.

Proof. Setting $v=-u$ in (2.3.2) and using (2.3.3a) we find

$$R^{12}(u)R^{21}(-u)=R^{23}(-u)R^{32}(u).$$

This implies that the l.h.s. of (D.1) commutes with matrixes of the form $X\otimes I$, $I\otimes X$ ($X\in \mathrm{End}(V)$). Hence it must be a scalar. □

The function $\rho(u)$ in (D.1) can be determined by comparing a particular matrix element. For the original eight vertex weight (2.3.1) we have

(D.2) $$R(u)R(-u)=\begin{bmatrix}1+u\\1\end{bmatrix}\begin{bmatrix}1-u\\1\end{bmatrix}I.$$

(Note that in this case $R(u)=CR(u)C$.) The unitarity relation for the (M,N)-weight $R_{MN}(u)$ in section 2.3 reads as follows. We set $R^{12}_{MN}(u)=R_{MN}(u)$, $R^{21}_{NM}(u)=C_{NM}R_{MN}(u-M+N)C_{NM}$, and regard both as acting on the same space $\underbrace{V\otimes\cdots\otimes V}_{M+N}$ (see section 2.3 for C_{MN}).

Lemma D.2. *Assuming $M\geq N$ we have*

(D.3) $$R^{12}_{MN}(u)R^{21}_{NM}(-u)=\begin{bmatrix}M+u\\N\end{bmatrix}\begin{bmatrix}N-u\\N\end{bmatrix}I.$$

Proof. For definiteness, let $M=3$ and $N=2$. Then by the definition

$$R^{12}_{32}(u)=\sigma(u)P_{123}P_{\bar{1}\bar{2}}R^{1\bar{2}}(u+2)R^{2\bar{2}}(u+1)R^{3\bar{2}}(u)R^{1\bar{1}}(u+1)R^{2\bar{1}}(u)R^{3\bar{1}}(u-1),$$

$$R^{21}_{23}(-u)=\sigma(-u-1)P_{123}P_{\bar{1}\bar{2}}$$
$$\times R^{\bar{1}3}(-u+1)R^{\bar{1}2}(-u)R^{\bar{1}1}(-u-1)R^{\bar{2}3}(-u)R^{\bar{2}2}(-u-1)R^{\bar{2}1}(-u-2),$$

where $\sigma(u)=[1]^6/([2]_2[u+2][u+1]^2[u])$. In the second line we have reshaffled the superfixes $(\bar{1}, \bar{2})$ and $(1, 2, 3)$ under the symmetrizers (cf. (2.3.6)). Using (2.3.6) and (D.2) repeatedly we obtain (D.3) (Fig. D.2). The general case is similar. □

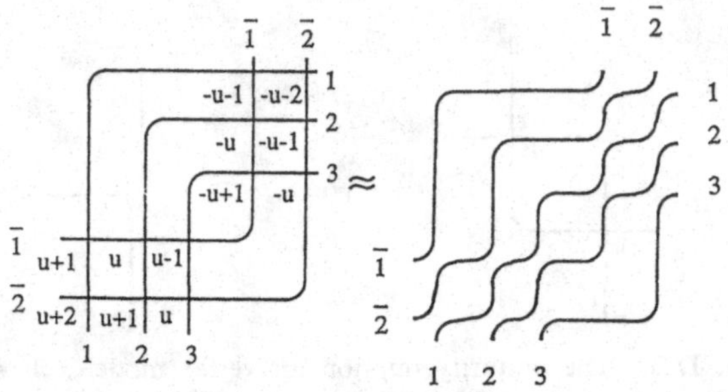

Fig. D. 2 The unitarity relation for fusion vertex models.

Hereafter we shall be concerned with the case $M=N$. Eq. (D.3), to be also called the "first inversion relation", then reads

(D.4) $$R_{NN}(u)CR_{NN}(-u)C=\begin{bmatrix}N+u\\N\end{bmatrix}\begin{bmatrix}N-u\\N\end{bmatrix}I,$$

where $C=C_{NN}$.

Let $\tilde{R}(u)$ denote the matrix of the eight vertex weight obtained by rotating the lattice through 90°, *i.e.*

(D.5) $$\tilde{R}^{\alpha\beta}_{\gamma\delta}(u)=R^{\beta\gamma}_{\delta\alpha}(u).$$

The following crossing symmetry holds:

$$R(-1-u)=-(\sigma^y\otimes I)\tilde{R}(u)(\sigma^y\otimes I), \qquad \sigma^y=\begin{pmatrix} & -i\\ i & \end{pmatrix}.$$

This implies together with (D.2) and the symmetry $R(u)=CR(u)C$ (Fig. D.3)

$$\tilde{R}(u)\tilde{R}(-2-u)=\begin{bmatrix}2+u\\1\end{bmatrix}\begin{bmatrix}-u\\1\end{bmatrix}I.$$

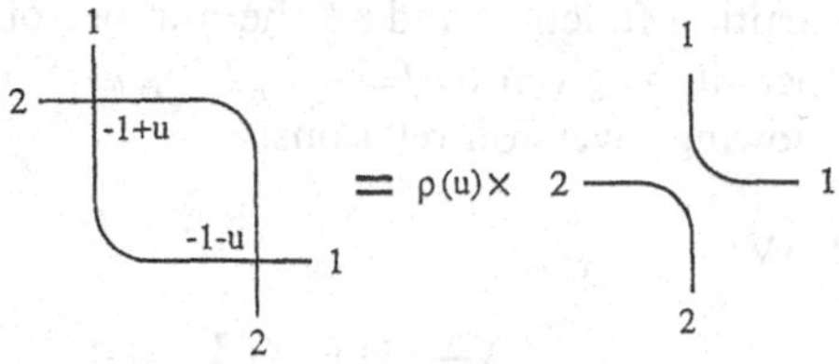

Fig. D. 3 The second inversion relation.

More generally, for the (N, N)-weight $R_{NN}(u)$ define $\tilde{R}_{NN}(u)$ by (D.5). In the same way as Lemma D.2 we have the "second inversion relation" for $R_{NN}(u)$:

Lemma D.3.

(D.6) $$\tilde{R}_{NN}(u)C\tilde{R}_{NN}(-2-u)C=\begin{bmatrix} N+1+u \\ N \end{bmatrix}\begin{bmatrix} N-1-u \\ N \end{bmatrix} I.$$

D.2. The free energy of the fusion vertex models and the restricted SOS models

In order to discuss the free energy for the fusion vertex models, we must specify the regimes to consider. Following the case of the restricted SOS models, we deal with the four cases below.

(D.7)
$$\begin{aligned} &\text{Regime I:} && -1<p<0, && 0<u<K/\lambda-1, \\ &\text{Regime II:} && 0<p<1, && 0<u<K/\lambda-1, \\ &\text{Regime III:} && 0<p<1, && -1<u<0, \\ &\text{Regime IV:} && -1<p<0, && -1<u<0. \end{aligned}$$

In regimes II and III, K' is real and positive, while in regimes I and IV so is $\tilde{K}'=K'-iK$ rather than K'. The end points $u=-1, 0$ are "inversion points" and $K/\lambda-1$ is a "virtual inversion point" in the terminology of [12]. We define w by

(D.8)
$$\begin{aligned} w&=e^{-2\pi\lambda u/K'} && \text{in regime II, III } (p=e^{-\pi K'/K}), \\ &=e^{-\pi\lambda u/\tilde{K}'} && \text{in regime I, IV } (p=-e^{-\pi\tilde{K}'/K}), \end{aligned}$$

and set

(D.9) $$\Lambda=2K/\lambda.$$

Now let κ denote the partition function per site

$$\kappa=\lim_{\mathcal{M}\to\infty} Z^{1/\mathcal{M}},$$

where Z is the partition function and $\mathcal{M}$ the number of sites of the lattice. The free energy per site is given by $f=-k_BT\log\kappa$. From (D.4) and (D.6) we obtain the following inversion relations.

Regime III, IV:

(D.10)
$$\kappa(u)\kappa(-u)=\begin{bmatrix} N+u \\ N \end{bmatrix}\begin{bmatrix} N-u \\ N \end{bmatrix},$$
$$\kappa(u)\kappa(-2-u)=\begin{bmatrix} N+1+u \\ N \end{bmatrix}\begin{bmatrix} N-1-u \\ N \end{bmatrix}.$$

Regime I, II:

(D.11)
$$\kappa(u)\kappa(-u)=\begin{bmatrix} N+u \\ N \end{bmatrix}\begin{bmatrix} N-u \\ N \end{bmatrix},$$
$$\kappa(u)\kappa(\Lambda-2-u)=\begin{bmatrix} N+1+u \\ N \end{bmatrix}\begin{bmatrix} N-1-u \\ N \end{bmatrix}.$$

The inversion relations for the restricted SOS models follow from (2.1.25) and (2.2.19). Identifying λ with $2K/L$ (and Λ with L) we find that they have the same form as (D.10–11). The definitions of regimes (D.7) and of the parameter w (D.8) agree with those for the SOS models.

It is straightforward to compute the free energy. We apply the conjugate modulus transformation (3.1.5–6) (wherein L is to be read as $2K/\lambda$) and solve (D.10–11) for $\kappa(u)$. The result is expressed as follows.

Regime I: $(\nu=\pi\lambda/\tilde{K}',\ \Lambda>2N)$

(D.12)
$$\log\kappa(u)=\sum_{j\in 2Z}F(j)+\sum_{j\in 2Z+1}G(j),$$
$$F(j)=F(-j)=\frac{1}{j}\times\frac{\operatorname{sh}(N\nu j/2)\operatorname{ch}((\Lambda-2N-2)\nu j/4)}{\operatorname{sh}(\nu j/2)\operatorname{ch}((\Lambda-2)\nu j/4)\operatorname{sh}(\Lambda\nu j/4)}$$
$$\times\operatorname{sh}(u\nu j/2)\operatorname{sh}((\Lambda-2-2u)\nu j/4),$$
$$G(j)=G(-j)=\frac{1}{j}\times\frac{\operatorname{sh}(N\nu j/2)\operatorname{sh}((\Lambda-2N-2)\nu j/4)}{\operatorname{sh}(\nu j/2)\operatorname{sh}((\Lambda-2)\nu j/4)\operatorname{ch}(\Lambda\nu j/4)}$$
$$\times\operatorname{sh}(u\nu j/2)\operatorname{ch}((\Lambda-2-2u)\nu j/4).$$

Regime II: $(\nu=2\pi\lambda/K',\ \Lambda>2N)$

$$\log\kappa(u)=\sum_{j\in Z}F(j),$$

$$F(j)=F(-j)=\frac{1}{j}\times\frac{\operatorname{sh}(N\nu j/2)\operatorname{sh}(u\nu\, j/2)}{\operatorname{sh}(\nu j/2)\operatorname{sh}((\Lambda-2)\nu j/2)\operatorname{sh}(\Lambda\nu j/2)}$$

(D.13)

$$\times(\operatorname{sh}((\Lambda-N-2)\nu j/2)\operatorname{sh}((\Lambda-1-u)\nu j/2)$$
$$-\operatorname{sh}(N\nu j/2)\operatorname{sh}((u+1)\nu j/2)).$$

Regime III: $(\nu=2\pi\lambda/K',\ \Lambda>N)$

$$\log\kappa(u)=\sum_{j\in Z}F(j),$$

$$F(j)=F(-j)=-\frac{2}{j}\times\frac{\operatorname{sh}(N\nu j/2)\operatorname{ch}((\Lambda-N-1)\nu j/2)}{\operatorname{sh}(\nu j)\operatorname{sh}(\Lambda\nu j/2)}$$

(D.14)

$$\times\operatorname{sh}(u\nu j/2)\operatorname{sh}((u+1)\nu j/2).$$

Regime IV: $(\nu=\pi\lambda/\tilde{K}',\ \Lambda>2N)$

$$\log\kappa(u)=\sum_{j\in 2Z}F(j)+\sum_{j\in 2Z+1}G(j),$$

$$F(j)=F(-j)=-\frac{2}{j}\times\frac{\operatorname{sh}(N\nu j/2)\operatorname{ch}((\Lambda-2N-2)\nu j/4)}{\operatorname{sh}(\nu j)\operatorname{sh}(\Lambda\nu j/4)}$$
$$\times\operatorname{sh}(u\nu j/2)\operatorname{sh}((u+1)\nu j/2),$$

(D.15)

$$G(j)=G(-j)=-\frac{2}{j}\times\frac{\operatorname{sh}(N\nu j/2)\operatorname{sh}((\Lambda-2N-2)\nu j/4)}{\operatorname{sh}(\nu j)\operatorname{ch}(\Lambda\nu j/4)}$$
$$\times\operatorname{sh}(u\nu j/2)\operatorname{sh}((u+1)\nu j/2).$$

Here we have assumed that $\Lambda>N$ or $2N$ for simplicity. Otherwise the expression should be modified. (This comes from the difference of the behavior of the factors $E(x^u, x^L)$ or $E(x^u, -x^{L/2})$ appearing in the r.h.s. of the inversion relations.)

As we remarked before, the free energy for the restricted SOS models are obtained simply by replacing λ by $2K/L$ (Λ by L).

D.3. Critical behavior

The critical behavior of the free energy can be studied in the following way. For a function $f(x)$, let

$$\hat{f}(\xi)=\int_{-\infty}^{\infty}f(x)\exp(2\pi ix\xi)dx$$

denote it Fourier transform. Poisson's summation formula asserts that in the case $f(-x)=f(x)$

$$\sum_{j\in Z}f(j)=\hat{f}(0)+2\sum_{\xi\geq 1}\hat{f}(\xi).$$

The free energy results (D.12–15) are rewritten as

Regime II, III:

$$\sum_{j\in Z} F(j)=\hat{F}(0)+2\sum_{\xi\geq 1}\hat{F}(\xi),$$

Regime I, IV:

$$\sum_{j\in 2Z} F(j)+\sum_{j\in 2Z+1} G(j)=\frac{1}{2}(\hat{F}(0)+\hat{G}(0))+\sum_{\xi\geq 1}(\hat{F}(\xi/2)+(-)^{\xi}\hat{G}(\xi/2)).$$

The 0-th term represents the critical value $\log\kappa^{(c)}(u)$. It is the same for regime I/II or regime III/IV, and is given by

Regime I, II:

$$\log\kappa^{(c)}(u)=\int_{-\infty}^{\infty}\frac{dt}{t}\frac{\operatorname{sh}(Nt/2)\operatorname{sh}(ut/2)}{\operatorname{sh}(t/2)\operatorname{sh}((\Lambda-2)t/2)\operatorname{sh}(\Lambda t/2)}$$
$$\times(\operatorname{sh}((\Lambda-N-2)t/2)\operatorname{sh}((\Lambda-1-u)t/2)$$
$$-\operatorname{sh}(Nt/2)\operatorname{sh}((u+1)t/2)).$$

Regime III, IV:

$$\log\kappa^{(c)}(u)=-2\int_{-\infty}^{\infty}\frac{dt}{t}\frac{\operatorname{sh}(Nt/2)\operatorname{ch}((\Lambda-N-1)t/2)}{\operatorname{sh}(t)\operatorname{sh}(\Lambda t/2)}$$
$$\times\operatorname{sh}(ut/2)\operatorname{sh}((u+1)t/2).$$

The above formula for regime III–IV agrees with the result of [13] (eq. (2.23) there).

To study the behavior of $\hat{F}(\xi)$ (or $\hat{F}(\xi/2)$, $\hat{G}(\xi/2)$) for $\xi>0$, we deform the contour of integration to surround the upper half plane and pick up the residues. Thus we get the following results for general Λ (the case of the fusion vertex models). Here $(\log\kappa(u))_{\text{sing}}$ denotes the non-analytic part of $\log\kappa(u)$ in $|p|$.

Regime I, II:

$$(\log\kappa(u))_{\text{sing}}=\frac{4\sin^2(N\pi/(\Lambda-2))}{\sin(2\pi/(\Lambda-2))}\sin(2\pi u/(\Lambda-2))\times|p|^{\Lambda/(\Lambda-2)}$$
$$+O(|p|^{2\Lambda/(\Lambda-2)}).$$

Regime III:

$$(\log\kappa(u))_{\text{sing}}=4\frac{\cos(\Lambda\pi/2)}{\sin(\Lambda\pi/2)}\sin\pi u\times|p|^{\Lambda/2}+O(|p|^{\Lambda})\qquad\text{if } N \text{ is odd,}$$

$\log\kappa(u)$ is regular if N is even.

Regime IV:

$$(\log \kappa(u))_{\rm sing} = 4\frac{\sin \pi u}{\sin (\Lambda\pi/2)} \times |p|^{\Lambda/2} + O(|p|^{\Lambda}) \qquad \text{if } N \text{ is odd},$$

$$\log \kappa(u) \quad \text{is regular} \qquad \text{if } N \text{ is even.}$$

Complications occur when Λ is an integer because of the double poles in $F(j)$ and $G(j)$. This is the case of the restricted SOS models. Careful examination shows that the results above should then be modified as follows.

Regime I, II:

$$(\log \kappa(u))_{\rm sing} = \frac{4 \sin^2(N\pi/(L-2))}{\sin (2\pi/(L-2))} \sin (2\pi u/(L-2)) \times |p|^{L/(L-2)} + O(|p|^{2L/(L-2)}).$$

The only exception occurs when $L=4$ and $N=1$ (Ising model). In this case we have

$$(\log \kappa(u))_{\rm sing} = -\frac{4}{\pi} \sin \pi u \times |p|^2 \log |p| + O(|p|^4 \log |p|).$$

Regime III:

$$(\log \kappa(u))_{\rm sing} = \frac{4}{\pi} \sin \pi u \times |p|^{L/2} \log |p| + O(|p|^L \log |p|)$$

if L is even and N is odd,

$\log \kappa(u)$ is regular otherwise.

Regime IV:

$$(\log \kappa(u))_{\rm sing} = (-)^{L/2}\frac{4}{\pi} \sin \pi u \times |p|^{L/2} \log |p| + O(|p|^L \log |p|)$$

if L is even and N is odd,

$$(\log \kappa(u))_{\rm sing} = (-)^{(L-1)/2} 4 \sin \pi u \times |p|^{L/2} + O(|p|^{3L/2})$$

if L is odd and N is odd,

$\log \kappa(u)$ is regular otherwise.

References

[1] E. Date, M. Jimbo, A. Kuniba, T. Miwa and M. Okado, Exactly Solvable SOS Models: Local Height Probabilities and Theta Function Identities, Nucl. Phys., **B290** [FS 20] (1987), 231–273.

[2] E. Date, M. Jimbo, T. Miwa and M. Okado, Fusion of the Eight Vertex SOS Model, Lett. Math. Phys., **12** (1986), 209–215; Erratum and Addendum, Lett. Math. Phys., **14** (1987), 97.

[3] ——, Automorphic Properties of Local Height Probabilities for Integrable Solid-on-Solid Models, Phys. Rev., **B35** (1987), 2105–2107.

[4] R. J. Baxter, Eight-Vertex Model in Lattice Statistics and One-dimensional Anisotropic Heisenberg Chain. II. Equivalence to a Generalized Ice-type Model, Ann. of Phys., **76** (1973), 25–47.

[5] G. E. Andrews, R. J. Baxter and P. J. Forrester, Eight-Vertex SOS Model and Generalized Rogers-Ramanujan identities, J. Stat. Phys., **35** (1984), 193–266.

[6] P. P. Kulish, N. Yu. Reshetikhin and E. K. Skylyanin, Yang-Baxter Equation and Representation Theory: I. Lett. Math. Phys., **5** (1981), 393–403.
I. V. Cherednik, Some Finite Dimensional Representations of Generalized Sklyanin Algebra, Funct. Anal. and Appl., **19** (1985), 77–79.

[7] M. Jimbo and T. Miwa, Irreducible Decomposition of Fundamental Modules for $A_\ell^{(1)}$ and $C_\ell^{(1)}$ and Hecke modular forms, Adv. Stud. Pure Math., **4** (1984), 97–119.

[8] R. J. Baxter, Exactly solved models in statistical mechanics, Academic, London, 1982.

[9] G. E. Andrews, The theory of partitions, Addison-Wesley, Massachusetts, 1976.

[10] V. G. Kac and D. H. Peterson, Infinite-Dimensional Lie Algebras, Theta Functions and Modular Forms, Advances in Math., **53** (1984), 125–264.

[11] R. J. Baxter and G. E. Andrews, Lattice Gas Generalization of the Hard Hexagon Model. I. Star-Triangle Relation and Local Density, J. Stat. Phys., **44** (1986), 249–271.
G. E. Andrews and R. J. Baxter, Lattice Gas Generalization of the Hard Hexagon Model. II. The Local Densities as Elliptic Functions, ibid., (1986), 713–728.
P. J. Forrester and G. E. Andrews, Height Probabilities in Solid-on-Solid Models II, preprint ITP-SB-86-94, 1986.

[12] R. J. Baxter, The Inversion Relation Method for Some Two-Dimensional Exactly Solved Models in Lattice Statistics, J. Stat. Phys., **28** (1982), 1–41.

[13] K. Sogo, Y. Akutsu and T. Abe, New Factorized S-Matrix and Its Application to Exactly Solvable q-State Model. II, Prog. Theoret. Phys., **70** (1983), 739–746.

E. Date
Department of Mathematics
College of General Education
Kyoto University
Kyoto 606, Japan

M. Jimbo, T. Miwa and M. Okado
Research Institute for Mathematical Sciences
Kyoto University
Kyoto 606, Japan

A. Kuniba
Institute of Physics
College of Arts and Sciences
University of Tokyo
Meguro-ku, Tokyo 153, Japan

Докл. Акад. Наук СССР
Том 287 (1986), № 5

Soviet Math. Dokl.
Vol. 33 (1986), No. 2

ON "QUANTUM" DEFORMATIONS OF IRREDUCIBLE FINITE-DIMENSIONAL REPRESENTATIONS OF $\mathfrak{gl}_N$

UDC 519.46+517.43

I. V. CHEREDNIK

Suppose that $\{I_\alpha\}$ is a base of a matrix algebra M_N of order N, I_0 is the identity matrix, ${}^1I_\alpha = I_\alpha \otimes I_0 \otimes \cdots \otimes I_0$, ${}^2I_\alpha = I_0 \otimes I_\alpha \otimes I_0 \otimes \cdots \otimes I_0$, etc. For $R = \sum_{\alpha,\beta} w_{\alpha\beta}(u) I_\alpha \otimes I_\beta$, where $\{w_{\alpha\beta}\}$ is a family of functions of a parameter $u \in \mathbf{C}$, set ${}^{ij}R = \sum_{\alpha,\beta} w_{\alpha\beta}(u_{ij}) {}^iI_\alpha {}^jI_\beta$, where $u_{ij} \stackrel{\text{def}}{=} u_i - u_j$, and the $u_i \in \mathbf{C}$ are parameters. Assume that R satisfies the equation

$$(1) \qquad {}^{12}R\,{}^{13}R\,{}^{23}R = {}^{23}R\,{}^{13}R\,{}^{12}R$$

for arbitrary $u_1, u_2, u_3 \in \mathbf{C}$. For fixed R, a *quantum analog* of a d-dimensional representation of $\mathfrak{gl}_N$ is defined to be a family of matrices $\{J_\alpha\} \subset M_d$ for which the Yang-Baxter-Faddeev relation (see [1] and [2])

$$(2) \qquad {}^{12}R\,{}^1L\,{}^2L = {}^2L\,{}^1L\,{}^{12}R,$$

holds, where ${}^iL = \sum_{\alpha,\beta} w_{\alpha\beta}(u_i) {}^iI_\alpha J_\beta$. The problem of describing such representations for a Baxter R ($N = 2$) was reformulated in [3] as a problem of constructing finite-dimensional representations of an associative algebra $\mathcal{A}$ generated by abstract A_β with defining relations following from (2) for arbitrary $u_{1,2,3}$ and for $L = \sum_{\alpha,\beta} w_{\alpha\beta}(u) I_\alpha A_\beta$. In [5] and the present paper $\mathcal{A}$ is a deformation of the universal enveloping algebra $U(\mathfrak{gl}_N)$ (in fact, of its graded version). In [4] representations of $\mathcal{A}$ for Baxter R are found which extend irreducible finite-dimensional representations of $\mathfrak{gl}_2$. In [5] and [6], by a different method, deformations of symmetric and exterior powers of the identity representation of $\mathfrak{gl}_N$ in the case of Belavin R-matrices are constructed. A number of results about representations of $\mathcal{A}$ for rational degeneracies (i.e. for Yang R-matrices) can be obtained from [7]. From the viewpoint of the concept of R-matrix quantification suggested by L. D. Faddeev and extended by V. G. Drinfel'd in his definition of "quantum" groups (see [8]), $\mathcal{A}$ is an algebra of a "quantum" group of currents $\mathrm{GL}_N(\mathbf{C}(u))$ "observed" on the simplest finite-dimensional submanifold (on a stratum).

In this note we suggest a method for constructing quantum representations of $\mathfrak{gl}_N$ (representations of $\mathcal{A}$) corresponding to any Young schemes for arbitrary R-matrices with rational $w_{\alpha\beta}(u)$ which have only one (modulo symmetries) pole of the first order for u. The principal attention is given to Belavin elliptic R-matrices. The structure of the corresponding $\mathcal{A}$ at a generic point is studied (cf. the hypotheses raised in [3]).

1. R-matrices. The group $\mathfrak{S}_n$ of permutations is generated by transpositions s_i, $1 \le i < n$, which transpose in any set of n elements indexed by $1, \dots, n$ the elements with indices i and $i+1$. Let s_0 be the identity of $\mathfrak{S}_n$. Assume that $\mathfrak{S}_n$ acts as automorphisms on some associative algebra $\mathcal{E}$ with identity 1. Elements of $\mathfrak{S}_n$ act by the formula $({}^wf)(u_1, \dots, u_n) = w(f(w^{-1}(u_1, \dots, u_n)))$ on functions f of $u_1, \dots, u_n \in \mathbf{C}$ defined in a neighborhood of zeros with values in $\mathcal{E}$.

1980 *Mathematics Subject Classification* (1985 *Revision*). Primary 20G05; Secondary 81E13.

Suppose that a function $R(u) \in \mathcal{E}$, $u \in \mathbf{C}$, satisfies the "locality" property ${}^{ij}R \stackrel{\text{def}}{=} w^{-1}(R(u_{12})) = w^{-1}(R(u_{ij}))$, does not depend on the choice of w: $(1, \ldots, n) \to (i, j, \ldots)$. Let the relations (1) hold. Then the function $R_{s_i} \stackrel{\text{def}}{=} {}^{i\,i+1}R, R_{s_0} = 1$, is uniquely extendable to a function $R_w(u_1, \ldots, u_n) \in \mathcal{E}$, $w \in \mathfrak{S}_n$, which satisfies the condition (of cocyclicity, see [9]) $R_{xy} = R_y\, {}^{y^{-1}}R_x$, where $l(xy) = l(x) + l(y)$, and $l(x)$ is the length of the reduced decomposition of $x \in \mathfrak{S}_n$ into $\{s_i\}$. Let ${}^{i\tilde{3}}R = {}^{i3}R\,{}^{i4}R \cdots {}^{in}R$, $i = 1, 2$. Then ${}^{12}R\,{}^{1\tilde{3}}R\,{}^{2\tilde{3}}R = {}^{2\tilde{3}}R\,{}^{1\tilde{3}}R\,{}^{12}R$. Let there exist $\eta \in \mathbf{C}$, $\eta \neq 0$, for which both parts of (1) vanish for $u_{13} = 0 = u_{12} \pm \eta$.

2. Young projections. Suppose that $m_i \in \mathbf{Z}_+$, $n - 2 = \sum_1^l m_k$, $m_1 \geq \cdots \geq m_l \geq 0$, and $\mu = (m_1, \ldots, m_l)$ is a Young scheme of l rows and m_1 columns. Filling the rows consecutively, we write a sequence $\tilde{3} = (3, \ldots, n)$ of numbers in μ. If $1 \leq i_k \leq l$ and $1 \leq j_k \leq m_{i_k}$ are the indices of a row and column of an entry k, then either $k > k' \Leftrightarrow i_k > i_{k'}$, or $i_k = i_{k'}$ and $j_k > j_{k'}$. From the array thus obtained write down all the numbers, column after column, in the reverse order: $\underset{\sim}{3} = (t_3, \ldots, t_n)$, where either $k > k' \Leftrightarrow j_{t_k} = j_{t_{k'}}, i_{t_k} < i_{t_{k'}}$, or $j_{t_k} < j_{t_{k'}}$. Let g denote the substitution g: $\tilde{3} \to \underset{\sim}{3}$, $g(1, 2) = 1, 2$. Let $u_k = (i_k - 1)v + (i_k - j_k)\eta$, where $v \in \mathbf{C}$ and $k \geq 3$.

THEOREM 1. *At $v = 0$ the function $R_g(v)$ has a zero of order no less than the number p of pairs $\{k, k'\}$, $3 \leq k' < k \leq n$, for which $i_k - j_k = i_{k'} - j_{k'}$.*

Let $v^{-p}R_g(v = 0) \stackrel{\text{def}}{=} R_\mu = \tilde{P}_\mu \tilde{F}_\mu = \underset{\sim}{F}_\mu \underset{\sim}{P}_\mu$ for projections $\tilde{P}_\mu$ and $\underset{\sim}{P}_\mu$ and for invertible $\tilde{F}_\mu, \underset{\sim}{F}_\mu \in \mathcal{E}$. At $v = 0$ set ${}^{i\underset{\sim}{3}}R = {}^{it_3}R\,{}^{it_4}R \cdots {}^{it_n}R$, $i = 1, 2$,

$$ {}^{i}\tilde{L} = {}^{i\tilde{3}}R\tilde{P}_\mu = \tilde{P}_\mu\, {}^{i\tilde{3}}R\tilde{P}_\mu, $$

$$ {}^{i}\underset{\sim}{L} = \underset{\sim}{P}_\mu {}^{i\underset{\sim}{3}}R = \underset{\sim}{P}_\mu\, {}^{i\underset{\sim}{3}}R\underset{\sim}{P}_\mu. $$

THEOREM 2. (1) ${}^{12}R\,{}^{1}L\,{}^{3}L = {}^{2}L\,{}^{1}L\,{}^{12}R$ *for* $L = \tilde{L}, \underset{\sim}{L}$.

(2) *At $u = q\eta$, $q \in \mathbf{Z}$ (the index i is omitted) L has a zero of order no less than the number of indices $n \geq k > 3$ with $i_k - j_k = q$ (cf. [5] and [6]).*

3. *Examples.* Let $\mathcal{E} = \mathbf{C}[\mathfrak{S}_n]$ be the group algebra of $\mathfrak{S}_n$, and let $R = (u/\eta)s_0 + s_1$ be the Yang matrix. Then $c_1 R_\mu$ is a Young projection $c_2 P_\mu Q_\mu$, where P_μ and Q_μ are, respectively, a row symmetrizer and a column antisymmetrizer of the table μ ($c_1, c_2 \neq 0$ are constants) and ${}^{w}(\cdot) = w(\cdot)w^{-1}$.

Baxter-Belavin R-matrices. Let $\theta_\alpha^0(u) = \theta_\alpha^0(u; t)$ be the theta-function corresponding to $\tau \in \mathbf{C}$ ($\operatorname{Im}\tau > 0$) with characteristics $\alpha_1/N + 1/2$ and $\alpha_2/N + 1/2$, where $\alpha = (\alpha_1, \alpha_2)$. Let

$$ w_\alpha(u) = \theta_\alpha^0(u + \eta/N)\theta_0^0(\eta/N)/\theta_\alpha^0(\eta/N)\theta_0^0(u + \eta/N), $$

where $\eta \in \mathbf{C}$ and $\alpha \in \mathbf{Z}_N^2$. Let $I_\alpha = g^{\alpha_1} h^{\sigma\alpha_2}$ for $\alpha \in \mathbf{Z}_N^2$, $\sigma \in \mathbf{Z}_N^*$, and matrices g and h with the properties $g^N = h^N = 1$ and $hg = \omega gh$, where $\omega \stackrel{\text{def}}{=} \exp(2\pi i/N)$. Then $R = \sum_{\alpha \in \mathbf{Z}_N^2} w_\alpha(u) I_\alpha \otimes I_\alpha^{-1}$ satisfies the conditions of §1 for $\mathcal{E} = M_N^{\otimes n}$ and $\mathfrak{S}_n$ acting by permutations of components of the product (see, for example, [6]). As above, using μ we construct $\tilde{L}$ choosing $\tilde{P}_\mu$ with the property ${}^{\tilde{3}}I_\alpha \tilde{P}_\mu\, {}^{\tilde{3}}I_\alpha^{-1} = \tilde{P}_\mu$, where ${}^{\tilde{3}}I_\alpha \stackrel{\text{def}}{=} {}^{3}I_\alpha \cdots {}^{n}I_\alpha$ and $\alpha \in \mathbf{Z}_N^2$. Let

$$ f_\mu(u) = \theta_0^0\left(u + \frac{(n-2)\eta}{N}\right) \prod_{k=4}^{n} \theta_0^0(u - u_k) \prod_{k=3}^{n} \theta_0^0(u - u_k + \eta/N)^{-1}. $$

Then

$$ \tilde{L}' \stackrel{\text{def}}{=} f_\mu^{-1}(u)\tilde{L}(u) = \sum_{\alpha \in \mathbf{Z}_N^2} w_\alpha\left(u + \frac{n-3}{N}\eta\right) I_\alpha \otimes \tilde{J}_\alpha, $$

where the $\tilde{J}_\alpha \in {}^{\bar 3}M_N = {}^{3}M_N \cdots {}^{n}M_n \cong M_N^{\otimes(n-2)}$ depend only on η and ${}^{\bar 3}I_\alpha \tilde{J}_\beta {}^{\bar 3}I_\alpha^{-1} = \omega^{\sigma\langle\alpha,\beta\rangle}\tilde{J}_\beta$, $\langle\alpha,\beta\rangle = \alpha_1\beta_2 - \alpha_2\beta_1$. The analogous result is true for $\underset{\sim}{L}'$ and $\underset{\sim}{J}_\alpha$ which are determined by $\underset{\sim}{L}$. Note that $R_\mu \to cP_\mu Q_\mu$ as $\eta \to 0$, where elements of $\mathfrak{S}_n$ are identified with corresponding permutation matrices and $c \neq 0$ (see [6]). Relations (2) hold for $\tilde{L}'$ and $\underset{\sim}{L}'$.

4. Deformation of $U(\mathfrak{gl}_N)$. For a Baxter-Belavin R-matrix define an algebra $\mathcal{A}_\eta$ as the quotient of a tensor algebra $\mathcal{T}$ of the vector space $\bigoplus_\alpha \mathbf{C}A_\alpha$ by the ideal $\mathcal{K}_\eta$ of all relations on A_α which follow from (2) and depend only on η for $L = \sum_{\alpha\in\mathbf{Z}_N^2} w_\alpha(u) I_\alpha A_\alpha$ (see [3] and [5]). The ideal $\mathcal{K}_\eta$ is generaeted by defining relations whose coefficients are rational functions of η over the elliptic curve $E = \mathbf{C}/N\mathbf{Z}+N\mathbf{Z}\tau$, and it is transformed into itself by actions of $\alpha \in \mathbf{Z}_N^2$ on $\{A_\beta\}$ defined by the formula $A_\beta \to w^{\sigma\langle\alpha,\beta\rangle}A_\beta$. Therefore, the quotient of the trivial bundle $\mathcal{T}_E$ over E with fiber $\mathcal{T}$ modulo the subbundle $\mathcal{K}$ generated by $\mathcal{K}_n$ at a generic point is a fibration of algebras $\mathcal{A}$ over E with an induced action of the group $\mathbf{Z}_N^2$. Setting $\deg A_\alpha = 1$ and $\deg 1 = 0$ we can $\mathbf{Z}_+$-grade $\mathcal{A}$, where the A_α are considered as global sections of $\mathcal{A}$.

The equality (2) at $\eta = 0$ leads to the relations (see, for example, [6])

$$(3) \qquad [A_\alpha, A_\beta] = (\omega^{-\sigma\alpha_1\beta_2} - \omega^{-\sigma\alpha_2\beta_1})A_{\alpha+\beta}A_0, \qquad [A_0, A_\beta] = 0.$$

Let $\tilde{A}$ denote the quotient of $\mathcal{T}_E$ by $\mathcal{K}$ at $\eta \neq 0$ and by (3) in a neighborhood of $\eta = 0$. The fiber $\tilde{\mathcal{A}}$ at $\eta = 0$ is determined by (3) and, after substitution of $A_0 = 1$, turns into $U(\mathfrak{gl}_N)$. The mapping $A_\alpha \to \tilde{J}_\alpha$ (or $A_\alpha \to \underset{\sim}{J}_\alpha$, as explained above) can be extended to a homomorphism of the generic fiber $\mathcal{A}$ into the algebra of ${}^{3}M_N$-valued rational functions on E which agrees with the action of $\alpha \in \mathbf{Z}_N^2$ on $\mathcal{A}$ and the action of ${}^{\bar 3}I_\alpha$ on ${}^{\bar 3}M_N$ by inner automorphisms.

THEOERM 3. *$\tilde{\mathcal{A}} = \mathcal{A}$, and the dimension of the subbundle $\mathcal{A}$ with $\deg = i$ equals the dimension of the space of homogeneous polynomials in N^2 variables of degree $i \in \mathbf{Z}_+$.*

For a proof it suffices, for every $i \in \mathbf{Z}_+$, to find a scheme μ for which the corresponding homomorphism $U(\mathfrak{gl}_N) \to M_d$ is injective in the component of degree at most i and use results of §3.

5. Maximal commutative subalgebra of $\mathcal{A}$. For $r \in \mathbf{Z}_+$, $r > 1$, let $\mathcal{Q}_1$ denote the "antisymmetrizer" of $\tilde{P}_\mu$ for a column of μ ($m_1 = 1$) constructed for the indices $(1,\dots,r)$ (instead of $(3,\dots,n)$). Then $\dim_{\mathbf{C}} \operatorname{Im}\mathcal{Q}_1 = \dim_{\mathbf{C}}(\Lambda^r\mathbf{C}^N)$ at a generic point η (cf. [5]). Let $L(u) = \sum_{\alpha\in\mathbf{Z}_N^2} w_\alpha I_\alpha A_\alpha$, where the A_α are sections of A and ${}^{\hat 1}L = \mathcal{Q}_1 {}^{r}L(u_1 + (r-1)\eta) \cdots {}^{1}L(u_1)\mathcal{Q}_1$. For uniformity, let $\mathcal{Q}_1 = I_0$ and ${}^{\hat 1}L = {}^{1}L(u_1)$ for $r = 1$. Analogously, for u_2 and the indices $(r+1,\dots,r+s)$, $s \geq 1$, define $\mathcal{Q}_2$ and ${}^{\hat 2}L$. Set

$$\begin{aligned} {}^{\hat 1\hat 2}R &= \mathcal{Q}_1\mathcal{Q}_2 \prod_{j=r}^{1}\left(\prod_{j=1}^{s} {}^{j(i+r)}R(u_{12} + (j-i)\eta)\right)\mathcal{Q}_1\mathcal{Q}_2 \\ &= \mathcal{Q}_1\mathcal{Q}_2({}^{r\,r+1}R(u_{12} + (r-1)\eta)\cdots {}^{1\,r+s}R(u_{12} + (1-s)\eta))\mathcal{Q}_1\mathcal{Q}_2. \end{aligned}$$

Then ${}^{\hat 1\hat 2}R\,{}^{\hat 1}L\,{}^{\hat 2}L = {}^{\hat 2}L\,{}^{\hat 1}L\,{}^{\hat 1\hat 2}R$ (cf. [5] and [6]), and $[H_r(u_1), H_s(u_2)] = 0$, where $H_r \overset{\text{def}}{=} \operatorname{tr} {}^{\hat 1}L$, tr is the matrix trace, and $1 \leq r$, $s \leq N$. Let H_r^k denote the coefficient of $H_r(u)$ at $(u + k\eta)^{-1}$ for $1 \leq r \leq N$ and $0 \leq k \leq r-1$.

From Chapter 4 of [6] we obtain

THEOREM 4. *At a generic point η the elements H_r^k generate the maximal commutative subalgebra in $\mathcal{A}_\eta$ and are algebraically independent, while the H_N^k generate the center of $\mathcal{A}_\eta$.*

Elements H_N^k belong to the center of $\mathcal{A}_\eta$, since ${}^{1\hat{2}}R = c(\eta)^{r+1} I_0 \mathcal{Q}_1 \varphi(u_{12})$ for $r = N$ and $s = 1$, where

$$\varphi(u) = \theta_0^0(u + N\eta) \prod_{k=1}^{N-1} \theta_0^0(u + (k-1)\eta) \left(\prod_{k=1}^{N} \theta_0^0(u + (k-1)\eta + \eta/N) \right)^{-1},$$

and c is a scalar function of η (cf. [3]).

THEOREM 5. *For $A_\alpha = \tilde{J}_\alpha$ constructed for a scheme μ (see* §3*),*

$$H_N\left(u + \frac{n-3}{N}\eta\right) = c^{n-2}(\eta)\psi(u)\tilde{P}_\mu,$$

where

$$\psi(u) = \prod_{i=1}^{N} \left(\theta_0^0(u + (m_i + N - i)\eta) / \theta_0^0\left(u + \frac{n-2}{N}\eta + (N-i)\eta\right)\right).$$

For degenerating $\eta \to 0$, $u \to 0$, $u = \lambda\eta$, and $c(\eta) \to 1$ one obtains a known formula for the action of Casimir elements on irreducible finite-dimensional representations.

REMARKS. Analogs of H_r can be defined for arbitrary schemes μ (they commute with H_r^k and can be expressed in their terms). The constructions in §§1 and 2 are compatible with the embedding of Young schemes, one into the other; they enable one to construct deformations of not necessarily irreducible representations of $\mathfrak{gl}_N$. The results of §§3, 4, and 5 can be transferred to degeneracies of Belavin R-matrices and may be extended to L which have a pole of arbitrary order in u. In a particular case of a Yang R-matrix (see above) irreducible representations of an algebra $\mathcal{A}_\eta$ constructed for L with a pole of indeterminate order correspond to representations of the Yangian (for $\mathfrak{gl}_N$) introduced by Drinfel'd.

The author thanks I. M. Gel'fand for attention to this work, and V. G. Drinfel'd for a useful discussion.

Interfaculty Problem Scientific-Research Laboratory
of Molecular Biology and Bioorganic Chemistry
Moscow State University

Received 2/APR/85

BIBLIOGRAPHY

1. L. A. Takhtadzhyan and L. D. Faddeev, Uspekhi Mat. Nauk **34** (1979), no. 5 (209), 13–63; English transl. in Russian Math. Surveys **34** (1979).

2. P. P. Kulish and E. K. Sklyanin, Zap. Nauchn. Sem. Leningrad. Otdel. Mat. Inst. Steklov. (LOMI) **95** (1980), 129–160; English transl. in J. Soviet Math. **19** (1982), no. 5.

3. E. K. Sklyanin, Funktsional. Anal. i Prilozhen. **16** (1982), no. 4, 27–34; English transl. in Functional Anal. Appl. **16** (1982).

4. ——, Funktsional. Anal. i Prilozhen. **17** (1983), no. 4, 34–48; English transl. in Functional Anal. Appl. **17** (1983).

5. I. V. Cherednik, Funktsional. Anal. i Prilozhen. **19** (1985), no. 1, 81–90; English transl. in Functional Anal. Appl. **19** (1985).

6. ——, Itogi Nauki: Algebra, Topologiya, Geometriya, vol. 22, VINITI, Moscow, 1984, pp. 205–265; English transl., to appear in J. Soviet Math.

7. P. P. Kulish and E. K. Sklyanin, Integrable Quantum Field Theories (Proc. Sympos., Tvärminne, Finland, 1981), Lecture Notes in Phys., vol. 151, Springer-Verlag, 1982, pp. 61–119.

8. V. G. Drinfel'd, Dokl. Akad. Nauk SSSR **273** (1983), 531–535; English transl. in Soviet Math. Dokl. **28** (1983).

9. I. V. Cherednik, Teoret. i Mat. Fiz. **61** (1984), 35–44; English transl. in Theoret. Math. Phys. **61** (1984).

Translated by B. M. SCHEIN

Volume 189, number 1,2 PHYSICS LETTERS B

TOWARDS THE CLASSIFICATION OF COMPLETELY INTEGRABLE QUANTUM FIELD THEORIES (THE BETHE-ANSATZ ASSOCIATED WITH DYNKIN DIAGRAMS AND THEIR AUTOMORPHISMS)

N. Yu. RESHETIKHIN

Leningrad Branch of the Steklov Mathematical Institute, Fontanka 27, Leningrad, USSR

and

P.B. WIEGMANN [1]

Institut Laue–Langevin, 156X, F-38042 Grenoble Cedex, France
and Laboratoire de Physique Théorique de L'Ecole Normale Supérieure [2], 24, rue Lhomond, F-75231 Paris Cedex 05, France

Received 13 February 1987

The eigenvalues of the unitarizable and factorized S-matrices and their transfer matrices are presented under the form of the Bethe-ansatz equations expressed in terms of the simple root system and of the automorphisms of the corresponding simple Lie algebra.

1. The basic concept of the theory of quantum integrability is the factorized two-particle S-matrix. Each of such unitarizable matrices creates a class of integrable (1 + 1)-quantum field theories. The hidden symmetry of the integrable models is coded by the consistency conditions of the factorization of scattering [1]:

$$S_{12}(\theta)S_{13}(\theta+\theta')S_{23}(\theta') = S_{23}(\theta')S_{13}(\theta+\theta')S_{12}(\theta). \tag{1}$$

Here $S_{12}(\theta_1-\theta_2) \equiv S_{kl}{}^{ij}(\theta)$ is the S-matrix of the two particles with rapidities θ_1, θ_2 which belong to the isotopic multiplets $V_1(i,\ k=1,\dots,n_1)$, and $V_2(j,\ l=1,\dots,n_2)$.

One can consider the condition (1) as an equation (known as the Yang–Baxter or triangle equation). The general solution of it would give the classification of quantum integrable systems.

The two-particle S-matrix is related to a many-body problem by the transfer matrix (or the monodromy matrix) [2–4].

$$\hat{T} \equiv {}_j^{\,i}T^{(i_1\dots i_{\mathcal{N}})}_{(j_1\dots j_{\mathcal{N}})}(\theta\,|\,\theta_1\dots\theta_{\mathcal{N}}) = S^{i i_1}_{j_1 k_1}(\theta-\theta_1)\dots S^{K_{\mathcal{N}-1} i_{\mathcal{N}}}_{j_{\mathcal{N}} j}(\theta-\theta_{\mathcal{N}}). \tag{2}$$

The diagonalization of the transfer matrix enables one to solve a relevant class of the quantum field theories.

Up to now a rich collection of factorized S-matrices has been found (see for example refs. [5–13]). Each of them was found by methods which are highly specific for each concrete S-matrix. The eigenvalues of the

[1] On leave from the Landau Institute for Theoretical Physics, Kosygina 2, 117 334 Moscow, USSR.
[2] Laboratoire Propre du Centre National de la Recherche Scientifique, associé à l'Ecole Normale Supérieure et à l'Université de Paris-Sud.

Volume 189, number 1,2 PHYSICS LETTERS B 30 April 1987

transfer matrices are computed specifically for each known S-matrix by different approaches (i.e. traditional coordinate [3,13], algebraic [14,7,15] and analytic [16] Bethe-ansatze) and are parametrized by the solutions of the so-called Bethe-ansatz equations.

Despite the diversity of the S-matrices their Bethe-ansatz equations look very universal.

Recently in refs. [8,9], the general expressions for the eigenvalues of the S-matrices as well as for the transfer matrices, invariant with respect to any simple Lie group, were presented in terms of the Bethe-ansatz equations which depend only on the highest weight of the representation and on the simple roots of the Lie algebra. These S-matrices are known as "rational".

In this letter we give, without proof, the generalization of this result for the trigonometrical and elliptic S-matrices, thus presenting the eigenvalues of almost all factorized S-matrices [‡1].

2. The factorized S-matrices are classified by the simple Lie algebras, their irreducible representations, and automorphisms [18] [‡2]. They naturally divide into three classes:

(i) Rational S-matrices. They are rational functions of θ. The rational S-matrices are invariant with respect to Lie group G.

$$\rho_a(g)\rho_b(g)S_{ab}(\theta)\rho_a(g)\rho_b(g^{-1})=S_{ab}(\theta),\quad g\in G, \tag{3}$$

where $\rho_a(g)$ are the representations of the Lie group acting in the space of a-particle states V_a. Under the proper normalization

$$S_{ab}(\theta)=\mathbb{1}+I_\mu^{(a)}\otimes I_\mu^{(b)}/\theta+O(1/\theta^2),\quad \text{at } \theta\to\infty, \tag{4}$$

where the I are the generators of the Lie algebra.

The irreducible multiplet of the factorized particle, i.e. the space V_a is the irreducible space of the "quantum group". It is unambiguously constructed starting from the irreducible representation of the Lie group v, however, V_a being considered as a representation of the Lie group generally speaking is reducible [19–21,8,9]. (See refs. [8,20] for the list of their irreducible moduli.)

The second class is the one-parameter family $S(\theta;\ \gamma)$ formed by:

(ii) Trigonometrical S-matrices. Their matrix elements are constructed starting with the irreducible representations of the Lie algebra and the automorphism of the Dynkin diagram σ [18]. The matrix possesses automorphicity and periodicity

$$S(\theta+i\pi/\gamma)=(U_a\otimes\mathbb{1}_b)S_{ab}(\theta)(U_a^{-1}\otimes\mathbb{1}_b),\quad S(\theta+g\cdot\pi i/\gamma)=S_{ab}(\theta), \tag{5}$$

where U is the matrix connected with the Coxeter automorphism $U^g=1$ (g is the order of the Coxeter automorphism) [18,10]. The rational S-matrix can be considered as the limit $\gamma\to 0$ of the trigonometrical ones.

Finally there is the two-parameter family $S(\theta;\ \gamma;\ \mu)$ formed by:

(iii) Elliptical S-matrices. They are connected only with $G=GL(N)$. The automorphicity has the form

$$S_{ab}(\theta+i\pi/\gamma+i\pi/\mu)=U_aS_{ab}(\theta+i\pi/\mu)U_a^{-1}=V_bS_{ab}(\theta+i\pi/\gamma)V_b^{-1},$$
$$S_{ab}(\theta+NK+NK')=S(\theta), \tag{6}$$

‡1 Actually, we consider unitarizable S-matrices having the "quasiclassical" property: $S(\theta,\eta)\to\mathbb{1}+0(\eta)$ where η is a parameter. Some strange and badly investigated S-matrices [17] possess this property at first sight. However there are some reasons to think that they are some special points of the quasiclassical family $S(\theta,\eta)$.

‡2 In the framework recently proposed by Drienfeld [19] the S-matrix is the representation of the so-called "quantum group" G which can be considered as a generalization of the Lie algebra. The quantum group is the noncommutative and noncocommutative Hopf algebra unambiguously determined by the Lie algebra and its automorphism and coincides with the Lie algebra in the quasiclassical limit $\eta\to 0$.

Volume 189, number 1,2 PHYSICS LETTERS B 30 April 1987

Table 1

Dynkin diagrams	Automorphism	Restricted Dynkin diagrams
$A^{(2)}_{2n}$: 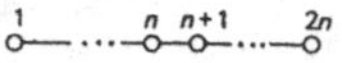 ;	$\sigma\alpha_i = \alpha_{2n-i}$ $i = 1,\dots,n \equiv \bar{r}$	$\bar{\Delta}(A^{(2)}_{2n}) = B_n$; 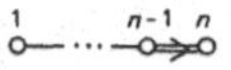
$A^{(2)}_{2n-1}$: 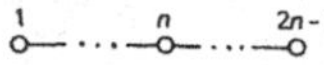 ;	$\sigma\alpha_i = \alpha_{2n-1-i}$ $\sigma\alpha_{n-1} = \alpha_{n-1}$ $\bar{i} = 1,\dots,n$	$\bar{\Delta}(A^{(2)}_{2n-1}) = C_n$; 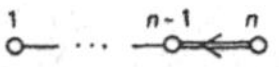
$D^{(2)}_n$: 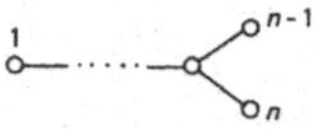 ;	$\sigma\alpha_i = \alpha_i$ $i = 1,\dots,n-2$ $\sigma\alpha_{n-1} = \alpha_n$ $\bar{r} = n-1$	$\bar{\Delta}(D^{(2)}_n) = B_{n-1}$; 
$E^{(2)}_6$: 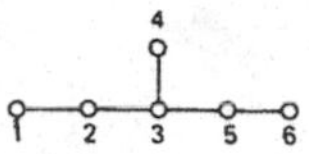 ;	$\sigma\alpha_1 = \alpha_6$ $\sigma\alpha_2 = \alpha_5$ $\sigma\alpha_3 = \alpha_3$ $\sigma\alpha_4 = \alpha_4$. $\bar{r} = 4$	$\bar{\Delta}(E^{(2)}_6) = F_4$; 
$D^{(3)}_4$: 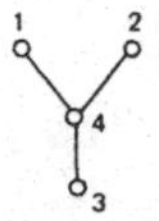 ;	$\sigma\alpha_1 = \alpha_2$ $\sigma\alpha_2 = \alpha_3$ $\sigma\alpha_3 = \alpha_1$ $\sigma\alpha_4 = \alpha_4$ $\bar{r} = 2$	$\bar{\Delta}(D^{(3)}_4) = G_2$; 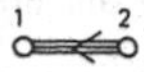

where U and V are Coxeter automorphisms of the $GL(N)\cdot(V^N = U^N = 1)$. The matrix elements of such an S-matrix are elliptic functions with periods K and K'.

Let us stress that the trigonometric and elliptic S-matrices do not possess Lie group invariance (3). It is broken up to $U(1)\dots U(1)$ for the trigonometric case and up to Z_N for the elliptic one. However, the matrices act in the same space as the one where their rational limits act and therefore they are characterized by the highest weight of the corresponding representation.

3. Before presenting the main expressions, let us introduce the following notation.

Let $\Delta = \{\alpha_i\}$ and $\{\omega_i\}$ $(i = 1,\dots,\text{rank G})$ be the simple roots and highest weights of the fundamental representations of the simple Lie algebra G: $(\omega_i, \alpha_j)/(\alpha_i, \alpha_j) = 2\delta_{ij}$. Let σ be the automorphism of the Dynkin diagram, and t is the order of the automorphism $\sigma^t = 1$. Let $\bar{\Delta} = \{\bar{\alpha}_i\}$ is a so-called restricted root system which is invariant under the automorphism

$$\bar{\alpha}_i = \frac{1}{t}\sum_{s=0}^{t-1} \sigma^s \alpha_i, \quad \bar{i} = 1,\dots,\bar{r}. \tag{7}$$

There are four simple Lie algebras which possess second-order automorphisms and only one which possesses the third-order automorphism, which is shown in table 1.

Proposition. The eigenvalues t of the trigonometrical transfer matrix $\hat{T}$ acting on the space

$$V\{\Omega_0\}\otimes\prod_{a=1}^{\mathscr{N}}V\{\Omega_a\}$$

of the Lie algebra G are parametrized by any set of integers m_i fulfilling the conditions

$$\sum_{a=0,\ldots,\mathscr{N}}(\Omega_a,\alpha_j)-\sum_{i=1}^{r}m_i(\alpha_i,\alpha_j)\geqslant 0,\quad i=1,\ldots,\text{rank G}=r, \tag{8}$$

and by a set

$$\{\lambda_\alpha^{(i)}\},\quad \alpha=1,\ldots,m_i,\qquad |\text{Im}\,\gamma\lambda|\leqslant\pi/2t \tag{9}$$

of the solutions of the Bethe-ansatz equations

$$t(\theta|\{\theta_a\}|\{m_i\})=\prod_{s=0}^{t-1}\prod_{i=1}^{r}\prod_{\alpha=1}^{m_i}\frac{\sinh(-\theta+\lambda_\alpha^{(i)}+i\pi s/\gamma t+2\pi i(\Omega_0,\sigma^s\alpha_i))}{\sinh(-\theta+\lambda_\alpha^{(i)}-i\pi s/\gamma t-2\pi i(\Omega_0,\sigma^s\alpha_i))}. \tag{10}$$

The Bethe-ansatz equations are

$$\prod_{a=0}^{\mathscr{N}}\prod_{s=0}^{t}\frac{\sinh(\lambda_\alpha^{(i)}-\theta_a+2\pi i(\Omega_a,\sigma^s\alpha_i)+i\pi s/t\gamma)}{\sinh(\lambda_\alpha^{(i)}-\theta_a-2\pi i(\Omega_a;\sigma^s\alpha_i)-i\pi s/\gamma t)}$$

$$=\prod_{s=0}^{t-1}\prod_{j=1}^{r}\prod_{\beta=1}^{m_j}\frac{\sinh(\lambda_\alpha^{(i)}-\lambda_\beta^{(j)}+2\pi i(\alpha_i,\sigma^s\alpha_j)+i\pi s/t\gamma)}{\sinh(\lambda_\alpha^{(i)}-\lambda_\beta^{(j)}-2\pi i(\alpha_i,\sigma^s\alpha_j)-i\pi s/\gamma t)}. \tag{11}$$

5. Let us consider the limit $\gamma\to 0$. At finite γ the eigenstates are nondegenerate. To make the limit $\gamma\to 0$ clear, let us extract the subset or λ's such as $|\text{Im}\,\gamma\lambda|=\pi/2t$. In the limit, these λ's come out from the game giving the degeneracy of the eigenstates. The solutions with $|\text{Im}\,\gamma\lambda|<\pi/2t$ have a regular limit. The number of all these solutions is denoted by m_i'. They form a space of a representation of the Lie algebra with the highest weight

$$\Omega=\sum_{a=0}^{\mathscr{N}}\Omega_a-\sum_{i=1}^{r}m_i'\alpha_i. \tag{12}$$

Thus to go to the limit $\gamma\to 0$, one should replace $\sinh\gamma x$ by x and m_i by m_i' and one is lead to the expression of ref. [8]. The limit does not depend on the automorphism.

6. There is another form of eqs. (10), (11) which reflects the restricted Dynkin diagram which operates with a set $\{\Lambda_1^{(i)}\ldots\Lambda_{M_i}^{(i)}\}$ combining the following sets of λ's:

$$\bigcup_{s=0}^{t-1}\left\{\lambda_1^{\sigma^s i}+i\pi s/t,\ldots,\lambda_{m_{\sigma^s i}}^{\sigma^s i}+i\pi s/2t\right\},$$

where $M_i=\sum_{s=0}^{t-1}m_{\sigma^s i}$. The numbers Λ are considered in the whole interval $(-\pi/2\gamma,\pi/2\gamma)$.

Volume 189, number 1,2 PHYSICS LETTERS B 30 April 1987

Then eqs. (10), (11) for the second-order automorphism can be rewritten in the form

$(t=2)$

$$t(\theta\,|\{\theta_a\}\,|\{m_i\}) = \prod_{i=1}^{\bar{r}} \prod_{\alpha=1}^{M_i} \frac{\sinh\gamma\big(2\theta - 2\Lambda_\alpha^{(i)} + 2\pi \mathrm{i}(\bar{\Omega}_0,\, \bar{\alpha}_i)\big) - \sinh 2\pi \mathrm{i}\gamma(\bar{\Omega}_0^* \alpha_i^*)}{\sinh\gamma\big(2\theta - 2\Lambda_\alpha^{(i)} - 2\pi \mathrm{i}(\bar{\Omega}_0,\, \bar{\alpha}_i)\big) + \sinh 2\pi \mathrm{i}\gamma(\bar{\Omega}_0^*,\, \alpha_i^*)}, \tag{13}$$

$$\prod_{a=0}^{\mathcal{N}} \frac{\sinh\gamma\big(2\Lambda_\alpha^{(i)} - 2\theta_a + 2\pi \mathrm{i}(\bar{\Omega}_a,\, \bar{\alpha}_i)\big) - \sinh 2\pi \mathrm{i}\gamma(\bar{\Omega}_a^*,\, \alpha_i^*)}{\sinh\gamma\big(2\Lambda_\alpha^{(i)} - 2\theta_a - 2\pi \mathrm{i}(\bar{\Omega}_a,\, \bar{\alpha}_i)\big) + \sinh 2\pi \mathrm{i}\gamma(\bar{\Omega}_a^*,\, \alpha_i^*)}$$

$$= \prod_{i=1}^{\bar{r}} \prod_{\beta=1}^{M_i} \frac{\sinh\gamma\big(2\Lambda^i_\alpha - 2\Lambda^j_\beta + 2\pi \mathrm{i}(\bar{\alpha}_i \bar{\alpha}_j)\big) - \sinh 2\pi \mathrm{i}\gamma(\bar{\alpha}_i^*,\, \bar{\alpha}_j^*)}{\sinh\gamma\big(2\Lambda^i_\alpha - 2\Lambda^j_\beta - 2\pi \mathrm{i}(\bar{\alpha}_i \bar{\alpha}_j)\big) + \sinh 2\pi \mathrm{i}\gamma(\bar{\alpha}_i^*,\, \bar{\alpha}_j^*)}, \tag{14}$$

where we introduced the restricted highest weights

$\bar{\Omega}_i = \frac{1}{2}(1+\sigma)\Omega_i$,

and additional systems

$\bar{\alpha}_i^* = \frac{1}{2}(1-\sigma)\alpha_i$, $\bar{\Omega}_i^* = \frac{1}{2}(1-\sigma)\Omega_i$.

Finally we present the explicit equations for $D_4^{(3)}$ possessing the third-order automorphism. These equations correspond to the G_2 Dynkin diagram.

$(t=3)$

$$t(\theta\,|\{\theta_a\}\,|\{m_i\}) = \prod_{\alpha=1}^{M_1} \frac{\sinh\gamma\big(-\theta + \Lambda_\alpha^{(1)} + 2\pi \mathrm{i}(\Omega_0,\, \bar{\alpha}_1)\big)}{\sin\gamma\big(-\theta + \Lambda_\alpha^{(1)} - 2\pi \mathrm{i}(\Omega_0,\, \bar{\alpha}_1)\big)} \prod_{\alpha=1}^{M_2} \frac{\sinh 3\gamma\big(-\theta + \Lambda_\alpha^{(2)} + 2\pi \mathrm{i}(\Omega_0,\, \bar{\alpha}_2)\big)}{\sinh 3\gamma\big(-\theta + \Lambda_\alpha^{(2)} - 2\Pi \mathrm{i}(\Omega_0,\, \bar{\alpha}_2)\big)},$$

$$\prod_{a=0}^{\mathcal{N}} \frac{\sinh\gamma\big(\Lambda_\alpha^{(1)} - \theta_a + 2\pi \mathrm{i}(\Omega_a,\, \bar{\alpha}_1)\big)}{\sinh\gamma\big(\Lambda_\alpha^{(1)} - \theta_a - 2\pi \mathrm{i}(\Omega_a,\, \bar{\alpha}_1)\big)}$$

$$= \prod_{\beta=1}^{M_1} \frac{\sinh\gamma\big(\Lambda_\alpha^{(1)} - \Lambda_\beta^{(1)} + 2\pi \mathrm{i}(\bar{\alpha}_1,\, \bar{\alpha}_1)\big)}{\sin\gamma\big(\Lambda_\alpha^{(1)} - \Lambda_\beta^{(1)} - 2\pi \mathrm{i}(\bar{\alpha}_1,\, \bar{\alpha}_1)\big)} \prod_{\beta=1}^{M_2} \frac{\sinh 3\gamma\big(\Lambda_\alpha^{(1)} - \Lambda_\beta^{(2)} + 2\pi \mathrm{i}(\bar{\alpha}_1,\, \bar{\alpha}_2)\big)}{\sinh 3\gamma\big(\Lambda_\alpha^{(1)} - \Lambda_\beta^{(2)} - 2\pi \mathrm{i}(\bar{\alpha}_1,\, \bar{\alpha}_2)\big)},$$

$$\prod_{a=0}^{\mathcal{N}} \frac{\sinh\gamma\big(\Lambda_\alpha^{(2)} - \theta_a + 2\pi \mathrm{i}(\Omega_a,\, \alpha_4)\big)}{\sinh\gamma\big(\lambda_\alpha^{(2)} - \theta_a - 2\pi \mathrm{i}(\Omega_a,\, \alpha_4)\big)}$$

$$= \prod_{\beta=1}^{M_2} \frac{\sinh 3\gamma\big(\Lambda_\alpha^{(2)} - \Lambda_\beta^{(2)} + 2\pi \mathrm{i}(\bar{\alpha}_2,\, \bar{\alpha}_2)\big)}{\sinh 3\gamma\big(\Lambda_\alpha^{(2)} - \Lambda_\beta^{(2)} - 2\pi \mathrm{i}(\bar{\alpha}_2,\, \bar{\alpha}_2)\big)} \prod_{\beta=1}^{M_1} \frac{\sinh 3\gamma\big(\Lambda_\alpha^{(2)} - \Lambda_\beta^{(1)} + 2\pi \mathrm{i}(\bar{\alpha}_2,\, \bar{\alpha}_1)\big)}{\sinh 3\gamma\big(\Lambda_\alpha^{(2)} - \Lambda_\beta^{(1)} - 2\pi \mathrm{i}(\bar{\alpha}_2,\, \bar{\alpha}_1)\big)}.$$

6. Eqs. (10), (11) can be interpreted as an averaging of the rational S-matrix over all automorphisms, which goes back to ref. [22]. In fact they can be rewritten in the form

$$t(\theta\,|\{\theta_i\}\,|\{m_i\}) = \prod_{n=-\infty}^{+\infty} \prod_{i=1}^{r} \prod_{\alpha=1}^{m_i} \frac{-\theta + \lambda_\alpha^{(i)} + \mathrm{i}n\pi/t\gamma + 2\pi \mathrm{i}(\Omega_0,\, \sigma^n\alpha_i)}{-\theta + \lambda_\alpha^{(i)} - \mathrm{i}n\pi/t\gamma - 2\pi \mathrm{i}(\Omega_0,\, \sigma^n\alpha_i)}, \tag{15}$$

$$\prod_{n=-\infty}^{+\infty} \prod_{a=0}^{\mathcal{N}} \frac{\lambda_\alpha^{(i)} - \theta_a + 2\pi \mathrm{i}(\Omega_a,\, \sigma^n\alpha_i) + \mathrm{i}m\pi/\gamma t}{\lambda_\alpha^{(i)} - \theta_a - 2\pi \mathrm{i}(\Omega_0,\, \sigma^n\alpha_i) - \mathrm{i}m\pi/\gamma t} = \prod_{n=-\infty}^{+\infty} \prod_{j=1}^{r} \prod_{\beta=1}^{m_j} \frac{\lambda_\alpha^{(i)} - \lambda_\beta^{(j)} + \mathrm{i}n\pi/\gamma t + 2\pi \mathrm{i}(\alpha_i,\, \sigma^n\alpha_j)}{\lambda_\alpha^{(i)} - \lambda_\beta^{(j)} - \mathrm{i}n\pi/\gamma t - 2\pi \mathrm{i}(\alpha_i,\, \sigma^n\alpha_j)}. \tag{16}$$

7. The elliptic transfer matrix which exists only for the series $A^{(1)}_{N-1}$ [18] is just an averaging of the rational matrices over the second automorphism on the physical sheet:

$$t(\theta|\{\theta_i\}) = \prod_{l=-\infty}^{+\infty} \prod_{n=-\infty}^{+\infty} \prod_{i=1}^{N-1} \prod_{\alpha} \frac{\theta - \lambda^{(i)}_\alpha + 2\pi i(\Omega_0, \alpha_i) + iKn + iK'l}{\theta - \lambda^{(i)}_\alpha - 2\pi i(\Omega_0, \alpha_i) + Kn + iK'l}, \tag{17}$$

$$\prod_{n,l=-\infty}^{+\infty} \prod_{a=0}^{\mathcal{N}} \frac{\lambda^{(i)}_\alpha - \theta_a + 2\pi i(\Omega_a, \alpha_i) + iKn + iK'l}{\lambda^{(i)}_\alpha - \theta_a - 2\pi i(\Omega_a, \alpha_i) + iKn + iK'l}$$

$$= \prod_{n,l=-\infty}^{+\infty} \prod_{j=1}^{N-1} \prod_{\beta} \frac{\lambda^{(i)}_\alpha - \lambda^{(j)}_\beta + 2\pi i(\alpha_i \alpha_j) + iKn + iK'l}{\lambda^{(i)}_\alpha - \lambda^{(j)}_\beta - 2\pi i(\alpha_i \alpha_j) + iKn + iK'l}, \tag{18}$$

where

$$(\alpha_i, \alpha_j) = (1/\mathcal{N})\left[\delta_{ij} - \tfrac{1}{2}(\delta_{i,j+1} + \delta_{j,i+1})\right].$$

8. Setting $\mathcal{N} = 1$ in eqs. (10)–(18) one gets the eigenvalues of the factorized S-matrix. The unitarity and CP-symmetry immediately follows from (10)–(18).

$$S_{ab}(\theta) = S^{-1}_{ba}(-\theta) = S^{*}_{ba}(-\theta). \tag{19}$$

These equations reproduce, unify and generalize all the results of the factorized theory of scattering known up to now. At $S_a = 1$ $(\Omega_a = \omega_a)$ the S-matrices should be crossing-invariant. This means

$$S^{t_b}_{ab}(i\pi - \theta) = f_{ab}(\theta) S_{ab}(\theta), \tag{20}$$

where the operation t_b means transposition and transfer to the contragradient representation in the space of the second particle, say b, and where $f_{ab}(\theta)$ is a scalar factor [‡3]. We hope that one can derive this property directly from eqs. (10)–(18). Such an S-matrix can be associated with a theory of relativistic particles. If the representation and its contragradient representation are conjugated by the automorphism (i.e. $\sigma\alpha_a = \alpha_{\bar{a}}$) there are some additional relations

$$S_{ab}(\theta) = p_{ab}(\theta) S_{a\bar{b}}(\theta + i\pi/t\gamma),$$

where $p_{ab}(\theta)$ is also a scalar factor.

9. Do the S-matrices presented conclude all possible unitarizable factorized S-matrices? We think they do. At least they include all quasiclassical S-matrices and therefore all physically interesting solutions. We also believe that the equations presented lead one into the scope of new mathematics and point to the existence of an alternative view and new results in the classical theory of representations.

Finally we should note that we have omitted all proofs not so much out of brevity but because of the absence of direct convincing proofs for the non-specialist. The results presented resulted from the analysis of known examples i.e. the series $A^{(1)}_n$ [15] and $A^{(2)}_n$, $B^{(1)}_n$, $C^{(1)}_n$, $D^{(1)}_n$, $D^{(2)}_n$ [23] and moreover, from the persistent idea to find the Lie algebra structure everywhere.

We should like to acknowledge conversations with E.I. Ogievetski and H.J. de Vega, and to thank I.M. Gelfand for his interest. One of us (P.W.) is grateful for the kind hospitality extended to him by Professor P. Nozieres at the Institut Laue–Langevin (Grenoble), and by Professor E. Brézin and the members of the Laboratoire de Physique Théorique at the Ecole Normale Supérieure (Paris).

‡3 To have conventional normalization of the physical sheet in (20) $0 < \mathrm{Im}\,\theta < \pi$, one should normalize the root system by the condition $B(E_\alpha, E_{-\alpha}) = -1$ [8] where $B(x, y)$ is the Killing form and $E_{\pm\alpha}$ form the nilpotent root spaces.

Volume 189, number 1,2 PHYSICS LETTERS B 30 April 1987

References

[1] A.B. Zamolodchikov and Al.B., Zamolochikov, Ann. Phys. (NY) 120 (1979) 253.
[2] L.D. Faddeev, Integrable models in 1+1 dimensional quantum field theory, in: Les Houches Lectures (1982) (Elsevier Science Publishers, Amsterdam, 1984).
[3] R.J. Baxter, Exactly solved models in statistical mechanics (Academic Press, New York, 1982).
[4] A.B. Zamolodchikov, Sov. Sci. Rev. Ser. Phys. 2 (1980) 2.
[5] P.P. Kulish and E.K. Sklyanin, Lecture Notes in Physics, Vol. 151 (Springer, Berlin, 1982) p. 61.
[6] P.P. Kulish, N.Yu. Reshetikhin and E.K. Sklyanin, Lett. Math. Phys. 5 (1985) 393.
[7] N.Yu. Reshetikhin, Theor. Math. Phys. 63 (1985) 347.
[8] E.I. Ogievetsky and P.B. Wiegmann, Phys. Lett. B 160 (1986) 360.
[9] E.I. Ogievetsky, N.Yu. Reshetikhin and P.B. Wiegmann, Nucl. Phys. B 280 (1987) 45.
[10] V.V. Bazhanov, Phys. Lett. B 159 (1985) 321.
[11] M. Jimbo, Lett. Math. Phys., to be published.
[12] A.A. Belavin, Nucl. Phys. B 180 [FS2] (1981) 189.
[13] M. Gaudin, La fonction d'onde de Bethe.
[14] L.D. Faddeev and L.A. Takhtajan, Usp. Math. Nauk. 34 (1979) 13;
P.P. Kulish and N.Yu. Reshetikhin, J. Phys. A 16 (1983) L591;
H.J. de Vega and M. Karowski, LPTHE preprint 86/21 (1986), Nucl. Phys. B, to be published.
[15] O. Babelon, H.J. de Vega and C.M. Viallet, Nucl. Phys. B 220 (1983) 283.
[16] N.Yu. Reshetikhin, Lett. Math. Phys. 7 (1983) 205.
[17] I. Cherednik, unpublished.
[18] AA. Belavin and V.G. Drienfield, Funkt. Anal. ego Pril. 16 (1982) 1; Sov. Sci. Rev. Ser. Math. 3 (1982) 1.
[19] V.G. Drienfeld, Dokl. Acad. Nauk. 283 (1985) 1060.
[20] A.N. Kirillov and N.Yu. Reshetikhin, J. Phys. A, to be published.
[21] M. Karowski, Nucl. Phys. B 153 (1979) 244.
[22] L.D. Faddeev and N.Yu. Reshetikhin, Theor. Math. Phys. 56 (1983) 323.
[23] N.Yu. Reshetikhin, Lett. Math. Phys., to be published.

GENERALIZATIONS

7. Generalizations

This last section concerns two important directions of generalizing YBE.

In the reprint [28] Zamolodchikov introduced the three dimensional analog of YBE, called tetrahedron equations, and postulated an explicit solution to it in trigonometric parametrization. That it indeed solves the tetrahedron equaitons was later proved in the reprint by Baxter [29], who also computed its free energy[23)]. Bazhanov and Stroganov[32)] formulated higher dimensional analogs of YBE (called d-simplex equations) and showed that they imply the commutativity of the transfer matrices. They also studied the free-fermion model in higher dimensions[33)]. See also [166–167)]. Related discussions on the higher dimensional geometry can be found in[173)]. Because of the complexity of the equations, it remains a challenge to find non-trivial solutions other than the examples mentioned above.

In all cases we have seen so far (including the three dimensional models), the solutions are parametrized in terms of rational, trigonometric or elliptic functions. As the spectral parameters enter additively in the ordinary YBE, this phenomenom is understandable in view of the addition theorem for these functions. However it is not clear why one does not see spectral parameters living on essentially higher dimensional manifolds such as Abelian varieties (cf. [55)]). Au-Yang et al.[6–7),181)] discovered that, if one drops the additivity of spectral parameters, there are indeed solutions to YBE whose specral parameters live on curves of genus greater than 1. The resulting chiral Potts model is an extention of the Zamolodchikov-Fateev model [23]. The reprint of Au-Yang, Baxter and Perk [30] affords a proof of YBE for its Boltzmann weights. Apparently the chiral Potts model belonged to an entirely new class distinct from any other solutions known so far. Very recently, however, Bazhanov and Stroganov [31] revealed that it emerges as a descendant of the 6 vertex model.

Commun. Math. Phys. 79, 489–505 (1981)

Communications in Mathematical Physics

Tetrahedron Equations and the Relativistic S-Matrix of Straight-Strings in $2+1$-Dimensions

A. B. Zamolodchikov

Landau Institute of Theoretical Physics, Vorobyevskoe Shosse, 2, Moscow V-334, USSR

Abstract. The quantum S-matrix theory of straight-strings (infinite one-dimensional objects like straight domain walls) in $2+1$-dimensions is considered. The S-matrix is supposed to be "purely elastic" and factorized. The tetrahedron equations (which are the factorization conditions) are investigated for the special "two-colour" model. The relativistic three-string S-matrix, which apparently satisfies this tetrahedron equation, is proposed.

1. Introduction

The progress of the last decade in studying two-dimensional exactly solvable models of quantum field theory and lattice statistical physics was motivated to some extent by using *the triangle equations.* These equations were first discovered by Yang [1]; they appeared in the problem of non-relativistic $1+1$-dimensional particles with δ-function interaction, as the self-consistency condition for Bethe's ansatz. Analogous (at least formally) relations were derived by Baxter [2], who had investigated the eight-vertex lattice model. These relations restrict the vertex weights and are of great importance for exact solvability. In particular, for the rectangular-lattice model they guarantee the commutativity of transfer-matrices with different values of the anisotropy parameter v. In the case of Baxter's general nonregular lattice $\mathscr{L}$ [3], the triangle relations for the vertex weights ensure the remarkable symmetry of the statistical system (the so-called Z-invariance): the partition function is unchanged under the deformations of the lattice, generated by the arbitrary shifts of the lattice axes. Z-invariant model on the lattice $\mathscr{L}$ is exactly solvable [3] (see also [4]).

Recently Faddeev, Sklyanin, and Takhtadjyan [5, 6] have developed a new general method of studying the exactly solvable models in $1+1$-dimensions – the quantum inverse scattering method. The triangle equations are the significant constituent of this method; they are to be satisfied by the elements of the R-matrix which determine the commutation relations between the elements of the monodromy matrix.

The triangle equations are also the central part of the theory of the relativistic purely elastic ("factorized") S-matrix in $1+1$-dimensions (for a review, see [7] and references therein). These equations (the "factorization equations") connect the elements of the two-particle S-matrix; they represent the conditions which are necessary for the factorization of the multiparticle S-matrix into two-particle ones. For the scattering theory including n different kinds of particles A_i; $i=1,2,\ldots,n$ the factorization equations have the form [7,4]

$$S_{i_1 i_2}^{k_1 k_2}(\theta) S_{i_3 k_1}^{k_3 j_1}(\theta+\theta') S_{k_2 k_3}^{j_2 j_3}(\theta') \\ = S_{i_2 i_3}^{k_2 k_3}(\theta') S_{i_1 k_3}^{k_1 j_3}(\theta+\theta') S_{k_1 k_2}^{j_1 j_2}(\theta), \tag{1.1}$$

where, for instance, $S_{i_1 i_2}^{k_1 k_2}(\theta)$ is the two-particle S-matrix, (i_1, i_2), (k_1, k_2) are the kinds of the initial (final) particles having the rapidities[1] θ_1 and θ_2, respectively; $\theta=(\theta_1-\theta_2)$. This equation has the following meaning. In the purely elastic scattering theory the three-particle S-matrix is factorized into three two-particle ones, as if the three-particle scattering were the sequence of successive pair collisions. If the rapidities $\theta_1, \theta_2, \theta_3$ of the initial particles are given, the two alternatives for the successions of these pair collisions are possible. The two different (in general) formal expressions for the three-particle S-matrix in terms of two-particle ones [the right- and left-hand sides of (1.1)] correspond to these alternatives. The conservation of the individual particle momenta requires the two "semifronts" of outgoing wave, which correspond to these two alternatives, to be coherent. The Eq. (1.1) expresses this requirement. The diagrammatic representation of the triangle Eq. (1.1) is given in Fig. 1, where the straight lines represent the "world lines" of three particles moving with the rapidities $\theta_1, \theta_2, \theta_3$. The two-particle S-matrices correspond to the intersection points of the lines; $i_a(j_a)$; $a=1,2,3$ are the kinds of the initial (final) particles; the summing over the kinds k_a of the "intermediate" particles is implied.

In [8] the version of factorized scattering theory in $2+1$-dimensions was proposed. In this theory the scattered objects are not the particles but one-dimensional formations like infinite straight-lined domain walls, which are characteristic of some models of $2+1$-dimensional field theory. We shall consider the quantum objects of this type and call them *the straight strings*. The stationary state of a moving straight string is characterized by the uniform momentum distribution along its length; its kinematics can be described completely by the direction of the string and by the transversal velocity. We assume also that the stationary states of any number of arbitrarily directed (intersecting, in general) moving straight strings are realizable[2]. The intersection points divide each string into segments, each being assumed to carry some internal quantum number i which will be called "colour". The relativistic case of the straight-string kinematics will be implied.

1 The rapidity of the relativistic $1+1$-dimensional particle is defined by the formulae

$$p_a^0 = m\cosh\theta_a\,; \quad p_a^1 = m\sinh\theta_a\,,$$

where p_a^μ is the two-momentum; $p^2=m^2$

2 Solutions of this type are likely in some completely integrable classical models in $2+1$-dimensions (S. Manakov, private communication)

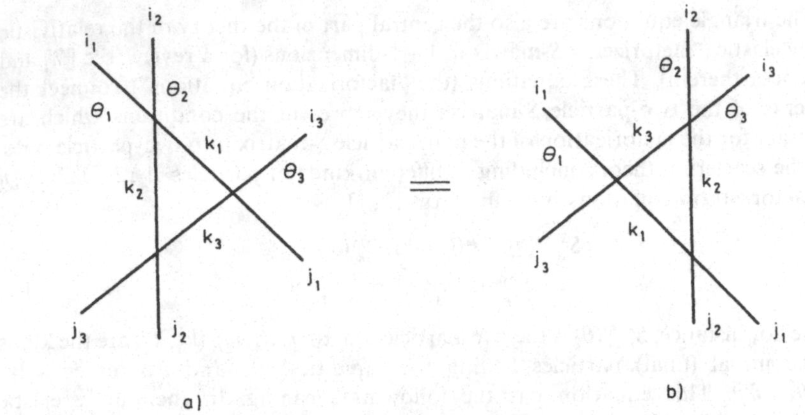

Fig. 1a and b. Diagrammatic representation of the triangle Eq. (1.1)

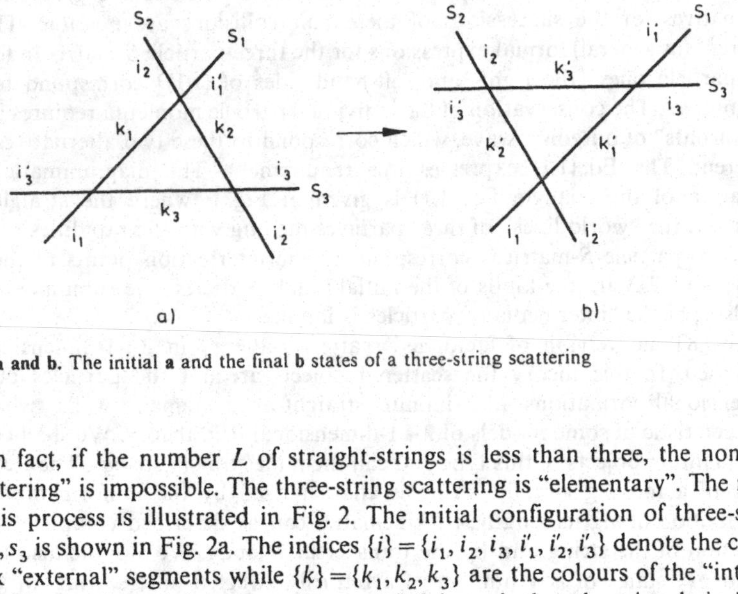

Fig. 2a and b. The initial **a** and the final **b** states of a three-string scattering

In fact, if the number L of straight-strings is less than three, the nontrivial "scattering" is impossible. The three-string scattering is "elementary". The nature of this process is illustrated in Fig. 2. The initial configuration of three-strings s_1, s_2, s_3 is shown in Fig. 2a. The indices $\{i\} = \{i_1, i_2, i_3, i'_1, i'_2, i'_3\}$ denote the colours of six "external" segments while $\{k\} = \{k_1, k_2, k_3\}$ are the colours of the "internal" ones. The motion of the strings S_a; $a = 1, 2, 3$ is such that the triangle in Fig. 2a shrinks with time. Shrinking and then "turning inside out" this triangle is the three-string scattering. After scattering, only states of the type shown in Fig. 2b appear. (This is, essentially, the meaning of the assumption of the "purely elastic" character of scattering.) The directions and velocities of the outgoing strings s_1, s_2, s_3 (Fig. 2b) coincide with those of the initial ones. The "internal" segments of strings, however, can be recoloured (in general $\{k'\} \neq \{k\}$).

In quantum theory the process shown in Fig. 2 is described by *the three-string scattering amplitude*

$$S^{\substack{i_1 k_1 i'_1 k'_1 \\ i_2 k_2 i'_2 k'_2 \\ i_3 k_3 i'_3 k'_3}}(\theta_1, \theta_2, \theta_3), \qquad (1.2)$$

where the variables $\theta_1, \theta_2, \theta_3$ ("interfacial angles", see below) describe the scattering kinematics.

One can imagine the three-string scattering as the intersection of three planes in 2+1-dimensional space-time. These planes represent the "world sheets" swept out by the moving straight-strings. Let n_1, n_2, n_3 $[n_a^2 = (n_a^1)^2 + (n_a^2)^2 - (n_a^0)^2 = 1]$ be the normal unit vectors of the planes corresponding to the strings s_1, s_2, s_3, respectively. The mutual orientation of three planes, and, hence, the kinematics of the three-string scattering, is described completely by three invariants

$$n_1 n_2 = -\cos\theta_3\,; \qquad n_1 n_3 = -\cos\theta_2\,; \qquad n_2 n_3 = -\cos\theta_1\,. \tag{1.3}$$

The two-plane intersection lines divide every plane s_a $(a=1,2,3)$ into four parts which will be called plaquettes. The colours of string segments, denoted by the indices i_a, k_a, i'_a, k'_a in (1.2) can be obviously attached to twelve plaquettes joined to the three-plane intersection point. In what follows this point will be called the vertex while the angles $\theta_1, \theta_2, \theta_3$, defined by (1.3) – the vertex variables.

The L-string scattering for $L>3$ has similar properties: the directions and velocities of all the strings s_a; $a=1, 2, \ldots, L$ remain unchanged after the scattering, the "internal" segments being, in general, recoloured. We assume the factorization of the multistring S-matrix: the L-string S-matrix is the product of $L(L-1)(L-2)/6$ three-string ones (1, 2), according to the idea that the L-string scattering can be thought of as the sequence of three-string collisions. The succession of this three-string collision is not determined uniquely by the directions and velocities of all the strings s_a but depends also on their "initial positions". Like the 1+1-dimensional case, the self-consistency of the factorization condition for the straight-string S-matrix requires the equality of different formal expressions for the L-string S-matrix in terms of three-string amplitudes, corresponding to the different successions of three-string collisions. It is easy to note that this requirement is equivalent to *the tetrahedron equation* shown in Fig. 3. In this figure the "world planes" of four strings s_a, $a=1,2,3,4$ (undergoing the four-string scattering) are shown. These planes form the tetrahedron in 2+1 space-time. The vertices of the tetrahedron represent the "elementary" three-string collisions; the corresponding S-matrices (1.2) are the multipliers in the expression for the 4-string S-matrix. The tetrahedra shown in Fig. 3a and 3b (which differ from each other by some parallel shift of the planes s_a) represent two possible successions of three-string collisions constituting the same four-string scattering process. The colours of the "external" plaquettes are fixed and respectively equal in the right- and the left-hand sides of the equality in Fig. 3; the summing over all possible colourings of the "internal" plaquettes (which are the faces of the tetrahedra) is implied. This tetrahedron equation should be satisfied at any mutual orientations of the planes s_1, s_2, s_3, s_4.

The 1+1-dimensional factorized S-matrix can be interpreted, after euclidean continuation, as the Z-invariant statistical model on the planar Baxter's lattice $\mathscr{L}$ (see [4]). The 2+1-dimensional factorized S-matrix of straight-strings admits similar interpretation [8]. The natural three-dimensional analog of Baxter's lattice $\mathscr{L}$ is the lattice formed by a large number L of arbitrarily directed intersecting planes in three-dimensional euclidean space. The fluctuating variables ("colours")

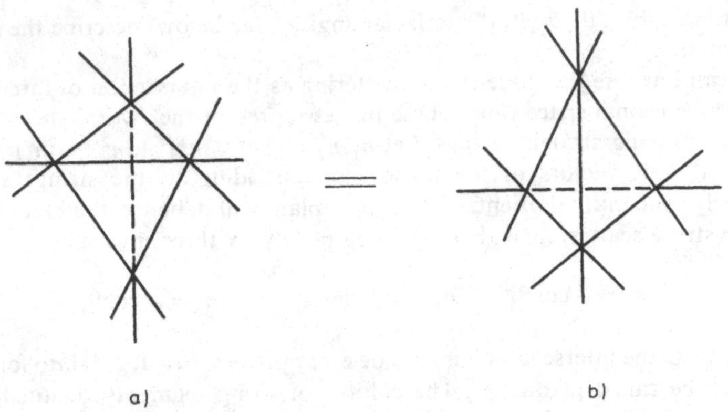

Fig. 3a and b. Diagrammatic representation of the tetrahedron equations

are attached to the lattice plaquettes. The partition function is defined as the sum over all possible colourings of all the plaquettes, each colour configuration being taken with the weight equal to the product of the vertex weights over all the vertices (the vertices are the points of triple intersections of the planes). The vertex weights are assumed to be the functions (common for all the vertices) of the mutual orientation of three planes intersecting in a given vertex. Identifying the vertex weights with the elements of the three-string S-matrix (1.2) (continued to the euclidean domain), one can note that, due to the tetrahedron equation (Fig. 3), the statistical system thus defined possesses Z-invariance.

The tetrahedron equation (Fig. 3) turns out to be highly overdefined system of functional equations; even in the simplest models the independent equations outnumber (by several hundredfold) the independent elements of the three-string S-matrix (1.2). Therefore, the compatibility of these equations is extremely crucial for the scattering theory of straight-strings. In [8] the *two-colour model* of straight-string scattering theory was proposed, and the explicit solution of the corresponding tetrahedron equations was found in the special "static limit" which corresponds to the case $v_a \to 0$, where v_a are the velocities of all the strings. In this paper we construct the relativistic three-string S-matrix for the two-colour model, which is apparently the solution of the "complete" tetrahedron equations. Although the complete evidence of the last statement is unknown we present some nontrivial checks.

The qualitative aspects of the factorized straight-string scattering theory have been described briefly in this Introduction; more detailed discussion can be found in [8]. In Sect. 2 the formulation of the two-colour model is given for the relativistic case. The corresponding tetrahedron equations are discussed in Sect. 3. In Sect. 4 the explicit formulae for the elements of the three-string S-matrix are proposed and the arguments that this S-matrix satisfies the tetrahedron equations are presented. In Sect. 5 it is shown that the obtained S-matrix is in agreement with the unitarity condition for the straight-string S-matrix.

2. Two-colour Model of Straight-strings Scattering Theory

Consider the relativistic scattering theory of straight-strings (see the Introduction) in which the strings' segments can carry only two colours – "white" or "black". Further, let us allow only the states satisfying the following requirement: the even number (i.e., 0, 2 or 4) of black segments can join in each point of two-string intersection. In other words, in any allowed state the black segments form continuous polygonal lines (which may intersect) without ends. Certainly, all the elements of the three-string S-matrix converting the allowed states into unallowed ones (and vice versa) are implied to be zero.

As it is explained in the Introduction, the three-string scattering kinematics can be represented by means of three intersecting "world planes" s_1, s_2, s_3 in $2+1$ space time, the vertex being the "place of collision". In the two-colour model each of the twelve plaquettes joining the vertex can be coloured into black or white so that the black plaquettes form the continuous broken surfaces without boundaries.

Each allowed coluring of these twelve plaquettes corresponds to some nonvanishing element of the three-string S-matrix.

It is convenient to perform the considerations in terms of the euclidean space-time: the "world planes" s_a can be treated as imbedded in the 3-dimensional euclidean space; each of the variables θ, defined by (1.3) being some interfacial angle. The "physical" amplitudes of scattering in the Minkowski space-time can be obtained from the euclidean formulae by means of analytical continuation.

Let us picture the "colour configuration" of the twelve plaquettes joining the vertex as follows. Consider the sphere with the vertex as its centre. The planes s_1, s_2, s_3 draw three great circles on this sphere; the variables θ_1, θ_2, θ_3 [see (1.3)] are exactly the intersection angles of these circles. The intersection points divide each of the circles into four segments; the colours of the plaquettes can be obviously attached to these segments. Performing the stereographic projection one can map these three circles on the plane as shown in Fig. 4. This picture can be interpreted as follows. The spherical triangle I_1 in Fig. 4 corresponds to the triangle in Fig. 2a and represents the initial state of some three-string scattering process. The final state of this process (shown in Fig. 2b) is represented by the spherical triangle F_1. The variables θ_1, θ_2, θ_3 are the interior angles of the triangles I_1 and F_1 (obviously, these triangles are equal on the sphere). Alternatively, one could consider, for instance, the triangle I_2 as representing the initial (and F_2 as the final) state of some other three-string process. This is just the cross-channel. Evidently, the transfer to this cross-channel is associated with the variable transformation

$$\theta_1 \to \pi - \theta_1; \quad \theta_2 \to \pi - \theta_2; \quad \theta_3 \to \theta_3 . \tag{2.1}$$

As it is clear from Fig. 4, each three-string scattering process has four cross-channels $I_1 \to F_1$, $I_2 \to F_2$, $I_3 \to F_3$, $I_4 \to F_4$.

We shall assume the P and T invariances of the straight-string scattering theory [8], and also its symmetry under the simultaneous recolouring of all black segments into white and vice versa ("colour symmetry"). Then the three-string S-matrix contains 8 independent amplitudes which are shown (together with the adopted notations) in Fig. 5.

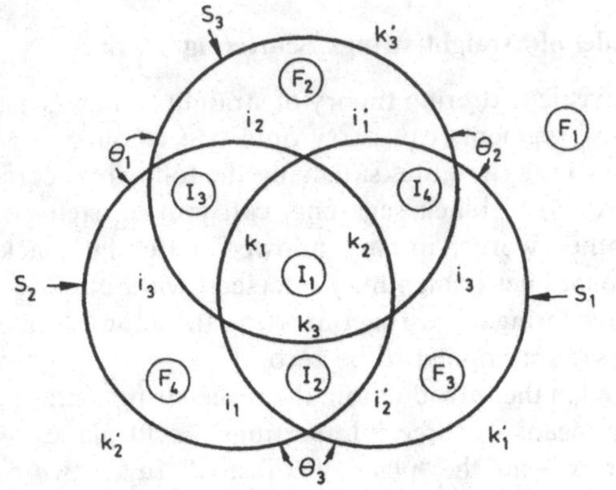

Fig. 4. Stereographic projection of the sphere surrounding the vertex

a) $S(\theta_1,\theta_2,\theta_3)$

b) $a(\theta_1,\theta_2,\theta_3)$

c) $V(\theta_1,\theta_2,\theta_3)$

d) $R(\theta_1,\theta_2,\theta_3)$

e) $K(\theta_1,\theta_2,\theta_3)$

f) $\omega(\theta_1,\theta_2,\theta_3)$

g) $d(\theta_1,\theta_2,\theta_3)$

h) $T(\theta_1,\theta_2,\theta_3)$

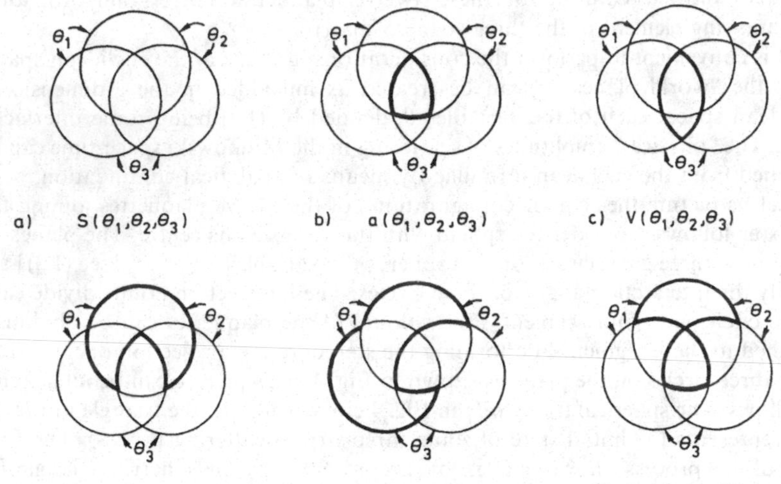

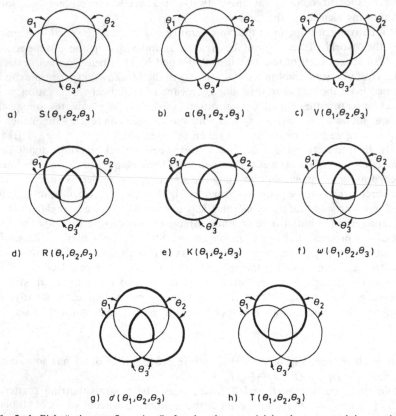

Fig. 5a–h. Eight "colour configurations" of twelve plaquettes joining the vertex, and the notations for corresponding three-string scattering amplitudes. The white (black) circular segments are represented by the ordinary (solid) lines

It is convenient to introduce, apart from the amplitudes defined in Fig. 5 the following 5 functions

$$U(\theta_1,\theta_2,\theta_3)=\mathcal{A}(\pi-\theta_1,\pi-\theta_2,\theta_3), \tag{2.2a}$$
$$L(\theta_1,\theta_2,\theta_3)=V(\pi-\theta_1,\theta_2,\pi-\theta_3), \tag{2.2b}$$
$$\Omega(\theta_1,\theta_2,\theta_3)=R(\pi-\theta_1,\pi-\theta_2,\theta_3), \tag{2.2c}$$
$$H(\theta_1,\theta_2,\theta_3)=R(\pi-\theta_1,\theta_3,\pi-\theta_2), \tag{2.2d}$$
$$W(\theta_1,\theta_2,\theta_3)=\sigma(\pi-\theta_1,\pi-\theta_2,\theta_3), \tag{2.2e}$$

which describe the cross-channels of the processes shown in Fig. 5. The three-string amplitudes should possess the following symmetries, which are the consequences of P, T, "colour" and crossing symmetries

$$S(\theta_1,\theta_2,\theta_3)=S(\theta_2,\theta_1,\theta_3)=S(\theta_1,\theta_3,\theta_2)$$
$$=S(\pi-\theta_1,\pi-\theta_2,\theta_3). \tag{2.3a}$$
$$\mathcal{A}(\theta_1,\theta_2,\theta_3)=\mathcal{A}(\theta_2,\theta_1,\theta_3)=\mathcal{A}(\theta_1,\theta_3,\theta_2),$$
$$U(\theta_1,\theta_2,\theta_3)=U(\theta_2,\theta_1,\theta_3). \tag{2.3b}$$
$$V(\theta_1,\theta_2,\theta_3)=V(\theta_2,\theta_1,\theta_3)=V(\pi-\theta_1,\pi-\theta_2,\theta_3),$$
$$L(\theta_1,\theta_2,\theta_3)=L(\theta_2,\theta_1,\theta_3). \tag{2.3c}$$
$$R(\theta_1,\theta_2,\theta_3)=R(\theta_2,\theta_1,\theta_3), \qquad H(\theta_1,\theta_2,\theta_3)=H(\theta_2,\theta_1,\theta_3),$$
$$\Omega(\theta_1,\theta_2,\theta_3)=\Omega(\theta_2,\theta_1,\theta_3). \tag{2.3d}$$
$$\omega(\theta_1,\theta_2,\theta_3)=\omega(\theta_2,\theta_1,\theta_3)=\omega(\pi-\theta_1,\pi-\theta_2,\theta_3)$$
$$=\omega(\pi-\theta_1,\theta_2,\pi-\theta_3). \tag{2.3e}$$
$$K(\theta_1,\theta_2,\theta_3)=K(\theta_2,\theta_1,\theta_3)=K(\pi-\theta_2,\pi-\theta_1,\theta_3)$$
$$=K(\pi-\theta_1,\theta_2,\pi-\theta_3). \tag{2.3f}$$
$$\sigma(\theta_1,\theta_2,\theta_3)=\sigma(\theta_2,\theta_1,\theta_3)=\sigma(\theta_1,\theta_3,\theta_2),$$
$$W(\theta_1,\theta_2,\theta_3)=W(\theta_2,\theta_1,\theta_3). \tag{2.3g}$$
$$T(\theta_1,\theta_2,\theta_3)=T(\theta_2,\theta_1,\theta_3)=T(\pi-\theta_1,\pi-\theta_2,\theta_3)$$
$$=T(\pi-\theta_1,\theta_2,\pi-\theta_3). \tag{2.3h}$$

The analytic properties of the three-string amplitudes will be considered in Sects. 3 and 4.

3. The Tetrahedron Equations

The hardest restrictions for the three-string S-matrix come from the tetrahedron equations, which are shown schematically in Fig. 3. Here we shall choose the four "world planes" s_1, s_2, s_3, s_4 shown in this figure to be placed into the euclidean space (see Sect. 2). The three-string S-matrices associated with the vertices of the tetrahedra in Fig. 3 are the functions of corresponding vertex variables. In the two-

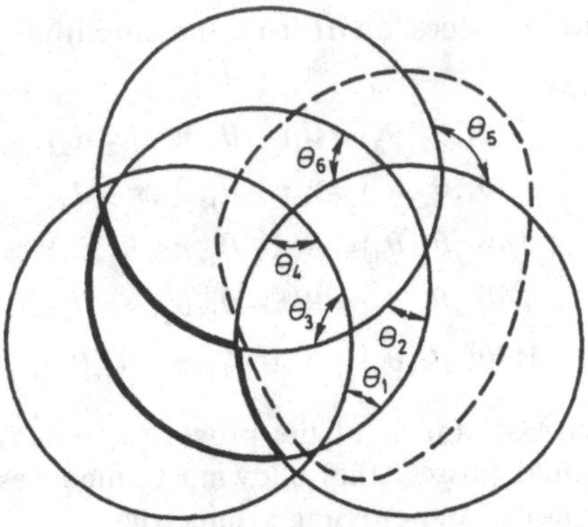

Fig. 6. Stereographic projection of large sphere surrounding the tetrahedron

colour model each plaquette in Fig. 3 can be black or white. Recall that the colours of 24 "external" plaquettes are fixed and equal in the right- and left-hand sides of the tetrahedron equation while the independent summing is performed over all colourings of "internal" plaquettes. Obviously, each allowed colouring of the "external" plaquettes gives rise to some functional equation connecting the three-string amplitudes.

To describe the colourings of the "external" plaquettes it is convenient to introduce again the large sphere (its radius is much larger than the size of the tetrahedra), taking some point near the vertices as the centre, and consider 4 great circles on the sphere corresponding to the planes s_1, s_2, s_3, s_4 (certainly, the tetrahedra in Figs. 3a and 3b are indistinguishable from this point of view). Stereographic projection of this sphere is shown in Fig. 6. The angles θ_1, θ_2, θ_3, θ_4, θ_5, θ_6, shown in this figure are just the interior interfacial angles (i.e., the angles between the planes s_a) of the tetrahedra.

Any allowed colouring of the "external" plaquettes in Fig. 3 corresponds in obvious manner to some allowed colouring of the 24 circular segments in Fig. 6 into black and white. In Fig. 6 some colouring of this type is shown as an example. This colouring gives rise, as it is evident from simple consideration, to the following functional equation

$$\begin{aligned} &S(\theta_1,\theta_2,\theta_3)S(\theta_1,\theta_4,\theta_6)S(\theta_5,\theta_4,\theta_3)\mathcal{a}(\theta_2,\theta_5,\theta_6) \\ &+\mathcal{a}(\theta_1,\theta_2,\theta_3)\mathcal{a}(\theta_1,\theta_4,\theta_6)\mathcal{a}(\theta_4,\theta_3,\theta_5)\sigma(\theta_2,\theta_5,\theta_6) \\ &=U(\theta_1,\theta_3,\theta_2)U(\theta_1,\theta_4,\theta_6)U(\theta_4,\theta_3,\theta_5)S(\theta_2,\theta_5,\theta_6) \\ &+V(\theta_1,\theta_3,\theta_2)V(\theta_1,\theta_4,\theta_6)V(\theta_3,\theta_4,\theta_5)\mathcal{a}(\theta_2,\theta_5,\theta_6)\,. \end{aligned} \tag{3.1}$$

The equation (3.1) is only one representative of the system of functional tetrahedron equations which arises if one considers all possible allowed colourings of the circular segments in Fig. 6. This system includes hundreds of independent equations and we are not able to present it here; the equation (3.1) is written down mainly for illustration.

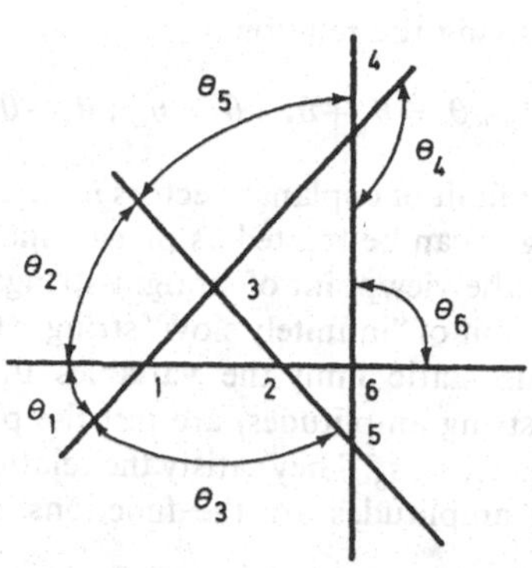

Fig. 7. Fragment of the diagram in Fig. 6 which is enclosed with the dotted curve

It is essential that the variables $\theta_1, \theta_2, \theta_3, \theta_4, \theta_5, \theta_6$ in the tetrahedron equations are not completely independent. Since the mutual orientation of four planes in three-dimensional space is determined completely by only five parameters, there is one relation between these six angles[3]. This relation can be derived, for instance, from the spherical trigonometry. To do so, concentrate on the fragment of Fig. 6, surrounded by dotted curve. This fragment is shown separately in Fig. 7, where the circular arcs are drawn schematically as the straight lines. Using the formulae of spherical trigonometry one can express the lengths of segments ℓ_{32} and ℓ_{25} in terms of the interior angles of the spherical triangles (123) and (256), respectively. On the other hand, the length of the segment $\ell_{35}=\ell_{32}+\ell_{25}$ may be expressed independently in terms of the interior angles of the triangle (354). This allows one to write

$$\left[\cos\frac{\theta_1+\theta_3-\theta_2}{2}\cos\frac{\theta_1+\theta_2-\theta_3}{2}\cos\frac{\theta_6+\theta_5-\theta_2}{2}\cos\frac{\theta_6+\theta_2-\theta_5}{2}\right]^{1/2}$$
$$-\left[\cos\frac{\theta_1+\theta_2+\theta_3}{2}\cos\frac{\theta_2+\theta_3-\theta_1}{2}\cos\frac{\theta_2+\theta_5+\theta_6}{2}\cos\frac{\theta_2+\theta_5-\theta_6}{2}\right]^{1/2}$$
$$=\sin\theta_2\left[\cos\frac{\theta_3+\theta_5-\theta_4}{2}\cos\frac{2\pi-\theta_3-\theta_4-\theta_5}{2}\right]^{1/2}. \tag{3.2}$$

Equation (3.2) is a variant of the desired relation.

4. The Solution of the Tetrahedron Equations

The relation (3.2) connecting the interior angles of the tetrahedron essentially complicates the direct investigation of the tetrahedron equations. However, one can concentrate at first on the special limiting case. Namely, consider the variables

3 Certainly, this relation is the imbedding condition of four vectors n_1, n_2, n_3, n_4 into the three-dimensional space. Its general form is $\det|n_a^\mu|=0$, where four vectors n_a^μ are treated formally as four-dimensional; $a, \mu=1,2,3,4$

$\theta_1, \theta_2, \theta_3, \theta_4, \theta_5, \theta_6$ satisfying the relation

$$\theta_1+\theta_2+\theta_3=\theta_2+\theta_5+\theta_6=\theta_4+\theta_3-\theta_5=\pi, \tag{4.1}$$

which corresponds to the limit of coplanar vectors n_1, n_2, n_3, n_4. In this case all the spherical triangles in Fig. 7 can be treated as planar ones and the relation (3.2) is certainly satisfied. From the viewpoint of straight-strings kinematics, the relation (4.1) corresponds to the limit of "infinitely slow" strings; therefore we call this case the "static limit". In the static limit the variables θ_1, θ_2, θ_3 (which are the arguments of the three-string amplitudes) are just the planar angles between the directions of three strings s_1, s_2, s_3. They satisfy the relation $\theta_1+\theta_2+\theta_3=\pi$. Hence the "static" three-string amplitudes are the functions, not of three, but of two variables, θ_1, θ_2.

Most of the tetrahedron equations do not become identities in the static limit [as it happens for the Eq. (3.1)]. Actually, considering the static limit, the number of the independent tetrahedron equations even increase, since the different cross-channels of the same "complete" equations give rise to the different "static" tetrahedron equations.

In [8] the solution of the static-limit tetrahedron equation was constructed; it has the form

$$\begin{aligned}
&S^{st}(\theta_1,\theta_2)=\sigma^{st}(\theta_1,\theta_2)=T^{st}(\theta_1,\theta_2)=W^{st}(\theta_1,\theta_2)=1;\\
&\alpha^{st}(\theta_1,\theta_2)=R^{st}(\theta_1,\theta_2)=0;\\
&L^{st}(\theta_1,\theta_2)=\omega^{st}(\theta_1,\theta_2)=-K^{st}(\theta_1,\theta_2)=\varepsilon_1 V^{st}(\theta_1,\theta_2)=\left[\operatorname{tg}\frac{\theta_1}{2}\operatorname{tg}\frac{\theta_2}{2}\right]^{1/2};\\
&H^{st}(\theta_1,\theta_2)=U^{st}(\theta_1,\theta_2)=-\varepsilon_1\Omega^{st}(\theta_1,\theta_2)=\varepsilon_2\left[\frac{\cos\left(\frac{\theta_1}{2}+\frac{\theta_2}{2}\right)}{\cos\frac{\theta_1}{2}\cos\frac{\theta_2}{2}}\right]^{1/2},
\end{aligned} \tag{4.2}$$

where the notations for the three-string amplitudes are the same as in Fig. 5 and in (2.2), the θ_3 being set equal to $\pi-\theta_1-\theta_2$; for instance,

$$U^{st}(\theta_1,\theta_2)\equiv U(\theta_1,\theta_2,\pi-\theta_1-\theta_2).$$

In (4.2) ε_1 and ε_2 are arbitrary signs; $\varepsilon_1^2=\varepsilon_2^2=1$. Expressions (4.2) satisfy all the "static" tetrahedron equations. We do not insist that (4.2) is the general solution; rather we think that it is not so.

Let us search for the solution of the "complete" tetrahedron equations which corresponds to the static limit (4.2). First consider the power expansion around the static limit. Namely, let the velocities of the scattered strings be not exactly zero but small. In this case the three-string scattering amplitudes can be conveniently considered as the functions of two angles θ_1, θ_2 (which determine the space directions of the strings s_1, s_2, s_3, see Fig. 2) and "symmetrical velocity" $w=\frac{1}{2}dr/dt$ where r is the radius of the circle inscribed in the triangle in Fig. 2. At small velocities of the strings s_1, s_2, s_3, s_4 the nonrelativistic kinematics is valid, and the "velocities" w, corresponding to four triangles (123), (256), (146), (453) in Fig. 7, are connected as follows:

$$w_{146}\frac{\sin\frac{\theta_{125}}{2}\sin\frac{\theta_2}{2}}{\sin\frac{\theta_1}{2}\sin\frac{\theta_{25}}{2}}=w_{123}\frac{\cos\frac{\theta_5}{2}\ \text{?}}{}$$

$$w_{146}\frac{\sin\frac{\theta_{125}}{2}\sin\frac{\theta_2}{2}}{\sin\frac{\theta_1}{2}\sin\frac{\theta_{25}}{2}}=w_{123}\frac{\sin\frac{\theta_{12}}{2}}{\sin\frac{\theta_1}{2}}+w_{256}\frac{\cos\frac{\theta_5}{2}}{\cos\frac{\theta_{25}}{2}};$$

$$w_{453}\frac{\sin\frac{\theta_{125}}{2}\sin\frac{\theta_2}{2}}{\sin\frac{\theta_5}{2}\sin\frac{\theta_{12}}{2}}=w_{123}\frac{\cos\frac{\theta_1}{2}}{\cos\frac{\theta_{12}}{2}}+w_{256}\frac{\sin\frac{\theta_{25}}{2}}{\sin\frac{\theta_5}{2}}, \tag{4.3}$$

where the notations $\theta_{12}=\theta_1+\theta_2$; $\theta_{25}=\theta_2+\theta_5$; $\theta_{125}=\theta_1+\theta_2+\theta_5$ are used. The investigation of the tetrahedron equations in the linear approximation (in w) leads to the result

$$\mathcal{A}(\theta_1,\theta_2,w)=-\varepsilon_1 R(\theta_1,\theta_2,w)=\varepsilon_2\lambda w\left[\sin\frac{\theta_1}{2}\sin\frac{\theta_2}{2}\sin\frac{\theta_3}{2}\right]^{-1/2}+O(w^2), \tag{4.4a}$$

$$S(\theta_1,\theta_2,w)=T(\theta_1,\theta_2,w)=1-\lambda w+O(w^2), \tag{4.4b}$$

$$\mathcal{O}(\theta_1,\theta_2,w)=W(\theta_1,\theta_2,w)=1+\lambda w+O(w^2), \tag{4.4c}$$

$$U(\theta_1,\theta_2,w)=H(\theta_1,\theta_2,w)=-\varepsilon_1\Omega(\theta_1,\theta_2,w)$$
$$=U^{st}(\theta_1,\theta_2)(1+O(w^2)), \tag{4.4d}$$

$$\mathcal{W}(\theta_1,\theta_2,w)=-K(\theta_1,\theta_2,w)$$
$$=L^{st}(\theta_1,\theta_2)\left(1+\lambda w\,\mathrm{ctg}\frac{\theta_1}{2}\mathrm{ctg}\frac{\theta_2}{2}+O(w^2)\right) \tag{4.4e}$$

$$L(\theta_1,\theta_2,w)=\varepsilon_1 V(\theta_1,\theta_2,w)$$
$$=L^{st}(\theta_1,\theta_2)\left(1-\lambda w\,\mathrm{ctg}\frac{\theta_1}{2}\mathrm{ctg}\frac{\theta_2}{2}+O(w^2)\right), \tag{4.4f}$$

where $\theta_3=\pi-\theta_1-\theta_2$, L^{st} and U^{st} are given by Eqs. (4.2), and λ is an arbitrary constant.

In studying the complete relativistic tetrahedron equations it is convenient to introduce the variables (spherical excesses)

$$\begin{aligned}2\alpha&=\theta_1+\theta_2+\theta_3-\pi,\\ 2\beta&=\pi+\theta_3-\theta_1-\theta_2,\\ 2\gamma&=\pi+\theta_1-\theta_2-\theta_3,\\ 2\delta&=\pi+\theta_2-\theta_1-\theta_3,\end{aligned} \tag{4.5}$$

obeying the relation

$$\alpha+\beta+\gamma+\delta=\pi. \tag{4.6}$$

Any transmutations of θ_1, θ_2, θ_3, and also any crossing transformations of the type of (2.1) lead, as one can easily verify, to some transmutations among the variables α, β, γ, δ. In fact, the quantity 2α is the area of the spherical triangle I_1 (and F_1) in Fig. 4, while 2β, 2γ, 2δ are the areas of the triangles I_2, I_3, I_4, respectively. Therefore the four cross-channels $I_1 \to F_1$, $I_2 \to F_2$, $I_3 \to F_3$, $I_4 \to F_4$ of the three-string scattering will be called α, β, γ, δ-channels, respectively.

In the static limit $\theta_1 + \theta_2 + \theta_3 \to \pi$, and we have

$$\alpha \to 0; \quad \beta \to \theta_3; \quad \gamma \to \theta_1; \quad \delta \to \theta_2. \tag{4.7}$$

The following relation is valid up to the main order in w

$$w = \sqrt{\frac{\alpha}{2}} \left[\operatorname{tg}\frac{\theta_1}{2} \operatorname{tg}\frac{\theta_2}{2} \operatorname{tg}\frac{\theta_3}{2} \right]^{1/2}. \tag{4.8}$$

Therefore, as it is seen from (4.4), the three-string amplitudes have the square-root branching plane $\alpha = 0$, which will be called the α-channel threshold. The crossing symmetry requires the amplitudes to possess the branching planes (also square-root) $\beta = 0$; $\gamma = 0$; $\delta = 0$, which are the thresholds of the β, γ, δ-channels.

These reasons allow one to write down the following formulae

$$a(\theta_1, \theta_2, \theta_3) = R(\theta_1, \theta_2, \theta_3) = \varepsilon_2 \left[\frac{\sin\frac{\alpha}{2}}{\cos\frac{\beta}{2}\cos\frac{\gamma}{2}\cos\frac{\delta}{2}} \right]^{1/2}; \tag{4.9a}$$

$$-V(\theta_1, \theta_2, \theta_3) = \left[\operatorname{tg}\frac{\gamma}{2} \operatorname{tg}\frac{\delta}{2} \right]^{1/2} - \left[\operatorname{tg}\frac{\alpha}{2} \operatorname{tg}\frac{\beta}{2} \right]^{1/2}; \tag{4.9b}$$

$$\omega(\theta_1, \theta_2, \theta_3) = -K(\theta_1, \theta_2, \theta_3) = \left[\operatorname{tg}\frac{\gamma}{2} \operatorname{tg}\frac{\delta}{2} \right]^{1/2} + \left[\operatorname{tg}\frac{\alpha}{2} \operatorname{tg}\frac{\beta}{2} \right]^{1/2}; \tag{4.9c}$$

$$S(\theta_1, \theta_2, \theta_3) = T(\theta_1, \theta_2, \theta_3) = 1 - \left[\operatorname{tg}\frac{\alpha}{2} \operatorname{tg}\frac{\beta}{2} \operatorname{tg}\frac{\gamma}{2} \operatorname{tg}\frac{\delta}{2} \right]^{1/2}; \tag{4.9d}$$

$$\sigma(\theta_1, \theta_2, \theta_3) = 1 + \left[\operatorname{tg}\frac{\alpha}{2} \operatorname{tg}\frac{\beta}{2} \operatorname{tg}\frac{\gamma}{2} \operatorname{tg}\frac{\delta}{2} \right]^{1/2}, \tag{4.9e}$$

which are in accordance with the expansion (4.4) provided $\varepsilon_1 = -1$ and $\lambda = 1$, and entirely satisfy the crossing relations (2.3). Therefore we suppose that the expressions (4.9) give the exact solution of the "complete" tetrahedron equations for the two-colour model.

Unfortunately, rigorous verification of this supposition is rather difficult. Direct substitution of (4.9) into the tetrahedron equation is complicated because of the relation (3.2), not to speak of the large number of equations to be verified. However, we have performed some simplified verifications; an example is given in the Appendix. Moreover, our supposition has been confirmed in various numerical checks.

Note that, like the triangle equations (1.1), the tetrahedron equations are homogeneous; the three-string S-matrix is determined by the equations only up to the overall factor which can be a function of the variables θ. The formulae (4.9) should be considered as expressions giving the ratios of different elements of the three-string S-matrix; the right-hand sides of all the equalities (4.9) are implied to be multiplied by some function

$$[Z(\alpha,\beta,\gamma,\delta)]^{-1}, \tag{4.10}$$

which is symmetric under arbitrary transmutations of the variables α, β, γ, δ [this is forced by the crossing symmetry requirements (2.3)]. This function will be determined by the unitarity condition for the straight-strings S-matrix, studied in the next section.

5. Unitarity Condition

In the euclidean domain the variables α, β, γ, δ [connected by (4.6)] are real and non-negative, and all the amplitudes (4.9) are real. The "physical" scattering of the strings s_a in Minkowski space-time corresponds to real negative values of α (provided the velocities of the strings s_a are not too large[4]. Here the amplitudes acquire the imaginary parts. Let us introduce the cutting hyperplane $\mathrm{Im}\,\alpha=\mathrm{Im}\,\beta=\mathrm{Im}\,\gamma=\mathrm{Im}\,\delta=0$; $\mathrm{Re}\,\alpha<0$ (corresponding to the branching plane $\alpha=0$) in the three-dimensional complex space of the variables α, β, γ, δ. Then the "upper" edge ($\mathrm{Im}\,\alpha=+0$; $\mathrm{Re}\,\beta>0$; $\mathrm{Re}\,\gamma>0$; $\mathrm{Re}\,\delta>0$) of this hyperplane represents the "physical" domain of α-channel. Continuing some amplitude to the "lower" edge $\mathrm{Im}\,\alpha=-0$, one obtains the complex-conjugated amplitude of reversed process (here we imply the T-invariance so that the amplitudes of direct and reversed processes are equal).

In the physical domain of α-channel the three-string unitarity condition should be satisfied, i.e.,

$$\sum_{k_1'k_2'k_3'} S^{i_1k_1i_1'k_1'}_{\substack{i_2k_2i_2'k_2'\\ i_3k_3i_3'k_3'}}(\theta_1,\theta_2,\theta_3)\,S^{i_1\ell_1i_1'k_1'}_{\substack{i_2\ell_2i_2'k_2'\\ i_3\ell_3i_3'k_3'}}(\theta_1,\theta_2,\theta_3)^*=\delta^{\ell_1}_{k_1}\delta^{\ell_2}_{k_2}\delta^{\ell_3}_{k_3}, \tag{5.1}$$

where S is the amplitude of the process shown in Fig. 2 and the star denotes the complex conjugation. If the second multiplier in the left-hand side of (5.1) is treated not as the complex-conjugated amplitude but the result of analytical continuation around the branching plane $\alpha=0$, the relation (5.1) becomes valid at any complex θ.

The requirement (5.1) for the two-colour string model leads, using (4.9), to the single equation for the "unitarizing factor" (4.10)

$$Z(\alpha,\beta,\gamma,\delta)Z^{(\alpha)}(\alpha,\beta,\gamma,\delta)=\frac{\cos\dfrac{\alpha+\beta}{2}\cos\dfrac{\alpha+\gamma}{2}\cos\dfrac{\alpha+\delta}{2}}{\cos\dfrac{\alpha}{2}\cos\dfrac{\beta}{2}\cos\dfrac{\gamma}{2}\cos\dfrac{\delta}{2}}, \tag{5.2}$$

where the suffix (α) denotes the continuation around the branching plane $\alpha=0$. This equation together with the requirement of symmetry under arbitrary

4 Actually, this is true unless the velocities of two-string intersection points exceed the speed of light

transmutation of the variables, determines the factor (4.10). The investigation of this equation is the subject of our further work.

It can be shown that, due to the factorization of the multistring S-matrix into three-string ones, the three-string unitarity condition (5.1) guarantees the unitarity of the total S-matrix of straight-strings.

6. Discussion

In a recent paper by Belavin [10] the remarkable symmetry of the triangle Eqs. (1.1) was discovered. This symmetry reveals the reasons for the compatibility of the overdefined system of functional Eqs. (1.1) and throws some light upon the nature of the general solution of these equations. It would be extremely interesting to find something like this symmetry in the tetrahedron equations.

As explained in the Introduction, the factorized S-matrix of straight-strings in the euclidean domain can be interpreted as the three-dimensional lattice statistical model which possesses Z-invariance and apparently is exactly solvable. Unfortunately, for the solution found in this paper some of the vertex weights turn out to be negative; therefore the existence of the thermodynamic limit of the corresponding lattice system becomes problematic. We suppose that there are solutions of tetrahedron equations which are free from this trouble. On the other hand, if the thermodynamic limit exists, there is the hypothesis that the partition function of a Z-invariant statistical system on the infinite lattice is simply connected to the "unitarizing factor" (4.10). (The two-dimensional analog of this hypothesis is discussed in [4].)

Appendix

Consider the Eq. (3.1) under the following condition

$$\theta_2+\theta_5+\theta_6=\pi. \tag{A.1}$$

Since $\mathcal{a}^{st}=0$, $S^{st}=\sigma^{st}=1$, the Eq. (3.1) acquires the form

$$\begin{aligned}&\mathcal{a}(\theta_1,\theta_2,\theta_3)\mathcal{a}(\theta_4,\pi-\theta_2-\theta_5,\theta_1)\mathcal{a}(\theta_4,\theta_3,\theta_5)\\&\quad=U(\theta_4,\theta_3,\theta_5)U(\theta_1,\theta_4,\pi-\theta_2-\theta_5)U(\theta_1,\theta_3,\theta_2).\end{aligned}\tag{A.2}$$

After the substitution of the explicit expressions (4.9) into (A.2) it can be rewritten

$$\begin{aligned}&\cos\frac{\theta_1+\theta_2+\theta_3}{2}\sin\frac{\theta_1+\theta_4-\theta_2-\theta_5}{2}\cos\frac{\theta_3+\theta_4+\theta_5}{2}\\&\quad=\cos\frac{\theta_4+\theta_3-\theta_5}{2}\sin\frac{\theta_1+\theta_2+\theta_4+\theta_5}{2}\cos\frac{\theta_1+\theta_3-\theta_2}{2}.\end{aligned}\tag{A.3}$$

The validity of this equality, assuming (3.2), remains to be proved.

The degeneration of the diagram in Fig. 7, corresponding to the case (A.1), is shown in Fig. 8. The length of the circular segment can be expressed independently in terms of the interior angles of two triangles in Fig. 8: either (123) or (124).

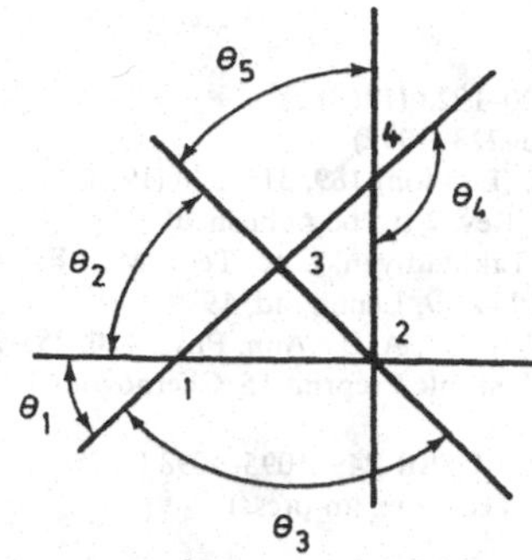

Fig. 8. Degeneration of diagram shown in Fig. 7, corresponding to the relation (A.1)

Comparing the results one obtains the relation

$$\left[\cos\frac{\theta_1+\theta_2+\theta_3}{2}\cos\frac{\theta_1+\theta_2-\theta_3}{2}\sin\frac{\theta_2+\theta_5-\theta_1-\theta_4}{2}\sin\frac{\theta_2+\theta_5+\theta_4-\theta_1}{2}\right]^{1/2}$$
$$=\left[\cos\frac{\theta_1+\theta_3-\theta_2}{2}\cos\frac{\theta_2+\theta_3-\theta_1}{2}\sin\frac{\theta_1+\theta_2+\theta_5-\theta_4}{2}\sin\frac{\theta_1+\theta_2+\theta_5+\theta_4}{2}\right]^{1/2}. \tag{A.4}$$

Doing the same with the segments ℓ_{34} and ℓ_{42} one gets two more relations

$$\left[\cos\frac{\theta_1+\theta_2+\theta_3}{2}\cos\frac{\theta_2+\theta_3-\theta_1}{2}\cos\frac{\theta_5+\theta_3-\theta_1}{2}\cos\frac{\theta_3+\theta_4+\theta_5}{2}\right]^{1/2}$$
$$=\left[\cos\frac{\theta_1+\theta_2-\theta_3}{2}\cos\frac{\theta_1+\theta_3-\theta_2}{2}\cos\frac{\theta_4+\theta_3-\theta_5}{2}\cos\frac{\theta_5+\theta_4-\theta_3}{2}\right]^{1/2}; \tag{A.5}$$

$$\left[\cos\frac{\theta_5+\theta_4-\theta_3}{2}\cos\frac{\theta_5+\theta_4+\theta_3}{2}\sin\frac{\theta_1+\theta_2+\theta_5-\theta_4}{2}\sin\frac{\theta_1+\theta_4-\theta_2-\theta_5}{2}\right]^{1/2}$$
$$=\left[\cos\frac{\theta_4+\theta_3-\theta_5}{2}\cos\frac{\theta_3+\theta_5-\theta_4}{2}\sin\frac{\theta_1+\theta_2+\theta_4+\theta_5}{2}\sin\frac{\theta_1-\theta_2-\theta_4-\theta_5}{2}\right]^{1/2}, \tag{A.6}$$

which are certainly equivalent to (A.4). Taking the products of the right- and left-hand sides of (A.4), (A.5), (A.6), one obtains exactly the equality (A.3).

Acknowledgements. I thank A. Belavin, V. Fateev, and Al. Zamolodchikov for many useful discussions. I am obliged to L. Shur, who has performed the computer verifications of the solution and to L. Agibalova for her invaluable help in preparing the English version of this article.

References

1. Yang, S.N.: Phys. Rev. **168**, 1920–1923 (1968)
2. Baxter, R.J.: Ann. Phys. **70**, 193–228 (1972)
3. Baxter, R.J.: Philos. Trans. Soc. (London) **189**, 315–346 (1978)
4. Zamolodchikov, A.B.: Sov. Sci. Rev. **2** (to be published)
5. Faddeev, L.D., Sklyanin, V.K., Takhtadjyan, L.A.: Teor. Mat. Fiz. **40**, 194–212 (1979)
6. Faddeev, L.D.: Preprint LOMI P-2-79, Leningrad, 1979
7. Zamolodchikov, A.B., Zamolodchikov, Al.B.: Ann. Phys. **120**, 253–291 (1979)
8. Zamolodchikov, A.B.: Landau Institute Preprint 15, Chernogolovka 1980, Zh. Eksp. Teor. Fiz. (in press)
9. Cherednik, I.: Dokl. Akad. Nauk USSR **249**, 1095–1098 (1979)
10. Belavin, A.A.: Pisma Zh. Eksp. Teor. Fiz. (in press)

Communicated by Ya. G. Sinai

Received July 10, 1980

Commun. Math. Phys. 88, 185–205 (1983)

Communications in Mathematical Physics

On Zamolodchikov's Solution of the Tetrahedron Equations

R. J. Baxter

Research School of Physical Sciences, Australian National University, Canberra, Australia 2600

Abstract. The tetrahedron equations arise in field theory as the condition for the S-matrix in $2+1$-dimensions to be factorizable, and in statistical mechanics as the condition that the transfer matrices of three-dimensional models commute. Zamolodchikov has proposed what appear (from numerical evidence and special cases) to be non-trivial particular solutions of these equations, but has not fully verified them. Here it is proved that they are indeed solutions.

1. Introduction

A number of two-dimensional models in statistical mechanics have been exactly solved [1–4] by using the "star-triangle equations" (or simply "triangle equations") [5–7], which are generalizations of the star-triangle relation of the Ising model [8, 9]. These equations are the conditions for two row-to-row transfer matrices to commute.

Alternatively, these models can be put into field-theoretic form by considering the transfer matrix that adds a single face to the lattice [10], and regarding this as an S-matrix. The star-triangle relations then become the condition for the S-matrix to factorize [11].

These equations can be generalized to three-dimensional models in statistical mechanics, corresponding to a $1+2$-dimensional field theory. Unfortunately, the resulting "tetrahedron" equations are immensely more complicated, the main problem being that there are 2^{14} individual equations to satisfy for an Ising-type model, as against 2^6 in two-dimensions. Symmetries reduce this number somewhat, but there are still apparently many more equations than unknowns and until recently there was little reason to suppose that the equations permitted any interesting solutions at all.

However, by what appears to be an extraordinary feat of intuition, Zamolodchikov [12, 13] has written down particular possible solutions and has shown that they satisfy some of the tetrahedron equations in various limiting cases. Extensive numerical tests have also been made by V. Bajanov and Yu.

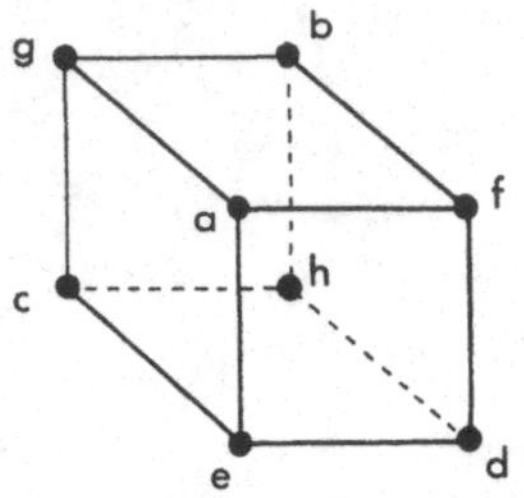

Fig. 1. Arrangement of the spins $a, \ldots, h$ on the corner sites of a cube

Stroganov, but a complete algebraic proof has hitherto not been obtained. Here (in Sects. 3–5) I give the required proof, which depends very much on classical nineteenth (and eighteenth) century mathematics, in particular on spherical trigonometry.

Let me refer to Zamolodchikov's two papers [12] and [13] as ZI and ZII, respectively. In ZI he considers the "static limit," when his solution simplifies considerably. Some of the Boltzmann weights then vanish, and the others satisfy various symmetry and anti-symmetry relations. In Sect. 6 I show that it is the only solution of the tetrahedron equations with these zero elements, symmetries and anti-symmetries.

From the statistical mechanical point of view, the anti-symmetry relations are rather unsatisfactory as they require that some of the Boltzmann weights be negative. We should like to replace them by strict symmetry relations, but unfortunately the tetrahedron equations then no longer admit a solution.

2. Interactions-Round-a-Cube Model and the Tetrahedron Equations

Just as the two-dimensional star-triangle relations can be obtained in convenient generality for an "Interactions-Round-a-Face" (IRF) model [7, 10], so can the three-dimensional tetrahedron relations be obtained for an "Interactions-Round-a-Cube" (IRC) model.

Consider a simple cubic lattice $\mathscr{L}$ of N sites. At each site i there is a "spin" σ_i, free to take some set of values. Each cube of the lattice has eight corner sites: let the spins thereon be $a, b, \ldots, h$, arranged as in Fig. 1, and allow all possible interactions between them. Then the Boltzmann weight of the cube will be some function of $a, b, \ldots, h$: let us write it (omitting commas) as $W(a|efg|bcd|h)$. The partition function is

$$Z = \sum \prod W(\sigma_i|\sigma_m\sigma_n\sigma_p|\sigma_j\sigma_k\sigma_l|\sigma_q), \tag{2.1}$$

where the product is over all N cubes of the lattice; for each cube $i, j, \ldots, p$ are the eight corner sites; the summation is over all values of all the N spins.

Let T be the layer-to-layer transfer matrix of the model. It depends on W, so can be written as $T(W)$. Consider another model, with a different weight function W'. Then, as has been shown by Jaekel and Maillard [14], the two-dimensional argument [7, 10] can be generalized to establish that $T(W)$ and $T(W')$ commute if

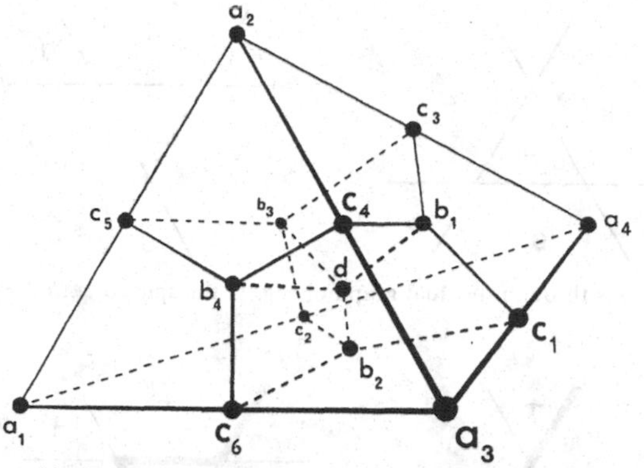

Fig. 2. The graph (a rhombic dodecahedron) whose partition function is the left hand side of (2.2). Some edges are shown by broken lines: this is merely to help visualisation

there exist two other weight functions W'' and W''' such that

$$\begin{aligned}&\sum_d W(a_4|c_2c_1c_3|b_1b_3b_2|d)\,W'(c_1|b_2a_3b_1|c_4dc_6|b_4)\\&W''(b_1|dc_4c_3|a_2b_3b_4|c_5)\,W'''(d|b_2b_4b_3|c_5c_2c_6|a_1)\\&\quad=\sum_b W'''(b_1|c_1c_4c_3|a_2a_4a_3|d)\,W''(c_1|b_2a_3a_4|dc_2c_6|a_1)\\&W'(a_4|c_2dc_3|a_2b_3a_1|c_5)\,W(d|a_1a_3a_2|c_4c_5c_6|b_4)\end{aligned}\tag{2.2}$$

for all values of the 14 spins $a_1, a_2, a_3, a_4, b_1, b_2, b_3, b_4, c_1, c_2, \ldots, c_6$. I shall refer to these 14 spins as "external," and to d as the "internal" spin.

We can think of each side of (2.2) as the partition function of four skewed cubes joined together, with a common interior spin d. This graph is a rhombic dodecahedron. For the left hand side of (2.2), the graph can be drawn (by distorting the angles) as in Fig. 2; for the right hand side, the centre spin d is to be connected to $a_1, \ldots, a_4$, instead of $b_1, \ldots, b_4$. In either case the lattice is bipartite: $a_1, \ldots, a_4$, $b_1, \ldots, b_4$ lie on one sub-lattice; $c_1, \ldots, c_6$, d on the other.

Now suppose that each spin σ_i can only take values $+1$ and -1, and that W is unchanged by negating all its eight arguments, i.e.

$$W(-a|-e,-f,-g|-b,-c,-d|-h) = W(a|efg|bcd|h)\,. \tag{2.3}$$

The Eq. (2.2) are then the tetrahedron equations used by Zamolodchikov. To see this, work with the duals $\mathscr{L}_D$ of the lattices discussed above. The cube shown in Fig. 1 is then replaced by three planes intersecting at a point, dividing three-dimensional space into eight volumes associated with the spins $a, \ldots, h$. Two parallel cross-sections of this diagram (one above the point of intersection, the other below) are shown in Fig. 3.

Adjacent spins are separated by faces of $\mathscr{L}_D$ (shown as lines in Fig. 3). Colour each face white if the spins on either side of it are equal, black if they are different. Then by letting the spins $a, \ldots, h$ take all possible values, one obtains the allowed colourings shown in Eq. (6.1) of ZI, a typical example being shown in Fig. 4.

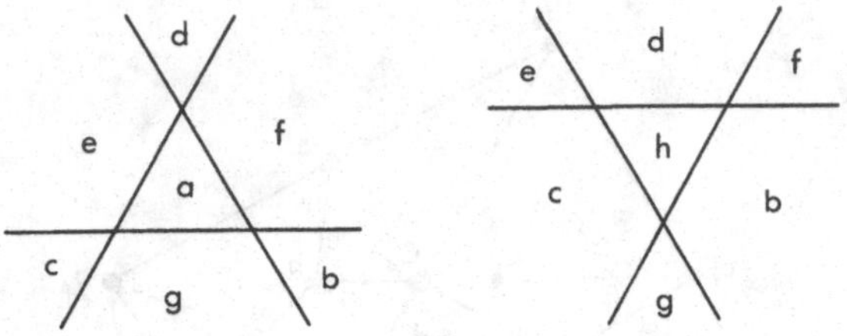

Fig. 3. Two cross-sections through the dual graph of Fig. 1: the spins $a, \ldots, h$ are here associated with volumes

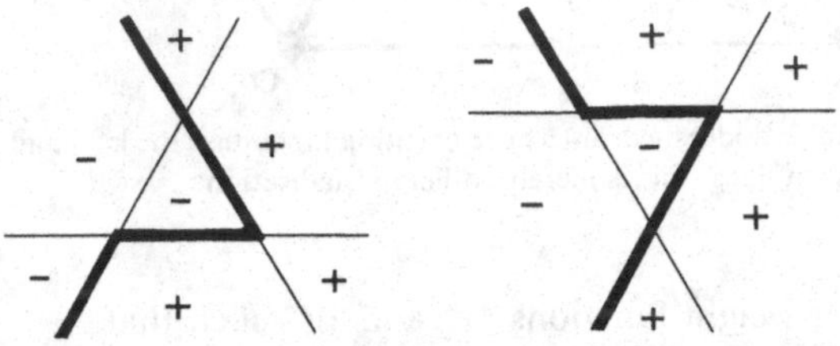

Fig. 4. A typical set of values of $a, \ldots, h$, and Zamolodchikov's corresponding plaquette colouring (heavy lines denote "black," lighter lines "white")

Conversely, any allowed colouring corresponds to just two sets of values of the spins $a, \ldots, h$, one set being obtained from the other by negating all of them. From (1.3), this negation leaves W unchanged, so the Boltzmann weight of the configuration is uniquely determined by the colouring of the faces. We can therefore replace the function W of the eight spins by a function S of the colours of the 12 faces. If we write "white" and "black" simply as "+" and "−", then the colour on the face between two spins c and g is simply the product cg of the spins. We can therefore explicitly define S by

$$S_{\substack{\mathrm{de,af,bg,ch} \\ \mathrm{bf,ag,ce,dh}}}^{\mathrm{cg,ae,df,bh}} = W(a|efg|bcd|h). \tag{2.4}$$

Define S', S'', S''' similarly, with W replaced repsectively by W', W'', W'''. Then (2.2) becomes precisely Eq. (3.9) of ZI, except only that in Zamolodchikov's notation S, S', S'', S''' become

$$S(z^{(23)}, z^{(12)}, z^{(13)}), \quad S(z^{(24)}, z^{(12)}, z^{(14)}),$$
$$S(z^{(34)}, z^{(13)}, z^{(14)}), \quad S(z^{(34)}, z^{(23)}, z^{(24)}),$$

respectively.

3. Zamolodchikov's Model

Zamolodchikov's model has the "black-white" symmetry property that S is unchanged by reversing the colours of all 12 faces. This means that W not only satisfies (2.3), but has the stronger sub-lattice symmetry properties

$$\begin{aligned} W(-a|efg|-b,-c,-d|h) &= W(a|-e,-f,-g|bcd|-h) \\ &= W(a|efg|bcd|h). \end{aligned} \tag{3.1}$$

Table 1. Values of the function W: the spin products λ, μ, ν, are defined by (3.3), the parameters $P_0, \ldots, R_3$ by (3.13)

λ	μ	ν	$W(a\|efg\|bcd\|h)$
+	+	+	$P_0 - abcd\,Q_0$
−	+	+	R_1
+	−	+	R_2
+	+	−	R_3
+	−	−	$ab\,P_1 + cd\,Q_1$
−	+	−	$ac\,P_2 + bd\,Q_2$
−	−	+	$ad\,P_3 + bc\,Q_3$
−	−	−	R_0

To express the function W in terms of Zamolodchikov's matrix elements $\sigma, S, \alpha, \ldots, K, H, V$, we simply note that $W(a|efg|bcd|h)$ corresponds to the spins as arranged in Fig. 3, translate each set of spin values to a face colouring, and look up the corresponding matrix element in Eq. (6.1) of ZI (using if necessary the black-white colour symmetry). For instance, using Fig. 4, considering also the effect of negating h:

$$W(-|-++|+-+|-)=K(\theta_1,\theta_2),$$
$$W(-|-++|+-+|+)=H(\theta_2,\theta_1). \tag{3.2}$$

Doing this, using Eqs. (2.2) and (4.9) of ZII, we find that W is "almost determined" by the values of the three spin products

$$\lambda = abeh\,, \quad \mu = acfh\,, \quad \nu = adgh\,. \tag{3.3}$$

More precisely, W has the values given in Table 1, where $P_0, \ldots, P_3, Q_0, \ldots, Q_3$, and $R_0, \ldots, R_3$ are constants, independent of the spins $a, \ldots, h$. The functions W', W'', W''' are also given by Table 1, but with different values of $P_0, \ldots R_3$.

Spin Symmetries

Before specifying these constants, it is worth considering the symmetries of W and the tetrahedron equations (2.2). Since (2.2) has to be valid for all values of all the 14 spins $a_1, a_2, \ldots c_6$, it consists of 2^{14} individual equations: a dauntingly large number! The spin reversal symmetry (2.3) helps: it reduces the number to 2^{13}. The stronger symmetry (3.1), together with the fact that the graph in Fig. 2 is bi-partite (the a_i and b_i lie on one sub-lattice, the c_i and d on the other), implies that (2.2) is unchanged not only by reversing all spins, but also by reversing all those on one sub-lattice. This reduces the number of distinct equations in (2.2) to 2^{12}.

This is still a very large number, but fortunately when we use the specific form of W given in Table 1 we find some dramatic simplifications. If W (and W', W'', W''') depended only on λ, μ, ν, then it would be true that (2.2) involved the 14 external spins $a_1, \ldots, c_6$ only via the 10 products a_1b_1, a_2b_2, a_3b_3, a_4b_4, $a_3a_4c_1$, $a_1a_4c_2$, $a_2a_4c_3$, $a_2a_3c_4$, $a_1a_2c_5$, $a_1a_3c_6$. Further, negating either the first four of these products, or the last six, would merely be equivalent to negating d. This would mean that (2.2) depended only on eight combinations of the external spin,s e.g.

$$\begin{array}{cccc} a_1a_2b_1b_2, & a_1a_3b_1b_3, & a_1a_4b_1b_4, & a_1a_3c_1c_2, \\ a_2a_3c_1c_3, & a_2a_4c_1c_4, & a_1a_4c_1c_6, & a_1a_2a_3a_4c_1c_5. \end{array} \tag{3.4}$$

Hence there would only be $2^8=256$ distinct equations. In fact W depends not only on λ, μ, ν: from Table 1 it has the form

$$W(a|efg|bcd|h)=\tau F(\lambda,\mu,\nu,abcd), \tag{3.5}$$

where F is a function of four spin products and τ is a sign factor, equal to either 1, ab, ac or ad. Also, the terms involving $abcd$ have $\lambda\mu\nu=1$, i.e. $abcd=efgh$. It follows that $abcd$ in (3.5) can be replaced by $efgh$ if desired.

Using the form (3.5) (and corresponding forms for W', W'', W''') in (2.2), the eight $abcd$ (or $efgh$) arguments that occur can be taken to be $b_1b_2b_3a_4$, $b_1b_2a_3b_4$, $b_1a_2b_3b_4$, $a_1b_2b_3b_4$, $b_1a_2a_3a_4$, $a_1b_2a_3a_4$, $a_1a_2b_3a_4$, $a_1a_2a_3b_4$. The ratios of these depend only on the first three products in (3.4), so they are determined by the products (3.4), apart from a single overall sign factor. This means that each of the previously mentioned 2^8 equations is in fact a pair of equations, one being obtained from the other by negating (say) a_1, a_2, a_3, a_4. Equivalently, one equation can be obtained from the other by negating R_0, Q_0, Q_1, Q_2, Q_3 in Table 1, for all four functions W, W', W'', W'''.

It still remains to examine the contributions to (2.2) of the τ sign factor in (3.5). I have done this (aided by a computer) for each of the 2^8 pairs of equations. In every case it is true that the multiplied contributions also depend only on $b_1b_2b_3a_4$ and the eight spin products in (3.4) [apart possibly from an overall sign factor multiplying both sides of (2.2)].

Thus there are just $2\times 2^8=512$ distinct equations: a great reduction on the original 2^{14}!

There are still further simplifications: the function W has the "diagonal reversal" property:

$$W(a|efg|bcd|h)=W(h|bcd|efg|a), \tag{3.6}$$

and similarly for W', W'', W'''. It follows that the two sides of (2.2) are interchanged by the transformation

$$\begin{array}{c} a_i \leftrightarrow b_i, \quad i=1,\dots,4, \\ c_1 \leftrightarrow c_5, \quad c_2 \leftrightarrow c_4, \quad c_3 \leftrightarrow c_6. \end{array} \tag{3.7}$$

This means that 64 of the 512 equations are satisfied identically, the right hand side being the same as the left hand side. The remaining 448 occur in pairs of type $B=A$ and $A=B$, so there are only 224 distinct equations remaining. These can conveniently be regarded as 112 pairs, one being obtained from the other by negating R_0, Q_0, …, Q_3 for all four functions W, W', W'', and W'''.

Set

$$X_j=P_j+Q_j, \qquad Y_j=P_j-Q_j \tag{3.8}$$

for $j=0,1,2,3$, and take W' to be also given by Table 1 and (3.8), but with P_j, Q_j, R_j, X_j, Y_j replaced by P'_j, Q'_j, R'_j, X'_j, Y'_j. Similarly for W'', W'''. Then two typical equations [obtained from (2.2) by taking all the external spins positive except a_1, and except a_1 and a_2] are

$$Y_0 Y_0' Y_0'' R_0''' + R_0 R_2' R_1'' X_0''' = Y_0''' R_0'' R_3' R_1 + R_0''' Y_1'' Y_1' Y_1 \,, \tag{3.9}$$

$$Y_0 Y_0' R_1'' R_0''' + R_0 R_2' X_0'' X_0''' = R_1''' R_0'' X_2' X_2 + Y_1''' Y_1'' R_0' R_2 \,. \tag{3.10}$$

Two other equations can be obtained immediately from these by negating $R_0, \ldots, R_0'''$ and $Q_j, \ldots, Q_j'''$, i.e. by negating each R_0 and replacing every X by a Y, and every Y by an X.

The Constants $P_0, \ldots, R_3$

Now let us return to the procedure described before (3.2), so as to obtain the constants $P_0, \ldots, R_3$ in Table 1, and hence the X_j, Y_j in (3.8), from Eqs. (2.2) and (4.9) of ZII. To do this we need various functions of the angles ϕ_1, ϕ_2, ϕ_3 of a spherical triangle, namely the spherical excesses

$$\begin{aligned} 2\alpha_0 &= \phi_1 + \phi_2 + \phi_3 - \pi \,, \\ 2\alpha_1 &= \pi + \phi_1 - \phi_2 - \phi_3 \,, \\ 2\alpha_2 &= \pi + \phi_2 - \phi_3 - \phi_1 \,, \\ 2\alpha_3 &= \pi + \phi_3 - \phi_1 - \phi_2 \,, \end{aligned} \tag{3.11}$$

and the quantities

$$t_i = [\tan(\alpha_i/2)]^{1/2} \,, \quad s_i = [\sin(\alpha_i/2)]^{1/2} \,, \quad c_i = [\cos(\alpha_i/2)]^{1/2} \,, \tag{3.12}$$

for $i = 0, 1, 2, 3$. We then find that $P_0, \ldots, R_3$ are given by

$$\begin{aligned} P_0 = 1 \,, \quad & Q_0 = t_0 t_1 t_2 t_3 \,, \quad R_0 = s_0/(c_1 c_2 c_3) \,, \\ P_i = t_j t_k \,, \quad & Q_i = t_0 t_i \,, \quad R_i = s_i/(c_0 c_j c_k) \,, \end{aligned} \tag{3.13}$$

for all permutations (i, j, k) of $(1, 2, 3)$. (Here ϕ_1, ϕ_2, ϕ_3 are Zamolodchikov's angles $\theta_2, \theta_1, \theta_3$.)

It is convenient to regard Table 1 and Eqs. (3.11)–(3.13) as defining W as a function of the angles ϕ_1, ϕ_2, ϕ_3, as well as of the spins $a, \ldots, h$. We can write it as

$$W[\phi_1, \phi_2, \phi_3 \,; a|efg|bcd|h] \,, \tag{3.14}$$

or, if the explicit spin dependence is not required, as $W[\phi_1, \phi_2, \phi_3]$. The other weights W', W'', W''' are also given by this function, but with different values of the arguments ϕ_1, ϕ_2, ϕ_3. Zamolodchikov's assertion is that (2.2) is satisfied if W, W', W'', W''' therein are given by

$$\begin{aligned} W = W[\theta_2, \theta_1, \theta_3] \,, \quad & W' = W[\pi - \theta_6, \theta_1, \pi - \theta_4] \,, \\ W'' = W[\theta_5, \pi - \theta_3, \pi - \theta_4] \,, \quad & W''' = W[\theta_5, \theta_2, \theta_6] \,, \end{aligned} \tag{3.15}$$

where $\theta_1, \ldots, \theta_6$ are the six angles of a spherical quadrilateral, as shown in Fig. 5, and equivalently in Fig. 7 of ZII. These angles are not independent: they necessarily satisfy the relation (3.2) of ZII.

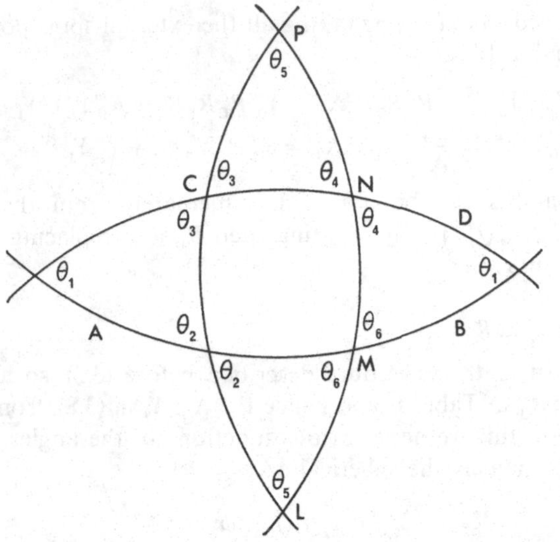

Fig. 5. A segment of a spherical quadrilateral, showing its associated four triangles, with interior angles $\theta_1, \ldots, \theta_6$

The parameters $P'_i, P''_i, P'''_i, Q'_i, \ldots, R'''_3$ are of course given by adding primes to P, Q, R in (3.13) and substituting the appropriate values of ϕ_1, ϕ_2, ϕ_3 into (3.11), and thence into (3.12) and (3.13). For instance, the double-primed parameters $P''_0, \ldots, R''_3$ are obtained by taking ϕ_1, ϕ_2, ϕ_3 to be θ_5, $\pi-\theta_3$, $\pi-\theta_4$.

4. Angle Symmetries

We want to verify Zamolodchikov's assertion that (2.2) is satisfied by (3.15). Fortunately we do not have to prove each of the 224 equations individually. We can regard each of them as an identity, to be verified for all values of $\theta_1, \ldots, \theta_6$ satisfying the spherical quadrilateral constraint (3.2) of ZII. It turns out that many of these identities are simple corollaries of one another.

Q Negation

We can regard $\theta_1, \theta_2, \theta_3$ as determined by $\theta_4, \theta_5, \theta_6$ and the arc lengths LM, MN in Fig. 5. These parameters can be varied so as to shift the line AB in Fig. 5 upwards through the point C. The $(\theta_1, \theta_2, \theta_3)$ triangle first shrinks to a point, and then reappears in an inverted configuration, as shown in Fig. 6. This gives a new spherical quadrilateral, with angles θ_1, θ_2, θ_3, $\pi-\theta_4$, $\pi-\theta_5$, $\pi-\theta_6$.

Since $\theta_1+\theta_2+\theta_3-\pi$ is the area of the $(\theta_1, \theta_2, \theta_3)$ triangle [15], it cannot become negative during this process: it has a double zero when the triangle shrinks to a point. This means that $[\tan(\theta_1+\theta_2+\theta_3-\pi)/4]^{1/2}$ has a simple zero, so is negated when analytically continued from Fig. 5 to Fig. 6. The same is true of $[\sin(\theta_1+\theta_2+\theta_3-\pi)/4]^{1/2}$. All other square roots retain their original positive sign.

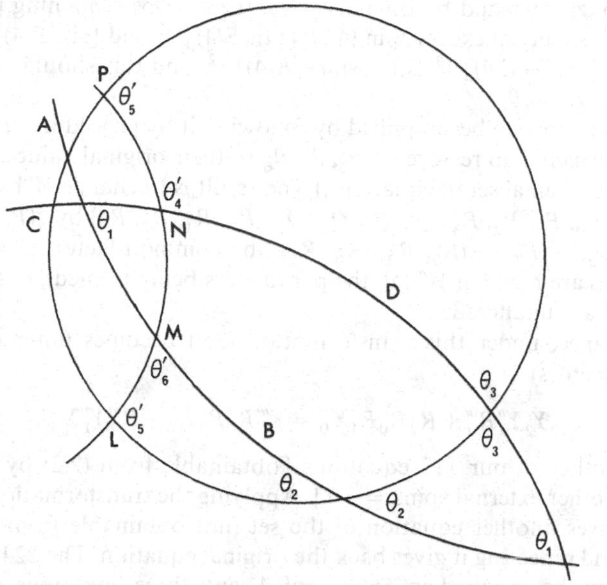

Fig. 6. The spherical quadrilateral of Fig. 5, after the line AB has been shifted through C. Here θ'_j denotes the supplement $\pi-\theta_j$ of the angle θ_j

It follows that for each of our equations, we can obtain another by the following procedure:

Use (3.15) and (3.11)–(3.13) to write the equation explicitly in terms of $\theta_1, \ldots \theta_6$; negate $[\tan(\theta_1+\theta_2+\theta_3-\pi)/4]^{1/2}$ and $[\sin(\theta_1+\theta_2+\theta_3-\pi)/4]^{1/2}$ throughout; replace $\theta_4, \theta_5, \theta_6$ by $\pi-\theta_4$, $\pi-\theta_5$, $\pi-\theta_6$.

Let us call this procedure P_{123}, and define P_{146}, P_{345}, P_{256} similarly. (Each corresponds to shrinking one of the triangles in Fig. 5 through a point.) By itself, each such procedure gives an equation which is not in our original set, but if we perform all four sequentially, then the result is to return $\theta_1, \ldots, \theta_6$ to their original values, having negated R_0 and Q_j for $j=0,1,2,3$ and for all four functions W, W', W'', W'''. (This corresponds to inverting each S matrix.) This is the (R_0, Q)-negation pair symmetry discussed between (3.5) and (3.11).

The 224 equations therefore occur in 112 pairs, each equation of a pair being a corollary of the other.

Negation of θ_1

Another way to analytically continue $\theta_1, \ldots, \theta_6$ is to allow the great circles AB and CD in Fig. 5 to first become coincident and then cross one another. The result (after vertically mirror inverting) is to replace $\theta_1, \ldots, \theta_6$ in Fig. 5 by $-\theta_1$, $\pi-\theta_2$, $\pi-\theta_3$, $\pi-\theta_4$, θ_5, $\pi-\theta_6$, respectively; i.e. to negate θ_1 and supplement $\theta_2, \theta_3, \theta_4, \theta_6$. In this process the spherical excesses (triangle areas)

$$\begin{aligned} &\theta_1+\theta_2+\theta_3-\pi, \quad \pi+\theta_1-\theta_2-\theta_3, \\ &\theta_1+\theta_4+\theta_6-\pi, \quad \pi+\theta_1-\theta_4-\theta_6, \end{aligned} \tag{4.1}$$

all pass through zero and become negative, their ratios remaining positive. Thus for each such spherical excess E in (4.1), $[\tan(E/4)]^{1/2}$ and $[\sin(E/4)]^{1/2}$ should be replaced by $i[\tan(-E/4)]^{1/2}$ and $i[\sin(-E/4)]^{1/2}$, and this should be done before transforming $\theta_1, \ldots, \theta_6$.

This procedure can be simplified by following it by procedures P_{123} and P_{146}, the effect of which is to restore $\theta_2, \theta_3, \theta_4, \theta_6$ to their original values. (It also keeps us within our original set of equations.) The result is to change W by replacing the parameters $(P_0, P_1, P_2, P_3; Q_0, Q_1, Q_2, Q_3; R_0, R_1, R_2, R_3)$ by $\xi(P_2, iQ_3, P_0, iQ_1; Q_2, -iP_3, Q_0, -iP_1; -iR_2, R_3, iR_0, R_1)$, the common factor ξ being P_2^{-1}. The same changes are made in W' (all the parameters being primed), but the functions W'' and W''' are unaltered.

For instance, under this transformation (3.9) becomes (after cancelling the common ξ factors)

$$Y_2 Y_2' Y_0'' R_0''' + R_2 R_0' R_1'' X_0''' = Y_0''' R_0'' R_1' R_3 - R_0''' Y_1'' X_3' X_3 , \tag{4.2}$$

which is another of our 112 equations [obtainable from (2.2) by taking b_2, c_2, $c_6 = -1$, all other external spins $= +1$]. Applying the transformation to any of the equations gives another equation of the set (not obtainable from the first by Q negation), and repeating it gives back the original equation. The 224 equations can therefore now be grouped in 56 sets of 4, any three equations of a set being corollaries of the fourth.

Permutation Symmetry

The weight function (3.14) has various symmetry properties, in particular

$$W(\phi_i, \phi_j, \phi_k; a|e_i e_j e_k|b_i b_j b_k|h) = \text{unchanged by permuting } i, j, k , \tag{4.3}$$

and

$$\begin{aligned} W[\phi_1, \phi_2, \phi_3; a|efg|bcd|h] &= W[\pi-\phi_1, \pi-\phi_3, \phi_2; f|dba|geh|c] \\ &= W[\phi_1, \pi-\phi_3, \pi-\phi_2; b|hfg|acd|e] \\ &= W[\phi_1, \phi_2, \phi_3; h|bcd|efg|a] . \end{aligned} \tag{4.4}$$

Using these, (2.2) can be put into the more obviously symmetric form

$$\begin{aligned} &\sum_d W[\theta_2, \theta_3, \theta_1; a_4|c_2 c_3 c_1|b_1 b_2 b_3|d]\, W[\theta_1, \theta_6, \theta_4; a_3|c_1 c_6 c_4|b_4 b_1 b_2|d] \\ &\quad \cdot W[\theta_4, \theta_3, \theta_5; a_2|c_4 c_3 c_5|b_3 b_4 b_1|d]\, W[\theta_5, \theta_6, \theta_2; a_1|c_5 c_6 c_2|b_2 b_3 b_4|d] \\ &= \sum_d W[\theta_5, \theta_6, \theta_2; b_1|c_1 c_3 c_4|a_2 a_3 a_4|d]\, W[\theta_4, \theta_3, \theta_5; b_2|c_2 c_6 c_1|a_3 a_4 a_1|d] \\ &\quad \cdot W[\theta_1, \theta_6, \theta_4; b_3|c_5 c_3 c_2|a_4 a_1 a_2|d]\, W[\theta_2, \theta_3, \theta_1; b_4|c_4 c_6 c_5|a_1 a_2 a_3|d] . \end{aligned} \tag{4.5}$$

For some purposes it is convenient to replace the angles θ_r and the spins c_r by ψ_{ij} and e_{ij}, where

$$\begin{aligned} &\theta_1 = \psi_{34}, \quad \theta_2 = \psi_{14}, \quad \theta_3 = \psi_{24}, \\ &\theta_4 = \psi_{23}, \quad \theta_5 = \psi_{12}, \quad \theta_6 = \psi_{13}, \\ &c_1 = e_{34}, \quad c_2 = e_{14}, \quad c_3 = e_{24}, \\ &c_4 = e_{23}, \quad c_5 = e_{12}, \quad c_6 = e_{13}. \end{aligned} \tag{4.6}$$

Making these substitutions into (4.5), all indices are integers between 1 and 4, and it is easy to check by using (4.3) that the equation is unchanged by permuting the integers 1, 2, 3, 4. Further, any such permutation is equivalent to merely re-drawing Fig. 5, by extending the quadrilateral to another part of the spherical surface and possibly rotating or reflecting it. Thus the permutation takes one set of quadrilateral angles $\theta_1, \ldots \theta_6$ to another.

There are 24 such permutations (we can think of them as the permutations of the four vertices of the tetrahedron in Fig. 2, $\theta_1, \ldots, \theta_6$ being associated with the edges). Each takes one of the 224 equations to itself or another, so this is a very strong symmetry. In fact, when use is also made of the Q-negation and θ_1-negation symmetries, it turns out that all the 224 equations are corollaries of just two archetypical equations, which we can take to be Eqs. (3.9) and (3.10) above. Thus our original 2^{14} equations have finally reduced to two!

[We cannot get down to just one equation with the above transformations: they all take R-parameters to R-parameters, and X or Y-parameters to X or Y-parameters. Since each term in (3.9) has an odd number of each, and each term in (3.10) has an even number, one equation cannot be transformed to the other. The 224 equations fall into two distinct classes: an "odd" class with 128 members, and an "even" class with 96.]

5. Proof of the two Archetypal Identities

To verify that Zamolodchikov's solution does indeed satisfy the tetrahedron equations, it remains only to prove that (3.9) and (3.10) are satisfied for all sets $(\theta_1, \ldots, \theta_6)$ of spherical quadrilateral angles.

First let us follow Zamolodchikov's notation [Eqs. (4.9) and (2.2) of ZII] and define functions σ, S, a, U, ω, V of ϕ_1, ϕ_2, ϕ_3 by

$$\begin{aligned} \sigma(\phi_1, \phi_2, \phi_3) &= 1 + t_0 t_1 t_2 t_3, \\ S(\phi_1, \phi_2, \phi_3) &= 1 - t_0 t_1 t_2 t_3, \\ a(\phi_1, \phi_2, \phi_3) &= s_0/(c_1 c_2 c_3), \\ U(\phi_1, \phi_2 | \phi_3) &= s_3/(c_0 c_1 c_2), \\ \omega(\phi_1, \phi_2 | \phi_3) &= t_0 t_3 + t_1 t_2, \\ V(\phi_1, \phi_2 | \phi_3) &= t_0 t_3 - t_1 t_2, \end{aligned} \tag{5.1}$$

where t_i, s_i, c_i are defined in terms of ϕ_1, ϕ_2, ϕ_3 by (3.12) and (3.11). (The functions σ, S, a are symmetric in ϕ_1, ϕ_2, ϕ_3; U, ω, V are symmetric only in ϕ_1 and ϕ_2.)

The 48 parameters X_j, Y_j, $R_j, \ldots, R_j'''$ are given by (3.11)–(3.13) and (3.8), with arguments ϕ_1, ϕ_2, ϕ_3 determined by (3.15). They are equal to particular values of the functions $\sigma, \ldots, V$ (possibly negated). For instance, setting ϕ_1, ϕ_2, ϕ_3 to be θ_5, $\pi - \theta_3$, $\pi - \theta_4$, we can verify that $Y_1'' = -V(\theta_3, \theta_4 | \theta_5)$. Doing this for all the parameters in (3.9) and (3.10), these equations can be written more explicitly as

$$\begin{aligned} & S(\theta_1,\theta_2,\theta_3)\,S(\theta_1,\theta_4,\theta_6)\,S(\theta_3,\theta_4,\theta_5)\,\alpha(\theta_2,\theta_5,\theta_6) \\ &\quad +\alpha(\theta_1,\theta_2,\theta_3)\,\alpha(\theta_1,\theta_4,\theta_6)\,\alpha(\theta_3,\theta_4,\theta_5)\,\sigma(\theta_2,\theta_5,\theta_6) \\ &\quad = U(\theta_1,\theta_3|\theta_2)\,U(\theta_1,\theta_4|\theta_6)\,U(\theta_3,\theta_4|\theta_5)\,S(\theta_2,\theta_5,\theta_6) \\ &\quad + V(\theta_1,\theta_3|\theta_2)\,V(\theta_1,\theta_4|\theta_6)\,V(\theta_3,\theta_4|\theta_5)\,\alpha(\theta_2,\theta_5,\theta_6), \end{aligned} \tag{5.2}$$

$$\begin{aligned} & S(\theta_1,\theta_2,\theta_3)\,S(\theta_1,\theta_4,\theta_6)\,\alpha(\theta_3,\theta_4,\theta_5)\,\alpha(\theta_2,\theta_5,\theta_6) \\ &\quad +\alpha(\theta_1,\theta_2,\theta_3)\,\alpha(\theta_1,\theta_4,\theta_6)\,\sigma(\theta_3,\theta_4,\theta_5)\,\sigma(\theta_2,\theta_5,\theta_6) \\ &\quad =\omega(\theta_2,\theta_3|\theta_1)\,\omega(\theta_4,\theta_6|\theta_1)\,U(\theta_3,\theta_4|\theta_5)\,U(\theta_2,\theta_6|\theta_5) \\ &\quad + U(\theta_2,\theta_3|\theta_1)\,U(\theta_4,\theta_6|\theta_1)\,V(\theta_3,\theta_4|\theta_5)\,V(\theta_2,\theta_6|\theta_5). \end{aligned} \tag{5.3}$$

[Equation (5.2) is precisely Eq. (3.1) of ZII.]

Simplification of $\sigma, \ldots, V$

An irritating feature of these equations is the proliferation of square roots that enter via (3.12). We can remove these by introducing the lengths of the sides of the spherical triangles, as well as their angles.

To do this, we need some basic formulae of spherical trigonometry, which are given by Todhunter and Leathem [15], here referred to as TL. Let A, B, C be the interior angles of a spherical triangle, and a, b, c the lengths of the corresponding (opposite) sides. Let

$$E=A+B+C-\pi, \tag{5.4}$$

$$s=\tfrac{1}{2}(a+b+c) \tag{5.5}$$

(E is the "spherical excess" of the triangle, $2s$ is the perimeter). Then, from (5.1), (3.12), and (3.11),

$$\sigma(A,B,C)=1+\left[\tan\frac{E}{4}\tan\frac{2A-E}{4}\tan\frac{2B-E}{4}\tan\frac{2C-E}{4}\right]^{1/2}. \tag{5.6}$$

The square-root expression in (5.6) is the "Lhuilierian" of the triangle (Sect. 137 of TL) and is equal to $\tan(E/4)\cot(s/2)$. It follows at once that

$$\begin{aligned}\sigma(A,B,C) &= 1+\tan(E/4)/\cot(s/2) \\ &= \sin[(2s+E)/4]/\cos\frac{E}{4}\sin\frac{s}{2}. \end{aligned} \tag{5.7}$$

Similarly,

$$S(A,B,C)=\sin[(2s-E)/4]/\cos\frac{E}{4}\sin\frac{s}{2}. \tag{5.8}$$

As is shown in Sect. 28 of TL, any relation in spherical trigonometry remains true if the angles are changed into the supplements of the corresponding sides, and vice-versa. Applying this duality principle to Eq. (32) of Sect. 139 of TL, we obtain

$$\sin^2\frac{1}{2}s=\frac{\sin\frac{E}{4}\cos\frac{2A-E}{4}\cos\frac{2B-E}{4}\cos\frac{2C-E}{4}}{\sin\frac{1}{2}A\sin\frac{1}{2}B\sin\frac{1}{2}C}, \tag{5.9}$$

and from this and (5.1), (3.11), and (3.12) we can verify that

$$a(A,B,C)=\frac{\sin(E/4)}{\sin\frac{s}{2}\left[\sin\frac{A}{2}\sin\frac{B}{2}\sin\frac{C}{2}\right]^{1/2}}. \tag{5.10}$$

Using Cagnoli's theorem (Sect. 132 of TL), we can establish that

$$\sin\tfrac{1}{2}E\sin(C-\tfrac{1}{2}E)=\sin^2\frac{c}{2}\sin A\sin B, \tag{5.11}$$

while from (5.1), (3.11), and (3.12),

$$\frac{U(A,B|C)}{a(A,B|C)}=\left[\frac{\sin(C-\frac{1}{2}E)}{\sin\frac{1}{2}E}\right]^{1/2}. \tag{5.12}$$

Eliminating $\sin(C-\frac{1}{2}E)$ between (5.11) and (5.12), then using (5.10), we obtain

$$U(A,B|C)=\frac{\sin\frac{c}{2}\left[\cos\frac{A}{2}\cos\frac{B}{2}\right]^{1/2}}{\cos\frac{E}{4}\sin\frac{s}{2}\left[\sin\frac{C}{2}\right]^{1/2}}. \tag{5.13}$$

Taking the duals of Eqs. (34) and (35) of Sect. 140 of TL, then dividing by (5.9), gives the formulae

$$\frac{\sin\frac{1}{2}(s-c)}{\sin\frac{1}{2}s}=\left\{\tan\frac{2A-E}{4}\tan\frac{2B-E}{4}\tan\frac{A}{2}\tan\frac{B}{2}\right\}^{1/2}, \tag{5.14}$$

$$\frac{\cos\frac{1}{2}(s-c)}{\sin\frac{1}{2}s}=\left\{\cot\frac{E}{4}\tan\frac{2C-E}{4}\tan\frac{A}{2}\tan\frac{B}{2}\right\}^{1/2}. \tag{5.15}$$

Using these, it follows from (5.1), (3.11), and (3.12) that

$$\omega(A,B|C)=\frac{\tan\frac{E}{4}\cos\frac{s-c}{2}+\sin\frac{s-c}{2}}{\sin\frac{s}{2}\left[\tan\frac{A}{2}\tan\frac{B}{2}\right]^{1/2}}=\frac{\sin[(E+2s-2c)/4]}{\cos\frac{E}{4}\sin\frac{s}{2}\left[\tan\frac{A}{2}\tan\frac{B}{2}\right]^{1/2}}. \tag{5.16}$$

Similarly,

$$V(A,B|C)=\frac{\sin[(E-2s+2c)/4]}{\cos\frac{E}{4}\sin\frac{s}{2}\left[\tan\frac{A}{2}\tan\frac{B}{2}\right]^{1/2}}. \tag{5.17}$$

For the purpose of verifying (5.2) and (5.3), these expressions for σ, S, a, U, ω, V are more convenient than the original definitions (5.1). They contain square roots only of multiplicative functions of individual angles, and these cancel out of Eqs. (5.2) and (5.3).

To use these forms, we need a notation for the lengths of the sides of the four triangles $(\theta_1, \theta_3, \theta_2)$, $(\theta_4, \theta_1, \theta_6)$, $(\theta_3, \theta_4, \theta_5)$, $(\theta_2, \theta_6, \theta_5)$ in Fig. 5. If θ_i and θ_j are two interior angles of a triangle, let $r_{ij}(=r_{ji})$ be the length of the side joining them. Thus $r_{65}=r_{56}=LM$, $r_{64}=MN$ and $r_{45}=NP$. Since every great circle has length 2π, and any two great circles bisect each other, the length LP (along either circle) is π. The lengths r_{ij} therefore satisfy the four relations

$$r_{56}+r_{64}+r_{45}=r_{52}+r_{23}+r_{35}=r_{12}+r_{26}+r_{61}=\pi, \tag{5.18}$$

$$r_{13}+r_{41}+r_{34}=\pi. \tag{5.19}$$

We shall also need following quantities associated with a triangle $(\theta_i, \theta_j, \theta_k)$:

$$2\alpha_{ijk}=\theta_i+\theta_j+\theta_k-\pi, \tag{5.20}$$

$$2s_{ijk}=r_{ij}+r_{jk}+r_{ki}, \tag{5.21}$$

$$e_i=\exp\tfrac{1}{2}\theta_i, \quad x_i=\cos\tfrac{1}{2}\theta_i, \quad y_i=\sin\tfrac{1}{2}\theta_i, \tag{5.22}$$

$$\begin{aligned} F_{ijk}&=\sin\tfrac{1}{2}(s_{ijk}+\alpha_{ijk}), \quad G_{ijk}=\sin\tfrac{1}{2}(s_{ijk}-\alpha_{ijk}) \\ H_{ijk}&=\tfrac{1}{2}\sin\alpha_{ijk}, \quad L_{ijk}=\sin\tfrac{1}{2}r_{ij}, \\ M_{ijk}&=\sin\tfrac{1}{2}(\alpha_{ijk}+s_{ijk}-r_{ij}), \\ N_{ijk}&=\sin\tfrac{1}{2}(\alpha_{ijk}-s_{ijk}+r_{ij}). \end{aligned} \tag{5.23}$$

From (5.18). (5.19), and (5.21) it is apparent that

$$s_{132}+s_{416}+s_{345}+s_{265}=2\pi. \tag{5.24}$$

First Identity

Substituting the forms given in (5.7)–(5.17) for the functions $\sigma, \ldots, V$ into the first identity (5.2), cancelling common factors, and using the definitions (5.22) and (5.23), the identity becomes

$$\begin{aligned} &\mu G_{132}G_{416}G_{345}H_{265}+H_{132}H_{416}H_{345}F_{265} \\ &=\lambda\mu L_{132}L_{416}L_{345}G_{265}+\lambda N_{132}N_{416}N_{345}H_{265}, \end{aligned} \tag{5.25}$$

where

$$\lambda=x_1x_4x_3, \quad \mu=y_1y_4y_3. \tag{5.26}$$

Each of H_{265}, F_{265}, G_{265} is defined by (5.23) in terms of a sine function. For these quantities, it is convenient to write $\sin u$ as $\mathrm{Im}[\exp iu]$ (or, for G_{265}, as $-\mathrm{Im}[\exp(-iu)]$. The resulting expressions can be factored, using (5.20) and (5.24), into a product of 3 terms associated with the other triangles, e.g.

$$F_{265}=\mathrm{Im}(A_{132}A_{416}A_{345}), \tag{5.27}$$

where

$$A_{ijk}=\exp[i(\pi+\theta_k-2s_{ijk})/4]. \tag{5.28}$$

(I use the convention that where i occurs as an index, it is an integer between 1 and 6; elsewhere it is the square root of -1.)

Subtracting the right hand side from the left hand side, it follows that (5.25) can be written as

$$\mathrm{Im}(J)=0\,, \tag{5.29}$$

where

$$\begin{aligned} J=&-\tfrac{1}{2}i\mu\hat{G}_{132}\hat{G}_{416}\hat{G}_{345}-e_1e_3e_4\hat{H}_{132}H_{416}\hat{H}_{345}\\ &-i\lambda\mu(e_1e_3e_4)^{-1}\hat{L}_{132}\hat{L}_{416}\hat{L}_{345}-\tfrac{1}{2}\lambda\hat{N}_{132}\hat{N}_{416}\hat{N}_{345}\,, \end{aligned} \tag{5.30}$$

and

$$\begin{aligned} \hat{G}_{ijk}&=G_{ijk}\exp[\tfrac{1}{2}i(\pi+\theta_k+r_{ij})]\,,\\ \hat{H}_{ijk}&=H_{ijk}\exp[i(\theta_k-\theta_i-\theta_j-\pi+2r_{ij}-2s_{ijk})/4]\\ \hat{L}_{ijk}&=L_{ijk}\exp[\tfrac{1}{2}i(\alpha_{ijk}+s_{ijk})]\,,\\ \hat{N}_{ijk}&=N_{ijk}\exp[\tfrac{1}{2}i(\pi+\theta_k)] \end{aligned} \tag{5.31}$$

Each of $\hat{G}_{ijk},\ldots,\hat{N}_{ijk}$ is a property of a single spherical triangle, with angles θ_i, θ_j, θ_k. We can express them in terms of two angles and an included side; in particular of θ_i, θ_j, and r_{ij}. This will give J as a function only of θ_1, θ_4, θ_3 and r_{13}, r_{41}, r_{34}. These variables are independent except for the simple relation (5.19), so we should be able to verify explicitly that (5.29) is satisfied.

In fact we can simplify this procedure. Using only (5.20), (5.23), and (5.31), it is readily seen that each of $\hat{G}_{ijk}$, $\hat{H}_{ijk}$, $\hat{L}_{ijk}$, $\hat{N}_{ijk}$ is a linear combination of the expressions

$$\begin{gathered} \exp[i(2s_{ijk}+\theta_k)/4]\,,\qquad \exp[-i(2s_{ijk}+\theta_k)/4]\,,\\ \exp[i(-2s_{ijk}+3\theta_k)/4]\,, \end{gathered} \tag{5.32}$$

with coefficients that are simple explicit functions of θ_i, θ_j, and r_{ij}.

We can relate the three expressions (5.32) by using spherical trigonometry. The duals of Eqs. (25) and (26) of Sect. 138 of TL are

$$\begin{aligned} \sin s&=\sin c\cos\tfrac{1}{2}A\cos\tfrac{1}{2}B\,\mathrm{cosec}\tfrac{1}{2}C\,,\\ \cos s&=[-\sin\tfrac{1}{2}A\sin\tfrac{1}{2}B+\cos\tfrac{1}{2}A\cos\tfrac{1}{2}B\cos c]\,\mathrm{cosec}\tfrac{1}{2}C\,. \end{aligned} \tag{5.33}$$

From these it follows that

$$e^{-is}\sin\tfrac{1}{2}C=[\cos\tfrac{1}{2}A\cos\tfrac{1}{2}B\,e^{-ic}-\sin\tfrac{1}{2}A\sin\tfrac{1}{2}B]\,. \tag{5.34}$$

Expanding $\sin\frac{1}{2}C$ on the left hand side in terms of $\exp(\pm\frac{1}{2}iC)$, then multiplying by $\exp[i(2s+C)/4]$ and replacing A, B, C, s, c by θ_i, θ_j, θ_k, s_{ijk}, r_{ij}, we obtain a linear relation between the expressions (5.32). Thus we can write $\hat{G}_{ijk},\ldots,\hat{N}_{ijk}$ as linear combinations of any two of the expressions (5.32), with coefficients that are simple explicit functions of θ_i, θ_j, and r_{ij}.

More conveniently, we can write them as linear combinations of

$$\begin{aligned} p_{ijk}&=\sin[\tfrac{1}{2}(s_{ijk}+\alpha_{ijk}-r_{ijk})]\,.\\ q_{ijk}&=\exp(\tfrac{1}{2}ir_{ij})\sin[\tfrac{1}{2}(s_{ijk}+\alpha_{ijk})] \end{aligned} \tag{5.35}$$

[which are themselves linear combinations of the expressions (5.32)]. We find that

$$\begin{aligned}
\hat{G}_{ijk} &= 2x_i x_j p_{ijk} + [-\cos\tfrac{1}{2}(\theta_i-\theta_j) + i\sin\tfrac{1}{2}(\theta_i+\theta_j)]q_{ijk}\,,\\
\hat{H}_{ijk} &= x_i x_j p_{ijk} - y_i y_j q_{ijk}\,,\\
\hat{L}_{ijk} &= -p_{ijk} + q_{ijk}\,,\\
\hat{N}_{ijk} &= -[\cos\tfrac{1}{2}(\theta_i-\theta_j) + i\sin\tfrac{1}{2}(\theta_i+\theta_j)]p_{ijk} + 2y_i y_j q_{ijk}\,.
\end{aligned} \tag{5.36}$$

Substituting these expressions into (5.30) and re-arranging, we obtain (after many cancellations)

$$2J = -x_1x_3x_4p_{132}p_{416}p_{345} - iy_1y_3y_4q_{132}q_{416}q_{345}\,. \tag{5.37}$$

Using (5.19), we find from (5.35) that $q_{132}q_{416}q_{345}$ is pure imaginary. Since x_j, y_j, p_{ijk} are real, it follows that J is real. We have therefore verified (5.29), and hence the identities (5.2) and (3.9).

It is interesting to note that after (5.36) we have only used the definitions (5.35) to note that $p_{132}p_{416}p_{345}$ and $iq_{132}q_{416}q_{345}$ are real. Thus (5.29) is true for any expression J given by (5.30) and (5.36), the only restrictions on p_{ijk} and q_{ijk} being these two reality conditions.

Second Identity

The same general techniques can be used to verify the second identity (5.3), but there is no longer such a simple symmetry between triangles (132), (416), and (345).

Substituting into (5.3) the expressions for σ, $S, \ldots, V$ given in (5.7)–(5.17), cancelling common factors and using the definitions (5.23), we obtain

$$\begin{aligned}
&y_iG_{132}G_{146}H_{345}H_{265} + y_5H_{132}H_{146}F_{345}F_{265}\\
&\quad = \varrho y_1M_{231}M_{641}L_{345}L_{265} + \varrho y_5L_{231}L_{641}N_{345}N_{265}\,,
\end{aligned} \tag{5.38}$$

where

$$\varrho = x_2x_3x_4x_6\,. \tag{5.39}$$

Writing the sine functions associated with triangle (256) as imaginary parts of exponentials, and using (5.18)–(5.20) and (5.24) to share out these exponentials between the other three triangles, (5.38) can be written as

$$\mathrm{Im}(K) = 0\,, \tag{5.40}$$

where

$$\begin{aligned}
K = {}& \tfrac{1}{2}iy_1\hat{G}_{132}\hat{G}_{146}\hat{\hat{H}}_{345} - e_1\hat{H}_{132}\hat{H}_{146}\hat{\hat{F}}_{345}\\
&+ iy_1x_3x_4\hat{M}_{132}\hat{M}_{146}\hat{\hat{L}}_{345} + e_1^{-1}x_3x_4\hat{Z}_{132}\hat{Z}_{146}\hat{\hat{N}}_{345}\,,
\end{aligned} \tag{5.41}$$

$\hat{G}$, $\hat{H}$ being defined by (5.36), and $\hat{M}, \ldots, \hat{\hat{N}}$ by

$$\begin{aligned}
\hat{M}_{ijk} &= x_kM_{kji}\exp[-\tfrac{1}{2}i(\pi + r_{ik})]\,,\\
\hat{Z}_{ijk} &= x_kL_{kji}\exp[i(\pi + \theta_i + \theta_k - \theta_j + 2s_{ijk} - 2r_{ik})/4]\,,\\
\hat{\hat{H}}_{345} &= H_{345}\exp[\tfrac{1}{2}i(\theta_5 + r_{34} - \pi)]\,,\\
\hat{\hat{F}}_{345} &= y_5F_{345}\exp[\tfrac{1}{2}i(\alpha_{345} - s_{345} + r_{34})]\,,\\
\hat{\hat{N}}_{345} &= y_5N_{345}\exp[\tfrac{1}{2}i(\alpha_{345} + s_{345})]\,.
\end{aligned} \tag{5.42}$$

Again we look for linear relations between the functions associated with each triangle, the coefficients being explicit functions of θ_1, θ_3, θ_4, r_{13}, r_{14}, r_{34}. Using (5.34), we find that

$$\begin{aligned}
\hat{M}_{ijk} &= x_i e_j p_{ijk} - y_i y_j q_{ijk} - i x_i y_j q^*_{ijk}, \\
\hat{Z}_{ijk} &= x_i p_{ijk} + i y_i q_{ijk}, \\
\hat{\hat{F}}_{345} &= \hat{\hat{H}}_{345} + x_3 x_4 L_{345}, \\
\hat{\hat{N}}_{345} &= \hat{\hat{H}}_{345} - y_3 y_4 L_{345},
\end{aligned} \tag{5.43}$$

q^*_{ijk} being the complex conjugate of q_{ijk}.

Substituting these expressions, and the expressions (5.36) for $\hat{G}$ and $\hat{H}$, into (5.41), one finds that K can be written as

$$K = K_1 + K_2 + K^*_2, \tag{5.44}$$

where

$$K_1 = i y_1 q_{132} q_{146} [\tfrac{1}{2} e_3^{-1} e_4^{-1} \hat{\hat{H}}_{345} - x_3 x_4 y_3 y_4 L_{345}], \tag{5.45}$$

$$\begin{aligned}
K_2 = \{ &\tfrac{1}{2} x_1 (y_1 x_3 y_4 + y_1 y_3 x_4 - x_1 x_3 x_4 - x_1 y_3 y_4) p_{132} p_{146} \\
&+ i x_1 y_1 (y_3 q_{132} - i e_3^{-1} p_{132})(y_4 q_{146} - i e_4^{-1} p_{146}) - y_1^2 y_3 y_4 q_{132} q^*_{146} \} x_1 x_3 x_4 L_{345}
\end{aligned} \tag{5.46}$$

Plainly $K_2 + K^*_2$ is real, so to verify (5.40) we have only to show that K_1 is real. Using (5.42), (5.23), (5.35), (5.19) and the spherical trigonometric formula

$$\sin\theta_3 \sin\theta_4 \cos r_{34} = \cos\theta_5 + \cos\theta_3 \cos\theta_4$$

(Sect. 54 of TL), we can establish that

$$K_1 = \tfrac{1}{8} y_1 F_{132} F_{146} [\sin\theta_5 - \sin(\theta_3 + \theta_4) + \sin\theta_3 \sin\theta_4 \sin r_{34}]. \tag{5.47}$$

Thus K_1 is real; we have verified (5.40) and hence the identities (5.3) and (3.10).

I have assumed that $\theta_1, \ldots, \theta_6$ and $r_{12}, \ldots, r_{56}$ are real numbers: this is really just a notational device to avoid writing (5.30) and (5.41) twice, once as given and once with i replaced by $-i$. The identities (5.2) and (5.3) are basically algebraic, so must of course be true for complex values of the parameters as well as real ones (so long as consistent choices are made of the branches of multi-valued functions).

6. The Static Solution

In the first of his two papers, i.e. in ZI, Zamolodchikov considers the "static limit" of the tetrahedron equations. This can be thought of as the limit when the $(\theta_1, \theta_2, \theta_3)$ and $(\theta_2, \theta_5, \theta_6)$ triangles in Fig. 5 are infinitesimally small, in which case

$$\begin{aligned}
\theta_1 + \theta_2 + \theta_3 &= -\theta_1 + \theta_4 + \theta_6 = \theta_3 + \theta_4 - \theta_5 \\
&= \theta_2 + \theta_5 + \theta_6 = \pi.
\end{aligned} \tag{6.1}$$

The spherical quadrilateral becomes planar, as in Fig. 7 of ZII.

In this case the parameters Q_0, Q_1, Q_2, Q_3, R_0 in Table 1 vanish, for all four functions W, W', W'', W'''. Thus each W is determined by just seven parameters: P_0, P_1, P_2, P_3, R_1, R_2, R_3.

It is interesting to see if the tetrahedron equations (2.2) admit some more general solution than that found by Zamolodchikov, containing Zamolodchikov's as a special case. In general this is a very difficult problem, but one possible start is to attempt to generalize the static limit solution, i.e. to look for solutions of (2.2) such that the weight functions W, W', W'', W''' all have the form given in Table 1 (different functions having different values of the constants), with Q_0, Q_1, Q_2, Q_3, R_0 all equal to zero.

Many simplifications arise in this limit. For arbitrary $P_0, \ldots, R_3'''$ we can still use the spin symmetries of Sect. 3 to reduce the number of equations to 224, occurring in 112 pairs, each equation of a pair being obtained from the other by negating every Q_j. However, since we are taking every Q_j to be zero, this means that the two equations of a pair are identical, so there are only 112 distinct equations.

Each of these equations is of the form

$$\pm A \pm B = \pm C \pm D, \tag{6.2}$$

where each of A, B, C, D is a product of four of the parameters $P_0, \ldots, R_3'''$ (one for each of the four weight functions W, W', W'', W'''). Since R_0 in Table 1 is zero, some of A, B, C, D may vanish. Indeed, 18 of the 112 equations are simply

$$0+0=0+0. \tag{6.3}$$

This leaves us with 94 non-trivial distinct equations, which break up into the following four main sets:

(i) 28 equations of the form

$$A+0=C+0 \tag{6.4}$$

(i.e. one non-zero product on each side),

(ii) 12 equations of the form

$$A-B=0+0, \tag{6.5}$$

(iii) 36 equations of the form

$$A \pm B = C+0, \tag{6.6}$$

(iv) 18 equations of the form

$$A \pm B = C \pm D. \tag{6.7}$$

Obviously (2.2) is unchanged by multplying any of the four weight functions by a constant. Assuming that P_0 is non-zero, we can without further loss of generality choose it to be unity, for each of the functions W, W', W'', W'''. Thus

$$P_0 = P_0' = P_0'' = P_0''' = 1. \tag{6.8}$$

This leaves us with six available parameters for each function, giving us 24 in all.

We now seek to systematically solve the 94 equations for these 24 unknowns. The 40 equations of types (i) and (ii) involve only two products, so are quite easy to examine. Assuming that our remaining 24 parameters are all non-zero, we could linearize these equations by taking logarithms. It turns out that they are satisfied if and only if there exist 12 parameters $x_1, \ldots, x_6, y_1, \ldots, y_6$ such that

$$\begin{aligned}
&P_1=(z_1z_3)^{1/2}, \quad P_2=(z_2z_3)^{1/2}, \quad P_3=(z_1z_2)^{1/2},\\
&R_1=(y_2/x_1x_3)^{1/2}, \quad R_2=(y_1/x_2x_3)^{1/2}, \quad R_3=(y_3/x_1x_2)^{1/2},\\
&P'_1=(z_1/z_4)^{1/2}, \quad P'_2=1/(z_4z_6)^{1/2}, \quad P'_3=(z_1/z_6)^{1/2},\\
&R'_1=(x_6/x_1y_4)^{1/2}, \quad R'_2=(y_1/y_4y_6)^{1/2}, \quad R'_3=(x_4/x_1y_6)^{1/2},\\
&P''_1=1/(z_3z_4)^{1/2}, \quad P''_2=(z_5/z_4)^{1/2}, \quad P''_3=(z_5/z_3)^{1/2},\\
&R''_1=(y_5/y_3y_4)^{1/2}, \quad R''_2=(x_3/y_4x_5)^{1/2}, \quad R''_3=(x_4/y_3x_5)^{1/2},\\
&P'''=(z_2z_6)^{1/2}, \quad P'''_2=(z_5z_6)^{1/2}, \quad P'''_3=(z_2z_5)^{1/2},\\
&R'''_1=(y_5/x_2x_6)^{1/2}, \quad R'''_2=(y_2/x_5x_6)^{1/2}, \quad R'''=(y_6/x_2x_5)^{1/2},
\end{aligned} \tag{6.9}$$

where for brevity I have introduced $z_1, \ldots, z_6$ such that

$$z_j=y_j/x_j, \tag{6.10}$$

and the $\frac{1}{2}$ powers are introduced for later convenience.

At this stage it may be noted that if the equations of type (ii) are changed from $A-B=0$ to $A+B=0$, then the combined set of 40 equations has no solutions with $P_1, \ldots, R'''_3$ all non-zero. This means that the sign factors *ab*, *ac*, *ad* in Table 1 are essential and that W cannot be chosen to have all its values non-negative. This is unfortunate from the viewpoint of statistical mechanics.

Now we substitute these expressions for $P_1, \ldots, R'''_3$ into the 36 equations of type (iii), and obtain

$$\begin{aligned}
y_1+y_2y_3&=x_2x_3\\
y_2+y_3y_1&=x_3x_1\\
y_3+y_1y_2&=x_1x_2\\
-y_1+y_4y_6&=x_4x_6\\
y_4-y_6y_1&=x_6x_1\\
y_6-y_1y_4&=x_1x_4\\
y_3-y_4y_5&=x_4x_5\\
y_4-y_5y_3&=x_5x_3\\
-y_5+y_3y_4&=x_3x_4\\
y_2+y_5y_6&=x_5x_6\\
y_5+y_6y_2&=x_6x_2\\
y_6+y_2y_5&=x_2x_5\\
y_2y_4+y_3y_6&=x_1x_5\\
y_3y_6-y_1y_5&=x_2x_4\\
y_1y_5+y_2y_4&=x_3x_6\,.
\end{aligned} \tag{6.11}$$

(The first 12 equations occur twice, the last 3 occur four times.)

Eliminating x_3 and y_3 between the first three equations gives

$$\Delta_1=\Delta_2, \tag{6.12}$$

where

$$\Delta_j=(x_j^2+y_j^2-1)/(x_jy_j). \tag{6.13}$$

By symmetry, it follows from the first three equations that $\Delta_1=\Delta_2=\Delta_3$. Similarly, the next three sets of three give $-\Delta_1=\Delta_4=\Delta_6$, $\Delta_3=\Delta_4=-\Delta_5$, $\Delta_2=\Delta_5=\Delta_6$.

The only solution of these equations is

$$\Delta_j=0, \quad j=1,\ldots,6, \tag{6.14}$$

so that

$$x_j^2+y_j^2=1. \tag{6.15}$$

We can therefore choose six unknowns $\theta_1,\ldots,\theta_6$ such that

$$x_j=\cos\tfrac{1}{2}\theta_j, \quad y_j=\sin\tfrac{1}{2}\theta_j \tag{6.16}$$

for $j=1,\ldots,6$. The first twelve equations in (6.11) then reduce to

$$\begin{aligned}\theta_1+\theta_2+\theta_3&=\pi,\\ -\theta_1+\theta_4+\theta_6&=\pi,\\ \theta_3+\theta_4-\theta_5&=\pi,\\ \theta_2+\theta_5+\theta_6&=\pi,\end{aligned} \tag{6.17}$$

(apart from additive multiples of 2π which can be absorbed into $\theta_1,\ldots,\theta_6$). These are precisely Eqs. (6.1), so we can regard $\theta_1,\ldots,\theta_6$ as the angles of a plane quadrilateral. Comparing these results with (3.11)–(3.15), we find that we have regained Zamolodchikov's solution in the static limit. Thus this is the only solution of (2.2) in which the weights W, W', W'', W''' have the form given in Table 1, with Q_0, Q_1, Q_2, Q_3, R_0 all zero.

The last three equations in (6.11), as well as all the type (iv) equations, are now satisfied automatically: indeed they have to be, as these are just special cases of the general equations which have been verified in Sects. 3–5.

7. Summary

The tetrahedron equations are given in (2.2) and Zamolodchikov's solution in Table 1 and Eqs. (3.11)–(3.15). For this solution, the 2^{14} tetrahedron equations reduce, first to 224 non-trivial distinct equations, and then to just two identities. In Sect. 5 I have proved these identities, thereby verifying Zamolodchikov's solution.

This solution is very special: it does contain three adjustable parameters for any particular Boltzmann weight function W, namely the three angles θ_1, θ_2, θ_3 of a spherical triangle. However, these parameters are probably "irrelevant" (in the language of renormalization group theory), just as the corresponding elliptic function argument u (or v) for the two-dimensional eight-vertex model is irrelevant [1]. If so, then no critical behaviour can be observed by varying these parameters. The solution also has the property that some weights occur in anti-symmetric pairs of opposite sign, which is unfortunate from the point of view of statistical mechanics.

Even so, it is remarkable that the tetrahedron equations have any non-trivial solutions at all, and this leads one to hope that other three-dimensional solutions may be found, perhaps by generalizing Zamolodchikov's result. In Sect. 6 I have attempted to do this in a modest way by restricting the weight functions to have the same symmetries, anti-symmetries and zero elements as Zamolodchikov's "static limit" solution. Unfortunately it turns out that no such generalization is possible: Zamolodchikov's is the only solution of this form. This is disappointing, but one can still hope that other, less restricted, generalizations or alternative solutions remain to be found.

Ultimately, of course, one is interested in statistical mechanics in calculating the partition-function per site $Z^{1/N}$. Zamolodchikov calls this the "unitarizing factor" (dropping the superfix $1/N$), and writes down the inversion equation for it in (5.2) of ZII. Unfortunately this determines $Z^{1/N}$ only if appropriate analyticity assumptions are made [16, 17], and it is not obvious what these are, or precisely how to use them. Again, one would like to generalize W to include a temperature-like variable. The analyticity assumptions could then be checked against low- or high-temperature series expansions, as can be done for two-dimensional exactly solved models [10].

References

1. Baxter, R.J.: Partition function of the eight-vertex Lattice model. Ann. Phys. **70**, 193–228 (1972)
2. Baxter, R.J., Enting, I.G.: 399th solution of the Ising model. J. Phys. A **11**, 2463–2473 (1978)
3. Baxter, R.J.: Hard hexagons: exact solution. J. Phys. A **13**, L 61–L 70 (1980)
4. Baxter, R.J., Pearce, P.A.: Hard hexagons: interfacial tension and correlation length. J. Phys. A **15**, 897–910 (1982)
5. Yang, C.N.: S-matrix for one-dimensional N-body problem with repulsive or attractive δ-function interaction. Phys. Rev. **168**, 1920–1923 (1968)
6. Baxter, R.J.: Solvable eight-vertex model on an arbitrary planar lattice. Philos. Trans. R. Soc. (London) **289**, 315–346 (1978)
7. Baxter, R.J.: Exactly solved models in statistical mechanics. London: Academic Press 1982
8. Wannier, G.H.: The statistical problem in cooperative phenomena. Rev. Mod. Phys. **17**, 50–60 (1945)
9. Onsager, L.: In: Critical phenomena in alloys, magnets and superconductors. Mills, R.E., Ascher, E., Jaffee, R.I. (eds.) New York: McGraw-Hill 1971
10. Baxter, R.J.: In: Fundamental problems in statistical mechanics. V. Cohen, E.G.D. (ed.). Amsterdam: North-Holland 1980.
11. Chudnovsky, D.V., Chudnovsky, G.V.: Characterization of completely X-symmetric factorized S-matrices for a special type of interaction. Applications to multicomponent field theories. Phys. Lett. **79** A, 36–38 (1980)
12. Zamolodchikov, A.B.: Zh. Eksp. Teor. Fiz. **79**, 641–664 (1980); J.E.T.P. **52**, 325–336 (1980)
13. Zamolodchikov, A.B.: Tetrahedron equations and the relativistic S-matrix of straight strings in 2+1 dimensions. Commun. Math. Phys. **79**, 489–505 (1981)
14. Jaekel, M.T., Maillard, J.M.: Symmetry relations in exactly soluble models. J. Phys. A **15**. 1309–1325 (1982)
15. Todhunter, I., Leathem, J.G.: Spherical trigonometry. London: MacMillan 1949
16. Baxter, R.J.: J. Stat. Phys. **28**, 1–41 (1982)
17. Pokrovsky, S.V., Bashilov, Yu.A.: Star-triangle relations in the exactly solvable statistical models. Commun. Math. Phys. **84**, 103–132 (1982)

Communicated by J. Fröhlich

Received October 28, 1982

Volume 128, number 3,4 PHYSICS LETTERS A 28 March 1988

NEW SOLUTIONS OF THE STAR-TRIANGLE RELATIONS FOR THE CHIRAL POTTS MODEL

R.J. BAXTER [1], J.H.H. PERK [2]
Research School of Physical Sciences, Australian National University, G.P.O. Box 4, Canberra, ACT 2601, Australia

and

H. AU-YANG
Institute for Theoretical Physics, State University of New York at Stony Brook, Stony Brook, NY 11794-3840, USA

Received 11 December 1987; accepted for publication 28 January 1988
Communicated by A.A. Maradudin

We present new explicit N-state solutions of the star-triangle relations for a nearest-neighbour two-spin interaction model. The solutions include families with real and positive Boltzmann weights. They are given in terms of two rapidities associated with two lines, which cross through each edge. The rapidities are 4-vectors, restricted to lie on the intersection of two Fermat surfaces. The usual difference property is not present.

In his 1944 solution of the two-dimensional Ising model, Onsager [1] noted the importance of a star-triangle relation. Many years later, this relation (and its generalisation to vertex models and interaction-round-a-face models) has become the cornerstone within a general theory of exactly solvable models of statistical mechanics [2]. Many solutions of the star-triangle relations have been found so far. Usually, with the model can be associated lines such that the Boltzmann weights are associated with their intersections and such that with each line corresponds a line variable (or rapidity variable). The weights are then uniformized in terms of elementary functions or Jacobi elliptic functions of fixed modulus, where the arguments are differences of the two rapidities [2].

Very recently, solutions of the star-triangle equations have been found, with N spin states per site, that were shown not to be of this form [3–6], and that are related to the theory of commensurate-incommensurate phase transitions, see e.g. ref. [7] for review. In particular a nonselfdual $N=3$ state model, uniformised by a genus 10 curve, was shown to exist [3], and $N=3$, 4 and 5 selfdual models were constructed in terms of Fermat curves [4,6]. Here the fact that the genus is larger than one implies that we cannot have difference variables [8–10], which also follows from Reshetikhin's criterion [11] when applied to the associated quantum spin chain hamiltonian. This criterion was written down to explain why the one-dimensional Hubbard model [12] is special. But this Hubbard case is different from the higher genus case of refs. [3–6], as Shastry [13] has shown that the corresponding two-dimensional model is uniformised in terms of just trigonometric functions, but now with sums and differences of rapidities as their arguments.

In the present note we shall give a general solution of the star-triangle equation, which includes the results of refs. [3–6] as special cases and in much more explicit form. The Z-invariant [2,14,15] formulation in terms of rapidities on certain lines turns out

[1] Currently visiting: Mathematics Department, The Faculties, Australian National University, G.P.O. Box 4, Canberra, ACT 2601, Australia.
[2] Visiting from: Institute for Theoretical Physics, State University of New York at Stony Brook, Stony Brook, NY 11794-3840, USA.

Volume 128, number 3,4 PHYSICS LETTERS A 28 March 1988

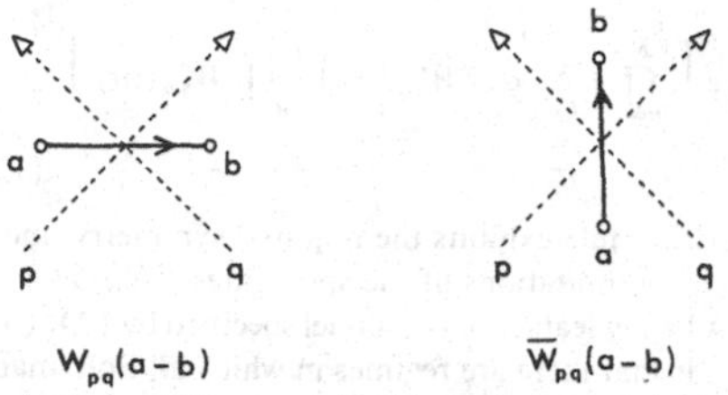

Fig. 1. The two Boltzmann weights depending on the orientation of the spin pair with respect to the rapidity lines p and q.

to be the natural framework for our model. So, in order to keep our presentation brief, we shall give each Boltzmann weight W_{pq} two subscripts p and q, corresponding to two rapidity variables, right from the start.

As in refs. [14,15], we shall start with a collection of straight lines in the plane, such that no three lines intersect in the same point. The lines (represented by broken lines in our figures) carry each a rapidity variable and a direction (indicated by an open arrow) away from a baseline [14], i.e. all arrows point in the same halfplane. The pth line carries a rapidity variable represented by a 4-vector (a_p, b_p, c_p, d_p) restricted to lie on a curve. We can colour the faces alternatingly black and white and on the black faces we put spin variables $\sigma=1, 2, ..., N \pmod N$. Then there are two kinds of neighbouring spin pairs, as indicated in fig. 1, with states a and b, and Boltzmann weigths $W_{pq}(a-b)$ and $\bar{W}_{pq}(a-b)$ on the edges (indicated by solid lines). Here the arrow from a to b indicates that the argument is $a-b \pmod N$, rather than $b-a$. This arrow corresponds to the chirality of our model.

We can now write down the star–triangle equation both graphically, as in figs. 2a and 2b, and algebraically, i.e.

$$\sum_{d=1}^{N} \bar{W}_{qr}(b-d)\, W_{pr}(a-d)\, \bar{W}_{pq}(d-c)$$

$$=R_{pqr}\, W_{pq}(a-b)\, \bar{W}_{pr}(b-c)\, W_{qr}(a-c)\,. \tag{1}$$

We have discovered a general solution of (1) of the form

$$\frac{W_{pq}(n)}{W_{pq}(0)} = \prod_{j=1}^{n} \frac{d_p b_q - a_p c_q \omega^j}{b_p d_q - c_p a_q \omega^j}\,, \tag{2}$$

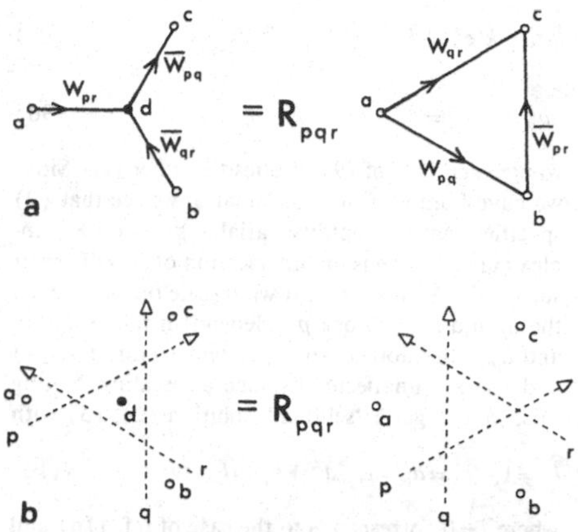

Fig. 2. (a) The star–triangle equation in spin language. (b) The same star–triangle equation, but now with rapidity lines exhibited.

$$\frac{\bar{W}_{pq}(n)}{\bar{W}_{pq}(0)} = \prod_{j=1}^{n} \frac{\omega a_p d_q - d_p a_q \omega^j}{c_p b_q - b_p c_q \omega^j}\,, \tag{3}$$

where

$$\omega \equiv e^{2\pi i/N}\,. \tag{4}$$

Requiring periodicity, i.e.

$$W_{pq}(n+N) = W_{pq}(n)\,,$$

$$\bar{W}_{pq}(n+N) = \bar{W}_{pq}(n)\,, \tag{5}$$

we are led to the conditions

$$\frac{a_p^N \pm b_p^N}{c_p^N \pm d_p^N} = \lambda_\pm = \text{independent of } p\,, \tag{6}$$

for the rapidity 4-vector

$$\boldsymbol{x}_p \equiv (a_p, b_p, c_p, d_p)\,, \tag{7}$$

associated with each line p. Therefore, we have two numbers k and k', which we can restrict to satisfy

$$k^2 + k'^2 = 1\,, \tag{8}$$

such that

$$a_p^N + k' b_p^N = k d_p^N\,, \tag{9a}$$

$$k' a_p^N + b_p^N = k c_p^N\,, \tag{9b}$$

$$ka_p^N + k' c_p^N = d_p^N \,, \tag{9c}$$

$$kb_p^N + k' d_p^N = c_p^N \,, \tag{9d}$$

where each pair of (9) implies the other two. Since we have homogeneous coordinates, we see that (9) specifies that the rapidity variable p lies on a complex curve, which is the intersection of two "Fermat surfaces". We note that, if we rescale on each line p the a_p and b_p with one p-independent factor and c_p and d_p with another p-independent factor, eqs. (2) and (3) are unaffected by such a rescaling. Noting this, we can get a "selfdual" solution (2), (3) with

$$k' = 1, \quad c_p = d_p = 1, \quad a_p^N + b_p^N = I \,, \tag{10}$$

where $I = 0$ corresponds to the case of ref. [16] and $I \neq 0$ can be reduced to the conjectured solution of refs. [4,6] which hitherto was verified only for $N \leqslant 5$.

The proof that (2) and (3), with (9), satisfy the star–triangle equation (1) is amazingly simple, but shall be presented elsewhere. The main feature is that we, as in ref. [3], Fourier transform (1) with respect to c, introducing

$$V_{ab} \equiv \sum_{k=1}^{N} \omega^{bk} W_{pr}(a+k) \bar{W}_{qr}(k) \,, \tag{11a}$$

and

$$\bar{V}_{ab} \equiv \sum_{k=1}^{N} \omega^{ak} \bar{W}_{pr}(k) W_{qr}(b+k) \,, \tag{11b}$$

which follows from (11a) by interchanging a with b and p with q. The proof of (1) follows from the observation that the product formulae (2) and (3) are equivalent to a pair of linear recurrence relations for each of V_{ab} and $\bar{V}_{ab}$.

The normalisation factor R_{pqr} does not explicitly enter this proof of (1). It is certainly given by (1) for any set of values a, b, c (e.g. $a = b = c = 0$). We conjecture (from the selfdual and $N = 2$ Ising cases, and from numerical tests) that it factors into the form

$$R_{pqr} = \frac{f_{pq} f_{qr}}{f_{pr}} \,, \tag{12}$$

where

$$f_{pq} \equiv \left[\prod_{m=1}^{N} \left(\sum_{k=1}^{N} \omega^{mk} \bar{W}_{pq}(k) \right) \Big/ \prod_{m=1}^{N} W_{pq}(m) \right]^{1/N} \tag{13}$$

This formula exhibits the required symmetry under cyclic permutations of the spin states 1, ..., N.

A happy feature of the model specified by (2), (3), (9) is that there are regimes in which all Boltzmann weights are real and positive. This happens, if for each line p

$$a_p^* c_p = \omega^{1/2} b_p^* d_p \,, \tag{14a}$$

$$|a_p| = |d_p| \,, \tag{14b}$$

$$|b_p| = |c_p| \,, \tag{14c}$$

where the asterisk denotes complex conjugation. Each of the three conditions (14a), (14b), (14c) implies the other two (to within choices of Nth roots of unity). If we parametrise our reality conditions by

$$b_p / c_p = \omega^{1/2} e^{i\theta_p} \,, \tag{15}$$

we find that the Boltzmann weigths (2), (3) are real and positive, as long as

$$0 < \theta_p < \theta_q < \pi/N, \quad (0 < k, k' < 1) \,. \tag{16}$$

The symmetries of our model are closely related to the $4N^2$ automorphisms of the curve (9), (14) generated by

$$p \to \mathrm{R}p \,,$$

$$(a_p, b_p, c_p, d_p) \to (b_p, \omega a_p, d_p, c_p) \,, \tag{17a}$$

$$p \to \mathrm{S}p \,,$$

$$(a_p, b_p, c_p, d_p) \to (\omega^{-1/2} c_p, d_p, a_p, \omega^{-1/2} b_p) \,, \tag{17b}$$

$$p \to \mathrm{T}p \,,$$

$$(a_p, b_p, c_p, d_p) \to (\omega a_p, b_p, \omega c_p, d_p) \,, \tag{17c}$$

with

$$\mathrm{R}^{2N} = \mathrm{S}^2 = \mathrm{T}^N = (\mathrm{RS})^2 = \mathrm{R}^{-1}\mathrm{TRT} = \mathrm{T}^{-1}\mathrm{STS} = 1 \tag{17d}$$

(up to normalisation). From these we find, e.g.,

$$\bar{W}_{pq}(n)=W_{q,Rp}(n)\,, \tag{18a}$$

$$W_{pq}(-n)=W_{Rp,Rq}(n)\,, \tag{18b}$$

$$W_{Sq,Sp}(n)=W_{pq}(n)\,, \tag{18c}$$

$$\bar{W}_{Sq,Sp}(n)=\bar{W}_{pq}(-n)\,. \tag{18d}$$

Furthermore, we note

$$W_{pp}(n)=W_{pp}(0)\,, \tag{19a}$$

$$\bar{W}_{pp}(n)=\bar{W}_{pp}(0)\delta_{n,0}\,, \tag{19b}$$

$$W_{pq}(n)W_{qp}(n)=W_{pq}(0)W_{qp}(0)\,. \tag{19c}$$

Therefore, if two rapidities become equal, the corresponding spins either decouple or become identical. This demonstrates the usefulness of the Z-invariant formulation, as many special lattices are contained as special cases. If we take a checkerboard lattice, such that we have vertical rapidities p_1 and p_2 alternatingly and horizontal rapidities q_1 and q_2, the spin model is also a checkerboard lattice diagonally oriented, see fig. 3a. By making q_1 equal to either p_1 or p_2, we reduce the "rapidity lattice" to a Kagomé lattice, and the spin model to a honeycomb or triangular lattice.

With the general checkerboard model is associated a uniform vertex model, as in fig. 3b, with Boltzmann weights

$$R_{\alpha\beta|\lambda\mu}=\bar{W}_{p_1q_1}(\alpha-\lambda)\bar{W}_{p_2q_2}(\mu-\beta)$$
$$\times W_{p_2q_1}(\alpha-\mu)W_{p_1q_2}(\lambda-\beta)\,. \tag{20}$$

After Fourier similarity transformation

$$\hat{R}_{\alpha\beta|\lambda\mu}\equiv N^{-2}$$
$$\times\sum_{\alpha'=1}^{N}\sum_{\beta'=1}^{N}\sum_{\lambda'=1}^{N}\sum_{\mu'=1}^{N}\omega^{-\alpha\alpha'-\mu\mu'+\beta\beta'+\lambda\lambda'}R_{\alpha'\beta'|\lambda'\mu'}\,, \tag{21a}$$

we find that the transformed weights vanish, unless

$$\alpha+\mu=\beta+\lambda\,. \tag{21b}$$

This model (21) is an N-state generalisation of the

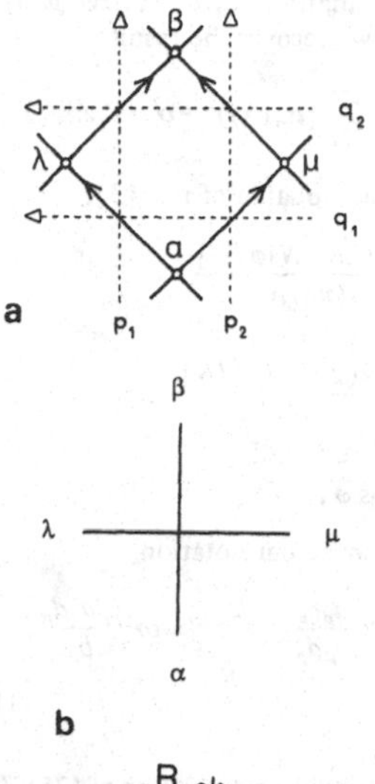

Fig. 3. (a) Unit cell of checkerboard lattice, with spins α, β, λ, μ and rapidity lines p_1, p_2, q_1, q_2. (b) Corresponding uniform vertex model weight $R_{\alpha\beta|\lambda\mu}$.

free-fermion model, but the mapping here differs from the one in ref. [15].

Assuming the normalisation

$$W_{pq}(0)=\bar{W}_{pq}(0)=1\,, \tag{22}$$

we find

$$R_{\alpha\beta|\lambda\mu}=\delta_{\alpha\lambda}\delta_{\beta\mu},\quad\text{for}\quad p_1=q_1\,,\; p_2=q_2\,. \tag{23}$$

So the shift operator is in the class of commuting transfer matrices of the vertex model and using the standard method [17] (see also ref. [2], section 10.14), we can derive two associated hamiltonians taking logarithmic derivatives with respect to q_1 and q_2 at the shift point (23), generalising the XY-model spin chain hamiltonian for $N=2$.

We shall present here only the result for the spin model on a uniform square lattice, that is in fig. 3a we take

$$p_1=p_2=p,\quad q_1=q_2=q\,. \tag{24}$$

The model now has a commuting family of row-to-row transfer matrices $T_p(q)$ at fixed p. By taking the limit $q \to p$, we recover the result

$$\mathscr{H}_p = -\sum_{j=1}^{L}\sum_{n=1}^{N-1} [\bar{\alpha}_n (X_j)^n + \alpha_n (Z_j Z_{j+1}^{\dagger})^n] \,. \quad (25)$$

where, in the notation of ref. [3],

$$\alpha_n = \frac{\exp[i(2n-N)\phi/N]}{\sin(\pi n/N)} ,$$

$$\bar{\alpha}_n = \lambda \frac{\exp[i(2n-N)\bar{\phi}/N]}{\sin(\pi n/N)} ,$$

$$\cos\phi = \lambda \cos\bar{\phi} \,, \quad (26)$$

but now also, in our notation,

$$e^{2i\phi/N} = \omega^{1/2} \frac{a_p c_p}{b_p d_p} , \quad e^{2i\bar{\phi}/N} = \omega^{1/2} \frac{a_p d_p}{b_p c_p} ,$$

$$\lambda = k' \,. \quad (27)$$

The above provides a derivation of (25), (26), which was previously [3,5] postulated on the basis of a perturbation expansion of the star–triangle equation. Special cases were introduced earlier and studied numerically [18,19].

The hamiltonian $\mathscr{H}_p$ is proportional to the logarithmic derivative of $T_p(q)$, evaluated at $q=p$. It depends explicitly on the p variable. This feature is unlike what happens in previous solutions of the star–triangle relation (e.g. the eight-vertex model and its associated XYZ hamiltonian [17]), where the difference property $T_p(q)=T(q-p)$ ensures that $\mathscr{H}$ is independent of p.

From a physical point of view, the variable k' in (9) is a temperature-like variable. The limit $k' \to 0$ corresponds to the zero-temperature limit of extreme order. We expect $k'=1$ to represent a critical line between an ordered and a disordered state, containing (for $N \geqslant 3$) the Lifshitz point, where the incommensurate phase begins [7].

One of us (JHHP) thanks his colleagues at the Australian National University for their hospitality. He also acknowledges support from the National Science Foundation under Grant DMR-8505419.

References

[1] L. Onsager, Phys. Rev. 65 (1944) 117.
[2] R.J. Baxter, Exactly solved models in statistical mechanics (Academic Press, London, 1982).
[3] H. Au-Yang, B.M. McCoy, J.H.H. Perk, S. Tang and M.L. Yan, Phys. Lett. A 123 (1987) 219.
[4] B.M. McCoy, J.H.H. Perk, S. Tang and C.H. Sah, Phys. Lett. A 125 (1987) 9.
[5] J.H.H. Perk, Preprint ITP-SB-87-57, to be published in Proc. 1987 Summer Institute on Theta functions (Am. Math. Soc.).
[6] H. Au-Yang, B.M. McCoy, J.H.H. Perk and S. Tang, Preprint ITP-SB-87-54, to be published in Prospect of Algebraic Analysis.
[7] M. den Nijs, in: Phase transitions and critical phenomena, Vol. 12, eds. C. Domb and J.L. Lebowitz (Academic Press, London), to be published.
[8] J.M. Maillard, J. Math. Phys. 27 (1986) 2776.
[9] J.B. McGuire and J.M. Freeman, Florida Atlantic University Preprint.
[10] J.B. McGuire and C.A. Hurst, Florida Atlantic University Preprint.
[11] P.P. Kulish and E.K. Sklyanin, in: Lecture notes in physics, Vol. 151. Integrable quantum field theories, eds. J. Hietarinta and C. Montonen (Springer, Berlin, 1982) p. 61.
[12] E.H. Lieb and F.Y. Wu, Phys. Rev. Lett. 20 (1968) 1445.
[13] B. Sriram Shastry, Tata Institute Preprint FIFR/TH/87-23.
[14] R.J. Baxter, Philos. Trans. R. Soc. A 289 (1978) 315.
[15] R.J. Baxter, Proc. R. Soc. A 404 (1986) 1.
[16] V.A. Fateev and A.B. Zamolodchikov, Phys. Lett. A 92 (1982) 37.
[17] R.J. Baxter, Ann. Phys. 70 (1972) 323.
[18] S. Howes, L.P. Kadanoff and M. den Nijs, Nucl. Phys. B 215 [FS7] (1983) 169.
[19] G. von Gehlen and V. Rittenberg, Nucl. Phys. B 257 [FS14] (1985) 351.

CHIRAL POTTS MODEL AS A DESCENDANT OF THE SIX-VERTEX MODEL

V.V Bazhanov, Yu.G. Stroganov

Centre for Mathematical Analysis
Australian National University
GPO Box 4
Canberra ACT 2601
Australia

Permanent Address:
Institute for High Energy Physics
Serpukhov
Moscow Region, USSR

June, 1989

AMS Nos: 82A67; 81E99.

Abstract

We observe that N-state integrable chiral Potts model can be considered as a part of some new algebraic structure related to six-vertex model. As a result we obtain a functional equation which determine all the eigenvalues of the chiral Potts model transfer matrix.

1.Introduction

The star-triangle (or Yang-Baxter) relation and its generalization play a central role in the theory of exactly solvable models in statistical mechanics [1,2] and field theory [3]. Many solutions of the Yang-Baxter equations have been found so far. Usually they are uniformized in terms of elementary functions or Jacobi elliptic functions of fixed modulus, where the argument is the difference of the "rapidities" of the two lines through that vertex.

Very recently, solutions of the star-triangle equations have been found [4-6] for the restricted class of N-state chiral Potts models, that were shown not to be of this form. In fact they should be uniformized by genus $g > 1$ curves and, hence, cannot have the difference property.

Due to its unusual properties chiral Potts model seems to have no close relation to the "conventional" integrable models, such as, e.g., six-vertex model by Lieb [7]. In the present paper we show, that it is not the case. In fact, the integrable chiral Potts model can be considered as a part of new algebraic structure related namely to the six-vertex model.

The organization of this paper is as follows. In Sect. 2 we find some new L-operators which are intertwined by six-vertex model R-matrix. Sect. 3 contains the basic definitions of the chiral Potts model. In Sect. 4 we establish a relationship between the six-vertex and chiral Potts models. In Sect. 5 we obtain the functional equations which determine all the eigenvalues of the transfer matrix of the chiral Potts model.

2. L-operators related to the six-vertex model R-matrix.

In this section we consider some solutions of the Yang-Baxter equations (YBE) related to the six-vertex model R-matrix. The latter is a four-index matrix function $R_{i_1 i_2}^{j_1 j_2}(x)$ (the indices run over the two values 0 and 1) with the following non-vanishing matrix elements

$$\begin{aligned} R_{00}^{00} = R_{11}^{11} &= \rho\sin(\theta+\eta), \quad & R_{01}^{01} = R_{10}^{10} &= \rho\sin\theta, \\ R_{01}^{10} &= \rho\sin\eta e^{i\theta}, & R_{10}^{01} &= \rho\sin\eta e^{-i\theta}, \end{aligned} \tag{2.1}$$

where $\theta = -i\log x$ is a variable, while ρ, η are considered as constants. $R(x)$ satisfies the YBE

$$R_{i_3 i_1}^{j_3 j_1}(x) R_{j_3 i_2}^{k_3 j_2}(y) R_{j_1 j_2}^{k_1 k_2}(xy^{-1}) = R_{i_1 i_2}^{j_1 j_2}(xy^{-1}) R_{i_3 j_2}^{j_3 k_2}(y) R_{j_3 j_1}^{k_3 k_1}(x), \tag{2.2}$$

where summation over repeated indices is assumed (see Fig. 1).

Let $L(x)$ be an operator in $C^2 \otimes C^N$, $N \geq 2$, satisfying the following equations (shown in Fig.2):

$$L_{i_1\alpha}^{j_1\beta}(x) L_{i_2\beta}^{j_2\gamma}(y) R_{j_1 j_2}^{k_1 k_2}(yx^{-1}) = R_{i_1 i_2}^{j_1 j_2}(yx^{-1}) L_{j_2\alpha}^{k_2\beta}(y) L_{j_1\beta}^{k_1\gamma}(x), \tag{2.3}$$

where $L_{i\alpha}^{j\beta}(x), i,j = 0,1; \quad \alpha,\beta = 0,1,\ldots,N-1$, denote the matrix elements of $L(x)$. The operator $L(x)$ is called a quantum L-operator related to a given R-matrix. It can conveniently be viewed as a two by two matrix with operator matrix elements acting in C^N. Then, one can rewrite (2.3) as

$$(L(x)\otimes L(y))\check{R}(xy^{-1}) = \check{R}(xy^{-1})(L(y)\otimes L(x)), \tag{2.3'}$$

where

$$\check{R}(x) = R(x)P, \tag{2.4}$$

P being a permutation operator in $C^2 \otimes C^2$, $P(x\otimes y) = (y \otimes x)$.

Discarding an interesting question about the most general solution of (2.3) let us search for an L-operator of the form

$$L(x) = xL_+ + x^{-1}L_-, \tag{2.5}$$

where $L_+(L_-)$ is independent of x and has an upper (lower) triangular form. The most obvious non-trivial solution of this form for $N = 2$ is the R-matrix itself. From (2.1), (2.4) it follows, that

$$R(x) = \frac{\rho}{2}(xR_+ + x^{-1}R_-), \tag{2.6}$$

where $R_\pm$ are independent of x. Introducing $\check{R}_\pm = R_\pm P$ as in (2.4), we have

$$\begin{aligned} \check{R}_+\check{R}_- &= 1 \\ \check{R}_+ + \check{R}_- &= 2\sin\eta 1 \end{aligned} \tag{2.7}$$

By using of (2.5)-(2.7) Eq. (2.3) reduces to only three independent relations

$$\begin{cases} (L_\pm \otimes L_\pm)\check{R}_+ = \check{R}_+(L_\pm \otimes L_\pm) \\ (L_- \otimes L_+)\check{R}_+ = \check{R}_+(L_+ \otimes L_-). \end{cases} \tag{2.8}$$

Explicitly, we have

$$\begin{cases} [(L_\sigma)_{ii}, (L_{\sigma'})_{jj}] = 0, & \sigma, \sigma' = \pm, \quad i,j = 0,1, \\ (L_\sigma)_{ii}(L_+)_{01} = \omega_1^{-\sigma\epsilon(i)}(L_+)_{01}(L_\sigma)_{ii}, & \sigma = \pm, \quad i = 0,1, \\ (L_\sigma)_{ii}(L_-)_{10} = \omega_1^{\sigma\epsilon(i)}(L_-)_{10}(L_\sigma)_{ii}, & \sigma = \pm, \quad i = 0,1, \\ [(L_+)_{01}, (L_-)_{10}] = (\omega_1 - \omega_1^{-1})\{(L_-)_{11}(L_+)_{00} - (L_+)_{11}(L_-)_{00}\}, \end{cases} \tag{2.8'}$$

where $\epsilon(0) = 1$, $\epsilon(1) = -1$, $\omega_1 = \exp(i\eta)$. These relations can be considered as a definition of some quadratic Hopf algebra [8] with six generating elements $(L_\sigma)_{ii}$, $i = 0,1$, $\sigma = \pm$ and $(L_+)_{01}, (L_-)_{10}$, which generalizes the $U_q(sl(2))$ algebra [9]. The latter arises if we set, e.g.,

$$(L_\sigma)_{00} = (L_{-\sigma})_{11}, \quad \sigma = \pm. \tag{2.9}$$

We are interested in representations of the algebra (2.8) which, in general, do not match the above constraints. Moreover, let us require that

$$\det{}_{C^N}(L_\sigma)_{ij} \neq 0 \tag{2.10}$$

for all values of σ, i, j. From the relations (2.8) it follows, that this property can be achieved only if $\omega_1^N = 1$, i.e.,

$$\eta = 2\pi k/N, \quad k = 0,1,\ldots,N-1 \tag{2.11}$$

Here we restrict ourselves to the case, when

$$N = \text{prime} \quad \text{number}. \tag{2.11'}$$

Then, one can show,that the most general* solution of (2.8), (2.10), (2.11) can be written as

$$\begin{aligned}
D_+ &= (L_+)_{00} = d_+A \\
D_- &= (L_-)_{00} = d_-B \\
F_+ &= (L_+)_{11} = f_+B \\
F_- &= (L_-)_{11} = f_-A \\
G &= (L_+)_{01} = (g_+B + g_-A)C \\
H &= (L_-)_{10} = (h_+B + h_-A)C^{-1}
\end{aligned} \tag{2.12}$$

where A, B, C are N by N matrices satisfying the relations

$$[A, B] = 0; \qquad CA = \omega_1 AC; \qquad CB = \omega_1^{-1} BC. \tag{2.13}$$

The eight parameters $d_+, d_-, f_+, f_-, g_+, g_-, h_+, h_-$ are arbitrary *modulo* the constraints

$$g_-h_- = f_-d_+, \quad g_+h_+ = f_+d_- \tag{2.14}$$

So, we can choose a set of six parameters

$$\mathcal{X} = \{d_+, d_-, f_+, f_-, g_+, g_-\} \tag{2.15}$$

as the independent ones.

Thus, Eqs. (2.5), (2.12)-(2.14) define a six-parameter solution of the YBE (2.3) for the case (2.11). A particular choice of the matrices A, B, C in (2.12) convenient for subsequent calculations is as follows

$$A = X^\rho, \qquad B = X^{-\rho}, \qquad C = Z. \tag{2.16}$$

---

* When N is not prime eqs. (2.12), (2.13) still give a solution of (2.8), (2.10), but, apparently, not the most general one.

where $\rho = (N-1)/2$ and X, Z are the N by N matrices.

$$\begin{aligned} X_{\alpha\beta} &= \delta_{\alpha,\beta+1}, \qquad Z_{\alpha\beta} = \delta_{\alpha\beta}\omega^{\alpha}, \\ ZX &= \omega XZ \end{aligned} \tag{2.17}$$

$$\delta_{\alpha\beta} = \begin{cases} 1, & \alpha = \beta \quad (\text{mod } N) \\ 0, & \alpha \neq \beta \quad (\text{mod } N) \end{cases}, \tag{2.18}$$

where $\omega = \omega_1^{-2}$ and $\alpha, \beta = 0, \ldots, N-1$.

Let us discuss some properties of the transfer matrices associated with $L(x)$ given by (2.5), (2.12)–(2.14). For the lattice of M by M' sites the column to column and row to row transfer matrices have the form (see Fig. 3 and Fig. 4 resp.).

$$T_{col} = \mathcal{T}(x, \mathcal{X})_{i_1,\ldots,i_N}^{j_1,\ldots,j_N} = \sum_{\{\alpha\}} \prod_{k=1}^{M'} (L_{i_k \alpha_{k+1}}^{j_k \alpha_k}) \tag{2.19}$$

$$T_{row} = \mathcal{T}(x, \mathcal{X})_{\alpha_1,\ldots,\alpha_N}^{\beta_1,\ldots,\beta_N} = \sum_{\{i\}} \prod_{k=1}^{M} (L_{i_k \quad \alpha_k}^{i_{k+1} \beta_k}) \tag{2.20}$$

The first one acts in $(C^2 \otimes)^{M'}$, while the second one acts in $(C^N \otimes)^M$. In addition introduce a six-vertex model transfer matrix acting in $(C^2 \otimes)^{M'}$

$$T_{6v}(x)_{i_1,\ldots,i_N}^{j_1,\ldots,j_N} = \sum_{\{k\}} \prod_{s=1}^{M'} (R_{i_s k_s}^{j_s k_{s+1}}(x)) \tag{2.21}$$

with R given by (2.1). It follows from (2.3) that

$$[\mathcal{T}(x, \mathcal{X}), T_{6v}(y)] = 0. \tag{2.22}$$

It is well known, that T_{6v} commutes with an arrow number operator $\mathcal{N}$

$$[T_{6v}, \mathcal{N}] = 0, \quad \mathcal{N} = \sum_{k=1}^{N} 1 \otimes \cdots \otimes \underset{k-\text{th}}{\sigma_z} \otimes \cdots 1 \tag{2.23}$$

Contrary to this, the transfer matrix $\mathcal{T}(x, \mathcal{X})$ does not commute with $\mathcal{N}$. This intriguing phenomenon is possible due to degeneracy of the spectrum of T_{6v} among sectors with the values of $\mathcal{N}$ differing by multiples of N. Thus, Eq. (2.22) implies the existence of a family of new arrow non-preserving integrals, commuting with the 6v-model transfer matrix (as we shall see in Sect. 4, these integrals, in general, do not commute among themselves).

Another interesting feature of $\mathcal{T}(x, \mathcal{X})$ is that it possesses the properties of Baxter's Q-matrix [10] for the 6v-model. Namely, one can show that

$$T_{6v}(\theta)\mathcal{T}(\theta) = \sin^N \theta \mathcal{T}(\theta+\eta) + \sin^N(\theta+\eta)\mathcal{T}(\theta-\eta),$$
$$\mathcal{T}(\theta) = \mathcal{T}(e^{i\theta}, \mathcal{X}(\theta)),$$
$$\mathcal{X}(\theta) = \{a, b, ce^{-i\theta}, de^{i\theta}, \lambda b, \lambda a\},$$

where $\theta = -i \log x$ and a, b, c, d, λ are arbitrary parameters.

The properties of $T(x, \mathcal{X})$ are interesting as well. In particular, we shall show in Sect. 4, that it commutes with transfer-matrix of integrable checkerboard N-state chiral Potts model.

3. Chiral Potts Model.

Following ref. [6] let us recall the basic definitions of the checkerboad integrable chiral Potts model. Consider an oriented square lattice $\mathcal{L}$ and its dual $\mathcal{L}'$ (shown in Fig. 5 by solid and dashed lines, resp.).

The vertical lines of $\mathcal{L}'$ carry rapidity variables q, q' in alternating order. Each rapidity variable q is represented by a 4-vector (a_q, b_q, c_q, d_q) restricted to lie on a curve. Similarly the horizontal lines carry the rapidity variables p, p'. Place spin variables $\sigma = 0, \ldots, n-1$ on sites of the original lattice $\mathcal{L}$. Then, there are two kinds of neighbouring spin pairs, as indicated in Fig. 6, with states a and b, and Boltzmann weights $W_{pq}(a-b), \overline{W}_{pq}(a-b)$ on the edges of $\mathcal{L}$. Here the arrow from a to b indicates that the argument is $a - b \pmod n$, rather than $b - a$. This arrow corresponds to the chirality of the model.

We can write down the star-triangle equation both graphically, as in Fig. 7, and algebraically, i.e.,

$$\sum_{e=1}^{N} \overline{W}_{qp}(a-e)W_{rp}(c-e)\overline{W}_{rq}(e-b) = R_{rqp}W_{rq}(c-a)\overline{W}_{rp}(a-c)W_{qp}(c-b) \quad (3.1)$$

where R_{rqp} is independent of the spins. The solution of (3.1) found in [6] has the form

$$\frac{W_{pq}(k)}{W_{pq}(0)} = \prod_{j=1}^{k} \frac{d_p b_q - a_p c_q \omega^j}{b_p d_q - c_p a_q \omega^j} \tag{3.2}$$

$$\frac{\overline{W}_{pq}(k)}{\overline{W}_{pq}(0)} = \prod_{j=1}^{k} \frac{\varepsilon a_p d_q - d_p a_q \omega^j}{c_p b_q - b_p c_q \omega^j} \tag{3.3}$$

where

$$\omega = \exp(2\pi i/N) \tag{3.4}$$

Each rapidity 4-vector

$$x_p = (a_p, b_p, c_p, d_p)$$

associated with a line p is restricted to satisfy the following relations

$$a_p^N + k' b_p^N = k d_p^N \tag{3.5a}$$

$$k' a_p^N + b_p^N = k c_p^N \tag{3.5b}$$

$$k a_p^N + k' c_p^N = d_p^N \tag{3.5c}$$

$$k b_p^N + k' d_p^N = c_p^N \tag{3.5d}$$

where $k'^2 = 1 - k^2$. Note that each pair of (3.5) implies the other two. The constant k is a parameter of the model. In the particular self-dual case $k = 0$ the model is reduced to the Fateev-Zamolodchikov model [10] (Z_N-model). The latter model is critical.

Let $\sigma_1, \ldots, \sigma_M$ and $\sigma'_1, \ldots, \sigma'_M$ be the spins of two adjacent N-site rows of $\mathcal{L}$. Then, for cylindrical boundary conditions one can define two transfer matrices

$$(U_{p,q,q'})_{\sigma\sigma'} = \prod_{i=1}^{N} \overline{W}_{pq}(\sigma_i - \sigma'_i) W_{pq'}(\sigma'_i - \sigma_{i+1}) \tag{3.6a}$$

$$(\hat{U}_{p,q,q'})_{\sigma\sigma'} = \prod_{i=1}^{N} \overline{W}_{pq'}(\sigma_i - \sigma'_{i+1}) W_{pq}(\sigma'_i - \sigma_i) \tag{3.6b}$$

Clearly

$$\hat{U}_{p,q,q'} = U_{p,q',q} \hat{P}, \tag{3.7}$$

where $\hat{P}$ shifts the spins one site

$$\hat{P}_{\sigma\sigma'} = \prod_{i=1} \delta_{\sigma_{i-1},\sigma_i} \tag{3.8}$$

For the homogeneous case when $q' = q$ the transfer matrices $U_{q,q,p}, \hat{U}_{q,q,p}$ with the same value of q but with different values of p commute among each other and with the operator P

$$[U_{pq}, V_{p'q}] = [U_{pq}, P] = [V_{pq}, P] = 0 \tag{3.9}$$

This is the consequence of star-triangle relation (3.1).

4. Chiral Potts model as a descendant of the 6-vertex model.

As it was noted in Sect. 2 the transfer matrix (2.19) commutes with the 6-vertex model transfer matrix (2.21). Nevertheless two different transfer matrices (2.19) not necessary commute among themselves because of the degeneracy of the spectrum of the 6-vertex model.

Let L and $\tilde{L}$ be two L-operators of the form (2.5), (2.12) with different sets of parameters $\mathcal{X}, \tilde{\mathcal{X}}$, (2.15). It is convenient to set $x = 1$ in (2.5) because it can be absorbed into the other parameters. Then we have

$$L = \begin{pmatrix} D & G \\ H & F \end{pmatrix}, \tag{4.1}$$

where $D = D_+ + D_-, F = F_+ + F_-$, the other notations being defined by (2.12).

Clearly, the transfer matrices corresponding to L and $\tilde{L}$ will commute if there exist an intertwining matrix S satisfying the equation

$$\sum_{i_2=0}^{1} \sum_{\alpha_2,\beta_2=0}^{n-1} L_{i_1\alpha_1}^{i_2\alpha_2} \tilde{L}_{i_2\beta_1}^{i_3\beta_2} S_{\alpha_2\beta_2}^{\alpha_3\beta_3} = \sum_{i_2=0}^{1} \sum_{\alpha_2,\beta_2=0}^{n-1} S_{\alpha_1\beta_1}^{\alpha_2\beta_2} \tilde{L}_{i_1\beta_2}^{i_2\beta_3} L_{i_2\alpha_2}^{i_3\alpha_3} \tag{4.2}$$

or using a matrix notations

$$(L_i^j \otimes \tilde{L}_j^k)\check{S} = \check{S}(\tilde{L}_i^j \otimes L_j^k) \tag{4.3}$$

where $(L_i^j \otimes \tilde{L}_j^k)$ denotes now a matrix product in C^2 and a direct product in $C^N \otimes C^N$ (the summation over repeated indices is assumed).

The last equation implies (in the case that S exists) that any of the four pairs of matrices $(L_i^j \otimes \tilde{L}_j^k)$ and $(\tilde{L}_i^j \otimes L_j^k)$, for $i,k = 0,1$, should have the same spectra. Requiring this one can show that the three quantities

$$\Gamma_1 = \frac{(d_+^n - f_+^n)(d_-^n - f_-^n)}{(h_+^n + h_-^n)(g_+^n + g_-^n)}; \quad \Gamma_2 = \frac{d_-^n - f_-^n}{d_+^n - f_+^n} \tag{4.4}$$

$$\Gamma_3 = \frac{h_+^n + h_-^n}{g_+^n + g_-^n} \tag{4.5}$$

should be the same for L and $\tilde{L}$.

At this stage it is convenient to introduce a parametrization for the coefficients in (2.12). Obviously, the matrix $\check{S}$ in (4.3) is unaffected by a simultaneous similarity transformation of L and $\tilde{L}$ considered as matrices in C^2

$$L_i^j \to \Lambda_{ik} L_k^l (\Lambda^{-1})_{lj}, \quad \tilde{L}_i^j \to \Lambda_{ik} \tilde{L}_k^l (\Lambda^{-1})_{lj} \tag{4.6}$$

If we choose $\Lambda = \mathrm{diag}(\lambda^{1/2}, \lambda^{-1/2})$ then (4.6) results only in a rescaling of the parameters g_+, g_-, h_+, h_- in (2.12). The values of Γ_1, Γ_2 remain unchanged, while Γ_3 rescales as

$$\Gamma_3 \to \Gamma_3 \lambda^{-2}. \tag{4.7}$$

Now, choose the modulus k and three points p, q, q' on the curve (3.5) such, that

$$\Gamma_1 = k^2, \quad \Gamma_2 = \frac{a_p^n b_p^n}{d_p^n c_p^n} \tag{4.8}$$

Then , using the transformation (4.6), (4.7) one can adjust Γ_3 so that

$$\Gamma_3 = -\frac{a_p^n c_p^n}{d_p^n b_p^n} \tag{4.9}$$

Note, that in this gauge we have the relation

$$\det F - \det D + \det H - \det G = 0, \tag{4.10}$$

where D, F, G, H are defined by (4.1). The coefficients in (2.12) can be parametrized as

$$\begin{aligned} d_+ &= -\rho_1 c_p d_p b_q b_{q'}, & h_+ &= \omega \rho_1 a_p c_p d_q a_{q'} \\ d_- &= \omega \rho_1 a_p b_p d_q d_{q'}, & h_- &= -\omega \rho_1 a_p c_p b_q c_{q'} \\ f_+ &= -\omega \rho_1 c_p d_p a_q a_{q'}, & g_+ &= -\omega \rho_1 b_p d_p a_q d_{q'} \\ f_- &= \omega \rho_1 a_p b_p c_q c_{q'}, & g_- &= \rho_1 b_p d_p c_q b_{q'} \end{aligned} \tag{4.11}$$

where $\omega = \omega_1^{-2}$. Eqs. (4.11) contain five independent parameters $\rho_1, k; p, q, q'$ instead of the six ones in (2.15). One parameter was absorbed by the gauge transformation (4.6), (4.7). Note, that the transfer matrix (2.20) does not depend on this gauge degree of freedom. In particular, applying transformation (4.6) with $\lambda = c_p/b_p$ to the coefficients (4.11), one can show that transfer matrix (2.20) can be written as

$$T^{(2)}(p; q, q') = T(1, x) = (c_p d_p c_q c_{q'})^M P(t_p) \tag{4.12}$$

where $P(t_p)$ is an M-th degree polynomial in the variable

$$t_p = (a_p b_p)/(c_p d_p)$$

Let us turn to the calculation of S in (4.3). Parametrize the coefficients of $\tilde{L}$ by the same formulae (4.11) with q, q' replaced by r, r' respectively. Solving now the linear system (4.2) for the elements S we obtain the following unexpected result

$$S_{\alpha\gamma}^{\beta\delta}(q, q', r, r') = W_{qr'}(\alpha, \gamma)\overline{W}_{q',r'}(\alpha, \beta)\overline{W}_{q,r}(\gamma, \delta)W_{q',r}(\beta, \delta) \tag{4.13}$$

where W and $\overline{W}$ are the Boltzmann weights of the chiral Potts model defined by (3.2), (3.3). Note, that matrix (4.13) was used [6] for the vertex formulation (i.e. with spins placed on the edges of lattice) of the checkerboard chiral Potts model. Using star-triangle relation (3.1) one can show that it satisfies the Yang-Baxter equations [6].

Recall now, how we came to the chiral Potts model. We started from six-vertex model R-matrix (2.1), satisfying the YBE (2.2). Then we solved another more general YBE (2.3) [with the Ansatz (2.5), (2.10), (2.11)], which includes this R-matrix as an "input". As a result we obtained the L-operators (2.12). Finally, the R-matrix S, (4.13) was found as a solution of the third YBE (4.2), which in turn includes as an input these L-operators. Thus, the chiral Potts model appeared here as the result of some unambiguous procedure, which exhibits some new algebraic structure related to the six-vertex model.

Before ending this section let us give two remarks about the structure of L-operator (4.1), (2.12). R.J. Baxter observed that our $T^{(2)}(p; q, q')$ coincide with T_{col}^{-1} in his recent paper [11] on superintegrable chiral Potts model (eqs. (8.7), (8.13) of ref. [11]). In fact, using (2.12), (4.1), (4.11) one can easily rewrite the matrix elements of L as

$$L_{i\alpha}^{j\beta} = f_i(\alpha - \beta) g_j(\alpha - \beta)$$

where $f_i(\alpha - \beta)$ depends only on p, q, k, while $g_j(\alpha - \beta)$ depends only on p, q', k). These connections are discussed in [12].

V.E. Korepin and V.O. Tarasov noticed that L-operator (4.1) can be decomposed in a product of two more elementary L-operators of massless lattice Sine-Gordon model [13]. Let X_1, Z_1 and $X_2 Z_2$ be the two sets of matrices (2.13) acting in different N-dimensional spaces. Then

$$L = \begin{pmatrix} aX_1 & bZ_1 \\ cZ_1^{-1} & dX_1^{-1} \end{pmatrix} \begin{pmatrix} \tilde{a}X_2 & \tilde{b}Z_2 \\ \tilde{c}Z_2^{-1} & \tilde{d}X_2^{-1} \end{pmatrix}$$

if we set $X_1 X_2 Z_2^{-1} Z_1 = 1$ and identify $X_2^{-1} Z_2, \quad Z_2^{-1} X_2, \quad X_1 X_2$ with A, B, C in (2.12) respectively.

5.Functional relations.

We wish to find the eigenvalues of the matrix $\mathbf{T}$ defined by (2.20). Just as in [14] let us first seach for vectors whose elements are products of single spin functions

$$\mathbf{Q}_{\{\alpha\}} = \phi_{\alpha_1}^1 \phi_{\alpha_2}^2 \cdots \phi_{\alpha_M}^M \tag{5.1}$$

which obeys the following relations

$$\mathbf{TQ} = \Phi_1 \mathbf{Q}' + \Phi_2 \mathbf{Q}'', \tag{5.2}$$

where Φ_1, Φ_2 are scalars and $\mathbf{Q}', \mathbf{Q}''$ are vectors of the same form as (5.1).

The calculations are closely parallel to those of ref. [11]. Rewrite $\mathbf{T}$ as

$$\mathbf{T}_{\alpha}^{\beta} = \mathrm{Tr}[\mathbf{L}(\alpha_1, \beta_1)\mathbf{L}(\alpha_2, \beta_2) \cdots \mathbf{L}(\alpha_M, \beta_M)], \tag{5.3}$$

where $\mathbf{L}(\alpha, \beta)$ is a two by two matrix

$$L(\alpha, \beta) = \begin{pmatrix} D_{\alpha\beta} & G_{\alpha p} \\ H_{\alpha\beta} & F_{\alpha\beta} \end{pmatrix}$$

whose elements are given by (4.1), (2.12). The product $\mathbf{TQ}$ is a vector which can be written as

$$(\mathbf{TQ})_\alpha = \mathrm{Tr}[\mathbf{K}_1(\alpha_1)\mathbf{K}_2(\alpha_2)\cdots\mathbf{K}_M(\alpha_N)], \tag{5.4}$$

$$\mathbf{K}_J(\alpha) = \sum_\beta \mathbf{L}(\alpha,\beta)\phi^J_\beta.$$

Note, that (5.4) is unaffected if we replace each $\mathbf{K}_J(\alpha)$ by

$$\mathbf{K}^*_J(\alpha) = \mathbf{O}_J^{-1}\mathbf{K}_J(\alpha)\mathbf{O}_{J+1} \tag{5.5}$$

for $J = 1,\ldots,M$, provided

$$\mathbf{O}_{M+1} = \mathbf{O}_M. \tag{5.6}$$

Choose $\mathbf{O}_J$ of the form

$$\mathbf{O}_J = \frac{1}{m_J}\begin{pmatrix} 1 & -x_J \\ x_J & 1 \end{pmatrix}, \tag{5.7}$$

where $m_J = \sqrt{1+x_J^2}$. The trace in (5.4) will simplify to the sum of two products if we can choose the ϕ^J_α and $\mathbf{M}_J$, so that all $K^*_J(\alpha)$ are upper triagular matrices, i.e., their bottom left elements vanish. This is so if

$$(x_J x_{J+1} G + x_J D - x_{J+1} F - H)\phi^J = 0, \tag{5.8}$$

where D, F, G, H are defined by (4.1).

Equating the determinant of the coefficients of this linear system to zero, we get

$$x_J^N x_{J+1}^N \det G + x_J^N \det D - x_{J+1}^N \det F - \det H = 0 \tag{5.9}$$

for $J = 1,\ldots,M$. Given x_J, (5.9) is an N-th degree equation for x_{J+1}. Thus, if we take x_1 be given, we can construct the entire sequence $x_1,\ldots,x_{M+1}$, having a choice of N alternatives at each stage. To satisfy (5.6) we require that $x_1 = x_{M+1}$. Comparing (5.9) with (4.10) one can readily find solutions for which $x_J^N = 1, \quad M$. In fact they are the only solutions satisfying (5.6). Hence, we have the N^M solutions of (5.9)

$$x_J = \omega^{\alpha_J}, \quad \alpha_J = 0,\ldots,N-1, \\ J = 1,\ldots,M, \tag{5.10}$$

where $\omega = \exp(2\pi/n)$, for all N^M choices of $\{\alpha\} = \{\alpha_1, \ldots, \alpha_J\}$. If we use the parametrization (4.11) then $\mathbf{T}, \mathbf{Q}, \phi$ become functions of the rapidities p, q, q'. When necessary we shall write this dependence explicitly. Solving now (5.8) for ϕ^J, we obtain

$$\phi^J_{\beta_J}(p; q, q') = C_1 W_{pq'}(\alpha_J - \beta_J)\overline{W}_{pq}(\beta_J - \alpha_{J+1}) \tag{5.11}$$

where W_{pq} and $\overline{W}_{pq}$ are given by (3.3), C_1 is a normalization factor. Then the vector $\mathbf{Q}$, (5.1), corresponding to the sequence $\{\alpha\}$ in (5.1) has the form

$$\mathbf{Q}^{\{\alpha\}}_{\{\beta\}}(p; q, q') = C_1^M \prod_{J=1}^{M} \left(W_{pq'}(\alpha_J - \beta_J)\overline{W}_{pq}(\beta_J - \alpha_{J+1})\right) \tag{5.12}$$

Obviously, one can view $\mathbf{Q}^{\{\alpha\}}_{\{\beta\}}$ as matrix elements of some N^M by N^M matrix $\mathbf{Q}$, which is nothing but the transfer matrix (3.6a) of the chiral Potts model for the inhomogeneous chain with alternating rapidities q and q'.

We can now calculate the diagonal elements of $\mathbf{K}^*_J$, (5.5), using (5.11) and the relations

$$\begin{aligned} (\mathbf{K}^*_J)_{00} &= \frac{m_J}{m_{J+1}}(D + x_{J+1}G)\phi^J \\ (\mathbf{K}^*_J)_{11} &= \frac{m_{J+1}}{m_J}(F - x_J G)\phi^J \end{aligned} \tag{5.13}$$

Substituting the resulting expression into (5.4) and introducing an index R for $\mathbf{Q}$ to emphasize that $\mathbf{Q}$ is multiplied by $\mathbf{T}$ from the right, we get (for normalization $W_{pq}(0) = \overline{W}_{pq}(0) = 1$)

$$\begin{aligned} \mathbf{T}(p; q, q')\mathbf{Q}_R(p; q, q') = \Phi_1(p; q, q')\mathbf{Q}_R(R^{N-1}p; q, q') + \\ + \Phi_2(p; q, q')\mathbf{Q}_R(R^{1-N}p; q, q') \end{aligned} \tag{5.14}$$

where R denotes one of the automorphisms of the curve (3.6) [6]:

$$\begin{aligned} &p \to Rp \\ &(a_p, b_p, c_p, d_p) \to (b_p, \omega a_p, d_p, c_p) \end{aligned} \tag{5.15}$$

and

$$\Phi_1(p; q, q') = \Phi(p; q, q') \left[\frac{\omega(x_p - x_q\omega^\rho)(t_{q'} - t_p)}{(y_{q'}\omega\rho - x_p)}\right]^M \tag{5.16}$$

$$\Phi_2(p; q, q') = \Phi(p; q, q') \left[\frac{(y_p - \varepsilon_{q'}^{\rho+1})(t_q - \varepsilon t_p)}{(\varepsilon^\rho y_q - y_p)}\right]^M \tag{5.17}$$

$$\Phi(p;q,q') = \left[\rho_1 c_p d_p d_q d_{q'} \overline{W}_{pq}(\rho) W_{pq'}(-\rho)\right]^M \tag{5.18}$$

where $\rho = (N-1)/2 \quad x_p = a_p/d_p, \quad y_p = b_p/c_p, \quad t_p = q_p b_p/c_p d_p$. Next, one can find a matrix $\mathbf{Q}_L$ with the similar properties. Repeating the calculations and using the fact that

$$\mathbf{T}(R^N p;q,q') = \mathbf{T}(p;q,q') \tag{5.19}$$

we obtain

$$\mathbf{Q}_L(p;q,q')\mathbf{T}(p;q,q') = \Phi_1(p;q',q)\mathbf{Q}_L(R^{N-1}p;q,q') + \\ + \Phi_2(p;q',q)\mathbf{Q}_L(R^{1-N}p;q,q'), \tag{5.20}$$

where

$$\mathbf{Q}_L(p;q,q') = \mathbf{Q}_R(p;q',q)\hat{\mathbf{P}}^{-1} \tag{5.21}$$

with $\hat{\mathbf{P}}$ given by (3.8). Combining (5.14), (5.20), (5.21) we have

$$\mathbf{T}(p;q,q')\mathbf{Q}_R(p;q,q') = \mathbf{Q}_R(p;q,q')\mathbf{T}(p;q',q). \tag{5.22}$$

Setting now $q = q'$, we have

$$\mathbf{Q}_L(p;q,q) = U_{pqq}, \quad \mathbf{Q}_R(p;q,q) = \hat{U}_{pqq} \tag{5.23}$$

$$[\mathbf{T}(p;q,q), \mathbf{Q}(p;q,q)] = 0, \tag{5.24}$$

where $\mathbf{Q} = \mathbf{Q}_L$ or $\mathbf{Q} = \mathbf{Q}_R$, while U, V are defined by (3.6). Moreover, the YBE (2.3) and the star-triangle equation (3.1) imply

$$[\mathbf{T}(p;q,q'), \mathbf{T}(p';q,q')] = 0 \tag{5.25}$$

$$[\mathbf{Q}(p;q,q)\mathbf{Q}(p';q,q)] = 0 \tag{5.26}$$

Uniformizing the curve (3.5) the weights (3.2), (4.10) and, hence, the matrices $\mathbf{T}, \mathbf{Q}$ (for the finite length of the chain, N) become meromorphic functions of p. Due to commutativity (5.24)-(5.26), $\mathbf{T}$ and $\mathbf{Q}$ have common eigenvectors which are independent of p. Therefore the eigenvalues will be also meromorphic functions of p. Their poles are, of cause, those of the weights.

Consider an action of (5.14) on some eigenvector. Let p^* be any (non-trivial) zero of the corresponding eigenvalue $Q(p;q,q)$, then from (5.14) we have the equations

$$\frac{Q(R^{N-1}p^*)}{Q(R^{1-N}p^*)} = -\frac{\Phi_2(p^*)}{\Phi_1(p^*)}, \tag{5.27}$$

which, in principle, fixes of all the zeros of Q and, hence, contains sufficient data to reconstruct Q. The corresponding eigenvalue of T can be then calculated from (5.14).

The main problem now is to find suitable uniformization of the curve (3.5).

In fact, one can derive additional relations among $\mathrm{T}(p,q,q')$, $U_{p,q,q'}$ and $\hat{U}_{p,q,q'}$. Here we only state the result emphasizing the most important points of the calculations. The detailed proof will be published in [12].

The first relation is:

$$(L^{(2)\beta}_{\alpha}(p,q,q'))_{ij} = \Lambda^{-1}_{ppq'} \sum_{\gamma\gamma'\delta} \omega^{\rho(\delta-\gamma)+j\delta-i\gamma} P^{(2)}_{\gamma\gamma'} \times$$
$$\times S^{\beta\delta}_{\gamma'\alpha}(p, R^{N+1}p, q, q') \tag{5.28}$$

where $S^{\gamma\delta}_{\alpha\beta}$ is given by (4.13), $L^{(2)\beta}_{\alpha}(p,q,q')$ denotes the matrix elements of L-operator (4.1), (2.12), $i,j=0,1$,

$$P^{(n)}_{\alpha\beta} = \frac{1}{N}\sum_{j=0}^{n-1} \omega^{\rho(n-1-2j)(\alpha-\beta)}, \qquad n=1,\ldots,N \tag{5.29}$$

$$P^{(N)}_{\alpha\beta} = \delta_{\alpha\beta}, \tag{5.30}$$

$$\Lambda_{pqq'} = \left[\frac{-S_{q',R_p}(x_{q'}-\omega y_p)W_{q,R^{N+1}p}(\rho)\overline{W}_{q',R^{N+1}p}(\rho)}{(t_{q'}-\omega^2 t_p)(y_q-\omega y_p)c_p d_p d_q c_{q'}}\right]^M \tag{5.31}$$

the variables x_p, y_p, t_p being defined after eq. (5.18),

$$S_{p,q} = N\frac{(t_p^N - t_q^N)(x_p-x_q)(y_p-y_q)}{(t_p-t_q)(x_p^N-x_q^N)(y_p^N-x_q^N)}, \tag{5.32}$$

Now consider the fusion procedure for the matrix T, (5.3). For our case this procedure is essentially the same as for 6v-model [15], because it is determined by the degeneracy points structure of the 6-vertex R-matrix (2.1). Introduce the notations

$$\tau^{(2)}_k = \mathrm{T}(R^{\Delta k}p,q,q'), \tag{5.33}$$

$$\phi_k = S_{R^{\Delta k+1}p,q} \tag{5.34}$$

$$\mu_k = \Phi_2(p,q,q')\Phi_1(R^2p,q,q') \tag{5.35}$$

where Φ_1, Φ_2 are defined by eqs (5.16), (5.17), $\Delta = N+1$. Note, that $\mu_k(c_p d_p d_q c_{q'})^{-2M}$ is a polynomial in t_p.

The fusion procedure leads to the following relations

$$\tau_0^{(k)} \tau_{(k-1)}^{(2)} = \mu_{k-2} \tau_0^{k-1} + \tau_0^{(k+1)}, \qquad k = 2, \dots, N-1 \tag{5.36}$$

where $\tau_k^{(1)} \equiv 1$ and

$$(\tau_k^{(n)})_{\alpha_1 \dots \alpha_M}^{\beta_1 \dots \beta_M} = T_r \left[L^{(n)\beta_1}_{\alpha_1}(R^{\Delta k} p, q, q') \quad \dots \quad L^{(n)\beta_M}_{\alpha_M}(R^{\Delta k} p, q, q') \right]$$

where $L^{(n)\beta_1}_{\alpha_1}$ are n by n matrices with the elements

$$\left(L^{(n+1)\alpha_{n+1}}_{\alpha_1}(p, q, q') \right)_{ab} = \sum_{\substack{i_1, \dots, i_n = 0 \\ j_1, \dots, j_n = 0}}^{1} \frac{1}{a!} \delta_{a, i_1 + i_2 + \dots + i_n} \delta_{b, j_1 + j_2 + \dots + j_n} \times$$
$$\times \sum_{\alpha_2, \dots, \alpha_n = 0}^{N-1} \prod_{m=1}^{n} \left(\mathrm{L}^{(2) j_m \alpha_{m+1}}_{i_m \alpha_m} (R^{m-1)\Delta} p, q, q') \right) \tag{5.38}$$

where $a, b = 0, \dots, n$.

Substituting now eq. (5.28) into (5.38) we obtain

$$\left(\prod_{m=0}^{n-2} \Lambda_m \right) \left(\prod_{m=1}^{n-2} \phi_m \right)^{-1} \left(\mathrm{L}^{(n)\beta}_{\alpha}(p, q, q') \right)_{a,b} =$$
$$= \frac{1}{N} \sum_{\gamma, \gamma', \delta} \omega^{\rho(n-1)(\delta-\gamma)+b\delta-a\gamma} P^{(n)}_{\gamma\gamma'} S^{\beta}_{\gamma'\alpha}(p, R^{(n-1)\Delta+1} p, q, q') \tag{5.39}$$

Setting now $n = N$, using (5.30) and remembering the definitions (4.13) and (4.6) we get

$$U_{p,q,q'} \hat{U}_{R^N p,q,q'} = \left(\prod_{m=0}^{n-2} \Lambda_m \right) \left(\prod_{m=1}^{n-2} \phi_m \right)^{-1} \tau_1^{(N)} \tag{5.40}$$

Now recall that in our notations $Q_R(p, q, q')$ coincide with $U_{p,q,q'}$. So, using eq. (5.14) and recurrence relations (5.36) one can express $\tau^{(N)}$ through U. Substituting the resulting expression into (5.40) we obtain closed relation which contains only U and $\hat{U}$. In the case $N = 3$, $q = q'$, it coincides with the relation conjectured in [16] on the basis of numerical calculations.

Acknowledgements.

The authors thank R.J. Baxter, M.N. Barber, M.T. Batchelor, V.A. Fatteev, V.E. Drinfeld, M. Jimbo, V.E. Korepin, R. Kashaev, T. Miwa, P.A. Pearce, G.R.W. Quispel, N. Reshetikhin and A.B. Zamolodchikov for the fruitful discussions.

This work was completed during the visit of one of the authors (VVB) to the Australian National University. He thanks the Australian National University and the Institute for High Energy Physics for their support and the Centre for Mathematical Analysis (especially Prof. R.J. Baxter) for their hospitality during the visit.

References

[1] L. Onsager. Phys. Rev. 65 (1944) 117.

[2] R.J. Baxter. "Exactly Solved Models in Statistical Mechanics". Academic, London, 1982.

[3] L.D. Faddeev. Sov. Sci.Rev. C1 (1980) 107-155.

[4] H. Au-Yang, B.M. McCoy, J.H.H. Perk, S. Tang and M. Yan. Phys. Lett. A 123 (1987) 219.

[5] B.M. McCoy, J.H.H. Perk, S. Tang and C.H. Sah. Phys. Lett. A 125 (1987) 9.

[6] R.J. Baxter, J.H.H. Perk and H. Au-Yang. Phys. Lett. A 128 (1988) 138.

[7] E.H. Lieb. Phys. Rev. 162 (1967) 162.

[8] V.G. Drinfeld. Dokl. Akad. Nauk. SSSR, 283 (1985) 1060.

[9] E.K. Sklyanin. Funk. Anal. Prilozh. 16(4) (1982) 27.

[10] V.A. Fateev, A.B. Zamolodchikov. Phys. Lett. A 92 (1982) 37.

[11] R.J. Baxter,"Superintegrable Chiral Potts Model, Thermodynamic Properties, an "Inverse" Model, and a Simple Associated Hamiltonian". Preprint CMA-R12-89, Canberra, 1989, accepted for J. Stat. Phys.

[12] R.J. Baxter, V.V. Bazhanov, J.H.H. Perk, "Functional Relations for Transfer Matrices of the Chiral Potts Model", in Proc. of Australian National University Workshop on Yang-Baxter equations, Canberra, 1989 (to be published in Int. J. Mod. Phys. B).

[13] A.G. Izergin, V.E. Korepin. Nucl. Phys. B205 (1982) 401.
V.O. Tarasov, Teor. Mat. Fiz. 63(2)(1985) 175.

[14] R.J. Baxter, Ann. Phys. 76 (1973) 1.

[15] A.N. Kirillov, N.Y. Reshetikhin, J. Phys. A: Math. Gen. 19(1986) 565.

[16] G. Albertini, B.M. McCoy, J.H.H. Perk. Phys. Lett. A135 (1989) 159.

Figure Captions

1. Graphical representation of the Yang-Baxter equation (2.2).
2. Graphical representation of the Yang-Baxter equation (2.3).
3. Graphical representation of the column to column transfer matrix (2.19).
4. Graphical representation of the row to row transfer matrix (2.20).
5. Part of the lattice for the formulation of checkerboard Chiral Potts model. The open arrows show the rapidities directions.
6. The two types of Boltzmann Weights.
7. Graphical representation of the star-triangle relation (3.1).
8. Graphical representation of the Yang-Baxter equations (4.2).

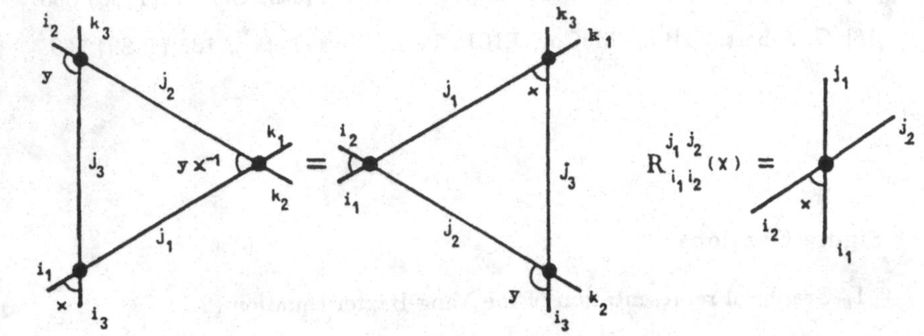

1. Graphical representation of the Yang-Baxter equation (2.2).

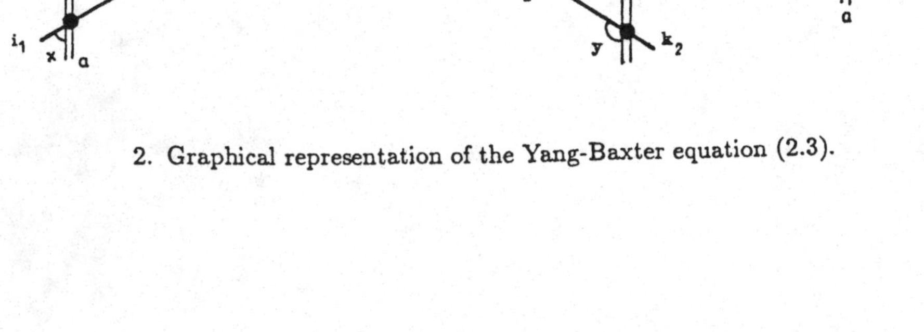

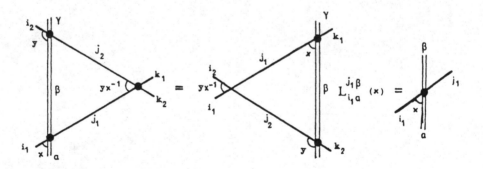

2. Graphical representation of the Yang-Baxter equation (2.3).

$$\mathcal{T}\,{}^{j_1\,,\,\ldots\,,\,j_{M'}}_{\,i_1\,,\,\ldots\,,\,i_{M'}} =$$

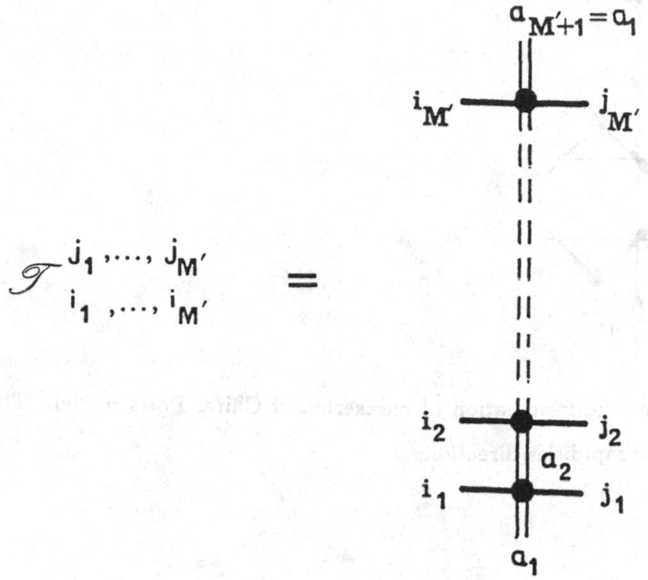

3. Graphical representation of the column to column transfer matrix (2.19).

(2) $\;\beta_1\,,\ldots,\beta_M$

$\quad a_1\,,\ldots,a_M$

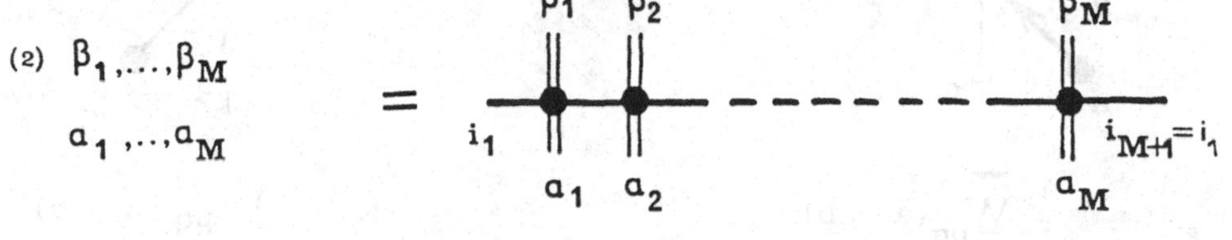

4. Graphical representation of the row to row transfer matrix (2.20).

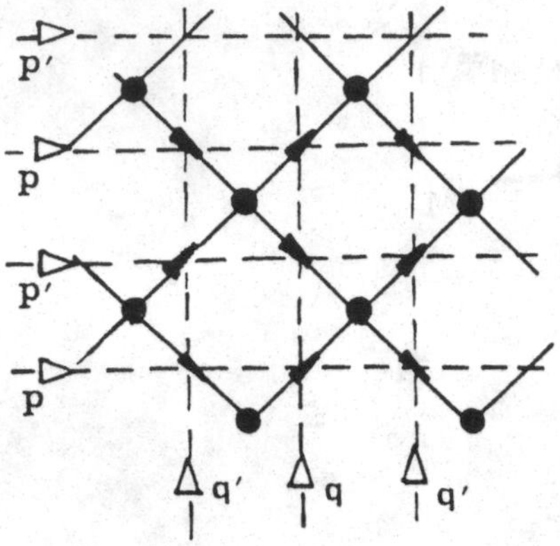

5. Part of the lattice for the formulation of checkerboard Chiral Potts model. The open arrows show the rapidities directions.

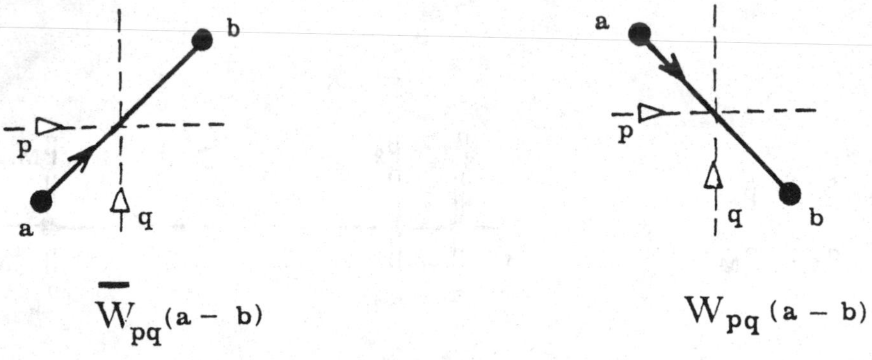

6. The two types of Boltzmann Weights.

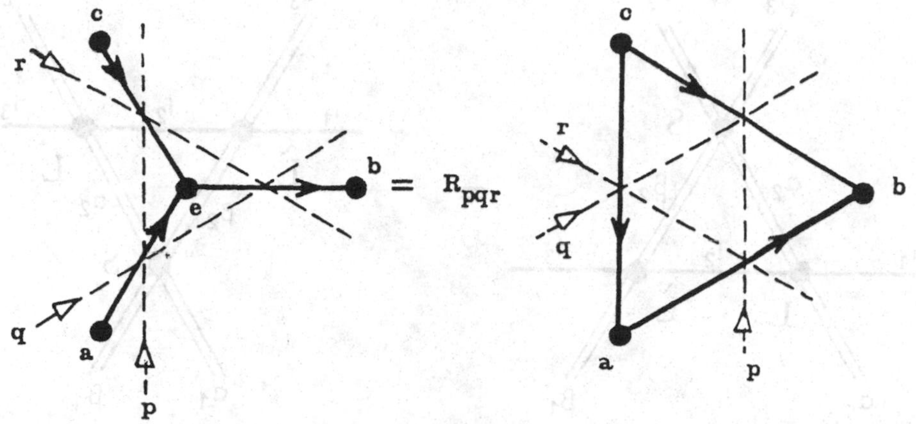

7. Graphical representation of the star-triangle relation (3.1).

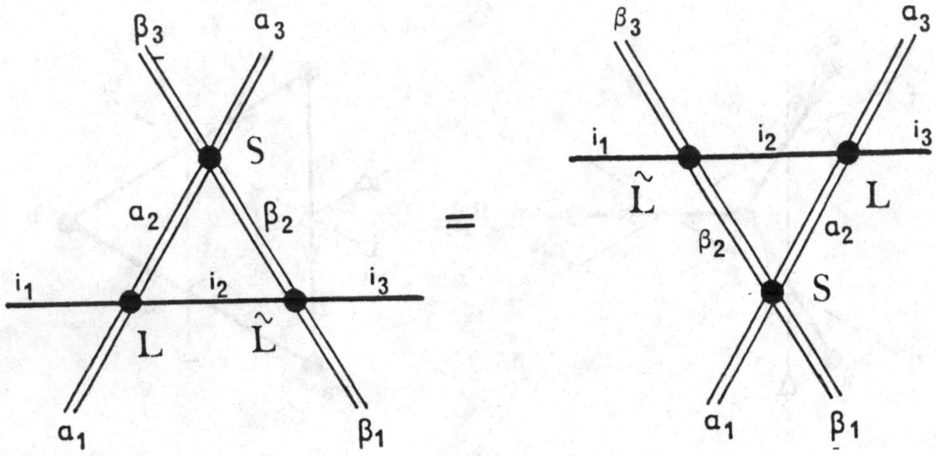

8. Graphical representation of the Yang-Baxter equations (4.2).

LIST OF REFERENCES

1. Y.Akutsu, A.Kuniba and M.Wadati, Exactly solvable IRF models, II S_N generalizations, *J. Phys. Soc. Japan* **55** (1986), 1466-1474
2. Y.Akutsu, A.Kuniba and M.Wadati, Exactly solvable IRF models, III A new hierarchy of solvable models, *J. Phys. Soc. Japan* **55** (1986), 1880-1886
3. S.I.Alishauskas and P.P.Kulish, Spectral decomposition of $SU(3)$-invariant solutions of the Yang-Baxter equation, *Zapiski Nauch. Sem. LOMI* **145** (1985), 3-21 [*J. Soviet Math.* (1986), 2563-2574]
4. H.H.Anderson, The linkage principle and the sum formula for quantum groups, *preprint* (1989),
5. G.E.Andrews, R.J.Baxter and P.J.Forrester, Eight-vertex *SOS* model and generalized Rogers-Ramanujan-type identities, *J. Stat. Phys.* **35** (1984), 193-266
6. H.Au-Yang, B.M.McCoy, J.H.H.Perk, S.Tang and M.-L.Yan, Commuting transfer matrices in the chiral Potts models: solutions of star-triangle equations with genus > 1, *Phys. Lett. A* **123** (1987), 219-223
7. H.Au-Yang, B.M.McCoy, J.H.H.Perk, and S.Tang , Solvable models in statistical mechanics and Riemann surfaces of genus greater than one, in *Algebraic Analysis* vol.1, Eds. M.Kashiwara and T.Kawai, Academic, (1989), 29-39
8. H.Au-Yang and J.H.H.Perk, Onsager's star-triangle equation: master key to integrability, *Adv. Stud. Pure Math.* **19** (1989), 57-94
9. O.Babelon, Representation of the Yang-Baxter algebra associated to Toda field theory, *Nucl. Phys. B* **230** [FS10] (1984), 241-249
10. O.Babelon, Jimbo's q-analogues and current algebras, *Lett. Math. Phys.* **15** (1988), 11-117
11. O.Babelon, H.J.de Vega and C.M.Viallet, Solutions of the factorization equations from Toda field theory, *Nucl. Phys. B* **190** [FS3] (1981), 542-552
12. R.J.Baxter, Eight-vertex model in lattice statistics, *Phys. Rev. Lett.* **26** (1971), 832-833
13. R.J.Baxter, Partition function of the eight-vertex lattice model, *Ann. Phys.* **70** (1972), 193-228
14. R.J.Baxter, Eight-vertex model in lattice statistics and one-dimensional anisotropic Heisenberg chain I. Some fundamental eigenvectors, *Ann. Phys.* **76** (1973), 1-24
15. R.J.Baxter, Eight-vertex model in lattice statistics and one-dimensional anisotropic Heisenberg chain II. Equivalence to a generalized ice-type model, *Ann. Phys.* **76** (1973), 25-47
16. R.J.Baxter, Eight-vertex model in lattice statistics and one-dimensional anisotropic Heisenberg chain III. Eigenvectors of the transfer matrix and Hamiltonian, *Ann. Phys.* **76** (1973), 48-71
17. R.J.Baxter, Solvable eight-vertex model on an arbitrary planar lattice, *Phil. Trans. Royal Soc. London* **289** (1978), 315-346
18. R.J.Baxter, Hard hexagons: exact solution, *J. Phys. A* **13** (1980), L61-L70

19. R.J.Baxter, Exactly solved models, in *Fundamental Problems in Statistical Mechanics,* Ed. E.G.D.Cohen **5** North Holland, Amsterdam, (1980), 109-141

20. R.J.Baxter, *Exactly Solved Models in Statistical Mechanics,* Academic, London 1982,

21. R.J.Baxter, The inversion relation method for some two-dimensional exactly solved models in lattice statistics, *J. Stat. Phys.* **28** (1982), 1-41

22. R.J.Baxter, On Zamolodchikov's solution of the tetrahedron equations, *Comm. Math. Phys.* **88** (1983), 185-205

23. R.J.Baxter, The Yang-Baxter equations and the Zamolodchikov model, *Physica D* **18** (1986), 321-347

24. R.J.Baxter and G.E.Andrews, Lattice gas generalization of the hard hexagon model I, Star-triangle relation and local densities, *J. Stat. Phys.* **44** (1986), 249-271

25. R.J.Baxter, J.H.H.Perk and H.Au-Yang, New solutions of the star-triangle relations for the chiral Potts model, *Phys. Lett. A* **128** (1988), 138-142

26. V.V.Bazhanov, Trigonometric solutions of triangle equations and classical Lie algebras, *Phys. Lett. B* **159** (1985), 321-324

27. V.V.Bazhanov, Quantum R matrices and matrix generalizations of trigonometric functions, *Teoret. Mat. Fiz.* **73** (1987), 26-32 [*Theoret. Math. Phys.* **73** (1987), 1035-1039]

28. V.V.Bazhanov, Integrable quantum systems and classical Lie algebras, *Comm. Math. Phys.* **113** (1987), 471-503

29. V.V.Bazhanov and A. G. Shadrikov, Trigonometric solutions of triangle equations for simple Lie superalgebras, *Teoret. Mat. Fiz.* **73** (1987), 402-419 [*Theoret. Math. Phys.* **73** (1987), 1302-1312]

30. V.V.Bazhanov and Yu.G.Stroganov, A new class of factorized S-matrices and triangle equations, *Phys. Lett. B* **105** (1981), 278-280

31. V.V.Bazhanov and Yu.G.Stroganov, Trigonometric and S_n symmetric solutions of triangle equations with variables on the faces, *Nucl. Phys. B* **205** [FS5] (1982), 505-526

32. V.V.Bazhanov and Yu.G.Stroganov, Condition of commutativity of transfer matrices on a multidimensional lattice, *Teoret. Mat. Fiz.* **52** (1982), 105-113 [*Theoret. Math. Phys.* **52** (1982), 685-691]

33. V.V.Bazhanov and Yu.G.Stroganov, Free fermions on a three-dimensional lattice and tetrahedron equations, *Nucl. Phys. B* **230** [FS10] (1984), 435-454

34. V.V.Bazhanov and Yu.G.Stroganov, On connection between the solutions of the quantum and classical triangle equations, Proc. VI-th Int. Seminar on High Energy Physics and Field Theory, **1**, 51-53 Protvino 1983

35. V.V.Bazhanov and Yu. G. Stroganov, Hidden symmetry of free fermion model. I. Triangle

equation and symmetric parametrization, *Teoret. Mat. Fiz.* **62** (1985), 377-387 [*Theoret. Math. Phys.* **62** (1985), 253-260]

36. V.V.Bazhanov and Yu.G.Stroganov, Hidden symmetry of free fermion model. II. Partition function, *Teoret. Mat. Fiz.* **63** (1985), 291-302 [*Theoret. Math. Phys.* **63** (1985), 519-527]
37. V.V.Bazhanov and Yu. G. Stroganov, Hidden symmetry of free fermion model. III. Inversion relations, *Teoret. Mat. Fiz.* **63** (1985), 417-427 [*Theoret. Math. Phys.* **63** (1985), 604-611]
38. V.V.Bazhanov and Yu.G.Stroganov, Chiral Potts model as a descendant of the six-vertex model, preprint 1989
39. A.A.Belavin, Dynamical symmetry of integrable quantum systems, *Nucl. Phys. B* **180** [FS2] (1981), 189-200
40. A.A.Belavin, Discrete groups and the integrability of quantum systems, *Funct. Anal. Appl.* **14** (1981), 260-267
41. A.A.Belavin and V.G.Drinfel'd, Solutions of the classical Yang-Baxter equation for simple Lie algebras, *Funct. Anal. Appl.* **16** (1983), 159-180
42. A.A.Belavin and V.G.Drinfel'd, Classical Yang-Baxter equation for simple Lie algebras, *Funkt. Anal. Priloz.* **17** (1983), 69-70 [*Funct. Anal. Appl.* **17** (1984), 220-221]
43. A.A.Belavin and V.G.Drinfel'd, Triangle equations and simple Lie algebras, *Soviet Scientific Reviews* section C, ed. S.P.Novikov, Harwood Academic Press, vol.4 (1984), 93-165
44. F.A.Berezin, C.P.Pokhil and V.M.Finkelberg, The Schrödinger equation for a system of one-dimensional particles with point interaction (Russian), *Vestn. Mosk. Gos. Univ.* **1** (1964), 21-28
45. F.A.Berezin and V.N.Sushko, Relativistic two-dimensional model of a self-interacting fermion field with nonvanishing rest mass, *Zh.Eksp.Theor.Fiz.* **48** (1965), 1293-1306 [*Sov. Phys. JETP* **21** (1965), 865-873]
46. B.Berg, M.Karowski, P.Weisz and V.Kurak, Factorized $U(n)$ symmetric S-matrices in two dimensions, *Nucl. Phys. B* **134** (1978), 125-132
47. D.Bernard, Vertex operator representations of the quantum affine algebra $U_q(B_r^{(1)})$, *Lett. Math. Phys.* **17** (1989), 239-245
48. A.Bovier, Factorized S matrices and generalized Baxter models, *J. Math. Phys.* **24** (1983), 631-641
49. E.Brezin and J.Zinn-Justin, Un problème à N corps soluble, *C. R. Acad. Sci. Paris* **B263** (1966), 670-673
50. N.Burrough, The universal R-matrix for $U_q\mathfrak{sl}(3)$ and beyond !, preprint DAMTP/R-89/4 1989
51. N.Burrough, Relating the approaches to quantised algebras and quantum groups, preprint DAMTP/R-89/11 1989

52. V.Chari and A.Pressley, Strings, qunatum Vandermonde determinants, and representations of Yangians, preprint Tata Institute 1989

53. V.Chari and A.Pressley, Fundamental representations of Yangians, preprint Tata Institute 1989

54. V.Chari and A.Pressley, Quantum R matrices and intertwining operators, preprint Tata Institute 1989

55. I.V.Cherednik, Certain S-matrices associated with Abelian manifolds, *Dokl. Akad. Nauk. CCCP* **249** (1979), 1095-1098 [*Soviet Phys. Doklady* **24** (1979), 974-976]

56. I.V.Cherednik, On a method of constructing factorized S matrices in elementary functions, *Teoret. Mat. Fiz.* **43** (1980), 117-119 [*Theoret. Math. Phys.* **43** (1980), 356-358]

57. I.V.Cherednik, On properties of factorized S-matrices in elliptic functions, *Yad. Fiz.* **32** (1982), 549-557

58. I.V.Cherednik, Bäcklund-Darboux transformations for classical Yang-Baxter bundles, *Usp. Math. Nauk* **38** (1983), 3-21

59. I.V.Cherednik, Factorizing particles on a half-line and root systems, *Teoret. Mat. Fiz.* **61** (1984), 35-44 [*Theoret. Math. Phys.* **61** (1984), 977-983]

60. I.V.Cherednik, Some finite-dimensional representations of generalized Sklyanin algebras, *Funkt. Anal. Priloz.* **19** (1985), 89-90 [*Funct. Anal. Appl.* **19** (1985), 77-79]

61. I.V.Cherednik, On $R-$matrix quantization of formal loop groups, *"Group Theoretical Methods in Physics"* (Proceedings of III Seminare Yurmala, USSR, May 1985), Nauka, VNU Publ.B.V.

62. I.V.Cherednik, Special bases of irreducible representations of a degenerate affine Hecke algebra, *Funkt. Anal. Priloz.* **20** (1986), 87-88 [*Funct. Anal. Appl.* **20** (1986), 76-78]

63. I.V.Cherednik, On "quantum" deformations of irreducible finite-dimensional representations of $\mathfrak{gl}_N$, *Dokl. Akad. Nauk. CCCP* **287** (1986), 1076-1079 [*Soviet Math. Doklady* **33** (1986), 507-510]

64. I.V.Cherednik, On irreducible representations of elliptic quantum $R-$algebras, *Dokl. Akad. Nauk. CCCP* **291** (1986), 49-53 [*Soviet Math. Doklady* **34** (1987), 446-450]

65. I.V.Cherednik, A new interpretation of Gel'fand-Tzetlin bases, *Duke Math. J.* **54** (1987), 563-577

66. I.V.Cherednik, An analogue of character formula for Hecke algebras, *Funkt. Anal. Priloz.* **21** (1987), 94-95 [*Funct. Anal. Appl.* **21** (1987), 172-174]

67. I.V.Cherednik, $q-$analogues of Gel'fand-Tzetlin bases, *Funct. Anal. Appl.* **22** (1988), 78-79

68. I.V.Cherednik, Quantum groups as hidden symmetries of classic representation theory, preprint 1989 (to appear in the Proceedings of 17-th DGM conference, Chester, August 1988)

69. D.V.Chudnovsky and G.V.Chudnovsky, Characterization of completely X−symmetric factorized S-matrices for a special type of interaction. Applications to multicomponent field theories, *Phys. Lett. A* **79** (1980), 36-38

70. D.V.Chudnovsky and G.V.Chudnovsky, Completely X−symmetric S−matrices corresponding to theta functions, *Phys. Lett. A* **81** (1981), 105-110

71. T.L.Curtright and C.K.Zachos, Deforming maps for quantum algebras, preprint ANL-HEP-PR-89-105 1989

72. E.Date, M.Jimbo, A.Kuniba, T.Miwa and M.Okado, Exactly solvable SOS models I :Local height probabilities and theta function identities, *Nucl. Phys. B* **290**[FS20] (1987), 231-273

73. E.Date, M.Jimbo, A.Kuniba, T.Miwa and M.Okado, Exactly solvable SOS models II : Proof of the star-triangle relation and combinatorial identities, *Adv. Stud. Pure Math.* **16** (1988), 17-122

74. E.Date, M.Jimbo, T.Miwa, Representations of $U_q(\mathfrak{gl}(n, \mathbf{C}))$ at $q = 0$ and the Robinson-Schensted correspondence, preprint RIMS 656 (1989), to appear in *Physics and Mathematics of Strings*, Memorial volume for Vadim Knizhnik, Eds. L.Brink, D.Friedan and A.M.Polyakov, World Scientific.

75. E.Date, M.Jimbo, T.Miwa and M.Okado, Fusion of the eight vertex SOS model, *Lett. Math. Phys.* **12** (1986), 209-215

76. R.Dipper and S.Donkin, Quantum GL_n, preprint 1989

77. V.G.Drinfel'd, Hamiltonian structures on Lie groups, Lie bialgebras and the geometric meaning of the classical Yang-Baxter equations, *Dokl. Akad. Nauk. CCCP* **268** (1983), 285-287 [*Soviet Math. Doklady* **27** (1983), 68-71]

78. V.G.Drinfel'd, On constant, quasi-classical solutions of the Yang-Baxter quantum equation, *Dokl. Akad. Nauk. CCCP* **273** (1983), 531-534 [*Soviet Math. Doklady* **28** (1983), 667-671]

79. V.G.Drinfel'd, Hopf algebras and the quantum Yang-Baxter equation, *Dokl. Akad. Nauk. CCCP* **283** (1985), 1060-1064 [*Soviet Math. Doklady* **32** (1985), 254-258]

80. V.G.Drinfel'd, Degenerate affine Hecke algebras and Yangians, *Funkt. Anal. Priloz.* **20** (1986), 69-70 [*Funct. Anal. Appl.* **20** (1986), 58-60]

81. V.G.Drinfel'd, Quantum groups, Proceedings of the International Congress of Mathematicians, Berkeley, California USA 1986, 798-820; *Zapiski Nauch. Sem. LOMI* **155** (1986), 18-49 [*J. Soviet Math.* **41** (1988), 898-915]

82. V.G.Drinfeld, Quasi-Hopf algebras and Knizhnik-Zamolodchikov equations, preprint ITP-89-43E Kiev 1989

83. L.D.Faddeev, Quantum completely integrable models in field theory, *Soviet Scientific Reviews*

Section C, vol.**1** (1980), 107-155

84. L.D.Faddeev, Integrable models in (1 + 1) dimensional quantum field theory, in *Les Houches Lectures XXXIX* Elsevier, Amsterdam (1982), 563-608
85. L.D.Faddeev, N.Yu Rheshetikhin and L.A.Takhtajan, Quantization of Lie groups and Lie algebras, in *Braid groups, knot theory and statistical mechanics* Eds. C.N.Yang and M.L.Ge, 97-110, World Scientific, Singapore, 1989
86. L.D.Faddeev, N.Yu Rheshetikhin and L.A.Takhtajan, Quantum groups, in *Algebraic Analysis* vol.**1** Eds. M.Kashiwara and T.Kawai, Academic, Boston, (1989), 129-139
87. V.A.Fateev, A factorized S matrix for particles of opposite partities and an integrable 21–vertex statistical model, *Soviet J. Nuclear Phys.* **33** (1981), 761-766
88. V.A.Fateev and A.B.Zamolodchikov, The exactly solvable case of a $2D$ lattice of plane rotators, *Phys. Lett. A* **92** (1982), 35-36
89. V.A.Fateev and A.B.Zamolodchikov, Self-dual solutions of the star-triangle relations in $\mathbf{Z}_N$–models, *Phys. Lett. A* **92** (1982), 37-39
90. B.U.Felderhof, Direct diagonalization of the transfer matrix of the zero-field free-fermion model, *Physica* **65** (1973), 421-451
91. B.U.Felderhof, Diagonalization of the transfer matrix of the zero-field free-fermion model II, *Physica* **66** (1973), 279-297
92. B.U.Felderhof, Diagonalization of the transfer matrix of the zero-field free-fermion model III, *Physica* **66** (1973), 509-526
93. P.Fendley and P.Ginsparg, Non-critical orbifolds, *Nucl. Phys. B* [FS] **324** (1989), 549-580
94. P.Di Francesco and J.-B. Zuber, $SU(N)$ lattice integrable models associated with graphs, preprint S. Ph-T/89/92 1989
95. I.B.Frenkel and N.Jing, Vertex representation of quantum affine algebras, *Proc. Nat. Acad. Sci. U.S.A.* **85** (1988), 9373-9377
96. M.Gaudin, *La fonction d'onde de Bethe,* Masson, Paris 1983,
97. M.Gaudin, Matrices R de dimension infinie, *J. Phys. France* **49** (1988), 1857-1865
98. M.Gaudin, Sur la nouvelle solution à deux rapidités de la relation de Yang-Baxter. Cas self-dual, preprint 1989
99. I.M.Gel'fand and I.V. Cherednik, Abstract Hamiltonian formalism for classical Yang-Baxter bundles, *Usp. Math. Nauk* **38** (1983), 3-21 [*Russian Math. Surveys* **38** (1983), 1-22]
100. I.M.Gel'fand and I.Ya.Dorfman, Hamiltonian operators and the classical Yang-Baxter equation, *Funct. Anal. Appl.* **16** (1983), 241-248
101. J.-L.Gervais and A.Neveu, Novel triangle relation and absence of tachyons in Liouville string field theory, *Nucl. Phys. B* **238** (1984), 125-141

102. V.A.Groza, I.I.Kachurik and A.U.Klymik, The quantum algebra $U_q(SU_2)$ and basic hypergeometric functions, preprint ITP-89-51E Kiev 1989

103. D.I.Gurevich, Poisson brackets associated with the classical Yang-Baxter equaiton, *Funkt. Anal. Priloz.* **23** (1989), 68-69 [*Funct. Anal. Appl.* **23** (1989), 57-59]

104. T.Hayashi, $q-$analogues of Clifford and Weyl algebras —Spinor and oscillator representations of quantum enveloping algebras—, preprint Nagoya Univ. 1989

105. M. Hashimoto and T.Hayashi, Quantum multilinear algebra, preprint Nagoya Univ. 1989

106. B.-Y.Hou, B.-Y.Hou and Z.-Q.Ma, Clebsch-Gordan coefficients, Racah coefficients, and braiding fusion of quantum $sl(2)$ enveloping algebra I, preprint BIHEP-TH-89-7,8 1989

107. B.-Y.Hou, B.-Y.Hou and Z.-Q.Ma, Clebsch-Gordan coefficients, Racah coefficients, and braiding fusion of quantum $sl(2)$ enveloping algebra II, preprint NWU-IMP-89-11,12 1989

108. A.G.Izergin and V.E.Korepin, The lattice quantum sine-Gordon model, *Lett. Math. Phys.* **5** (1981), 199-205

109. A.G.Izergin and V.E.Korepin, Lattice version of quantum field theory models in two dimensions, *Nucl. Phys.* **B205** [FS5] (1982), 401-413

110. A.G.Izergin and V.E.Korepin, The inverse scattering method approach to the quantum Shabat -Mikhailov model, *Comm. Math. Phys.* **79** (1981), 303-316

111. A.G.Izergin and V.E.Korepin, Quantum inverse scattering method, *Fisika Elementarnykh Chasits i Atomnogo Yadra* **13** (1982), 501-541 (Dubna edition); [*Sov. J. Particles and Nucleus* **13** (1982), 207-223]

112. A.G.Izergin and V.E.Korepin, The most general $L-$operator for the $R-$matrix of the $XXX-$model, *Lett. Math. Phys.* **8** (1984), 259-265

113. M. Jimbo, Quantum R matrix for the generalized Toda system, *Comm. Math. Phys.* **102** (1986), 537-547

114. M. Jimbo, A $q-$difference analogue of $U(\mathfrak{g})$ and the Yang-Baxter equation, *Lett. Math. Phys.* **10** (1985), 63-69

115. M. Jimbo, A $q-$analogue of $U(\mathfrak{gl}\ (N+1))$, Hecke algebra, and the Yang- Baxter equation, *Lett. Math. Phys.* **11** (1986), 247-252

116. M.Jimbo, Introduction to the Yang-Baxter equation, *Int. J. Mod. Phys. A* **4** (1989), 3759-3777; in *Braid groups, knot theory and statistical mechanics* Eds. C.N.Yang and M.L.Ge, World Scientific, Singapore, 1989

117. M.Jimbo, A.Kuniba, T.Miwa and M.Okado, The $A_n^{(1)}$ face models, *Comm. Math. Phys.* **119** (1988), 543-565

118. M.Jimbo and T.Miwa, Some remarks on the differential approach to the star-triangle relation, *Lett. Math. Phys.* **8** (1984), 529-534

119. M.Jimbo and T.Miwa, A solvable lattice model and related Rogers-Ramanujan type identities, *Physica D* **15** (1985), 335-353

120. M.Jimbo and T.Miwa, Classification of solutions to the star-triangle relation for a class of 3-and 4-state IRF models, *Nucl. Phys. B* **257** [FS14] (1985), 1-18

121. M.Jimbo, T.Miwa and M.Okado, An $A^{(1)}_{n-1}$ family of solvable lattice models, *Mod. Phys. Lett. B* **1** (1987), 73-79

122. M.Jimbo, T.Miwa and M.Okado, Solvable lattice models whose states are dominant integral weights of $A^{(1)}_{n-1}$, *Lett. Math. Phys.* **14** (1987), 123-131

123. M.Jimbo, T.Miwa and M.Okado, Local state probabilities of solvable lattice models: an $A^{(1)}_{n-1}$ family, *Nucl. Phys. B* **300**[FS22] (1988), 74-108

124. M.Jimbo, T.Miwa and M.Okado, Symmetric tensors of the $A^{(1)}_{n-1}$ family, in *Algebraic Analysis* vol.**1** Eds. M.Kashiwara and T.Kawai, Academic, Boston, (1989), 253-256

125. M.Jimbo, T.Miwa and M.Okado, Solvable lattice models related to the vector representation of classical simple Lie algebras, *Comm. Math. Phys.* **116** (1988), 507-525

126. I.I.Kachurik and A.U.Klimyk, On Clebshc-Gordan coefficients of quantum algebra $U_q(SU_2))$, preprint ITP-89-48E Kiev 1989

127. M.Karowski, Exact $S-$matrices and form factors in $1+1$ dimensional field theoretic models with soliton behavior, *Physics Reports* **49** (1979), 229-237

128. M.Karowski and H.J.Thun, Complete S-matrix of the massive Thirring model, *Nucl. Phys. B* **130** (1977), 295

129. M.Karowski, H.J.Thun, T.T.Truong and P.Weisz, On the uniqueness of a purely elastic S-matrix in $1+1$ dimensions, *Phys. Lett. B* **67** (1977), 321-322

130. M.Karowski and P.Weisz, Exact form factors in $1+1$ dimensional field theoretic models with soliton behavior, *Nucl. Phys. B* **139** (1978), 455-476

131. M.Kashiwara and T.Miwa, A class of elliptic solutions to the star-triangle relation, *Nucl. Phys.* **B275** [FS17] (1986), 121-134

132. A.N.Kirillov and N.Yu.Reshetikhin, Representations of the algebra $U_q(sl(2))$, $q-$orthogonal polynomials and invariants of links, in *Infinite-dimensional Lie algebras and groups,* Ed. V.G.Kac,. World Scientific, 1989, 285-339.

133. A.U.Klimyk and V.A.Groza, Representations of the quantum pseudounitary algebras, preprint ITF-89-37P 1989

134. H.T.Koelink and T.H.Koornwinder, The Clebsch-Gordan coefficients for the quantum group $S_\mu U(2)$ and $q-$Hahn polynomials, preprint 1988, to appear in Nederl. Akad. Wetensch. Proc. Ser A

135. T.H.Koornwinder, Representations of the twisted $SU(2)$ quantum group and some

q-hypergeometric orthogonal polynomials, Nederl. Akad. Wetensch. Proc. Ser A, **92** (1989), 97-117

136. T.H.Koornwinder, The addition formula for little q-Legendre polynomials and the $SU(2)$ quantum group, preprint AM-R8906 1989

137. T.H.Koornwinder, Continuous q-Legendre polynomials are spherical matrix elements of irreducible representations of the quantum $SU(2)$ group, *CWI Quarterly* **2** (1989), 171-173

138. I.M.Krichever, Baxter's equations and algebraic geometry, *Funkt. Anal. Priloz.* **15** (1981), 22-35 [*Funct. Anal. Appl.* **15** (1981), 92-103]

139. P.P.Kulish, Representation of the Zamolodchikov-Faddeev algebra, *Zapiski Nauch. Sem. LOMI* **109** (1981), 83-92 [*J. Soviet Math.* **24** (1984), 208-215]

140. P.P.Kulish, Quantum superalgebra $osp(2|1)$, *Zapiski Nauch. Sem. LOMI* **169** (1988), 95-106

141. P.P.Kulish and N.Yu.Reshetikhin, Quantum linear problem for the sine-Gordon equation and higher representations, *Zapiski Nauch. Sem. LOMI* **101** (1981), 101-110 [*J. Soviet Math.* **23** (1983), 2435-2441]

142. P.P.Kulish and N.Yu.Reshetikhin, Integrable fermion chiral models connected with the classical Lie algebras, *Zapiski Nauch. Sem. LOMI* **133** (1984), 146-159 [*J. Soviet Math.* (1985), 3352-3361]

143. P.P.Kulish, N.Yu.Reshetikhin and E.K.Sklyanin, Yang-Baxter equation and representation theory I, *Lett. Math. Phys.* **5** (1981), 393-403

144. P.P.Kulish and E.K.Sklyanin, Solutions of the Yang-Baxter equation, *Zapiski Nauch. Sem. LOMI* **95** (1980), 129-160 [*J. Soviet Math.* **19** (1982), 1596-1620]

145. P.P.Kulish and E.K.Sklyanin, Quantum spectral transform method. Recent developments, *Lecture Notes in Physics* **151** (Springer, 1982) 61-119

146. A.Kuniba, A new family of solvable lattice models associated with $A_n^{(1)}$, *Adv. Stud. Pure Math.* **19** (1989), 367-398

147. A.Kuniba, Quantum R matrix for G_2 and a solvable 173 vertex model, preprint 1989

148. A.Kuniba and T.Yajima, Local state probabilities for an infinite sequence of solvable lattice models, *J. Phys. A* **21** (1988), 519-527

149. A.Kuniba, Y.Akutsu and M.Wadati, Exactly solvable IRF models, I. A three state model, *J. Phys. Soc. Japan I* **55** (1986), 1092-1101

150. A.Kuniba, Y.Akutsu and M.Wadati, Exactly solvable IRF models, IV. Generalized Rogers-Ramanijan identities and solvable hierarchy, *J. Phys. Soc. Japan IV* **55** (1986), 2166-2176

151. A.Kuniba, Y.Akutsu and M.Wadati, Exactly solvable IRF models, V. A further new hierarchy, *J. Phys. Soc. Japan V* **55** (1986), 2605-2617

152. A.Kuniba, Y.Akutsu and M.Wadati, The Gordon generalization hierarchy of exactly solvable IRF models, *J. Phys. Soc. Japan* **55** (1986), 3338-3353

153. A.Kuniba, Y.Akutsu and M.Wadati, An exactly solvable 4-state IRF model, *Phys. Lett. A* **116** (1986), 382-386

154. A.Kuniba, Y.Akutsu and M.Wadati, An exactly solvable 5-state IRF model, *Phys. Lett. A* **117** (1986), 358-364

155. D.A.Leites and V.V.Serganova, Solutions of the classical Yang-Baxter equation for simple superalgebras, *Theoret. Math. Phys.* **58** (1984), 16-24

156. G.Lusztig, Quantum deformations of certain simple modules over enveloping algebras, *Adv. Math.* **70** (1988), 237-249

157. G.Lusztig, Modular representations and quantum groups, *Contemporary Mathematics* **82** (1989), 59-77

158. G.Lusztig, On quantum groups, preprint 1989

159. G.Lusztig, Finite dimensional Hopf algebras arising from quantum groups, preprint 1989

160. G.Lusztig, Quantum groups at roots of 1, preprint 1989

161. J.-M.Maillard, Automorphisms of algebraic varieties and Yang-Baxter equations, *LPTHE* 86/01

162. J.-M.Maillard and T.Garel, Towards an exhaustive classification of the star-triangle relation : I, *J. Phys. A* **17** (1984), 1251-1256

163. J.-M.Maillard and T.Garel, Towards an exhaustive classification of the star-triangle relation : II, *J. Phys. A* **17** (1984), 1257-1265

164. J.-M.Maillet, Kac-Moody algebra and extended Yang-Baxter relations in the $O(N)$ non-linear $\sigma-$model, *Phys. Lett. B* **162** (1985), 137-142

165. J.-M.Maillet, New integrable canonical structures in two dimensional models, *Nucl. Phys. B* **269** (1986), 54-76

166. J.-M.Maillet and F.Nijhoff, The tetrahedron equation and the four simplex equation, *Phys. Lett. A* **134** (1989), 221-228

167. J.M.Maillet and F.Nijhoff, Integrability for multidimensional lattice models, preprint CERN-TH-5332/89 *Phys. Lett. A*

168. J.-M.Maillet and F.Nijhoff, Gauging the quantum group, preprint CERN TH-5449/89 1989

169. S.Majid, Construction of non-commutative non commutative groups by Hopf algebra bicrossproduct, preprint 1988

170. S.Majid, Quasitriangular Hopf algebras and Yang-Baxter equations, preprint 1989 to appear in *Int. J. Mod. Phys. A*

171. Yu.I.Manin, Some remarks on Koszul algebras and quantum groups, *Ann. Inst. Fourier* **37** (1987), 191-205

172. Yu.I.Manin, *Quantum groups and non-commutative geometry, Publ. C. R. M., Univ. Montréal 1988*

173. Yu.I.Manin and V.V.Schechtman, Arrangements of hyperplanes, higher braid groups and higher Bruhat orders, *Adv. Stud. Pure Math.* **17** (1989), 289-308

174. T.Masuda, K.Mimachi, Y.Nakagami, M.Noumi and K.Ueno, Representations of quantum groups and a $q-$analogue of orthogonal polynomials, *C. R. Acad. Sci. Paris t.* **307** (1988), 559-564

175. T.Masuda, K.Mimachi, Y.Nakagami, M.Noumi and K.Ueno, Representations of the quantum group $SU_q(2)$ and the little $q-$Jacobi polynomials, preprint 1988, to appear in J. Functional Analysis

176. T.Masuda, K.Mimachi, Y.Nakagami, M.Noumi and K.Ueno, Representation of quantum groups, preprint 1988

177. T. Masuda, K. Mimachi, Y. Nakagami, M. Noumi, Y. Saburi and K. Ueno, Unitary representation of the quantum group $SU_q(1,1)$, I, II, preprint 1989, to appear in *Lett. Math. Phys.*

178. T. Masuda, Y. Nakagami and J. Watanabe, Non commutative differential geometry on the quantum $SU(2)$ I — an algebraic viewpoint, preprint 1989, to appear in *J. K-theory*

179. T. Masuda, Y. Nakagami and J. Watanabe, Non commutative differential geometry on the quantum 2-spheres of Podleś, preprint IHES/M/89/62 1989

180. T. Masuda and J. Watanabe, Sur les espaces vectoriels topologiques associés aux groupes quantiques $SU_q(2)$ et $SU_q(1,1)$, preprint 1989

181. B.M.McCoy, J.H.H.Perk, S.Tang and C.-H.Sah, Commuting transfer matrices for the four-state self-dual chiral Potts model with a genus-three uniformizing Fermat curve, *Phys. Lett. A* **125** (1987), 9-14

182. J.B.McGuire, Study of exactly solvable one-dimensional $N-$body problem, *J. Math. Phys.* **5** (1964), 622-636

183. T.Miwa, Multi-state solutions to the star-triangle relation with abelian symmetries, *Nucl. Phys. B* **270** [FS16] (1986), 50-60

184. M.Noumi and K.Mimachi, Quantum 2-spheres and big $q-$Jacobi polynomials, preprint 1989, to appear in *Comm. Math. Phys.*

185. M.Noumi and K. Mimachi, Big q-Jacobi polynomials, q-Hahn polynomials and a family of quantum 3-spheres, preprint 1989, to appear in *Lett. Math. Phys.*

186. M.Noumi, H.Yamada and K.Mimachi, Zonal spherical functions on the quantum homogeneous

space $SU_q(n+1)/SU_q(n)$, *Proc. Japan Acad.* **65** (1989), 169-171

187. E.Ogievetsky and P.Wiegmann, Factorized $S-$matrix and the Bethe Ansatz for simple Lie groups, *Phys. Lett. B* **168** (1986), 360-366

188. E.Ogievetsky, N.Reshetikhin and P.Wiegmann, The principal chiral field in two dimensions on classical Lie algebras : The Bethe-Ansatz solution and factorized theory of scattering, *Nucl. Phys. B* **280** [FS18] (1987), 45-96

189. D.I.Olive and N.Turok, Algebraic structure of Toda systems, *Nucl. Phys. B* **220** [FS8] (1983), 491-507

190. G.I.Ol'shanskii, Extension of the algebra $U(\mathfrak{g})$ for infinite-dimensional classical Lie algebras $\mathfrak{g}$ and the Yangians $Y(\mathfrak{gl}(m))$, *Dokl. Akad. Nauk. CCCP* **297** (1987), 1050-1053 [*Soviet Math. Doklady* **36** (1988), 569-573]

191. R.N.Onody and M.Karowski, Exact solution of a ten-vertex model in two dimensions, *J. Phys. A* **16** (1983), L31-L35

192. L.Onsager, Cristal statistics I. A two dimensional model with an order-disorder transition, *Physics Reports* **65** (1944), 117-149

193. L.Onsager, in Critical Phenomena in Alloys, Magnets and Supercondutors, Eds. R.E.Mills, E.Ascher and R.J.Jaffe McGraw-Hill, New York, 1971

194. H.C.Öttinger and J.Honerkamp, Note on the Yang-Baxter equations for generalized Baxter models, *Phys. Lett. A* **88** (1982), 339-343

195. B.Parshall and J.-P. Wang, Quantum linear groups I, preprint 1989

196. B.Parshall and J.-P. Wang, Quantum linear groups II, preprint 1989

197. V.Pasquier, Exact solubility of the D_n series, *J. Phys. A* **20** (1987), L217-L220

198. V.Pasquier, Two-dimensional critical systems labelled by Dynkin diagrams, *Nucl. Phys. B* **285** [FS19] (1987), 162-172

199. V.Pasquier, Etiology of IRF models, *Comm. Math. Phys.* **118** (1988), 335-364

200. V.Pasquier and H.Saleur, Common structures between finite systems and conformal field theories through quantum groups, preprint 1989

201. P.A.Pearce and K.A.Seaton, Solvable hierarchy of cyclic solid-on-solid lattice models, *Phys. Rev. Lett.* **60** (1988), 1347

202. J.H.H.Perk and C.L.Schultz, New families of commuting transfer matrices in $q-$state vertex models, *Phys. Lett. A* **84** (1981), 407-410

203. J.H.H.Perk and C.L.Schultz, Families of commuting transfer matrices in $q-$state vertex models, in *Non-linear Integrable systems —Classical Theory and Quantum Theory* Eds. M.Jimbo and T.Miwa, World Scientific, Singapore, 1983

204. P.Podleš, Quantum spheres, *Lett. Math. Phys.* **14** (1987), 193-202

205. S.V.Pokrovsky and Yu.A.Bashilov, Star-triangle relations in exactly solvable statistical models, *Comm. Math. Phys.* **84** (1982), 103-132

206 W.Pusz and S.L.Woronowicz, Twisted second quantization, preprint 1988

207. N.Yu.Reshetikhin, Integrable models of quantum one-dimensional magnets with $O(n)$ and $Sp(2k)$ symmetry, *Theoret. Math. Phys.* **63** (1985), 555-569

208. N.Yu.Reshetikhin, The spectrum of the transfer matrices connected with Kac-Moody algebras, *Lett. Math. Phys.* **14** (1987), 235-246

209. N.Yu.Reshetikhin, Quantized universal enveloping algebras, the Yang-Baxter equation and invariants of links I, preprint E-4-87 LOMI 1988

210. N.Yu.Reshetikhin, Quantized universal enveloping algebras, the Yang-Baxter equation and invariants of links II, preprint E-17-87 LOMI 1988

211. N.Yu.Reshetikhin, Algebraic Bethe Ansatz for the $SO(n)$ invariant transfer-matrices, *Zapiski Nauch. Sem. LOMI* **169** (1988), 122-140

212. N.Yu.Reshetikhin and M.A.Semenov-Tian-Shansky, Quantum R matrices and factorization problems in quantum groups, *J. Diff. Geom. and Phys.* **5** (1988),

213. N.Yu.Reshetikhin and P.Wiegmann, Towards the classification of completely integrable quantum field theories (the Bethe-Ansatz associated with Dynkin diagrams and their automorphisms), *Phys. Lett. B* **189** (1987), 125-131

214. M.P.Richey and C.A.Tracy, $\mathbf{Z}_n$ Baxter model : symmetries and the Belavin Parametrization, *J. Stat. Phys.* **42** (1986), 311-348

215. M.P.Richey and C.A.Tracy, Symmetry group for a completely symmetric vertex model, *J. Phys. A* **20** (1987), 2667-2677

216. M.Rosso, Comparaison des groupes $SU(2)$ quantiques de Drinfeld et de Woronowicz, *C. R. Acad. Sci. Paris t.* **304** (1987), 323-326

217. M.Rosso, Repésentations irréductibles de dimension finie du $q-$analogue de l'algèbre enveloppante d'une algèbre de Lie simple, *C. R. Acad. Sci. Paris t.* **305** (1987), 587-590

218. M.Rosso, Finite-dimensional representations of quantum analog of the enveloping algebra of a complex simple Lie algebra, *Comm. Math. Phys.* **117** (1988), 581-593

219. M.Rosso, An analogue of P.B.W. theorem and the universal $R-$matrix for $U_h\mathfrak{sl}(N+1)$, *Comm. Math. Phys.* **124** (1989), 307-318

220. M.Rosso, Algèbres enveloppantes quantifiées, groupes quantiques de matrices et calcul differentiel non commutatif, preprint 1989

221. C.L.Schultz, Solvable $q-$state models in lattice statistics and quantum field theory, *Phys. Rev. Lett.* **46** (1981), 629-632

222. M.A.Semenov-Tyan-Shanskii, What is a classical $R-$matrix ?, *Funkt. Anal. Priloz.* **17** (1983), 17-33 [*Funct. Anal. Appl.* **17** (1984), 259-272]

223. M.A.Semenov-Tyan-Shanskii, Classical $r-$matrices and quantization, *Zapiski Nauch. Sem. LOMI* **133** (1984), 228-235 [*J. Soviet Math.* **31** (1985), 3411-3416]

224. M.A.Semenov-Tyan-Shanskii, Dressing transformations and Poisson group actions, *Publ. RIMS* **21** (1985), 1237-1260

225. M.A.Semenov-Tyan-Shanskii, Poisson groups and Dressing transformations, *Zapiski Nauch. Sem. LOMI* **46** (1986), 119-142 [*J. Soviet Math.* **46** (1989), 1641-1656]

226. R.Shankar and E.Witten, S matrix of the supersymmetric nonlinear σ model, *Phys. Rev.* D **17** (1978), 2134

227. B.S. Shastry, Decorated star-triangle relations and exact integrability of the one-dimensional Hubbard model, *J. Stat. Phys.* **50** (1988), 57-79

228. E.K.Sklyanin, On complete integrability of the Landau-Lifschitz equation, preprint E-3-1979 1979

229. E.K.Sklyanin, The quantum version of the inverse scattering method, *Zapiski Nauch. Sem. LOMI* **95** (1980), 55-128 [*J. Soviet Math.* **19** (1982), 1546-1596]

230. E.K.Sklyanin, Some algebraic structures connected with the Yang-Baxter equation, *Funkt. Anal. Priloz.* **16** (1982), 27-34 [*Funct. Anal. Appl.* **16** (1983), 263-270]

231. E.K.Sklyanin, Some algebraic structures connected with the Yang-Baxter equation. Representations of quantum algebras, *Funkt. Anal. Priloz.* **17** (1983), 34-48 [*Funct. Anal. Appl.* **17** (1984), 273-284]

232. E.K.Sklyanin, An algebra generated by quadratic relations, *Usp. Math. Nauk* **40** (1985), 214

233. E.K.Sklyanin, Classical limits of the $su(2)-$invariant solutions of the Yang-Baxter equation, Zapiski Nauch. Sem. LOMI**146** (1985), 119 [*J. Soviet Math.* **40** (1988), 93-107]

234. E.K.Sklyanin and L.D.Faddeev, Quantum-mechanical approach to completely integrable field theory models, *Dokl. Akad. Nauk. CCCP* **243** (1978), 1430-1433 [*Soviet Phys. Doklady* **23** (1978), 902-904]

235. E.K.Sklyanin, L.A.Takhtajan and L.D.Faddeev, Quantum inverse problem method I, *Teoret. Mat. Fiz.* **40** (1979), 194-220 [*Theoret. Math. Phys.* **40** (1980), 688-706]

236. D.-J.Smit, The quantum group structure in a class of $d = 2$ conformal field theories, preprint 1988

237. K.Sogo, Y.Akutsu and T.Abe, New factorized $S-$matrix and its application to exactly solvable $q-$state model I, *Prog. Theoret. Phys.* **70** (1983), 730-738

238. K.Sogo, Y.Akutsu and T.Abe, New factorized $S-$matrix and its application to exactly solvable $q-$state model II, *Prog. Theoret. Phys.* **70** (1983), 739-746

239. Ja. S. Soibelman, Irreducible representations of the algebra of functions on quantum group $SU(n)$ and Schubert cells, preprint 1988

240. Yu.G.Stroganov, A new calculation method for partition functions in some lattice models, *Phys. Lett. A* **74** (1979), 116-118

241. E.Taft and J.Towber, Quantum deformation of flag schemes and Grassmann schemes I. A q-deformation of the shape-algebra for $GL(n)$, preprint 1989

242. M.Takeuchi, Quantum orthogonal and symplectic groups and their embedding into quantum GL, preprint Tsukuba Univ. 1988

243. M.Takeuchi, Matric bialgebras and quantum groups, preprint Tsukuba Univ. 1989

244. M.Takeuchi, The q-bracket product and the P-B-W theorem for quantum enveloping algebras of classical types (A_n), (B_n),(C_n), and (D_n), preprint Tsukuba Univ. 1989

245. L.A.Takhtadzhyan, Solutions of the triangle equations with $\mathbf{Z}_n \times \mathbf{Z}_n$−symmetry and matrix analogs of the Weierstrass Zeta-and Sigma-functions, *Zapiski Nauch. Sem. LOMI* **133** (1984), 258-276 [*J. Soviet Math.* **31** (1985), 3432-3444]

246. L.A.Takhtadzhyan and L.D.Faddeev, The quantum method of the inverse problem and the Heisenberg XYZ model, Usp. Math. Nauk**34** (1979), 13-63 [*Russian Math. Surveys* **34** (1979), 11-68]

247. L.A.Takhtadzhyan, Quantum groups and integrable models, *Adv. Stud. Pure Math.* **19** (1989), 435-457

248. T.Tanisaki, Harish-Chandra isomorphism for quantum algebras, preprint Osaka Univ. 1989

249. T.Tanisaki, Finite dimensional representations of quantum groups, preprint Osaka Univ. 1989

250. V.O.Tarasov, Irreducible monodromy matrices for the R matrix of the XXZ model, and local quantum lattice Hamiltonians, *Dokl. Akad. Nauk. CCCP* **278** (1984), 1102-1105 [*Soviet Phys. Doklady* **29** (1984), 804-806]

251. V.O.Tarasov, On the structure of quantum L−operators for the R−matrix of XXZ−model, *Teoret. Mat. Fiz.* **61** (1984), 163-173 [*Theoret. Math. Phys.* **61** (1984), 1065-1072]

252. V.O.Tarasov, Irreducible monodromy matrices for the R matrix of the XXZ model and local lattice quantum Hamiltonian, *Teoret. Mat. Fiz.* **63** (1985), 175-196 [*Theoret. Math. Phys.* **63** (1985), 440-454]

253. C.A.Tracy, Embedded elliptic curves and the Yang-Baxter equation, *Physica D* **16** (1985), 203-220

254. C.A.Tracy, Complete integrability in statistical mechanics and the Yang-Baxter equations, *Physica D* **14** (1985), 253-264

255. K.Ueno, T.Takebayashi and Y.Shibukawa, Gel'fand-Tsetlin basis for $U_q(gl(N+1))$ modules, preprint 1989, to appear in *Lett. Math. Phys.*

256. K.Ueno, T.Takebayashi and Y.Shibukawa, Gel'fand-Tsetlin basis for representations of the quantum group $GL_q(N+1)$ and the basic quantum affine space, preprint 1989

257. L.L.Vaksman and Ya.S.Soibelman, Algebra of functions on the quantum group $SU(2)$, *Funkt. Anal. Priloz.* **22** (1988), 1-14 [*Funct. Anal. Appl.* **22** (1989), 170-181]

258. J.-L.Verdier, Groupes quantiques, *Sém. Bourbaki* **685** (1986-87), n° 685

259. H. Wenzl, Quantum groups and subfactors of type B, C and D, preprint 1989

260. S.L.Woronowicz, Tannaka-Krein duality for compact matrix pseudogroups. Twisted $SU(N)$ groups, *Invent. Math.* **93** (1988), 35-76

261. S.L.Woronowicz, Differential calculus on compact matrix pseudogroups (quantum groups), preprint 1988

262. S.L.Woronowicz, Twisted $SU(2)$ group. An example of a non-commutative differential calculus, *Publ. RIMS, Kyoto Univ.* **23** (1987), 117-181

263. S.L.Woronowicz, Compact matrix pseudogroups, *Comm. Math. Phys.* **111** (1987), 613-665

264. H.Yamane, A Poincaré-Birkhoff-Witt theorem for quantized universal enveloping algebras of type A_N, *Publ. RIMS, Kyoto Univ.* **25** (1989), 503-520

265. C.N.Yang, Some exact results for the many-body problem in one dimension with repulsive delta-function interaction, *Phys. Rev. Lett.* **19** (1967), 1312-1315

266. C.N.Yang, $S-$matrix for one-dimensional $N-$body problem with repulsive or attractive $\delta-$function interaction, *Phys. Rev.* **168** (1968), 1920-1923

267. A.B.Zamolodchikov, Exact two particle $S-$matrix of quantum sine-Gordon solitons, *Comm. Math. Phys.* **55** (1977), 183-186

268 A.B.Zamolodchikov, $\mathbf{Z}_4-$symmetric factorized $S-$matrix in two space-time dimensions, *Comm. Math. Phys.* **69** (1979), 165-178

269. A.B.Zamolodchikov, Factorized S matrices and lattice statistical systems, *Soviet Scientific Reviews* section A, Ed. Khalatnikov, Harwood Academic Press, vol.**2** (1980), 1-40

270. A.B.Zamolodchikov, Tetrahedron equations and integrable systems in three dimensional space, *Zh.Eksp.Theor.Fiz.* **79** (1980), 641-664 [*Sov. Phys. JETP* **52** (1980), 325-336]

271. A.B.Zamolodchikov, 'Fishing-net' diagrams as a completely integrable system, *Phys. Lett. B* **97** (1980), 63-66

272. A.B.Zamolodchikov, Tetrahedron equations and the relativistic $S-$matrix of straight strings in $2+1$ dimensions, *Comm. Math. Phys.* **79** (1981), 489-505

273. A.B.Zamolodchikov and V.A.Fateev, A model factorized $S-$matrix and an integrable spin-1 Heisenberg chain, *Yad. Fiz.* **32** (1980), 581-590 [*Soviet J. Nuclear Phys.* **32** (1980), 298-303]

274. A.B.Zamolodchikov and Al.B.Zamolodchikov, Relativistic factorized $S-$matrix in two dimensions having $O(N)$ isotopic symmetry, *Nucl. Phys. B* **133** (1978), 525-535

275. A.B.Zamolodchikov and Al.B.Zamolodchikov, Exact S matrix of Gross-Neveu "elementary" fermions, *Phys. Lett. B* **72** (1978), 481-483

276. A.B.Zamolodchikov and Al.B.Zamolodchikov, Factorized $S-$matrices in two dimensions as the exact solutions of certain relativistic quantum field theory models, *Ann. Phys.* **120** (1979), 253-291